Animal Diversity

NINTH EDITION

Cleveland P. Hickman, Jr.
Professor Emeritus
Washington and Lee University

Susan L. Keen
University of California, Davis

Allan Larson
Washington University

David J. Eisenhour
Morehead State University

Original Artwork by
William C. Ober, M.D., and Claire W. Garrison, R.N.

ANIMAL DIVERSITY, NINTH EDITION

Published by McGraw Hill LLC, 1325 Avenue of the Americas, New York, NY 10121.

This book is printed on acid-free paper.

1 2 3 4 5 6 7 8 9 LWI 25 24 23 22 21 20

ISBN 978-1-260-24088-7 (bound edition)
MHID 1-260-24088-6 (bound edition)
ISBN 978-1-260-44315-8 (loose-leaf edition)
MHID 1-260-44315-9 (loose-leaf edition)

Portfolio Director: *Michelle Vogler*
Product Developer: *Erin DeHeck*
Senior Marketing Manager: *Kelly Brown*
Lead Content Project Managers: *Susan Trentacosti, Tammy Juran*
Senior Buyer: *Susan K. Culbertson*
Designer: *Beth Blech*
Senior Content Licensing Specialist: *Melissa Homer*
Cover Image: *©rhonny dayusasono/Shutterstock*
Compositor: *MPS Limited*

Library of Congress Cataloging-in-Publication Data

Names: Hickman, Cleveland P., Jr., author. | Keen, Susan L., author. | Larson, Allan, author. | Eisenhour, David J., author.
Title: Animal diversity / Cleveland P. Hickman, Jr., Professor Emeritus, Washington and Lee University, Susan L. Keen, University of California, Davis, Allan Larson. Washington University. David J. Eisenhour. Morehead State University; original artwork by William C. Ober, M.D., and Claire W. Garrison, R.N.
Description: Ninth edition. | New York : McGraw Hill LLC, 2021. | Includes bibliographical references and index.
Identifiers: LCCN 2020017423 | ISBN 9781260240887 (hardcover) | ISBN 9781260443158 (spiral bound)
Subjects: LCSH: Zoology. | Animal diversity.
Classification: LCC QL47.2 .H527 2021 | DDC 590–dc23
LC record available at https://lccn.loc.gov/2020017423

mheducation.com/highered

Brief Contents

Contents

Chapter 7

Cnidarians and Ctenophores 151

Chapter 8

Xenacoelomorpha, Platyhelminthes, Gastrotricha, Gnathifera, including Chaetognatha, and Mesozoa 177

Chapter 9

Polyzoa and Trochozoa: *Cycliophora, Entoprocta, Ectoprocta, Brachiopoda, Phoronida, and Nemertea* 201

Chapter 10

Molluscs 213

Chapter 11

Annelids 240

Chapter 12

Smaller Ecdysozoans 260

Chapter 13

Arthropods 273

Chapter 14

Echinoderms and Hemichordates 319

Chapter 15

Vertebrate Beginnings: *The Chordates* 338

Chapter 16

Fishes 354

Chapter 17

The Early Tetrapods and Modern Amphibians 378

Chapter 18

Amniote Origins and Nonavian Reptiles 394

Chapter 19

Birds 414

Chapter 20

Mammals 439

About the Authors

Cleveland P. Hickman, Jr.

Cleveland P. Hickman, Jr., Professor Emeritus of Biology at Washington and Lee University in Lexington, Virginia, has taught zoology and animal physiology for more than 30 years. He received his PhD in comparative physiology from the University of British Columbia, Vancouver, B.C., in 1958 and taught animal physiology at the University of Alberta before moving to Washington and Lee University in 1967. He has published numerous articles and research papers on fish physiology, in addition to co-authoring these highly successful texts: *Integrated Principles of Zoology, Biology of Animals, Animal Diversity, Laboratory Studies in Animal Diversity,* and *Laboratory Studies in Integrated Principles of Zoology.*

Over the years, Dr. Hickman has led many field trips to the Galápagos Islands. His current research is on intertidal zonation and marine invertebrate systematics in the Galápagos. He has published four field guides in the Galápagos Marine Life Series for the identification of echinoderms, marine molluscs, marine crustaceans, and corals and other radiates. (To read more about these field guides, visit *http://www.galapagosmarine.com.*)

His interests include scuba diving, woodworking, and participating in chamber music ensembles.

Dr. Hickman can be contacted at
hickman.c@rockbridge.net

Susan Keen

Susan Keen is a professor in the Department of Evolution and Ecology at the University of California at Davis. She was Associate Dean for Undergraduate Academic Programs, College of Biological Sciences, at the same institution from 2012 to 2018. She received her PhD in zoology from the University of California at Davis, following an MSc from the University of Michigan at Ann Arbor. She is a native of Canada and obtained her undergraduate education at the University of British Columbia in Vancouver.

Dr. Keen is an invertebrate zoologist fascinated with sea jelly life histories and life cycles where both asexual and sexual phases of organisms are present. Her other research has included work on sessile marine invertebrate communities, spiders, and Andean potato evolution.

Dr. Keen has been teaching evolution and animal diversity within the Introductory Biology series for 27 years. She enjoys all facets of the teaching process, from lectures and discussions to the design of effective laboratory exercises. In addition to her work with introductory biology, she offers seminars on teaching methods and studies the use of visualizations in how students learn. She has won awards for Excellence in Education from the Associated Students group at Davis. She attended the National Academies Summer Institute on Undergraduate Education in Biology, and was a National Academies Education Fellow in the Life Sciences for 2005–2006. Her interests include horseback riding, gardening, travel, and mystery novels.

Dr. Keen can be contacted at:
slkeen@ucdavis.edu.

Allan Larson

Allan Larson is a professor at Washington University, St. Louis, MO. He received his PhD in genetics at the University of California, Berkeley, following BS and MS degrees in zoology from the University of Maryland. His fields of specialization include evolutionary biology, molecular population genetics and systematics, and amphibian systematics. He teaches courses in zoology, genetics, macroevolution, molecular evolution, and the history of evolutionary theory, and has organized and taught a special course in evolutionary biology for high-school teachers.

Dr. Larson's research uses DNA sequences to examine evolutionary relationships among vertebrate species, especially salamanders and lizards. The students in Dr. Larson's laboratory have participated in zoological field studies around the world, including in Africa, Asia, Australia, Madagascar, North America, South America, and the Caribbean Islands. Dr. Larson has authored numerous scientific publications and has edited for the journals *The American Naturalist, Evolution, Journal of Experimental Zoology, Molecular Phylogenetics and Evolution, and Systematic Biology*. Dr. Larson serves as an academic advisor to undergraduate students and supervises the undergraduate biology curriculum at Washington University.

Dr. Larson can be contacted at
larson@wustl.edu

David J. Eisenhour

David J. Eisenhour is a professor of biology at Morehead State University in Morehead, Kentucky. He received his PhD in zoology from Southern Illinois University, Carbondale. He teaches courses in environmental science, general biology, mammalogy, human anatomy, general zoology, comparative anatomy, ichthyology, and vertebrate zoology. Dr. Eisenhour has an active research program that focuses on systematics, conservation biology, and natural history of North American freshwater fishes. He has a particular interest in the diversity of Kentucky's fishes and is writing a book about that subject. He and his undergraduate and graduate students have authored several publications. In 2001 he was given a Master Teacher Award by a student group at MSU, and in 2016 he was given the Distinguished Researcher Award at MSU. David serves as an academic advisor to prepharmacy students.

His interests include fishing, landscaping, softball, hiking, and entertaining his three children, who, along with his wife Lynn, are enthusiastic participants in fieldwork.

Dr. Eisenhour can be contacted at
d.eisenhour@morehead-st.edu

Preface

Animal Diversity is tailored for the restrictive requirements of a one-semester or one-quarter course in zoology, and is appropriate for both nonscience and science majors of varying backgrounds. This ninth edition of *Animal Diversity* presents a survey of the animal kingdom with emphasis on diversity, evolutionary relationships, functional adaptations, and environmental interactions.

Organization and Coverage

The sixteen survey chapters of animal diversity are prefaced by four chapters presenting the principles of evolution, ecology, animal architecture, and taxonomy. Throughout this revision, we updated references and worked to improve pedagogy.

Chapter 1 begins with a brief explanation of the scientific method—what science is (and what it is not)—and then introduces evolutionary principles. Following a historical account of Charles Darwin's life and discoveries, we present the five major components of Darwin's evolutionary theory, the important challenges and revisions to his theory, and an assessment of its current scientific status. This approach reflects our understanding that Darwinism is a composite theory whose component parts guide active research and can be modified by new discoveries. It also prepares the student to dismiss the arguments of creationists who misconstrue scientific challenges to Darwinism as contradictions to the validity of organic evolution. The chapter summarizes the major principles of molecular genetics, population genetics, and macroevolution.

Chapter 2 explains the principles of ecology, with emphasis on populations, community ecology, and variations in the life-history strategies of natural populations. The treatment includes discussions of niche, population growth and its regulation, limits to growth, competition, energy flow, nutrient cycles, and extinction.

Chapter 3, on animal architecture, is a short but important chapter that describes the organization and development of the body plans that distinguish major groups of animals. This chapter includes a picture essay of tissue types and a section explaining important developmental processes responsible for the evolutionary diversification of the bilateral animals.

Chapter 4 treats taxonomy and phylogeny of animals. We present a brief history of how animal diversity has been organized for systematic study, emphasizing current use of Darwin's theory of common descent as the major principle underlying animal taxonomy. Our summary of continuing controversies over concepts of species and higher taxa includes discussion of how alternative taxonomic philosophies guide our study of evolution. We give special attention to phylogenetic systematics (cladistics) and the interpretation of cladograms. Chapter 4 also emphasizes that current issues in ecology and conservation biology depend upon our taxonomic system.

The sixteen survey chapters provide comprehensive, current, and thoroughly researched coverage of the animal phyla in the context of eukaryotic diversity and evolution. We emphasize the unifying phylogenetic, architectural, and functional themes of each group, and illustrate them with detailed coverage of representative forms. Each chapter includes succinct statements of the diagnostic characteristics and major subgroups of the focal taxa. Discussions of phylogenetic relationships take a cladistic viewpoint, with cladograms showing the structure of each group's history and the origin of the principal shared derived characters. Phylogenetic trees add temporal evolutionary hypotheses to the cladistic analyses. A cladogram depicting phylogenetic relationships among animal taxa appears in the inside front cover and serves to order our coverage of animal diversity in Chapters 5–20. A smaller version of this cladogram appears at the start of each taxonomic chapter, with the chapter's taxonomic coverage highlighted on it.

Changes in the Ninth Edition

Starting with this edition, the Learning Objectives listed at the start of each chapter correspond to the chapter's main sections. Chapter summaries, many of which are expanded in content, are restructured in tabular form to list the main highlights of each section of the chapter. This correspondence between the learning objectives, chapter sections, and summary should help students to grasp the main lessons of each chapter. Our extensive cross-referencing of material among the different parts of the text now uses section numbers, with live links available in electronic versions of the text.

Notes and essays separate from the main text are now numbered for reference and organized according to six Key Themes: (1) Adaptation and Physiology, (2) Ecology, (3) Evolution, (4) Genetics and Development, (5) Human Connections, and (6) Science Explained. **Adaptation and Physiology** connects the proximate and ultimate causes underlying organismal functioning. **Ecology** addresses the interactions of animal populations with their environments, including factors that influence their geographic distribution and abundance. **Evolution** highlights common descent of animal species and the historical processes that modify organismal characteristics in natural populations. **Genetics and Development** covers the mechanisms of heredity and the ways that an organism uses genetic information to progress from a zygote all the way to an adult animal. **Human Connections** highlights ways that zoological findings influence human welfare, including medical applications and maintaining environmental health, or how human activities have affected animal species. And **Science Explained** covers scientific methodology and the history of scientific discovery.

In addition to these organizational revisions, we have replaced many photographs and diagrams throughout the book to improve clarity and vibrancy. We have also updated geological periods on our phylogenetic trees throughout the book; for example, the former Tertiary period is replaced with the Paleogene and Neogene periods. Next, we highlight some additional revisions in the order of the numbered chapters.

Chapter 1 includes more information on the last universal common ancestor of life (LUCA). We expand coverage of August Weismann to introduce his concepts of germ plasm and soma. We introduce polymorphism as a population-genetic concept and explicitly contrast this usage with an alternative meaning of polymorphism used in describing some colonial invertebrates.

Chapter 2 opens with a section titled "The Hierarchy of Ecology." Mutualism is added to competition, parasitism, and predation in forming the set of interactions among species that structure a community. We update our coverage of demography by contrasting profiles for Mexico in 1975 and 2019—the growth curve for the human population now shows information through 2019. A new Key Theme box introduces the neutral theory of community assembly. Our boxed essay "Life without the Sun" updates the taxonomy of siboglinid worms to place them within polychaete annelids. A new boxed essay covers the United Nations 2019 report on biodiversity and extinction, which is the subject of two new review questions. The periodic cycling of snowshoe hare and lynx populations of Canada as caused by predator-prey dynamics now appears in Chapter 2 rather than Chapter 20.

Chapter 3 features a revised opening essay and numerous revised figures. We improve discussion of the basal lamina.

Chapter 4 includes a new boxed essay instructing students how to read a phylogenetic tree. We revise the indented taxonomies of fishes and of Bilateria in light of new phylogenetic results. We improve presentation of the blocks of aligned DNA sequences in the boxed essay "Phylogenies from DNA Sequences."

Chapter 5 is extensively revised to incorporate descriptions of eukaryotic cellular components and how these components serve to diagnose clades or to explain cell function. We add four new figures and revise others.

Chapter 6 features a revised opening essay, a new Key Theme box on bioactive chemicals from sponges, descriptions of two newly described water-flow patterns, and a new image of a glass sponge. A section on phylum Placozoa concludes the chapter.

We divide Chapter 7 into a larger number of sections. A new Key Theme box with an accompanying photograph covers the origin of the term "medusa." Another new Key Theme box explains oceanic dead zones.

Chapter 8 now includes Chaetognatha because taxonomic revision places Chaetognatha within clade Gnathifera. Taxonomic revision removes clade Platyzoa due to lack of evidence supporting the group. We update the phylogenetic figures to reflect these changes. We present parasitic infection rates for many taxa using recent statistics. We add a new Key Theme box on schistosomiasis, and a new figure on mesozoans.

In Chapter 9, we remove the group Kryptrochozoa due to lack of evidence supporting it, and add clade Trochozoa. We revise the phylogenetic figures and text accordingly. A new Key Theme box introduces invasive bryozoans.

Chapter 10 features a revised description of torsion; a new Key Theme box on invasive slugs and their spreading of disease; a new Key Theme box on invasive bivalves with an accompanying photograph; and a revised Key Theme box on squid. An improved discussion of two hypotheses for mollusc diversification is included.

Chapter 11 features extensive new discussion of recent taxonomic revisions to the annelids, covering the groups added and the rationales for these additions. We divide Chapter 11 into a greater number of sections, and revise usage of the terms "polychaete" and "oligochaete" throughout the chapter.

Chapter 12 features a new Key Theme box on cryptobiosis, updated statistics on infection rates for nematode parasites, and many revised images. We add discussion of neglected tropical diseases, with some examples of successful control of these diseases.

Chapter 13 has a new opening essay focusing on recent data showing a global decline in insect populations. Other changes include greater emphasis on beneficial insects, revised section heads throughout, a new Key Theme box on the evolution of bee behavior in response to Colony Collapse Disorder, an expanded Key Theme box on human responses to barnacles fouling ship bottoms, and a new discussion of *Wolbachia* as a way to control the spread of mosquito-borne diseases. Numerous figures are revised.

Chapter 14 no longer covers chaetognaths, whose taxonomic revision now places them in Gnathifera (Chapter 8). General discussion of relationships among deuterostome phyla is moved forward to section 14.1, followed by "Phylum Echinodermata" (section 14.2) and "Phylogeny and Adaptive Diversification of Echinoderms" (section 14.3). Revised text explains that some, but not all, sea cucumbers eviscerate. A new, small section covers feeding ecology of sea urchins.

We reorganize Chapter 15 to place the related topics of evolution and classification of chordates first as a framework for the rest of the chapter. We improve our explanation of the problem of using Linnaean ranks in higher vertebrate taxonomy, particularly for the Class level. Protochordates are more clearly defined. Feeding and respiration in amphioxus get new descriptions. We reduce coverage of Garstang's hypothesis of paedomorphosis in the origin of chordates to avoid an impression that this is the currently favored explanation. We heavily revise the text on ostracoderms, and shift its position to link it to the origin of jawed vertebrates. We add new artwork to illustrate ostracoderms. Along the same lines, we move discussion of the first jawed fishes, placoderms and acanthodians, forward and modestly revise the text. A beautiful new art piece illustrates a typical placoderm, with clearly visible jaws and pectoral and pelvic fins. We substantially revise discussion of the evolutionary origin of jaws, including a new figure illustrating evolution of jaws from pharyngeal arches in an agnathan fish.

Chapter 16 now consolidates all characteristics of fishes in a single table for easy comparison. Revised text in the first section provides a better framework for the chapter as a whole.

Chapter 17 includes a renumbering of sections, giving the three major groups of amphibians their own separate learning objectives and summary. The new section "Temnospondyls and Modern Amphibians" provides new information on temnospondyls. We introduce the perennibranchiate condition of some salamanders with new information on experimental thyroid hormone treatment. We present updated information for the oldest fossil amphibian, as well as for the numbers of taxonomic families, genera, and species of living amphibians.

In Chapter 18, the special section on dinosaurs appears earlier and closer to the text discussion of dinosaurs. We add coverage of green sea turtles following a reviewer's suggestion. We replace a figure illustrating crocodilian anatomy with two figures, one illustrating changes to the lungs of amniotes, and another showing blood flow patterns through the heart in turtles/squamates. New figures and text cover amniote adaptations, especially for respiration and circulation. A new subsection under the Squamata heading describes the reproductive biology of squamates. We update and clarify the status of pythons in Florida. Numerous minor updates include the diet of glass lizards, taxonomic revision to recognize only a single species of tuatara, and the survival and reproduction of captive tuataras.

Chapter 19 features a completely rewritten section on avian evolution as well as new information on the conservation status of birds, using some new (2018-2019) information about sharp declines in the world's birds. We replace the art in several figures, including the illustration of a feather and the contrast between precocial and altricial development. Our explanation of the dynamics of bird flight is extensively rewritten, reorganized, and consolidated. Two figures formerly presented as line drawings are now full illustrations. A new section examines hypotheses on the origin of flight.

Chapter 20 includes a revised and expanded section on mammary glands, a new section on olfaction and the mammalian brain, and a new section on torpor and hibernation as strategies to deal with northern winters. In addition, there are many small updates, including the number of mammal species on the IUCN list, calcium content of moose antlers, ability of some bats to fly upward from horizontal surfaces, and consolidation of material on human evolution. We revise the spelling of Neanderthals to reflect more common recent usage. Material on the predator-prey dynamics of mammalian populations is now relocated to Chapter 2.

Teaching and Learning Aids

Vocabulary Development

The glossary provides a definition and often a pronunciation and derivation for many of the key terms that appear in boldface in the chapters. Derivations of generic names of animals are given where they first appear in the text. In addition, derivations of many technical and zoological terms are provided, allowing students to recognize the more common roots that recur in many technical terms.

Learning Objectives

Each chapter begins with a list of learning objectives that identify the major organizing principles of each section of the chapter. Chapter summaries are likewise arranged to correspond to each section of the chapter. Students enhance their understanding by using these guiding principles as they read a chapter's detailed material.

Chapter Prologues

A distinctive feature of this text is an opening essay at the beginning of each chapter. Each essay presents a theme or topic relating to the subject of the chapter to stimulate interest. Some present biological—particularly evolutionary—principles; others illuminate distinguishing characteristics of the animal group treated in the chapter.

Key Theme Boxes

Key Theme boxes, which appear throughout the book, augment the text material and offer interesting sidelights without interrupting the narrative.

Instructors: Student Success Starts with You

Tools to enhance your unique voice

Want to build your own course? No problem. Prefer to use our turnkey, prebuilt course? Easy. Want to make changes throughout the semester? Sure. And you'll save time with Connect's auto-grading too.

65%
Less Time Grading

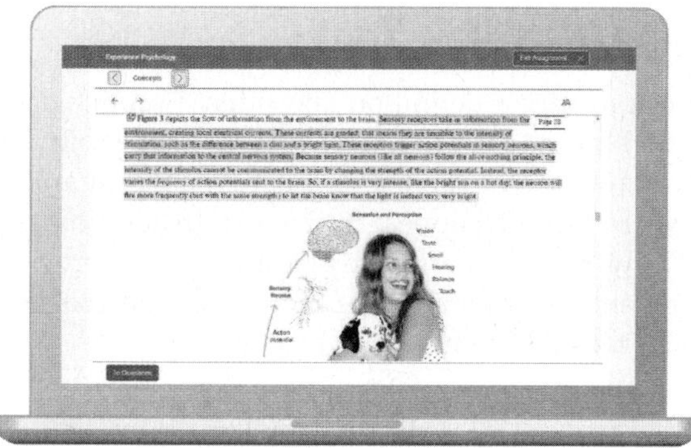

Laptop: McGraw Hill; Woman/dog: George Doyle/Getty Images

Study made personal

Incorporate adaptive study resources like SmartBook® 2.0 into your course and help your students be better prepared in less time. Learn more about the powerful personalized learning experience available in SmartBook 2.0 at **www.mheducation.com/highered/connect/smartbook**

Affordable solutions, added value

Make technology work for you with LMS integration for single sign-on access, mobile access to the digital textbook, and reports to quickly show you how each of your students is doing. And with our Inclusive Access program you can provide all these tools at a discount to your students. Ask your McGraw Hill representative for more information.

Padlock: Jobalou/Getty Images

Solutions for your challenges

A product isn't a solution. Real solutions are affordable, reliable, and come with training and ongoing support when you need it and how you want it. Visit **www.supportateverystep.com** for videos and resources both you and your students can use throughout the semester.

Checkmark: Jobalou/Getty Images

Students: Get Learning that Fits You

Effective tools for efficient studying

Connect is designed to make you more productive with simple, flexible, intuitive tools that maximize your study time and meet your individual learning needs. Get learning that works for you with Connect.

Study anytime, anywhere

Download the free ReadAnywhere app and access your online eBook or SmartBook 2.0 assignments when it's convenient, even if you're offline. And since the app automatically syncs with your eBook and SmartBook 2.0 assignments in Connect, all of your work is available every time you open it. Find out more at **www.mheducation.com/readanywhere**

"I really liked this app—it made it easy to study when you don't have your textbook in front of you."

- Jordan Cunningham, Eastern Washington University

Calendar: owattaphotos/Getty Images

Everything you need in one place

Your Connect course has everything you need—whether reading on your digital eBook or completing assignments for class, Connect makes it easy to get your work done.

Learning for everyone

McGraw Hill works directly with Accessibility Services Departments and faculty to meet the learning needs of all students. Please contact your Accessibility Services Office and ask them to email accessibility@mheducation.com, or visit **www.mheducation.com/about/accessibility** for more information.

Top: Jenner Images/Getty Images, Left: Hero Images/Getty Images, Right: Hero Images/Getty Images

For Review

Each chapter ends with a concise summary and review questions. The review questions enable students to test themselves for retention and understanding of the more important chapter material. A list of annotated selected references is available online for each of the chapters, followed by a list of general references.

Art Program

The appearance and usefulness of this text are much enhanced by numerous full-color paintings by William C. Ober and Claire W. Garrison. Bill's artistic skills, knowledge of biology, and experience gained from an earlier career as a practicing physician have enriched the authors' zoology texts through many editions. Claire practiced pediatric and obstetric nursing before turning to scientific illustration as a full-time career. Texts illustrated by Bill and Claire have received national recognition and won awards from the Association of Medical Illustrators, American Institute of Graphic Arts, Chicago Book Clinic, Printing Industries of America, and Bookbuilders West. Bill and Claire also are recipients of the Art Directors Award.

Supplements

Instructor's Manual

Each chapter of the Instructor's Manual provides a detailed chapter outline, lecture enrichment suggestions, a commentary, and critical thinking questions. This material should be particularly helpful for first-time users of the text, although experienced teachers also may find much of value. The Instructor's Manual is available on this text's Online Learning Center through Connect.

Acknowledgments

The authors express their gratitude to the able and conscientious staff of McGraw Hill Education, who brought this book to its present form. We extend special thanks to Portfolio Director Michelle Vogler, Product Developer Erin DeHeck, Content Project Manager Susan Trentacosti, and Marketing Manager Kelly Brown. All played essential roles in shaping the ninth edition.

1

Science of Zoology and Evolution of Animal Diversity

LEARNING OBJECTIVES

Readers will be able to:

1.1 Explain that science consists in testing, possibly rejecting, and improving our simplest and best explanations using data, not in proving the correctness of a conjecture.

1.2 Describe the historical context of and major influences on Darwin's evolutionary theory.

1.3 Describe the five major conjectures of Darwin's evolutionary theory: perpetual change, common descent, multiplication of species, gradualism, and natural selection.

1.4 Explain how Darwin's theories of perpetual change, common descent, and multiplication of species are supported by all relevant data, and why continuing controversies about the roles of gradualism and natural selection do not challenge these first three theories.

1.5 Describe neo-Darwinism as a change to the hereditary component of Darwin's theory.

1.6 Explain allele frequency as the population-genetic basis for measuring evolutionary change, and the major forces of evolution at the population level.

1.7 Explain how paleontological studies of speciation and extinction have expanded evolutionary theory to include species selection and catastrophic species selection.

Evolutionary diversification of Hawaiian honeycreepers.

A Legacy of Change

Life's history is a legacy of perpetual change. Despite the apparent permanence of the natural world, change characterizes all things on earth and in the universe. Countless kinds of animals and plants have flourished and disappeared, leaving behind an imperfect fossil record of their existence. Many, but not all, have left living descendants that resemble them to varying degrees.

We observe and measure life's changes in many ways. On a short evolutionary timescale, we see changes in the frequencies of different genetic traits within populations. For example, evolutionary changes in the relative frequencies of light- and dark-colored moths occurred within a single human lifetime in polluted areas of industrial England. On the other hand, formation of new species and dramatic changes in organismal appearance, as shown by evolutionary diversification of Hawaiian birds, require longer timescales covering 100,000 to 1 million years. Major evolutionary trends and episodic mass extinctions occur on even larger timescales, covering tens of millions of years. The fossil record of horses through the past 50 million years shows a series of different species replacing older ones. The fossil record of marine invertebrates shows episodic mass extinctions separated by intervals of approximately 26 million years.

Organic evolution is the irreversible, historical change that we observe in living populations and in the earth's fossil record. Because every feature of life is a product of evolutionary processes, biologists consider organic evolution the keystone of all biological knowledge.

Zoology (Gr. *zōon*, animal, + *logos*, discourse on, study of) is the scientific study of animals. It is part of biology (Gr. *bios*, life, + *logos*), the study of all life. Explaining the panorama of animal diversity–how animals function, live, reproduce, and interact–is exciting and challenging.

To explain the diversity of animal life, we must study its long history, whose fossil evidence spans more than 540 million years. From the earliest animals to the millions of animal species living today, this history demonstrates extensive and ongoing change, which we call **evolution**. We depict the history of animal life as a branching genealogical tree, called a **phylogeny** or **phylogenetic tree**. We place the earliest species ancestral to all animals at the trunk; all living animal species fall at the growing tips of the branches. Each successive branching event represents the historical splitting of an ancestral species to form new ones. Newly formed species inherit many characteristics from their immediate ancestor, but they also evolve new features that appear for the first time in the history of animal life. Each branch therefore has its own unique combination of characteristics and contributes a new dimension to the spectrum of animal diversity.

The scientific study of animal diversity has two major goals. The first is to reconstruct a phylogeny of animal life and to find where in evolutionary history we can locate the origins of major characteristics–multicellularity, a **coelom, spiral cleavage,** vertebrae, **homeothermy**–and all other dimensions of animal diversity as we know it. The second major goal is to understand historical processes that generate and maintain diverse species and adaptations throughout evolutionary history. Darwin's theory of evolution allows us to apply scientific principles to attain both goals.

1.1 Principles of Science

A basic understanding of zoology requires understanding what science is, what it excludes, and how one gains knowledge using the scientific method. In this section we examine the methodology that zoology shares with science as a whole. These features distinguish the sciences from other disciplines, such as art and religion.

Despite the enormous impact of science on our lives, many people have only a minimal understanding of science. Public misunderstanding of scientific principles as applied to animal diversity revealed itself to us on March 19, 1981, when the governor of Arkansas signed into law the Balanced Treatment for Creation-Science and Evolution-Science Act (Act 590 of 1981). This act falsely presented creation-science as a valid scientific endeavor. Further legal scrutiny revealed that creation-science was not science, but rather a religious position advocated by a minority of America's religious community.

Enactment of this law incited a historic lawsuit tried in December 1981 in the court of Judge William R. Overton, U.S. District Court, Eastern District of Arkansas. The American Civil Liberties Union filed the suit on behalf of 23 plaintiffs, including religious leaders and groups representing several denominations, individual parents, and educational associations. Plaintiffs contended that this law violated the First Amendment to the U.S. Constitution, which prohibits establishment of religion by government. This amendment prohibits passing a law that would favor one religious position over another one. On January 5, 1982, Judge Overton permanently prohibited Arkansas from enforcing Act 590.

Considerable testimony during the trial clarified the nature of science. On the basis of testimony by scientists, Judge Overton stated explicitly these essential characteristics of science:

1. It is guided by natural law.
2. It must be explanatory by reference to natural law.
3. Its conjectures are testable against the empirical world.
4. Its conclusions are tentative and not necessarily the final word.
5. It is falsifiable.

Pursuit of scientific knowledge is guided by physical and chemical laws that govern the state of existence. Scientific knowledge must explain observations by reference to natural law without intervention of any supernatural being or force. We must record observations that directly or indirectly test hypotheses about nature. We must discard or modify any conclusion if further observations contradict it. As Judge Overton stated, "While anybody is free to approach a scientific inquiry in any fashion they choose, they cannot properly describe the methodology used as scientific, if they start with a conclusion and refuse to change it regardless of the evidence developed during the course of the investigation." Science lies outside religion, and scientific knowledge does not favor one religious position over another.

Unfortunately, the religious position formerly called creation-science later reappeared in American politics with the name "intelligent design theory." We once again defended science education against this scientifically meaningless doctrine. On December 20, 2005, Judge John E. Jones III of the U.S. District Court for the Middle District of Pennsylvania ruled unconstitutional the teaching of intelligent design, which had been mandated by the Dover school board. The local voters already had rejected the eight board members who supported the intelligent-design requirement, replacing them with candidates who actively opposed teaching intelligent design as science.

Scientific Method

The essential criteria of science form the **hypothetico-deductive method**. One begins this process by generating **hypotheses**, or potential explanations of a phenomenon of nature. These hypotheses are usually based on prior observations of nature (figure 1.1) or on theories derived from such observations. Scientific hypotheses often constitute general statements that might explain a large number of diverse observations. The hypothesis of natural selection, for example, explains our observations that many different species have accumulated favorable characteristics that adapt them to their environments. Based on a hypothesis, a scientist must say, "If my hypothesis

A

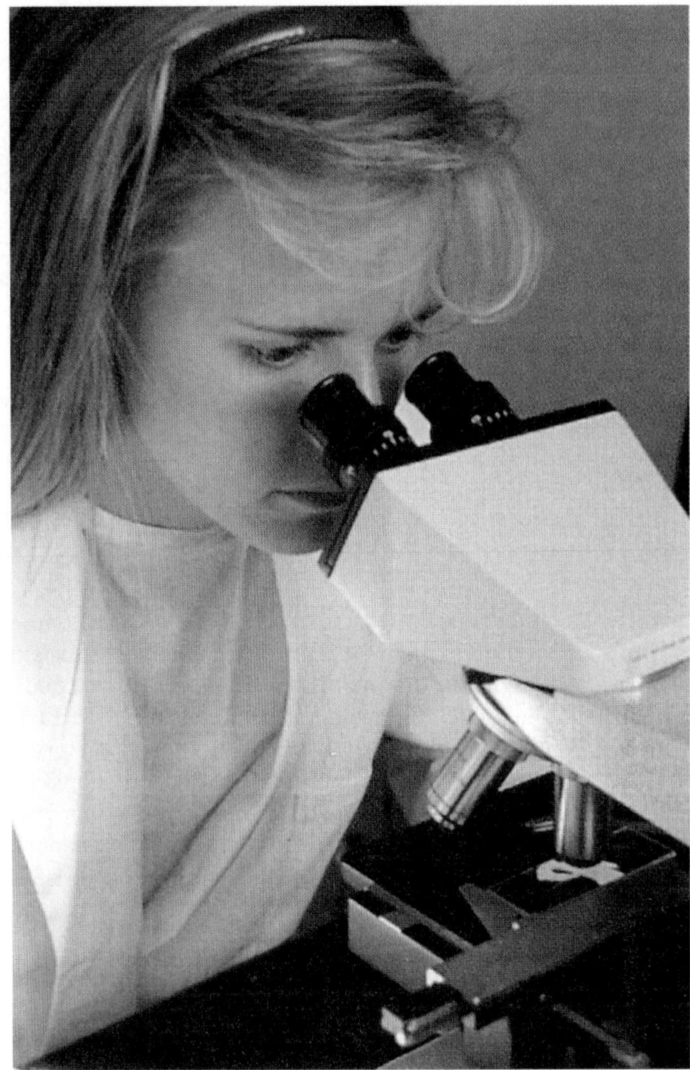

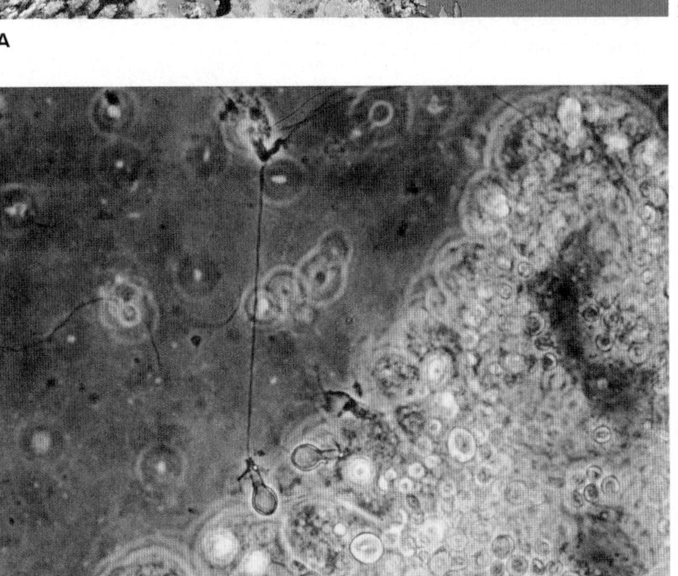

C

B

figure 1.1

Examples of observation in zoological research. **A,** Observing a coral reef. **B,** Observing nematocyst discharge, **C,** from cnidarian tentacles (see section 7.1).

correctly explains past observations, then future observations must match specific expectations."

The scientific method comprises six steps:

1. Observation
2. Question
3. Hypothesis
4. Empirical test
5. Conclusions
6. Publication

Observations are a critical first step in evaluating the biological characteristics and evolutionary histories of animal populations. For example, observations of moth populations in industrial areas of England for more than a century have revealed that moths in polluted areas mostly have darkly colored wings and body, whereas moths of the same species in unpolluted areas are more lightly colored. This observation pertains to multiple moth species, but we focus here on *Biston betularia* (figure 1.2).

Our question is, Why do pigmentation patterns vary according to habitat? With no prior knowledge of the biology of these moth populations, one might hypothesize that coloration is influenced somehow by a direct action of the environment. Does consumption of soot by caterpillars somehow darken pigmentation of the adult moth? One could test this hypothesis by rearing moths under artificial conditions. If darkly pigmented moths and lightly pigmented moths are allowed to reproduce in unpolluted conditions, our hypothesis predicts that offspring of both will be lightly pigmented; by contrast, offspring of both groups would be darkly pigmented if raised in polluted conditions.

To test our hypothesis, we construct a null hypothesis. A null hypothesis is one that permits a statistical test of our data

figure 1.2

Light and melanic forms of peppered moths, *Biston betularia*, on **A,** an unpolluted lichen-covered tree and **B,** a soot-covered tree near industrial Birmingham, England. These color variants have a simple genetic basis. **C,** Recent decline in frequency of the melanic form due to diminished air pollution in industrial areas of England. Frequency of the melanic form exceeded 90% in 1960, when smoke and sulfur dioxide emissions were still high. Later, as emissions fell and light-colored lichens began to grow again on tree trunks, the melanic form became more conspicuous to predators. By 1986, only 50% of the moths were melanic, the rest having been replaced by the light form.

to reject its predictions if the hypothesis is false. We can choose as our null hypothesis the prediction that population of origin has no effect on moth color: moths reared in unpolluted conditions should be lightly pigmented regardless of whether their parents were from light or dark populations, and offspring from both populations reared in polluted conditions should be dark. This experiment is a special case of a "common garden" experiment, as used in agriculture. Do contrasting populations from different habitats retain their contrasting characteristics when reared in a common garden?

For *Biston betularia,* a common garden experiment reveals that the contrasting wing colors of populations from polluted and unpolluted environments are maintained in the common garden. Offspring of moths from polluted populations retain the dark pigmentation of their parents, whereas offspring of lightly pigmented moths are lightly colored like their parents. We thereby reject the hypothesis that the color contrasts represent a direct action of environmental conditions.

We have gained important knowledge by rejecting our initial hypothesis, and we now test an alternative hypothesis, that pigmentation is a genetic trait in *Biston betularia*. Using standard genetic methodology, we cross the darkly and lightly colored populations and trace the inheritance of pigmentation in subsequent generations. Experimental results reveal that the offspring produced by crossing light and dark populations have dark pigmentation, and that the second-generation progeny include both dark and light moths in the 3:1 ratio predicted by the null hypothesis for a single-gene trait with dark pigmentation being genetically dominant.

We still have not answered our initial question—why pigmentation differs between populations in polluted versus unpolluted environments. We have learned, however, that the critical question is why different forms of a single gene have contrasting frequencies in these two areas. We know that the moth populations have inhabited England since well before the introduction of industrial pollution. The lightly pigmented populations most likely resemble the ancestral condition, so why have darkly pigmented moths accumulated in the polluted environments? The simplest hypothesis is that darkly pigmented individuals are more likely to survive and to reproduce in polluted environments.

Further observations reveal that *Biston betularia* is typical of moths in being active at night and inactive during the day, resting on the bark of trees. Contrasting photographs of light and dark moths resting on unpolluted, lichen-covered tree bark versus sooty tree bark lead us to a hypothesis that might explain why dark moths predominate in polluted areas. Figure 1.2 shows that the lightly colored moth is camouflaged against the unpolluted substrate, whereas the dark moth is highly visible; by contrast, the dark moth is camouflaged against the sooty bark, whereas the light moth is highly visible. Camouflage suggests that a predator using vision to find its prey preferentially kills moths that contrast with the background color of their diurnal resting place. How can we test this hypothesis?

Many birds are diurnal predators guided to their prey by vision. Many experiments have revealed that birds will attack clay models that closely resemble their favored prey items. We can test our hypothesis by constructing clay models of light and dark moths. We place equal numbers of the light and dark models against the bark of unpolluted trees and equal numbers of light and dark models against sooty tree bark. When a bird attacks a clay model, it typically leaves an imprint of its beak in the clay. Because beak shape varies among bird species, the beak shape marked in the clay often reveals which species attacked the model. Our null hypothesis is that equal numbers of dark and light models have beak impressions on both the unpolluted and the polluted substrates. We reject this hypothesis if we find a large excess of beak marks in the uncamouflaged models relative to the camouflaged ones; dark models should be attacked preferentially in unpolluted conditions and light models attacked preferentially in polluted conditions. Note that in this case, we use a null hypothesis that is the *opposite* of our favored explanation, that birds preferentially destroy uncamouflaged moths. In this case, data that reject the null hypothesis serve to verify our favored explanation.

Experiments of this kind have rejected the null hypothesis as expected, verifying our explanation that dark moths prevail in polluted environments because the dark color protects them from predation by birds during the day. Note that our experiments led us to a strong, specific explanation for the initial observations. It is a strong working hypothesis, but our experiments have not proven the correctness of this hypothesis. We can test it further in various ways. For example, we might raise light and dark moths in equal numbers in an outdoor enclosure that excludes birds; our null hypothesis is then that the dark and light forms should persist in equal numbers regardless of whether the tree bark is polluted or unpolluted. Rejection of this null hypothesis would tell us that our favored explanation was not the full answer to our original question.

We publish our results and conclusions to guide other researchers further to test our hypotheses. Over the past century, many research papers have reported results and conclusions to explain "industrial melanism" in moths. With some ambiguities, the favored explanation is that differential bird predation on uncamouflaged moths best explains industrial melanism. These studies have drawn much attention because this explanation illustrates Darwin's theory of natural selection (see section 1.3).

Experimental Versus Comparative Methods

One can group the many questions raised about animal life into two major categories. The first category seeks to explain **proximate causes** (also called immediate causes) that guide biological systems at all levels of complexity. It includes explaining how animals perform their metabolic, physiological, and behavioral functions at molecular, cellular, organismal, and even population levels. For example, how is genetic information expressed to guide the synthesis of proteins? What signal causes cells to divide to produce new cells? How does population density affect the physiology and behavior of organisms?

We test hypotheses of proximate causes using the **experimental method**. This method has three steps: (1) predicting from a tentative explanation how a system being studied would respond to a treatment, (2) making the treatment, and (3) comparing observed results to predicted ones. An investigator repeats the experiment multiple times to eliminate chance occurrences that might produce errors. **Controls** (repetitions of an experimental procedure that lack the treatment) eliminate any unperceived conditions that might bias an experiment's outcome.

Our example in the preceding section of using clay models of moths to test avian predation on differently colored forms illustrates experimental testing of a hypothesis. By placing darkly colored models on both light and dark backgrounds, we see that birds attack the ones on light backgrounds much more frequently than they do dark models on dark backgrounds. Our interpretation that dark moths on dark backgrounds avoid predation by camouflage requires a control. Perhaps birds choose to feed only on light, unpolluted branches. Our control is to place light moths on both light and dark backgrounds. When we observe that birds preferentially attack the light models placed on dark backgrounds, we reject the hypothesis that birds choose not to feed on dark, polluted substrates. The simplest interpretation of the results as described here is that birds will eat both dark and light moths that fail to match their backgrounds, and that camouflage conceals potential prey items from avian predators.

Processes by which animals maintain their body temperature under different environmental conditions, digest food, migrate to new habitats, or store energy are additional examples of phenomena studied by experimentation. Experimental sciences in biology include molecular biology, cell biology, endocrinology, immunology, physiology, developmental biology, and community ecology.

In contrast to proximate causes, ultimate causes are the processes that have produced biological systems and their properties through evolutionary time. For example, what evolutionary factors cause some birds to make complex seasonal migrations between temperate and tropical regions? Why do different species of animals have different numbers of chromosomes in their cells? Why do some animal species maintain complex social systems, whereas individuals of other species remain largely solitary?

Key Theme 1.1

SCIENCE EXPLAINED

Ultimate Causes and Teleology

A scientist's use of the phrase "ultimate cause," unlike Aristotle's usage, does not imply a preconceived goal for natural phenomena. An argument that nature has a predetermined goal, such as evolution of the human mind, is termed teleological. **Teleology** is the mistaken notion that the evolution of living organisms is guided by purpose toward an optimal design. A major success of Darwinian evolutionary theory is its rejection of teleology in explaining biological diversification.

Tests of hypotheses of ultimate causality require the **comparative method**. Characteristics of molecular biology, cell biology, organismal structure, development, and ecology are compared among species to identify patterns of variation. Scientists then use patterns of similarity and dissimilarity to test hypotheses of relatedness and thereby to reconstruct the phylogenetic tree that relates the species being compared. Systematics is the ordering of organisms according to their inferred evolutionary relationships for comparative study. Recent advances in DNA sequencing technology permit precise tests of evolutionary relationships among all animal species. Comparative studies also serve to test hypotheses of evolutionary processes that have molded diverse animal species.

We use the evolutionary tree to examine hypotheses of the evolutionary origins of the diverse molecular, cellular, organismal, and populational characteristics observed in the

The Power of a Theory

Darwin's theory of common descent (see section 1.3) illustrates the scientific importance of general theories that give unified explanations to diverse kinds of data. Darwin proposed his theory of descent with modification of all living forms because it explained the patterns of similarity and dissimilarity among organisms in anatomical structures and cellular organization.

Anatomical similarities between humans and apes led Darwin to propose that humans and apes share more recent common ancestry with each other than they do with any other species. Darwin was unaware that his theory, a century later, would provide the primary explanation for similarities and dissimilarities among species in the structures of their chromosomes, sequences of amino acids in homologous proteins, and sequences of bases in homologous genomic DNA.

The accompanying figure shows photographs of a complete haploid set of chromosomes from four selected ape species: human (*Homo sapiens*), bonobo (the pygmy chimpanzee, *Pan paniscus*), western gorilla (*Gorilla gorilla*), and Sumatran orangutan (*Pongo abelii*). Each chromosome in the human genome has a corresponding chromosome with similar structure and gene content in the genomes of other ape species. The most obvious difference between human and ape chromosomes is that the large second chromosome in the human nuclear genome was formed evolutionarily by a fusion of two smaller chromosomes characteristic of the ape genomes. Detailed study of the human and other ape chromosomes shows remarkable correspondence between them in genic content and organization. Ape chromosomes are more similar to each other than they are to chromosomes of any other animals.

Comparison of DNA and protein sequences among apes likewise confirms their close genetic relationships, with humans and the two chimpanzee species being closer to each other than any of these species are to other apes. DNA sequences from the nuclear and mitochondrial genomes independently support the close relationships among ape species and especially the grouping of humans and chimpanzees as close relatives. Homologous DNA sequences of humans and chimpanzees are approximately 99% similar in base sequence.

Studies of variation in chromosomal structure, mitochondrial DNA sequences, and nuclear DNA sequences produced multiple independent data sets, each one potentially capable of rejecting Darwin's theory of common descent. Darwin's theory would be rejected, for example, if the chromosomal structures and DNA sequences of apes were no more similar to each other than to those of other animals. The data in this case support rather than reject predictions of Darwin's theory. The ability of Darwin's theory of common descent to make precise predictions of genetic similarities among these and other species, and to have those predictions confirmed by numerous empirical studies, illustrates its great strength. As new kinds of biological data have become available, the scope and strength of Darwin's theory of common descent have increased enormously. Indeed, nothing in biology makes sense in the absence of this powerful explanatory theory.

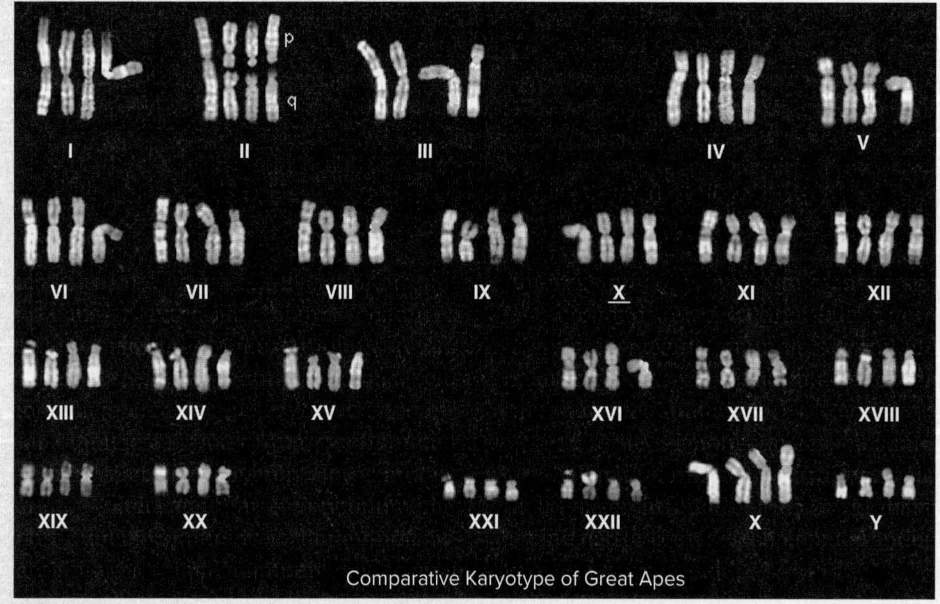

Dr. Mariano Rocchi

The human haploid genome contains 22 autosomes (I–XXII) and a sex chromosome (X or Y). The human chromosome is shown first in each group of four, followed by the corresponding chromosomes of bonobo, gorilla, and orangutan, in that order. Note that the chromatin of human chromosome II corresponds to that of two smaller chromosomes (marked p and q) in other apes.

animal world. For example, comparative methodology rejects the hypothesis of a common origin for flight in bats and birds. Comparative morphology of vertebrates and comparisons of DNA sequences from living species clearly place bats within the mammals (Chapter 20) and birds within a separate group that also includes crocodilians, lizards, snakes, and turtles (see figure 18.2). The most recent common ancestor of these vertebrates clearly could not fly, and close inspection reveals that bats and birds evolved flight via very different modifications of their bodies and forelimbs. The ultimate causes of flight in bats and birds thus require separate explanations, not a shared one. The comparative method likewise reveals that **homeothermy** evolved in a lineage ancestral to birds and separately in a lineage ancestral to mammals. Furthermore, comparative studies of fossil birds reject the hypothesis that feathers arose for the purpose of flight, because feathers preceded evolution of the flight apparatus in avian ancestry. Initially, feathers most likely served primarily for insulation and only later acquired a role in aerodynamics. It should be clear that none of these important historical questions could have been answered by experiment.

Clearly, the comparative method often relies on results of experimental sciences to reveal the characteristics being compared among animals. The comparative method utilizes all levels of biological complexity, as illustrated by the fields of comparative biochemistry, molecular evolution, comparative cell biology, comparative anatomy, comparative physiology and behavior, and phylogenetic systematics.

Source: U.S. National Library of Medicine

A

New York Public Library/Science Source

B

figure 1.3

Founders of the theory of evolution by natural selection. **A,** Charles Robert Darwin (1809–1882). **B,** Alfred Russel Wallace (1823–1913) in 1895. Darwin and Wallace independently developed the same theory. A letter and essay from Wallace written to Darwin in 1858 spurred Darwin into writing *On the Origin of Species*, published in 1859.

1.2 Origins of Darwinian Evolutionary Theory

Charles Robert Darwin and Alfred Russel Wallace (figure 1.3) were the first to establish evolution as a powerful scientific theory. Today, evolution can be denied only by abandoning reason. As the English biologist Sir Julian Huxley wrote, "Charles Darwin effected the greatest of all revolutions in human thought, greater than Einstein's or Freud's or even Newton's, by simultaneously establishing the fact and discovering the mechanism of organic evolution." Darwinian theory allows us to explain both the genetics of populations and long-term trends in the fossil record. Darwin and Wallace did not originate the basic idea of organic evolution, which has an ancient history. We review first the history of evolutionary thinking as it led to Darwin's theory and then discuss critical evidence supporting Darwin's theory.

Pre-Darwinian Evolutionary Ideas

Early Greek philosophers, notably Xenophanes, Empedocles, and Aristotle, recorded the idea that life has a long history of evolutionary change. They recognized fossils as evidence for former life, which they thought had been destroyed by natural catastrophe. Despite their inquiry, ancient Greeks failed to establish an evolutionary concept that could guide a meaningful study of life's history. Evolutionary thinking declined as the metaphorical biblical account of earth's creation became accepted as requiring no mechanistic explanation. The year 4004 B.C. was fixed by Archbishop James Ussher (mid-seventeenth century) as the time of life's creation. Evolutionary views were considered heretical, but they refused to die. The French naturalist Georges Louis Buffon (1707–1788) stressed environmental influences on modifications of animal form and extended the earth's age to 70,000 years.

Lamarckism: The First Scientific Hypothesis for Evolution

The first complete hypothesis for evolution was authored by the French biologist Jean Baptiste de Lamarck (1744–1829) (figure 1.4) in 1809, the year of Darwin's birth. He made the first convincing argument that fossils were remains of extinct animals. Lamarck's evolutionary mechanism, **inheritance of acquired characteristics**, tentatively answered the challenging question of how evolution could construct biological characteristics that seemed designed to serve their possessors' needs: By striving to make best use of their environmental resources, organisms would acquire adaptations and pass them by heredity to their offspring. According to Lamarck, giraffes evolved a long neck because their ancestors lengthened their necks by stretching to obtain food and then passed the lengthened neck to their offspring. Lamarck proposed that over many

New York Public Library/Science Source

figure 1.4

Jean Baptiste de Lamarck (1744–1829), French naturalist who offered the first scientific explanation of evolution. Lamarck's hypothesis that evolution proceeds by inheritance of acquired characteristics was rejected by genetic research.

Hulton Archive/Getty Images

figure 1.5

Sir Charles Lyell (1797–1875), English geologist and friend of Darwin. His book *Principles of Geology* greatly influenced Darwin during Darwin's formative period.

generations, these changes accumulated to produce the long necks of modern giraffes.

We call Lamarck's concept of evolution *transformational,* because as individual organisms transform their characteristics through the use and disuse of body parts, heredity makes corresponding adjustments to produce evolution. We now reject transformational theories because genetic studies show that traits acquired during an organism's lifetime, such as strengthened muscles, are not transmitted to offspring.

Darwin's evolutionary theory differs from Lamarck's in being a *variational* theory based in genetic differences that occur among organisms within a population. Evolution occurs at the level of the **population,** and it includes changes across generations in the organismal characteristics that prevail in the population. Darwin argued that organisms whose hereditary characteristics conferred an advantage for survival or reproduction would contribute the greatest numbers of offspring to future generations. Populations would thus accumulate, across generations, the characteristics most favorable for the organisms possessing them. Any less favorable alternative characteristics would decline in frequency and eventually disappear.

Charles Lyell and Uniformitarianism

The geologist Sir Charles Lyell (1797–1875) (figure 1.5) established in his *Principles of Geology* (1830–1833) the principle of **uniformitarianism**. Uniformitarianism encompasses two important assumptions that guide scientific study of the history of nature. These assumptions are (1) that the laws of physics and chemistry have not changed throughout earth's history, and (2) that past geological events occurred by natural processes similar to those that we observe in action today. Lyell showed that natural forces, acting over long periods of time, could explain the formation of fossil-bearing rocks. For example, as skeletal remains of corals (see figure 7.25), foraminiferans (see figure 5.24), and molluscs (see figure 10.1) accumulate on the sea floor, they form sediments of calcium carbonate that eventually become compressed into limestone. Lyell's geological studies convinced him that earth's age must be measured in hundreds of millions of years. Measured rates of sedimentation are much too slow to have produced earth's sedimentary rock formations in a shorter period of time. These principles were important because they discredited miraculous and supernatural explanations of the history of nature and replaced them with scientific explanations. Lyell also stressed the gradual nature of geological changes that occur through time, and he argued that such changes have no inherent directionality. Both of these claims left important marks on Darwin's evolutionary theory.

Darwin's Great Voyage of Discovery

"After having been twice driven back by heavy southwestern gales, Her Majesty's ship *Beagle,* a ten-gun brig, under the command of Captain Robert FitzRoy, R.N., sailed from Devonport on the 27th of December, 1831." Thus began Charles Darwin's account of the historic five-year voyage of the *Beagle* around the world (figure 1.6). Darwin, not quite 23 years old, had asked to accompany Captain FitzRoy on the *Beagle,* a small vessel only 90 feet in length, which was about to make an extensive surveying voyage to South America and the Pacific (figure 1.7). It was the beginning of the most important scientific voyage of the nineteenth century.

During this voyage (1831–1836), Darwin endured seasickness and erratic companionship from Captain FitzRoy, but his endurance and early training as a naturalist equipped him for his work. The *Beagle* made many stops along the coasts of South America and adjacent islands. Darwin made extensive collections and observations of the faunas and floras of these regions. He unearthed numerous fossils of animals long extinct and noted a resemblance between fossils of South American pampas and known fossils of North America. In the Andes, he encountered seashells embedded in rocks at 13,000 feet. He experienced a severe earthquake and watched mountain torrents that relentlessly wore away the earth. These observations and his reading of Lyell's *Principles of Geology* during the voyage strengthened Darwin's conviction that natural forces could explain geological features of the earth.

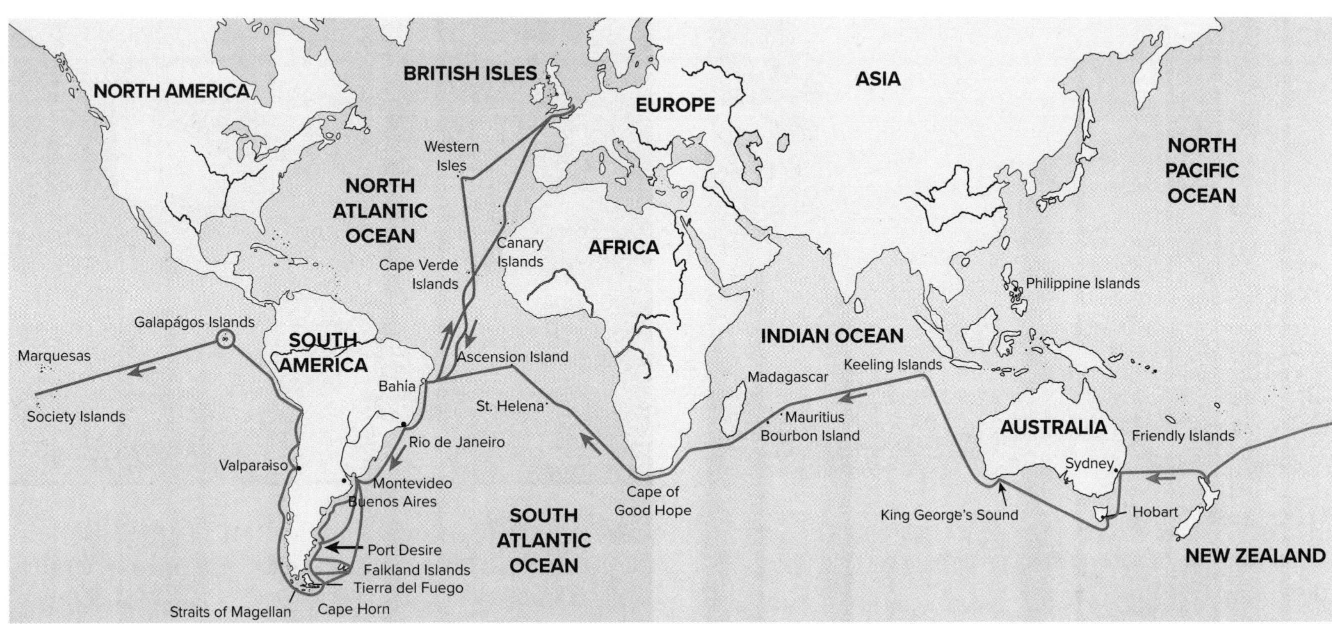

figure 1.6

Five-year voyage of H.M.S. *Beagle*.

figure 1.7

Charles Darwin and H.M.S. *Beagle*. **A,** Darwin in 1840, four years after the *Beagle* returned to England, and a year after his marriage to his cousin, Emma Wedgwood. **B,** H.M.S. *Beagle* sails in Beagle Channel, Tierra del Fuego, on the southern tip of South America in 1833. This watercolor was painted by Conrad Martens, one of two official artists during the voyage of the *Beagle*.

In mid-September of 1835, the *Beagle* arrived at the Galápagos Islands, a volcanic archipelago straddling the equator 600 miles west of Ecuador (figure 1.8). The fame of these islands stems from their oceanic isolation and rugged volcanic terrain. Circled by capricious currents, surrounded by shores of twisted lava, bearing skeletal brushwood baked by equatorial sunshine, almost devoid of vegetation, inhabited by strange reptiles and by convicts stranded by the Ecuadorian government, these islands had few admirers among mariners. By the middle of the seventeenth century, Spaniards called these islands "Las Islas Galápagos"—the tortoise islands. The giant tortoises, used for food first by buccaneers and later by American and British whalers, sealers, and ships of war, were the islands' principal attraction. At the time of Darwin's visit, these tortoises already were heavily exploited.

Cleveland P. Hickman, Jr.

figure 1.8

The Galápagos Islands viewed from the rim of a volcano.

Cleveland P. Hickman, Jr.

figure 1.9

Darwin's study at Down House in Kent, England, is preserved today much as it was when Darwin wrote *On the Origin of Species.*

During the *Beagle*'s five-week visit to the Galápagos, Darwin documented the unique character of the Galápagos plants and animals, including the giant tortoises, marine iguanas, mockingbirds, and ground finches. Darwin later described these studies as the "origin of all my views."

Darwin discovered that although the Galápagos Islands and Cape Verde Islands (visited earlier in this voyage) were similar in climate and topography, Galápagos plants and animals resembled most closely those of the South American mainland, and they were entirely different from the African-derived forms of the Cape Verde Islands. Each Galápagos island often contained a unique species that nonetheless resembled forms on other Galápagos islands. In short, Galápagos life must have originated in continental South America, colonized islands in rare events of trans-oceanic dispersal, and then undergone modification in various environmental conditions of different islands. He concluded that these species were neither divinely created nor immutable; they were products of a long history of evolutionary change.

Key Theme 1.2

EVOLUTION

Darwin's Thoughts on His Work

"Whenever I have found that I have blundered, or that my work has been imperfect, and when I have been contemptuously criticized, and even when I have been overpraised, so that I have felt mortified, it has been my greatest comfort to say hundreds of times to myself that 'I have worked as hard and as well as I could, and no man can do more than this.' "

—Charles Darwin, in his autobiography, 1876.

On October 2, 1836, the *Beagle* returned to England, where Darwin conducted most of his scientific work (figure 1.9). Most of Darwin's extensive collections had preceded him there, as had his notebooks and diaries kept during the cruise. Darwin's journal, published three years after the *Beagle*'s return to England, was an instant success and required two additional printings within its first year. *The Voyage of the Beagle* would become one of the most popular travel books of all time.

The main product of Darwin's voyage, his theory of evolution, would continue to develop for more than 20 years after the *Beagle*'s return. In 1838, he "happened to read for amusement" an essay on populations by T. R. Malthus (1766–1834), who stated that animal and plant populations, including human populations, have the reproductive capacity to increase beyond the capacity of their environment to support them. Darwin already had been gathering information on artificial selection of animals under domestication. He was especially fascinated by artificial breeds of pigeons. Many pigeon breeds differed so much in appearance and behavior that they would be considered different species if found in nature. All clearly had been derived from a single wild species, the rock dove (*Columbia livia*). After reading Malthus's article, Darwin realized that a process of selection in nature, driven by a "struggle for existence" because of overpopulation, could be a powerful force for evolution of wild species.

Darwin allowed the idea to develop in his own mind, writing private, trial essays in 1844 and 1846. In 1856, he began to assemble his voluminous data into a work on origins of species. He expected to write four volumes, a very big book, "as perfect as I can make it." However, his plans took an unexpected turn.

In 1858, he received a manuscript from Alfred Russel Wallace (1823–1913), an English naturalist in Malaya with whom he corresponded. Darwin was stunned to find that in a few pages, Wallace summarized the main points of the natural selection theory on which Darwin had worked for two decades. Rather than withhold his own work in favor of Wallace as he was inclined to do, Darwin took the advice of two close friends, Lyell and a botanist, Hooker, to publish his views in a brief statement that would appear together with Wallace's paper in the *Journal of the Linnean Society.* Portions of both papers were read before an unresponsive audience on July 1, 1858.

For the next year, Darwin worked urgently to prepare an "abstract" of the planned four-volume work. This book was published in November 1859, with the title *On the Origin of Species by Means of Natural Selection, or the Preservation of Favoured Races in the Struggle for Life.* The 1250 copies of the first printing sold out the first day! The book instantly generated a storm that has never abated. Darwin's views were to have extraordinary consequences on scientific and religious beliefs, and they remain among the greatest intellectual achievements of all time.

Once Darwin's caution had been swept away by publication of *On the Origin of Species,* he entered an incredibly productive period of evolutionary thinking for the next 23 years, producing five revisions of *On the Origin of Species* and a dozen new books. He died on April 19, 1882, and was buried in Westminster Abbey. The *Beagle* had already disappeared, having been retired in 1870 and sold for scrap.

1.3 Darwin's Theory of Evolution

Darwin's theory of evolution is now over 160 years old. Biologists frequently are asked, "What is Darwinism?" and "Do biologists still accept Darwin's theory of evolution?" These questions do not have simple answers because Darwinism encompasses several different, although mutually connected, theories. Professor Ernst Mayr of Harvard University argued that Darwinism should be viewed as five major theories. These five theories have somewhat different origins and fates and cannot be treated as only a single hypothesis. The theories are (1) **perpetual change,** (2) **common descent,** (3) **multiplication of species,** (4) **gradualism,** and (5) **natural selection.** The first three theories are generally accepted as having universal application throughout the living world. Gradualism and natural selection remain somewhat controversial among evolutionists; they are clearly important evolutionary processes, but they might not explain as much of animal evolution as Darwin thought. Creationists often misrepresent legitimate controversies regarding gradualism and natural selection as challenges to the first three theories, whose validity is strongly supported by all relevant facts.

1. **Perpetual change.** This is the basic theory of evolution on which the others depend. It states that the living world has a long history of ongoing change, with hereditary continuity from past to present life. Organismal characteristics undergo modification across generations throughout time. This theory originated in antiquity but did not gain widespread acceptance until Darwin advocated it in the context of his other four theories. Perpetual change is documented by the fossil record, which clearly refutes any claims for a recent origin of all living forms. Because it has withstood repeated testing and is supported by an overwhelming number of observations, we now regard perpetual change as fact.
2. **Common descent.** The second Darwinian theory, common descent, states that all forms of life propagated from a common ancestor through a branching of lineages (figure 1.10). An opposing argument, that different forms of life arose independently and descended to the present in linear, unbranched genealogies, is refuted by comparative studies of

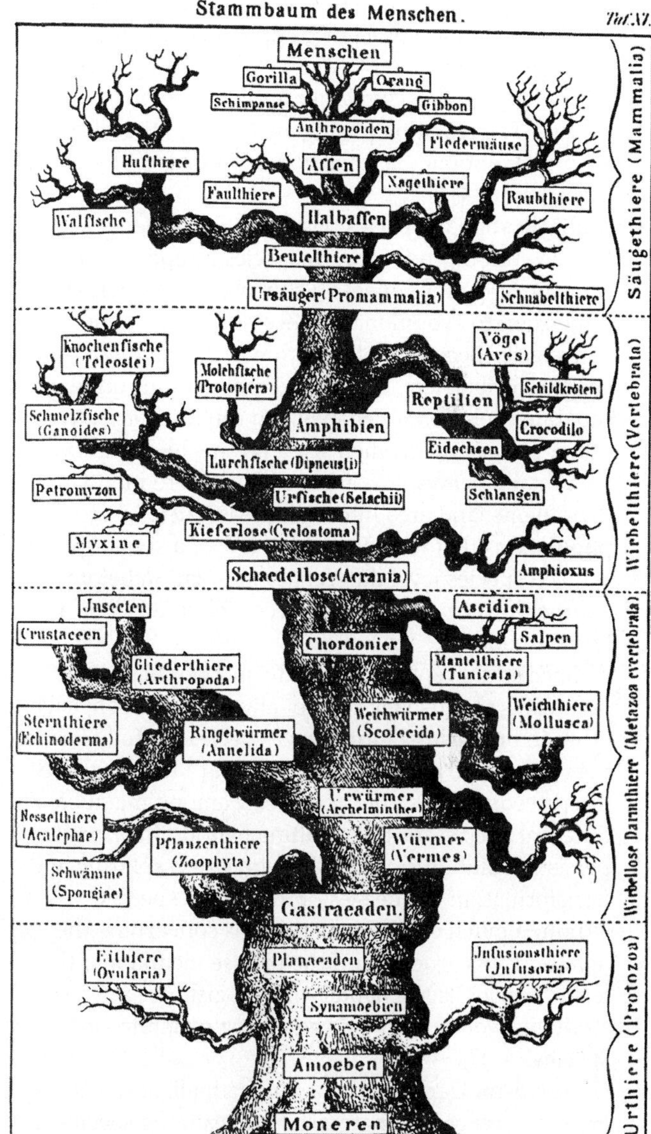

figure 1.10

An early tree of animal life drawn in 1874 by the German biologist Ernst Haeckel, who was strongly influenced by Darwin's theory of common descent. Some hypotheses, including the grouping of humans (*Menschen*) with anthropoid apes, have been verified by subsequent testing. Other hypotheses have been rejected in favor of contrasting hypotheses; for example, humans and chimpanzees are more closely related to each other than either is to gorillas, and orangs are more closely related to the combined chimpanzees, gorillas, and humans than they are to gibbons.

Source: Haeckel, Ernst, *The Evolution of Man,* New York, NY: D. Appleton, 1886.

organismal form, cellular structure, and macromolecular structures (including those of the genetic material, DNA). All of these studies confirm the theory that life's history has the structure of a branching evolutionary tree, called a phylogeny. Species that share relatively recent common ancestry (within the past several million years) have more similar features at all levels than do species whose most recent common ancestor occurred tens or hundreds of millions of years ago. Darwin's theory of common descent guides much ongoing research to reconstruct life's phylogeny using patterns of similarity and dissimilarity observed among species. The resulting phylogeny provides the basis for our taxonomic classification of animals (see Chapter 4).

3. **Multiplication of species.** Darwin's third theory states that evolution produces new species by splitting and transforming older ones. This theory adds a spatial dimension to evolutionary processes. When populations of a species become isolated from each other by geographic barriers, the isolated populations undergo separate evolutionary change and can diverge from each other. For example, when sea level was higher in the past than it is now, low areas of Cuba were inundated, dividing its land area into multiple isolates. Lizard populations that were formerly parts of a single species evolved species-level differences in isolation before another lowering of sea level reconsolidated Cuba as we know it today.

   Species are reproductively distinct populations of organisms that usually but not always differ from each other in organismal form. Once species are fully formed, they propagate as separate evolutionary lineages, and interbreeding does not occur freely among members of different species, or the resulting hybrid offspring do not persist. Evolutionists generally agree that splitting and transformation of lineages produce new species, although much controversy remains concerning the details of this process and the precise meaning of the term "species"(see Chapter 4). Biologists are actively studying evolutionary processes that generate new species.

4. **Gradualism.** Darwin's theory of gradualism states that large differences in anatomical traits among species originate by accumulation of many small incremental changes over very long periods of time. This theory opposes the notion that large anatomical differences arise by sudden genetic changes within a generation. This theory is important because genetic changes having very large effects on organismal form are usually harmful to an organism. It is possible, however, that some genetic changes of large effect are nonetheless sufficiently beneficial to be favored by natural selection. For example, the genetic mutation that produced dark pigmentation in *Biston betularia* (see figure 1.2) and thus permitted camouflage on polluted substrates was favored in polluted environments despite causing an abrupt change. Gradual evolution of industrial melanism would have involved accumulation of slightly darker forms over many generations to produce the melanic moth, and the genetic data contradict the gradual interpretation. Therefore, although we know that gradual evolution occurs, it does not necessarily explain the origins of all structural differences among species. Scientists are studying this question actively.

5. **Natural selection.** Natural selection explains why organisms are constructed to meet the demands of their environments, a phenomenon called **adaptation.** This theory describes a natural process by which populations accumulate favorable characteristics throughout long periods of evolutionary time. Adaptation formerly was considered strong evidence against evolution. Darwin's theory of natural selection was therefore important for convincing people that a natural process, amenable to scientific study, could produce new adaptations and new species. Demonstration that natural processes could produce adaptation was important to the eventual acceptance of all five Darwinian theories. Darwin developed his theory of natural selection as a series of five observations, and he made three inferences from them (see box Darwin's Explanatory Model of Evolution by Natural Selection):

Observation 1: Organisms have great potential fertility. All populations produce large numbers of gametes and potentially large numbers of offspring each generation. Population size would increase exponentially at an enormous rate if all fertilized eggs produced each generation successfully completed their development to form reproductively active adult individuals. Darwin calculated that, even for slow-breeding organisms such as elephants, a single pair breeding from age 30 to 90 and having only six offspring could produce 19 million descendants in 750 years.

Observation 2: Natural populations normally remain constant in size, except for minor fluctuations. Natural populations fluctuate in size across generations and sometimes go extinct, but no natural populations show the continued exponential growth that their reproductive capacity theoretically could sustain.

Observation 3: Natural resources are limited. Exponential growth of a natural population would require unlimited natural resources to provide food and habitat for an expanding population, but natural resources are finite.

Inference 1: A continuing *struggle for existence* occurs among members of a population. Survivors represent only a portion, often a very small portion, of all individuals produced each

Darwin's Explanatory Model of Evolution by Natural Selection

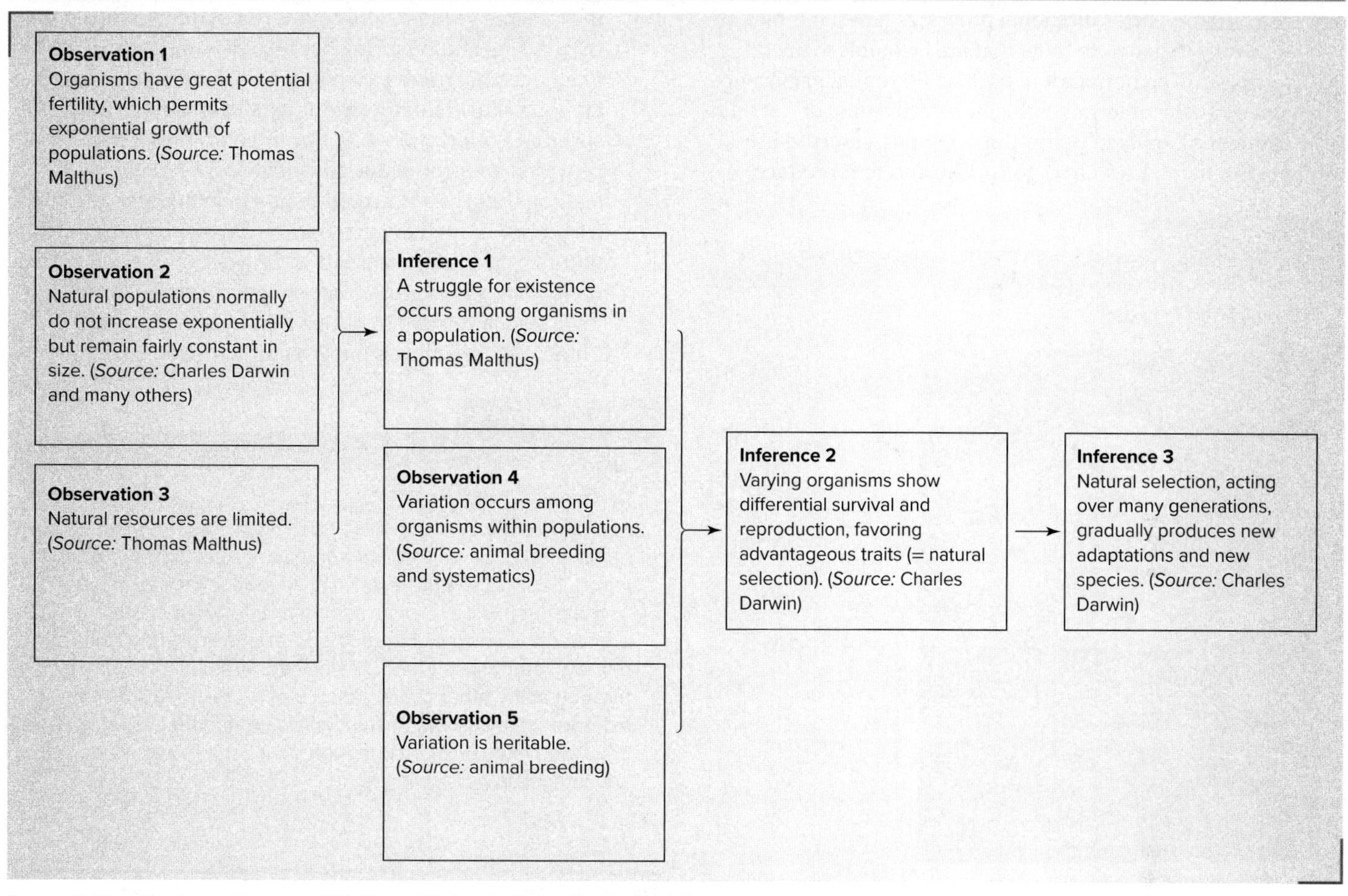

Source: *E. Mayr,* One Long Argument, *1991, Harvard University Press, Cambridge, MA.*

generation. Darwin wrote in *On the Origin of Species* that "it is the doctrine of Malthus applied with manifold force to the whole animal and vegetable kingdoms." Struggle for food, shelter, and space becomes increasingly severe as overpopulation develops.

Observation 4: All populations show organismal *variation*. No two individuals are exactly alike. They differ in size, color, physiology, behavior, and many other ways.

Observation 5: Variation is heritable. Darwin noted that offspring tend to resemble their parents, although he did not understand how. Many years later, the hereditary mechanism discovered by Gregor Mendel would be applied to Darwin's theory.

Inference 2: Varying organisms show *differential survival and reproduction* favoring advantageous traits (= natural selection). Survival in a struggle for existence is not random with respect to contrasting hereditary traits present in a population. Some traits give their possessors an advantage in using their environmental resources for effective survival and reproduction. Survivors transmit their favored traits to offspring, thereby causing those traits to accumulate in the population.

Inference 3: Over many generations, natural selection gradually produces new adaptations and new species. The preferential propagation of favorable traits across generations gradually transforms species and causes their long-term "improvement." Darwin knew that people often use hereditary variation to produce useful new breeds of livestock and plants. *Natural* selection acting over millions of years should be even more effective in producing new types than *artificial* selection imposed during a human lifetime. Natural selection acting independently on geographically separated populations would cause them to diverge from each other, thereby generating reproductive barriers that lead to speciation.

Natural selection can be considered a two-step process with a random component and a nonrandom component. Production of variation among organisms is the random part. Mutational processes have no inherent tendency to generate traits that are favorable to an organism. The nonrandom part is differential persistence among traits, determined by the effectiveness of contrasting traits in permitting their possessors to use environmental resources to survive and to reproduce.

Key Theme 1.3

EVOLUTION

Thomas Henry Huxley

Time Life Pictures/The LIFE Picture Collection/Getty Images

Thomas Henry Huxley (1825–1895), one of England's greatest zoologists, on first reading the convincing evidence of natural selection in Darwin's *On the Origin of Species*, is said to have exclaimed, "How extremely stupid not to have thought of that!" He became Darwin's foremost advocate and engaged in often bitter debates with Darwin's critics. Darwin, who disliked publicly defending his own work, was glad to leave such encounters to his "bulldog," as Huxley called himself.

Differential survival and reproduction among varying organisms is called **sorting** and should not be equated with natural selection. We now know that even random processes (genetic drift, see section 1.6) can produce sorting. If a garden planted with equal numbers of red- and white-flowered plants suffers severe damage from a hurricane, it is unlikely that equal numbers of red- and white-flowered plants will survive to produce seeds. If red-flowered plants constitute 70% of the survivors, sorting has occurred in favor of the red-flowered plants. In this case, flower color provided no advantage in withstanding the hurricane damage. Most likely, a larger number of red-flowered plants happened to be growing in better-protected locations, permitting their differential survival. This sorting therefore cannot be attributed to natural selection because the character being sorted had no causal influence on the outcome. If in the same garden, white-flowered plants produced more seeds and offspring because they were more visible to a nocturnal moth pollinator, we would observe sorting favoring the white flowers and could attribute this sorting to selection; white flower color in this case provided a reproductive advantage over red color, leading the white-flowered plants to increase in frequency in the next generation. Darwin's theory of natural selection states that sorting occurs *because certain traits give their possessors advantages in survival and reproduction* relative to others that lack those traits. Therefore, selection is one specific cause of sorting.

Key Theme 1.4

EVOLUTION

"Survival of the Fittest"

The popular phrase "survival of the fittest" predates the publication of Darwin's theory of natural selection; it was introduced by the British social philosopher Herbert Spencer, and later applied to Darwin's theory. Unfortunately, this phrase often implies unbridled aggression and violence in a bloody, competitive world. In fact, natural selection operates through many other characteristics of living organisms. For example, many Russian evolutionists argued that animals practicing mutual aid enjoyed the greatest survival advantages in harsh climates.

1.4 Evidence for Darwin's Five Theories of Evolution

Perpetual Change

Perpetual change in the form and diversity of animal life throughout its history reveals itself most directly in the fossil record of the past 540 million years. A **fossil** is a remnant of past life uncovered from the earth's crust (figure 1.11). Some fossils constitute complete remains (insects in amber and mammoths), actual hard parts (teeth and bones), or petrified skeletal parts infiltrated with silica or other minerals (ostracoderms and molluscs). Other fossils include footprints or other impressions, burrows of marine worms in sediment on the sea floor, and fossil excrement (coprolites). In addition to documenting organismal evolution, fossils reveal profound changes in the earth's physical environments, including major changes in the locations of lands and seas. Fossils formed on the floors of ancient seas can be quarried high atop current mountains (Burgess Shale, figure 1.12). Discovery of new fossils and reinterpretation of familiar ones expand our knowledge of how the forms and diversity of animals changed through geological time. Evolutionary study of the fossil record is called paleontology.

Alan Morgan

A

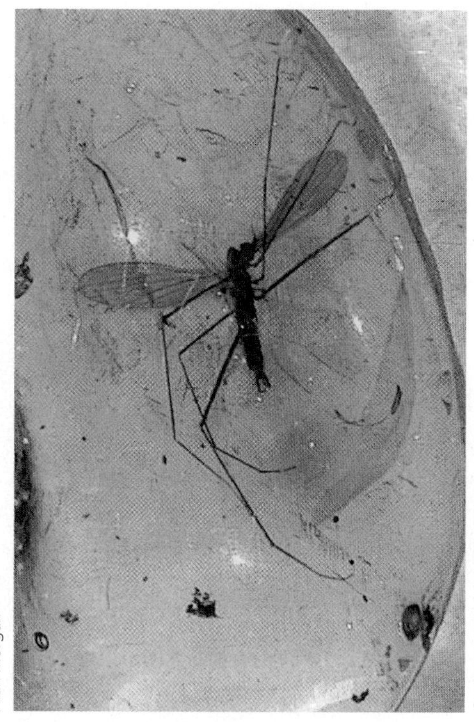
Carlyn Iverson/McGraw Hill Education

B

Alan Morgan

C

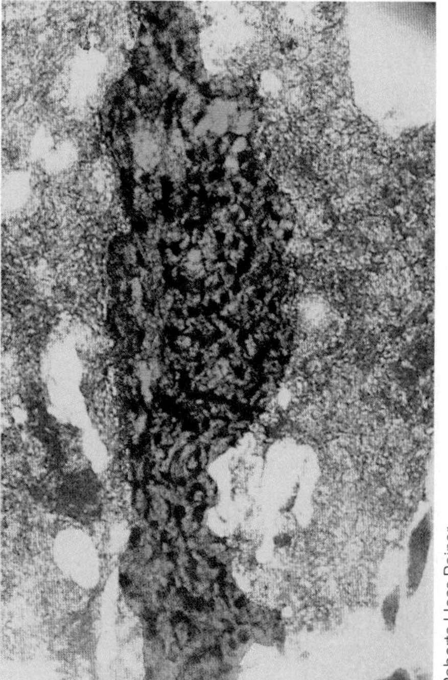
Roberta Hess Poinar

D

figure 1.11

Four examples of fossil material. **A,** Stalked crinoids (sea lilies, class Crinoidea, phylum Echinodermata; see figure 14.21) from Devonian rocks. The fossil record shows that these echinoderms reached their greatest diversity millions of years earlier and began a slow decline to the present. **B,** The fossil of an insect that got stuck in the resin of a tree approximately 25 million years ago, after which the resin hardened into amber. **C,** Fish fossil of the perciform genus *Priscacara* from rocks of the Green River Formation, Wyoming. Such fish swam here during the Eocene epoch, approximately 50 million years ago. **D,** Electron micrograph of tissue from a fly fossilized as shown in **B;** the nucleus of a cell is marked in red.

Key Theme 1.5
EVOLUTION

Fossils of Soft Tissues

On rare occasions, fossil remains include soft tissues preserved so well that electron microscopy reveals recognizable cellular organelles! Insects and even small vertebrates, such as lizards, can be entombed in amber, the fossilized resin of trees. One study of a fly entombed in 40-million-year-old amber revealed structures corresponding to muscle fibers, nuclei, ribosomes, lipid droplets, endoplasmic reticulum, and mitochondria (figure 1.11D). This extreme case of mummification probably occurred because chemicals in the plant sap diffused into the embalmed insect's tissues. A fictional extraction and cloning of DNA from embalmed insects that had bitten and then sucked the blood of dinosaurs was the technical basis for Michael Crichton's best-seller *Jurassic Park.*

Interpreting the Fossil Record

The fossil record is biased because preservation is selective. Vertebrate skeletal parts and invertebrates with shells and other hard structures left the best record (figure 1.11). Soft-bodied animals, including jellyfishes and most worms, are fossilized only under very unusual circumstances, such as those that formed the Burgess Shale of British Columbia (figure 1.12). Exceptionally favorable conditions for fossilization produced a Precambrian fossil bed in South Australia, tar pits at Rancho La Brea (Hancock Park, Los Angeles), great dinosaur beds (Alberta, Canada, and Jensen, Utah; figure 1.13), the Olduvai Gorge of Tanzania, and the early Cambrian Chengjiang beds of China.

Fossil deposits form stratified layers, with new deposits forming above older ones. If left undisturbed, which is rare, the ages of fossils in a preserved sequence are directly proportional to their depth in stratified layers. Stratigraphy is the study of fossil-bearing rocks. Characteristic fossils often serve to identify particular layers. Certain widespread marine invertebrate fossils, including various foraminiferans (see figure 5.24) and echinoderms (see section 14.2), are such good indicators of specific geological periods that we call them "index," or "guide," fossils. Unfortunately, layers are usually tilted or folded or show faults (cracks). Old deposits exposed by erosion might be covered by new deposits in a different plane. When exposed to tremendous pressures or heat, stratified sedimentary rock metamorphoses into crystalline quartzite, slate, or marble, thereby destroying fossils.

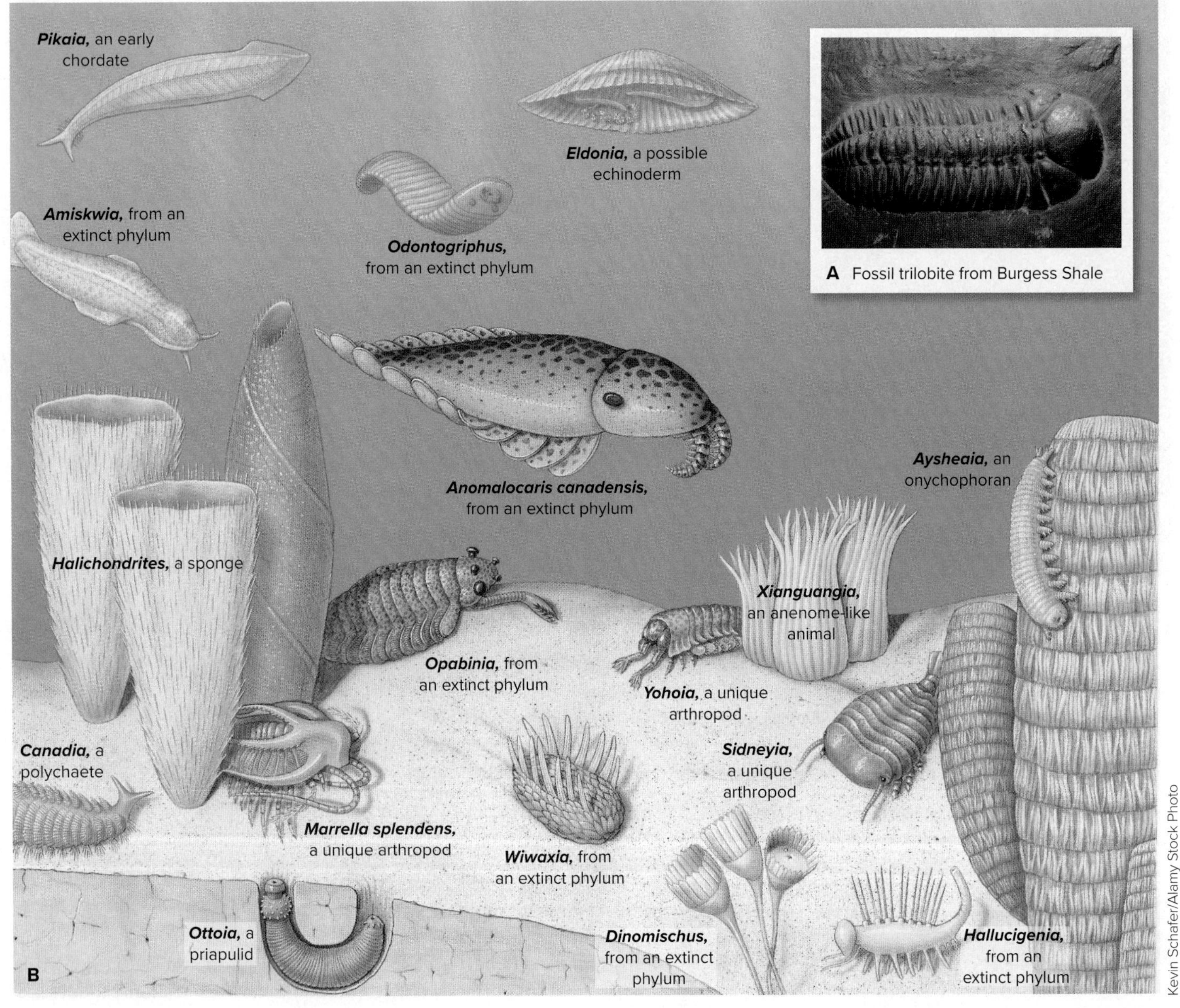

figure 1.12

A, Fossil trilobite. **B,** Animals of the Cambrian period, beginning approximately 540 million years ago, as reconstructed from fossils preserved in the Burgess Shale of British Columbia, Canada. The major body plans of living animals appeared rather abruptly at this time.

Stratigraphy of fossils for two major groups of African antelopes is summarized in figure 1.14. These antelope species have different characteristic sizes and shapes of their horns, which form much of the fossil record of this group. Solid vertical lines in figure 1.14 show the temporal distributions of species as determined by the presence of their characteristic horns in rock strata of various ages. Red lines denote the fossil records of living species, and gray lines denote the fossil records of extinct species. The dotted gray lines show the inferred relationships among living and fossil species based upon their sharing of homologous structural features.

Geological Time

Long before the earth's age was known, geologists divided its history into a table of succeeding events based on ordered layers of sedimentary rock. The "law of stratigraphy" produced a relative dating, with the oldest layers at the bottom and the most recent at the top of a sequence. Time was divided into eons, eras, periods, and epochs as shown on the endpapers inside the back cover of this book. Time during the last eon (Phanerozoic) is expressed in eras (for example, Cenozoic), periods (for example, Cambrian), epochs (for example, Paleocene), and sometimes smaller divisions of an epoch.

figure 1.13

A fossil skeleton from Dinosaur Provincial Park, Alberta, Canada.

In the late 1940s, radiometric dating methods were developed for determining the absolute ages (in years) of rock formations. Several independent methods are now used, all based on radioactive decay of naturally occurring elements into other elements. These "radioactive clocks" are independent of pressure and temperature changes and therefore not affected by often violent earth-building activities.

One method, potassium-argon dating, uses the decay of potassium-40 (^{40}K) to argon-40 (^{40}Ar) (12%) and calcium-40 (^{40}Ca) (88%). Argon is a noble gas that evaporates from liquid media. It accumulates in the crystal structure of rock only after the rock has solidified and the nuclear decay of potassium-40 produces a trapped atom of argon. The half-life of potassium-40 is 1.3 billion years, meaning that half of the original atoms will decay in 1.3 billion years, and half of the remaining atoms will be gone at the end of the next 1.3 billion years. This decay continues until all radioactive potassium-40 atoms are gone. To measure the age of a rock, one calculates the ratio of remaining potassium-40 atoms to the amount of potassium-40 originally there (the remaining potassium-40 atoms plus the argon-40 and calcium-40 into which they have decayed).

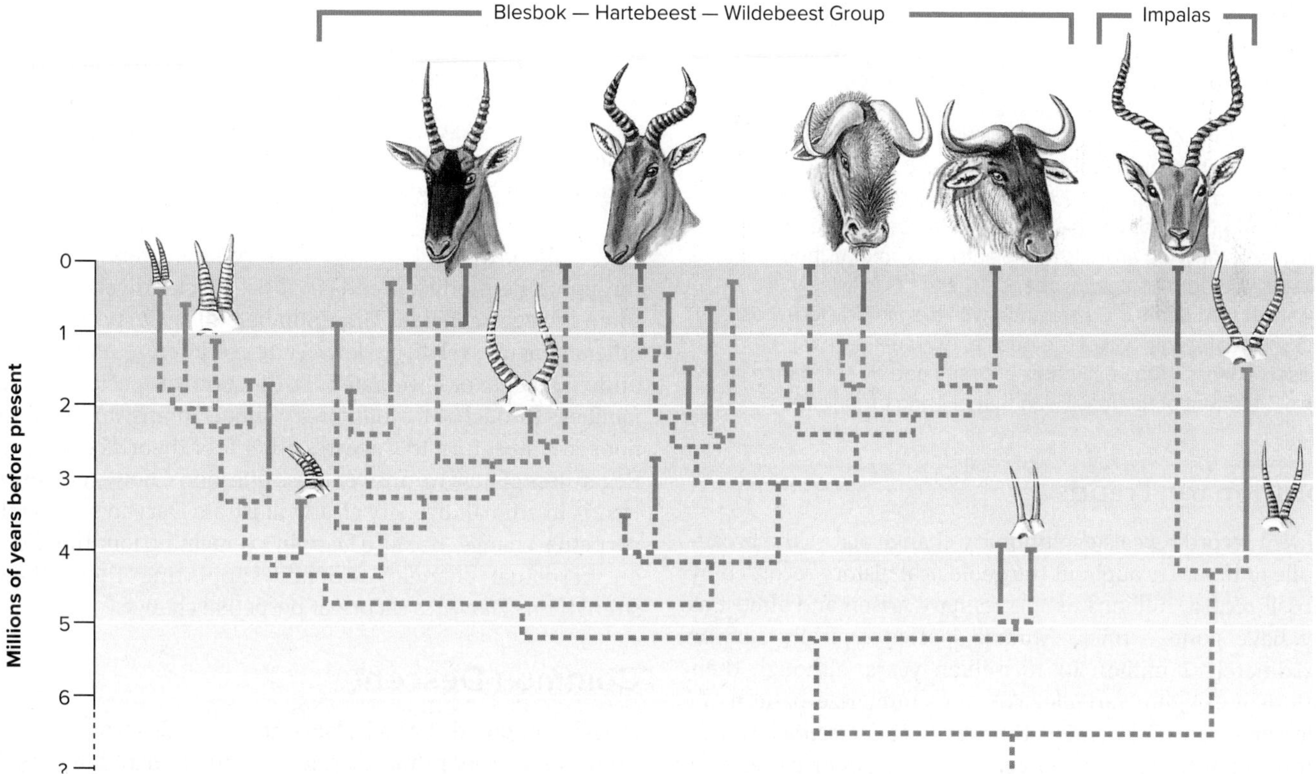

figure 1.14

Stratigraphic record and inferred evolutionary relationships among alcelaphine (blesboks, hartebeests, wildebeests) and aepycerotine (impalas) antelopes in Africa. Species in this group are identified by characteristic sizes and shapes of horns found in rock strata of various ages. Solid vertical lines show the temporal distributions of species in rock strata whose ages are shown on the scale at the left side of the figure. Red lines show the temporal distributions of living species, and gray lines show the temporal distributions of extinct species in rock strata. Dotted gray lines show the inferred relationships among species based on their sharing of homologous structural features. The relative constancy of horn structure within species through geological time is consistent with the theory of punctuated equilibrium (see figure 1.27). This fossil record shows that rates of speciation and extinction are higher for alcelaphine antelopes than for impalas.

A standard equation converts these data to the time elapsed since the formation of the rock as a function of the half life of potassium-40. Several other isotopes can be used in a similar manner to date the ages of rock formations, some for dating the age of the earth itself. One of the most useful radioactive clocks depends on decay of uranium into lead. With this method, rocks over 2 billion years old can be dated with a probable error of less than 1%.

The fossil record of macroscopic organisms begins near the start of the Cambrian period of the Paleozoic era, approximately 540 million years before present (BP). Geological time before the Cambrian period is called the Precambrian era or the Proterozoic eon. Although the Precambrian era occupies 85% of all geological time, it receives much less attention than do later eras, partly because oil, which provides a commercial incentive for much geological work, seldom exists in Precambrian formations. The Precambrian era contains well-preserved fossils of bacteria and algae, and casts of jellyfishes, sponge spicules, soft corals, segmented flatworms, and worm trails. Most, but not all, are microscopic fossils.

Key Theme 1.6
EVOLUTION

Carbon Dating

The more well-known carbon-14 (^{14}C) dating method is of little help in estimating ages of geological formations because its short half-life restricts the use of ^{14}C to quite recent events (less than about 40,000 years). It is especially useful, however, for archaeological studies. This method is based on the production of radioactive ^{14}C (half-life of approximately 5570 years) in the upper atmosphere by bombardment of nitrogen-14 (^{14}N) with cosmic radiation. Radioactive ^{14}C enters the tissues of living animals and plants, and an equilibrium is established between atmospheric ^{14}C and ^{14}N in living organisms. At death, ^{14}C exchange with the atmosphere stops. In 5570 years, only half of the original ^{14}C remains in a preserved fossil. One estimates a fossil's age by comparing the ^{14}C content of the fossil with that of living organisms.

Evolutionary Trends

The fossil record reveals evolutionary change across the broadest scale of time. Throughout the geological history recorded by the fossil record, millions of species have arisen and almost as many have gone extinct. Animal species typically survive approximately 1 million to 10 million years, although their durations are highly variable. We can summarize patterns of species or taxon replacement through time as **trends.** Trends are directional changes in characteristic features or patterns of diversity in a group of organisms. Fossil trends clearly demonstrate Darwin's principle of perpetual change.

A well-studied fossil trend is the evolution of horses from the Eocene epoch to the present. Looking back to the Eocene epoch, we see many different genera and species of horses that replaced each other through time (figure 1.15). George Gaylord Simpson (see figure 4.2) showed that this trend is compatible with Darwinian evolutionary theory. Three characteristics that show the clearest trends in horse evolution are body size, tooth structure, and foot structure. Compared to modern horses, those of extinct genera were small, their teeth had a relatively small grinding surface, and their feet had a relatively large number of toes (four). Throughout the subsequent Oligocene, Miocene, Pliocene, and Pleistocene epochs, new genera arose and old ones went extinct. In each case, there was a net increase in body size, expansion of the grinding surface of teeth, and reduction in number of toes. As the number of toes declined, the central digit became increasingly more prominent in the foot, and eventually only this central digit remained.

The fossil record shows a net change not only in the characteristics of horses but also in the numbers of different horse genera (and numbers of species) that exist through time. Many horse genera of past epochs have been lost to extinction, leaving only a single survivor. Evolutionary trends in diversity are observed in fossils of many different groups of animals (figure 1.16).

Key Theme 1.7
EVOLUTION

Evolutionary Trends

Our use of "evolutionary trend" does not imply that more recent forms are superior to older ones or that the changes represent progress in adaptation or organismal complexity. Although Darwin predicted that such trends would show progressive adaptation, many contemporary paleontologists consider progressive adaptation rare among evolutionary trends. Observed trends in the evolution of horses do not imply that contemporary horses are superior in any general sense to their Eocene predecessors.

Different rates of species formation versus extinction through time produce trends in fossil species diversity. Why do some lineages generate large numbers of new species whereas others generate relatively few? Why do different animal groups undergo higher or lower rates of extinction (of species, genera, or families) throughout evolutionary time? To answer these questions, we must turn to Darwin's other four theories of evolution. Regardless of how we answer these questions, however, observed trends in animal diversity clearly illustrate Darwin's principle of perpetual change. Because Darwin's remaining four theories rely on perpetual change, evidence supporting these theories strengthens Darwin's theory of perpetual change.

Common Descent

Darwin proposed that all plants and animals "have descended from some one primordial form, into which life was first breathed." Life's history forms a branching tree, called a phylogeny, that gives all of life a unified evolutionary history. Current estimates place the last universal common ancestor (LUCA) of living forms at over 4 billion years ago. It would have been a unicellular form resembling living bacteria. Pre-Darwinian evolutionists, including Lamarck, advocated multiple independent origins of life, each of which gave rise to lineages that changed through time without the extensive branching required by Darwin's theory. Like all good scientific theories, common descent makes

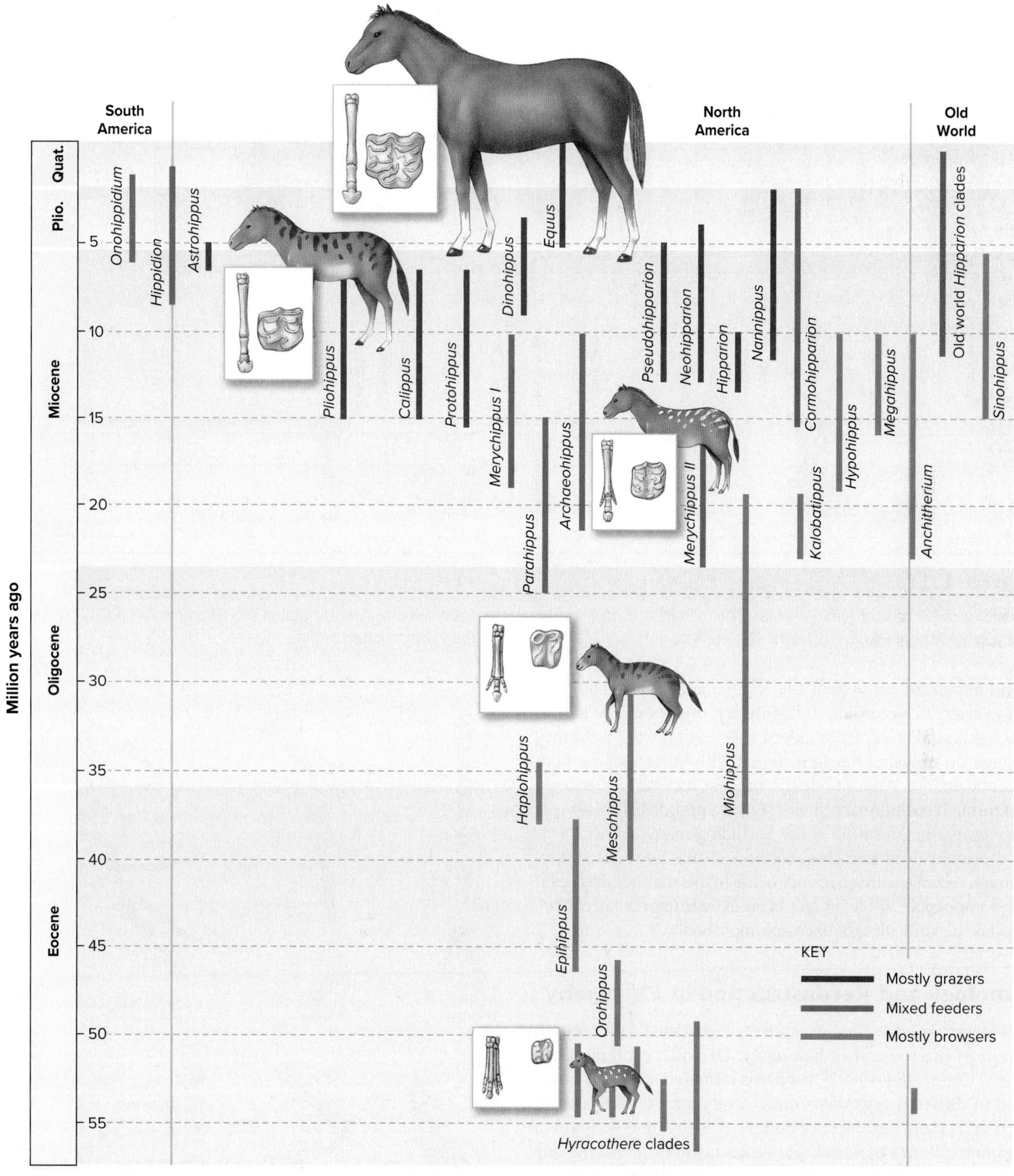

figure 1.15

Stratigraphy of genera of horses from the Eocene epoch to the present. Evolutionary trends toward increased size, elaboration of molars, and loss of toes are shown along with bars denoting temporal durations and continental locations of genera.

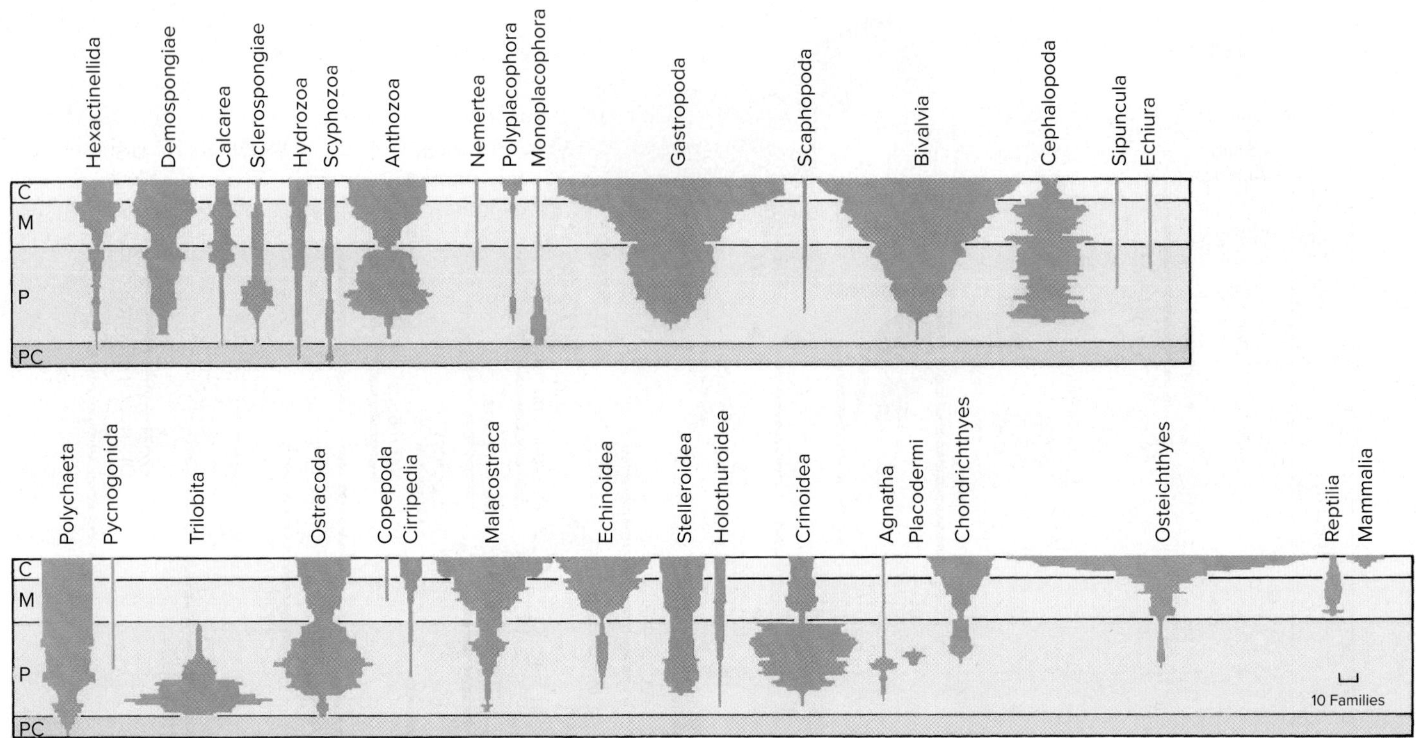

figure 1.16

Diversity profiles of taxonomic families from different animal groups in the fossil record. A scale marks the Precambrian (PC), Paleozoic (P), Mesozoic (M), and Cenozoic (C) eras. The number of families is indicated by the width of each profile.

several important predictions that can be tested and potentially used to reject it. According to this theory, one should be able to trace the genealogies of all modern species backward until they converge on ancestral lineages shared with other species, both living and extinct. We continue this process, moving farther backward through evolutionary time, to reach a primordial ancestor of all life on earth. All forms of life, including many extinct forms that represent dead branches, connect to this tree somewhere. Although reconstructing a history of life in this manner may seem almost impossible, it has in fact been extraordinarily successful. How has this difficult task been accomplished?

Homology and Reconstruction of Phylogeny

Darwin recognized a major source of evidence for common descent in the concept of **homology.** Darwin's contemporary, Richard Owen (1804–1892), used this term to denote "the same organ in different organisms under every variety of form and function."A classic example of homology is the limb skeleton of vertebrates. Bones of vertebrate limbs maintain characteristic structures and patterns of connection despite diverse modifications for different functions (figure 1.17). According to Darwin's theory of common descent, structures that we call homologies represent characteristics inherited with some modification from a corresponding feature in a common ancestor.

Darwin devoted an entire book, *The Descent of Man and Selection in Relation to Sex,* largely to the idea that humans share common descent with apes and other animals. This idea was repugnant to many Victorians, who responded with outrage

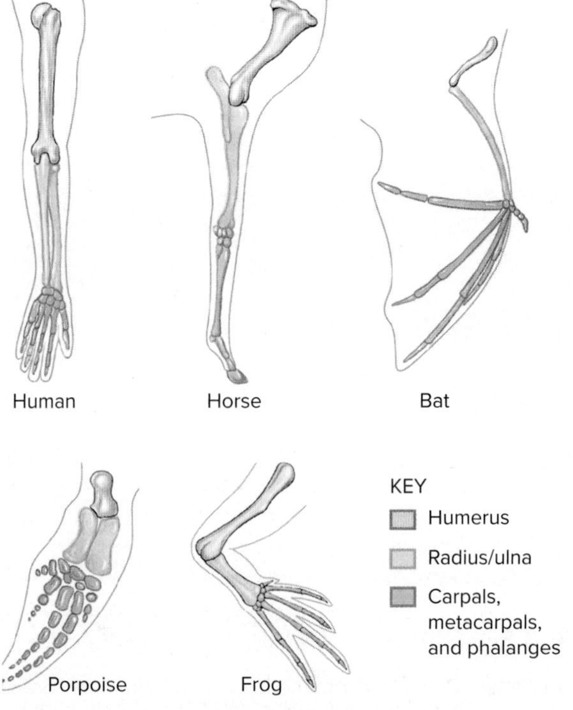

figure 1.17

Forelimbs of five vertebrates show skeletal homologies: brown, humerus; orange, radius and ulna; purple, "hand" (carpals, metacarpals, and phalanges). Homologies of bones and patterns of connection are evident despite evolutionary modification for various uses.

Source: Library of Congress Prints and Photographs Division, [LC-USZ62-48534]

figure 1.18

This 1873 advertisement for Merchant's Gargling Oil ridicules Darwin's theory of the common descent of humans and apes, which was widely doubted by the general public during Darwin's lifetime.

(figure 1.18). Darwin built his case mostly on anatomical comparisons that revealed homology between humans and apes. To Darwin, the close resemblances between apes and humans could be explained only by common descent.

Throughout the history of all forms of life, evolutionary processes generate new characteristics that are transmitted across generations. Every time a new feature becomes established in a lineage destined to be ancestral to others, a new homology originates. The sharing of these homologies among species provides evidence for common descent and allows us to reconstruct a branching evolutionary history of life. We illustrate such evidence using a phylogenetic tree of ground-dwelling flightless birds (figure 1.19). A new skeletal homology arises on each of the lineages shown (descriptions of these homologies are not included because they are highly technical). The different groups of species located at the tips of the branches contain different combinations of these homologies, thereby revealing common ancestry. For example, ostriches show homologies 1 through 5 and 8, whereas kiwis show homologies 1, 2, 13, and 15.

The tree structure inferred from analysis of skeletal structures of flightless birds can be tested by data gathered independently from DNA-sequence information (see Chapter 4). The phylogeny of flightless birds inferred from DNA-sequence data does not agree completely with the one inferred from skeletal structures (figure 1.19); if we choose the hypothesis favored by DNA-sequence data, then we must hypothesize that some of the skeletal structures either arose multiple times or were lost on some lineages as shown in figure 1.19B. Conflict between the phylogenetic hypotheses derived from skeletal structures and from DNA sequences guides systematists to examine their phylogenetic characters and analyses for sources of error in inferring the detailed phylogenetic relationships among these species. All of the phylogenetic data support the hypothesis that these flightless birds are more closely related to each other than they are to any other living species.

Branches of an evolutionary tree combine species into a **nested hierarchy** of groups within groups (see Chapter 4). Smaller groups (species grouped near terminal branches) are contained within larger ones (species grouped by basal branches, including the trunk of the tree). If we erase the tree structure but retain the patterns of homology observed in the terminal groups of species, we can reconstruct the branching structure of the entire tree. Evolutionists test the theory of common descent by observing patterns of homology present in all groups of organisms. The pattern formed by all homologies taken together should specify a single branching tree that represents the evolutionary genealogy of all living organisms.

The nested hierarchical structure of homology is so pervasive in animals that it forms the basis for our systematic groupings (genera grouped into families, families grouped into orders, orders into classes, classes into phyla, and phyla into the animal kingdom). Plants and fungi show similar hierarchical patterns of homology and corresponding systematic groupings. Hierarchical classification even preceded Darwin's theory because this pattern was so evident, but it was not explained scientifically before Darwin. Once common descent was understood, biologists began investigating structural, molecular, and/or chromosomal homologies of animal groups to infer evolutionary relationships. Taken together, the nested hierarchical patterns uncovered by these studies permit us to reconstruct the evolutionary trees of many groups and to continue testing our phylogenetic hypotheses with new data. Methods presented in Chapter 4 show how we use Darwin's theory of common descent to reconstruct the evolutionary history of life and to construct a taxonomic system that summarizes evolutionary relationships among species.

Key Theme 1.8
EVOLUTION

Homology versus Analogy

Characters of different organisms that perform similar functions are not necessarily homologous (inherited from a common ancestor). The wings of bats and birds, although homologous as vertebrate forelimbs, are not homologous as wings. The most recent common ancestor of bats and birds had forelimbs, but the forelimbs were not in the form of wings. The wings of bats and birds evolved independently and have only superficial similarity in their flight structures.

Bat wings are formed by skin stretched over elongated digits, whereas bird wings are formed by feathers attached along the forelimb. Such functionally similar but nonhomologous structures are often called *analogues*.

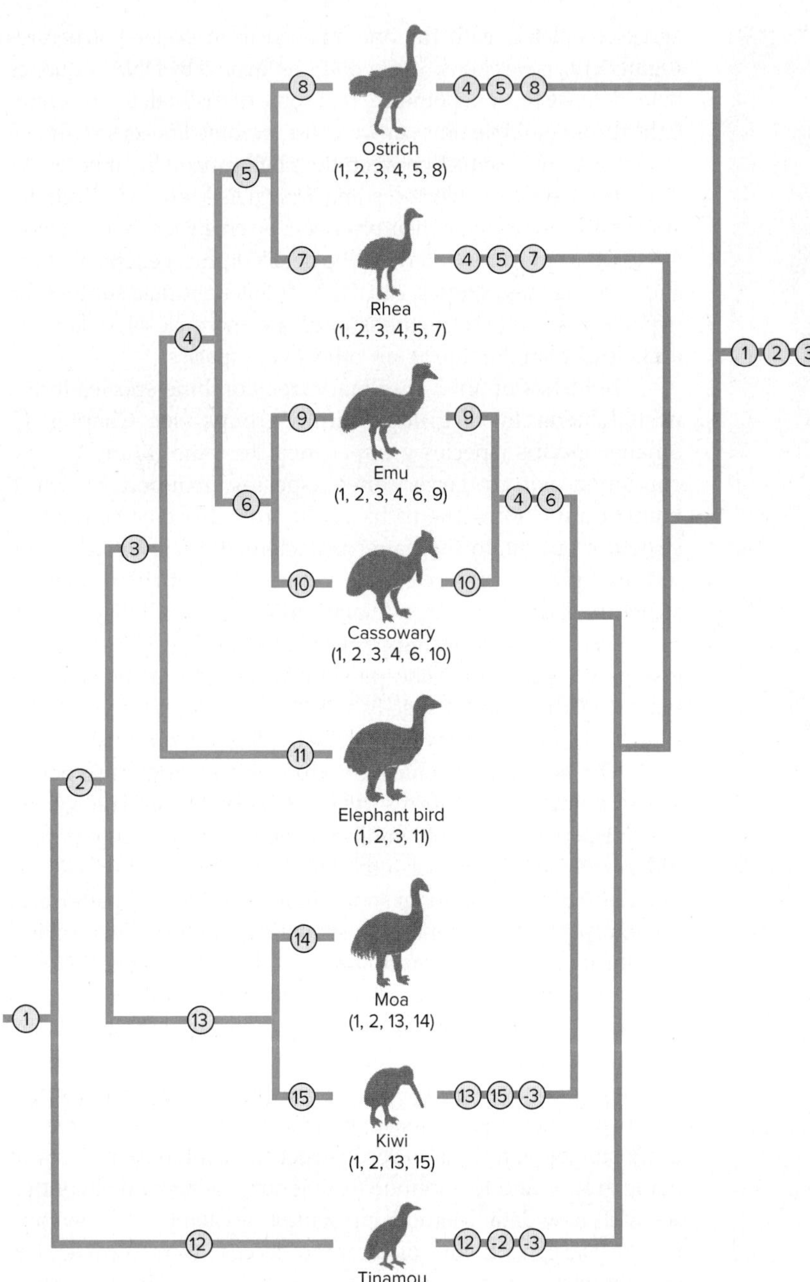

figure 1.19

A, The phylogenetic pattern specified by 15 homologous structures in the skeletons of a group of flightless birds. Homologous features are numbered 1 through 15 and are marked both on the branches of the tree on which they arose and on the birds that have them. If you were to erase the tree structure, you would be able to reconstruct it without error from the distributions of homologous features shown for the birds at the terminal branches. **B,** A phylogenetic analysis of molecular data suggests a different pattern of relationships among the living flightless birds (all except moas and elephant birds). If the molecular analysis is correct, then one must reinterpret evolution of skeletal characters 2, 3, 4, and 5. In the interpretation shown, character 2 is lost by tinamous (−2), and character 3 is lost by both kiwis and tinamous (−3). Character 4 arises independently in ostriches, rheas, and a common ancestor of cassowaries and emus. Character 5 originates separately in ostriches and rheas. Multiple origins and losses complicate phylogenetic analysis, as explained in Chapter 4.

Note that the earlier evolutionary hypothesis that life arose many times, forming unbranched lineages, predicts linear sequences of evolutionary change with no nested hierarchy of homologies among species. Because we do observe nested hierarchies of homologies, that hypothesis is rejected. Note also that a supernatural creationist argument can make no testable predictions about any pattern of homology and therefore fails the criteria of a scientific theory of animal diversity.

Ontogeny, Phylogeny, and Recapitulation

Zoologists find in animal development important clues to an animal's evolutionary history. **Ontogeny** is the development of an organism through its entire life, from its origin as a fertilized egg or bud throughout adulthood to death. Early developmental and embryological features contribute greatly to our knowledge of homology and common descent. Comparative studies of ontogeny show how evolutionary changes in developmental timing generate new body forms and structures.

Comparisons of gene expression among animals show that in forms as dissimilar as insects and humans, homologous genes may guide developmental differentiation of anterior versus posterior body segments. For example, mutations of such genes, termed **homeotic genes,** in fruit flies can cause awkward developmental changes such as legs in the place of antennae or an extra pair of wings. Such genes provide an evolutionary "tool kit" that can be used to construct new body parts by relocating patterns of gene expression to different parts of a developing embryo. Perhaps the most famous homeotic genes are those containing a sequence of 180 base pairs, called the **homeobox,** which encodes a protein sequence that binds to other genes, thereby altering their expression. For example, evolution of the paired limbs of terrestrial vertebrates occurred by activating a caudal "tool kit" of homeobox genes at the paired pectoral and pelvic regions of the body where limbs develop. Genetic and developmental tools evolved for constructing a tail were thus available for the new role of constructing paired limbs.

The German zoologist Ernst Haeckel, a contemporary of Darwin, proposed the influential hypothesis that each successive stage in an organism's development represented an adult form present in its evolutionary history. For example, a human embryo with gill-like depressions in its neck was considered to signify a fishlike ancestor. On this basis, Haeckel gave his generalization: *ontogeny (individual development) recapitulates (repeats) phylogeny (evolutionary descent).* This notion later was simply called **recapitulation**, or the **biogenetic law.** Haeckel based his biogenetic law on the premise that evolutionary change often occurs by successively adding stages onto the end of an unaltered ancestral ontogeny, condensing the ancestral ontogeny into earlier developmental stages. This notion was based on Lamarck's concept of inheritance of acquired characteristics (see section 1.2).

A nineteenth-century embryologist, K. E. von Baer, gave an alternative explanation of the relationship between ontogeny and phylogeny. He argued that early developmental features were simply more widely shared among different animal groups than were later ones. For example, figure 1.20 shows early embryological similarities between organisms whose adult forms are very different. Adults of animals with relatively short and simple ontogenies often resemble the pre-adult stages of other animals whose ontogeny is more elaborate, but the embryos of descendants do not necessarily resemble the adults of their ancestors. However, even early development undergoes evolutionary divergence among groups, and it is not as stable as von Baer thought.

We now know many parallels between ontogeny and phylogeny, but features of an ancestral ontogeny can be shifted to either earlier or later stages in descendant ontogenies. Evolutionary change in the timing of development is called **heterochrony,** a term initially used by Haeckel to denote exceptions to recapitulation. Because the lengthening or shortening of ontogeny can change different parts of an organism independently, we often see a mosaic of different kinds of developmental evolutionary change in a single lineage. Therefore, cases in which an entire ontogeny recapitulates phylogeny are rare.

Despite many changes in scientific thinking about the relationships between ontogeny and phylogeny, one important fact remains clear. Darwin's theory of common descent is strengthened enormously by the many homologies found among developmental stages of organisms belonging to different species.

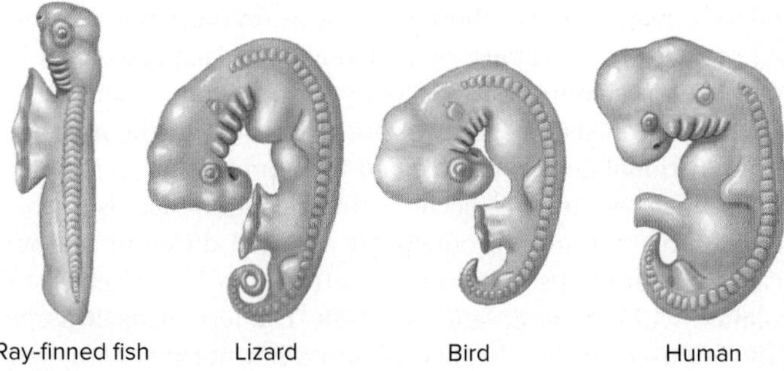

figure 1.20

Comparison of gill arches (purple) of different embryos. All are shown separated from the yolk sac. Note the remarkable similarity of these four embryos at this early stage in development.

Multiplication of Species

Multiplication of species through time is a logical corollary to Darwin's theory of common descent. A branch point on a phylogenetic tree means that an ancestral species has split into two different ones. Darwin's theory postulates that variation present within a species, especially variation between geographically separated populations, provides material from which new species are produced. Because evolution is a branching process, the total number of species produced by evolution increases through time, although most of these species eventually go extinct without leaving descendant species. A major challenge for evolutionists is to discover the processes by which an ancestral species "branches" to form two or more descendant species.

Before we explore multiplication of species, we must decide what we mean by "species." No consensus exists regarding the definition of species (see Chapter 4). Most biologists would agree, however, on three important criteria for recognizing a species:

1. Individuals of the same species descend from a common ancestral population and form an unbranched lineage of ancestral-descendant populations.
2. Individuals of the same species exhibit reproductive compatibility (ability to interbreed) among individuals and reproductive incompatibility between species (for sexually reproducing organisms).
3. Individuals within the same species maintain genotypic and phenotypic cohesion (lack of abrupt differences among populations in allelic frequencies [see section 1.6] and in organismal appearance).

The criterion of reproductive compatibility has received the greatest attention in studies of species formation, also called **speciation**.

Biological factors that prevent different species from interbreeding are called **reproductive barriers**. A primary problem of speciation is to discover how two initially compatible populations evolve reproductive barriers that cause them to become distinct, separately evolving lineages. How do populations diverge from each other in their reproductive properties while maintaining complete reproductive compatibility within each population?

Key Theme 1.9
EVOLUTION

Reproductive Barriers versus Geographic Barriers

Geographical barriers between populations are not equivalent to reproductive barriers. Geographical barriers refer to spatial separation of two populations. They prevent gene exchange and are usually a precondition for speciation. What constitutes a geographic barrier varies with species' dispersal abilities; a barrier to a land snail is not a barrier to a hawk. Reproductive barriers are various morphological, physiological, ecological, and behavioral factors that prevent interbreeding between different species. Geographical barriers do not guarantee that reproductive barriers will evolve. Reproductive barriers are most likely to evolve under conditions that include a generation of small population size, a favorable combination of selective factors, and long periods of geographical isolation. One or both of a pair of geographically isolated populations may become extinct prior to evolution of reproductive barriers between them. Over the vast span of geological time, however, conditions sufficient for speciation have occurred many millions of times.

figure 1.21

Geographic formation of species in the sea urchin (*Eucidaris*). Formation of the Isthmus of Panama separated an ancestral population, leading to evolution of separate Caribbean (*E. tribuloides*) and Pacific (*E. thouarsi*) species.

Reproductive barriers between populations usually evolve gradually. Evolution of reproductive barriers requires that diverging populations be kept physically separate ("isolated") for long periods of time. If the diverging populations were reunited before reproductive barriers were completely formed, interbreeding would occur between the populations and they would merge. Speciation by gradual divergence in animals usually requires perhaps 10,000–100,000 years or more. Geographical isolation followed by gradual divergence is the most effective way for reproductive barriers to evolve, and many evolutionists consider geographical separation a prerequisite for branching speciation. Speciation that results from evolution of reproductive barriers between geographically separated populations is called **allopatric speciation**, or geographical speciation. For example, geological formation of the Isthmus of Panama fragmented an ancestral population of *Eucidaris* sea urchins into separate Caribbean and Pacific Ocean isolates, leading to formation of the pair of species shown in figure 1.21.

Evidence for allopatric ("in another land") speciation occurs in many forms, but perhaps most convincing is an occurrence of geographically separated but adjoining, closely related populations that illustrate the gradual origin of reproductive barriers. Populations of a salamander, *Ensatina eschscholtzii,* in California are a particularly clear example (figure 1.22). These populations show evolutionary divergence in color pattern and collectively form a geographical ring around California's central valley. Genetic exchange between differentiated, geographically adjoining populations is evident through the formation of hybrids and occasionally regions of extensive genetic exchange (called zones of introgression). Two populations at the southern tip of the geographical range (called *E. e. eschscholtzii* and *E. e. klauberi*) make contact but do not interbreed. A gradual accumulation of reproductive differences among contiguous populations around the ring is visible, with the two southernmost populations separated by strong reproductive barriers.

Additional evidence for allopatric speciation comes from observations of animal diversification on islands. Oceanic islands formed by volcanoes are initially devoid of life. Plants and animals from a continent or from other islands occasionally colonize new islands in separate invasions. Invaders often encounter situations ideal for evolutionary diversification, because environmental resources that were exploited heavily by other species on the mainland are free for colonization on a sparsely populated island. Because colonization of oceanic islands is rare, populations established on islands are effectively isolated geographically from their parental populations and can undergo divergent evolution, leading to reproductive barriers and speciation. Archipelagoes, such as the Galápagos Islands, greatly increase opportunities for speciation in this manner.

Evolutionists have often wondered whether the isolation of populational gene pools needed for reproductive barriers to evolve might sometimes occur without geographical isolation. Populations that are reproductively active at different seasons or on different substrates could, in theory, achieve gene-pool isolation without geographic isolation. The term *sympatric speciation* is used to denote species formation not involving geographical isolation. Intermediate between allopatric and sympatric speciation is a condition in which the diverging populations are geographically separate but make contact along a borderline; speciation arising in this manner is called *parapatric speciation*. Many zoologists doubt that sympatric and parapatric speciation have been important in animal evolution.

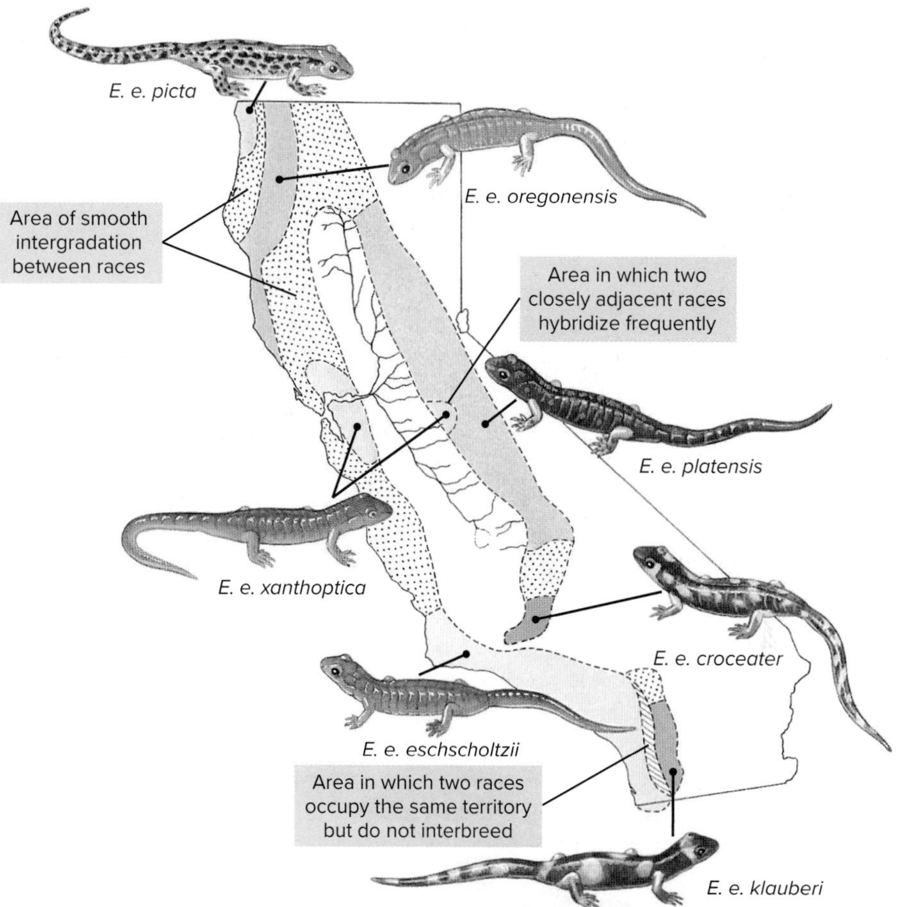

figure 1.22

Speciation in progress: geographic variation of color patterns in the salamander genus *Ensatina*. Populations of *Ensatina eschscholtzii* form a geographic ring around the Central Valley of California. Adjacent, differentiated populations throughout the ring can exchange genes except at the bottom of the ring where the subspecies *E. e. eschscholtzii* and *E. e. klauberi* overlap without interbreeding. These two subspecies would be recognized as distinct species if intermediate populations linking them across the ring were extinct. This example demonstrates that reproductive barriers between populations can evolve gradually.

Production of many ecologically diverse species from a common ancestral species is called **adaptive radiation**, especially when these species arise within a short interval of geological time (a few million years). Some of our best examples of adaptive radiation occur in lakes or on young islands, which provide new evolutionary opportunities for aquatic and terrestrial organisms, respectively. Oceanic islands formed by volcanoes are initially devoid of life. They are gradually colonized by plants and animals from a continent or from other islands in separate founder events. These colonizations involve floating of seeds across an expanse of ocean followed by germination on the island. Birds or insects might be carried by wind from a continent to the island, or animals might be carried by flotation in sea-swept debris. The probability that a particular species will make a successful colonization is low, but given large numbers of continental species and millions of years of evolutionary time, such colonizations occur and eventually establish an oceanic island biota. Archipelagoes, such as the Galápagos Islands, greatly increase opportunities for both founder events and ecological diversification. The entire archipelago is isolated from the continent, and each island is geographically isolated from the others by sea; moreover, each island is different from every other one in its physical, climatic, and biotic features. Galápagos finches clearly illustrate adaptive radiation on an oceanic archipelago (figures 1.23 and 1.24).

Galápagos finches (the name "Darwin's finches" was popularized in the 1940s by the British ornithologist David Lack) are close relatives, but each species differs from others in the size and shape of its beak and in its feeding habits. Darwin's finches descended from a single ancestral population that arrived from South America and subsequently colonized different islands of the Galápagos archipelago. These finches underwent adaptive radiation, occupying habitats that in South America were denied to them by other species better able to exploit those habitats. Galápagos finches thus acquired characteristics of mainland birds as diverse and unfinchlike as warblers and woodpeckers (figure 1.24B). The founding of new island populations by a small number of migrants may have accelerated evolutionary divergence among these island finches. A fourteenth finch population, found on isolated Cocos Island far north of the Galápagos archipelago, resembles the Galápagos finches and almost certainly descends from the same ancestral founder. Despite their great adaptive diversification, the different forms of Galápagos finches do not form clearly distinct species lineages. Molecular genetic data indicate genetic compatibility and gene exchange among the adaptively diverse forms.

Gradualism

Darwin's theory of gradualism opposed arguments for a sudden origin of species. Small differences, resembling those that we observe among organisms within populations today, are the raw material from which different major forms of life evolved. This theory shares with Lyell's uniformitarianism a notion that we must not explain past changes by invoking unusual catastrophic events that are not observable today. If new species

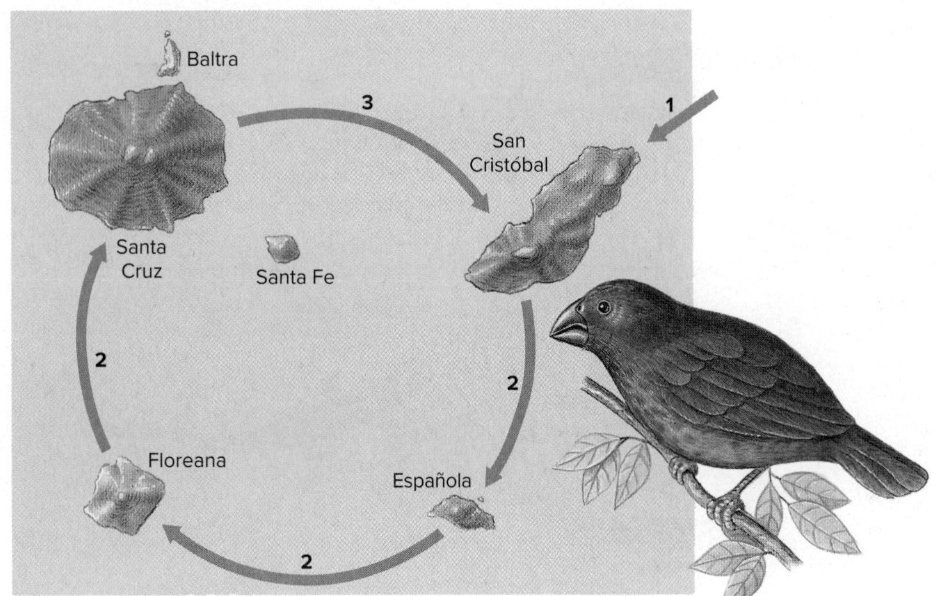

figure 1.23

Tentative model for evolution of the 13 forms of Darwin's finches on the Galápagos Islands. This model postulates three steps: (1) Immigrant finches from South America reach the Galápagos and colonize an island; (2) after a population becomes established, finches disperse to other islands where they adapt to new conditions and change genetically; and (3) after a period of isolation, secondary contact is established between different populations. Different populations would be recognized as separate species if they cannot interbreed successfully, although reproductive isolation apparently is not strong among the different forms of Darwin's finches.

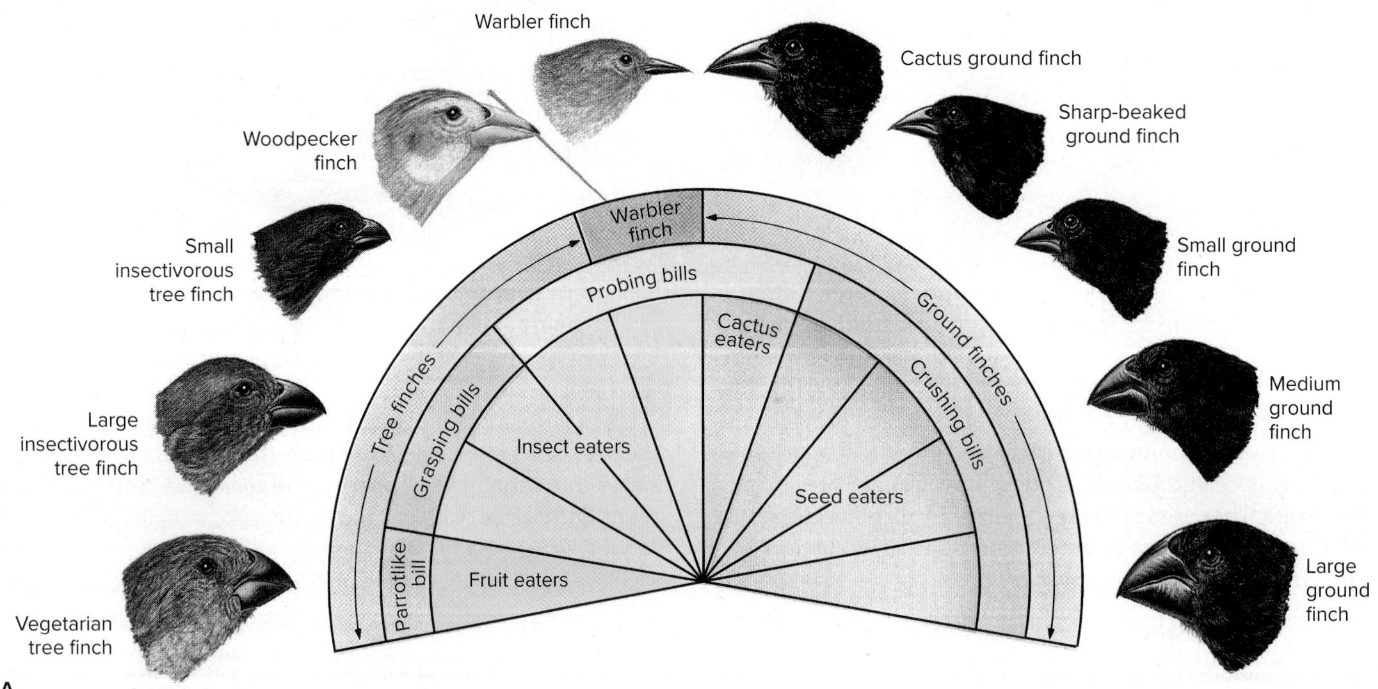

Cleveland P. Hickman, Jr.

figure 1.24

A, Adaptive radiation in 10 contrasting forms of Darwin's finches from Santa Cruz, one of the Galápagos Islands. Differences in beaks and feeding habits are shown. All apparently descended from a single common ancestral finch from South America. **B,** Woodpecker finch, one of 13 forms of Galápagos Islands finches, using a slender twig as a tool for feeding. This finch worked for about 15 minutes before spearing and removing a wood roach from a break in the tree.

originated in single, catastrophic events, we should be able to see these events happening today, and we do not. What we observe instead are small, continuous changes in phenotypes occurring in natural populations. Such continuous changes can produce major differences among species only by accumulating over many thousands to millions of years. A simple statement of Darwin's theory of gradualism is that accumulation of quantitative changes leads to qualitative change.

Phenotypic gradualism was controversial when Darwin first proposed it, and it is still controversial. Not all phenotypic changes are small, incremental ones. For example, some mutations that appear during artificial breeding, traditionally called "sports," change a phenotype substantially in a single mutational step. Sports that produce dwarfing occur in many species, including humans, dogs, and sheep, and have been used by animal breeders to achieve desired results; for example, a sport that deforms limbs was used to produce ancon sheep, which cannot jump hedges and are therefore easily contained (figure 1.25). Many colleagues of Darwin who accepted his other theories considered phenotypic gradualism too extreme. If sports can be used in animal breeding, why must we exclude them from our evolutionary theory? In favor of gradualism, some have replied that sports always have negative side effects that would prevent affected organisms from surviving in natural populations. Indeed, it is questionable whether the ancon sheep in figure 1.25, despite its attractiveness to farmers, would propagate successfully in the presence of its long-legged relatives without human intervention. Naturalists nonetheless report mutations of large effect that appear adaptive in natural populations; a mutation responsible for a large difference in beak size in the African finch species *Pyrenestes ostrinus* allows large-beaked birds to eat hard seeds, whereas small-beaked forms eat softer seeds.

When we view Darwinian gradualism on a geological timescale, we may expect to find in the fossil record a long series of intermediate forms bridging phenotypes of ancestral and descendant populations (figure 1.26). This predicted pattern is called **phyletic gradualism**. Darwin recognized that phyletic gradualism is not often revealed by the fossil record. Studies conducted since Darwin's time likewise have failed to produce a continuous series of fossils as predicted by phyletic gradualism. Is the theory of gradualism therefore refuted by the fossil record? Darwin and others have claimed that it is not, because the fossil record is too imperfect to preserve each minor change in animal form that occurs across generations in a species lineage. Although evolution is a slow process by our standards, it is rapid relative to the rate at which good fossil deposits accumulate. Others have argued, however, that abrupt origins and extinctions of species in the fossil record force us to conclude that phyletic gradualism is rare.

Niles Eldredge and Stephen Jay Gould proposed a model called **punctuated equilibrium** in 1972 to explain discontinuous evolutionary changes observed throughout geological time. Punctuated equilibrium states that phenotypic evolution is concentrated in relatively brief events of branching speciation, followed by much longer intervals of morphological evolutionary stasis (figure 1.27). Speciation is an episodic event,

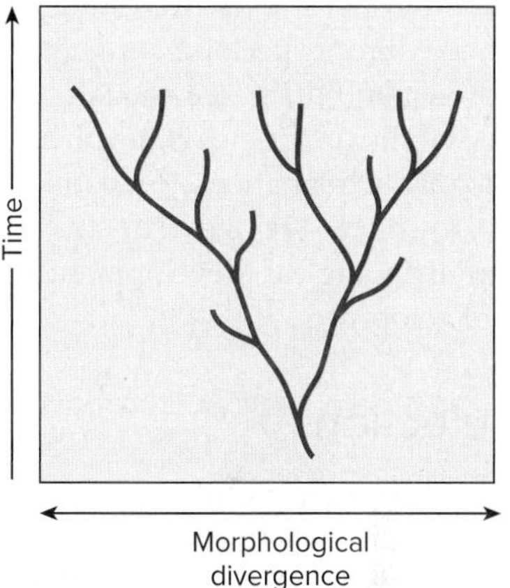

figure 1.26

A gradualist model of evolutionary change in morphology, viewed as proceeding more or less steadily through geological time (millions of years). Bifurcations followed by gradual divergence led to speciation.

figure 1.25

The ancon breed of sheep (center) arose from a "sport" mutation that caused dwarfing of their legs. Many of Darwin's contemporaries criticized him for claiming that such large mutations are not important for evolution by natural selection.

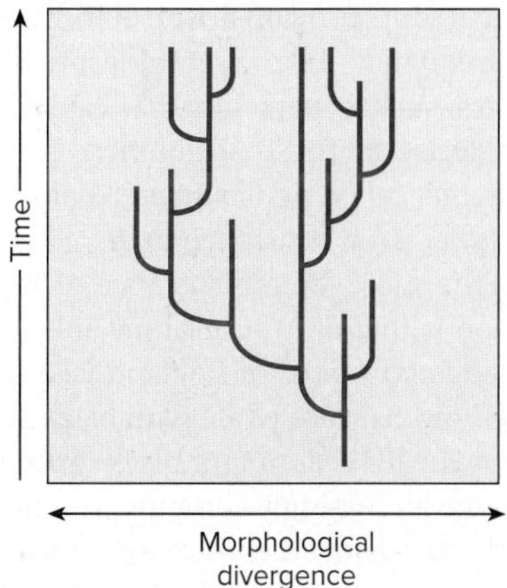

figure 1.27

A punctuated equilibrium model shows evolutionary change concentrated in relatively rapid bursts of branching speciation (*lateral lines*) followed by prolonged periods of little change throughout geological time (millions of years).

occurring over a period of approximately 10,000–100,000 years. Because species may survive for 5 million to 10 million years, the speciation event is a "geological instant," representing 1% or less of a species' existence. Ten thousand years is plenty of time, however, for Darwinian evolution to accomplish dramatic changes. Therefore, a small fraction of the total evolutionary history of a group accounts for most of the morphological evolutionary change that occurs.

Probably the best-documented case of punctuated equilibrium is for ectoprocts (see section 9.4) of the genera *Metrarabdotos* and *Stylopoma* of the Caribbean Sea. These genera originated 15–25 million years ago and have an excellent fossil record as well as living species. Durations of morphologically static species range from 2 to 16 million years, whereas events of branching speciation can be narrowed to no more than 0.86 million years and often less than 0.16 million years in the fossil record. Molecular genetic studies of living populations confirm that the morphological data accurately diagnose biological species. Morphological change is episodic and concentrated in events of branching speciation. Interestingly, rates of morphological divergence during speciation are no greater than would be expected by genetic drift (see section 1.6). The persistent morphological stability of species is the phenomenon that challenges evolutionary explanations. Evolution by natural selection appears to have created effective mechanisms for maintaining morphological developmental stability within these species of ectoprocts.

Natural Selection

Many examples show how natural selection alters populations in nature. Sometimes selection can proceed very rapidly as happened, for example, when flies and mosquitoes quickly evolved high resistance to insecticides. Doses that at first killed almost all pests later failed to control them. As more insects were exposed to insecticides, those most sensitive were killed, leaving more space and less competition for resistant strains to multiply. Thus, as a result of selection, mutations bestowing high resistance, but previously rare in these populations, increased in frequency. Likewise, widespread medical use of some antibiotics, such as penicillin, has caused bacterial strains to evolve resistance to these antibiotics. Some staphylococci secrete an enzyme, called penicillinase, that inactivates penicillin and confers bacterial resistance to it.

Perhaps the most famous instance of rapid evolution by natural selection is that of industrial melanism (dark pigmentation) in the peppered moths of England (see figure 1.2). Before 1850, peppered moths were white with black speckling on their wings and body. In 1849, a mutant black form of the species appeared. It became increasingly common, reaching frequencies of 98% in Manchester and other heavily industrialized areas by 1900. We illustrated scientific methodology (see section 1.1) with the kinds of tests that ultimately favored the hypothesis that the moths least visible to avian predators were most likely to survive and to reproduce: white moths in unpolluted environments, and black moths in polluted ones. When pollution diminished, the frequency of lightly pigmented individuals increased in moth populations (see figure 1.2C), just as the hypothesis of natural selection predicted.

A recurring criticism of natural selection is that it cannot generate new structures or species but can only modify existing ones. Most structures in their early evolutionary stages could not have performed roles that the fully formed structures perform, and therefore it is unclear how natural selection could have favored them. What use is half a wing or the rudiment of a feather for a flying bird? To answer this criticism, Darwin proposed that many structures evolved initially for purposes different from the uses that they have today. Rudimentary feathers could have been useful in thermoregulation, for example. Their role in flying would have evolved later after they incidentally acquired aerodynamic properties that subjected them to selection for improved flying. **Exaptation** denotes the utility of a structure for a biological role that was not part of the structure's evolutionary origin. Exaptation contrasts with adaptation, which implies that a structure arose by natural selection for the utility in question. Bird feathers are therefore adaptations for thermoregulation but exaptations for flight. Because the anatomical differences observed among organisms from different, closely related species resemble variation observed within species, it is unreasonable to propose that selection cannot lead beyond a species boundary.

1.5 Revisions of Darwinian Evolutionary Theory

Neo-Darwinism

The most serious weakness in Darwin's argument was his failure to identify correctly a mechanism for inheritance. Darwin saw heredity as a blending phenomenon in which the hereditary factors of parents melded in their offspring. Darwin also invoked the Lamarckian hypothesis that an organism could alter its heredity through use and disuse of body parts and through direct environmental influence. Darwin did not realize that hereditary factors could be discrete and nonblending, and that a new genetic variant therefore could persist unaltered from one generation to the next. The German biologist August Weismann (1834–1914) rejected Lamarckian inheritance. His microscopic studies of animal development showed that cells destined to become gametes (the "germ plasm") separated from body cells (the "soma") early in development; there is thus no means by which the use and disuse of body parts could alter heredity. He showed experimentally that modifications of an organism during its lifetime do not change its heredity, and he revised Darwinian evolutionary theory accordingly. We now use the term **neo-Darwinism** to denote Darwinian evolutionary theory as revised by Weismann. The genetic basis of neo-Darwinism eventually became what we now call the **chromosomal theory of inheritance**, a synthesis of Mendelian genetics and cytological studies of the segregation of chromosomes into gametes.

Key Theme 1.10
EVOLUTION

Darwin and Mendel

Gregor Mendel published his theories of inheritance in 1868, 14 years before Darwin died. Darwin presumably never read this work, although it was found in Darwin's extensive library after his death. Had Mendel written to Darwin about his results, it is possible that Darwin would have modified his theory accordingly. However, it is also possible that Darwin could not have seen the importance of hereditary mechanisms to the continuous variations and gradual changes that represent the hub of Darwinian evolution. It took other scientists many years to establish the relationship between Mendelian genetics and Darwin's theory of natural selection.

Emergence of Modern Darwinism: A Synthetic Theory

In the 1930s, a new generation of geneticists reevaluated Darwinian evolutionary theory from a mathematical perspective. These were population geneticists, scientists who studied variation in natural populations of animals and plants and who had a sound knowledge of statistics. Gradually, a new comprehensive theory emerged that brought together population genetics, paleontology, biogeography, embryology, systematics, and animal behavior in a Darwinian framework.

Population geneticists study evolution as change in the genetic compositions of populations. With the establishment of population genetics, evolutionary biology became divided into two subfields. **Microevolution** pertains to evolutionary changes in frequencies of variant forms of genes within populations. **Macroevolution** refers to evolution on a grand scale, encompassing origins of new organismal structures and designs, evolutionary trends, adaptive radiation, phylogenetic relationships of species, and mass extinction. Macroevolutionary research is based in systematics and the comparative method (see section 1.7). Following this evolutionary synthesis, both macroevolution and microevolution have operated firmly within a tradition of neo-Darwinism, and both have expanded Darwinian theory in important ways.

1.6 Microevolution: Genetic Variation and Change within Species

Microevolution is the study of genetic change occurring within natural populations. A population is a reproductively cohesive group of organisms of the same species; populations of sexually reproducing forms show interbreeding among their members. Variant forms of a single gene are called **alleles**. For example, in human populations, three different allelic forms occur for the gene encoding ABO blood types: I^A, I^B, and *i*. A person has two copies of this gene, one copy inherited from each parent. A person's **genotype** denotes the pair of allelic forms present for a particular gene. For the ABO blood type, the possible genotypes are I^AI^A, I^AI^B, I^Ai, I^BI^B, I^Bi, *ii*. A genotype is *homozygous* if it contains two copies of one allele (for example, I^AI^A, I^BI^B, *ii*) and *heterozygous* if it contains copies of two alleles (for example, I^AI^B, I^Ai, I^Bi). An organismal trait associated with a particular genotype is called the *phenotypic effect* of that genotype. A capital letter in the allelic designation indicates that an allele is genetically dominant, being expressed in both homozygous and heterozygous genotypes. A lower-case letter indicates that an allele is genetically recessive, revealing its characteristics only in its homozygous genotype. Therefore, two genotypes specify blood types A (I^AI^A, I^Ai) and B (I^BI^B, I^Bi), but blood types AB (I^AI^B) and O (*ii*) have single genotypes. Many genetic traits show this standard pattern of inheritance, called Mendelian inheritance.

The flow of information from the sequence of bases in a gene to the sequence of amino acids in a protein, sometimes called the "central dogma" of molecular biology, explains the relationship between genotype and phenotype. A protein-coding gene is expressed when a molecule of messenger RNA (mRNA) is copied from one of its two DNA strands, a process called transcription. The mRNA is complementary in base sequence to the DNA strand that serves as the template for its transcription. Messenger RNA is transported to ribosomes in the cell's cytoplasm, where it guides synthesis of a protein. Each triplet of RNA bases, called a codon, specifies an amino acid in the protein being synthesized. The sequence of amino acids in the protein matches the sequence of codons in the messenger RNA. Alleles I^A and I^B encode proteins that differ from each other in amino acid sequence. Antibodies to protein A do not recognize protein B, and vice versa. The recessive *i* allele does not encode a protein product, and thus represents a "loss of function." Genotypes I^AI^A and I^Ai both have protein A present but not B. Genotypes I^BI^B and I^Bi both have protein B present but not A. Genotype I^AI^B has proteins A and B. Many recessive genetic traits share in common the loss of a protein function. They are recessive because when paired with an allele that encodes a functional protein, the heterozygous individual produces functional protein from its functional allele. For an enzymatic protein, sometimes loss of function occurs from change of a single amino acid in the protein sequence; although protein synthesis occurs normally, the resulting protein lacks its normal enzymatic function, a genetically recessive phenotype.

Occurrence of different alleles of a gene in a population is called **polymorphism**. All alleles of all genes possessed by members of a population collectively form its **gene pool**. Polymorphism is potentially enormous in large populations because at observed mutation rates, many different alleles are expected for all genes. This population-genetic meaning of polymorphism contrasts with zoological usage of the same term to describe colonial forms whose members are differentiated to perform various different functions, as in colonial cnidarians (see section 7.1).

Population geneticists study polymorphism by identifying allelic forms of a gene present in a population and then measuring their relative frequencies in that population. The relative frequency of a particular allele of a gene in a population is called its **allelic frequency**. Because each individual genotype contains two copies of this gene, the total number of copies present in a population is twice the number of individuals. What fraction of this total is represented by each different allelic form? In France, we find the following allelic frequencies: $I^A = 0.46$, $I^B = 0.14$, and $i = 0.40$. In Russia, the corresponding allelic frequencies differ ($I^A = 0.38$, $I^B = 0.28$, and $i = 0.34$), demonstrating microevolutionary divergence between these populations (figure 1.28). Genetically, alleles I^A and I^B are dominant to i, but i is nearly as frequent as I^A and exceeds the frequency of I^B in both populations. Dominance describes the *phenotypic effect* of an allele in heterozygous individuals, not its relative abundance in a population of individuals. In many human populations, genetically recessive traits, including the O blood type, blond hair, and blue eyes, are very common. In the following discussion, we demonstrate that Mendelian inheritance and dominance do not alter allelic frequencies directly or produce evolutionary change in a population.

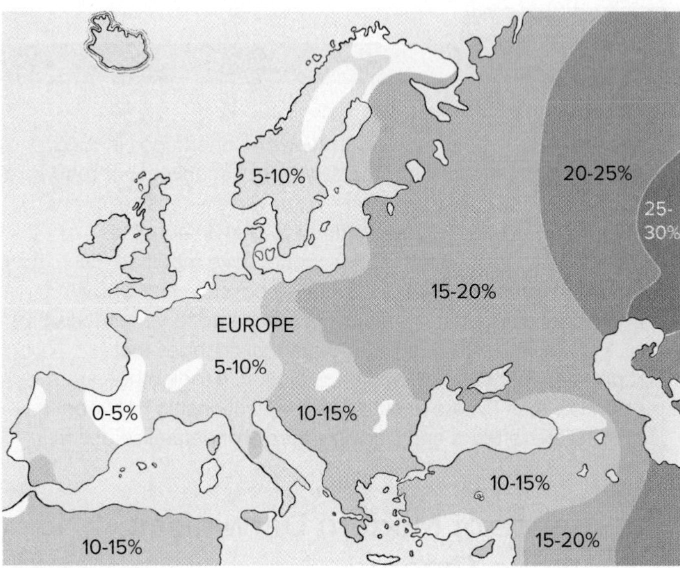

figure 1.28

Frequencies of the blood-type B allele (I^B) among humans in Europe. This allele is more common in the east and rarer in the west. The allele may have arisen in eastern Europe and gradually diffused westward through genetically continuous populations. This allele has no known selective advantage, and its changing frequency probably represents random genetic drift.

Key Theme 1.11
GENETICS AND DEVELOPMENT

Gene Expression and Development

Most genes express themselves only at specific times and locations in a developing organism. The problem in development is to explain how, if every cell has a full gene complement, certain genes are "turned on" to produce proteins required at a particular developmental stage while other genes remain silent. Transcription of a gene requires that the gene contain a sequence of bases, called a promoter, that binds RNA polymerase and appropriate **transcription factors.** A transcription factor is a DNA-binding protein that typically can activate over 100 target genes whose products collectively guide a particular developmental process. Genes that encode a transcription factor thus influence expression of many other genes (see section 3.3) and serve to recruit various gene products as needed for a particular developmental event. A particular target gene might serve multiple developmental processes, each one activated by a different transcription factor. A gene whose protein product is needed for separate developmental events in a fly's brain and eyes would contain in its promoter a DNA sequence that binds a brain-specific transcription factor and another DNA sequence that binds an eye-specific transcription factor. Evolutionary change in the tissue-specific expression of a gene thus occurs by insertion or deletion of promoter sequences specific to the various transcription factors that regulate gene expression in animal development.

Protein Polymorphism

Beginning in the 1960s, studies of protein variation provided the first unambiguous evidence that animal populations typically contain large amounts of genetic variation. Studies of protein polymorphism are now largely superseded by studies of variation in DNA sequences sampled from both the nuclear and mitochondrial genomes. DNA studies reveal even larger amounts of variation than did studies of proteins. We focus here on protein variation, for both its historical importance and its more straightforward interpretation using the principles of Hardy-Weinberg equilibrium.

Different allelic forms of genes encode proteins that often differ slightly in their amino acid sequence. This phenomenon is called **protein polymorphism.** If these differences affect the protein's net electric charge, the different allelic forms can be separated in the laboratory using protein electrophoresis (figure 1.29). We can identify the genotypes of particular individuals for protein-coding genes and measure allelic frequencies in a population.

Geneticists using this approach have discovered far more variation than was previously expected. Despite the high levels of polymorphism discovered using protein electrophoresis, these studies underestimate both protein polymorphism and the total genetic variation present in a population. For example, protein polymorphism that does not involve charge differences is not detected. Furthermore, because the genetic code is degenerate (more than one codon for most amino acids), protein polymorphism does not reveal all of the genetic variation present in protein-coding genes. Genetic changes that do not alter protein structure sometimes alter patterns of protein synthesis during development and can be very important to an organism. When all kinds of variation are considered, it is evident that most species have an enormous potential for further evolutionary change.

Genetic Equilibrium

A mathematical theorem called **Hardy-Weinberg equilibrium** (see box, Hardy-Weinberg Equilibrium: Why Mendelian Heredity Does Not Change Allelic Frequencies) permits us to estimate the relationship between the frequencies of alleles at a genetic locus and the frequencies of genotypes formed by random fertilization of gametes with respect to this variation. We apply the

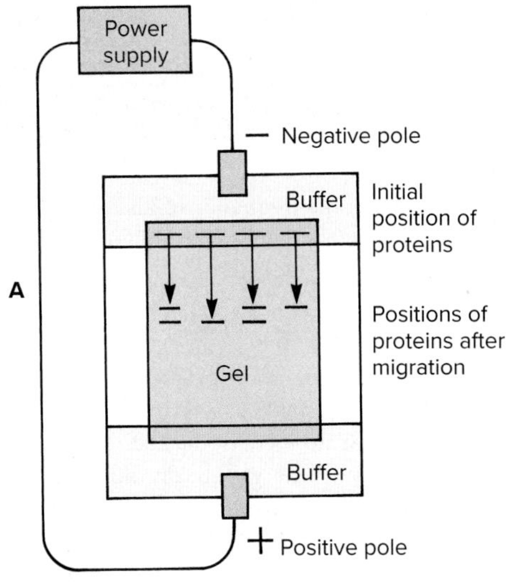

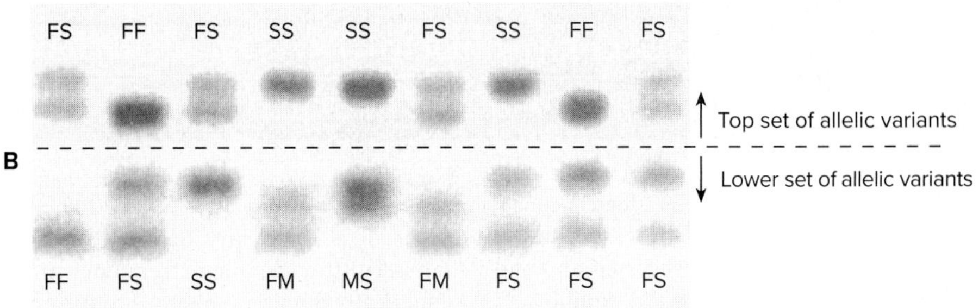

figure 1.29

Study of genetic variation in proteins using gel electrophoresis. **A,** An electrophoretic apparatus separates allelic variants of proteins that differ in charge because of differences in their sequence of amino acids. **B,** Genetic variation in the protein leucine aminopeptidase for nine brown snails, *Helix aspersa*. Two different sets of allelic variants are revealed. The top set contains two alleles [denoted fast (F) and slow (S) according to their relative movement in the electric field]. Individuals homozygous for the fast allele show only a single fast band on the gel (FF), those homozygous for the slow allele show only a single slow band (SS), and heterozygous individuals have both bands (FS). The lower set contains three different alleles denoted fast (F), medium (M), and slow (S). Note that no individuals shown are homozygous for the medium (M) allele.

theorem to one locus at a time, tracing the alleles through the formation of gametes followed by fertilization of gametes to produce individuals of the next generation. The frequency of an allele in the gametes produced by a population equals its frequency in the gene pool as defined near the start of this section. Random mating consists mathematically of drawing pairs of gametes at random from the gene pool, uniting each random pair, and then quantifying the frequencies of the resulting organismal genotypes. Mating is usually random with respect to most molecular genetic traits, such as blood types, because they are unlikely to influence mate choice or recognition. We can estimate from Hardy-Weinberg equilibrium the ratios of genotypes and phenotypes that should occur in the following generation.

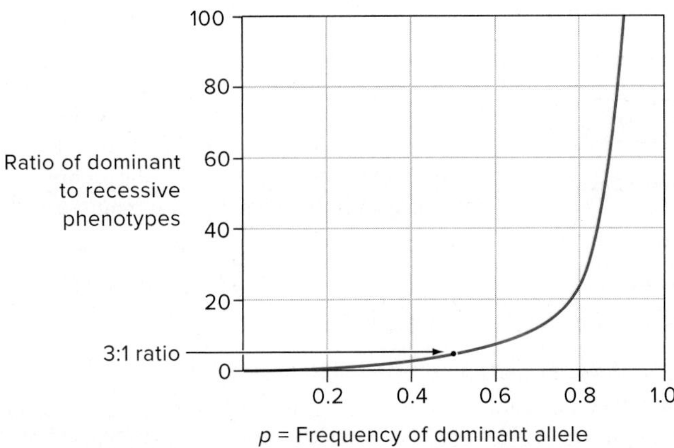

figure 1.30

The ratio of dominant to recessive phenotypes in a randomly mating population (in Hardy-Weinberg equilibrium) plotted as a function of the frequency of the dominant allele. Frequency of the recessive allele equals $1 - p$. The 3:1 ratio of dominant to recessive phenotypes in the second generation of Mendel's crosses occurred because all individuals of the preceding generation were heterozygous, thus $p = q = 0.5$, a condition not common in natural populations. As the dominant allele approaches a frequency of one, virtually no individuals express the recessive phenotype, thus permitting recessive lethal alleles to persist in a population at very low frequencies (0.001).

Consider a human population that contains a dominant allele for normal pigmentation (A) and a recessive allele for albinism (a), which is rare in the gene pool. Figure 1.30 shows the expected frequency of the dominant phenotype in a population as a function of the frequency of the dominant allele in the gene pool. When an allele is rare, copies of that allele occur almost entirely in heterozygous genotypes (Aa in our example), which express the dominant allele phenotypically; only a tiny fraction of the copies of a rare allele occurs in homozygous form (aa). The frequency of the recessive phenotype in the population is thus much less than the frequency of the recessive allele in the gene pool. Using conventional symbols, p denotes the frequency of allele A and $q = (1 - p)$ denotes the frequency of allele a. Under random mating, the genotypic frequencies are p^2 (AA), $2pq$ (Aa) and q^2 (aa). The frequency of albinism in humans is approximately 1/20,000. Assuming that mating is random with respect to genotype at this locus, we calculate using Hardy-Weinberg equilibrium:

$$q^2 = \frac{1}{20{,}000}$$

$$q = \sqrt{\frac{1}{20{,}000}} = \frac{1}{141}$$

$$p = 1 - q$$

$$= \frac{141}{141} - \frac{1}{141} = \frac{140}{141}$$

Hardy-Weinberg Equilibrium: Why Mendelian Heredity Does Not Change Allelic Frequencies

The Hardy-Weinberg law is a logical consequence of Mendel's first law of inheritance, also called the law of segregation, and expresses a tendency toward equilibrium inherent in Mendelian heredity. Mendel's first law of inheritance states that each organism contains a pair of genetic factors for each variable trait. Different forms of these factors are called alleles. The paired factors in a given organism may be copies of the same allele (homozygous) or different alleles (heterozygous). In either case, the formation of gametes involves "segregation" of the paired factors so that each gamete receives only one factor for a given trait. Each gamete from a homozygous organism contains a copy of the same allele. A gamete from a heterozygous organism contains one of its two alleles, and that organism produces its two kinds of gametes in equal frequency.

Let us select for our example a population having a single locus bearing just two alleles, *T* and *t*. The phenotypic expression of this gene might be, for example, ability to taste a chemical compound called phenylthiocarbamide. Individuals in a population can be of three genotypes for this locus, *T/T*, *T/t* (both tasters), and *t/t* (non-tasters). In a sample of 100 individuals, suppose that 20 have the genotype *T/T*, 40 are *T/t*, and 40 are *t/t*. The following table shows the allelic frequencies (remember that every individual's genotype has two copies of a gene):

Genotype	Number of Individuals	Copies of the *T* Allele	Copies of the *t* Allele
T/T	20	40	
T/t	40	40	40
t/t	40		80
TOTAL	100	80	120

Of the 200 copies, the proportion of the *T* allele is 80/200 = 0.4 (40%), and the proportion of the *t* allele is 120/200 = 0.6 (60%). It is customary to use *p* and *q* to represent the two allelic frequencies. The genetically dominant allele is represented by *p*, and the genetically recessive by *q*. Thus:

$$p = \text{frequency of } T = 0.4$$

$$q = \text{frequency of } t = 0.6$$

$$\text{Therefore, } p + q = 1$$

Having calculated allelic frequencies in this sample, let us determine whether these frequencies change spontaneously in a new generation of the population. Assuming that mating is random (gametes are sampled independently in pairs), each individual is expected to contribute an equal number of gametes to a common pool from which the next generation is formed. Frequencies of gametes in the pool are proportional to allelic frequencies in the sample: 40% of the gametes are *T*, and 60% are *t* (ratio of 0.4:0.6). Both ova and sperm will, of course, show similar frequencies. The next generation is formed as shown here:

Sperm	Ova: $T = 0.4$	Ova: $t = 0.6$
$T = 0.4$	$T/T = 0.16$	$T/t = 0.24$
$t = 0.6$	$T/t = 0.24$	$t/t = 0.36$

Collecting genotypes, we have:

$$\text{frequency of } T/T = 0.16$$

$$\text{frequency of } T/t = 0.48$$

$$\text{frequency of } t/t = 0.36$$

Next, we determine the values of *p* and *q* from randomly mated populations. From the table, we see that the frequency of *T* is the sum of genotypes *T/T*, which is 0.16, and one-half of the genotype *T/t*, which is 0.24:

$$T\,(p) = 0.16 + .5(0.48) = 0.4$$

Similarly, the frequency of *t* is the sum of genotypes *t/t*, which is 0.36, and one-half the genotype *T/t*, which is 0.24:

$$t\,(q) = 0.36 + .5(0.48) = 0.6$$

The new generation bears exactly the same allelic frequencies as its parent population! Note that no increase has occurred in the frequency of the genetically dominant allele *T*. Thus *in a freely interbreeding, sexually reproducing population, the frequency of each allele would remain constant generation after generation in the absence of natural selection, migration, recurring mutation, and genetic drift.* A mathematically minded reader would recognize that the genotype frequencies *T/T*, *T/t*, and *t/t* are actually a binomial expansion of $(p + q)^2$:

$$(p + q)^2 = p^2 + 2pq + q^2 = 1$$

Note that the equilibrium calculations give *expected* frequencies, which are unlikely to be realized exactly in a population of limited size. For this reason, limits to population size cause evolutionary change.

Most genes have more than just a single pair of alleles, especially when we measure genetic variation at the DNA sequence level. The binomial expansion shown above can be used for any number of alleles. Suppose that we have three alleles (T_1, T_2, T_3) whose frequencies are denoted *p*, *q*, and *r*, respectively. We now have six possible genotypes with the following Hardy-Weinberg equilibrium frequencies:

$$(p + q + r)^2 = \overset{T_1T_1}{p^2} + \overset{T_1T_2}{2pq} + \overset{T_2T_2}{q^2} + \overset{T_1T_3}{2pr} + \overset{T_2T_3}{2qr} + \overset{T_3T_3}{r^2} = 1$$

As the number of alleles at a gene increases, the proportion of the population having heterozygous genotypes also increases.

The frequency of carriers is:

$$A/a = 2pq = 2 \times \frac{140}{141} \times \frac{1}{141} = \frac{1}{70}$$

One person in every 70 is a carrier!

Tay-Sachs disease in humans is associated with homozygosity for a recessive lethal allele; individuals homozygous for the lethal allele die early in childhood. Natural selection keeps recessive lethal alleles rare in the population, because individuals homozygous for such alleles never reproduce. Natural selection does not eliminate recessive lethal alleles from the population, however, because virtually all copies of those alleles occur in heterozygous genotypes, which are phenotypically normal. Mating is thus random with respect to whether individuals carry the lethal allele. For a recessive lethal allele present in 2 of every 100 persons (but homozygous in only 1 in 1000 fertilizations), 50 generations of selection are required to reduce the frequency of the allele to 1 of every 100 persons.

Key Theme 1.12
GENETICS AND DEVELOPMENT

Eugenics

Eugenics deals with improvement of hereditary qualities in humans by social control of mating and reproduction. While the genetic argument against eugenics in the preceding paragraph is compelling, a eugenics program lingered in the United States (and elsewhere) until it was finally dispatched following Adolf Hitler's attempt to "purify" the races in Europe by genocide.

Nonrandom Mating

If mating is nonrandom, genotypic frequencies deviate from Hardy-Weinberg expectations. For example, if two different alleles of a gene are equally frequent ($p = q = .5$), we expect half of the genotypes to be heterozygous ($2pq = 2$ [.5] [.5] $= .5$) and one-quarter to be homozygous for each allele ($p^2 = q^2$ $[.5]^2 = .25$). In **positive assortative mating,** individuals mate preferentially with others of the same genotype, as when albinos mate with other albinos, for example. Matings among homozygous parents generate offspring that are homozygous like themselves. Matings among heterozygous parents produce on average 50% heterozygous offspring and 50% homozygous offspring (25% of each alternative type) each generation. Positive assortative mating increases the frequency of homozygous genotypes and decreases the frequency of heterozygous genotypes in a population but does not change allelic frequencies.

Preferential mating among close relatives also increases homozygosity and is called **inbreeding.** Whereas positive assortative mating usually affects one or a few traits, inbreeding simultaneously affects all variable traits. Strong inbreeding greatly increases the chances that rare recessive alleles become homozygous and thereby are expressed.

Key Theme 1.13
EVOLUTION

Genetic Inbreeding in Zoo Populations

Inbreeding is a serious problem in zoos holding small populations of rare animals. Matings of close relatives tend to bring together genes from a common ancestor and increase the probability that two copies of a deleterious gene come together in the same organism. The result is "inbreeding depression." The management solution is to enlarge genetic diversity by bringing together captive animals from different zoos or by introducing new stock from wild populations if possible. Paradoxically, where zoo populations are extremely small and no wild stock can be obtained, deliberate inbreeding is recommended. Individuals homozygous for deleterious genes die, thereby removing those genes from the population. The surviving offspring, often few in number, will not carry deleterious genes.

Forces of Evolutionary Change

Population geneticists measure evolutionary change as a change in the frequency of an allele in the gene pool of a population. "Forces" of evolutionary change are factors that can alter the frequency of an allele in the gene pool, and include (1) recurring mutation, (2) genetic drift, (3) migration, (4) natural selection, and (5) interactions among these factors. Recurring mutation is the ultimate source of variability in all populations, but it usually requires interaction with one or more of the other factors to cause a noteworthy change in allelic frequencies. For diploid individuals, a new mutation occurs in one copy in heterozygous condition in a single individual. Its frequency is 1/2N where N is the number of individuals in the population. The total change in allelic frequencies caused by a single mutation is vanishingly small unless the new allele increases in frequency by one of the other forces of evolutionary change. We consider these other forces individually.

Genetic Drift

Some species, such as the cheetah (figure 1.31), contain very little genetic variation, probably because their ancestral lineages were sometimes restricted to very small populations. A small population clearly cannot contain large amounts of

hilton kotze/iStock/Getty Images

figure 1.31

Cheetahs, a species whose genetic variability has been depleted to very low levels because of small population size in the past.

genetic variation. Each individual organism has at most two different allelic forms of each gene, and a single breeding pair contains at most four different allelic forms of a gene. Suppose that we have such a breeding pair. We know from Mendelian genetics that chance decides which allelic forms of a gene get passed to offspring. It is therefore possible by chance alone that one or two parental alleles in this example are not passed to any offspring. It is highly unlikely that the different alleles present in a small ancestral population are all passed to descendants without any change in allelic frequency. This chance fluctuation in allelic frequency from one generation to the next, including loss of alleles from a population, is called **genetic drift.**

Genetic drift occurs to some degree in all populations of finite size. Perfect constancy of allelic frequencies, as predicted by Hardy-Weinberg equilibrium, occurs only in populations whose size approaches infinity, and such populations occur only in mathematical models. All populations of animals are finite and therefore experience some effect of genetic drift, which becomes greater, on average, as population size declines. Genetic drift erodes the genetic variability of a population. If population size remains small for many generations in a row, genetic variation can be greatly depleted. Alleles are lost by chance at a much higher rate than new ones are introduced by mutation. This loss is harmful to the evolutionary success of a species because it restricts potential genetic responses to environmental change. Indeed, biologists are concerned that cheetah populations may have insufficient variation for continued survival.

The term *population bottleneck* denotes a large reduction in the size of a population, causing loss of genetic variation, by genetic drift, followed by an increase in population size. The loss of variation is proportional to the number of generations for which the number of individuals in the population remains small. A bottleneck associated with the founding of a new geographic population is called a founder effect and sometimes initiates the formation of a new species.

Migration

Migration is the movement of individuals from one population to another one prior to mating. Movement of alleles between populations in this manner is called "gene flow." Migration is an evolutionary force only in species that are geographically subdivided into two or more separate populations; otherwise, the species evolves as a single undivided gene pool. Migration prevents different populations of a species from becoming genetically dissimilar. If a species is divided into many small populations, genetic drift and selection acting separately in different populations can produce evolutionary divergence among them. A small amount of migration among populations each generation keeps different populations from diverging strongly. For example, the French and Russian populations show some genetic divergence (see figure 1.28), but continuing migration prevents them from becoming completely distinct.

Natural Selection

Natural selection can change both allelic frequencies and genotypic frequencies in a population. Although effects of selection are often reported for particular polymorphic genes, it is important to remember that natural selection acts on whole animals, not on isolated traits. The organism that possesses a combination of traits most successful in a particular environment is favored. An animal may have some traits that confer no advantage or even a disadvantage, but the animal is successful overall if its combination of traits is favorable. When we claim that a particular genotype has a higher relative **fitness** than do alternative genotypes, we state that the genotype on average confers an advantage in survival and reproduction in a population. If alternative genotypes have unequal probabilities of survival and reproduction, Hardy-Weinberg equilibrium is upset.

Using the genetic theory of natural selection, one can measure relative fitness values associated with different genotypes in a population. Geneticists often use W to denote the expected average fitness of a genotype in a population, with the genotype of highest fitness given a value of one and fitnesses of other genotypes indicated as fractions.

We illustrate measurement of fitness using genetic variation associated with the disease sickle cell anemia in human populations. Considering only the alleles for normal hemoglobin (*A*) and sickle cell hemoglobin (*S*) for the beta-hemoglobin gene in human populations, the possible genotypes are *AA*, *AS*, and *SS*. Measurements of viability of individuals of these three genotypes in nonmalarial environments give a fitness value of 1 to genotypes *AA* and *AS* and a fitness of 0.2 to genotype *SS*. People having the *SS* genotype, who are susceptible to severe anemia, are expected to contribute only 20% as many offspring to the next generation on average as are individuals having the *AA* or *AS* genotypes. In malarial environments, genotype *AS* has the highest fitness (= 1); genotype *AA* has a slightly decreased fitness (= 0.9) because these individuals have a greater incidence of malaria than do *AS* individuals, and *SS* has a low fitness (= 0.2) because of anemia. From these measured fitness values and knowledge of the frequencies of alleles in a population and its system of mating, one can calculate the **average effect** that an allele has on the phenotype of relative fitness in that population. In the example of sickle cell anemia, the average effect of allele *S* on fitness in a malarial environment is a balance between the strongly negative effect it has when homozygous and the positive effect that it has when heterozygous with allele *A*.

The average effect of an allele on fitness is expressed not only by its direct contribution to the fitness of its possessors but also by aid that its possessors give to close relatives, who are likely to contain copies of the allele. The term "inclusive fitness" pertains to cases where the average effect of an allele on its own propagation in the gene pool would be calculated incorrectly if its effects on survival of its possessors' relatives were ignored. For example, natural selection might favor an allele for a behavior through which an individual dies in saving the lives of many relatives.

figure 1.32

A pair of wood ducks. Brightly colored feathers of male birds probably confer no survival advantage and might even be harmful by alerting predators. Such colors nonetheless confer advantage in attracting mates, which overcomes, on average, any negative consequences of these colors for survival. Darwin used the term "sexual selection" to denote traits that give an individual an advantage in attracting mates, even if these traits are neutral or harmful for survival.

Some traits and combinations of traits are advantageous for certain aspects of an organism's survival or reproduction and disadvantageous for others. Darwin used the term **sexual selection** to denote selection of traits that are advantageous for obtaining mates but perhaps harmful for survival. For example, bright colors and elaborate feathers might enhance a male bird's competitive ability to obtain mates while simultaneously increasing his vulnerability to predators (figure 1.32). Environmental changes can alter the selective values of different traits.

Selection is often studied using quantitative traits, those that show continuous variation in a population with no obvious pattern of Mendelian segregation in their inheritance. The values of a trait in offspring often are intermediate between the values of their parents. Such traits are influenced by variation at many genes, each of which follows Mendelian inheritance and contributes a small, incremental amount to a phenotype. Traits that show quantitative variation include tail length in mice, length of a leg segment in grasshoppers, number of gill rakers in sunfishes, number of peas in pods, and height of adult humans. When values are graphed with respect to frequency distribution, they often approximate a normal, bell-shaped curve (figure 1.33A). Most individuals fall near the average, fewer fall well above or below the average, and extreme values form "tails" of the frequency curve with increasing rarity.

Selection can act on quantitative traits to produce three different kinds of evolutionary response. **Stabilizing selection** favors average values of a trait and disfavors extreme ones (figure 1.33B). **Directional selection** favors an extreme value of a phenotype and causes a population average to shift toward it

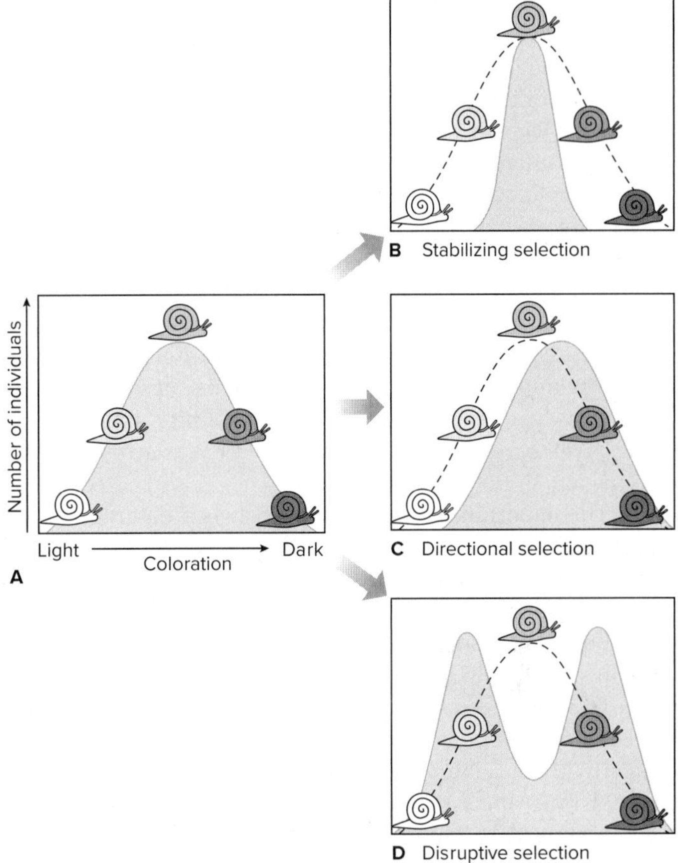

figure 1.33

Responses to selection on a continuous (polygenic) character, coloration in a snail. **A,** Frequency distribution of coloration before selection. **B,** Stabilizing selection culls extreme variants from the population, in this case eliminating unusually light or dark individuals and thereby stabilizing the mean. **C,** Directional selection shifts the population mean, in this case by favoring darkly colored variants. **D,** Disruptive selection favors both extremes but not the mean; the mean is unchanged, but the population no longer has a bell-shaped distribution of phenotypes.

(figure 1.33C). When we think about natural selection producing evolutionary change, we usually have directional selection in mind, although that is not the only possibility. A third alternative is **disruptive selection,** in which two different extreme phenotypes are favored simultaneously, but their average is disfavored (figure 1.33D). If inbreeding or positive assortative mating accompanies disruptive selection, this can cause a population to become bimodal, meaning that two very different phenotypes predominate.

Interactions of Selection, Drift, and Migration

Subdivision of a species geographically into small populations that exchange migrants is an optimal way to promote rapid adaptive evolution of a species. Interactions of genetic drift and selection in different populations permit many different genetic combinations of many polymorphic genes to be tested against natural selection. Migration among populations permits particularly favorable new genetic combinations to spread throughout the species as a whole. Interactions of selection, genetic drift, and migration in this example produce evolutionary change qualitatively different from what would result if any of these three factors acted alone. The geneticist Sewall Wright called this interaction *shifting balance* because it permits a population to switch from an initial adaptive combination of traits to a new and possibly better one. Genetic drift, nonrandom mating, migration, and natural selection interact in natural populations to create an enormous opportunity for evolutionary change; the perpetual stability predicted by Hardy-Weinberg equilibrium almost never lasts across any substantial amount of evolutionary time.

The importance of interactions between natural selection and genetic drift in adaptive evolution is illustrated by the beta-hemoglobin variation discussed in the preceding sub section. We noted that in malarial environments, natural selection retains both the *A* and *S* alleles in the population because individuals with the *AS* genotype uniquely combine the benefits of malarial resistance and normal respiration. A rare third allele called hemoglobin *C* also occurs in some malarial regions of western Africa. Like hemoglobin *S*, allele *C* was derived from allele *A* by a single mutation changing the sixth codon; in allele *C*, lysine (AAG) substitutes for glutamic acid (GAG). In malarial areas, natural selection acts against allele *C* in heterozygous genotypes because *AC* individuals are susceptible to malaria (as are *AA* individuals), and *CS* individuals have severe anemia. Natural selection favors the *C* allele in homozygous form; *CC* individuals have malarial resistance, normal respiration, and much higher fitness than do *AS* individuals in malarial areas. If the population were fixed for the *C* allele, all individuals would benefit from both malarial resistance and normal respiration.

Why has the *C* allele not become the most frequent one in malarial regions of western Africa? We know from Hardy-Weinberg equilibrium that in a randomly mating population, a rare allele occurs almost exclusively in heterozygous genotypes with the more common alleles. Because selection favors *AS* individuals over *AC* and *CS* individuals in malarial areas, selection acts to eliminate allele *C* from those populations.

In a few local populations of western Africa, genetic drift caused the *C* allele to reach a relatively high frequency prior to the onset of malaria. In these populations alone, the *CC* genotype occurs with sufficient frequency that the positive action of natural selection on these individuals exceeds the negative effects of selection on the *AC* and *CS* individuals in the same population. The *C* allele has increased in frequency by natural selection in these local populations. One expects that natural selection eventually would cause the *C* allele to become fixed in these local populations, and that gene flow from them to others would allow the *C* allele to become fixed across malarial Africa. One hopes, of course, that effective treatment of malaria will overcome the need for an evolutionary solution to this problem, as the latter would require many generations and much intervening illness. Nonetheless, this example shows how interactions between genetic drift, natural selection, and gene flow can shift a population's response to malarial selection from maintaining a polymorphism of the *A* and *S* alleles to fixing the *C* allele.

1.7 Macroevolution: Major Evolutionary Events and Processes

Macroevolution describes large-scale events and processes in organic evolution. Speciation links macroevolution and microevolution. The major trends in the fossil record described earlier (see figures 1.14 through 1.16) fall clearly within the realm of macroevolution. Patterns and processes of macroevolutionary change emerge from those of microevolution, but they acquire some degree of autonomy in doing so. The emergence of new adaptations and species, and the varying rates of speciation and extinction observed in the fossil record, transcend the fluctuations of allelic frequencies within populations.

Speciation and Extinction Through Geological Time

Evolutionary change at the macroevolutionary level provides a new perspective on Darwin's theory of natural selection. Although a species can persist for many millions of years, it ultimately has two possible evolutionary fates: it may give rise to new species or become extinct without leaving descendants. Rates of speciation and extinction vary among lineages, and lineages that have the highest speciation rates and the lowest extinction rates produce the greatest diversity of living forms. The characteristics of a species might make it more or less likely than others to undergo speciation or extinction. Because many characteristics are passed from ancestral to descendant species analogous to heredity at the organismal level, lineages whose properties enhance the probability of speciation and confer resistance to extinction should dominate the living world. This species-level process that produces differential rates of speciation and extinction among lineages is analogous in many ways to natural selection. It represents an expansion of Darwin's theory of natural selection.

Species selection denotes the differential survival and multiplication of species through geological time based on variation among lineages in species-level properties. These species-level properties include mating rituals, social structuring, migration patterns, and geographic distribution. Descendant species usually resemble their ancestors for these properties. For example, a "harem" system of mating in which a single male and several females compose a breeding unit characterizes some mammalian lineages but not others. We expect speciation rates to be enhanced by social systems that promote the founding of new populations by small numbers of individuals. Certain social systems might increase the probability that a

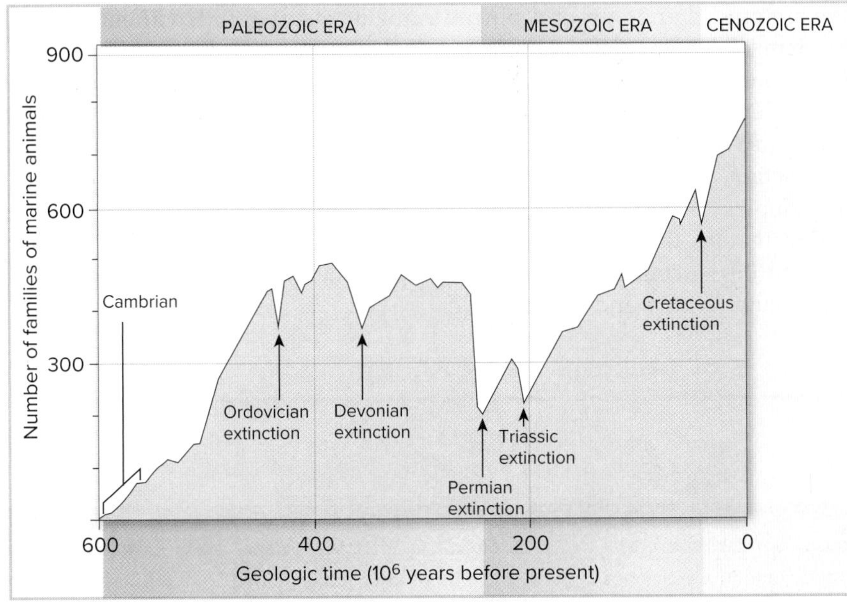

figure 1.34

Changes in numbers of taxonomic families (see section 4.1) of marine animals through time, from the Cambrian period to the present. Sharp drops represent five major extinctions of skeletonized marine animals. Note that despite these extinctions, the overall number of marine families has increased over the past 540 million years. (Results follow the analysis of David Raup and J. John Sepkoski, Jr., as reported in 1982. Subsequent revisions to geological dating move the start of the Cambrian forward from 600 million years ago as shown to 542 million years ago, requiring compression of the left 60 million years of the figure.)

species survives environmental challenges through cooperative action. Such properties would be favored over geological time by species selection.

Mass Extinctions

When we study evolutionary change on an even larger timescale, we observe episodic events in which large numbers of taxa become extinct nearly simultaneously. These events are called **mass extinctions** (figure 1.34). The most cataclysmic of these extinction episodes happened about 250 million years ago, when at least half of the families of shallow-water marine invertebrates and fully 90% of marine invertebrate species disappeared within a few million years. This was the **Permian extinction.** The **Cretaceous extinction,** which occurred about 65 million years ago, marked the end of dinosaurs as well as numerous marine invertebrates and many small reptilian species.

Causes of mass extinctions and their occurrence at intervals of approximately 26 million years throughout the past 250 million years are difficult to explain. Some researchers have proposed biological explanations for these mass extinctions, and others consider them artifacts of our statistical and taxonomic analyses. Walter Alvarez proposed that the earth was occasionally bombarded by asteroids, causing the mass extinction at the end of the Cretaceous and perhaps others (figure 1.35). The drastic effects of such a bombardment were observed in July 1994 when fragments of Comet Shoemaker-Levy 9 bombarded Jupiter. The first fragment to hit Jupiter was estimated to have the force of 10 million hydrogen bombs. Twenty additional fragments hit Jupiter within the following week; one of them was 25 times more powerful than the first fragment. This bombardment was the most violent event in the recorded history of our solar system. Bombardments by asteroids or comets could change the earth's climate drastically, sending debris into the atmosphere and blocking sunlight. Temperature changes would challenge ecological tolerances of many species. This hypothesis is being tested in several ways, including searches for impact craters left by asteroids and for altered mineral content of rock strata where mass extinctions occurred. Unusual concentrations of the rare-earth element iridium in strata at the end of the Cretaceous indicate that this element entered the earth's atmosphere through asteroid bombardment. So far, the end-Cretaceous mass extinction is the only one that coincides with strong geological evidence for asteroid bombardment.

Source: D. Roddy(LPI)/NASA

figure 1.35

Meteor crater of the Arizona desert is among the youngest and best-preserved impact craters. Geologists estimate that it formed approximately 50,000 years ago when a meteor measuring 30 meters wide and weighing 100,000 tons landed at a speed of 20 kilometers per second. Asteroid impacts associated with the end-Cretaceous mass extinctions would have greatly exceeded the power of this one, but geological changes accumulating over the past 65 million years have lessened the preservation of those craters.

Sometimes lineages favored by species selection are unusually at risk during a mass extinction. Climatic changes caused by hypothesized asteroid bombardments could produce selective challenges very different from those encountered at other times in the earth's history. Selective discrimination of particular biological traits by mass extinction events is termed **catastrophic species selection.** For example, mammals survived the end-Cretaceous mass extinction that destroyed dinosaurs and other prominent vertebrate and invertebrate groups. The ability of small mammals to occupy underground burrows might have permitted them to withstand climatic stresses that killed more exposed animals. Following this event, mammals were able to use environmental resources that previously had been denied them, leading to their adaptive radiation.

Natural selection, species selection, and catastrophic species selection interact to produce macroevolutionary trends that we see in the fossil record. Intensive study of these interacting causal processes has made modern paleontology an active and exciting field.

■ SUMMARY

Sections	Key Concepts
1.1 Principles of Science	• Zoology is the scientific study of animals. Science is the acquisition of knowledge guided by hypotheses that are explanatory with reference to natural law, and that are testable, tentative, and falsifiable. These criteria form the basis for the scientific method, which is called hypothetico-deductive. • On the basis of prior observations, a scientist formulates an explanatory hypothesis. Predictions about future observations based on the hypothesis guide researchers to collect data that can support or falsify the hypothesis. • A testable hypothesis for which there is a large amount of supporting data, particularly a hypothesis that explains a very large number of observations, is called a theory. • Causal factors that influence or regulate the operation of a biological system are called proximate or immediate causes. We test hypotheses of proximate causes using controlled experiments, which reveal the mechanisms underlying animal form and function. • By contrast, ultimate causes denote the factors in evolutionary history that produced the biological system being studied. We test hypotheses of ultimate causality using the comparative method, guided by knowledge of the phylogenetic relationships of the species being studied.
1.2 Origins of Darwinian Evolutionary Theory	• Charles Darwin's theory of organic evolution explains life's diversity as the historical outcome of gradual change from previously existing forms. • The idea of evolution came from ancient Greek thought and from Lamarck's theory, published the year of Darwin's birth. • Darwin derived much of the material used to construct his theory from his experiences on a five-year voyage around the world aboard H.M.S. *Beagle*. He read Charles Lyell's geological theory and applied Lyell's uniformitarian gradualism to living systems. • Darwin transcended Lamarck's theory in proposing that the population is the unit of evolutionary change; variation among organisms in a population is the basis for evolutionary change.
1.3 Darwin's Theory of Evolution	• Darwin's evolutionary theory has five major components. Its most basic proposition is *perpetual change,* the theory that life has a long history of irreversible change with continuity from past to present life. • Darwin's theory of *common descent* states that all organisms descend from a common ancestor through branching genealogical lineages. • A corollary of common descent is *multiplication of species* through evolutionary time. Species are reproductively distinct populations that form independent evolutionary lineages. Splitting of an ancestral species into geographically isolated populations, followed by evolutionary divergence of the separated populations, leads to formation of new species. • Darwin's theory of *gradualism* states that large phenotypic differences between species are produced by accumulation through evolutionary time of many small incremental changes. • Darwin's fifth major statement is that *natural selection* is the major guiding force of evolution. This principle is founded on observations that all species overproduce their kind, causing a struggle for limited resources. Because no two organisms are exactly alike, and because variable traits are at least partially heritable, those organisms whose hereditary endowment enhances their use of resources for survival and reproduction contribute disproportionately to the next generation. Over many generations, the sorting of variation by selection produces new species and new adaptations. • Darwin's theories of perpetual change, common descent and multiplication of species do not depend on Darwin's more controversial theories of gradualism and natural selection.

Sections	Key Concepts
1.4 Evidence for Darwin's Five Theories of Evolution	• The fossil record amply demonstrates perpetual change of animal form and diversity following the Cambrian explosion about 540 million years ago. • Morphological homologies among organisms are characteristics inherited with modification from a corresponding feature of their common evolutionary ancestor. A nested hierarchy of homologies in animal diversity strongly supports a history of phylogenetic branching of population lineages from a common ancestor. • Allopatric speciation is the evolution of reproductive barriers between geographically separated populations to generate new species. Adaptive radiation denotes proliferation of many adaptively disparate species from a single ancestral lineage, especially when the process occurs within a relatively short interval of geological time (for example, within one million years). Oceanic archipelagoes, such as the Galápagos Islands, are particularly conducive to adaptive radiation of terrestrial organisms. • Gradualism remains controversial in population genetics because mutations that have large effects on phenotype have been used in animal breeding and occur in some natural populations. • At the macroevolutionary level, punctuated equilibrium states that most evolutionary change occurs in relatively brief events of branching speciation, separated by long intervals in which a species shows morphological stasis. • An adaptation is a character that evolved by natural selection for a particular biological role, whereas exaptation denotes the utility of a character for a biological role that was not part of the character's evolutionary origin.
1.5 Revisions of Darwinian Evolutionary Theory	• Darwin's theory of natural selection was modified around 1900 and in subsequent decades by correcting his genetic errors. August Weismann's studies of animal development and heredity rejected the Lamarckian notion that use and disuse of body parts could alter their heredity. Darwin's theory as modified by Weismann is called neo-Darwinism. Mendelian inheritance and the chromosomal theory further clarified the particulate and non-blending nature of genetic factors. • In the mid-twentieth century, a synthetic neo-Darwinian evolutionary theory unified the fields of population genetics, paleontology, biogeography, embryology, systematics, and animal behavior.
1.6 Microevolution: Genetic Variation and Change Within Species	• Mutations are the ultimate source of variation on which selection acts to produce evolution. • Darwin's theory postulates that variation arises at random with respect to an organism's needs, and that differential survival and reproduction direct evolutionary change. • Population geneticists discovered the principles by which populations undergo genetic change through time. Hardy-Weinberg equilibrium shows that the hereditary process itself does not cause evolution. • Sources of evolutionary change include mutation, genetic drift, nonrandom mating, migration, natural selection, and their interactions. • Genetic drift is random change in the frequencies of allelic variants caused by finite population size and the resulting sampling error that occurs from one generation to the next. • Mutation and genetic drift together suffice to produce evolutionary change. • Gene flow is the movement of genes from one population to another by migration of individuals between populations. • Polymorphism for the beta-hemoglobin gene in human populations in malarial regions of Africa illustrates natural selection and its interactions with genetic drift and gene flow.
1.7 Macroevolution: Major Evolutionary Events and Processes	• Macroevolution constitutes evolutionary change over large scales of geological time, encompassing many millions of years. • Evolutionary processes emerging above the level of the population include species selection: species-level properties, such as geographical range and amount of gene flow among populations, can cause some lineages to be more prone than others to extinction or to speciation. • Episodic mass extinctions, such as the end-Cretaceous extinction associated with an asteroid impact, may select for characteristics different from ones that prospered between extinction peaks. This is catastrophic species selection and may reset evolution—as occurred when mammals survived the end-Cretaceous mass extinction, but dinosaur lineages did not.

■ Review Questions

1. What are the essential characteristics of science? Describe how evolutionary studies fit these characteristics. Why are arguments based on supernatural or miraculous explanations, such as "creation-science" or "intelligent design," not valid scientific hypotheses?
2. What is the relationship between a hypothesis and a theory?
3. Explain how biologists distinguish between experimental and comparative methods.
4. Briefly summarize Lamarck's hypothesis for evolution. What evidence rejects this hypothesis?
5. What is "uniformitarianism"? How did it influence Darwin's evolutionary theory?
6. Explain how Darwin's journey on the *Beagle* and his study of pigeon breeding affected his thinking.
7. What key idea, contained in Malthus's essay on populations, helped Darwin to formulate his theory of natural selection?
8. Explain how each of the following contributes to Darwin's evolutionary theory: fossils, geographic distributions of closely related animals, homology, and animal classification.
9. How do modern evolutionists view the relationship between ontogeny and phylogeny?
10. What major evolutionary lesson is illustrated by Darwin's finches on the Galápagos Islands?
11. How does observation of "sports" in animal breeding challenge Darwin's theory of gradualism? Why did Darwin consider such mutations to have little evolutionary importance?
12. What does the theory of punctuated equilibrium state about the occurrence and importance of speciation throughout geological time?
13. Describe the observations and inferences that compose Darwin's theory of natural selection.
14. Identify the random and nonrandom components of Darwin's theory of natural selection.
15. Describe some recurring criticisms of Darwin's theory of natural selection. How are these criticisms refuted?
16. A common but mistaken notion is that because some alleles are genetically dominant and others are recessive, dominants eventually replace all recessives in a population. How does the Hardy-Weinberg equilibrium refute this notion?
17. Assume that you are measuring frequencies of alternative conditions of a variable trait from two animal populations. The trait is controlled by a single allelic pair, *A* and *a*, and you can distinguish all three phenotypes, *AA*, *Aa*, and *aa* (intermediate inheritance). Your sample includes:

Population	*AA*	*Aa*	*aa*	TOTAL
I	300	500	200	1000
II	400	400	200	1000

Calculate the distribution of phenotypes in each population as expected under Hardy-Weinberg equilibrium. Is population I in equilibrium? Is population II in equilibrium?

18. After studying a population for a trait determined by a single pair of alleles, you find that this population departs from predictions of Hardy-Weinberg equilibrium in having higher than expected frequencies of homozygous genotypes and lower than expected frequencies of heterozygous genotypes. What possible reasons might explain its departure from equilibrium?
19. Explain why genetic drift is more powerful in small populations.
20. Is it easier for selection to remove a deleterious recessive allele from a randomly mating population or from a highly inbred population? Why?
21. Contrast microevolution and macroevolution.

For Further Thought Why will our knowledge of the origin of life always be more tentative than our knowledge of the subsequent evolution of life's diversity?

2

Animal Ecology

LEARNING OBJECTIVES

Readers will be able to:

2.1 Describe the hierarchical structure of ecological systems including the levels of organisms, populations, ecological communities, and ecosystems.

2.2 Explain the critical concept of niche.

2.3 Explain population demography and how population growth is regulated.

2.4 Describe how parasitism, competition, and predation influence the species composition of an ecological community.

2.5 Define an ecosystem and explain the flow of matter and energy through an ecosystem.

2.6 Explain how species extinction rates have varied through geological time, including the contrast between low levels of extinction and mass extinction.

Cleveland P. Hickman, Jr.

Talkeetna Mountain Range, Alaska.

Ecological Niche and Animal Species Diversity

Life's diversity is organized into a hierarchy of interacting units: an individual organism, a population, a community, and most inclusively an ecosystem. Central to ecological study of animal diversity is habitat, the spatial location where an animal lives. What an animal does in its habitat is its niche: how it gets its food, how it achieves reproductive perpetuity—in short, how it survives and stays adapted in a Darwinian sense.

As a species evolves, so too does its niche, and no two species in a community can evolve to exploit exactly the same resources at the same place and time. This illustrates the "competitive exclusion principle": no two species can occupy the same niche if they are to live together in a stable ecological community. Different species are therefore able to form an ecological community in which each has a different role in their shared environment.

In the mid-nineteenth century, German zoologist Ernst Haeckel introduced the term **ecology**, defined as the relation of the animal to its organic as well as inorganic environment. Environment here includes everything external to the animal but, most important, its niche. Although we no longer restrict ecology to animals alone, Haeckel's definition remains basically sound. Animal ecology is now a highly synthetic science that incorporates the behavior, physiology, genetics, and evolution of animals to study interactions between populations of animals and their environments. The major goal of ecological studies is to explain how these diverse interactions influence the geographical distribution and abundance of animal populations. Measuring the niche requirements of each species is crucial for understanding animal diversity and for ensuring the continued survival of many populations.

2.1 The Hierarchy of Ecology

Ecology encompasses a hierarchy of biological systems in interaction with their environments. At the base of the ecological hierarchy is an **organism** such as an individual animal. To understand why animals live where they do, ecologists must examine the varied physiological and behavioral mechanisms that animals use to survive, to grow, and to reproduce. For example, a near-perfect physiological balance between production and loss of heat is required for success in certain endothermic species (birds and mammals, whose body temperature is regulated internally) under extreme temperatures that occur in the Arctic or a desert. Other species succeed in these situations by escaping the most extreme conditions through migration or hibernation. Insects, fishes, and other ectotherms (animals whose body temperature is determined by environmental heat) compensate for fluctuating temperatures through behavioral responses (figure 2.1) or by altering biochemical and cellular processes involving enzymes, lipid organization, and the neuroendocrine system. Thus an animal's physiological capacities permit it to occupy changing and often adverse environments. Behavioral responses are important also for obtaining food, finding shelter, escaping enemies and unfavorable environments, finding a mate, courting, and tending to offspring. Physiological mechanisms and behaviors that improve adaptability to environments help organisms to survive. Ecologists who focus their studies at the organismal level are called physiological ecologists or behavioral ecologists.

Animals in nature coexist with others of the same species as reproductive units called **populations**. Measurable properties of populations include genetic variability among individuals (polymorphism), a gene pool (see section 1.6), growth in numbers over time, and factors that limit the density of individuals in a given area. Ecological studies of these properties help us to predict the future success of endangered species and to discover controls for pest species.

Just as individuals are not alone in nature, populations of different species co-occur in more complex associations called **ecological communities**. One measure of an ecological community's complexity is the number of species present, called **species richness**. The populations of species in a community interact with one another in many ways, the most prevalent ones being **predation**, **parasitism**, **competition**, and **mutualism**. **Predators** obtain energy and nutrients by killing and eating prey. **Parasites** derive similar benefits from their host organism, but they live on or in the host and usually do not kill the host. A **parasitoid** lives on or in a host but eventually kills its host organism. Competition occurs when food and/or space are limited and members of different species interfere with each other's use of their shared resources. Mutualism occurs when two different species both benefit from their association in a community. Communities are complex because all of these interactions occur simultaneously, and their individual effects on the community often cannot be isolated.

Ecological communities are biological components of even larger, more complex entities called **ecosystems**. An ecosystem

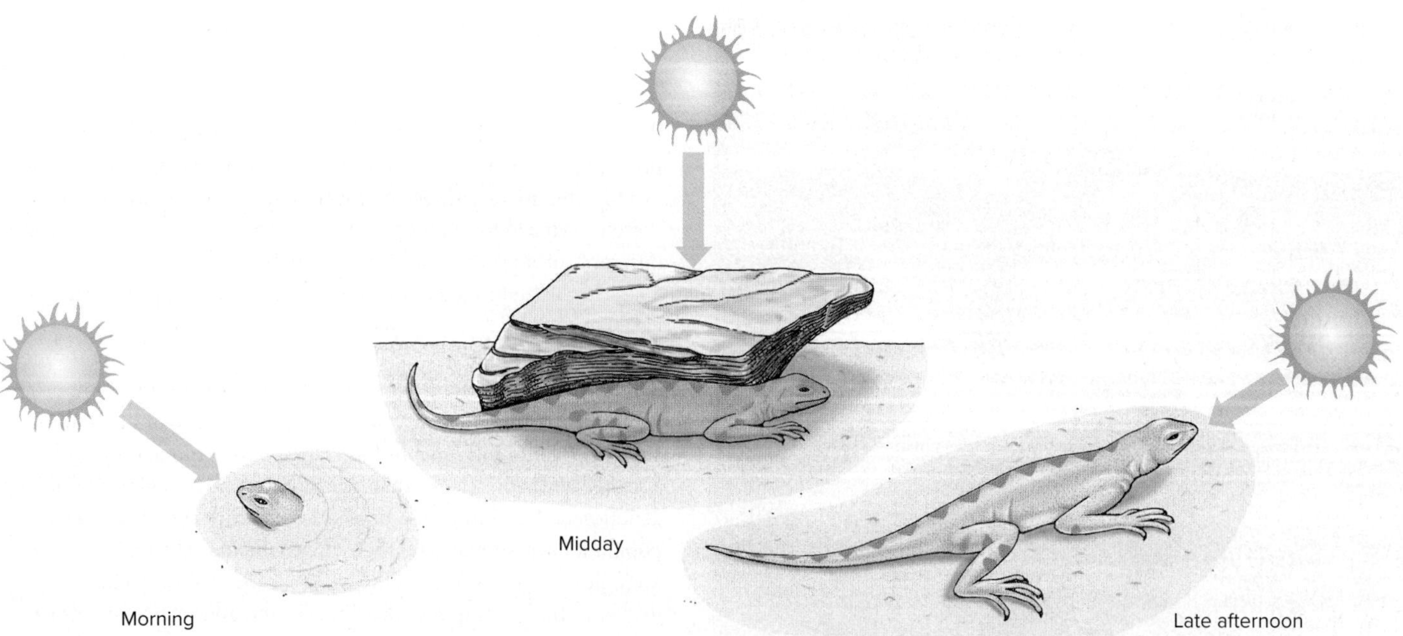

figure 2.1

How a lizard (an ectotherm) regulates its body temperature behaviorally. In the morning, the lizard absorbs the sun's heat through its head while keeping the rest of its body protected from cool morning air. Later it emerges to bask. At noon, with its body temperature high, it seeks shade from the hot sun. When the air temperature drops in the late afternoon, it emerges and lies parallel to the sun's rays.

consists of all populations in an ecological community and their physical environment. The study of ecosystems reveals two key processes in nature: the flow of energy and the cycling of materials through biological channels. The largest ecosystem is the **biosphere**, the thin veneer of land, water, and atmosphere that envelops the earth and supports all life.

2.2 Environment and the Niche

An animal's *environment* comprises all conditions that directly affect its chances for survival and reproduction. These factors include space; forms of energy, such as sunlight, heat, wind, and water currents; and materials, such as soil, air, water, and numerous chemicals. Environment also includes other organisms, which can be an animal's food or its predators, competitors, hosts, mutualists, or parasites. Environment thus includes both abiotic (nonliving) and biotic (living) factors. **Resources**, such as space and food, are the factors that an animal uses directly.

An investigator categorizes a particular resource as being either expendable or nonexpendable, depending on how an animal uses it. Food is expendable, because once eaten it is no longer available. Food therefore must be continuously replenished. Space, whether total living area or a subset, such as the number of suitable nesting sites, is not exhausted by being used, and thus is nonexpendable.

The physical space where an animal lives is its **habitat**. Size of habitat varies. A rotten log is a normal habitat for carpenter ants. Such logs occur in larger habitats called forests where deer also live. However, deer forage in open meadows, so their habitat is larger than the forest. On a larger scale, some migratory birds occupy forests of the north temperate region during summer and move to the tropics during winter. Thus, habitat is defined by an animal's normal activity rather than by arbitrary physical boundaries.

Animals of any species demonstrate environmental limits of temperature, moisture, and food within which they can grow, reproduce, and survive. A suitable environment therefore must meet all requirements for life. A freshwater clam inhabiting a tropical lake could tolerate the temperature of a tropical ocean, but the ocean's salinity would be fatal. A brittle sea star living in the Arctic Ocean could tolerate the salinity of a tropical ocean but not its temperature. Thus temperature and salinity are two separate dimensions of an animal's environmental limits. If we add another variable, such as pH, we increase our description to three dimensions (figure 2.2). If we consider all environmental conditions that permit members of a species to survive and multiply, we define a role for that species in nature distinct from those of other species. This unique, multidimensional fingerprint of a species is called its **niche**. Dimensions of niche vary among members of a species, making the niche subject to evolution by natural selection. The niche of a species undergoes evolutionary changes over successive generations.

One can classify particular animals as either generalists or specialists with respect to a particular environmental resource or condition. For example, most fish species are adapted to either freshwater or seawater, but not both. However, those that live in salt marshes, such as the killifish, *Fundulus heteroclitus,* are generalists regarding salinity, and easily tolerate changes in salinity as freshwater from land mixes with seawater during tidal cycles. Similarly, although most snakes are capable of eating a wide variety of animal prey, others are narrow dietary specialists; for example, the African snake *Dasypeltis inornata* is specialized for eating bird eggs (see figure 18.19).

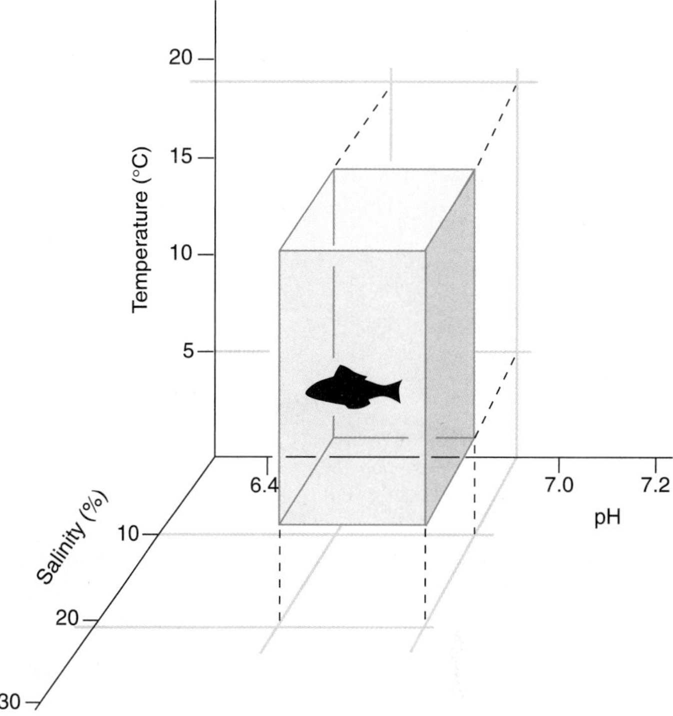

figure 2.2

Three-dimensional niche volume of a hypothetical animal showing three tolerance ranges. This graphic representation is one way to show the multidimensional nature of environmental relations. This representation is incomplete, however, because additional environmental factors also influence growth, reproduction, and survival.

However broad its tolerance limits, an animal experiences only a single set of conditions at a time. An animal probably does not experience in its lifetime all of the environmental conditions that it potentially can tolerate. Thus, we must distinguish an animal's **fundamental niche**, which describes its potential role, and its **realized niche**, the subset of potentially suitable environments that an animal actually experiences. Likewise, we must distinguish fundamental niche from realized niche at the population and species levels. For example, competition within a community may limit the realized niche of a species to a much smaller range of conditions than those predicted by its fundamental niche. For example, the lungless slimy salamander *Plethodon teyahalee* occupies forest-floor habitats in the southern Appalachian Mountains. *Plethodon teyahalee* occupies the coolest, mountaintop forests only where species of the *Plethodon jordani* group are absent (figure 2.3). Climatic requirements restrict the *P. jordani*–group species to

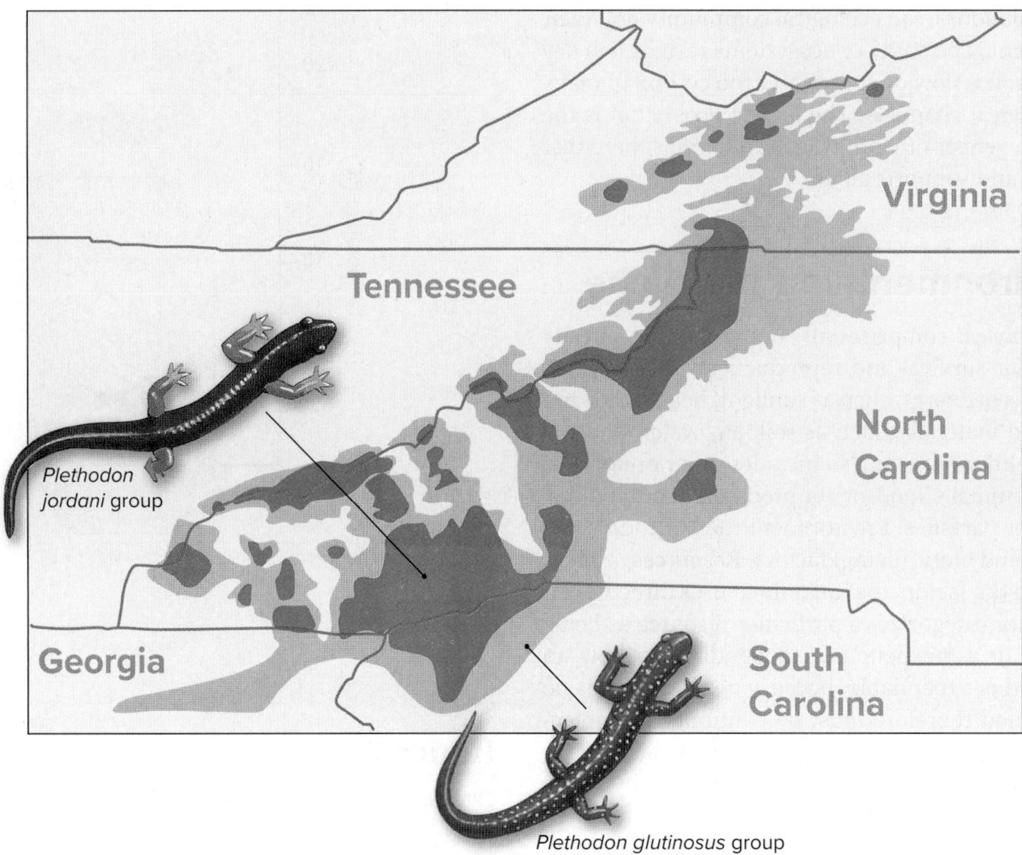

figure 2.3

Geographic consequences of competitive exclusion. The fundamental niches of woodland salamanders of the *Plethodon jordani* group restrict them to the coldest environments located at mountaintops in the southern Appalachian Mountains (red shading). Salamanders of the *Plethodon glutinosus* group (which includes the species *P. teyahalee* mentioned in the text) have a broader temperature tolerance, usually occupying warmer woodlands at lower elevations (including areas shown in gray on the map), but they can occupy the colder habitats favored by the *P. jordani* group provided that members of the *P. jordani* group are absent. Where they occur, populations of the *P. jordani* group aggressively exclude members of the *P. glutinosus* group from their geographic territory. Climatic cycles occurring on tens of thousands of years alter the geography of this competitive exclusion; the gray areas on the map would be cold enough to support *P. jordani*–group populations at times of maximum glaciation, and *P. glutinosus*–group populations thus would be driven from these areas during colder parts of the climatic cycle. Competition from the *P. jordani* group thus limits the realized niche of many *P. glutinosus*–group populations to a subset of their fundamental niche in the dimensions of temperature and elevation.

Source: Weisrock, D. W., and A. Larson. "Testing hypotheses of speciation in the *Plethodon jordani* species complex with allozymes and mitochondrial DNA sequences," *Biological Journal of the Linnean Society 89,* 2006, 25–51.

cool mountaintop isolates, from which they exclude *P. teyahalee* through aggressive encounters. Competition with the *P. jordani*–group species thereby constrains the realized niche of some *P. teyahalee* populations to a smaller and higher average range of temperatures than their fundamental niche would permit.

2.3 Populations

Animals exist in nature as members of populations. A population is a reproductively interactive group of animals of a single species. A species might comprise a single, cohesive population or many geographically disjunct populations. A geographically and genetically cohesive population that is separable from other such populations is called a **deme**. Because members of a deme regularly interbreed, they share a common gene pool (see section 1.6). A deme of cichlid fishes in Cuatro Cienegas, Mexico, was mistakenly diagnosed as two separate species because some individuals have strong "molariform" jaws capable of crushing snails, whereas others have weaker "papilliform" jaws capable of processing only soft food items. Molecular genetic studies show that despite their morphological differences, these fishes interbreed and share a common gene pool, thereby forming a single deme of the species *Cichlasoma minckleyi*.

To study the spatial ecology of a species that has multiple geographic populations, ecologists use a model called the **metapopulation**. A metapopulation comprises two or more geographically distinct demes, each of which maintains continuity in time and space. Most interactions among individual organisms, including reproduction, occur within a deme, but some individuals move from one deme to a different one. This movement between demes is called migration if the individual's change of demes is a permanent one.

Movement of individuals among demes within a species can impart some evolutionary cohesion to the species as a whole. Local environments can change unpredictably, sometimes depleting or eliminating a local deme. Migration is therefore a crucial source of replacement among demes within a region. A species can avoid extinction if risk of extinction is spread among many demes, because simultaneous destruction of all demes is unlikely unless a catastrophe is geographically widespread. Interaction among demes in this manner is called **metapopulation dynamics**. In some species, migration and recolonization among demes are nearly symmetrical. If some demes are stable and others are more susceptible to extinction, the more stable ones, termed **source demes**, differentially supply emigrants to the less stable ones, called **sink demes**.

Suppose that the freshwater fish *Cichlasoma minckley* is subdivided geographically into multiple demes that differ from each other in the relative proportions of molariform and papilliform individuals. A deme that contains mostly molariform individuals should be more stable in times of food scarcity than demes having mostly "papilliform" individuals, because the molariform jaws permit a fish to exploit snails as food when the preferred softer food items are depleted. Such a deme could function as a source deme following a decline in the fish's favored prey species.

Demography

Each population or deme has a characteristic **age structure**, **sex ratio**, and **growth rate**. **Demography** is the study of these properties and the factors that influence them. Demographic characteristics (individual size, age, number of offspring) vary according to the lifestyles of the species under study. For example, some animals (and many plants) are **modular**. Modular animals, such as sponges, corals, and ectoprocts (see section 9.4), form colonies of genetically identical organisms. Reproduction is by asexual **cloning**, as described for hydrozoans in Chapter 7 (see section 7.3). Colonies propagate also by fragmentation, as seen on coral reefs during severe storms. Pieces of coral can be scattered by wave action on a reef, forming propagules for new reefs. For these modular animals, age structure and sex ratio are difficult to measure. Changes in colony size are used to measure growth rate, but counting individuals is difficult and less meaningful than in **unitary** animals, which live independently.

Populations of unitary animals that reproduce by **parthenogenesis** share some characteristics with asexual clones. Parthenogenetic species occur in many animal taxa, including insects, fishes, salamanders, and lizards. Usually such groups contain only females, which lay unfertilized eggs that hatch into daughters genotypically identical to their mothers. The praying mantis *Bruneria borealis*, common in the southeastern United States, is a unitary parthenogenetic animal.

Key Theme 2.1

GENETICS AND DEVELOPMENT

Parthenogenesis

Parthenogenesis ("virgin origin") is the development of an embryo from an unfertilized egg or from one in which the male and female nuclei fail to unite following fertilization. Parthenogenesis takes many forms and is surprisingly widespread ecologically and taxonomically in the animal kingdom.

Most animals are biparental, and reproduction follows a period of organismal growth and maturation. Each new generation begins with a **cohort** of individuals born at the same time. Of course, individuals of any cohort do not all survive to reproduce. For a population to retain constant size from generation to generation, each adult female must replace herself on average with one daughter that survives to reproduce. If females produce on average more than one viable daughter, the population grows; if fewer than one, the population declines.

Animal species have different characteristic patterns of **survivorship** from birth until death of the last member of a cohort. The three principal types of survivorship are illustrated in figure 2.4. In curve I, most individuals die at old age through senescence. Human populations that have the resources to support expanding population size and to prevent infant mortality approximate curve I. Curve II, in which rate of mortality as a proportion of survivors is constant over all ages, characterizes some animals that care for their young, as do many birds, and thus avoid high infant mortality. In contrast to curve I, however, populations at curve II have important sources of mortality other than old age. Predation or lack of food, for example, would kill individuals at any age. Curve III characterizes populations whose infant or juvenile mortality is very high relative to that of young adults.

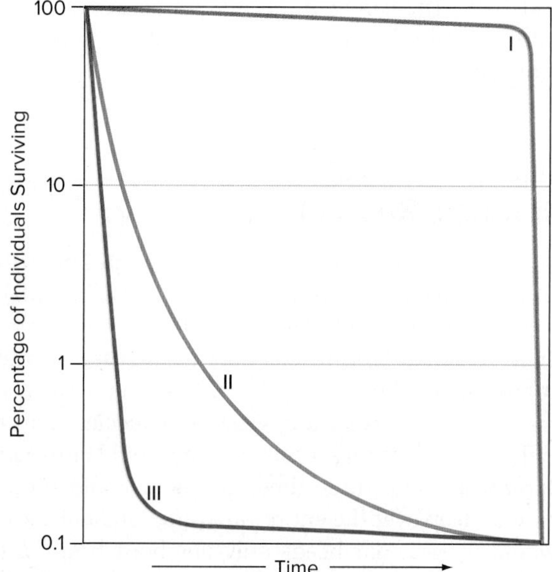

figure 2.4

Three types of theoretical survivorship curves. See text for explanation.

Survivorship of most invertebrates, and of vertebrates such as fish that produce great numbers of offspring, resembles curve III. For example, a mature female marine snail, *Ilyanassa obsoleta,* produces thousands of eggs each reproductive period. Zygotes become free-floating planktonic veliger larvae, which travel far from the mother's location in oceanic currents. They experience high mortality from numerous animals that consume plankton. Furthermore, larvae require a specific, sandy substrate on which to settle and then to metamorphose into an adult snail. The probability of a larva surviving long enough to find a suitable habitat is very low, and most of the cohort dies during the veliger stage. One therefore sees a rapid drop in survivorship in the first part of the curve. The few larvae that do survive to become snails have better odds of surviving further, as shown by the more gentle slope of the curve for older snails. Thus, high reproductive output balances high juvenile mortality.

Most animals do not survive to reach reproductive age, and those that do might typically reproduce only once before they die. In many insect species of the temperate zone, adults reproduce once before the onset of winter and die, leaving only their eggs to overwinter and repopulate their habitat the following spring. Similarly, Pacific salmon after several years return from the ocean to fresh water to spawn only once, after which all the adults of a cohort die. This condition in which an organism reproduces only once during its life history is termed **semelparity**. However, other animals survive long enough to produce multiple cohorts of offspring that may mature and reproduce while their parents are still alive and reproductively active. **Iteroparity** denotes the occurrence of more than one reproductive cycle in an organism's life.

Populations of animals containing multiple cohorts, such as robins, box turtles, and humans, exhibit **age structure**. Analysis of age structure reveals whether a population is actively growing, stable, or declining. Figure 2.5 shows age-structure profiles for the people of Mexico as measured in 1975 and 2019. The earlier profile shows an actively growing population, whereas the later one shows a population maintaining a stable size.

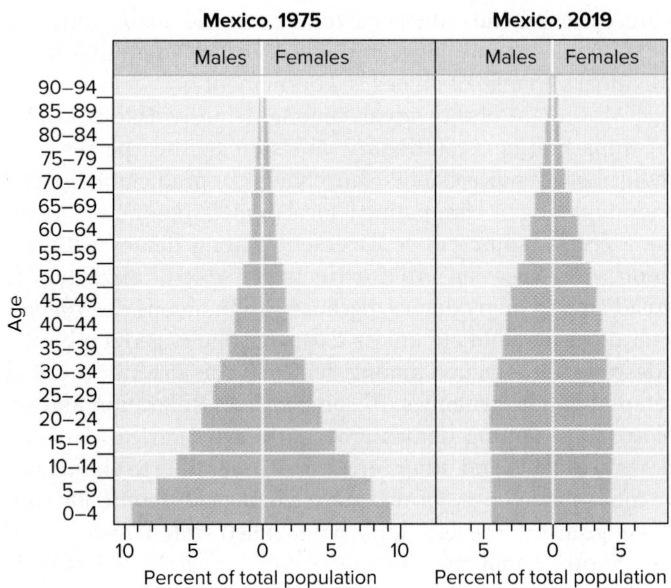

figure 2.5

Age structure profiles of the human populations of Mexico in 1975 and 2019 contrast the rapidly growing, youthful population of 1975 with the stable population of 2019, where the fertility rate approximates replacement. This demographic transition from high birth and death rates to low ones often occurs as a country becomes increasingly industrial in its economic development. Popular references to "the demographic transition" in human societies are specifically to this kind of change.

Population Growth and Intrinsic Regulation

Population growth is the difference between rates of birth and death. As Darwin recognized from an essay by Thomas Malthus, all populations have an inherent ability to grow exponentially. This ability is called the intrinsic rate of increase or **intrinsic growth rate**, denoted by the symbol **r**. The steeply rising curve in figure 2.6 shows this kind of growth. If species continually grew in this fashion, earth's resources soon would be exhausted. A bacterium dividing three times per hour could produce a colony a foot deep over the entire earth after 36 hours, and this mass would be over our heads only one hour later. Potential growth rates of bacterial populations greatly exceed those of animals, but animal populations could achieve the same kind of result over a longer period of time, given unlimited resources. Many insects lay thousands of eggs each year. A single Atlantic codfish, *Gadus morhua,* can spawn 6 million eggs in a season, and a field mouse can produce 17 litters of five to seven offspring each year.

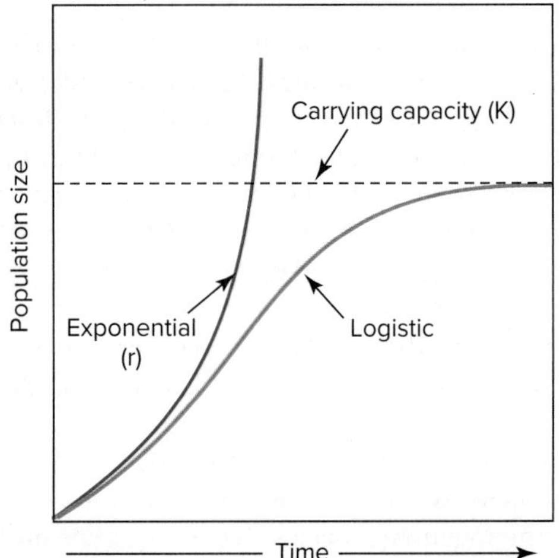

figure 2.6

Population growth, showing exponential growth of a species in an unlimited environment, and logistic growth in a limited environment. K = population size at carrying capacity; r = intrinsic rate of population growth.

Exponential and Logistic Growth

We describe the sigmoid growth curve (see figure 2.6) by a simple model called the logistic equation. The slope at any point on the growth curve is the growth rate, how rapidly the population size changes with time. If N represents the number of organisms and t the time, we can express growth in the language of calculus as an instantaneous rate:

dN/dt = the rate of change in number of organisms per time at a particular instant in time.

When populations experience unlimited resources (unlimited food and space, and no competition from other organisms), growth is limited only by the inherent capacity of the population to reproduce itself. Under these ideal conditions, growth is expressed by the symbol r, defined as the intrinsic rate of population growth per capita. The index r is actually the difference between birth rate and death rate per individual in the population at any instant. The growth rate of the population as a whole is then

$$dN/dt = rN$$

This expression describes the rapid, exponential growth illustrated by the early upward-curving portion of the sigmoid growth curve (see figure 2.6).

Growth rate for a natural population slows as the upper limit is approached, and eventually stops. At this point, N has reached its maximum density because the space being studied has become "saturated" with animals. This limit is called the carrying capacity of the environment and is expressed by the symbol K. The sigmoid population growth curve can now be described by the logistic equation, written as

$$dN/dt = rN([K - N]/K)$$

This equation states the rate of increase per unit of time (dN/dt = rate of growth per capita [r] × population size [N] × unutilized freedom for growth [K − N]/K). One can see from the equation that when the population approaches carrying capacity, K − N approaches 0, dN/dt also approaches 0, and the curve flattens.

Populations occasionally overshoot the carrying capacity of the environment so that N exceeds K. The population then exhausts some resource (usually food or shelter). The rate of growth, dN/dt, then becomes negative, and the population must decline.

Obviously, unrestricted growth is not prevalent in nature. Even in the most benign environment, a growing population eventually exhausts food or space. Exponential increases such as locust outbreaks or planktonic blooms in lakes must end when a population expends food or space. Among all the resources that could limit a population, the one in shortest supply relative to the population's needs is the first one that the population depletes. This one is termed the **limiting resource**. The largest population that the limiting resource can support in a habitat is called the **carrying capacity** of that environment, symbolized by **K.** Ideally, a population slows its growth rate in response to diminishing resources until it just reaches K, as represented by the sigmoid (S-shaped) curve (blue line) in figure 2.6. The box on this page compares the mathematical expressions of exponential and sigmoid (or logistic) growth curves.

Sigmoid growth occurs when population density exerts negative feedback on growth rate. Intrinsic regulation of a population by negative feedback from population size is **density-dependent.** If the limiting resource is expendable, as is food, carrying capacity is reached when the rate of resource replenishment equals the rate of depletion by the population; the population is then at K for that limiting resource. According to the sigmoid model, when population density reaches K, rates of birth and death are equal, and growth of the population ceases. If food is replaced at a rate that supports the current population but no more, a population of grasshoppers in a green meadow, for example, may be at carrying capacity even though we see plenty of unconsumed food.

Although experimental populations of unicellular organisms often fit a sigmoid growth curve closely, most populations in nature tend to fluctuate above and below carrying capacity. For example, after sheep were introduced to Tasmania around 1800, their numbers changed logistically, with small oscillations around an average population size of about 1.7 million; we thereby infer the carrying capacity of the environment to be 1.7 million sheep (figure 2.7A). Ring-necked pheasants introduced on an island in Ontario, Canada, exhibited wider oscillations (figure 2.7B).

Why do intrinsically regulated populations oscillate this way? First, the carrying capacity of an environment can change over time, causing a change in population density as dictated by a limiting resource. Second, animals always experience a lag between the time that a resource becomes limiting and the time that the population responds by reducing its rate of growth. Third, **extrinsic factors** occasionally limit a population's growth below carrying capacity.

On a global scale, humans have the longest record of exponential population growth (figure 2.7C). Although famine and war have restrained growth of populations locally, global human growth declined only when bubonic plague ("black death") decimated much of Europe during the fourteenth century. What, then, is the carrying capacity for the human population? The answer is unclear, because technological advances have increased our ability to extract useful resources from our environments.

Agriculture increased the carrying capacity of the environment, and the human population grew steadily from 5 million around 8000 B.C., when agriculture was introduced, to 16 million around 4000 B.C. Despite the toll of terrible famines, disease, and war, the population reached 500 million by 1650. With the Industrial Revolution in Europe and England in the

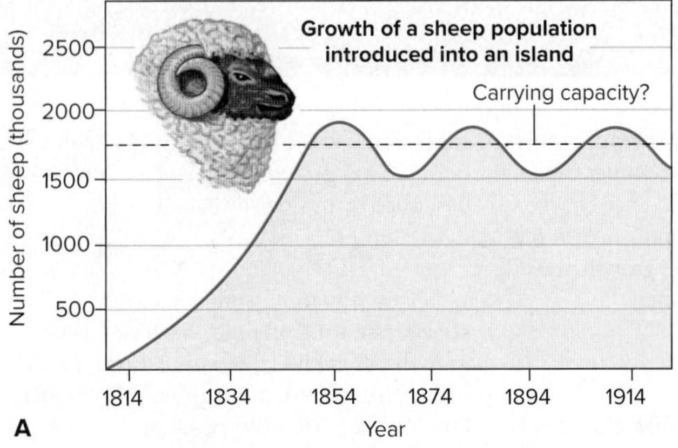

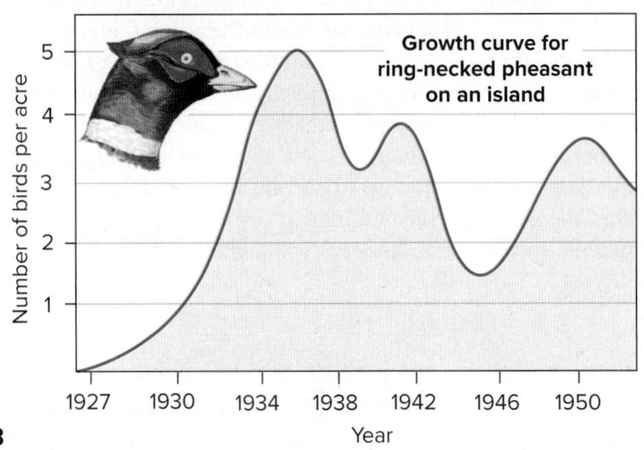

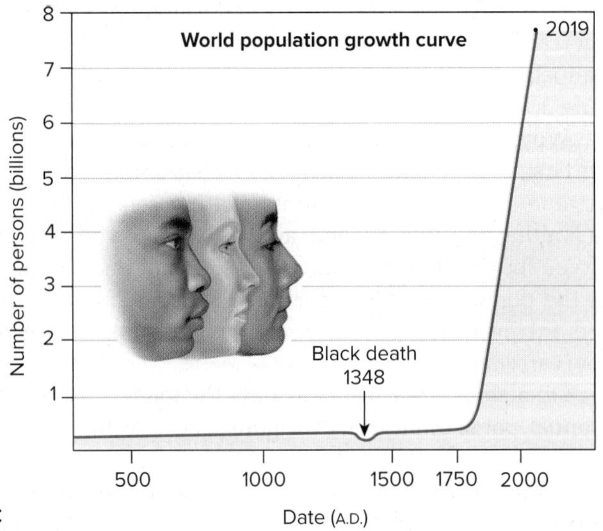

figure 2.7

Growth curves for **A,** sheep *Ovis aries*, **B,** ring-necked pheasant, *Phasianus colchicus*, and **C,** world human populations throughout history. Note that the sheep population on an island is stable because people control the population, but the ring-necked pheasant population oscillates greatly, probably due to large changes in carrying capacity. Where would you place the carrying capacity for the human population?

eighteenth century, followed by a medical revolution and better agricultural practices, the human carrying capacity increased dramatically. The population doubled to 1 billion around 1850. It doubled again to 2 billion by 1927, reached 4 billion in 1974, passed 6 billion in October 1999, and is expected to reach 8.9 billion by 2040. Thus, growth has been exponential and remains high (figure 2.7C). What would constitute the maximum sustainable agricultural output is uncertain, but food production cannot sustain population growth indefinitely. The United Nations has taken the stance that K for humans is approximately 7–11 billion people.

Key Theme 2.2

HUMAN CONNECTIONS

Human Population Growth

Between 1970 and 2019 the annual growth of the human population decreased from 2.0% to 1.1%, but the total size of the population doubled. The decrease in growth rate is credited to better education for women and family planning. The United Nations predicts that the human population will increase from 7.7 billion people in 2019 to 11.2 billion by the end of the century.

Extrinsic Limits to Growth

We have seen that the intrinsic carrying capacity of a population for an environment prevents unlimited exponential growth of the population. Population growth also can be limited by extrinsic biotic factors, such as predation, parasitism (including disease-causing pathogens), and interspecific competition, or by abiotic influences, such as floods, fires, and storms. Although abiotic factors certainly reduce populations in nature, they cannot truly regulate population growth because their effect is wholly independent of population size; abiotic limiting factors are thus **density-independent**. A single hailstorm might kill most of the young wading birds, or a forest fire might eliminate entire populations of many animals, regardless of how many individuals exist.

In contrast, biotic factors can and do act in a density-dependent manner. Predators and parasites respond to changes in density of their prey and host populations, respectively, to maintain those populations at fairly constant sizes. These sizes are below carrying capacity, because populations regulated by predation or parasitism are not limited by their resources. Competition between species for a common limiting resource lowers the effective carrying capacity for each species below that of either one alone. Salamanders of the species *Plethodon teyahalee* occur together with single species of the *Plethodon jordani* group at some scattered intermediate elevations in the southern Appalachian Mountains; where the species overlap, both species have lower densities than are typical for them outside the small area of co-occurrence. Experimental removal of either species from a site of co-occurrence increases population density of the remaining species.

2.4 Community Ecology

Interactions Among Populations in Communities

Populations of animals that form an ecological community interact in various ways that can be detrimental (−), beneficial (+), or neutral (0) to each species, depending on the nature of the interaction. For instance, a predator's effect on prey is (−), because survival of the prey animal is reduced. However, the same interaction benefits the predator (+) because the food obtained from prey increases the predator's ability to survive and to reproduce. Thus, the predator-prey interaction is + −. Ecologists use this shorthand notation to characterize interspecific interactions because it shows the direction in which the interaction affects each species.

We see other kinds of + − interactions. One of these is **parasitism**, in which the parasite benefits by using the host as a home and a source of nutrition, and the host is harmed. **Herbivory**, in which an animal eats a plant, is another + − relationship. **Commensalism** is an interaction that benefits one species and neither harms nor benefits the other (0 +). For example, most bacteria that normally inhabit our intestinal tracts do not affect us (0), but the bacteria benefit (+) by having food and a place to live. Another example of commensalism is the association of pilot fishes and remoras with sharks (figure 2.8). These fishes get the "crumbs" remaining when the host shark makes its kill.

Organisms engaged in mutualism (+ +) both benefit from their ecological interaction (figure 2.9). A distinction between commensalism and mutalism is sometimes difficult to make in practice. For example, if the harmless bacteria residing in the human gut prevent entry of harmful bacteria, the apparent commensalism becomes a mutualism. Likewise, if remoras remove parasites from their shark hosts (see figure 2.8), then this commensalism likewise becomes a mutualism.

Some mutualistic relationships are not only beneficial but necessary for the survival of one or both species. Such relationships are termed *obligatory mutualisms*; by contrast, *facultative mutualisms* are not required for a species' survival. The shark-remora example is a facultative mutualism. The relationship between a termite and parabasalids (a kind of eukaryotic unicellular organism, described in section 5.5) inhabiting its gut illustrates

Noble Proctor/Science Source

figure 2.8

Four remoras, *Remora* sp., attached to a shark. In this classic example of commensalism, remoras feed on fragments of food left by their shark host, as well as on pelagic invertebrates and small fishes. Although they actually are good swimmers, remoras prefer to be pulled through the water by marine creatures or boats. The shark host may benefit by having embedded copepod skin parasites removed by remoras.

Cleveland P. Hickman, Jr.

Cleveland P. Hickman, Jr.

figure 2.9

Among the many examples of mutualism that abound in nature is that of the whistling thorn acacia, *Acacia drepanolobium*, of the African savanna and the ants, *Crematogaster mimosae* and other species, that make their homes in the acacia's swollen galls. The acacia provides both protection for the ants' larvae (*lower photograph of opened gall*) and honeylike secretions used by ants as food. In turn, ants protect the tree from herbivores by swarming out as soon as the tree is touched. Giraffes, *Giraffa camelopardalis*, which love the tender acacia leaves, seem immune to the ants' fiery stings.

an obligatory mutualism. Bacterial symbionts of the parabasalids can digest wood eaten by the termite because the bacteria produce an enzyme, lacking in the termite, that digests cellulose; the termite lives on waste products of parabasalid-bacterial metabolism. In return, the parabasalids and their bacteria gain a habitat and food supply. Such absolute interdependence among species is a liability if one of the participants is lost.

Competition between species reduces abundances of both (– –). Many biologists, including Darwin, considered competition the most common and important interaction in nature. Ecologists have constructed most of their theories of community structure based on the premise that competition is the chief organizing factor in species assemblages. Sometimes the effect on one species in a competitive relationship is negligible. This condition is called **amensalism**, or **asymmetric competition** (0 –). For example, two species of barnacles, *Chthamalus stellatus* and *Balanus balanoides,* compete for space on rocky intertidal habitats. A famous experiment by Joseph Connell[1] demonstrated that *B. balanoides* excluded *C. stellatus* from a portion of the habitat, whereas *C. stellatus* had no effect on *B. balanoides*.

We have treated interactions as occurring between pairs of species. However, in natural communities containing populations of many species, a predator often has more than one prey species, and several animal species often compete for the same resource. Thus, ecological communities are complex and dynamic, a challenge to ecologists who study this level of natural organization.

Competition and Resource Partitioning

Competition occurs when two or more species share a **limiting resource**. Simply sharing food or space with another species does not produce competition unless the resource is in short supply relative to the needs of the species that share it. Thus, we cannot assume that competition occurs in nature simply by showing that two species share a resource. However, we can find evidence of competition by investigating the different ways that species exploit a resource.

Competing species reduce conflict by reducing overlap of their niches. **Niche overlap** is the portion of resources shared by the niches of two or more species. For example, if two species of birds eat seeds of exactly the same size, competition eventually excludes the species less able to exploit the resource. This example illustrates the principle of **competitive exclusion**: strongly competing species cannot coexist indefinitely. This principle was discovered in 1932 by the Soviet microbiologist G. F. Gause, who performed experiments on the mechanism of competition between two ciliate species (*Paramecium aurelia* and *P. caudatum*) and two yeast species (*Saccharomyces cerevisiae* and *S. kefir*). To coexist in the same habitat, species must specialize by partitioning a shared resource and using different portions of it. We call ecological specialization of this kind **resource partitioning**.

Resource partitioning usually appears as differences in organismal morphology or behavior related to exploitation of a resource. For example, in his classic study of the Galápagos finches (see figure 1.24), the English ornithologist David Lack noticed that beak sizes of these birds depended on whether they occurred together on the same island (figure 2.10). On the islands Daphne and Los Hermanos, where *Geospiza fuliginosa* and *G. fortis* occur separately and therefore do not compete with each other, beak sizes are nearly identical; on the island Santa Cruz, where both *G. fuliginosa* and *G. fortis* coexist, their beak sizes do not overlap. These results suggest resource partitioning, because beak size determines the size of seeds selected for food. Later work by the American ornithologist Peter Grant confirms what Lack suspected: *G. fuliginosa* with its smaller beak selects smaller seeds than does *G. fortis* with its larger beak. Where the two forms coexist, competition between them led to evolutionary displacement of beak sizes to diminish competition. Absence of competition today has been called "the ghost of competition past."

Ecological character displacement is the mechanism of resource partitioning illustrated by these Galápagos finches; co-occurrence in the same community causes species to evolve

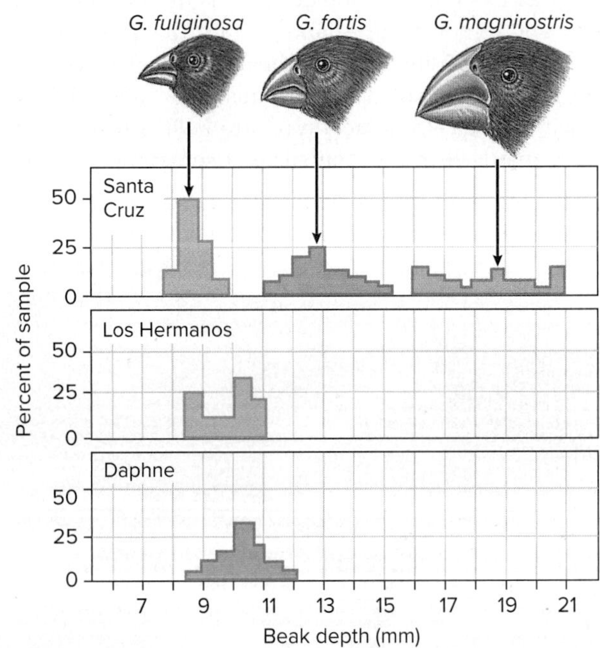

figure 2.10

Resource partitioning occurs by character displacement of beak sizes in Darwin's finches from the Galápagos Islands. Beak depths are given for the ground finches *Geospiza fuliginosa* and *G. fortis* where they occur together (sympatric) on Santa Cruz Island and where they occur alone on the islands Daphne and Los Hermanos. *G. magnirostris* is another large ground finch that lives on Santa Cruz. Although traditionally considered separate species as shown, molecular genetic results suggest that these forms are more appropriately considered ecomorphs of a single species.

[1]Connell, J. H. 1961. The influence of interspecific competition and other factors on the distribution of the barnacle *Chthamalus stellatus*. *Ecology* **42**:710–723.

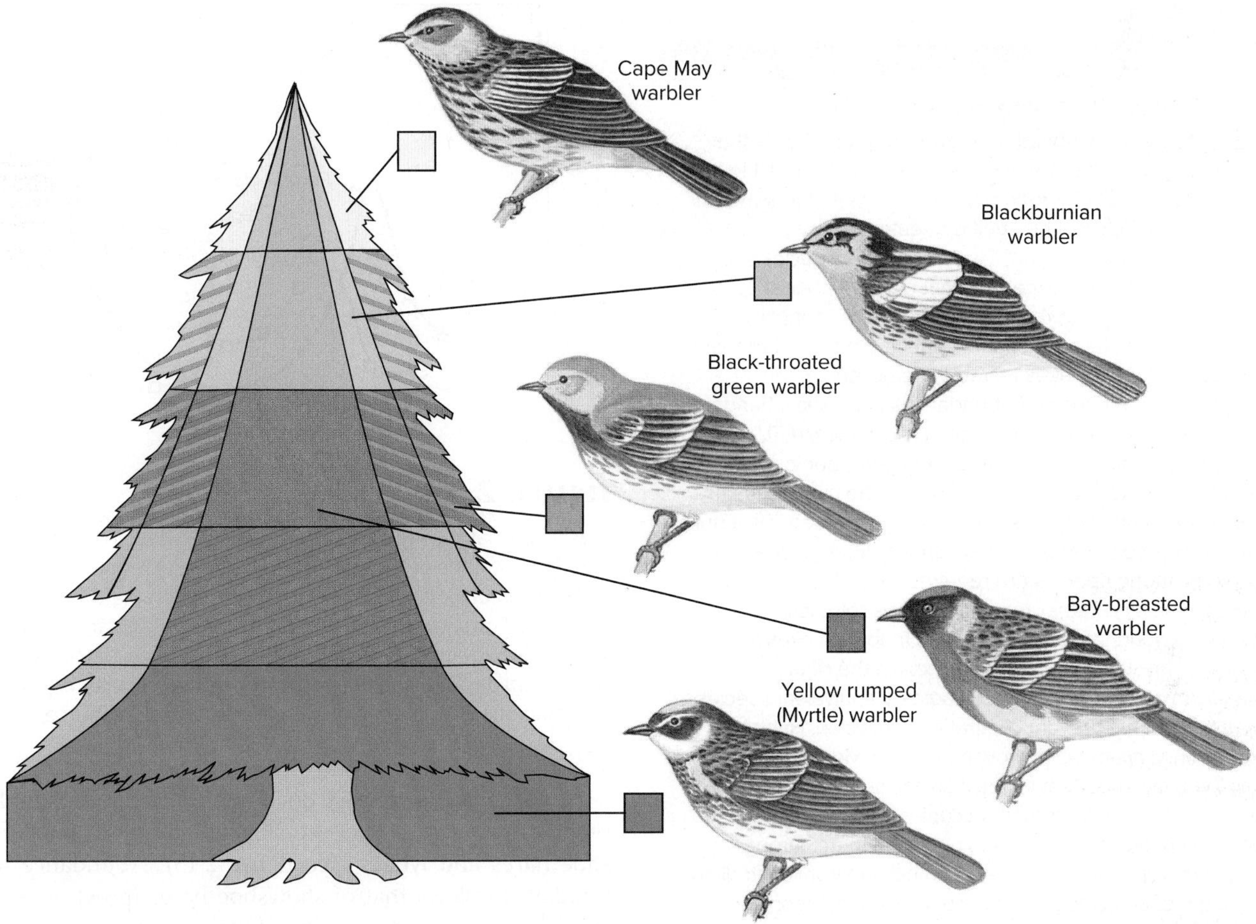

figure 2.11

Distribution of foraging effort among five species of wood warblers in a northeastern spruce forest. The warblers form a feeding guild. Resource partitioning by selection of different microhabitats among species permits them to coexist in the same trees despite sharing a limited number of insects as prey.

niche dimensions more restricted than those of their ancestors. An alternative and perhaps more common means of evolving resource partitioning is through **microhabitat selection**, the formation of a community by species whose preferred microhabitats differ sufficiently to prevent them from competing for what otherwise would be a shared limiting resource.

When several species share the same general resources, they form a **guild**. Just as a guild in medieval times constituted a brotherhood of men sharing a common trade, species in an ecological guild share a common livelihood. The term guild was introduced to ecology by Richard Root in his 1967 paper on niche patterns of the blue-gray gnatcatcher *Polioptila caerulea*.[2]

The theory of competitive exclusion predicts that when different species of a guild occur in the same geographic area, detailed study will reveal resource partitioning among the guild's species. A classic example of resource partitioning among species in a bird guild is Robert MacArthur's study of a feeding guild comprising five species of warblers in spruce woods of the northeastern United States.[3]

At first glance, one might ask how five birds, very similar in size and appearance, could coexist by feeding on insects in the same tree. However, on close inspection, MacArthur found subtle differences in sites of foraging among these birds (figure 2.11). One species searched only on the outer branches of spruce crowns; another species used the top 60% of the tree's outer and inner branches away from the trunk; another species concentrated on inner branches closer to the trunk; another species used the midsection from the periphery to the trunk; and still another species foraged in the bottom 20% of the tree. These observations suggest that structural differences in habitat separate each warbler's niche within this guild.

---

[2]Root, R. B. 1967. The niche exploitation pattern of the blue-gray gnatcatcher. *Ecological Monographs* **37**:317–350.

[3]MacArthur, R. H. 1958. Population ecology of some warblers of northeastern coniferous forests. *Ecology* **39**:599–619.

Key Theme 2.3
ECOLOGY

Neutral Theory in Community Ecology

Ecologist Stephen P. Hubbell has challenged the theory that species must have different ecological niches to coexist in an ecological community for long periods. The most fundamental and controversial idea of his "unified neutral theory of biodiversity and biogeography" is that multiple species in a community are ecologically equivalent and can coexist indefinitely. By assuming that individuals of all potentially competing species have the same expected rates of birth and death, neutral theory makes testable predictions of the geographic distributions and abundances of species. Random changes in species abundance, called "ecological drift," and random movement of individuals among geographic communities are, together with speciation, the major causal processes of community assembly. The neutral theory is a null hypothesis whose rejection by data shows where niche differences among species are required to explain a pattern of species diversity. Rather than assuming ecological differences among species, the neutral theory invokes such differences only when necessary to explain the data.

Long-term co-occurrence of ecologically equivalent species requires factors that prevent competitive exclusion. Hubbell invokes strong limitations to movement of individuals among sites as the most important factor. If dispersal is highly restricted, an inferior competitor occupies many sites because the best competitor for the site did not arrive there. Strong limitations to dispersal can prevent competition from controlling the structure of an ecological community. Neutral theory has predicted the diversity of tree species in tropical forests.

Predators and Parasites

The ecological warfare waged by predators against their prey causes coevolution: predators get better at catching prey, and prey get better at escaping predators. This is an evolutionary race that a predator cannot afford to win. If a predator became so efficient that it exterminated its prey, the predator species would become extinct. Because most predators feed on more than a single species, specialization on a single prey to the point of extermination is uncommon.

When a predator does rely primarily on a single prey species, both populations tend to fluctuate cyclically. First, prey density increases; then predator density increases until prey become scarce. At that point, predators must adjust their population size downward by leaving the area, lowering reproduction, or dying. When the density of a predator population falls enough to allow reproduction by prey to outpace mortality from predation, the cycle begins again. Thus, populations of both predators and prey show cycles of abundance, but increases and decreases in predator abundance are slightly delayed relative to those of prey because of the time lag in a predator's response to changing prey density (figure 2.12). Laboratory experiments with ciliates (see section 5.5: Ciliophora) reveal this process (figure 2.12).

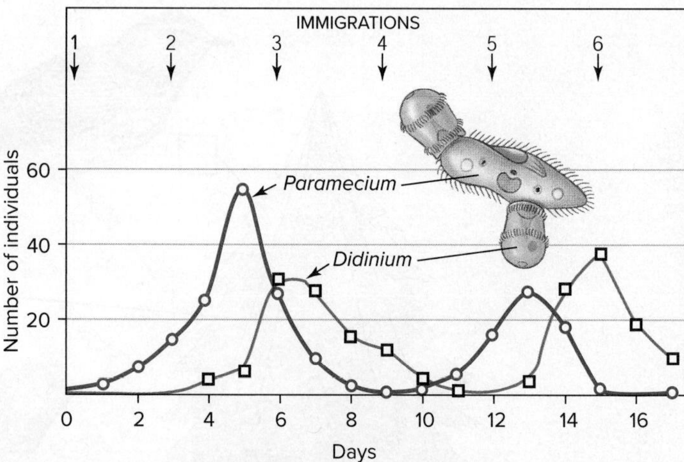

figure 2.12

Classic predator-prey experiment by the Soviet biologist G. F. Gause in 1934 shows the cyclic interaction between predator (*Didinium*) and prey (*Paramecium*) in laboratory culture. When the *Didinium* find and eat all the *Paramecium*, the *Didinium* themselves starve. Gause could keep the two species coexisting only by occasionally introducing one *Didinium* and one *Paramecium* to the culture (*arrows*); these introductions simulated migration from an outside source.

Perhaps the longest documented natural example of a predator-prey cycle is between Canadian populations of snowshoe hares and lynxes (see figure 2.13). Abundance of lynx (predator) follows that of showshoe hares (prey) on a 10-year cycle. Interestingly, the timing of these predator-prey cycles appears to track 10-year cycles in abundance of sunspots, which increase the solar energy that reaches earth and possibly also growth of vegetable matter on which hares feed. A global climatic variable thus influences timing of the density-dependent predator-prey cycles.

The war between predators and prey reaches high art in the evolution of defenses by potential prey. Potential prey can escape detection because their bodies match their background or resemble some inedible feature of the environment (such as

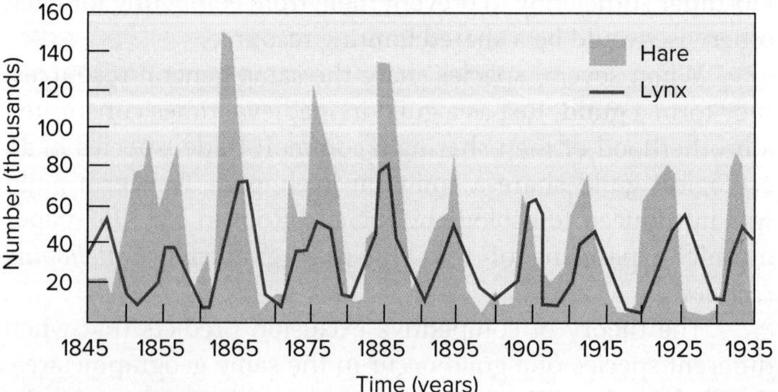

figure 2.13

Changes in population size of snowshoe hare and lynx in Canada as indicated by pelts received by the Hudson's Bay Company. Abundance of lynx (predator) follows that of the hare (prey).

Nuridsany et Perennou/Science Source

John Keates/Alamy Stock Photo

Mark Newman/Lonely Planet Images/Getty Images

B

Bill Brooks/Alamy Stock Photo

Bill Brooks/Alamy Stock Photo

C

Arthur C. Twomey/Science Source

A

figure 2.14

Artful guises abound. **A,** Batesian mimicry: A harmless yellowjacket clearwing moth *Pennisetia marginata* (*top*) mimics a yellowjacket wasp *Vespula maculifrons* (*bottom*), which has a powerful stinger. **B,** Müllerian Mimicry: Neogropical butterflies *Heliconius ismenius* (*top*) and *Lycorea cleobaea* (*bottom*) share distastefulness and matching coloration to warn common predators. **C,** Mimicry between the viceroy butterfly *Limenitis archippus* (*top*) and monarch butterfly *Danaus plexippus* (*bottom*) is often presented as Batesian mimicry in which the presumably tasteful viceroy mimics warning coloration of the distasteful monarch; evidence that the viceroy also is distasteful would make this a case of Müllerian mimicry.

a stick). Such defenses are called *cryptic*. In contrast to cryptic defenses, animals that are toxic or distasteful to predators actually advertise their strategy with bright colors and conspicuous behavior, a type of defense called *aposematic*. These species are protected because predators learn to avoid them after distasteful encounters. A predator can use mimicry aggressively by matching another organism that attracts its prey. For example, the crab spider shown in figure 13.9A mimics flower petals among which it sits to draw as prey insects that visit the flowers in search of nectar or pollen.

When distasteful prey adopt warning coloration, advantages arise for palatable prey. Palatable prey can deceive potential predators by mimicking distasteful prey, a phenomenon called *Batesian mimicry*. Coral snakes and yellowjacket wasps (figure 2.14A) are both brightly colored, harmful prey. Coral snakes have a venomous bite, and yellowjacket wasps have a powerful sting. Species of both groups serve as **models** for other species, called **mimics**, which are harmless but resemble the harmful model species. A Batesian mimic essentially parasitizes the model population; a predator that encounters the edible mimic first might later harm models, expecting them to be edible (figure 2.14A).

In another form of mimicry, termed *Müllerian mimicry*, two or more species that are toxic or otherwise harmful resemble each other (figure 2.14B). What might an animal that has its own poison gain by evolving resemblance to another poisonous animal? The answer is that a predator needs only to experience the toxicity of one species to avoid all similar prey. A predator can learn one warning signal more easily than many! The different species participating in a case of Müllerian mimicry often are not equally distasteful. If a mildly distasteful species shares warning coloration with a highly toxic species, the mildly distasteful species gains more advantage from the mimicry than does the highly toxic species. Batesian mimicry is thus an extreme case in a continuum of differential toxicity among species that share warning coloration. For example, mimicry between viceroy and monarch butterflies (figure 2.14C) is often presented as Batesian mimicry of the distasteful monarch by the presumably tasteful viceroy; however, some data suggest that the viceroy also is distasteful, which would make the shared warning coloration of monarchs and viceroys a case of Müllerian mimicry.

Sometimes one population's influence on others is so pervasive that its absence drastically changes the entire

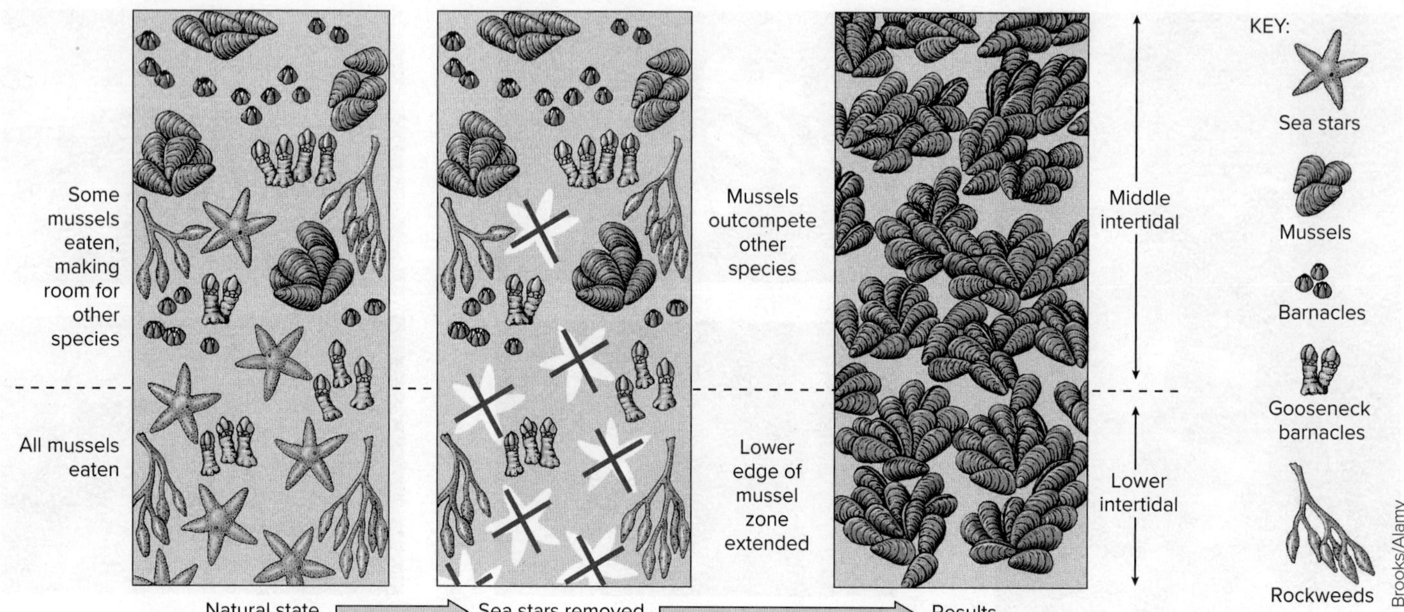

figure 2.15

Experimental removal of a keystone species, the predatory sea star *Pisaster ochraceus*, from an intertidal community completely changes the structure of the community. With their principal predator missing, mussels form dense beds by outcompeting and replacing other intertidal species.

community. We call such a population a **keystone species**.[4] On rocky intertidal shores of western North America, the sea star *Pisaster ochraceous* is a keystone species. Sea stars are a major predator of the mussel *Mytilus californianus*. When sea stars were removed experimentally from a patch of Washington State coastline, mussel populations expanded, occupying all of the space previously used by 25 other invertebrate and algal species (figure 2.15). Keystone predators act by reducing prey populations below a level where resources, such as space, are limiting. The original notion that all keystone species are predators has been broadened to include any species whose removal causes the extinction of others.

By reducing competition, keystone species allow more species to coexist on a resource. Consequently, they contribute to maintaining diversity in a community. Keystone species illustrate a more general phenomenon, disturbance. Periodic natural disturbances, such as fires and hurricanes, also can prevent monopolization of resources and competitive exclusion by a few broadly adapted competitors. Disturbances can permit more species to coexist in such highly diverse communities as coral reefs and rain forests.

Parasites are often considered freeloaders because they appear to benefit from their hosts at no expense. **Ectoparasites**, such as ticks and lice, infest many different kinds of animals. The host provides nutrition from its body and aids dispersal of the parasite. However, we must consider that the evolutionary pathway to parasitism from free-living forms often has costs as well as benefits. **Endoparasites**, such as tapeworms (see Chapter 8), have lost their ability to choose habitats. Also, because they must move among hosts to complete their life cycle, the chance that a single individual will live to reproduce is very low. The more intermediate hosts involved in a parasite's life cycle, the lower the likelihood of success, and the greater reproductive output must be to balance mortality.

Biologists often are puzzled by the complexity of parasite-host relationships. For example, a trematode parasite (see section 8.3: Class Trematoda) of the marine gastropod *Ilyanassa obsoleta* actually changes its host's behavior to complete its life cycle. These snails live in sandy intertidal substrates in eastern North America. If the snails are exposed to air when the tide recedes, they normally burrow into sand to avoid desiccation. If, however, a snail is infected with the trematode *Gynaecotyla adunca,* it moves shoreward on high tides preceding low night tides to be left on the beach on the receding tide. Then, as in the legend of the Trojan Horse, the snail sheds larval forms called cercariae into the sand, where they can infect the next intermediate host, a beach-living crustacean. The crustacean can be eaten by a gull or other shorebird, the definitive hosts for this trematode. The life cycle is completed when the bird defecates into water, releasing eggs from which hatch larvae that will infect more snails.

Coevolution between parasite and host is expected to generate an increasingly benign, less virulent relationship if host organisms are uncommon and/or difficult for a parasite to infest. Selection favors a benign relationship, because a parasite's chances of survival are diminished if its host dies.

[4]Paine, R. T. 1969. A note on trophic complexity and community stability. *American Naturalist* **103**:91–93.

Virulence of a parasite is correlated, at least in part, with availability of new hosts. A highly virulent parasite should evolve only if the parasite can easily move to a new host after killing a previous one.

■ 2.5 Ecosystems

Transfer of energy and materials among organisms within ecosystems is the ultimate level of organization in nature. Energy and materials are required to construct and to maintain life, and their incorporation into biological systems is called **productivity**. Ecologists subdivide productivity into component **trophic levels** based on how organisms obtain energy and materials. Trophic levels are linked together into **food chains**, which denote movement of energy from plant compounds to organisms that eat plants, then to other organisms that eat the plant feeders, and possibly further through a linear series of organisms that feed and are then eaten by others (figure 2.16). A **food web** shows the branching pathways for transfer of energy and materials among species in an ecosystem. Figure 2.16 summarizes a food web for a salt marsh, with each group of species ordered vertically according to its level in the corresponding food chain, shown at the left of the figure.

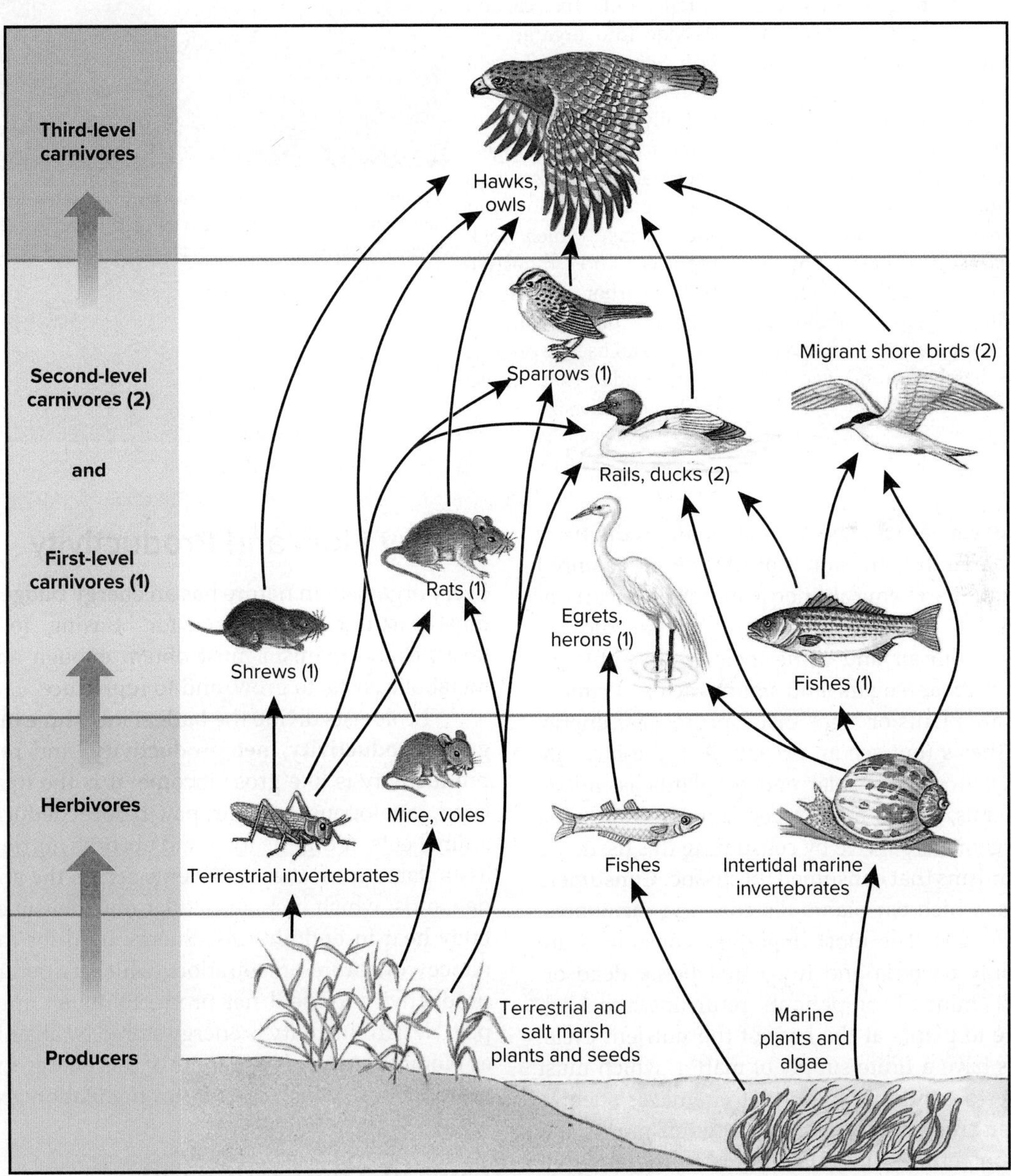

figure 2.16

Primary producers and consumers (herbivores, carnivores) in a midwinter food web in *Salicornia* salt marsh of San Francisco Bay area. The connected trophic levels shown at the left constitute a food chain for this community. The connections between the species that occupy these trophic levels, summarized in the rest of the figure, constitute a food web for this community.

Life Without the Sun

For many years, ecologists thought that all animals depended directly or indirectly on primary production from solar energy. However, in 1977 and 1979, dense communities of animals were discovered living on the sea floor adjacent to vents of hot water issuing from rifts (such as the Galápagos Rift and East Pacific Rise) where tectonic plates on the sea floor are slowly spreading apart. Such communities were subsequently found on numerous ocean ridges (Juan de Fuca, East Pacific, Southeast Pacific, mid-Atlantic, and Indian Ocean), back-arc basins (mostly West Pacific), hydrocarbon or hypersaline cold seeps (Gulf of Mexico, West Coast of Africa, Asia), and subduction zones (Mediterranean, west coasts of North and South America). These communities (see photo) include several species of molluscs, some crabs, enteropneusts (acorn worms), and polychaete worms, including giant siboglinid tubeworms. The temperature of seawater above and immediately around vents is 7–23°C where it is heated by basaltic intrusions, whereas the surrounding normal seawater is 2°C.

The producers in these vent communities are chemoautotrophic bacteria that derive energy from oxidation of large amounts of hydrogen sulfide or methane in vent water and fix carbon dioxide into organic carbon. Some animals in vent communities—for example, bivalve molluscs—are filter feeders that ingest bacteria. Others, such as the giant siboglinid tubeworms, which lack mouths and digestive tracts, harbor colonies of symbiotic bacteria in their tissues and use the organic compounds that these bacteria synthesize.

A population of giant siboglinid tubeworms grows in dense profusion near Sully Vent, a hydrothermal vent in the Main Endeavour Vent Field of the Juan de Fuca Ridge, NE Pacific Ocean, photographed at 2250 m (about 7400 feet) below sea level.

Primary producers are organisms that begin productivity by fixing carbon and nitrogen, and storing energy acquired usually from sunlight as chemical energy in covalent carbon-carbon bonds. "Fixing" consists in acquiring carbon dioxide and nitrogen gases from air and using them to assemble organic molecules, such as sugars and amino acids. Primary producers usually are plants or algae that capture solar energy through **photosynthesis** (but see an exception in the box Life Without the Sun). Powered by solar energy, plants assimilate and organize minerals, water, and carbon dioxide into living tissue. All other organisms survive by consuming this tissue, or by consuming organisms that consumed this tissue. **Consumers** include **herbivores**, which eat plants directly, and **carnivores**, which eat other animals. The most important consumers are **decomposers**, mainly bacteria and fungi that break dead organic matter into its mineral components, returning it to a soluble form available to plants at the base of the nutrient cycle. Earth's ecosystems have a finite supply of matter, which must be recycled for life to continue. Important chemicals such as carbon and nitrogen are endlessly reused as organisms die, and their decomposition returns these chemicals to the environment, from which plants construct new living matter. Unlike matter, energy ultimately dissipates from the ecosystem as heat and is not recycled. Thus, no ecosystems are truly closed; all require input of new energy from the sun or from hydrothermal vents (see box Life Without the Sun).

Energy Flow and Productivity

Every organism in nature has an **energy budget**. Just as we each must partition our income for housing, food, utilities, and taxes, each organism must obtain enough energy to meet its metabolic costs, to grow, and to reproduce.

Ecologists divide the budget into three main components: **gross productivity**, **net productivity**, and **respiration**. Gross productivity is like gross income; it is the total energy assimilated, analogous to your pay before deductions. When an animal eats, it digests food and absorbs nutrients. Most energy assimilated from these nutrients serves the animal's metabolic demands, which include cellular metabolism and regulation of body heat in endotherms. Energy used for metabolic maintenance constitutes respiration, which is deducted from gross productivity to yield net productivity, an animal's take-home pay. Net productivity is energy stored by an animal in its tissues as **biomass**. Animals use some of this energy for growth and for reproduction, which constitutes population growth.

Laws of Thermodynamics

Scientists' observations of how heat and energy interact within a system revealed the laws of thermodynamics. The first law of thermodynamics states that energy cannot be created or destroyed. Although energy can change from one form to another, the total amount of energy in an isolated system remains unchanged.

In the context of an animal's energy budget, the first law of thermodynamics can be expressed by the following equation, in which gross and net productivity are represented by $\mathbf{P_g}$ and $\mathbf{P_n}$, respectively, and respiration is **R:**

$$P_n = P_g - R$$

This equation conveys the important messages that (1) the energy budget of every animal is finite and may be limiting and (2) energy is available for growth of individuals and populations only after satisfying maintenance. Energy for maintenance, **R**, usually constitutes more than 90% of the assimilated energy ($\mathbf{P_g}$) for animal consumers.

When we study energy transfers between trophic levels in food webs, we must recall from the second law of thermodynamics that a molecular system tends toward greater disorder, and that a predator thus will not recover all of the chemical-bond energy of its prey in a useful form. More than 90% of the energy in an animal's food is lost as heat, and less than 10% is stored as biomass. Each succeeding trophic level therefore contains only 10% of the energy in the next lower trophic level. Most ecosystems are thereby limited to five or fewer trophic levels.

Our ability to feed a growing human population is influenced profoundly by the second law of thermodynamics. Humans, who are at the end of the food chain, may eat grain, fruits, and vegetables of plants that fix the sun's energy in chemical bonds; this very short chain represents an efficient use of potential energy. When humans eat beef from cattle, which in turn eat grass that fixes the sun's energy, the addition of a trophic level decreases available energy by a factor of 10. Nonetheless, cattle and other agricultural animals can convert to meat and/or milk or eggs vegetable matter, such as grass, that is unsuited for direct human consumption. Because grass is abundant and otherwise inaccessible for human nutrition, farm animals that convert it to meat or other food give us access to an important source of energy despite adding a trophic level to the food chain.

Ecological Pyramids

When we examine the food chain in terms of biomass at each level, we can construct **ecological pyramids** either of numbers or of biomass. A pyramid of numbers (figure 2.17A), also called an **Eltonian pyramid**, depicts numbers of organisms that are transferred between each trophic level. This pyramid provides a vivid impression of the great difference in numbers of organisms involved in each step of the chain, and it supports the observation that large predatory animals are rarer than the small animals on which they feed. However, a pyramid of numbers does not indicate actual mass of organisms at each level.

More instructive are pyramids of biomass (figure 2.17B), which depict the total bulk, or "standing crop," of organisms at each trophic level. Such pyramids usually slope upward because energy is lost at each transfer, and there is thus less energy available to sustain production of biomass at each successively higher trophic level. However, in aquatic ecosystems whose producers are algae, which have short life spans and rapid turnover, the pyramid inverts. Algae tolerate heavy exploitation by zooplankton consumers. Therefore, the base of the pyramid (biomass of phytoplankton) is smaller than the biomass of zooplankton that it supports. This inverted pyramid is analogous to a person who weighs far more than the food in a refrigerator, but who can be sustained from the refrigerator because food is constantly replenished.

Key Theme 2.4
ECOLOGY

Charles Elton

Oxford ecologist Charles Elton invented the concepts of food chains and ecological pyramids in 1923. Working for a summer on a treeless arctic island, Elton watched arctic foxes as they roamed, noting what they ate and, in turn, what their prey had eaten, until he traced the complex cycling of nitrogen in food throughout the animal community. Elton realized that life in a food chain comes in discrete sizes, because each form has evolved to be much bigger than what it eats. He thus explained the common observation that large animals are rare while their smaller prey are common; the ecological pyramids illustrating this phenomenon now bear Elton's name.

A third type of pyramid is a pyramid of energy, which shows rate of energy flow between levels (figure 2.17C). An energy pyramid is never inverted because the amount of energy transferred from each level is less than the amount that entered it. A pyramid of energy gives the best overall picture of community structure because it depicts the loss of energy from a community as energy flows from primary producers through the food web. In the English Channel, energy of phytoplankton exceeds that of zooplankton, even though the biomass of phytoplankton is less than the biomass of zooplankton (because grazing by zooplankton consumers is heavy).

Nutrient Cycles

All elements essential for life come from environmental air, soil, rocks, and water. When plants and animals die and their bodies decay, or when organic substances are burned or oxidized, elements and inorganic compounds essential for life (nutrients) return to the environment. Decomposers fulfill an essential role in this process by feeding on the remains of plants and animals and on fecal material. The result is that nutrients flow in a perpetual cycle between biotic and abiotic components of the ecosystem. Nutrient cycles are often called **biogeochemical cycles** because they involve exchanges between living organisms (*bio-*) and the rocks, air, and water of the earth's crust (*geo-*). Continuous input of energy from the sun keeps nutrients flowing and the ecosystem functioning (figure 2.18).

We measure biogeochemical cycles in terms of naturally occurring elements, such as carbon, phosphorus, and nitrogen.

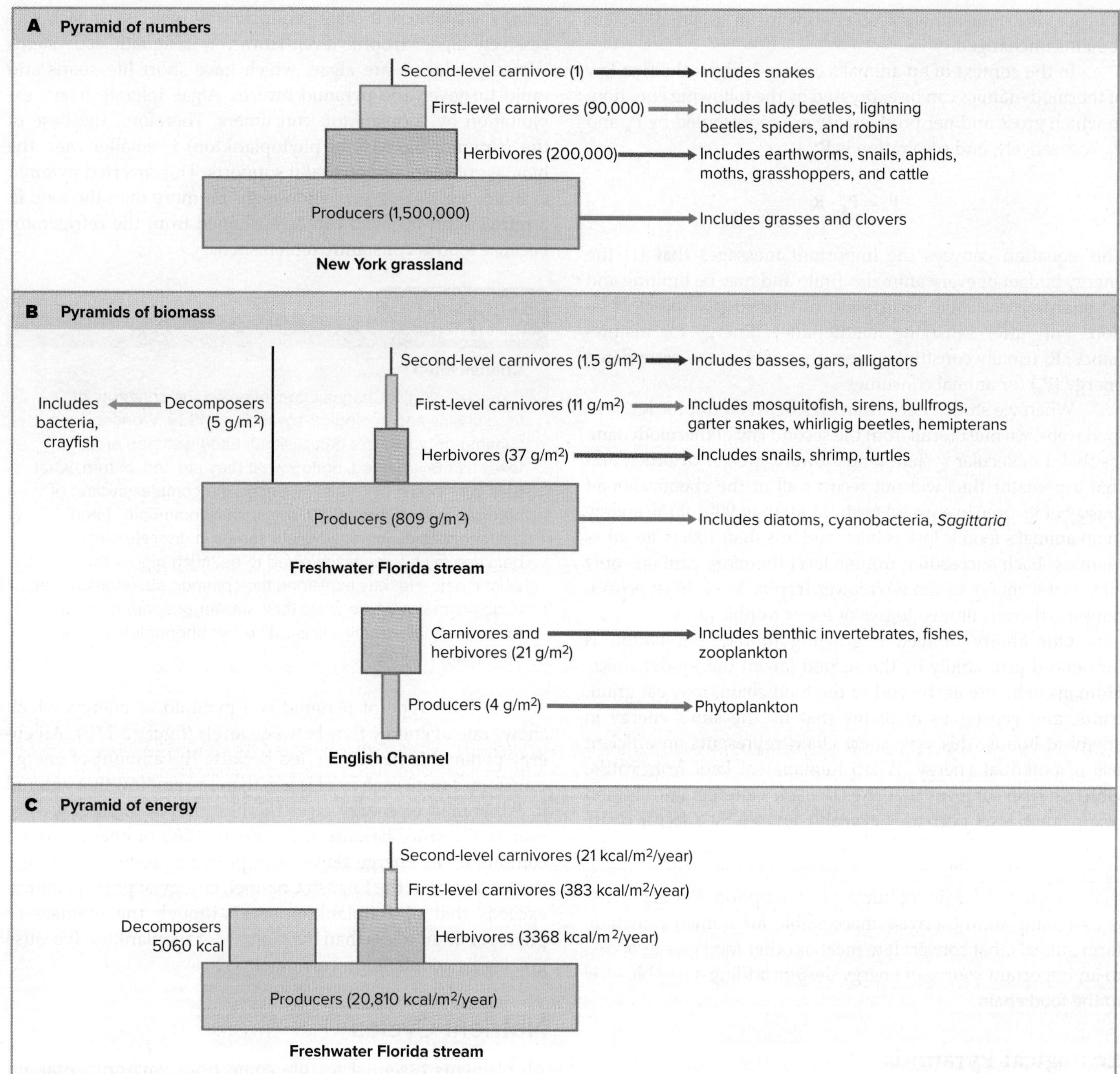

figure 2.17

Ecological pyramids with representative organisms from each trophic level. **A,** Pyramid of numbers of organisms counted for 100 square feet of grassland in New York. **B,** Pyramids of biomass for a freshwater Florida stream (*top*) and plankton of the English Channel (*inverted pyramid at bottom*). **C,** Pyramid of energy for the same Florida stream shown in **B.** (*From the work of E. P. Odum and H. T. Odum.*)

For example, the nutrient pool of the soil shown in figure 2.18 contains a store of phosphorus, some of which is in compounds that can be extracted by living plants. Phosphorus obtained by plants can be stored in the plants' tissues, returned to the soil, or consumed by herbivorous animals. The herbivores likewise store phosphorus in their bodies, return it to the soil via waste products, or provide it to carnivores that consume them. These carnivores likewise retain some phosphorus, return some to the soil via waste products, and may themselves be the prey of other carnivores. The amount of phosphorus available in the nutrient pool can be a limiting factor in the nutrient cycle of a local community.

Many industrially produced organic chemicals that enter the ecosystem as waste challenge nature's nutrient cycling

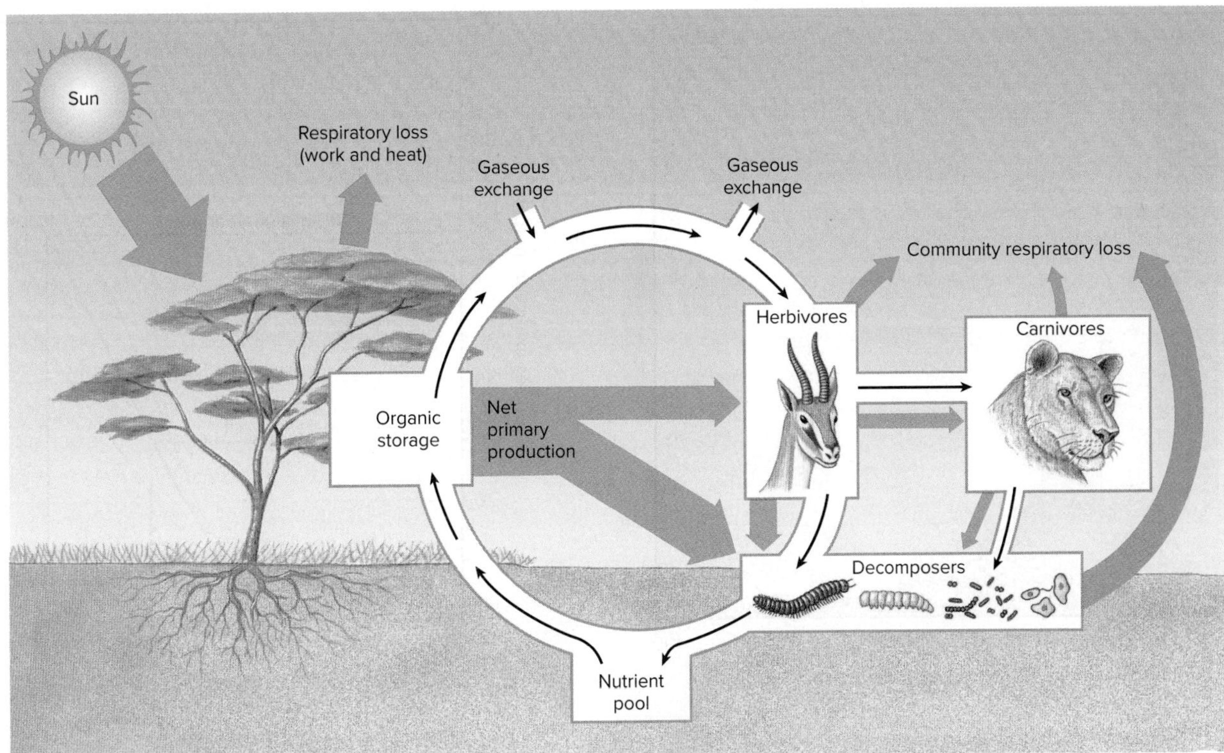

figure 2.18

Nutrient cycles and energy flow in a terrestrial ecosystem. Note that nutrients are recycled (black arrows), whereas energy flow (*orange*) is one way.

because decomposers have not evolved ways to degrade the industrial chemicals. Among the most harmful synthetic compounds are pesticides. Pesticides in natural food webs can be insidious for three reasons. First, many pesticides become concentrated as they travel up succeeding trophic levels. The highest concentrations occur in top carnivores, such as hawks and owls, diminishing their ability to reproduce. Second, many species that are killed by pesticides are not pests, but merely innocent bystanders, called nontarget species. Nontarget effects happen when pesticides move out of the agricultural field to which they were applied, through rainwater runoff, leaching through soil, or dispersal by wind. The third problem is persistence; some chemicals used as pesticides have a long life span in the environment, so that nontarget effects persist long after the pesticides have been applied. Genetic engineering of crop plants aims to improve their resistance to pests to lessen the need for chemical pesticides.

2.6 Biodiversity and Extinction

Biodiversity exists because rates of speciation on average slightly exceed rates of extinction in earth's evolutionary history. Approximately 99% of all species that have lived are extinct. Speciation rates represent an ongoing process of geographic expansion of populations by dispersal followed by geographic fragmentation (see section 1.4: Multiplication of Species), producing a multiplication of species. Speciation rates vary enormously among animal taxa and geographic area. Typical rates include 0.2–0.4 speciation events per species per million years, as measured for Cretaceous marine gastropods of the Atlantic coast; average duration of these gastropod species was 2–6 million years.

Extinction rates show episodic peaks and valleys throughout earth's evolutionary history. Paleontologist David Raup analyzed episodicity of extinction peaks by dividing the 600-million-year history[5] of the marine fossil record into successive intervals of 1 million years duration; he then measured the percentage of existing species that suffered extinction during each interval. Rates of species extinction in the 600 intervals range from near zero to 96%, averaging about 25% (figure 2.19A). We measure episodicity of extinction by asking, How long must we wait, on average, for an extinction peak killing at least 30% of existing species, or perhaps 65%? Answers to these questions are shown by Raup's "kill curve" (figure 2.19B). At least 5% of the existing species go extinct during any million-year interval of geological time. Extinction events killing at least 30% of

[5]Following Raup's analysis, the estimated age of the Phanerozoic fossil record (see inside back cover) was decreased from 600 million years to 542 million years. This revision should not alter the major conclusions of Raup's analysis, but the numbers are no longer consistent with geological dating. When describing Raup's results, we continue to refer to the 600-million-year number used in his analyses while using the newer estimates elsewhere in the book.

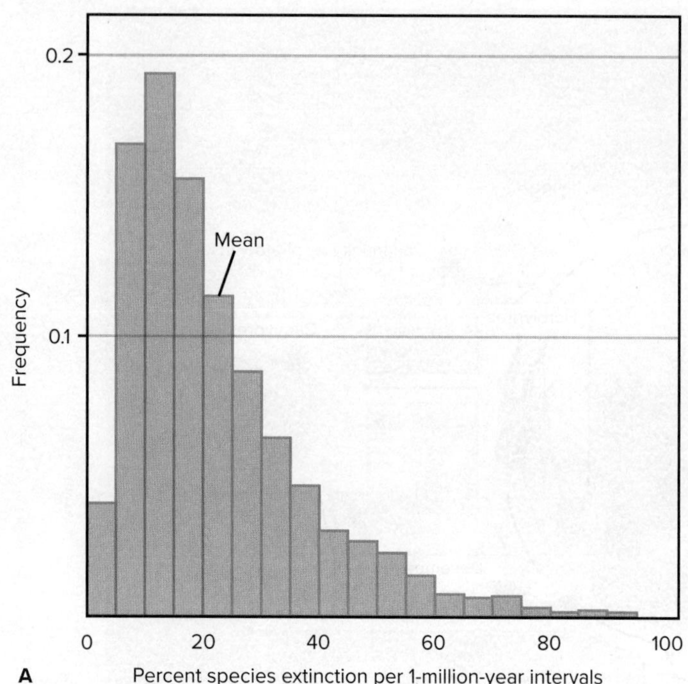

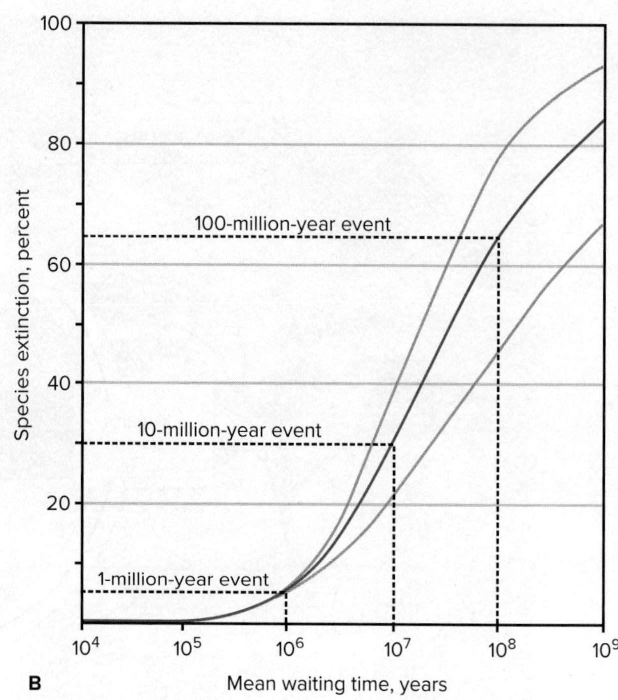

figure 2.19

A, Variation in extinction rates for species in the marine fossil record. In this analysis, David Raup divided the fossil record into 600 consecutive intervals, each one having a duration of 1 million years (Myr), beginning 600 Myr ago. Percent species extinction was measured for each interval. Almost 20% of the intervals had extinction rates of 10–15%, the highest bar on this graph. The mean species extinction rate is 25% per million years, and the mean species duration is 4 Myr. **B,** Species "kill curve" for the data summarized in **A.** Waiting time is the average interval between events equal to or greater than a given intensity of species extinction. The kill curve demonstrates an episodic distribution of extinction peaks over the past 600 Myr. If extinction events were distributed evenly through time, the kill curve would be indistinguishable from the *x* axis. The lighter curves on either side of the dark one show the statistical measurement error; the actual curve could lie anywhere between these lighter curves, but the dark line provides the best estimate.

After David Raup (1995).

existing species occur every 10 million years on average, whereas extinction events killing at least 65% of existing species occur on average every 100 million years; the latter episodes clearly qualify as mass extinctions (see section 1.7: Mass Extinctions).

Figure 2.19A reveals a continuous distribution of extinction rates from very high to very low. Extinction rates intermediate between mass extinction and background extinction occur primarily in the Paleozoic era, with the contrast between extinction peaks and valleys being more pronounced in the post-Paleozoic fossil record. The contrast between "mass extinction" and "background extinction" thus remains useful despite the apparent continuity of high and low extinction rates in Raup's analysis.

Fossil studies show that species whose geographic ranges are large have lower average rates of extinction than do those with small geographic ranges, although mass extinctions can erase this contrast. Species of Atlantic Cretaceous gastropods differed greatly in geographic range depending on modes of larval feeding. Some species had pelagic, plankton-feeding ("planktotrophic") larvae, which were carried long distances by oceanic currents; these species maintained large geographic ranges averaging 2000 km along the Atlantic Coast. Other species had heavy larvae that settled to the ocean floor as benthic feeders immediately upon hatching; these nonplanktotrophic species average less than one-fourth the geographic range of their plankton-feeding counterparts. A nonplanktotrophic species is about three times more likely to suffer extinction than a plankton-feeding one, but their geographic fragmentation also makes nonplanktotrophic species twice as likely to undergo speciation.

A paradox of biodiversity is that geographic **habitat fragmentation** of a species simultaneously increases rates of both local extinction and speciation. The African antelopes shown in figure 1.14 illustrate a similar contrast; in the past 6 million years, one group (hartebeests, blesboks, and wildebeests) experienced multiple events of speciation and extinction with seven species alive today, whereas the other lineage (impalas) persisted as a single species through the same time period. Impalas are about equal to the sum of the other seven species in number of animals alive today. This contrast shows that increased species diversity evolves at the peril of greater extinction risk for each individual species.

Higher taxa, such as orders, families, and genera (see section 4.1: Linnaeus and Taxonomy), also gain some protection

Table 2.1 Comparison of Species Extinction Levels for the Big Five Mass Extinctions*

Extinction Episode	Age, Myr Before Present	Percent Extinction
Cretaceous	66	76
Triassic	200	76
Permian	251	96
Devonian	359	82
Ordivician	443	85

After David Raup (1995), with revised dates for these extinctions.

from extinction by having large geographic ranges. Raup notes that higher taxa containing many species collectively distributed across a large geographic area are unlikely to go extinct. When such extinctions do happen, as seen for many dinosaur and marine ammonoid taxa at the end of the Cretaceous era (table 2.1), unusually catastrophic conditions appear responsible. An asteroid bombardment and the resulting global dust cloud at the end-Cretaceous extinction (see section 1.7: Mass Extinctions) appears to have caused within a short time widespread fires and extreme darkness and cold followed by extreme heat, all conditions well outside the evolutionary tolerances of many animal taxa whose members were formerly abundant. Only by chance would a taxon likely contain species able to withstand a challenge unprecedented in the group's evolutionary history.

Darwin explained extinctions of higher taxa by interspecific competition, but paleontological studies now refute this claim. Paleontologist Michael Benton estimates that less than 15% of the extinct tetrapod families could have been killed by competition with other families. Ecological and fossil studies of ectoprocts (see section 9.4) demonstrate that species of one order (Cheilostomatida) have outcompeted those of another order (Tubuliporata) ecologically by overgrowing their colonies for many millions of years without driving Tubuliporata to extinction.

Decline or extinction of one taxon often frees resources to which another taxon becomes adapted at a much later time, leading to evolutionary proliferation of ecologically diverse species in the latter. Resources freed by extinction of dinosaur

An Extinction Crisis and Conservation

The United Nations reported in 2019 that nature is declining globally at rates unprecedented in human history, and that the rate of species extinctions is accelerating. Around 1 million animal and plant species are considered threatened with extinction, many within decades, from an estimated total of 8 million species (5.5 million being insects). More than 40% of amphibian species, almost 33% of reef-forming corals and more than a third of all marine mammals are threatened. The estimate for threatened insect species is 10%. On average, extinction threatens about 25% of animal and plant species. The rate of global extinction is estimated to be tens to hundreds of times higher than the average for the past 10 million years. More than 500,000 of the world's estimated 5.9 million terrestrial species have insufficient habitat for continued survival without restoration. At least 680 vertebrate species have gone extinct by human actions since the sixteenth century. Currently, more than 6 species of hoofed mammals would be extinct or surviving only in captivity without conservation measures.

The five major causes of extinction are, in descending order, (1) changes in land and sea use, (2) direct exploitation of organisms, (3) climate change, (4) pollution, and (5) invasive alien species. Approximately 75% of terrestrial and 66% of marine environments have been severely altered by human actions. Ecosystems show a 47% reduction in geographic range relative to baseline conditions, with a continuing decline of 4% per decade. More than 85% of the wetlands present in the year 1700 had been lost by 2000. Average abundance of native species declined by more than 20% in most major terrestrial biomes since 1900. Habitat loss and deterioration have reduced the integrity of terrestrial habitats globally by more than 30%. Over a third of the earth's terrestrial surface is devoted to crop or livestock production. Global forest cover is 68% of pre-industrial levels. Commercial fishing covers more than half of the earth's oceanic area, with 33% of marine fish stocks being harvested at unsustainable levels and another 60% being harvested at the maximally sustainable levels.

Climatic warming is projected to cause a 3-25% decrease in fish biomass by the end of this century. A 100% increase since 1980 in greenhouse gas emissions has raised the average global temperature by at least 0.7°C. An estimated 5% of species would risk extinction from 2°C warming, rising to 16% for 4.3°C warming. Even a global warming of 1.5 to 2 degrees would profoundly shrink the ranges of most terrestrial species. Plastic pollution has increased tenfold since 1980. Approximately 300-400 million tons of heavy metals, solvents, toxic sludge, and other industrial wastes enter the world's waters annually. Fertilizers entering coastal ecosystems have produced more than 400 oceanic "dead zones" with a combined area of over 245,000 km^2, exceeding that of the United Kingdom. Detailed records from 21 countries show an average 70% increase in the number of invasive alien species since 1970.

The UN report's authors argue that conservation of nature requires a fundamental, system-wide reorganization of technological, economic, and social factors at all levels from local to global. Conservation investments in 109 countries from 1996 to 2008 reduced the extinction risk for mammals and birds by 29%. Eradication of invasive mammals on islands has benefited more than 107 highly threatened species of birds, mammals, and reptiles. The United Nations Convention on Biological Diversity sets targets for biodiversity and supports policy actions and societal initiatives to expand and strengthen conservation measures.

taxa at the end of the Cretaceous period are considered important for the later proliferation of mammalian adaptive and species diversity in the Cenozoic era (see figure 20.1).

Fossil studies of extinction help us to put into evolutionary perspective the consequences of human-mediated ecological changes for biodiversity. Fragmentation of populations as seen especially on islands produces locally high rates of species formation and endemism (see section 4.2), but these young species are unusually prone to extinction because their geographic ranges are small. About half of all areas in the world that contain at least two endemic bird species are islands, even though islands form less than 10% of the earth's terrestrial habitats. Island species are often particularly prone to destruction by introduction of invasive exotic species. For example, land snails of the genus *Partula* on the Tahitian island of Moorea were a major study system for island speciation until introduction of exotic snails displaced these native species. Mainland habitats, such as forests, are fragmented into virtual islands when development clears vast areas of habitat and when introduced species invade these areas. Because tropical regions have high species endemism, human fragmentation of these environments is particularly likely to produce species extinction.

A major challenge for animal species conservation is to obtain an inventory of earth's species diversity. Estimates of earth's total number of species are typically as large as 8 million, and this number may be too low by an order of magnitude. Taxonomists make these estimates by calculating ratios of described versus undescribed species in a sample from a geographic region, from rates of description of new species through time, and by judging how thoroughly species diversity has been measured in various taxa. Current estimates of species diversity are least satisfactory for microorganisms. Vagueness of these estimates reflects both practical and conceptual problems. A thorough survey of geographic genetic variation among natural populations requires expensive molecular-genetic analyses (see Chapter 1) and is critical for applying any of the current species concepts (see Chapter 4). This analysis is feasible only for taxa with relatively small numbers of large individuals, such as those at the top of a food chain or ecological pyramid. Beetles and nematodes constitute two taxa whose large numbers of small organisms challenge hopes for a comprehensive taxonomic survey. Even with appropriate data in hand, conflicting perspectives on what should constitute a species (discussed in Chapter 4) preclude an unambiguous tally of species numbers. Such conflicts would be particularly intense in groups of animals that do not show primarily a simple bisexual mode of reproduction. Clearly, conservation efforts cannot await a thorough taxonomic inventory of all animal populations. Maintenance of diverse ecosystems is an initial priority for preventing widespread species extinction.

One hopes that human disturbance of natural environments will not make the present time a competitor with the big five mass extinctions of table 2.1. Human activity clearly has induced numerous species extinctions, and we must avoid making the present time one that climbs too high on Raup's kill curve (figure 2.19B). Evolutionary studies suggest, however, that widely distributed higher taxa are unlikely to go extinct even during episodes of high species extinction. Lacking an inventory of species diversity in animals, we must avoid creating conditions that would selectively destroy any particular higher taxon.

■ SUMMARY

Sections	Key Concepts
2.1 The Hierarchy of Ecology	• Ecology is the study of relationships between organisms and their environments to explain the distribution and abundance of species on earth. • Ecology encompasses a hierarchy of biological systems in interaction with their environments: organism, population/species, ecological community, ecosystem, and the biosphere.
2.2 Environment and the Niche	• The physical space that contains an animal's environment is its habitat. • Within the habitat are physical and biological conditions suitable to survival and reproduction, which constitute the niche of an animal, population, or species.
2.3 Populations	• Animal populations consist of demes of interbreeding members sharing a common gene pool. A metapopulation comprises two or more demes connected by migration of individuals between demes. • Birth cohorts of animals have characteristic patterns of survivorship that represent adaptive trade-offs between parental care and numbers of offspring. • Animal populations consisting of overlapping birth cohorts have age structure that indicates whether they are growing, declining, or at equilibrium. • Every species in nature has an intrinsic rate of increase that provides a potential for exponential growth. • Population growth may be regulated intrinsically, by the carrying capacity of the environment; or extrinsically, by competition between species for a limiting resource, or by predators or parasites. • Density-independent abiotic factors can limit, but not truly regulate, population growth.

Sections	Key Concepts
2.4 Community Ecology	• Communities consist of populations that interact with one another through competition, predation, parasitism, and mutualism. • These relationships produce coevolution among populations within communities. • Guilds of species avoid competitive exclusion by partitioning limited resources, often featuring morphological specialization. • Keystone predators are those that control community structure and reduce competition among prey, which increases species diversity. • Parasites and their hosts can evolve a benign relationship that enables their coexistence.
2.5 Ecosystems	• Ecosystems consist of communities and their abiotic environments. • Animals occupy the trophic levels of herbivorous and carnivorous consumers within ecosystems. • All organisms have an energy budget consisting of gross and net productivity as well as respiration. • For animals, respiration usually is at least 90% of this budget. Thus, transfer of energy from one trophic level to another is limited to about 10%, which in turn limits the number of trophic levels in an ecosystem. • Ecological pyramids of energy depict how productivity decreases in successively higher trophic levels of food webs. • Ecosystem productivity is described by measuring energy flow and material cycles within ecosystems. All energy is lost as heat, but nutrients and other materials are recycled. • No ecosystem is closed because all depend upon acquiring energy from outside sources.
2.6 Extinction and Biodiversity	• Biodiversity exists because rates of speciation on average slightly exceed rates of extinction in earth's evolutionary history. • Approximately 99% of all species that have lived are now extinct. • Extinction rates in the geologic past are highly episodic, with species extinction ranging from near zero to 96% in different million-year intervals. • Species having large geographic ranges experience lower average rates of extinction than do species with smaller ranges, and the same relationship holds for higher taxa. • Paleontological studies of extinction provide an important perspective for evaluating potential evolutionary consequences of human-mediated species extinction.

■ Review Questions

1. The term *ecology* is derived from a Greek word meaning "house" or "place to live." However, as used by scientists, the terms ecology and environment do not mean the same thing. How do these terms differ?
2. How would you distinguish the concepts of ecosystem, community, and population?
3. What is the distinction between habitat and environment?
4. Define a niche. How does the *realized niche* of a population differ from its *fundamental niche?* How does the concept of niche differ from the concept of guild?
5. Populations of independently living (unitary) animals have a characteristic age structure, sex ratio, and growth rate. However, these properties are difficult to measure for modular animals. Why?
6. Explain which of the three survivorship curves in figure 2.4 best fits the following: (a) a population in which mortality as a proportion of survivors is constant; (b) a population in which few individuals die early and most individuals live to old age; (c) a population in which the very young experience heavy mortality but the survivors live to old age. Offer an example from the real world of each survivorship pattern.
7. Contrast exponential and logistic growth of a population. Under what conditions might you expect a population to exhibit exponential growth? Why cannot exponential growth be perpetuated indefinitely?
8. Growth of a population can be limited by either density-dependent or density-independent mechanisms. Define and contrast these two mechanisms. Offer examples of how growth of the human population might be curbed by either agent.
9. Herbivory is beneficial for an animal (+) but harmful to the plant it eats (–). What are some + – interactions among animal populations? What is the difference between commensalism and mutualism?
10. Explain how character displacement can ease competition between coexisting species.
11. Define predation. How does the predator-prey relationship differ from the parasite-host relationship? Why is the evolutionary race between predator and prey one that the predator cannot afford to win?
12. Mimicry of yellowjacket wasps by clearwing moths illustrates a harmless and potentially edible species resembling a hazardous one.

What is the advantage to the moth of this form of mimicry? What is the advantage to a toxic species of mimicking another toxic species?

13. A keystone species is one whose removal from a community causes extinction of other species. How does this extinction happen?
14. What is a food chain? How does a food chain differ from a food web?
15. What is a trophic level, and how does it relate to a food chain?
16. Define *productivity* as the word is used in ecology. What is a primary producer? What is the distinction between gross productivity, net productivity, and respiration? How is net productivity related to biomass?
17. What conditions produce an inverted pyramid of biomass in which the consumers have a greater biomass than the producers? Describe an example of an inverted pyramid of numbers of organisms, in which there are, for example, more herbivores than plants on which they feed.
18. The pyramid of energy has been offered as an example of the second law of thermodynamics. Why?
19. Animal communities surrounding deep-sea thermal vents apparently exist in total independence of solar energy. How is this existence possible?
20. What are the implications of Raup's kill curve for the contrast between background extinction and mass extinction in geological history?
21. What do paleontological studies show about the relationship between the geographic range of a species and its probability of extinction? How does geographic habitat fragmentation produce a paradox for biodiversity?

For Further Thought Natural processes of species formation and extinction emerge on a timescale measured in millions of years, whereas human-mediated extinctions are measured on a scale of years to decades. How does this discrepancy complicate our attempts to conserve biodiversity?

3

Animal Architecture

LEARNING OBJECTIVES

Readers will be able to:

3.1 Distinguish five levels of body complexity in unicellular and multicellular organisms.

3.2 Identify the body symmetry of any organism.

3.3 Compare and contrast cytoplasmic, conditional, and syncytial cell specifications and their roles in pattern formation.

3.4 Explain how the body plans of protostome and deuterostome animals are produced from embryonic germ layers and body cavities.

3.5 Distinguish the basic body plans that occur across the animal phyla.

3.6 Recognize the tissues and extracellular components of the animal body.

3.7 Discuss the ways in which surface-area-to-volume relationships and metabolic costs change with animal body size.

Larry S. Roberts

Cnidarian polyps have radial symmetry and the cell-tissue level of organization (*Dendronephthya* sp.).

New Designs for Living

Zoologists today recognize 32 animal phyla, each phylum characterized by a distinctive body plan and array of biological properties that set it apart from all other phyla. Nearly all are the survivors of perhaps 100 phyla that existed 541 million years ago during the Cambrian period. How long it took for these taxa to evolve is the subject of active debate, but virtually all of the major body plans that we see today, together with many other novel plans that we know only from the fossil record, were established by the Cambrian. A recently discovered fossil of *Yilingia spiciformis* shows that bilaterally symmetrical animals predate the Cambrian. The themes in animal architecture that evolved so long ago persist today. Although we cannot identify the group to which *Yilingia* belongs, we recognize segments in its tiny body. Segmentation evolved independently in three very ancient lineages: annelids, arthropods, and chordates. The speciation and extinction events that followed the Cambrian period produced only variations on the body plans already established by this time. In most animal groups, we recognize the ancestral body forms in modern descendant lineages.

Some animals, like the cnidarian polyps at left, have only one opening to the digestive tract, but most animals have separate anterior and posterior openings. Animal body plans vary in terms of morphological features such as tissue layers, body openings, and body cavities. Studies of gene expression during development have given us surprising insights into how some of these structures are made.

Humans share fundamental similarities in building materials and body plan with most other animals, despite vast differences in structural complexity. This essential uniformity of biological organization derives from the common ancestry of animals and from their basic cellular construction. In this chapter, we consider the limited number of body plans that underlie the diversity of animal form and examine some of the common architectural themes that animals share.

3.1 The Hierarchical Organization of Animal Complexity

The eukaryotic cell is the basic building block of animals. There are five major levels of organization within the eukaryotes. Our description of levels begins with a single cell and encompasses four types of cell-cell interactions in multicellular bodies (table 3.1).

At the first level of organization, the *cytoplasmic level,* cells vary in terms of their subcellular structures, including organelles. Organisms at the next level of organization, *the cellular level,* may be colonial or multicellular. Cells at this level communicate with other cells in the aggregation and may secrete an extracellular matrix, called an **ECM,** to which cells adhere. Collagen is a familiar component of ECM found in animals. The distinction between coloniality and multicellularity lies in the degree of specialization. Colonial organisms consist of aggregations of unspecialized cells, all able to reproduce. In multicellular organisms, cells are specialized for particular functions and reproduction is limited to some cells in the aggregation. The evolution of multicellularity occurred independently in perhaps as many as 25 lineages of unicellular eukaryotes (see Chapter 5 for some examples). One such event within a lineage called Opisthokonta led to the animals.

Animals evolved greater structural complexity by combining cells into larger units. The simplest animals show the cellular level of organization, in which cells demonstrate

TABLE 3.1 Grades of Organization in Organismal Complexity

Grade of organization	Description	Examples
Protoplasmic	*Protoplasmic grade of organization*–characterizes unicellular organisms. All life functions are confined within the boundaries of a single cell, the fundamental unit of life. Within a cell, protoplasm is differentiated into organelles capable of performing specialized functions.	Unicellular organisms *Paramecium*
Cellular	*Cellular grade of organization*–an aggregation of cells that are functionally differentiated. A division of labor is evident, so that some cells are concerned with, for example, reproduction, and others with nutrition. Some flagellates, such as *Volvox*, that have distinct somatic and reproductive cells are placed at the cellular level of organization. Many authorities also place sponges at this level.	Aggregations of cells *Volvox*
Cell–tissue	*Cell–tissue grade of organization*–an aggregation of similar cells into definite patterns or layers and organized to perform a common function, to form a tissue. Sponges are considered by some authorities to belong to this grade, although sea jellies and their relatives (Cnidaria) more clearly demonstrate the tissue plan. Both groups are still largely of the cellular grade of organization because most cells are scattered and not organized into tissues. An excellent example of a tissue in cnidarians is the nerve net, in which nerve cells and their processes form a definite tissue structure, with the function of coordination.	Aggregations of cells that form tissues Sea jellies (Cnidarians)
Tissue–organ	*Tissue–organ grade of organization*–an aggregation of tissues that form organs in a further step in complexity. Organs are usually composed of more than one kind of tissue and have a more specialized function than tissues. This is the the overall organizational level of flatworms (Platyhelminthes), that possess well-defined organs such as eyespots, proboscis, and excretory organs. In flatworms, the reproductive structures are organized into a system that is characteristic of the next level of organizational complexity.	Aggregations of tissues that form organs *Planaria*
Organ–system	*Organ–system grade of organization*–organs working together to perform some function, producing the highest level of organization—an organ system. Systems are associated with basic body functions such as circulation, respiration, and digestion. The simplest animals having this type of organization are nemertean worms, which have a complete dilgestive system distinct from the circulatory system. Most animal phyla demonstrate this type of organization.	Aggregations of organs that form systems Nemertean worms Arthropods

division of labor but are not strongly associated to perform a specific collective function (table 3.1). In the more complex **cell-tissue** level, similar cells are grouped together and perform their common functions as a highly coordinated unit. In animals of the *tissue-organ level* of organization, tissues are assembled into still larger functional units called **organs**. Usually one type of tissue carries the burden of an organ's chief function, as muscle tissue does in the heart; other tissues—epithelial, connective, and nervous—perform supportive roles. The chief functional cells of an organ are called its **parenchyma** (pa-ren'ka-ma; Gr. *para,* beside, + *enchyma,* infusion). The supportive tissues are its **stroma** (Gr. bedding). For instance, in the vertebrate pancreas the secreting cells are the parenchyma; the capsule and connective tissue framework represent the stroma.

Most animals have an additional level of complexity in which different organs operate together; this is the **organ-system level** of organization. Eleven different organ systems occur in animals: skeletal, muscular, integumentary, digestive, respiratory, circulatory, excretory, nervous, endocrine, immune, and reproductive. The great evolutionary diversity of these organ systems is covered in Chapters 8–20.

3.2 Animal Body Plans

Animal body plans differ in level of organization, in number of embryonic germ layers, in form and number of body cavities, and in body symmetry. Symmetry can generally be determined from the external appearance of an animal, but other features require more detailed study.

Animal Symmetry

The term **symmetry** refers to balanced proportions, or correspondence in size and shape of parts on opposite sides of a median plane. *Asymmetrical* bodies are not balanced; there is no plane through which they are divided into identical halves (figure 3.1A). **Spherical symmetry** means that any plane passing through the center divides the body into equivalent, or mirrored, halves (figure 3.1B). This type of symmetry occurs chiefly among some unicellular eukaryote groups and is rare in animals. Spherical forms are best suited for floating and rolling.

Bodies with **radial symmetry** can be divided into similar halves by more than two planes passing through the longitudinal axis (figure 3.1C). These are tubular-, vase-, or bowl-shaped animals, in which one end of the longitudinal axis is usually the mouth (the oral surface). Examples are the hydras, and jellyfishes in phylum Cnidaria, sea urchins (phylum Echinodermata), and some sponges; in sessile forms, the aboral surface is used in attachment. A variant form is biradial symmetry, in which, because of some part that is single or paired rather than radial, only one or two planes passing through the longitudinal axis produce mirrored halves. Sea walnuts or comb jellies (phylum Ctenophora), which are roughly globular in form but have a pair of tentacles, are an example. Radial and biradial animals are usually sessile, freely floating, or weakly swimming. Echinoderms (sea stars and their kin) are primarily bilateral animals (their larvae are bilateral) that have become secondarily radial as adults.

Bilateral symmetry applies to animals that are divided along a sagittal plane into two mirrored portions, right and left halves (figure 3.1D). Some convenient terms used for locating regions of animal bodies are **anterior**, to designate the head end; **posterior**, the opposite or tail end; **dorsal**, the back side; and **ventral**, the front or belly side (figure 3.2). **Medial** refers to the midline of the body; **lateral** refers to the sides. **Distal** parts are farther from the middle of the body; **proximal** parts are nearer. Figure 3.2 also illustrates three planes of the body. A **frontal plane** (also sometimes called a coronal plane) divides a bilateral body into dorsal and ventral halves by running through the anteroposterior axis and the right-left axis at right angles to the **sagittal plane**, the plane dividing an animal into right and left halves. A **transverse plane** (also called a cross section) would cut through a dorsoventral and a right-left axis at right angles to both the sagittal and frontal planes, separating the body into anterior and posterior portions.

A Asymmetry

B Spherical symmetry

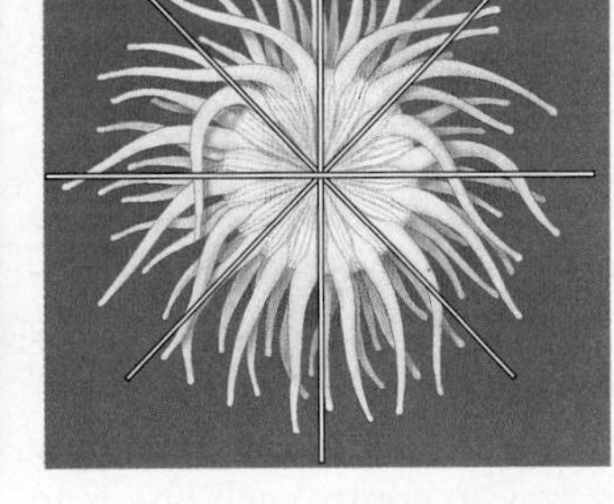

C Radial symmetry

D Bilateral symmetry

figure 3.1

Asymmetrical bodies **A**, lack planes of symmetry. Planes of symmetry are illustrated by spherically **B**, radially **C**, and bilaterally **D**, symmetrical animals.

figure 3.2

Terms used to describe positions on the body of a bilaterally symmetrical animal.

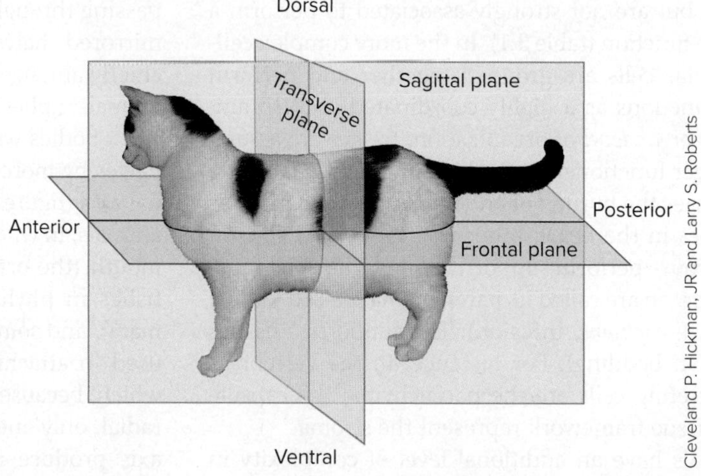

The evolution of bilateral symmetry was a major innovation because bilateral animals are much better fitted for directional (forward) movement than are radially symmetrical animals. Bilateral animals are collectively called **Bilateria**. Bilateral symmetry is strongly associated with **cephalization**, the differentiation of a head end. Concentration of nervous tissue and sense organs in a head bestows obvious advantages to an animal moving through its environment head first. This is the most efficient positioning of instruments for sensing the environment and responding to it. An animal's mouth usually occurs on the head where sensory structures guide forward locomotion toward food.

3.3 Development and Genetics

The body plan of an animal forms through an inherited developmental sequence (figure 3.3). The sequence begins after a sperm fertilizes an egg to create a **zygote**. The zygote is a single very large cell that will be divided into a large number of smaller cells called blastomeres by a process called **cleavage**. Cleavage is an orderly sequence of cell divisions where the zygote divides into two cells, the two cells divide to make four cells, the four make eight cells, and so on until there are hundreds of cells in an embryo.

How does an embryo generate a multitude of cell types from a single zygotic nucleus? We know that cleavage is a mitotic process, so an organism's cells carry the same set of genes, yet cells have many possible fates. What causes some cells to develop into neurons while others develop into skeletal muscle? One means of controlling cell fate is by determining the contents of the cytoplasm; as cleavage proceeds, certain messenger RNAs or proteins present in cytoplasm are unevenly partitioned among cells. Another way to control cell fate is to regulate the timing of gene expression. Genes to be expressed are transcribed into messenger RNA (mRNA) which is then translated into protein products. Transcription factors and enhancers that bind to promoter regions of genes during transcription regulate gene expression. Enhancers determine where, when, and how much protein is made—by acting on transcription. A single gene may have more than one enhancer, so that one enhancer works in one kind of cell and another works in a different cell type. For example, there is one enhancer for the *Pax6* gene in mouse pancreas cells and another for this gene in mouse eye cells. By controlling the enhancers, the organism controls gene activation in different cell types. In certain cell types, a combination of enhancers may be required; both of the mouse *Pax6* examples just given require more than one enhancer. Enhancers may activate a promoter or repress it, so some enhancers may be "negative enhancers," or silencers.

Transcription factors act on a larger scale than enhancers and may bind to enhancers or directly to promoters to control transcription. Like enhancers, transcription factors may repress or activate transcription. The fates of cells may be determined by multiple interacting factors, such as promoters, enhancers, and transcription factors, turning on and off the production of particular proteins in highly structured patterns. The proteins produced determine cell function.

In most animals, excluding insects, there are two major ways by which the early cells become committed to particular developmental fates: (1) cytoplasmic specification in which determinative molecules deposited in the egg cytoplasm by the mother are partitioned among cleaving cells, and (2) conditional specification in which cell fates are determined by interactions with neighboring cells (induction). All animals use both of these mechanisms to specify different cell types, but in some animals cytoplasmic specification is important in early development, whereas others rely predominantly on inductive interactions.

Cytoplasmic Specification

The cytoplasm of a fertilized egg is not homogeneous. It contains unequally distributed morphogenetic components such as mRNAs and proteins, which act as enzymes or as transcription factors. These components may be tethered to the cytoskeleton. As cleavage proceeds, the mRNAs and proteins are unequally partitioned among the resulting new cells (blastomeres) and thus influence cell fate. In animals with cytoplasmic

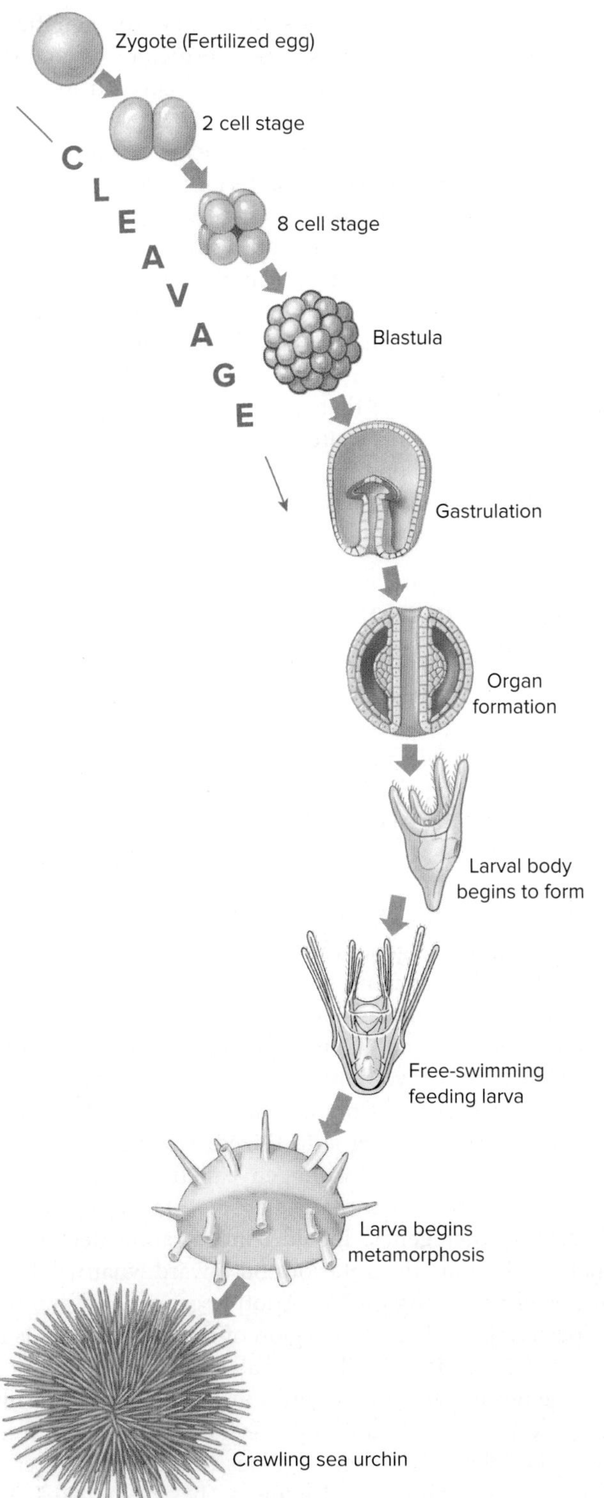

figure 3.3

Development in a sea urchin. Fertilization of an urchin egg is followed by cleavage, producing a mass of cells. This mass rearranges to form a single cell layer surrounding a cavity (blastula stage). In the next stage, a gut and more tissue layers form (gastrulation). Once a gut is present, the urchin embryo develops a larval body. The free-swimming larva feeds and grows in ocean surface waters. The larva metamorphoses into a tiny bottom-dwelling sea urchin; the urchin feeds and grows, reaching sexual maturity in this body form.

specification, such as snails and annelid worms, single cells isolated from the rest of the embryo will often continue to differentiate for a time along the path dictated by their cytoplasmic components. This results in mosaic development of the embryo. The term "mosaic" is used because the embryo appears to be a composite of independently developing parts rather than of interacting parts (see figure 3.11B). Cytoplasmic specification is also called autonomous specification.

Cytoplasmic specification is especially striking (and easily visualized) in some sessile marine animals called tunicates. In some species the fertilized egg contains as many as five differently colored types of cytoplasm (figure 3.4). During cleavage, these differently pigmented cytoplasms are segregated into different blastomeres that then proceed to form distinct tissues or organs. For example, yellow cytoplasm gives rise to muscle cells, whereas gray equatorial (midline) cytoplasm produces the notochord and neural tube (see section 15.2). Clear cytoplasm produces larval epidermis, and gray vegetal cytoplasm (not visible in figure 3.4) gives rise to the gut.

We know now that yellow cytoplasm gives rise to muscle cells because it contains macho-1 mRNA. Macho-1 mRNA encodes a transcription factor that activates other genes leading to muscle formation in cells descended from yellow cytoplasm cells. However, not all descendant cells make muscles; those in contact with cells whose descendants will make the gut are directed or induced to a different fate by a process called conditional specification.

Conditional Specification

Conditional specification differs from cytoplasmic specification because the fate of a particular cell is not fixed until the cell receives positional information from its neighbors. The cell is induced to a particular fate by the interaction between populations of cells. **Induction** is the capacity of some cells to evoke a developmental response, such as a change in cell shape or cell fate, in other cells. For example, formation of hair or feathers occurs when the epidermis, derived from an embryonic germ layer called ectoderm, responds to inducers from the underlying dermis, which itself is derived from another germ layer called mesoderm.

Although the interaction requires an inducer and a responder, the interactions may be reciprocal in that the tissues influence each other. Inducers act through contacts between the surface proteins of adjacent cells, or through diffusible molecules that travel between cells. Such molecules are called growth and differentiating factors. Diffusion of the molecules away from the source cell population produces a gradient in the strength of the signal available to the responding cell. In some cases, a threshold level of signal strength is required to change the fate of a cell receiving the signal.

Syncytial Specification

Not all embryos are multicellular early in development. In insects, nuclear division proceeds without cytoplasmic division, so the zygote is syncytial. A syncytium occurs when a

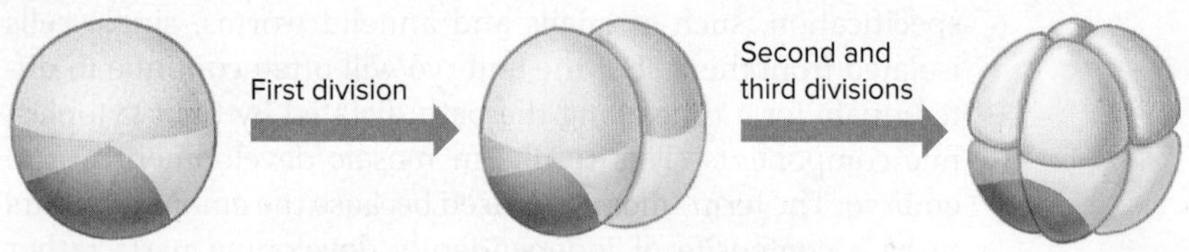

figure 3.4

Bilateral cleavage in tunicate embryos. The first cleavage division divides the asymmetrically distributed cytoplasm evenly between the first two blastomeres, establishing the future right and left sides of the adult animal. Bilateral symmetry of the embryo is maintained through subsequent cleavage divisions.

single cell membrane surrounds multiple nuclei. A familiar example of a syncytium is the egg of the fruit fly genus *Drosophila* (figure 3.5). In insect development, the syncytium is eventually divided into many cells, but some developmental processes occur before this point. Syncytial specification is similar to conditional specification, but the molecules that influence cell fate diffuse *within* the cytoplasm of a single large cell and not among cells. Species of *Drosophila* are model organisms for studies of genetics and development, so many principles of embryonic pattern formation were discovered in syncytial embryos.

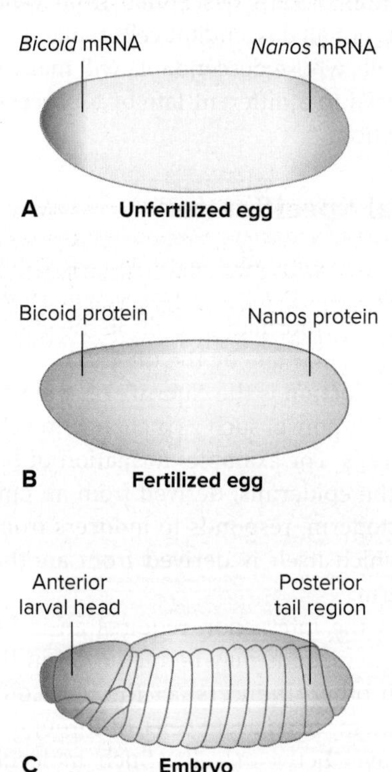

figure 3.5

The anterior-posterior axis in a fruit fly embryo is determined before fertilization by maternally positioned mRNAs. **A,** *Bicoid* and *Nanos* mRNAs occur at opposite ends of an egg. **B,** After fertilization, mRNAs are translated into proteins that diffuse outward, resulting in concentration gradients. Bicoid protein is most concentrated at what will become the anterior end, whereas nanos protein is most concentrated at what will become the posterior end. **C,** Anterior and posterior cells fates are determined by position within the concentration gradient.

One important principle discovered in *Drosophila* is how the anterior-posterior axis is determined. There are three general stages in the formation of the body: pattern formation, determination of position in the body, and induction of limbs and organs appropriate for that position. Each stage is guided by gradients of gene products that function as morphogens. Morphogens are diffusible molecules that create a concentration gradient as they move away from a source. A cell's position in the concentration gradient of one or more morphogens determines its fate. Cells respond only to morphogen concentrations above a threshold level.

Pattern Formation

The first step in pattern formation during embryo development is specification of the three body axes: front-to-rear (anteroposterior), left-to-right, and back-to-front (dorsoventral). In *Drosophila* the anteroposterior axis is determined even before an egg is fertilized. Christiane Nüsslein-Volhard and her colleagues in Germany found that this determination is due to a gradient of mRNA that is secreted into the egg by nurse cells in the mother. To make nurse cells, the oocyte divides mitotically several times, all prior to fertilization. The cells produced are genetically identical to the oocyte and to each other. They remain interconnected and eventually contribute all their cytoplasm and its constituents to the oocyte. In nurse cells, one mRNA involved in specifying the anterior region of the embryo, which will eventually form the head region, is transcribed from a gene called *bicoid* (pronounced BICK-oyd). After an egg is fertilized, *bicoid* mRNA is translated into a protein morphogen called bicoid (unlike the names of genes, protein names are not italicized) that binds to certain other genes. Bicoid concentration declines from what will become the anterior end toward what will become the posterior end (figure 3.5). Another gene, *nanos,* is involved in specifying the posterior region of the embryo that is fated to become the larva's abdominal region. The products of these genes activate other genes in a cascade that results in the production of an anteroposterior axis. *Bicoid* and *nanos* are two of about 30 maternal genes that control pattern formation in an embryo and allow a fly's head to be distinguished from its tail.

Bicoid-mutant embryos that do not produce the morphogen develop without a head; those that do not produce *nanos* lack an abdomen. Some of the maternal genes also specify a second dorsoventral axis that allows the fly's back and belly to be distinguished. For instance, the gene *short gastrulation* leads to development of ventral structures, such as the nerve cord. Clearly, the heterogeneous nature of the egg is critical for development: the gradients in morphogens provided by

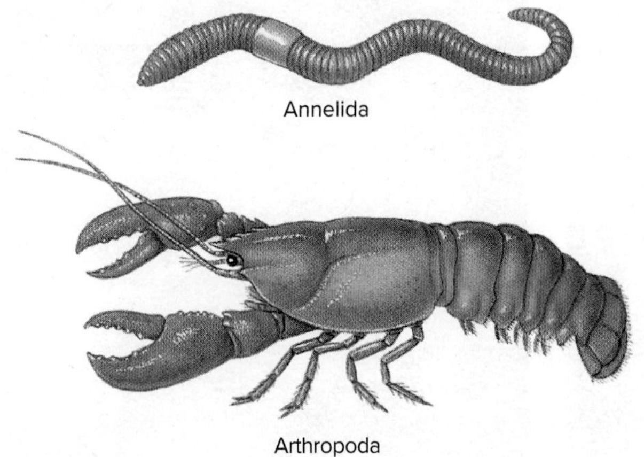

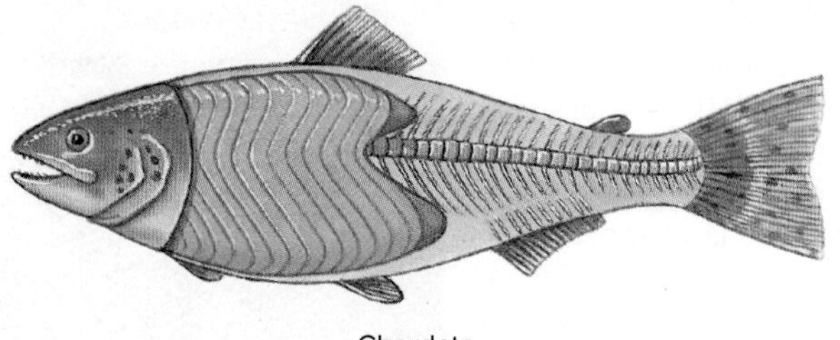

figure 3.6

Segmented phyla. These three phyla illustrate an important principle in nature—metamerism, or repetition of structural units. Segmentation permits specialization because segments, especially in arthropods, have become modified for different functions.

the mother define the axes for subsequent development, at which time the zygotic genes in the embryonic nuclei are activated.

One of the most exciting discoveries in developmental genetics has been that the developmental genes of vertebrates and many other animals are similar to those of *Drosophila;* they are conserved over a wide range of animals. For instance, a gene similar to *bicoid* is important in pattern formation in vertebrates. In vertebrates, however, the gene, called *Pitx2,* determines positioning of certain internal organs to either the left or the right side of the body. Mutations in *Pitx2* in frogs, chicks, and mice can place the heart and stomach on the right instead of the left side. Such mutations may explain a reversal of organ position that sometimes occurs in humans. *Pitx2* is activated by a protein produced by the gene *sonic hedgehog* (*Shh*), which is similar to a *Drosophila* gene called *hedgehog.* (The name hedgehog refers to the bristly appearance of fruit flies lacking the gene. The "sonic" comes from the video-game character Sonic the Hedgehog.) In vertebrates, *sonic hedgehog* is active in the left side only at the anterior end of the embryo, which determines the anteroposterior axis. Genes critical for development in a wide range of organisms are sometimes called "toolkit genes."

In *Drosophila,* as well as other arthropods, annelid worms, and chordates, one important aspect of pattern formation along the anteroposterior axis is **segmentation**, also called metamerism. Segmentation is the serial repetition of similar body regions along the anteroposterior axis of the body (figure 3.6). Segments are obvious in developing insects such as caterpillars or maggots, but in fishes, segmentation is apparent only in somites that produce structures such as vertebrae and repeated muscle bands.

The segments are identical early in development, but later activation of different combinations of genes causes each segment to differentiate. For example, the anterior segment of insect embryos forms antennae, eyes, and mouthparts, while segments farther back form legs. In *Drosophila,* segmentation genes control the number and orientation of segments. There are three classes of segmentation genes: gap, pair-rule, and segment-polarity. Gap genes are activated first and divide an embryo into regions such as head, thorax, and abdomen. Pair-rule genes divide these regions into segments. Finally, segment-polarity genes, such as *hedgehog,* organize the anterior posterior structures within each segment.

Homeotic and *Hox* Genes

Segmentation genes apparently regulate expression of other genes, ensuring that they are active only in appropriate segments. Such segment-specific genes are called homeotic genes. Mutations in homeotic genes, called homeotic mutations, result in appendages or other structures appearing in the wrong part of the body. For example, in *Drosophila* the homeotic gene *Antennapedia,* which helps trigger development of legs, is normally active only in the thorax. If *Antennapedia* is activated by a homeotic mutation in the head of a larval fly (a maggot), the adult fly will have legs in place of antennae (figure 3.7). *Antennapedia* and some other homeotic genes, as well as many other genes involved in development, include a sequence of 180 DNA base pairs, called the homeobox. The homeobox produces the part of a protein that attaches to the DNA of other genes, activating or blocking their expression.

In *Drosophila,* several other homeotic and nonhomeotic genes that are clustered on the same chromosome close to *Antennapedia* also include a homeobox. They specify a location within the body along the anterior-posterior axis. All such genes are usually called *Hox* genes. Most *Hox* genes occur in a cluster on one chromosome. Mammals have four clusters, each on a different chromosome, with 9–11 *Hox* genes each (figure 3.8). The sequence of *Hox* genes within a cluster is the same as the front-to-rear order in which they are expressed in the body. *Hox* genes occur in insects and chordates, as well as unsegmented animals such as hydra and nematode worms. They occur in plants, in yeasts, and perhaps in all eukaryotes.

figure 3.7

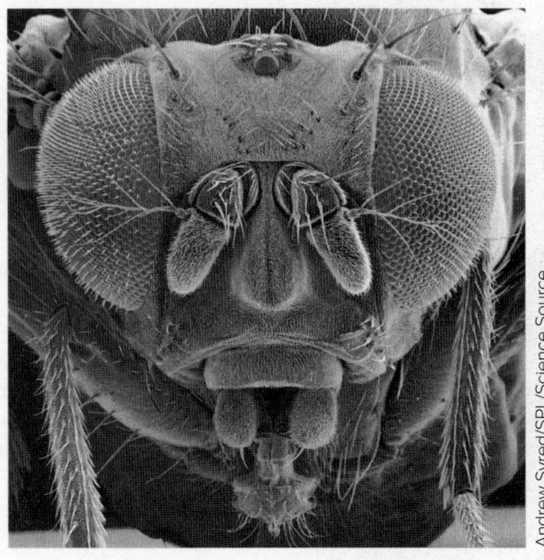

A, Head of a normal fruit fly with two antennae. **B,** Head of a fruit fly with a pair of legs growing out of head sockets where antennae normally grow. The *ntennapedia* homeotic gene normally specifies the second thoracic segment (with legs), but the dominant mutation of this gene leads to this bizarre phenotype.

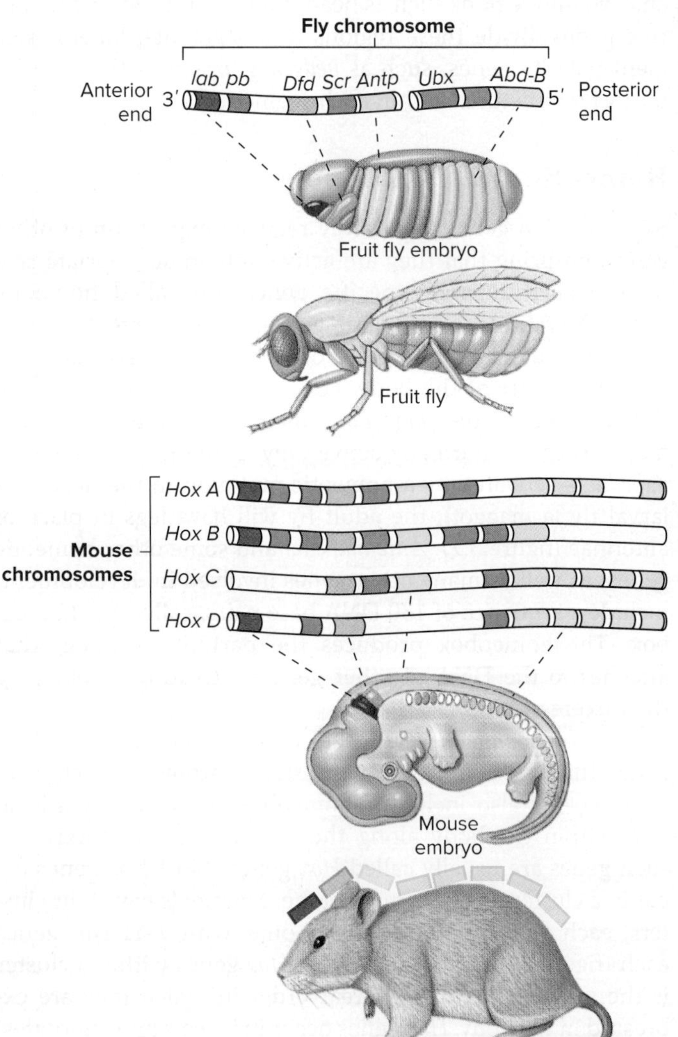

Morphogenesis of Limbs and Organs

Hox and other homeobox genes also play a role in shaping individual organs and limbs. Cheryll Tickle and her co-workers at University College London examined formation and development of limb buds in chicks (figure 3.9). They found that a new limb bud can be induced to grow from the side of a chick, where it would not normally occur, by implanting a bead soaked in fibroblast growth factor (FGF). This result implies that limbs are normally induced to develop by activation of the gene for FGF in appropriate parts of the body. Whether the limb bud develops into a wing or a leg depends on whether the FGF is applied toward the front or the rear of the chick. FGF also plays a role in shaping the limb. It is secreted by cells in an apical ectodermal ridge at the end of the limb bud. FGF acts as a morphogen that forms a gradient from the apical ectodermal ridge to the base of a limb bud. This gradient helps

figure 3.8

Homology of *Hox* genes in insects and mammals. These genes in both insects (fruit fly) and mammals (mouse) control the subdivision of the embryo into regions of different developmental fates along the anterior-posterior axis. The homeobox-containing genes lie on a single chromosome of the fruit fly and on four separate chromosomes in the mouse. Clearly defined homologies between the two, and the parts of the body in which they are expressed, are shown in color. The open boxes denote areas where it is difficult to identify specific homologies between the two. The *Hox* genes shown here are only a small subset of all the homeobox genes.

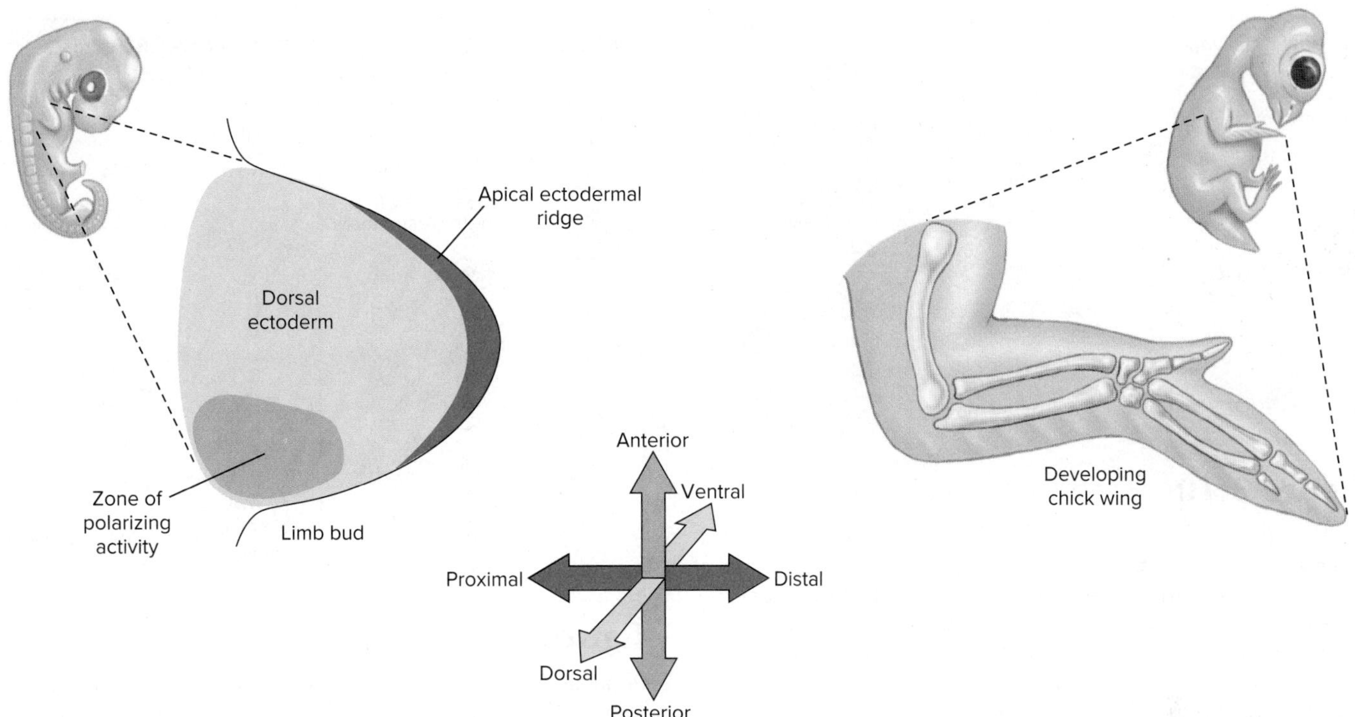

figure 3.9

Morphogenesis in a vertebrate limb bud. The skeleton of a mature chicken limb is shown for orientation. Three axes are established in the limb bud: a proximal-distal axis by fibroblast growth factor (FGF) from the apical ectodermal ridge; an anterior-posterior axis by sonic hedgehog protein from the zone of polarizing activity; and a dorsal-ventral axis by Wnt7a protein from dorsal ectoderm.

establish a proximodistal axis–one of three axes that guide development of a limb (figure 3.9). Fingers or toes develop at the end of the proximodistal axis with the highest level of FGF. An anteroposterior axis is established by a gradient of sonic hedgehog and ensures that fingers or toes develop in the appropriate order. Finally, the protein Wnt7a helps determine the dorsal-ventral axis.

There are fundamental similarities in the ways that animals develop. The fate of most cells is determined sequentially: following cleavage, the embryo forms blastula and gastrula stages with two or three embryonic germ layers: ectoderm, endoderm, and mesoderm. Later each of these germ layers makes particular derivatives. For example, skin, eye, and nerve cells are derived only from ectodermal cells. What emerges is a sequential pattern of development involving not only inductions but cell movement, changes in adhesive properties of cells, and cell proliferation. There is no "hardwired" master control panel directing development, but rather a sequence of local patterns in which one step in development is a subunit of another. However, there are groups of animals that follow the same developmental pathway because they inherited this pathway from a common ancestor. These pathways result in similar body plans. There are only a few major body plans, as described in the following section.

3.4 Development of Animal Body Plans

As shown in figure 3.3, there are many stages in development, but cleavage of the fertilized egg is usually the first step. Cleavage occurs in several different ways: sponges and cnidarians (sea anemones and their kin) lack a distinct cleavage pattern, but bilateral animals typically exhibit either radial or spiral cleavage (figure 3.10). Although there is much debate about which type of cleavage occurred first, biologists assume that these two types of cleavage each evolved only once. Animals with spiral cleavage form a monophyletic group, as do those with radial cleavage (see section 3.5).

In **radial cleavage**, cleavage produces tiers, or layers, of cells on top of one another in an early embryo (figure 3.10A). Radial cleavage typically occurs with **regulative development** where each blastomere of the early embryo, if separated from the others, can adjust or "regulate" its development into a complete and well-proportioned (though possibly smaller) embryo (figure 3.11A).

Spiral cleavage, found in several phyla, differs from radial in several ways. By the eight-cell stage cleavage produces two quartets of cells that come to lie not on top of each other but in furrows between the cells (see figure 3.10B). In addition,

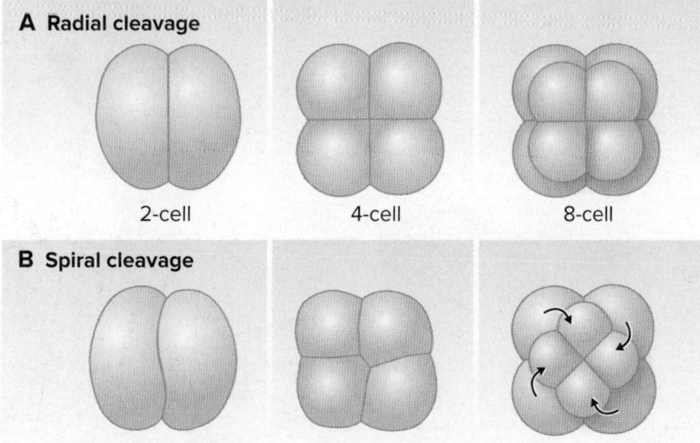

figure 3.10

Radial and spiral cleavage patterns shown at two-, four-, and eight-cell stages; all diagrams are top views. **A,** Radial cleavage, typical of echinoderms, chordates, and hemichordates. **B,** Spiral cleavage, typical of molluscs, annelids, and other protostomes. Arrows indicate clockwise movements of small cells following division of large cells in spiral cleavage. Compare the position of the top set of cells in spiral cleavage with the position of these cells following radial cleavage.

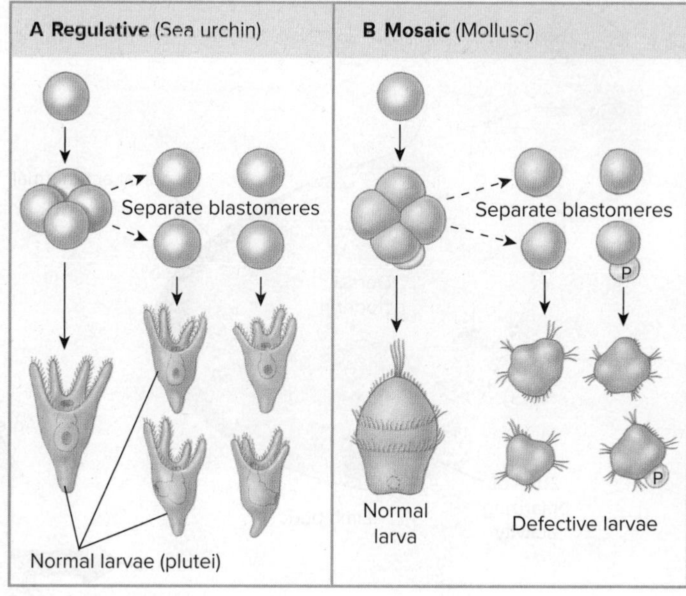

figure 3.11

Regulative and mosaic development. **A,** Regulative development. If the early blastomeres of a sea urchin embryo are separated, each develops into a complete larva. **B,** Mosaic development. When the blastomeres of a mollusc embryo are separated, each gives rise to a partial, defective larva. P indicates polar bodies, which are shed.

spirally cleaving eggs tend to pack their cells tightly together much like a cluster of soap bubbles, rather than just lightly contacting each other as do those in many radially cleaving embryos. Spirally cleaving embryos also differ from radial embryos in that most have a form of **mosaic development**, in which the organ-forming determinants in the egg cytoplasm are positioned within the egg, even before the first cleavage division. The result is that if the early blastomeres are separated, each will continue to develop for a time as though it were still part of the whole. Each forms a defective, partial embryo (figure 3.11B).

A special feature of most spirally cleaving embryos is that, at about the 29-cell stage, a blastomere called the 4d cell is formed that will give rise to all mesoderm of the embryo. **Mesoderm** is one of three germ layers usually present in an embryo; the other two germ layers are **ectoderm** and **endoderm**. Ultimately, all features of an adult animal are derived from one of these three germ layers, but adult structures do not form until well after cleavage.

Cleavage proceeds until the zygote is divided into many small cells, typically surrounding a fluid-filled cavity (figure 3.12). At this point, the embryo is called a **blastula**, and the fluid-filled cavity is a **blastocoel**. In animals other than most sponges, the blastula becomes a two-layered stage called a **gastrula** (figure 3.12A). The gastrula has two germ cell layers, ectoderm and endoderm. The outer germ cell layer, ectoderm, surrounds the blastocoel. The endoderm surrounds and defines an inner body cavity called the **gastrocoel**. The gastrocoel will become the gut cavity in an adult animal. Some body cavities, such as the gut, persist into adulthood, but other cavities are lost, and still others appear only later in development. Adult sea anemones and their kin are two-layered animals, so they do not develop beyond the gastrula stage, but most animals develop three embryonic germ layers and additional body cavities.

Body Cavities

The most obvious internal space or body cavity is a gut developing from an embryonic gastrocoel (figure 3.12A). The gut always has at least one opening, the blastopore. In adult sea anemones and their kin, this opening ingests food and releases digestive wastes. A gut with only one opening is called a two-way, blind, or incomplete gut. Most animals develop a second opening to the gut, creating a tube (figure 3.12B). Tubular guts permit sequential movement of food from mouth to anus and are called one-way or complete guts. Complete guts may evolve into regionalized processing guts, as seen in an earthworm (see figure 11.17A) or a vertebrate (see figure 20.10). It is not possible to process food sequentially if wastes must be ejected via the mouth.

In many animals, the gut is surrounded by a fluid-filled cavity. This cavity can form in several ways. For animals with only two germ layers, such as cnidarians (including sea anemones, corals, and jellyfishes), the embryonic blastocoel persists into adulthood as a cavity surrounding the blind gut (figure 3.12A). The blastocoel may also persist in animals with a third germ layer, mesoderm, but the addition of a layer of mesoderm inside the blastocoel causes structural changes.

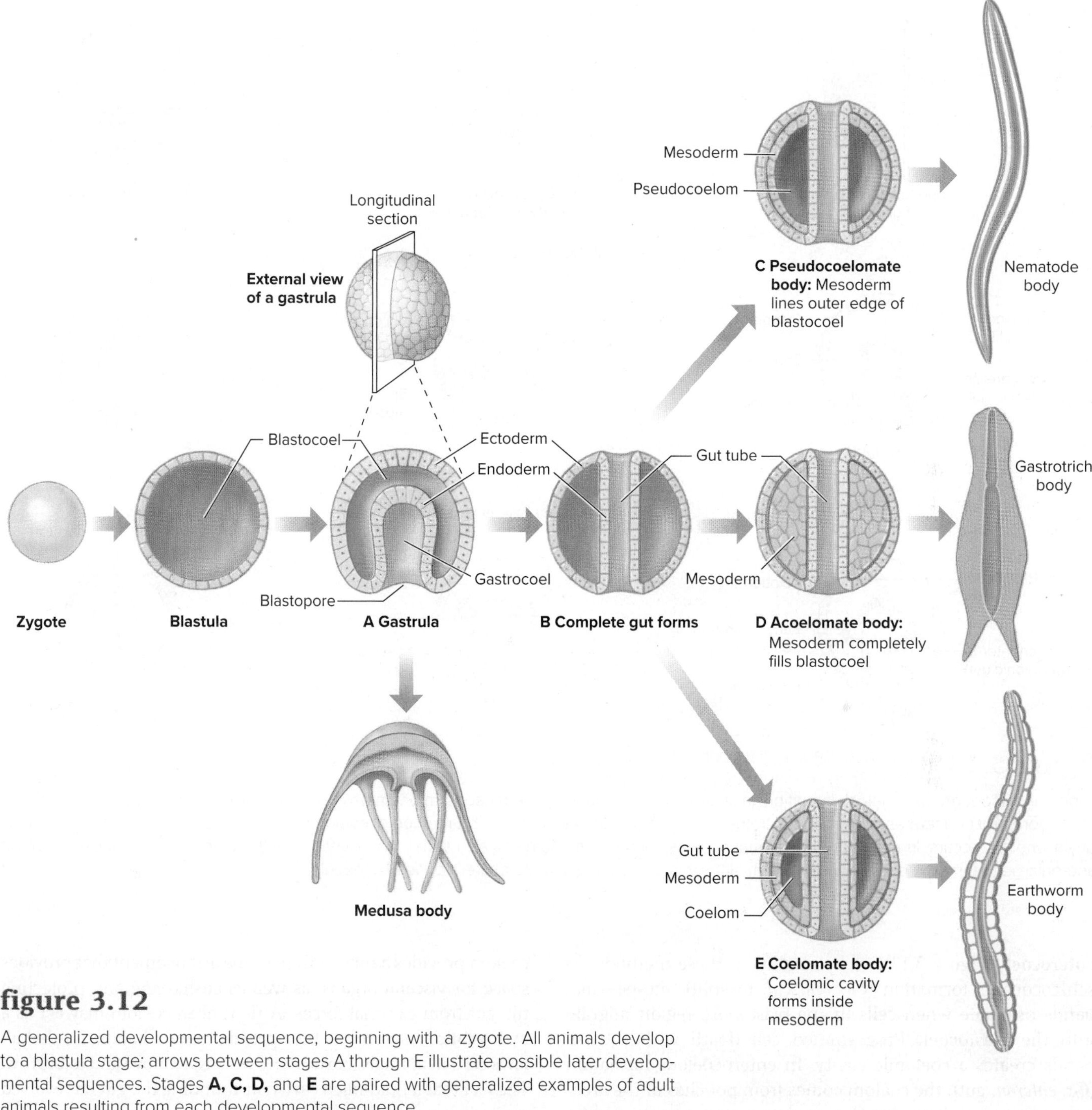

figure 3.12

A generalized developmental sequence, beginning with a zygote. All animals develop to a blastula stage; arrows between stages A through E illustrate possible later developmental sequences. Stages **A, C, D,** and **E** are paired with generalized examples of adult animals resulting from each developmental sequence.

In some animals, mesoderm lines the outer edge of the blastocoel, lying next to the ectoderm (figure 3.12C); when this occurs, the blastocoel is renamed the **pseudocoelom** in adult animals. "Pseudocoelom" means "false coelom," and the descriptive nature of this name will be clear once a coelom is understood.

In other animals, mesoderm completely fills the blastocoel (figure 3.12D). These animals are **acoelomate**, meaning that the body is without a coelom. The only body cavity is the space inside the gut tube, the gut being surrounded by a mass of tissue derived from mesoderm.

In the development of most bilaterally symmetrical animals, the blastocoel fills with mesoderm, and then a new cavity forms *inside* the mesoderm (figure 3.12E). This new cavity, completely surrounded by mesoderm, is a **coelom** (Gr. *koilos*, hollow or cavity). A coelom differs from a pseudocoelom, which has mesoderm on only the outer edge of the cavity. A coelom may form by one of two methods: **schizocoely** or

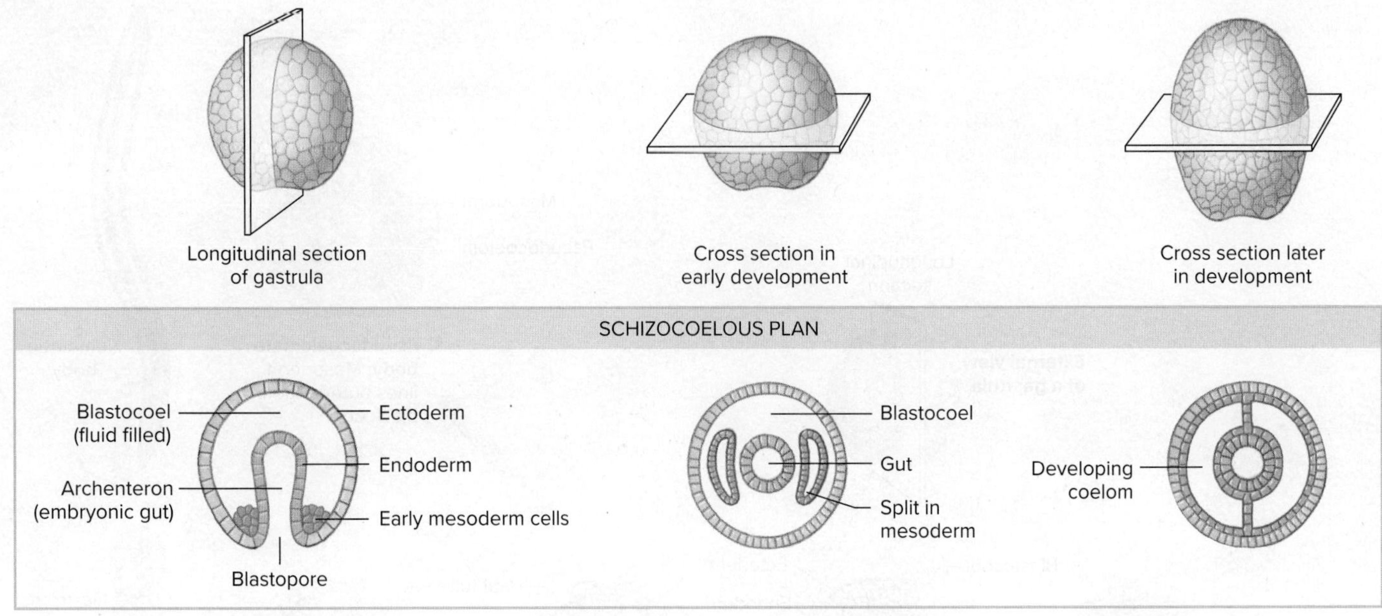

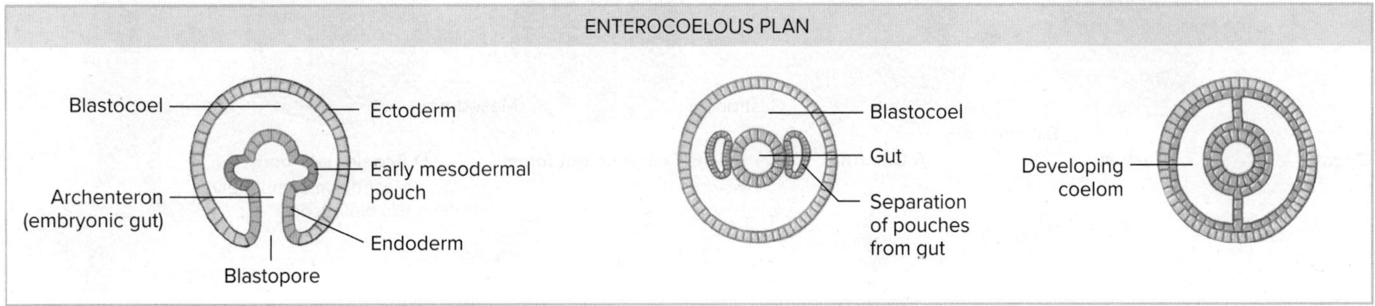

figure 3.13

Types of mesoderm and coelom formation. In schizocoelous formation, the mesoderm originates from the wall of the embryonic gut near the blastopore and proliferates into a band of tissue that splits to form the coelom. This method of coelom formation may not be homologous in all phyla where it occurs. In enterocoelous formation, most mesoderm originates as a series of pouches from the embryonic gut; these pinch off and enlarge to form the coelom. In both formations, the coeloms expand to obliterate the blastocoel.

enterocoely (figure 3.13), or by modifying these methods. In schizocoelous formation (Gr. *schizein,* to split), mesodermal bands originate when cells in the blastopore region migrate into the blastocoel. Programmed cell death within these bands creates a coelomic cavity. In enterocoelous formation (Gr. *enteron,* gut), the coelom comes from pouches of the archenteron, or primitive gut, that push outward into the blastocoel (figure 3.13). In this case, the mesoderm and cavity within it are created at the same time.

Once development is complete, the results of schizocoelous and enterocoelous formations are indistinguishable. Both produce a true coelom lined with a mesodermal peritoneum (Gr. *peritonaios,* stretched around) and having mesenteries in which the visceral organs are suspended.

The evolution of a coelom, a fluid-filled body cavity between the outer body wall and the gut, was a major developmental event affecting the body plans of many bilateral animals. A coelom provides a tube-within-a-tube arrangement that provides space for visceral organs, as well as cushioning and protecting the gut from external forces. A fluid-filled coelom may act as a hydrostatic skeleton (see the opening essay, Chapter 11) that allows certain animals, particularly worms, to move and to burrow. The fact that a fluid-filled cavity surrounding the gut has evolved multiple times, as in a pseudocoelom, or a true coelom formed by enterocoely or by schizocoely, attests to its functional value.

3.5 How Many Body Plans Are There?

Animals comprise 32 phyla as shown on the cladogram inside the front cover. Members of all phyla develop from cleavage through the blastula stage, but the number of embryonic germ layers formed after the blastula varies among phyla. The

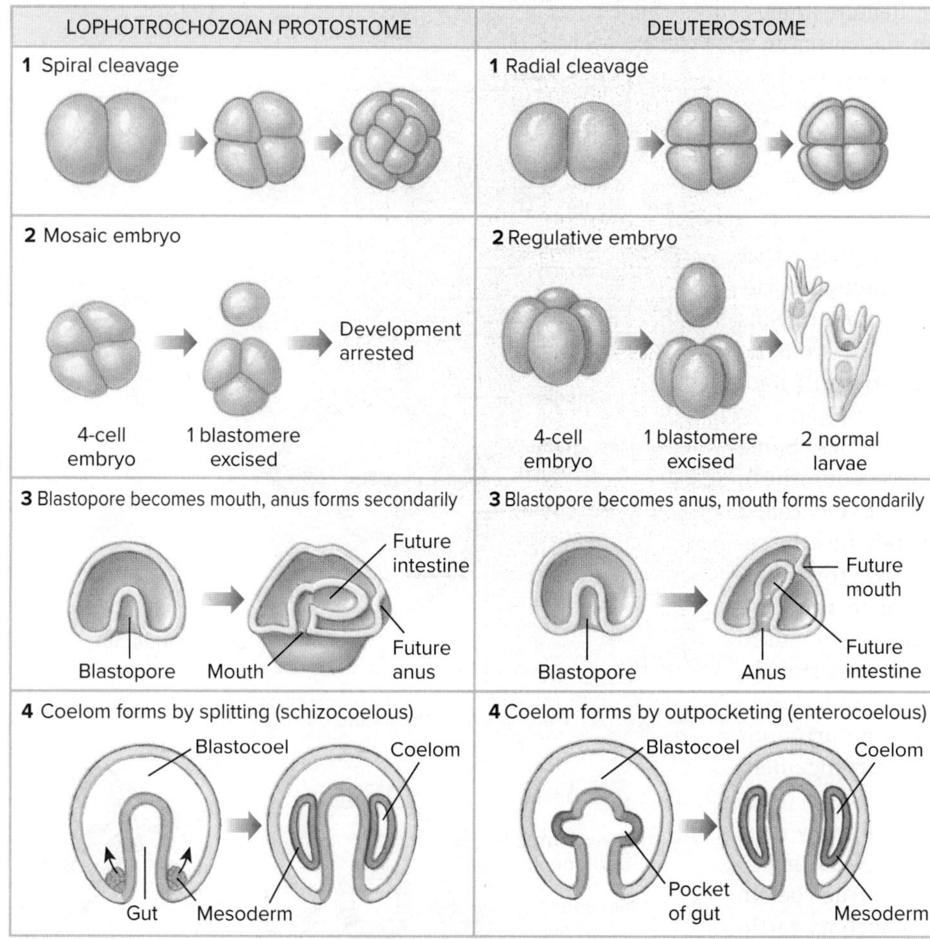

figure 3.14

Developmental tendencies of protostomes and deuterostomes. These tendencies are much modified in some groups—for example, the vertebrates. Cleavage in mammals, reptiles, birds, and many fishes is neither radial nor spiral in form. Within deuterostomes, vertebrates have evolved a mode of coelom formation that is schizocoelous, but not homologous to protostome schizocoely. Coelom formation may have evolved several times within the protostomes.

number of germ layers affects the tissues present in adults. Sponges, in phylum Porifera, have no tissue layers in the adult body, and were originally thought to develop only to a single-layered blastula. However, a detailed study has demonstrated gastrula formation and the presence of two tissue layers in the larval stages of a few species. Reorganization of the body after larval settlement obscures tissue layers. Members of a few other phyla, such as Cnidaria (sea anemones and their kin) develop two embryonic germ layers and have two adult body layers. These animals are typically radially symmetrical.

A vast majority of phyla contain bilaterally symmetrical animals developing from the three embryonic germ layers mentioned previously: ectoderm, endoderm, and mesoderm. The presence of three germ layers is called "triploblasty," so these animals are collectively the triploblastic Bilateria. Within the Bilateria, biologists recognize two groups of animals: **Protostomia** (called protostomes) and **Deuterostomia** (called deuterostomes) (find these groups on the cladogram inside front cover). These two evolutionary groups, or clades, differ in the pattern of cleavage, in the formation of mesoderm, and in the way that a coelom is made, if a coelom is present.

Deuterostome Body Plans

Deuterostomes share an inherited common developmental sequence that begins with radial regulative cleavage and progresses from a blastula to a gastrula. The blastopore becomes the anus (figure 3.14). The name Deuterostomia means "second mouth" and refers to formation of the mouth from a second opening in the embryo. Other distinguishing hallmarks of the deuterostomes are summarized in figure 3.14 and include formation of a coelom by enterocoely in most group members. A notable exception is the vertebrate chordates. Invertebrate chordates form the coelom by enterocoely as expected, but vertebrates develop via a modified, and independently derived, version of schizocoely.

There are three deuterostome phyla. Members of phylum Echinodermata, which includes the sea stars and sea urchins, undergo typical deuterostome development to form a free-swimming larval stage (figure 3.3). This larval stage metamorphoses into a small version of an adult echinoderm. Other deuterostomes include acorn worms and related animals in phylum Hemichordata, as well as tunicates, lancelets (see sections 15.4 and 15.5), and vertebrates in phylum Chordata.

The Xenoturbellida, sometimes considered deuterostomes, have recently been placed with acoelomorph flatworms in a taxon now called Xenacoelomorpha. Metamerism occurs in the chordates.

Protostome Body Plans

Protostomes share an inherited developmental sequence that begins with spiral mosaic cleavage and proceeds through a blastula and gastrula stage (see figure 3.12). Protostome means "mouth first" and refers to the formation of the mouth from the embryonic blastopore. The anus forms secondarily in protostomes (see figure 3.14).

There is no body plan common to protostomes. Some protostomes, such as flatworms (phylum Platyhelminthes), have acoelomate bodies where a spongy mass of cells called parenchyma, derived from mesoderm, completely fills the region between the epidermis and the digestive tract (figure 3.15, top panel). Readers should note that in this case "parenchyma" does not have the same meaning as when used for cells within an organ in section 3.1. Other protostomes, such as roundworms in phylum Nematoda, are pseudocoelomate (figure 3.15, middle panel). A pseudocoelom surrounds the gut, giving the body a tube-within-a-tube arrangement similar to that in animals with a true coelom. However, a pseudocoelom is a persistent blastocoel and is not lined with a peritoneum, the thin cellular membrane derived from mesoderm that lines the body cavity in animals with a true coelom (figure 3.15, lower panel). **Coelomate** animals such as earthworms (phylum Annelida) and insects (phylum Arthropoda) are also protostomes; when a coelom is present, it forms by schizocoely in protostomes.

Protostome Subgroups

Protostomia comprises two subgroups that were initially identified using molecular characters. One subgroup, the ecdysozoan protostomes or **Ecdysozoa**, includes all animals that molt their cuticles. The name comes from the root *ekdysis*, meaning "to strip off." This group comprises arthropods, nematodes, and members of six other phyla (see cladogram inside front cover).

The other protostome subgroup, **Lophotrochozoa**, includes a very diverse collection of animals. Part of the name comes from *lophos*, meaning "crest or tuft," in reference to a horseshoe-shaped whorl of tentacles called a **lophophore** (see Chapter 9 introduction) that occurs in some members of the group, such as lampshells (phylum Brachiopoda). Another part of the name comes from *trochos*, meaning "wheel," in reference to the band of cilia on a larval form called a **trochophore** (see figure 10.6). A trochophore larval stage occurs in the life cycles of many lophotrochozoans, including snails (phylum Mollusca) and worms (phylum Annelida). There are 17 lophotrochozoan phyla (see cladogram inside front cover). Metamerism occurs in phylum Annelida.

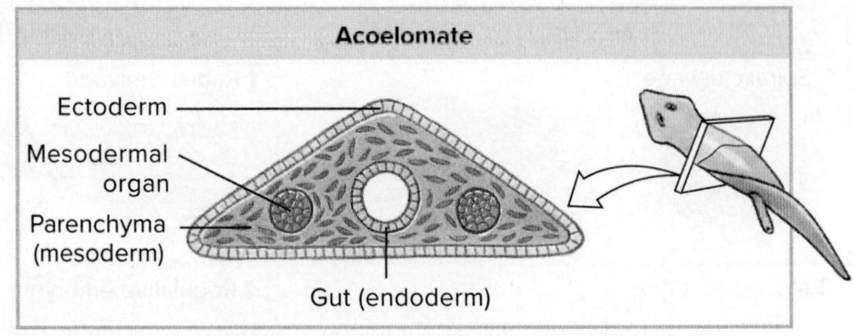

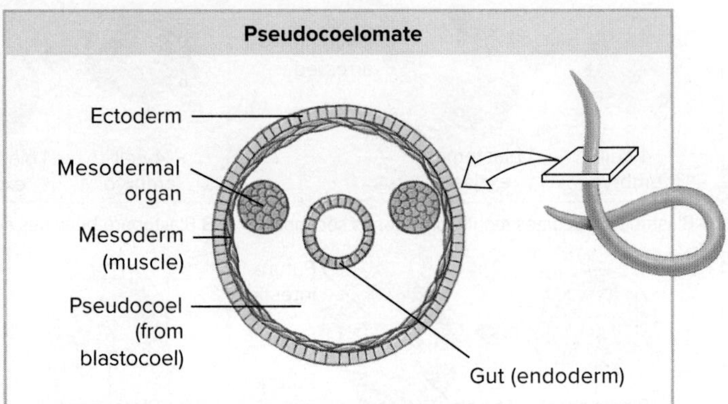

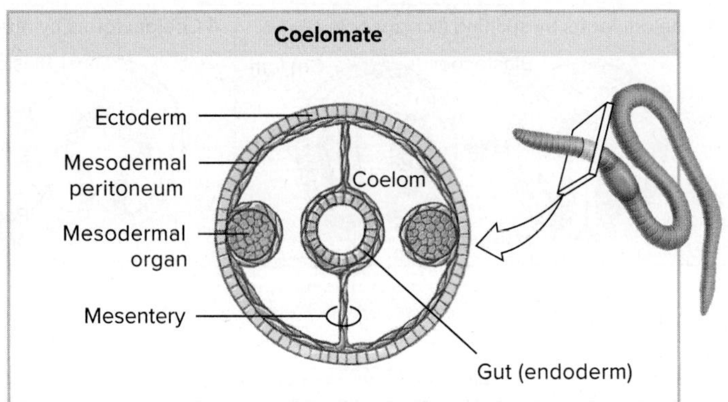

figure 3.15

Acoelomate, pseudocoelomate, and coelomate body plans are shown as cross sections of representative animals. Note the relative positions of mesodermally derived parenchyma, peritoneum, and body organs.

3.6 Components of Animal Bodies

Animal bodies consist of cellular components (the multicellular tissues and organs derived from the three embryonic germ layers) and extracellular components (the fluids and structures that cells deposit outside their cell membranes). Extracellular components include an ECM in epithelial and connective tissues, serum that carries blood cells through a vertebrate body, and the bone matrix of vertebrates, among others.

Cellular Components: Tissues

A tissue is a group of cells specialized for performing a common function. The study of tissues is called **histology** (Gr. *histos,*

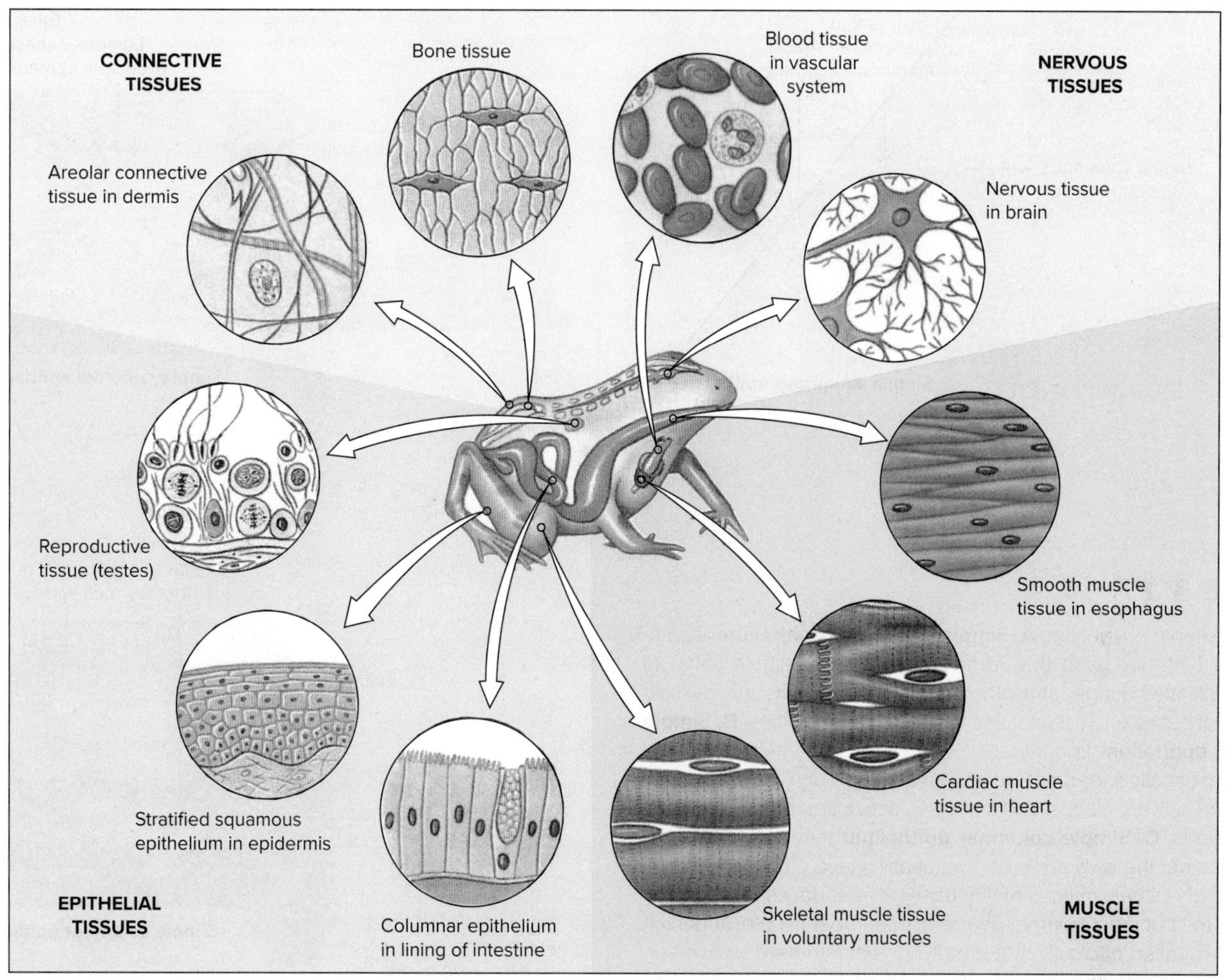

figure 3.16

Animal tissue types illustrated within the body of a frog.

tissue, + *logos*, discourse). During embryonic development, the three germ layers become differentiated into different tissues. In animals at or above the tissue level of organization, all cells exist as part of a tissue. Some tissues are composed of several kinds of cells, and some tissues, such as connective tissue, include extracellular materials made by tissue cells, as well as noncellular fluids.

There are four kinds of tissues: epithelial, connective (including vascular), muscular, and nervous (figure 3.16). This surprisingly short list of basic tissue types meets the diverse requirements of animal life.

Epithelial Tissue

An **epithelium** (pl., epithelia) is a sheet of cells that covers an external or internal surface. Outside the body, epithelium forms a protective covering. Inside, an epithelium lines all the organs of the body cavity, as well as ducts and passageways through which various materials and secretions move. On many surfaces, epithelial cells are often modified into glands that produce lubricating mucus or specialized products such as hormones or enzymes.

Epithelia are classified by cell form and number of cell layers. Simple epithelia (figure 3.17) occur in all animals that have tissues, whereas stratified epithelia (figure 3.18) are mostly restricted to vertebrates. All types of epithelia are supported an ECM called a basal lamina or a basement membrane. The basal lamina is a thin mat containing **collagen** (Gr. *kolla*, glue, + *genos*, descent), a protein material of great tensile strength. Blood vessels never enter epithelial tissues, so they depend on diffusion of oxygen and nutrients from underlying tissues.

Connective Tissue

Connective tissues are a diverse group of tissues that serve various binding and supportive functions. They are so widespread in the body that removal of other tissues would still leave the complete form of the body clearly apparent. Connective tissue is composed of relatively few cells, a great many extracellular fibers, and a fluid, gel-like or rigid **ground**

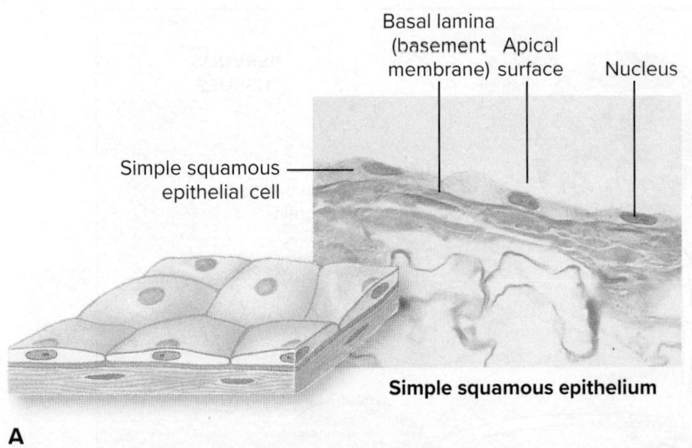

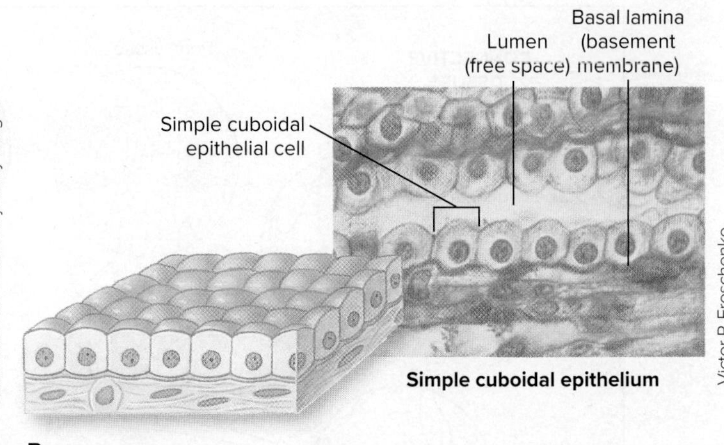

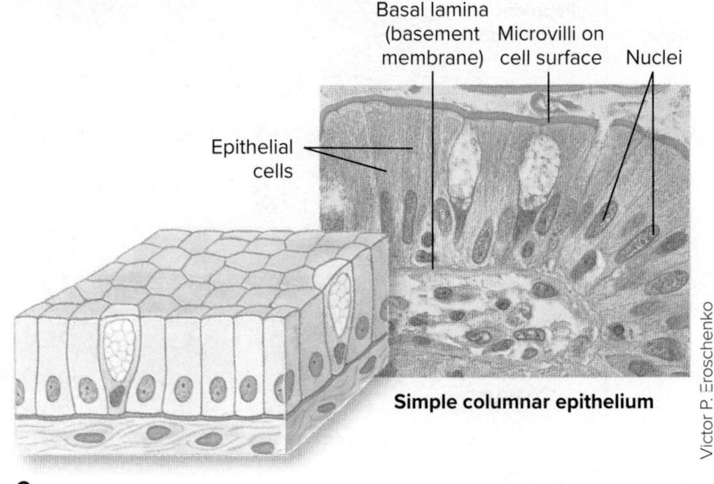

figure 3.17

Types of simple epithelium. **A, Simple squamous epithelium** is composed of flattened cells that form a continuous delicate lining of blood capillaries, lungs, and other surfaces. It permits the passive diffusion of gases and tissue fluids into and out of cavities. **B, Simple cuboidal epithelium** is composed of short, boxlike cells. Cuboidal epithelium usually lines small ducts and tubules, such as those of the kidney and salivary glands, and may have active secretory or absorptive functions. **C, Simple columnar epithelium** resembles cuboidal epithelium, but the cells are taller and usually have elongated nuclei. This type of epithelium lines highly absorptive surfaces, such as the intestinal tract of most animals. The cells often bear minute, fingerlike projections called microvilli that greatly increase the absorptive surface. In some organs, such as the female reproductive tract, the cells may be ciliated.

substance (also called **matrix**), in which the fibers are embedded. Fibers and ground substance together form an ECM. We recognize several different types of connective tissue. Two kinds of connective tissue proper occur in vertebrates. **Loose connective tissue** is composed of fibers and both fixed and wandering cells suspended in a syrupy ground substance (figure 3.19A). **Dense connective tissue**, which includes tendons and ligaments, is composed largely of densely packed fibers (figure 3.19B). Much of the fibrous component of connective tissue is composed of collagen, which is the most abundant protein in animals. It occurs in animal bodies wherever both flexibility and resistance to stretching are required. The connective tissue of invertebrates, as in vertebrates, consists of cells, fibers, and extracellular matrix, but usually it is not as elaborately developed.

Other types of connective tissue include **blood**, **lymph**, and **interstitial fluid**, composed of distinctive cells in a watery ground substance, the plasma. As plasma leaks from capillaries into the spaces between cells, it becomes interstitial fluid. Excess interstitial fluid is collected in lymphatic ducts, where it joins lymph. Lymph consists of lymphocytes and fluids. Some lymph is returned to the circulatory system when lymph ducts meet veins going to the heart. Blood, lymph, and interstitial fluid are collectively considered vascular tissue. Vascular tissue lacks fibers under normal conditions. **Cartilage** is a semirigid form of connective tissue with closely packed fibers embedded in a gel-like ground substance (figure 3.19C). **Bone** is a calcified connective tissue containing calcium salts organized around collagen fibers (figure 3.19D).

Key Theme 3.1

SCIENCE EXPLAINED

What Do Inter, Intra, and Extra Mean?

The term "intercellular," meaning "between cells," should not be confused with the term "intracellular," meaning "within cells." "Extracellular" means outside of cells.

Stratified squamous epithelium consists of two to many layers of cells adapted to withstand mild mechanical abrasion. The basal layer of cells undergoes continuous mitotic divisions, producing cells that are pushed toward the surface, where they are shed and replaced by new cells beneath them. This type of epithelium lines the oral cavity, esophagus, and anal canal of many vertebrates, and the vagina of mammals.

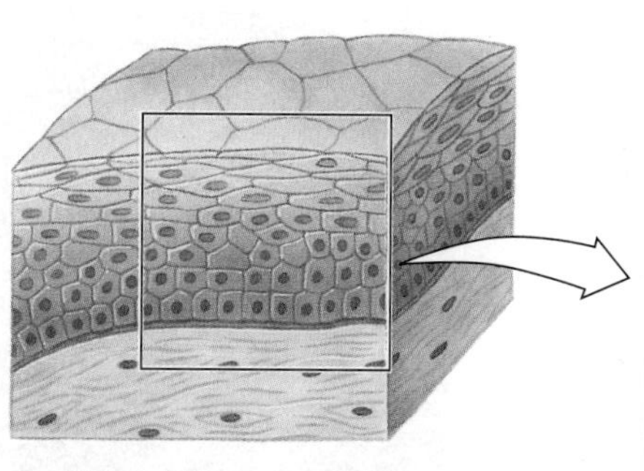

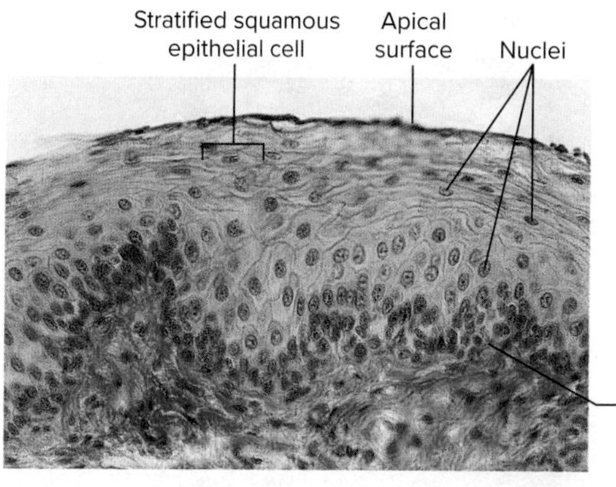

Transitional epithelium is a type of stratified epithelium specialized to accommodate great stretching. This type of epithelium lines the urinary tract and bladder of vertebrates. In the relaxed state, it appears to be four or five cell layers thick, but when stretched out, it has only two or three layers of extremely flattened cells.

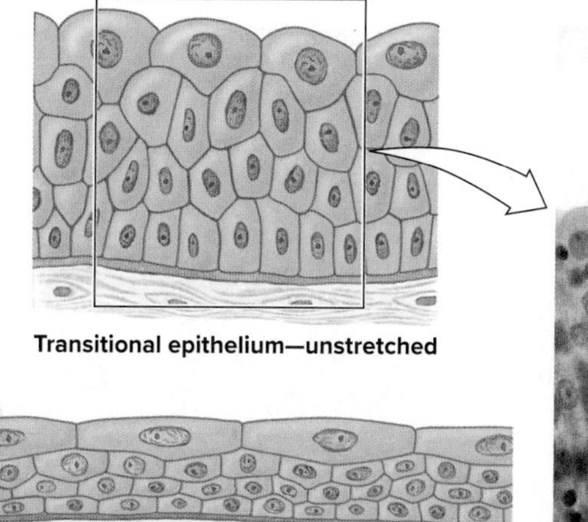

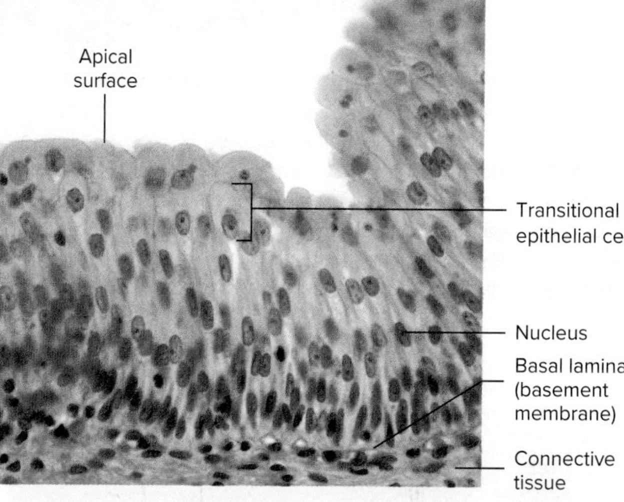

figure 3.18

Types of stratified epithelium.

Muscular Tissue

Muscle is the most common tissue of most animals. It originates (with few exceptions) from mesoderm, and its unit is the cell, or **muscle fiber**, which is specialized for contraction. When viewed with a light microscope, **striated muscle** appears transversely striped (striated), with alternating dark and light bands (figure 3.20A, B). In vertebrates, we recognize two types of striated muscle: **skeletal** and **cardiac muscle**. A third kind of muscle is **smooth** (or visceral) **muscle**, which lacks the characteristic alternating bands of the striated type (figure 3.20C). Unspecialized cytoplasm of muscles is called **sarcoplasm**, contractile elements within the fiber are **myofibrils**.

Nervous Tissue

Nervous tissue is specialized for receiving stimuli and conducting impulses from one region to another. Two basic types of cells in nervous tissue are **neurons** (Gr. nerve), the basic functional unit of the nervous system, and **neuroglia** (nurŏ-gle ə, Gr.nerve + *glia,* glue), a variety of nonnervous cells that insulate neuron membranes and serve various supportive functions. Figure 3.21 shows the functional anatomy of a typical neuron.

Extracellular Components

In addition to cellular structures, animals contain two important noncellular components: body fluids and extracellular structural elements. Body fluids are subdivided into two fluid "compartments": those that occupy **intracellular** space, within the body's cells, and those that occupy **extracellular** space, outside the cells. In animals with closed vascular systems (such as segmented worms and vertebrates), extracellular fluids are subdivided further into **blood plasma, lymph**, and **interstitial fluid**, as described in the previous section on connective tissues. Interstitial fluid, sometimes called tissue fluid, occupies the space surrounding cells. However, many invertebrates have open vascular systems, with no true separation of blood plasma from interstitial fluid.

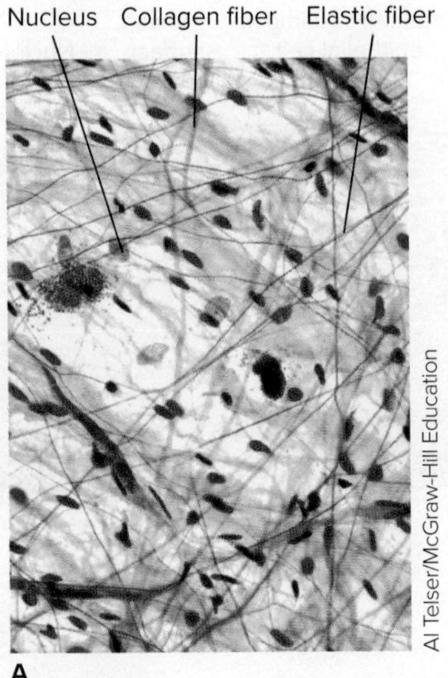

Loose connective tissue, also called areolar connective tissue, is the "packing material" of the body that anchors blood vessels, nerves, and organs. It contains fibroblasts that synthesize the fibers and a gel-like ground substance with wandering macrophages that phagocytize pathogens or damaged cells. The different fiber types include strong collagen fibers (thick and violet in micrograph) and elastic fibers (thin and branching in micrograph) formed of the protein elastin. Adipose (fat) tissue is considered a type of loose connective tissue.

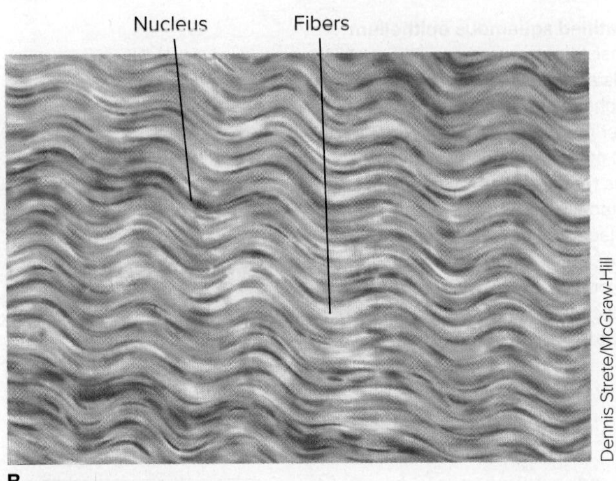

Dense connective tissue forms tendons, ligaments, and fasciae, the latter arranged as sheets or bands of tissue surrounding skeletal muscle. In a tendon (shown here), the collagenous fibers are extremely long and tightly packed together.

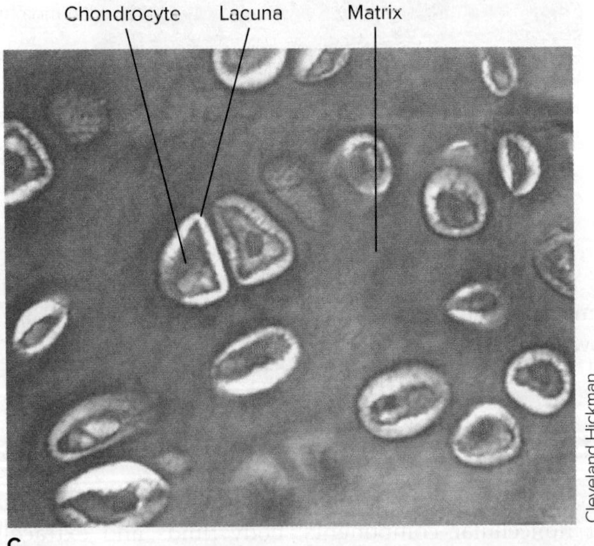

Cartilage is a vertebrate connective tissue composed of a firm gel matrix containing cells (chondrocytes) living in small pockets called lacunae, and collagen fibers or elastic fibers (depending on the type of cartilage). In the hyaline cartilage shown here, both collagen fibers and ground substance are stained uniformly purple, and cannot be distinguished from one another. Because cartilage lacks a blood supply, all nutrients and waste materials must diffuse through the ground substance from surrounding tissues.

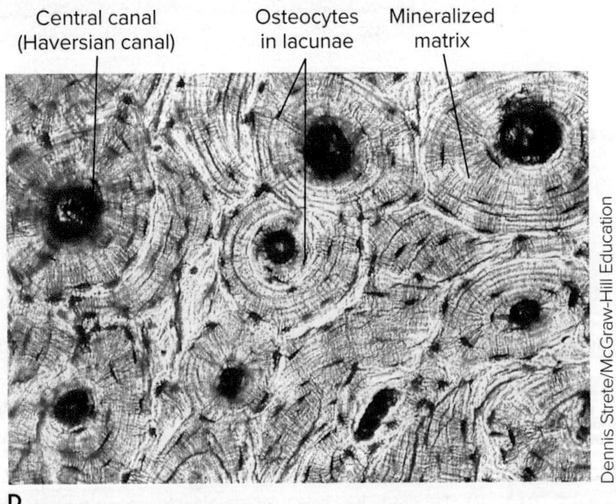

Bone, strongest of the vertebrate connective tissues, contains mineralized collagen fibers. Small pockets (lacunae) within the matrix contain bone cells, called osteocytes. The osteocytes communicate with blood vessels that penetrate bone by means of a tiny network of channels called canaliculi. Unlike cartilage, bone undergoes extensive remodeling during an animal's life, and can repair itself following even extensive damage.

figure 3.19

Types of connective tissue.

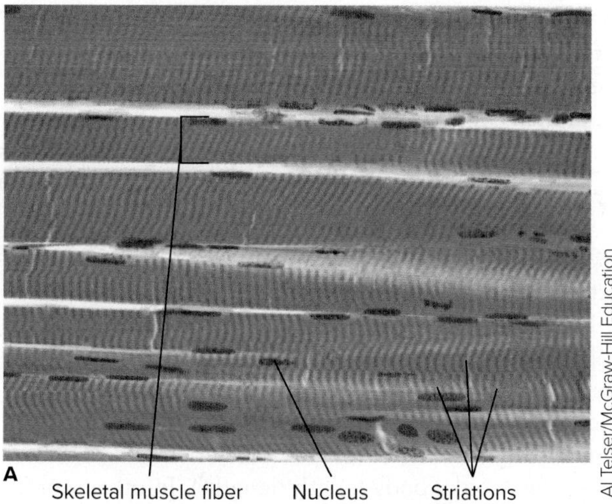

Skeletal muscle is a type of striated muscle found in both invertebrates and vertebrates. It is composed of extremely long, cylindrical fibers, which are multinucleate cells that may reach from one end of the muscle to the other. Viewed through the light microscope, the cells appear to have a series of stripes, called striations, running across them. Skeletal muscle is called voluntary muscle (in vertebrates) because it contracts when stimulated by nerves under conscious cerebral control.

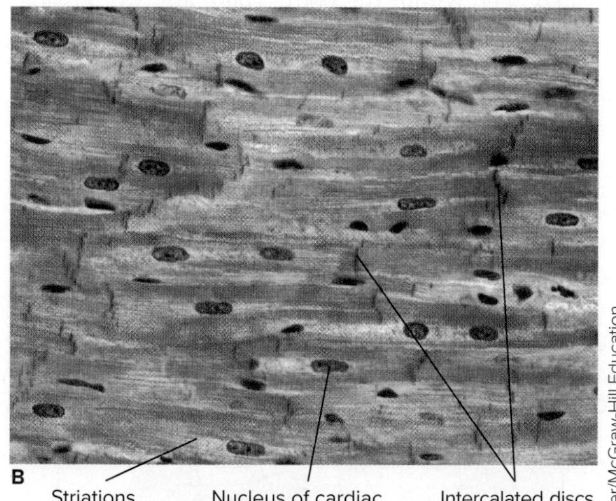

Cardiac muscle is another type of striated muscle found only in the vertebrate heart. The cells are much shorter than those of skeletal muscle and have only one nucleus per cell (uninucleate). Cardiac muscle tissue is a branching network of fibers with individual cells interconnected by junctional complexes called intercalated discs. Cardiac muscle is called involuntary muscle because it is not under conscious control. Instead, heart rate is controlled by specialized pacemaker cells located in the heart itself. However, autonomic nerves from the brain may alter pacemaker activity.

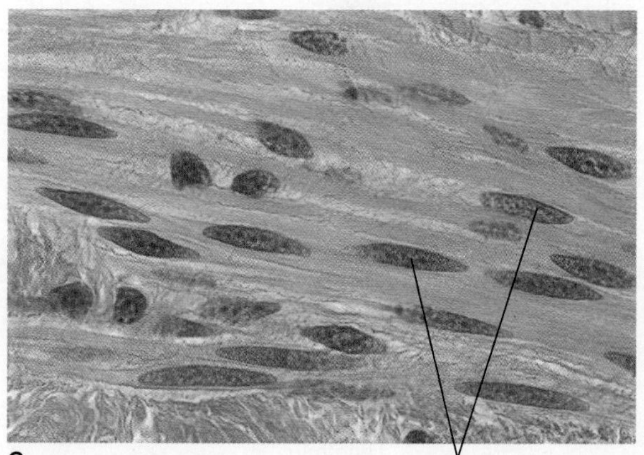

Smooth muscle is nonstriated muscle found in both invertebrates and vertebrates. Smooth muscle cells are long, tapering strands, each containing a single nucleus. Smooth muscle is the most common type of muscle in invertebrates, where it serves as body wall musculature and lines ducts and sphincters. In vertebrates, smooth muscle cells are organized into sheets of muscle circling the walls of the gut tube, blood vessels, respiratory passages, and urinary and genital ducts. Smooth muscle is typically slow acting and can maintain prolonged contractions with very little energy expenditure. Its contractions are involuntary and unconscious. The principal functions of smooth muscles are to push the material in a tube, such as the intestine, along its way by active contractions or to regulate the diameter of a tube, such as a blood vessel, by sustained contraction.

figure 3.20

Types of muscle tissue.

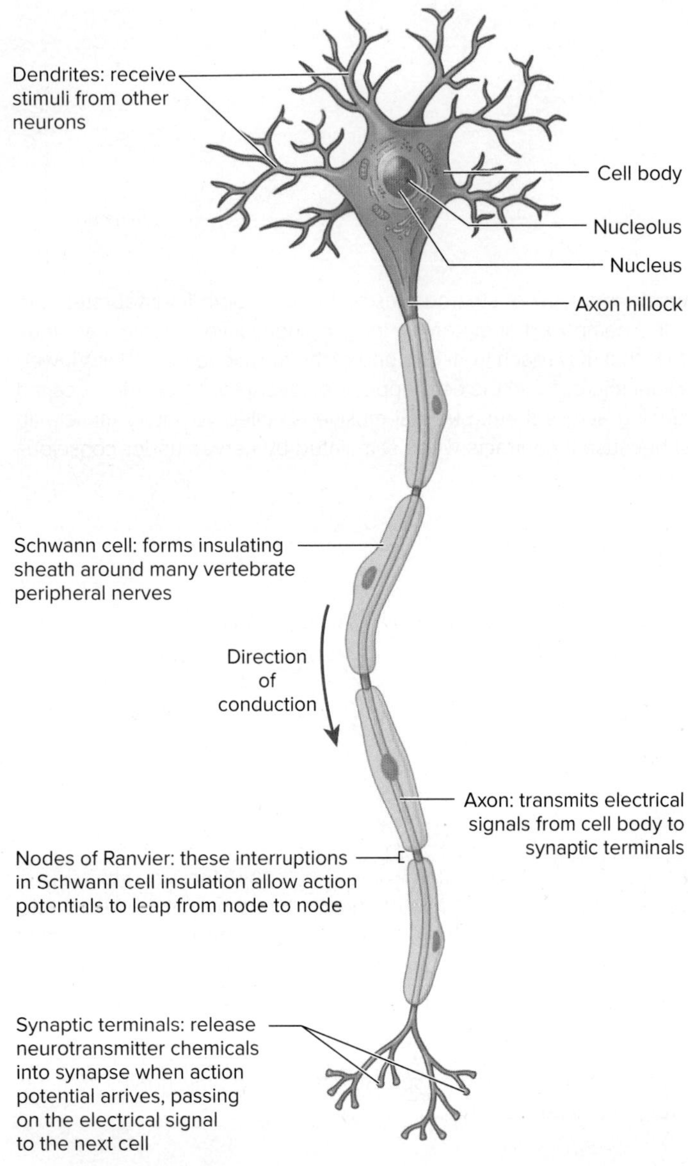

figure 3.21

Functional anatomy of a neuron. From the nucleated body, or soma, extend one or more **dendrites** (Gr. *dendron*, tree), which receive electrical impulses from receptors or other nerve cells, and a single axon that carries impulses away from the cell body to other nerve cells or to an effector organ. The axon is often called a **nerve fiber**. Schwann cells are examples of neuroglia. Neurons are separated from other neurons or from effector organs by specialized junctions called synapses.

If we were to remove all specialized cells and body fluids from the interior of the body, we would be left with the third element of the animal body: extracellular structural elements. This is the supportive material of the organism, forming the noncellular parts of loose connective tissue (especially well developed in vertebrates but present in all animals), cartilage (molluscs and chordates), bone (vertebrates), and cuticle (arthropods, nematodes, annelids, and others). These elements provide mechanical stability and protection. In some instances, they also act as a depot for exchange of materials and serve as a medium for extracellular reactions. We describe the diversity of extracellular skeletal elements characteristic of the different groups of animals in Chapters 10–20.

3.7 Complexity and Body Size

The most complex levels of animal organization permit, and to some extent even promote, evolution of large body size (figure 3.22). Large size confers several important physical and ecological consequences for an organism. As animals become larger, the body surface increases much more slowly than does body volume because surface area increases as the square of body length ($length^2$), whereas volume (and usually mass) increases as the cube of body length ($length^3$). In other words, a large animal will have less surface area relative to its volume than will a small animal of the same shape. The surface area of a large animal may be inadequate for respiration and nutrition by cells located deep within the body. There are two possible solutions to this problem. One solution is to fold or invaginate the body surface to increase surface area or, as exploited by flatworms, to flatten the body into a ribbon or disc so that no

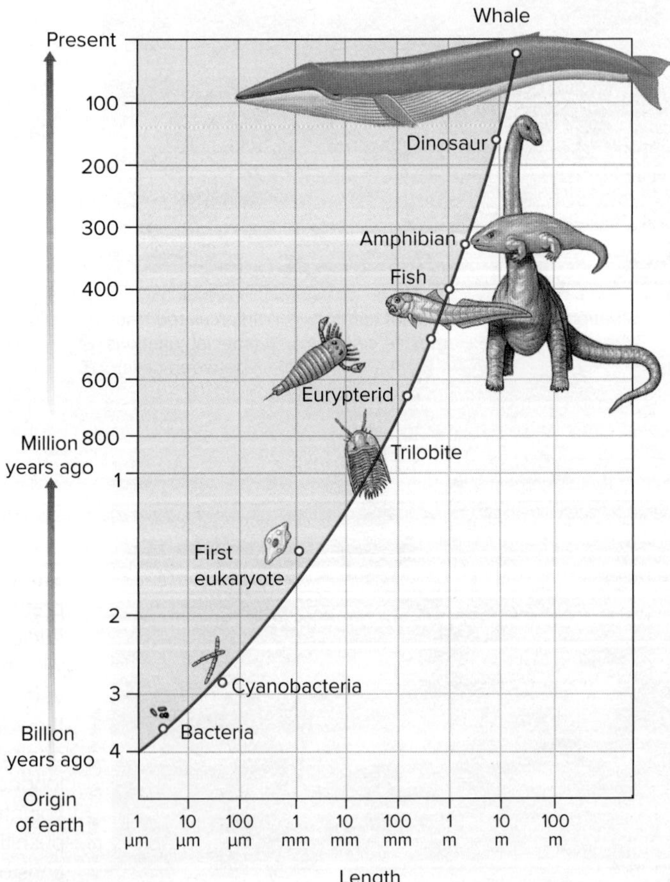

figure 3.22

Graph showing the largest organisms (absent trees and fungi) present at different periods of life on earth. Note that both scales are logarithmic.

internal space is far from the surface. This solution allows the body to become large without internal complexity. However, most large animals evolved a second solution; they developed internal transport systems to shuttle nutrients, gases, and waste products between the cells and the external environment.

Larger size buffers an animal against environmental fluctuations, provides greater protection against predation and enhances offensive tactics, and permits more efficient use of metabolic energy. A large mammal uses more oxygen than a small mammal, but the cost of maintaining its body temperature is less per gram of weight for a large mammal than for a small one. Large animals also can move at less energy cost than small animals can. A large mammal uses more oxygen in running than does a small mammal, but the energy cost of moving 1 g of its body over a given distance is much less for a large mammal than for a small one (figure 3.23). For all of these reasons, the ecological opportunities of larger animals are very different from those of small ones. The earth is home to organisms across a wide range of sizes and types. The next chapter describes the astonishing diversity of unicellular forms.

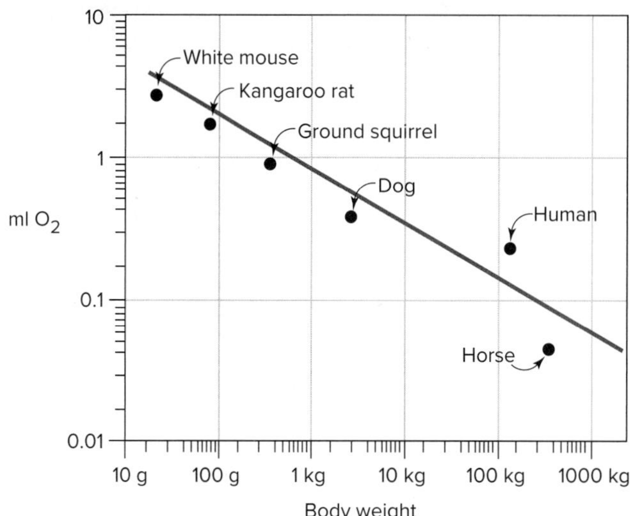

figure 3.23

Net cost of running for mammals of various sizes. Each point represents the cost (measured in rate of oxygen consumption) of moving 1 g of body over 1 km. The cost decreases with increasing body size.

■ SUMMARY

Sections	Key Concepts
3.1 Hierarchical Organization of Body Complexity	• Unicellular organisms perform all life functions in cytoplasm and organelles. • Colonial organisms are cellular aggregations where all cells reproduce. • Multicellarity requires a division of labor where only some cells reproduce. • Colonial and multicellular organisms are comprised of communicating cells, both may secrete an extracellular matrix, commonly called an ECM. • An animal has a multicellular body with varying degrees of organization and interaction among cells. • In most animals, cells combine into tissues that perform particular functions. • Tissues are organized into larger functional units called organs. • Organs are associated to form systems in the most complex bodies.
3.2 Animal Body Plans	• Every organism has an inherited body plan characterized by particular symmetry; presence or absence of body cavities; presence or absence of segmentation; degree of cephalization; type of nervous system. • Symmetrical bodies may be spherical, radial, or bilateral. • Bilateral symmetry tends to be correlated with cephalization and movement in one direction. • Bilaterally symmetrical animals typically have three tissue layers. • Most bilaterally symmetrical animals have a complete gut with a mouth and anus. • Many radially symmetrical animals have incomplete guts. • The gut is the primary body cavity of an animal, but it may be surrounded by another cavity such as a coelom or pseudocoelom.
3.3 Development and Genetics	• After fertilization, an egg becomes a zygote. • A zygote undergoes cleavage to form a blastula and then a gastrula stage. • Prior to the gastrula stage, development is determined by cytoplasmic, conditional, or syncytial specification. • Pattern formation is controlled by networks of regulatory genes that determine the primary body axes, segments, limbs, and organs.
3.4 Development of Animal Body Plans	• Sponges, cnidarians, and a few other groups of animals are not bilaterally symmetrical; adult animals in these groups exhibit fewer than three embryonic tissue layers. • Bilaterally symmetrical animals possess three embryonic tissue layers. • Bilaterally symmetrical animals cleave radially or spirally; their first developmental opening (blastopore) forms the mouth or the anus; development is either mosaic or regulative.

(*continued*)

Sections	Key Concepts
3.5 How Many Body Plans Are There?	• Sponges, cnidarians, and a few other groups of animals have non-layered or two-layered adult bodies. • Bilaterally symmetrical animals are triploblastic and are divided into two major groups based on developmental characteristics. • Deuterostomes have radial cleavage and regulative development; the mouth forms secondarily to the anus rather than from the blastopore. • Protostomes are characterized by spiral cleavage and mosaic development; the mouth forms at or near the embryonic blastopore. • Protostomia contains two subgroups: the lophotrochozoan protostomes and the ecdysozoan protostomes.
3.6 Components of Animal Bodies	• Animal cells are organized into embryonic and adult tissues. • The basic adult tissue types are nervous, connective, epithelial, and muscular. • In addition to cells, an animal body contains fluids and extracellular structural elements. • Fluids such as plasma, lymph, and interstitial fluid are divided into intracellular and extracellular compartments. • Extracellular structural elements lie in the spaces outside cells; they may be fibrous or formless elements and serve various structural functions.
3.7 Complexity and Body Size	• One correlate of increased body complexity in animals is increased body size. • Surface-area-to-body-volume ratios change with animal body size. • Metabolic rates and energetic costs change with body size. • Body size offers certain advantages, such as more effective predation, reduced energy cost of locomotion, and improved homeostasis. • The oldest, most species-rich, and abundant types of organisms are not large.

Review Questions

1. Name the five levels of organization in organismal complexity, and explain how each successive level differs from the one preceding it.
2. Distinguish between an asymmetrical and a symmetrical body, and contrast spherical, radial, and bilateral symmetries.
3. Explain the nature of an ECM and why it might account for the toughness of some epithelia and skin. (Hint: in these situations the ECM is sometimes called a basement membrane.)
4. Use the following terms to identify regions on your body and on the body of a frog anterior, posterior, dorsal, ventral, lateral, distal, proximal.
5. How would frontal, sagittal, and transverse planes divide your body?
6. What is the difference between radial cleavage and spiral cleavage?
7. Aside from the gut cavity, what types of body cavities exist and where do these cavities lie, relative to the gut cavity?
8. Match each description with at least one illustrative organism:

2 embryonic tissues _____	A. Sea star
3 embryonic tissues _____	B. Sea anemone
Protostome _____	C. Flatworm
Deuterostome _____	D. Earthworm
Metameric body _____	E. Lobster
	F. Fish
	G. Frog
	H. Insect

9. What is segmentation and how is it controlled in *Drosophila*?
10. What are the distinguishing developmental hallmarks of the two major groups of bilateral animals, the Protostomia and the Deuterostomia?
11. What are the four major types of tissues in the body of an animal?
12. How would you distinguish simple and stratified epithelium? What characteristic of stratified epithelium might explain why it, rather than simple epithelium, lines the oral cavity, esophagus, and vagina?
13. What three elements are present in all connective tissue? Give some examples of the different types of connective tissue.
14. What are the three kinds of muscle found in animals? Explain how each is specialized for particular functions.
15. Describe the principal structural and functional features of a neuron.
16. Explain why plasma, lymph, and interstitial fluid are considered extracellular.
17. Describe the molecular basis of mosaic cleavage.
18. Explain how the genes *bicoid* and *nanos* influence development of the anterioposterior axis of *Drosophila*.

For Further Thought Calculate the surface area and volume for two spherical organisms, one with a radius of 1 mm and the other with a radius of 10 mm. Determine the surface-area-to-volume ratio for each organism. What problems might occur for each organism because of its size? How would these problems be solved?

4

Taxonomy and Phylogeny of Animals

LEARNING OBJECTIVES

Readers will be able to:

4.1 Explain and distinguish the concepts of taxonomy, classification, systematization, and binomial nomenclature.

4.2 Explain the biological, evolutionary, cohesion, and phylogenetic species concepts, and the differences among them in defining the species category.

4.3 Explain the concept of homology and how the nested hierarchy of homologous characters reveals phylogeny.

4.4 Explain the taxonomic principles of evolutionary taxonomy and phylogenetic systematics (= cladistics) and how they differ.

4.5 Explain the basis for recognizing three domains of life.

4.6 Explain the major subdivisions of the animal kingdom above the phylum level and the characters used to identify them.

Molluscan shells from the collection of Jean Baptiste de Lamarck (1744–1829).

Cleveland P. Hickman, Jr.

Order in Diversity

Evolution has produced a great diversity of species in the animal kingdom. Zoologists have named more than 1.5 million species of animals, and thousands more are described each year. Some zoologists estimate that the species named so far constitute less than 20% of all living animals and less than 1% of all those that have existed.

Despite its magnitude, the diversity of animals is not without limits. Many conceivable forms do not exist in nature, as our myths of minotaurs and winged horses show. Animal diversity is not random but has a definite order. Characteristic features of humans and cattle never occur together in a single organism as they do in mythical minotaurs; nor do the characteristic wings of a bird and body of a horse occur together naturally as they do in the mythical horse Pegasus. Humans, cattle, birds, and horses are distinct groups of animals, and yet they do share some important features, including vertebrae and a skull, that separate them from even more dissimilar forms, such as insects and flatworms.

All human cultures classify familiar animals according to patterns in animal diversity. These classifications have many criteria. Some societies classify animals according to their usefulness or destructiveness to human endeavors; others may group animals according to their roles in mythology. Biologists organize animal diversity in a nested hierarchy of groups within groups according to their evolutionary relationships. Ordered patterns in the sharing of homologous features among species reveal these evolutionary relationships to us. This ordering is called a "natural system" because it reflects relationships that exist among animals in nature, outside the context of human activity. A systematic zoologist has three major goals: to discover all species of animals, to reconstruct their evolutionary relationships, and to communicate those relationships by a precise taxonomic system.

Darwin's theory of common descent (see Chapter 1) is the underlying principle that guides our search for order in the diversity of animal life. Our science of **taxonomy** ("arrangement law") produces a formal system for naming and grouping species to communicate this order. Animals that have very recent common ancestry share many features in common and are grouped most closely in our taxonomic system; dissimilar animals that share only very ancient common ancestry are placed in different taxonomic groups except at the "highest" or most inclusive levels.

Taxonomy is part of a broader science of **systematics,** which studies variation among animal populations to reveal their evolutionary relationships. Taxonomy predates evolutionary biology, however, and thus many taxonomic practices are relics of pre-evolutionary worldviews. Adjusting our taxonomic system to accommodate evolution has produced many problems. Taxonomy has reached an unusually active and controversial point in its development, with several alternative taxonomic systems competing for use. To explain this controversy, we must review the history of animal taxonomy.

4.1 Linnaeus and Taxonomy

The Greek philosopher and biologist Aristotle was the first to group organisms based on their structural similarities. Following the Renaissance in Europe, the English naturalist John Ray (1627–1705) introduced a more comprehensive system of classification and a new concept of species. Rapid growth of systematics in the eighteenth century culminated in the work of Carolus Linnaeus (1707–1778; figure 4.1), who produced our current scheme of classification.

Linnaeus was a Swedish botanist at the University of Uppsala. Linnaeus produced an extensive system of taxonomy for both plants and animals. This scheme, published in his great work, *Systema Naturae,* used **morphology** (the comparative study of organismal form) for arranging specimens in collections. He divided the animal kingdom into species and gave each one a distinctive name. He grouped species into genera, genera into orders, and orders into "classes." (We use quotation marks or a capital letter to distinguish "class" as a formal taxonomic rank from its broader meaning as a group of organisms that share an essential property.) Because Linnaeus's knowledge of animals was limited, his lower categories, such as genera, were much more inclusive than are the groupings currently given these ranks. Much of his classification has been drastically altered, but its basic principles remain.

figure 4.1

Carolus Linnaeus (1707–1778). This portrait was made of Linnaeus at age 68, three years before his death.

Linnaeus's scheme of arranging organisms into an ascending series of groups of increasing inclusiveness is a **hierarchical system** of taxonomy. Each major grouping of organisms, called a **taxon** (pl., **taxa**), is given one of several standard taxonomic ranks to indicate its general inclusiveness. The hierarchy of taxonomic ranks has been expanded considerably since Linnaeus's time (table 4.1). It now includes seven mandatory ranks for the animal kingdom, in descending series: kingdom, phylum, "class," order, family, genus, and species. Any organism being classified must be placed into at least seven taxa, one at each of these mandatory ranks. Taxonomists have the option of subdividing these seven ranks even further to recognize more than seven taxa (superfamily, subfamily, superorder, suborder, and others) for any particular group of organisms. More than 30 taxonomic ranks are used. For very large and complex groups, such as fishes and insects, these additional ranks are needed to express different degrees of evolutionary divergence.

A taxonomist's choice of a group of species for recognition as a formally ranked taxon always has some arbitrariness. For example, should the taxonomic family Hominidae be restricted to genus *Homo* (humans) and all fossil genera that are closer to *Homo* than to genus *Pan* (bonobos and chimpanzees), or should it comprise the genera *Homo, Pan, Gorilla,* and *Pongo* (orangutans) plus fossils closer to these genera than to gibbons? Within the last few decades, anthropologists have shifted Hominidae from the former usage primarily to the latter one. Does the confusing arbitrariness of ranking taxa outweigh the usefulness of ranks for reminding us which taxa are more inclusive than others? Could we devise a rank-free taxonomy that encodes the positions of species on an evolutionary tree of common descent? As we write, taxonomists are actively constructing both ranked and rank-free taxonomies of animals. Rank-free taxonomies often use indentations to specify levels of inclusiveness of each taxon as illustrated by the rank-free taxonomic ordering of bilaterally symmetrical animals shown in the last section of this chapter.

Systematization Versus Classification

Introduction of evolutionary theory into animal taxonomy changed the taxonomist's task from classification to **systematization.** Classification denotes the construction of classes, groupings of organisms possessing a common feature, called an essence, that is used to define the class. Organisms that possess the essential feature are members of the class by definition, and those that lack it are excluded. Because evolving species are always subject to change, the static nature of classes makes them a poor basis for a taxonomy of living systems. In systematization, on the other hand, the taxonomist will group species that form units of common evolutionary descent. Species placed into a taxonomic group include the most recent common ancestor of the group and its descendants and thus form a branch of the phylogenetic tree of life. The species of such a group thus constitute a system of common descent, not a class defined by possession of an essential characteristic. It remains common, although technically erroneous, for systematists to call their taxonomic systems "classifications."

Because organismal characteristics are inherited from ancestral to descendant species, character variation is used to diagnose systems of common descent, but there is no requirement that an essential character be maintained throughout the system for its recognition as a taxon. The role of morphological or other features in systematization is therefore fundamentally different from the role of such characters in classification. In classification, a taxonomist asks whether a species being classified contains the defining feature(s) of a particular taxonomic class; in systematization, a taxonomist asks whether the characteristics of a species confirm or reject the hypothesis that it descends from the most recent common ancestor of a particular taxon. For example, tetrapod vertebrates descend from a common ancestor that had four limbs, a condition retained in most but not all of its descendants. Although they lack limbs, caecilians (see section 17.3) and snakes (see section 18.2: Snakes: Serpentes) are tetrapods because they are descendants of the most recent common ancestor of all tetrapods; other morphological and molecular characters group them respectively with living amphibians and lizards.

Although the hierarchical structure of Linnean classification is retained in current taxonomy, the taxa are groupings of species related by evolutionary descent with modification, as diagnosed by sharing of homologous characters. Moving up the taxonomic hierarchy from a species toward more inclusive groups, each taxon represents the descendants of an earlier ancestor, a larger branch of the tree of life.

Binomial Species Nomenclature

Linnaeus's system for naming species is called **binomial nomenclature.** Each species has a Latinized name composed of two words (hence binomial) written in italics (underlined if handwritten or typed). The first word is the name of the **genus,** written with a capital initial letter; the second word is the **species epithet,** which identifies the species within the genus and is written with a lowercase initial letter (table 4.1). The great communicative value of Latin species names is that they are used consistently by scientists in all countries and languages; they are much more precise than are "common names," which vary culturally and geographically.

The name of a genus is always a noun, which is either feminine or masculine. The species epithet is usually an adjective that must agree in gender with the genus. For instance, the scientific name of a common robin is *Turdus migratorius* (L. *turdus,* thrush; *migratorius,* of the migratory habit). A species epithet never stands alone; the complete binomial must be used to name a species. Names of genera must refer only to single

Table 4.1 Examples of Taxonomic Categories to Which Representative Animals Belong

	Human	Gorilla	Southern Leopard Frog	Fork-Tailed Bush Katydid
Kingdom	Animalia	Animalia	Animalia	Animalia
Phylum	Chordata	Chordata	Chordata	Arthropoda
Subphylum	Vertebrata	Vertebrata	Vertebrata	Hexapoda
Class	Mammalia	Mammalia	Amphibia	Insecta
Subclass	Theria	Theria	—	Pterygota
Order	Primates	Primates	Anura	Orthoptera
Suborder	Haplorhini	Haplorhini	Neobatrachia	Ensifera
Family	Hominidae	Hominidae	Ranidae	Tettigoniidae
Subfamily	Homininae	Homininae	Raninae	Phaneropterinae
Genus	*Homo*	*Gorilla*	*Lithobates*	*Scudderia*
Species	*Homo sapiens*	*Gorilla gorilla*	*Lithobates sphenocephalus*	*Scudderia furcata*
Subspecies	—	—	—	*Scudderia furcata furcata*

Hierarchical taxonomy of four species (human, gorilla, southern leopard frog, and fork-tailed bush katydid). Higher taxa generally are more inclusive than lower-level taxa, although taxa at two different levels may be equivalent in content. Closely related species are united at a lower point in the hierarchy than are distantly related species. For example, humans and gorillas are united at the subfamily (Homininae) and above; they are united with southern leopard frogs at the subphylum level (Vertebrata) and with bush katydids at the kingdom (Animalia) level. Mandatory Linnean ranks are shown in bold type.

groups of organisms; a single name cannot be given to two different genera of animals. However, the same species epithet may be used in different genera to denote different species. For example, the scientific name of a white-breasted nuthatch is *Sitta carolinensis.* The species epithet *"carolinensis"* is used in other genera, including *Poecile carolinensis* (Carolina chickadee) and *Anolis carolinensis* (green anole, a lizard) to mean "of Carolina." All ranks above species are designated using a capitalized noun, such as Iguanidae (the lizard family that contains *Anolis*).

Key Theme 4.1

SCIENCE EXPLAINED

Authority on Species

The person who first names and describes a species is called the authority. The authority deposits in a museum a formally designated type specimen, which carries the name of the species. The authority's name and date of publication of the species description often appear after the species name. Thus, *Didelphis marsupialis* Linnaeus, 1758, tells us that Linnaeus was the first person to publish the species name of opossums. The authority citation is not part of the scientific name but rather an abbreviated bibliographical reference. If the genus name of a species is changed following its initial description, the authority's name appears in parentheses.

4.2 Species

While discussing Darwin's book *On the Origin of Species* in 1859, Thomas Henry Huxley (see Key Theme 1.3) asked, "In the first place, what is a species? The question is a simple one, but the right answer to it is hard to find, even if we appeal to those who should know most about it." So far in this chapter, we have used the term "species" as if it had a simple and unambiguous meaning, but actually, Huxley's commentary is as valid today as it was over 140 years ago. Our concepts of species have become more sophisticated, but the different concepts and disagreements surrounding their use are as evident now as in Darwin's time.

Criteria for Recognition of Species

Despite widespread disagreement about what constitutes a species, biologists agree that certain criteria are important for recognizing species. First, **common descent** is central to nearly all modern concepts of species. Members of a species must trace their ancestry to a common ancestral population, although not necessarily to a single pair of parents. Species are thus historical entities. Variation among organisms serves to test the hypothesis that a particular grouping of organisms forms a population evolving separately from other such populations.

A second criterion is that species must be the *smallest distinct groupings* of organisms sharing patterns of ancestry and descent; otherwise, it would be difficult to separate species from higher taxa whose members also share common descent. Morphological characters traditionally have been used to identify such groupings, but chromosomal and molecular characters are now extensively used for this purpose.

A third important criterion is the *reproductive community,* which pertains to sexually reproducing organisms; members of a species must form a reproductive community that excludes members of other species. This criterion is very important to many modern concepts of species. For organisms whose reproduction is strictly **asexual,** reproductive community entails occupation of a particular ecological niche in a particular place so that a reproducing population responds as a unit to evolutionary processes such as natural selection and genetic drift (see section 1.6: Forces of Evolutionary Change).

Any species has a distribution through space, called its *geographic range,* and a distribution through time, called its *evolutionary duration.* Species differ greatly from each other in both of these dimensions. Species having very large geographic ranges or worldwide distributions are called **cosmopolitan,** whereas those with very restricted geographic distributions are called **endemic**. If a species were restricted to a single point in space and time, we would have little difficulty recognizing it, and nearly every species concept would lead us to the same decision. We have little difficulty distinguishing from each other the different species of animals that inhabit our local parks or woods. However, when we compare a local population to similar but not identical populations located hundreds of miles away, it may be hard to determine whether these populations represent a single species or multiple species (see figure 1.22).

Throughout the evolutionary duration of a species, its geographic range might change many times. A geographic range could be either continuous or disjunct, having breaks within it where the species is absent. Suppose that we find two similar but not identical populations living 300 miles apart with no closely related populations between them. Are we observing a single species with a disjunct distribution, or two different but closely related species? Suppose that these populations have been separated historically for 50,000 years. Is this enough time for them to have evolved separate reproductive communities, or do they form parts of a single reproductive community? Species concepts differ in how they answer these questions.

Concepts of Species

Before Darwin, a species was considered a distinct and immutable entity. The concept that species are defined by fixed, essential features (usually morphological) is called the **typological species concept.** Taxonomists discarded this concept following establishment of Darwinian evolutionary theory.

The most influential concept of species inspired by Darwinian evolutionary theory is the **biological species concept** formulated by Theodosius Dobzhansky and Ernst Mayr. In 1983, Mayr stated the biological species concept as follows: *"A species is a reproductive community of populations (reproductively isolated from others) that occupies a specific niche in nature."* Note that a species is identified here according to reproductive properties of populations, groups of related organisms inhabiting a particular geographic area, not according to organismal morphology. This definition establishes that a species is an *interbreeding* population of individuals having *common descent.* By adding the criterion of **niche,** an ecological concept denoting an organism's role in its ecological community (see section 2.2), we recognize that members of a reproductive community constitute an ecological entity in nature. Because a reproductive community should maintain genetic cohesiveness, organismal variation should be relatively smooth and continuous within species and discontinuous between them. Although a biological species is based on reproductive properties of populations rather than on organismal morphology, morphology nonetheless can help us to diagnose biological species.

The biological species concept has been strongly criticized for several reasons. First, it refers to contemporary populations and provides little guidance for tracing the temporal duration of a species lineage through its past history. How far back must we trace a species lineage before we cross its historical boundary with its ancestral species? For example, humans constitute one biological species relative to other living populations. Human populations worldwide form a single reproductive community, reproductively isolated from all other populations. We have no guidance from the biological species concept regarding whether we should place human fossils of increasing evolutionary age into the current human species or into separate species. Should all fossil populations that are closer to us than to other living apes be judged part of our species or a series of temporal species that replaced one another through the past 5–10 million years of human evolution?

Second, according to the biological species concept, species do not exist in groups of organisms that reproduce only asexually. It is common taxonomic practice, however, to describe species in all groups of organisms.

A third problem is that systematists using the biological species concept often disagree on the amount of reproductive divergence necessary for considering two populations separate species. For example, should occurrence of limited hybridization between populations in a small geographic area cause them to be considered a single species despite evolutionary differences between them? Hybrids between populations of salamanders *Ensatina eschscholtzii platensis* and *E. e. xanthoptica* (see figure 1.22) occur geographically only at their point of contact in the Sierra Nevada mountains; molecular genetic distinctness between these taxonomic subspecies is otherwise so large that many authorities consider them separate reproductive communities and thus separate species under the biological species concept. Different authors invoking the biological species concept have interpreted the same data as demonstrating that the *Ensatina* populations of figure 1.22 constitute one species versus demonstrating at least 11 distinct species.

The **evolutionary species concept** was proposed by the mammalian paleontologist George Gaylord Simpson (figure 4.2) in the 1940s to add an evolutionary time dimension to the biological species concept. This concept persists in a modified form today. A current definition of the evolutionary species is *a single lineage of ancestor-descendant populations that maintains its identity from other such lineages and that has its own evolutionary tendencies and historical fate.* Note that the criterion of common descent is retained here in the need for a species to have a distinct historical identity. Unlike the biological species concept, the evolutionary species concept applies both to sexually and asexually reproducing forms. As long as continuity of diagnostic features is maintained by an evolving lineage, it is recognized as a single species. Abrupt changes in diagnostic features mark a boundary between different species in evolutionary time.

SPL/Science Source

figure 4.2

Mammalian paleontologist George Gaylord Simpson (1902–1984) formulated the evolutionary species concept and principles of evolutionary taxonomy.

An ability of geographic populations to evolve collectively as a single, genetically cohesive unit through evolutionary time is critical to the evolutionary species concept. Population geneticist Alan Templeton updated this concept in 1989 to make explicit the expectation that populations of a species evolve as a genetically cohesive unit by natural selection and genetic drift. Using his **cohesion species concept**, Templeton defined a species as follows: *the most inclusive population of individuals having the potential for phenotypic cohesion through intrinsic cohesion mechanisms.* The cohesion mechanisms include gene flow across the geographic expanse of the species, shared loss of alleles through genetic drift, and shared genetic changes caused by natural selection. Another way of stating the cohesion criterion is that any individual in a species is a possible common ancestor of the entire species at some future time. For example, a new allele that arises by mutation in a single person could spread throughout the human population over many generations, ultimately to become shared by all members of the species at some future time. New mutations arising in other species, even in our closest relatives of the genus *Pan*, could not enter the human gene pool.

The last concept that we present is the **phylogenetic species concept**. Ornithologist Joel Cracraft defines the phylogenetic species concept as an *irreducible (basal) grouping of organisms diagnosably distinct from other such groupings and* *within which there is a parental pattern of ancestry and descent.* This concept also emphasizes common descent, and both asexual and sexual groups are covered. Any population that has become spatially separated from others and has undergone character evolution that distinguishes it would be recognized as a species. The criterion of irreducibility requires that no more than one such population be placed in a single species. The phylogenetic species concept emphasizes recognizing as species the smallest groupings of organisms that have undergone independent evolutionary change. The evolutionary and cohesion species concepts place greater emphasis on whether historically separated populations have the biological potential to merge into a single lineage in the future. In general, the phylogenetic species concept would describe a larger number of species than would any other concept. The phylogenetic species concept is intended to encourage us to reconstruct patterns of evolutionary common descent on the finest scale possible. How would the phylogenetic species concept resolve the controversy over how many species exist in *Ensatina eschscholtzii* (see figure 1.22)? The molecular genetic data reveal at least 11 diagnosably distinct lineages, each of which would constitute a separate phylogenetic species. Grouping these lineages as a single species because they retain a potential for future genetic exchanges is not an option under the phylogenetic species concept.

Herpetologist Kevin de Queiroz argues that the various competing concepts of species have a common underlying principle despite their differences. In each case, a species constitutes a segment of a population-level lineage, what de Queiroz calls the **general lineage concept** of species. For the biological species concept, the segment is a temporally short one, with reproductive community among sexually reproducing populations being the critical secondary attribute that separates the biological species concept from alternatives. For the phylogenetic species concept, a population lineage diagnosable as having evolved independently since its evolutionary separation from another such lineage provides the secondary attribute that distinguishes this concept from others. The general lineage concept of species has gained popularity among systematists because it emphasizes the common goal of identifying the phylogenetic history of population-level lineages in detail. It does not solve the problem, however, that taxonomists using contrasting species concepts may differ greatly in how many species they judge worthy of a Linnean Latin binomial (see boxed essay, Species Concepts in Practice).

Taxonomists agree that historically distinct population lineages, the species of the phylogenetic species concept, are real entities in nature. Such entities exist as the indivisible units of evolutionary process and change independent of our knowledge of them. Advocates of the other species concepts do not deny these claims, but they consider such lineages too numerous and too ephemeral for each one to deserve recognition with a Latin species binomial, especially when the biological differences among lineages are judged superficial. Given the power of molecular genetic data to diagnose species lineages, it is perhaps impractical to expect each one to be given formal species status. A taxonomic system must be practical to serve us well, but when we defer to practicality we risk making our recognized species arbitrary constructs that lose their integrity as natural individuals.

Current disagreements concerning concepts of species should not be considered trivial or discouraging. Whenever a field of science enters a phase of dynamic growth, old concepts are reevaluated and either refined or replaced with newer, more progressive ones. Active debate among systematists shows that this field has acquired unprecedented activity and importance in biology. Just as Thomas Henry Huxley's time was one of enormous advances in biology, so is the present time. Both times are marked by fundamental reconsiderations of the meaning of species.

Key Theme 4.2

SCIENCE EXPLAINED

Subspecies

Some species are divided into subspecies, in which case a trinomial nomenclature is employed (see katydid example, table 4.1, and salamander example, figure 1.22); such species are called **polytypic**. Generic, specific, and subspecific names are printed in italics (underlined if handwritten or typed). A polytypic species contains one subspecies whose subspecific name is a repetition of the species epithet and one or more additional subspecies whose names differ. Thus, to distinguish geographic variants of *Ensatina eschscholtzii*, one subspecies is named *Ensatina eschscholtzii eschscholtzii*, and different subspecies names are used for each of the six other subspecies (see figure 1.22). Both the genus name and the species epithet may be abbreviated as shown in figure 1.22. Formal recognition of subspecies has lost popularity among taxonomists in part because the increasing influence of the phylogenetic species concept discourages systematists from grouping multiple separately evolving lineages as subspecies of one biological species. Recognition of subspecies in other cases often errs by highlighting one or a few superficial characters that do not always diagnose an evolutionarily distinct unit. Subspecies, therefore, should be viewed as tentative statements indicating that the species status of the populations needs further investigation.

DNA Barcoding of Species

DNA barcoding is a technique for diagnosing organisms to species using sequence information from a standard gene present in all animals. The mitochondrial gene encoding cytochrome *c* oxidase subunit 1 (*COI*), which contains about 650 nucleotide base pairs, is a standard "barcode" region for animals. DNA sequences of *COI* usually vary among individuals of the same species but not extensively, so that variation within a species is much smaller than differences among species. DNA barcoding is applied to specimens in nature by taking a small DNA sample from blood or another expendable tissue. The method is useful also for specimens in natural history museums, zoos, aquaria, and frozen tissue collections. DNA sequences from such sources are checked against a public reference library of species identifiers to assign unknown specimens to known species. DNA

Species Concepts in Practice

These hypothetical examples illustrate conditions that challenge taxonomists in judging the numbers of species represented by a group of populations being studied. I. Two geographic populations of sexually reproducing forms make geographic contact along a borderline at which hybrids (offspring whose parents come from different geographic populations) occur (I. A. natural hybrids are fertile; I. B. natural hybrids are infertile). II. Two geographically allopatric populations of sexually reproducing forms appear morphologically and ecologically equivalent (I. A. artificial crosses produce fertile hybrids; I. B. artificial crosses produce infertile hybrids). III. Six genetically divergent populations of sexually reproducing forms have geographic and phylogenetic relationships as shown with hybrids produced between populations B and D. Populations A–C (group 1) are morphologically indistinguishable from each other but differ from populations D–F (group 2), which are morphologically indistinguishable from each other (III. A. groups 1 and 2 are ecologically equivalent; III. B. groups 1 and 2 are ecologically distinct). IV. Sexually reproducing populations A–E are genetically divergent but similar in ecology and morphology; population F was derived from population E by a founder event and is reproductively isolated from the others. V. Two allopatric populations of animals that reproduce only asexually are ecologically and morphologically distinct from each other. Each box indicates the number of different species that a taxonomist would recognize using the species concept(s) at the head of the column. Some concepts have inherent ambiguities permitting some room for individual judgment separating taxonomic "lumpers" from "splitters"; such cases are indicated by "1-2" in the appropriate column with the source of ambiguity indicated in column 5. Assume that the individual populations discussed are internally genetically homogeneous based on molecular genetic data, but that they are diagnosably distinct from others using those same data.

Geography / Phylogeny		Biological Species	Evolutionary and Cohesion Species	Phylogenetic Species	Source of Ambiguity
I. 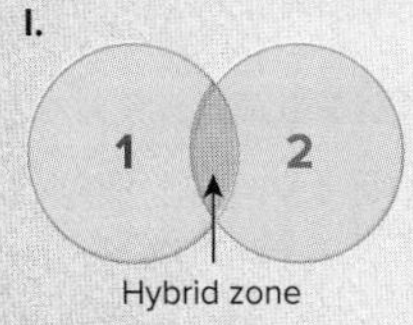	A. hybrids fertile	1 - 2	1 - 2	2	Will 1 & 2 eventually merge or remain distinct with a small hybrid zone?
	B. hybrids infertile	2	2	2	None
II. Artificial cross 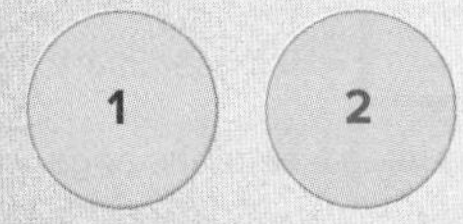	A. hybrids fertile	1	1	2	None
	B. hybrids infertile	2	2	2	None
III. 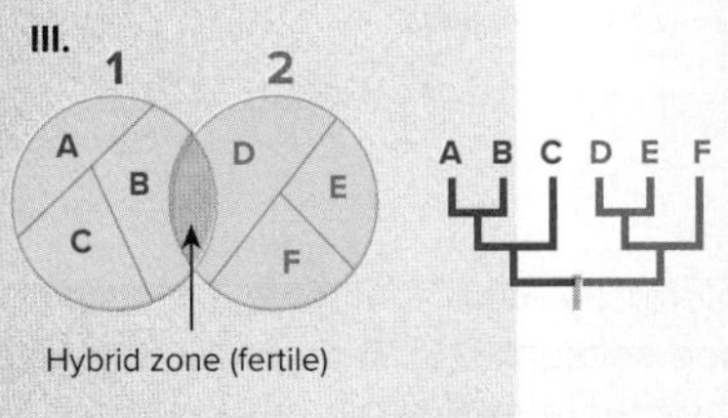	A. 1 & 2 are ecologically similar	1 - 2	1 - 2	6	Will 1 & 2 eventually merge or remain distinct with a small hybrid zone?
	B. 1 & 2 are ecologically distinct	1 - 2	2	6	Is ecological isolation sufficient for separate species status without reproductive isolation?
IV. 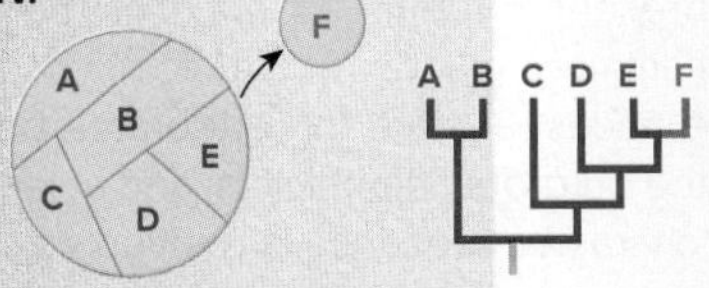		2	2	6	None
V. 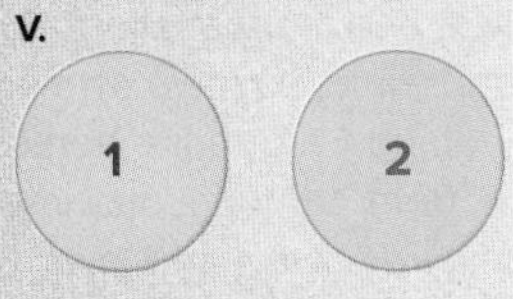		0	2	2	None

barcoding does not solve the controversies regarding use of different species concepts, but it often permits the origin of a specimen to be identified to a particular local population, which is valuable information regardless of the species status that a taxonomist assigns to that population.

4.3 Taxonomic Characters and Reconstruction of Phylogeny

A major goal of systematics is to infer an evolutionary tree or **phylogeny** that relates all extant and extinct species. One constructs a phylogenetic tree by studying organismal features, formally called **characters**, that vary among species. A character is any feature that a taxonomist uses to study variation within or among species. Potentially useful taxonomic characters include morphological, chromosomal, and molecular features. Taxonomists find characters by observing patterns of similarity among organisms. If two organisms possess similar features, they may have inherited these features from an equivalent one in a common ancestor.

Character similarity that results from common ancestry is called **homology** (see section 1.4: Homology and Reconstruction of Phylogeny). The character of feathers in birds illustrates homology; all birds have feathers that were inherited with various modifications from those of their most recent common ancestor. However, similarity does not always reflect common ancestry. Independent evolutionary origins of similar features on different lineages produce patterns of similarity among organisms that misrepresent common descent; this occurrence complicates the work of taxonomists. Character similarity that does not accurately represent patterns of common descent is called nonhomologous similarity, or **homoplasy.** For example, independent evolution of image-forming eyes by cephalopod molluscs and by vertebrates illustrates homoplasy; the most recent common ancestor of molluscs and vertebrates did not have such eyes. For an example of molecular homoplasy, see the interpretation of character 41 in the boxed essay, Phylogenies from DNA Sequences (see section 4.4).

Using Character Variation to Reconstruct Phylogeny

To reconstruct the phylogeny of a taxon using characters that vary among its members, the first step is to ask, for each character, which character state was present in the most recent common ancestor of the entire taxon. This form is called the **ancestral character state** for the taxon as a whole. We presume that all contrasting states of the character arose later within the group, and these forms are called evolutionarily **derived character states.** The ancestral/descendant relationships among the alternative states of a character in the taxon of interest constitute the **polarity** of the character.

For example, suppose that our study taxon is the tetrapod vertebrates, which includes all amphibians, reptiles (including birds), and mammals. Two contrasting character states of egg structure are presence versus absence of an extraembryonic membrane called the **amnion** in the egg. An amnion occurs in the eggs of all reptiles and mammals, but it is absent from all eggs of amphibians. Which of these alternative states most likely characterized the most recent common ancestor of all tetrapod vertebrates?

The method used to examine polarity of a variable character is called **outgroup comparison.** We begin by selecting an additional group of organisms, called an **outgroup,** that is phylogenetically close but not within the taxon being studied, which we call the "ingroup." A character state shared by ingroup and outgroup species is judged ancestral for the ingroup. Bony fishes constitute an appropriate outgroup for polarizing variation in egg structure of tetrapods. Next, we infer that any character state found within *both* the study group and its outgroup most likely characterized the most recent common ancestor of the study group. All bony fishes lack an amnion in their eggs; therefore, we infer that the absence of an amnion is ancestral for the tetrapods and presence of an amnion is derived. Polarity of this character indicates that an amnion first evolved in an ancestral lineage of all modern reptiles and mammals. Polarity of characters is evaluated most effectively when several different outgroups are used. All character states found in a study group that are absent from appropriate outgroups are considered derived within the study group (see figure 4.3 for additional derived characters of vertebrates).

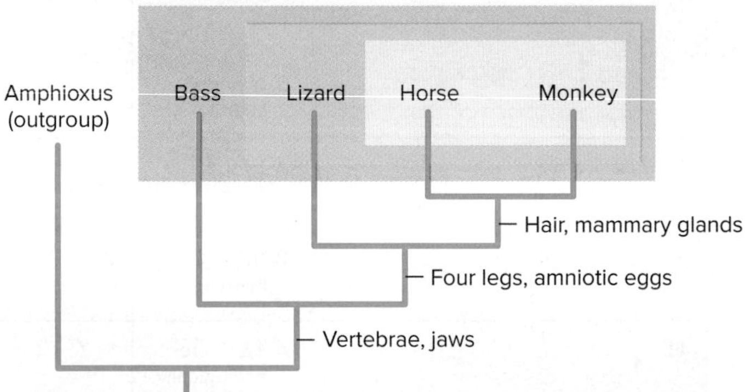

figure 4.3

A cladogram as a nested hierarchy of taxa. Amphioxus, a small limbless swimming chordate (see section 15.5), is the outgroup, and the study group comprises four vertebrates (bass, lizard, horse, and monkey). Four characters that vary among vertebrates are used to generate a simple cladogram: presence versus absence of four limbs, amniotic eggs, hair, and mammary glands. For all four characters, absence is considered the ancestral state in vertebrates because this condition is shared by the outgroup, amphioxus, and at least one ingroup member; for each character, presence is derived within vertebrates. Because they share presence of four limbs and amniotic eggs as synapomorphies, lizards, horses, and monkeys form a clade relative to bass. This clade is subdivided further by two synapomorphies (presence of hair and mammary glands) that unite horses and monkeys relative to lizards. We know from comparisons involving even more distantly related animals that vertebrae and jaws constitute synapomorphies of vertebrates and that amphioxus, which lacks these features, falls outside the vertebrate clade.

A **clade** (Gr. *klados*, branch) is the fundamental unit of phylogenetic grouping of species; it comprises an ancestral lineage and all species descended from that lineage. A derived character shared uniquely by all members of a clade is formally called a **synapomorphy** (Gr. *synapsis*, joining together, + *morphē*, form) of that clade. The derived character arose in the common ancestral lineage of the clade and was inherited by all descendants of that ancestor. Taxonomists use shared derived characters as evidence of homology to infer that a particular group of species forms a clade. Among living amniotes, jaws forming a beak without teeth and presence of feathers are synapomorphies that identify birds as a clade. Within birds, the derived morphological character of a flexible palate marks Neognathae as a nested clade separate from Paleognathae. Within Neognathae, derived, swimming-adapted wings mark the penguin order Sphenisciformes as a nested clade. The pattern formed by the derived states of all characters within our ingroup reveals a **nested hierarchy** of clades within clades. Our goal is to identify all of the different clades nested within the ingroup, which would reveal the structure of common descent among its species.

A nested hierarchy of clades is presented as a branching diagram called a **cladogram** (figure 4.3; see also figure 1.19). The branches on a cladogram are a formal device for indicating a nested hierarchy of clades within clades. The same nested hierarchy is sometimes presented as an indented list of taxon names; for example, the cladogram for major taxa of living gnathostome vertebrates shown in figure 16.2 could be represented as:

Chondrichthyes
  Holocephali
  Elasmobranchii

Osteichthyes
  Actinopterygii
  Sarcopterygii
    Actinistia
    Dipnoi (lungfishes)
    Tetrapoda

Taxonomists often make a technical distinction between a cladogram and a **phylogenetic tree.** A cladogram is not strictly equivalent to a phylogenetic tree, whose branches represent real lineages that occurred in the evolutionary past. To obtain a phylogenetic tree, one must add to a cladogram information concerning ancestors, durations of evolutionary lineages, or amounts of evolutionary change that occurred on lineages. However, because the branching order of a cladogram matches that of the corresponding phylogenetic tree, a cladogram often serves as a first approximation of the structure of the phylogenetic tree.

Sources of Phylogenetic Information

We find characters that are useful for constructing cladograms in comparative morphology (including embryology), comparative cytology, and comparative biochemistry. **Comparative morphology** examines varying shapes and sizes of organismal structures, including their developmental origins. As we discuss in later chapters, variable structures of skull bones, limb bones, and integument (scales, hair, feathers) are particularly important for reconstructing the phylogeny of vertebrates. Comparative morphology uses specimens obtained from both living organisms and fossilized remains. **Comparative biochemistry** uses the sequences of amino acids in proteins and the sequences of nucleotides in nucleic acids to identify variable characters for constructing a cladogram or phylogenetic tree (figure 4.4). Recent work shows that some fossils retain enough DNA for comparative biochemical studies. **Comparative cytology** (also called **karyology**) uses variation in the numbers, shapes, and sizes of chromosomes and their parts to obtain variable characters for constructing cladograms. Comparative cytology is used almost exclusively on living rather than

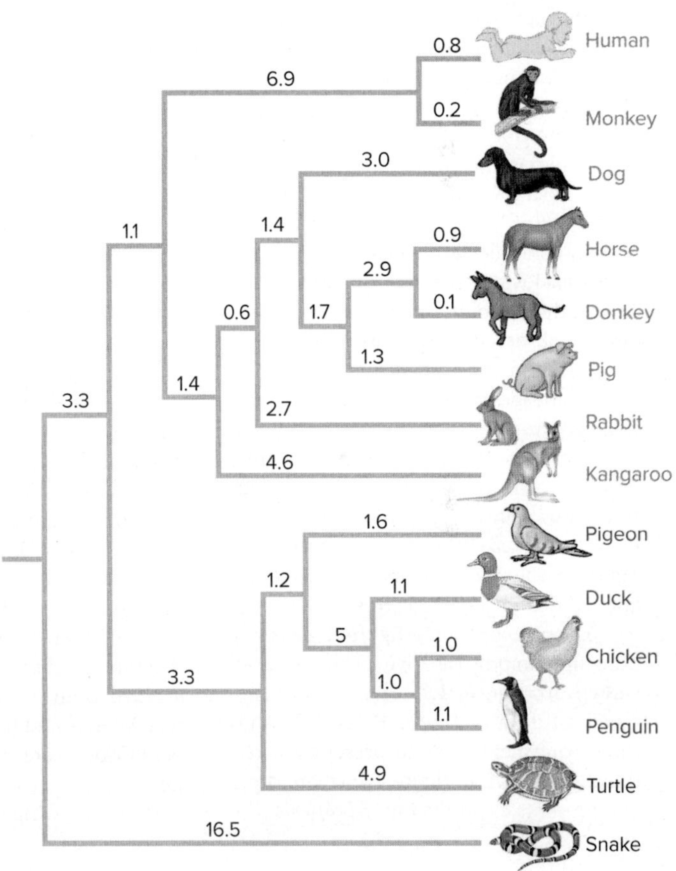

figure 4.4

An early phylogenetic tree of representative amniotes based on amino acid sequences of the respiratory protein, cytochrome *c*. The numbers on the branches are estimated minimum numbers of mutational changes needed to explain amino acid substitutions along different evolutionary lineages. Publication of this tree by Fitch and Margoliash in 1967 convinced systematists that molecular sequences contain phylogenetic information. Subsequent work confirms some hypotheses, including the hypothesis that mammals (blue) and birds (red) form non-overlapping clades, while rejecting others; the kangaroo, for example, should be outside a branch containing all other mammals shown.

Source: Fitch, W. M. and E. Margoliash. 1967. Construction of phylogenetic trees. *Science* **155**:279–284.

How to Read a Phylogenetic Tree or Cladogram

The critical information in a phylogenetic tree is contained in the pattern of branching. It is a visual representation of the evolutionary history of lineages hypothesized to connect the species or higher taxa being studied to their most recent common ancestor, represented by the root of the tree. The history of common descent moves forward in time from the root to the tips, each of which is a species or higher taxon used to construct the tree. Each point at which an ancestral lineage splits to form two or more descendant lineages is called a **node**. A node represents the most recent common ancestor of all species that descend from it. An ancestral species or lineage and all of its descendants collectively form a clade. The branching pattern thus constitutes a hierarchy of less inclusive, younger clades nested within older, more inclusive ones.

To construct a phylogenetic tree, one works backward from the tips to the root. Taxa at the tips of the tree are the only ones directly observed; they provide the character variation used to construct the tree. Ancestral lineages and nodes are hypotheses inferred from analyzing the variation in characters present across the tip taxa. Interpreted as a cladogram, the branching structure shows only the nested hierarchy of less inclusive clades within more inclusive ones, omitting the evolutionary interpretations applied to a phylogenetic tree.

The accompanying figure depicts relationships among the five primate genera shown in figure 4.7: *Hylobates, Pongo, Gorilla, Pan,* and *Homo*. Parts A–C show three equivalent ways to present the relationships shown in figure 4.7, whereas part D represents a contrasting hypothesis.

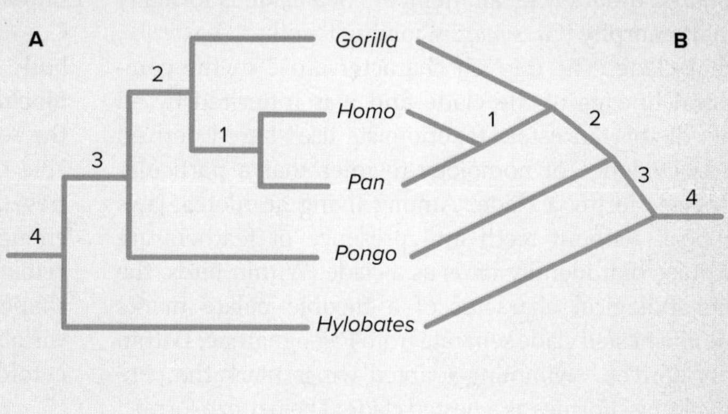

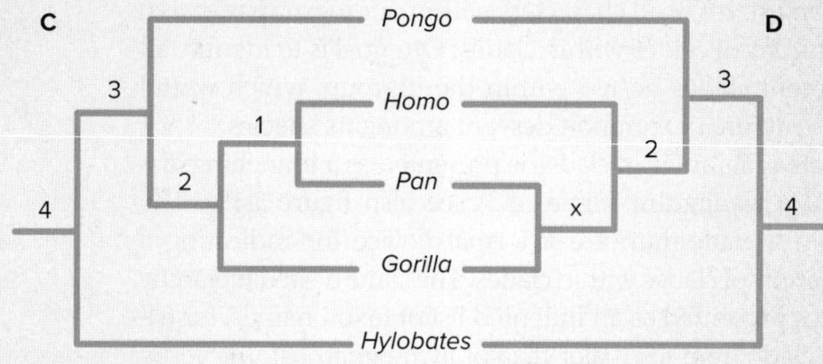

Parts A and B show two different ways to draw the branches, either in a rectangular form as in A or as angled branches in B. The four numbered clades are the same in the two trees. Clade 1 comprises *Homo* and *Pan* alone; clade 2 comprises *Gorilla, Homo,* and *Pan*; clade 3 comprises *Gorilla, Homo, Pan,* and *Pongo*; and clade 4 comprises *Gorilla, Homo, Hylobates, Pan,* and *Pongo*.

Any branch can be rotated on the tree diagram *without changing the relationships depicted*. Branches 2 and 3 in C are rotated relative to A and B, but the same four numbered clades are present. A common mistake in reading a tree diagram is to interpret part A as showing *Homo* close to *Gorilla* and part C as showing *Homo* close to *Pongo*. This error comes from reading the diagram from side to side rather than by tracing the paths of common descent between the root and tips.

Cutting a branch and moving it to a new location produces a contrasting phylogenetic hypothesis. In D, the branch for genus *Pan* is removed from its pairing with *Homo* and attached to the branch for *Gorilla*. A clade comprising *Gorilla* and *Pan* is marked by x in part D. Molecular phylogenetic analyses reject the hypothesis in part D in favor of the one in parts A–C. Part D nonetheless contains clades 2–4 in common with the hypothesis shown in A–C.

fossilized organisms because chromosomal structure is not well preserved in fossils (see figure 1.11D).

To add the evolutionary timescale needed for producing a phylogenetic tree, we must consult the fossil record. We look for the earliest appearance of derived morphological characters in fossils to estimate ages of clades distinguished by those characters. The ages of fossils showing derived characters of a particular clade are determined by radioactive dating (see section 1.4: Geological Time) to estimate the age of the clade.

4.4 Theories of Taxonomy

A theory of taxonomy establishes principles that we use to recognize and to rank taxonomic groups. There are two currently popular theories of taxonomy: (1) evolutionary taxonomy and (2) phylogenetic systematics (cladistics). Both are based on evolutionary and phylogenetic principles but differ on how those evolutionary principles are used. These differences have important implications for how we use a taxonomy to study evolutionary processes. Evolutionary taxonomy predates phylogenetic systematics and retains many aspects of Linnean taxonomy; for this reason, it is sometimes called "traditional evolutionary taxonomy." Evolutionary taxonomy was well established by the 1940s; phylogenetic systematics arose in the 1960s as a replacement for evolutionary taxonomy, which some systematists considered too subjective and misleading.

The relationship between a taxonomic group and a phylogenetic tree or cladogram is important for both theories. This relationship takes one of three forms: **monophyly, paraphyly,** or **polyphyly.** A taxon is monophyletic if it includes the most recent common ancestor of all members of a group and all descendants of that ancestor (figure 4.5A). The terms monophyletic group and clade are synonymous. A taxon is paraphyletic if it includes the most recent common ancestor of all members of a group and some but not all descendants of that ancestor (figure 4.5B). A taxon is polyphyletic if it does not include the most recent common ancestor of all members of a group; this situation requires the group to have had at least two separate evolutionary origins, usually requiring independent evolutionary acquisition of a diagnostic feature (figure 4.5C). For example, if birds and mammals were grouped in a taxon called Homeothermia, it would be a polyphyletic taxon because birds and mammals descend from two quite separate amniotic lineages that have evolved homeothermy independently. The most recent common ancestor of birds and mammals is not homeothermic and does not occur in the polyphyletic Homeothermia just described.

Monophyletic and paraphyletic groups share the property of **convexity,** which distinguishes them from polyphyletic groups. A group is convex if you can trace a path between any two members of the group on a cladogram or phylogenetic tree without leaving the group. For example, on figure 4.5 you could trace a connection between any pair of points in the blue areas of parts A or B without leaving the blue area. For the polyphyletic grouping in part C of figure 4.5, one cannot trace the path between species C and E without leaving the group designated by blue shading. In figure 4.5C, if a systematist added the full path connecting species C and E to the group shown but continued to omit the paths leading to species A, B, and H, then the new grouping thus formed would be convex and paraphyletic rather than polyphyletic. Demonstration that a group is not convex is the formal criterion for considering the group polyphyletic. Both evolutionary and cladistic taxonomies accept monophyletic groups and reject polyphyletic groups. They differ regarding acceptance of paraphyletic groups.

Evolutionary Taxonomy

Evolutionary taxonomy retains the basic structure of Linnean taxonomy; species are grouped in a nested hierarchy of increasingly more inclusive higher taxa, each of which receives one of the mandatory or optional Linnean ranks. The taxonomy is evolutionary because all taxa must have a single evolutionary origin, must include the most recent common ancestor of all members of the taxon, and must be convex on the phylogenetic tree. Unlike pre-evolutionary uses of Linnean taxonomy, evolutionary taxonomy explicitly excludes polyphyletic groups.

George Gaylord Simpson (see figure 4.2) was highly influential in developing and formalizing principles of evolutionary taxonomy. According to Simpson, a particular branch on an evolutionary tree is given the status of a higher taxon if it represents a distinct **adaptive zone.** Simpson describes an adaptive zone as "a characteristic reaction and mutual relationship between environment and organism, a way of life and not a place where life is led." By entering a new adaptive zone through a fundamental change in organismal structure and behavior, an evolving population can use environmental resources in a completely new way.

A taxon forming a distinct adaptive zone is termed a **grade.** Simpson gives the example of penguins as a distinct adaptive zone within birds. The lineage immediately ancestral to all penguins underwent fundamental changes in form of the body and wings to switch from aerial to aquatic locomotion (figure 4.6). Aquatic birds that can fly both in air and underwater are

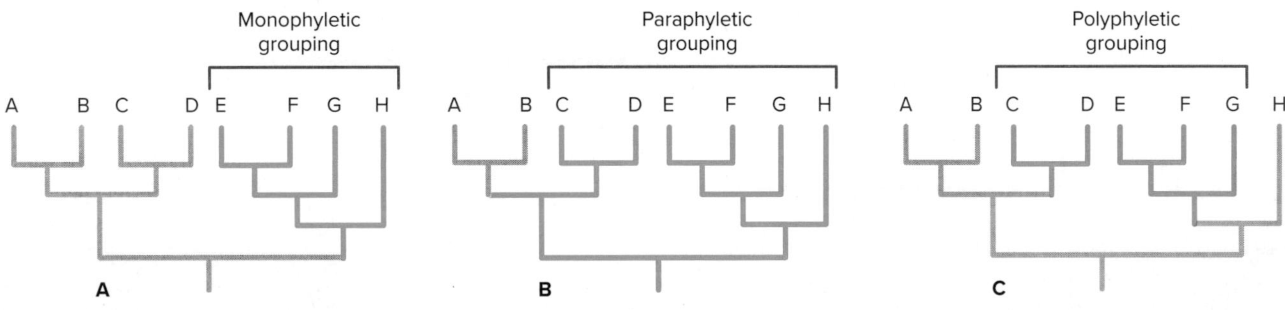

figure 4.5

Relationships between phylogeny and taxonomic groups illustrated for a hypothetical phylogeny of eight species (A through H). **A,** A monophyletic group contains the most recent common ancestor of all members of the group and all of its descendants. **B,** A paraphyletic group contains the most recent common ancestor of all members of the group and some but not all of its descendants. **C,** A polyphyletic group does not contain the most recent common ancestor of all members of the group, thereby requiring the group to have at least two separate phylogenetic origins. Monophyletic and paraphyletic groups are *convex*, meaning that one can trace a path from any member of the group to any other member without leaving the group; any group that fails the convexity criterion is considered polyphyletic.

figure 4.6

A, Penguin. **B,** Diving petrel. Penguins (avian family Spheniscidae) were considered by George G. Simpson a distinct adaptive zone within birds because of their adaptations for submarine flight. Simpson believed that the adaptive zone ancestral to penguins resembled that of diving petrels, which display adaptations for combined aerial and aquatic flight. Adaptive zones of penguins and diving petrels are distinct enough to be recognized taxonomically as different families within a common order (Ciconiiformes).

somewhat intermediate in habitat, morphology, and behavior between aerial and aquatic adaptive zones. Nonetheless, obvious modifications of a penguin's wings and body for swimming represent a new grade of organization. Penguins are therefore recognized as a distinct taxon within birds, family Spheniscidae. The Linnean rank of a taxon depends upon the breadth of its adaptive zone: the broader the adaptive zone when fully realized by a group of organisms, the higher the taxon is ranked.

Evolutionary taxa may be either monophyletic or paraphyletic. Recognition of paraphyletic taxa requires, however, that taxonomies distort patterns of common descent. An evolutionary taxonomy of anthropoid primates provides a good example (figure 4.7). This taxonomy places humans (genus *Homo*) and their closest fossil relatives in family Hominidae. It groups bonobos and chimpanzees (genus *Pan*), gorillas (genus *Gorilla*), and orangutans (genus *Pongo*) in family Pongidae. However, the pongid genera *Pan* and *Gorilla* share more recent common ancestry with Hominidae than they do with the remaining pongid genus, *Pongo*. Family Pongidae is therefore paraphyletic because it excludes humans, who also descend from the most recent common ancestor of all pongids (figure 4.7). Evolutionary taxonomists nonetheless consider the pongid genera a single, family-level grade of **arboreal** (tree-dwelling), herbivorous primates having no language; they show a family-level adaptive zone. Humans are terrestrial, omnivorous primates who have language and culture, thereby forming

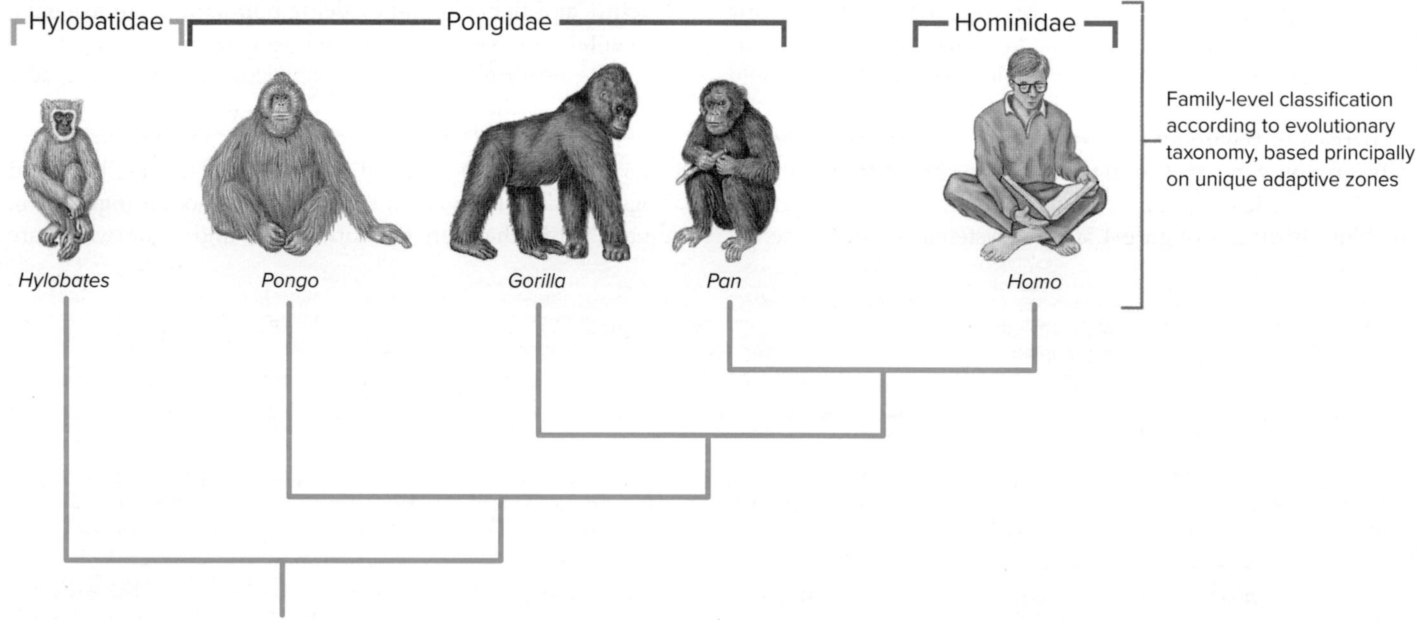

figure 4.7

Phylogeny and family-level classification of anthropoid primates. Evolutionary taxonomy groups the genera *Gorilla*, *Pan*, and *Pongo* into the paraphyletic family Pongidae because they share the same adaptive zone or grade of organization. Humans (genus *Homo*) are phylogenetically closer to *Gorilla* and *Pan* than any of these genera are to *Pongo*, but humans are placed in a separate family (Hominidae) because they represent a different grade of organization. Cladistic taxonomy eliminates the paraphyletic family Pongidae and groups *Pongo*, *Gorilla*, *Pan*, and *Homo* into a single monophyletic family, Hominidae. Gibbons (genus *Hylobates* and three others not shown) form the monophyletic family Hylobatidae, which is compatible with both evolutionary and cladistic classifications.

a distinct adaptive zone at the taxonomic level of a family. Unfortunately, if we want our taxa to constitute such adaptive zones, our taxonomy will distort common descent.

Evolutionary taxonomy faced two opposite challenges. **Phenetic taxonomy** sought to group species into higher taxa according to quantitative measurements of overall similarity among species, without regard to phylogeny. Phylogeny was considered too difficult to measure to be the basis of our taxonomic system. Later work showed that phylogeny could be measured precisely and that the notion of overall similarity was highly ambiguous. Phenetic taxonomy contributed some useful analytical methods but did not have a strong impact on animal taxonomy, and scientific interest in this approach has declined.

Phylogenetic Systematics/Cladistics

A second and stronger challenge to evolutionary taxonomy is called **phylogenetic systematics,** or **cladistics.** As the first name implies, this approach emphasizes the criterion of common descent, and as the second name implies, it is based on the cladogram of a group being classified. This approach to taxonomy was first proposed in 1950 by the German entomologist Willi Hennig (figure 4.8) and therefore is sometimes called "Hennigian systematics." All taxa recognized by Hennig's cladistic system must be monophyletic. We saw previously how evolutionary taxonomists' recognition of the primate families Hominidae and Pongidae distorts genealogical relationships to emphasize the adaptive uniqueness of the Hominidae. Because the most recent common ancestor of the paraphyletic family Pongidae is also an ancestor of family Hominidae, recognition of Pongidae is incompatible with cladistic taxonomy. To avoid paraphyly, cladistic taxonomists have discontinued use of the traditional family Pongidae, placing chimpanzees, gorillas, and orangutans with humans in the family Hominidae (as shown for *Gorilla gorilla* in table 4.1).

Disagreement regarding the validity of paraphyletic groups may seem trivial at first, but its important consequences become clear when we discuss evolution. For example, claims that amphibians evolved from bony fish, that birds evolved from reptiles, or that humans evolved from apes might be made by an evolutionary taxonomist but are meaningless to a cladist. We imply by these statements that a descendant group (amphibians, birds, or humans) evolved from part of an ancestral group (bony fish, reptiles, or apes, respectively) from which the descendant is excluded. This usage automatically makes the ancestral group paraphyletic, and indeed bony fish, reptiles, and apes as traditionally recognized are paraphyletic groups. How are such paraphyletic groups recognized? Do they share distinguishing features not shared by a descendant group?

Paraphyletic groups are usually defined in a negative manner. They are distinguished only by the absence of features found in a particular descendant group, because any traits shared from their common ancestry are present also in the excluded descendants (unless secondarily lost). For example, apes are those "higher" primates that are not humans. Likewise, fish are those vertebrates that lack the distinguishing

Dr. Wolfgang Hennig

figure 4.8

Willi Hennig (1913–1976), German entomologist who formulated the principles of phylogenetic systematics/cladistics in 1950.

characteristics of tetrapods (amphibians and amniotes). What does it mean then to say that humans evolved from apes? To an evolutionary taxonomist, apes and humans are different adaptive zones or grades of organization; to say that humans evolved from apes states that bipedal organisms of large brain capacity evolved from arboreal, quadrupedal organisms of smaller brain capacity. To a cladist, however, the statement that humans evolved from apes says essentially that humans evolved from an arbitrary grouping of species that lack the distinctive characteristics of humans; such a statement contains no useful information. An extinct ancestral group is always paraphyletic because it excludes a descendant that shares its most recent common ancestor. Although many such groups have been recognized by evolutionary taxonomists, none are recognized by cladists.

Cladists denote the common descent of different taxa by identifying **sister taxa.** Sister taxa share more recent common ancestry with each other than either one does with any other taxon. For example, the sister group of humans appears to be the species pair bonobos and chimpanzees (genus *Pan*), with gorillas forming the sister group to humans, bonobos, and chimpanzees combined. Orangutans are the sister group of the clade that comprises humans, bonobos, chimpanzees, and gorillas; gibbons form the sister group of the clade that comprises orangutans, bonobos, chimpanzees, gorillas, and humans (see figure 4.7). These apes and humans share a common ancestor that is not shared with any other living species.

Evolutionary paleontologists traditionally used the family-level adjective **hominid** to denote humans and fossil species closer to humans than to other apes; however, a cladist's

Phylogenies From DNA Sequences

Most phylogenetic reconstruction currently underway comes from collecting genomic DNA sequences for the species of interest and analyzing those data using statistical approaches, especially maximum-likelihood and Bayesian methods. Such analysis requires sophisticated computer algorithms and cannot be done by hand. Nonetheless, statistical inference of phylogenies is designed to separate patterns of homology from homoplasy in the data, and to use the former to trace common descent of the species being studied. To explain statistical phylogenetics, we first must examine the basic phylogenetic principles of character analysis that underlie statistical phylogenetics.

A simple example illustrates cladistic analysis of DNA sequence data to examine phylogenetic relationships among species. Note that this procedure of character analysis follows the general principles described in section 4.3. The study group in this example contains three species of chameleons, two from the island of Madagascar (*Brookesia theili* and *B. brygooi*) and one from Equatorial Guinea (*Chamaeleo feae*). The outgroup is a lizard of the genus *Uromastyx,* which is a distant relative of chameleons. Do the molecular data in this example confirm or reject the prior taxonomic hypothesis that the two Madagascan chameleons are more closely related to each other than either one is to the Equatorial Guinean species?

The molecular information in this example comes from a piece of the mitochondrial DNA sequence (57 bases) for each species. Each sequence encodes amino acids 221–239 of a protein called "NADH dehydrogenase subunit 2" in the species from which it was obtained. These DNA base sequences are aligned and numbered as

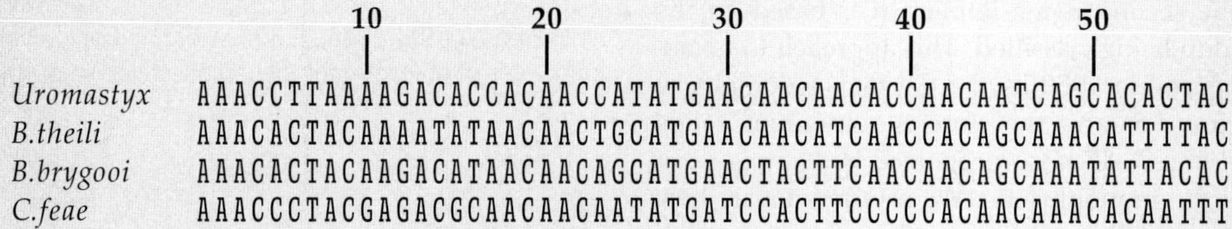

Each column in the aligned sequences constitutes a character that takes one of four states: A, C, G, or T (a fifth possible state, absence of the base, is not observed in this example). Only characters that vary among the three chameleon species potentially contain information on which pair of species is most closely related. Twenty-three of the 57 aligned bases show variation *among chameleons,* as shown here in bold letters:

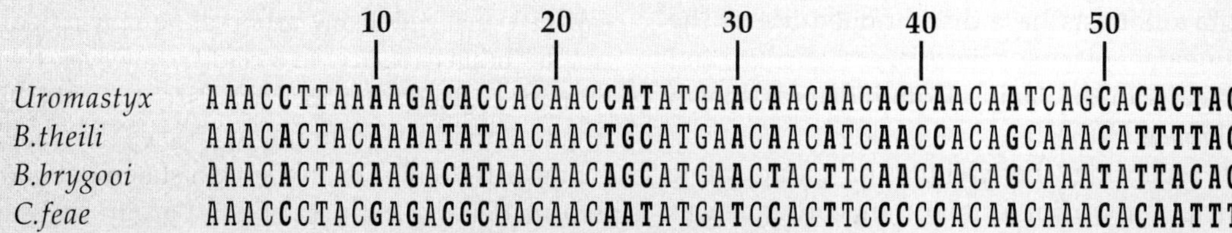

To be useful for constructing a cladogram, a character must demonstrate sharing of derived states (= synapomorphy). Which of these 23 characters demonstrate synapomorphies for chameleons? For each of the 23 variable characters, we must ask whether one of the states observed in chameleons is shared with the outgroup, *Uromastyx.* If so, this state is judged ancestral for chameleons and the alternative state(s) derived. Derived states are identified for 21 of the 23 characters just identified; derived states are shown here in blue:

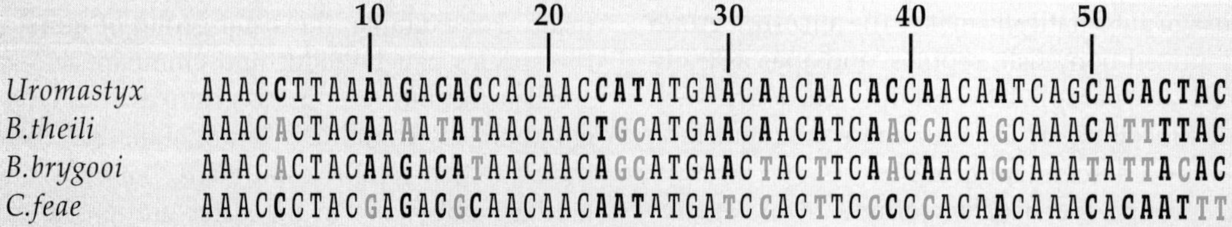

Note that polarity is ambiguous for two variable characters (at positions 23 and 54) whose alternative states in chameleons are not observed in the outgroup.

Of the characters showing derived states, 10 show synapomorphies among chameleons. These characters are marked here with numbers 1, 2, or 3 below the appropriate column.

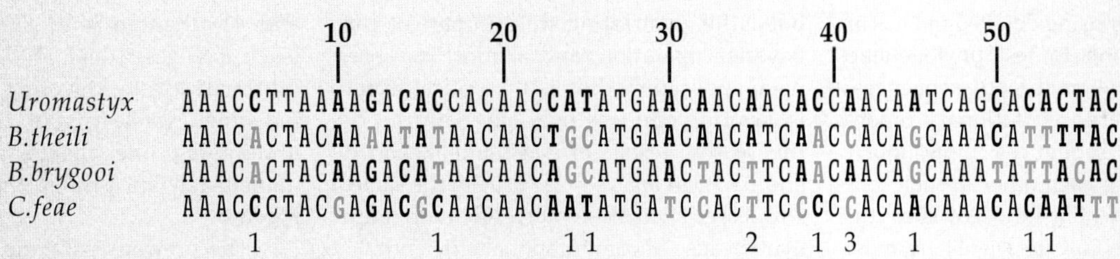

The eight characters marked "1" show synapomorphies grouping the two Madagascan species (*Brookesia theili* and *B. brygooi*) to the exclusion of the Equatorial Guinean species, *Chamaeleo feae.* We can represent these relationships as a cladogram:

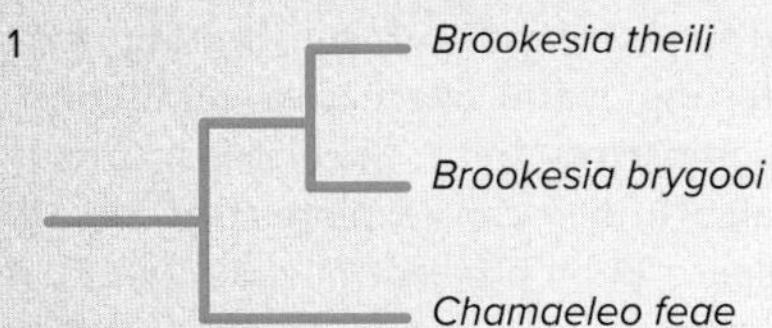

We can explain evolution of all characters favoring this cladogram by placing a single mutational change on the branch ancestral to the two *Brookesia* species. This is the simplest explanation for evolutionary changes of these characters.

Characters marked "2" and "3" disagree with our cladogram and favor alternative relationships:

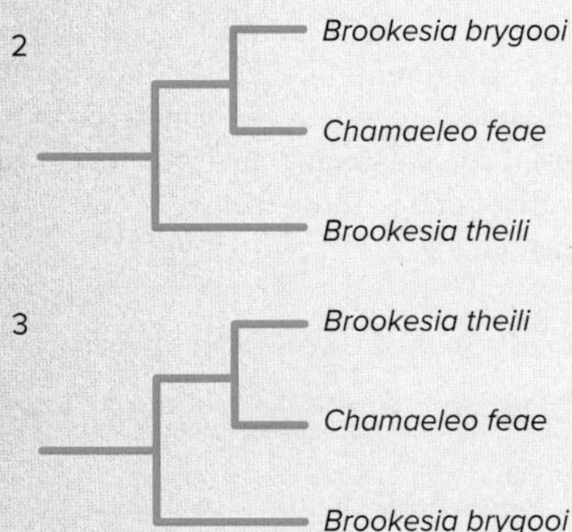

To explain evolutionary changes in characters favoring cladograms 2 or 3 using cladogram 1, we need at least two changes per character. Likewise, if we try to explain evolution of characters favoring cladogram 1 on cladogram 2 or 3, we need at least two changes for each of these characters. These two diagrams show the minimum numbers of changes required for character 5 (which favors cladogram 1) and character 41 (which favors cladogram 3) on cladogram 1; the ancestral state of each character appears at the root of the tree and the nucleotide states observed in each species at the tips of the branches:

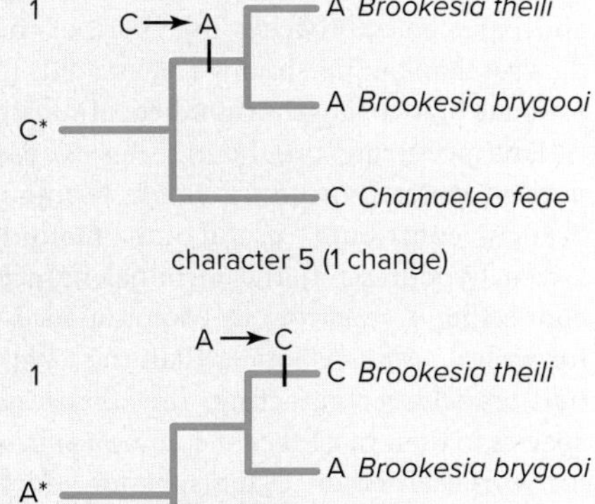

character 5 (1 change)

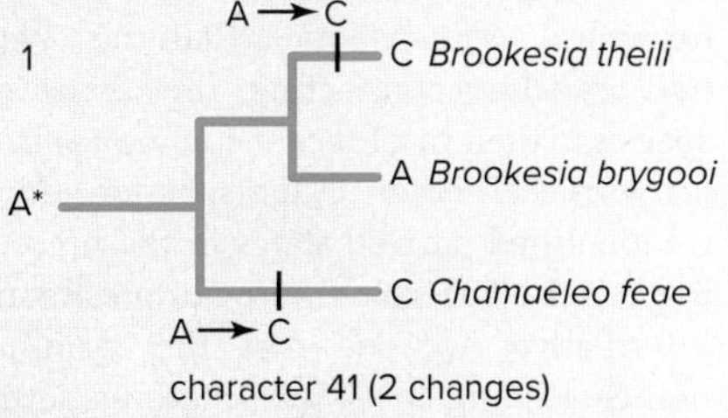

character 41 (2 changes)

A principle called **parsimony** can be used to resolve conflicts among taxonomic characters, as seen here. Parsimony represents the simplest limiting case of the maximum-likelihood and Bayesian methods for statistical inference of phylogeny, so we present it first and then show how the statistical methods build on this framework. We choose as our best working hypothesis the cladogram that requires the smallest total amount of character change. In our example, parsimony favors cladogram 1. For all 10 phylogenetically informative characters, cladogram 1 requires a total of 12 changes of character state (one for each of the 8 characters favoring it and two for each of the other 2 characters); cladograms 2 and 3 each require at least 19 character-state changes, 7 steps longer than cladogram 1. By choosing cladogram 1, we claim that characters favoring cladograms 2 and 3 show homoplasy in their evolution.

The molecular sequences shown in this example therefore confirm predictions of the prior hypothesis, based on the appearance and geography of these chameleons, that the *Brookesia* species shared a common ancestor with each other more recently than either one did with *Chamaeleo feae.*

As a further exercise, you should convince yourself that the 12 characters that vary among chameleons but do not demonstrate unambiguous sharing of derived states are equally compatible with each of the three possible cladograms shown. For each character, find the minimum total number of changes that must occur to explain its evolution on each cladogram. You will see, if you do this exercise correctly, that the three cladograms do not differ in minimum numbers of changes required for each of these characters. For this reason, the characters are phylogenetically uninformative by the parsimony criterion.

The parsimony method just illustrated makes some assumptions: base substitutions are equally likely to occur at any of the 57 sites, all 3 possible substitutions are equally likely at any site (for example, at site 5 one assumes that C would be equally likely to change to A, G, or T given that a substitution occurs), and the expected amount of molecular evolution on any branch is proportional to its temporal duration. Suppose that site 16 is unusually subject to mutational change, that a change from C to T is ten times more likely than one from C to A or G, and that the internal branch on the tree (a branch is called "internal" if it groups two or more of the species being studied to the exclusion of other species) is very short compared to the tip branches (those that terminate in a living species whose DNA sequence was determined). The likelihood of getting the observed data for that site by parallel changes C to T in the tip branches of the two *Brookesia* species might equal or exceed the likelihood of a single change C to T in the internal branch of hypothesis 1. The data at site 16 then would not necessarily favor hypothesis 1 over the alternatives. If we can obtain detailed knowledge of the evolutionary properties of the sites in this DNA sequence (for example, by studying its evolution in a large number of lizards), we might favor a phylogenetic method that incorporates a detailed model of DNA-sequence evolution.

Maximum-likelihood and Bayesian approaches to phylogenetic inference are

---

* Ancestral nucleotide state inferred by outgroup comparison.

(continued)

efficient means for using detailed models of molecular evolution to test phylogenetic hypotheses from aligned DNA sequences. First, we estimate an evolutionary model for the aligned sequences: how much do sites differ from each other in their tendencies to vary, which kinds of substitutions (C to A, C to G, C to T, etc.) are most likely to occur. Second, each site is evaluated with respect to each of the alternative possible trees to determine which tree has the highest likelihood of producing the observed data (such as the CTTC pattern shown for site 16). In this second step, branches of the tree can vary in length, relaxing the assumption that expected amounts of substitution are proportional to the temporal duration of the branch. For our data, the maximum-likelihood method would evaluate the probability of observing the results in each of the 57 columns considering all possible trees, and measure the likelihood of the entire data set for each contrasting tree. The tree with the highest likelihood of producing the observed data is the favored phylogenetic hypothesis. We can reject contrasting trees whose likelihoods are judged much lower than would be expected by chance alone. The Bayesian approaches operate in a similar manner, but they permit an investigator to evaluate the contributions of a new data set relative to prior results. For example, we could evaluate the probability that the data favor hypothesis 1 after incorporating phylogenetic results for these same species obtained from another data set. Calculating likelihoods for contrasting trees is part of the Bayesian operation, and in most phylogenetic applications this is the critical factor in favoring one tree over alternatives. For this reason, results of maximum-likelihood and Bayesian analyses are usually the same, with Bayesian analyses currently more popular because of computational efficiency.

Note that the maximum-likelihood and Bayesian approaches use more of the data set than our parsimony analysis did. To estimate branch lengths (numbers of substitutions occurring on them), sites whose derived states arose on a single terminal branch contribute useful information. Given hypothesis 1, the terminal branch connecting *C. feae* from its common ancestor with *Brookesia* is longer than the terminal branches connecting the *Brookesia* species to their most recent common ancestor with each other. Using sites for which we identified derived states in the preceding exercise but which were not parsimony informative, we see that the terminal branch leading to *C. feae* requires seven substitutions (sites 10, 15, 30, 32, 38, 56, 57), whereas the terminal branch leading to *B. brygooi* requires only three changes (sites 32, 50, 55), and the terminal branch leading to *B. theili* requires only two changes (sites 12, 14). Using the likelihood method, these sites collectively would favor hypothesis 1 even though none of the sites is parsimony informative. We thus expect parallel substitutions to occur more frequently in the *C. feae* lineage and one of the two *Brookesia* lineages (as probably occurred at sites 35 and 41) than in both *Brookesia* lineages (such changes would produce parsimony-informative sites favoring hypothesis 1, indistinguishable in our data from sites undergoing one change in the lineage immediately ancestral to the two *Brookesia* species).

The previous paragraph should make clear why inferring phylogenies using maximum-likelihood and Bayesian approaches would be very hard to do by hand, even for the data in our example. Finding the optimal tree topologies, branch lengths, relative probabilities of substitution at different sites, and relative probabilities of different kinds of substitution requires testing many alternative conditions and comparing their likelihoods. Computer algorithms can explore this parameter space in an efficient manner, but it would be prohibitively tedious to do by hand. Nonetheless, sharing of derived states at a site as predicted by hypotheses of homology remains the primary reason why the tree that requires only a single evolutionary change to explain the observed variation has a higher likelihood than do trees that require two or more parallel substitutions. Statistical inference of phylogeny using maximum-likelihood and Bayesian methods thus remains grounded in the basic cladistic principles. When parameters of the maximum-likelihood and Bayesian analyses are set to the simple conditions used for parsimony analysis, their results correspond to those that we obtained using parsimony.

Data from Townsend, T. and A. Larson. 2002. Molecular phylogenetics and mitochondrial genomic evolution in the Chamaeleonidae (Reptilia, Squamata). Molecular Phylogenetics and Evolution 23:22–36.

use of "hominid" refers to any member of the expanded family Hominidae, which includes humans, bonobos, chimpanzees, gorillas, orangutans, and fossil forms that are phylogenetically closer to these species than to other living primates.

Current State of Animal Taxonomy

The formal taxonomy of animals that we use today was established using principles of evolutionary systematics and has been revised recently in part using principles of cladistics. Introduction of cladistic principles initially replaces paraphyletic groups with monophyletic subgroups while leaving the remaining taxonomy mostly unchanged. A thorough revision of taxonomy along cladistic principles, however, will require profound changes. A new taxonomic system called PhyloCode is being developed as an alternative to Linnean taxonomy; this system replaces Linnean ranks with codes that denote the nested hierarchy of monophyletic groups conveyed by a cladogram. In our coverage of animal taxonomy, we emphasize taxa that are monophyletic and therefore consistent with criteria of both evolutionary and cladistic taxonomy. We use Linnean ranks where this system still prevails in the current taxonomy of a group of animals. Where unranked taxa have gained common usage, we present a rank-free taxonomy. Some new groups, especially ones comprising single-celled organisms, encompass multiple phyla and are therefore designated as named clades without a Linnean rank. For familiar taxa that are clearly paraphyletic grades, we note this fact and suggest alternative taxonomic schemes that contain only monophyletic taxa.

When discussing patterns of descent, we avoid statements such as "mammals evolved from reptiles" that imply paraphyly and instead specify appropriate sister-group relationships to convey the order of branching in a phylogeny. We avoid calling groups of organisms primitive, advanced, specialized, or generalized because all groups of animals contain combinations of primitive, advanced, specialized, and generalized features; these terms are best restricted to describing specific characteristics and not an entire group.

Revision of taxonomy according to cladistic principles can cause confusion. In addition to new taxonomic names, we see old ones used in unfamiliar ways. For example, cladistic use of "bony fishes" includes amphibians and amniotes (including reptilian groups, birds, and mammals) in addition to the finned, aquatic animals that evolutionary taxonomists normally group under the term "fish." Cladistic use of "reptiles" includes birds in addition to snakes, lizards, turtles, and crocodilians; however, it excludes some fossil forms, such as extinct therapsid groups, that evolutionary taxonomists traditionally placed in Reptilia (see figure 18.1). When using these seemingly familiar terms, taxonomists must be very careful to specify whether they are referencing traditional evolutionary taxa or newer cladistic taxa.

■ 4.5 Major Divisions of Life

The oldest phylogenetic events in the history of life have been obscure because the different forms of life share very few characters that can be compared among them to reconstruct phylogeny. More recently, however, a cladistic classification of all life-forms has been proposed based on phylogenetic information obtained from molecular data (the nucleotide base sequence of DNA encoding ribosomal RNA). According to this tree (see figure 4.9), Woese, Kandler, and Wheelis (1990) recognized three monophyletic **domains** above the kingdom level: Eucarya (all eukaryotes), Bacteria (the true bacteria), and Archaea (prokaryotes differing from bacteria in membrane structure and ribosomal RNA sequences, also called Archaebacteria). Multicellular animals, plants, and fungi are each traditionally given the Linnean rank of kingdom. Because each of these taxa is a monophyletic group, continued recognition of these three kingdoms is compatible with phylogenetic systematics. Prior to the 1990s, it was common also to group unicellular eukaryotes as kingdom Protista and all prokaryotes as kingdom Monera in a five-kingdom Linnean classification. Monera and Protista are paraphyletic groups, and their recognition as taxa is now discontinued. Unicellular eukaryotes were sometimes placed within the animal kingdom as phylum Protozoa, a hypothesis clearly rejected by all current phylogenetic analyses of eukaryotes (Chapter 5). Calling unicellular eukaryotes "protozoans" or "protistans" is archaic and should be resisted. Likewise, "metazoan" is archaic because this term formerly denoted all nonprotozoan members of the animal kingdom. Under current taxonomy, "metazoan" is a redundant synonym for "animal."

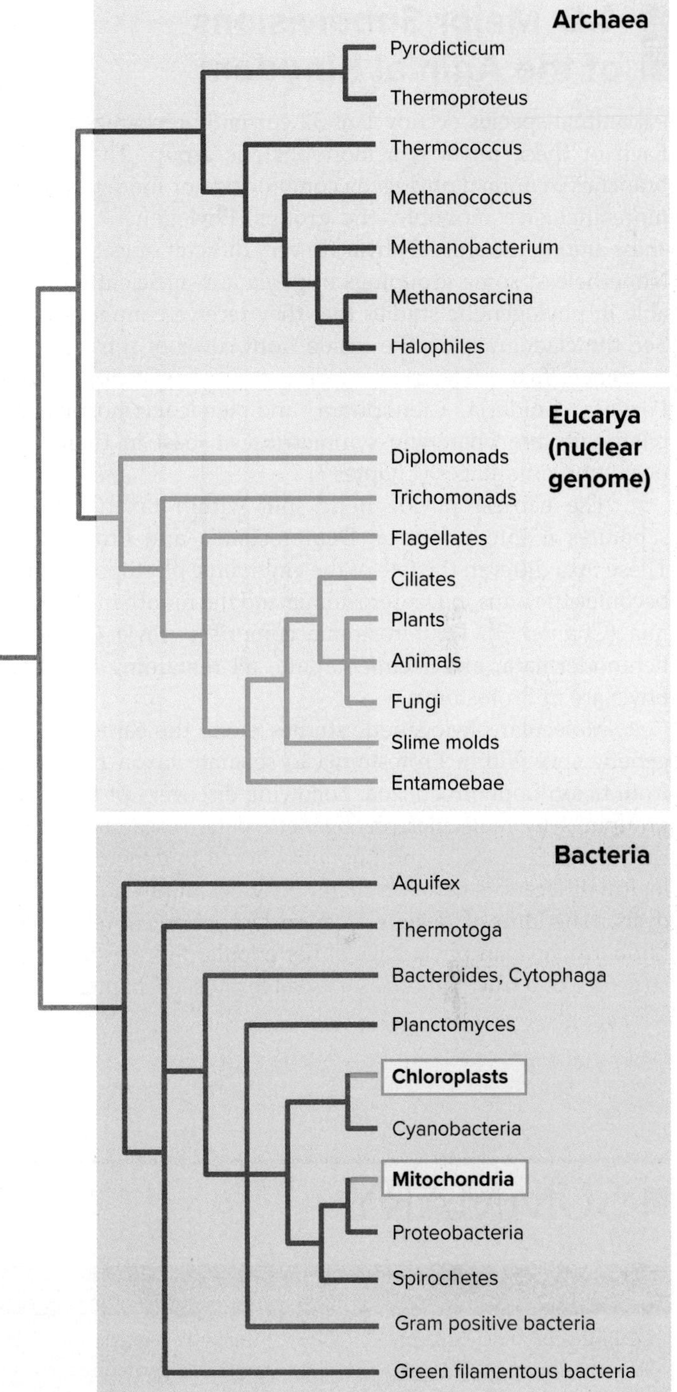

figure 4.9

Phylogenetic overview of the three domains of life—Archaea, Bacteria, and Eucarya—based on analysis of genes encoding ribosomal RNA. Because of their endosymbiotic origins, organellar genomes of domain Eucarya (mitochondria, chloroplasts) are phylogenetically within the Bacteria rather than the clade that includes all eucaryotic nuclear genomes. Organisms of domain Eucarya therefore include cellular components of disparate evolutionary origins. See figure 5.2 for a more detailed current summary of phylogenetic relationships among eukaryotes. Although animals, plants, and fungi form a monophyletic group within the limited taxon sampling shown here, figure 5.2 reveals that these taxa originated separately from different unicellular lineages.

4.6 Major Subdivisions of the Animal Kingdom

All animal species occupy 1 of 32 currently recognized phyla. Each of these phyla is a monophyletic group. The deepest branches in animal phylogeny combine two or more phyla into more inclusive monophyletic groups. Phylogenetic relationships among the animal phyla are very difficult ones to resolve. Nonetheless, some groupings of phyla are sufficiently repeatable in phylogenetic studies that they receive common usage. See the cladogram on the inside front cover of this book for reference. Taxon Bilateria comprises all phyla except Placozoa, Porifera, Cnidaria, Ctenophora, and Xenacoelomorpha. All bilaterians are bilaterally symmetrical at least in their early postembryonic stages (Chapter 3).

The earliest phylogenetic split within taxon Bilateria separates it into two taxa, Deuterostomia and Protostomia. These taxa differ in the fate of the embryonic blastopore, which becomes the anus in Deuterostomia and the mouth in Protostomia (Chapter 3). Deuterostomia comprises phyla Chordata, Echinodermata, and Hemichordata; all remaining bilaterian phyla are in Protostomia.

Molecular phylogenetic studies reveal the earliest phylogenetic split within Protostomia to separate taxon Ecdysozoa from taxon Lophotrochozoa. Following discovery of these two groupings by molecular phylogenetic data, researchers chose the names Ecdysozoa and Lophotrochozoa to denote morphological differences between them. Ecdysozoan phyla feature ecdysis, a shedding of the outer body surface, during development. Lophotrochozoan phyla have either a lophophore as a feeding structure or a trochophore as a larval form (see Chapter 9).

Taxonomic ordering of Bilateria can be shown using an indented system, with the leftmost taxon including all taxa listed below it farther to the right, up to any subsequently listed taxon equal in indentation to the first one. Bilateria is the leftmost taxon in this listing and includes Protostomia and Deuterostomia as mutually exclusive taxa at the next rightward level of indentation. Within Protostomia, taxon Ecdysozoa and taxon Lophotrochozoa appear equivalent to each other at the most rightward nesting.

Bilateria
- Protostomia:
  - Lophotrochozoa: phyla Acanthocephala, Annelida, Brachiopoda, Chaetognatha, Echiurida, Ectoprocta, Entoprocta, Gastrotricha, Gnathostomulida, Mesozoa, Micrognathozoa, Mollusca, Nemertea, Phoronida, Platyhelminthes, Rotifera
  - Ecdysozoa: phyla Arthropoda, Kinorhyncha, Loricifera, Nematoda, Nematomorpha, Onychophora, Priapulida, Tardigrada
- Deuterostomia: phyla Chordata, Echinodermata, Hemichordata

In summary, there are five unranked taxa above the level of phylum whose usage is widespread and should be familiar to students of zoology: Bilateria, Deuterostomia, Protostomia, Ecdysozoa, and Lophotrochozoa. Although further study is needed to test these new groupings, we use them to organize our survey of animal diversity. We present further details of animal taxonomy in Chapters 5–20.

SUMMARY

Sections	Key Concepts
4.1 Linnaeus and Taxonomy	• Animal systematics has three major goals: (1) to identify all species of animals, (2) to evaluate evolutionary relationships among animal species, and (3) to produce a taxonomy that groups animal species in a hierarchy of taxonomic groups (taxa). • Taxa are traditionally ranked to denote increasing inclusiveness as follows: species, genus, family, order, "class," phylum, and kingdom. All of these ranks can be subdivided to signify taxa intermediate between them. • In a classification, taxa are defined by possession of essential characters. In systematization, taxa are defined based on common ancestry, with characters serving to diagnose common ancestry rather than to define the taxon. • Names of species are binomial, with the first name designating the genus to which the species belongs (capitalized) followed by a species epithet (lowercase), both italicized. Taxa at all higher ranks have single capitalized but non-italicized names.

Sections	Key Concepts
4.2 Species	• A biological species is defined as a reproductive community of populations (reproductively isolated from others) that occupies a specific niche in nature. • Because the biological species concept may be difficult to apply in spatial and temporal dimensions, and because it excludes asexually reproducing forms, alternative concepts are proposed. These alternatives include the evolutionary species concept, the cohesion species concept, and the phylogenetic species concept. • An evolutionary species is a single lineage of ancestral-descendant populations that maintains its identity from other such lineages and has its own evolutionary tendencies and historical fate. • The cohesion concept of species is a revision of the evolutionary species concept that emphasizes population-genetic mechanisms underlying the evolutionary cohesion of a lineage: the most inclusive population of individuals having the potential for phenotypic cohesion through intrinsic cohesion mechanisms. • By emphasizing the most inclusive units that manifest or have the potential for evolutionary cohesion, the evolutionary and cohesion concepts sometimes group into one species population lineages that have diagnosably distinct evolutionary histories. • The phylogenetic species concept treats each of these diagnosably distinct population lineages as separate species. • A common feature of these species concepts is that a species is a segment of a population-level lineage, which is the general lineage concept of species.
4.3 Taxonomic Characters and Reconstruction of Phylogeny	• Characters are homologous if they descend, with or without modification, from an equivalent character of a common ancestor. • Phylogeny is the structure of common descent of species, with each internal lineage of the tree of life being ancestral to a unique set of descendants. • A clade comprises the most recent common ancestor of a group of species and all descendants of that ancestor. • A cladogram shows the nested hierarchy of clades within clades diagnosed by shared derived characters called synapomorphies. • We hypothesize that synapomorphies represent homologies that arose in the clade's most recent common ancestor. • Reconstruction of phylogeny consists in finding morphological, chromosomal, and molecular characters that mark each branch on the tree of life. • Using a simple example, we illustrate how a systematist infers a phylogenetic tree from aligned DNA sequences using the principles of parsimony, maximum-likelihood, and Bayesian statistics. • The fossil record provides estimates of the ages of evolutionary lineages.
4.4 Theories of Taxonomy	• Evolutionary taxonomy groups species into higher taxa according to the joint criteria of common descent and adaptive evolution; such taxa have a single evolutionary origin and occupy a distinctive adaptive zone. • A second approach, called phylogenetic systematics or cladistics, emphasizes common descent exclusively in grouping species into higher taxa. Only monophyletic taxa (= clades, those having a single evolutionary origin and containing all descendants of the group's most recent common ancestor) are used in cladistics. • In addition to monophyletic taxa, evolutionary taxonomy recognizes some taxa that are paraphyletic (having a single evolutionary origin, but excluding some descendants of the most recent common ancestor of the group because they have evolved a distinct adaptive zone). • Both schools of taxonomy exclude polyphyletic taxa (those having more than one evolutionary origin).
4.5 Major Divisions of Life	• The three most inclusive taxa of living organisms consistent with cladistic taxonomy are domains Archaea, Bacteria, and Eukarya, which includes animals. • The phylogeny of life's domains was constructed using sequences of genes encoding ribosomal RNA. • Although placement of the root of the tree of life is difficult, Archaea and Eukarya are often grouped as a clade to the exclusion of Bacteria.
4.6 Major Subdivisions of the Animal Kingdom	• The most inclusive formal taxon within animals is the phylum, but zoologists commonly use some more inclusive taxa above the phylum level. • Taxon Bilateria, diagnosed by primary bilateral symmetry, comprises clades Deuterostomia and Protostomia, distinguished by the fate of the embryonic blastopore to become the anus versus the mouth, respectively. • Protostomia comprises clades Ecdysozoa and Lophotrochozoa, diagnosed respectively by the presence of molting versus a trochophore larva or lophophore.

■ Review Questions

1. List in order, from most inclusive to least inclusive, the principal categories (ranks of taxa) in Carolus Linnaeus's system of classification.
2. Explain why the system for naming species that originated with Linnaeus is binomial.
3. How does the biological species concept differ from earlier typological concepts of species? Why do evolutionary biologists prefer it to typological species concepts?
4. What problems have been identified with the biological species concept? How do other concepts of species attempt to overcome these problems? What problems do these other concepts entail?
5. What are the properties of phylogenetically useful (or informative) characters? How are such characters used to construct a cladogram?
6. What is the difference between a cladogram and a phylogenetic tree? Given a cladogram for a group of species, what additional information is needed to obtain a phylogenetic tree?
7. How do monophyletic, paraphyletic, and polyphyletic taxa differ? How do these differences affect the validity of such taxa for both evolutionary and cladistic taxonomies?
8. How do cladists and evolutionary taxonomists differ in their classifications of humans and apes? Contrast their respective interpretations of the statement that humans evolved from apes, which evolved from monkeys.
9. What are the three major domains of life consistent with the principles of cladistic taxonomy?
10. For the human, gorilla, and leopard-frog taxa in table 4.1, construct a rank-free ordering using indentations to denote nesting of groups within groups, as shown in section 4.6.
11. What are the five major unranked taxa above the phylum level in animal taxonomy?

For Further Thought If a taxonomist constructs a rooted phylogentic tree for a group of living species, the structure of the tree alone can be used to distinguish hypotheses of monophyly versus nonmonophyly of a particular subgroup. If monophyly is rejected for a particular subgroup, tree topology alone cannot distinguish paraphyly from polyphyly. What additional information is needed to distinguish paraphyly from polyphyly?

5

Unicellular Eukaryotes

LEARNING OBJECTIVES

Readers will be able to:

5.1 Describe the components of a eukaryotic cell and explain their functions; make special reference to structures used in cell movement.

5.2 Compare and contrast feeding mechanisms used in autotrophy, heterotrophy, and mixotrophy.

5.3 Explain why osmoregulation is more important for freshwater organisms than for marine organisms.

5.4 Explain the roles of gamete formation and sexual and asexual reproduction, including fission and cyst formation, in a complex life cycle such as that of *Plasmodium* or another parasite.

5.5 Exemplify the diversity of habits and lifestyles used by unicellular eukaryotes and illustrate the impact of these organisms on humans as endoparasites and as key members of food chains.

5.6 Explain how using molecular characters has changed our understanding of taxonomic relationships among unicellular eukaryotes.

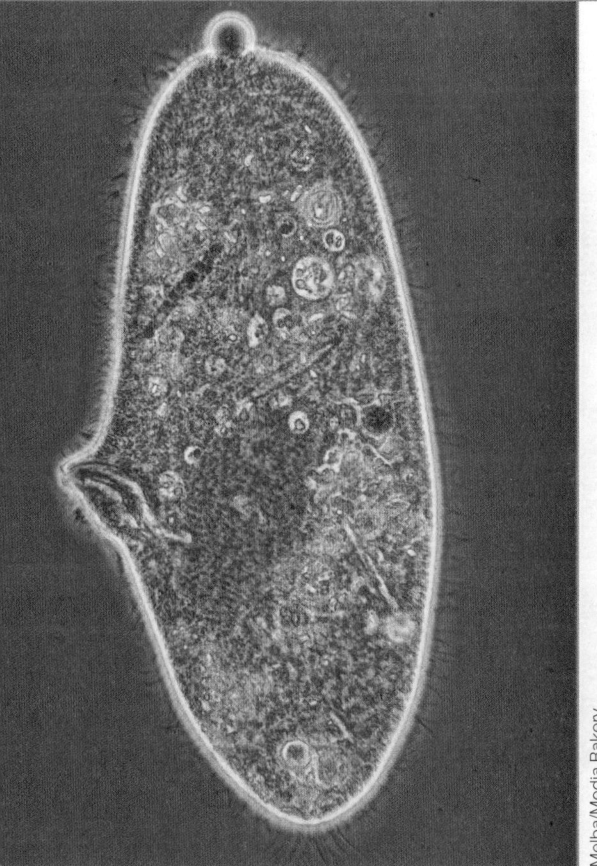
Melba/Media Bakery

A paramecium.

Emergence of Eukaryotes and a New Life Pattern

The first reasonable evidence for life on earth dates from approximately 3.5 billion years ago. The first cells were, like living Archaea (archaeons) and Bacteria, "prokaryotic" in being relatively small and having no nucleus or mitochondria. They evolved on an earth with little atmospheric oxygen. From the first cells, the Bacteria and Archaea diversified over an enormous time span prior to the origin of a eukaryotic cell. Although several bacterial lineages evolved photosynthetic abilities, only cyanobacteria release oxygen gas as a by-product of photosynthesis. As cyanobacteria increased in abundance, they enriched the atmosphere with oxygen. A separate bacterial cell lineage, alpha-proteobacteria, evolved an ability to use the atmospheric oxygen released by cyanobacteria for a highly efficient oxygen-based respiration. The symbiotic merger of this alpha-proteobacterium with an archaeon-like cell produced the first eukaryote; the nucleus came from the archaeon-like cell. The alpha-proteobacterial symbiont became the first eukaryotic organelle, the mitochondrion, and the site of aerobic respiration in eukaryotic cells. There are at least two hypotheses for how and why these cells merged, but their product, the "eukaryotic" cell, characterizes many unicellular organisms and all animals, plants, and fungi. DNA is contained in both the nucleus and in mitochondria.

Plant cells and some unicellular eukaryotes also contain DNA in their photosynthetic plastids. Plastids, called chloroplasts in plants, are descendants of cyanobacteria, the same group whose production of atmospheric oxygen was critical for the earlier origin of eukaryotic cells. A cyanobacterium merged with a eukaryotic cell, eventually being reduced to the organelle we call a chloroplast.

When a prokaryote is engulfed and modified to become a eukaryotic organelle, the organelle is said to develop by primary endosymbiosis. Chloroplasts in clade Plantae (red algae, green algae, and land plants, among others) arose in this way. However, a eukaryotic cell may also obtain plastids from another eukaryote. This is secondary endosymbiosis. Two similar cells with plastids may have formed very differently, so it can be hard to untangle the evolutionary relationships among the diverse array of unicellular eukaryotes alive today.

The assemblage of eukaryotic unicellular organisms was formerly called protista or protozoa. The inclusion of "zoa" in the name originally referred to two animal-like features: the absence of a cell wall and the presence of at least one motile stage in the life cycle. However, these are not homologous features and we now recognize unicellular eukaryotes as belonging to many different taxonomic groups whose ages exceed those of animals, fungi, and plants. As you will see here, many unicellular eukaryotes impact humans; they feed on corneas or brain tissue, cause diseases such as malaria or dysentery, and infect livestock. Our ability to treat these disease agents depends on our ability to classify them correctly because susceptibility to particular drugs is often lineage-dependent. This is an exciting time to study these tiny organisms.

In a unicellular eukaryote, all life activities proceed within the limits of a single plasma membrane. Over 64,000 species have been named, and more than half of these are known only from fossils. Some researchers estimate that there may be 250,000 species of unicellular eukaryotes; many taxa are identified from environmental DNA samples. They are highly adaptable and easily distributed from place to place. They require moisture, whether they live in marine or freshwater habitats, soil, decaying organic matter, or plants and animals. They may be sessile or free-swimming, and they form a large part of the floating plankton. Many make external skeletons, called shells or tests, of calcium, silica, sand grains, or other materials. These skeletons form gigantic oceanic and soil deposits (see figures 5.24 and 5.25). Some species may have spanned geological eras of more than 100 million years.

Unicellular eukaryotes play an enormous role in the economy of nature. About 10,000 species are symbiotic in or on animals or plants, or other unicells. The symbiotic relationship may be **mutualistic** (both partners benefit), **commensalistic** (one partner benefits without affecting the other), or **parasitic** (one partner benefits at the expense of the other) (see Chapter 2). Some of the most important diseases of humans and domestic animals are caused by parasitic forms.

Unicellular eukaryotes are often described by their body type: ciliates have numerous cilia covering the plasma membrane; flagellates have one or more flagella used to propel the cell; amebas have irregular shapes caused by flowing cytoplasm inside the plasma membrane (figure 5.1). Amebas may be testate (shelled) or naked. When present, tests may be simple coverings with only one opening or elaborate and beautiful structures pierced with many holes through which the cell protrudes.

Unicellular eukaryotes form a paraphyletic group because their most recent common ancestor is also an

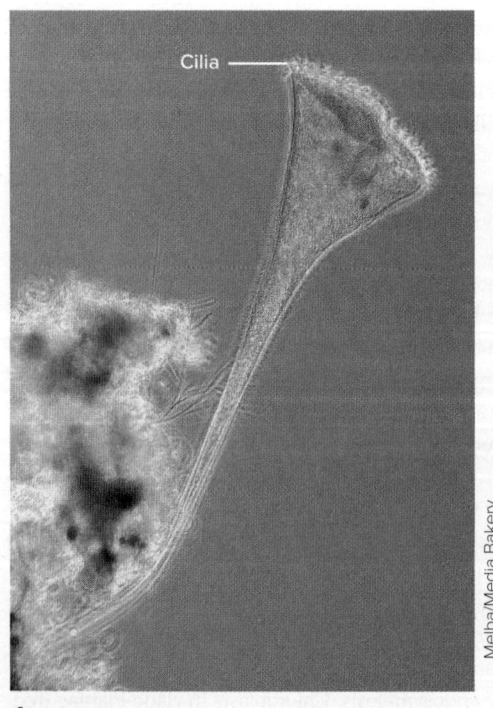

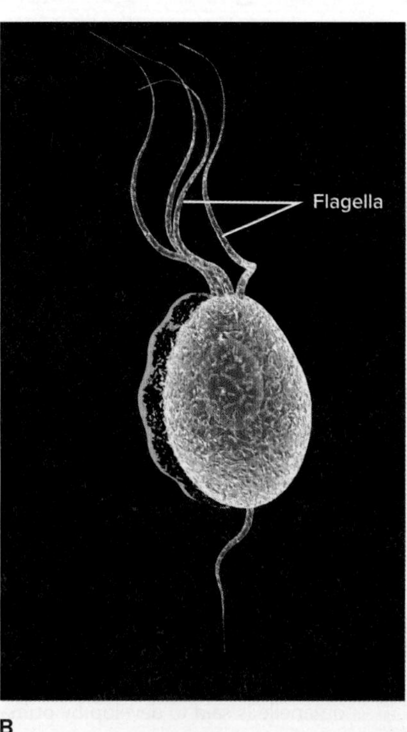

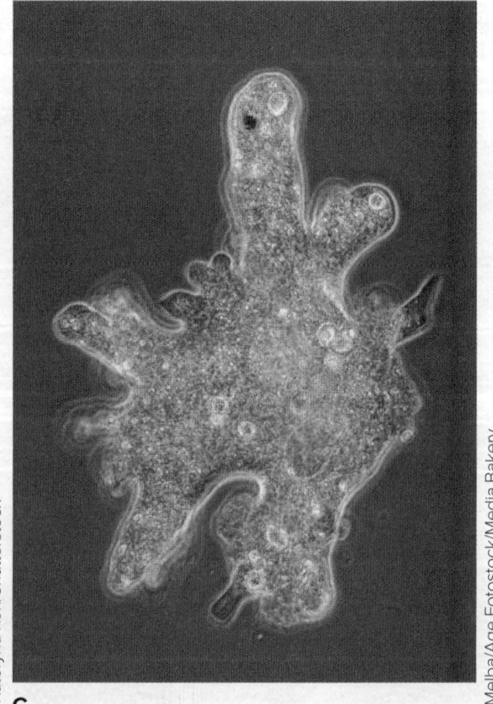

figure 5.1

Three body forms of the eukaryotic cell. **A,** a ciliate; **B,** a flagellate (*Giardia lamblia*); **C,** an ameba.

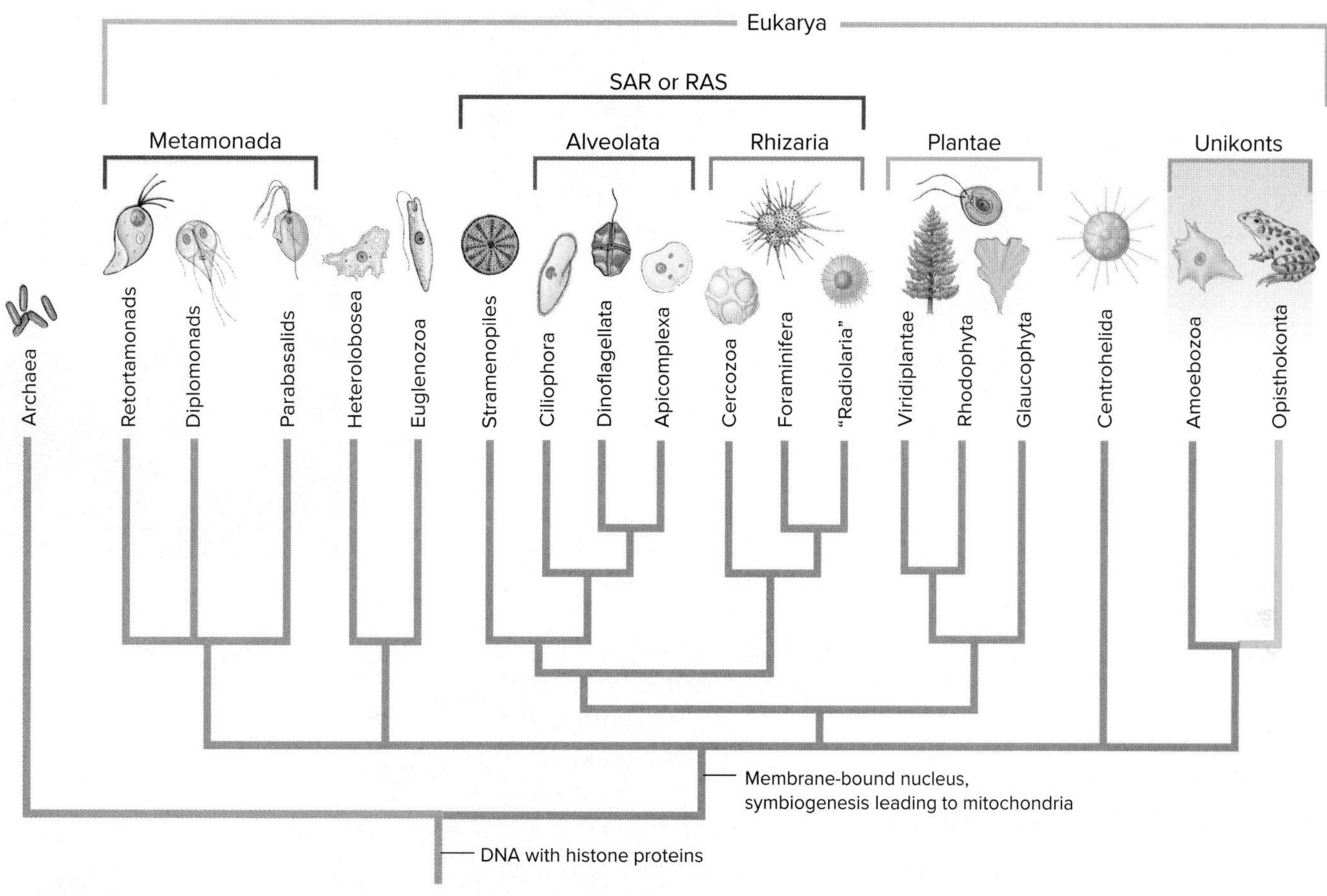

figure 5.2

Cladogram showing the major eukaryotic clades; in most cases the order of branching remains to be determined. Opisthokonta is a very large clade comprising choanoflagellates, fungi, and multicellular animals. The terms "SAR" and "RAS" are interchangeable; SAR represents the first letter of the names Stramenopiles, Alveolates, and Rhizaria, whereas RAS represents the same taxa listed in the reverse order.

ancestor of animals, plants, and fungi. Analyses of sequences of bases in genes, primarily those encoding the small subunit of ribosomal RNA, but also those encoding several proteins, have revolutionized our concepts of phylogenetic affinities. The origin of the first eukaryote, as described in the opening essay, was followed by diversification into the many unicellular and multicellular taxa observed today (figure 5.2). In this chapter we briefly describe 18 major eukaryotic clades (see figure 5.2) from the more than 30 such clades currently recognized. Only the ciliated body type diagnoses a clade. The ameboid and flagellated forms appear to have evolved independently several times, so these forms occur in many different taxonomic groups. Some traditional group names, such as Foraminifera, are still in use, but other names have been newly created. For example, the taxon Opisthokonta was formed when choanoflagellates, fungi, and animals were found to form a clade. A description of the major clades that contain unicellular eukaryotes is in section 5.5, following a description of the features of a eukaryotic cell.

5.1 Form and Function

Unicellular eukaryotes live within a single plasma membrane. They use organelles and sub-cellular processes to solve life's problems, such as supporting the body, moving, gathering food, and accessing genetic information. For example, unicellular eukaryotes have a cytoskeleton, cilia, flagella, or pseudopodia to move and to capture food, as well as an endomembrane system and a nucleus containing DNA. Some unicellular eukaryotic lineages have unique features specialized for lifestyles as diverse as predation in the ocean plankton or intracellular parasitism. Taxonomists group unicellular eukaryotes into more than 30 major taxa whose contents are mutually exclusive; we cover the 18 most-studied ones. It is not practical to describe all specializations in the 18 taxa discussed in this chapter; however, we describe the major organelles within a generalized eukaryotic cell and note aspects of these organelles that occur in unicellular groups of particular interest, frequently those with relevance to human health.

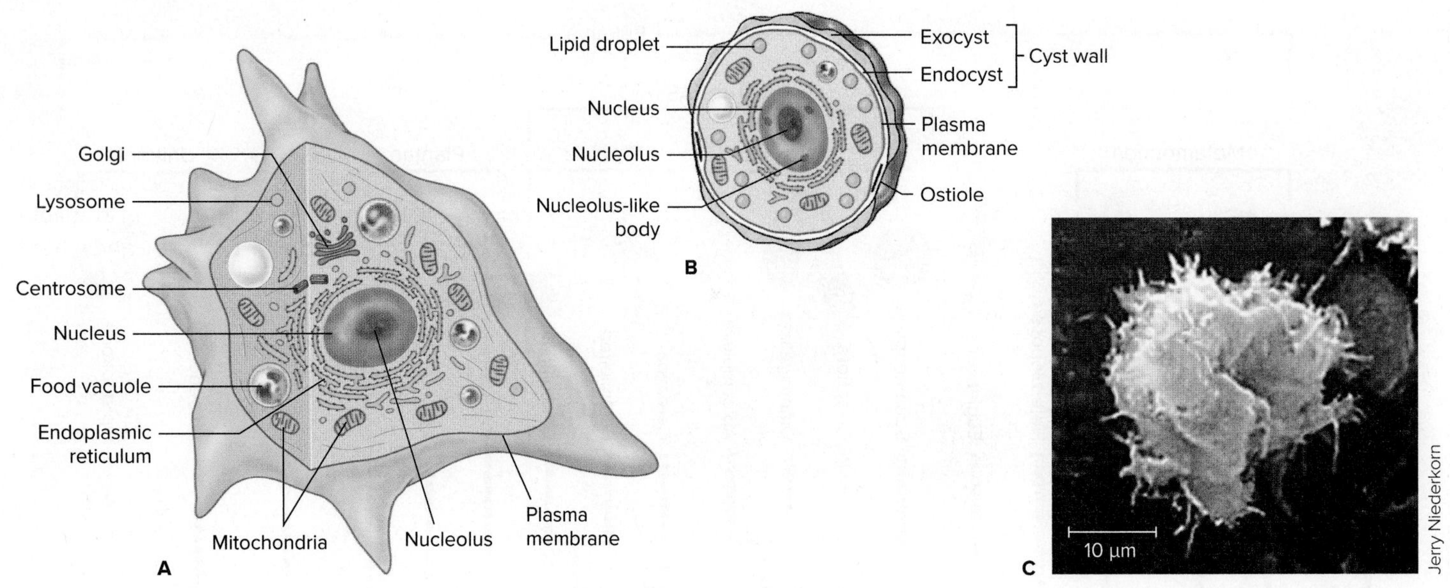

figure 5.3

Structure of *Acanthamoeba palestinensis*, an ameboid unicellular eukaryote. **A,** Active, feeding form. **B,** Cyst. **C,** *Acanthamoeba castellanii* kills cells of the human cornea; it is spread by contact lenses that have not been correctly disinfected.

Membrane and Cytoskeleton

A plasma membrane separates a unicellular eukaryote from its surrounding medium. The membrane is a selectively permeable phospholipid bilayer able to control the traffic of a wide range of chemical substances across it. Immediately beneath the plasma membrane, and bound to it, is an actin cortex with associated actin-binding proteins. The cell cortex influences shape and other properties of the plasma membrane. Moving inward from the cortex, cell cytoplasm consists of a fluid portion, called cytosol, and the numerous organelles and membranes that exist outside the nuclear envelope (figure 5.3A). The nucleus, bounded by a porous membrane, is not part of the cytoplasm.

It would be a serious error to imagine the cell as a simple fluid-filled sac with a few inclusions inside it. In fact, the cytoplasm is highly structured. It is densely packed with molecules needed for chemical reactions and crossed by cytoskeletal fibers that form and dissociate as needed. There are three types of cytoskeletal fibers: microtubules, intermediate filaments, and microfilaments. Microtubules are composed of tubulin molecules attached together to make 13 linear strands arranged in the form of a cylinder (figure 5.4). Microtubules may associate with motor proteins and function in transport within the cell, moving chromosomes during cell division, moving messenger RNA to particular areas of the cell, and transporting vesicles between the plasma membrane and organelles such as the endoplasmic reticulum or the Golgi complex. Microtubules also give internal structure to cilia and flagella (see figure 5.9). In certain unicellular eukaryotes, they stiffen other cell extensions such as axopodia and axostyles, as discussed below. They are also responsible for stiffening the plasma membrane in euglenids and alveolates.

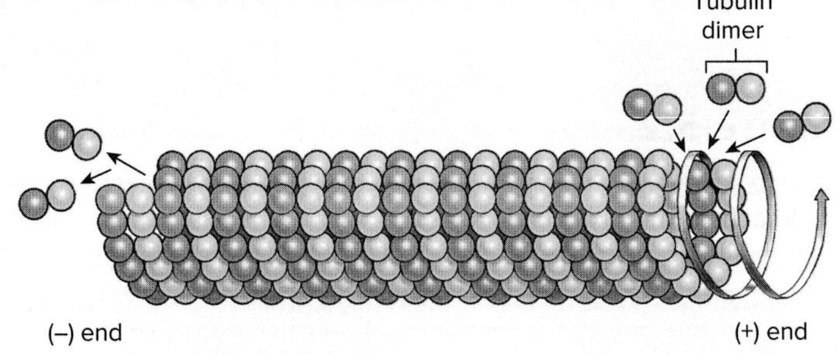

figure 5.4

A microtubule is composed of 13 strands of tubulin molecules. Each molecule is a dimer. Tubulin dimers are added to and removed from the plus end of the microtubule more rapidly than at the minus end.

Intermediate filaments are smaller than microtubules, but larger than microfilaments. There are six biochemical variants of intermediate filaments. Functions are largely structural—supporting the plasma membrane when it contacts other cells or the ECM and limiting cell stretching. In animals and other multicellular eukaryotes, they have a key role in connecting cells together, but this function is not important in unicellular taxa. Although there is some controversy, the general assumption is that intermediate filaments do not occur in unicellular eukaryotes. A recently discovered family of proteins called alveolins in the super-group Alveolata (see section 5.5: Alveolata) has been described structurally as "intermediate-filament-like." Research indicates that alveolins create rigidity in the cell; they are often found associated with microtubules.

Microfilaments are composed of actin subunits that bind together reversibly to make long chains. Microfilaments, often called actin filaments, form the cell cortex immediately interior to the plasma membrane. Because the cortex underlies the plasma membrane, it affects properties of the membrane and interactions between the cell surface and surfaces over which a cell moves. Actin filaments are very important in cell movement and shape change, particularly in ameboid cells.

Ameboid Cell Extension via Actin Filaments

Ameboid cells assume a wide variety of shapes and forms (figure 5.5). The plasma membrane is flexible and responds to changes in the dynamic cytoskeleton beneath it. Actin subunits in the cytosol undergo rapid chain formation (polymerization) to form microfilaments, which may push against the plasma membrane and deform it (figure 5.6). The plasma membrane takes on different shapes due to the underlying microfilament skeleton as the cell moves forward. General cell extensions are called **pseudopodia** (Gr. *pseudēs*, false, + *podion*, small foot) and may be distinguished by their shapes. Large, blunt extensions are called **lobopodia**, whereas thin pointed extensions are called **filopodia** (figure 5.6).

Connecting the actin subunits together to make a filament requires energy in the form of ATP (adenosine triphosphate), so filaments extend and shorten depending on available energy. A filament has two different ends: a growing end where new subunits are being added, and the opposite end where subunits dissociate and the filament shortens. To create a narrow pointed filopodium, a bundle of parallel actin filaments forms to push the plasma membrane outward. Cross-linking proteins connect the filaments to stabilize the membrane extension (figure 5.6). If energy is not continually provided to add new actin subunits to the growing ends of the filaments, then the filopodium will dissipate. In a case where new actin subunits are added to the growing end of the filament at the same rate as actin subunits are lost from the other end of the filament, the filopodium persists. This balanced rate of subunit addition and dissociation is called "actin treadmilling."

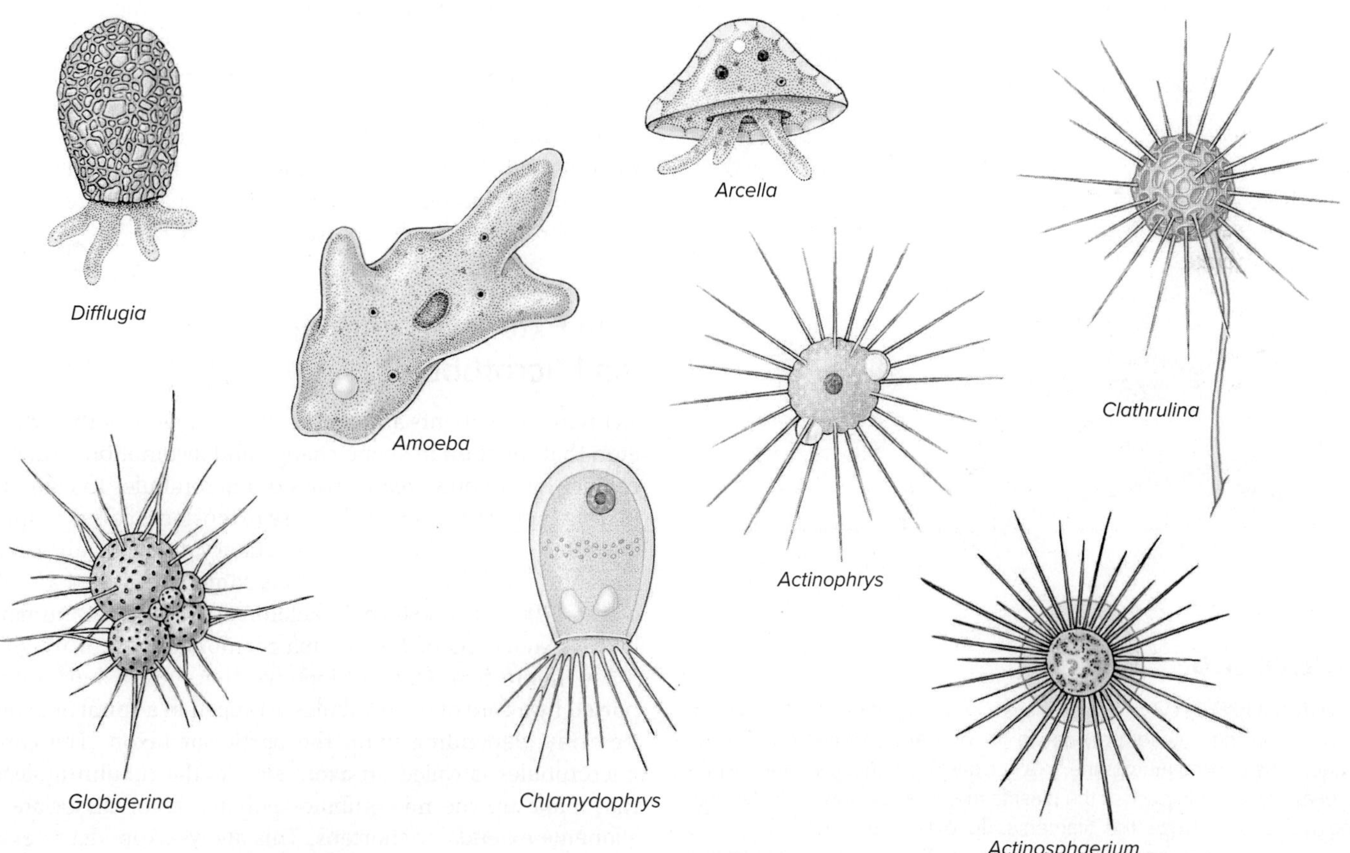

figure 5.5

The amebas are a very diverse group distributed across many clades: *Arcella*, *Difflugia*, and *Amoeba* have lobopodia and are amoebozoans; *Globigerina* has a testate body with net-like reticulopodia and is a foraminiferan; *Chlamydophrys* and *Clathrulina* are cercozoans, whereas *Actinosphaerium* and *Actinophrys* are stramenopiles. Axopodia are present in *Actinophrys*, *Actinosphaerium*, and *Clathrulina*.

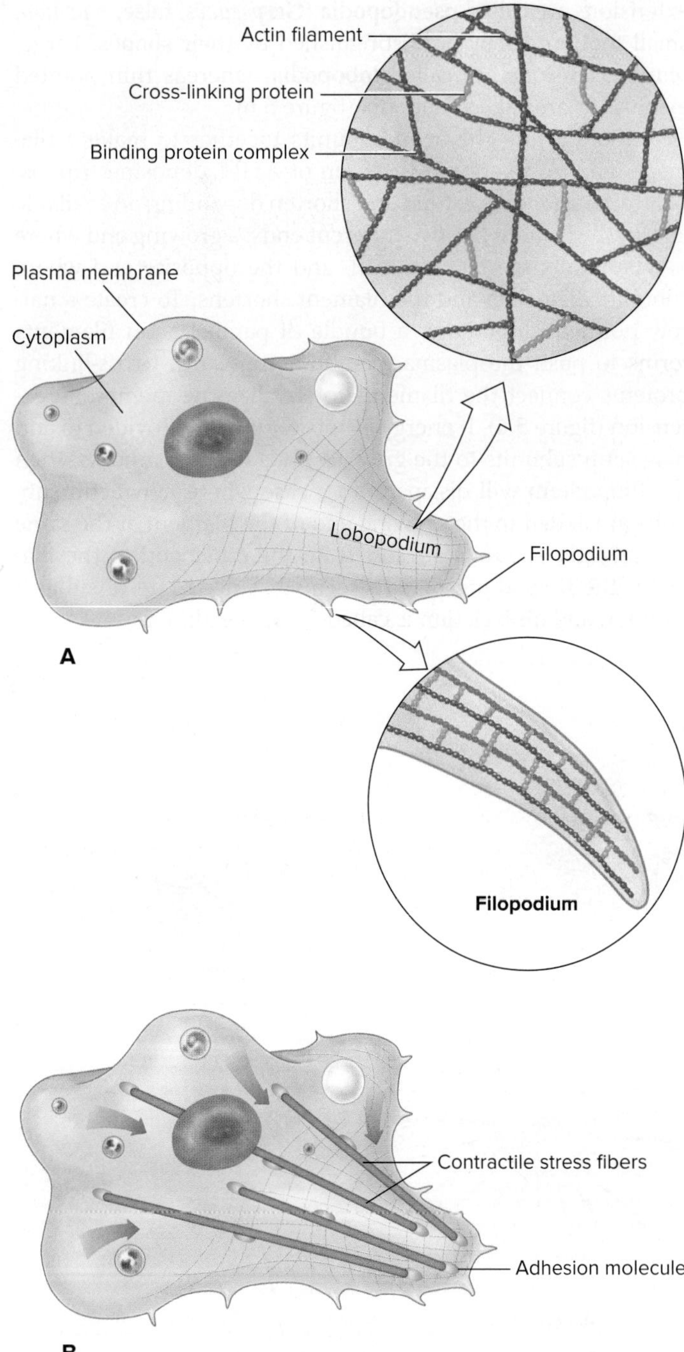

figure 5.6

The dynamic actin cytoskeleton plays a large role in cell shape. **A,** Actin subunits in the cytosol polymerize to form microfilaments that push against the plasma membrane. In a narrow filopodium, a bundle of parallel actin filaments pushes the plasma membrane outward. Cross-linking proteins stabilize the filaments. To extend the cell membrane broadly, as in a lobopodium, actin filaments are stabilized by branching. Branches occur at sites with actin-binding proteins within the filament. **B,** Adhesion proteins bind to the surfaces over which cells move, creating temporary anchors. New adhesion sites are made at the front of the plasma membrane as the cell moves forward. Contractile stress fibers that run between adhesion sites contract to bring the trailing edge of the cell forward and then adhesion sites are released.

To extend the plasma membrane over a much larger area than in a filopodium, as in a large blunt lobopodium (figure 5.6), actin filaments pushing against the membrane must be stabilized by branching. Filaments branch due to the addition of specialized actin-binding proteins at various sites within the chain—these proteins allow new filaments to form on the sides of existing filaments. Branched filaments with cross-linking proteins underlie large extensions of the plasma membrane (figure 5.6).

Extensions of the plasma membrane and movement of the accompanying cytoplasm push the cell forward along a leading side. If the cell is to move ahead, how is the trailing edge of the cell brought forward? Imagine a cell moving over a surface—it will need anchor points. Adhesion proteins within the plasma membrane bind to the surfaces over which cells move, creating temporary anchor points. As the cell moves ahead, new adhesion sites are made at the front of the plasma membrane. Actin filaments, now bound with myosin motor proteins, create contractile stress fibers that run between adhesion sites. These fibers contract to bring the trailing edge of the cell forward and then adhesion sites are released. All the processes just described take place continuously, in fractions of a second, as amebas change shape.

Cell shape change produces cell movement, but extensions of the plasma membrane also allow the cell to engulf food particles during feeding by **phagocytosis** (figure 5.7). As the plasma membrane surrounds a food particle in 3-dimensions, it creates an internal membrane-bound vacuole. Following this, the outer edges of the plasma membrane fuse. Food particles, once outside the plasma, are now internal and there has been no opportunity for cell rupture (figure 5.6C).

Cell Extension and Movement via Microtubules

Actin microfilaments are not the only elements of the cytoskeleton that function in shape change and locomotion. Some unicellular eukaryotes create arrays of microtubules to extend the plasma membrane outward. These extensions, called **axopodia**, are often present in unicellular eukaryotes with an external skeleton; recall that the skeleton is sometimes called a shell or a test. When the external skeleton is pierced by numerous holes, extensions of the plasma membrane pass through the holes (figure 5.8). Unlike filopodia, these extensions are supported by a core of microtubules arranged in a spiral or geometric array, depending upon the particular taxon. The core of microtubules is called an **axoneme**. As the tubulin molecules that make up the microtubules polymerize or dissociate, the axoneme extends or shortens. This allows axopodia to extend or retract quickly. Because the cytoskeleton is beneath the plasma membrane, cytoplasm flows along the axoneme, moving outward on one side of the axopod and returning to the cell on the other side.

Reticulopodia add a new dimension to the beautiful and fascinating cell extensions discussed in the chapter introduction.

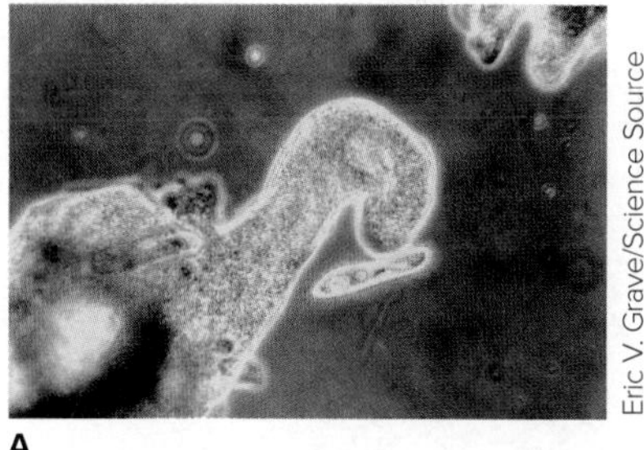

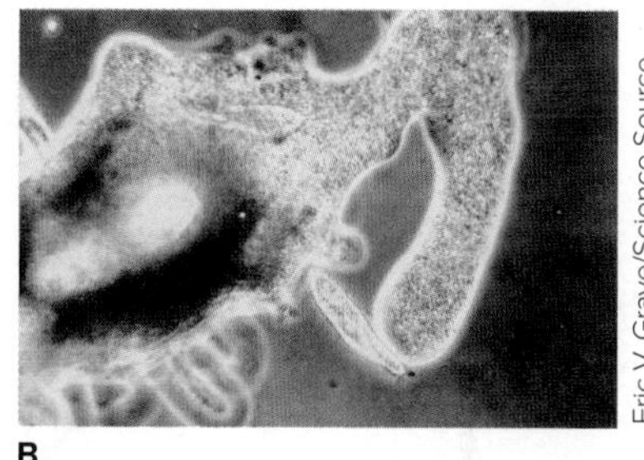

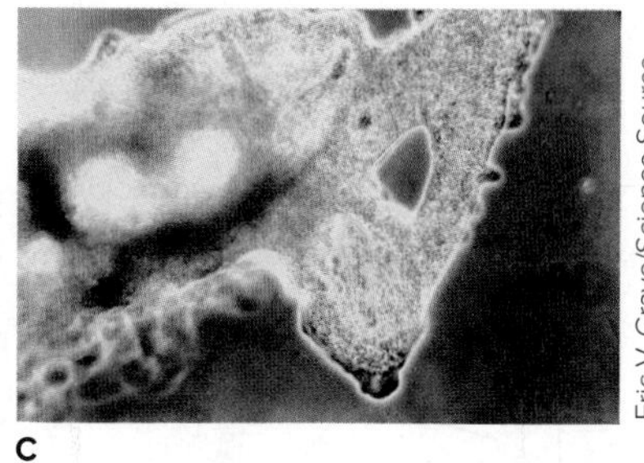

figure 5.7

Large blunt pseudopodia, called lobopodia, are used to engulf food items in a process called phagocytosis. **A, B,** the ameba extends a lobopodium toward a food item. **C,** the ameba engulfs the food item by phagocytosis.

In unicellular eukaryotes called foraminferans, axopodia extend outward, contacting and merging with other axopodia from the same cell. As the plasma membranes atop the axonemes merge, a branched net-like mesh called a reticulum develops (see *Globigerina* in Figure 5.5). The mesh is used in prey capture, formation of the exoskeleton (shell), cell attachment, and cell movement. Cytoplasm moves bi-directionally throughout the reticulum.

Two groups of unicellular eukaryotes, oxymonads (not discussed here) and parabasalids (important human parasites; see section 5.5: Metamonada) have an organelle called an **axostyle**. The axostyle is a flexible ribbon-like structure. However, "ribbon-like" does not imply that the axostyle is thin or sheathlike–but it is about 10 times wider than it is thick. It is composed of microtubules cross-linked in parallel rows. It creates the longitudinal axis in a parabasalid cell (see *Trichomonas* in figure 5.13), and is physically associated with the nucleus, and in some cases with the Golgi apparatus. In living organisms, it maintains a constant rapid undulation and appears to move the cell. The anterior end of the axostyle associates with a short rod, and sometimes a sheath surrounds the axostyle. Details of axostyle structure differ in oxymonads and parabasalids, so biologists assume that they evolved independently in each group.

In Euglenozoa, longitudinal microtubules just beneath the plasma membrane stiffen the membrane creating the characteristic "**pellicle**" (see section 5.5: Euglenozoa). Alveolates, a supergroup of three phyla (see section 5.5: Alveolata) also have a microtubule layer beneath the plasma membrane forming a pellicle, but the exact position of this layer varies among phyla (see section on Cell Movement by Glideosome and section 5.5: Alveolata and Dinoflagellata).

Again, a central core of microtubules is called an *axoneme*, and readers will see this term used to describe the core of microtubules at the center of a cilium or a flagellum. Details of the core of the microtubules that support cell extensions differ. In the axostyle and in the axoneme of axopodia, the microtubules are singlets; this means that they occur unbound to their neighbors. In cilia and flagella, the microtubules in the axoneme are paired as "doublets" (compare the images in figure 5.8 with figure 5.9). At the base of the cilium or flagellum, microtubules occur in groups of three as "triplets."

Microtubules are very important for cell structure. What determines where they form? Microtubule formation is determined by "microtubule organizing centers" commonly called "MTOCs." At MTOCs, free tubulin molecules come together to form strands in much the same way as actin molecules form filaments. Thirteen strands form a microtubule (figure 5.4), and depending on the location, microtubules may be singlets, doublets, or triplets. There are many different kinds of MTOCs in the cell.

A centrosome is an MTOC; microtubules assemble between two centrosomes to make a spindle that aligns chromosomes during cell division (mitosis and meiosis). A centrosome is composed of two centrioles as well as a protein mass that guides tubulin strand formation and anchors the microtubules. Centrioles are short cylinders made of nine interconnected microtubule triplets. A modified centriole at the base of a cilium or flagellum is called a **basal body**; this structure is also called a kinetosome, a basal granule, or a blepharoplast. The basal body determines where a cilium or a flagellum forms in the cell by creating conditions for the growth of the axoneme microtubules. Recall that an axoneme is the structural center of the cilium or flagellum. The basal body has nine *triplets* of microtubules, but in the axoneme above it, there are only nine *pairs* of microtubules and they surround two central microtubules (figure 5.9).

Cell Movement by Glideosome

The Alveolata is a super-group composed of three phyla: Apicomplexa, Dinoflagellata, and Ciliophora. The two diagnostic features of this group are membranous sacs called **alveoli** that underly the plasma membrane and the presence of a family of proteins called **alveolins**, described structurally as similar to intermediate filaments. Across the Alveolata, the position and function of the alveoli vary among groups, but in apicomplexans, alveoli and the cytoskeleton work together to create a **glideosome** (see figure 5.22). Apicomplexans are parasites with multistage life cycles. Some of the stages are motile and glide across the surface of host cells for more than 10 minutes. They do not have pseudopodia, cilia, or flagella, so how they moved

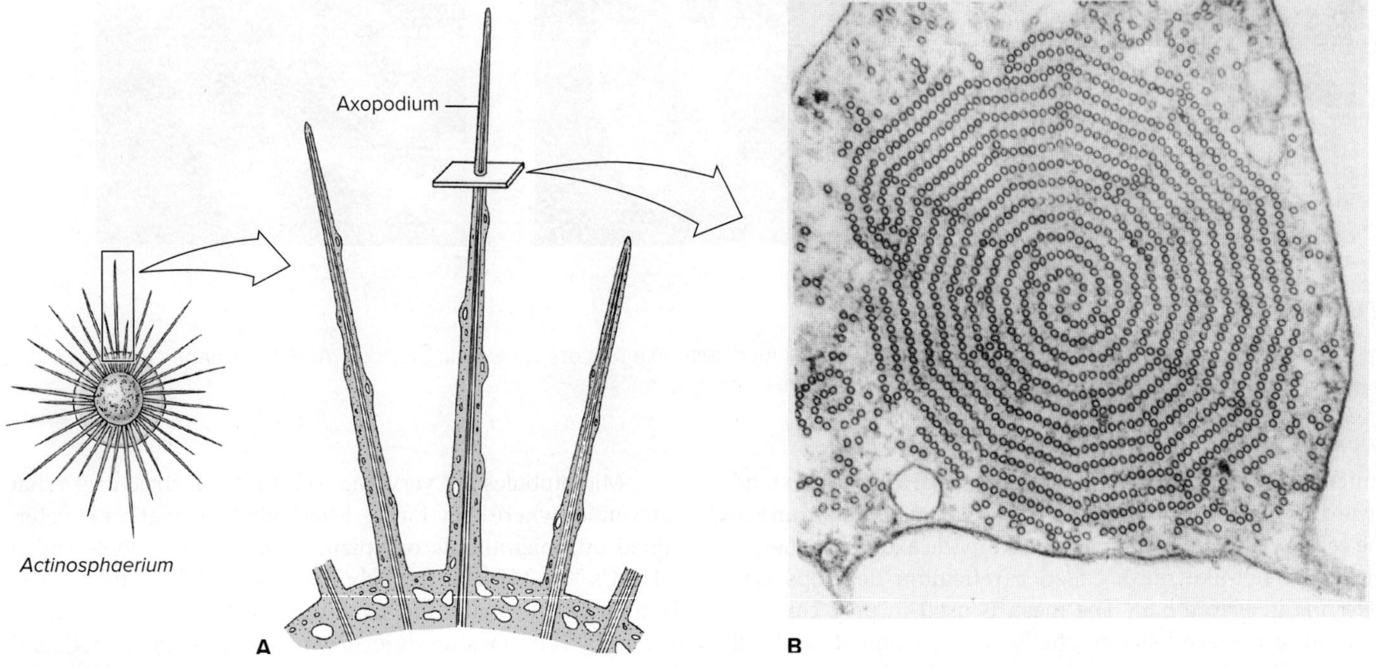

figure 5.8

A, Diagram of axopodium microtubules. **B,** An electron micrograph of axopodium (from *Actinosphaerium nucleofilum*) in cross section. The axoneme of the axopodium is composed of an array of microtubules, which may vary from three to many in number, depending on the species. Some species can extend or retract their axopodia quite rapidly (× 99,000).

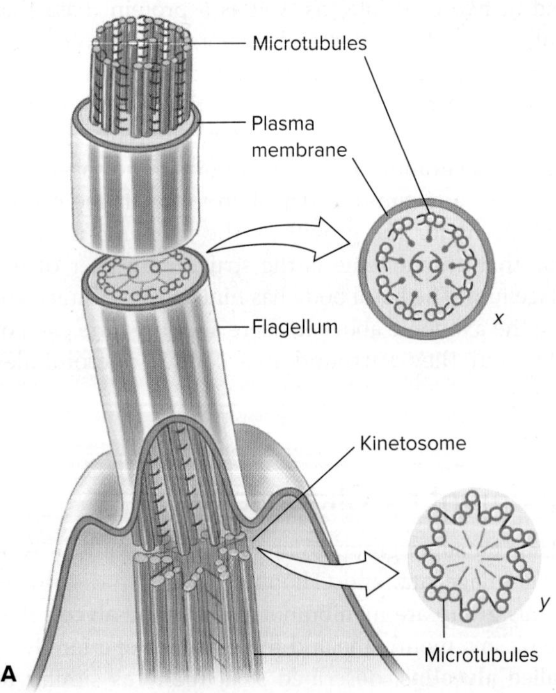

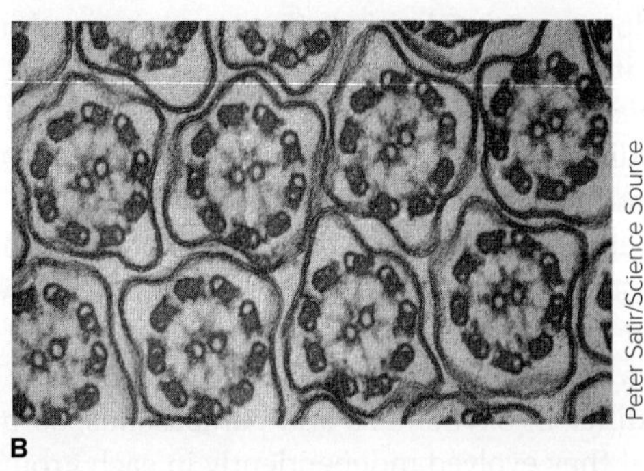

figure 5.9

Internal structure of a flagellum. **A,** The axoneme is composed of nine pairs of microtubules plus a central pair, and it is enclosed within the plasma membrane. The central pair ends at about the level of the cell surface in a basal plate (axosome). The peripheral microtubules continue inward for a short distance to compose two of each of the triplets in the kinetosome (at level *y* in **A**). **B,** Electron micrograph of a section through several cilia, corresponding to section at *x* in **A** (× 133,000).

was not understood until recently. We now know that the plasma membrane of this stage of the parasite binds to the plasma membrane of the host cell in a number of places. These binding sites are linked to the actin cytoskeleton just inside the plasma membrane of the parasite and then firmly anchored to the alveoli deeper inside the parasite cell (see figure 5.22 for a detailed description). The parasite actin cytoskeleton is connected to myosin motors that "walk" the parasite forward. In 3-D, the parasites glide in a corkscrew motion pushing the host cell membrane inward. The parasite pushes into the host cell without penetrating the host plasma membrane; instead, the *host* membrane forms a vacuole around the parasite cell. The parasite is now inside the host cell within a protective membrane that the host does not identify as foreign. The parasite cell later copies itself through cell division.

Cell Movement by Cilia and Flagella

No sharp morphological distinction exists between cilia and flagella, and some investigators have preferred to call them both *undulipodia* (L. dim. of *unda,* a wave, + Gr. *podos,* a foot). However, their actions are somewhat different. A cilium propels water parallel to the surface to which the cilium is attached, whereas a flagellum propels water parallel to the main axis of the flagellum. Cilia are typically present in large numbers and are relatively short, whereas flagella are longer and less numerous. Each flagellum or cilium contains nine pairs of longitudinal microtubules arranged in a circle around a central pair (figure 5.9). This tube of microtubules in a flagellum or cilium is its **axoneme;** an axoneme is covered by a membrane continuous with the plasma membrane covering the rest of the organism. At about the point where an axoneme enters the cell proper, the central pair of microtubules ends at a small plate within the circle of nine pairs. Also at about that point, another microtubule joins each of the nine pairs, so that these form a **basal body**. All typical flagella and cilia have a basal body at their base, whether borne by a unicellular eukaryote or by an animal cell.

The current explanation for ciliary and flagellar movement is the *sliding-microtubule hypothesis.* Movement is powered by the release of chemical bond energy in ATP. Two little arms are visible in electron micrographs on each of the pairs of peripheral tubules in the axoneme (level *x* in figure 5.9), and these bear the enzyme adenosine triphosphatase (ATPase), which cleaves the ATP. When bond energy in ATP is released, the arms "walk along" one of the microtubules in the adjacent pair, causing it to slide relative to the other tubule. Shear resistance, causing the axoneme to bend when the microtubules slide past each other, is provided by "spokes" from one of the tubules in each doublet projecting toward the central pair of microtubules. These spokes are also visible in electron micrographs.

Key Theme 5.1
ADAPTATION AND PHYSIOLOGY

Cell Movement in Animals

Unicellular eukaryotes use cilia, flagella, and extensions of the cytoskeleton as locomotory structures, but animal cells also use these structures. Many small animals use cilia for locomotion, as exemplified by rotifers and ctenophores. Larger animals such as clams and polychaete worms use cilia to create water currents for their feeding and respiration. Ciliary movement is also vital to many animals in functions such as ingestion of food, reproduction, excretion, and osmoregulation (see flatworm flame cells). Ameboid motion of cells is critical for animal macrophages and other cells in the immune system. Almost all animal sperm rely on a flagellum to reach the egg.

CHARACTERISTICS of Unicellular Eukaryotes

1. **Unicellular;** some colonial, and some with multicellular stages in their life cycles.
2. **Mostly microscopic,** although some are large enough to be seen with the unaided eye; they are sometimes called microbial eukaryotes in reference to their size.
3. All symmetries represented; shape variable or constant (oval, spherical, or other).
4. **No germ layer present**.
5. No organs or tissues, but **specialized organelles** are found; nucleus single or multiple.
6. Free-living, mutualism, commensalism, parasitism all represented in the groups.
7. Locomotion by **pseudopodia, axopodia, axostyle, glideosome, flagella, cilia,** and direct cell movements; some sessile.
8. Some have a **simple endoskeleton** or **exoskeleton**, but most are naked.
9. **Nutrition of all types:** autotrophic (manufacturing own nutrients by photosynthesis), heterotrophic (depending on other plants or animals for food), mixotrophic (combining heterotrophy and autotrophy), saprozoic (using nutrients dissolved in the surrounding medium).
10. Aquatic or terrestrial habitat; free-living or symbiotic mode of life.
11. Reproduce **asexually** by fission, budding, and cysts and **sexually** by conjugation or by syngamy (union of male and female gametes to form a zygote).
12. Multicellularity has evolved in at least 25 lineages with unicellular ancestors. By definition, multicellular forms have distinct reproductive cells and nonreproductive (somatic) cells. *Volvox carteri* is a multicellular chlorophyte.

Key Theme 5.2
SCIENCE EXPLAINED

Colloids

Colloidal systems are permanent suspensions of finely divided particles that do not precipitate, such as milk, blood, starch, soap, ink, and gelatin. Colloids in living systems are commonly proteins, lipids, and polysaccharides suspended in the cell cytoplasm. Such systems may undergo sol-gel transformations, depending on whether the fluid or particulate components become continuous. In the sol state of cytoplasm, solids are suspended in a liquid, and in the semisolid gel state, liquid is suspended in a solid.

Exoskeletons

The actin cortex, microtubules, actin filaments, and alveoli form a flexible endoskeleton. Many unicellular eukaryotes make exoskeletons, variously called tests or shells, from collected particles or by secreting silica or calcium. Agglutinated tests are made from particles, but these vary from smooth to rough depending on the size and shape of particles selected (see figure 5.5). Tests made from glassy silica occur in diatoms (see section 5.5: Stramenopiles); phaeodarians (formerly radiolarians, now in Cercozoa) make a skeleton with hollow spines using amorphous silica plus magnesium, calcium, and copper; and radiolarians (see figure 5.25 and section 5.5: Radiolaria) make a test of amorphous silica or strontium sulfate. Centrohelids (see section 5.5: Centrohelida) form an external coat from silica scales. One genus of Foramanifera (see section 5.5: Foraminifera) also makes a silica test; however, most foraminiferans have tests of calcium carbonate (see figure 5.24B). Calcium crystal size and shape determine whether the tests appear glassy or milky. In some, grains of calcite form the tests. Tests are usually pierced by numerous holes through which axopodia protrude (see figure 5.24A).

Cell Nucleus

The **nucleus** encloses the eukaryotic chromosomes. It is bound by two membranes with a space between them that together form the nuclear envelope. Each nuclear membrane is a phospholipid bilayer, but the chemical composition differs somewhat from the plasma membrane. The nuclear envelope has pores that allow communication between the nucleus and the cytoplasm. Eukaryotic chromosomes are linear strands of chromatin composed of DNA and binding proteins. The **nucleolus** contains DNA that controls the synthesis of ribosomal RNA; ribosomal RNA combines with proteins to make ribosomes. Ribosomes are the sites of protein synthesis. Protein synthesis occurs outside the nucleus, so once made, the ribosomes leave the nucleus through nuclear pores open to the cytoplasm.

Many unicellular eukaryotes have a typical nucleus, but in the ciliate clade (see section 5.5: Ciliophora) two kinds of nuclei are present: a germ line nucleus called the **micronucleus** and a somatic nucleus called the **macronucleus**. Only genes of the macronucleus are transcribed, so this nucleus regulates cell function (see section 5.5: Ciliophora, as well as figures 5.19 and 5.20). Another atypical nuclear feature occurs in Euglenozoa. In taxa outside this group, nucleoli are not visible during mitosis; here, nucleoli persist during mitosis, so this is potentially a diagnostic feature for Euglenozoa.

Endomembrane System

The endomembrane system is a set of membranes that creates compartments within the cytoplasm. Different compartments house different chemical reactions. The endomembrane system comprises the outer nuclear membrane, the endoplasmic reticulum (ER), the Golgi apparatus, lysosomes, and the plasma membrane (figure 5.3). It also includes vesicles within the cytoplasm. Recall that ribosomes leave the nucleus to begin protein synthesis in the cytoplasm—they join the ER to make "rough endoplasmic reticulum," a site for protein synthesis. Smooth endoplasmic reticulum lacks ribosomes and is a site for lipid synthesis, among other functions. Proteins and polypeptides made in the ER are encapsulated by small sections of ER membrane as transport vesicles. Transport vesicles move the polypeptides and proteins to the stacked disc membrane system (called cisternae or dictyosomes) of the Golgi apparatus, where they are modified for particular functions. Products of the Golgi apparatus are then repackaged in membrane vesicles and moved through the cytoplasm. Some vesicles might contain substances for incorporation into the plasma membrane or for export from the cell. When cysts are being formed, wall material is made in the Golgi apparatus and sent to the plasma membrane. Lysosomes are membrane-bound organelles containing enzymes used for digestion.

How can enzymes within membrane-bound structures function in digestion? Recall learning about phagocytosis in section 5.2: Nutrition and Digestion, and figure 5.7. Food particles outside the cell were engulfed in the plasma membrane and brought into the cell in food vacuoles. When the membrane of a food vacuole fuses with the membrane of a digestive lysosome, the food is digested, but the enzymes and their products are contained and not released into the cytoplasm, where they would be toxic.

If a vesicle from the Golgi apparatus contains proteins to be incorporated into the plasma membrane, the vesicle membrane fuses with the plasma membrane. When proteins or waste products contained in a vesicle are to be exported from the cell, the vesicle membrane also fuses with the plasma membrane positioning the vesicle contents outside the cell. The membrane's ability both to create compartments and eliminate them through fusion makes possible a myriad of cell functions.

Parabasalids (see section 5.5: Metamonada) are endoparasitic or symbiotic unicellular eukaryotes. *Trichomonas vaginalis* is a well-known sexually transmitted example for humans. Parabasalids are characterized by a structure called the "**parabasal apparatus,**" in which striated fibers (called root fibers) run from a basal body—or another MTOC—to a very well-developed system of stacked discs (cisternae or dictyosomes) of the Golgi apparatus. In animal cells, microtubules connect to the cisternae of the Golgi apparatus to hold them in position near the nucleus, so biologists hypothesize that the striated fibers of parabasalids fulfill the same role in the absence of microtubules. The chemical composition of the striated fibers is not known, but preliminary studies indicate that it is neither tubulin nor actin.

Extrusomes

The general term "extrusome" refers to a membrane-bound organelle used to extrude something from the cell. The wide variety of structures suggests that extrusomes are not all homologous. The ciliate trichocyst is a well-known extrusome (see section 5.5: Ciliophora and figure 5.17). Some ciliates also produce toxicysts that contain a poison used to paralyze prey. A new type of extrusome, the **ancoracyst**, was recently discovered inside a new unicellular eukaryote, *Ancoracysta twista*. The ancoracysts seem to be used to immobilize prey.

Mitochondria

The mitochondrion (figure 5.3) is an extremely interesting organelle. Energy required for cellular metabolism is recovered from digested food and transferred within the cell as a high-energy phosphodiester bond in ATP (adenosine triphosphate). Enzymatic reactions that require energy are often coupled with the conversion of ATP to ADP (adenosine diphosphate) and inorganic phosphate. In the classical mitochondrion, oxidative phosphorylation of ADP occurs; energy from the carbon bonds of pyruvate (a product of food digestion) is used to convert ADP to ATP, with oxygen as the terminal electron acceptor in the electron transport chain. This pathway works in oxygen-rich environments, but biologists have discovered modified mitochondria within unicellular eukaryotes living in environments with little or no oxygen, such as deep anaerobic sediments. Anaerobic niches for parasitic forms occur inside the bodies of hosts or inside host cells for intracellular parasites.

When there is no oxygen to serve as the terminal electron acceptor, how do mitochondria make ATP? In some cases, mitochondria maintain the electron transport chain, but use organic (e.g., fumerate) or environmental compounds as terminal electron acceptors. These cells do not release hydrogen gas. More highly modified mitochondria are called "mitochondria-related organelles," or MROs. There are three general types: (1) in some MROs the electron transport chain is functional, but protons are the terminal electron acceptor, and hydrogen gas is produced; (2) in cells in anaerobic environments or those with very low oxygen levels, MROs called **hydrogenosomes** use substrate-level phosphorylation to produce ATP and hydrogen gas is released; and (3) MROs called **mitosomes** do not produce ATP and most have lost their genomes. Many different mutations have produced the MROs we see now—the assembly of iron-sulphur (Fe-S) cofactors is the only function shared by mitochondria and all their descendant MROs.

The modifications in ATP production among mitochondria are strongly related to the abundance of oxygen in the habitat. Mitochondrial type is not a good phylogenetic character. Within the clade Heterolobosea (see section 5.5: Heterolobosea), some organisms such as *Naegleria* have classical mitochondria, but others in low oxygen environments have hydrogenosomes. The parasite *Giardia* (see section 5.5: Metamonada), which lives in humans and other mammals, inhabits anaerobic gut environments and has a mitosome.

The one situation where mitochondrial structure is unique and characterizes a clade of unicellular eukaryotes occurs in the parasitic Kinetoplasta (see section 5.5: Euglenozoa). As explained in the chapter opening essay, mitochondria arose through symbiogenesis. Their ancestors were alpha-proteobacteria, so they possess a circular bacterial genome, generally called mitochondrial DNA, or mtDNA. In Kinetoplasta, the mtDNA occurs as a large set of interconnected DNA circlets. The circlets are of two types; "maxicircles," which comprise the protein-coding genes, and "minicircles," which edit the products of transcription from maxicircles. This unique set of circlets is called kinetoplast DNA or kDNA. Kinetoplastans have only one mitochondrion situated with a basal body at the base of the flagellum. The part of the mitochondrion that connects to a basal body and houses the kDNA is called a **kinetoplast**.

Some unicellular eukaryotes lack both mitochondria and MROs. When first discovered, biologists disagreed about whether these unicells could have descended from the first eukaryote that existed before the symbiosis with an alpha-proteobacterium began. Some biologists argued that a mitochondrion had been present in the ancestors of these cells, but that it was lost later in evolution. To resolve the controversy, we have to learn more about the mitochondrial genome. The ancestral mitochondrion inherited genes on a circular DNA strand from its bacterial ancestor. Over time, many of the genetic instructions for protein synthesis migrated out of the mitochondrion and are now encoded by DNA of the eukaryotic cell nucleus. Even though the mitochondrial genome has been lost in hydrogenosomes and mitosomes, this does not mean that all proteins normally made by mitochondria do not exist in the cell. Some proteins normally made in the mitochondrion are now made in the cytosol using instructions from nuclear DNA. These proteins are translocated into the mitochondrion. If production in the cytosol and translocation into mitochondria takes place, it means that the ancestors of these cells possessed a mitochondrion. When researchers studied the nuclear DNA of eukaryotes without mitochondria, they discovered mitochondrial genes embedded in the nuclear genome; all the evidence indicates that the most recent common ancestor of all known eukaryotes contained mitochondria.

Plastids

Plastids are organelles containing photosynthetic pigments. The original addition of a plastid to eukaryotic cells occurred when a cyanobacterium was engulfed and not digested. **Chloroplasts** (see figure 5.15) contain different versions of chlorophylls (*a, b,* or *c*), but other kinds of plastids contain other pigments. For example, red algal plastids contain phycobilins. Particular pigments shared among lineages may indicate ancestry, but plastids could also have been gained by secondary endosymbiosis rather than direct inheritance from a common ancestor (see the chapter opening essay).

■ 5.2 Nutrition and Digestion

Unicellular eukaryotes can be categorized broadly into two modes of feeding: autotrophs, which synthesize their own organic constituents from inorganic substrates, and heterotrophs, which must have organic molecules synthesized by other organisms. Within heterotrophs, those that ingest visible particles of food (**phagotrophs,** or holozoic feeders) contrast with those ingesting food in a soluble form (**osmotrophs,** or saprozoic feeders). Mixotrophs make some of their own nutrients, but also eat other organisms, combining autotrophy and heterotrophy in a new mode called **mixotrophy**.

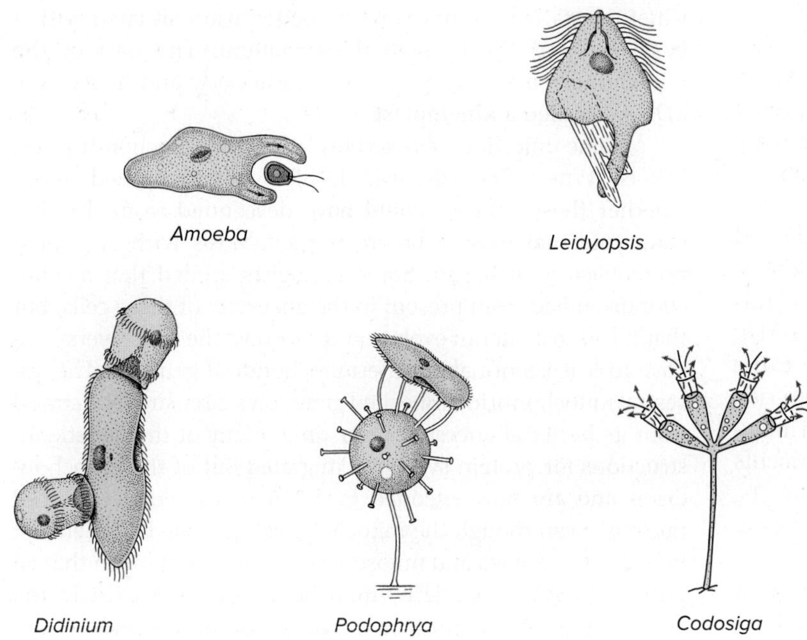

figure 5.10

Types of phagocytosis. *Amoeba* surrounds a small flagellate with pseudopodia. *Leidyopsis*, a flagellate living in the intestine of termites, forms pseudopodia and ingests wood chips. *Didinium*, a ciliate, feeds only on *Paramecium*, which it swallows through a temporary cytostome in its anterior end. Sometimes more than one *Didinium* feeds on the same *Paramecium*. *Podophrya* is a suctorian ciliophoran. Its tentacles attach to its prey and suck prey cytoplasm into its body, where the cytoplasm is pinched off to form food vacuoles. *Codosiga*, a sessile choanoflagellate with a collar of microvilli, feeds on particles suspended in the water drawn through its collar by the beat of its flagellum. The particles are moved to the cell body and ingested (surrounded by small pseudopodia).

Holozoic nutrition implies **phagocytosis** (figure 5.10; see figure 5.5), in which there is an infolding or invagination of the plasma membrane around the food particle. As the invagination extends farther into the cell, it is pinched off at the surface. A food particle is thus contained in an intracellular, membrane-bound vesicle, the **food vacuole** or **phagosome.** Lysosomes, small vesicles containing digestive enzymes, fuse with the phagosome and pour their contents into it, where digestion begins. As digested products are absorbed across the vacuolar membrane, the phagosome becomes smaller. Any undigestible material may be released to the outside by exocytosis, the vacuole again fusing with the plasma membrane. In most ciliates, many flagellates, and many apicomplexans, the site of phagocytosis is a definite mouth structure, the **cytostome** (see figure 5.18). In amebas, phagocytosis can occur at almost any point by envelopment of the particle with pseudopodia. Many ciliates have a characteristic structure for expulsion of waste matter, called a **cytoproct** (see figure 5.18) or **cytopyge,** occupying a characteristic location.

A new feeding process, called **trogocytosis** was discovered in *Entamoeba histolytica* (see section 5.5: Amoebozoa), an important human intestinal parasite that causes amebic dysentery. In trogocytosis, the unicellular parasite bites and consumes sections of the host cell membrane, cytoplasm, and organelles. The same process is used when *Naegleria fowleri* ingests brain tissue (see section 5.5: Heterolobosea).

■ 5.3 Excretion and Osmoregulation

Water balance, or osmoregulation, is a function of the one or more **contractile vacuoles** (see figures 5.11 and 5.18) typically present in freshwater forms, which live in a hypoosmotic environment. These vacuoles are often absent in marine or parasitic taxa, which live in a nearly isosmotic medium.

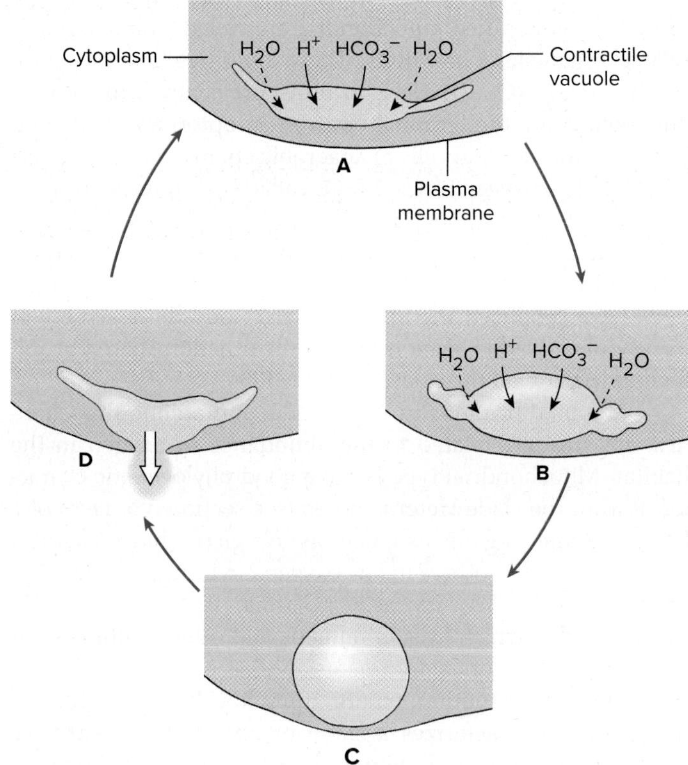

figure 5.11

Proposed mechanism for operation of contractile vacuoles composed of a system of cisternae and tubules. **A, B,** Proton pumps in their membranes transport H^+ and cotransport HCO_3^- into the vacuoles. Water passively diffuses inward to maintain an osmotic pressure equal to that in the cytoplasm. **C,** Once the vacuole has filled, **D,** its membrane fuses with the cell's plasma membrane, expelling water, H^+, and HCO_3^-. Protons and bicarbonate ions are replaced readily by action of carbonic anhydrase on carbon dioxide and water.

Contractile vacuoles act as pumps to remove excess water from the cytoplasm. One hypothesis suggests that proton pumps on the vacuolar surface and on tubules radiating from it actively transport H^+ and cotransport bicarbonate (HCO_3^-) (figure 5.11), which are osmotically active particles. According to this hypothesis, as these particles accumulate within a vacuole, water is drawn into the vacuole. Fluid within the vacuole remains isosmotic to the cytoplasm. Then, as the vacuole finally joins its membrane to the surface membrane and empties its contents to the outside, it expels water, H^+, and HCO_3^-. These ions can be replaced readily by action of the enzyme carbonic anhydrase on CO_2 and H_2O. Carbonic anhydrase occurs in the cytoplasm.

Nitrogenous wastes from metabolism diffuse through the plasma membrane, but some are emptied also by contractile vacuoles.

5.4 Reproduction and Life Cycles

Asexual Reproduction: Fission

The cell-multiplication process that produces genetically identical individuals in unicellular eukaryotes is called **fission.** The most common type is **binary fission,** which yields two essentially identical individuals (figure 5.12; see figure 5.19). When a progeny cell is considerably smaller than the parent and then grows to adult size, the process is called **budding.** This process occurs in some ciliates. In **multiple fission,** several nuclear divisions precede division of the cytoplasm (cytokinesis) so that numerous individuals are produced almost simultaneously (see figure 5.23B). Multiple fission, or **schizogony,** is common among the Apicomplexa and some amebas. When multiple fission leads to spore or sporozoite formation, it is called **sporogony.**

Sexual Processes

Although all unicelllular eukaryotes reproduce asexually, and some are apparently exclusively asexual, the widespread occurrence of sexual reproduction testifies to the value of genetic recombination. Sexual processes may precede certain phases of asexual reproduction, but embryonic development does not occur; unicellular eukaryotes do not have embryos. The essential features of sexual processes include a reduction division of the chromosome number to half (diploid number to haploid number), development of sex cells (gametes) or at least gamete nuclei, and usually fusion of the gamete nuclei (see figure 5.20).

The gamete nuclei, or pronuclei, which fuse in fertilization to restore the diploid number of chromosomes, are usually borne in special gamete cells. Gametes may be similar in appearance or unlike.

In animals, meiosis usually occurs during or just before gamete formation (called **gametic meiosis**). This is the case in the Ciliophora and some flagellated and ameboid groups. However, in other flagellated groups and in the Apicomplexa, the first divisions *after* fertilization (zygote formation) are meiotic (*zygotic meiosis*) (see figure 5.23A), and all individuals produced asexually (mitotically) in the life cycle up to the next zygote are haploid. In some amebas (foraminiferans), haploid and diploid generations alternate (**intermediary meiosis**), a widespread phenomenon among plants.

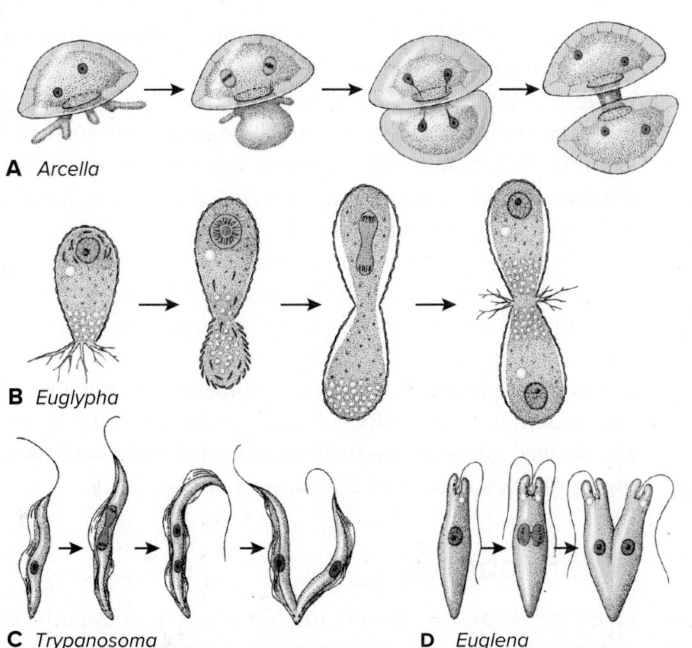

figure 5.12

Binary fission in some amebas and Euglenozoa. **A,** The two nuclei of *Arcella* divide as some of its cytoplasm is extruded and begins to secrete a new shell for the daughter cell. **B,** The shell of another ameba, *Euglypha,* is constructed of secreted platelets. Secretion of the platelets for the daughter cell begins before the cytoplasm enters it. As these form the shell of the daughter cell, the nucleus divides. **C,** *Trypanosoma* has a kinetoplast near the basal body of its flagellum close to its posterior end in the stage shown. All of these parts must replicate before the cell divides. **D,** Division of *Euglena.*

Fertilization of an individual gamete by another is called **syngamy,** but some sexual phenomena in unicellular taxa do not involve syngamy. Examples are **autogamy,** in which gametic nuclei arise by meiosis and fuse to form a zygote within the same organism that produced them, and **conjugation,** in which an exchange of gametic nuclei occurs between paired organisms (conjugants). Conjugation is described further in the discussion of *Paramecium.*

Encystment and Excystment

Separated as they are from their external environment only by their delicate external plasma membrane, unicellular eukaryotes are surprisingly successful in habitats frequently subjected to extreme conditions, such as heat, cold, or dryness. Survival under harsh conditions surely is related to an ability to form cysts, dormant forms that have resistant external coverings and undergo a more or less complete shutdown of metabolic machinery. Cyst formation is also important to many parasitic forms that must survive a harsh environment between hosts (see figure 5.3). However, some parasites do not form cysts, apparently depending on direct transfer from one host to another. Reproductive phases such as fission, budding, and syngamy may occur in cysts of some species. Encystment has not been found in *Paramecium,* and it is rare or absent in marine forms.

Key Theme 5.3

ADAPTATION AND PHYSIOLOGY

Remarkable Survival of Cysts

Cysts of some soil-inhabiting and freshwater unicells have amazing durability. For instance, cysts of the soil ciliate *Colpoda* can survive 7 days in liquid air and 3 hours at 100°C. Survival of *Colpoda* cysts in dried soil has been shown for up to 38 years, and those of a certain small flagellate (*Podo*) can survive up to 49 years! Not all cysts are so sturdy, however. Those of *Entamoeba histolytica*, an agent for amebic dysentery, tolerate gastric acidity but not desiccation, temperature above 50°C, or sunlight.

The conditions stimulating encystment are incompletely understood, although in some cases cyst formation is cyclic, occurring at a certain stage in the life cycle. In most free-living forms, adverse environmental change favors encystment. Such conditions may include food deficiency, desiccation, increased environmental osmotic pressure, decreased oxygen concentration, or change in pH or temperature.

During encystment, a number of organelles, such as cilia or flagella, are resorbed, and the Golgi apparatus secretes cyst wall material, which is carried to the surface in vesicles and extruded.

Although the exact stimulus for excystation (escape from cysts) is usually unknown, a return of favorable conditions initiates excystment when cysts are at a resistant stage. In parasitic forms, an excystment stimulus may be more specific, requiring conditions similar to those found in the host.

5.5 Unicellular Eukaryotic Taxa

Characteristic cellular structures permit easy diagnosis of some major clades, whereas only genomic sequence information can effectively diagnose others. For example, the presence of branching, netlike pseudopodia in a shelled ameba identifies the organism as belonging to Foraminifera (see figure 5.2). We provide identifying morphological features for those clades discussed, if such features are present.

Metamonada

The Metamonada comprises three clades: retortamonads, diplomonads, and parabasalids. Many of these organisms are parasites or endosymbionts.

Retortamonads include commensal and parasitic unicells, such as *Chilomastix* and *Retortamonas. Giardia* (see figures 5.13 and 5.14), a representative diplomonad, is a common intestinal parasite of humans and other animals. It uses a ventral adhesive disc, resembling a large suction cup, to attach to the host intestine (figure 5.14). In some infected animals, the entire intestinal lining appears to have scales because there are so many attached parasites. *Giardia* is often asymptomatic but sometimes causes an unpleasant, but not fatal, diarrhea. Cysts are passed in the feces, and new hosts are infected by ingestion of cysts, often in contaminated water.

Parabasalids have a sheet of microtubules called an **axostyle** that originates at the base of a flagellum and is used in locomotion (see section 5.1: Cell Extension and Movement via Microtubules). The group is named for a modified region of the Golgi apparatus called a **parabasal** apparatus (see section 5.1: Endomembrane System) connected to a basal body (see section 5.1: Cell Extension and Movement via Microtubules).

Among parabasalids, some trichomonads (for example, *Trichonympha* and *Trichomonas* in figure 5.13) are of medical or veterinary importance. *Trichomonas vaginalis* infects the urogenital tract of humans and is sexually transmitted. It produces no symptoms in males but is the most common cause of vaginitis in females. *Trichonympha* is a termite gut symbiont that houses cellulose-digesting bacteria. Without *Trichonympha,* termites could not feed on wood.

In general, metamonads live in anoxic or low oxygen regions and lack mitochondria entirely or have mitochondrial derivatives (see section 5.1: Mitochondria). *Trichomonas* and other parabasalids have hydrogenosomes. Diplomonads, such as *Giardia,* have mitosomes, as do some retortamonads. Mitosomes do not contain DNA, but the presence of mitochondrially derived genes in the nucleus indicates mitochondria were once present (see section 5.1: Mitochondria).

Heterolobosea

Heteroloboseans are naked amebas whose pseudopodia form abruptly, in what is sometimes called an "eruptive" manner. The life cycle of many heteroloboseans includes both amebic

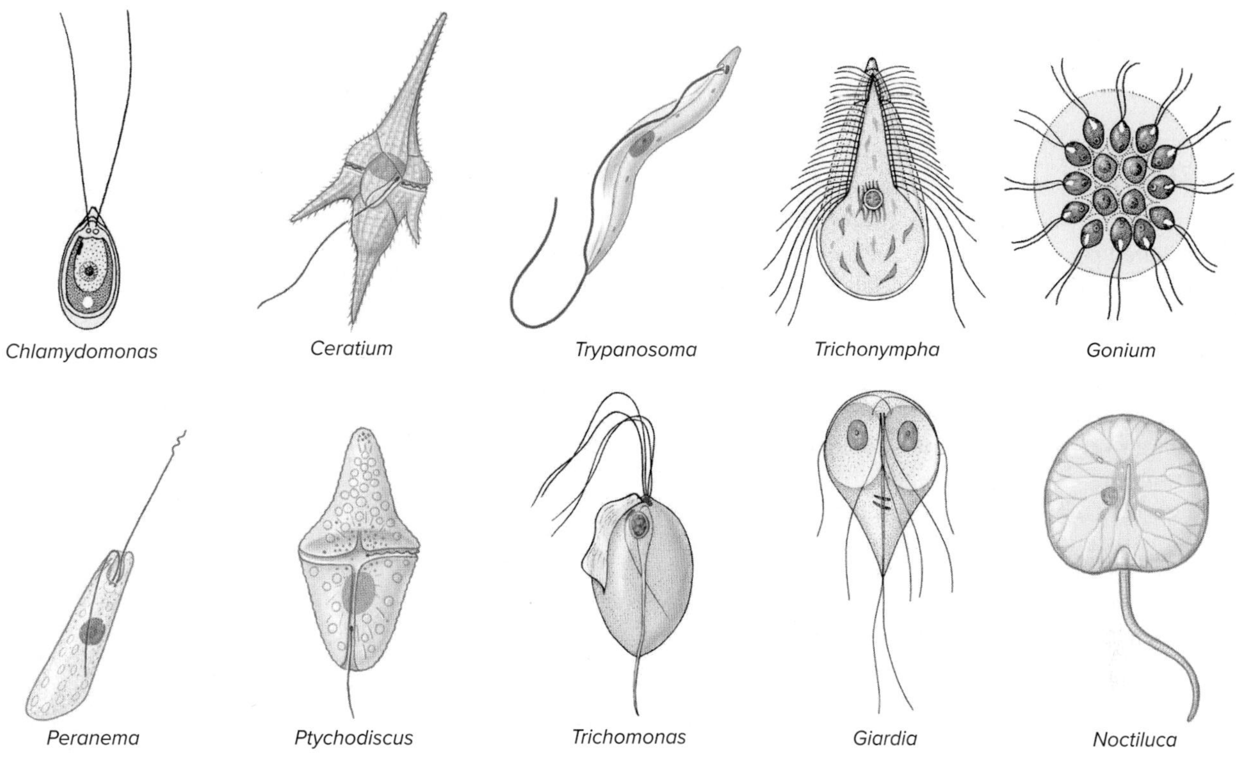

figure 5.13

Representatives of several unicellular taxa having flagella: Diplomonads (*Giardia*); Parabasalids (*Trichomonas, Trichonympha*); Plantae (*Chlamydomonas, Gonium*); Euglenozoa (*Peranema, Trypanosoma*); Dinoflagellata (*Ceratium, Ptychodiscus, Noctiluca*).

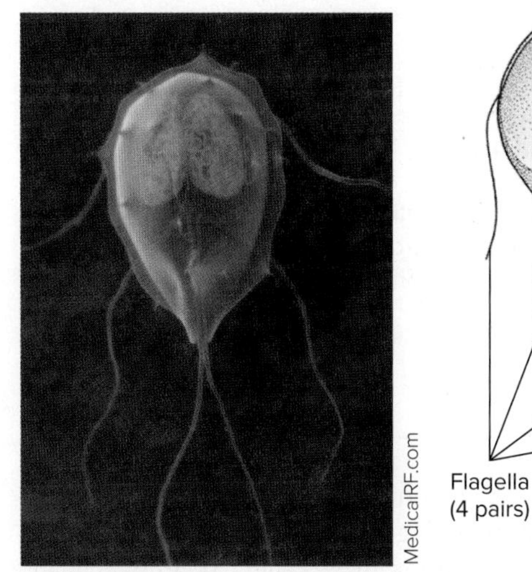

A 3D structure of *Giardia lambia*

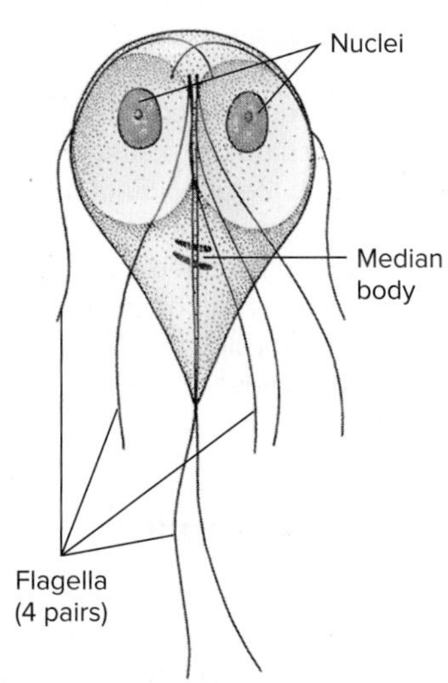

B Illustration of *Giardia lambia*

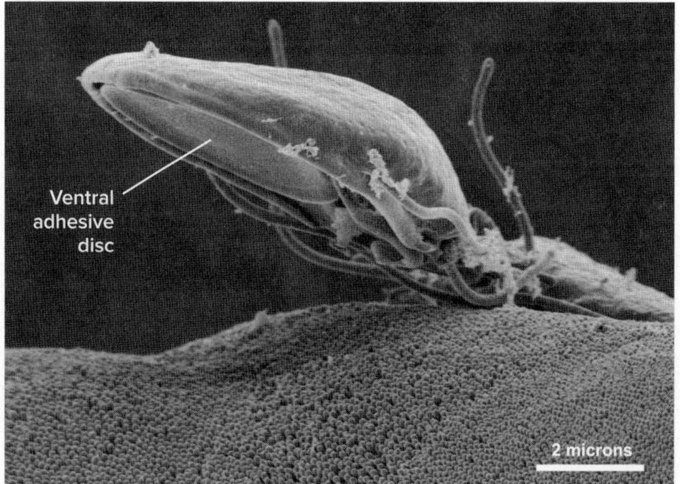

C *Giardia* attaching to intestinal wall

figure 5.14

Giardia lamblia causes diarrhea in humans infected via drinking water. **A** and **B,** Photograph and labeled illustration of *Giardia* in top view. **C,** Photograph showing the ventral disc used to attach each cell to the intestinal wall of the host.

and flagellated stages, so group members are sometimes called amoeboflagellates or schizopyrenids. In *Naegleria gruberi,* the amebic stage feeds on bacteria, but once all local food is exhausted, the ameba completely alters its cytoskeleton, transforming into a flagellated cell within 90 minutes. The flagellated stage is better able than the amebic stage to seek distant food sources. Most heteroloboseans feed on bacteria and are harmless, but *Naegleria fowleri* causes a few deaths each year. It lives in hot pools and causes primary amebic meningoencephalitis in humans when water containing amebas is inhaled. The amebas enter through the nasal passages and migrate along olfactory nerves to the brain, where tissue is destroyed.

Euglenozoa

Euglenozoa is generally recognized as a monophyletic group, based on the shared persistence of the nucleoli during mitosis, and the presence of discoid mitochondrial cristae. Cristae of this form occur also in a clade of amebas (Heterolobosea), so these groups are shown as sister taxa in figure 5.2 and are united as "Discobids." Members of Euglenozoa have a series of longitudinal microtubules just beneath the plasma membrane that help to stiffen the membrane into a **pellicle** (figure 5.15). There are two subgroups: Euglenida and Kinetoplasta.

Euglenids have chloroplasts with chlorophyll *b.* These chloroplasts are surrounded by a double membrane and are likely to have arisen via secondary endosymbiosis. *Euglena* is a representative of this group. It has a light-sensitive **stigma,** or eyespot, which is a shallow pigment cup that allows light from only one direction to strike a light-sensitive receptor (figure 5.15).

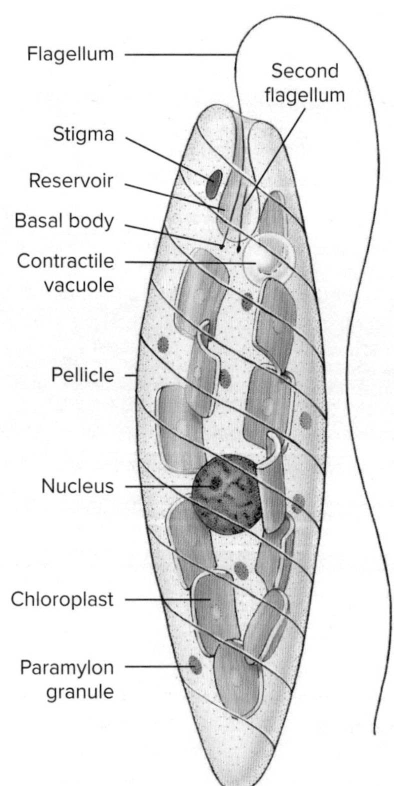

figure 5.15

Euglena. Features shown are a combination of those visible in living and stained preparations.

Kinetoplastans are all parasitic in plants and animals. They have a unique organelle called a **kinetoplast** (see section 5.1: Mitochondria). In kinetoplastids, the cellular basal body is connected to the kDNA inside the mitochondria by a filamentous attachment complex that spans the mitochondrial membrane. Because there is only one mitochondrion in the cell, it is important that nuclear division occurs at the same time as mitochondrial division. Both cells produced by division need a nucleus and a mitochondrion. Microtubules from the basal body position the kDNA correctly within the dividing mitochondrion.

Various species of *Trypanosoma* (Gr. *trypanon,* auger, + *soma,* body) (see figure 5.13) cause serious diseases in humans and other animals. Two subspecies of *Trypanosoma brucei* (*T. b. rhodesiense* and *T. b. gambiense*) cause clinically distinct forms of African sleeping sickness in humans. Another subspecies, *T. b. brucei,* along with several other trypanosomes, causes a similar disease in domestic animals, which makes agriculture very difficult in large areas of Africa. *T. brucei* is spread by tsetse flies of the genus *Glossina*. *Trypanosoma cruzi* causes American trypanosomiasis, or Chagas disease, a serious disease of humans in Mexico, South and Central America that may infect up to 8 million people. *T. cruzi* is spread by true bugs of the genus *Triatoma.*

Key Theme 5.4
HUMAN CONNECTIONS

Novel Targets for Drugs

The presence of unique cellular structures such as the kinetoplast in parasitic trypanosomes might seem like a trivial detail, but in fact these structures are vital for developing medicines. Because both the parasite and the human host are eukaryotes, there are few targets for drug action that would not harm the host as much as the parasite. However, because only the parasite has a kinetoplast, novel proteins needed for kDNA synthesis or other unique functions can become drug targets. Treating or preventing sleeping sickness in humans and other animals would save many lives.

Stramenopiles

Members of the clade Stramenopiles have tubular mitochondrial cristae. They may have flagellated cells, but stramenopiles are heterokont (Gr. *hetero,* different, + *kontos,* pole) flagellates. They have two different flagella, both inserted at the cell anterior, instead of the posterior as in opisthokonts (Gr. *opisth,* posterior). In heterokonts, the forward-directed flagellum is long and hairy, whereas the other is short and smooth, and trails behind the cell. This clade is sometimes called Heterokonta; the name stramenopile (L. *stramen,* straw, + *pile,* hair) refers to three-part tubular hairs covering the flagellum. This clade contains brown algae, yellow algae, and diatoms, all autotrophic forms collecting energy with plastids, but heterotrophic forms are also present. The opalinids, commensals in animals such as frogs, and once thought to be modified ciliates, as well as plant pathogens such as slime nets and oomycetes, are among the organisms in this clade. Perhaps the most well-known oomycetes belong to the genus *Phytophthora. Phytophthora infestans* causes late potato blight and was responsible for the Irish potato famine. A related species causes "sudden oak death" which is currently damaging large areas of California. Stramenopiles also includes some amebas with axopodia, such as *Actinosphaerium* (see figure 5.5) and *Actinophrys.*

Alveolata

Based on molecular and morphological evidence, three taxa (Ciliophora, Dinoflagellata, Apicomplexa) are often grouped together in Alveolata. Alveolates possess **alveoli,** membrane-bound sacs that lie beneath the plasma membrane. Recently, a new family of structural proteins called alveolins was discovered in all Alveolates (see section 5.1: Cell Movement by Glideosome). Alveolins co-occur with alveoli, but these proteins and membranous sacs lie in different relative positions across the Alveolata, as described in the sections for each group.

Ciliophora

Ciliates are a large and interesting group, represented in all types of freshwater and marine habitats. They are the most structurally complex and diversely specialized of all unicells. Most are free-living, although some are commensal or parasitic. They are usually solitary and motile, but some are colonial and others are sessile. There is a great diversity of shape and size. In general, ciliates are larger than most other unicells, ranging from 10 μm to 3 mm long. At some stage, all have cilia that beat in a coordinated rhythmic manner, although the arrangement of the cilia varies. In suctorians, the young possess cilia and are free-swimming, but the adults grow an attachment stalk, become sessile, and lose their cilia.

In ciliates, a pair of alveoli underlie each cilium and alveolins occur at a ridge where the two alveoli meet (figure 5.17). Alveoli co-occur with basal bodies, extrusomes (see section 5.1: Extrusomes), and the cell cortex. In many species, the alveoli and the proteins thicken the plasma membrane, creating a stiff pellicle. Cilia are short and usually arranged in longitudinal or diagonal rows. Cilia may cover the surface of the animal or be restricted to the oral region or to certain bands. In some forms, cilia are fused into a sheet called an **undulating membrane** or into smaller **membranelles;** both structures propel food into the **cytopharynx** (gullet). Other forms may have fused cilia forming stiffened tufts called **cirri,** which are often used in locomotion by the creeping ciliates, such as *Euplotes* in figure 5.16.

The basal bodies and structural fibers just beneath the pellicle form the **infraciliature** in ciliates (figure 5.16). The infraciliature apparently does not coordinate ciliary beat, as formerly hypothesized. Ciliary movement seems to be coordinated by waves of depolarization of the plasma membrane moving down the animal, similar to conduction of a nerve impulse.

Some ciliates have **trichocysts** or **toxicysts** between the bases of the cilia (figures 5.17 and 5.18). Upon mechanical or chemical stimulation, these small bodies explosively expel a long, threadlike structure. The mechanism of expulsion is unknown. The function of trichocysts is probably defensive. When a paramecium is attacked by a predatory *Didinium,* it expels its trichocysts but to no avail. Toxicysts, however, release a poison that paralyzes the prey of carnivorous ciliates. Toxicysts are structurally quite distinct from trichocysts. Many dinoflagellates have structures very similar to trichocysts.

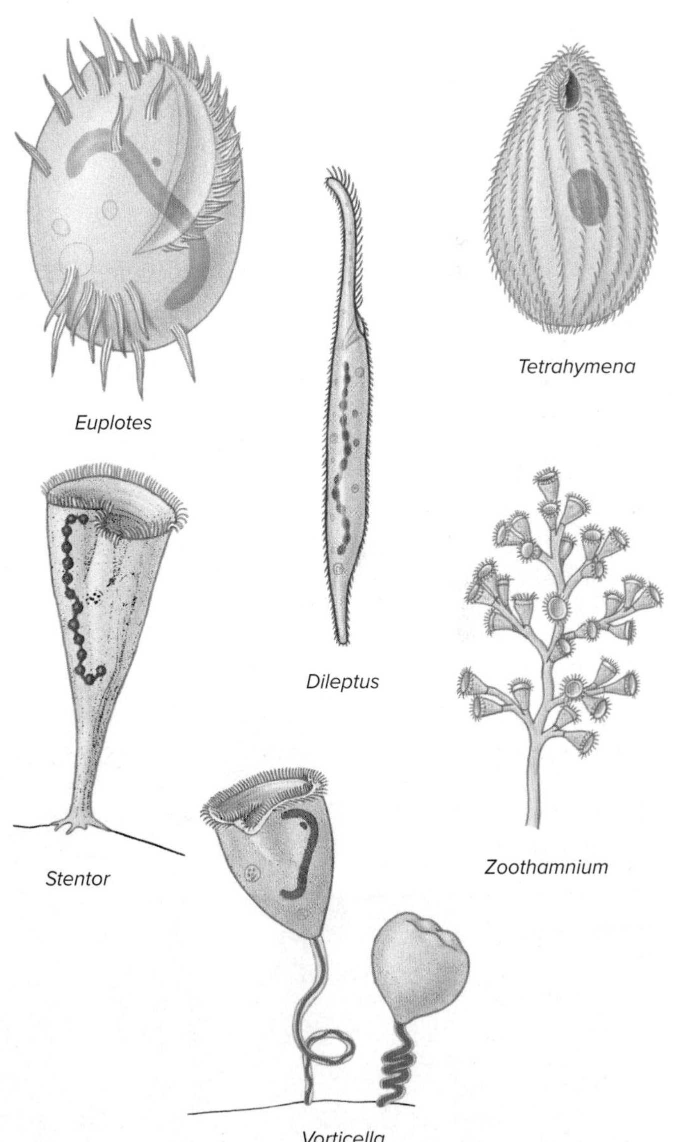

figure 5.16

Some representative ciliates. *Euplotes* have stiff cirri used for walking on the surface of an object or an air bubble. Contractile *myonemes* in the ectoplasm of *Stentor* and in the stalks of *Vorticella* allow great expansion and contraction. Note the macronuclei—long and curved in *Euplotes* and *Vorticella*, and shaped like a string of beads in *Stentor.*

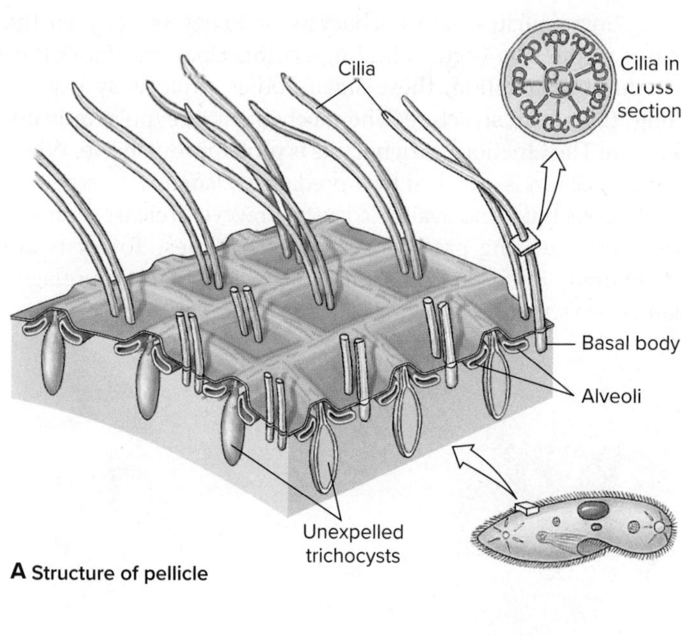

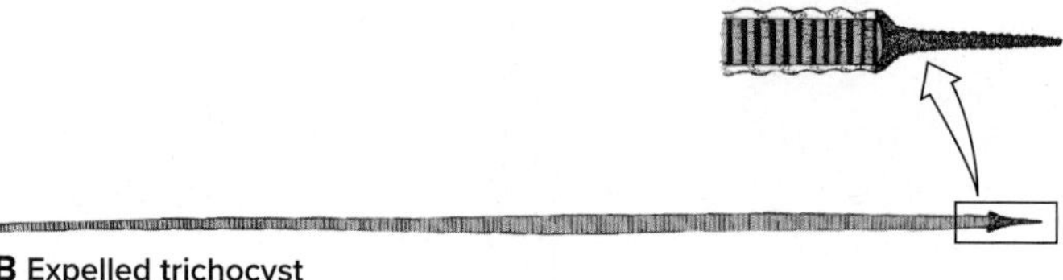

figure 5.17

A, The ciliate pellicle comprises the plasma membrane, the cell cortex, a pair of alveoli underlying each cilium, alveolins, microtubules, and basal bodies. **B,** Expelled trichocyst.

Most ciliates are holozoic, possessing a cytostome (mouth) that in some forms is a simple opening and in others is connected to a gullet or ciliated groove. The mouth in some is strengthened by stiff, rod-like structures for swallowing larger prey; in others, such as paramecia, ciliary water currents carry microscopic food particles toward the mouth. *Didinium* has a proboscis for engulfing the paramecia on which it feeds (see figure 5.10). Suctorians paralyze their prey and then ingest their contents through tubelike tentacles by a complex feeding mechanism that apparently combines phagocytosis with a sliding filament action of microtubules in the tentacles. An example is *Podophrya* in figure 5.10.

Contractile vacuoles occur in all freshwater and some marine ciliates; the number varies from one to many among the different species. In most ciliates, the vacuole is fed by one or more collecting canals (figure 5.18). The vacuoles occupy a fixed position, and each discharges through a more-or-less permanent pore.

Ciliates are always multinucleate, possessing at least one **macronucleus** and one **micronucleus,** but varying from one to many of either type. The macronuclei are responsible for metabolic, synthetic, and developmental functions. Macronuclei are varied in shape among the different species (see figures 5.16 and 5.18). Micronuclei participate in sexual reproduction and give rise to macronuclei after exchange of micronuclear material between individuals. Micronuclei divide mitotically, and macronuclei divide amitotically.

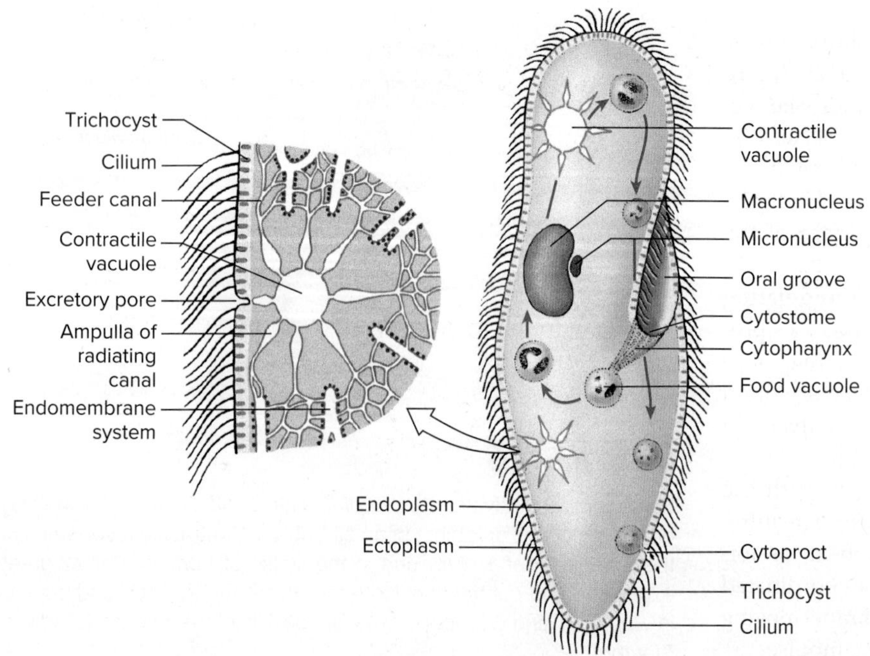

figure 5.18

Left, Enlarged section of a contractile vacuole of *Paramecium.* Water is collected by the endoplasmic reticulum, emptied into feeder canals, and then emptied into the vesicle. The vesicle contracts to empty its contents to the outside, thus serving as an osmoregulatory organelle. *Right, Paramecium,* showing gullet (cytopharynx), food vacuoles, macronucleus and micronucleus.

Reproduction and Life Cycles

Life cycles of ciliates usually involve both asexual binary fission (figure 5.19) and a type of sexual reproduction called **conjugation** (figure 5.20). Conjugation is a temporary union of two individuals for the purpose of exchanging chromosomal material. During union, the macronucleus of each cell degenerates, and the micronucleus of each individual undergoes meiosis, giving rise to four haploid micronuclei, three of which degenerate (figure 5.20A and B). The remaining micronucleus then divides into haploid pronuclei (figure 5.20C), one of which is exchanged for a pronucleus of the conjugant partner. When the exchanged pronucleus unites with the pronucleus of the partner, the diploid number of chromosomes is restored. The two partners, each with fused pronuclei (now comparable to a zygote), separate, and each divides twice by mitosis, thereby giving rise to four daughter paramecia each (figure 5.20D–G).

Conjugation always involves two individuals of different mating types. This prevents inbreeding. Most ciliate species are divided into several varieties, each variety containing two *mating types.* Mating occurs only between the differing mating types within each variety.

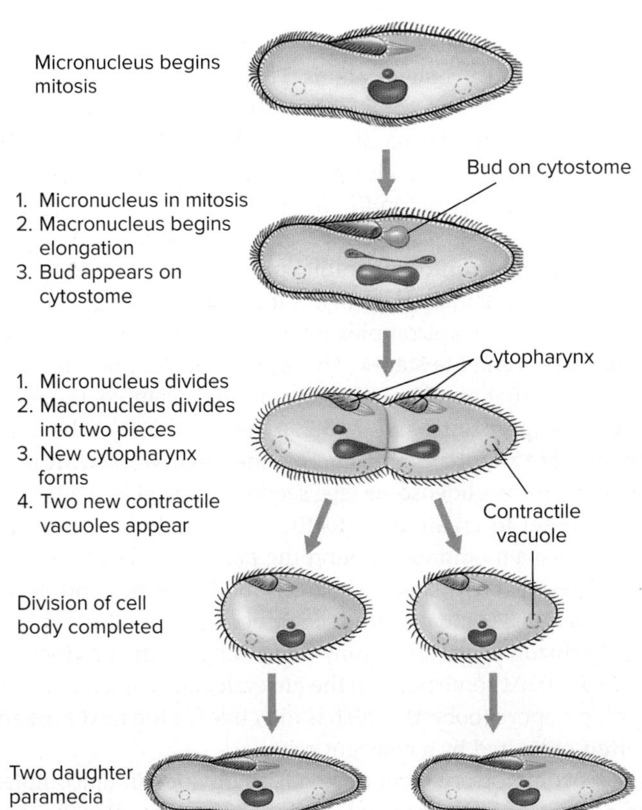

figure 5.19

Binary fission in a ciliophoran (*Paramecium*). Division is transverse, across rows of cilia.

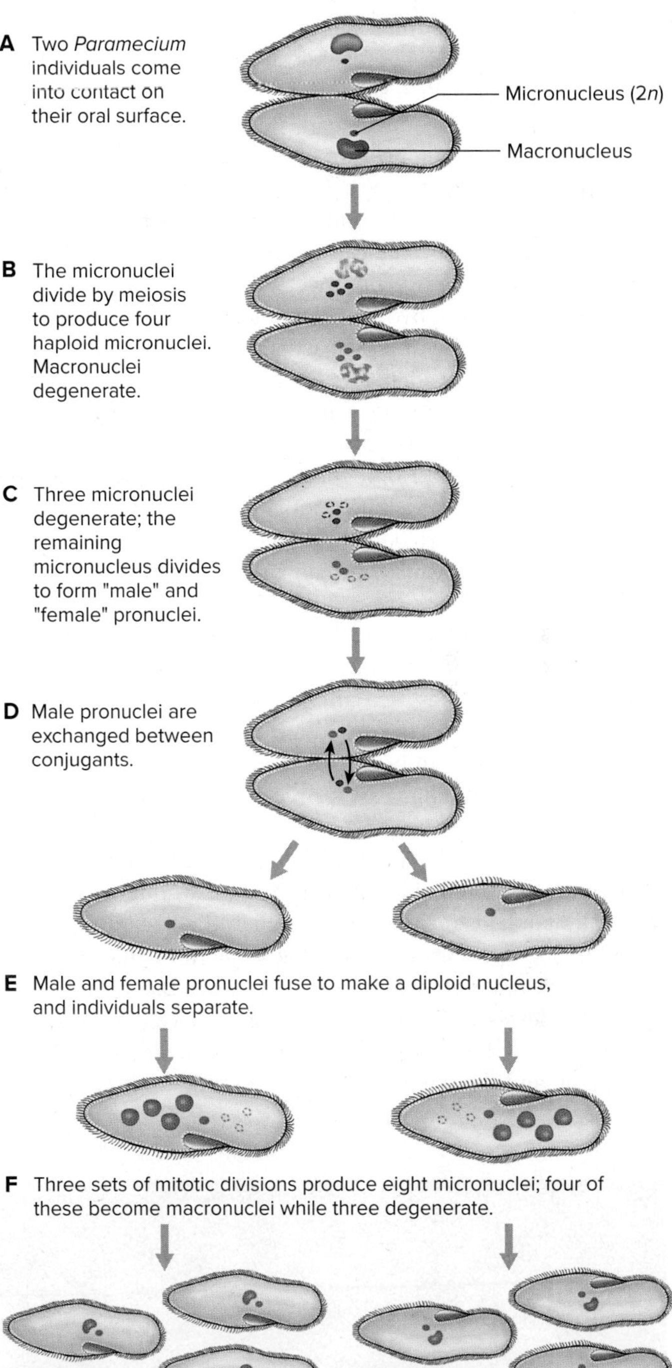

figure 5.20

Conjugation in *Paramecium.*

Dinoflagellata

As members of clade Alveolata, dinoflagellates have alveoli and alveolins. Alveolins and cortical microtubules lie between the plasma membrane and alveoli; all these structures form the pellicle. In armored dinoflagellates such as *Ceratium* (figure 5.13), the armor consists of cellulose plates within the alveoli. In naked dinoflagellates, such as *Noctiluca* (figure 5.13), the alveolar sacs are empty of plates.

Many dinoflagellates are photoautotrophic, with chromatophores bearing chlorophylls *a* and *c* but not *b*. Most dinoflagellates have brown or yellow chromatophores, although some are colorless. Ecologically, some species are among the most important primary producers in marine environments. They commonly have two flagella, one equatorial and one longitudinal, each borne at least partially in grooves on the body (for example, see *Ceratium*, *Ptychodiscus*, and *Noctiluca* in figure 5.13). Heterotrophic species can ingest prey through a mouth region between the plates near the posterior area of the body. *Noctiluca* (see figure 5.13), a colorless dinoflagellate, is a voracious predator and has a long, motile tentacle, near the base of which its single, short flagellum emerges. *Noctiluca* is one of many marine organisms that can produce light (bioluminescence).

Zooxanthellae are dinoflagellates that live in mutualistic association in tissues of certain invertebrates, including other unicellular eukaryotes, sea anemones, horny and stony corals, and clams. The association with stony corals is of ecological and economic importance because only corals with symbiotic zooxanthellae can form coral reefs (see section 7.8: Coral Reefs).

Dinoflagellates can damage other organisms, as when they produce a "red tide" (figure 5.21). Although this name originally applied to situations in which the organisms reproduced in such profusion (producing a "bloom") that the water turned red from their color, any instance of a bloom producing detectable levels of toxic substances is now called a red tide. The water may be red, brown, yellow, or not remarkably colored at all, and the phenomenon has nothing to do with tides!

Mark Conlin/Getty Images

figure 5.21

"Red tide" makes a striking appearance on the coastline. The water appears colored because it contains a high concentration of dinoflagellates.

Dinoflagellates may produce a toxin, such as saxitoxin, inside their cells. When they are consumed by suspension feeders including clams, mussels, or other bivalve molluscs (see Figure 10.23), this toxin accumulates in the mollusc tissues. No harm comes to the dinoflagellates or to the molluscs, but when vertebrate predators, including humans, eat the bivalves, the neurotoxin causes numbness, paralysis, and often death. When a dinoflagellate bloom is detected, fisheries officials prohibit harvesting of bivalves until the dinoflagellates have dispersed and the toxin is no longer present in the molluscan tissues. Red tides have caused considerable economic losses to the shellfish industry.

Pfiesteria piscicida is one of several related dinoflagellate species that may affect fish in brackish waters along the Atlantic Coast, south of North Carolina. Much of the time, *Pfiesteria* feeds on algae and bacteria, but something in the excreta of large schools of fishes causes it to release a powerful, short-lived toxin. The toxin may stun or kill fishes, often creating skin lesions. *Pfiesteria* has flagellated and ameboid forms among its more than 20 body types; some forms feed on fish tissues and blood. Although it does not have chloroplasts, it may sequester chloroplasts from its algal prey and gain energy from them over the short term. This fascinating group of species was discovered in 1988.

Apicomplexa

As members of clade Alveolata, apicomplexans make membranous alveoli and alveolin proteins. The alveoli, sometimes called an "inner membrane complex" in this group, occur immediately beneath the plasma membrane. Microtubules and alveolins are internal to the alveoli; all the structures support the actin-myosin cytoskeleton (figure 5.22).

Apicomplexans are almost all endoparasites. They use hosts from many different animal phyla and their complex life cycles frequently include an insect intermediate host. The diagnostic feature for apicomplexans is the presence of an **apical complex** in some life stages. This apical complex occurs in life cycle stages that enter or exit host cells, for example, the **merozoites** and **sporozoites** (figure 5.23). Two key parts of the apical complex are **rhoptries** and **micronemes** that work with other elements of the glideosome (see section 5.1: Cell Movement by Glideosome) to create a protective vacuole derived from the host's plasma membrane around the parasite cell (figure 5.22).

The life cycle usually includes both asexual and sexual reproduction. Asexual reproduction occurs by multiple fission, called schizogony or sporogony depending on the product (see figure 5.23). At some point in the life cycle, most apicomplexans develop a spore (oocyst), which is infective for the next host and is often protected by a resistant coat.

Locomotor organelles are not as obvious in apicomplexans as they are in other unicellular eukaryotes. Pseudopodia occur in some intracellular stages, and gametes of some species are flagellated. Tiny contractile fibrils can form waves of contraction across the body surfaces to propel the organism through a liquid medium.

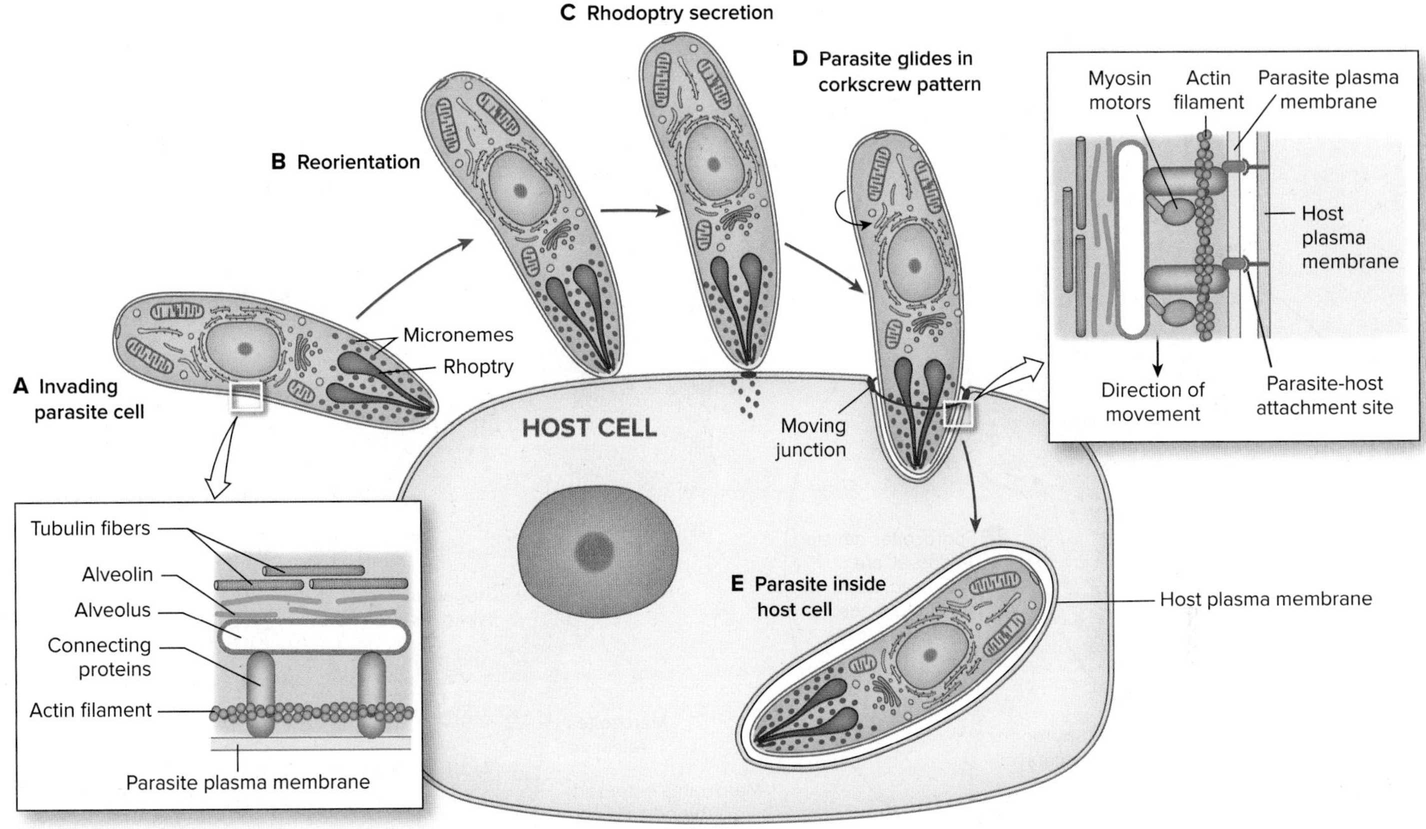

figure 5.22

A generalized description of glideosome function during host cell invasion in an apicomplexan. **A,** The invasive stage of the parasite is initially outside the host cell. The inset shows the parasite plasma membrane, actin filament, membranous sacs that make up the alveoli, and proteins connecting these structures together. **B,** The parasite secretes microneme proteins to begin forming a junction between parasite and host and then reorients to adhere to the host cell plasma membrane. The number of micronemes present determines the time over which gliding can occur. **C,** Secretions from the micronemes and rhoptries create the moving junction between the parasite and hosts; rhoptery proteins aide in creating traction. The inset shows proteins linking the parasite plasma membrane to the host plasma membrane and the addition of myosin motors to the parasite structures described in the inset from part A. Myosin motor proteins allow the parasite to move. **D,** As the parasite glides over the host in a corkscrew fashion, the moving junction between the parasite cell and the host cell causes the host membrane to indent. **E,** At the end of the process, the parasite has moved inside the host cell, surrounded by a membrane collected from the host during invasion.

Apicomplexan parasites belong to two groups: **coccidians** (Coccidia [Gr. *kokkos,* kernel or berry]), which infect both vertebrates and invertebrates, and **gregarines,** which live mainly in the digestive tract and body cavity of certain invertebrates. We discuss two examples of medical importance from class Coccidea: *Plasmodium* and *Toxoplasma.*

Plasmodium causes **malaria.** The vectors (carriers) of the parasites are female *Anopheles* mosquitos, and the life cycle is depicted in figure 5.23. Malaria is one of the most important diseases in the world: 41% of the earth's people live in malarial regions. Global estimates of deaths caused by malaria range from 700,000 to over 2 million per year, with 75% being African children. Although these statistics represent a decline from the situation 45 years ago, some resurgence is caused by increased resistance of mosquitos to insecticides, rising numbers of *Plasmodium* strains that are resistant to drugs, and socioeconomic conditions and civil strife that interfere with control of malaria in tropical countries.

Key Theme 5.5

HUMAN CONNECTION

Understanding Malaria

Although malaria has been recognized as a disease and a scourge of humanity since antiquity, its unicellular eukaryote cause was discovered only 100 years ago by Charles Louis Alphonse Laveran, a French army physician. At that time, the mode of transmission was still mysterious, and "bad air" (hence the name malaria) was a popular suspect. Ronald Ross, an English physician in the Indian Medical Service, determined some years later that the malarial organism was carried by *Anopheles* mosquitos. Because Ross knew nothing about mosquitos or their normal parasites, his efforts were long frustrated by studying the wrong mosquitos and parasites. That he persisted is a tribute to his determination. The momentous discovery earned Ross the Nobel Prize in 1902 and knighthood in 1911.

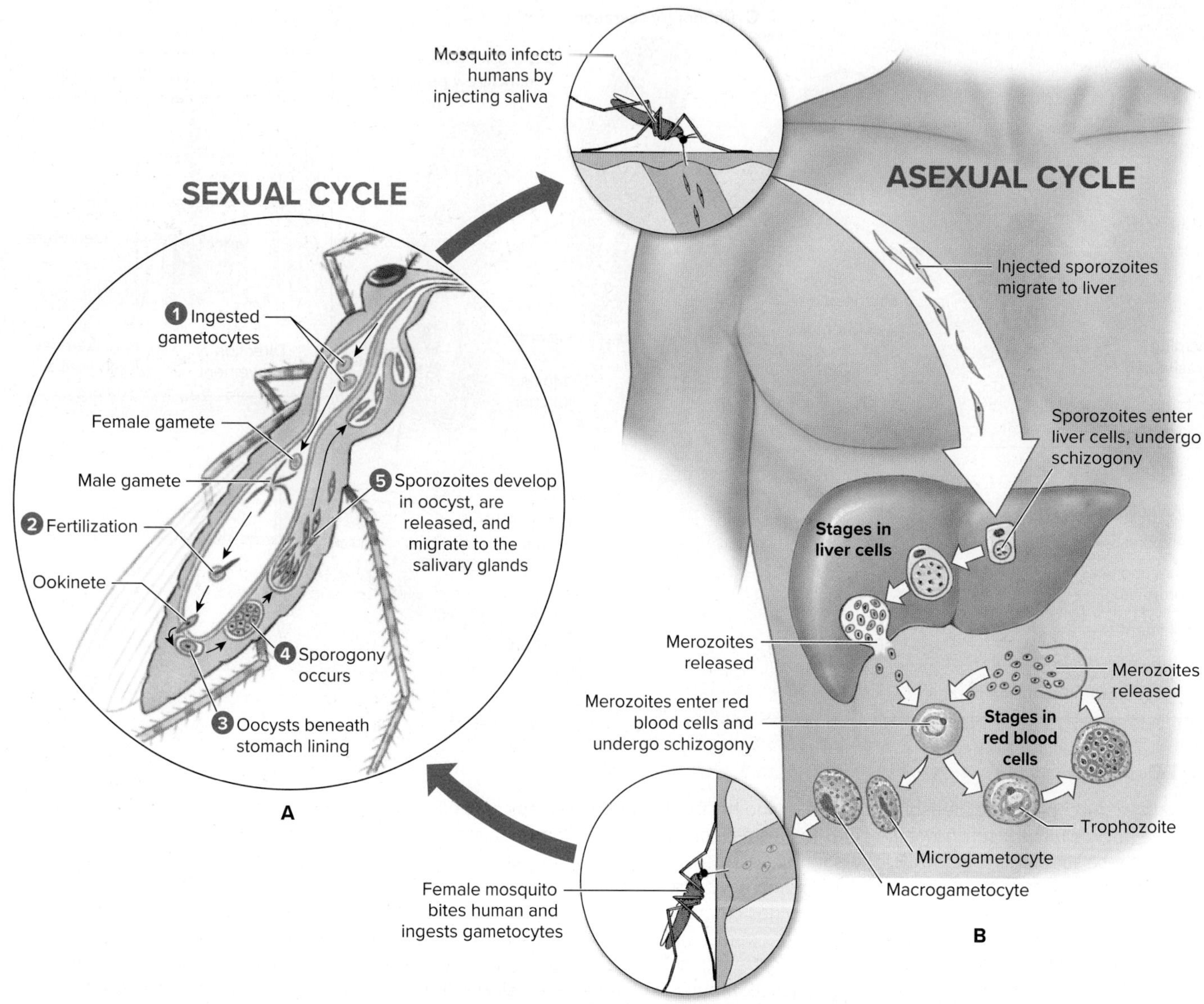

figure 5.23

Life cycle of *Plasmodium vivax,* one of the unicellular eukaryotes (class Coccidea) that causes malaria in humans. **A,** The sexual cycle produces sporozoites in the body of the mosquito. **B,** Sporozoites infect humans and reproduce asexually, first in liver cells, and then in red blood cells. Malaria is spread by the *Anopheles* mosquito, which ingests gametocytes along with human blood. Then, when biting another victim, the mosquito leaves sporozoites in the new wound.

Toxoplasma (Gr. *toxo,* a bow, + *plasma,* molded) is a common parasite in the intestinal tissues of cats. However, this parasite can produce extraintestinal stages in a wide variety of other animals as well—for example, rodents, cattle, and humans. The extraintestinal forms do not produce gametes and oocysts, but they can initiate the intestinal cycle in a cat if the cat eats infected prey. In humans, *Toxoplasma* causes few or no ill effects except in AIDS patients or in women infected during pregnancy. Prenatal infection greatly increases the chances of a birth defect in the baby; about 2% of all mental retardation in the United States is a result of congenital toxoplasmosis. Humans can become infected by eating insufficiently cooked beef, pork, or lamb or by accidentally ingesting oocysts from the feces of cats. For these reasons, pregnant women should not eat raw meat or empty cats' litterboxes.

Key Theme 5.6
HUMAN CONNECTIONS

Infection by *Toxoplasma* and *Cryptosporidium*

Some 16% or more of adults in the United States are infected with *Toxoplasma gondii* but have no symptoms because the parasite is held in check by the immune system. However, *T. gondii* is one of the most important opportunistic infections in AIDS patients. The latent infection is activated in 5–15% of AIDS patients, often in the brain, with serious consequences. Another coccidian, *Cryptosporidium parvum*, was first reported in humans in 1976. We now recognize it as a major cause of diarrheal disease worldwide, especially in children in tropical countries. Waterborne outbreaks have occurred in the United States, and the diarrhea can be life-threatening in immunocompromised patients (such as those with AIDS). The latest coccidian pathogen to emerge has been *Cyclospora cayetanensis*. U.S. infection rates for 2005 were about 0.2 cases per 100,000 persons, with diarrhea being the most common symptom of infection. Infection usually occurs by ingestion of contaminated food or water.

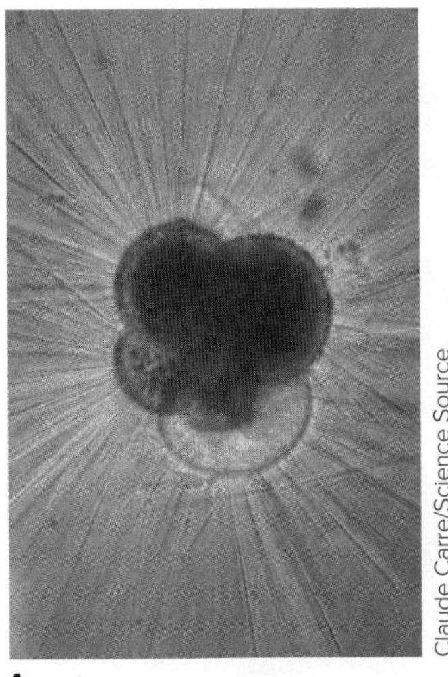
Claude Carre/Science Source
A

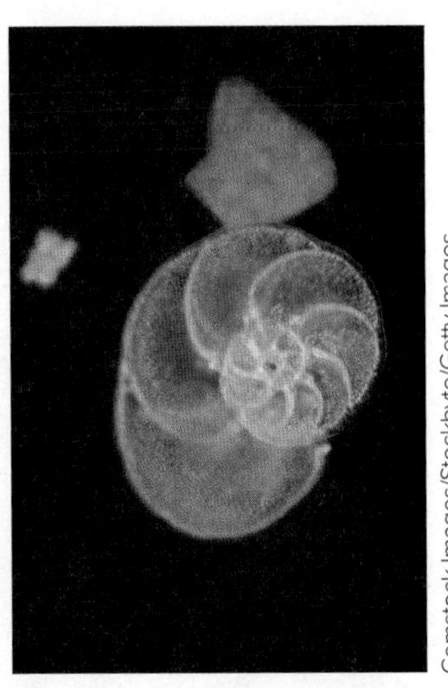
Comstock Images/Stockbyte/Getty Images
B

figure 5.24

A, Air-dried foraminiferan, showing spines extending from the test. **B,** A test of the foraminiferan. Foraminiferans are ameboid marine unicellular eukaryotes that secrete a calcareous, many-chambered test in which to live, and then extrude protoplasm through pores to form a layer over the outside. The animal begins with one chamber, and as it grows, it secretes a succession of new and larger chambers, continuing this process throughout life. Many foraminiferans are planktonic, and when they die, their shells are added to sediments on the sea floor.

Cercozoa

Cercozoans do not share a common body plan, but form a clade in molecular phylogenetic studies. There are flagellated and ameboid members. The ameboid members of the group may be naked or testate. The testate amebas include *Euglypha*, which makes a test from collected particles (see figure 5.12); *Clathrulina*, which has a siliceous capsule (see figure 5.5); and phaeodarians with an amorphous silica skeleton with magnesium, calcium, and copper added. The skeleton of the latter group contains hollow spines and is known from microfossils dating back to the Cambrian period.

Foraminifera

The foraminiferans (L. *foramen*, hole, + *fero*, bear) are mostly marine forms that secrete complex, many-chambered tests of calcium carbonate (figure 5.24). They usually add sand grains to the secreted material, using great selectivity in choosing colors. Slender pseudopodia extend through openings in the test and then run together to form a protoplasmic net (reticulum) in which they ensnare their prey. They are beautiful little creatures with many pseudopodia radiating out from a central test.

The foraminiferan test is a very useful structure because it is distinctive enough to identify particular species and because it preserves very well in ocean sediments, lasting long after the ameba that made it has died. Add to this the fact that different planktonic foraminiferan species thrive only at certain temperatures, and one has the basis to reconstruct paleo-climates by reading history in deep-sea sediments. Such sediments may be 600–3600 m deep (approximately 2000–12,000 feet) and contain as many as 50,000 foraminiferans per gram. Changing percentages of different species from oldest to youngest sediments tell the story of temperature change in oceanic surface waters.

Some fossil foraminiferans reached 100 mm in diameter, but they are small in comparison to the deep-sea xenophyophorans. These giant, multinucleate benthic forms reach 20 cm in diameter. They cement a wide variety of particles into a fragile test. They may be filter or deposit feeders, and are unusual in that their cytoplasm often contains barium sulfate crystals and their fecal pellets, retained in the body, concentrate heavy metals such as lead or mercury.

Radiolaria

Radiolarians (L. *radiolus*, small ray) are marine forms, mostly living as plankton, that have intricate and beautiful siliceous skeletons (figure 5.25). A central capsule that separates the inner and outer cytoplasm is perforated to allow cytoplasmic continuity. Around the capsule is a frothy mass of cytoplasm from which axopodia arise (see section 5.2: Cell Extension and Movement via Microtubules). Sticky axopodia capture prey items, which are carried by the streaming cytoplasm through the capsule. The cytoplasm on one side of the axial rod moves outward toward the axial tip, while that on the other side moves

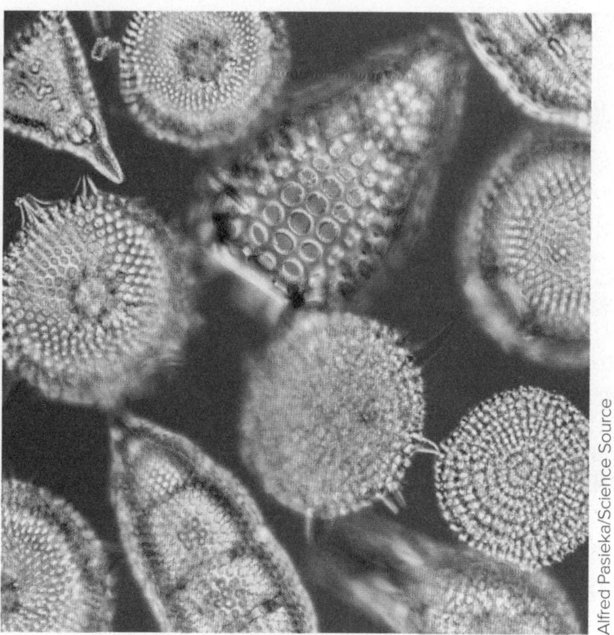

figure 5.25

Types of radiolarian tests. In his study of these beautiful forms collected on the famous *Challenger* expedition of 1872 to 1876, Ernst Haeckel worked out our present concepts of symmetry.

inward toward the test. Solitary radiolarians feed on bacteria, micro-algae, and micro-flagellates, whereas colonial radiolarians collect nutrients from symbiotic algae. With the exception of one subgroup, which is now placed in Cercozoa, organisms described as radiolarians before the advent of molecular phylogenetics are still considered radiolarians.

Radiolarians are among the oldest known unicellular eukaryotes, and they and foraminiferans have left excellent fossil records. For millions of years, tests of dead unicellular eukaryotes have been dropping to the sea floor. Many limestone and chalk deposits on land were formed by foraminiferans when the land was covered by sea. In contrast, the silica skeletons of radiolarians decay to an ooze that lithifies to form chert.

Plantae

The clade Plantae, sometimes called Archaeplastida, contains glaucophytes, red algae, green algae, bryophytes, and vascular plants, among others. Within Plantae, members of the clade Viridiplantae have chlorophylls *a* and *b* and include unicellular, colonial, and multicellular green algae. The volvocine green algae are exemplified by unicellular *Chlamydomonas* (see figure 5.13), colonial *Gonium* (see figure 5.13), and multicellular *Volvox carteri* (figure 5.26). As discussed in Chapter 3

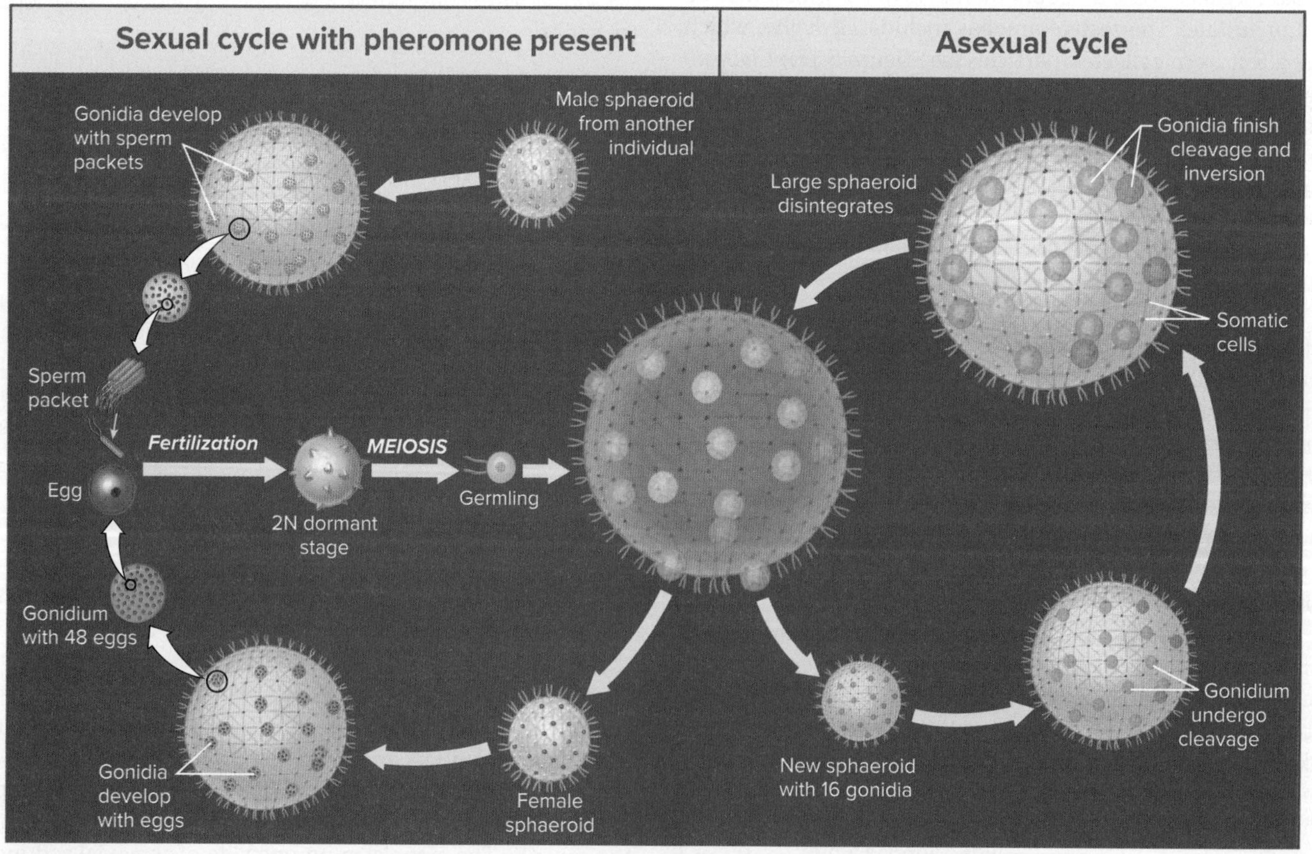

figure 5.26

The life cycle of *Volvox carteri* showing both the asexual and the sexual phases of the life cycle. New sphaeroids develop from specialized reproductive cells called gonidia. Somatic cells in sphaeroids live only 48 hours.

(see section 3.1), both colonial and multicellular organisms are embedded in an extracellular matrix (ECM). However, in colonial forms, all cells can reproduce sexually, whereas this capacity is limited to cells specialized for that purpose in multicellular organisms. *Volvox carteri* has a complete division of labor: small, haploid, motile, photosynthetic, somatic cells embed in an ECM to form a sphaeroid body and 16 large, haploid, nonmotile reproductive cells called gonidia embed beneath the somatic cells. Gonidia undergo cleavage and inversion to form new sphaeroids with both somatic and reproductive cells present. As the 16 new sphaeroids that developed from gonidia leave by digesting through the ECM, the sphaeroids of somatic cells that once housed them die (figure 5.26).

Male and female strains of *V. carteri* are not distinguishable in the asexual phase, but when exposed to a particular pheromone, the sexes become distinct. In female sphaeroids, gonidia develop into sphaeroids containing about 48 eggs, whereas the gonidia of male sphaeroids develop into organisms containing 64–128 sperm packets. Sperm packets are motile. Fertilization produces a dormant diploid resting stage that tolerates freezing and dessication. Upon germination, the resting stage undergoes meiosis to produce a single viable haploid "germling" that undergoes cleavage to make a new sphaeroid. Sexual reproduction occurs only sporadically.

The persistence of unicellular, colonial, and multicellular forms within one lineage permits biologists to make inferences about steps in the evolution of multicellularity. In the volvocine algae it appears that the ECM evolved through modifications of the inner and outer plant cell walls. A diverse array of cell wall proteins is important for ECM function, but these proteins are very different from the ECM proteins of animals or other multicellur forms.[1]

Centrohelida

Prior to the widespread use of molecular methods in phylogenetics, the Heliozoa was a large group of testate amebas forming axopodia. All species formerly placed in Heliozoa, except members of the order Centrohelida, have been moved to other taxa, hence Heliozoa has been discontinued. Centrohelids are amebas with flattened mitochondrial cristae. The structure of their axopodia is distinctive: microtubules (visible in figure 5.8) within the axoneme are arranged in hexagons or triangles in this group. In most centrohelids, the axopodia extend through a coat of variously shaped silica scales, but this coat may be absent and is replaced in some taxa by a mucous layer. Most centrohelids live in freshwater, but a few clades have colonized brackish or marine habitats. These lovely unicells are predators. Some former heliozoans are now in Stramenopiles (for example, *Actinophrys*) or Cercozoa (for example, *Clathrulina*).

Amoebozoa

Amoebozoans include naked and testate amebas, as well as amebas with flagellated stages in the life cycle. Amoebozoans typically have branched tubular mitochondrial cristae, but this feature is not unique to the group. Ameboid forms include the fascinating plasmodial and cellular slime molds, sometimes called social amebas (for example, *Dictyostelium*), testate amebas with lobose pseudopodia such as *Arcella* (see figures 5.5 and 5.12), and naked lobose amebas such as *Chaos carolinense, Amoeba proteus*, or members of genus *Acanthamoeba* (see figure 5.3). *Chaos carolinense* or *A. proteus* are sometimes used in biology laboratories, but *Acanthamoeba castellani* (see figure 5.3) has gained notoriety because of its impact on human health. *Acanthamoeba castellani* kills cells of the human cornea and is spread by contact lenses that are not properly disinfected. Amoebozoa also houses entozoic amebas—those that live inside humans or other animals. *Entamoeba histolytica* causes amebic dysentery in infected people.

Opisthokonta

Opisthokonta is a clade characterized by a combination of flattened mitochondrial cristae and one posterior flagellum on flagellated cells, if such cells are present. Opisthokonta contains animals and fungi as well as microsporidians, choanoflagellates, and some little-known unicells (figure 5.27). The little-known forms include ichthyosporeans (animal parasites sometimes called DRIPs; one type (*Ichthyophonus*) causes skin lesions in fishes), nucleariid amebas, corallochytreans, and ministeriid amebas. Microsporidians are intracellular parasites now recognized as specialized fungi; some cause disease in humans. Choanoflagellates are solitary or colonial unicellular eukaryotes currently placed as the sister group to animals (metazoans).

Choanoflagellates possess genes, such as tyrosine kinase and cadherin genes, involved in cell-cell signaling and in cell adhesion that are homologous to those of animals. Choanoflagellate cells are distinctive; they are ovoid in shape, and at one end a collar of microvilli surrounds a flagellum. Beating of the flagellum moves water through the collar where food particles carried by the water are collected. Choanoflagellate cells are very similar to the choanocytes of sponges (see Chapter 6). In fact, such cells may be an ancestral characteristic of animals because in addition to sponges, they occur in certain corals and in some echinoderms. If they are ancestral, they have been lost in the evolutionary ancestry of most animal phyla.

[1]Abedin, M., and N. King. 2010. Diverse evolutionary paths to cell adhesion. *Trends in Cell Biology* **20:**734–742.

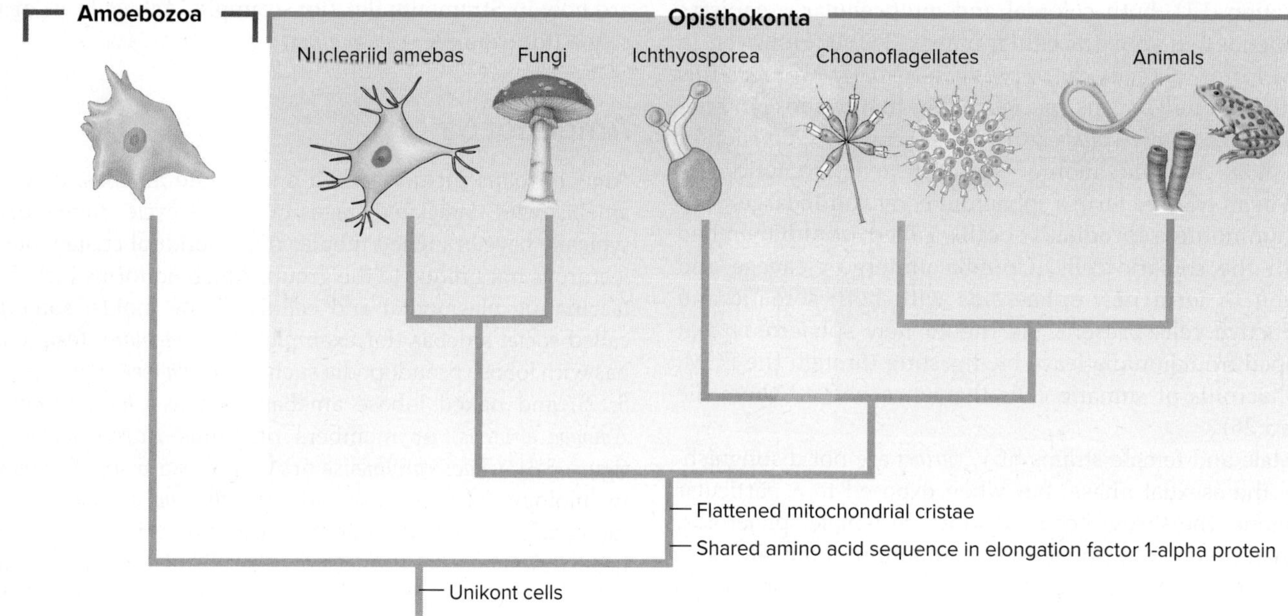

figure 5.27

One hypothesis regarding relationships among some members of Opisthokonta: Choanoflagellates are shown as the sister taxon to Animals. Choanoflagellates shown are *Codonosiga* on the left and *Proterospongia* on the right.

5.6 Phylogeny

The phylogeny of eukaryotes is important to evolutionary biologists and to those with more practical goals. One reason to find phylogenetic homes for disease-causing organisms is that treatments effective against particular pathogens often work on related organisms. For example, there were no effective treatments against microsporidians until it was discovered that microsporidians were highly specialized fungi and thus anti fungal drugs could kill them.

Relationships among unicellular eukaryotes have been difficult to discover because the "single-cell" body plan has relatively few unique morphological features. There are many unicells with an ameboid body–how can we be sure they inherited this body plan from a common ancestor rather than evolving it independently? There are unicells that switch between ameboid and flagellated cell bodies within one life cycle–how will we know which cell body was the ancestral form? Most cells share the same basic suite of organelles, so we cannot use presence of an organelle like a mitochondrion to identify sister taxa. Using electron microscopy, we have been able to look for details of organelles that might indicate evolutionary relationships–details such as the shape of the internal membranes in mitochondria–but these differences often relate more to function than to ancestry. It was once presumed that reduction or loss of mitochondria could indicate evolutionary history, but we know now that organisms from environments with little or no oxygen will have reduced mitochondria.

The advent of gene sequencing and the use of multiple genes to infer phylogeny show more promise; figure 5.2 illustrates hypothesized relationships among 18 groups of unicellular eukaryotes. These 18 groups do not represent the full diversity of unicellular eukaryotes but are the most well-known of more than 70 groups discovered so far. Characters that unify some of the 18 taxa have emerged, as in the case of the Alveolata, where all members of the group have membranous sacs called alveoli beneath the plasma membrane. Recent work also identified a unique family of structural proteins called alveolins present in all group members surveyed so far. Most research places Alveolata as the sister taxon to the Stramenopiles, with this pair as the sister taxon to the Rhizaria (see figure 5.2).

The supergroup Rhizaria unites Cercozoa, Foraminifera, Radiolaria, and some other taxa to form a clade whose many members are testate amebas, some with rhizopodia. However, some groups, such as Radiolaria, comprise different sets of species than when they were originally formulated. Many species of amebas were removed from Radiolaria based on their positions in molecular phylogenies. Rhizaria includes *Gromia sphaerica*, which made headlines when it was discovered rolling across the seafloor off the Bahamas at a depth of 2000 feet. This ameba is several centimeters in diameter, so the size is striking, but the really exciting part is that its track on the sea floor is very like fossil tracks attributed to bilaterally symmetrical organisms. Paleontologists are now reevaluating some trace fossil evidence, wondering about the role of giant amebas.

The union of Stramenopiles, Alveolata, and Rhizaria into a supergroup called SAR (sometimes called RAS when the order of taxa is reversed) emerged from several different phylogenetic studies, and other work indicated that this large group is the sister taxon to Plantae (also called Archaeplastida). Plantae is a group of photosynthetic organisms descended from the lineage that evolved the primary symbiosis between a cyanobacterium and a eukaryotic cell. The endosymbiotic cyanobacterium evolved into a chloroplast. Photosynthetic eukaryotes outside of Plantae, of which there are many in Euglenozoa, Stramenopiles, and Dinoflagellata, to name just a few, gained their chloroplasts through secondary endosymbiosis. They merged with another eukaryotic cell, not with a cyanobacterium. Examining the cladogram in figure 5.2, readers will see that Apicomplexa, where almost all taxa are endoparasites, is the sister taxon to the Dinoflagellata, which are photosynthetic organisms. After studying this relationship, biologists wondered if there was any evidence of a chloroplast in apicomplexans. This led to the discovery of a chloroplast derivative, called an apicoplast, in apicomplexans. The apicoplast has taken on new biochemical functions, but it still retains some features of the ancestral chloroplast. There is an apicoplast in *Plasmodium*, the organism that causes malaria (see figure 5.23), a human scourge. The truly exciting aspect of this discovery is that the chemistry of the apicoplast can be used as a drug target for antimalarials. Drugs that attack the apicoplast will distinguish between the cells of the parasite and the cells of the host; as explained in Key Theme 5.4, this is very important because both host and parasite are eukaryotes whose cells share a fundamentally similar structure and biochemistry.

The Unikonts comprise two clades: Amoebozoa and Opisthokonta, united on the grounds that when flagellated cells are present, there is only one flagellum. In Opisthokonta, the single flagellum is posterior, as it is a typical animal sperm cell. As readers will see in Chapter 6, Opisthokonta comprises fungi, choanoflagellates, and all the animals. Many readers are surprised that fungi are more closely related to animals than to plants because in traditional biology courses they are often taught in the botany component. However, features such as the shared presence of chitin in fungi and animals make sense in light of the phylogenetic relationship. The remaining chapters of this book focus on animals, but animal interactions with unicellular eukaryotes will appear again in the context of food webs, symbioses, disease agents, and host-parasite interactions.

Classification of Unicellular Eukaryotes

The classification below is not exhaustive, and with few exceptions, we list only taxa discussed in this chapter.

Metamonada (met′-a-mon-a′-da) (Gr. *meta*, after + L. *monas*, unit). Basal bodies and flagella typically in groups of four, often associated with a nucleus.

Phylum Retortamonada (re-tor′ta-mo′nad-a) (L. *retorqueo*, to twist back, + *monas*, single, unit). Golgi apparatus lacking; mitochondria reduced or absent; three anterior and one recurrent (running toward posterior) flagellum lying in a groove; intestinal parasites or free-living in anoxic environments. Example: *Retortamonas*.

Phylum Diplomonada (di′plo-mon-a′da) (Gr. *diploos*, double, + L. *monas*, unit). One or two basal bodies associated with a nucleus; individual flagellated life stages with one to four flagella; mitotic spindle within nucleus; cysts present; free-living or parasitic. Example: *Giardia*.

Phylum Parabasala (par′a-bas′a-la) (Gr. *para*, beside, + *basis*, base). With very large Golgi apparatus supported by unique parabasal fibers; up to thousands of flagella. *Trichomonas* and two other forms comprise clade Parabasalids.

Order Trichomonadida (tri′ko-mon-a′di-da) (Gr. *trichos*, hair, + *monas*, unit). Typically at least some basal bodies associated with rootlet filaments characteristic of trichomonads; parabasal apparatus present; division spindle extranuclear; hydrogenosomes present; no sexual reproduction; true cysts rare; all parasitic. Examples: *Dientamoeba, Trichomonas*.

Phylum Heterolobosea (het′ə-rō-lōbō′se-ə) (Gr. *heteros*, different, + *lobos*, lobe). Naked amebas with eruptive pseudopodia; a typical life cycle includes both amebic and flagellated stages. Group members sometimes called amoeboflagellates or schizopyrenids. Examples: *Naegleria fowleri, Naegleria gruberi*.

Phylum Euglenozoa (yu-glen-a-zo′a) (Gr. *eu-*, good, true, + *glēnē*, cavity, socket, + *zōon*, animal). With cortical microtubules; flagella often with paraxial rod (rodlike structure accompanying axoneme in flagellum); mitochondria with discoid cristae; nucleoli persist during mitosis. This phylum is synonomous with clade Euglenozoa.

Subphylum Euglenida (yu-glen′i-da) With pellicular microtubules that stiffen pellicle.

Class Euglenoidea (yu-glen-oyd′e-a) (Gr. *eu-*, good, true, + *glēnē*, cavity, socket, + *-ōideos*, form of, type of). Two heterokont flagella (flagella with different structures) arising from apical reservoir; some species with light-sensitive stigma and chloroplasts. Example: *Euglena*.

Subphylum Kinetoplasta (ky-neet′o-plas′ta) (Gr. *kinētos*, to move, + *plastos*, molded, formed). Cells have a single mitochondrion with a special region called a kinetoplast that contains kDNA and connects to a basal body; paraxial rod.

Class Trypanosomatidea (try-pan′o-som-a-tid′e-a) (Gr. *trypanon*, a borer, + *sōma*, the body). One or two flagella arising from pocket; flagella typically with paraxial rod that parallels axoneme; single mitochondrion (nonfunctional in some forms) extending length of body

(*continued*)

as tube, hoop, or network of branching tubes, usually with single conspicuous DNA-containing kinetoplast located near flagellar kinetosomes; Golgi body typically in region of flagellar pocket, not connected to kinetosomes and flagella; all parasitic. Examples: *Leishmania, Trypanosoma.*

Phylum Stramenopiles (stra-men′ō-piles) (L. *stramen*, straw,+ *pile*, hair). Flagellates with two different flagella, one long and one short; forward flagellum is covered with tubular three-part hairs; mitochondria with tubular cristae. Free-living and parasitic plantlike and animal-like forms. Examples: *Phytophthora infestans, Actinosphaerium, Actinophrys.*

Phylum Ciliophora (sil-i-of′-or-a) (L. *cilium*, eyelash, + Gr. *phora*, bearing). Cilia or ciliary organelles in at least one stage of life cycle; two types of nuclei, with rare exception; binary fission across rows of cilia, budding and multiple fission also occur; sexuality involving conjugation, autogamy, and cytogamy; nutrition heterotrophic; contractile vacuole typically present; most species free-living, but many commensal, some parasitic. (This is a very large group, now divided by the Society of Protozoologists into three classes and numerous orders and suborders. The classes are separated on the basis of technical characteristics of the ciliary patterns, especially around the cytostome, the development of the cytostome, and other characteristics.) Examples: *Paramecium, Colpoda, Tetrahymena, Balantidium, Stentor, Blepharisma, Epidinium, Euplotes, Vorticella, Carchesium, Trichodina, Podophrya, Ephelota.* This phylum is within clade Alveolata.

Phylum Dinoflagellata (dy′no-fla-jel-at′a) (Gr. *dinos*, whirling, + *flagellum*, little whip). Typically with two flagella, one transverse and one trailing; body usually grooved transversely and longitudinally, each groove containing a flagellum; chromoplasts usually yellow or dark brown, occasionally green or blue-green, bearing chlorophylls *a* and *c*; nucleus unique among eukaryotes in having chromosomes that lack or have low levels of histones; mitosis intranuclear; body form sometimes of spherical unicells, colonies, or simple filaments; sexual reproduction present; members free-living, planktonic, parasitic, or mutualistic. Examples: *Zooxanthella, Ceratium, Noctiluca, Ptychodiscus.* This phylum is within clade Alveolata.

Phylum Apicomplexa (ap′i-compleks′a) (L. *apex*, tip or summit, + *complex*, twisted around). Characteristic set of organelles (apical complex) associated with anterior end present in some developmental stages; cilia and flagella absent except for flagellated microgametes in some groups; cysts often present; all parasitic. This phylum is within clade Alveolata.

Class Gregarinea (gre-ga-ryn′e-a) (L. *gregarius*, belonging to a herd or flock). Mature gamonts (individuals that produce gametes) large, extracellular; gametes usually alike in shape and size; zygotes forming oocysts within gametocysts; parasites of digestive tract or body cavity of invertebrates; life cycle usually with one host. Examples: *Monocystis, Gregarina.*

Class Coccidea (kok-sid′e-a) (Gr. *kokkos*, kernel, grain). Mature gamonts small, typically intracellular; life cycle typically with merogony, gametogony, and sporogony; most species live inside vertebrates. Examples: *Cryptosporidium, Cyclospora, Eimeria, Toxoplasma, Plasmodium, Babesia.*

Phylum Cercozoa (ser-kō-zo′a) (Gr. *kerkos*, tail, + *zōon*, animal). A diverse group of unicells, heterogeneous in lifestyle and morphology; monophyly supported by molecular data. Most are free-living, some parasitic. Examples: *Euglypha, Clathrulina.*

Phylum Foraminifera (for′a-min-if′er-a) (L. *foramin*, hole, + *fero*, to bear). Shelled amebas bearing slender pseudopodia that extend through many openings in the test, forming a net that ensnares prey; includes xenophyophores. Examples: *Vertebralina, Globigerina.*

Phylum Radiolaria (rade-a-la′re-a) (L. *radiolus*, small sunbeam). Most are amebas with a well-developed internal skeleton of strontium sulfate or silica, forming beautiful tests. Axopodia are present. Examples: *Tetrapyle, Pterocorys.*

Phylum Viridiplantae (vir′i-di-plan′tee) (L. *viridis*, green, + *planto*, to set, plant). Unicellular and multicellular photoautotrophs with chlorophylls *a* and *b*; reserve material is starch. Examples: *Chlamydomonas, Volvox, Zea mays.*

Phylum Centrohelida (cen-tro-hel′i-da) (Gr. *kentron*, center of a circle, + *hēlios*, the sun). Amebas with flattened mitochondrial cristae; axoneme of axopodia with microtubules arranged in hexagons or triangles; in most, axopodia extend through a coat of variously shaped silica scales. Mostly freshwater, some marine. Examples: *Acanthocystis, Pterocystis, Heterophrys.*

Phylum Amoebozoa (uh-mee′bo-zo′a) (Gr. *amoibē*, to change, + *zōon*, animal). Naked and shelled amebas, many with flagellated stages in life cycle; mitochondria, when present, have tubular and branched cristae. Free-living and parasitic. Examples: *Entamoeba, Dictyostelium, Chaos.*

Phylum Opisthokonta (o-pist′tho-kon′ta) (Gr., *opisthen*, behind, at the back, + *kontos*, a pole, referring to a flagellum). Many are flagellates with one posterior flagellum; the group includes nucleariid amebas, choanoflagellates, fungi, and the animals. Examples: *Codonosiga, Penicillium,* animals.

■ SUMMARY

Sections	Key Concepts
5.1 Form and Function	• Unicellular eukaryotes are surrounded by a plasma membrane, and most have organelles, such as the nucleus and endomembrane system, similar to those of other eukaryotic cells. Ciliates have multiple nuclei. • Movement is accomplished by dynamic cytoskeletal elements such as actin filaments and microtubules that form pseudopodia, lobopodia, filopodia, rhizopodia, axopodia, and axostyles. • Flagella and cilia are structurally similar locomotory structures with an internal microtubular structure arising from a basal body. • Glideosomes are unique locomotory structures in apicomplexans that allow parasites to invade host cells. • Extrusomes are unique organelles that extrude substances, often toxic or defensive, from cells. • Mitochondria are often modified in unicellular eukaryotes and are then called MROs (mitochondria-related organelles). There are three basic types of MROs present among organisms that occupy low-oxygen or anoxic environments. • Kinetoplastans are unicells with one mitochondrion that contains a special region called a kinetoplast. The kinetoplast connects the mitochondrion to a basal body and contains the uniquely structured DNA called kDNA. • Photosynthetic organelles called plastids may contain various pigments including chlorophyll.
5.2 Nutrition and Digestion	• Various unicellular eukaryotes feed by holophytic, holozoic, or saprozoic means. • Phagocytosis is the mechanism for holophytic and holozoic feeding. • Trogocytosis is a recently described feeding mechanism that occurs in parasitic taxa such as *Entamoeba* or *Naegleria*. Parasite cells bite host cells, consuming organelles. • Some taxa, such as ciliates, possess a distinct oral region or gullet where food is ingested.
5.3 Excretion and Osmoregulation	• Oxygen uptake and nitrogenous waste elimination are accomplished across the plasma membrane in unicellular eukaryotes. Some wastes may be eliminated using contractile vacuoles. • Unicells living in isosmotic environments do not accumulate excess water. • Unicells living in freshwater habitats expel excess water using contractile vacuoles.
5.4 Reproduction and Life Cycles	• Asexual reproduction is accomplished by binary fission, multiple fission (schizogony), and budding. • Sexual processes, such as conjugation in ciliates, are common, but patterns of ploidy differ from those in animals. Some unicells exhibit alternation of generations in life cycles similar to those of plants. • Meiosis may occur before gametes are made or after zygotes are formed, depending upon the taxon. • Cyst formation is an adaptation to withstand adverse environmental conditions.
5.5 Unicellular Eukaryotic Taxa	• Photoautotrophic species occur in several taxa, including Viridiplantae, Euglenozoa, and Dinoflagellata; many are important primary producers in oceans. • Kinetoplasta is a subgroup of Euglenozoa that comprises many non-photosynthetic species that are important parasites of humans and other animals; serious illnesses include sleeping sickness and Chagas disease. • Metamonada comprises human parasites such as *Giardia* and *Trichomonas*, as well as nonparasitic forms such as *Trichonympha*, a symbiont of termites that houses cellulose-digesting bacteria. • Heteroloboseans have both ameboid and flagellated life stages; a few species cause diseases such as amebic meningoencephalitis. • Stramenopiles is a diverse group of unicells that includes autotrophs and heterotrophs with a variety of body forms. The group includes plant parasites called oomycetes that cause blights and kill oak trees; an oomycete caused the Irish potato famine. • Apicomplexa are almost all parasitic, including the malarial agent, *Plasmodium*. • Ciliophora is a large and diverse group, whose members move by cilia or ciliary organelles. Many are complex, possessing micronuclei and macronuclei with different functions, as well as extrusomes. • Amebas are assigned to a number of phyla, including Cercozoa, Heterolobosea, Foraminifera, Radiolaria, and Amoebozoa, among others. • Rhizaria comprises the morphologically diverse Cercozoa, and testate amebas with tests (shells) of glass or calcium. • Testate amebas have an excellent fossil record and can be used to date sediments. • Centrohelids are testate amebas forming axopodia with a distinctive pattern of microtubules. • Ameobozoans are free-living and testate amebas; one species causes human amebic dysentery. • Opisthokonta have a single posterior flagellum when a flagellum is present; this group comprises fungi, choanoflagellates, and animals, among others.

(*continued*)

Sections	Key Concepts
5.6 Phylogeny	• The evolution of a eukaryotic cell was followed by diversification of lineages to form morphologically disparate clades, some of which contain both unicellular and multicellular forms. • Molecular phylogenies using multiple genes have led to a better understanding of evolutionary relationships among the unicellular eukaryotes. • Knowledge of evolutionary relationships allows humans to design drugs to attack parasites. • Choanoflagellates are the closest living relatives of animals.

Review Questions

1. Before the advent of modern phylogenetic methods, biologists thought that all unicellular eukaryotes were closely related to animals. Two terms, protozoan and metazoan, were used to describe the unicellular and multicellular animals, respectively. Why, in light of recent phylogeny, can we replace the term metazoan with animal?
2. Distinguish among the following phyla: Euglenozoa, Apicomplexa, Ciliophora, Dinoflagellata.
3. Explain the role of actin in ameboid movement.
4. Distinguish between cell extensions formed by actin filaments and those formed by microtubules.
5. Explain the morphological features that unite ciliates, dinoflagellates, and apicomplexans as members of Alveolata.
6. Multicellularity is hypothesized to have evolved independently about 25 times. Distinguish coloniality from multicellularity and use the volvocine algae to illustrate steps in the evolution of a multicellular body plan.
7. Use taxa described in this chapter to illustrate the importance of insect vectors to the spread of diseases caused by unicellular eukaryotes.
8. Distinguish the following: sexual and asexual reproduction; binary fission, budding, and multiple fission.
9. What is the survival value of encystment?
10. Contrast and give examples of autotrophic and heterotrophic unicellular eukaryotes.
11. If you needed to know the age of particular deep-sea sediments, which unicellular eukaryotes could prove useful?
12. Outline the general life cycle of malarial organisms. How do you account for the resurgence of malaria in recent years?
13. What is the public health importance of *Toxoplasma,* and how do humans become infected with it? What is the public health importance of *Cryptosporidium* and *Cyclospora?*
14. How do dinoflagellate taxa harm humans?
15. Distinguish primary endosymbiosis from secondary endosymbiosis.
16. Explain how surface temperatures in ancient oceans may be reconstructed.

For Further Thought Mark the distribution of photoautotrophic taxa on figure 5.2. Recent research suggests that members of Ciliophora contain genes of algal origin. From a "tree-thinking" perspective, how would you explain the origin of these genes?

6

Sponges
Phylum Porifera

LEARNING OBJECTIVES

Readers will be able to:

6.1 Outline the role of sponges in a marine ecosystem and explain their value to human health.

6.2 Explain how choanocytes and other types of cells function within a living sponge.

6.3 Compare and contrast the organization and features of members of the four sponge classes.

6.4 Discuss the evolution of complexity in suspension-feeding sponges and in carnivorous forms.

6.5 Describe the body plan of a placozoan.

A Caribbean demosponge, *Aplysina fistularis*.

Larry S. Roberts

The Advent of Multicellularity

Organisms may be unicellular, colonial, or multicellular. The body of colonial organisms, such as choanoflagellates, is made of multiple cells, but there is no specialization among cells. To be "multicellular," an organism must specialize cell function and limit sexual reproduction to certain cells. Multicellularity has arisen in unicellular lineages at least 25 times. Independent evolutionary events produced multicellularity in brown algae (kelp), plants, and of course, animals. Before molecular characters were used to follow the pathway of evolution, biologists placed some unicellular taxa within animals, using the terms "protozoan" and "metazoan," respectively, for the unicellular and multicellular parts of this kingdom. We now know that "protozoans" do not form a monophyletic group with each other or with "metazoans." "Metazoan" is redundant with "animal." This chapter describes two enigmatic animal groups, sponges and placozoans. Sponges are the sister taxon to all other animals, but their body organization is not typical of other animals. Most sponge bodies are exquisitely structured to bring the surrounding water into their bodies and to capture suspended food particles from it. They have a range of cell types that move water, collect food, stiffen the body, and create a skeleton of calcium or glass. They vary in the complexity of the internal channels that control water flow and in the organization of the skeleton. There is one highly specialized branch of the sponge lineage that inhabits very deep oceans where there are few particles to collect—these sponges have become carnivores, trapping small prey items on their body surface and using a combination of phagocytosis and extracellular digestion to consume them. Because sponges have no mouth and no limbs or tentacles, it might appear to the casual observer that sponges are not doing anything, but in fact, their collection of cells allows them to do almost everything one expects of more obviously complex animals. Placozoans have plate-like bodies without a mouth, but they also do extracellular digestion. They glide over surfaces and when they encounter food, typically cyanobacteria and algae, they secrete digestive enzymes onto it. They are able to mimic a digestive cavity by mounding up the body and using the space between the substrate and the body wall to digest food. Placozoans and sponges solve life's problems through unique arrangements of cells.

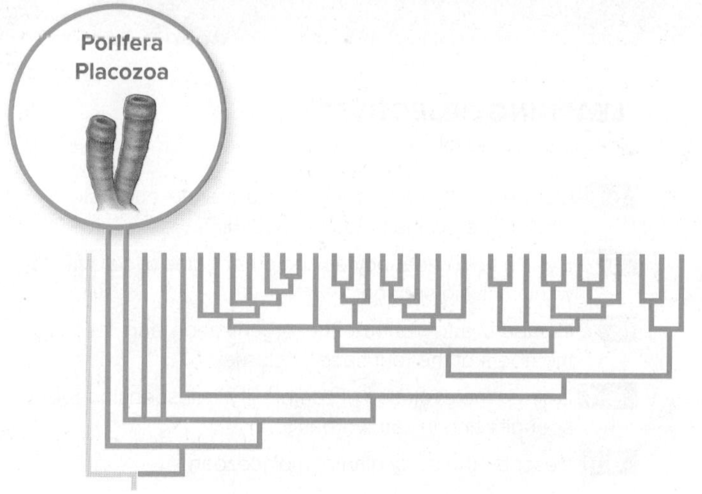

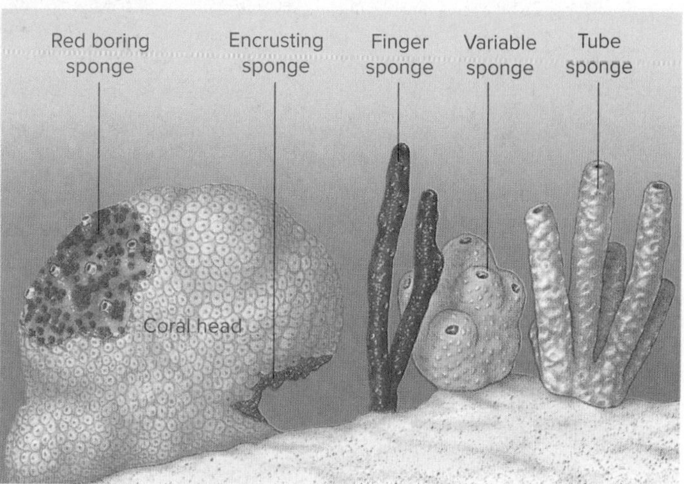

figure 6.1

Some growth habits and forms of sponges.

Sponges belong to phylum Porifera (po-rif′-er-a) (L. *porus,* pore, + *fera,* bearing). Sponges bear myriads of tiny pores and canals that constitute a filter-feeding system. Water flow through the unique canal systems in these sessile animals brings in food and oxygen and carries away their body wastes. They have two layers of cells: an outer pinacoderm and an inner choanoderm. Between these layers lies mesohyl, an extracellular matrix (ECM; see section 3.1), home to up to ten different cell types and stiffened by a skeleton of minute **spicules** of calcium carbonate or silica and collagen. An extracellular matrix is a substance secreted by cells for structural support. It is usually made of collagen and proteoglycans. Most sponges have no organs or true tissues, and their cells show a certain degree of independence. For example, dissociated sponge cells can re-form a new sponge body. There is no nervous system or sense organs, but sponges do have some contractile elements.

Sponges vary in size from a few millimeters to the great loggerhead sponges, which may reach 2 m or more across. Many sponge species are brightly colored because of pigments in their dermal cells. Red, yellow, orange, green, and purple sponges are not uncommon. However, their color fades quickly when they are removed from water. Many sponges are quite irregular in shape, but vase-like and columnar forms also occur. Some stand erect, some are branched or lobed, and others are low, even encrusting, in form (figure 6.1). Some bore holes into shells or rocks.

Sponges are an ancient group, with an abundant fossil record extending back to the early Cambrian period and even, according to some claims, the Precambrian. Living poriferans

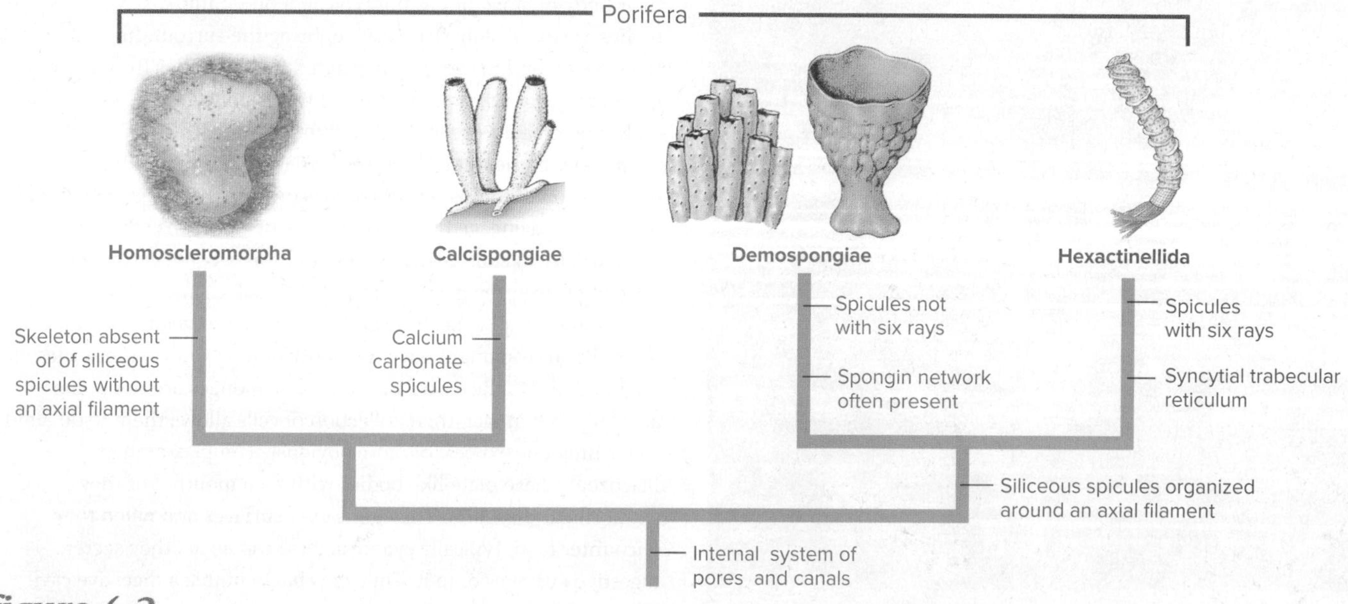

figure 6.2

Cladogram depicting evolutionary relationships among the four classes of sponges with living representatives.

were assigned to three classes: Calcarea (having calcareous spicules), Demospongiae (having a skeleton of siliceous spicules or spongin or both), and Hexactinellida (having six-rayed siliceous spicules) (figure 6.2). A fourth class (Homoscleromorpha) was created from a subset of the Demospongiae. Homoscleromorphs are thin encrusting sponges that have simple spicules or lack spicules entirely.

6.1 Ecological Relationships

Most of the 5000 or more sponge species are marine, although some 150 species occupy fresh water. Marine sponges abound in all seas and at all depths, and a few occupy brackish water. Although their embryos are free-swimming, adults are always attached, usually to rocks, shells, corals, or other submerged objects (figure 6.3). Some bottom-dwelling forms even grow on sand or mud. Their growth patterns often depend on shape of the substratum, direction and speed of water currents, and availability of space, so that the same species may differ markedly in appearance under different environmental conditions. Sponges in calm waters may grow taller and straighter than those in rapidly moving waters.

Many animals (crabs, nudibranchs, mites, bryozoans, and fish) live as commensals or parasites in or on sponges. Larger sponges particularly tend to harbor a large variety of invertebrate commensals. On the other hand, sponges grow on many other living animals, such as molluscs, barnacles, brachiopods, corals, and hydroids. One sponge has been described that preys on shrimp. Some crabs attach pieces of sponge to their carapace for camouflage and protection, since most predators seem to find sponges distasteful. Certainly one reason for the success of sponges as a group is that they have few enemies. Because of a sponge's elaborate skeletal framework and often noxious odor, most potential predators find sampling a sponge about as pleasant as eating a mouthful of glass splinters embedded in evil-smelling gristle. Some reef fishes, however, do graze on shallow-water sponges.

Larry S. Roberts

figure 6.3

This orange demosponge, *Mycale laevis*, often grows beneath platelike colonies of the stony coral, *Montastrea annularis*. The large oscula of the leuconoid sponge are seen at the edges of the plates. Unlike some other sponges, *Mycale* does not burrow into the coral skeleton and may actually protect the coral from invasion by more destructive species. Pinkish radioles of a Christmas tree worm, *Spirobranchus giganteus* (phylum Annelida, class Polychaeta), also project from the coral colony. An unidentified reddish sponge is seen farther to the right of and below the Christmas tree worm.

Key Theme 6.1

HUMAN CONNECTIONS

Bioactive Chemicals from Sponges

Sponges, and the microorganisms that live in or on them, produce a wide variety of bioactive chemicals. The sessile lifestyle of sponges makes them vulnerable to predators and competitors for space, so many have a diverse repertoire of secondary metabolites. An extract from a marine sponge appears effective against leishmaniasis (see section 5.5: Euglenozoa), and another shows promise for treating herpetic infections. Many bacteria isolated from marine taxa also have antimicrobial or antiviral effects; for example, some inhibit *Staphylococcus aureus* infections, and others are active against *Escherichia coli*, some strains of which cause food poisoning. Anti-cancer, anti-fungal, anti-malarial, cytotoxic, and immunosuppressive substances have also been found. The discovery of a fibrous chitin skeleton in some sponges (see section 6.2: Types of Skeletons) is exciting because it has potential as a scaffold for human tissue engineering. One major challenge for commercial use of materials from sponges lies in finding sufficient quantities of particular sponge tissues in nature. Sea farming is being explored for several sponge species, including *Sarcotragus spinosulus*, so perhaps large-scale culturing may provide a source for these valuable materials.

6.2 Form and Function

Sponges are sessile. Their only body openings are pores, usually many tiny ones called **dermal ostia** for incoming water, and one to a few large ones called **oscula** (sing., **osculum**) that serve as water outlets. These openings are connected by a system of canals, some of which are lined with peculiar flagellated collar cells called **choanocytes,** whose flagella maintain a current of environmental water through the canals. The choanocyte lining forms the choanoderm. Choanocytes not only keep the water moving but also trap and phagocytize food particles from the water. Collapse of the canals is prevented by the skeleton, which, depending on the species, may be composed of needlelike calcareous or siliceous spicules, a meshwork of organic spongin fibers, or a combination of the two. **Chitin** has also been identified in the skeletons of some members of Demospongaie and Hexactinellida.

CHARACTERISTICS of Phylum Porifera

1. Multicellular, body, an aggregation of several types of cells differentiated for various functions, some of which are organized into **incipient tissues** of a low level of integration. However, the pinacoderm approaches a true tissue epithelium in homoscleromorph sponges.
2. Body with pores (ostia), canals, and chambers that form a unique system of **water currents** on which sponges depend for food and oxygen.
3. Mostly marine; all aquatic.
4. Body is asymmetrical in most groups, but radial in Hexactinellida.
5. Outer surface of flat pinacocytes creates a pinacoderm layer; most interior surfaces have a choanoderm layer of flagellated collar cells (choanocytes) that create water currents; a gelatinous protein ECM called mesohyl contains amebocytes of various types and skeletal elements.
6. Skeletal structure of fibrillar collagen (a protein) and calcareous or siliceous crystalline spicules, often combined with variously modified collagen (spongin); type IV collagen, characteristic of other animals, occurs only in homoscleromorph sponges; fibrous chitin support structures occur in one class.
7. No organs or true tissues; digestion intracellular in most, but extracellular in carnivorous forms; excretion and respiration by diffusion.
8. Reactions to stimuli apparently local and independent in cellular sponges, but electrical signals occur in syncytial glass sponges; nervous system probably absent.
9. All adults sessile and attached to, or embedded in, substratum.
10. Asexual reproduction by buds or gemmules and sexual reproduction by eggs and sperm; free-swimming flagellated larvae in most.

Types of Canal Systems

Traditionally, sponges were described as having three types of canal systems—asconoid, syconoid, or leuconoid (figure 6.4), but some biologists use a five-category system, adding sylleibids and solenoids.

Asconoids–Flagellated Spongocoels

Asconoid sponges have the simplest organization. They are small and tube-shaped. Water enters through microscopic dermal pores into a large cavity called a **spongocoel,** which is lined with choanocytes. The beating of choanocyte flagella pull water through the pores and expel it through a single large osculum (figure 6.4). *Leucosolenia* (Gr. *leukos,* white, + *solen,* pipe) is an asconoid type of sponge. Its slender, tubular individuals grow in groups attached by a common **stolon,** or stem, to objects in shallow seawater. *Clathrina* (L. *clathri,* latticework) is an asconoid with bright yellow, intertwined tubes (figure 6.5). A presence of many small tubes, rather than one larger body, is a consequence of the asconoid design. Choanocytes can only collect food from a small area around them, so each tube is narrow, and the body grows by making new tubes. All asconoids are in the Calcarea.

Syconoids–Flagellated Canals

Syconoid sponges look somewhat like larger editions of asconoids, from which they were derived. They have a tubular body and a single osculum, but instead of a simple choanocyte layer lining the spongocoel, as in asconoids, this layer in syconoids is folded back and forth to make canals. Readers might recognize this folding as a classic method of increasing the surface area for food capture. The choanocytes line certain folds called **radial canals** (see figure 6.4). Water, entering the body through dermal pores, moves first to incurrent canals and then into radial canals via small lateral openings called prosopyles (figure 6.6). From the radial canals, filtered water moves through apopyles into the spongocoel, finally exiting by the osculum. The spongocoel in syconoids is lined with epithelial-type cells rather than with the choanocytes found in asconoids. Syconoids are in the Calcarea. *Sycon* (Gr. *sykon,* a fig) is a commonly studied example of the syconoid type of sponge (figure 6.6).

Leuconoids–Flagellated Chambers

Leuconoid organization is the most complex of the sponge types and, because internal feeding surface area is maximized through a myriad of tiny chambers, it permits an increase in sponge size. Most leuconoids form large masses with numerous oscula (see figure 6.3).

Clusters of flagellated chambers are filled from incurrent canals and discharge water into excurrent canals that eventually lead to the osculum (see figure 6.4). There is no spongocoel. Most sponges are of the leuconoid type, which occurs in most Calcarea and in all other classes.

Other Canal Systems

The three types of canal systems—asconoid, syconoid, and leuconoid—demonstrate an increase in the complexity and efficiency of the water-pumping system, but they do not imply an evolutionary or developmental sequence. Two new canal systems were described by sponge biologists: sylleibid and solenoid. The sylleibid canal system lies between the syconoid and leuconoid system in complexity because water does not enter the feeding canals directly from the outside. Instead, there is a tube without choanocytes leading into a radially arranged set of feeding canals in sylleibids. This body plan occurs in Calcarea.

The leuconoid grade of construction has evolved independently many times in sponges. Possession of a leuconoid plan is of clear adaptive value; it increases the proportion of flagellated surfaces compared with the volume, thus providing more collar cells to meet food demands. In the solenoid body plan, the cham-

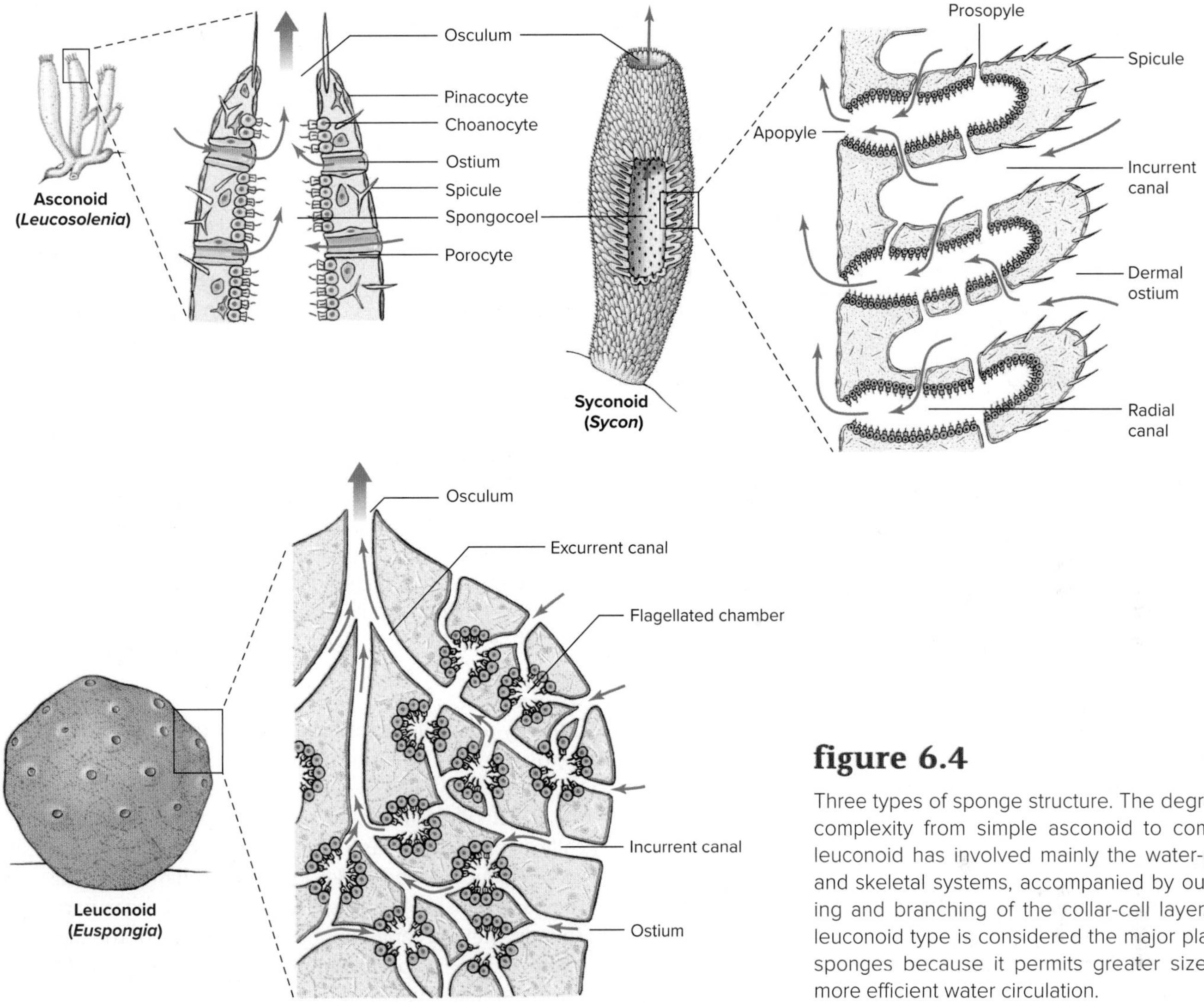

figure 6.4

Three types of sponge structure. The degree of complexity from simple asconoid to complex leuconoid has involved mainly the water-canal and skeletal systems, accompanied by outfolding and branching of the collar-cell layer. The leuconoid type is considered the major plan for sponges because it permits greater size and more efficient water circulation.

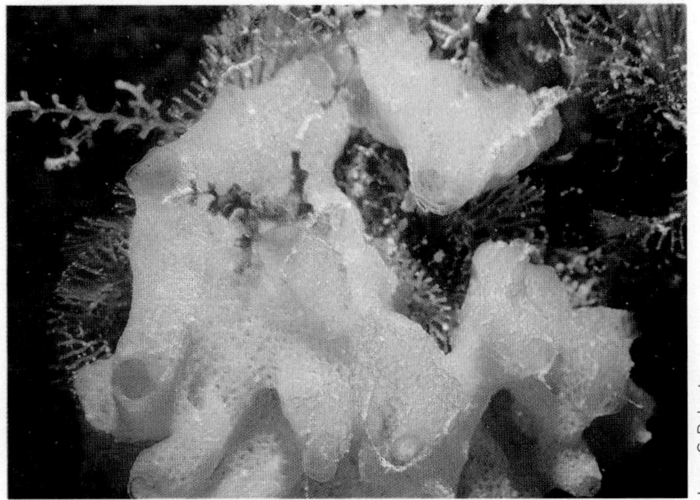

Larry S. Roberts

figure 6.5

Clathrina canariensis (class Calcarea) is common in caves and under ledges on Caribbean reefs.

bered arrangement of a leuconoid system is modified into a set of internal interconnected feeding tubes. Only the tube walls have choanocytes. This body plan occurs in Calcarea.

Types of Cells

Sponge cells make layers, called epithelia, as in the internal choanoderm and external pinacoderm. In most animals, an **epithelium** is strengthened by an underlying non-living protein basal lamina. The basal lamina is a type of ECM. In an epithelium, there are different kinds of junctions between adjacent cells and between the cells and the ECM. For a sheet of cells to be a "true tissue epithelium," as seen in more complex animal bodies, there must be two types of cell junctions present, along with a basal lamina (see section 6.3: Class Homoscleromorpha). In contrast to cells in epithelia, cells in the sponge **mesohyl** are loosely arranged (figure 6.7). The mesohyl is the "connective tissue" of the sponges; in it are found various ameboid cells, fibrils, and skeletal elements. Several types of cells occur in sponges.

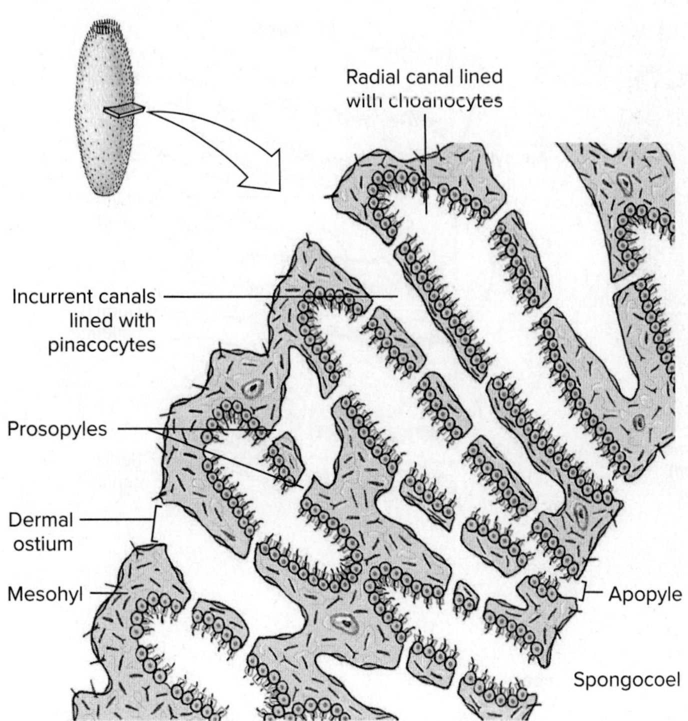

figure 6.6

Cross section through the body wall of the sponge *Sycon*, showing the canal system.

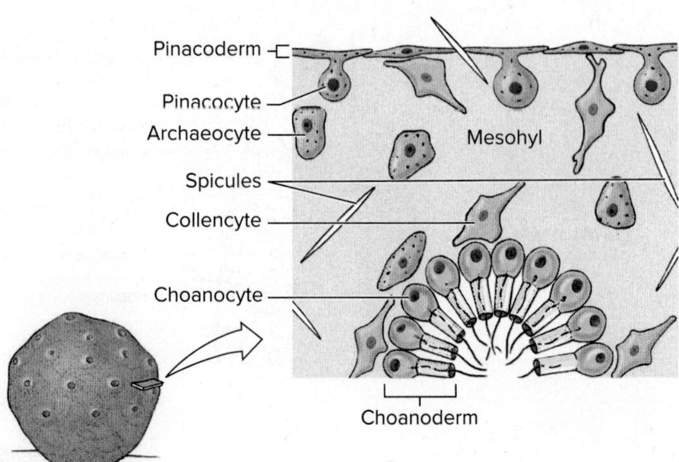

figure 6.7

Small section through the body wall of a sponge, showing four types of sponge cells. Pinacocytes are protective and contractile; choanocytes create water currents and engulf food particles; archaeocytes have a variety of functions, including phagocytosis of food particles and differentiation into other cell types; collencytes secrete collagen.

Pinacocytes

The nearest approach to a true tissue epithelium in sponges occurs in **pinacocyte** cells (figure 6.7) of the external pinacoderm layer. These thin, flat, epithelial-type cells cover the exterior surface and some interior surfaces, but only in one group do they rest on a basal lamina ECM. Some are T-shaped, with their cell bodies extending into the mesohyl. Pinacocytes are somewhat contractile and help to regulate the surface area of the sponge. Some pinacocytes are modified as contractile **myocytes,** which are usually arranged in circular bands around the oscula or pores, where they help to regulate the rate of water flow. Myocytes contain microfilaments similar to those found in muscle cells of other animals.

Porocytes

Tubular cells that pierce the body wall of asconoid sponges, through which water flows, are called **porocytes** (see figure 6.4).

Choanocytes

Choanocytes, which line flagellated canals and chambers, are ovoid cells with one end embedded in mesohyl and the other exposed. The exposed end bears a flagellum surrounded by a collar (figures 6.7 and 6.8). Electron microscopy shows that the collar is composed of adjacent microvilli, connected to each other by delicate microfibrils, so that the collar forms a fine filtering device for straining food particles from the water (figure 6.8B and C). The beat of a flagellum pulls water through the sievelike collar and forces it out through the open top of the collar. Particles too large to enter the collar become trapped in secreted mucus and slide down the collar to the base, where they are phagocytized by the cell body. Larger particles have already been screened out by the small size of the dermal pores and prosopyles. Food engulfed by the cells is passed to a neighboring archaeocyte for digestion.

Archaeocytes

Archaeocytes are ameboid cells that move through the mesohyl (see figure 6.7) and perform a number of functions. They can phagocytize particles at the external epithelium and receive particles for digestion from choanocytes. Archaeocytes apparently can differentiate into any of the other types of more specialized cells in the sponge. Some, called **sclerocytes,** secrete spicules. Others, called **spongocytes,** secrete the spongin fibers of the skeleton, and **collencytes** secrete fibrillar collagen.

Types of Skeletons

The skeleton gives support to a sponge, preventing collapse of canals and chambers. The major structural protein in the animal kingdom is collagen, and fibrils of collagen occur throughout the ECM of all sponges. Collagen occurs in many different types and only one group of sponges, homoscleromorphs, makes the type IV collagen characteristic of other animals.

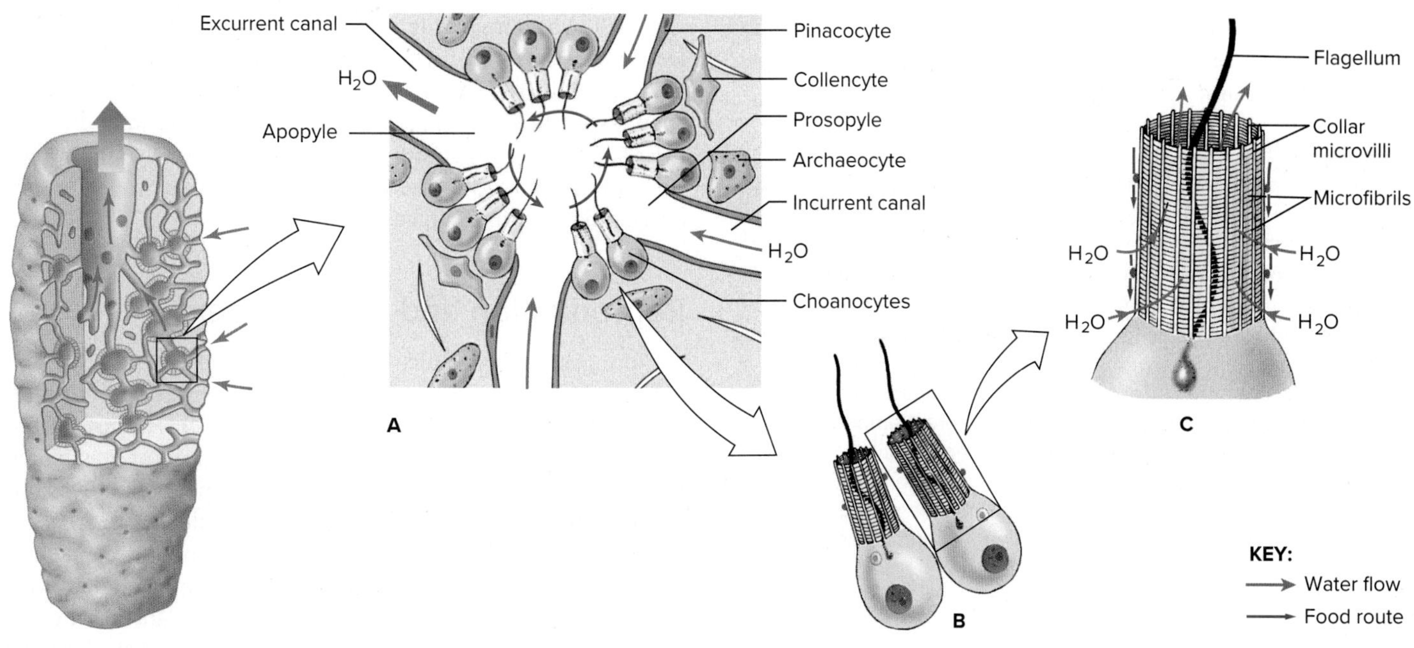

figure 6.8

Food trapping by sponge cells. **A,** Cutaway section of canals showing cellular structure and direction of water flow. **B,** Two choanocytes, and **C,** structure of the collar. In part C, the small red arrows indicate movement of food particles.

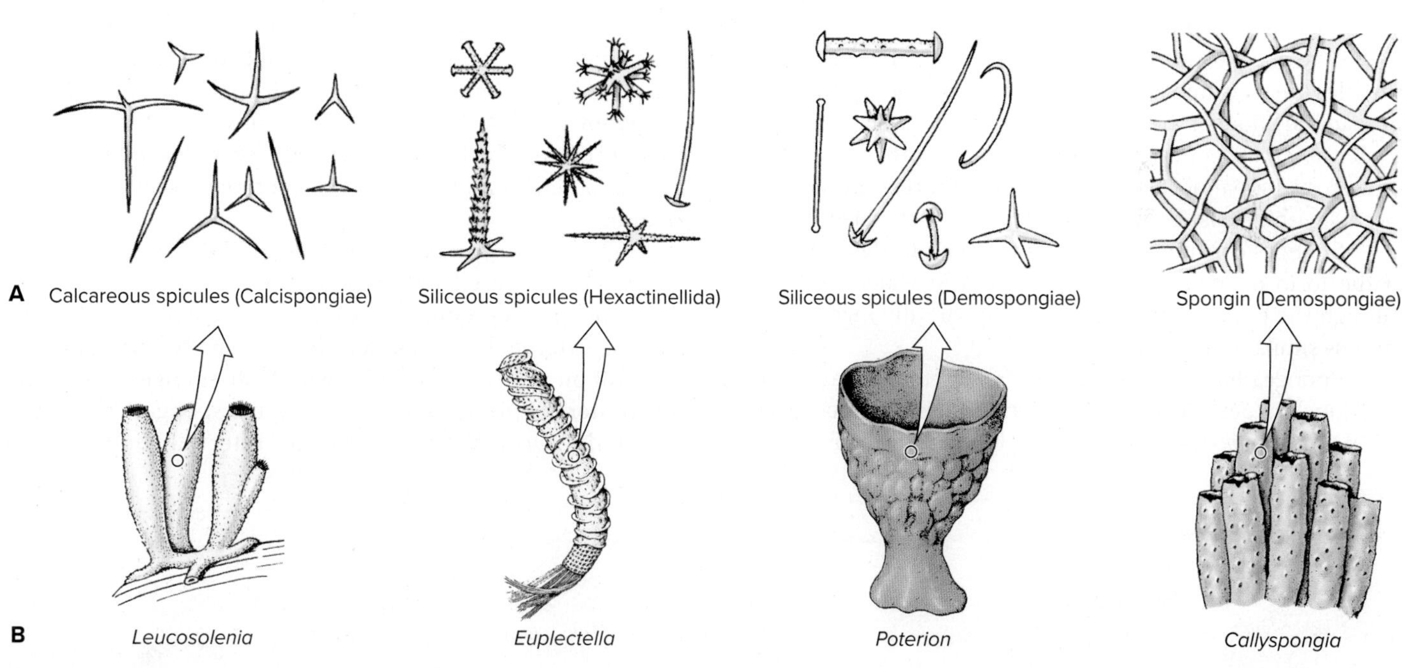

figure 6.9

A, The many types of spicules in sponges provide amazing diversity, beauty, and complexity of form. **B,** Some examples of sponge body forms possessing the support structure shown above in panel A.

Demospongiae secrete a form of collagen traditionally called spongin (figure 6.9A). Demospongiae also secrete siliceous spicules, and calcareous sponges secrete spicules composed mostly of crystalline calcium carbonate that have one, three, or four rays (figure 6.9A). Glass sponges (Hexactinellida) have siliceous spicules with six rays arranged in three planes at right angles to each other. There are many variations in the shape of spicules, and these structural variations are of taxonomic importance. Sponge bodies possessing each type of support in figure 6.9A are shown in figure 6.9B. A support system using

chitin was discovered in Demospongaie and Hexactinallida; this chitin has a unique 3-D fibrous structure that may have medical applications (see Key Theme 6.1).

Key Theme 6.2
HUMAN CONNECTIONS

Glass Spicules as Fiber-Optic Cables

Sponges harbor microalgae and cyanobacteria on the body surface and deep inside the body. The presence of photosynthetic organisms inside the sponge led some to propose that spicules were able to transmit light into the body. The fiber-optic properties of siliceous spicules have now been confirmed. This has sparked interest among materials scientists and engineers in the enzymatic machinery needed to form silica nanoparticles and to fuse these particles into spicules inside and outside sponge cells. In many cases, the exterior simplicity of a sponge masks chemical and functional sophistication.

Sponge Physiology

Sponges feed primarily on particles suspended in water pumped through their canal systems. They consume detritus particles, planktonic organisms, and bacteria in sizes ranging from 50 μm (average diameter of ostia) to 0.1 μm (width of spaces between the microvilli of the choanocyte collar). Pinacocytes may phagocytize particles at the surface, but most larger particles are consumed in the canals by archaeocytes that move close to the lining of the canals. The smallest particles, accounting for about 80% of the particulate organic carbon, are phagocytized by choanocytes. Digestion is entirely **intracellular** (occurs within cells), a chore performed by the archaeocytes.

Sponges consume a significant portion of their nutrients in the form of organic matter dissolved in water circulating through the system. Such material is apparently ingested by a process similar to phagocytosis.

Sponges have no respiratory or excretory organs; these functions are performed by diffusion. Contractile vacuoles occur in the archaeocytes and choanocytes of freshwater sponges.

All of a sponge's life activities depend on a current of water flowing through its body. A sponge pumps a remarkable amount of water, up to 1500 liters of water a day in large sponges. At least some sponges can crawl (move laterally over their supporting substratum) at speeds of up to 4 mm per day. This ability may give them an advantage over more sessile encrusting organisms in competition for space.

Reproduction and Development

All sponges can reproduce both sexually and asexually. In sexual reproduction, most sponges are **monoecious** (having both male and female sex cells in one individual). Sperm arise from transformation of choanocytes. In Calcarea and at least some Demospongiae, oocytes also develop from choanocytes; in other demosponges, oocytes apparently develop from archaeocytes. Sperm are released into the water by one individual and taken into the canal system of another. There, choanocytes phagocytize them, transform into carrier cells, and then carry the sperm through the mesohyl to the oocytes, where fertilization occurs.

Zygotes develop into flagellated larvae, which break loose and are carried away by water currents. Some sponges are oviparous and expel both oocytes and flagellated sperm into the water. Fertilization takes place in the water, but there are a few oviparous species where ova are released in a mucous bundle that attaches to the parent body and is fertilzed *in situ*. The free-swimming larva of most sponges is a solid-bodied **parenchymula** (figure 6.10). The outwardly directed, flagellated cells on the larval surface migrate to the interior after the larva settles and become choanocytes in the flagellated chambers.

The loose organization of sponges is ideally suited for regeneration of injured and lost parts, and for asexual reproduction. Sponges reproduce asexually by fragmentation and by forming external buds that detach or remain to form colonies. In addition to external buds, which all sponges can form, freshwater sponges and some marine sponges reproduce asexually by regularly forming internal buds called

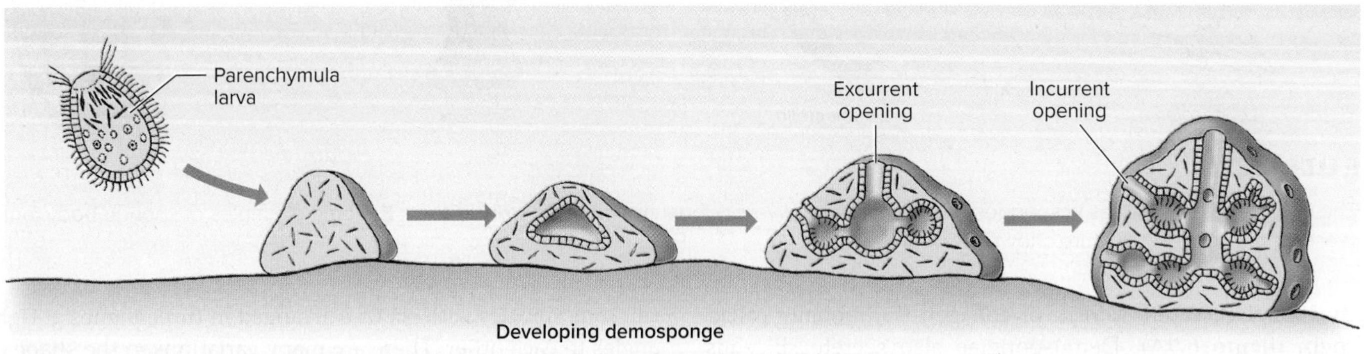

figure 6.10

Development of a demosponge.

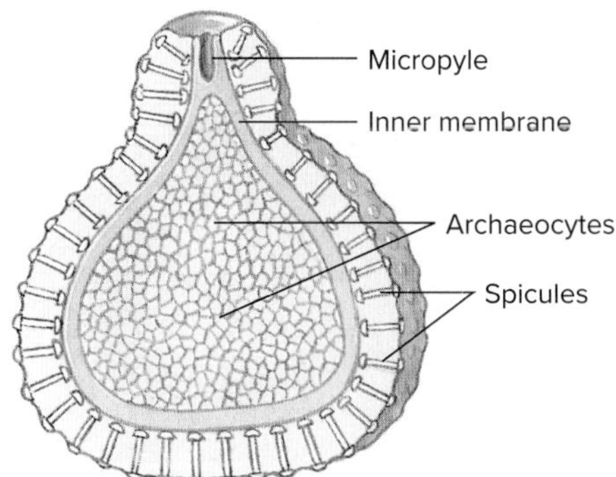

figure 6.11

Section through a gemmule of a freshwater sponge (Spongillidae). Gemmules are a mechanism for surviving the harsh conditions of winter. Upon the return of favorable conditions, the archaeocytes exit through the micropyle to form a new sponge. The archaeocytes of the gemmule give rise to all cell types of the new sponge.

gemmules (figure 6.11). These dormant masses of encapsulated archaeocytes form during unfavorable conditions. They can survive periods of drought and freezing and more than three months without oxygen. Later, with the return of favorable conditions for growth, archaeocytes in the gemmules escape and develop into new sponges.

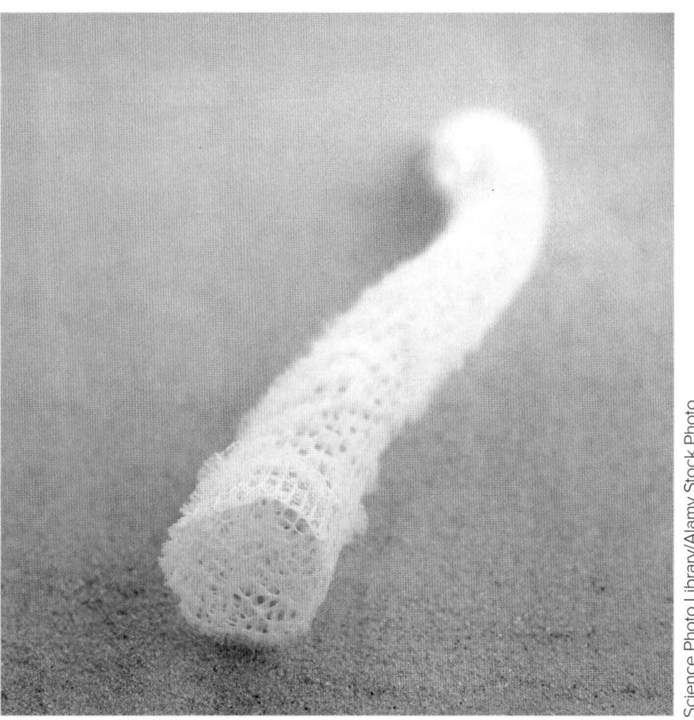
Science Photo Library/Alamy Stock Photo

figure 6.12

The complex skeleton of a glass sponge (Hexactinellida) remains after living tissue is removed. The skeleton has fiber optic properties.

6.3 Brief Survey of Sponges

Class Calcarea (Calcispongiae)

Calcarea are calcareous sponges, so called because their spicules are composed of calcium carbonate. Their spicules are straight monaxons or have three or four rays (see figure 6.9A). The sponges tend to be small–10 cm or less in height–and tubular or vase shaped. They may be asconoid, syconoid, or leuconoid in structure. Although many are drab, some are bright yellow, red, green, or lavender. *Leucosolenia, Clathrina* (see figure 6.5), and *Sycon* are common examples.

Class Hexactinellida (Hyalospongiae)

Hexactinellids are nearly all deep-sea forms. Most are radially symmetrical and range in length from 7 to 10 cm to more than 1 m. One distinguishing feature, reflected in the name Hexactinellida, is the skeleton of six-rayed siliceous spicules bound together in an exquisite glasslike latticework (see figure 6.12).

The tissue structure of hexactinellids differs so dramatically from that of other sponges that some scientists advocate placing them in a subphylum separate from other sponges. The body is syncytial–there are many nuclei inside a single very large plasma membrane. This single, continuous syncytial tissue is called a **trabecular reticulum.** A 1-m-diameter glass sponge makes the largest syncytium known within the animals.

The trabecular reticulum is bilayered and can be sheetlike or tubular. Between the layers of the sheet, or inside the tubes, is a thin collagenous mesohyl in which cells, such as archeocytes or choanoblasts, occur (figure 6.13). Choanoblasts and other cells are connected to each other, and to the trabecular reticulum, by cytoplasmic bridges. Choanoblasts are unusual cells that make two or more flagellated outgrowths called **collar bodies.** The flagellum on a collar body beats to drive water flow in the same way that it would on a choanocyte.

An assemblage of collar bodies forms a flagellated chamber. Here, the trabecular reticulum branches to become two distinct bilayered sheets: a primary reticulum and a thin secondary reticulum that lacks mesohyl. The two sheets make a sandwich around the center of a collar body (figure 6.13). Collar bodies extend through openings in both sheets, but the openings surround the collar bodies tightly. There is a space between the two sheets. To collect food, the incurrent water is directed to the primary reticulum, where pores enter the space between the primary and secondary reticular sheets. Water entering this space must leave by moving through the mesh of microvilli on collar bodies; water cannot go anywhere else because it is blocked by the secondary reticulum. Food particles captured on microvilli are shared throughout the syncytium.

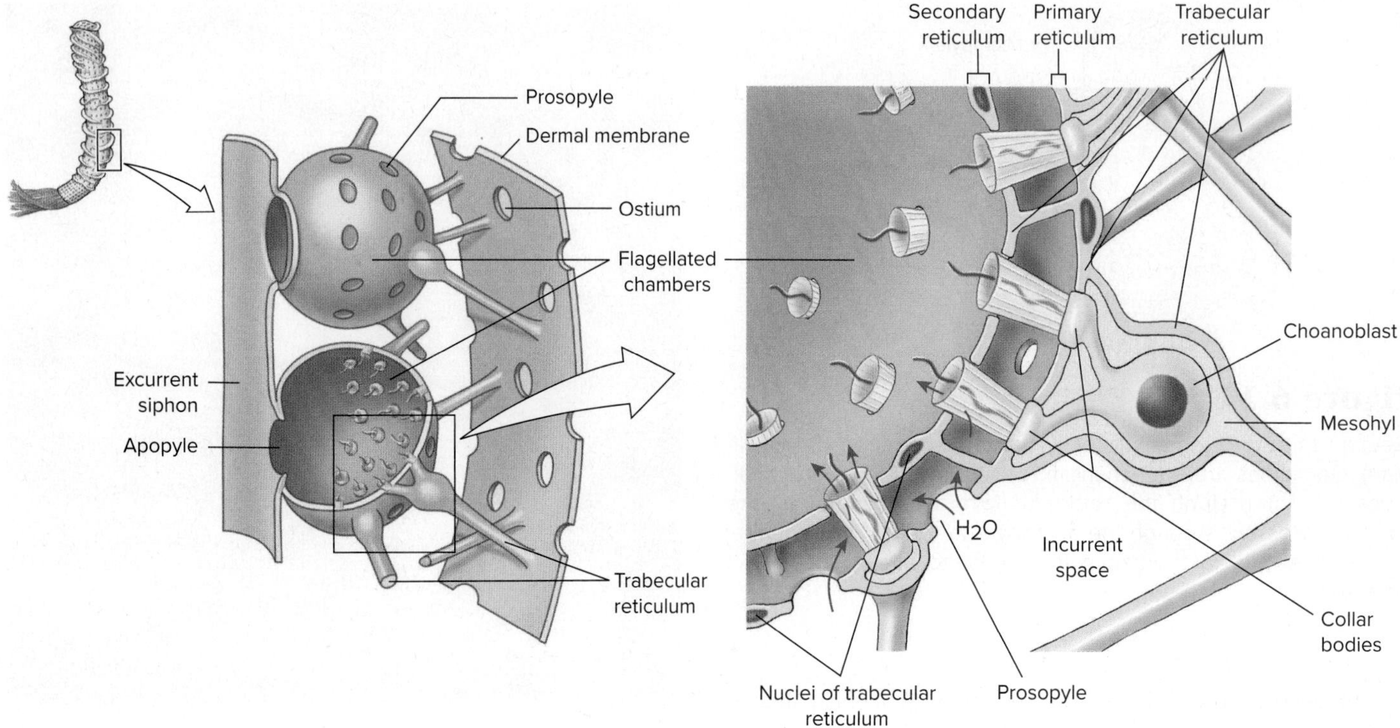

figure 6.13

Diagram of part of a flagellated chamber of hexactinellids. The primary and secondary reticula are branches of the trabecular reticulum, which is syncytial. Cell bodies of the choanoblasts and their processes are borne by the primary reticulum and are embedded in a thin, collagenous mesohyl. Processes of the choanoblasts end in collar bodies, whose collars extend up through the secondary reticulum. Flagellar action propels water (*arrows*) to be filtered through the mesh of collar microvilli (figure 6.8C).

The lattice-like network of spicules in many glass sponges is of exquisite beauty, as seen, for example, in *Euplectella* (NL. from Gr. *euplektos,* well-plaited), a classic example of Hexactinellida (see figure 6.12).

Class Demospongiae

Demospongiae comprise approximately 80% of all sponge species, including most larger sponges. Their skeletons may be composed of siliceous spicules, spongin fibers, or both (see figure 6.9A). All members of the class are leuconoid, and all are marine except one family, the freshwater Spongillidae. Freshwater sponges occur widely in well-oxygenated ponds and streams, where they encrust plant stems and old pieces of submerged wood. They resemble a bit of wrinkled scum, pitted with pores, and are brownish or greenish in color. Freshwater sponges die and disintegrate in late autumn, leaving gemmules to survive the winter.

Marine Demospongiae vary in both color and shape. Some are encrusting; some are tall and fingerlike; and others are shaped like fans, vases, cushions, or balls (figure 6.14). Some sponges bore into and excavate molluscan shells and coral skeletons. Loggerhead sponges may grow several meters in diameter. So-called bath sponges belong to the group called horny sponges, which have only spongin skeletons. They can be cultured by cutting out pieces of the individual sponges, fastening them to a weight, and dropping them into the proper water conditions. It takes many years for a horny sponge to grow to market size. Most commercial "sponges" now on the market are synthetic, but the harvest and use of bath sponges persists.

Class Homoscleromorpha

Homoscleromorphs are marine sponges that occur in a range of colors, but live in cryptic habitats, so they are often overlooked. They are more common in nearshore habitats, but they do occur in deep water. Sponges in this class were formerly placed in Class Demospongiae, but were separated because they possess unique features such as a basal lamina composed of ECM underlying the pinacoderm cell layer (see section 6.2: Types of Cells). Recall that a true tissue epithelium requires two distinct types of cell junctions and the presence of a basal lamina. Pinacoderm cells link to the ECM with adherens junctions but do not link to each other with desmosome junctions, so one type of junction is missing. The pinacoderm layer fails to meet the definition of a true tissue epithelium and is instead called an incipient epithelium. Representative genera are *Plakina*, *Oscarella*, and *Corticium*.

Taxonomy of Phylum Porifera

Class Calcarea (cal-cárē-ə) (L. *calcis,* lime, + Gr. *spongos,* sponge) **(Calcispongiae).** Have spicules of calcium carbonate that often form a fringe around the osculum; spicules needle-shaped or three- or four-rayed; all three types of canal systems (asconoid, syconoid, leuconoid) represented, plus two newly distinguished types (sylleibids and solenoids); all marine. Examples: *Sycon, Leucosolenia, Clathrina.*

Class Hexactinellida (hex-ak-tin-el'i-da) (Gr. *hex,* six, + *aktis,* ray) **(Hyalospongiae).** Have six-rayed, siliceous spicules extending at right angles from a central point; spicules often united to form a network; body often cylindrical or funnel-shaped. Flagellated chambers in simple syconoid or leuconoid arrangement. Habitat mostly deep water; all marine. Examples: Venus' flower basket *(Euplectella), Hyalonema.*

Class Demospongiae (de-mo-spun'jē) (tolerated misspelling of Gr. *desmos,* chain, tie, bond, + *spongos,* sponge). Have skeleton of spongin or siliceous spicules that are not six-rayed, or both; chitin fibers support the body in some. Leuconoid-type canal systems. One family found in fresh water; all others marine. Examples: *Thenea, Cliona, Spongilla, Myenia,* and all bath sponges.

Class Homoscleromorpha (hō-mō-skle'-rō-mor-fə) (Gr. *homos,* same, + *skleros,* hard, + *morphe,* form). Previously a subgroup of Demospongiae; spicules may be absent as in *Oscarella;* if present, spicules are small, simple in shape, and do not form around an axial filament; pinacoderm with a distinct basal lamina. Examples: *Oscarella, Corticium.*

A

B

C

figure 6.14

Marine Demospongiae on Caribbean coral reefs. **A,** *Pseudoceratina crassa* is a colorful sponge growing at moderate depths. **B,** *Ectyoplasia ferox* is irregular in shape, and its oscula form small, volcano-like cones. It is toxic and may cause skin irritation if touched. **C,** *Monanchora unguifera* with commensal brittle star, *Ophiothrix suensoni* (phylum Echinodermata, class Ophiuroidea).

6.4 Phylogeny and Adaptive Diversification

Phylogeny

Sponges are the sister taxon to a group comprising all other animal phyla, as shown on the "Cladogram of All Animal Phyla" inside the front cover. Sponges originated before the Cambrian period. Two groups of calcareous spongelike organisms occupied early Paleozoic reefs. The Devonian period saw rapid development of many glass sponges. The simple body plans of most sponges, aside from glass sponges, might suggest that sponges share few features with other animals, but this is not true. To form a multicellular body, some cells form layers adhering to other cells, to the ECM, or to both items. Some proteins used in cell adhesion and cell signaling in sponges are homologous to those in other animals; in fact many of these occur in choanoflagellates, evolving before the last common ancestor of all animals.[1] Sponge development includes the characteristic animal blastula stage, and some sponges actually develop to the two-layered gastrula stage before reorganizing their bodies into

[1]Abedin, M., and N. King. 2010. Diverse evolutionary paths to cell adhesion. *Trends in Cell Biology* **20**:734–742.

asymmetrical adults. It is possible that the sessile lifestyle of sponges favored a deceptively simple body in most species.

Adaptive Diversification in Porifera

Sponges are a highly successful group that includes several thousand species in a variety of marine and freshwater habitats. Their diversification centers largely on their unique water-current system, with multiple independent evolutionary events leading to body plans that maximize the surface area available for food capture.

One very novel way of feeding has evolved within a family of sponges, called cladorhizids, inhabiting nutrient-poor deep-water caves. Each sponge has a fine coating of tiny, hook-like spicules over its highly branched body. The spicule layer entangles the appendages of tiny crustaceans, often copepods (see section 13.6: Copepoda), swimming near the sponge surface. Later, sponge digestive cells and cells filled with cytoplasmic bacteria cover the prey, enveloping and digesting them. Bacteria aid in digestion, which is both extracellular and intracellular. Complete digestion takes 8 to 10 days for 8 mm prey items. These animals are carnivores, not suspension feeders; they lack choanocytes and internal canals but have siliceous spicules like typical members of class Demospongiae. In addition to capturing prey, some augment their diets with nutrients obtained from symbiotic methanotrophic bacteria. To colonize such a nutrient-poor habitat initially, the ancestors of this group must have had at least one alternative feeding system, either carnivory or chemoautotrophy, already in place. Presumably, after the alternative method of food capture was in use, the choanocytes and internal canals were no longer formed.

Further body modifications in this lineage might make it difficult to identify descendants as sponges. The discovery of the deep-sea harp sponge, *Chondrocladia lyra* (figure 6.15) shows us the complexity of a predatory sponge. This animal has multiple veins, each with a basal stolon close to 40 cm in length, anchored by rhizoids. Vertical branches extend nearly 20 cm upward from stolons. Terminal balls on each branch act as floats and may contain spermatophores; eggs are held midway up the branches. Body form varies from stalked and spherical to branching in the 36 other species in this genus.

©Monterey Bay Aquarium Research Institute (MBARI)

figure 6.15

The carnivorous sponge, *Chondrocladia lyra*, is commonly called a "harp sponge." It lives off the coast of California at depths down to 3500 meters, where it was discovered in 2012 by researchers using undersea robots operated by the Monterey Bay Aquarium Research Institute (MBARI). Image: © 2012 MBARI.

6.5 Placozoa

Placozoans are plate-like marine organisms 1–2 mm in diameter, but only 20 microns in thickness (figure 6.16). They have no fixed shape and no plane of symmetry. They have ciliated dorsal and ventral epithelia with a syncytial central layer of fiber cells. There are only six cell types present, but these include contractile cells with actin filaments, cells that secrete digestive enzymes, and cells that absorb nutrients. Many aspects of their cell biology are similar to those in sponges and choanoflagellates, but the sister taxon to this group remains obscure.

The simple body of placozoans appears to mask ancestral complexity. Placozoan epithelia have only one type of cell junction and do not have basal laminae. Placozoans do not secrete an ECM. However, genetic sequencing reveals the presence of genes that make proteins in the ECM of other animals, suggesting that they descend from an ancestral form that had an ECM. Ongoing research is likely to increase our understanding of sponge and placozoan evolution.

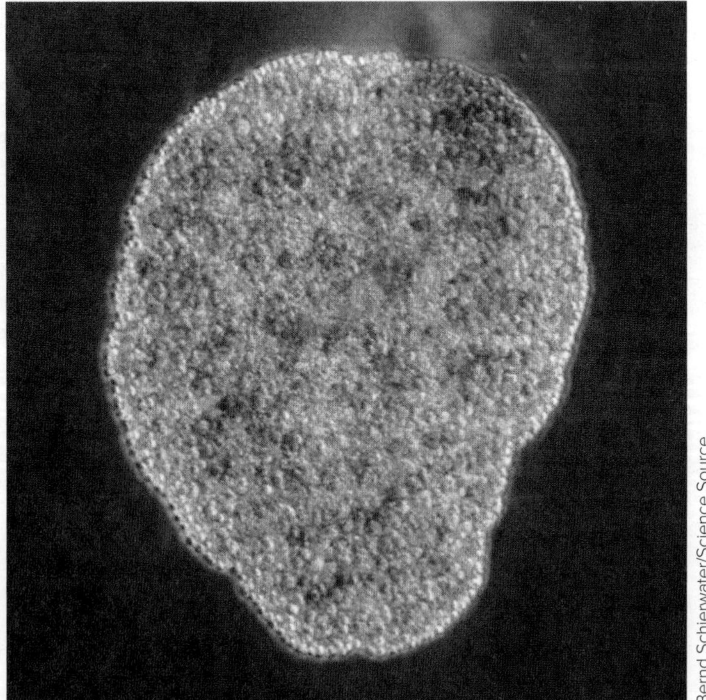

Bernd Schierwater/Science Source

figure 6.16

A photograph of the placozoan *Trichoplax adhaerens*.

SUMMARY

Sections	Key Concepts
6.1 Ecological Relationships	• Sponges are an abundant marine group with some freshwater representatives. • Sponge bodies serve as homes for many other animals. • Sponges and the microorganisms that inhabit them produce a wide range of bioactive substances with valuable medicinal properties for humans. Skeletal elements may be useful as scaffolds for engineered tissues.
6.2 Form and Function	• Sponges are sessile and almost all suspension-feed by drawing water into their bodies. Food particles less than 50 microns in size are captured from the water by choanocytes. Large sponges filter up to 1500 liters of water per day. • Sponges depend on the flagellar beat of their choanocytes to circulate water through their bodies for gathering food and exchanging respiratory gases. • The position of the choanocyte cell layer varies from simple to complex sponge bodies. In the simplest forms, choanocytes line a central chamber (asconoid), whereas choanocytes are restricted to internal chambers or canals in complex forms (leuconoid and solenoid). There are up to five arrangements of choanocytes. • Sponges have a range of cell types, but cells are not organized into tissues or organs. The external sponge epithelium, called pinacoderm, has some, but not all, features of true tissue epithelium in one class of sponges (Homoscleromorpha). • Sponges do not have a mouth or a gut cavity. The vast majority of sponges digest food intracellularly. • Sponges secrete skeletons of collagen in the form of large fibers or filaments (spongin). Calcareous or siliceous spicules are also skeletal elements. Spicules and spongin co-occur in most species, but one group of sponges makes a skeleton from chitin fibers. • Most sponges are monoecious but produce sperm and oocytes at different times. Embryogenesis is unusual because flagellated cells migrate from the embryo surface to the interior of the larva. The larva is free-swimming but settles to become a sessile adult. Sponges have great regenerative capabilities and may reproduce asexually by budding, fragmentation, or gemmules (internal buds).
6.3 Brief Survey of Sponges	• There are four classes of sponges. • Class Calcarea comprises sponges with a skeleton of calcareous spicules. Members of this group illustrate all the possible arrangements of choanocytes and internal water flow patterns. • Class Hexactinellida comprises the deep-sea glass sponges. The skeletal system is a complex grid of siliceous spicules. The living tissue is a syncytium. • Class Demospongiae comprises 80% of sponge species. Their skeletons are typically siliceous spicules and/or spongin, but some have skeletal fibers of chitin. • Class Homoscleromorpha comprises marine sponges with two features present in animals more complex than sponges: (1) the pinacoderm layer has a basal lamina, but only one of the two types of cell junctions present in most animals, and (2) the skeleton contains type IV collagen, a characteristic of other animals.
6.4 Phylogeny and Adaptive Diversification	• Sponges are the sister taxon of a group comprising all other animals. • Although their body plans differ from those of other animals, most aspects of sponge embryological development are typical of animals. • Sponges are an ancient group, present before the Cambrian period. • Adaptive diversification is centered on elaboration of the water circulation and suspension-feeding system. • One group of deep-water and cave-dwelling demosponges (cladorhizids) does not filter-feed. As carnivores, they trap small prey on a Velcro-like surface made from tiny spicules. Prey are digested extracellularly and intracellularly.
6.5 Placozoa	• Placozoans are thin, plate-like organisms able to assume a variety of shapes. There is no plane of symmetry. • Placozoans contain six different types of cells. There are two cellular epithelia, with a fibrous syncytial layer between them. • Placozoans mound their bodies over food and secrete digestive enzymes onto food. Digestion is extracellular.

Review Questions

1. Biologists used to talk about single-celled animals and multicellular animals. Why was this dichotomy incorrect?
2. Briefly describe asconoid, syconoid, and leuconoid body types in sponges. How do the two newly described body types, sylleibids and solenoids, differ from these three types?
3. What is a pinacoderm, and how is it distinct from a true tissue epithelium?
4. Explain how the body of a leuconoid sponge is designed for efficiency in feeding and respiration.
5. Define the following: pinacocytes, choanocytes, archaeocytes, sclerocytes, collencytes.
6. What material occurs in the skeleton of all sponges?
7. Describe the skeleton of each of the four classes of sponges.
8. Describe how sponges feed, respire, and excrete.
9. What is a gemmule?
10. Why are gemmules likely to occur in sponges in areas with harsh winters?
11. Describe how gametes are produced and how fertilization occurs in most sponges.
12. How have some sponges altered their morphology to function as predators instead of filter feeders?

For Further Thought Mutualisms are interactions between two species where both participants benefit. How might the relationship between a sponge with glassy spicules and microalgae or cyanobacteria be a mutualism?

7

Cnidarians and Ctenophores

LEARNING OBJECTIVES

Readers will be able to:

7.1 Describe the habitat and lifestyles of cnidarians.

7.2 Explain how the cnidocyte and other cells function in the polyp and medusa stages of cnidarians.

7.3 Describe the typical life cycle of Hydrozoa and give examples of variation within this group.

7.4 Describe the typical life cycle of Scyphozoa and give examples of variation within this group.

7.5 Describe the body plan of Staurozoa.

7.6 Describe the body plan and sensory systems of Cubozoa; explain why some are dangerous to humans.

7.7 Explain why Myxozoa is now placed in Cnidaria.

7.8 Describe the typical life cycle of Anthozoa and give examples of variation within this group, including sea anemones and corals.

7.9 Describe the body plan and lifestyle of ctenophores and contrast these animals with cnidarians.

7.10 Discuss muscle tissue as a key phylogenetic character for cnidarians and ctenophores.

Larry S. Roberts

Tentacles of a Caribbean sea anemone, *Condylactis gigantea.*

A Fearsome Tiny Weapon

Most cnidarians, aside from corals, have soft bodies. Sea anemones and many others are attached to hard surfaces and thus unable to move away from danger. The jellies that swim are not able to swim against ocean currents. Indeed, we might easily get the false impression that cnidarians provide easy meals for other animals. However, many cnidarians are very effective predators that kill and eat much more highly organized, swift, and intelligent animals. They manage these feats because they possess tentacles that bristle with tiny, remarkably sophisticated weapons called nematocysts.

As a nematocyst forms within a cell, it is endowed with potential energy to power its discharge. A nematocyst is like a factory-manufactured gun that rolls off the assembly line cocked and ready with a bullet in its chamber. Like the cocked gun, the completed nematocyst requires only a small stimulus to make it fire. Rather than a bullet, a tiny thread bursts from the nematocyst. Achieving a velocity of 2 m/sec and an acceleration of 40,000 × gravity, it instantly penetrates its prey and injects a paralyzing toxin. A small animal unlucky enough to brush against one of the tentacles is suddenly speared with hundreds or even thousands of nematocysts and quickly immobilized. Some nematocyst threads can penetrate human skin, causing sensations ranging from minor irritation to great pain, or even death, depending on the species. A nematocyst is a fearsome, but wondrous, tiny weapon.

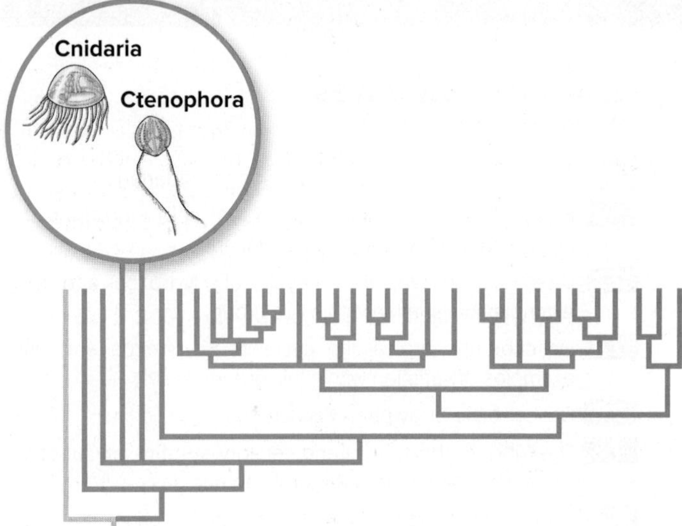

Phyla Cnidaria and Ctenophora exhibit primary radial or biradial symmetry. Radial symmetry, in which body parts are arranged concentrically around an oral-aboral axis, is particularly suitable for sessile or sedentary animals and for free-floating animals because they approach their environment (or it approaches them) from all sides equally. Biradial symmetry is basically a type of radial symmetry in which only two planes through the oral-aboral axis divide the animal into mirror images because the body contains a paired structure. All other animals, except sponges, have a primary bilateral symmetry; they are bilateral or were derived from an ancestor that was bilateral.

Before the advent of molecular phylogenies, biologists hypothesized that cnidarians and ctenophores were each other's closest relatives, but new phylogenies call this assumption into question. Radial symmetry may have evolved independently in cnidarians and ctenophores. Phylogenies using different kinds of data place ctenophores in different parts of the tree of life;[1] awaiting greater consensus, we do not depict a branching order for the basal nodes in animal phylogeny. The shared presence of true muscles was one character that seemed to group ctenophores and bilaterally symmetrical animals as a clade, but new analyses muddy the waters on this issue (see section 7.10).

CHARACTERISTICS of Phylum Cnidaria

1. **Cnidocytes** present, typically housing stinging organelles called **nematocysts**.
2. Entirely aquatic, some in freshwater, but most marine.
3. **Radial symmetry** or biradial symmetry around a longitudinal axis with oral and aboral ends; no definite head.
4. Two types of individuals: **polyps** and **medusae**.
5. Adult body two-layered **(diploblastic)** with epidermis and gastrodermis derived from embryonic ectoderm and endoderm, respectively.
6. Mesoglea, an extracellular matrix ("jelly") lies between body layers; amount of mesoglea varies; some have mesoglea with cells and connective tissue from ectoderm.
7. Incomplete gut called **gastrovascular cavity;** often branched or divided with septa.
8. Extracellular digestion in gastrovascular cavity and intracellular digestion in gastrodermal cells.
9. Extensible tentacles usually encircle mouth or oral region.
10. Muscular contractions via epitheliomuscular cells, which form an outer layer of longitudinal fibers at base of epidermis and an inner layer of circular fibers at base of gastrodermis; modifications of plan in hydrozoan medusae (independent ectodermal muscle fibers) and other complex cnidarians.
11. Sense organs include well-developed statocysts (organs of balance) and ocelli (photosensitive organs); complex eyes in members of Cubozoa.
12. Nerve net with symmetrical and asymmetrical synapses; diffuse conduction; two nerve rings in hydrozoan medusae.
13. Asexual reproduction by budding in polyps forms clones and colonies; some colonies exhibit **polymorphism.***
14. Sexual reproduction by gametes in all medusae and some polyps; monoecious or dioecious; holoblastic indeterminate cleavage; planula larval form.
15. No excretory or respiratory system.
16. No coelomic cavity.

*Note that polymorphism here refers to more than one structural form of individual within a species, as contrasted with the use of the word in genetics (see section 1.6), where it refers to different allelic forms of a gene in a population.

7.1 Phylum Cnidaria

Phylum Cnidaria (nī-dar′-ē-ə) (Gr. *knidē*, nettle, + L. *aria* [pl. suffix]; like or connected with) is a group of more than 9000 species. It takes its name from cells called **cnidocytes,** which contain the stinging organelles (cnidae) characteristic of the phylum. Cnidae come in several types, including the common **nematocysts.** Nematocysts are formed only by cnidarians.

Cnidarians are an ancient group with the longest fossil history of any animal phylum, reaching back more than 700 million years. They are widespread in marine habitats, and some occupy fresh water. Although cnidarians are mostly sessile or, at best, fairly slow moving or slow swimming, they are quite efficient predators of organisms that are much swifter and more complex.

The phylum Cnidaria includes some of nature's strangest and loveliest creatures: branching, plantlike hydroids; flowerlike sea anemones; sea jellies; and those architects of the ocean floor, gorgonian corals (sea whips, sea fans, and others) and stony corals, whose thousands of years of calcareous house-building have produced great reefs and coral islands. Algae frequently live as mutualists in the tissues of cnidarians,

[1]Dohrmann, M., and G. Wörheide. 2013. Novel scenarios of early animal evolution: Is it time to rewrite textbooks? *Integr. Comp. Biol.* **53**:503–511.

notably in some freshwater hydras and in reef-building corals. The presence of algae in reef-building corals limits coral reefs to relatively shallow, clear water where sunlight is sufficient for the photosynthetic requirements of the algae. These corals are an essential component of coral reefs, and reefs are extremely important habitats in tropical waters. Coral reefs are discussed further later in the chapter (see section 7.8: Coral Reefs).

Four classes of Cnidaria were traditionally recognized: Hydrozoa (the most variable class, including hydroids, fire corals, Portuguese man-of-war, and others), Scyphozoa ("true" jellies), Cubozoa (cube, or box jellies), and Anthozoa (the largest class, including sea anemones, stony corals, soft corals, and others) (figure 7.1). A fifth class, Staurozoa, has been proposed for a group of odd animals whose bodies resemble polyps with a medusa-like region on the top (see section 7.5). Molecular data now identifies a sixth cnidarian class of highly modified parasitic forms, the Myxozoa (figure 7.1).

Ecological Relationships

Cnidarians occur most abundantly in shallow marine habitats, especially in warm temperatures and tropical regions. There are some freshwater species, but no terrestrial species. Colonial hydroids are usually found attached to mollusc shells, rocks, wharves, and other animals in shallow coastal water, but some species occur at great depths. Floating and free-swimming medusae occur in open seas and lakes, often far from shore. Colonies such as the Portuguese man-of-war and *Velella* (L. *velum,* veil, + *ellus,* dim. suffix) have floats or sails by which the wind carries them.

Some ctenophores, molluscs, and flatworms eat hydroids bearing nematocysts and use these stinging structures for their own defense. Other animals, such as some molluscs and fishes, feed on cnidarians, but cnidarians rarely serve as food for humans. Planktonic medusae may

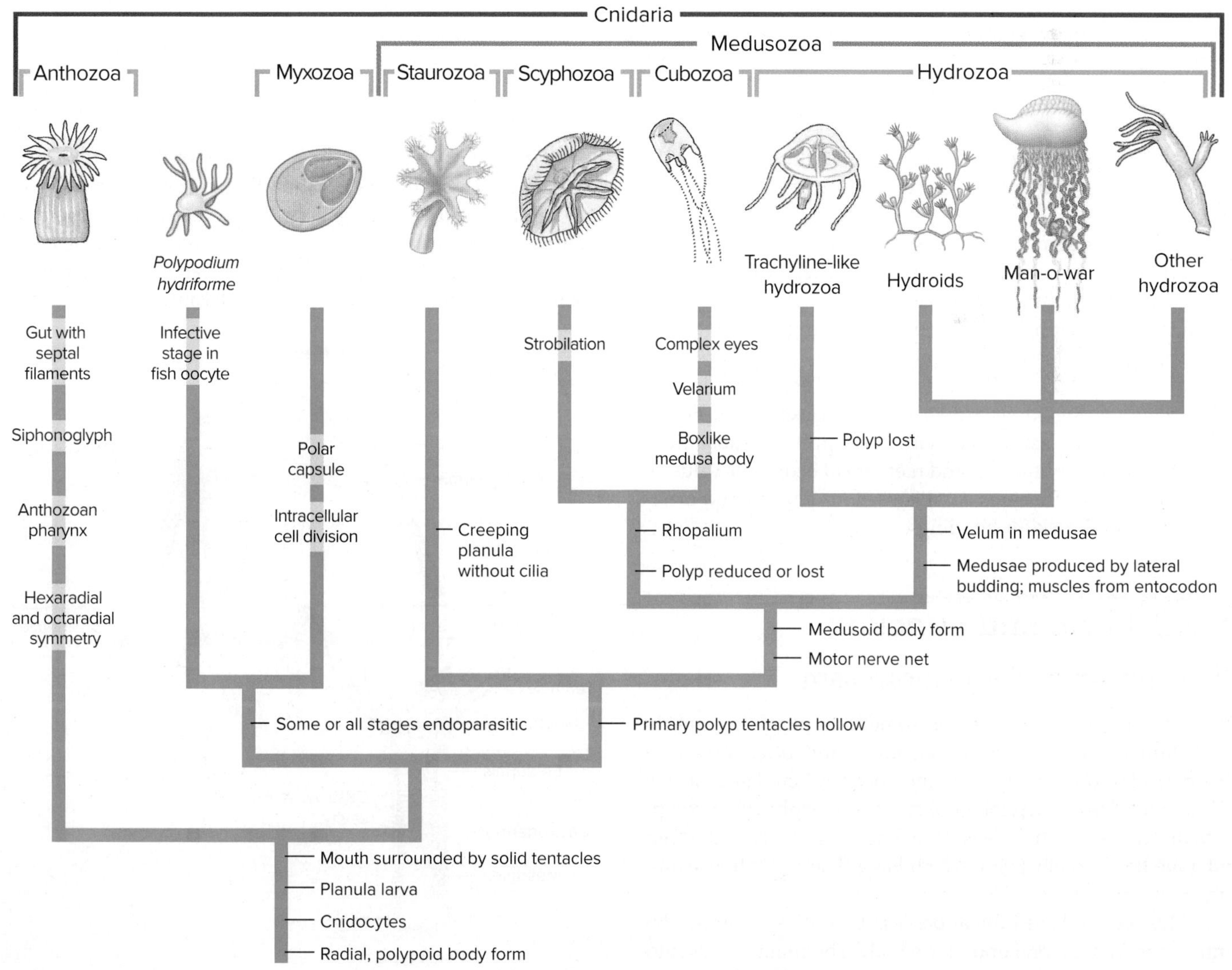

figure 7.1

Cladogram showing hypothetical relationships of cnidarian classes with some shared derived characters (synapomorphies) indicated.

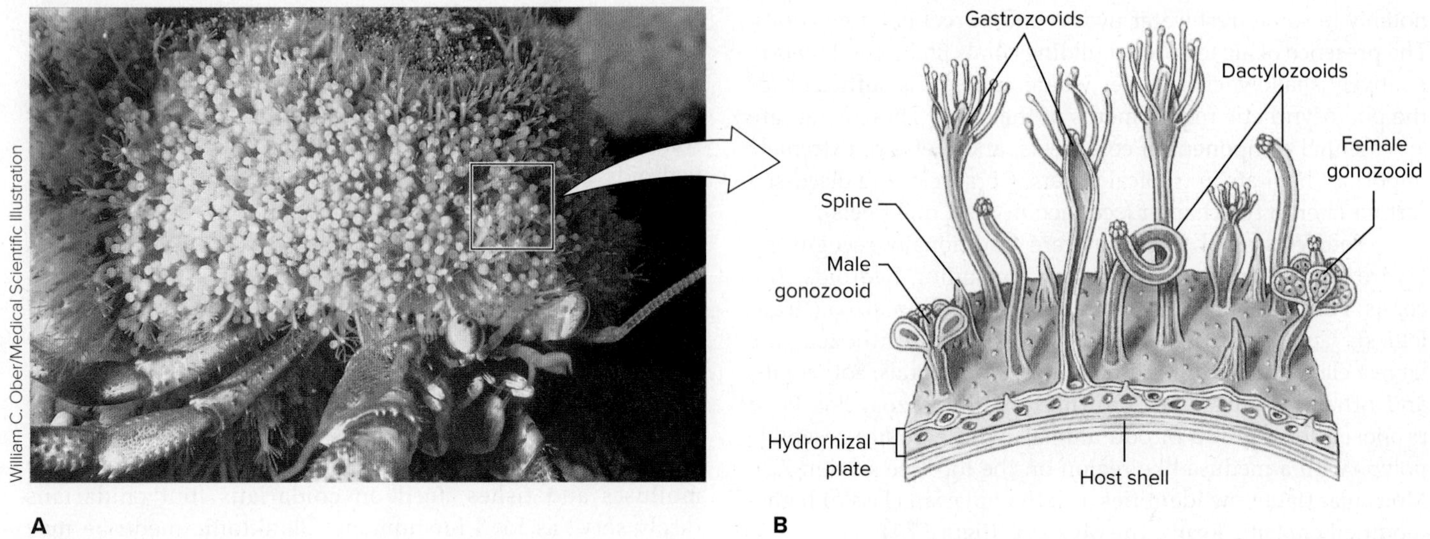

figure 7.2

A, A hermit crab with its cnidarian mutuals. The shell is blanketed with polyps of the hydrozoan *Hydractinia symbiolongicarpus.* The crab gets some protection from predation by the presence of cnidarians, and the cnidarians get a free ride and bits of food from their host's meals. **B,** Portion of a colony of *Hydractinia,* showing the types of zooids and the stolon (hydrorhiza) from which they grow. Note that male and female gonozooids do not naturally occur on the same colony.

be of some importance as food for commercially valuable fishes, but the reverse is also true–young fish fall prey to cnidarians.

Cnidarians sometimes live symbiotically with other animals, often as commensals on the shell or other surface of their host. Certain hydroids (figure 7.1) and sea anemones commonly live on snail shells inhabited by hermit crabs (figure 7.2), giving the crabs some protection from predators.

Although many cnidarians have little economic importance, reef-building corals are an important exception. Fishes and other animals associated with reefs provide substantial amounts of food for humans, and reefs are of economic value as tourist attractions. Precious coral is used for jewelry and ornaments, and coral rock for buildings.

7.2 Form and Function

Dimorphism and Polymorphism

One of the most interesting–and sometimes puzzling–aspects of phylum Cnidaria is the dimorphism and often polymorphism displayed by many of its members. All cnidarian forms fit into one of two morphological types (**dimorphism**): a **polyp,** or hydroid form, which has a sedentary or sessile lifestyle, and a **medusa,** or jelly form, which has a floating or free-swimming existence (figure 7.3).

Most polyps have tubular bodies. A mouth surrounded by tentacles marks the oral end of the body. The mouth leads into a blind gut or gastrovascular cavity (figure 7.3). The aboral end of the polyp is usually attached to a substratum by a pedal disc or other device.

Polyps may reproduce asexually by budding, fission, or laceration of the pedal disc. In **budding,** a knob of tissue forms on the side of an existing polyp and develops a functional mouth and tentacles (see figure 7.8). A bud that detaches from the polyp that made it is a clone. Clones can also be formed by **fission,** in which one-half of a polyp pulls away from the other,

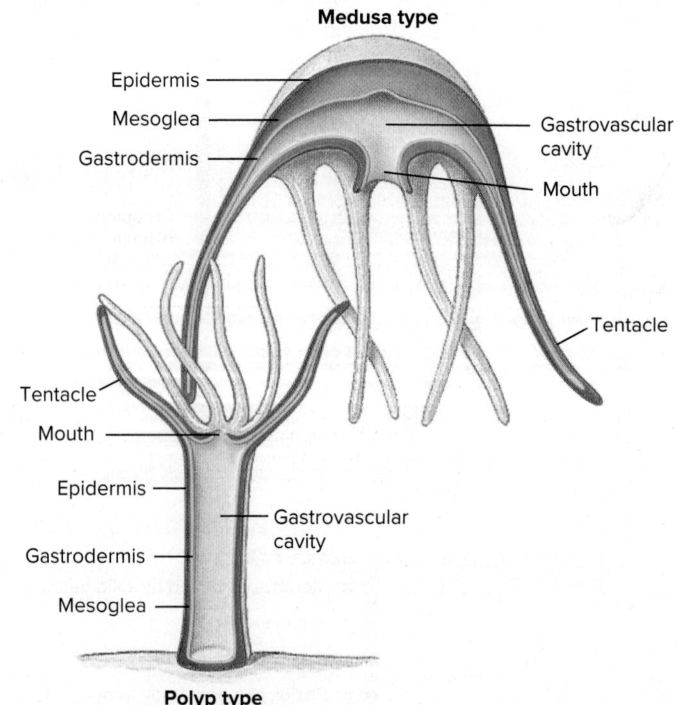

figure 7.3

Comparison between the polyp and medusa types of individuals.

or by laceration of the pedal disc, in which tissue torn from the pedal disc (see figure 7.22) develops into tiny new polyps. Polyps that do not bud are solitary; others form clones or colonies.[2] The distinction between clones and colonies is blurred when a colony fragments.

When buds stay attached to the polyp that made them, a colony forms, and food may be shared through a common gastrovascular cavity (see figure 7.2). A shared gastrovascular cavity permits some polyps (often called zooids) to specialize for particular functions, such as feeding (gastrozooids), defense (dactylozooids), or making sexually reproducing stages (gonozooids). Gastrozooids and dactylozooids do not collect food, but are fed through the shared gastrovascular cavity. A colony that contains several body forms is **polymorphic**; such colonies are common in class Hydrozoa.

Key Theme 7.1

SCIENCE EXPLAINED

Why is a Sea Jelly Called a Medusa?

The name "medusa" was suggested by a fancied resemblance to the Gorgon Medusa, a mythological creature with snaky tresses that turned to stone any who gazed upon her. Can you see how some might think the medusa in the photograph resembles a sea jelly?

Digital image courtesy of the Getty's Open Content Program.

Medusae are usually free-swimming and have bell-shaped or umbrella-shaped bodies and tetramerous symmetry (body parts arranged in fours). The mouth is usually centered on the concave side, and tentacles extend from the rim of the umbrella.

Superficially, polyps and medusae seem very different, but each has retained the saclike body plan basic to the phylum (see figure 7.3). A medusa is essentially an unattached polyp with the tubular portion widened and flattened into the bell shape.

Both a polyp and a medusa possess two tissue layers with an extracellular matrix (ECM) called mesoglea between them, but the jellylike layer of mesoglea is much thicker in a medusa, forming the bulk of the animal and making it more buoyant. Because of this mass of mesoglea ("jelly"), medusae are commonly called jellyfishes, although the more correct term "jellies" is preferred because they are not "fishes."

Locomotion

Colonial polyps are permanently attached, but hydras can move freely across a substrate by gliding on their pedal disc, aided by mucous secretions. Sea anemones can move similarly on their pedal discs. Hydras can also use a "measuring worm" movement, looping along a surface by bending over and attaching their tentacles to the substratum. They may even turn handsprings or detach and, by forming a gas bubble on the pedal disc, float to the surface.

Most medusae can move freely, and they swim by contracting the bell, which expels water from the concave, oral side. The muscular contractions are antagonized by the compressed mesoglea and elastic fibers within it. Usually, medusae contract several times and move generally upward, then sink slowly. Cubozoan medusae, however, can swim strongly.

Life Cycles

In a cnidarian life cycle, polyps and medusae play different roles. The exact life cycle varies among the classes, but in general a zygote develops into a free-swimming **planula** larva. The planula settles and metamorphoses into a polyp. A polyp may make other polyps asexually, but polyps in classes Hydrozoa and Scyphozoa eventually make medusae. Although medusae are made asexually from the polyp body, they reproduce sexually. Medusae develop into either male or female individuals (dioecious) and produce gametes. Fertilization typically occurs in open water, although some species brood their zygotes until the planula stage.

Sea anemones and corals (class Anthozoa) are all polyps; hence, both sexual and asexual reproduction occur in the polyp phase. The sea jellies (class Scyphozoa) have a conspicuous medusoid form, although most also have a polypoid larval stage. Colonial hydroids of class Hydrozoa typically have life histories that feature both a polyp stage and a free-swimming medusa stage. A species that has both an attached polyp and a floating medusa within its life history can take advantage of the feeding and distribution possibilities of both pelagic (open-water) and benthic (bottom) environments.

Feeding and Digestion

Cnidarians prey on a variety of organisms of appropriate size; larger species are usually capable of killing and eating larger prey. The mouth opens into the gastrovascular cavity, which communicates with cavities in the tentacles. The mouth may be surrounded by an elevated manubrium or by elongated

[2] In Chapter 3, distinctions were made among unicellular, colonial, and multicellular organisms. However, the term "colonial" is also used for multicellular animals and plants where the body is made of repeated units. A colonial cnidarian consists of multiple polyps; polyps may be identical or specialized for different functions.

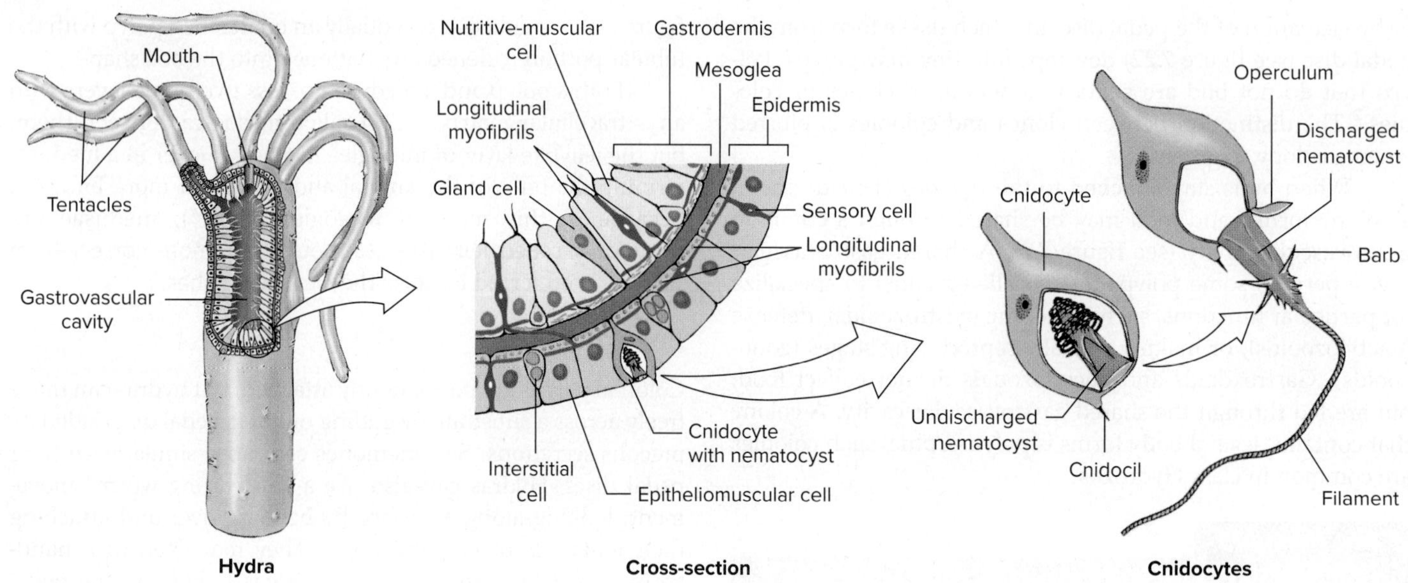

figure 7.4

A Cnidarian polyp, illustrated by *Hydra* (*left*), has specialized cells in the body wall (*center*). The epidermal layer of the body wall contains specialized cnidocytes with nematocysts. Nematocysts (*right*) arise from epidermal interstitial cells and are used to sting prey.

oral lobes in the medusa stage. Normally, prey organisms are drawn into the gastrovascular cavity into which gland cells discharge enzymes that initiate **extracellular digestion.** In addition, nutritive-muscular cells phagocytize many food particles for **intracellular digestion.** Ameboid cells can carry undigested particles to the gastrovascular cavity, where they are eventually expelled with other indigestible matter.

Body Wall

In a cnidarian polyp, such as *Hydra,* the body wall surrounding the gastrovascular cavity consists of an outer **epidermis** (ectodermal) and an inner **gastrodermis** (endodermal) with **mesoglea** between them (figure 7.4). Each layer and the cells it contains are discussed briefly.

Mesoglea Mesoglea lies between the epidermis and the gastrodermis and adheres to both layers (see figures 7.3 and 7.4). It is a gelatinous ECM without fibers or cellular elements in hydrozoan polyps. It is thicker in medusae and has elastic fibers; in scyphozoan medusae, it has ameboid cells. The mesoglea of anthozoans contains ameboid cells and epitheliomuscular cells (discussed in section 7.2: Epidermis).

Gastrodermis The gastrodermis, a layer of cells lining the gastrovascular cavity, consists chiefly of large, ciliated, columnar epithelial cells with irregular flat bases (see figure 7.4, *center*). Cells of the gastrodermis include nutritive-muscular, interstitial, and gland cells and, in classes other than Hydrozoa, cnidocytes. Gonads are gastrodermal in most cnidarians.

Nutritive-muscular cells are usually tall columnar cells that have laterally extended bases containing myofibrils. In hydrozoans, the myofibrils run at right angles to the body or tentacle axis and so form a circular muscle layer. However, this muscle layer is very weak, and longitudinal extension of the body and tentacles occurs mostly by increasing the volume of water in the gastrovascular cavity. Water is brought into the cavity through the mouth by the beating of cilia on the nutritive-muscular cells in hydrozoans or by ciliated cells in the pharynx of anthozoans. Thus, water in the gastrovascular cavity serves as a **hydrostatic skeleton.** The two cilia on the free end of each cell also serve to circulate food and fluids in the digestive cavity. The cells often contain large numbers of food vacuoles. Gastrodermal cells of green hydras (*Chlorohydra* [Gr. *chlōros,* green, + *hydra,* a mythical nine-headed monster slain by Hercules]) bear green algae (zoochlorellae), but in marine cnidarians they bear a type of dinoflagellate (see section 5.5: Dinoflagellata: zooxanthellae). Both are examples of mutualism, with the algae furnishing organic compounds to their cnidarian hosts.

Interstitial cells scattered among the bases of the nutritive cells can transform into other cell types. Gland cells are tall cells that secrete digestive enzymes.

Epidermis The epidermal layer contains epitheliomuscular, interstitial, gland, cnidocyte, and sensory and nerve cells. Gonads are epidermal in Hydrozoa.

Epitheliomuscular cells form most of the epidermis (figure 7.5) and serve both for covering and for muscular contraction. The bases of most of these cells are extended parallel to the tentacle or body axis and contain myofibrils, thus forming a layer of longitudinal muscle next to the mesoglea. Contraction of these fibrils shortens the body or tentacles.

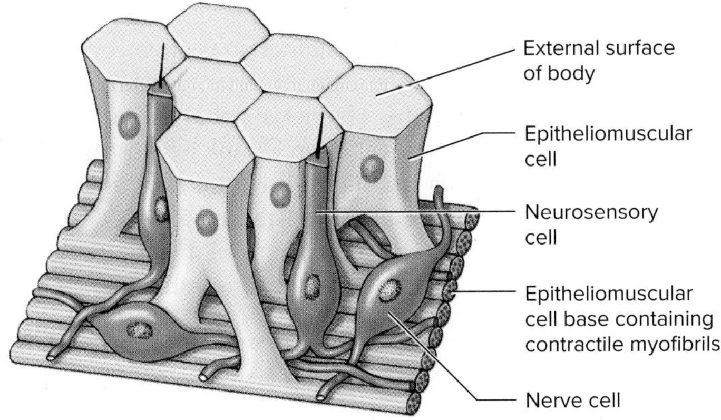

figure 7.5

Epitheliomuscular and nerve cells in hydra.

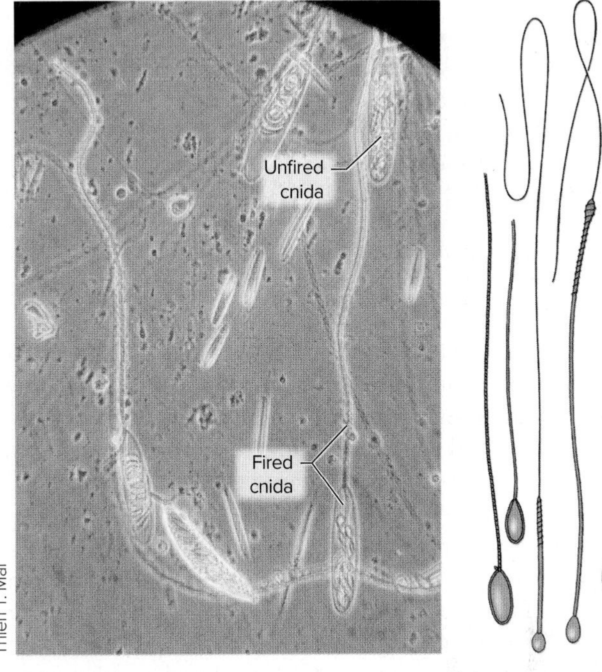

figure 7.6

A, Fired and unfired cnidae from *Corynactis californica.*
B, Several types of cnidae shown after discharge.

Interstitial cells are undifferentiated stem cells found among the bases of the epitheliomuscular cells. Differentiation of interstitial cells produces cnidoblasts, sex cells, buds, nerve cells, and others, but generally not epitheliomuscular cells (which reproduce themselves).

Gland cells are particularly abundant around the mouth and in the pedal disc of hydra. They secrete mucus or adhesive material.

Cnidocytes (containing cnidae) occur throughout the epidermis (see figure 7.4). They may be between the epitheliomuscular cells or housed in invaginations of these cells, and they are most abundant on the tentacles.

Sensory cells are scattered among the other epidermal cells, especially around the mouth and tentacles. The free end of each sensory cell bears a flagellum, which is the sensory receptor for chemical and tactile stimuli. The other end branches into fine processes, which synapse with nerve cells.

Nerve cells of the epidermis are often multipolar (have many processes), although in more highly organized cnidarians, the cells may be bipolar (with two processes). Their processes (axons) form synapses with sensory cells and other nerve cells, and junctions with epitheliomuscular cells and cnidocytes. Both one-way and two-way synapses with other nerve cells are present (figure 7.5).

Cnidae Over 20 different types of cnidae have been described, and they are important taxonomic characters. **Cnidae** are tiny capsules contained within cnidocytes. They are composed of material similar to chitin and contain a coiled tubular "thread" or filament, which is a continuation of the narrowed end of the capsule (figure 7.6). This end of the capsule is covered by a little lid, or **operculum.** The inside of the undischarged thread may bear tiny barbs, or spines, as in the most common cnida, the **nematocyst**. There are three functional types of cnidae in hydras: those that penetrate prey and inject venom (penetrants, see figure 7.4); those that recoil and entangle prey (volvents, see figure 7.6); and those that secrete an adhesive substance used in locomotion and attachment (glutinants).

A cnida is enclosed in the cell that produced it. (During its development, a cnidocyte is properly called a **cnidoblast.**) Except in Anthozoa, cnidocytes are equipped with a trigger-like **cnidocil,** which is a modified cilium. Anthozoan cnidocytes have a somewhat different ciliary mechanoreceptor. In some sea anemones and perhaps other cnidarians, small organic molecules from a prey "tune" the mechanoreceptors, sensitizing them to the frequency of vibration caused by the prey swimming. Tactile stimulation causes a nematocyst to discharge. Cnidocytes are borne in invaginations of ectodermal cells and, in some forms, in gastrodermal cells, and they are especially abundant on the tentacles. When a nematocyst has discharged, its cnidocyte is absorbed and a new one replaces it.

The mechanism of nematocyst discharge is remarkable. Present evidence indicates that discharge is caused by a combination of tensional forces generated during nematocyst formation and also by an astonishingly high osmotic pressure within the nematocyst: 140 atmospheres. When stimulated to discharge, the permeability of the nematocyst changes, and the high internal osmotic pressure causes water to rush into the capsule. The operculum opens, and the rapidly increasing *hydrostatic pressure* within the capsule forces the thread outward with great force, the thread turning inside out as it goes. At the everting end of the thread, the barbs flick to the outside like tiny switchblades. This minute but awesome weapon then injects venom when it penetrates prey.

Key Theme 7.2

SCIENCE EXPLAINED

Osmotic versus Hydrostatic Pressure

Note the distinction between osmotic and hydrostatic pressure. A nematocyst is never actually required to contain 140 atmospheres of hydrostatic pressure within itself; such a hydrostatic pressure would doubtless cause it to explode. As water rushes in during discharge, osmotic pressure falls rapidly, while hydrostatic pressure increases rapidly.

The nematocysts of most cnidarians are not harmful to humans and are a nuisance at worst. However, stings of the Portuguese man-of-war (see figure 7.13) and certain jellies are quite painful and sometimes dangerous. The venom of box jellies (Class Cubozoa, Section 7.6) may be lethal.

Nerve Net The nerve net of the cnidarians is one of the best examples of a diffuse nervous network in the animal kingdom. This plexus of nerve cells occurs both at the base of the epidermis and at the base of the gastrodermis, forming two interconnected nerve nets. Nerve processes (axons) end on other nerve cells at synapses or at junctions with sensory cells or effector organs (nematocysts or epitheliomuscular cells). Nerve impulses are transmitted from one cell to another by release of a neurotransmitter from small vesicles on one side of the synapse or junction. One-way transmission between nerve cells in higher animals is ensured because the vesicles are located on only one side of the synapse. However, cnidarian nerve nets are peculiar in that many of the synapses have vesicles of neurotransmitters on both sides, allowing transmission across the synapse in either direction. Another peculiarity of cnidarian nerves is the absence of any sheathing material (myelin) on the axons.

Cnidarians have no concentrated grouping of nerve cells to suggest a "central nervous system." However, nerves are grouped in the "ring nerves" of hydrozoan medusae and in the marginal sense organs of scyphozoan medusae. In some cnidarians, the nerve nets form two or more systems: Scyphozoa has a fast conducting system to coordinate swimming movements and a slower one to coordinate movements of tentacles.

Key Theme 7.3

ADAPTATION AND PHYSIOLOGY

Radial Symmetry and Nerve Nets

Note that there is little adaptive value in pairing a radially symmetrical animal with a central nervous system with a brain. Predators and prey may approach from any direction in a watery 3-D world. A nerve net makes possible an immediate response toward or away from a stimulus in any direction. Sending signals and information to a single central brain is much more useful for bilaterally symmetrical animals that typically move in one plane.

Nerve cells of the net have synapses with slender sensory cells that receive external stimuli, and the nerve cells have junctions with epitheliomuscular cells and nematocysts. Together with the contractile fibers of epitheliomuscular cells, the sensory cell and nerve-net combination is often considered a cellular neuromuscular system, an important landmark in the evolution of nervous systems. The nerve net arose early in animal evolution, and it has never been completely lost phylogenetically. Annelids, such as the earthworm, have it in their digestive systems. In the human digestive system, it occurs as nerve plexuses in the musculature controlling the rhythmic peristaltic movements of the stomach and intestine.

7.3 Class Hydrozoa

Most Hydrozoa are marine and colonial in form, and the typical life cycle includes both an asexual polyp and a sexual medusa stage. Some, however, such as the freshwater hydras, have no medusa stage. Certain marine hydroids do not have free medusae (figure 7.7), whereas some hydrozoans occur only as medusae and have no polyp stage.

Hydras, although not typical hydrozoans, have become favorites as an introduction to Cnidaria because of their size and ready availability. Combining study of a hydra with that of a representative colonial marine hydroid such as *Obelia* (Gr. *obelias*, round cake) gives an excellent idea of class Hydrozoa.

Hydra: A Freshwater Hydrozoan The common freshwater hydra (figure 7.8) is a solitary polyp and one of the few cnidarians found in fresh water. Its normal habitat is the underside of aquatic leaves and lily pads in cool, clean freshwater pools and streams. The hydra family occurs throughout the world, with 16 species in North America.

Key Theme 7.4

ADAPTATION AND PHYSIOLOGY

Hydra is a Model Organism for Regeneration

Over 230 years ago, Abraham Trembley was astonished to discover that isolated sections of the stalk of hydra could regenerate and each become a complete animal. Since then, over 2000 investigations of hydra have been published, and the organism has become a classic model for the study of morphological differentiation. The mechanisms governing morphogenesis have great practical importance, and the simplicity of hydra lends itself to these investigations. Substances controlling development (morphogens), such as those determining which end of a cut stalk will develop a mouth and tentacles, can function at extremely low concentrations (10^{-10} M).

The body of a hydra can extend to a length of 25–30 mm or can contract to a tiny, jellylike mass. It is a cylindrical tube with the lower (aboral) end drawn out into a slender stalk, ending in a pedal disc for attachment. This pedal disc has gland cells that enable a hydra to adhere to a substratum and also to secrete a gas bubble for floating. In the center of the disc may be an excretory pore. The mouth, located on a conical elevation called the **hypostome,** is encircled by six to ten hollow tentacles

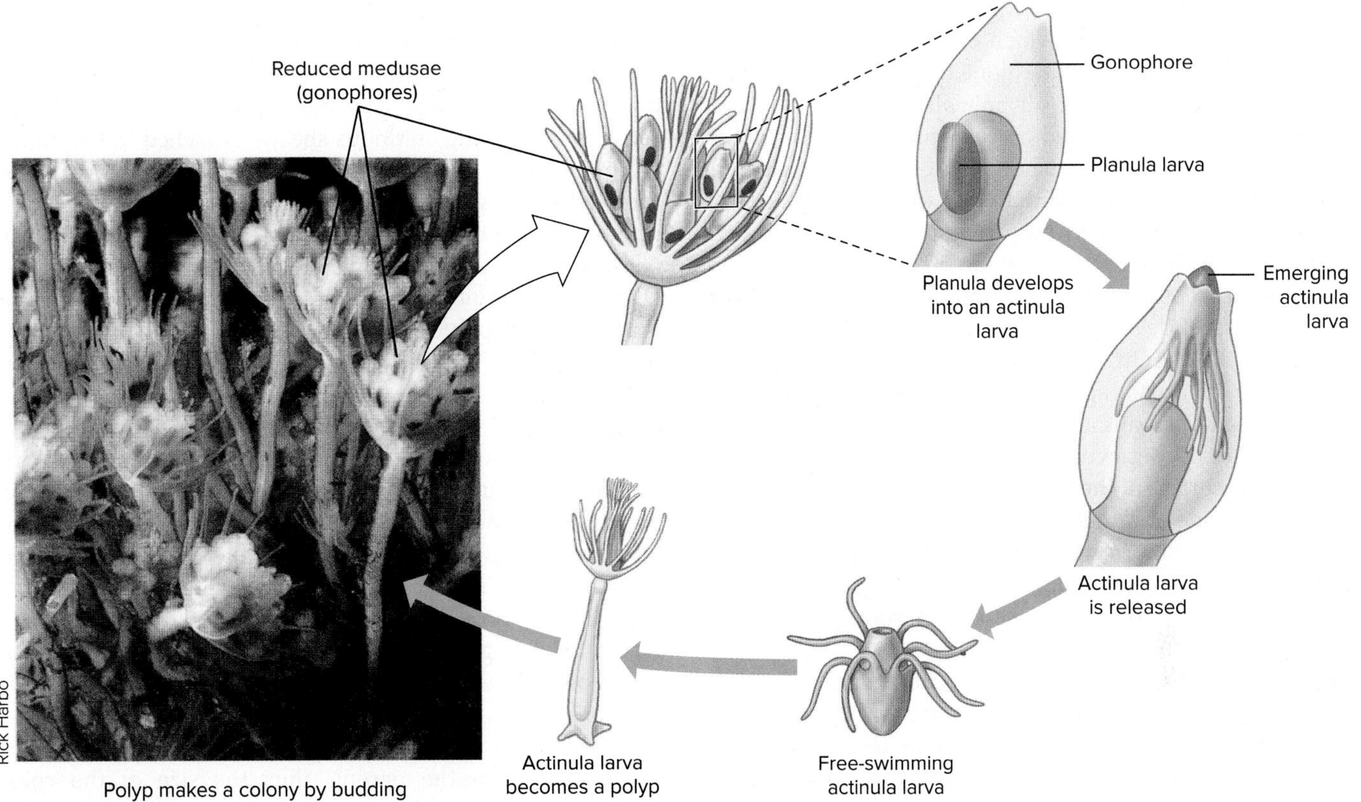

figure 7.7

In some hydroids, such as this *Tubularia crocea*, medusae are reduced to gonadal tissue and do not detach. These reduced medusae are known as gonophores.

that, like the body, can greatly extend when the animal is hungry. The mouth opens into the gastrovascular cavity, which communicates with the cavities in the tentacles.

Hydras feed on a variety of small crustaceans, insect larvae, and annelid worms. The hydra awaits its prey with tentacles extended (figure 7.9). A food organism that brushes against its tentacles may find itself harpooned by scores of nematocysts that render it helpless, even though it may be larger than the hydra. The tentacles move the prey toward the mouth, which slowly widens to completely engulf it.

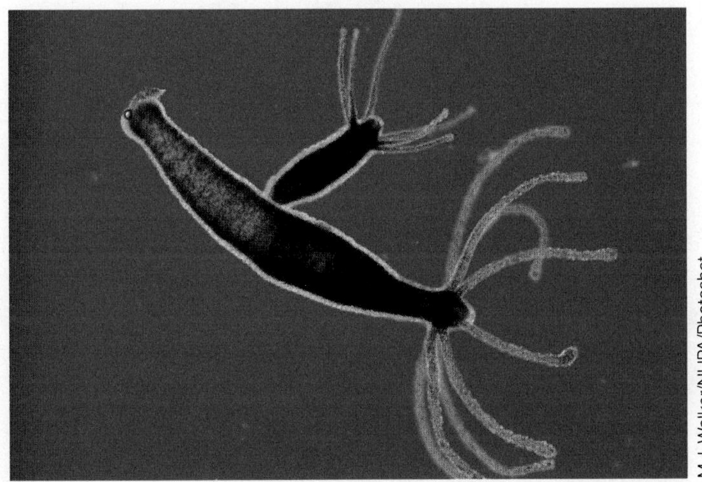

figure 7.8

This green hydra has symbiotic algae inside its gastrodermal cells. A bud with tentacles is attached to one side of the body.

figure 7.9

Hydra catches an unwary crustacean with the nematocysts on its tentacles. Notice the buds attached to hydra near its base, and developing gonads above the buds.

The chemical activator causing the mouth to open is the reduced form of glutathione, which occurs to some extent in all living cells. Prey release glutathione through wounds made by the nematocysts, but not all animals release enough of the chemical to activate a feeding response by a hydra. This explains how a hydra distinguishes between *Daphnia,* which it relishes, and some other forms that it refuses. If glutathione is added to water containing hydras, each hydra will go through the motions of feeding, even though no prey is present.

In asexual reproduction, buds appear as outpocketings of the body wall and develop into young hydras that eventually detach from the parent. In sexual reproduction, temporary gonads (figure 7.9) usually appear in autumn, stimulated by lower temperatures and perhaps also by reduced aeration of stagnant waters. Testes or ovaries, when present, appear as rounded projections on the surface of the body (figure 7.9). Eggs in the ovary usually mature one at a time and are fertilized by sperm shed into the water. A cyst forms around the embryo before it breaks loose from the parent, enabling it to survive the winter. Young hydras hatch in spring when the weather is favorable.

Hydroid Colonies Far more representative of class Hydrozoa than hydras are those hydroids that have a medusa stage in their life cycle. *Obelia* is often used in laboratory exercises for beginning students to illustrate the hydroid type (figure 7.10).

A typical hydroid has a base, a stalk, and one or more terminal polyps (zooids). The base by which colonial hydroids attach to the substratum is a rootlike stolon, which gives rise to one or more stalks. The living cellular part of the stalks secretes a nonliving chitinous sheath. Attached to the ends of the branches of the stalks are individual zooids. Most zooids are feeding polyps called **hydranths,** or **gastrozooids.** They may be tubular, bottle shaped, or vaselike, but all have a terminal mouth and a circlet of tentacles. In some forms, such as *Obelia,* the chitinous sheath continues as a protective cup around the polyp into which the polyp can withdraw for protection (see figure 7.10). In others, the polyp is naked. Dactylozooids are polyps specialized for defense in some species (see figure 7.2B).

Hydranths, much like hydras, capture and ingest prey, such as tiny crustaceans, worms, and larvae, thus providing nutrition for the entire colony. After partial digestion in a hydranth, the digestive broth passes into the common gastrovascular cavity where intracellular digestion occurs.

Circulation within the gastrovascular cavity is a function of the ciliated gastrodermis, but rhythmic contractions and pulsations of the body, which occur in many hydroids, also aid circulation.

In contrast to hydras, new individuals that bud do not detach from the parent; thus the size of the colony increases. New polyps may be hydranths or reproductive

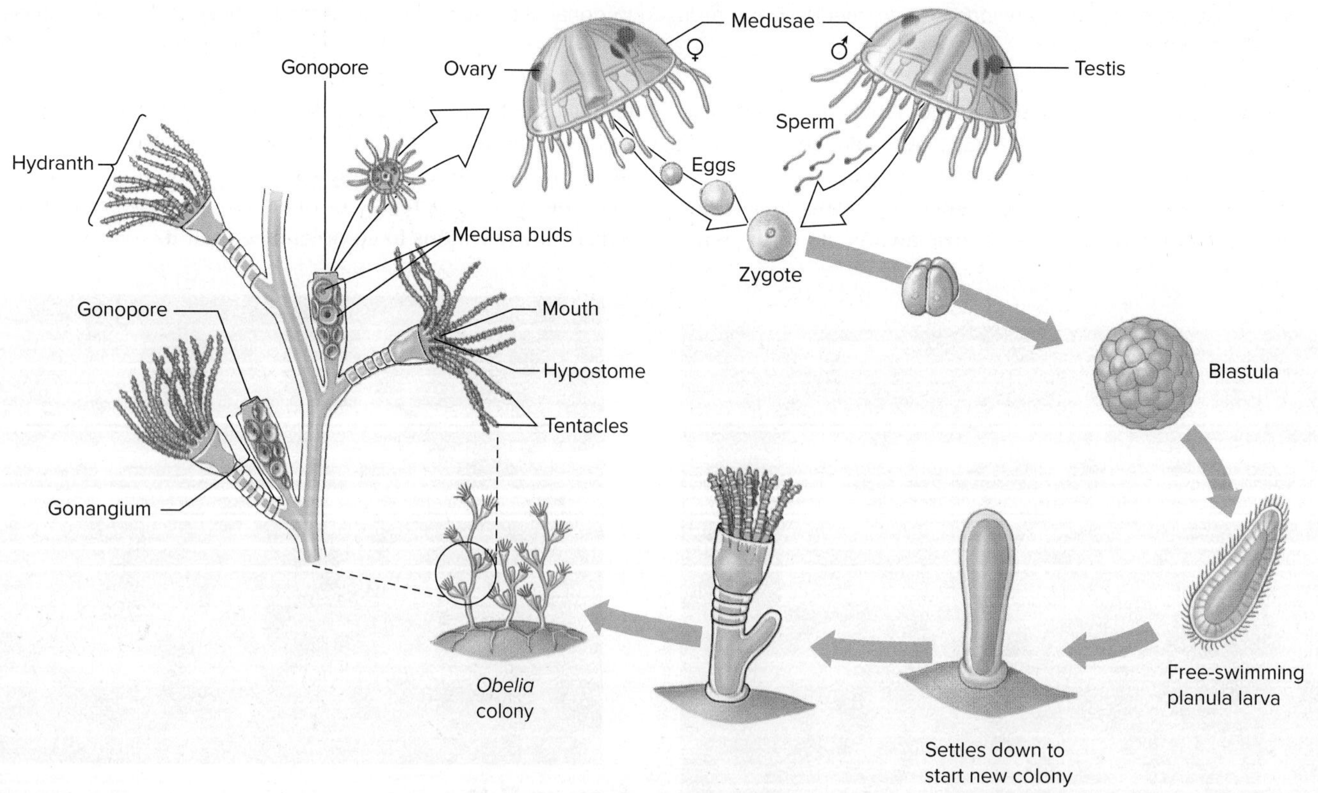

figure 7.10

Life cycle of *Obelia,* showing alternation of polyp (asexual) and medusa (sexual) stages. *Obelia* is a calyptoblastic hydroid, meaning that a nonliving covering protects its polyps as well as its stem.

polyps called **gonangia.** Medusae are produced by budding within the gonangia. Young medusae leave the colony as free-swimming individuals that mature and produce gametes (eggs and sperm) (figure 7.10). In some species, medusae remain attached to the colony and shed their gametes there. In other species, medusae never develop, gametes being shed by male and female gonophores. Development of a zygote produces a ciliated planula larva that swims for a time. Then it settles onto a substratum, becoming a minute polyp that gives rise, by asexual budding, to the hydroid colony, thus completing the life cycle.

Hydroid medusae are usually smaller than their scyphozoan counterparts, ranging from 2 or 3 mm to several centimeters in diameter (figure 7.11). The margin of the bell projects inward as a shelflike **velum,** which partly closes the open side of the bell (figure 7.12). Muscular pulsations that alternately fill and empty the bell propel the animal forward, aboral side first, with a sort of "jet propulsion." The velum projects outward as the bell empties, reducing the cross-sectional area of the exit stream and increasing the power of the jet. Tentacles attached to the bell margin are richly supplied with nematocysts.

The mouth opening at the end of a suspended **manubrium** leads to a stomach and four radial canals that connect with a ring canal around the margin. This in turn connects with the hollow tentacles. Thus, the gastrovascular cavity is continuous from mouth to tentacles, and the entire system is lined with gastrodermis. Nutrition is similar to that of hydranths.

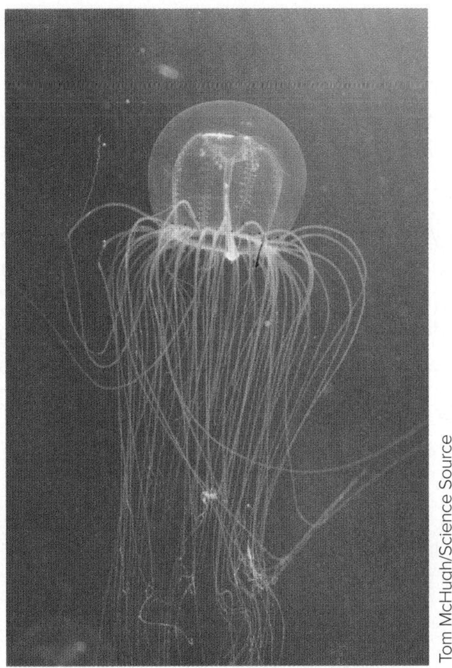

Tom McHugh/Science Source

figure 7.11

The medusa of *Polyorchis penicillatus* is thumb-sized and commonly encountered on the Pacific Coast of North America, although the polyp stage has never been identified.

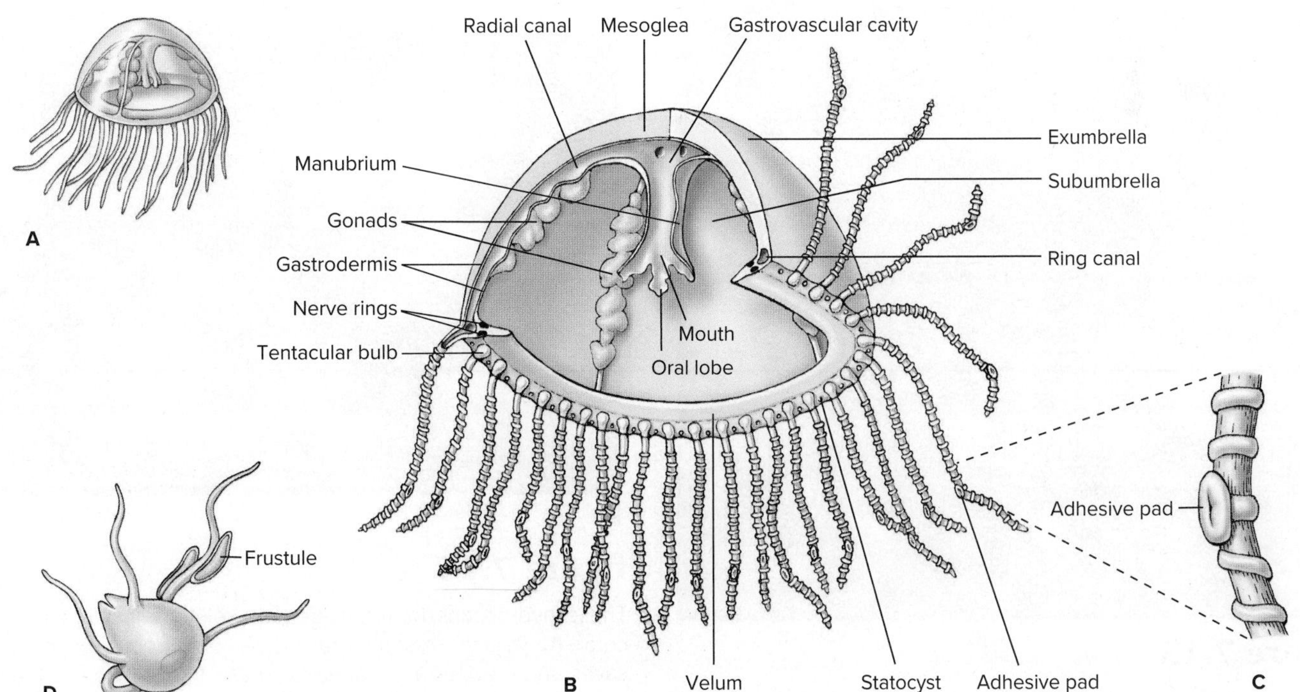

figure 7.12

Structure of *Gonionemus*. **A,** Medusa with typical tetramerous arrangement. **B,** Cutaway view showing morphology. **C,** Portion of a tentacle with its adhesive pad and ridges of nematocysts. **D,** Tiny polyp, or hydroid stage, that develops from the planula larva. It can produce more polyps by budding (frustules) or produce medusa buds.

The nerve net is usually concentrated into two nerve rings at the base of the velum. The bell margin is liberally supplied with sensory cells. It usually also bears two kinds of specialized sense organs: **statocysts,** which are small organs of equilibrium (see figure 7.12), and **ocelli,** which are light-sensitive organs.

Other Hydrozoans Siphonophores are hydrozoans that form floating colonies and include *Physalia* (Gr. *physallis,* bladder), the Portuguese man-of-war (figure 7.13). These colonies include several types of modified medusae and polyps. *Physalia* has a rainbow-hued float, probably a modified polyp, which carries it at the mercy of winds and currents. It contains an air sac filled with secreted gas and acts as a carrier for the generations of individuals that bud from it and hang suspended in the water. There are several types of individuals, including feeding polyps, reproductive polyps, long stinging tentacles, and so-called jelly polyps. Many swimmers have experienced the uncomfortable sting that these colonial floaters can inflict. The pain, along with the panic of the swimmer, can increase the danger of drowning.

Other hydrozoans secrete massive calcareous skeletons that resemble true corals (figure 7.14). They are sometimes called **hydrocorals.**

Wild Horizon/Getty Images

figure 7.13

A Portuguese man-of-war colony, *Physalia physalis* (order Siphonophora, class Hydrozoa). Colonies often drift onto southern ocean beaches, where they are a hazard to bathers. Each colony of medusa and polyp types is integrated to act as one individual. As many as a thousand zooids may be found in one colony. The nematocysts secrete a powerful neurotoxin.

Larry S. Roberts

A

Larry S. Roberts

B

figure 7.14

These hydrozoans form calcareous skeletons that resemble true coral. **A,** *Stylaster roseus* (order Stylasterina) occurs commonly in caves and crevices in coral reefs. These fragile colonies branch in only a single plane and may be white, pink, purple, red, or red with white tips. **B,** Species of *Millepora* (order Milleporina) form branching or platelike colonies and often grow over the skeletons of gorgonians. They have a generous supply of powerful nematocysts that produce a burning sensation on human skin, justly earning them the common name fire coral.

7.4 Class Scyphozoa

Class Scyphozoa (si-fō-zo′ə) (Gr. *skyphos,* cup) includes most of the larger jellies, or "cup animals." A few, such as *Cyanea* (Gr. *kyanos,* dark-blue substance), may attain a bell diameter exceeding 2 m and tentacles 60–70 m long (figure 7.15). Most scyphozoans, however, range from 2 to 40 cm in diameter. Most float in the open sea, some even at depths of 3000 m. Their coloring ranges from colorless to striking orange and pink hues.

The medusae of scyphozoans, unlike those of hydrozoans, have no velum. The bells of different species vary in depth from a shallow saucer shape to a deep helmet or goblet shape, and many have a scalloped margin, each notch bearing a sense organ called a **rhopalium** and a pair of lobelike projections called lappets. *Aurelia* (L. *aurum,* gold) has eight such notches (figures 7.16 and 7.17); others may have four or sixteen. Each rhopalium bears a statocyst for balance, two sensory pits containing concentrations of sensory cells, and sometimes an ocellus (simple photoreceptor). The mesoglea is thick and contains cells as well as fibers.

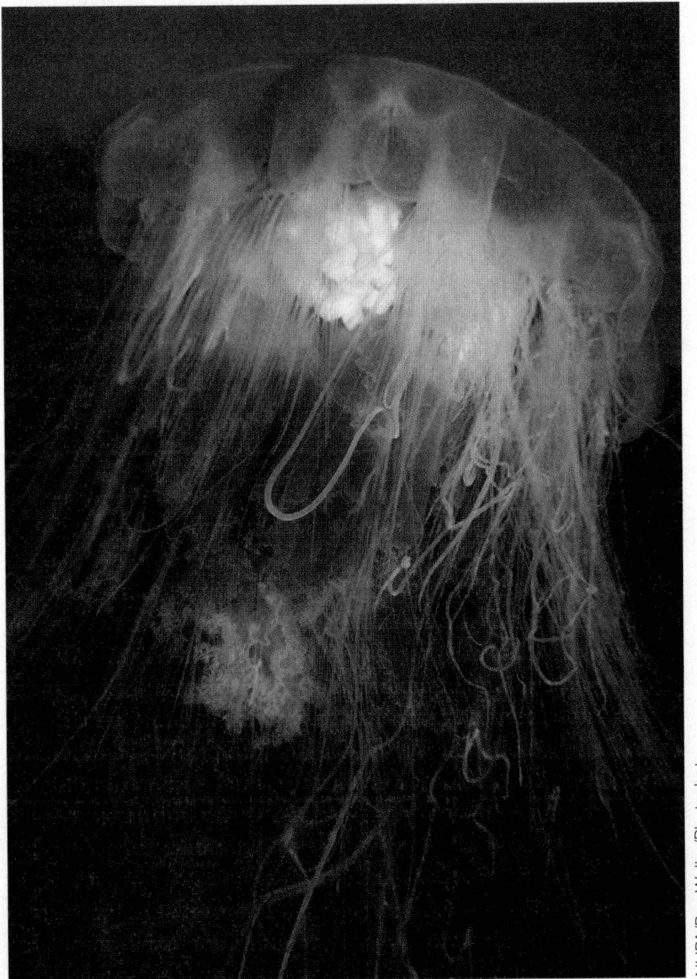

NHPA/Roy Waller/Photoshot

figure 7.15

Giant jelly, *Cyanea capillata* (order Semeaeostomeae, class Scyphozoa). A North Atlantic species of *Cyanea* reaches a bell diameter exceeding 2 m. Many fishermen know it as "sea blubber."

The mouth is centered on the subumbrellar side. The manubrium is usually drawn out into four frilly **oral lobes** used in food capture and ingestion. Marginal tentacles may be many or few; they may be short, as in *Aurelia,* or long, as in *Cyanea.* The tentacles, the manubrium, and often the entire body surface of scyphozoans are well supplied with nematocysts. Scyphozoans feed on small organisms, from unicellular eukaryotes to fishes. Capture of some prey involves stinging and manipulation with tentacles and oral lobes, but methods vary. *Aurelia* also feeds on small planktonic animals caught in mucus of the umbrella surface, carried to "food pockets" on the umbrella margin by cilia, and picked from the pockets by the oral lobes, whose cilia carry the food to the gastrovascular cavity. Cilia on the gastrodermis keep a current of water moving to bring food and oxygen into the gastrovascular cavity and to remove wastes.

Internally, four gastric pouches containing nematocysts connect a complex system of **radial canals** that branch from the pouches to the ring canal (see figure 7.16) forming the gastrovascular cavity, through which nutrients circulate.

The sexes are separate in scyphozoans; their gonads are located in the gastric pouches. Fertilization is internal, with sperm carried by ciliary currents into the gastric pouch of the female. Zygotes may develop in seawater or may be brooded in folds of the oral arms. The ciliated planula larva attaches and develops into a **scyphistoma,** a hydralike form (see figure 7.16) that may bud to make other polyps. By a process called **strobilation,** the scyphistoma of *Aurelia* forms a series of saucerlike buds (**ephyrae**) and becomes a **strobila** (see figure 7.16). When the ephyrae break free, they grow into mature jellies.

Key Theme 7.5

ADAPTATION AND PHYSIOLOGY

Jellies Abound in Ocean Dead Zones

An ocean dead zone is a very low-oxygen region where few animals survive, but marine scyphozoan jellies often predominate. These ancient animals evolved when oceanic oxygen levels were low, so their physiology may suit them to life in dead zones. The bulk of a jelly's body is mesoglea, with a thin layer of living tissue on the top of the animal and more tissue adjacent to the underside of the medusa bell. Oxygen diffuses into the body. Mesoglea is largely acellular, so it does not require oxygen, but instead acts as a reservoir. Metabolically active tissues rely on diffusion to bring in oxygen, but when levels are low, they draw oxygen from stores in mesoglea. In laboratory experiments at zero oxygen levels, stored oxygen allowed jellies to swim slowly. Jellies are often in the news as disrupting fisheries, clogging cooling tubes at nuclear power plants, and reducing beach access for swimming. Increases in abundance may have several causes: as humans harvest ever more fish, jellies have fewer competitors for food; as oceans warm, larvae develop faster; and new underwater structures built by humans provide more surface area for growth of the polyp stages that produce the medusae. Jellies seem well-suited to our changing world—in response, European researchers seek to increase commercial use of jellies, exploring jelly mucus as a way to bind ocean microplastics and jelly bodies as potential fertilizers or as human foods.

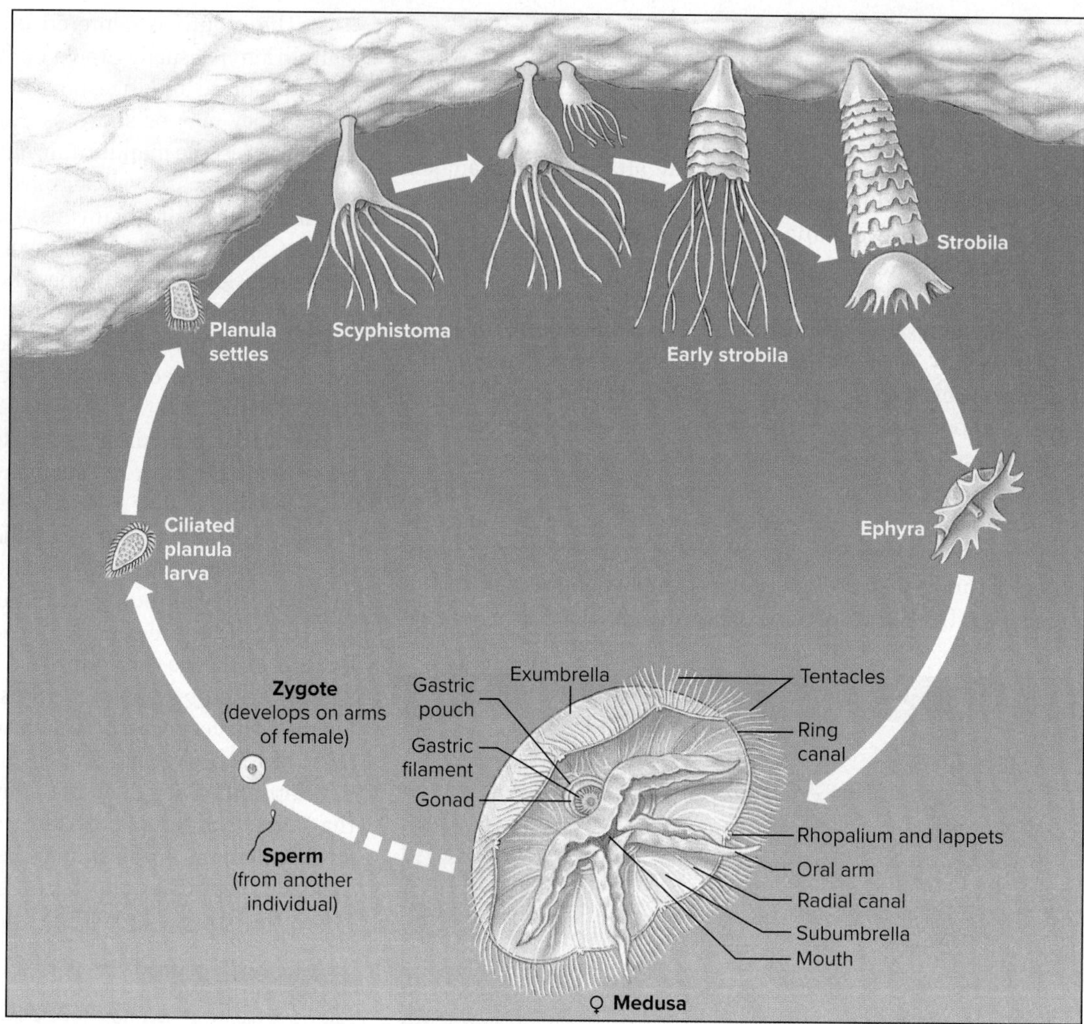

figure 7.16

Life cycle of *Aurelia*, a marine scyphozoan medusa.

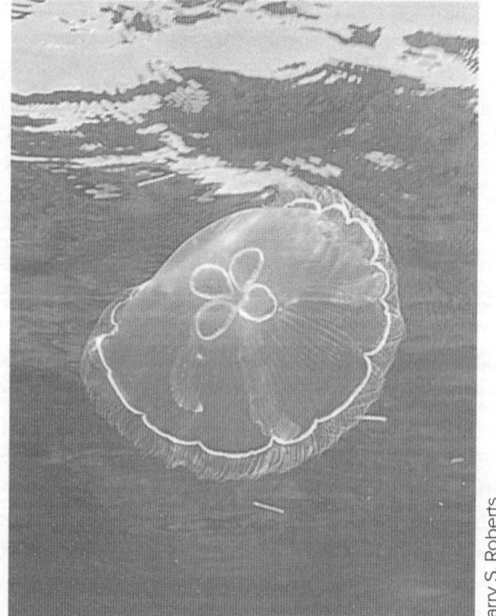

figure 7.17

Moon jelly *Aurelia aurita* (class Scyphozoa) is worldwide in distribution. It feeds on planktonic organisms caught in mucus on its umbrella.

7.5 Class Staurozoa

Animals in class Staurozoa are commonly called stauromedusans and were previously considered unusual scyphozoans, even though their life cycle does not include a medusa phase. The body is stalked (figure 7.18) and uses a basal adhesive disc to attach to seaweeds and other objects on the sea floor. The lower portion of the body is polyp-like in having longitudinal muscles and septa dividing the gastrovascular cavity. The top of the body resembles a medusa in that it has a circular coronal muscle, gonads, and gastric filaments. White spots on the body mark clusters of nematocysts positioned to defend the gonads. It has eight extensions ("arms"), ending in tentacle clusters, surrounding the mouth. The arms bear unusual adhesive pads used in temporary attachment. Polyps reproduce sexually. Other cnidarians make a free-swimming ciliated planula larva, but staurozoans have a creeping planula. The larva develops into a juvenile form, which later metamorphoses into a sessile adult. The creeping planula is a diagnostic feature and two other features, white spots with nematocysts and adhesive pads, may also be diagnostic.

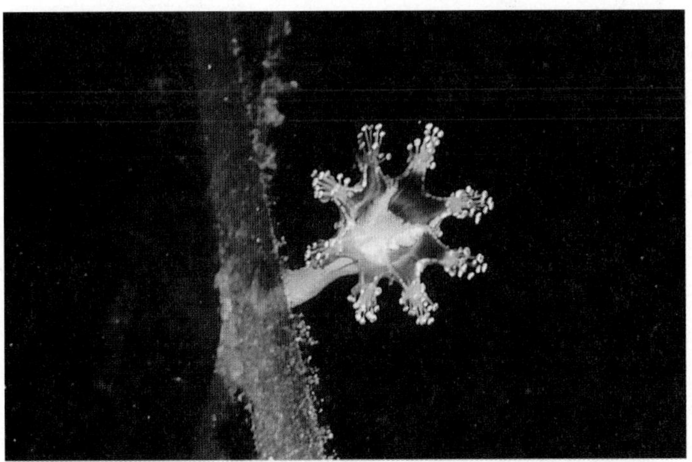

figure 7.18

Thaumatoscyphus hexaradiatus in class Staurozoa is an unusual cnidarian. The oral end of the animal has some features of a medusa stage.

7.6 Class Cubozoa

In Cubozoa the medusa is the predominant form (figure 7.19); the polyp is inconspicuous and in most cases unknown. In the few cases where they have been studied, a polyp buds to make other polyps, and each polyp transforms into a medusa. In transverse section, medusa bells are almost square, explaining the common name *box-jelly* as well as the term "cubomedusa." A tentacle or group of tentacles occurs at each corner of the square at the umbrella margin. The base of each tentacle is differentiated into a flattened, tough blade called a **pedalium** (figure 7.19). Rhopalia each house six eyes, of three different types, in addition to other sense organs. Some eyes are image-forming. The umbrella margin is not scalloped, and the subumbrella edge turns inward to form a **velarium.** The velarium functions as a velum does in hydrozoan medusae, increasing swimming efficiency, but it differs structurally. Cubomedusae are strong swimmers and voracious predators, feeding mostly on fish. Members of at least one species exhibit complex mating behaviors, including twined tentacles as the male passes a spermatophore to the female.

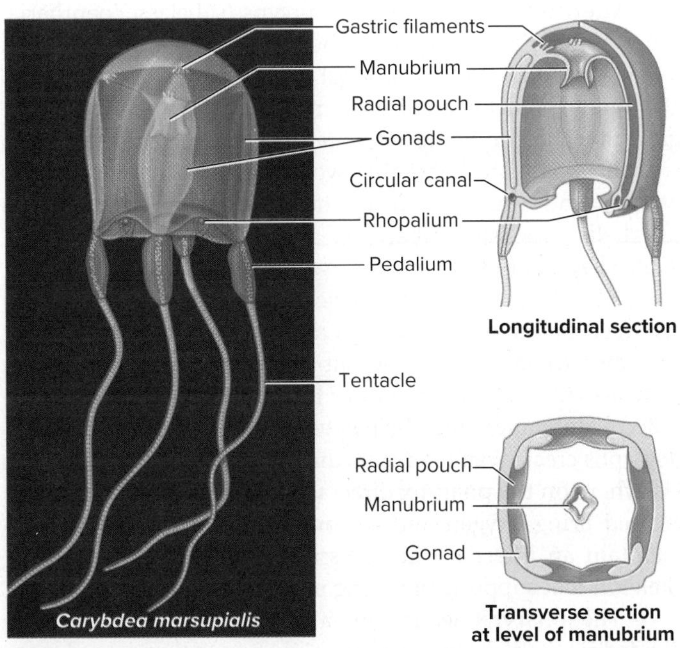

figure 7.19

Carybdea, a cubozoan medusa.

Key Theme 7.6

HUMAN CONNECTIONS

Fatal Stingers

Chironex fleckeri (Gr. *cheir,* hand, + *nexis,* swimming) is a large cubomedusa called the sea wasp. Its stings are quite dangerous and sometimes fatal. Most fatalities have been reported from tropical Australian waters, usually following massive stings. Witnesses have described victims as covered with "yards and yards of sticky wet string." Stings are very painful, and death, if it occurs, ensues within minutes due to cardiac arrest. Three mechanisms for heart damage are proposed: venom proteins attack heart muscle cells; ion flow across membranes is disrupted by channel blockers; and venom proteins damage red blood cells, causing a rapid increase of potassium. Research is proceeding in all three areas, and it is not clear why only some people are severely affected. If death does not occur within 20 minutes after stinging, complete recovery is likely.

7.7 Class Myxozoa

Myxozoans are obligate parasites whose typical life cycle has two aquatic hosts: a fish and an annelid worm (sometimes a bryozoan). Their economic impact results from damage to commercially valuable fishes such as salmon and trout. Whirling disease in trout develops when young fish eat an infected annelid or when fish skin is penetrated by spiny parasitic spores released from an annelid. Inside the trout, parasite cells divide and spread, colonizing cartilage in the skull and spine. Cell division is unique: new cells grow inside older cells. Eventually, very tough myxospores develop and leave the fish. These are eaten by *Tubifex* worms (see figure 11.16). In the worm gut, the polar capsule of the spore extrudes a long filament that attaches to the gut lining. Germ cells move into spaces between cells of the gut lining, where they multiply. Later, sexual stages develop, eventually producing spiny spores that exit the worm via the anus and find a new fish host.

Fish infected with whirling disease (*Myxobolus cerebralis*) exhibit skeletal deformities and odd behaviors, reducing survival. The spread of myxozoans has damaged commercial and sport fisheries around the world and provided a new threat to some endangered species. Whirling disease is common in the Western and Central United States, and in the northeastern states.

Myxozoans have been known since the 1880s, with more than 2180 species identified. However, only recently have molecular studies confirmed that myxozoans are highly reduced cnidarians. The sister taxon to Myxozoa is a freshwater cnidarian called *Polypodium hydriforme. Polypodium hydriforme* has a free-living phase that resembles a cluster of medusae with distinct tentacles, but produces an infective sexual stage that enters the oocytes of fishes such as paddlefish or sturgeon. Inside the oocytes, a planula-like larval stage feeds on yolk and develops into a stolon. When the fish host spawns and oocytes are released into freshwater, the stolon breaks into individual medusa-like sections. Each section multiples by longitudinal fission, eventually reproducing sexually to produce new infective stages. *Polypodium hydriforme* is an endoparasitic cnidarian that retains key cnidarian features including cnidocytes with nematocysts. The polar capsules of myxozoans are homologous to nematocysts, but there are few other myxozoan features that indicate their cnidarian affinities. Myxozoans have one of the smallest genomes in the animal kingdom.

figure 7.20

Sea anemones (order Actiniaria, subclass Zoantharia) are the familiar and colorful "flower animals" of tide pools, rocks, and pilings of the intertidal zone. Most, however, are subtidal, their beauty seldom revealed to human eyes. This is *Bunodosoma grandis*.

7.8 Class Anthozoa

Anthozoans, or "flower animals," are polyps with a flowerlike appearance (figure 7.20). There is no medusa stage. Anthozoa are all marine and occur in both deep and shallow water and in polar seas as well as tropical seas. They vary greatly in size and may be solitary or colonial. Many are supported by skeletons.

Class Anthozoa has three subclasses: **Zoantharia** (or **Hexacorallia**), composed of the sea anemones, hard corals, and others; **Ceriantipatharia,** which includes only tube anemones and thorny corals; and **Octocorallia** (or **Alcyonaria**), containing soft and horny corals, such as sea fans, sea pens, sea pansies, and others. Zoantharians and ceriantipatharians have a **hexamerous** body plan (based on six or multiples of six) or polymerous symmetry and have simple tubular tentacles arranged in one or more circlets on the oral disc. Octocorallians are **octomerous** (built on a plan of eight) and always have eight pinnate (featherlike) tentacles arranged around the margin of the oral disc (see figure 7.21).

The gastrovascular cavity is large and partitioned by septa, or mesenteries, that are inward extensions of the body wall. Where one septum extends into the gastrovascular cavity from the body wall, another extends from the diametrically opposite side; thus, these septa are *coupled.* In Zoantharia, septa are not only coupled, but also *paired* on each side (figure 7.22). The muscular arrangement varies among different groups, but the body wall usually has circular muscles, while the septa have longitudinal and transverse muscles.

Anthozoans tend to exhibit biradial symmetry in the septal arrangement and in the shape of the mouth and pharynx. They have no special organs for respiration or excretion.

Sea Anemones Sea anemone polyps (subclass Zoantharia, order Actiniaria) are larger and heavier than hydrozoan polyps (see figures 7.20 and 7.21). Most range from 5 mm or less to 100 mm in diameter, and from 5 mm to 200 mm long, but some grow much larger. Some are quite colorful. Anemones occur in coastal areas all over the world, especially in warmer waters, and they attach by means of their pedal discs to shells, rocks, timber, or whatever submerged substrata they can find. Some burrow in mud or sand.

Sea anemones are cylindrical, with a crown of tentacles arranged in one or more circles around the mouth on the flat **oral disc** (figure 7.22). The slit-shaped mouth leads into a **pharynx.** At one or both ends of the mouth is a ciliated groove called a **siphonoglyph,** which extends into the pharynx. Siphonoglyphs create water currents directed into the pharynx. Cilia elsewhere on the pharynx direct water outward. Currents thus created bring oxygen and remove wastes. They also help to maintain an internal fluid pressure, providing a hydrostatic skeleton that supports opposing muscles.

The pharynx leads into a large gastrovascular cavity divided into radial chambers by pairs of septa that extend vertically from the body wall toward the pharynx (see figure 7.22).

figure 7.21

A, Orange sea pen *Ptilosarcus gurneyi* (order Pennatulacea, subclass Octocorallia, class Anthozoa). Sea pens are colonial forms that inhabit soft sediment. The base of the fleshy body of the primary polyp is buried in sediment. It gives rise to numerous secondary, branching polyps. **B,** Close-up of a gorgonian. The pinnate tentacles characteristic of the subclass Octocorallia are apparent.

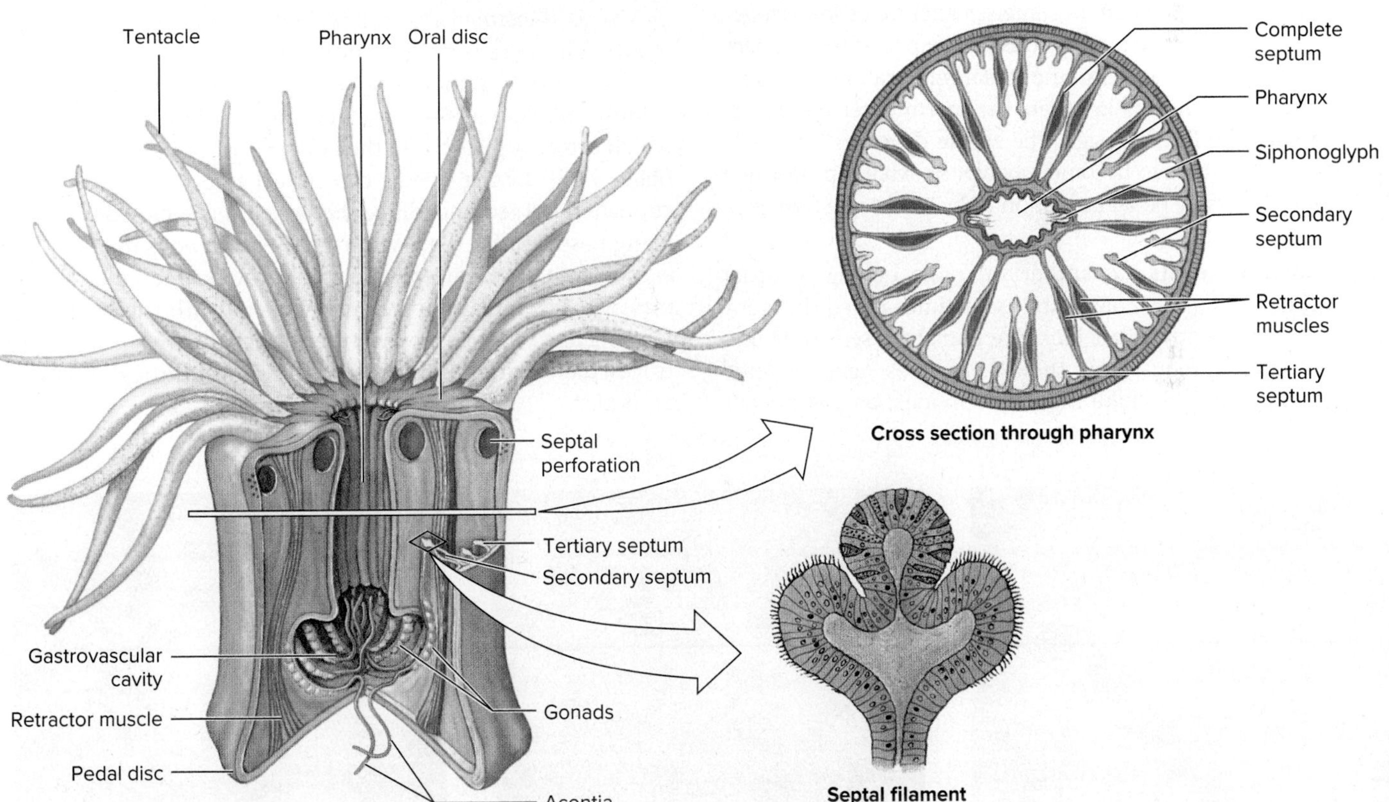

figure 7.22

Structure of the sea anemone. The free edges of the septa and the acontia threads are equipped with nematocysts to complete the paralyzation of prey begun by the tentacles.

These chambers communicate with each other and are open below the pharynx. In many anemones, the lower ends of the septal edges are prolonged into threads called **acontia,** which are also provided with nematocysts and gland cells, that can be protruded through the mouth or through pores in the body wall to overcome prey or to provide defense. The pores can rapidly discharge water from the body when the animal is endangered and contracts to a small size.

Key Theme 7.7

ECOLOGY

Mutualisms

Anemones form some interesting mutualistic relationships with other organisms. Many anemones house unicellular algae in their tissues (as do reef-building corals), from which they derive some nutrients. Some hermit crabs place anemones on the snail shells in which the crabs live, gaining some protection from predators by the presence of the anemone, while the anemone dines on particles of food dropped by the crab. Anemone fishes (figure 7.23) of the tropical Indo-Pacific form associations with large anemones. A property of the skin mucus of this fish prevents the anemone's nematocysts from discharging. By studying the mucus, biologist Amit Lotan was able to develop a protective lotion for swimmers.

Larry S. Roberts

figure 7.23

Orangefin anemone fish (*Amphiprion chrysopterus*) nestles in the tentacles of its sea anemone host. Anemone fishes do not elicit stings from their hosts but may lure unsuspecting other fish to become meals for the anemone.

Sea anemones are carnivorous, feeding on fishes or almost any live animal of suitable size. Some species live on minute forms caught by ciliary currents.

The sexes are separate in some sea anemones (dioecious), and other anemones are monoecious (hermaphroditic). Gonads are arranged on the margins of the septa. Fertilization is external in some species, whereas in others the sperm enter the gastrovascular cavity to fertilize eggs. The zygote develops into a ciliated larva that settles to become a polyp. Asexual reproduction commonly occurs by pedal laceration or by transverse fission.

Zoantharian Corals Zoantharian corals belong to order Scleractinia of subclass Zoantharia, sometimes called the stony (or true) corals. Stony corals look like miniature sea anemones that live in calcareous cups they themselves have secreted (figures 7.24 and 7.25). Like that of anemones, a coral polyp's gastrovascular cavity is subdivided by septa arranged in multiples of six (hexamerous), and its hollow tentacles surround the mouth, but there is no siphonoglyph.

Instead of a pedal disc, the epidermis at the base of the column secretes a limy skeletal cup, including sclerosepta, which project upward into the polyp between its true septa (see figure 7.25). Living polyps can retract into the safety of their cup when not feeding. Since the skeleton is secreted below the living tissue rather than within it, the calcareous material is an exoskeleton. In many colonial corals, the skeleton may become massive, accumulating over many years, with the living coral forming a sheet of tissue over the surface (figure 7.26). The gastrovascular cavities of the polyps all connect through this sheet of tissue.

Larry S. Roberts

A

Larry S. Roberts

B

Larry S. Roberts

C

figure 7.24

Zoantherian corals. **A,** Cup coral, *Tubastrea* sp. The polyps form clumps resembling groups of sea anemones. Although often found on coral reefs, *Tubastrea* is not a reef-building coral (ahermatypic) and has no symbiotic zooxanthellae in its tissues. **B,** The polyps of *Montastrea cavernosa* are tightly withdrawn in the daytime but open to feed at night, as in **C** (order Scleractinia, subclass Zoantharia).

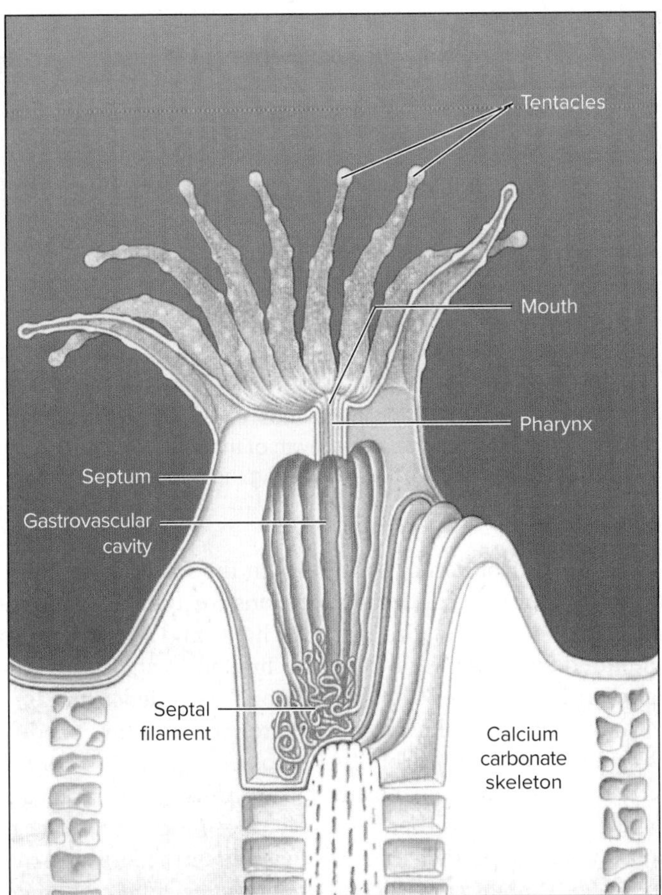

figure 7.25

Polyp of a zoantharian coral (order Scleractinia), showing calcareous cup (exoskeleton), gastrovascular cavity, septa, and septal filaments.

Octocorallian Corals Octocorals include soft corals, sea pansies, sea pens, and sea fans and other gorgonian corals (horny corals). They have strict octomerous symmetry, with eight pinnate tentacles and eight unpaired, complete septa (figure 7.27). Almost all are colonial, and the gastrovascular cavities of polyps in colonies communicate through a system of gastrodermal tubes called solenia. The tubes run through an extensive mesoglea in most octocorals, and epidermis covers the surface of the colony.

The graceful beauty of octocorals–in hues of yellow, red, orange, and purple–helps to create the "submarine gardens" of coral reefs (figure 7.28). Octocoral colonies may be thick encrusting mats or ribbons, tall forms resembling an antique quill pen, or whiplike clusters of long vertical branches. Soft corals have fleshy bodies with calcareous spicules in the mesoglea, but the skeletons of other octocorals typically combine fused or separate spicules with a stiffened yet flexible protein called gorgonin. Gorgonin has chemical similarities to keratin and to collagen; it is responsible for structural support in horny corals, including gorgonians. Because skeletal structures are secreted within the mesoglea, most octocorals have an endoskeleton.

Coral Reefs Coral reefs are among the most productive of all ecosystems, and their diversity of life forms is rivaled only by tropical rain forests. Coral reefs are large formations of calcium carbonate (limestone) in shallow tropical seas laid down by living organisms over thousands of years; living plants and animals are confined to the top layer of reefs, where they add more calcium carbonate to that deposited by their predecessors. The most important organisms that take dissolved calcium and carbonate ions from seawater and precipitate it as limestone to form reefs are reef-building corals and coralline algae. Reef-building corals have mutualistic algae (zooxanthellae) living in their tissues. Coralline algae include several types of red algae, and they may be encrusting or form upright, branching growths. Not only do they contribute to the total mass of calcium carbonate, but their deposits help hold the reef together. Some octocorals and hydrozoans (especially *Millepora* [L. *mille,* a thousand, + *porus,* pore] spp., the "fire coral," see figure 7.14B) contribute in some measure to the calcareous material, and an enormous variety of other organisms contribute small amounts. However, reef-building corals are essential to the formation of large reefs, since such reefs do not occur where these corals cannot live.

figure 7.26

Boulder star coral, *Montastrea annularis* (subclass Zoantharia, class Anthozoa). Colonies can grow up to 10 feet (3 m) high.

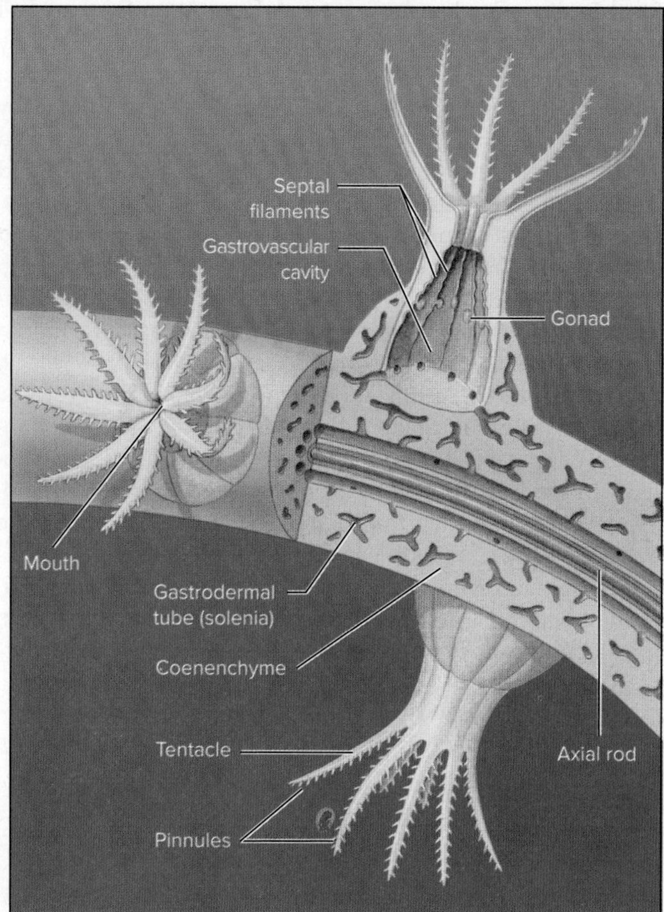

figure 7.27

Polyps of an octocorallian coral. Note the eight pinnate tentacles, coenenchyme, and solenia. These corals have an endoskeleton of limy spicules, often with a hard structural protein, which may form an axial rod.

figure 7.28

A soft coral, *Dendronephthya* sp. (order Alcyonacea, subclass Octocorallia, class Anthozoa), on a Pacific coral reef. The showy hues of this soft coral vary from pink and yellow to bright red and contribute much color to Indo-Pacific reefs.

Key Theme 7.8

SCIENCE EXPLAINED

Ancient Coral Reefs

Because zooxanthellae are vital to reef-building corals and water absorbs light, reef-building corals rarely live below a depth of 30 m (100 feet). Interestingly, some deposits of coral-reef limestone, particularly around Pacific islands and atolls, reach great thickness—even thousands of feet. Clearly, the corals and other organisms could not have grown from the bottom in the abyssal blackness of the deep sea and reached shallow water where light could penetrate. Charles Darwin was the first to realize that such reefs began their growth in *shallow* water around volcanic islands; then, as the islands slowly sank beneath the sea, growth of the reefs kept pace with the rate of sinking, thus explaining the deep deposits.

The distribution of coral reefs in the world is limited to locations that offer optimal conditions for their zooxanthellae. Because they require warmth, light, and the salinity of undiluted seawater, coral reefs are limited to shallow waters between 30° N and 30° S latitude and excluded from areas with upwelling of cold water or where river outflows produce low salinity and high turbidity. Photosynthesis and fixation of carbon dioxide by the zooxanthellae furnish food molecules for the coral. Zooxanthellae recycle phosphorus and nitrogenous waste compounds that otherwise would be lost, and they enhance the ability of the coral to deposit calcium carbonate.

Despite their great intrinsic and economic value, coral reefs in many areas are threatened by a variety of factors, mostly of human origin. These include overenrichment with nutrients (from sewage and runoff of agricultural fertilizer used on nearby land) and overfishing of herbivorous fishes, both of which contribute to overgrowth of multicellular algae. Agricultural pesticides, sediment from tilled fields and dredging, and oil spills contribute to reef degradation. When such environmental stresses do not kill corals directly, they may make the organisms more susceptible to the numerous coral diseases observed in recent years.

Coral reefs are apparently suffering from the effects of global warming. The highly beneficial symbiosis between corals and zooxanthellae is threatened by **coral bleaching** (figure 7.29), where corals become white and brittle after expelling their zooxanthellae. The loss of zooxanthellae correlates with global warming and the resultant increase in ocean temperature. As the water warms, heat damages part of the photosynthetic mechanism in the zooxanthellae, leading to the production of harmful oxidants. The oxidants diffuse into coral tissues, destroying the finely tuned mutualism. Zooxanthellae die or are expelled in what appears to be an immune response from the corals. An initial reduction in the numbers of symbionts worsens the problem because the strongly reflective coral skeleton brings more light into the damaged photosynthetic pathway. There are at least eight separate taxa of symbionts that vary in thermal tolerance, but none appears able to survive continued warming or to prevent bleaching. Bleaching occurred

Taxonomy of Phylum Cnidaria

Class Hydrozoa (hī-drō-zō'ə) (Gr. *hydra,* water serpent, + *zōon,* animal). Solitary or colonial; asexual polyps and sexual medusae, although one body type may be suppressed; hydranths with no mesenteries; medusae (when present) with a velum; both freshwater and marine. Examples: *Hydra, Obelia, Physalia, Tubularia.*

Class Scyphozoa (si-fō-zō'ə) (Gr. *skyphos,* cup, + *zōon,* animal). Solitary; polyp stage sometimes reduced or absent; bell-shaped medusae without velum; gelatinous mesoglea much enlarged; margin of bell or umbrella typically has eight notches containing sensory structures; all marine. Examples: *Aurelia, Cassiopeia, Cyanea, Rhizostoma.*

Class Staurozoa (stō-ro-zōá) (Gr. *strauros,* a cross, + *zōon,* animal). Solitary; polyps only; medusa absent; polyp surface extended into eight clusters of tentacles surrounding mouth; attachment via adhesive disc; all marine. Examples: *Haliclystis, Lucernaria.*

Class Cubozoa (kū' bō-zo-'ə) (Gr. *kybos,* a cube, + *zōon,* animal). Solitary; polyp stage reduced; bell-shaped medusae are square in cross section, with tentacle or group of tentacles hanging from a bladelike pedalium at each corner of the umbrella; margin of umbrella entire, without velum but with velarium; all marine. Examples: *Tripedalia, Carybdea, Chironex, Chiropsalmus.*

Class Myxozoa (mik'sō-zō'ă) (Gr. *myxa,* slime, + *zōon,* animal). Spore-producing aquatic endoparasites whose typical life cycle alternates between fish and annelid worm hosts. Body is reduced to a few cells with no obvious cnidarian features except a polar capsule with an extrusible filament that is homologous to the nematocyst. Examples: *Myxobolus, Buddenbrockia.*

Class Anthozoa (an-thō-zō'ə) (Gr. *anthos,* flower, + *zōon,* animal). All polyps; no medusae; solitary or colonial; enteron subdivided by mesenteries or septa bearing nematocysts; gonads endodermal; all marine.

Subclass Zoantharia (zo'an-tha're-ə) (N. L. from Gr. *zōon,* animal, + *anthos,* flower, + L. *aria,* like or connected with) (**Hexacorallia**). Have simple unbranched tentacles; mesenteries in pairs, in multiples of six; include sea anemones, hard corals, and others. Examples: *Metridium, Anthopleura, Tealia, Astrangia, Acropora, Montastrea, Tubastrea.*

Subclass Ceriantipatharia (se're-ant-ip'a-tha're-ə) (N. L. combination of Ceriantharia and Antipatharia). Have simple, unbranched tentacles; mesenteries unpaired, initially six; include tube anemones and black or thorny corals. Examples: *Cerianthus, Antipathes, Stichopathes.*

Subclass Octocorallia (ok'to-ko-ral'e-ə) (L. *octo,* eight, + Gr. *korallion,* coral) (**Alcyonaria**). Have eight pinnate tentacles; eight complete, unpaired mesenteries; include soft and horny corals. Examples: *Tubipora, Alcyonium, Gorgonia, Plexaura, Renilla, Ptilosarcus.*

Reinhard Dirscherl/Getty Images

figure 7.29

A comparison of healthy and bleached coral polyps in *Acropora.* Indonesia, Papua Province.

prior to global warming, but never at the intensity or on the scale seen now. Bleaching levels in 2002 on the Great Barrier Reef in Australia were the worst in recorded history, with 60% of the entire reef showing bleaching; in some areas bleaching was visible in 90% of the coral present. Caribbean reefs also showed 90% bleaching, which was accompanied by death of half of the affected corals. Furthermore, higher atmospheric concentrations of carbon dioxide (from burning hydrocarbon fuels) tend to acidify ocean water, which makes precipitation of $CaCO_3$ by corals more difficult metabolically.

7.9 Phylum Ctenophora

Ctenophora (te-nof'o-rə) (Gr. *kteis, ktenos,* comb, + *phora,* pl. of bearing) contains about 150 species. All are marine forms occurring in all seas but especially in warm waters. They take their name from eight rows of comblike plates, or ctenes, which they bear for locomotion. Common names for ctenophores are "sea walnuts" and "comb jellies." Ctenophores, along with cnidarians, represent the only two phyla having primary radial symmetry, in contrast to other animals, which have primary bilateral symmetry. In both phyla, the oral/aboral axis is used as a reference point because there is no concentration of sensory or nerve cells to form a head. Like cnidarians, ctenophores exhibit the tissue level of organization. They have a unique cleavage pattern and develop an ECM, mesoglea, between ectoderm and endoderm. They have no definite organ systems, but they do have a true muscle layer, developed from endoderm, that lies next to the ECM (see section 7.10).

A *Pleurobrachia*

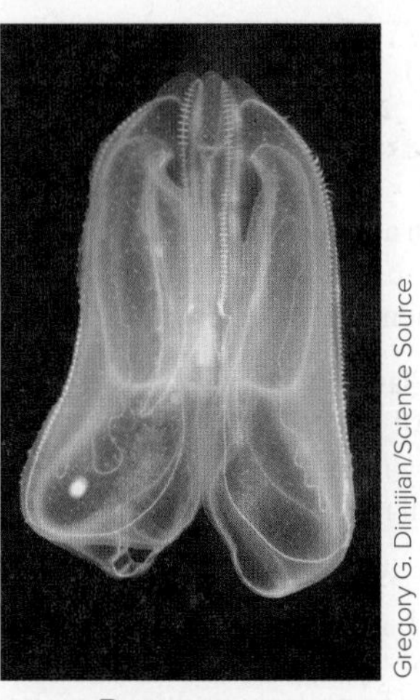

B *Mnemiopsis*

figure 7.30

A, Comb jelly *Pleurobrachia* sp. (order Cydippida, class Tentaculata). Its fragile beauty is especially evident at night when its comb rows luminesce. **B,** *Mnemiopsis* sp. (order Lobata, class Tentaculata).

Ctenophores may make adhesive cells called **colloblasts,** but they do not make nematocysts. However, one ctenophore species (*Haeckelia rubra,* named after Ernst Haeckel, nineteenth-century German zoologist) carries undischarged nematocysts on certain regions of its tentacles and lacks colloblasts. These nematocysts are appropriated from cnidarian medusae on which *H. rubra* feeds.

Except for a few creeping and sessile forms, ctenophores are free-swimming (figure 7.30). Although they are feeble swimmers and are more common in surface waters, ctenophores are sometimes found at considerable depths. Highly modified forms, such as *Cestum* (L. *cestus,* girdle), use sinuous body movements, as well as their comb plates, in locomotion (figure 7.31).

The fragile, transparent bodies of ctenophores are easily seen at night when they emit light (luminesce). Ctenophora has a greater proportion of luminescent species than any other phylum.

Form and Function

Pleurobrachia (Gr. *pleuron,* side, + L. *brachia,* arms), is a representative tentaculate ctenophore (figure 7.32). Its surface has eight longitudinal rows of transverse plates bearing long, fused cilia called **comb plates** or **ctenes.** The beating of the cilia in each row starts at the aboral end and proceeds along the rows to the oral end, thus propelling the animal forward. All rows beat in unison. A reversal of the wave direction drives the animal backward. Ctenophores may be the largest animals that swim exclusively by cilia.

Two long tentacles carried in a pair of tentacle sheaths (figure 7.32) can stretch to a length of perhaps 15 cm. The surface of the tentacles bears specialized glue cells called colloblasts, which secrete a sticky substance that facilitates catching

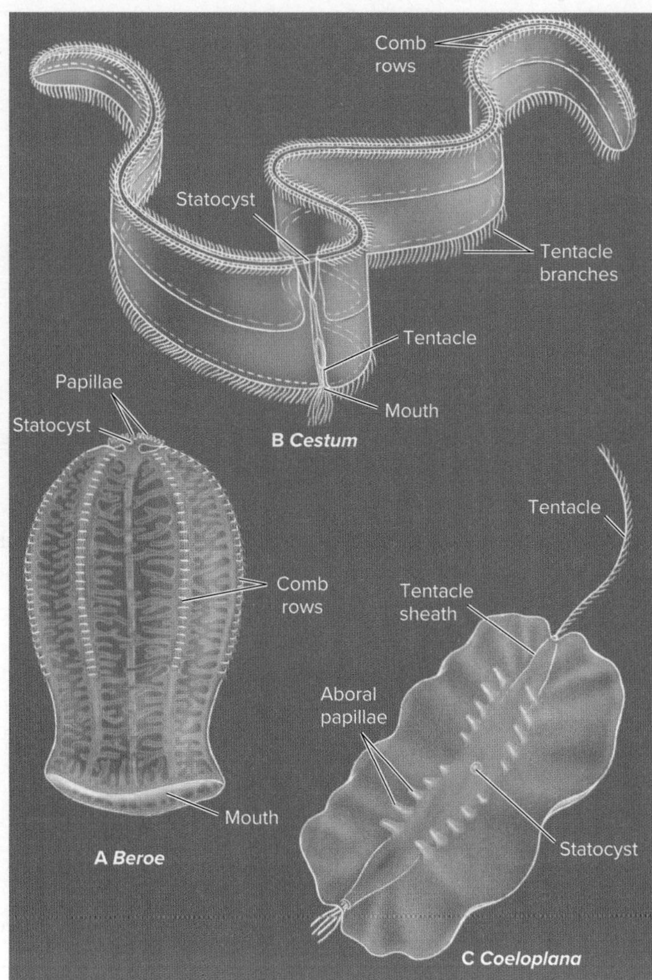

figure 7.31

Diversity among phylum Ctenophora. **A,** *Beroe* sp. (order Beroida). **B,** *Cestum* sp. (order Cestida). **C,** *Coeloplana* sp. (order Platyctenea).

small prey. When covered with food, the tentacles contract and food is wiped onto the mouth. Digestion occurs as food passes through a pharynx, a gastrovascular cavity, and a system of gastrovascular canals. Rapid digestion occurs in the pharynx; then, partly digested food circulates through the rest of the system, where digestion is completed intracellularly. Residues are expelled through anal pores in the aboral end. The presence of a functional complete gut has only recently been confirmed in ctenophores.

Lobate ctenophores, such as *Mnemiopsis* (figure 7.30B), do not have tentacles for prey capture. Instead they use cilia lining the lobes around the mouth to ingest large volumes of water from which prey are captured. The invasive *Mnemiopsis leidyi* is an extremely efficient generalist predator able to capture food items ranging in size from 50 μm to more than 3 mm. Prey are unable to detect the gentle influx of water in which they are carried into the ctenophore, making *M. leidyi* a serious threat to coastal plankton communities (see Key Theme 7.9).

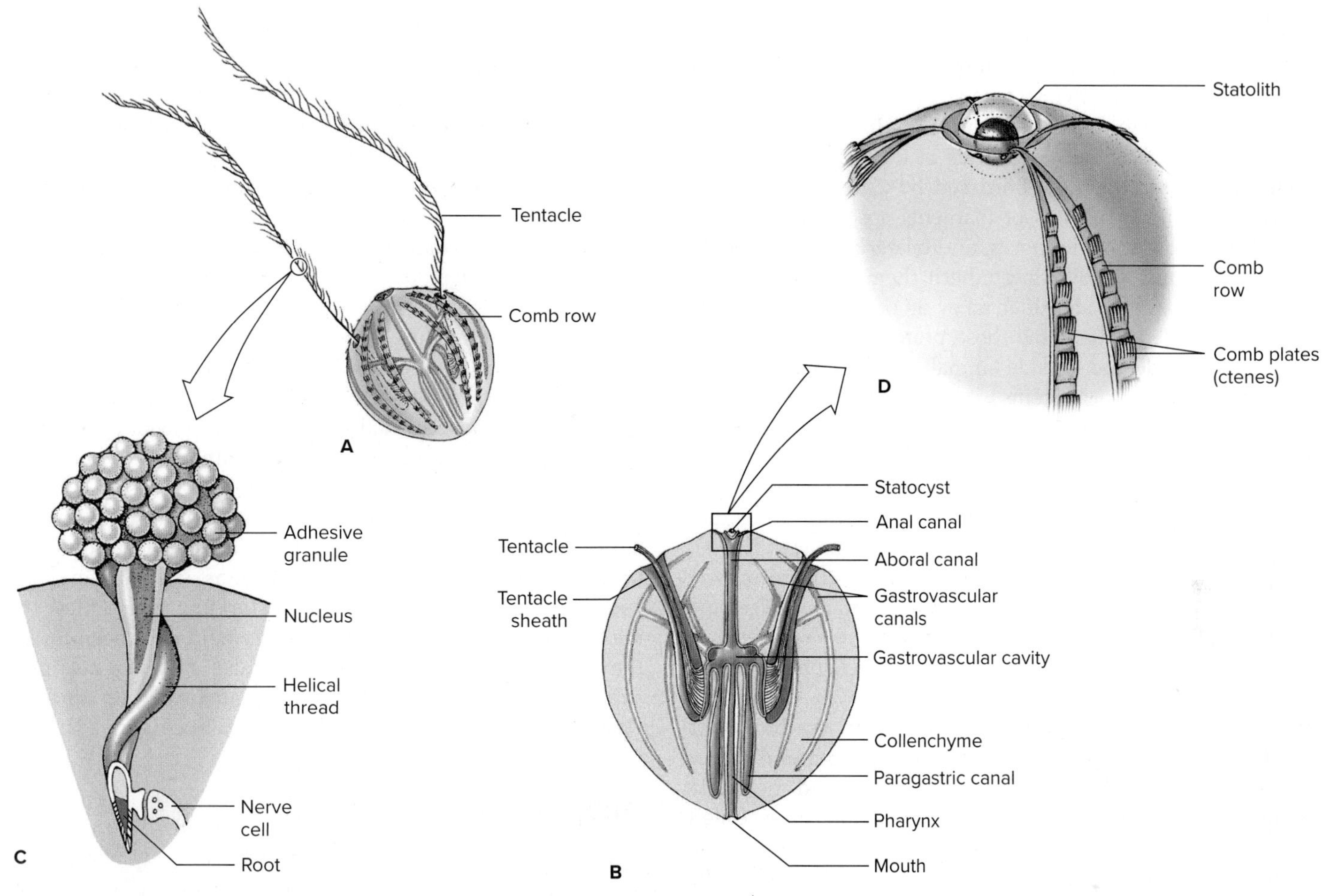

figure 7.32

Comb jelly *Pleurobrachia*, a ctenophore. **A,** External view. **B,** Hemisection. **C,** Colloblast, an adhesive cell characteristic of ctenophores. **D,** Portion of comb rows showing comb plates, each composed of transverse rows of long, fused cilia.

Key Theme 7.9

ECOLOGY

Invasive Ctenophores

In the 1980s, population explosions of *Mnemiopsis leidyi* in the Black and Azov Seas led to catastrophic declines in fisheries there. Inadvertently introduced from the coast of the Americas with ballast water of ships, the ctenophores fed on zooplankton, including small crustaceans and the eggs and larvae of fish. The normally inoffensive *M. leidyi* is kept in check in the Atlantic by certain specialized predators. Recently, the accidental introduction of a predatory ctenophore into the Black Sea caused a decline of *M. leidyi*, and then loss of the new predator. However, *M. leidyi* has since invaded the Caspian, Adriatic, Baltic, and North Seas.

A nerve net system similar to that of the cnidarians includes a subepidermal plexus concentrated under each comb plate.

The sense organ at the aboral pole is a **statocyst,** or organ of equilibrium. The statocyst is also concerned with the beating of the comb rows but does not trigger their beat. Other sensory cells are abundant in the epidermis.

All ctenophores are monoecious, bearing both an ovary and a testis. Gametes are shed into the water, except in a few species that brood their eggs, and there is a free-swimming larva.

7.10 Phylogeny and Adaptive Diversification

Phylogeny

Sponges, placozoans, cnidarians, and ctenophores are not bilaterally symmetrical. Which of these taxa are most closely related to members of the Bilateria? The answer is not yet clear, but if some of these four taxa shared features, such as muscles, with the bilaterians to the exclusion of other taxa, we could infer a branching order. Sponges and placozoans lack muscles, whereas cnidarians and ctenophores have a range of contractile elements that includes epitheliomuscular cells, smooth muscles, and striated muscles. Epitheliomuscular cells (see figure 7.5) contain myosin filaments and actin filaments, are contained within a tissue (epithelium), and anchor to the ECM. True muscle cells, whether smooth or striated, contain the same two filaments, but do not form a tissue epithelium. In striated muscles, striations result from repeated units of the two kinds of filaments (called sarcomeres), bordered by Z-discs. In smooth muscle cells, the two kinds of filaments are not in repeated units. If cnidarians, ctenophores, and bilaterians have smooth and striated muscles, did they inherit them from a common ancestor? Researchers identified a key set of muscle motor proteins and decided to see how these proteins were distributed among taxa inside and outside animals. To their surprise they found that these motor proteins were present outside animals—they occur in other members of Opisthokonta and in other unicellular eukaryotes such amoebozoans and heteroloboseans (see figures 5.2 and 5.26). In unicells, the contractile motor proteins function in cell movement and shape change. Thus, the essential parts of muscle function predate muscles. The researchers also found differences in the ways striated muscle cells were assembled in cnidarians and bilaterians, suggesting convergent evolution, not shared ancestry.

Bilaterians are triploblastic animals (see Chapter 3) whose muscles are derived from the mesodermal layer. Cnidarians and ctenophores lack a mesodermal layer, so biologists wondered whether their muscle cells arose from embryonic ectoderm or endoderm. Recall that mesoderm normally originates from endoderm, although a middle layer called ectomesoderm does sometimes occur. Most ctenophores have only smooth muscle, but one species has striated muscle in its tentacles. Ctenophore muscle cells arise from endoderm. The cells lie within the mesoglea and are not epithelial, so they are "true" muscle cells. Some biologists consider ctenophores triploblastic because of the presence of muscles. It appears that ctenophores independently evolved striated muscles, as did cnidarians. Cnidarian muscle cells develop from the ectoderm; in medusae the ectoderm gives rise to a tissue layer called the entocodon that is separated from the ectoderm and endoderm by ECM. The entocodon layer produces striated muscle cells, but there is no evidence that this layer is homologous to mesoderm. It appears only in the medusa phase of the life cycle. Striated muscle cells also appear in a highly derived group of swimming sea anemones (anthozoans), but in this case the cells are epithelial and lie within the epidermis. The most recent studies suggest that striated muscles evolved independently at least four times: in highly derived swimming anemones, in medusozoans, in ctenophores, and in triploblasts. Some results even suggest that the striated muscles of triploblasts evolved independently in protostomes and deuterostomes, but that is another story.

Within the Cnidaria, a central question remains: Which came first, the polyp or the medusa? Some researchers hypothesize that the life cycles of a specialized group of hydrozoans suggest that the medusa came first. In these hydrozoans, the planula develops to a second larva called an actinula and then directly into a medusa without an intervening polyp stage. If the ancestral cnidarian had this life cycle, polyps would have been added to the life cycle later, and anthozoans would have lost the medusa stage.

The alternative hypothesis is that the anthozoan life was ancestral and the medusoid form was added to the life cycle later. This would suggest that hydrozoans, scyphozoans, and cubozoans shared a common ancestor that had added the medusa to the life cycle. The medusa would have been lost in the lineage leading to *Hydra.* Molecular evidence currently supports this hypothesis, and some researchers group the three classes with medusae into the Medusozoa (see figure 7.1).

Adaptive Diversification

In Cnidaria, polyp and medusa have similar body plans. They are efficient predators, many feeding on prey quite large in relation to themselves. Some are adapted for feeding on small particles. There are many colonial species, with some colonies growing to great size among corals, and others, such as siphonophores, showing astonishing polymorphism and specialization of individuals within the colony. The myxozoans are obligate aquatic parasites with two hosts in the life cycle; they are highly modified cnidarians.

Ctenophores have retained the basic arrangement of eight comb plates, with or without tentacles, and biradial symmetry in most. However, some, like Venus' girdle, have an elongate body.

■ SUMMARY

Sections	Key Concepts
7.1 Phylum Cnidaria	• Cnidarians are an ancient group estimated at 700 million years old. • Cnidarians have a primary radial symmetry. • Radial symmetry serves sessile or free-floating organisms well because environmental stimuli are equally likely from all directions. • Cnidaria occur in all world oceans and in freshwater. • Cnidaria are surprisingly efficient predators because they possess unique stinging organelles called cnidae.
7.2 Form and Function	• Cnidarians are diploblastic, with a body wall composed of epidermis and gastrodermis and a mesoglea between these cell layers. • Cnidarians are at the tissue level of organization. • The digestive-respiratory (gastrovascular) cavity has a mouth and no anus. • The unique organelle called a cnida is produced by a cnidoblast (which becomes the cnidocyte) and is coiled within a capsule. There are 20 types of cnidae. • A common cnida is the nematocyt. When discharged, nematocysts penetrate prey and inject venom. Discharge occurs because of a change in the permeability of the capsule, creating high osmotic pressure within the capsule; this produces an increase in internal hydrostatic pressure. • Many cnidarians have two basic body types: polyp and medusa. • In many hydrozoans and scyphozoans, the life cycle involves both an asexually reproducing polyp and a sexually reproducing medusa. In anthozoans, sexual reproduction occurs in the polyp.
7.3 Class Hydrozoa	• Most hydrozoans are colonial and marine. • Solitary freshwater hydras, commonly used in biology courses, have a typical polyp form, are not colonial, and have no medusa stage. • Most marine hydrozoans exhibit a branching colony form with many polyps (hydranths). Most colonies are sessile, but a few drift or swim in the sea. • Most hydrozoans produce medusae; unlike scyphozoan medusae, those of hydrozoans tend to be small, bell-shaped, and possess an internal shelf called a velum. • Hydrozoan medusae may be free-swimming or remain attached to the colony.
7.4 Class Scyphozoa	• Scyphozoans are typical jellies, in which the medusa is the large and dominant body form, and most have an inconspicuous polyp stage. • Medusae may reach 2 meters in diameter; there is no velum. • Medusae are made asexually, but they reproduce sexually.
7.5 Class Staurozoa	• Staurozoans are odd, sessile cnidarians where the body has features of typical cnidarian polyps (longitudinal muscles and gastric septa) as well as features of medusae (circular coronal muscles, gastric filaments, and gonads). Most medusa-like features are at the oral end. • A diagnostic feature is the creeping planula larva stage, which replaces the typical ciliated free-swimming planula.
7.6 Class Cubozoa	• Cubozoans, also called box jellies or sea wasps, are predominantly medusoid. The medusae has a cubic form with a shelf-like velarium along the free edge of the medusa bell. • The polyp form is not well-studied but, in some species, polyps transform into medusae. • Medusae are strong swimmers and voracious predators. • Sea wasps produce powerful venom: stings may be fatal due to cardiac arrest. • Medusae have good sensory systems that include image-forming eyes.
7.7 Class Myxozoa	• Myxozoans, recently identified as cnidarians, are obligate parasites. • The life cycle has two hosts: typically a fish and an annelid worm. • They cause whirling disease in fish. • The myxozoan polar capsule is homologous to the nematocyst; it is the only shared morphological feature of myxozoans and other cnidarians.

(continued)

Sections	Key Concepts
7.8 Class Anthozoa	• Anthozoans are all marine with a polyp body form; medusa stage is absent. • The most important subclasses are Zoantharia (six-part or polymerous symmetry) and Octocorallia (8-part symmetry). • The largest zoantharian orders contain sea anemones, which are solitary and do not have a skeleton, and stony corals, which are mostly colonial and secrete a calcareous exoskeleton. • Stony corals are the critical component in coral reefs, which are habitats of great beauty, productivity, and ecological and economic value. • Octocorallia contain the soft and horny corals, many of which are important and beautiful components of coral reefs.
7.9 Phylum Ctenophora	• Ctenophora, called comb jellies or sea walnuts, are all marine. • Ctenophora have biradial symmetry. • They swim by means of eight comb rows of fused cilia (ctenes); a few forms are creeping or sessile. • Colloblasts are used to capture small prey; they characterize the phylum. • Ctenophores have a complete gut ending in functional anal pores and true muscles.
7.10 Phylogeny and Adaptive Diversification	• Evolutionary relationships among cnidarians, ctenophores, and bilaterally symmetrical organisms are not clear. • Although cnidarians and ctenophores share radial symmetry and are both diploblastic, there is no evidence that they are sister taxa. • Based on the shared presence of true muscles in ctenophores and bilaterians, some have inferred a sister-taxon relationship between these lineages, but the most recent evidence suggests that muscles evolved independently in these two taxa. • There is debate as to the first cnidarian body type, but evidence supports the polyp as the ancestral form with a medusa form evolving later in the common ancestor of scyphozoans, hydrozoans, and cubozoans.

Review Questions

1. Explain the utility of radial symmetry for sessile and free-floating animals.
2. What characteristics of phylum Cnidaria are most important in distinguishing it from other phyla?
3. Name and distinguish the classes in phylum Cnidaria.
4. Distinguish between polyp and medusa forms.
5. Explain the mechanism of nematocyst discharge. How can a hydrostatic pressure of one atmosphere be maintained within a nematocyst until it receives an expulsion stimulus?
6. What is an unusual feature of the nervous system of cnidarians?
7. Diagram a hydra, and label the main body parts.
8. Name and give the functions of the main cell types in the epidermis and the gastrodermis of hydra.
9. What stimulates feeding behavior in hydras?
10. Give an example of a highly polymorphic, floating, colonial hydrozoan.
11. Distinguish the following from each other: statocyst and rhopalium; scyphomedusae and hydromedusae; scyphistoma, strobila, and ephyrae; polyp forms in Zoantharia and in Octocorallia.
12. What are the causes and consequences of coral bleaching?
13. Describe three specific symbiotic interactions of anemones with other organisms.
14. Contrast the skeletons of zoantharian and octocorallian corals.
15. Why are coral reefs generally restricted to shallow marine waters?
16. Specifically, what kinds of organisms are most important in deposition of calcium carbonate on coral reefs?
17. How do zooxanthellae contribute to the welfare of reef-building corals?
18. What characteristics of Ctenophora do you consider most important in distinguishing it from other phyla?
19. How do ctenophores swim, and how do they obtain food?
20. What arguments favor the ideas that the first cnidarian was either a medusa or a polyp?

For Further Thought What would be the best way to make non-biologists aware of the ecological and economic costs of global warming as it affects coral reefs?

8

Xenacoelomorpha, Platyhelminthes, Gastrotricha, Gnathifera, including Chaetognatha, and Mesozoa

LEARNING OBJECTIVES

Readers will be able to:

8.1 Explain the features used to unite two taxa as Xenacoelomorpha and describe the lifestyle of each taxon.

8.2 Discuss the major clades within the Bilateria and those within Protostomia.

8.3 Describe the body plan in each class of Platyhelminthes and compare adaptations to ecto- and endo-parasitism, noting the ancestral state of each character; illustrate the impacts of trematodes and cestodes on humans.

8.4 Describe the form and habitat of gastrotrichs.

8.5 Explain the basis for grouping five phyla as taxon Gnathifera.

8.6 Describe the form and habitat of gnathostomulids.

8.7 Describe the form and habitat of chaetognaths.

8.8 Describe the form and habitat of micrognathozoans.

8.9 Describe the form, habitat, and reproductive strategies of rotifers.

8.10 Describe the form and habitat of acanthocephalans, noting adaptations for parasitism.

8.11 Explain why the body plan of mesozoans is unusual for bilaterians.

8.12 Describe the new evolutionary position of chaetognaths and explain what a polytomy means for our understanding of evolutionary relationships.

Larry S. Roberts

Thysanozoon nigropapillosum, a marine turbellarian (order Polycladida).

Getting Ahead

For animals that spend their lives sitting and waiting, as do most members of Cnidaria and Ctenophora, radial symmetry is ideal. One side of the animal is just as important as any other for snaring prey coming from any direction. But if an animal is active in seeking food, shelter, home sites, and reproductive mates, it develops a different set of strategies and a new body organization. An elongated body form with head (anterior) and tail (posterior) ends serves active, directional movement. In addition, one side of the body faces up (dorsal), and the other side, specialized for locomotion, faces down (ventral). What results is a bilaterally symmetrical animal in which the body could be divided along only one plane of symmetry to yield two halves that are mirror images of each other. Furthermore, because it is better to perceive where one is going than where one has been, sense organs and centers for nervous control have come to be located at the anterior end. This process is called cephalization. Thus, cephalization and primary bilateral symmetry occur together in almost all triploblastic animals.

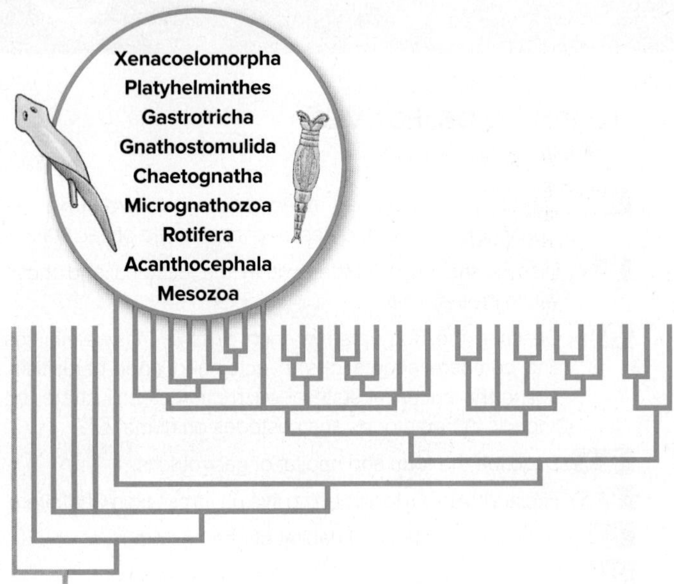

The term "worm" has been applied loosely to elongated, bilateral invertebrate animals without appendages. At one time, zoologists considered worms (Vermes) a taxon that included a highly diverse assortment of forms. This unnatural assemblage has been reclassified into various phyla. By tradition, however, zoologists still call various groups of these animals flatworms, ribbon worms, roundworms, segmented worms, and others. In this chapter we introduce two phyla of flatworms and six other phyla that lack a mesodermally lined body cavity. All but one phylum contains bilaterally symmetrical, **triploblastic** animals. Beyond bilateral symmetry and the presence of three tissue layers however, these animals differ in body plan.

Members of two phyla that we discuss here have **acoelomate** (Gr. *a,* not, + *koilōma,* cavity) bodies (see section 3.4: Body Cavities); in other words, they lack a coelom. Readers may recall that diploblasts also lack a coelom, but are not called acoelomate; the term applies only to animals possessing mesoderm.

Typical acoelomates have only one internal space, the digestive cavity (figure 8.1). The region between the epidermis and the digestive cavity lining is filled with a cellular, mesodermally derived parenchyma. Parenchyma is a form of packing tissue containing more cells and fibers and less extracellular matrix (ECM) than does the mesoglea of cnidarians. Organs are another derivative of mesoderm that increases internal complexity in triploblasts. We see some internal complexity in members of the acoelomate phyla Acoelomorpha and Platyhelminthes.

A **pseudocoelomate** (Gr. *pseudo,* false, + *koilōma,* cavity) body has an internal cavity, the pseudocoelom, surrounding the gut, but the cavity is not completely lined with mesoderm as it is in a coelomate animal (figure 8.1). The mesodermal layer lies only on the outside of the pseudocoelom; the inner lining of the cavity is formed by the endodermally derived gut lining. A pseudocoelom is an embryonic blastocoel that persists throughout development, so some biologists call animals with this body plan blastocoelomates rather than pseudocoelomates.

A pseudocoelom may be filled with fluid or with a gelatinous matrix containing some mesenchymal cells (derived from mesoderm). It may provide space for expansion of digestive, excretory, and reproductive systems or for storage of waste products, but when filled with fluid, it often functions as a hydrostatic support or permits circulation of body fluids. Many pseudocoelomate animals are quite small and lack true circulatory systems.

Animals sharing a particular body plan do not necessarily form a monophyletic group. Pseudocoelomate groups include rotifers and acanthocephalans, among others discussed in later chapters, whereas acoelomate taxa include xenacoelomorphs, platyhelminths, gastrotrichs, and gnathostomulids. Xenacoelomorpha is the sister taxon to the combined group of protostomes and deuterostomes sometimes called Bilateria or

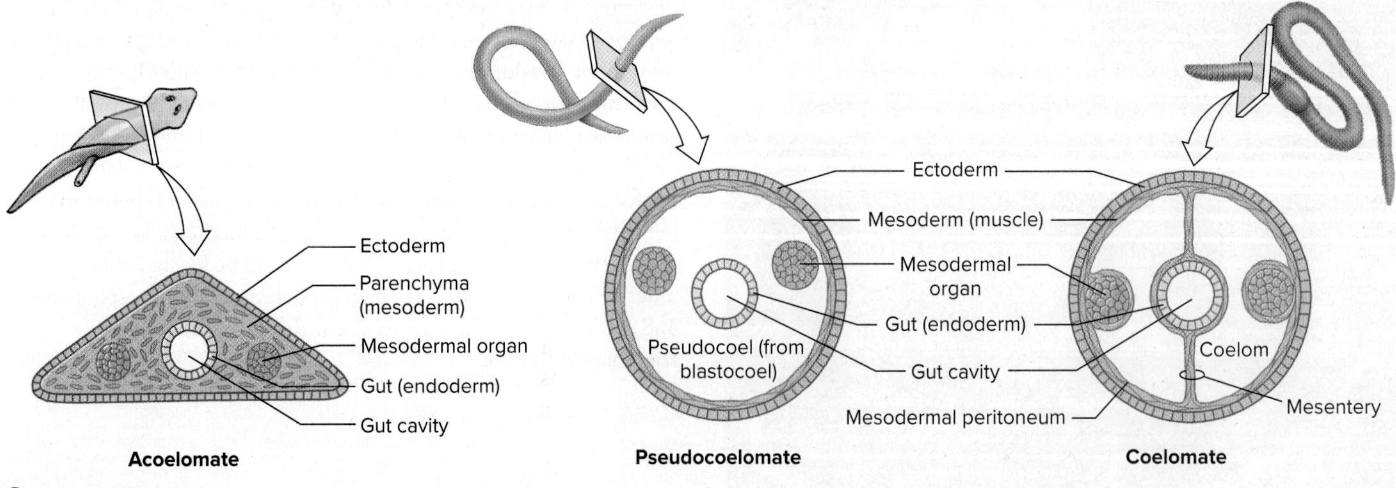

figure 8.1

Acoelomate, pseudocoelomate, and coelomate body plans for comparison. Notice that peritoneum and mesenteries are present only in coelomate animals.

Nephrozoa. Examine the Cladogram of All Animal Phyla inside the front cover to see this branching pattern. Notice that the other phyla discussed in this chapter belong to Protostomia and are placed in the subgroup Lophotrochozoa (see section 3.5: Protostome Subgroups).

8.1 Phylum Xenacoelomorpha

Xenacoelomorpha is new taxon housing two sister clades: Xenoturbellida and Acoelomorpha. Xenoturbellids (figure 8.2) are wormlike ciliated animals first identified in 1949. There is only one genus, *Xenoturbella,* but the recent discovery of four new species brings the species count to six. One recently discovered species, *Xenoturbella monstrosa,* reaches 20 cm in length. Another species, *X. profunda,* occurs near a hydrothermal vent off the coast of Mexico at a depth of 3700 m. Some new species are purple or pink, in contrast to the yellowish species first discovered.

Xenoturbellid bodies have an external ciliated furrow, called the ring furrow, and a second longitudinal furrow called the side furrow. They have a relatively thick epidermis supported by a basal lamina, a subepidermal nerve net, and a newly discovered ventral epithelial glandular network of unknown function. Circular and longitudinal muscles are present. There is a ventral mouth leading into a blind gut. Xenoturbellids feed on bivalve mollusc eggs and adults. They reproduce sexually with both simultaneous and sequential hermaphroditism being described. Zygotes undergo direct development without a feeding larval stage, in a manner similar to that of acoelomorph worms.

figure 8.2

This large colorful xenoturbellid, *Xenoturbella profunda* was discovered in Monterey Bay, California at a depth of 2890 meters. It is more than 20 centimeters long. It was photographed by MBARI's remotely operated vehicle Doc Ricketts.

Acoelomorpha (Gr. *a,* without, + *koilos,* hollow, + *morphe,* form) comprises worms that previously belonged to two orders within class Turbellaria in phylum Platyhelminthes (see section 8.3). These worms are less than 5 mm long and have a cellular, ciliated epidermis. Some have a saclike gut without an anus, and others entirely lack a digestive cavity. In gutless worms, food particles enter through the mouth and move into a tissue mass derived from endoderm; sometimes, a temporary digestive cavity forms around the food.

Acoelomorphs have a distinct anteroposterior axis, but their diffuse set of anterior nerve cells lacks the ganglia typical of a "true" brain. Acoelomorphs have a radial arrangement of nerves in the body, instead of the ladderlike pattern (see figure 8.9B) seen in flatworms of phylum Platyhelminthes. Potential synapomorphies include special cilia and a frontal organ.

Acoelomorphs typically inhabit marine sediments, although a few forms are pelagic. Some species live in brackish water. Most acoelomorphs are free-living, but some are symbiotic and others parasitic. The group contains approximately 350 species.

8.2 Clades within Protostomia

Examine the Cladogram of All Animal Phyla inside the front cover. Xenacoelomorpha is placed outside the clade of bilaterally symmetrical animals called Bilateria. All phyla described in the remaining chapters of this book are bilaterally symmetrical or had bilaterally symmetrical ancestors. All triploblastic animals within Bilateria are divided between two clades: Protostomia and Deuterostomia. Division into these two groups is based largely on features of development (see section 3.5), although the two groups are also recovered in most phylogenies using molecular characters. Protostomia is divided into two large clades: **Lophotrochozoa** and **Ecdysozoa**. Platyhelminthes is the first protostome phylum discussed; it and the remaining phyla included in this chapter belong to Lophotrochozoa.

Lophotrochozoans first appeared as a clade in molecular phylogenies. Prior to the construction of these phylogenies, the primary taxonomic division within protostomes was based on body plan: acoelomate phyla were placed in one group and all coelomate phyla in a separate group. Molecular phylogenies rejected the hypothesis that the acoelomates and the coelomates formed separate monophyletic groups, instead dividing protostomes into two subsets (Lophotrochoza and Ecdysozoa) with distinctive molecular signatures. Some morphological characters support each subset. Members of Ecdysozoa possess a cuticle that is molted as their bodies grow. Most members of Lophotrochozoa share either an odd horseshoe-shaped feeding structure, the **lophophore** (see figures 9.1 and 9.4), or a particular larval form called the **trochophore** (see figure 10.6).

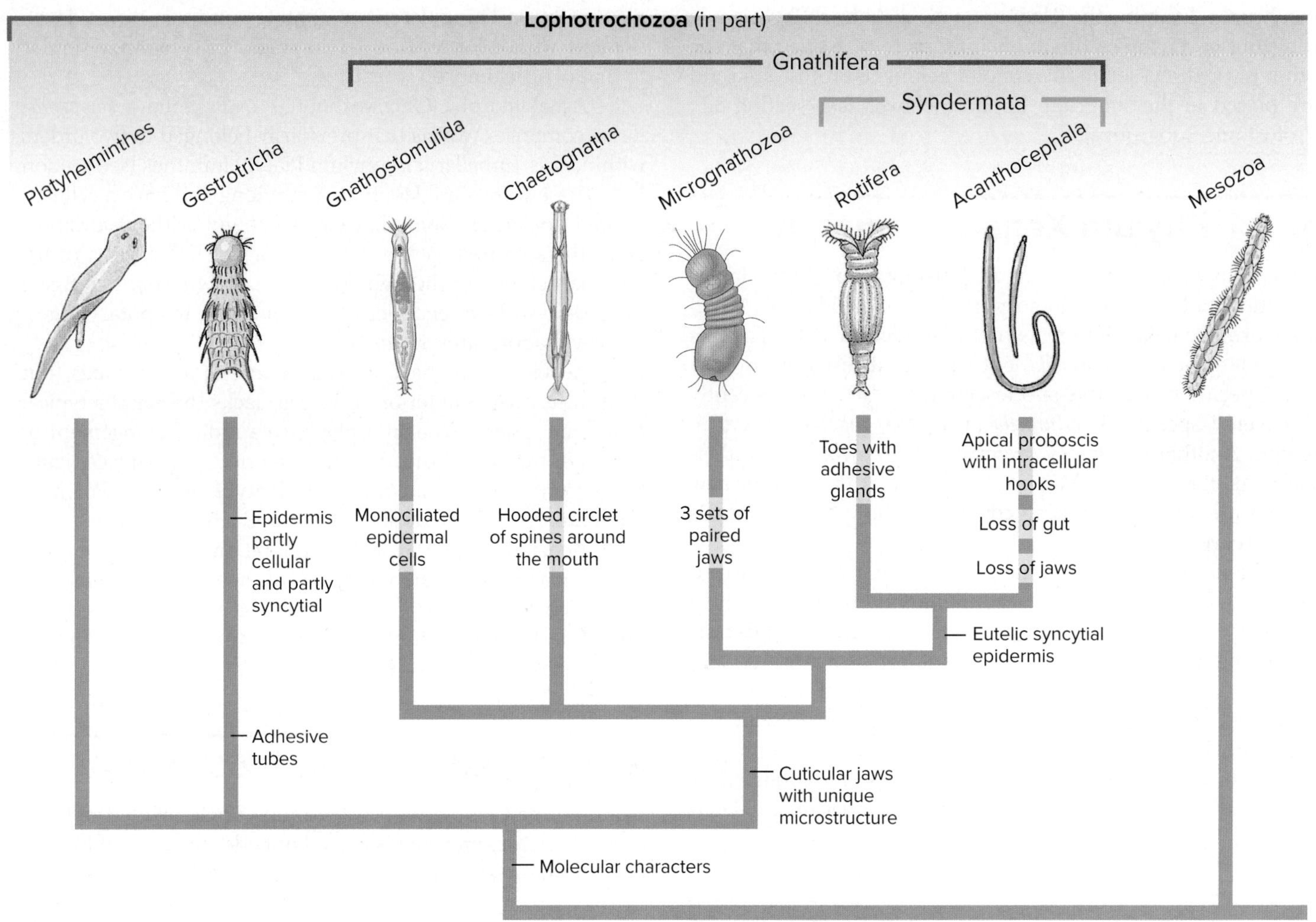

figure 8.3

Hypothetical relationships among members of Platyzoa. Characters are modified subsets of those in Kristensen, R. M. 2002. An introduction to Loricifera, Cycliophora, and Micrognathozoa. *Integrative and Comparative Biology* **42:**641–651; Brusca, R. C. and G. J. Brusca. 2003. *Invertebrates*. ed. 2. Sunderland, MA: Sinauer Associates, Inc.

Trochophore larvae are minute, translucent, and roughly top-shaped. They have a prominent circlet of cilia and sometimes one or two accessory circlets. Trochophores occur in the early development of many marine members of Annelida and Mollusca and are assumed ancestral for these groups. Trochophore-like larvae occur in some marine members of Platyhelminthes, Nemertea, Echiura, and Sipuncula, among others. However, Lophotrochozoa also includes some phyla that possess neither a trochophore larva nor a lophophore, so researchers have looked at other characters to unite the group. Some researchers substitute the name Spiralia for the clade that we call Lophotrochozoa because these phyla exhibit spiral cleavage early in development (see section 3.4) or are assumed to have ancestors with spiral cleavage. When Spiralia is used in this manner, the name Lophotrochozoa is often retained for various subgroups of Spiralia, but its content varies among researchers. Lophophorate phyla are discussed in Chapter 9. In the remainder of this chapter, we discuss flatworms in Phylum Platyhelminthes, the tiny aquatic animals in Gastrotricha, five phyla united in clade Gnathifera sharing a ciliated pharyngeal ganglion and jaws made from cuticle (Acanthocephala, Chaetognatha, Gnathostomulida, Micrognathozoa, and Rotifera), and a group of poorly understood worm-like animals called Mesozoa (figure 8.3).

■ 8.3 Phylum Platyhelminthes

Platyhelminthes (Gr. *platys*, flat, + *helmins*, worm) range in length from a millimeter or less to many meters (in some tapeworms). Their typically flattened bodies may be slender, broadly leaflike, or long and ribbonlike. There are four classes within Platyhelminthes. Class Turbellaria comprises nonparasitic forms, whereas members of classes Trematoda, Monogenea, and Cestoda are parasitic (figure 8.4). Although members of the parasitic classes all possess a syncytial tegument, there is no single feature yet discovered that diagnoses phylum Platyhelminthes.

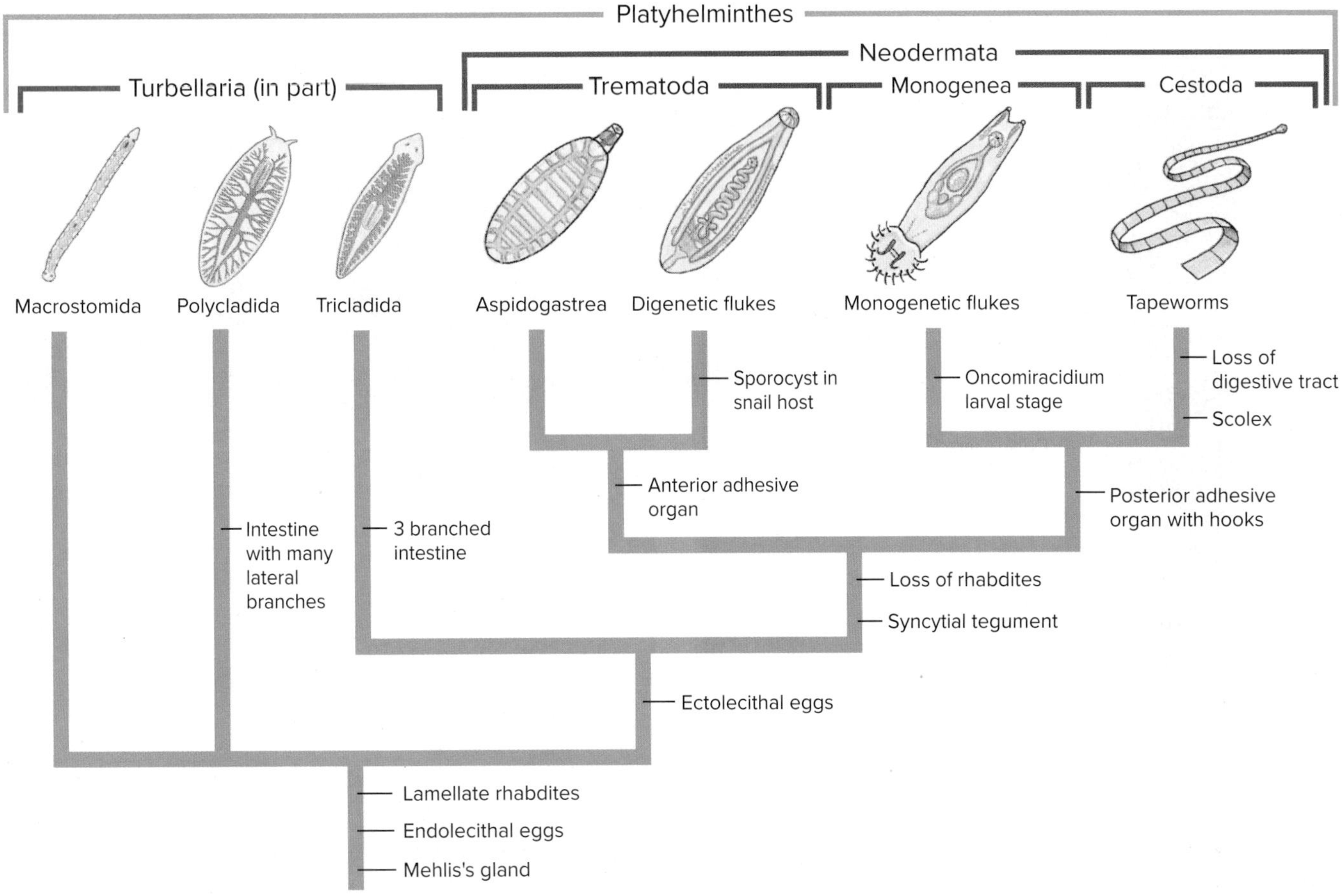

Platyhelminthes
Neodermata
Turbellaria (in part)
Trematoda
Monogenea
Cestoda
Macrostomida
Polycladida
Tricladida
Aspidogastrea
Digenetic flukes
Monogenetic flukes
Tapeworms
Sporocyst in snail host
Oncomiracidium larval stage
Loss of digestive tract
Scolex
Anterior adhesive organ
Posterior adhesive organ with hooks
Intestine with many lateral branches
3 branched intestine
Loss of rhabdites
Syncytial tegument
Ectolecithal eggs
Lamellate rhabdites
Endolecithal eggs
Mehlis's gland

figure 8.4

Hypothetical relationships among Platyhelminthes. The traditional class Turbellaria is paraphyletic. Some turbellarians have ectolecithal development and, together with the Trematoda, Monogenea, and Cestoda, form a clade and the sister group of the endolecithal turbellarians. For the sake of simplicity, the synapomorphies of those turbellarians and of the Aspidogastrea, as well as many others given by Brooks (1989), are omitted.

Source: Modified from D. R. Brooks. The phylogeny of the Cercomeria (Platyhelminthes: Rhabdocoela) and general evolutionary principles. Journal of Parasitology **75:**606–616, 1989.

Ecological Relationships

Free-living flatworms occur exclusively in class Turbellaria. A few turbellarians are symbiotic (commensals or parasites), but most are adapted as bottom-dwellers in marine or fresh water or live in moist places on land. Many, especially larger species, occur on the undersides of stones and other hard objects in freshwater streams or in littoral zones of the ocean.

Relatively few members of class Turbellaria inhabit fresh water, but freshwater planarians, such as *Dugesia* (order Tricladida), are used extensively in introductory laboratory courses. Planarians and some others frequent streams and spring pools; others prefer the flowing water of mountain streams. Some species occur in moderately hot springs. Terrestrial turbellarians inhabit moist places such as under stones and logs or on moist vegetation (figure 8.5).

All members of classes Monogenea and Trematoda (flukes) and class Cestoda (tapeworms) are parasitic. Most Monogenea are ectoparasites, but all trematodes and cestodes

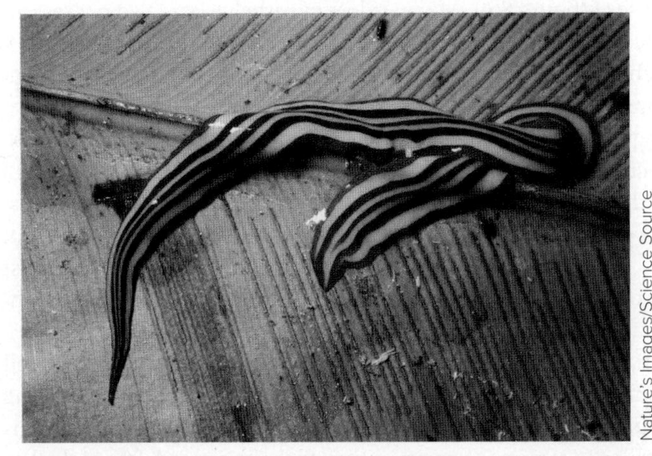

figure 8.5

A terrestrial turbellarian from the Peruvian Amazon.

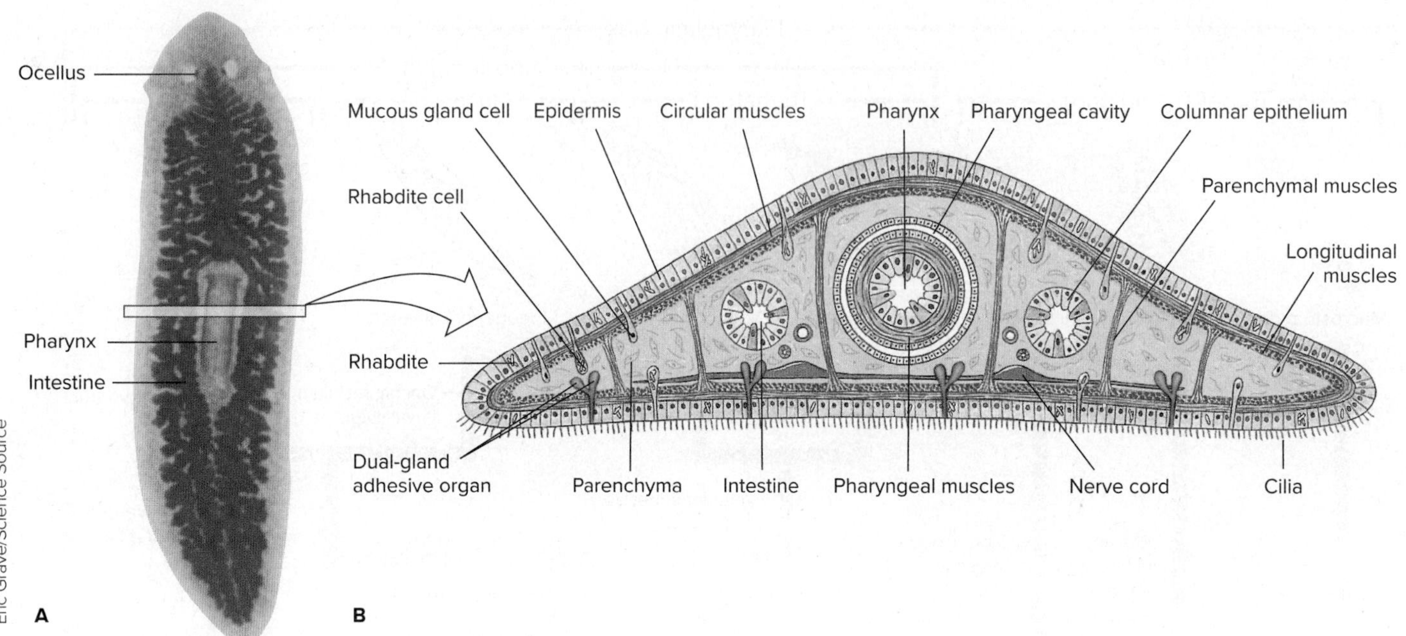

figure 8.6

Cross section through the pharyngeal region of a common planarian, showing relationships of body structures.

are endoparasitic. Many species have indirect life cycles with more than one host; the first host is often an invertebrate, and the final host is usually a vertebrate. Humans serve as hosts for some species. Certain larval stages may be free-living.

Form and Function

Tegument, Muscles

Most turbellarians have a ciliated cellular epidermis resting on a basal lamina. The epidermis typically contains rod-shaped **rhabdites,** composed of fused vesicles from the Golgi apparatus, that swell to form a protective mucous sheath around the body when discharged with water; single-celled mucous glands may also be present (figure 8.6). Many orders of turbellarians also possess an interesting locomotory system that allows them quickly to attach and detach from surfaces. In such cases, the epidermis contains **dual-gland adhesive organs** consisting of three cell types: viscid and releasing gland cells and anchor cells (figure 8.7). Secretions of the viscid gland cells apparently fasten microvilli of the anchor cells to the substrate, and secretions of the releasing gland cells provide a quick, chemical detaching mechanism.

In contrast to the ciliated cellular epidermis of most turbellarians, all members of the parasitic classes possess a nonciliated body covering called a syncytial tegument. The term syncytial means that many nuclei occur within a single cell membrane. It might seem that a completely new body covering appeared in parasites, but two atypical forms of the epidermis similar to tegument occur within turbellarians. A few turbellarian species have a syncytial epidermis, and at least one species has a syncytial *insunk* epidermis where cell bodies (containing

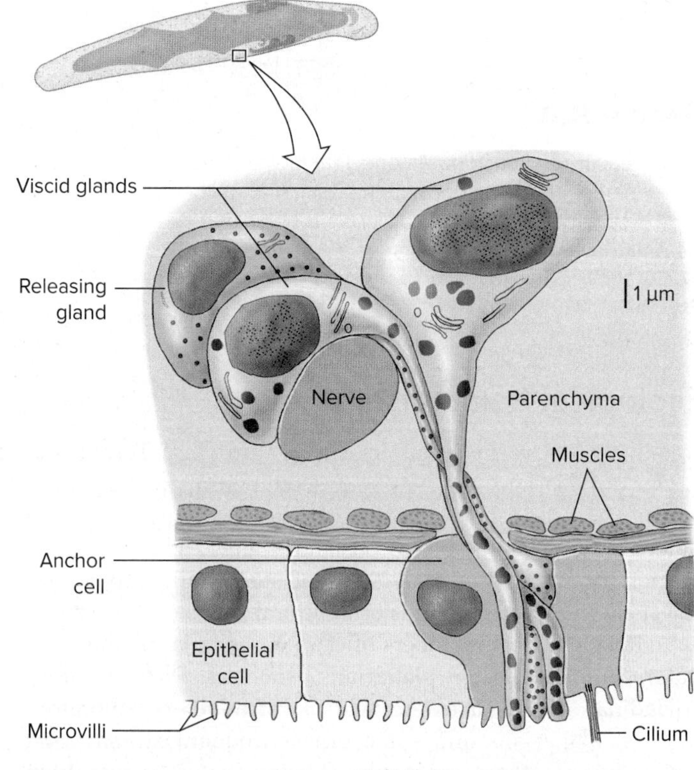

figure 8.7

Reconstruction of a dual-gland adhesive organ of a turbellarian, *Haplopharynx* sp. Two viscid glands and one releasing gland lie beneath the body wall. The anchor cell lies within the epidermis, and one of the viscid glands and the releasing gland contact a nerve.

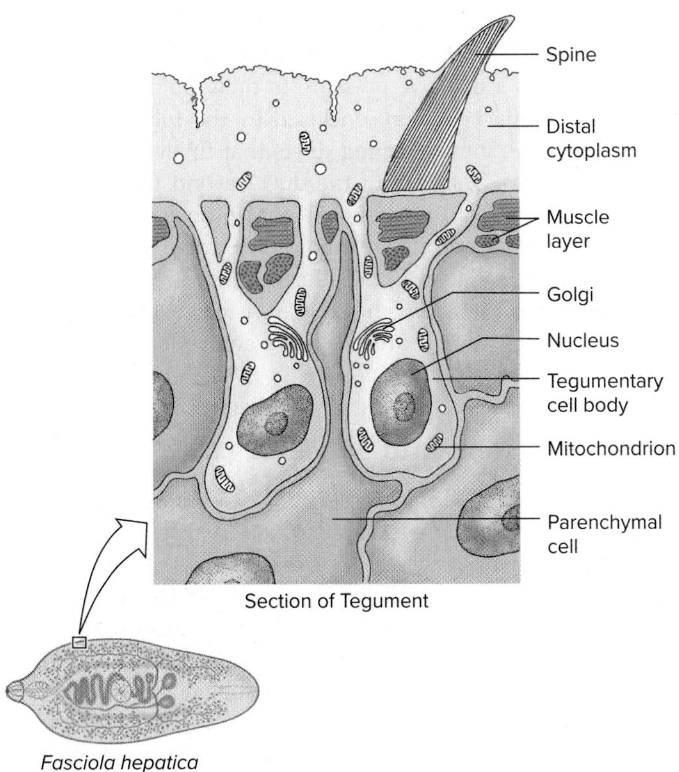

figure 8.8

Structure of the tegument of a trematode, *Fasciola hepatica*.

the nuclei) lie beneath the basal lamina of the epidermis and communicate with the distal (surface) cytoplasm through cytoplasmic channels. The term "insunk" is a misnomer because the cell bodies below the basal lamina send extensions upward to the epidermis, rather than the other way around. The extensions then fuse to form the syncytial covering, much as they do in the parasitic classes.

Adults of all members of Trematoda, Monogenea, and Cestoda share a syncytial covering that entirely lacks cilia and is designated a **tegument** (figure 8.8). This distinctive tegumental plan is the basis for uniting trematodes, monogeneans, and cestodes in a taxon called Neodermata. It is a peculiar epidermal arrangement and may be related to adaptations for parasitism in ways that are still unclear.

In the body wall below the basal lamina of flatworms are layers of muscle fibers that run circularly, longitudinally, and diagonally. A meshwork of parenchyma cells, developed from mesoderm, fills the spaces between muscles and visceral organs. Parenchyma cells in some, perhaps all, flatworms are not a separate cell type, but are noncontractile portions of muscle cells.

Nutrition and Digestion

The digestive system of most platyhelminthes includes a mouth, pharynx, and intestine. In turbellarians the muscular **pharynx** opens posteriorly just inside the mouth, through which it can extend (figure 8.9). The mouth is usually at the anterior end in

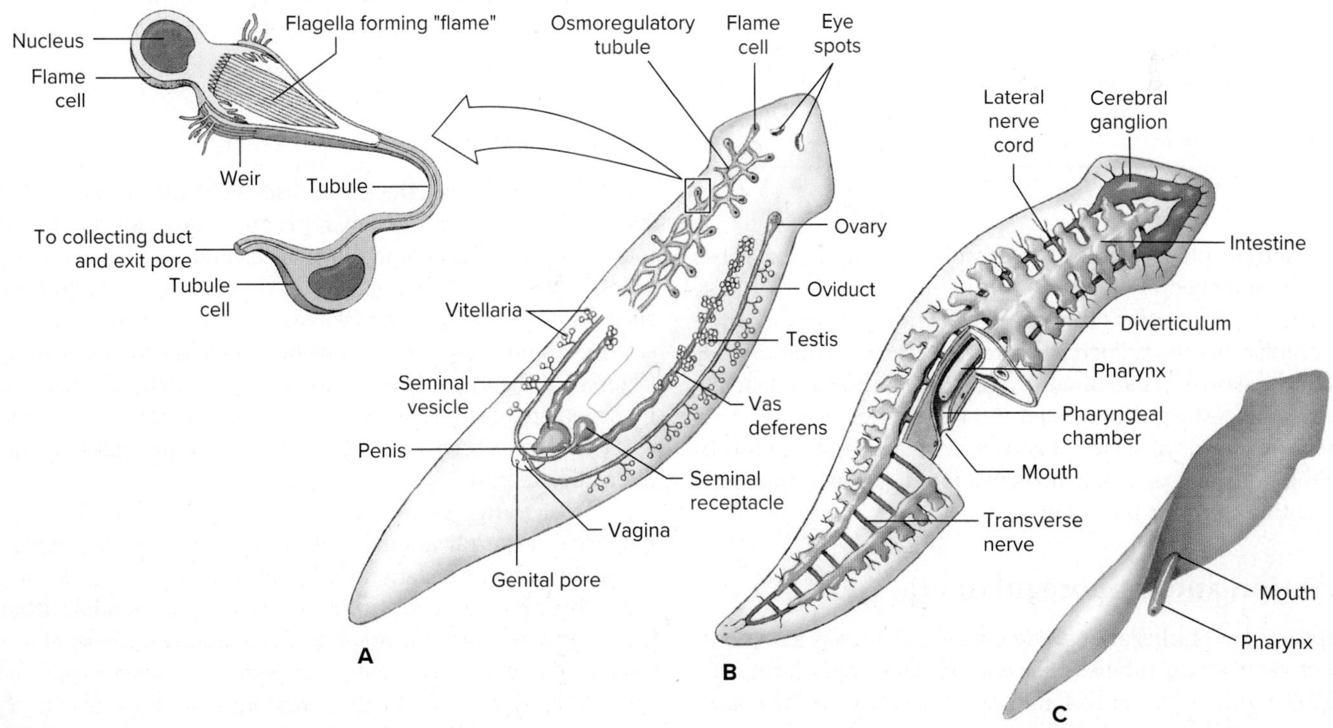

figure 8.9

Structure of a planarian. **A,** Reproductive and excretory systems, shown in part. Inset at left is enlargement of a flame cell. **B,** Digestive tract and ladder-type nervous system. The pharynx is shown in the resting position. **C,** Pharynx extended through ventral mouth.

CHARACTERISTICS of Phylum Platyhelminthes

1. No single diagnostic feature.
2. Marine, freshwater, and moist terrestrial habitats.
3. Turbellarian flatworms are mostly free-living; classes Monogenea, Trematoda, and Cestoda are entirely parasitic.
4. **Bilateral symmetry;** definite polarity of anterior and posterior ends; **body flattened dorsoventrally**.
5. Adult body three-layered **(triploblastic)**.
6. Body acoelomate.
7. Epidermis may be cellular or syncytial (ciliated in some); **rhabdites** in epidermis of most Turbellaria; epidermis a syncytial **tegument** in Monogenea, Trematoda, Cestoda, and some Turbellaria.
8. Gut incomplete; may be branched; absent in cestodes.
9. Muscular system primarily a sheath form of mesodermal origin; layers of circular, longitudinal, and sometimes oblique fibers beneath the epidermis.
10. Nervous system consisting of a **pair of anterior ganglia** with **longitudinal nerve cords** connected by transverse nerves and located in the mesenchyme in most forms.
11. Sense organs include statocysts (organs of balance) and ocelli, auricles, and rheoreceptors.
12. Asexual reproduction by fragmentation and other methods as part of complex parasitic life cycles.
13. Most forms monoecious; reproductive system complex, usually with well-developed gonads, ducts, and accessory organs; internal fertilization; development direct in free-swimming forms and those with single hosts; complicated life cycle often involving several hosts in many internal parasites.
14. Excretory system comprises two lateral canals with branches bearing **flame cells (protonephridia)**; lacking in some forms.
15. Respiratory, circulatory, and skeletal systems lacking; lymph channels with free cells in some trematodes.

flukes, and the pharynx is not protrusible. The intestine may be simple or branched.

Intestinal secretions contain proteolytic enzymes for some extracellular digestion. Food is sucked into the intestine, where cells of the gastrodermis often phagocytize it and complete intracellular digestion. Undigested food is egested through the pharynx. Tapeworms lack a digestive system and must absorb all of their nutrients as small molecules (predigested by the host) directly through their tegument.

Excretion and Osmoregulation

Except in some turbellarians, the osmoregulatory system consists of canals with tubules that end in **flame cells (protonephridia)** (figure 8.9A). Each flame cell surrounds a small space into which a tuft of flagella projects. In some turbellarians and in the other classes of flatworms, the protonephridia form a weir (Old English *wer,* a fence placed in a stream to catch fish). In a weir, the rim of the cup formed by the flame cell bears fingerlike projections that interdigitate with similar projections of a tubule cell. The beat of the flagella (resembling a flickering flame) provides a negative pressure to draw fluid through the weir into the space (lumen) enclosed by the tubule cell. The lumen continues into collecting ducts that finally open to the outside by pores. The wall of the duct beyond the flame cell commonly bears folds or microvilli that probably function in reabsorption of certain ions or molecules. It is likely that this system is osmoregulatory in most forms; it is reduced or absent in marine turbellarians, which do not have to expel excess water.

Metabolic wastes are largely removed by diffusion through the body wall.

Nervous System and Sense Organs

Flatworms are cephalized but vary in the complexity of their nervous system. The simplest system, found in some turbellarians, is a subepidermal nerve plexus resembling the nerve net of the cnidarians. Other flatworms have, in addition to a nerve plexus, one to five pairs of longitudinal nerve cords (figure 8.9B) lying under the muscle layer. Connecting nerves form a "ladder-type" pattern. Their brain is a mass of ganglion cells arising anteriorly from the nerve cords. The neurons are organized into sensory, motor, and association types—an important advance in the evolution of nervous systems.

Tactile cells and chemoreceptive cells are abundant over the body, and in planarians they form definitive organs on the **auricles** (the earlike lobes on the sides of the head). Some also have **statocysts** for equilibrium and **rheoreceptors** for sensing water current direction. **Ocelli,** or light-sensitive eyespots, are common in turbellarians (figure 8.8A), monogeneans, and larval trematodes.

Reproduction

Many flatworms reproduce both asexually and sexually. Many freshwater turbellarians can reproduce asexually by fission, merely constricting behind the pharynx and separating into two animals, each of which regenerates the missing parts. In forms such as *Stenostomum* and *Microstomum,* individuals do not separate at once but remain attached, forming chains of zooids (figure 8.10B and C). Flukes reproduce asexually in their snail intermediate host (see figure 8.11), and some tapeworms, such as *Echinococcus,* can bud off thousands of juveniles in their intermediate host.

Most flatworms are monoecious (hermaphroditic) and practice cross-fertilization. Fertilization is internal via a penis or cirrus. In some turbellarians, yolk for nutrition of the developing embryo is contained within the egg cell itself **(endolecithal),** just as it is normally in other phyla of animals. Possession of endolecithal eggs is considered ancestral for flatworms. Other turbellarians, as well as all trematodes, monogeneans, and cestodes, share a derived condition in which female gametes contain little or no yolk, and yolk is contributed by cells released from separate organs (yolk glands). Usually yolk cells surround the zygote within an eggshell **(ectolecithal).**

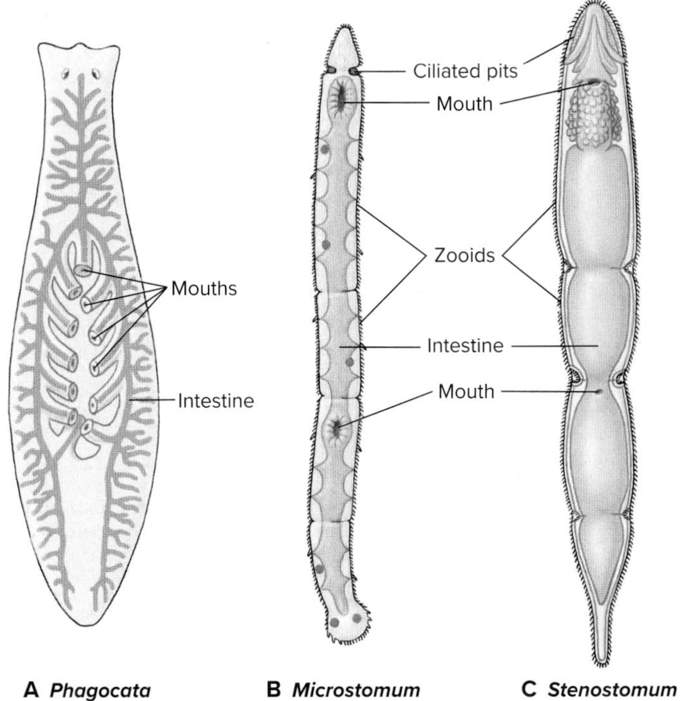

figure 8.10

Some small freshwater turbellarians. **A,** *Phagocata* has numerous mouths and pharynges. **B, C,** Incomplete fission yields a series of temporarily attached zooids.

Development following formation of the zygote may be direct or indirect. In some flatworms, an embryo becomes a larva, which may be ciliated or not, according to the group.

Class Turbellaria

Turbellarians are mostly free-living worms that range in length from 5 mm to nearly 50 cm. Their mouth is on the ventral side and leads into a gut with a variable form. The orders within class Turbellaria are distinguished by the form of the gut (present or absent; simple or branched; pattern of branching) and the pharynx (simple; folded; bulbous). Except for order Polycladida (Gr. *poly,* many, + *klados,* branch), turbellarians with endolecithal eggs have a simple gut and a simple pharynx. Polyclads have a folded pharynx and a gut with many branches. Polyclads include many marine forms of moderate to large size (3 to more than 40 mm) (see figure 8.4), and a highly branched intestine is correlated with larger size in turbellarians. Members of order Tricladida (Gr. *treis,* three, + *klados,* branch), which are ectolecithal and include freshwater planaria, have a three-branched intestine.

Turbellarians are typically creeping forms that combine muscular with ciliary movements to achieve locomotion. Very small planaria swim by means of their cilia. Others move by gliding, with the head slightly raised, over a slime track secreted by the marginal adhesive glands. Beating of epidermal cilia in the slime track moves the animal forward, while rhythmic muscular waves pass backward from the head. Large polyclads and terrestrial turbellarians crawl by muscular undulations, much in the manner of a snail.

Some turbellarians have simple life cycles without a distinct larval stage. In some freshwater planarians, egg capsules are attached by little stalks to the undersides of stones or plants, and embryos emerge as juveniles that resemble miniature adults. Marine turbellarians have a ciliated larva that is very similar to the trochophore (see figure 10.6) of some annelids, molluscs, and members of other phyla.

As traditionally recognized, turbellarians form a paraphyletic group. Several synapomorphies, such as insunk epidermis and ectolecithal development, show that some turbellarians are phylogenetically closer to Trematoda, Monogenea, and Cestoda than they are to other turbellarians. Ectolecithal turbellarians therefore appear to form a clade with trematodes, monogeneans, and cestodes to the exclusion of endolecithal turbellarians (see figure 8.4). Endolecithal turbellarians also are paraphyletic; the presence of a dual-gland adhesive system in some endolecithal turbellarians indicates a clade with ectolecithal flatworms to the exclusion of other endolecithal turbellarian lineages. Turbellaria therefore describes an artificial group, and the term is used here only because it is prevalent in zoological literature.

Class Trematoda

Trematodes are all parasitic flukes, and as adults they are almost all endoparasites of vertebrates. They are chiefly leaflike in form and are structurally similar in many respects to more complex Turbellaria. A major difference is the tegument (described earlier; see figure 8.8).

Other structural adaptations for parasitism are apparent: various penetration glands or glands that produce cyst material; organs for attachment, such as suckers and hooks; and increased reproductive capacity. Otherwise, trematodes share several characteristics with turbellarians, such as a well-developed gut tube (but with the mouth at the anterior, or cephalic, end) and similar reproductive, excretory, and nervous systems, as well as a musculature and parenchyma that differ only slightly from those of turbellarians. Sense organs are poorly developed.

Of the three subclasses of Trematoda, two are small and poorly known groups, but Digenea (Gr. *dis,* double, + *genos,* descent) is a large group that includes many species of medical and economic importance.

Subclass Digenea

With rare exceptions, digenetic trematodes have a complex life cycle, the first host **(intermediate host)** being a mollusc and the final host **(definitive host)** being a vertebrate. A definitive host is one in which the parasite reproduces sexually. In some species, a second, and sometimes even a third, intermediate host intervenes. The group has many species, and they can inhabit diverse sites in their hosts: all parts of the digestive tract, respiratory tract, circulatory system, urinary tract, and reproductive tract.

Among the world's most amazing phenomena are digenean life cycles. Although cycles of different species vary

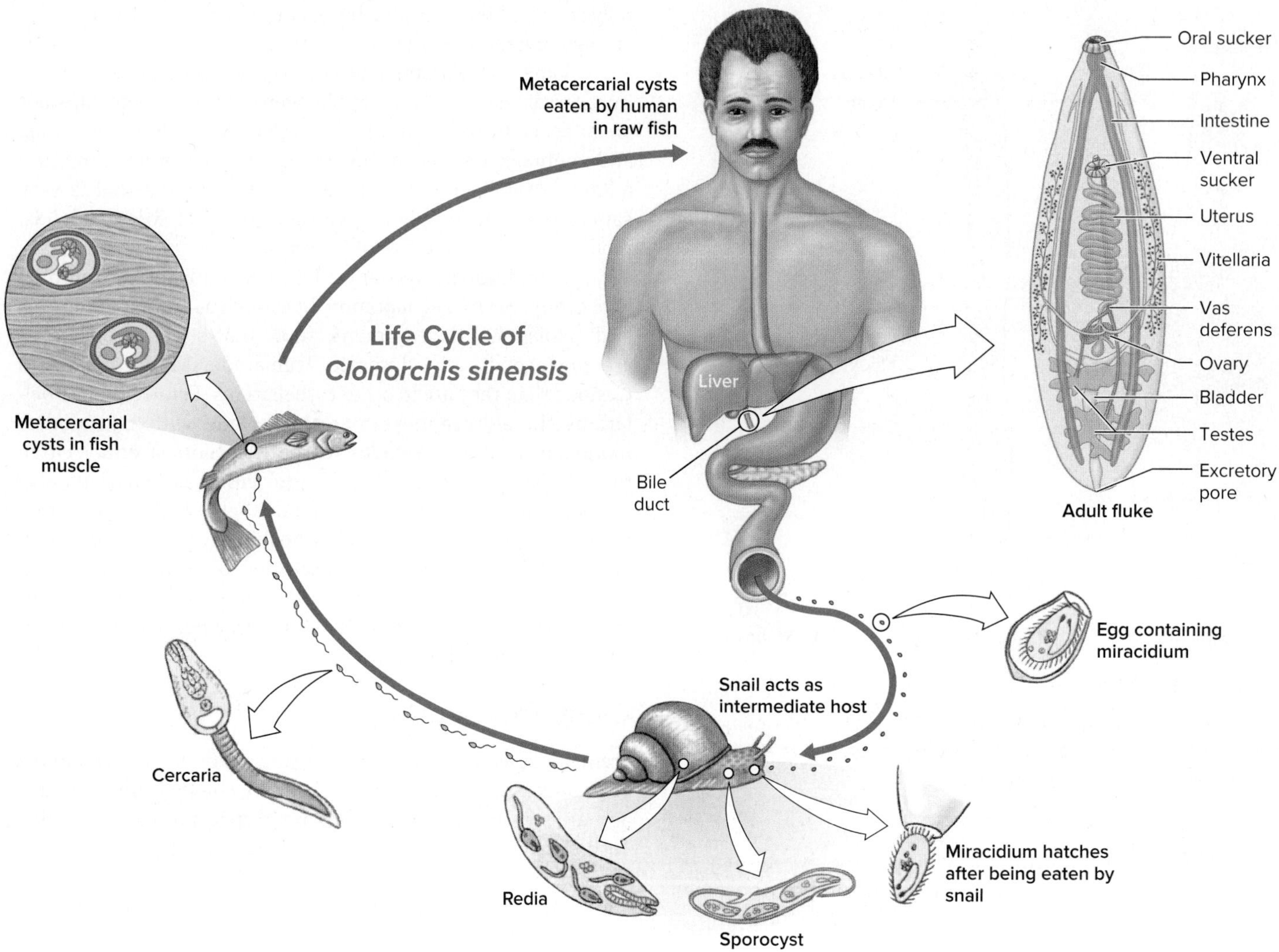

figure 8.11

Life cycle of *Clonorchis sinensis*.

widely in detail, typical stages include adult, egg (shelled embryo), miracidium, sporocyst, redia, cercaria, and metacercaria (figure 8.11). The shelled embryo or larva usually passes from the definitive host in excreta and must reach water to develop further. There, it hatches to a free-swimming, ciliated larva, the **miracidium.** The miracidium penetrates the tissues of a snail, where it transforms into a sacklike **sporocyst.** Sporocysts reproduce asexually to yield either more sporocysts or a number of **rediae.** Rediae, in turn, reproduce asexually to produce more rediae or to produce **cercariae.** In this way, a single zygote can give rise to an enormous number of progeny. Cercariae emerge from the snail and penetrate a second intermediate host or encyst on vegetation or other objects to become **metacercariae,** which are juvenile flukes. Adults grow from the metacercaria when that stage is eaten by a definitive host.

Some of the most serious parasites of humans and domestic animals belong to the Digenea (table 8.1). *Clonorchis* (Gr. *clon,* branch, + *orchis,* testis) (figure 8.11) is the most important liver fluke of humans and is common in China, southern Asia, and Japan. Cats, dogs, and pigs are also often infected. Adult flukes live in the bile passages, and shelled miracidia are passed in feces. If ingested by certain freshwater snails, sporocyst and redia stages develop, and free-swimming cercariae emerge. Cercariae that manage to find a suitable fish encyst in the skin or muscles as metacercariae. When the fish is eaten raw, the juveniles migrate up the bile duct to mature and may survive there for 15–30 years. The effect of the flukes on humans depends mainly on the extent of infection. A heavy infection may cause pronounced cirrhosis of the liver and death.

Schistosomiasis, an infection with blood flukes of genus *Schistosoma* (Gr. *schistos,* divided, + *soma,* body) (figure 8.12A), ranks as one of the major infectious diseases in the world, with 200 million people infected. The disease is widely prevalent over much of Africa and parts of South America, the West Indies, the Middle East, and the Far East. It is spread when shelled miracidia shed in human feces and urine enter water

Table 8.1 Examples of Digenetic Flukes Infecting Humans

Common and scientific names	Means of infection; distribution and prevalence in humans
Blood flukes (*Schistosoma* spp.); three widely prevalent species, others reported	Cercariae in water penetrate skin; 200 million people infected with one or more species
S. mansoni	Africa, South and Central America
S. haematobium	Africa
S. japonicum	Eastern Asia
Chinese liver flukes (*Clonorchis sinensis*)	Eating metacercariae in raw fish; about 30 million cases in Eastern Asia
Lung flukes (*Paragonimus* spp.); seven species, most prevalent is *P. westermani*	Eating metacercariae in raw freshwater crabs, crayfish; Asia and Oceania, sub-Saharan Africa, South and Central America; several million cases in Asia
Intestinal fluke (*Fasciolopsis buski*)	Eating metacercariae on aquatic vegetation; 10 million cases in Eastern Asia
Sheep liver fluke (*Fasciola hepatica*)	Eating metacercariae on aquatic vegetation; widely prevalent in sheep and cattle, occasional in humans

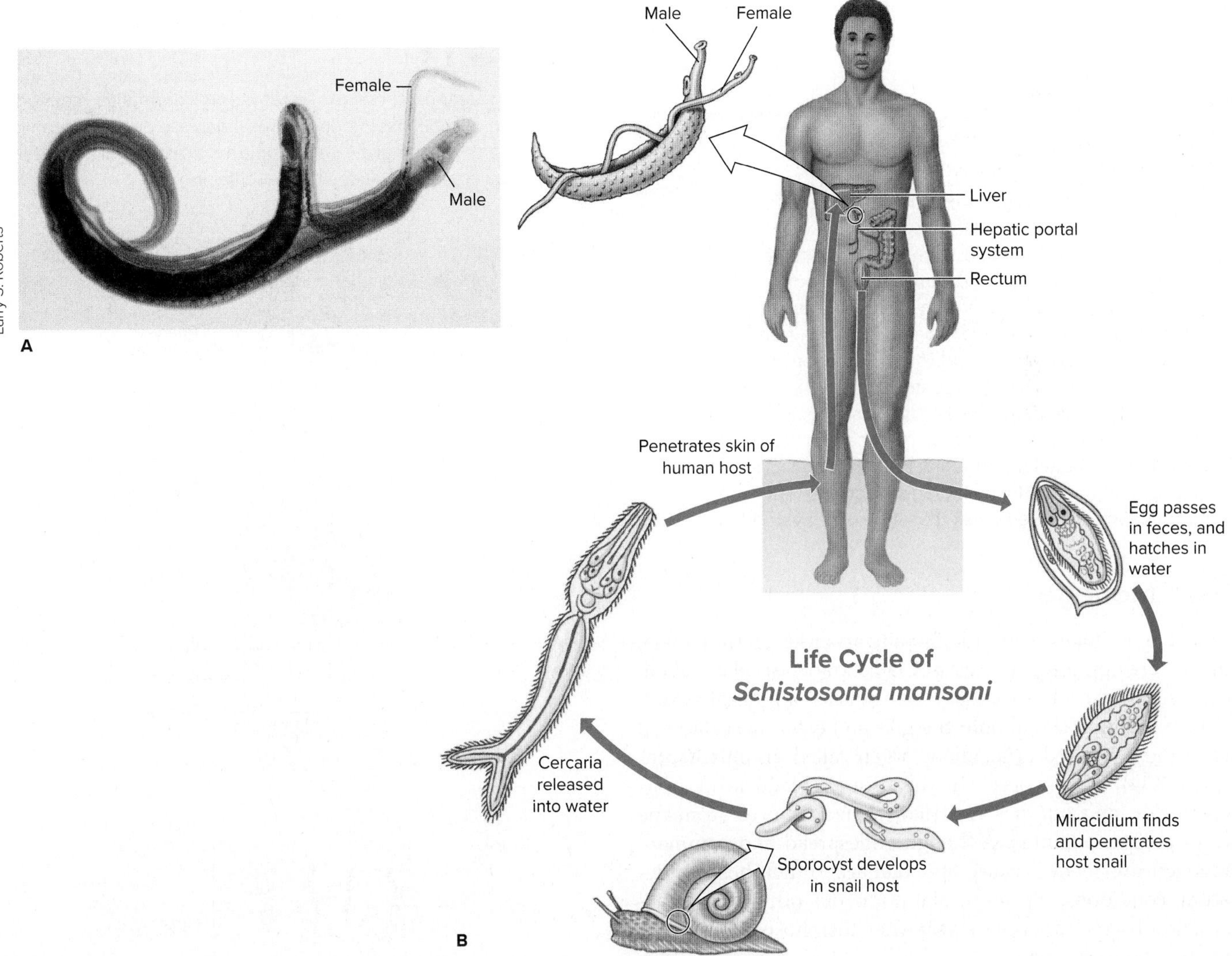

figure 8.12

A, Adult male and female *Schistosoma japonicum* in copulation. Blood flukes differ from most other flukes in being dioecious. Males are broader and heavier and have a large, ventral gynecophoric canal, posterior to the ventral sucker. The gynecophoric canal embraces the long, slender female (darkly stained individual) during insemination and oviposition. **B,** Life cycle of *Schistosoma* spp.

containing host snails (figure 8.12B). Cercariae that contact human skin penetrate the skin to enter blood vessels, which they follow to certain favorite regions depending on the type of fluke. *Schistosoma mansoni* lives in venules draining the large intestine; *S. japonicum* localizes more in venules draining the small intestine; and *S. haematobium* lives in venules draining the urinary bladder. Blood flukes differ from most other flukes in having separate males and females. In each species, many eggs released by female worms do not find their way out of the body but lodge in the liver or other organs, where they are sources of chronic inflammation. The inch-long adults may live for years in a human host, and their eggs cause such disturbances as severe dysentery, anemia, liver enlargement, bladder inflammation, and brain damage.

Cercariae of several genera whose normal hosts are birds often enter the skin of human bathers in their search for a suitable bird host, causing a skin irritation called "swimmer's itch" (figure 8.13). In this case, humans are a dead end in the fluke's life cycle because the fluke cannot develop further in the human body.

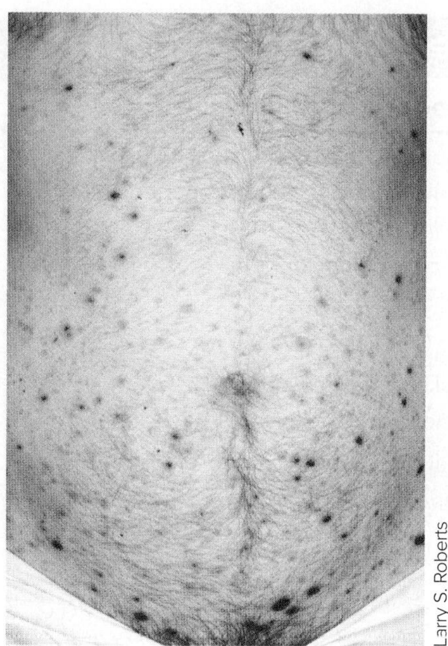

figure 8.13

Human abdomen, showing schistosome dermatitis caused by penetration of schistosome cercariae, which are unable to complete development in humans. Sensitization to allergenic substances released by cercariae causes rash and itching.

Key Theme 8.1

HUMAN CONNECTIONS

Controlling Schistosomiasis

Humans infected with schistosomiasis can be effectively treated with Praziquantel. Development of a vaccine is the subject of much research, but an effective vaccine is not yet available. Proper disposal of human wastes, snail eradication, and vector control have some promise for reducing infection. Biological controls include introduction of species of snails, crayfish, and fish that prey on the snail vectors. However, biological control attempts for other species have often been fraught with unexpected ecological impacts. In some cases, the biological control has been more of a problem in the long run than the pest species it was supposed to control. Many biologists consider such introductions an extreme risk that should be avoided.

Class Monogenea

Monogenetic flukes were traditionally an order of Trematoda, but they are now a separate class. Cladistic analysis places them as the sister taxon to Cestoda. Monogeneans are mostly external parasites that clamp onto the gills and external surfaces of fish using a hooked attachment organ called an **opisthaptor** (figure 8.14). A few occur in the urinary bladders of frogs and turtles, and one has reportedly been removed from the eye of a hippopotamus. Although widespread and common, monogeneans seem to cause little damage to their hosts under natural conditions. However, like numerous other fish pathogens, they become a serious threat when their hosts are crowded together, as in fish farming.

The life cycles of monogeneans are simple, having a single host, as suggested by the name of the group, which means "single descent." The egg hatches a ciliated larva that attaches to a host, sometimes following a free-swimming phase.

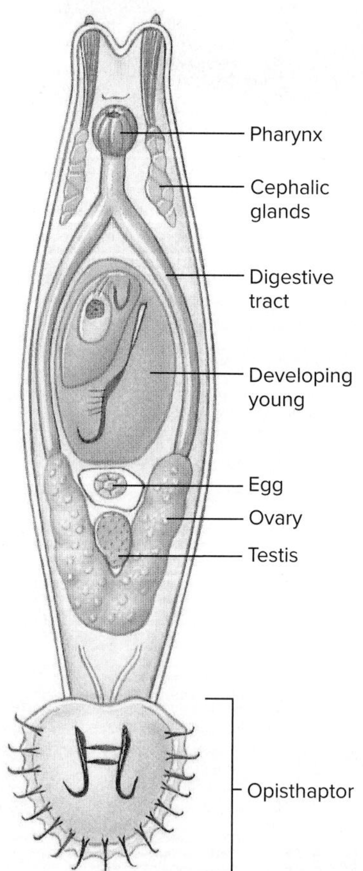

figure 8.14

Gyrodactylus sp. (class Monogenea), ventral view.

Class Cestoda

Cestoda, or tapeworms, differ in many respects from the preceding classes: they usually have long, flat bodies composed of a **scolex,** for attachment to the host, followed by many reproductive units or **proglottids** (figure 8.15). The scolex, or holdfast, is usually provided with suckers or suckerlike organs and often with hooks or spiny tentacles as well. Tapeworms entirely lack a digestive system, but they have well-developed muscles, and their excretory and nervous systems are somewhat similar to those of other flatworms. They have no special sense organs, but sensory endings in their tegument are modified cilia (figure 8.16).

As in Monogenea and Trematoda, adult cestodes have no external, motile cilia, and the tegument is composed of a distal cytoplasm with sunken cell bodies beneath the superficial layer of muscle (see figure 8.8). In contrast to monogeneans and trematodes, however, their entire surface is covered with minute projections called **microtriches** (sing., **microtrix**), which are similar in certain respects to microvilli of the vertebrate small intestine (figure 8.16). These microtriches greatly amplify the surface area of the tegument, which is a vital adaptation for a tapeworm, since it must absorb all its nutrients across the tegument.

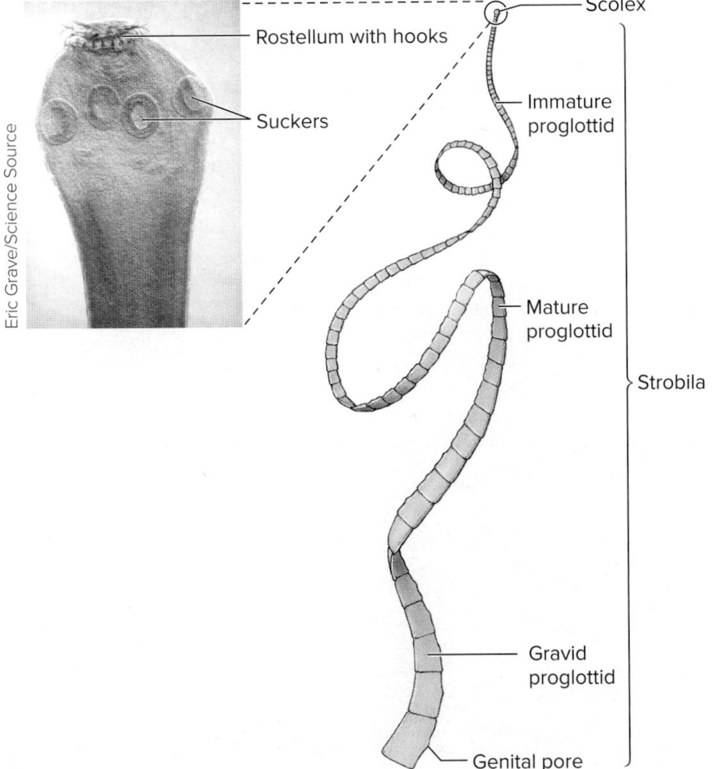

figure 8.15

A tapeworm, showing strobila and scolex. The scolex is the organ of attachment; it is hypothesized to be a posterior structure. Tapeworms do not have a head.

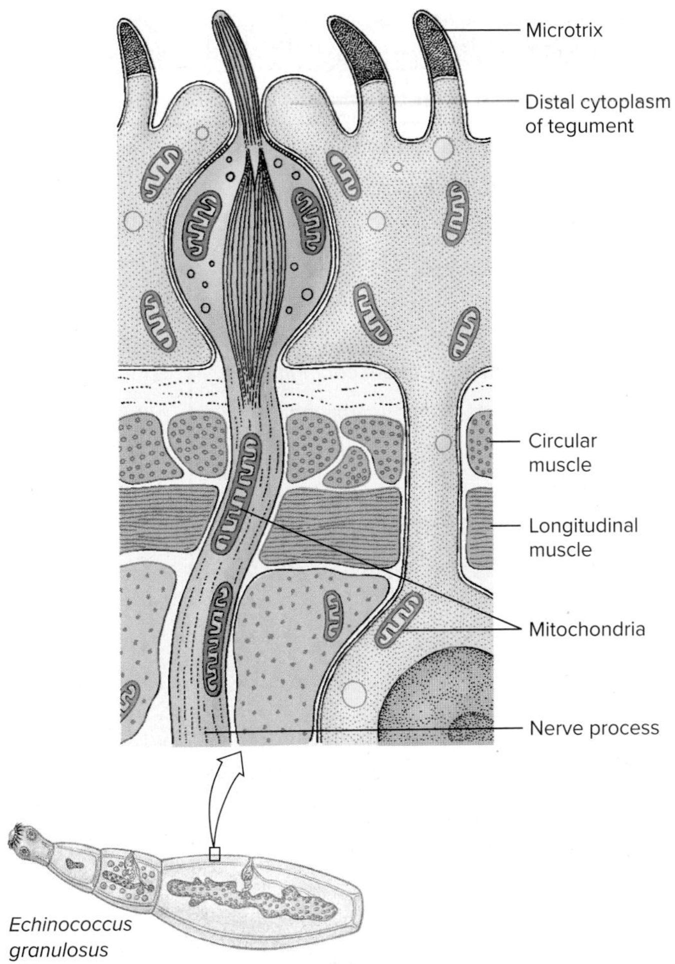

figure 8.16

Schematic drawing of a longitudinal section through a sensory ending in the tegument of *Echinococcus granulosus*, a tapeworm. Note the position of the nerve process.

Tapeworms are nearly all monoecious. The main body of a cestode is a chain of proglottids called a **strobila** (see figure 8.15). Typically, new proglottids form at a **germinative zone** just behind the scolex (proximal to it). As new proglottids differentiate in front of it, each previously formed proglottid moves distally in the strobila, and its gonads mature (figure 8.17). A proglottid is usually fertilized by another proglottid in the same or a different strobila. Shelled embryos form in the uterus of the proglottid, and either they are expelled through a uterine pore, or the entire proglottid detaches from the worm as it reaches the distal end.

About 4000 species of tapeworms are known. With rare exceptions, all cestodes require at least two hosts, and adults are parasitic in the digestive tract of vertebrates. Often one of the intermediate hosts is an invertebrate. Almost all vertebrate species can be infected. Normally, adult tapeworms do little harm to their hosts. Table 8.2 lists the most common tapeworms in humans.

Table 8.2 Common Cestodes Infecting Humans

Common and scientific names	Means of infection; prevalence in humans
Beef tapeworm (*Taenia saginata*)	Eating rare beef; most common of large tapeworms in humans
Pork tapeworm (*Taenia solium*)	Eating rare pork; less common than *T. saginata*; incidence of beef and pork tapeworms ca. 1000 reports annually in the United States
Fish tapeworm (*Diphyllobothrium latum*)	Eating rare or poorly cooked fish; fairly common in Great Lakes region of the United States and other areas of world where raw fish is eaten; incident reports vary from 100 to 2000 annually in Asia
Dog tapeworm (*Dipylidium caninum*)	Unhygienic habits of children (juveniles in flea and louse); moderate frequency
Dwarf tapeworm (*Hymenolepis nana*)	Juveniles in flour beetles; common
Unilocular hydatid (*Echinococcus granulosus*)	Cysts of juveniles in humans; infection by contact with dogs; common wherever humans are in close relationship with dogs and ruminants
Multilocular hydatid (*Echinococcus multilocularis*)	Cysts of juveniles in humans; infection by contact with foxes; less common than unilocular hydatid

In the beef tapeworm, *Taenia saginata* (Gr. *tainia*, band, ribbon), shelled larvae shed from the human host are ingested by cattle (figure 8.18). The six-hooked larvae (oncospheres) hatch, burrow into blood or lymph vessels, and migrate to skeletal muscle, where they encyst to become "bladder worms" (**cysticerci**). Each of these juveniles develops an invaginated scolex and remains quiescent until the uncooked muscle is eaten by humans. In the new host, the scolex evaginates, attaches to the intestine, and matures in 2 or 3 weeks; then ripe proglottids may be expelled daily for many years. Humans

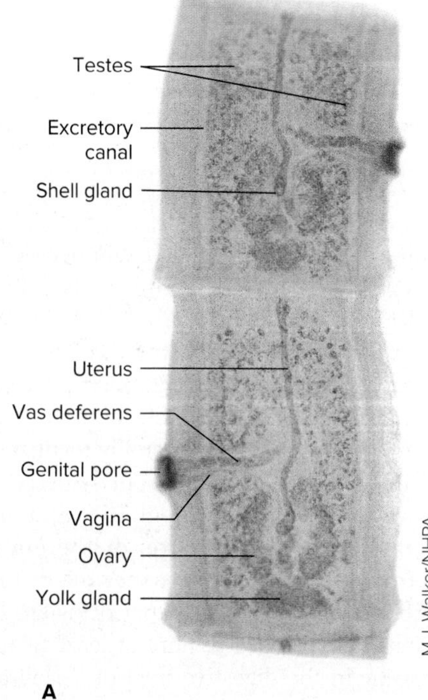

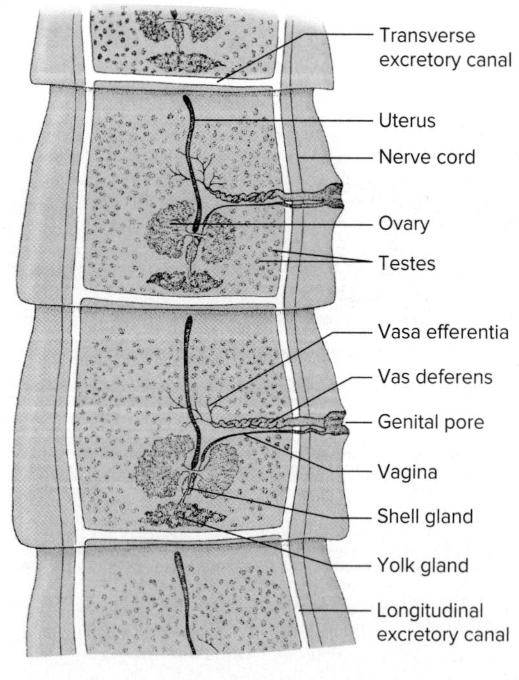

figure 8.17

A, Two mature proglottids of *Taenia pisiformis*, a dog tapeworm. **B,** More anatomical details are marked in a drawing of two complete proglottids, two other partial proglottids are shown.

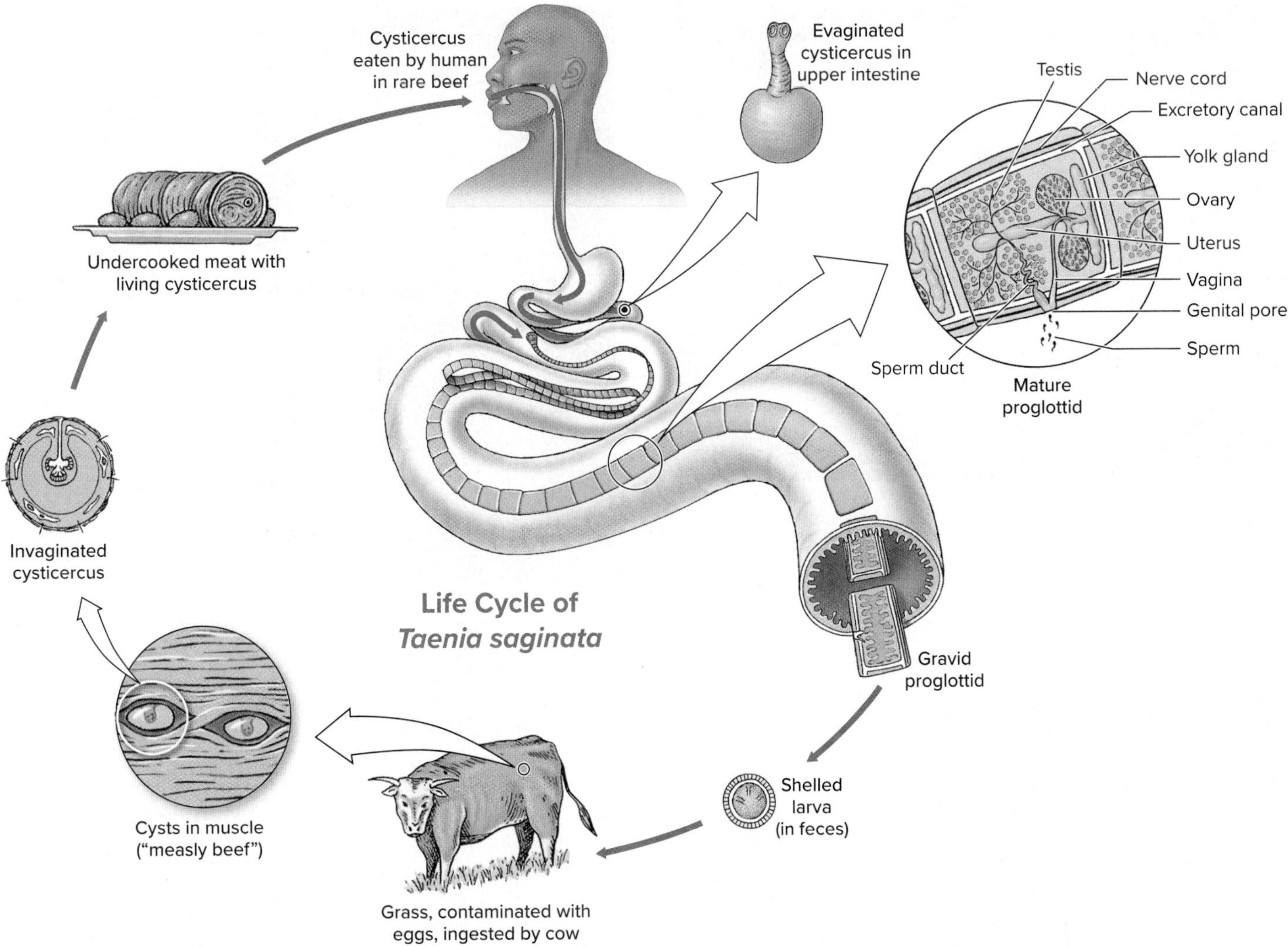

figure 8.18

Life cycle of beef tapeworm, *Taenia saginata*. Gravid proglottids detach in the human intestine, leave the body in feces, crawl onto grass, and are ingested by cattle. Eggs hatch in the cow's intestine, freeing oncospheres, which penetrate muscles and encyst, developing into "bladder worms." A human eats infected rare beef, and a cysticercus is freed in the intestine where it attaches to the intestinal wall, forms a strobila, and matures.

Taxonomy of Phylum Platyhelminthes

Class Turbellaria (tur′bel-lar′e-ə) (L. *turbellae* [pl.], stir, bustle, + *-aria*, like or connected with). **Turbellarians.** Usually free-living forms with soft, flattened bodies; covered with ciliated epidermis containing secreting cells and rodlike bodies (rhabdites); mouth usually on ventral surface sometimes near center of body; no body cavity other than the blind gut; mostly hermaphroditic; some undergo asexual fission. A paraphyletic taxon. Examples: *Dugesia* (planaria), *Microstomum, Planocera.*

Class Trematoda (trem′a-to′də) (Gr. *trēma,* a hole, + *eidos,* form). **Digenetic flukes.** Body covered with a syncytial tegument without cilia; leaflike or cylindrical in shape; usually have oral and ventral suckers, no hooks; gut usually has two main branches; mostly monoecious; life cycle complex, with first host a mollusc, final host usually a vertebrate; parasitic in all classes of vertebrates. Examples: *Fasciola, Clonorchis, Schistosoma.*

Class Monogenea (mon′o-gen′e-ə) (Gr. *mono,* single, + *genos,* descent). **Monogenetic flukes.** Body covered with a syncytial tegument without cilia; body usually leaflike to cylindrical in shape; posterior attachment organ with hooks, suckers, or clamps, usually in combination; monoecious; life cycle simple, with single host and usually with free-swimming, ciliated larva; all parasitic, mostly on skin or gills of fish. Examples: *Dactylogyrus, Polystoma, Gyrodactylus.*

Class Cestoda (ses-to′də) (Gr. *kestos,* girdle, + *eidos,* form). **Tapeworms.** Body covered with nonciliated, syncytial tegument; general form of body tapelike; have scolex with suckers or hooks, sometimes both, for attachment; body usually divided into series of proglottids; no gut or digestive organs; usually monoecious; parasitic in digestive tract of all classes of vertebrates; life cycle complex, with two or more hosts; first host may be vertebrate or invertebrate. Examples: *Diphyllobothrium, Hymenolepis, Taenia.*

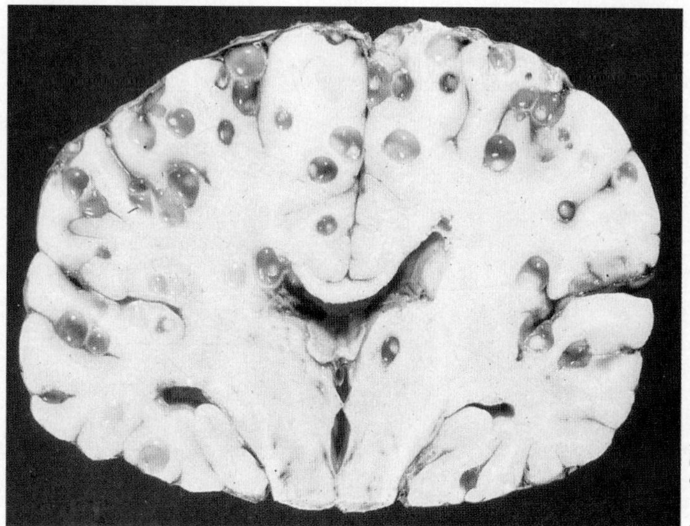

figure 8.19

Section through the brain of a person who died of cerebral cysticercosis, an infection with cysticerci of *Taenia solium*.

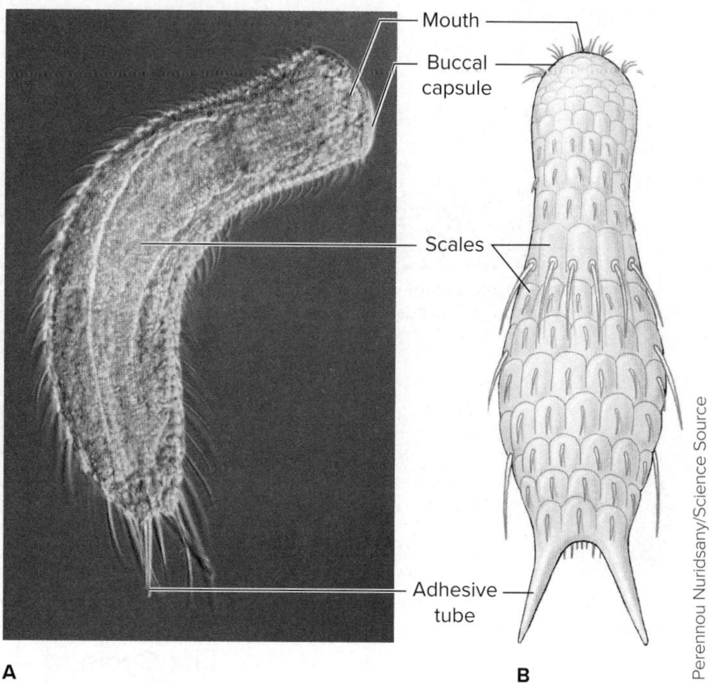

figure 8.20

A, A live *Chaetonotus simrothi*, a common gastrotrich. **B,** External structure of *Chaetonotus*.

become infected by eating raw or rare infested ("measly") beef. Adult worms may attain a length of 7 m or more, folded back and forth in the host intestine.

The pork tapeworm, *Taenia solium,* uses humans as definitive hosts and pigs as intermediate hosts. Adult tapeworms live in the human gut, and infected humans shed fertilized tapeworm eggs with their feces. Pigs are colonized by consuming infected human fecal material. Inside a pig, the tapeworm larvae form cysts in muscle tissue. When humans consume infected undercooked pork, the cysts hatch and develop into adult tapeworms; this is the most common mode of infection. However, there is another way that humans become infected: people who live with, or contact, other people infected with tapeworms may develop cysticercosis, a serious disease caused by ingesting the fertilized eggs directly, without any intermediate host. When the fertilized eggs are ingested by a human, instead of by a pig, the resulting larva migrates to tissues such as the brain, spinal cord, liver, muscles, or eyes. Infection of the brain or spinal cord is very serious and may cause death (figure 8.19). Infection of other organs may be treatable.

8.4 Phylum Gastrotricha

The phylum Gastrotricha (gas-trō-tri′kə) (Gr. *gaster,* belly, + *thrix,* hair) is a small group (about 460 species) of microscopic animals, approximately 65–500 μm long (figure 8.20). They are usually bristly or scaly in appearance and flattened on the ventral side; many move by gliding on ventral cilia. Others move in a leechlike fashion by briefly attaching the posterior end by means of adhesive glands. Both marine and freshwater species exist, and they are common in lakes, ponds, and seashore sands. They feed on bacteria, diatoms, and small unicellular eukaryotes. Gastrotrichs are hermaphroditic. However, the male system of many freshwater species is nonfunctional, and the female system produces offspring parthenogenetically.

8.5 Clade Gnathifera

Clade Gnathifera (nath-if′er-ə) traditionally comprised four phyla, but Chaetognatha is added here, so the group now comprises five phyla. Members of Gnathostomulida, Micrognathozoa, and Rotifera are tiny, free-living aquatic animals, whereas members of Acanthocephala are wormlike endoparasites of fishes or other vertebrates. Chaetognaths, commonly called arrow worms, reach up to 12 cm in length and are planktonic predators. Clade Gnathifera ("jaw-bearing") is so named because its members, other than the acanthocephalans, possess small, cuticular jaws with a homologous microstructure. The recently added chaetognaths have a somewhat larger complex jaw apparatus of similar microstructure. The number of pairs of such jaws varies within the clade, but jaws are absent from acanthocephalans. Recent work suggests that a ciliated pharyngeal ganglion is also a gnathiferan synapomorphy.

Acanthocephala and Rotifera are presumed sister taxa, together forming a clade called Syndermata. Their close relationship first appeared in molecular phylogenies and led morphologists to examine acanthocephalans anew, searching for evidence that these parasites were highly derived rotifers. There is little external similarity between free-swimming rotifers and endoparasitic worms, but members of both groups have a eutelic syncytial epidermis. **Eutely** denotes constancy in the numbers of nuclei present, as illustrated by the constant numbers of nuclei in various organs of one species of rotifer: A researcher, E. Martini (1912), always counted 183 nuclei in the brain, 39 in the stomach, and 172 in the coronal epithelium. Despite the shared syncytial epidermis, the union of two morphologically disparate taxa in clade Syndermata is still controversial.

8.6 Phylum Gnathostomulida

The Gnathostomulida (nath′ō-sto-mūlid′-a) (Gr. *gnatho*, jaw, + *stoma*, mouth, + *L-ulus*, diminutive) was first observed in 1928 in the Baltic, but its description was not published until 1956. Since then, these animals have been found in many parts of the world, including the Atlantic coast of the United States, and over 80 species in 18 genera have been described.

Gnathostomulids are delicate, wormlike animals measuring 0.5–1 mm long (figure 8.21). They occupy interstitial spaces of very fine sandy coastal sediments and silt and can endure conditions of very low oxygen.

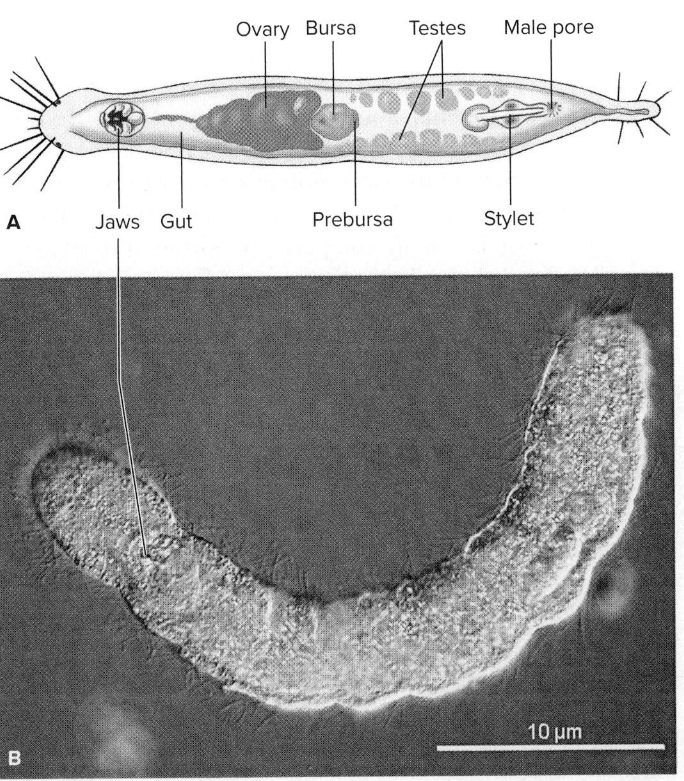

figure 8.21

A, *Gnathostomula jenneri* (phylum Gnathostomulida) is a tiny member of the interstitial fauna living between grains of sand or mud. Species in this family are among the most commonly encountered jaw worms, found in shallow water and down to depths of several hundred meters. **B,** *Gnathostomula paradoxa* is abundant in sediments near burrows of marine polychaetes in the North Sea. Its ecology is very similar to that of *G. jenneri* from the North American Atlantic coast.

Gnathostomulids scrape bacteria and fungi from a substratum using a pair of lateral jaws in the pharynx. The pharynx, structurally similar to the rotifer mastax (see figure 8.24), leads into a blind gut. The body is acoelomate with a poorly developed parenchyma layer. There is no circulatory system, and respiration relies on diffusion. The epidermis is ciliated, but each epidermal cell has only one cilium; this condition is rare among other bilateral animals, although it does occur in some gastrotrichs. Gnathostomulids are simultaneous hermaphrodites. Development proceeds via spiral cleavage.

8.7 Phylum Chaetognatha

Chaetognatha (kē-tog′na-thə) (Gr. *chaitē*, long hair, + *gnathos*, jaw) is a group of about 100 species of marine animals, sometimes called arrow worms, adapted for a planktonic existence. Their small, straight bodies resemble miniature torpedoes, or darts, ranging from 1 to 12 cm in length. Arrow worms are voracious feeders, consuming other zooplankton, especially copepods, and even small fishes. The vestibule, a depression surrounding the mouth, contains jaws, and is flanked by curved, chitinous spines (figure 8.22B). When an arrow worm captures prey, its jaws and raptorial spines spread apart and then snap closed with startling speed. Arrow worms usually

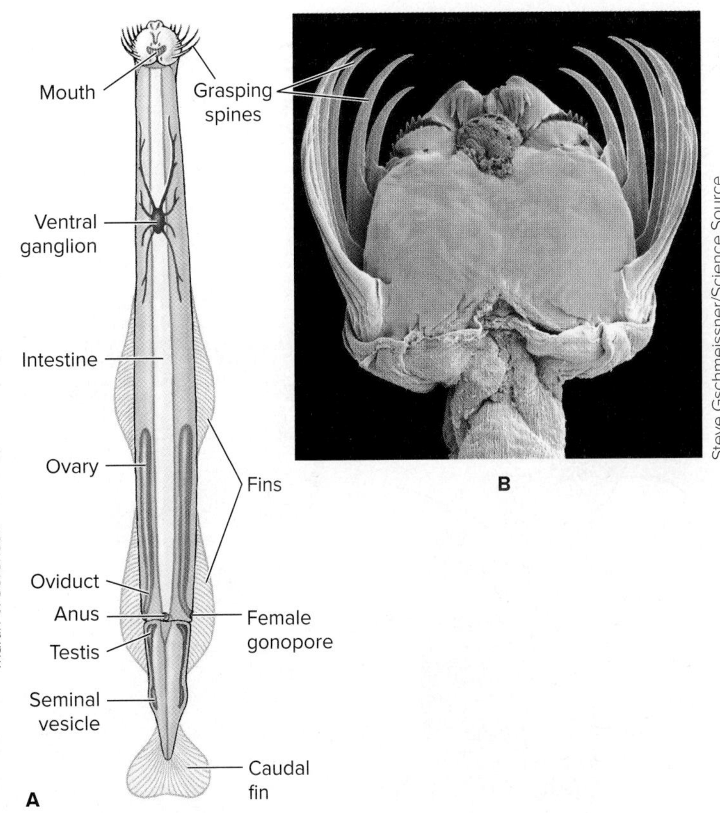

figure 8.22

Arrow worms. **A,** Internal structure of *Sagitta*. **B,** Scanning electron micrograph of the anterior end of an arrow worm.

swim to the surface at night and descend during the day. Much of the time, they drift passively, but they can dart forward in swift spurts, using their caudal fin and longitudinal muscles—a fact that no doubt contributes to their success as planktonic predators. Horizontal fins bordering the trunk function as stabilizers during swimming. The body of an arrow worm is unsegmented and is covered with a thin cuticle. They have a complete digestive system, a well-developed coelom, eyes and sensory bristles, two ventral nerve cords and a circumesophageal brain. Respiratory and excretory systems, however, are entirely absent. Arrow worms are hermaphroditic with either reciprocal fertilization or self-fertilization. Juveniles develop directly without metamorphosis.

Chaetognath embryogenesis suggests deuterostome affinities. The mouth does not arise from the blastopore, and the coelom develops by enterocoely. Early descriptions of chaetognath cleavage characterized it as radial, complete, and equal—but subsequent studies dispute this description, finding instead that the cleavage planes in four-cell embryos are similar to those of crustaceans and nematodes. The structure of the nervous system is similar to that of other protostomes. Some phylogenetic studies indicate that chaetognaths branched before the protostome-deuterostome split, and one study using mitochondrial gene order places chaetognaths with deuterostomes. However, a very recent study using newly generated molecular data supports the present placement of chaetognaths within Gnathifera.

8.8 Phylum Micrognathozoa

The only known micrognathozoan species, *Limnognathia maerski* (figure 8.23), was collected from Greenland in 1994 but not formally described until 2000. Micrognathozoans are tiny, interstitial animals (living between sand grains) about 142 μm long. The body consists of a two-part head, a thorax, and an abdomen with a short tail. The cellular epidermis has dorsal plates but lacks plates ventrally. These animals move using cilia and also possess a unique ventral ciliary adhesive pad that produces glue. They have a very complicated jaw system of plates and associated teeth; the mouth leads into a relatively simple gut. An anus opens to the outside only periodically. Only female reproductive organs have been found, so reproduction is not well understood.

8.9 Phylum Rotifera

Rotifera (rō -tif'e-rə) (L. *rota*, wheel, + *fero*, to bear) derive their name from their characteristic ciliated crown, or **corona,** which, when beating, often gives the impression of rotating wheels. Rotifers range in size from 40 μm to 3 mm in length, but most are between 100 and 500 μm long. Some have beautiful colors, although most are transparent. Some have odd or bizarre shapes, and their shapes are often correlated with their mode of life. Floaters are usually globular and saclike; creepers and swimmers are somewhat elongated and wormlike; and sessile types are commonly vaselike, with a cuticular envelope. Some are colonial.

Rotifers are a cosmopolitan group of about 2000 species, some of which occur throughout the world. Most species are freshwater inhabitants, a few are marine, some are terrestrial, and some are epizoic (living on the surface of other animals) or parasitic. Most often they are benthic, occurring on submerged vegetation in ponds or shallow margins of freshwater lakes.

The body is usually composed of a head, a trunk, and a foot. The head region bears the corona. The corona may form

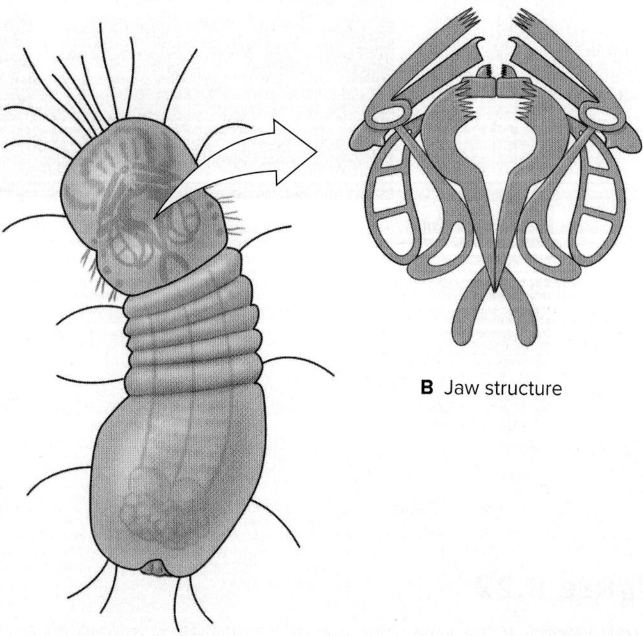

A *Limnognathia maerski*, a micrognathozoan

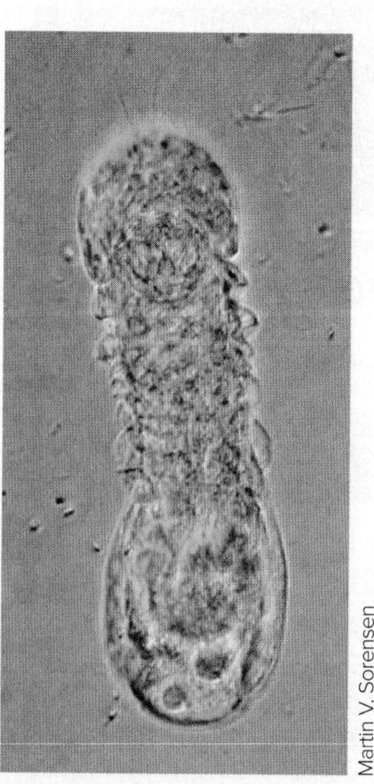

C

figure 8.23

A, *Limnognathia maerski*, a micrognathozoan. **B,** Detail of complex jaws. **C,** A living specimen. This animal was found on moss in a freshwater spring on Disko Island, Greenland. It swims, or crawls, consuming bacteria, blue-green algae, and diatoms.

a ciliated funnel, with its upper edges folded into lobes bearing bristles, or the corona may comprise a pair of ciliated discs (figure 8.24). The cilia create currents of water toward the mouth, bringing small planktonic forms for food. The corona may be retractile. The mouth, surrounded by some part of the corona, opens into a modified muscular pharynx called a **mastax,** which is unique to rotifers. The mastax has a set of intricate jaws used for grasping and chewing. The trunk contains visceral organs. A terminal foot, when present, is segmented and, in some, ringed with joints that can telescope to shorten. The one to four toes secrete a sticky substance from the pedal glands for attachment.

Rotifers have a pair of protonephridial tubules with flame bulbs, which eventually drain into a cloacal bladder that collects excretory and digestive wastes. They have a bilobed "brain" and sense organs that include eyespots, sensory pits, and papillae.

Female rotifers have one or two syncytial ovaries (**germovitellaria**) that produce yolk as well as oocytes. Although rotifers are dioecious, males are unknown in many species; in these, reproduction is entirely parthenogenetic. In parthenogenetic species, females produce only diploid eggs that have not undergone meiosis, cannot be fertilized, and develop only into females. Such eggs are called **amictic** eggs. Other species can produce two kinds of eggs—amictic eggs, which develop parthenogenetically into females, and **mictic** eggs, which have undergone meiosis and are haploid. Mictic eggs, if unfertilized, develop quickly and parthenogenetically into males. Males are haploid and make sperm via mitosis. If mictic eggs are fertilized, they secrete a thick shell and become dormant for several months before developing into females. Such dormant eggs can withstand desiccation and other adverse conditions and permit rotifers to inhabit temporary ponds.

Key Theme 8.2
SCIENCE EXPLAINED

Mictic and Amictic Eggs

Mictic (Gr., *miktos*, mixed, blended) refers to the capacity of haploid eggs to be fertilized (that is, "mixed") with the male's sperm nucleus to form a diploid embryo. Amictic ("without mixing") eggs are already diploid and can develop only parthenogenetically.

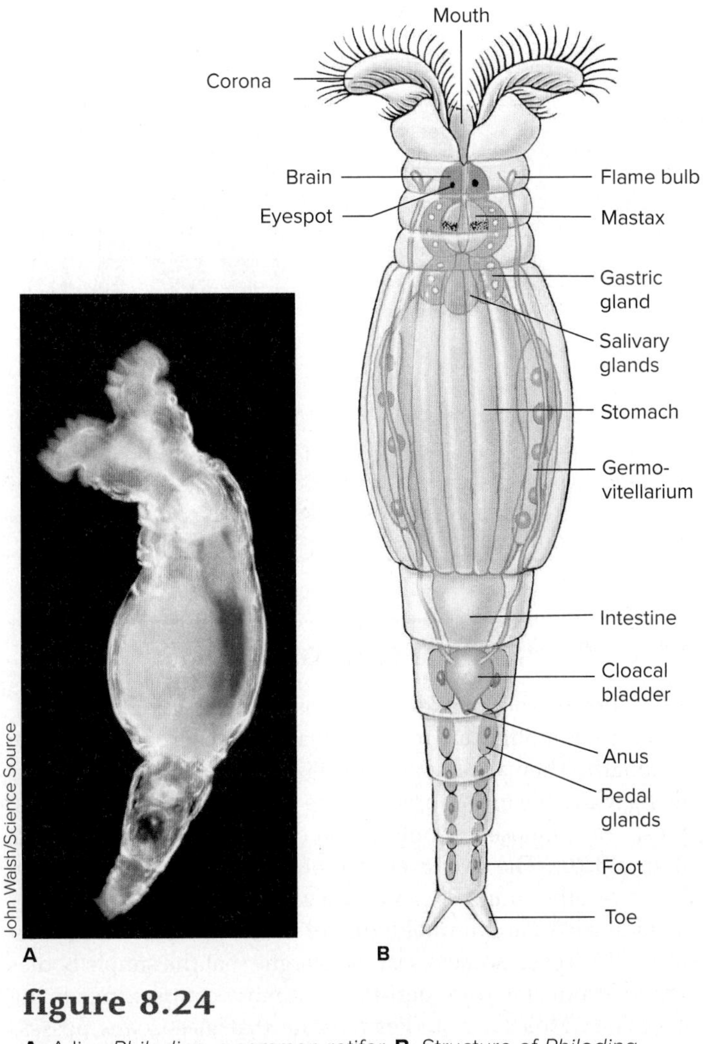

figure 8.24

A, A live *Philodina*, a common rotifer. **B,** Structure of *Philodina*.

8.10 Phylum Acanthocephala

Members of phylum Acanthocephala (a-kan′tho-sef′a-la) (Gr. *akantha,* spine or thorn, + *kephalē,* head) are commonly called "spiny-headed worms." The phylum derives its name from one of its most distinctive features, a cylindrical invaginable **proboscis** (figure 8.25) bearing rows of recurved spines, by which the worms attach themselves to the intestine of their host. All acanthocephalans are endoparasitic, living as adults in the intestines of vertebrates.

Over 1100 species of acanthocephalans are known. Most of them parasitize fishes, birds, and mammals, and the phylum is worldwide in distribution. No species is normally a parasite of humans, although rarely humans are infected with species that normally use other hosts.

Various species of acanthocephalans range in length from less than 2 mm to over 1 m, with the females of a species usually larger than the males. In life, the body is normally bilaterally flattened, with numerous transverse wrinkles. These worms are usually cream color, but may be yellowish or brown due to absorption of pigments from intestinal contents.

The body wall is syncytial, and its surface is punctuated by minute crypts 4–6 μm deep, which greatly increase the surface area of the tegument. About 80% of the thickness of the tegument is the radial fiber zone, which contains a **lacunar system** of ramifying, fluid-filled canals (figure 8.25). The function of the lacunar system is unclear, but it may help distribute nutrients to the peculiar, tubelike muscles in the body wall of these organisms.

Excretion is across the body wall in most species. One family has a pair of **protonephridia** with flame cells that unite to form a common tube opening into the sperm duct or uterus.

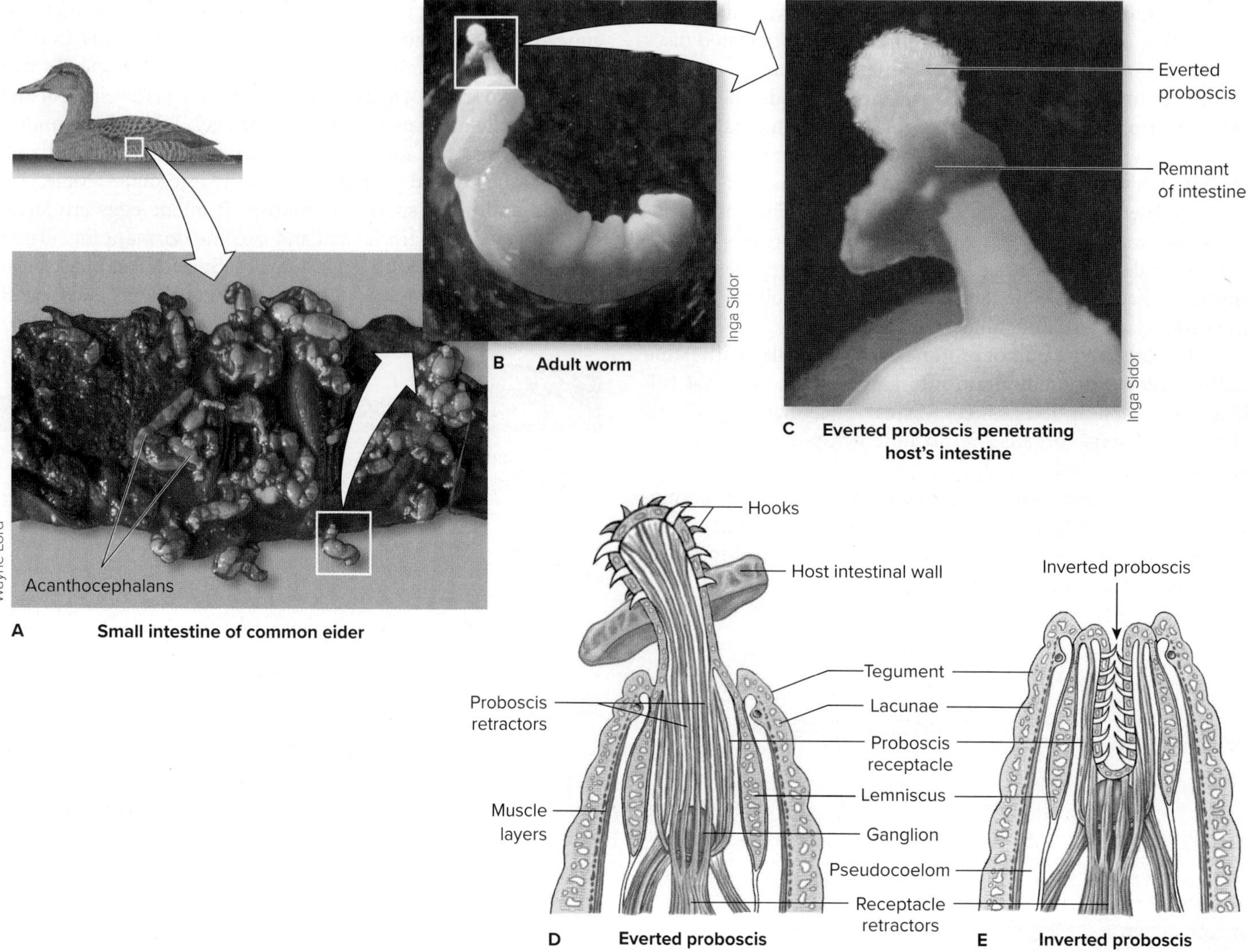

figure 8.25

A, Intestine of an eider duck filled with a lethal number of acanthocephalan worms, *Polymorphus botulus*. Prior to death, the duck fed on shore crabs, an intermediate host for the endoparasites, due to absence of blue mussels, the preferred prey for eider ducks on Cape Cod. **B,** An adult worm. **C,** An everted proboscis, showing the trait for which spiny-headed worms are named. The proboscis attaches to the intestinal wall; nutrients are absorbed across the tegument. **D,** Diagrammatic longitudinal section of an everted proboscis showing muscles. **E,** Diagrammatic longitudinal section of an inverted proboscis within the pseudocoelom.

Because acanthocephalans have no digestive tract, they must absorb all nutrients through their tegument. They can absorb various molecules by specific membrane transport mechanisms, and their tegument can perform pinocytosis (cell drinking).

Sexes are separate. Males have a protrusible penis, and at copulation sperm travel up the genital duct and escape into the pseudocoelom of females. Zygotes develop into shelled acanthor larvae. Shelled larvae escape from the vertebrate host in feces, and if eaten by a suitable arthropod, they hatch and work their way into the hemocoel, where they grow to become juvenile acanthocephalans. At that point, either their development ceases until the arthropod is eaten by a suitable host, or they pass through several transport hosts in which they encyst until eaten by the definitive host.

8.11 Phylum Mesozoa

Members of the phylum Mesozoa (Gr. *mesos,* in the middle, + *zōon,* animal) are minute, ciliated animals only 0.5–7 mm in length. They are highly specialized parasites or symbionts living in marine invertebrates; some live in cephalopod kidneys. Most are composed of only 20–30 cells arranged in two layers (figure 8.26). The two layers are not homologous to the germ layers of other animals, and mesozoan development does not include gastrulation, unlike that of other diploblasts or triploblasts. However, some researchers argue that the simple bodies in modern forms were derived from more complex free-living ancestors. Molecular studies indicate that mesozoans possess

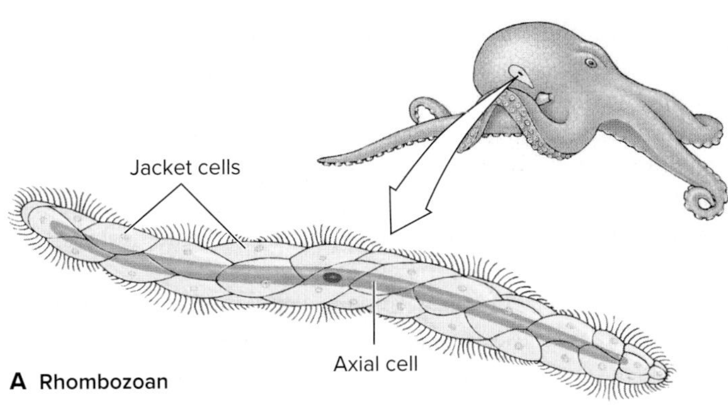

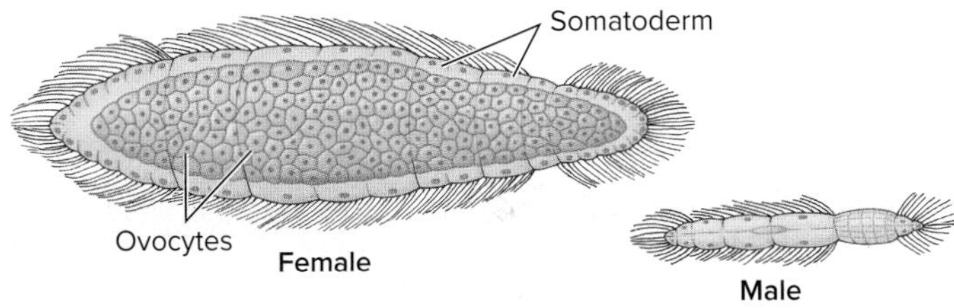

B Orthonectid (*Rhopalura*)

figure 8.26

A, Rhombozoan mesozoans live in cephalopod kidneys. **B,** Female and male orthonectid mesozoans are parasites of invertabrates such as flatworms, annelids, and molluscs. Ciliated epidermal cells surround a mass of sex cells.

some genetic and biochemical markers present in triploblastic animals. Their position on the Cladogram of All Animal Phyla inside the front cover reflects this evidence.

8.12 Phylogeny and Adaptive Diversification

Phylogeny

Examination of the Cladogram of All Animal Phyla inside the front cover shows a large clade called Bilateria (or Nephrozoa) that contains almost all the triploblastic animals. Xenacoelomorpha is the sister taxon to all other Bilateria. This taxon contains xenoturbellids and acoelomorphs, both of which share some features with flatworms and have some very unique features.[1] Acoelomorphs differ from flatworms in phylum Platyhelminthes in their embryonic cleavage patterns, in the way that their mesoderm forms, and in the structure of the nervous systems.[2] Xenoturbellida was previously placed with the deuterostome animals, but here it is united with Acoelomorpha in the phylum Xenacoelomorpha. However, both the present placement and the previous position are supported by new phylogenies that differ in the types of molecular data used (nuclear versus mitochondrial DNA). The phylogenetic position of Xenacoelomorpha remains uncertain.

Within Bilateria, evidence from sequence analysis of ribosomal genes suggests that ancestral protostomes split from ancestral deuterostomes in the Precambrian period. Protostomes later split into two large groups, sometimes called superphyla, Ecdysozoa and Lophotrochozoa. Some researchers prefer the name Spiralia for everything we depict as Lophotrochozoa; they argue that spiral cleavage is ancestral for the whole clade, whereas only some members of the group have a lophophore or trochophore larva. Evolutionary relationships within Lophotrochozoa are still in flux. The evolutionary relationship of Chaetognatha to other protostomes was not clear, but a recent study supports placement of chaetognaths within Gnathifera. Chaetognaths are much larger than other gnathiferans, but they do share a similar jaw structure with other clade members. We depict a polytomy for Gnathifera, Gastrotricha, and Platyhelminthes.

Within Platyhelminthes, class Turbellaria is clearly paraphyletic, but we are retaining the taxon because thorough cladistic analysis would require introducing many more taxa and characteristics that are beyond the scope of this book and not yet common in zoological literature. For example, ectolecithal turbellarians should be allied with trematodes, monogeneans, and cestodes in the sister group to endolecithal turbellarians. Some ectolecithal turbellarians share a number of other derived characters with trematodes and cestodes and have been placed by Brooks (1989) in a group designated Cercomeria (Gr. *kerkos,* tail, + *meros,* part) (see figure 8.4). Several synapomorphies, including the unique architecture of the tegument, indicate that neodermatans (trematodes, monogeneans, and cestodes) form a monophyletic group, and monophyly of Neodermata is supported by sequence data.

The group Gnathifera emerges in several different phylogenies, but most molecular studies do not include sequences from Micrognathozoa. Within Gnathifera, the sister-taxon relationship between Acanthocephala and Rotifera is surprising for reasons already discussed (see section 8.5). Both taxa are united as Syndermata, sharing a eutelic syncytial epidermis. Syndermata emerges from phylogenetic studies repeatedly, and several recent studies show that acanthocephalans belong within Rotifera. The classification of rotifer taxa is in flux, but a group called bdelloids is of interest. These rotifers are swimming or creeping forms, most with trochal discs at the top (see *Philodina,* figure 8.24). There are no known male bdelloids, and reproduction is by parthenogenesis. Several recent phylogenetic studies place acanthocephalans as the sister taxon to bdelloids, making the Rotifera paraphyletic as presently conceived. Acanthocephala will be subsumed within phylum Rotifera if this result is confirmed. The group name Syndermata would no longer be necessary.

[1]Achatz, J. G., M. Chiodin, W. Salvenmoser, S. Tyler, and P. Martinez. 2013. The Aceola: Their kind and kinships, especially with nemertodermatids and xenoturbellids (*Bilateria incertae sedis*). *Organisms, Diversity and Evolution* **13:**267–286.

[2]Ruiz-Trillo et al. 2004. *Molecular Phylogenetics and Evolution* **33:**321–332.

Mesozoans are identified as lophotrochozoan protostomes based on molecular data. Absence of the complex body plan typically associated with protostomes may be related to their lifestyles: they are parasites and endosymbionts.

Adaptive Diversification

The flatworm body plan, with its creeping adaptation, placed a selective advantage on bilateral symmetry and further development of cephalization, ventral and dorsal regions, and caudal differentiation. Xenacoelomorphs have simple triploblastic bodies; many entirely lack a gut. Free-living worms in phylum Platyhelminthes have an extensible pharynx, a blind gut, and thin, flattened bodies. Because of their body shape and metabolic requirements, early flatworms must have been well predisposed toward parasitism and gave rise to symbiotic descendants in Neodermata. These descendants diversified greatly as parasites, and many flatworms became highly specialized for that mode of existence. Specializations include loss of the head, reduced sensory organs, and loss of the gut in tapeworms.

An ancestral form with complex cuticular jaws may have diversified into the five phyla now placed within Gnathifera. This group includes the acanthocephalans whose jaw loss is assumed to have accompanied the transition to parasitism.

■ SUMMARY

Sections	Key Concepts
8.1 Phylum Xenacoelomorpha	• Recent phylogenies place Xenacoelomorpha as the sister taxon to a clade comprising all members of Bilateria. • Xenacoelomorpha comprises two groups: Acoelomorpha and Xenoturbellida. • Acoelomorphs are among the simplest bilaterally symmetrical animals at the organ-system level. Some entirely lack a gut. • Xenoturbellida comprises six species of wormlike animals with a blind gut and two distinct body furrows. They may be large and brightly colored. • Xenoturbellids once were placed with deuterostomes, and their current placement is still under debate.
8.2 Clades within Protostomia	• Most triploblastic animals are divided between two large clades, sometimes called superphyla: Protostomia and Deuterostomia. • Protostomia is divided into two clades: Lophotrochozoa and Ecdysozoa. • Animals in this chapter are lophotrochozoans; many but not all members of this group have a lophophore feeding structure or a trochophore larva. Because these features are not in every taxon, some biologists would like to rename the group Spiralia based on shared spiral cleavage. However, some taxa do not exhibit spiral cleavage, although they may have had spirally cleaving ancestors. • We describe Platyhelminthes, Gastrotricha, Mesozoa and the five phyla of clade Gnathifera in this chapter. Phylum Chaetognatha was recently placed within Gnathifera on the basis of molecular data and the shared presence of complex jaws, joining phyla Acanthocephala, Gnathostomulida, Micrognathozoa, and Rotifera.
8.3 Phylum Platyhelminthes	• Platyhelminthes contains free-living turbellarians and three classes of parasitic forms. • Body surface of turbellarians is a cellular epithelium. • Members of Platyhelminthes outside Turbellaria are ecto- or endoparasites covered by a nonciliated, syncytial tegument having a vesicular distal cytoplasm and cell bodies beneath superficial muscle layers. • Digestion is both extracellular and intracellular in most flatworms; cestodes lack a gut and instead absorb predigested nutrients across their tegument. • Osmoregulation is by flame-cell protonephridia, and removal of metabolic wastes and respiration occur across the body wall. Except for some turbellarians, flatworms have a ladder-type nervous system with motor, sensory, and association neurons. • Most flatworms are hermaphroditic, and asexual reproduction occurs in some groups. • Class Turbellaria is a paraphyletic group with mostly free-living and carnivorous members. • Trematodes (flukes) are of several types. Digenea includes a number of important parasites of humans and domestic animals. • Digenetic trematodes have mollusc intermediate hosts and almost always a vertebrate definitive host. • Monogeneans are important ectoparasites of fishes and have a direct life cycle (without intermediate hosts). • Cestodes (tapeworms) generally have a scolex at their posterior end, followed by a long chain of proglottids, each of which contains a complete set of reproductive organs of both sexes. There is no head. • Cestodes live as adults in the digestive tract of vertebrates. Juveniles develop in a vertebrate or invertebrate intermediate host.

Sections	Key Concepts
8.4 Phylum Gastrotricha	• Gastrotrichs are tiny aquatic animals. • They have ventrally flattened bodies with bristles or scales. • They move by cilia or adhesive glands.
8.5 Clade Gnathifera	• Clade Gnathifera contains five phyla whose common ancestor is hypothesized to possess cuticular jaws with a unique microstructure. • Presence of a ciliated pharyngeal ganglion has been proposed as a new synapomorphy for the clade. • Traditionally the clade included Gnathostomulida, Micrognathozoa, Rotifera, and Acanthocephala. The phylum Chaetognatha is a recent addition to this group.
8.6 Phylum Gnathostomulida	• Gnathostomulida contains tiny, wormlike animals living among sand grains and silt. • These animals feed on bacteria and fungi. They have a blind gut.
8.7 Phylum Chaetognatha	• Members of phylum Chaetognatha are commonly called arrow worms. • They are important and effective predators of zooplankton. • Prey are captured with jaws and chitinous spines that surround the mouth.
8.8 Phylum Micrognathozoa	• Micrognathozoa consists of a single species of tiny animals living between sand grains. • These animals have three pairs of complex jaws similar to those of rotifers and gnathostomulids.
8.9 Phylum Rotifera	• Phylum Rotifera is composed of small, mostly freshwater species with a ciliated corona that creates currents of water to draw planktonic food toward the mouth. • The mouth opens into a muscular pharynx, or mastax, which is equipped with jaws. • Rotifers have complex life cycles. • Bdelloid rotifers are obligate parthenogens and no males have been found. • Rotifers are grouped with spiny-headed endoparasites known as acanthocephalans in a clade called Syndermata. This grouping is based on a shared eutelic syncytial epidermis.
8.10 Phylum Acanthocephala	• Acanthocephalans are all parasitic in the intestine of vertebrates as adults, and their juvenile stages develop in arthropods. • They have an anterior, invaginable proboscis armed with spines, which they embed in the intestinal wall of their host. • They do not have a digestive tract and so must absorb all nutrients across their tegument. • They are placed as the sister taxon to Rotifera based on a shared eutelic syncytial epidermis. • Unlike other gnathiferans, they lack jaws; as the sister taxon to rotifers, their ancestors would have possessed jaws.
8.11 Phylum Mesozoa	• Mesozoans are wormlike parasites or symbionts of marine invertebrates. • Although they have only two body layers, mesozoans are presumed to have had triploblastic protostome ancestors.
8.12 Phylogeny and Adaptive Diversification	• Ancestral protostomes split from the deuterostome lineage in the Precambrian period. • Evolutionary relationships among protostomes are in flux. • Within Platyhelminthes, class Turbellaria is paraphyletic and some of its members share morphological features with worms in the three clades of parasitic forms. • Phylum Chaetognatha was recently placed with clade Gnathifera on the basis of molecular data, but chaetognaths share internal jaws with most other gnathiferans. Acanthocephalans lack jaws and represent highly derived rotifers, according to some molecular phylogenies.

■ Review Questions

1. Why does bilateral symmetry have adaptive value for actively motile animals?
2. How does an acoelomorph flatworm differ from a free-living member of phylum Platyhelminthes?
3. Match the terms in the right column with the classes in the left column:

_____ Turbellaria	a. Endoparasitic
_____ Monogenea	b. Free-living and commensal
_____ Trematoda	c. Ectoparasitic
_____ Cestoda	

4. There is no diagnostic feature for the Platyhelminthes. How many classes share the syncytial tegument?
5. Distinguish two mechanisms by which flatworms supply yolk for their embryos. Which system is evolutionarily ancestral for flatworms, and which one is derived?
6. How do flatworms digest their food?
7. Briefly describe the osmoregulatory system and the nervous system and sense organs of Platyhelminthes.
8. Contrast asexual reproduction in Turbellaria and Trematoda.
9. Contrast the typical life cycle of a monogenean with that of a digenetic trematode.
10. Independent evolutionary loss of what important feature causes acanthocephalans and tapeworms to rely on absorption of nutrients across the integument?
11. Answer the following questions with respect to both *Clonorchis* and *Schistosoma:* (a) How do humans become infected? (b) What is the general geographical distribution? (c) What main disease conditions are produced?
12. Define each of the following with reference to cestodes: scolex, microtriches, proglottids, strobila.
13. Why is *Taenia solium* a more dangerous infection than *Taenia saginata?*
14. What are some of the adaptations that permit flukes and tapeworms to have a parasitic lifestyle?
15. What evidence indicates that the traditional class Turbellaria is paraphyletic?
16. Explain the difference between an acoelomate body and one possessing a pseudocoelom.
17. What is the normal size of rotifers; where are they found; and what are their major body features?
18. Explain the difference between mictic and amictic eggs of rotifers.
19. Under what environmental conditions might amictic eggs be more advantageous than mictic eggs?
20. What are the approximate lengths of gastrotrichs, gnathostomulids, and micrognathozoans?
21. What habitat is shared by micrognathozoans and gnathostomulids?
22. What characters unite members of clade Gnathifera?
23. What characters are used to unite rotifers and acanthocephalans as members of clade Syndermata?
24. What is eutely?
25. The evolutionary ancestry of acanthocephalans is particularly obscure. Describe some characters of acanthocephalans that challenge the hypothesis that acanthocephalans are derived rotifers.
26. Where do chaetognaths live and how do they feed?
27. Where do mesozoans live?
28. Why do you think most species of xenoturbellids were discovered only recently?

For Further Thought How could an endoparasite evolve from a free-living ancestor, and what other ways of living (for example, symbiont, commensal, ectoparasite) would you predict as part of the transition?

9

Polyzoa and Trochozoa

Cycliophora, Entoprocta, Ectoprocta, Brachiopoda, Phoronida, and Nemertea

LEARNING OBJECTIVES

Readers will be able to:

9.1 Explain the basis for grouping three phyla as Polyzoa and which group would be removed if all lophophorate phyla are grouped together.

9.2 Describe the form and habitat of cycliophorans.

9.3 Describe the form and habitat of entoprocts.

9.4 Describe the form and habitat of ectoprocts, identifying the lophophore, and distinguishing the forms of individual zooids from that of the colony.

9.5 Explain the basis for grouping five phyla with Trochozoa.

9.6 Describe the morphological features that unite two phyla within Brachiozoa.

9.7 Describe the form and habitat of brachiopods.

9.8 Describe the form and habitat of phoronids.

9.9 Describe the form and habitat of nemertean worms, and distinguish them from flatworms.

9.10 Compare and contrast embryological development among the lophophorate taxa discussed.

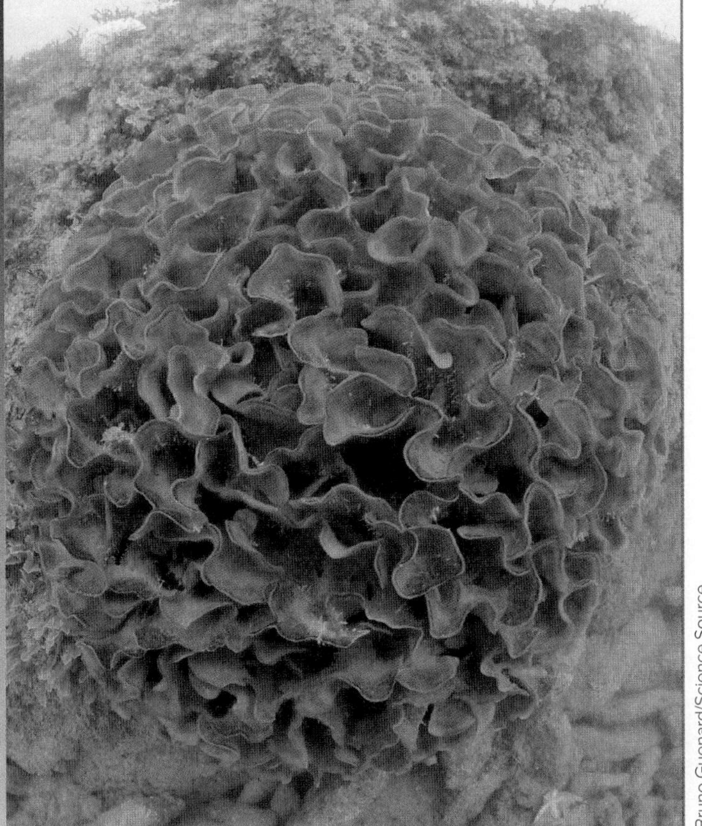

The common name of this flower-like bryozoan colony (*Pentapora foliacea*) is "rose of the sea" or "rose de mer" in its native Oléron, France.

Bruno Guenard/Science Source

Some Evolutionary Experiments

The early Cambrian period, about 570 million years ago, was the most fertile time in evolutionary history. For 3 billion years before this period, evolution had forged little more than bacteria and blue-green algae. Then, within the space of a few million years, all major phyla, and probably all smaller phyla, became established. This was the Cambrian explosion, the greatest evolutionary "bang" the world has known. In fact, the fossil record suggests that more phyla existed in the Paleozoic era than exist now, but some disappeared during major extinction events that punctuated evolution of life on earth. The greatest of these disruptions was the Permian extinction about 250 million years ago. Thus, evolution has led to many "experimental models." Some of these models failed because they were unable to survive in changing conditions. Others gave rise to abundant and dominant species and individuals that inhabit the world today. Some persisted with small numbers of species, living in very specialized habitats—for example, on the lips of lobsters. The odd shapes and unusual feeding structures of many such animals seem more in the realm of science fiction than reality.

Three phyla—Phoronida, Ectoprocta, and Brachiopoda—possess a crown of hollow ciliated tentacles, called a lophophore, that they use in food capture and respiration. Brachiopods were abundant in the Paleozoic era but began to decline thereafter. Phylum Ectoprocta arose in the Cambrian period, became widespread during the Paleozoic era, and remains a prevalent group today. Homology of the lophophore remains an intense area of research, with some molecular phylogenies supporting a return to a clade of lophophorate animals, and others supporting the relationships as we depict them.

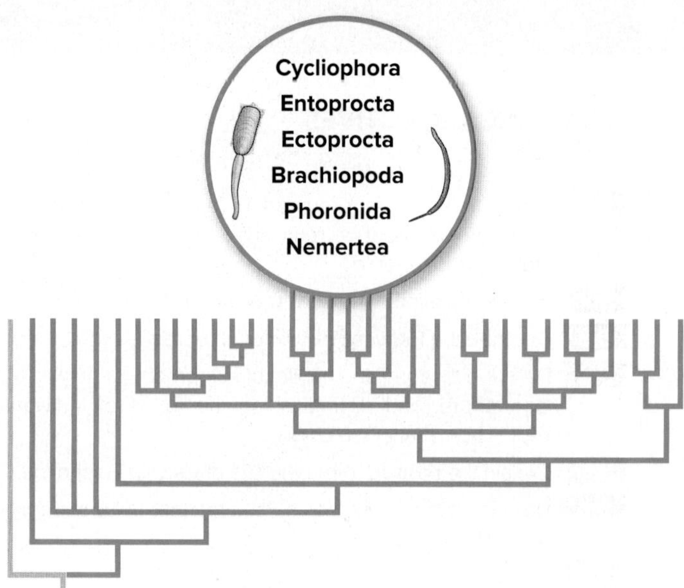

The six phyla described in this chapter are lophotrochozoan protostomes. The name "Lophotrochozoa" was developed by merging terms for two features present in many animals in this group: the lophophore and the trochophore (see figure 9.4). A **lophophore** (Gr. *lophas*, crest or tuft, + *phorein*, to bear) is a crown of tentacles covered with cilia that is borne on a ridge or fold of the body wall. It is an efficient feeding device. The cavity inside the lophophore is part of the coelom and is filled with coelomic fluid. The thin ciliated walls of the lophophore act as a respiratory surface for exchange of gases between environmental water and coelomic fluid. A lophophore can usually be extended for feeding and respiration and withdrawn for protection.

9.1 Lophophorate Phylogeny and Clade Polyzoa

Three animal phyla possess a lophophore: Ectoprocta, Brachiopoda, and Phoronida. Does the common possession of a lophophore indicate that these three phyla form a monophyletic group? Some evidence suggests not. Identifying homologous structures requires attention to detail and a detailed analysis of lophophore structure and function[1] indicates that the lophophore evolved twice, once in Ectoprocta and once in the common ancestor of Brachiopoda and Phoronida. These issues are discussed further in section 9.10.

Brachiopoda and Phoronida are united in a clade called Brachiozoa (figure 9.1). The sister taxon to Brachiozoa differs among phylogenetic analyses, but we depict Nemertea in this position. Nemerteans are very thin, stretchy, marine worms with an unusual way of capturing prey (see section 9.9). The three taxa are nested in a larger clade called Trochozoa (figure 9.1). The remaining members of Trochozoa, Mollusca, and Annelida are discussed in Chapters 10 and 11. The term "troch" in the name Trochozoa refers to part of a trochophore larva. A **trochophore** is a free-swimming, feeding, larval stage having a ring of large ciliated cells in front of the mouth (see figure 10.6). This ring of cells is called a prototroch and is used for locomotion primarily, although it may play a role in feeding. The trochophore stage is obvious in the development of molluscs, annelids, and sipunculans, among other taxa. The trochophore is assumed to be homologous, but the developmental patterns, molecular characteristics, and morphologies of the animals covered here are all under intense study, so the favored evolutionary relationships of lophotrochozoans may change.

The clade Polyzoa unites cycliophorans with entoprocts and ectoprocts, also called bryozoans (see figure 9.1). It is supported by some phylogenetic studies using multiple genes, but contrasting studies place ectoprocts with members of Brachiozoa. A close relationship between Ectoprocta and Entoprocta was proposed on morphological grounds 40 years ago. Cycliophora was not discovered until 1995, so it could not have been included in those early studies, but similarities to entoprocts were noted when the animals were first described. All three phyla have ciliated tentacles, but there are structural differences among them. The tentacles of ectoprocts contain a coelomic cavity and thus are hollow, whereas those of entoprocts are solid. The tentacles of cycliophorans are compound cilia. In a typical cilium or flagellum, there is one axoneme (column of microtubules; see figure 5.9) enclosed by an extension of the plasma membrane, but in compound cilia, multiple axonemes are enclosed within one extension of the plasma membrane. The stiff comb plates or ctenes in ctenophores (see figure 7.32D) are also compound cilia.

9.2 Phylum Cycliophora

In December 1995, P. Funch and R. M. Kristensen reported their discovery of some very strange little creatures clinging to the mouthparts of the Norway lobster (*Nephrops norvegicus*). The animals were tiny, only 0.35 mm long and 0.10 mm wide, and did not fit into any known phylum. They were named *Symbion pandora*, the first members of phylum Cycliophora (figure 9.2). Two other species have since been found on other species of lobsters, but they have not been formally described.

Cycliophorans have a very specialized habitat: they live on mouthparts of marine decapod crustaceans in the Northern Hemisphere. They attach to bristles on the mouthparts with an adhesive disc on the end of an acellular stalk. They feed by collecting bacteria, or bits of food dropped from their lobster host, on a ring of compound cilia that surrounds the mouth.

[1]Nielsen, C. 2002. The phylogenetic position of Entoprocta, Ectoprocta, Phoronida, and Brachiopoda. *Integrative and Comparative Biology* **42**:685–691.

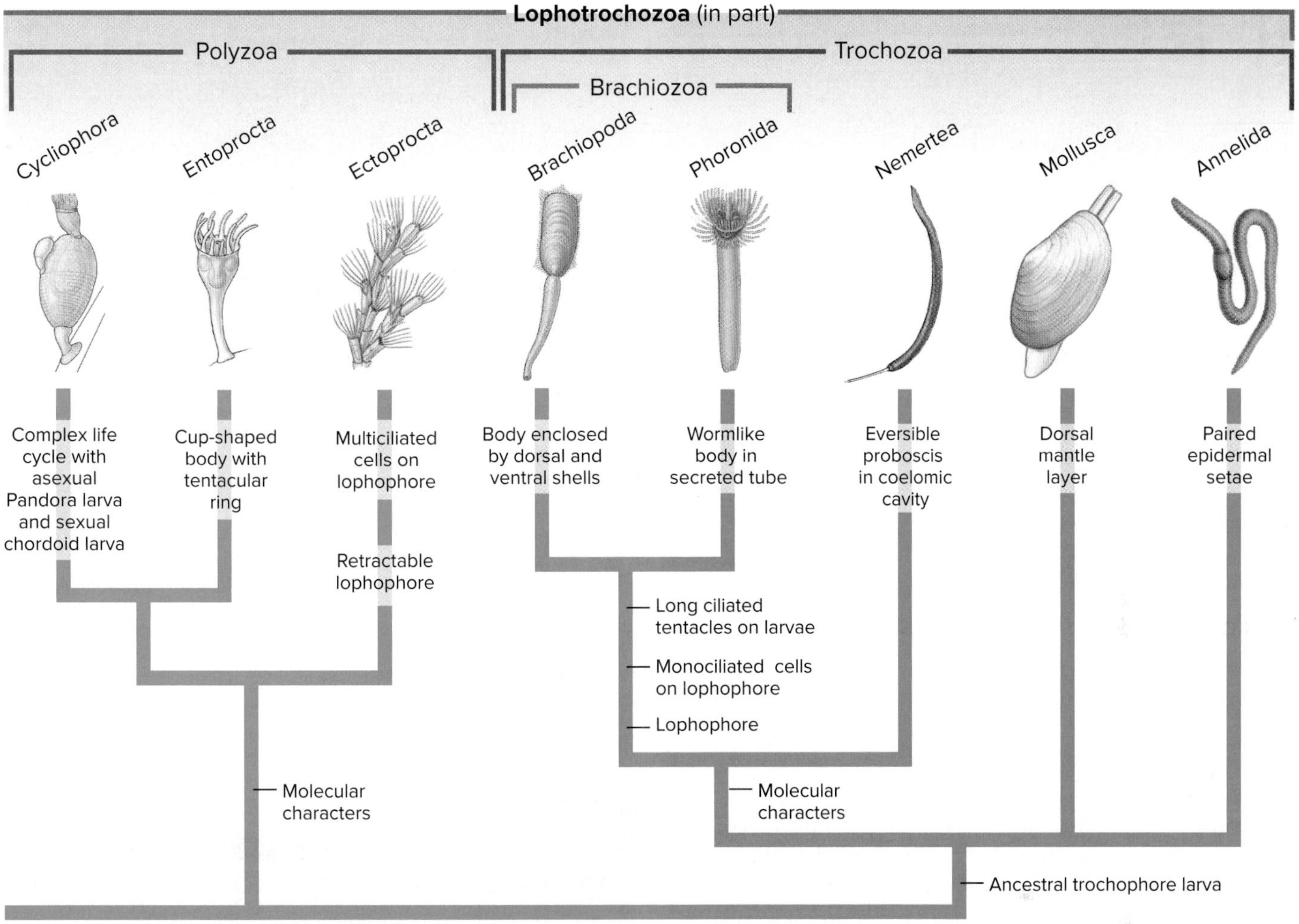

figure 9.1

Hypothetical relationships among polyzoans and trochozoans. Characters are modified subsets of those in Nielsen, C. 2002. The phylogenetic position of Entoprocta, Ectoprocta, Phoronida, and Brachiopoda. *Integrative and Comparative Biology* **42:**685–691; Brusca, R. C. and G. J. Brusca. 2003. *Invertebrates*. ed. 2. Sunderland, MA: Sinauer Associates, Inc.

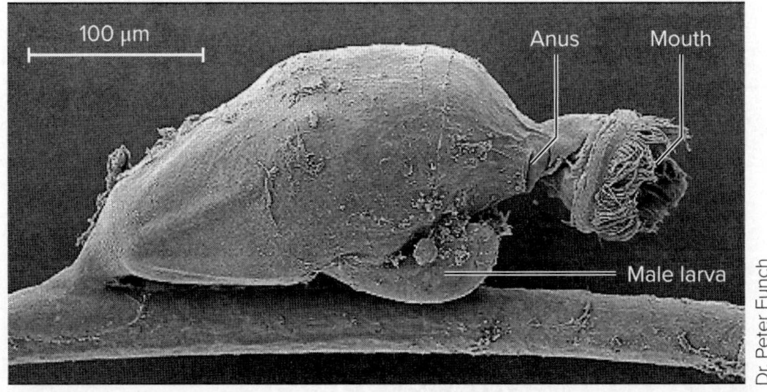

figure 9.2

Symbion pandora, a cycliophoran living on setae on the mouthparts of lobsters. A male larva is attached to the female stage.

The body plan is relatively simple. The mouth leads into a U-shaped gut ending with an anus that opens outside the ciliated ring (figure 9.2). The body is acoelomate. The epidermis is cellular and surrounded by a cuticle.

The cycliophoran life cycle has complex sexual and asexual phases. Asexual reproduction permits a clone quickly to fill vacant space on lobster mouthparts. As a prelude to sexual reproduction, male larvae are made. Male larvae attach to feeding stages that contain female larvae (figure 9.2). After further asexual reproduction of males, sexual reproduction occurs. Sexually produced larvae swim to new lobster hosts, where they form clones.

■ 9.3 Phylum Entoprocta

Entoprocta (en´to-prok´ta) (Gr. *entos*, within, + *proktos*, anus) is a small phylum containing less than 100 species of tiny, sessile animals that, superficially, look much like hydroid cnidarians, except that their tentacles are ciliated and tend

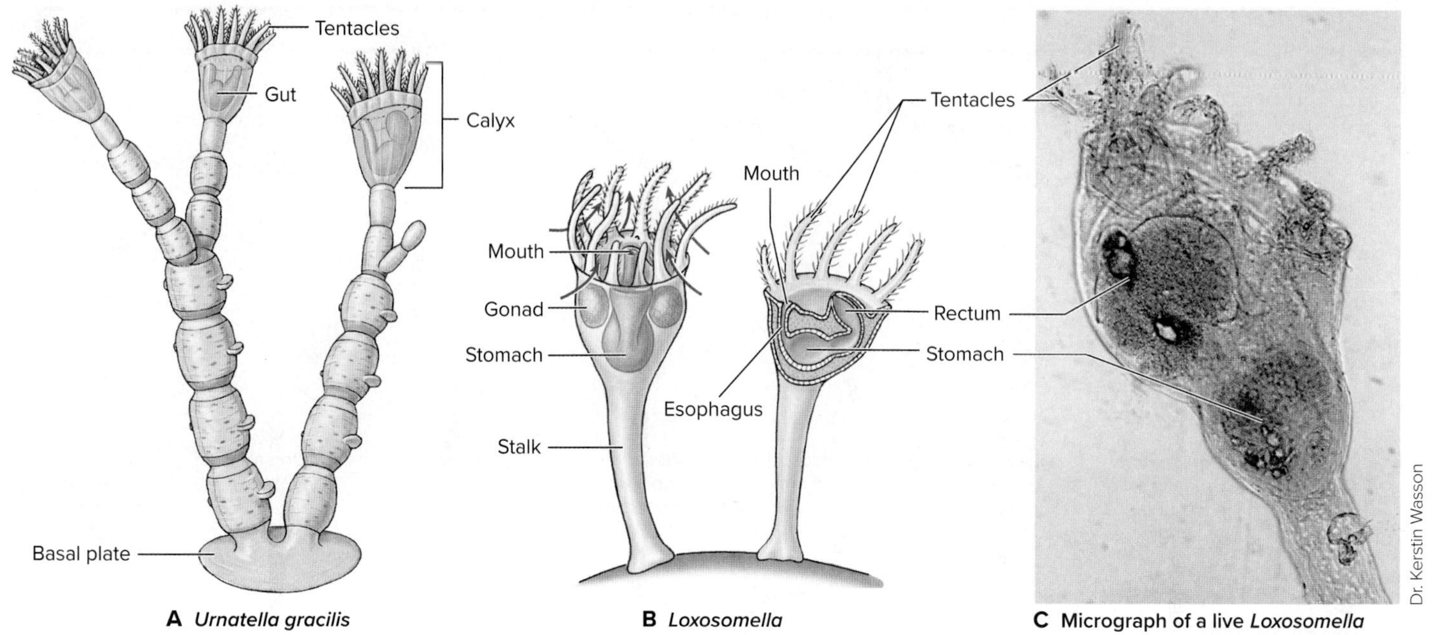

figure 9.3

A, *Urnatella*, a freshwater entoproct, forms small colonies of two or three stalks from a basal plate. **B,** *Loxosomella*, a solitary entoproct. Both solitary and colonial entoprocts can reproduce asexually by budding, as well as sexually. **C,** A live *Loxosomella*.

to roll inward (figure 9.3). Most entoprocts are microscopic, and none is more than 5 mm long. They are all stalked and sessile; some are colonial, and some are solitary. All are ciliary feeders.

With the exception of one genus, all entoprocts are marine forms that are widely distributed from polar regions to the tropics. Most marine species are restricted to coastal and brackish waters and often grow on shells and algae. Some are commensals on marine annelid worms. Freshwater entoprocts (figure 9.3) occur on the undersides of rocks in running water.

The body of an entoproct is cup-shaped, bears a crown or circle of ciliated tentacles, and attaches to a substrate by a stalk in solitary species. In colonial species, the stalks of individuals connect to the stolon of the colony. Both tentacles and stalk are continuations of the body wall.

The gut is U-shaped and ciliated, and both the mouth and anus open within the circle of tentacles, as expected from the meaning of the phylum name. The organism captures food particles in the current created by tentacular cilia and then passes these particles along the tentacles to the mouth. Entoprocts have a pair of protonephridia, but no circulatory or respiratory organs.

Some species are monoecious, some are dioecious, and others seem to be hermaphrodites in which the gonad at first produces sperm and then eggs (a condition described as *protandrous*). Cleavage is modified spiral and mosaic, and a trochophore-like larva is produced, similar to that of some molluscs and annelids (see figure 10.6).

Key Theme 9.1

SCIENCE EXPLAINED

A Pioneering Zoologist

Libbie Henrietta Hyman (1888–1969), author of the monumental multi-volume treatise, *The Invertebrates*, was a distinguished American zoologist. She earned all her academic degrees from the University of Chicago, where she received her Ph.D. in 1915 and remained doing research until 1931. She published a laboratory manual for elementary zoology in 1919 and one for comparative vertebrate anatomy in 1922. The latter remained popular for the next 50 years, allowing Hyman to live modestly on proceeds from her books while pursuing her first love, the study of invertebrates. She became a research associate at the American Museum of Natural History and published 135 research papers, chiefly on turbellarians, as well as the treatise on invertebrates for which she is best remembered. Her six-volume treatise covers the protozoan groups through molluscs to gastropods. Thousands of invertebrate zoologists owe Hyman a debt of gratitude.

9.4 Phylum Ectoprocta

Ectoprocta (ek´to-prok´ta) (Gr. *ektos*, outside, + *proktos*, anus) have long been called bryozoans (Gr. *bryon*, moss, + *zōōn*, animal), or moss animals, a term that originally included Entoprocta also.

Of the 4000 or so species of ectoprocts, few are more than 0.5 mm long. All are aquatic, both freshwater and marine, but they largely occur in shallow waters. With very few exceptions,

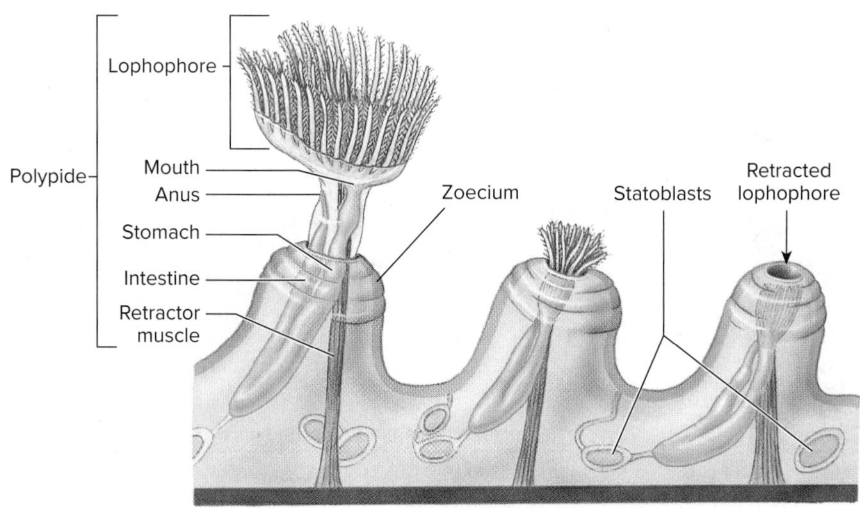

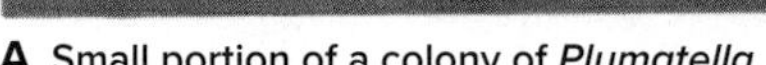
A Small portion of a colony of *Plumatella*

B Freshwater bryozoan *Plumatella repens*

figure 9.4

A, Small portion of freshwater colony of *Plumatella* (phylum Ectoprocta), which grows on the underside of rocks. These tiny individuals disappear into their chitinous zoecia when disturbed. **B,** *Plumatella repens*, a freshwater ectoproct. It grows on the undersides of rocks and vegetation in lakes, ponds, and streams.

most are colony builders. Ectoprocts, unlike most other phyla considered in this chapter, were abundant and widespread in the past and remain so today. They have left a rich fossil record since the Ordovician period. Modern marine forms exploit all kinds of firm surfaces, such as shells, rock, large brown algae, mangrove roots, and ship bottoms. As one of the most important groups of fouling organisms on boat hulls, they decrease the efficiency of a hull passing through the water and make periodic scraping of the hull necessary.

Key Theme 9.2
ECOLOGY

Invasive Bryozoans

One bryozoan species, *Tricellaria inopinate*, is spreading through the northeastern Atlantic Ocean and the Mediterranean Sea, displacing native bryozoans. Although it began to decline naturally in Venice, native byrozoans did not reestablish. Like many invasive species, *T. inopinate* reproduces throughout the year and disperses well. Cold temperatures appear to limit its northernmost distribution. In 2018, Alaskan citizens noted a new bryozoan, *Bugula neritina*, likely brought in by cruise ships. *B. neritina* is spreading rapidly and being monitored for effects on the local community.

Each member of a colony occupies a tiny chamber, called a **zoecium**, which is secreted by its epidermis (figure 9.4). Each individual, or **zooid**, comprises a feeding **polypide** and a case-forming **cystid**. Polypides include a lophophore, digestive tract, muscles, and nerve centers. Cystids are the body wall of the animal, together with its secreted exoskeleton, which is called a zoecium. A zoecium may, according to the species, be gelatinous, chitinous, or stiffened with calcium and possibly also impregnated with sand. Its shape may be boxlike, vaselike, oval, or tubular.

Some ectoproct colonies form limy encrustations on seaweed, shells, and rocks (figure 9.5); others form fuzzy or shrubby growths or erect, branching colonies that look like seaweed. Some ectoprocts might easily be mistaken for hydroids but can be distinguished under a microscope by the presence of ciliated tentacles (figure 9.4B). In some freshwater forms, individuals are borne on finely branching stolons that form delicate tracings on the underside of rocks or plants. Other freshwater ectoprocts are embedded in large masses of gelatinous material. Although zooids are minute, colonies may be several centimeters in diameter; some encrusting colonies may be a meter or more in width, and erect forms may reach 30 cm or more in height.

A polypide lives a type of jack-in-the-box existence, popping up to feed and then quickly withdrawing into its little chamber, which often has a tiny trapdoor (operculum) that shuts to conceal its inhabitant. To extend the tentacular crown, certain muscles contract, which increases hydrostatic pressure within the body cavity and pushes the lophophore out. Other muscles contract to withdraw the crown to safety with great speed.

Lophophore ridges tend to be circular in marine ectoprocts and U-shaped in freshwater species. When feeding, an animal extends its lophophore, and its tentacles spread out to a funnel shape (figure 9.4B). Cilia on the tentacles draw water into the funnel and out between tentacles. Food particles trapped in mucus in the funnel are drawn into the mouth by pumping of the muscular pharynx and by action of cilia in the pharynx. The digestive tract is U-shaped with the mouth opening inside the lophophore ring and the anus opening outside the ring (see figure 9.4).

Respiratory, vascular, and excretory organs are absent. Gas exchange occurs through the body surface, and since

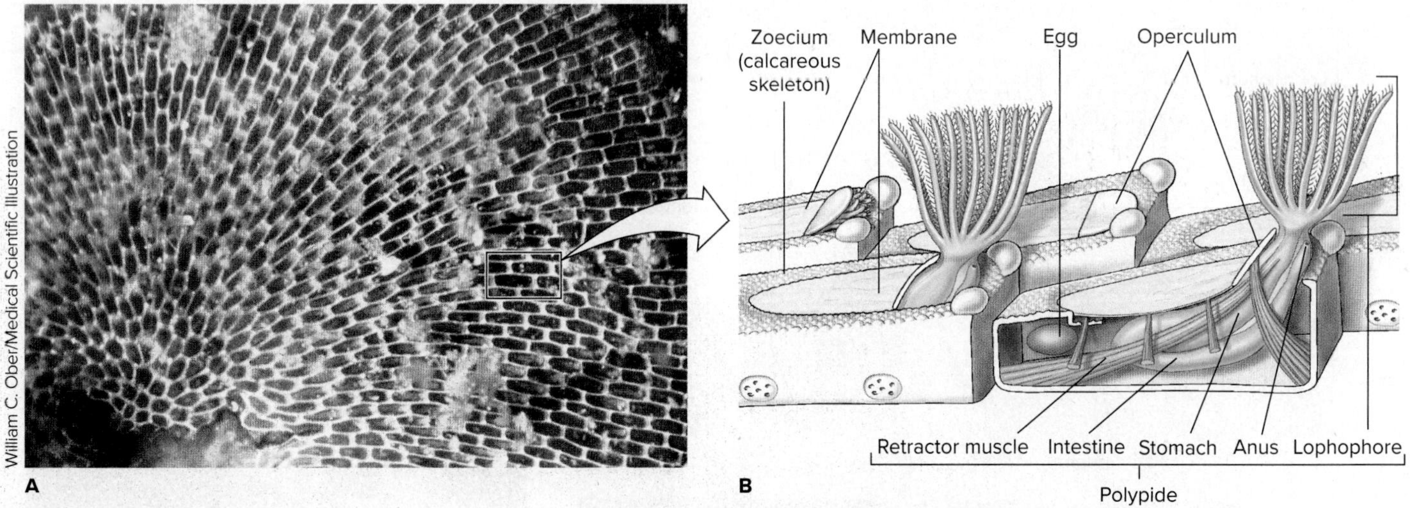

figure 9.5

A, A colony of *Membranipora*, a marine encrusting form of Ectoprocta. Each little oblong zoecium is the calcareous home of a tiny ectoproct. **B,** Each zoecium holds one polypide. To feed, each polypide lifts its operculum and extends its lophophore.

ectoprocts are small, coelomic fluid is adequate for internal transport. Coelomocytes (cells in coelomic fluid) engulf and store waste materials. A ganglionic mass and a nerve ring surround the pharynx, but no sense organs are present. A septum divides the coelom into an anterior region in the lophophore and a larger posterior area. A third coelomic region is present only in freshwater ectoprocts. Pores in the walls between adjoining zooids permit exchange of materials by way of coelomic fluid.

Feeding individuals dominate most colonies, but polymorphism also occurs. One type of modified zooid resembles a bird beak that snaps at small invading organisms that might foul a colony. Another type has a long bristle that sweeps away foreign particles.

Most ectoprocts are hermaphroditic. Some species shed eggs into seawater, but most brood their eggs, some within the coelom and some externally in a special zoecium in which embryos develop. Cleavage is radial but apparently determinate.

Brooding is often accompanied by degeneration of the lophophore and gut of adults, the remains of which contract into minute dark balls, or **brown bodies**. Later, new internal organs regenerate in old chambers. Brown bodies may remain in the body or be eliminated by the new digestive tract—an unusual kind of excretion.

Freshwater species reproduce both sexually and asexually. Asexual reproduction is by budding or by means of **statoblasts**, which are hard, resistant capsules containing a mass of germinative cells that form during the summer and fall (see figure 9.6). When a colony dies in late autumn, statoblasts are released, and in spring they can give rise to new polypides and eventually to new colonies.

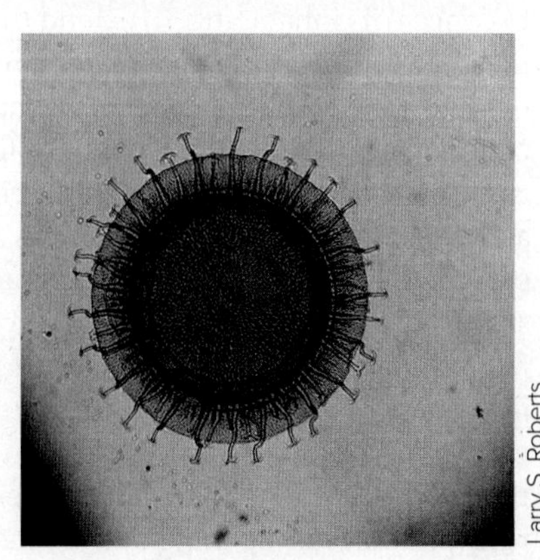

figure 9.6

Statoblast of a freshwater ectoproct, *Cristatella*. Statoblasts are a kind of bud that survives over winter when a colony dies in the autumn. This one is about 1 mm in diameter and bears hooked spines.

Key Theme 9.3

HUMAN CONNECTIONS

Anti-cancer Compounds from Marine Invertebrates

There are many ways to value biological diversity, but people looking for practical or economic value need go no further than the anti-cancer properties of compounds from a wide range of marine invertebrates. The bryozoan *Bugula neritina* is the source of bryostatin-1, a treatment for tumors and lymphoma. Bryostatin is actually made by a bacterial symbiont of *B. neritina* called *Candidatus endobugula sertula*, but the bacterium cannot be cultured outside the host. Other anti-cancer compounds are derived from sponges (e.g., discodermolide), corals (e.g., eleutherobin analogues), tunicates (e.g., ecteinascidin), clams (e.g., spisulosine), and sea cucumbers (e.g., calcigeroside B). The organisms just named are among the best-known marine invertebrates, so we wonder which unique compounds will be discovered within the animals discussed in this chapter. Known only to specialists, many animals covered here may have untapped potential uses. We know as little about their role in maintaining a healthy ocean (ecosystem services) as we do about their biochemical novelty.

9.5 Clade Trochozoa

Trochozoa comprises animals whose development includes a trochophore larva. Here we discuss the marine worms in Nemertea and two other taxa that form clade Brachiozoa, but molluscs and annelids, discussed in Chapters 10 and 11 respectively, are also trochozoans.

9.6 Clade Brachiozoa

The clade Brachiozoa comprises brachiopods and phoronids; both molecular characters and morphological features support this pairing (see figure 9.1). Both taxa possess a lophophore, and their larval stages have long, ciliated tentacles. Brachiopods occupy a two-part shell that superficially resembles a clam, whereas phoronids are wormlike and occupy a secreted submerged tube.

9.7 Phylum Brachiopoda

Brachiopoda (brak´ēō-podə) (Gr. *brachiōn*, arm, + *pous, podos*, foot), the lamp shells, is an ancient group. Compared with the fewer than 300 species now living, some 30,000 fossil species, which flourished in the Paleozoic and Mesozoic seas, have been described. Brachiopods were once very abundant, but they are now apparently in decline. Some modern forms have changed little from early ones. Genus *Lingula* (L., little tongue) (figure 9.7A) has existed virtually unchanged for over 400 million years. Most modern brachiopod shells range between 5 and 80 mm, but some fossil forms reached 30 cm in length.

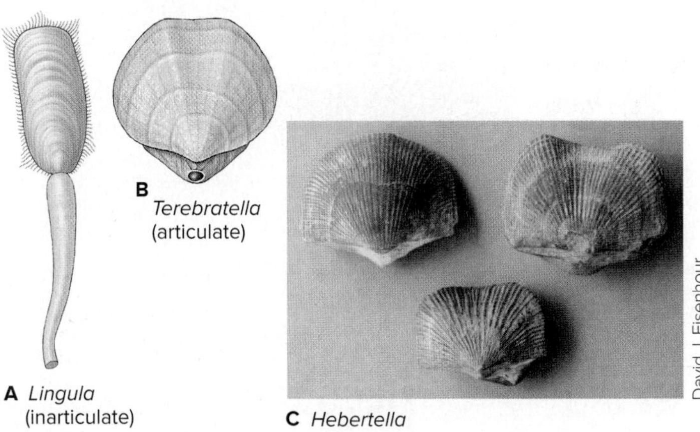

figure 9.7

Brachiopods. **A,** *Lingula*, an inarticulate brachiopod that normally occupies a burrow. Its contractile pedicel can withdraw the body into the burrow. **B,** An articulate brachiopod, *Terebratella*. Its valves have a tooth-and-socket articulation, and a short pedicel projects through one valve to attach to the substratum (pedicel shown in figure 9.8). **C,** Fossil brachiopods of the genus *Hebertella* from the late Ordovician period, northern Kentucky.

Brachiopods are all attached, bottom-dwelling, marine forms that mostly prefer shallow water. Externally, brachiopods resemble bivalve molluscs in having two calcareous shell valves secreted by a mantle. They were, in fact, grouped with molluscs until the middle of the nineteenth century, and their name refers to the arms of their lophophore, which were thought homologous to the mollusc foot. However, brachiopods have **dorsal** and **ventral valves** instead of right and left lateral valves that characterize bivalve molluscs, and also unlike bivalves, most of them are attached to a substrate either directly or by means of a fleshy stalk called a **pedicel** (or pedicle).

In most brachiopods, the ventral (pedicel) valve is slightly larger than the dorsal (brachial) valve, and one end projects in the form of a short, pointed beak that is perforated where the fleshy stalk passes through (figure 9.7B). In many, the shape of the pedicel valve is that of a classical oil lamp of Greek and Roman times, so that brachiopods came to be called "lamp shells."

There are two classes of brachiopods based on shell structure. The shell valves of Articulata are connected by a hinge with an interlocking tooth-and-socket arrangement (articular process); those of Inarticulata lack the hinge and are held together by muscles only.

The body occupies only the posterior part of the space between the valves (figure 9.8A), and extensions of the body wall form mantle lobes that line and secrete the shell. (Note that the name "mantle" reflects the fact that the brachiopods were once grouped with molluscs; this structure is not homologous to the mollusc mantle.) The large, horseshoe-shaped lophophore in the anterior mantle cavity bears long, ciliated tentacles used in respiration and feeding. Ciliary water currents carry food particles between the gaping valves and over the lophophore. Food is caught in mucus on the tentacles and carried in a ciliated food groove along an arm of the lophophore to the mouth (figure 9.8B). The digestive tract is U-shaped with the mouth inside the lophophore ring and the anus outside it (figure 9.8A).

There are only two coelomic cavities in articulates, but inarticulates have three coelomic cavities with the posterior cavity housing the viscera. One or two pairs of nephridia open into the coelom and empty into the mantle cavity. The circulatory system is open, with a contractile heart. A nerve ring with a small dorsal and a larger ventral ganglion is present.

Sexes are separate, and paired gonads discharge gametes through the nephridia. Development of brachiopods is similar in some ways to that of deuterostomes, with radial, mostly equal, holoblastic cleavage, and the coelom forms entercoely in articulates. Free-swimming larvae of articulates resemble a trochophore.

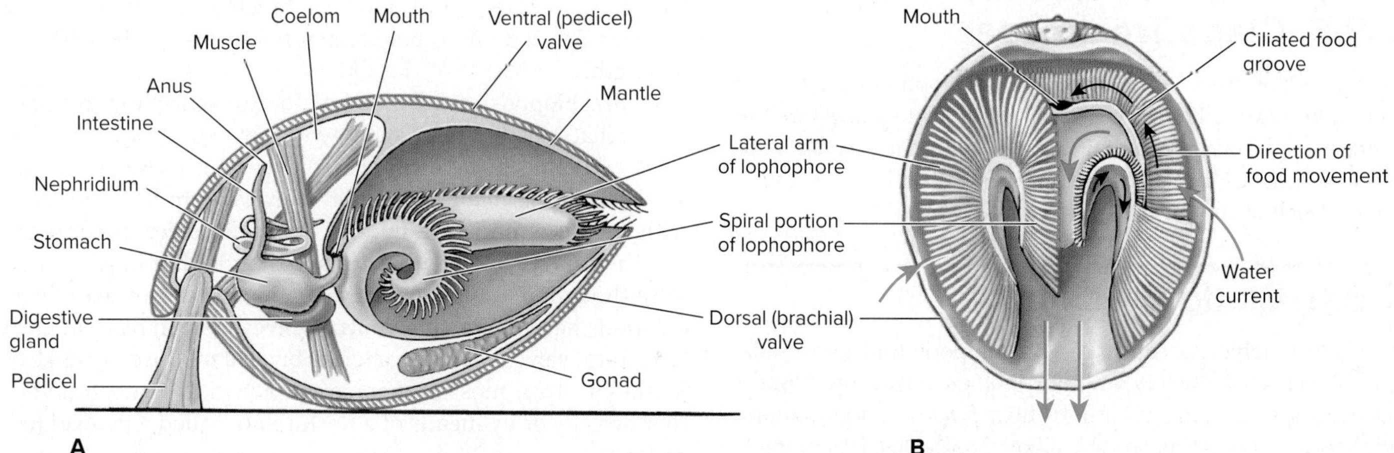

figure 9.8

Brachiopod anatomy. **A,** An articulate brachiopod (longitudinal section). **B,** Feeding and respiratory currents. Blue arrows show water flow over its lophophore; black arrows indicate food movement toward its mouth in the ciliated food groove.

9.8 Phylum Phoronida

Phylum Phoronida (fo-ron´i-də) (Gr. *phoros*, bearing, + L. *nidus*, nest) comprises approximately 10 species of small wormlike animals that live in benthic substrates under shallow coastal waters, especially in temperate seas. The phylum name refers to their tentacled lophophore. Phoronids range from a few millimeters to 30 cm in length. Each worm secretes a leathery or chitinous tube in which it lies free, but which it never leaves (figure 9.9). Their tubes may be anchored singly or in a tangled mass on rocks, shells, or pilings, or buried in the sand. Tentacles on the lophophore are thrust out for feeding, but if an animal is disturbed, it can withdraw completely into its tube.

Their lophophore has two parallel ridges curved in a horseshoe shape, the bend located ventrally and the mouth lying between the two ridges. Cilia on the tentacles direct a water current toward a groove between the two ridges, which leads toward their mouth. Plankton and detritus caught in this current become entangled in mucus and are carried by cilia to their mouth. As in brachiopoda, the mouth is inside the lophophore ring and the anus is outside.

The coelomic cavity is divided into two or three parts by mesenteries. Phoronids have a closed system of contractile blood vessels but no heart; their red blood contains hemoglobin. There is a pair of metanephridia. A nerve ring sends nerves to tentacles and the body wall.

There are both monoecious (the majority) and dioecious species of Phoronida, and at least one species reproduces asexually. Cleavage combines features of both spiral and radial types.

9.9 Phylum Nemertea (Rhynchocoela)

Nemerteans are thread- or ribbon-shaped predatory worms, often called ribbon worms (figure 9.10). Their name (Gr. *Nemertes*, one of the nereids, unerring one) refers to the unerring aim of the proboscis, a long muscular tube (figure 9.11) that can be thrust out swiftly to grasp the prey. The phylum is also called Rhynchocoela (Gr. *rhynchos*, beak, + *koilos*, hollow), which also refers to the proboscis. There are about 1300 species in the group; nearly all are marine.

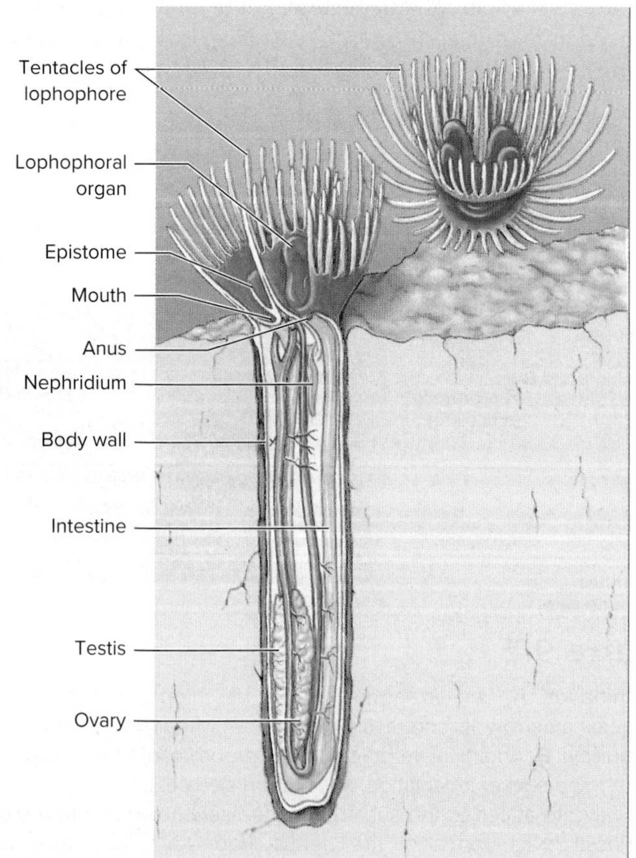

figure 9.9

Internal structure of *Phoronis* (phylum Phoronida), in diagrammatic vertical section.

figure 9.10

This nemertean, or ribbon worm, *Baseodiscus delineatus*, is from the Kewalo Basin, Honolulu, Oahu, in the Hawaiian Islands.

Form and Function

Many nemerteans are difficult to examine because they are so long and fragile. *Lineus longissimus* is the longest known animal on earth, reaching 60 m in length, although it is only 5 to 10 mm wide. *Amphiporus* (Gr. *amphi*, on both sides, + *poros*, pore), one of the smaller genera that ranges from 2 to 10 cm in length, is fairly typical of nemertean structure (figure 9.11). Its body wall consists of ciliated epidermis and layers of circular and longitudinal muscles. Most movement occurs by gliding over a slime track, although larger species move by muscular contractions.

The mouth is anterior and ventral, and the digestive tract is complete, extending the full length of the body and ending at an anus. The presence of an anus marks a significant advancement over the gastrovascular systems of flatworms and radiates, because ingestion and egestion can occur simultaneously. Cilia move food through the intestine. Digestion is largely extracellular.

Nemerteans are carnivorous, feeding primarily on annelids and other small invertebrates. They seize their prey with a

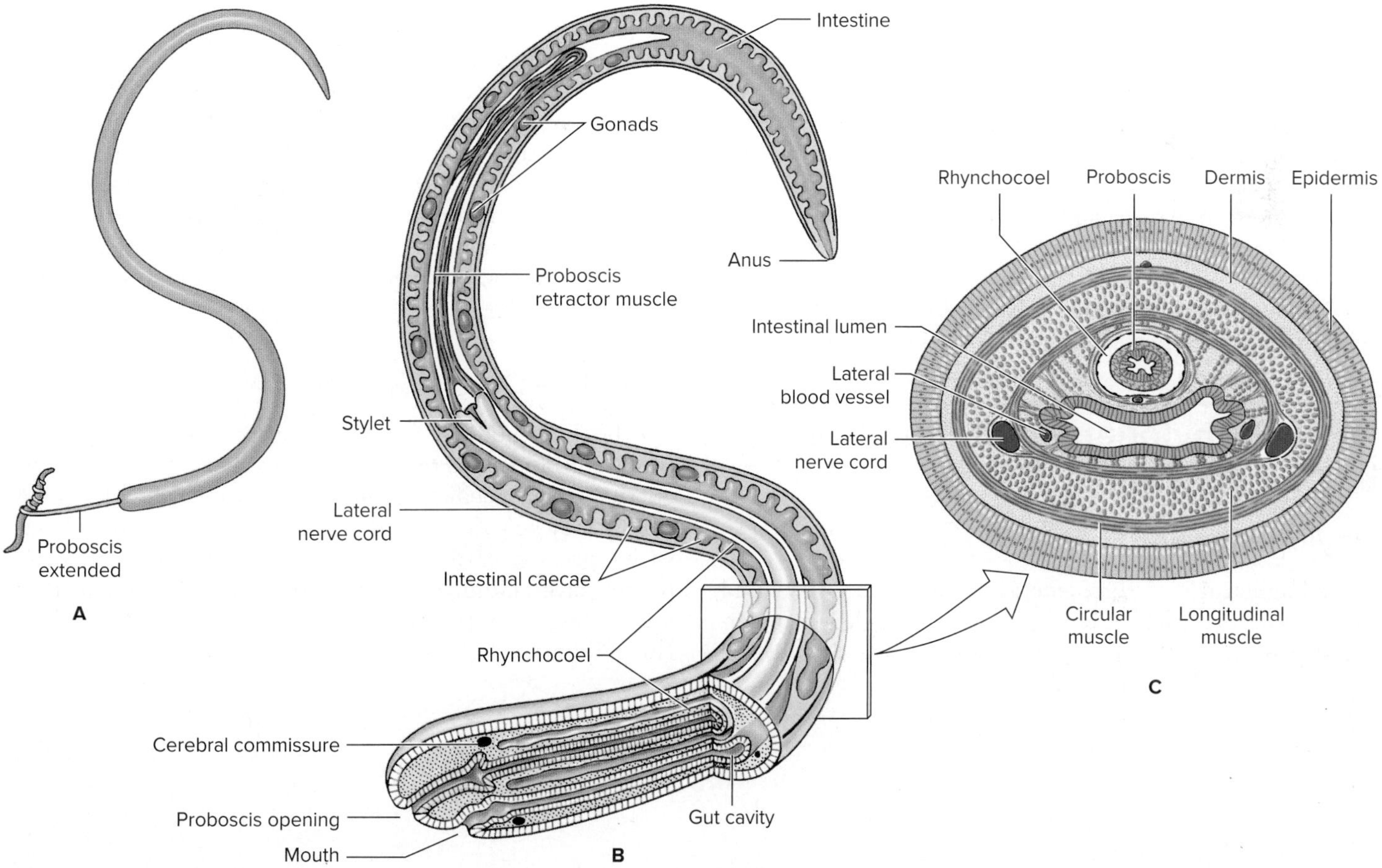

figure 9.11

A, *Amphiporus* with proboscis extended to catch prey. **B,** Internal structure of female ribbon worm (diagrammatic). Dorsal view to show proboscis. **C,** Diagrammatic cross section of female nemertean.

proboscis that lies in an interior cavity of its own, the **rhynchocoel**, above the digestive tract (but not connected with it). The proboscis itself is a long, blind muscular tube that opens at the anterior end at a proboscis pore above the mouth. (In a few nemerteans, the esophagus opens through the proboscis pore rather than through a separate mouth.) Muscular pressure on fluid in the rhynchocoel causes the long tubular proboscis to be everted rapidly through the proboscis pore. Eversion of the proboscis exposes a sharp barb, called a stylet (absent in some nemerteans). The sticky, slime-covered proboscis coils around the prey and stabs it repeatedly with the stylet, while pouring potent neurotoxins on the prey. Then, retracting its proboscis, a nemertean draws the prey near its mouth, through which the esophagus is thrust to engulf the prey.

A recent review highlights the wide range of pharmacologically active chemicals used by nemerteans, including tetrodotoxin, more commonly known from pufferfishes. Potential applications of these chemicals range from treatments for Alzheimer's disease to use in pesticides.

The rhynchocoel is a cavity entirely contained within mesoderm, so it is a true coelom. However, in all other organisms with a coelom, this fluid-filled cavity surrounds the gut. The nemertean rhynchocoel lies above the gut, rather than around it, and houses the eversible proboscis. It is not known whether the rhynchocoel is an independently derived cavity or some form of modified coelom, but most researchers support the former explanation.

Nemerteans have a true circulatory system, and an irregular flow of blood is maintained by the contractile walls of the vessels. Many flame-bulb protonephridia are closely associated with the circulatory system, so that their function appears to be truly excretory (for disposal of metabolic wastes), in contrast to their apparently osmoregulatory role in Platyhelminthes.

Nemerteans have a pair of nerve ganglia, and one or more pairs of longitudinal nerve cords are connected by transverse nerves.

Some species reproduce asexually by fragmentation and regeneration. Most nemerteans are dioecious.

9.10 Phylogeny and Adaptive Diversification

Phylogeny

Since Lophotrochozoa first emerged from phylogenies based on molecular characters, biologists have struggled to understand patterns of morphological change across this diverse group of phyla. The suite of developmental characters associated with spiral cleavage is presumed to be ancestral for the clade. This suite includes spirally cleaving embryos, mosaic cleavage where cell fates are initially determined by the cytoplasmic factors they contain, and formation of mesoderm from derivatives of a particular cell, called the 4d cell, that is present at about the 64-cell stage of cleavage. Mesoderm is derived from both endoderm (via the 4d cell) and ectoderm in most taxa with spiral cleavage. If some or all of these characters are ancestral, then there must be character loss in several taxa. For example, spiral cleavage does not occur in rotifers, gastrotrichs, bryozoans, phoronids, or brachiopods; nor does mesoderm come from the 4d cell in most of these taxa. The likelihood that these characters were lost over evolutionary time, as opposed to never having been present, cannot be determined until there is a good understanding of the branching pattern within Lophotrochozoa. Phylogenetic analysis has proved challenging because results vary depending on which genes are examined, even in a very large 2019 study. The next few years may yield resolution of this problem as large multigene datasets are developed for most or all taxa. Several studies recovered the lophophorates as a clade, and did not support Polyzoa. Details of tentacle structure differ within Polyzoa and recent studies showed significant differences between the patterns of nerves controlling the tentacles in entoprocts and ectoprocts.

Placement of Nemertea is a contentious issue. Nemerteans were once allied with Platyhelminthes because they share a ciliated epidermis and flame cell excretory structures, but nemerteans are unlike flatworms in that they have a complete digestive track and an eversible proboscis in a unique coelomic cavity. The position of the coelomic cavity above the digestive tract, rather than surrounding it, sets nemerteans apart from other coelomate animals. Recent analyses place nemerteans within Trochozoa, but their phylogenetic position within this group is unclear.

Adaptive Diversification

The boxlike modular unit of ectoprocts has proved very flexible. Zooids bud to form colonies, but a colony may be encrusting or treelike, to name just two extreme forms. A colony may be soft or calcified, and the shape of the zooids varies greatly, as does the surface ornamentation. Ectoprocts have colonized freshwater habitats, although they are far less widespread in freshwater as compared to marine habitats.

SUMMARY

Sections	Key Concepts
9.1 Lophophorate Phylogeny and Clade Polyzoa	• Polyzoa comprises three phyla of small animals that use cilia or ciliated tentacles to feed. Details of tentacle structure differ within Polyzoa. • Group members are cycliophorans, entoprocts, and ectoprocts. • Only ectoprocts have a lophophore. Tentacles in a lophophore are hollow and contain a coelomic cavity.
9.2 Phylum Cycliophora	• Cycliophorans are very tiny animals living on the setae of the mouthparts of lobsters. • Cycliophorans have complex life cycles with sexual and asexual phases.
9.3 Phylum Entoprocta	• Entoprocts are small, sessile aquatic animals with a cup-shaped body on a short stalk. • A single crown of ciliated solid feeding tentacles encircles both the mouth and anus.
9.4 Phylum Ectoprocta	• An ectoproct possesses a horseshoe-shaped crown of ciliated tentacles surrounding the mouth but not the anus. This is called a lophophore. • The lophophore functions as both a respiratory and feeding structure, its cilia creating water currents from which food particles are filtered. • Ectoprocts, also called bryozoans, are abundant in marine habitats, and a number of species are common in fresh water. • Ectoprocts are colonial, and although each individual is quite small, colonies are commonly several centimeters or more in width or height. • In a colony, each individual lives in a chamber (zoecium). A chamber is a secreted exoskeleton of chitinous, calcium carbonate, or gelatinous material. Individuals have a U-shaped digestive tract. • Ectoprocts are sessile as adults, but have a free-swimming larva.
9.5 Clade Trochozoa	• Trochozoans are animals with a trochophore larval stage in their development. • The group comprises five phyla, of which three—Brachiopoda, Phoronida, and Nemertea—are discussed in this chapter.
9.6 Clade Brachiozoa	• Brachiozoa unites phoronids and brachiopods in clade Brachiozoa. • Both taxa possess a lophophore. • Brachiopods are shelled and superficially resemble bivalve molluscs, whereas phoronids are wormlike and occupy a secreted tube.
9.7 Phylum Brachiopoda	• Brachiopods are shelled animals with an internal lophophore. • The body and lophophore are covered by a mantle, which secretes a dorsal and ventral valve (shell). Brachiopods are usually attached to the substrate directly or by a pedicel. • Brachiopods were misidentified as molluscs because of the superficial similarity of their shells, so terminology for parts such as the mantle do not imply homology with the corresponding molluscan features. • Brachiopods were very abundant in the Paleozoic era but have been declining in numbers and species since the early Mesozoic era.
9.8 Phylum Phoronida	• Phoronida are the least common lophophorates, living in tubes embedded in sediments. Most are in shallow coastal waters. • Phoronids extend the lophophore out of the tube for feeding.
9.9 Phylum Nemertea (Rhynchocoela)	• Members of Nemertea have a complete digestive system with an anus and a true circulatory system. • They are free-living and mostly marine; they ensnare prey with a long, eversible proboscis.
9.10 Phylogeny and Adaptive Diversification	• Many, but not all, lophotrochozoans share a suite of developmental characters associated with spiral cleavage. • The evolutionary branching patterns of taxa covered in this chapter are currently unclear because studies using different genes arrive at different outcomes. • There is debate over the homology of the lophophore. We depict two evolutionary events for this structure, but some phylogenies support a clade of lophophorates. • The nemerteans are predatory worms with an unusual position of the coelom. The sister taxon for this group is unclear. • Adaptive diversification is greatest in the ectoprocts, where the modular body plan may be encrusting, bushy, or tree-like. Colonies may be soft or hardened and occur in marine and freshwater habitats.

Review Questions

1. Where would you look if you had to find a cycliophoran?
2. How does an entoproct differ from an ectoproct?
3. Examine figure 9.1. How many times did the lophophore evolve? What evidence suggests that lophophorates do not form a clade?
4. Define each of the following for an ectoproct: lophophore, zoecium, zooid, polypide, cystid, brown bodies, and statoblasts.
5. Are lophophorates coelomate, acoelomate, or pseudocoelomate?
6. Brachiopods superficially resemble bivalve molluscs. How would you explain the difference to a layperson?
7. Lophophorates are sometimes placed in a phylogenetic position between protostomes and deuterostomes. How would you justify or oppose such placement?
8. In what ways are nemertean worms different from other coelomate taxa?
9. How does a nemertean catch and consume prey?

For Further Thought How would you advocate for more research into the animals discussed in this chapter? Consider economics, practical use, aesthetics, and the role of a diverse and healthy ecosystem in your answer.

10

Molluscs

LEARNING OBJECTIVES

Readers will be able to:

10.1 Describe the habitats and lifestyles typical of molluscs.

10.2 Explain the effects of ocean acidification on mollusc harvests and coastal economies.

10.3 Describe the key features of the molluscan body plan, referencing figure 10.3.

10.4 Explain in what ways caudofoveates and solenogasters are unusual molluscs.

10.5 Explain the concept of serial repetition with reference to Monoplacophora.

10.6 Describe the form and habitat of chitons.

10.7 Describe the form and habitat of scaphopods.

10.8 Describe the morphological and lifestyle variation across the gastropods.

10.9 Explain how the bivalve body plan is adapted for feeding, respiration, and reproduction.

10.10 Explain how the basic molluscan features are modified in a cephalopod.

10.11 Compare and contrast the body plan of the ancestral mollusc and the evolution of repeated structures under the Serialia and Conchifera hypotheses.

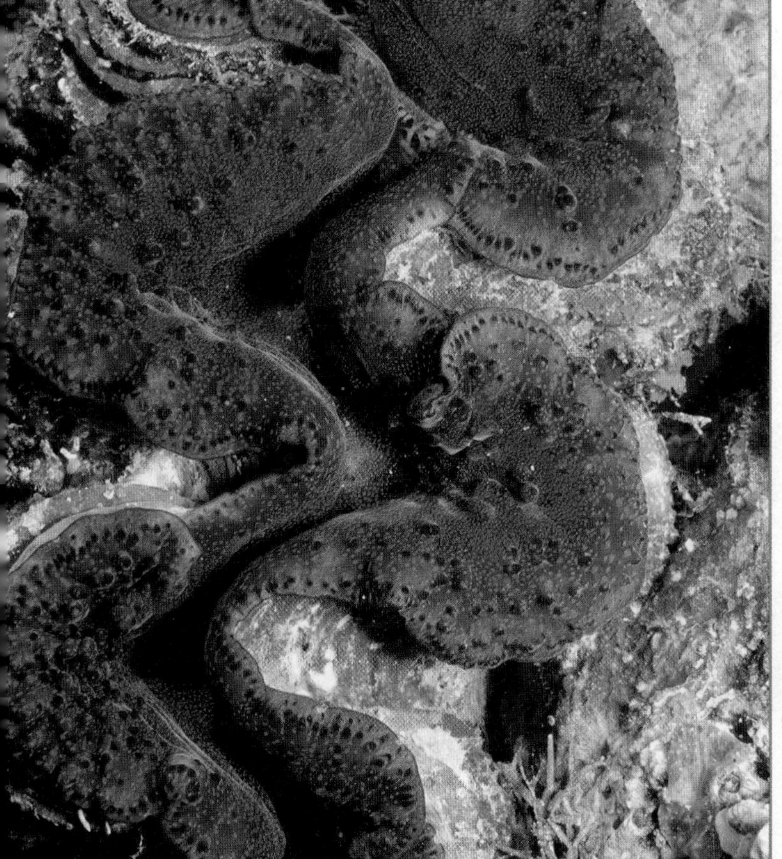

Giant clam, *Tridacna gigas*.

Comstock Images/Getty Images

A Shell Collector's Delight

Molluscs are astonishingly diverse. The group includes wormlike animals and giant squids, as well as animals with a single shell, two shells, eight shell plates, or no shell at all. Some researchers have wondered whether the molluscs really form a clade, but there are several features present in almost all molluscs that support monophyly of the group. Most molluscs have an unusual ribbon of teeth called a radula that they use to feed, often scraping algae from hard surfaces. Most also have a large muscular foot used in locomotion and a unique tissue layer called a mantle. The mantle secretes the shell and makes the respiratory and sensory organs, among other things. These common molluscan features are employed in very different ways across the eight classes of molluscs: for example, in some species the radular teeth are used to inject a paralyzing venom, whereas another group has lost the radula. The foot can be used for crawling in snails or divided into muscular arms in an octopus.

Humans have exploited this diversity in many ways. Shells were used as money on almost every continent—the money cowry is a snail whose shell was widely used as currency. There is debate as to whether the first use of shells was as currency or adornment, but shells are still used in jewelry, buttons, and decoration throughout the world. By far the most common use of molluscs is as food. We eat clams, oysters, scallops, mussels, snails, abalone, squid, and octopus, to name just a few molluscs exploited commercially. Sometimes we eat the mantle, sometimes the foot, and other times the entire body. Our dependence on these animals behooves us to be good stewards of the environment for molluscs by maintaining unpolluted riverine, coastal, and oceanic habitats and by developing sustainable harvest practices.

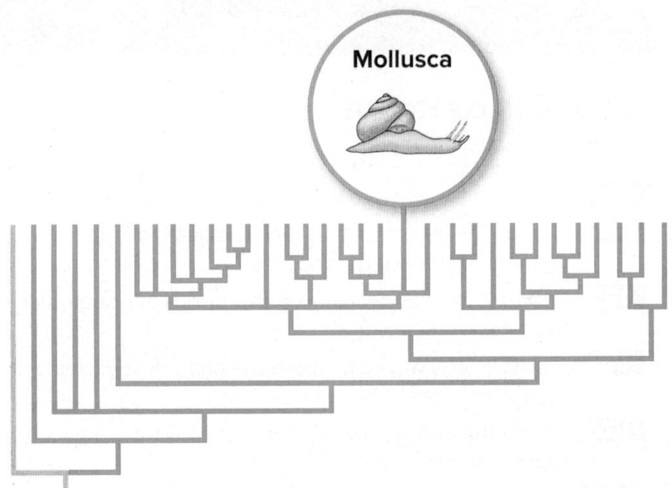

Mollusca (mol-lus´ka) (L. *molluscus,* soft) is among the largest animal phyla after Arthropoda. There are nearly 90,000 named living species and some 70,000 fossil species. Many more molluscs await formal description. The name Mollusca indicates one of their distinctive characteristics, a soft body.

This group includes organisms as disparate as chitons, snails, clams, and octopuses (figure 10.1). They range from fairly simple organisms to some of the most complex invertebrates, and from almost microscopic in size to the giant squid, *Architeuthis harveyi* (Gr. *archi,* primitive, + *teuthis,* squid). The body of this huge species may grow up to 18 m long with tentacles extended and may weigh up to 454 kg (1000 pounds). The shells of some giant clams, *Tridacna gigas* (Gr. *tridaknos,* eaten at three bites), which inhabit the Indo-Pacific coral reefs, reach 1.5 m in length and weigh over 225 kg (see figure 10.23). These are extremes, however, since probably 80% of all molluscs have a maximum shell size of less than 5 cm.

The enormous variety, great beauty, and availability of mollusc shells have made shell collecting a popular pastime. However, many amateur shell collectors, although able to name hundreds of the shells that grace our beaches, know very little about the living animals that created those shells and once occupied them. The largest classes of molluscs are Gastropoda (snails and their relatives), Bivalvia (clams, oysters, and others), Polyplacophora (chitons), and Cephalopoda (squids, octopuses, nautiluses). Monoplacophora, Scaphopoda (tusk shells), Caudofoveata, and Solenogastres are much smaller classes.

The evolutionary relationships among molluscan classes remain controversial. One hypothesis (figure 10.2) recognizes as clade Conchifera all molluscan classes that possess solid shells, although some gastropods have lost this feature. A contrasting hypothesis derived from molecular data unites Monoplacophora and Polyplacophora as a clade called Serialia based on the shared presence of serially repeated body parts. This hypothesis places Caudofoveata closest to the cephalopods. One problem with the latter hypothesis is that it requires additional gains and/or losses of the solid shell.

figure 10.1

Molluscs: a diversity of life forms. The basic body plan of this ancient group has become variously adapted for different habitats. **A,** Two chitons (*Katharina tunicata*), class Polyplacophora. **B,** A marine intertidal snail (*Littorina sitkana*), class Gastropoda. **C,** A nudibranch (*Chromodoris leopardus*), class Gastropoda. **D,** Pacific giant clams (*Panope abrupta*), with large siphons, class Bivalvia. **E,** An octopus (*Octopus briareus*), class Cephalopoda, forages at night on a Caribbean coral reef. **F,** A land snail (*Helix pomatia*).

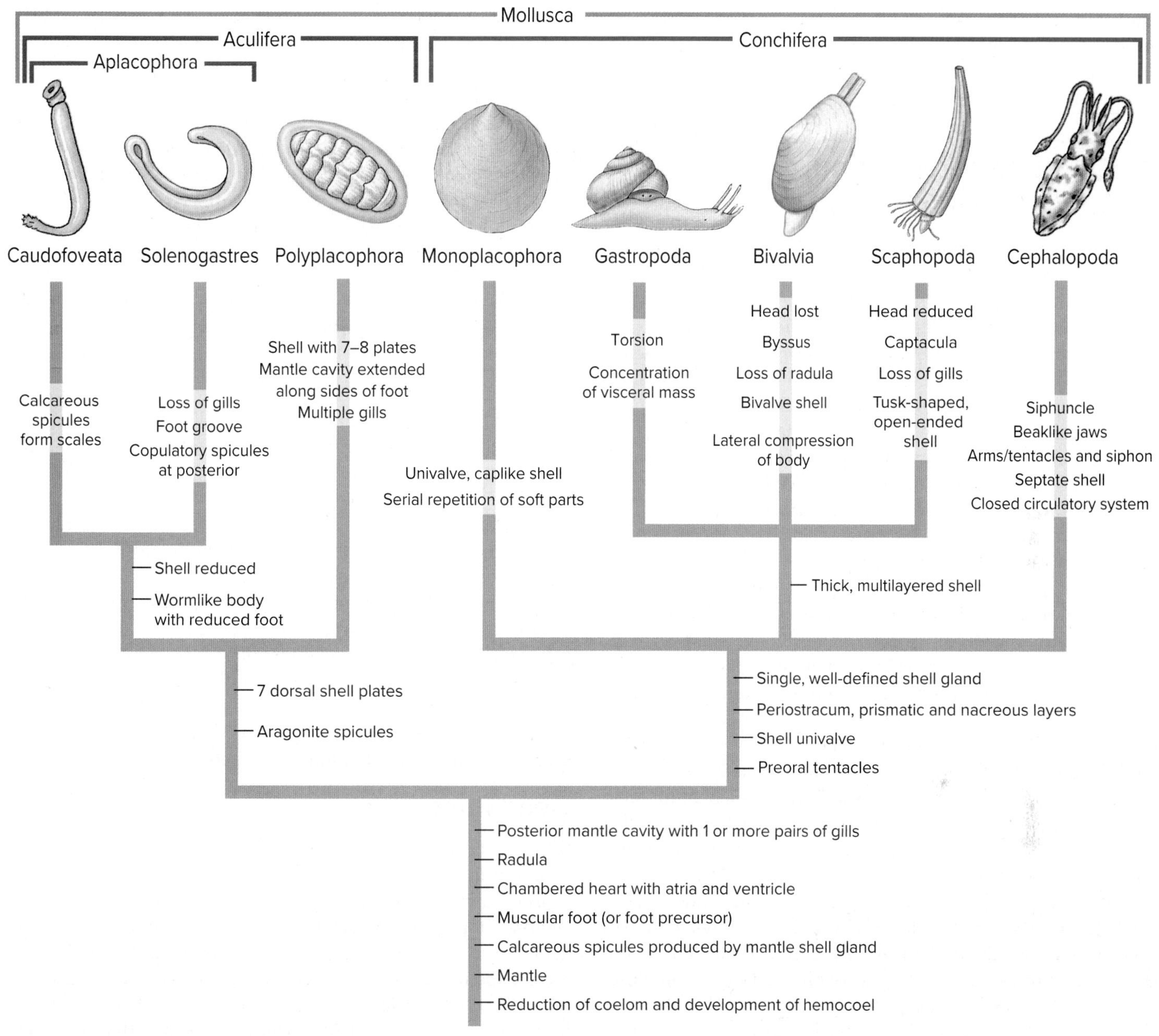

figure 10.2

Cladogram showing relationships among classes of Mollusca according to the Conchifera hypothesis. Synapomorphies that identify the various clades are shown, although a number of these have been modified or lost in some descendants. For example, the univalve shell (as well as shell coiling) has been reduced or lost in many gastropods and cephalopods, and many gastropods have undergone detorsion. The bivalve shell of Bivalvia was derived from an ancestral univalve shell. The byssus is lost in most adult bivalves but functions in larval attachment in many; therefore, the byssus is considered a synapomorphy of Bivalvia.

Source: Modified from R. C. Brusca and G. J. Brusca, ***Invertebrates,*** 2nd ed. Sunderland, MA: Sinauer Associates, Inc., 2003.

10.1 Ecological Relationships

Molluscs occupy a great range of habitats: from the tropics to polar seas; at altitudes exceeding 7000 m; in ponds, lakes, and streams; on mudflats; in pounding surf; and in the open ocean, from the surface to abyssal depths. Most live in the sea, and they represent a variety of lifestyles, including bottom-feeders, burrowers, borers, and pelagic forms. The phylum includes some of the most sluggish as well as some of the swiftest and most active invertebrates. It encompasses herbivorous grazers, predaceous carnivores, and ciliary filter feeders.

According to fossil evidence, molluscs originated in the sea, and most have remained there. Much of their evolution occurred along shores, where food was abundant and habitats were varied. Only bivalves and gastropods moved into brackish and freshwater habitats. As filter feeders, bivalves are unable to leave their aquatic surroundings; however, snails (gastropods) invaded land and may have been among the first animals to do so. Terrestrial snails are limited in range by their need for humidity, shelter, and calcium in the soil.

10.2 Economic Importance and Ocean Acidification

Molluscs are an extremely important food source for people around the world; 60,339,000 tonnes of molluscs were harvested commercially in 2014 on the eastern seaboard of the United States, with the western seaboard and the Gulf of Mexico adding 138,110 and 153,100 tonnes, respectively. The value of the eastern seaboard catch alone was nearly \$2.68 billion (source: http://www.seaaroundus.org), so the economic importance of molluscs cannot be overstated. A healthy mollusc "fishery" depends on healthy oceans, but recent threats to mollusc populations come from an unexpected direction: global climate change. While ocean warming does impact the distributions of many marine organisms, it is ocean acidification that is now causing concern for the future.

Increased amounts of CO_2 in the atmosphere initiate a set of chemical reactions in the oceans that lower pH. As the ocean becomes more acidic, levels of biologically available calcium decline, making it more difficult for marine organisms to form calcium skeletons. Calcium is an essential component for the foundation beneath living coral tissue (see section 7.8: Coral Reefs) and is critical to healthy mollusc shells. Clams, oysters, mussels, and scallops produce thinner and weaker shells when reared at low pH. Similarly, larval abalone in such conditions fail to develop normally, and many cannot create normal shells. Larval survivorship for oysters is reduced when ocean acidity increases, so the oyster industry in the Pacific Northwest is concerned about its \$278 million revenue (2009 data). One Oregon oyster farm adds calcium chloride and sodium carbonate to seawater to increase the amount of biologically available calcium for young animals. The one bright spot in the future of mollusc harvesting lies with cephalopods such as squid and octopus. They appear to be much less affected by ocean acidification than their kin, presumably because they do not have shells.

Research has shown direct adverse effects of ocean acidification on coral settlement and growth, and on the neurological development of larval tropical reef fishes, but indirect damage to the coral reef ecosystem is equally worrying. The complex food webs (see section 7.8: Coral Reefs) present on the reefs will change with the distribution and abundance of corals and coralline algae. Some researchers expect an increased abundance of algal turfs and sea grass beds as acidification proceeds, so community changes may act on those molluscs and other taxa that are not directly impacted by the acidification of our oceans.

10.3 Form and Function

Body Plan

Reduced to its simplest dimensions, a mollusc body consists of a head-foot portion and a visceral mass portion. The head-foot region contains feeding, cephalic sensory, and locomotor organs (see mouth and foot in figure 10.3). It depends primarily on muscular action for its function. The visceral mass contains digestive, circulatory, respiratory, and reproductive organs, and it depends primarily on ciliary tracts for its functioning. Two folds of skin, outgrowths of the dorsal body wall, form a protective *mantle,* which encloses a space between itself and the body wall called a mantle cavity. The mantle cavity houses gills or a lung, and in many molluscs the mantle secretes a protective shell over the head-foot and visceral mass. Modifications of structures that form the head-foot and the visceral mass produce the great diversity of molluscan body plans.

Head-Foot

Some molluscs have a well-developed head, but in other animals the head is identified as the region bearing the mouth, sensory organs (which often include photoreceptors), chemoreceptors, statocysts and tactile sensors, and a brain or central ganglion. Photosensory receptors range from fairly simple pigment cups to the complex eyes of cephalopods. Tentacles are often present. Within the mouth is a structure unique to molluscs, the radula, and usually posterior to the mouth is the chief locomotor organ, or foot (figure 10.3).

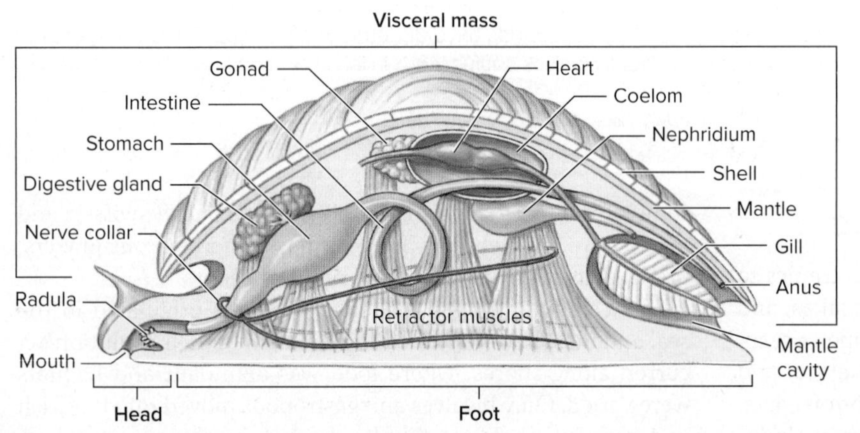

figure 10.3

This representation of a generalized mollusc has been called a "hypothetical ancestral mollusc," but the ancestral mollusc may not have had a distinct shell or a well-developed foot.

Radula

The **radula** is a rasping, protrusible, tonguelike organ found in all molluscs except bivalves and some gastropods and solenogasters. It is a ribbonlike membrane on which are mounted rows of tiny chitinous teeth that point backward (figure 10.4). Protractor and retractor muscles move the radula and its supporting cartilages **(odontophore)** outside and back into the mouth for feeling. A radula may have from a few to as many as 250,000 teeth that, when protruded, can scrape, pierce, tear, or cut particles of food. A radula may carry particles in a continuous stream toward the digestive tract.

Foot

The molluscan foot is variously adapted for locomotion, for attachment to a substrate, or for a combination of functions. The foot is usually a ventral, solelike structure that effects a creeping locomotion through waves of muscular contraction. However, there are many modifications, such as the attachment disc of limpets (which creates suction), the laterally compressed "hatchet foot" of bivalves, or the funnel for jet propulsion in squids and octopuses. Secreted mucus often aids adhesion or provides a slime track for small molluscs that glide on cilia.

Visceral Mass

Mantle and Mantle Cavity

The mantle is a sheath of skin, extending dorsally from the visceral mass, that wraps around each side of the body, protecting the soft parts and creating the mantle cavity between itself and the visceral mass. The outer surface of the mantle secretes the shell.

The mantle cavity plays an enormous role in the life of a mollusc. It usually houses respiratory organs (gills or a lung), which develop from the mantle, and the mantle's own exposed surface also serves gas exchange. Products from the digestive, excretory, and reproductive systems empty into the mantle cavity. In aquatic molluscs, a continuous current of water, kept moving by surface cilia or by muscular pumping, performs multiple functions: it supplies oxygen, and in some forms, food; flushes out wastes; and carries reproductive products out to the environment. In aquatic forms, the mantle is usually equipped with chemoreceptors. In cephalopods (squids and octopuses), the muscular mantle and its cavity create jet propulsion used in locomotion.

Shell

The shell of a mollusc, when present, is secreted and lined by the mantle. A shell typically has three layers (figure 10.5). The **periostracum,** the outer organic layer, is composed of resistant proteins called conchiolins. They protect the underlying calcareous layers from erosion by boring organisms. It is secreted by a fold of the mantle edge, and growth occurs only at the margin of the shell. On the older parts of the shell, the periostracum often wears away. The middle prismatic layer is composed of densely packed prisms of calcium carbonate laid down in a protein matrix. It is secreted by the glandular margin of the mantle, and increase in shell size occurs at the shell margin as the animal grows.

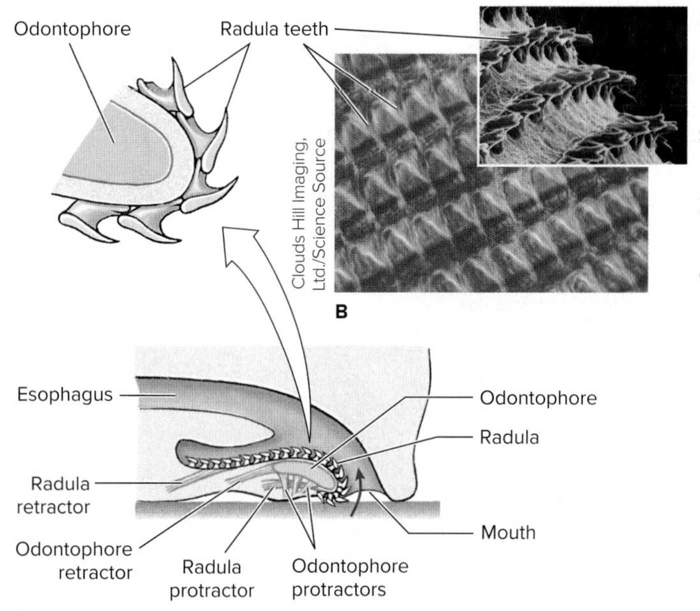

figure 10.4

A, Diagrammatic longitudinal section of gastropod mouth showing the radula and its scraping teeth. The radula wraps around the odontophore cartilage. As the animal grazes, the mouth opens, the odontophore is thrust forward, the radula scrapes backward to bring food into the pharynx, and the mouth closes. The sequence is repeated rhythmically. The radula ribbon wears out anteriorly, and it is continually replaced posteriorly. **B,** Microscope image of a radula showing a close-up view of teeth.

The inner **nacre,** or nacreous layer, of the shell is composed of calcium carbonate sheets laid down over a thin protein matrix. This layer is secreted continuously by the mantle surface, so that it becomes thicker during the animal's life.

Freshwater molluscs usually have a thick periostracum that gives some protection against acids produced in the water by decay of leaf litter. The periostracum varies from thick to absent in marine forms. Shell structure varies greatly. Calcium for the shell comes from environmental water or soil or from food. The first shell appears during the larval period and grows continuously throughout life.

Key Theme 10.1

HUMAN CONNECTIONS

Why Make Pearls?

Pearl production is a by-product of a protective device used by a mollusc when a foreign object, such as a grain of sand or a parasite, lodges between the shell and the mantle. The mantle secretes many layers of nacre around the irritating object (figure 10.5). Pearls are cultured by inserting small spheres, usually made from the shells of freshwater clams, in the mantle of a certain species of oyster and by maintaining the oysters in enclosures. The oyster deposits its own nacre around the "seed" in a much shorter time than would be required to form a pearl normally.

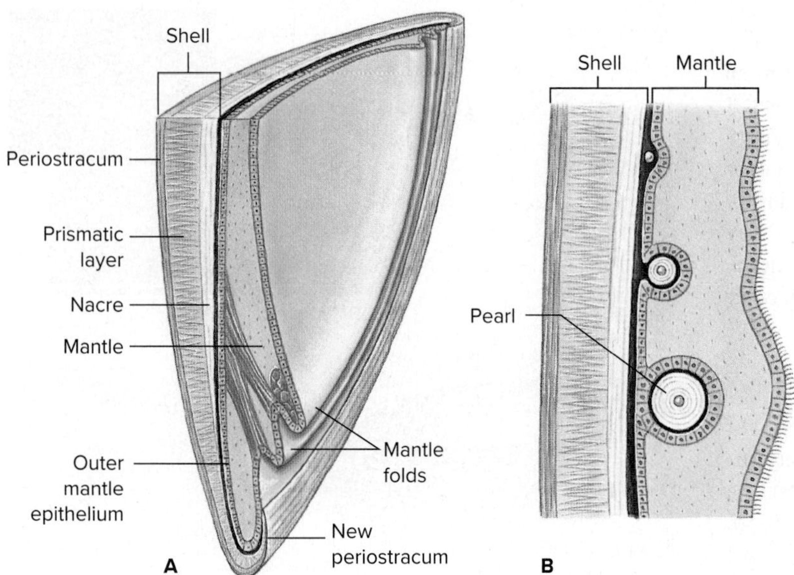

figure 10.5

A, Diagrammatic vertical section of shell and mantle of a bivalve. The outer mantle epithelium secretes the shell; the inner epithelium is usually ciliated. **B,** Formation of pearls between the mantle and shell as nacre accumulates around a parasite or bit of sand under the mantle.

Internal Structure and Function

Gas exchange occurs through the body surface, particularly the mantle, and in specialized respiratory organs such as gills or lungs. Most molluscs have an open circulatory system with a pumping heart, blood vessels, and blood sinuses. In an open circulatory system, blood is not entirely contained within blood vessels; rather, it flows through vessels in some parts of the body and enters open sinuses in other parts. An open circulatory system is less efficient at supplying oxygen to all tissues in the body, so it is common in slow-moving animals. Insects are a notable exception, but in these animals oxygen is distributed by the tracheal system, not by the circulatory system. In a closed circulatory system, blood moves to and from tissues within blood vessels. Most members of class Cephalopoda move rapidly and have a closed circulatory system with a heart, vessels, and capillaries.

The digestive tract is complex and highly specialized according to the feeding habits of the various molluscs. Most molluscs have a pair of kidneys (metanephridia), a type of nephridium in which the inner end opens into the coelom; ducts of the kidneys in many forms serve also for discharge of eggs and sperm. The nervous system consists of several pairs of ganglia with connecting nerve cords. Molluscs have various highly specialized sense organs.

Most molluscs are dioecious, although some gastropods are hermaphroditic. Following fertilization, embryos exhibit spiral cleavage, developing into a free-swimming **trochophore** larval stage (figure 10.6) in aquatic forms. In many taxa, the trochophore is followed by a second swimming larval stage, a **veliger** (figure 10.7). A veliger occurs in most marine snails, tusk shells, and bivalves. It has the beginning of a foot, shell, and mantle.

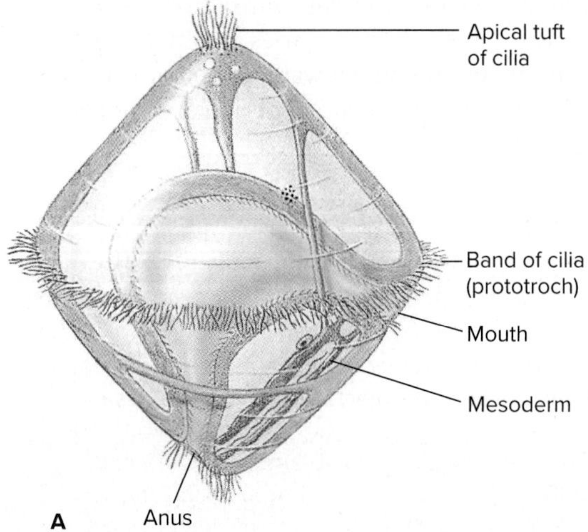

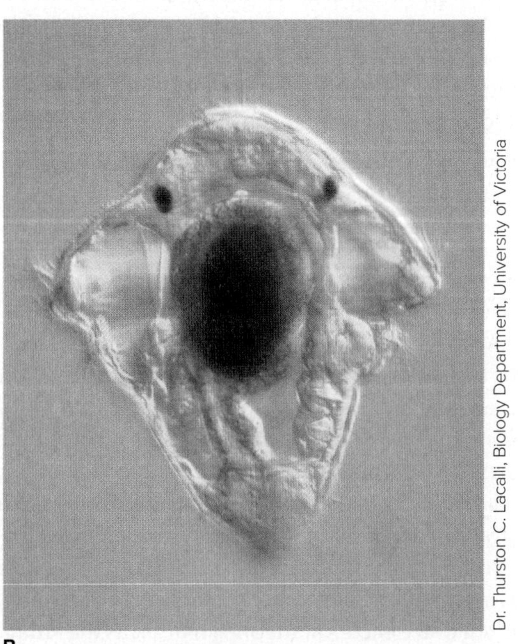

figure 10.6

A, Generalized trochophore larva. Molluscs and annelids with ancestral embryonic development have trochophore larvae, as do several other phyla. **B,** Trochophore of a Christmas tree worm, *Spirobranchus spinosus* (Annelida).

Key Theme 10.2

GENETICS AND DEVELOPMENT

What Is a Trochophore?

Trochophore larvae (figure 10.6) are minute, translucent, often top-shaped, and they have a prominent circlet of cilia (prototroch) and sometimes one or two accessory circlets. They are the ancestral larval form of molluscs and annelids. A trochophore-like larva occurs also in marine turbellarians, nemerteans, brachiopods, phoronids, sipunculids, and echiurans. Possession of a trochophore or trochophore-like larva supports assigning these phyla to superphylum Lophotrochozoa.

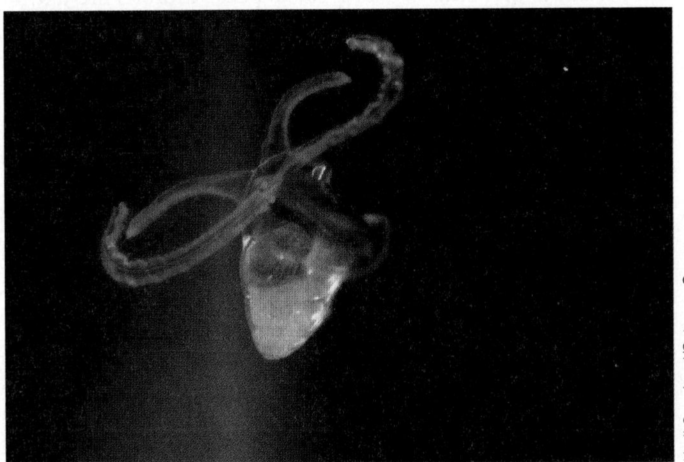

figure 10.7

Veliger of a snail, *Pedicularia,* swimming. The adults are parasitic on corals. The ciliated process (velum) develops from the prototroch of the trochophore (see figure 10.6A).

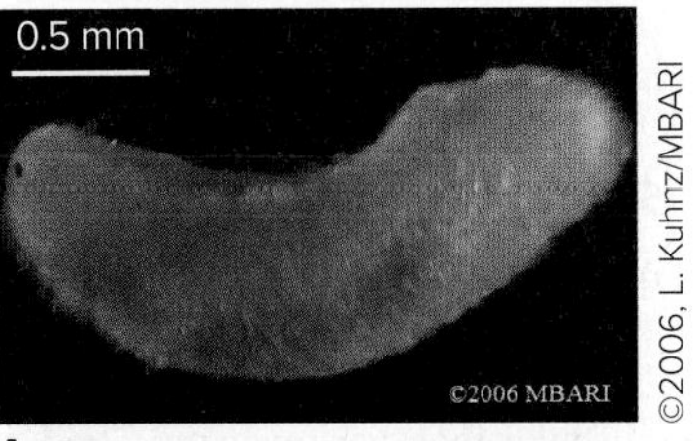

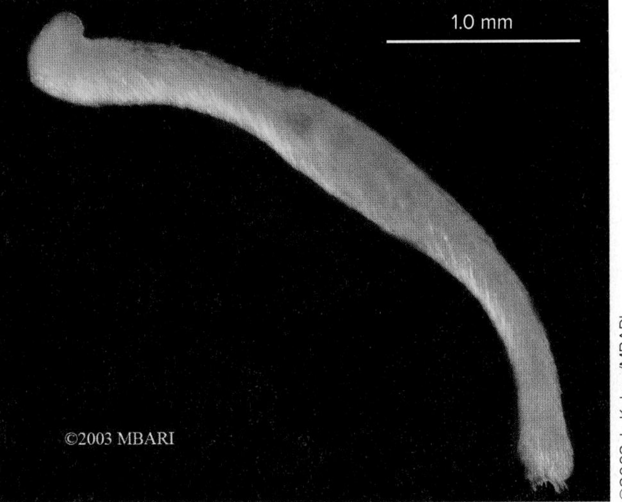

figure 10.8

Spicules are clearly visible on the skin of solenogasters and caudofoveates. **A,** *Neomeniomorpha,* a solenogaster; **B,** *Chaetoderma elegans,* a caudofoveate.

10.4 Classes Caudofoveata and Solenogastres

Caudofoveates and solenogasters (figure 10.8) both are wormlike and shell-less, with calcareous scales or spicules in their integument. Members of both groups have a reduced head and lack nephridia. Caudofoveates have one pair of gills and are dioecious. They burrow in marine sediments, feeding on microorganisms and detritus. Class Caudofoveata is sometimes called Chaetodermomorpha. In contrast to caudofoveates, solenogasters usually have no true gills, and they are hermaphroditic. Solenogasters live freely on the ocean floor and often feed on cnidarians. Class Solenogastres is sometimes called Neomeniomorpha.

10.5 Class Monoplacophora

Until 1952, Monoplacophora (mon-o-pla-kof´-o-ra) were known only from Paleozoic shells. However, in that year, living specimens of *Neopilina* (Gr. *neo,* new, + *pilos,* felt cap) were dredged up from the ocean floor near the west coast of Costa Rica. These molluscs are small and have a low, rounded shell and a creeping foot (figure 10.9). They superficially resemble

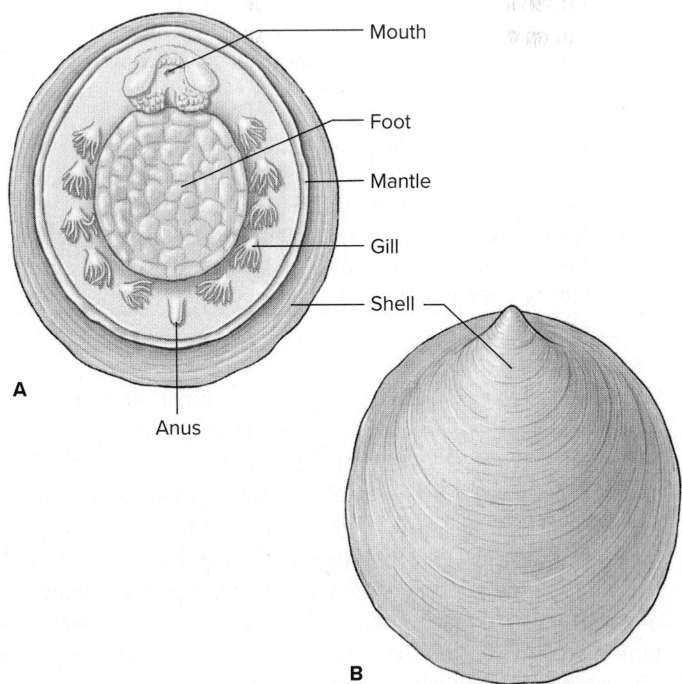

figure 10.9

Neopilina, class Monoplacophora. Living specimens range from 3 mm to about 3 cm in length. **A,** Ventral view. **B,** Dorsal view.

CHARACTERISTICS of Phylum Mollusca

1. Dorsal body wall forms a pair of folds called the **mantle,** which encloses the **mantle cavity,** is modified into **gills** or **lungs,** and secretes the **shell** (shell absent in some); ventral body wall specialized as a muscular **foot,** variously modified but used chiefly for locomotion; radula in mouth.
2. Live in marine, freshwater, and terrestrial habitats.
3. Free-living or occasionally parasitic.
4. Body bilaterally symmetrical (bilateral asymmetry in some); unsegmented; often with definite head.
5. Triploblastic body.
6. **Coelom** limited mainly to area around heart; and perhaps lumen of gonads, part of kidneys, and occasionally part of the intestine.
7. Surface epithelium usually ciliated and bearing mucous glands and sensory nerve endings.
8. Complex digestive system; rasping organ **(radula)** usually present; anus usually emptying into mantle cavity; internal and external **ciliary tracts** often of great functional importance.
9. Circular, diagonal, and longitudinal muscles in the body wall; mantle and foot highly muscular in some classes (for example, cephalopods and gastropods).
10. Nervous system composed of paired cerebral, pleural, pedal, and visceral ganglia, with nerve cords and subepidermal plexus; ganglia centralized in nerve ring in gastropods and cephalopods.
11. Sensory organs of touch, smell, taste, equilibrium, and vision (in some); the highly developed direct **eye** (photosensitive cells in retina face light source) of cephalopods is similar to the indirect eye (photosensitive cells face away from light source) of vertebrates but arises as a skin derivative in contrast to the brain eye of vertebrates.
12. No asexual reproduction.
13. Both **monoecious** and **dioecious** forms; **spiral cleavage;** ancestral larva a **trochophore,** many with a **veliger** larva, some with direct development.
14. One or two kidneys **(metanephridia)** opening into the pericardial cavity and usually emptying into the mantle cavity.
15. Gas exchange by **gills, lungs, mantle,** or **body surface.**
16. **Open circulatory system** (secondarily closed in cephalopods) of heart (usually three-chambered), blood vessels, and sinuses; respiratory pigments in blood.

limpets (see figure 10.20A), but unlike in most other molluscs, a number of organs, such as metanephridia, gonads, and gills, are serially repeated. Serial repetition occurs less extensively in chitons. Body structures repeat in each segment of an annelid worm (see Chapter 11 opening essay)—are these repeated structures an indication that molluscs had a segmented (metameric) ancestor? Some authors consider monoplacophorans truly segmented, but most argue that *Neopilina* shows only pseudometamerism and that molluscs did not have a metameric ancestor. The serialia hypothesis (see this chapter's opening essay and section 10.11) provides another possibility: repeated body parts arose in the common ancestor of monoplacophorans and chitons.

10.6 Class Polyplacophora: Chitons

Chitons are somewhat flattened and have a convex dorsal surface that bears eight (rarely seven) articulating calcareous plates (or valves), which give them their name (figures 10.10 and 10.11). Polyplacophora means "bearing many plates," in contrast to Monoplacophora, which bear one shell (*mono,* single). The plates overlap posteriorly and are usually dull in color, like the rocks to which chitons cling.

Most chitons are small (2–5 cm); the largest rarely exceeds 30 cm. They commonly occur on rocky surfaces in intertidal regions, although some live at great depths. Chitons are stay-at-home organisms, straying only very short distances for feeding. When they feed, a sensory subradular organ protrudes from their mouth to explore for algae or colonial organisms. The radula protrudes to scrape food from the rocks. A chiton can cling tenaciously to rock with its broad flat foot. If detached, it can roll up like an armadillo for protection.

The mantle forms a girdle around the margin of the plates (figure 10.11), and in some species mantle folds cover part or all of the plates (see figure 10.1A). On each side of the broad ventral foot and lying between the foot and the mantle is a row of gills suspended from the roof of the mantle cavity. With the foot and the mantle margin adhering tightly to the substrate, these grooves become closed chambers, open only at the ends. Water enters the grooves anteriorly, flows across the gills, and leaves posteriorly, thus bringing a continuous supply of oxygen to the gills.

Blood pumped by the three-chambered heart reaches gills by way of an aorta and sinuses. Two kidneys carry waste from the pericardial cavity to the exterior. Two pairs of longitudinal nerve cords connect in the buccal region. Sense organs include shell eyes on the surface of the shell (in some) and a pair of **osphradia** (chemosensory organs for sampling water).

figure 10.10

Dorsal view of *Chiton tuberculatus* on a rock surface. The upper surface of the mantle, or "girdle," has a beautiful "snakeskin" pattern of scales. The rigid scales and the flexible mantle together serve as armor for the chiton. The chiton's eight plates are clearly visible.

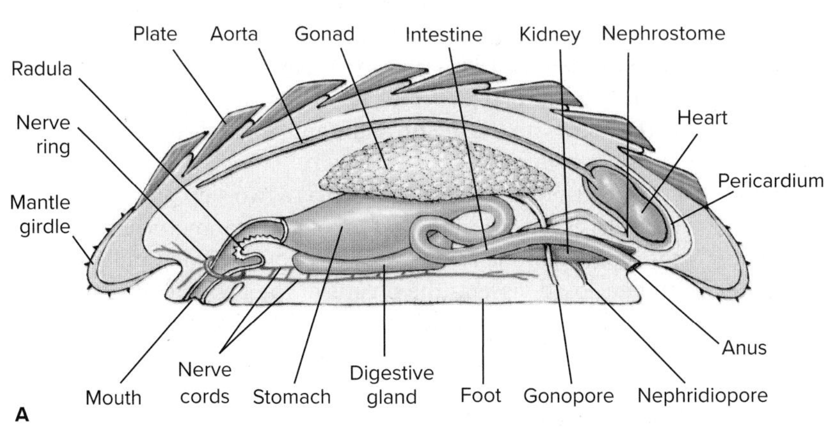

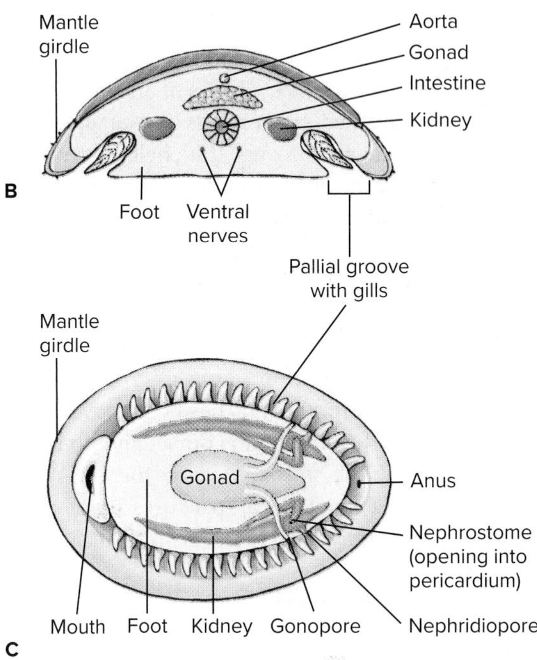

figure 10.11

Anatomy of a chiton (class Polyplacophora). **A,** Midsagittal section. **B,** Transverse section. **C,** Ventral view.

Sexes are separate in most chitons. Eggs are released singly or in strings or masses of jelly; they are shed into the female mantle cavity or into the sea. Sperm shed by males into the excurrent water may enter the gill grooves of females through incurrent openings. Trochophore larvae metamorphose directly into juveniles, without a second larval stage.

10.7 Class Scaphopoda

Scaphopoda (ska-fo-pod´a), commonly called tusk shells or tooth shells, are sedentary marine molluscs that have a slender tubular shell open at both ends. Inside the shell the mantle is wrapped around the viscera and fused to form a tube. Most scaphopods are 2.5–5 cm long, although they range from 4 mm to 25 cm.

The foot, which protrudes through the larger end of the shell, functions in burrowing into mud or sand, always leaving the small end of the shell exposed to water above (figure 10.12). Respiratory water circulates through the mantle cavity both by movements of the foot and by ciliary action. Food is caught on cilia of the scaphopod's foot or on the mucus-covered ciliated knobs of its long tentacles. Food is mainly detritus and microbial organisms from the substrate.

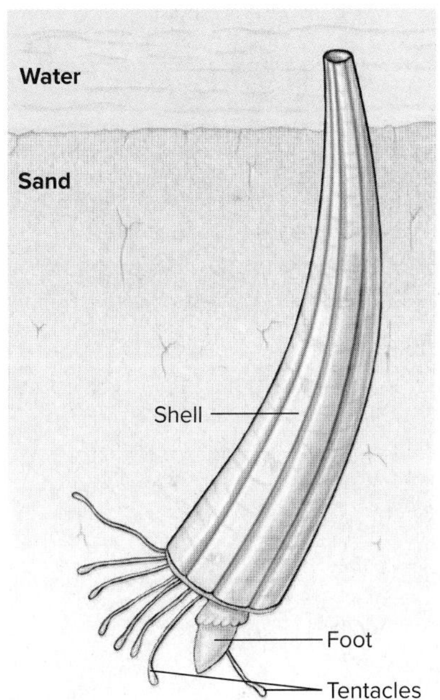

figure 10.12

The tusk shell, *Dentalium*, a scaphopod. It burrows into soft mud or sand and feeds by means of its prehensile tentacles. Respiratory currents of water are drawn in by ciliary action through the small open end of the shell, and then expelled through the same opening by muscular action.

10.8 Class Gastropoda

Among molluscs, class Gastropoda (gas-tro-pod´-a) (Gr. *gastér*, stomach, + *pous, podos,* foot) is by far the largest and most diverse, containing about 70,000 living and 15,000 fossil species. Its members differ so widely that no single general term in our language can apply to them as a group. They include snails, limpets, slugs, whelks, conchs, periwinkles, sea slugs, sea hares, sea butterflies, and others. They range from marine molluscs with many ancestral characters to air-breathing snails and slugs.

Gastropods are often sluggish, sedentary animals because most of them have heavy shells and slow locomotor organs. When present, the shell is almost always of one piece (univalve) and may be coiled or uncoiled. Some snails have an **operculum,** a hard proteinaceous plate that covers the shell aperture when the body withdraws into the shell. It protects the body and prevents water loss. Gastropods are primarily bilaterally symmetrical, but because of **torsion,** a twisting process that occurs during development, the visceral mass has become asymmetrical. To visualize gastropod torsion, compare the simple bilateral symmetry of an adult chiton (figure 10.11) or a generalized adult mollusc (figure 10.3) with the contorted body plan of an adult gastropod (see figure 10.19A). Note that the gastropod anus has been moved forward relative to its position on the chiton; it is dorsal and no longer opens at the posterior end of the body. Ontogenic torsion is the developmental process by which anatomical structures are repositioned in gastropod development relative to the primary bilateral symmetry.

Form and Function

Torsion

Gastropod development is highly variable, but in general a trochophore larval stage is followed by a veliger larval stage in which the shell arises. The veliger has two ciliated velar lobes, used in swimming, and the developing foot is present (see figure 10.7). Examine the images of veligers illustrated in lateral and dorsal views in figure 10.13: five stages in a developmental sequence are shown; the earliest stages are in the panels on the left. Development ends with an adult snail illustrated in the panel on the right. In the panel on the left side of figure 10.13, note the primary bilateral symmetry of the body with the mouth at the anterior end and the anus at the posterior end.

Torsion is usually described as a two-step process. The first step places a bend in the gut tube. An asymmetrical foot retractor muscle on one side of the body contracts and pulls the shell and enclosed viscera (the organs of the body) 90 degrees counterclockwise, relative to the head. This movement brings the anus from the posterior to the side of the body as shown in the panel second from the left in the dorsal view in figure 10.13. The shell rotates between 90 and 180 degrees counterclockwise into a position that persists into adulthood at this time. In the panel second from the left, note that the mantle cavity develops on the side of the body near, but initially separate from, the anus. In the second step of torsion, illustrated by the central panels in figure 10.13, the anus and mantle cavity usually move further in a counterclockwise direction and the mantle cavity is remodeled to encompass the anus. Toward the end of development, the mantle cavity and anus are within the forward part of the shell as illustrated in the fourth panel in figure 10.13. In a slower and more variable series of changes, the digestive tract moves both

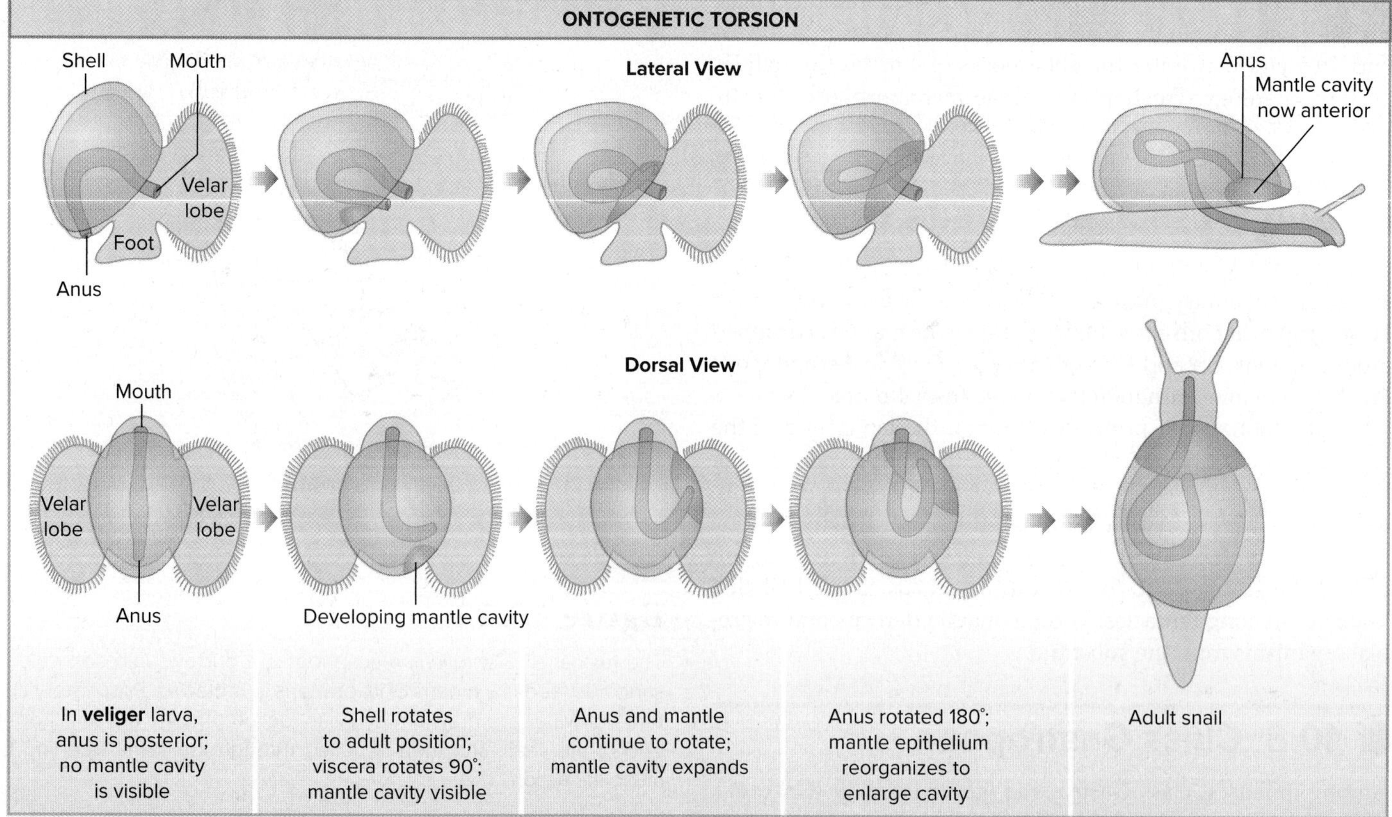

figure 10.13

Ontogenetic torsion in a gastropod veliger larva is illustrated in a time sequence of five panels, with an early stage in the left panel and the adult snail in the final panel on the right.

laterally and dorsally so that the anus lies above the head within the mantle cavity (see the adult snail in the fifth panel in figure 10.13; lateral view). Although not illustrated in figure 10.13, gastropod organs move with the digestive tract. In the lateral view of the adult snail (fifth panel), a twist in the gut is visible. This twist reverses the relative positions of the gills, kidney, and heart atria from one side to the other. The nerve cords form a figure eight.

Ontogenetic torsion contrasts with *evolutionary torsion,* the series of changes that produced the modern torted gastropod body from the ancestral untorted form. The hypothetical ancestral gastropod was assumed to have a posterior mantle cavity like that of the hypothetical ancestral mollusc (see figure 10.3). It had long been assumed that morphological changes in ontogenetic torsion represented the sequence of evolutionary changes. However, newer studies led researchers to hypothesize that the ancestral gastropod had two lateral mantle cavities, much like those in *Neopilina* (see figure 10.9) and in chitons (see figure 10.11). A single mantle cavity at the leading edge of the shell may have arisen when the left lateral mantle cavity was lost, and the right cavity expanded toward the middle of the body after the first 90 degrees of torsion.

The forward position of the anus when torsion is complete creates the possibility of wastes washing over the gills **(fouling)** and causes us to wonder what evolutionary factors favored such a strange realignment of the body. Several explanations have been proposed, none entirely satisfying. For example, sense organs of the mantle cavity (osphradia) would better sample water when turned in the direction of travel, and as mentioned already, the forward mantle cavity creates a space into which the head could be withdrawn. Certainly, the consequences of torsion and the resulting need to avoid fouling have been very important in the subsequent evolution of gastropods. We explore these consequences after describing an interacting feature of gastropods—coiling of the shell and visceral mass.

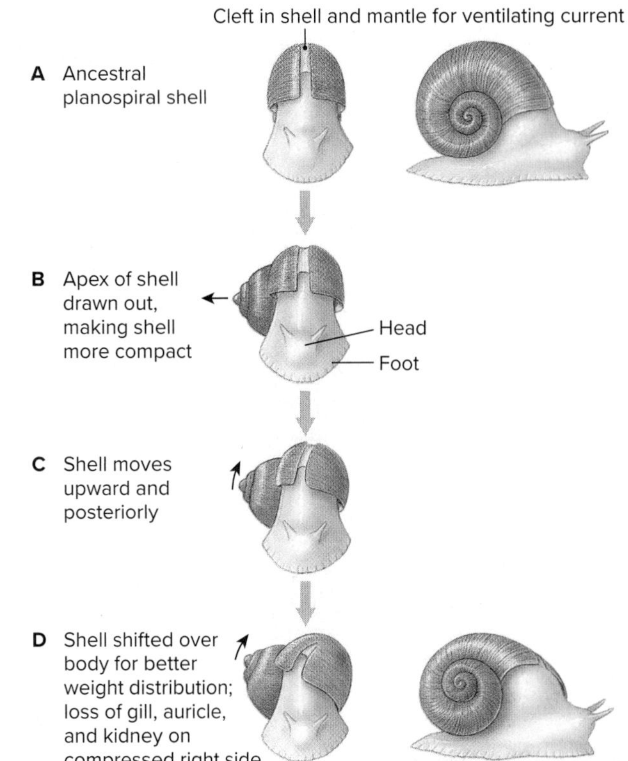

figure 10.14

Evolution of a shell in gastropods. **A,** The earliest coiled shells were planospiral, each whorl lying completely outside the preceding whorl. Interestingly, the shell has become planospiral secondarily in some living forms. **B,** Better compactness was achieved by snails in which each whorl lay partially to the side of the preceding whorl. **C, D,** Better weight distribution resulted when the shell was moved upward and posteriorly.

Coiling

Coiling, or spiral winding, of the shell and visceral mass may occur in the larval stage at the same time as torsion, but the fossil record shows that coiling was a separate evolutionary event and originated in gastropods earlier than torsion did. Coiling also occurs in cephalopods. Nevertheless, all living gastropods descend from coiled, torted ancestors, whether or not they now show these characteristics.

Early gastropods had a bilaterally symmetrical shell with all whorls lying in a single plane (figure 10.14A). Such a shell was not very compact, since each whorl had to lie completely outside the preceding one. A few modern species have secondarily returned to that form. The lack of compactness of the planospiral shell was resolved by a shape in which each succeeding whorl was at the side of the preceding one (figure 10.14B). However, this shape clearly was unbalanced, with much weight over to one side. Later gastropods achieved better weight distribution by shifting the shell upward and posteriorly, with the shell axis oblique to the longitudinal axis of the foot (figure 10.14C and D). However, the weight and bulk of the main body whorl, the largest whorl of the shell, pressed on the right side of the mantle cavity, and apparently interfered with the organs on that side. Accordingly, the gill, auricle, and kidney of the right side have been lost in all except a few living gastropods, leading to bilateral asymmetry.

Adaptations to Avoid Fouling

Although loss of the right gill was probably an adaptation to the mechanics of carrying a coiled shell in a "side-saddle" position, this loss made it possible to reduce the effects of fouling, where wastes leaving the anus are deposited onto gills and near the head. In most modern gastropods, water flows one-way; it is brought into the left side of the mantle cavity and out the right side, carrying with it the wastes from the anus and nephridiopore, which lie near the right side (figure 10.15). Some gastropods with the ancestral characteristic of two gills avoid fouling by venting excurrent water through a dorsal cleft above the anus (see figure 10.14), or through one or a series of holes in the shell (abalone, figure 10.16). Opisthobranchs (nudibranchs and others) have evolved an even more elaborate "twist"; after undergoing torsion as larvae, they develop various degrees of *detorsion* as adults.

Detorsion is a post-veliger change in the orientation of the digestive and nervous systems of opisthobranchs. After detorsion, the anus is posterior rather than anterior, and lies on the right side of the body. Compare the position of the anus in the shelled

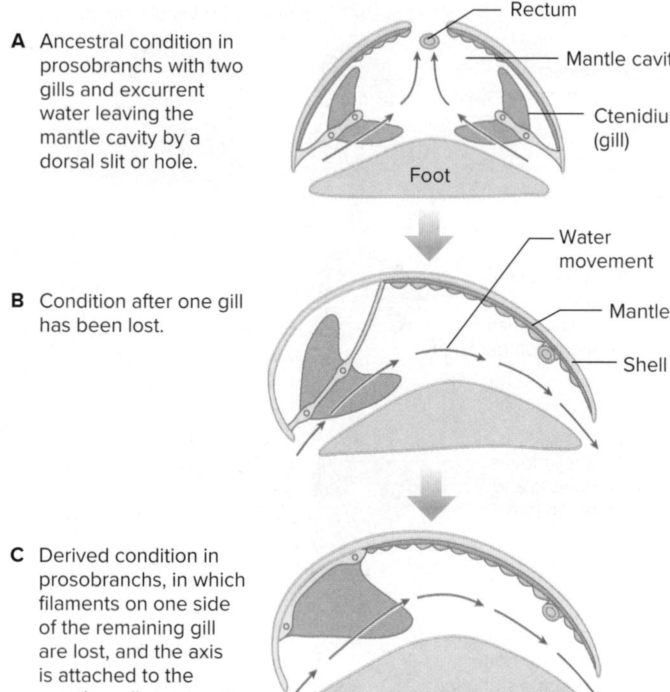

figure 10.15

Evolution of gills in gastropods. The diagrams represent cross sections of the mantle cavity viewed from the posterior. **A,** Ancestral condition with two gills. **B,** Loss of one gill allows unidirectional water flow. **C,** Filaments on one side of the gill are lost.

figure 10.16

A, Red abalone, *Haliotus rufescens.* Notice the series of "exit" holes in the shell that permit a one-way water flow over the sensory organs and gills. This huge, limpetlike snail is prized as food. Abalones are strict vegetarians, feeding especially on sea lettuce and kelp. **B,** Moon snail, *Naticarius orientalis* from Sulawesi, Indonesia, is similar to the North American moon snail, a predator of clams and mussels. It uses its radula to drill neat holes in its victim's shell, through which it then extends its proboscis to eat the bivalve's fleshy body.

gastropod after torsion (see figure 10.13) with that of the opisthobranch after torsion and detorsion (figure 10.17). The name suggests that torsion reverses, but in the sea hare (figure 10.17), the mantle complex migrates from the anterior of the body to the posterior right side accompanied by differential growth and folding of the digestive tract. In other taxa, torsion may cease earlier or be followed by different degrees of body reorganization.

Feeding Habits

The feeding habits of gastropods are as varied as their shapes and habitats, but all use some adaptation of the radula. Many gastropods are herbivorous, rasping algae from a substrate. Some herbivores are grazers, some are browsers, and others planktonic feeders. Abalones (see figure 10.16A) hold seaweed with their foot and break off pieces with their radula. Some snails are scavengers, living on dead and decayed flesh; others are carnivores, tearing prey with their radular teeth. Snails such as oyster borers and moon snails (see figure 10.16B) have an extensible proboscis for drilling holes in the shells of bivalves whose soft parts they find delectable. Some even have a spine for opening shells. Most pulmonates (air-breathing snails) (see figure 10.22) are herbivorous, but some eat earthworms and other snails. After maceration by the radula or by some grinding device, such as the so-called gizzard in sea hares, digestion is usually extracellular in the lumen of the stomach or digestive glands.

Some sessile gastropods, such as slipper shells, are ciliary feeders that use the gill cilia to collect particulate matter, which they roll into a mucous ball and carry to their mouth. Some sea butterflies secrete a mucous net to catch small planktonic forms and then draw the web into their mouth. In ciliary feeders, the stomachs are sorting regions, and most digestion is intracellular in the digestive gland.

Key Theme 10.3
ADAPTATION AND PHYSIOLOGY

Venomous Cone Shells

Among the most interesting predators are venomous cone shells (figure 10.18), which feed on vertebrates or other invertebrates, depending on the species. When *Conus* senses its prey, a single radular tooth slides into position at the tip of the proboscis. When the proboscis strikes prey, it expels the tooth like a harpoon, and the venom tranquilizes or kills the prey at once. Some species can deliver very painful stings, and the sting of several species is lethal to humans. The venom consists of a series of toxic peptides, and each *Conus* species carries peptides (conotoxins) specific for the neuroreceptors of its preferred prey. Studies of toxin chemistry have led to a synthetic version used to treat chronic pain as well as research into applications for Parkinson's disease.

Internal Form and Function

Respiration in most gastropods is performed by a gill (two gills in a few), although some aquatic forms lack gills and depend on the skin. Pulmonates (most freshwater and terrestrial snails) have lost their gill altogether, and the vascularized mantle wall has evolved to function as a lung. The lung opens to the outside through a pneumatostome (figure 10.19B and figure 10.22B).

figure 10.17

A, A sea hare, *Aplysia dactylomela,* crawls and swims across a coral reef, assisted by large, winglike parapodia, here curled above the body. **B,** When attacked, sea hares squirt a copious protective secretion derived from their red algal food source. Substances from the algae pass from the digestive gland to a "purple gland" where they are modified. Secretions from the purple gland exit the body through the mantle cavity, as do wastes from the anus. The anus and mantle cavity are posterior in sea hares because these animals undergo detorsion during development.

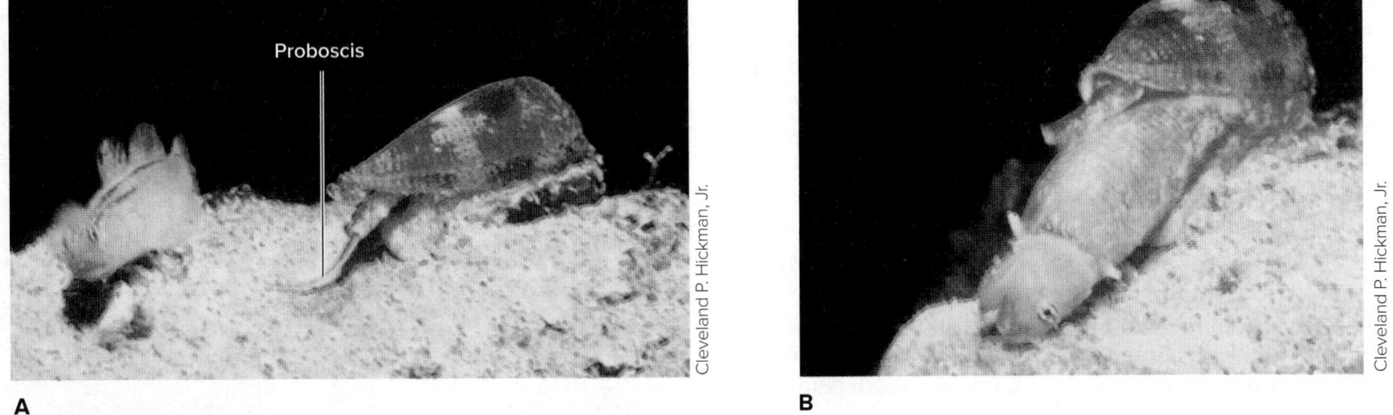

figure 10.18

A, *Conus* extends its long, wormlike proboscis. When the fish attempts to consume this tasty morsel, the *Conus* stings it in the mouth and kills it. **B,** The snail engulfs the fish with its distensible stomach, and then regurgitates the scales and bones some hours later.

Because the anus empties near the pneumatostome, waste is expelled forcibly with air or water from the lung. For the lung of a freshwater snail to take in air or to expel it, the animal must surface and expose the pneumatostome. The edge of the snail's mantle is rolled into a tube for air intake.

Most gastropods have a single nephridium (kidney). The circulatory and nervous systems are well developed (figure 10.19). The nervous system includes three pairs of ganglia connected by nerves. Sense organs include eyes, statocysts, tactile organs, and chemoreceptors.

There are both dioecious and hermaphroditic gastropods. Many perform elaborate courtship ceremonies. During copulation in hermaphroditic species, the two snails swap **spermatophores** (bundles of sperm), avoiding self-fertilization. Most land snails lay their eggs in holes in the ground or under logs. Some aquatic gastropods lay their eggs in gelatinous masses; others enclose them in gelatinous capsules or in parchment egg cases. Most marine gastropods go through a free-swimming veliger larval stage during which torsion and coiling occur. Others develop directly into a crawling juvenile within an egg capsule.

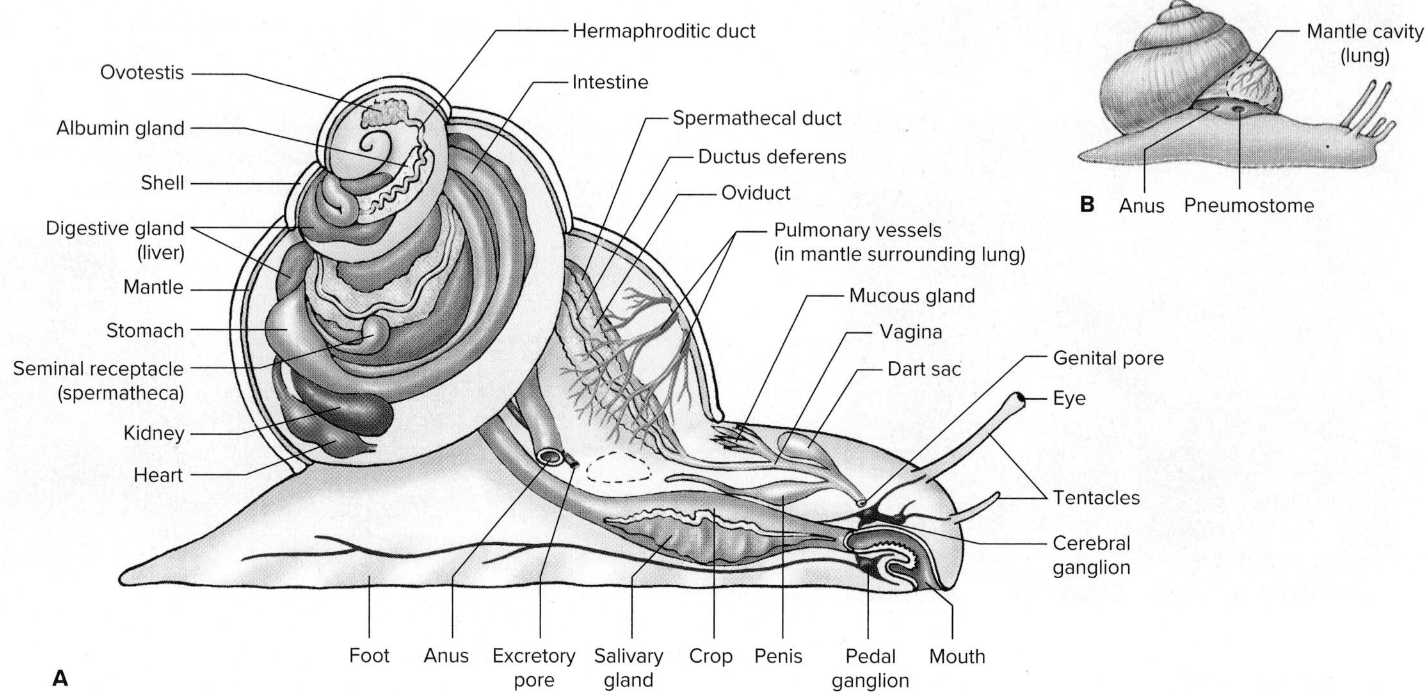

figure 10.19

A, Anatomy of a pulmonate snail. **B,** Position of the mantle cavity serving as a lung. Air enters and exits via the pneumostome.

Major Groups of Gastropods

Traditional classification of class Gastropoda recognized three subclasses: Prosobranchia, Opisthobranchia, and Pulmonata. Prosobranchia is by far the largest subclass, and its members are almost all marine. Familiar examples of marine prosobranchs are periwinkles, limpets (figure 10.20A), whelks, conchs, abalones (see figure 10.16A), slipper shells, oyster borers, rock shells, and cowries.

Opisthobranchia is an assemblage of marine gastropods including sea slugs, sea hares, nudibranchs, and canoe shells. At present, 8–12 groups of opisthobranchs are recognized. Some have a gill and a shell, although the latter may be vestigial, and some have no shell or true gill. Large sea hares, such as *Aplysia* (see figure 10.17), have large, earlike anterior tentacles, chemosensory rhinophores, and a vestigial shell. Nudibranchs have no shell as adults and rank among the most beautiful and colorful of molluscs (figure 10.21). Having lost the gill, the body surface of some nudibranchs is often increased for gas exchange by small projections **(cerata)** or by a ruffling of the mantle edge.

The third major group, Pulmonata, contains most land and freshwater snails and slugs. Usually lacking gills, their mantle cavity has become a lung, which fills with air by contraction of the mantle floor. Aquatic and a few terrestrial species have one pair of nonretractile tentacles, at the base of which are eyes; land forms usually have two pairs of tentacles, with the posterior pair bearing eyes (figure 10.22; see figure 10.19).

Currently, gastropod taxonomy is in flux, and some workers regard any attempt to present a classification as premature. A phylogeny of the gastropods based on morphological characters divided the class into two groups differing in form of the radula and other features. However, a preliminary molecular analysis did not support this grouping. The molecular data generally support a clade (Euthyneura) combining opisthobranchs and pulmonates, but Opisthobranchia appears to be paraphyletic.

Mandy Lindeberg, NOAA/NMFS/AKFSC

A

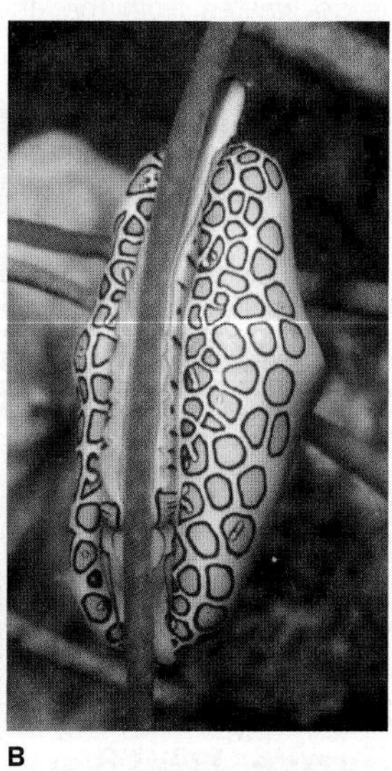
William C. Ober

B

figure 10.20

A, An Alaskan limpet grazes on algae, leaving behind it a cleared rock surface. **B,** A flamingo tongue snail, *Cyphoma gibbosum,* is a showy inhabitant of Caribbean coral reefs. These snails have a smooth, creamy orange-to-pink shell that is normally covered by the brightly marked mantle. Here the white flexible foot is visible as the snail crawls along a gorgonian coral.

figure 10.21

Phyllidia ocellata, a nudibranch. Like other *Phyllidia* spp., it has a hard body with dense calcareous spicules and bears its gills along the sides, between its mantle and foot.

A Pulmonate land snail

B Banana slug, *Ariolimax columbianus*

figure 10.22

A, Pulmonate land snail. Note two pairs of tentacles; the second, larger pair bears the eyes. **B,** Banana slug, *Ariolimax columbianus*. Note the pneumostome leading into the lung.

Key Theme 10.4

ECOLOGY

Invasive Slugs and Snails Spread Disease

Human infection with rat lung worm causes eosinophilic meningitis, a disease also called neuroangiostrongyliasis. Symptoms range from headache and nausea to paralysis, severe pain, and even death. Readers might wonder how rat lung worms interact with snails, but the nematode worm parasite (see section 12.1) has two hosts in its life cycle: rats and molluscs. Nematode worms mature in rats and their larvae are excreted with rat feces. Snails and slugs consume the feces and are infected. When rats eat molluscs, the life cycle is complete. Humans are not part of this life cycle, but may become infected with nematodes by handling the slugs or snails, drinking water or eating foods contaminated by them, or eating undercooked molluscan predators such as freshwater shrimp or frogs. Two invasive molluscs are implicated in disease spread: the giant African snail, *Lissachatina fulica*, and the semi-slug, *Parmarion martensi*. *L. fulica* is considered one of the most invasive species in the world—three animals were moved illegally to Florida as pets in 1966 and their descendants numbered 18,000 in 7 years. Rat lung worm in humans is known from tropical and subtropical regions of Asia, the Pacific, India, the Caribbean, South America, the southern United States, and Hawaii. Hawaiian monitoring by the Centers for Disease Control and Prevention confirmed 82 cases from 2007 to 2017.

10.9 Class Bivalvia (Pelecypoda)

Bivalvia (bi-val´ve-a) is also called Pelecypoda (pel-e-sip´o-da) (Gr. *pelekus*, hatchet, + *pous, podus*, foot). This class includes mussels, clams, scallops, oysters, and shipworms and range in size from tiny seed shells 1–2 mm in length to the giant South Pacific *Tridacna* (figure 10.23). Most bivalves are sedentary **suspension feeders** that depend on ciliary currents produced by the gills to collect their food. Unlike gastropods, they have no head, no radula, and very little cephalization.

Most bivalves are marine, but many live in brackish water and in streams, ponds, and lakes.

Form and Function

Shell

Bivalves are laterally compressed, and their two shells **(valves)** are held together dorsally by a hinge ligament that causes the valves to gape ventrally. **Adductor** muscles work in opposition to the hinge ligament and draw the valves together (figure 10.24B, C, and D). Projecting above the hinge ligament on each valve is the **umbo** (figure 10.24A), which is the oldest part of the shell. The valves function largely for protection, but those of shipworms (figure 10.25) have microscopic teeth for rasping wood, and rock borers use spiny valves for boring into rock.

figure 10.23

A giant clam (*Tridacna gigas*) lies buried in coral rock. The richly colored mantle tissue around the siphons bears enormous numbers of symbiotic dinoflagellates (zooxanthellae) that provide much of the clam's nutrition.

Body and Mantle

The visceral mass is suspended from the dorsal midline, and the muscular foot attaches to the visceral mass anteroventrally (figure 10.25). The name Pelecypoda, which translates to "hatchet foot," is a reference to shape of the foot, flattened between the two shells. The gills hang down on each side, each covered by a fold of the mantle. The posterior edges of the mantle folds form dorsal excurrent and ventral incurrent openings (see figure 10.24A). In some marine bivalves, part of the mantle is drawn out into extremely long, muscular siphons allowing the clam to burrow into the mud or sand and extend the siphons to the water above. Cilia on the gills and inner surface of the mantle direct the flow of water over the gills, bringing in food and oxygen (figure 10.25).

Bivalves have a three-chambered heart that pumps blood through the gills and mantle for oxygenation and to the kidneys for waste elimination (figure 10.26). They have three pairs of widely separated ganglia and generally poorly developed sense organs. A few bivalves have ocelli. The steely blue eyes of some scallops (see figure 10.29), located around the mantle edge, are remarkably complex, each being equipped with a cornea, lens, and retina.

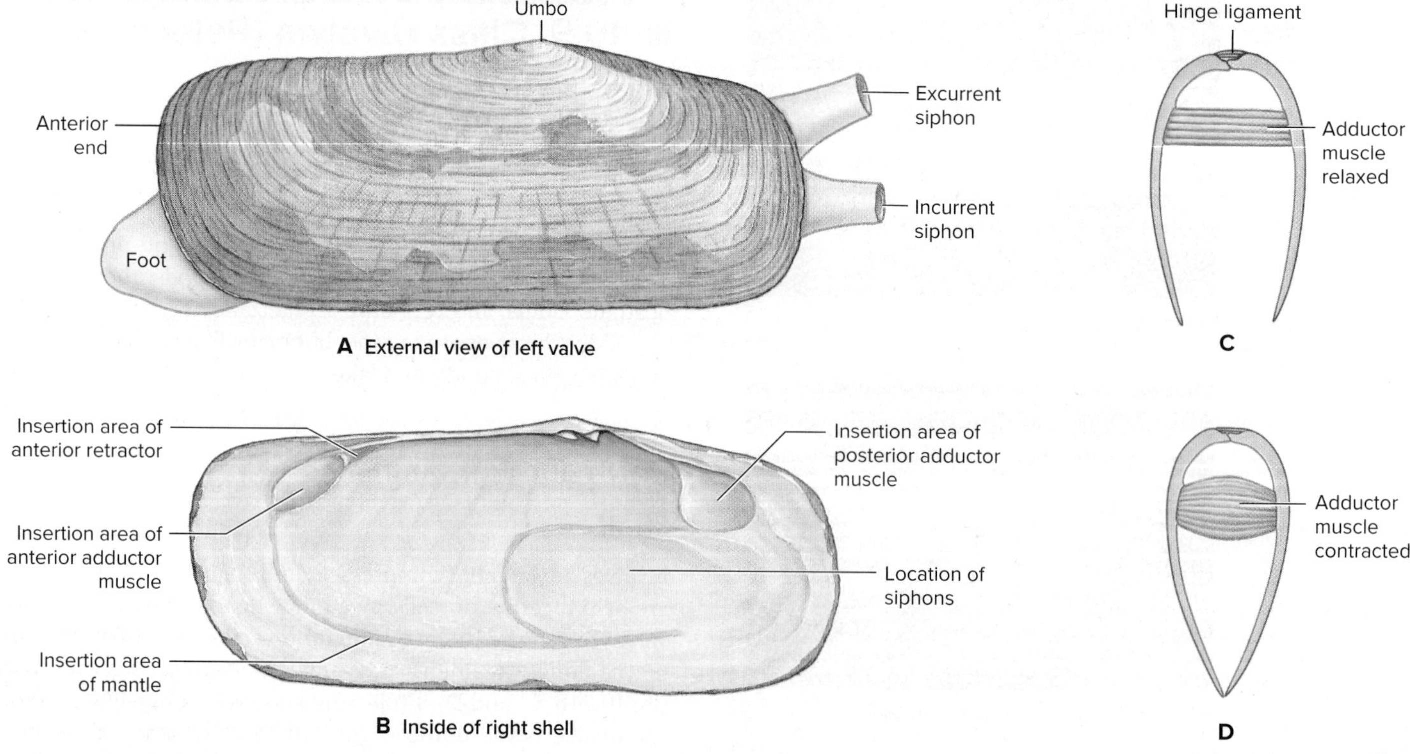

figure 10.24

Tagelus plebeius, the stubby razor clam (class Bivalvia). **A,** Left valve. **B,** Muscle scars inside the shell show the position of muscles and organs. **C, D,** Sections showing the function of adductor muscles and the hinge ligament as the shell opens and closes.

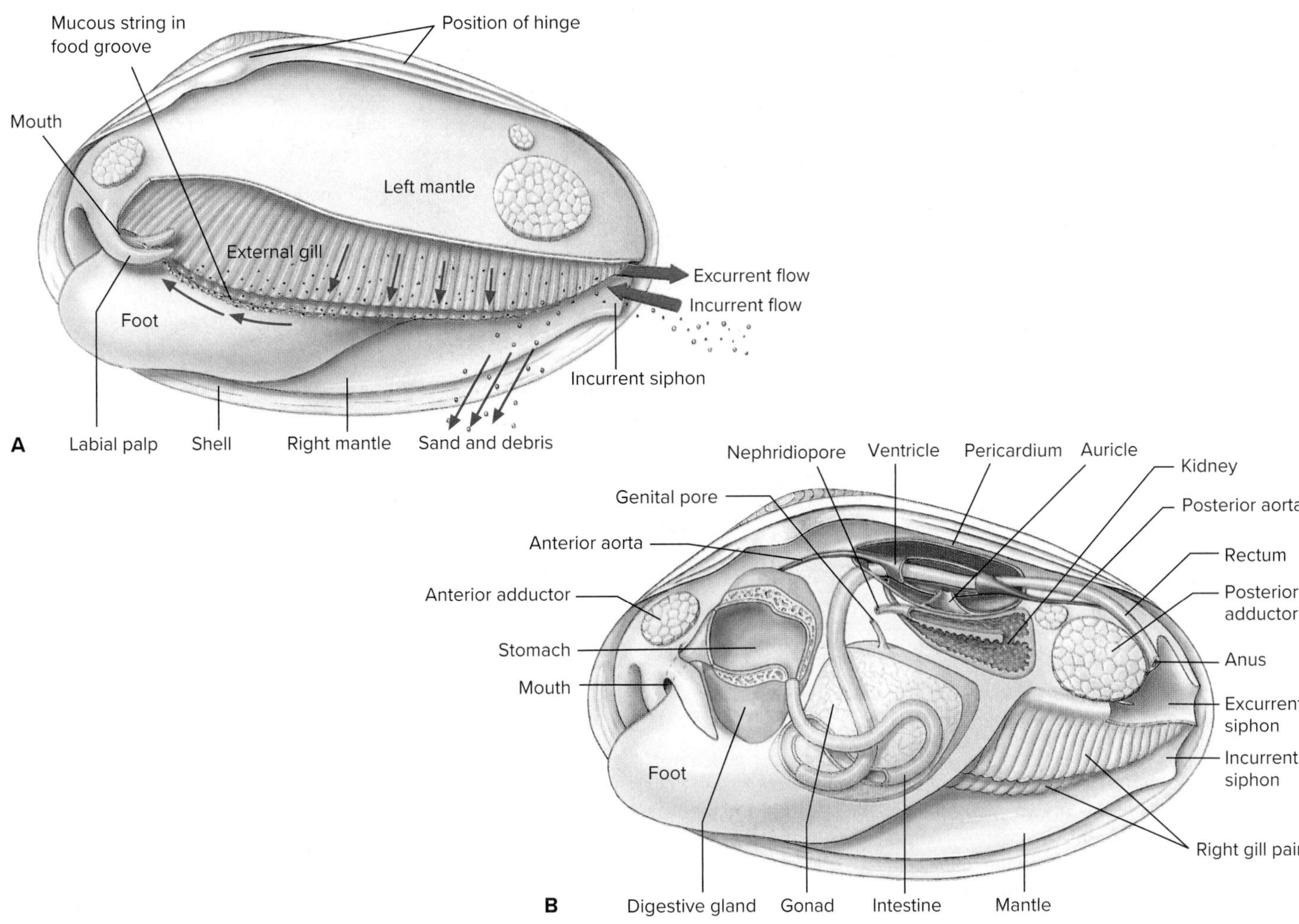

figure 10.25

A, Feeding mechanism of a freshwater clam, with the left valve and mantle removed. Water enters the mantle cavity posteriorly and is drawn forward by ciliary action to the gills and palps. As water enters the tiny openings of the gills, food particles are collected in strings of mucus, which are then carried by cilia to the palps and directed to the mouth. Sand and debris drop into the mantle cavity and are removed by cilia. **B,** Clam anatomy.

Feeding and Digestion

Most bivalves are suspension feeders. Their respiratory currents bring both oxygen and organic materials to their gills, where ciliary tracts direct them to the tiny gill pores. Gland cells on the gills and labial palps secrete copious amounts of mucus, which entangles food particles suspended in the water entering gill pores. Ciliary tracts move the particle-laden mucus to the mouth (see figure 10.25).

In the stomach, the mucus and food particles are kept whirling by a rotating gelatinous rod, called a **crystalline style.** As the style rotates, enzymes for extracellular digestion dissolve from it in layers. Ciliated ridges of the stomach sort food particles and direct suitable particles to the digestive gland for intracellular digestion.

Shipworms (figure 10.27) feed on the particles they excavate as they burrow in wood. Symbiotic bacteria occupy a special organ in these bivalves and produce cellulase to digest wood. Other bivalves, such as giant clams, gain much of their nutrition from the photosynthetic products of symbiotic algae (zooxanthellae, see section 5.5: Dinoflagellata) in their mantle tissue (see figure 10.23).

Locomotion

Most bivalves move by extending their slender muscular foot between the valves (see figure 10.24A). They pump blood into the foot, causing it to swell and to act as an anchor in mud or sand. Then longitudinal muscles contract to shorten the foot and pull the animal downward. Most bivalves use the foot for burrowing, but a few creep. Some bivalves are sessile: oysters attach their shells to a surface by secreting cement, and mussels (figure 10.28) attach to surfaces by secreting a number of slender stiff threads called byssal threads, or a byssus. The foot is often reduced in sessile bivalves. A few bivalves, such as scallops (figure 10.29), use their shells for locomotion by clapping the valves together to move in spurts.

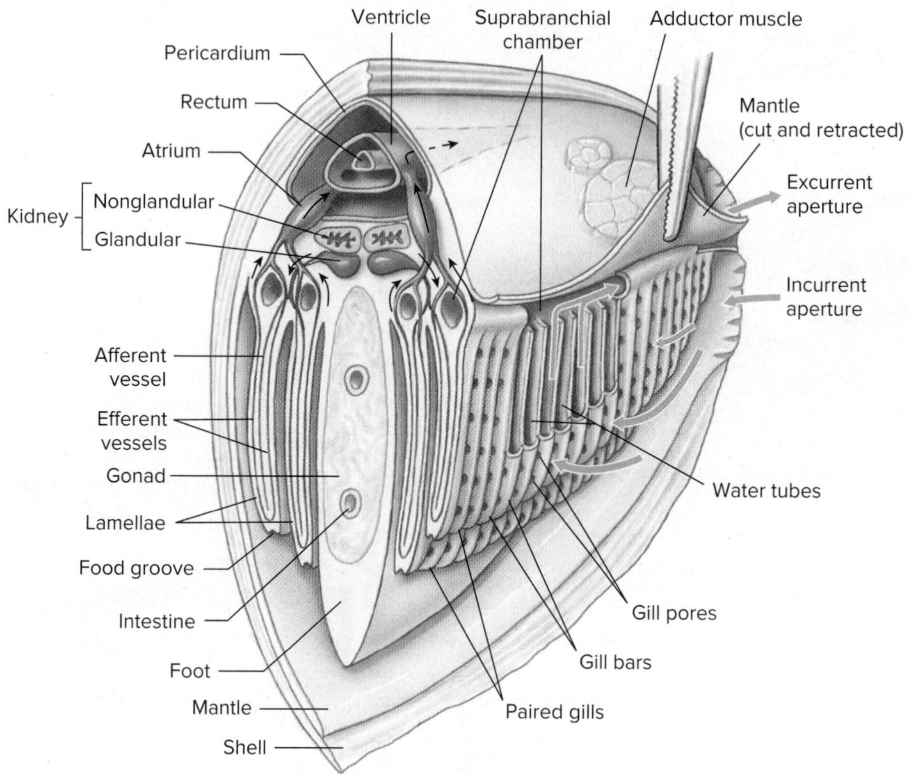

figure 10.26

Section through the heart region of a freshwater clam to show the circulatory and respiratory systems in relation to one another. Respiratory water currents: water is drawn in by cilia, enters gill pores, and then passes up water tubes to suprabranchial chambers and out the excurrent aperture. Blood in the gills exchanges carbon dioxide and oxygen. Blood circulation: the ventricle pumps blood forward to sinuses of their foot and viscera, and posteriorly to the mantle sinuses. Blood returns from the mantle to auricles; it returns from viscera to the kidney, and then goes to the gills, and finally to the auricles.

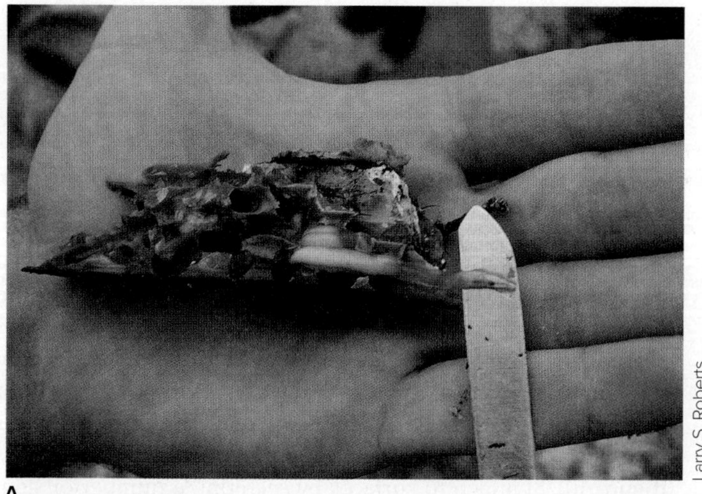

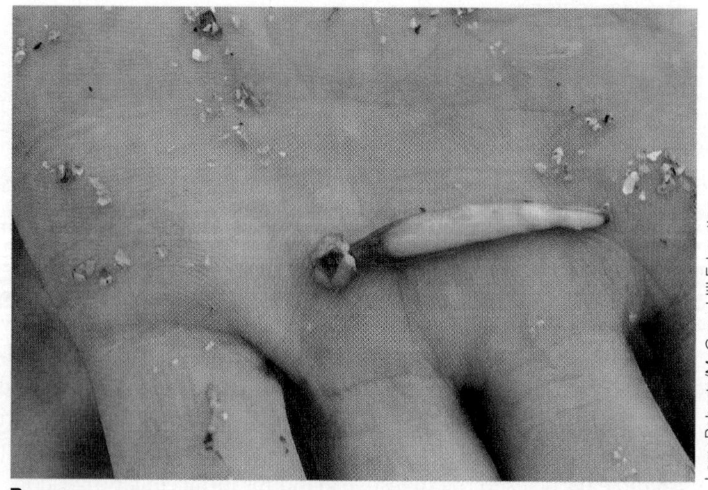

figure 10.27

A, Shipworms (*Teredo, Bankia,* and others) are bivalves that burrow in wood, causing great damage to unprotected wooden hulls and piers. **B,** The two small, anterior valves, seen at left, are used as rasping organs to extend the burrow.

NHPA/Matt Bain

figure 10.28

Mussels, *Mytilus edulis*, occur in northern oceans around the world; they form dense beds in the intertidal zone. A host of marine creatures live protected beneath attached mussels.

Larry S. Roberts

figure 10.29

Representing a group that has evolved from burrowing ancestors, the surface-dwelling bay scallop, *Aequipecten irradians*, has developed sensory tentacles and a series of blue eyes along its mantle edges.

Reproduction

Sexes are usually separate, and fertilization is usually external. Marine embryos typically go through three free-swimming larval stages—trochophore, veliger larva, and young spat—before reaching adulthood. In freshwater clams, fertilization is internal, and some gill tubes become temporary brood chambers. There, larvae develop into specialized veligers called **glochidia,** which are discharged with the excurrent flow (figure 10.30). If glochidia come in contact with a passing fish, they hitchhike a ride as parasites in the fish's gills for the next 20–70 days before sinking to the bottom to become sedentary adults. In lakes or streams, attachment to a fish prevents the small larvae from being moved out of the lake or farther downstream.

Key Theme 10.5
ECOLOGY

Endangered Bivalves

Freshwater clams were once abundant and diverse in streams throughout the eastern United States, but they are now easily the most jeopardized group of animals in the country. Of the more than 300 species once present, 24 are extinct, 160 are listed as threatened or endangered, and many more may be listed soon. A combination of causes is responsible, of which decline in water quality is among the most important. Pollution and sedimentation from mining, industry, and agriculture are among the culprits. Habitat destruction due to altered natural water courses and damming is an important factor. Poaching to supply the Japanese cultured pearl industry is partially to blame (see Key Theme 10.1). In addition, the prolific zebra mussels (see Key Theme 10.6) attach in great numbers to native clams, exhausting food supplies (phytoplankton) for native bivalves in the surrounding water.

Key Theme 10.6
ECOLOGY

Invasive Bivalves

Vitalii Hulai/123RF

Zebra mussels, *Dreissena polymorpha*, are a potentially disastrous biological introduction into North America. They were apparently picked up as veligers with ballast water by one or more ships in freshwater ports in northern Europe and then expelled between Lake Huron and Lake Erie in 1986. The 4 cm bivalve (see image) spread throughout the Great Lakes by 1990, and by 1994 it was as far south on the Mississippi River as New Orleans, as far north as Duluth, Minnesota, and as far east as the Hudson River in New York. It attaches to any firm surface and filter-feeds on phytoplankton. Large numbers accumulate rapidly. They foul water intake pipes of municipal and industrial plants, impede intake of water for municipal supplies, and have far-reaching effects on the ecosystem. Zebra mussel control costs approximately $500 million annually. However, on a slightly brighter note, some evidence suggests that a few species of native bivalves may survive in areas with zebra mussels, although their population sizes are severely reduced.

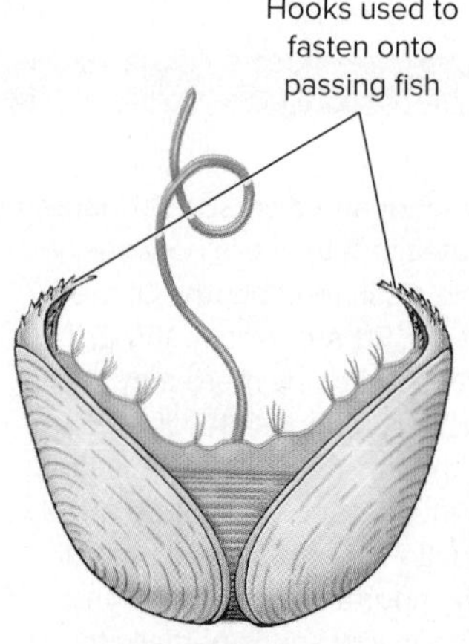

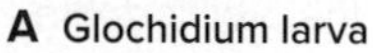

B Pocketbook mussel, *Lampsilis ovata*

figure 10.30

A, Glochidium, or larval form, for some freshwater clams. When 0.3 mm larvae are released from the mother's broad pouch, they may become attached to a fish's gill by clamping their valves closed. They remain as parasites on the fish for several weeks. **B,** Some clams have adaptations that help their glochidia find a host. The mantle edge of this female pocketbook mussel (*Lampsilis ovata*) mimics a small minnow, complete with eye. When a smallmouth bass comes to dine, it is doused with glochidia.

10.10 Class Cephalopoda

Cephalopoda (sef´-a-lo-pod´a) include squids, octopuses, nautiluses, and cuttlefishes. All are marine, and all are active predators.

Cephalopods (Gr. *kephalē,* head, + *pous, podos,* foot) have an odd body plan (figure 10.31) that develops as the embryonic head and foot become indistinguishable. The ring around the mouth bearing the arms and tentacles develops from the anterior margin of the head, whereas the circle of arms or tentacles itself is derived from the anterior margin of the foot. The foot also forms a **funnel** for expelling water from the mantle cavity.

Cephalopods range in size from 2 to 3 cm up to the giant squid, *Architeuthis,* which is the largest invertebrate known. The squid *Loligo* (L., cuttlefish) is about 30 cm long (figure 10.31A).

Cephalopods are predaceous, feeding chiefly on small fishes, molluscs, crustaceans, and worms. Their arms, used to capture and handle food, have a complex musculature and make delicately controlled movements. They are highly mobile and swiftly seize prey and bring it to the mouth. Octopuses and cuttlefishes have salivary glands that secrete a venom for immobilizing prey. Strong, beaklike jaws grasp prey, and the radula tears off pieces of flesh (figure 10.31A). However, not all of the prey is shredded by the beaks and radula—salivary gland secretions begin the digestive process so that partially liquefied food can be ingested. Digestion is extracellular and occurs in the stomach and cecum.

Key Theme 10.7
ECOLOGY

Sperm Whales Prey on Giant Squid

The giant squid, *Architeuthis,* is known only from stranded animals, those captured in nets of fishermen, and specimens found in stomachs of sperm whales. The mantle length is 5–6 m, and the head is up to one meter long. Squid are an important food item for sperm whales, who detect their prey using sonar. Squid have the largest eyes of any animal, reaching up to 27 cm in diameter with a 9 cm pupil. The eyes appear to be adapted for long-range vision, so although whale sonar would reveal the presence of the squid before the squid could see the whale, long-range vision might still allow the squid time for evasive action. The skin of sperm whales is often scarred by suckers from squid tentacles, evidence of deadly underwater battles.

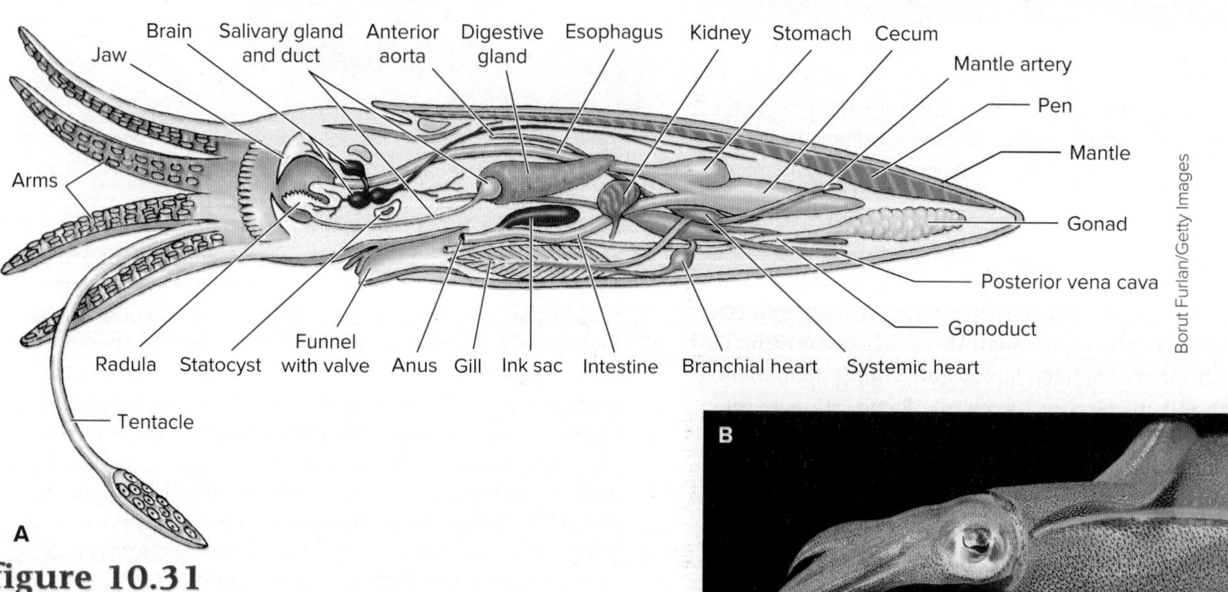

B

Todd J. Morris PhD

figure 10.31

A, Lateral view of anatomy of a typical squid, such *Loligo*, with the left half of the mantle removed. **B,** *Loligo vulgaris* from the Mediterranean Sea, Kas, Turkey.

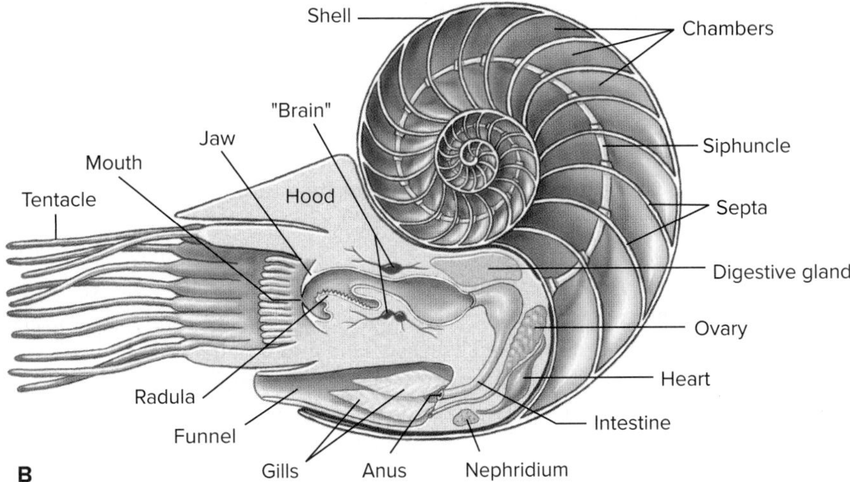

A B

figure 10.32

Nautilus, a cephalopod. **A,** Live *Nautilus,* feeding on a fish. **B,** Longitudinal section, showing the gas chambers of the shell and a diagram of the internal body structure.

figure 10.33

Cuttlefish, *Sepia latimanus*, has an internal shell familiar to keepers of caged birds as "cuttlebone."

Form and Function

Shell

Fossil records of cephalopods extend to the Cambrian period. The earliest shells were straight cones. Cephalopods without shells or with internal shells (such as octopuses and squids) probably evolved from a straight-shelled ancestor. Other fossil cephalopods had curved or coiled shells, as in ammonoids and nautiloids. The only remaining members of the once-flourishing nautiloids are the modern *Nautilus* species (Gr. *nautilos,* sailor) (figure 10.32). Ammonoids were widely prevalent in the Mesozoic era but became extinct by the end of the Cretaceous period. Reasons for their extinction remain a mystery. Present evidence suggests that they were gone before the asteroid bombardment at the end of the Cretaceous period (see section 1.7: Mass Extinctions).

Although early nautiloid and ammonoid shells were heavy, they were made buoyant by a series of gas chambers, as in of *Nautilus* (figure 10.32B), which enabled the animal to swim while carrying its shell. The shell of *Nautilus,* although coiled, is quite different from that of a gastropod. Transverse septa divide the shell into internal chambers (figure 10.32B). The living animal inhabits only the last chamber. As it grows, it moves forward, secreting behind it a new septum. The chambers are connected by a cord of living tissue called a **siphuncle,** which extends from the visceral mass. Ammonoids had chambered shells analogous to those of nautiloids, but the septa were more complex.

Cuttlefishes also have a small, coiled or curved shell, but it is entirely enclosed by the mantle (figure 10.33). In squids, most of the shell has disappeared, leaving only a thin, flexible strip called a **pen,** which the mantle encloses. *Octopus* (Gr. *oktos,* eight, + *pous, podos,* foot) has no shell.

Key Theme 10.8

ADAPTATION AND PHYSIOLOGY

Gas Chambers in *Nautilus*

After *Nautilus* secretes a new septum, the new chamber is filled with fluid similar in ionic composition to that of the *Nautilus's* blood (and of seawater). Fluid removal involves the active secretion of ions into tiny intercellular spaces in the siphuncular epithelium, so that a very high local osmotic pressure is produced, and the water is drawn from the chamber by osmosis. The gas in the chamber is only the respiratory gas from the siphuncle tissue that diffuses into the chamber as the fluid is removed. Thus the gas pressure in the chamber is 1 atmosphere or less because it is in equilibrium with the gases dissolved in the seawater surrounding the *Nautilus.* These dissolved gases are in turn in equilibrium with the air at the surface of the sea, despite the fact that the *Nautilus* may be swimming at 400 m beneath the surface. That the shell can withstand implosion by the surrounding 41 atmospheres (about 600 pounds per square inch), and that the siphuncle can remove water against this pressure, are marvelous feats of natural engineering!

Body and Mantle

In *Nautilus,* the head with its tentacles can be extruded from the opening of the body compartment of the shell (see figure 10.32). Its 60–90 or more tentacles have no suckers but adhere to prey by secretions. The tentacles search for, sense, and grasp food. Beneath the head is the funnel. The shell shelters the mantle, mantle cavity, and visceral mass. Two pairs of gills occupy the mantle cavity.

Cephalopods other than nautiloids have only one pair of gills. Octopuses have 8 arms with suckers. Squids and cuttlefishes (decapods) have 10 arms: 8 arms with suckers and a pair of long, retractile tentacles. The thick mantle covering the trunk fits loosely at the neck region, allowing water to be taken into the mantle cavity. When the mantle edges contract closely about the neck, water is forcefully expelled through the funnel. The water current thus created provides oxygen for gills in the mantle cavity, jet power for locomotion, and a means of carrying wastes and sexual products away from the body.

The active habits of cephalopods are reflected in their internal anatomy, particularly their respiratory and circulatory systems. Ciliary propulsion would not circulate enough water over the gills for an active animal, so cephalopods ventilate their gills by muscular action of the mantle wall. They have a closed circulatory system with a network of vessels, and blood flows through the gills via capillaries. Branchial (gill) hearts, or accessory hearts (see figure 10.31A), increase blood pressure in the gills, which occur at the end of the molluscan circulatory system. This increased blood pressure permits more rapid gas exchange.

Cephalopods have well-developed nervous systems, including the most complex brain among invertebrates (see figure 10.31A). Except for *Nautilus,* which has relatively simple eyes, cephalopods have elaborate eyes with a cornea, lens, chambers, and a retina (figure 10.34)—similar to the camera-type eye of vertebrates.

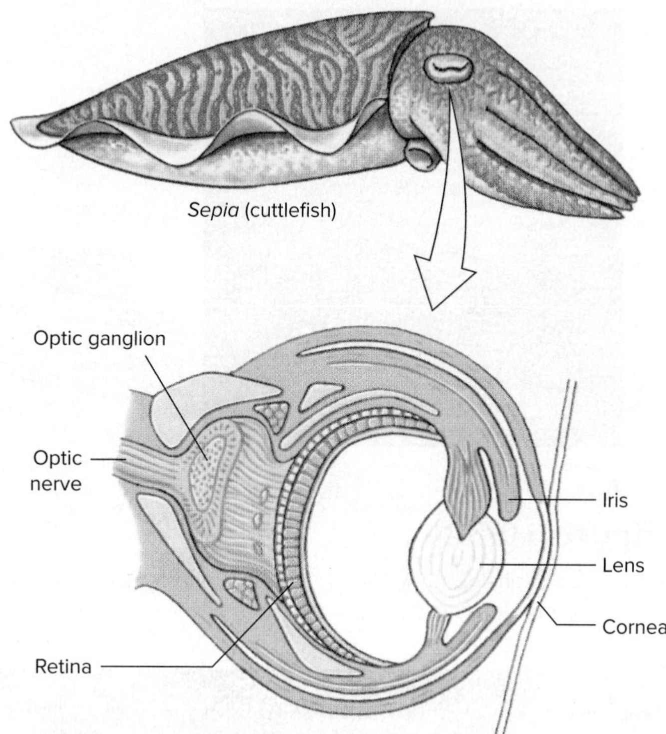

figure 10.34

Eye of a cuttlefish (*Sepia*). The structure of cephalopod eyes shows a high degree of convergent evolution with the eyes of vertebrates.

Color Changes

Special pigment cells called **chromatophores** in the skin of most cephalopods produce color changes by expanding and contracting. They are controlled by the nervous system and perhaps by hormones. Some color changes are protective to match background hues; most are behavioral and associated with alarm or courtship. The degree to which cephalopods are able to match colorful backgrounds made it difficult to believe that they lack color vision. However, in an experiment where background colors were equal in intensity, cephalopods were unable to match colors. This result, coupled with the physiology of their eyes, makes color vision unlikely. Cephalopod photoreceptors are rhabdomere cells, not rods and cones, as in vertebrates. Their ability to see color depends on the number of visual pigments present; almost all cephalopods have only one such pigment and cannot detect color. An exception is the bioluminescent firefly squid, which has three visual pigments, the same number found in humans. It appears that the wonderful color patterns we see on cephalopod skin are visible to other cephalopods as polarized light patterns; all cephalopods detect differences in polarized light. They also use this ability to enhance discrimination of translucent prey and fish with silvery reflective scales in much the same way that fishermen use polarized glasses to reduce glare from water.

Ink Production

Most cephalopods other than nautiloids have an ink sac that empties into the rectum. The sac contains an ink gland that secretes a dark fluid containing the pigment melanin. When the animal is alarmed, it releases a cloud of ink through the anus to form a "smokescreen" to confuse an enemy.

Locomotion

Most cephalopods swim by forcefully expelling water from the mantle cavity through a ventral funnel—a sort of jet-propulsion method. The funnel is mobile and can be pointed forward or backward to control direction; the force of water expulsion determines speed.

Squids and cuttlefishes are excellent swimmers. The squid body is streamlined and built for speed (see figure 10.31B). Cuttlefishes (see figure 10.33) swim more slowly. Both squids and cuttlefishes have lateral fins that can serve as stabilizers,

but they are held close to the body for rapid swimming. The gas-filled chambers of *Nautilus* keep the shell upright. Although not as fast as squids, nautiluses move surprisingly well.

Octopus has a rather globular body and no fins (see figure 10.1E). Octopuses can swim backward by spurting jets of water from their funnel, but they are better adapted to crawling over rocks and coral, using the suction discs on their arms to pull or to anchor themselves. Some deep-water octopods have fins and arms webbed like an umbrella; they swim in a medusa-like fashion. Very large octopuses may flatten their bodies and swim by undulation.

Reproduction

Sexes are separate in cephalopods. Before copulation, males often display changes in skin pigmentation and patterning, apparently directed toward females and against rival males. In the male seminal vesicle, spermatozoa are encased in spermatophores and stored in a sac that opens into the mantle cavity. During copulation, one arm of an adult male plucks a spermatophore from his own mantle cavity and inserts it into the female's mantle cavity near the oviduct opening (figure 10.35). Eggs are fertilized as they leave the oviduct and are usually attached to stones or other objects to develop. Some octopuses tend their eggs.

figure 10.35

Copulation in cephalopods. **A,** Mating cuttlefishes. **B,** A male octopus uses its modified arm to deposit spermatophores in the female mantle cavity to fertilize her eggs. Octopuses often tend their eggs during development.

10.11 Phylogeny and Adaptive Diversification

Phylogeny

Fossils attributed to Mollusca date from the early Cambrian period (figure 10.36). Some had shells and others had wormlike bodies without shells. Was the ancestral mollusc wormlike? Two classes of living molluscs have wormlike bodies: Caudofoveata and Solenogastres. According to the Conchifera hypothesis shown in figure 10.2, caudofoveates and solenogasters separate from a common ancestor of all other molluscs at the base of the molluscan tree. For many years, biologists assumed that the wormlike form was ancestral for both classes and that caudofoveates possessed many features of the ancestral mollusc. On this basis the ancestral mollusc was thought to be wormlike with a ventral gliding surface and a dorsal mantle with a chitinous cuticle and calcareous scales. There are two hypotheses for the evolution of molluscs: the Conchifera hypothesis and the Serialia hypothesis. We illustrate one version of Conchifera in figure 10.2 where the modern wormlike taxa, caudofoveates and solenogastres, form the sister taxon to chitons. The ancestor of this group had seven shell plates, which were greatly reduced in the wormlike molluscs.

However, according to the Serialia hypothesis, only the solenogasters have an ancestral wormlike body, and the wormlike body of caudofoveates is derived from a body plan resembling those of gastropods or cephalopods. Under this hypothesis caudofoveates, cephalopods, scaphopods, and gastropods form a clade; relationships among these four taxa vary depending on the method of tree construction used, but in some trees, caudofoveates are the sister taxon to cephalopods. Thus, the morphology of a caudofoveate must be interpreted in entirely different ways under the two main hypotheses for evolution of the molluscan classes.

The Serialia and Conchifera hypotheses also differ in the phylogenetic position of the two taxa with repeated body parts. In chitons, there are typically eight, and occasionally seven, calcareous shell plates along the back of the animal. Plates have their own musculature; the number of pairs of gills ranges from 6 to more than 80. In the monoplacophorans, there are three to six pairs of gills, two pairs of auricles, three to seven pairs of metanephridia, one or two pairs of gonads, and a ladderlike nervous system with ten pairs of pedal nerves. The body is not segmented in chitons or monoplacophorans—so what is the significance of the repeated parts? One possibility is that repeated parts are remnant features of a truly segmented ancestor; another is that they evolved independently in each group.

Under the Conchifera hypothesis, multiple foot retractor muscles are ancestral traits for all shelled molluscs, but multiple gills and other serially repeated features evolved independently in chitons and monoplacophorans (see figure 10.2). In strong contrast, under the Serialia hypothesis, these features arose in a common ancestor of the Serialia clade and were inherited by

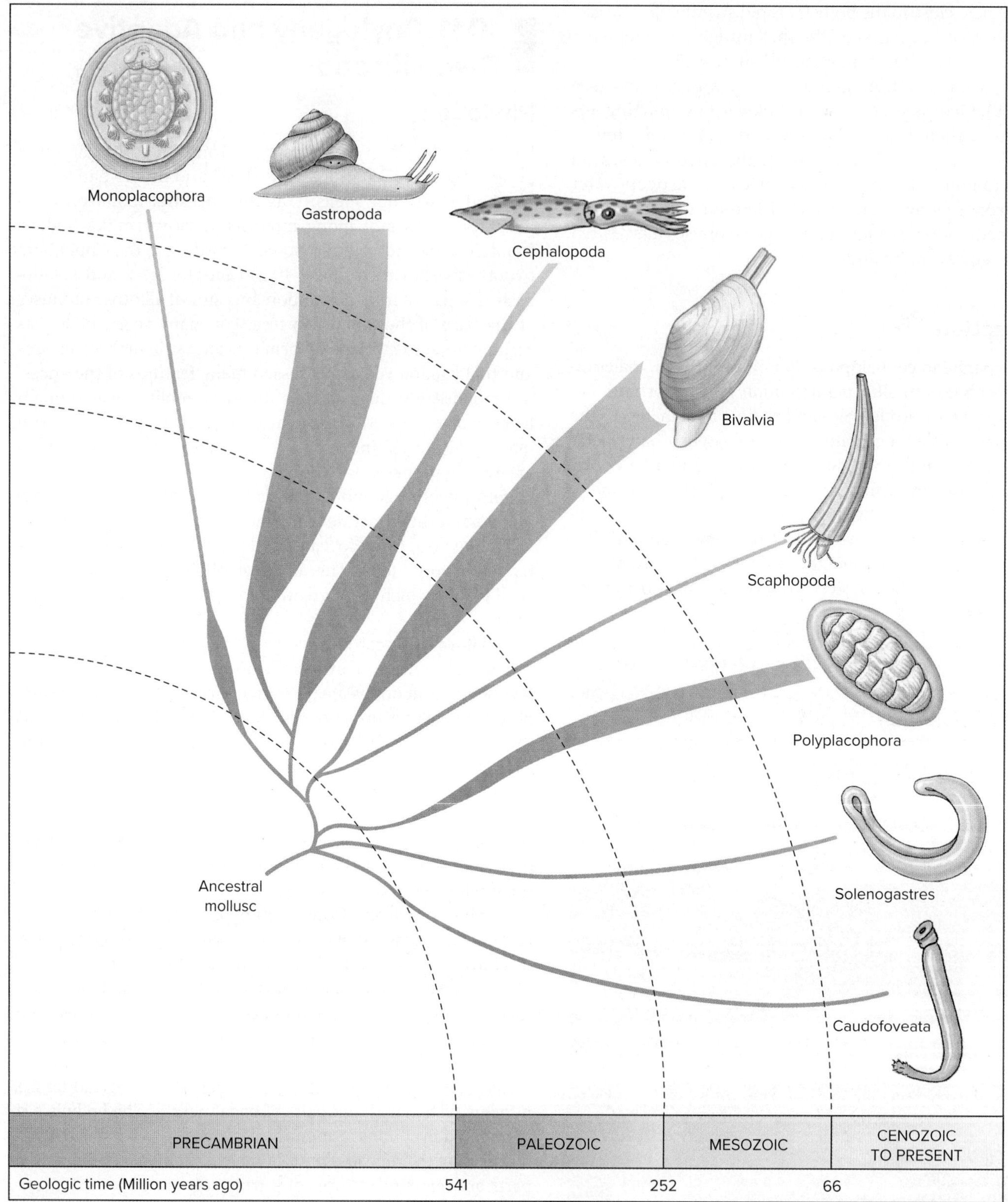

figure 10.36

Classes of Mollusca, showing their derivations and relative abundance.

the sister taxa: chitons and monoplacophorans. However, a 2019 analysis of current information from fossil forms, developmental patterns, and genetic data concluded that the modern wormlike forms were simplified from shelled ancestors and that the first mollusc was not wormlike. The Conchifera hypothesis (figure 10.2) is currently better supported than the Serialia hypothesis.

Adaptive Diversification

Both gastropods and cephalopods have a greatly expanded visceral mass. In gastropods, the mantle cavity was brought toward the head by torsion, but in cephalopods the mantle cavity was extended ventrally. Evolution of a chambered shell in cephalopods was a very important contribution to their freedom from the substrate and their ability to swim. Elaboration of their respiratory, circulatory, and nervous systems is correlated with their predatory and swimming habits.

Scaphopods and bivalves have an expanded mantle cavity that essentially envelops the body. Adaptations for burrowing characterize these taxa: a spatulate foot and reduction of the head and sense organs.

Most of the diversity among molluscs is related to their adaptation to different habitats and modes of life. Until recently, all terrestrial slugs were presumed to have evolved from terrestrial snail ancestors whose shells prevented dessication and permitted the colonization of land. These early snail colonists belonged to several different lineages, each independently losing the shell as they adapted to moist habitats. However, research in Palau, Western Pacific, has identified a new terrestrial slug, *Aiteng marefugitus,* whose intertidal marine ancestors did not have shells. Ancestral intertidal slugs were able to invade the rainforest habitat in Palau because of its high humidity and lack of terrestrial competitors. Molluscs exhibit a wide variety of feeding methods, ranging from sedentary filter feeding to active predation. Adaptations for food gathering include an enormous variety in radular structure and function, particularly among gastropods.

The versatile glandular mantle secretes the shell and forms the mantle cavity, but it is also variously modified into gills, lungs, siphons, and apertures, and it sometimes functions in locomotion, in feeding, or in a sensory capacity. The shell, too, has undergone a variety of evolutionary adaptations.

Taxonomy of Phylum Mollusca

Class Caudofoveata (kaw´do-fo-ve-at´a) (L. *cauda,* tail, + *fovea,* small pit): **caudofoveates.** Wormlike; shell, head, and excretory organs absent; radula usually present; mantle with chitinous cuticle and calcareous scales; oral pedal shield near anterior mouth; mantle cavity at posterior end with pair of gills; sexes separate; often united with solenogasters in class Aplacophora. Examples: *Chaetoderma, Limifossor* (see figure 10.8).

Class Solenogastres (so-len´o-gas´trez) (Gr. *solēn,* pipe, + *gastēr,* stomach): **solenogasters.** Wormlike; shell, head, and excretory organs absent; radula usually present; mantle usually covered with scales or spicules; mantle cavity posterior, without true gills, but sometimes with secondary respiratory structures; foot represented by long, narrow, ventral pedal groove; hermaphroditic. Example: *Neomenia* (see figure 10.8).

Class Monoplacophora (mon´o-pla-kof´o-ra) (Gr. *monos,* one, + *plax,* plate, + *phora,* bearing): **monoplacophorans.** Body bilaterally symmetrical with a broad, flat foot; a single limpetlike shell; mantle cavity with five or six pairs of gills; large coelomic cavities; radula present; six pairs of nephridia, two of which are gonoducts; separate sexes. Example: *Neopilina* (see figure 10.9).

Class Polyplacophora (pol´y-pla-kof´o-ra) (Gr. *polys,* many, several, + *plax,* plate, + *phora,* bearing): *chitons.* Elongated, dorsoventrally flattened body with reduced head; bilaterally symmetrical; radula present; shell of seven or eight dorsal plates; foot broad and flat; gills multiple, along sides of body between foot and mantle edge; sexes usually separate, with a trochophore but no veliger larva. Examples: *Chiton tuberculatus* (see figure 10.10), *Chaetopleura.*

Class Scaphopoda (ska-fo-pod´a) (Gr. *skaphē,* trough, boat, + *pous, podos,* foot): **tusk shells.** Body enclosed in a one-piece tubular shell open at both ends; conical foot; mouth with radula and tentacles; head absent; mantle for respiration; sexes separate; trochophore larva. Example: *Dentalium* (see figure 10.12).

Class Gastropoda (gas-tro-pod´a) (Gr. *gastēr,* belly, + *pous, podos,* foot): **snails and relatives.** Body asymmetrical; usually in a coiled shell (shell uncoiled or absent in some); head well developed, with radula; foot large and flat; dioecious or monoecious, some with trochophore, typically with veliger, some without larva. Examples: *Busycon, Polinices* (see figure 10.16B), *Physa, Helix, Aplysia* (see figure 10.17).

Class Bivalvia (bi-val´ve-a) (L. *bi,* two, + *valva,* folding door, valve) **(Pelecypoda): bivalves.** Body enclosed in a two-lobed mantle; shell of two lateral valves of variable size and form, with dorsal hinge; head greatly reduced but mouth with labial palps; no radula; no cephalic eyes; gills platelike; foot usually wedge shaped; sexes usually separate, typically with trochophore and veliger larvae. Examples: *Mytilus* (see figure 10.28), *Venus, Bankia* (see figure 10.27).

Class Cephalopoda (sef-a-lo-pod´-a) (Gr. *kephalē,* head, + *pous, podos,* foot): **squids and octopuses.** Shell often reduced or absent; head well developed with eyes and a radula; head with arms or tentacles; foot modified into a funnel; nervous system of well-developed ganglia, centralized to form a brain; sexes separate, with direct development. Examples: *Loligo, Sepioteuthis* (see figure 10.31), *Octopus* (see figure 10.1E), *Sepia* (see figure 10.33), *Nautilus* (see figure 10.32).

■ SUMMARY

Sections	Key Concepts
10.1 Ecological Relationships	• Mollusca is one of the largest and most diverse phyla. • Molluscs range in size from very small organisms to the largest invertebrates. • Their basic body divisions are the head-foot and the visceral mass, usually covered by a shell. • A majority are marine, but some are freshwater, and a few are terrestrial.
10.2 Economic Importance and Ocean Acidification	• Many molluscs are economically important as harvested food items. • Ocean acidification is a serious threat to molluscs because increased acidity makes it harder to secrete calcium, the key component of healthy shells. • Bivalves, such as oysters, are particularly affected, damaging economies in coastal communities.
10.3 Form and Function	• The molluscan body is divided into two regions: head-foot and visceral mass. • The body is usually covered by a shell. • Molluscs are coelomate (have a coelom), although their coelom is limited to the area around the heart. • The mantle secretes the shell and overlies a part of the visceral mass to form a cavity housing the gills. The mantle cavity has been modified into a lung in some molluscs. • The foot is usually a ventral, solelike, locomotory organ, but it may be variously modified, as in the cephalopods, where it has become a funnel and arms. • The radula is a protrusible, tongue-like organ with teeth used in feeding. Most molluscs, except bivalves and some solenogasters, have a radula. • The circulatory system of molluscs is open, with a heart and blood sinuses, except in cephalopods, which have a closed circulatory system. • Molluscs usually have a pair of nephridia connecting with the coelom, and a complex nervous system with a variety of sense organs. • The ancestral larva of molluscs is the trochophore, and most marine molluscs have a derived second larva, the veliger.
10.4 Classes Caudofoveata and Solenogastres	• Classes Caudofoveata and Solenogastres are small groups of wormlike molluscs with calcareous spines or scales, but there is no shell.
10.5 Class Monoplacophora	• Class Monoplacophora is a small group of marine molluscs with many serially repeated body parts beneath a single cap-like shell.
10.6 Class Polyplacophora: Chitons	• Polyplacophorans are commonly called chitons. • They have seven or eight shell plates along the dorsal surface, surrounded by a band of mantle, called the girdle. They have a row of gills along each side of their foot. • Chitons are slow-moving grazers.
10.7 Class Scaphopoda	• Scaphopoda is a small class whose members have a tubular shell, open at both ends, and the mantle wrapped around the body.
10.8 Class Gastropoda	• Gastropoda is the largest and most diverse class of molluscs. • Gastropods occur in marine and freshwater habitats and are the only molluscs to exploit terrestrial environments. • All gastropods exhibit torsion where, early in development, the posterior end rotates to the anterior so that the mantle cavity, gills, and anus are near the head. • Most gastropods also exhibit coiling, an elongation and spiraling of the visceral mass. • Torsion may cause fouling, where excreta are released onto the gills and near the head. • Fouling is avoided in various ways, including a unidirectional water flow system beginning on one side of the mantle cavity and exiting on the other side (seen in many gastropods), some degree of detorsion (opisthobranchs and pulmonates), or conversion of the mantle cavity into a lung (pulmonates).
10.9 Class Bivalvia (Pelecypoda)	• Class Bivalvia includes marine and freshwater forms. • Bivalves have two shells joined by a dorsal ligament and held together by an adductor muscle. • Most bivalves are suspension feeders, drawing water through their gills by ciliary action.

Sections	Key Concepts
10.10 Class Cephalopoda	• Members of class Cephalopoda are the most complex molluscs; all are predators. • Most cephalopods can swim rapidly. They swim by expelling a jet of water from their mantle cavity through a funnel (derived from the foot). • Their tentacles and arms capture prey by adhesive secretions or by suckers. • The cephalopod nervous system is very complex; color change via chromatophores is under nervous control and used in crypsis and communication with other members of their species.
10.11 Phylogeny and Adaptive Diversification	• Molluscs may be wormlike and shell-less, or have one, two, seven, or eight shell plates. • Some molluscan bodies have repeated body parts, which has been interpreted as evidence of a segmented ancestor. Current evidence indicates that molluscs with repeated body parts are likely derived from unsegmented molluscan ancestors. • Evolutionary relationships for molluscan classes are not yet clear. There are two hypotheses for relationships among the molluscan classes: Conchifera and Serialia. • The best supported hypothesis, Conchifera, proposes that the deepest phylogenetic split within molluscs produced two major clades: one clade comprises the two wormlike classes, Caudofoveata and Solenogastres, and the superficially segmented chitons. The second clade comprises the remaining five classes: Monoplacophora, Gastropoda, Bivalvia, Scaphopoda, and Cephalopoda.

Review Questions

1. Explain how ocean acidification is likely to affect molluscs. Describe the long-term effects of this process on human economies.
2. Members of phylum Mollusca are extremely diverse, and yet the phylum clearly constitutes a monophyletic group. What evidence can you cite in support of this statement?
3. Use examples to illustrate the various functions of the radula across the molluscs.
4. Distinguish among the following classes of molluscs: Polyplacophora, Gastropoda, Bivalvia, Cephalopoda.
5. Define the following: radula, odontophore, periostracum, prismatic layer, nacreous layer, trochophore, veliger.
6. Briefly describe the habitat and habits of a typical chiton.
7. Define torsion and fouling with respect to gastropods.
8. Torsion in gastropods created a selective disadvantage: fouling. Suggest one or more potential selective advantages that could have offset the disadvantage. How have gastropods evolved to avoid fouling?
9. Distinguish between opisthobranchs and pulmonates.
10. Briefly describe how a typical bivalve feeds and how it burrows.
11. What is the function of the siphuncle of *Nautilus*?
12. Describe how cephalopods swim and eat.
13. Cephalopods are actively swimming predators, but they evolved from a slow-moving, probably grazing ancestor. Describe evolutionary modifications of the ancestral plan that make the cephalopod lifestyle possible.
14. Briefly describe differences in the way that the morphology of caudofoveates is interpreted under the Conchifera and Serialia hypotheses.

For Further Thought Provide two potential evolutionary explanations for the presence of serially repeated body parts in monoplacophorans.

11

Annelids

LEARNING OBJECTIVES

Readers will be able to:

- **11.1** Describe the habitat and lifestyles typical of annelids.
- **11.2** Explain the major clades within phylum Annelida and list the recently added taxa.
- **11.3** Describe the key features of the annelid body plan, referencing figure 11.3.
- **11.4** Describe the habitat and lifestyle of *Chaetopterus*.
- **11.5** Describe the body plan of sipunculans and explain why they are atypical annelids.
- **11.6** List the clades nested within Pleistoannelida.
- **11.7** Describe the habitat and lifestyle of errantiate annelids.
- **11.8** Explain how sedentary annelids feed.
- **11.9** Describe the habitat and lifestyle of siboglinids; explain why they are atypical annelids.
- **11.10** Describe the habitat and lifestyle of echiurans; explain why they are atypical annelids.
- **11.11** Explain which taxa are placed with Clitellata.
- **11.12** Describe the key features of the clitellate body plan.
- **11.13** Describe the leech body plan and discuss adaptations to ectoparasitism and blood-feeding.
- **11.14** Predict the morphology of an ancestral annelid based on the fossil record.

Ethan Daniels/Getty Images

Cloeia sp., an annelid worm with a polychaete body plan (Class Errantia).

Annelids Illustrate Metamerism

Each body segment in an annelid is a serially repeated unit that contains components of circulatory, nervous, and excretory organ systems. The metameric body plan occurs in two other animal phyla: Arthropoda and Chordata. If our current understanding of animal phylogeny is correct, each of these phyla evolved metamerism independently. Why is metamerism a successful body plan?

A dominant feature of modern annelids is the fluid-filled coelomic cavity in every segment. The cavity is surrounded by longitudinal and circular muscles. Circular muscle contractions act on the fluid in the closed coelomic space to make the segment long and thin. In contrast, when longitudinal muscles contract, the segment becomes shorter and wider. Partitions, called septa, prevent fluid from moving from one segment to the next, so muscle contractions change the shape of a segment but not its volume. To understand why this shape change is important, one has only to watch an earthworm crawl or burrow. The worm uses short wide segments as anchors and extends its body forward from an anchor point by becoming long and thin. The front of the worm then anchors in its new position and the rest of the body is pulled up behind the anchor point, after which forward extension begins again. A burrower requires a compartmentalized coelom.

Was the ancestral annelid a burrower? Looking at an earthworm, one is tempted to picture the ancestral annelid as a streamlined metameric animal with minimal sensory structures. One might assume that enhanced sensory structures and expanded bodies are derived features. However, recent phylogenies indicate that the earthworm morphology is derived and that the ancestral annelid had flaplike extensions on the sides of its body and sensory palps on its head that were used also to collect food. It was a less-elaborate version of the marine worms you will see in this chapter.

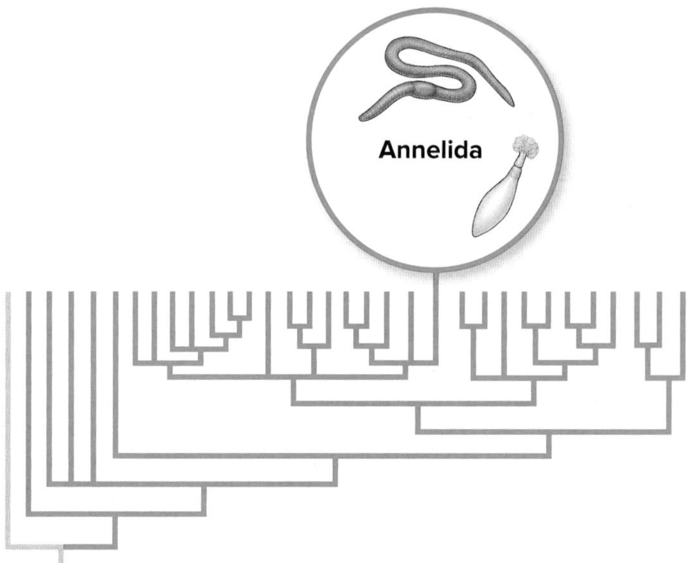

The worms described in this chapter are coelomate protostomes belonging to Lophotrochozoa. They develop by spiral mosaic cleavage, form mesoderm from derivatives of the 4d cell (see section 3.4) make a coelom by schizocoely (see section 3.4), and share a trochophore as the ancestral larval form.

Members of phylum Annelida are segmented worms living in marine, freshwater, and moist terrestrial habitats. Marine bristle worms, leeches, and the familiar earthworms belong to this group. Annelida also includes several recently added taxa that do not resemble typical annelids (see section 11.2).

11.1 Phylum Annelida

Annelida (an-nel′i-d) (L. *annellus,* little ring, + *-ida,* suffix) consists of worms with paired epidermal setae; almost all are segmented. It is a large phylum, numbering approximately 17,000 species, the most well known of which are earthworms, freshwater worms, marine bait worms, and leeches. However, the less familiar marine worms form approximately two-thirds of the phylum. Some are grotesque; others are graceful and beautiful. They include clam worms, plumed worms, parchment worms, scaleworms, lugworms, and many others.

Annelids are a highly developed group in which the nervous system is more centralized and the circulatory system more complex than those of the phyla we have studied so far. Annelids are sometimes called "bristle worms" because, except for leeches and sipunculans, most annelids bear tiny, chitinous bristles called **setae** (L. *seta,* hair or bristle). Short, needlelike setae help anchor segments during locomotion to prevent backward slipping; long, hairlike setae aid aquatic forms in swimming. Because many annelids burrow or live in secreted tubes, stiff setae also help prevent a worm from being pulled or washed out of its home.

Ecological Relationships

Annelids are worldwide in distribution, occurring in the sea, freshwater, and terrestrial soil. Some marine annelids live in tubes or burrow into bottom mud or sand. Some feed on organic matter in the mud through which they burrow; others feed on suspended particles with elaborate ciliary or mucous

CHARACTERISTICS of Phylum Annelida

1. Unique annelid head and paired epidermal setae present (lost in leeches and sipunculans); parapodia present in the ancestral condition.
2. Marine, freshwater, and terrestrial.
3. Most free-living, some symbiotic, some ectoparasitic.
4. Body bilaterally symmetrical, **metameric,** often with a distinct head; metamerism lost in echiurans and sipunculans.
5. Triploblastic body.
6. Coelom (schizocoel) well developed and divided by septa, except in leeches and echiurans; coelomic fluid functions as hydrostatic skeleton.
7. Epithelium secretes outer transparent, moist cuticle.
8. Digestive system complete and not segmentally arranged; replaced with a reliance on bacterial symbionts in some.
9. Body wall has outer circular and inner longitudinal muscle layers.
10. Nervous system has a double ventral nerve cord and a pair of ganglia with lateral nerves in each segment; brain is a pair of dorsal cerebral ganglia with connectives to ventral nerve cord.
11. Sensory system of tactile organs, taste buds, statocysts (in some), photoreceptor cells, and eyes with lenses (in some); specialization of head region into differentiated organs, such as tentacles, palps, and eyespots of polychaetes.
12. Asexual reproduction by fission and fragmentation; capable of complete regeneration; asexual reproduction by budding in some.
13. Hermaphroditic or separate sexes; larvae, if present, are trochophore type; spiral cleavage and mosaic development.
14. Excretory system typically a **pair of nephridia for each segment;** nephridia remove waste from blood as well as from coelom.
15. Respiratory gas exchange through skin, **gills,** or **parapodia**.
16. **Circulatory system closed** with muscular blood vessels and aortic arches ("hearts") for pumping blood, segmentally arranged; respiratory pigments (hemoglobin, hemerythrin, or chlorocruorin) often present; amebocytes in blood plasma.

devices for trapping food. A few annelids rely on bacterial symbionts for their nutrition. Many are predators, either swimming in the open sea or hiding in crevices of coral or rock except when hunting. Freshwater annelids burrow in mud or sand, live among vegetation, or swim freely. Some leeches are bloodsuckers, and others are carnivores; most of them live in freshwater. The terrestrial realm is occupied by earthworms.

Economic Importance

Much of the economic importance of annelids is indirect, deriving from their ecological roles. Many are members of grazing food chains or detritus food chains, serving as prey for organisms of more direct interest to humans, such as fishes. Consequently, there is a thriving market for some annelids as fish bait. Earthworm burrows increase the drainage and aeration of soils, and migrating worms help mix the soil and distribute organic matter to deeper layers (see section 11.12). Some marine annelids that burrow serve an analogous role; lugworms (*Arenicola,* see figure 11.11) are sometimes called "earthworms of the sea." Medical uses for leeches (see section 11.13) have revived the market in blood-sucking leeches and established "leech farms" where these organisms are raised in captivity.

■ 11.2 Taxa within Annelida

An earthworm with a segmented body is familiar to almost everyone. Earthworms, leeches, and marine bristle worms have long been considered members of phylum Annelida, but recently the phylum has been broadened to include several unusual forms. Molecular phylogenetic studies have expanded the taxa that are now considered annelids as well as rearranged the contents of many annelid subgroups. Figure 11.1 illustrates the current groups within phylum Annelida.

Historically annelids were divided among three taxonomic classes: Polychaeta, Oligochaeta, and Hirudinida. However, phylogenetic analyses demonstrated that classes Polychaeta and Oligochaeta were paraphyletic groups, so these classes were abolished (figure 11.1). However, the terms "oligochaete" and "polychaete" still have descriptive value, even if they no longer represent taxonomic classes. The term "oligochaete" (Gr. *oligos*, few, + *chaite-*, long hair) refers to a worm with few bristles, called **setae** or chaetae, along the sides of its body. An earthworm provides a useful image for an oligochaete body form. The earthworm body lacks the flaplike paired appendages, called **parapodia,** that are visible in figure 11.2 and present on many other annelids. The term "polychaete" (Gr. *polys*, many, + *chaite-*, long hair) applies to a worm with many bristles (setae) on the sides of the body, as in figure 11.2E. There are many flaplike parapodia on the sides of the worm in figure 11.2; each parapodium bears many setae. Setae are shown in detail in figure 11.2D. We now use the term "polychaete" to denote a body plan with parapodia and setae, as seen in 80 morphologically distinct families of annelids.

Most annelids now belong to a single very large clade called "Pleistoannelida." Five groups of annelids were placed *outside* subphylum Pleistoannelida, but of these excluded taxa we discuss only two (see figure 11.1): Chaetopteridae and Sipuncula. Chaetopterids (section 11.4) are segmented worms with three distinct body regions, but the worms in Sipuncula (section 11.5) do not resemble other annelids. Sipunculans lack setae, and their bodies are not segmented; in the past they have been allied with many different taxa, including molluscs. Molecular evidence supports their recent placement within Annelida and indicates that they lost the diagnostic annelid features. Figure 11.1 lists six features expected in the ancestral annelid: (1) bicellular eyes composed of only two cells, one for photoreception and a second cell that supports and shades the first cell; (2) nuchal organs that have a chemosensory function; (3) paired epidermal setae that are stiff bristles used for locomotory and sensory functions; (4) grooved palps that are used to sense the environment surrounding the head; (5) biramous parapodia with internal setae that are flaplike extensions of body wall with two lobes, each stiffened by an internal rod, used in locomotion, respiration and other functions; and (6) a metameric body composed of serially repeated segments, each with an internal coelomic cavity (figure 11.3). Descendant taxa without these features represent character losses.

Within subphylum Pleistoannelida, recent phylogenies based on molecular characters identified two classes of annelids (figure 11.1): Errantia (Latin *errare*, to wander) and Sedentaria (Latin *sedere*, to sit). The division of annelids into two such groups, one errant (freely moving; figure 11.2) and the other sedentary (spending most of their lives in tubes or burrows; figure 11.4) was proposed long ago on morphological grounds, but later rejected. Many biologists were surprised to see these groups reemerge. Class Errantia comprises worms with a polychaete body plan. Many of these, like the clam worm, *Nereis* (Greek, name of a sea nymph), are predatory. *Nereis* has an eversible muscular pharynx armed with jaws or teeth that can be thrust out with surprising speed to capture prey (figure 11.2A).

Class Sedentaria comprises worms with polychaete and oligochaete body plans. Sedentaria includes a large clade called Clitellata (figure 11.1). Members of Clitellata have a reproductive structure called a **clitellum.** We discuss this feature in detail in section 11.12, but anyone who has noticed a thick band around the body of an earthworm has seen a clitellum. Worms within Clitellata have an oligochaete body plan; the leeches (Hirudinidae) are clitellates with even fewer external features. Figure 11.1 depicts three families of worms within Sedentaria, but outside Clitellata. The Sabellidae are tube-dwelling fan worms with typical segmented bodies beneath a head of tentacles, but the Siboglinidae and the Echiuridae do not resemble other annelids at first glance.

Both Siboglinidae and Echiuridae are relatively recent additions to phylum Annelida. Readers may know the deep-ocean siboglinids as pogonophorans and vestimentiferans, formerly in phylum Pogonophora according to some classifications. These worms were first identified when deep ocean rift vent

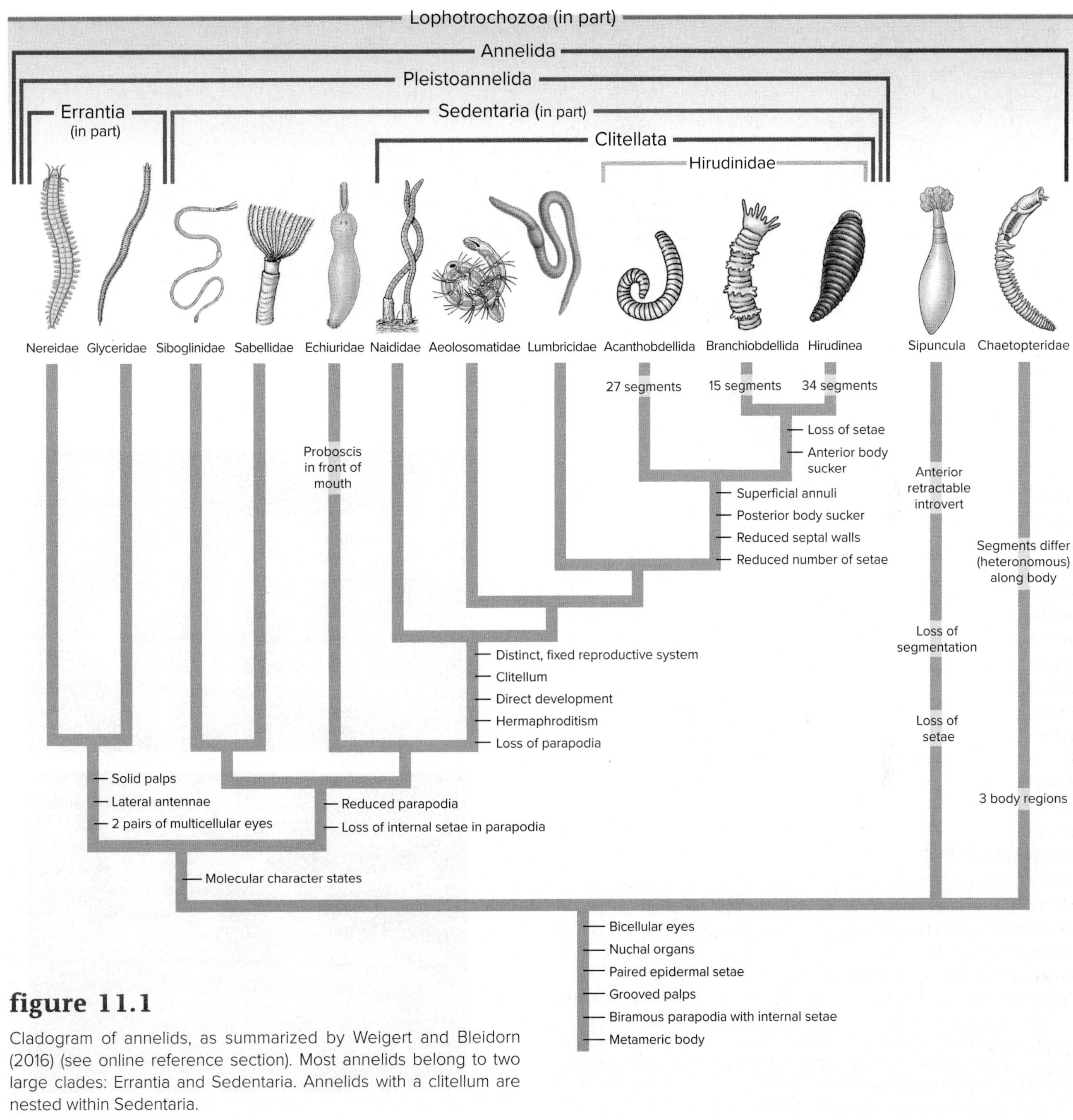

figure 11.1

Cladogram of annelids, as summarized by Weigert and Bleidorn (2016) (see online reference section). Most annelids belong to two large clades: Errantia and Sedentaria. Annelids with a clitellum are nested within Sedentaria.

communities were explored—they are large tube-dwellers that rely on bacterial symbionts to obtain food (see section 11.9). Members of Echiuridae are sausage-shaped "spoon worms" that do not have segmented bodies. They were once in phylum Echiura, but the presence of paired setae and repeated body parts in many species indicates an affinity to the annelids. It is likely that they had a segmented ancestor with a more typical annelid body plan.

■ 11.3 General Annelid Body Plan

An annelid body typically has a two-part head, composed of a **prostomium** and a **peristomium,** a series of segments (sometimes called metameres), and a terminal **pygidium** bearing an anus (figure 11.2). Neither the head nor the pygidium is considered a true segment. During growth, new segments form just in front of the pygidium; thus, the oldest segments are at the

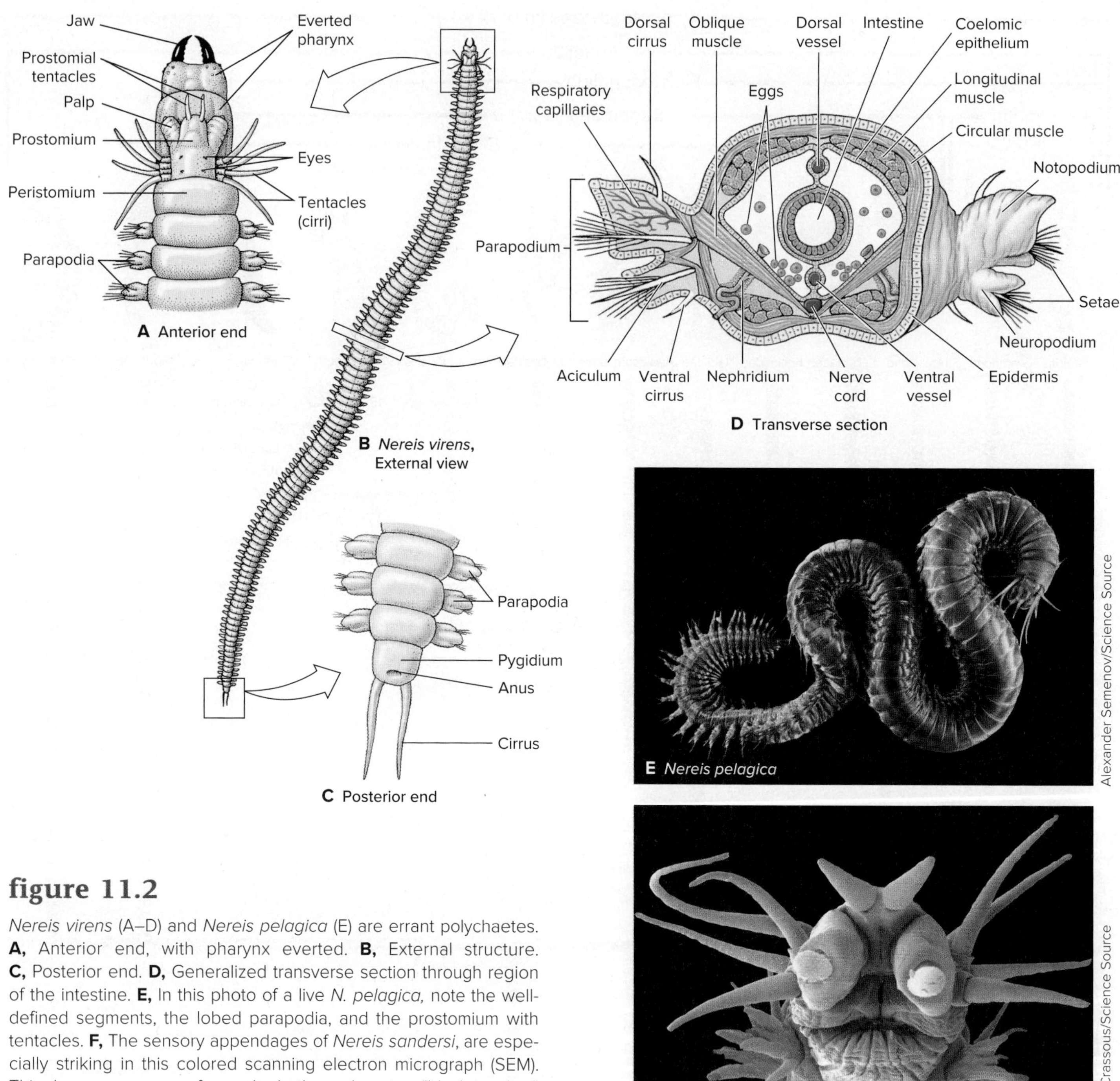

figure 11.2

Nereis virens (A–D) and *Nereis pelagica* (E) are errant polychaetes. **A,** Anterior end, with pharynx everted. **B,** External structure. **C,** Posterior end. **D,** Generalized transverse section through region of the intestine. **E,** In this photo of a live *N. pelagica,* note the well-defined segments, the lobed parapodia, and the prostomium with tentacles. **F,** The sensory appendages of *Nereis sandersi*, are especially striking in this colored scanning electron micrograph (SEM). This deep ocean worm from a hydrothermal vent or "black smoker" has a central pair of antennae, a pair of two-part palps, and four pairs of tentacles. Eyes are absent.

anterior end, and the youngest segments are at the posterior end. Each segment typically contains circulatory, respiratory, nervous, and excretory structures, as well as a coelom.

In most annelids, the coelom develops embryonically as a split in the mesoderm on each side of the gut **(schizocoel),** forming a pair of coelomic compartments in each segment. Each compartment is surrounded by **peritoneum** (a layer of mesodermal epithelium), which lines the body wall, forms dorsal and ventral mesenteries (double-membrane partitions that support the gut), and covers all organs (figure 11.3). Where peritonea of adjacent segments meet, **septa** (singular is **septum**) are formed. The body wall surrounding the peritoneum and coelom contains strong circular and longitudinal muscles adapted for swimming, crawling, and burrowing (figure 11.2D).

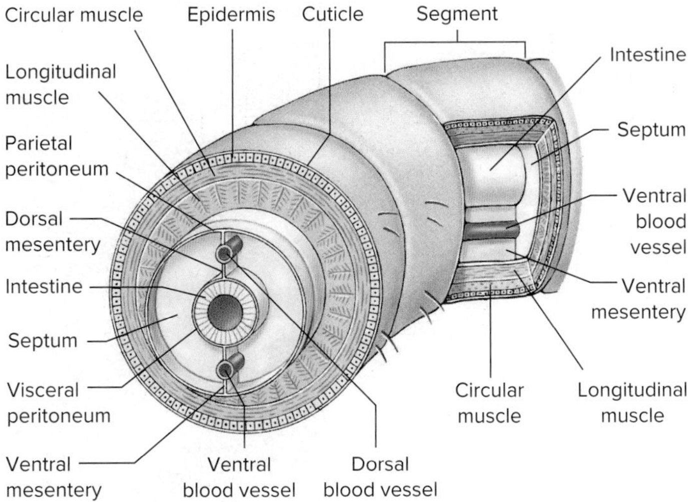

figure 11.3

Annelid body plan.

Except in leeches, the coelom is filled with fluid and serves as a **hydrostatic skeleton.** Because the volume of fluid in a coelomic compartment is essentially constant, contraction of longitudinal body-wall muscles causes a segment to shorten and become larger in diameter, whereas contraction of circular muscles causes it to lengthen and become thinner. The presence of septa means that widening or elongation occurs in restricted areas; crawling motions are effected by *alternating* waves of contraction by longitudinal and circular muscles passing down the body (peristaltic contractions). Longitudinal muscles contract in some segments, allowing those segments to widen and thus to anchor themselves against burrow walls or other substrates. In other segments, circular muscles contract, allowing the segments to elongate and stretch forward. Forces powerful enough for burrowing as well as locomotion can thus be generated. Swimming forms use undulatory rather than peristaltic movements in locomotion.

An annelid body has a thin, outer layer of nonchitinous cuticle surrounding the epidermis (figure 11.3). Paired epidermal setae (see figure 11.2D) are considered ancestral for annelids, although they have been reduced or lost in some. Parapodia are present in some taxa (figure 11.2) and absent in others (figure 11.3). The annelid digestive system is not segmented although it is regionalized: the gut runs the length of the body, perforating each septum (figure 11.3). Longitudinal dorsal and ventral blood vessels follow the same path, as does a ventral nerve cord.

The annelid head is rich with sensory organs such as eyes, tentacles, and palps in the errant annelids (figure 11.2), but is often modified to a crown of tentacles used in respiration and food capture in tube-dwelling forms (figure 11.4). The basic annelid body plan has proven highly adaptable, as is clear in the following sections on each major annelid clade.

The annelid body plan is modified in many annelid taxa. We first describe two groups of annelids that are placed outside subphylum Pleistoannelida (Chaetopteridae and Sipuncula) and then discuss the clades within Pleistoannelida (see figure 11.1).

Source: G. P. Schmahl, NOAA FGBNMS Manager

A

Larry S. Roberts

B

figure 11.4

Tube-dwelling sedentary annelids. **A,** Christmas tree "worms," *Spirobranchus giganteus*, live in a calcareous tube. On their heads are two whorls of modified tentacles (radioles) used to collect suspended food particles from the surrounding water. Notice the finely branched filters visible on the edge of one radiole. **B,** Sabellids such as *Bispira brunnea* live in leathery tubes. Because of their appearance, they are commonly called "feather duster" worms or fanworms.

11.4 Chaetopteridae

Members of Chaetopteridae are unusual tube-dwelling annelids with three distinct body regions (figure 11.5). The parchment worm, *Chaetopterus* (Gr. *chaitē*, long hair, pteron, wing), feeds on suspended particles. It lives in a U-shaped, parchment-like tube buried, except for the tapered ends, in sand or mud along the shore. The worm attaches to the side of the tube by ventral suckers. Fans (modified parapodia on segments 14–16) pump water through the tube by rhythmical movements. A pair of enlarged parapodia on segment 12 secretes a long mucous net that reaches back to a small food cup just in front of the fans. All water passing through the tube is filtered through this mucous net, the end of which is rolled into a ball by cilia in the cup. When the ball is about the size of a BB shot (about 3 mm in diameter), the fans stop beating and the ball of food and mucus is rolled forward by ciliary action to the mouth and swallowed.

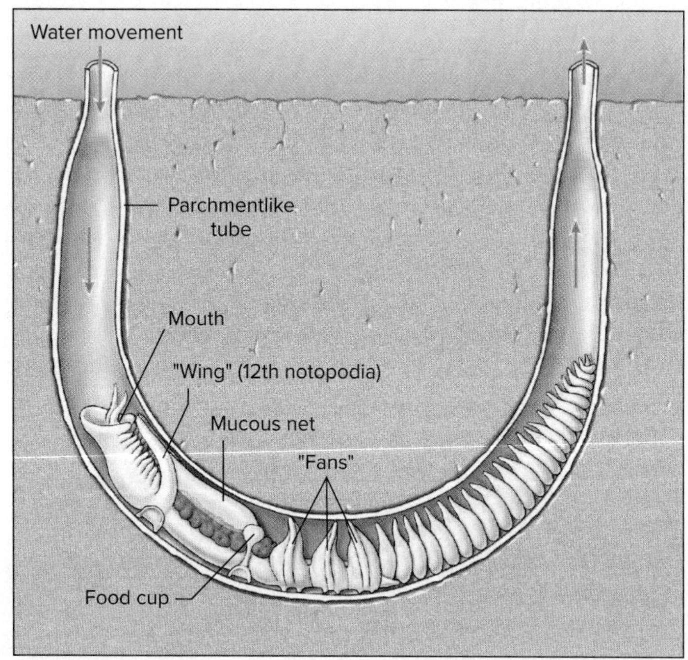

figure 11.5

Chaetopterus, a sedentary annelid, occupies a U-shaped tube in the sea floor. It pumps water through a parchment-like tube (of which one-half has been cut away here) with its three pistonlike fans. The fans beat 60 times per minute to keep water currents moving. Winglike notopodia of the twelfth segment continuously secrete a mucous net that strains food particles. As the net fills with food, the food cup rolls it into a ball, and when the ball is large enough (about 3 mm), the food cup bends forward and deposits the ball in a ciliated groove to be carried to the mouth and swallowed.

11.5 Sipuncula

Sipuncula (sī-pun´kū lə) (L. *sipunculus*, little siphon) consists of about 250 species of benthic marine worms, predominantly littoral or sublittoral. Sometimes called "peanut worms," they live

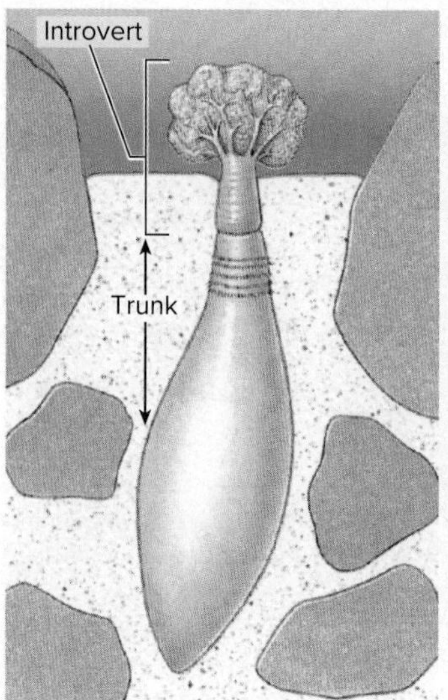

A *Themiste*, a sipunculan

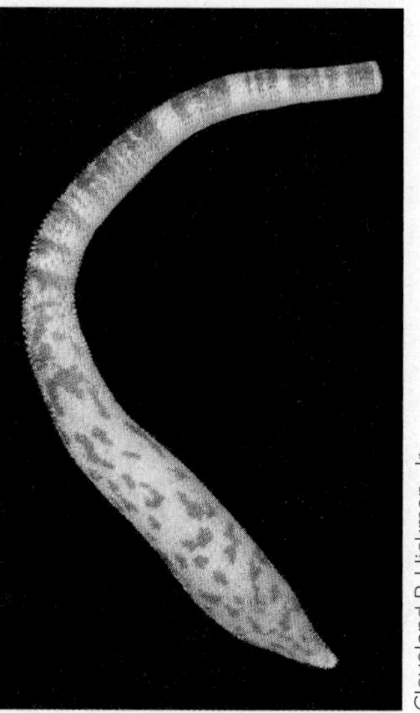

B *Phascolosoma*

figure 11.6

A, *Themiste*, a sipunculan, has extended its introvert. **B,** *Phascolosoma*, another sipunculan, has the introvert tucked into the trunk.

sedentary lives in burrows in mud or sand (figure 11.6), occupy borrowed snail shells, or live in coral crevices or among vegetation. Some species construct their own rock burrows by chemical and perhaps mechanical means. Most species are restricted to tropical zones. Some are tiny, slender worms, but most range from 15 to 30 cm in length.

Sipunculans are not metameric, nor do they possess setae. Their head forms an **introvert,** which is crowned by ciliated tentacles surrounding the mouth (figure 11.6). They are largely deposit feeders, extending their introvert and tentacles from their burrow to explore and to feed. They have a cerebral ganglion, nerve cord, and pair of nephridia. They have a U-shaped gut lying within a spacious coelom. The coelomic fluid contains red blood cells bearing a respiratory pigment called hemerythrin.

Sipunculan larvae are trochophores (see figure 10.6). Sexes are separate in most cases, and gonads develop seasonally.

11.6 Pleistoannelida

Pleistoannelida comprises two large clades: Errantia and Sedentaria. We will provide a brief description of the errant body plan and then discuss representative errant worms. Descriptions of the major types of sedentary annelids, including members of Clitellata, follow the discussion on errant taxa.

11.7 Class Errantia

Errantia comprises the motile annelids, most with a polychaete body plan. Most errant annelids are marine, and while most are 5–10 cm long, some are less than 1 mm, and others may be as long as 3 m. They may be brightly colored in reds and greens, iridescent, or dull. Many are euryhaline and can tolerate a wide range of environmental salinity. The freshwater fauna are more diversified in warmer regions than in temperate zones. Many errant annelids live under rocks, in coral crevices, or in abandoned shells. Some are planktonic. They play a significant part in marine food chains because fish, crustaceans, hydroids, and many other predators consume them.

An errant worm typically has a prostomium, which may or may not be retractile and often bears eyes, antennae, and sensory palps (see figure 11.2A and F). The peristomium surrounds the mouth and may bear setae, palps, or, in predatory forms, chitinous jaws (see figure 11.2).

The trunk is segmented, and most segments bear parapodia, which may have lobes, cirri, setae, and other structures on them (see figure 11.2D). Parapodia are composed of two main parts–a dorsal **notopodium** and a ventral **neuropodium**–either of which may be prominent or reduced in a given species. Parapodia are used for crawling, or swimming. They usually serve as the chief respiratory organs, although some also have gills.

Sense organs are highly developed in errant worms and include eyes and statocysts. Eyes, when present, may range from simple eyespots to well-developed organs. Usually eyes consist of retinal cups, with rodlike photoreceptor cells lining the cup wall and directed toward the lumen of the cup.

Errantiate worms have no permanent sex organs, possess no permanent ducts for their sex cells, and usually have separate sexes. Gonads appear as temporary swellings of the peritoneum and shed their gametes into the coelom. Gametes leave the body through gonoducts, through nephridia, or by rupture of the body wall. Fertilization is external, and development is indirect with a trochophore larva preceding the adult.

Some errant worms are free-moving pelagic forms, whereas others are active burrowers or crawlers. An example of an active, predatory worm is *Nereis* (from Greek mythology, a Nereid, or daughter of Nereus, ancient sea god), the clam worm (see figure 11.2). *Nereis* has a muscular, eversible pharynx equipped with jaws that can be thrust out with surprising speed and dexterity for capturing prey. Scale worms (figure 11.7) often live as commensals with other invertebrates, and fireworms (figure 11.8) feed on gorgonians and stony corals.

Key Theme 11.1

ADAPTATION AND PHYSIOLOGY

Swarming Sexual Stages

Some members of Errantia live most of the year as sexually unripe animals called atokes, but during the breeding season, a portion of the body develops into a sexually ripe worm called an *epitoke*, which is swollen with gametes (figure 11.9). For example, palolo worms live in burrows among coral reefs of the South Seas. During the reproductive cycle, their posterior segments become swollen with gametes. During the swarming period, which occurs at the beginning of the last quarter of the October-November moon, these epitokes break off and swim to the surface. Just before sunrise, the sea is literally covered with them, and at sunrise they burst, freeing eggs and sperm for fertilization. The anterior portions of the worms regenerate new posterior sections. A related form swarms in the Atlantic in the third quarter of the June-July moon. Swarming is of great adaptive value because synchronous maturation of all epitokes ensures the maximum number of fertilized eggs. However, it is very hazardous; many types of predators, including local residents who consider the worms a great delicacy, have a feast. In the meantime, the atoke remains safe in its burrow to produce another epitoke at the next cycle!

Cleveland P. Hickman, Jr.

figure 11.7

The scale worm, *Hesperonoe adventor*, normally lives as a commensal in the tubes of *Urechis*, commonly called the "fat inkeeper worm" (see section 11.10).

Larry S. Roberts

figure 11.8

A fireworm, *Hermodice carunculata* (Class Errantia), feeds on gorgonians and stony corals. Its needle-like setae repel predators by breaking off when brushed. Despite previous reports that setae are hollow and contain venom, new research shows that setae are solid. A combination of chitin and calcium carbonate makes them brittle.

11.8 Class Sedentaria

Sedentaria contains many worms with polychaete and oligochaete body plans that live in tubes or burrows. It also includes members of Clitellata (see figure 11.1). The body plan of sedentary annelids is much like that of errant forms, except that the head is often modified by the addition of tentacles used for food capture. Many parapodia are small, and some are modified to help anchor the worm in the tube; setae may be hooklike to attach to the tube wall. Parapodia may function in respiration, but many tube-dwellers also have gills.

Most sedentary tube and burrow dwellers are particle feeders, using ciliary or mucoid methods of obtaining food. Their principal food source is plankton and detritus. Some, such as *Amphitrite* (from Greek mythology, sea goddess) (figure 11.10), protrude their heads from the mud and send long, extensible tentacles over the surface. Cilia and mucus on the tentacles entrap particles found on the sea floor and move them toward the mouth, a practice called **deposit feeding.**

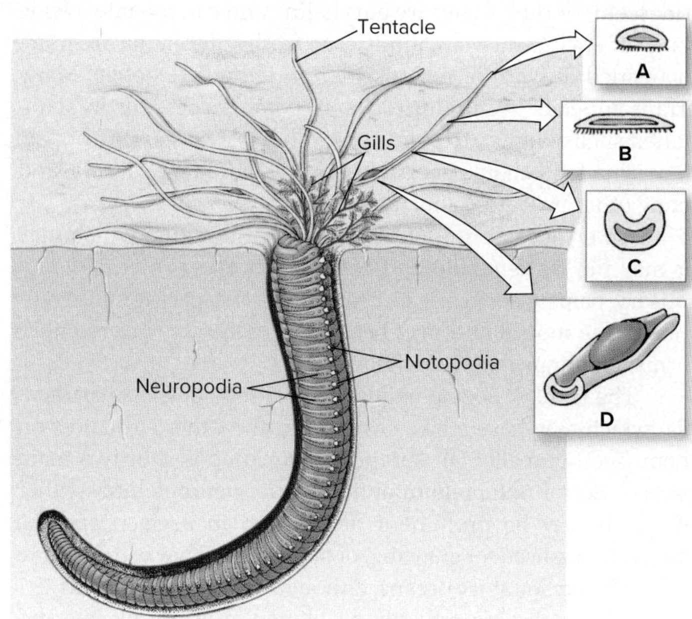

figure 11.10

Amphitrite, which builds its tubes in mud or sand, extends long, grooved tentacles out over the mud to gather bits of organic matter. The smallest particles are moved along food grooves by cilia, and larger particles are carried by peristaltic movement. The plumelike gills are blood red. **A,** Section through the exploratory end of a tentacle. **B,** Section through a tentacle in an area adhering to the substrate. **C,** Section showing a ciliary groove. **D,** A particle being carried toward the mouth.

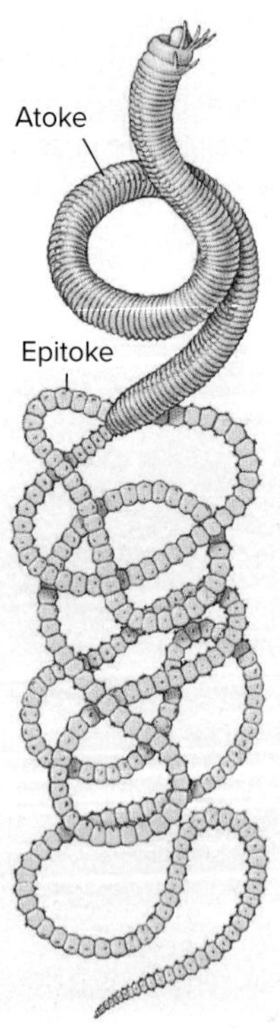

figure 11.9

Eunice viridis, the Samoan palolo worm. Posterior segments make up the epitokal region, consisting of segments packed with gametes. Each segment has one eyespot on the ventral side. Once a year, the worms swarm, and the epitokes detach, rise to the surface, and discharge their ripe gametes, leaving the water milky. By the next breeding season, epitokes are regenerated.

Fanworms, or "feather duster" worms, are beautiful tubeworms, fascinating to watch as they emerge from their secreted tubes and unfurl their lovely tentacular crowns to feed (see figure 11.4). A slight disturbance, sometimes even a passing shadow, causes them to retract quickly into the safety of the homes they have built. Food attracted to the feathery arms, or **radioles,** by ciliary action is trapped in mucus and carried down ciliated food grooves to the mouth (figure 11.11). Particles too large for the food grooves are carried along the margins and discarded. Further sorting may occur near the mouth, where only small particles of food can enter, and sand grains are stored in a sac to be used later in enlarging the tube.

Lugworms, *Arenicola* (L. *arena,* sand, + *colere,* to inhabit), live in L-shaped burrows in which, by peristaltic movements, they cause water to flow. Sand at the front of their burrows collects filtered food particles. The worms ingest the food-laden sand (figure 11.12).

Tube dwellers secrete many types of tubes. Some are parchment-like; some are firm, calcareous tubes attached to rocks or other surfaces (see figure 11.4A); and some are grains of sand or bits of shell or seaweed cemented together with mucous secretions. Many burrowers in sand and mudflats simply line their burrows with mucus (figure 11.12).

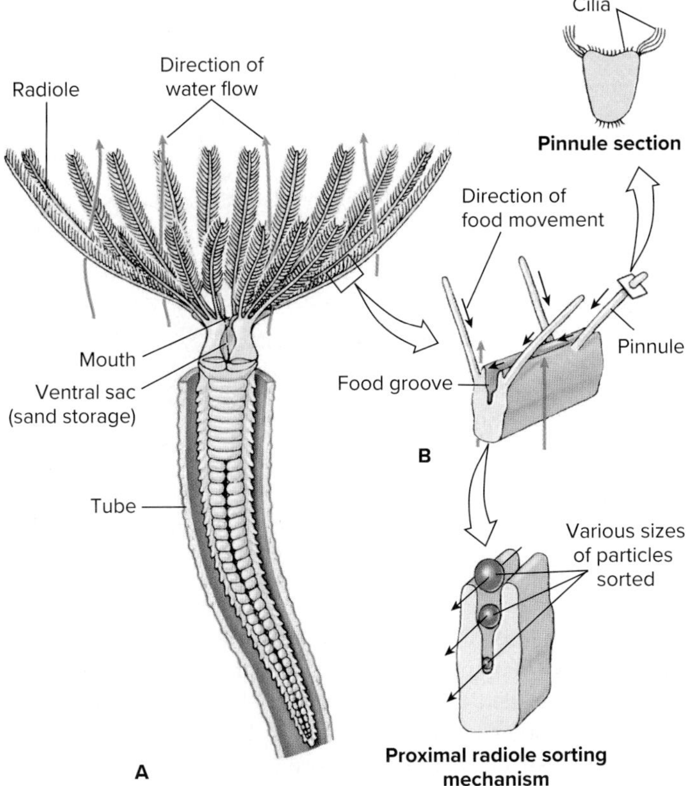

figure 11.11

Sabella, a suspension feeder. **A,** Anterior view of the crown. Cilia direct small food particles along grooved radioles to the mouth and discard larger particles. Sand grains are directed to storage sacs and later used in tube building. **B,** Distal portion of a radiole, showing ciliary tracts of pinnules and food grooves.

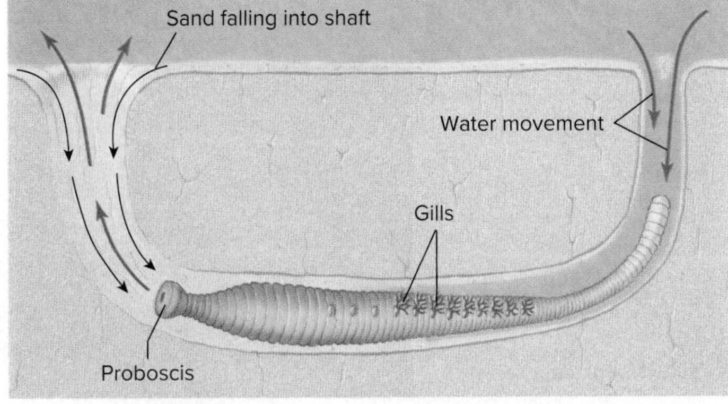

figure 11.12

Arenicola, the lugworm, lives in an L-shaped burrow in intertidal mudflats. It burrows by successive eversions and retractions of its proboscis. By peristaltic movements, it keeps water coming in the open end of the tube and filtering through the sand at the head end. The worm then ingests the food-laden sand.

Key Theme 11.2
ADAPTATION AND PHYSIOLOGY

Worms That Eat Whale Corpses

The 2004 discovery of bone-eating marine worms in genus *Osedax* was very exciting. These siboglinids bore into bones of whale carcasses through a "root" system and use endosymbiotic bacteria to digest the bone. The green, branching, vascularized "root" system develops from the posterior part of the egg sac and invades bone marrow. One might wonder if whale falls (carcasses) are sufficiently abundant for this lifestyle to succeed—surprisingly, they are typically only 5–15 km apart in coastal regions where the worms occur. Female worms take no chances in finding mates; dwarf males share the tubes of females. How do males and females find each other? Sex determination seems to be environmental, so larvae that land on females become male. Several species of these animals have been found, and surprising aspects of their biology appear regularly in the literature.

11.9 Siboglinidae (Pogonophorans)

Members of the former phylum Pogonophora (po-go-nof´e-ra) (Gr. *pōgōn*, beard, + *pherō*, to bear), or beardworms, were entirely unknown before the twentieth century. The first specimens to be described were collected from deep-sea dredgings in 1900 off the coast of Indonesia. They have since been discovered in several seas, including the western Atlantic off the eastern coast of the United States. Some 150 species have been described so far. Segmentation occurs only in one body region, but the cuticle and setae are homologous to those of other annelids.

Most siboglinids live in soft substrates on the ocean floor, usually at depths of more than 200 m. Their usual length varies from 5 to 85 cm, and their diameter is usually less than a millimeter. They are sessile, and secrete very long, chitinous tubes in which they live, extending their anterior end only for feeding.

The siboglinid body has a short forepart, a long, very slender trunk, and a small, segmented opisthosoma (figure 11.13). It is covered with a cuticle and bears setae on the trunk and opisthosoma. A series of coelomic compartments divides the body. The forepart bears from one to many tentacles.

Siboglinids are remarkable in having no mouth or digestive tract, making their mode of nutrition puzzling. They absorb some nutrients, such as glucose, amino acids, and fatty acids dissolved in seawater, through pinnules and microvilli of their tentacles. However, they apparently derive most of their energy from a mutualistic association with chemoautotrophic bacteria. These bacteria oxidize hydrogen sulfide to provide energy for producing organic compounds from carbon dioxide. An expanded region of the midgut fills with endoderm to form an organ called a **trophosome** that bears the bacteria. All traces of a foregut and hindgut are absent in adults.

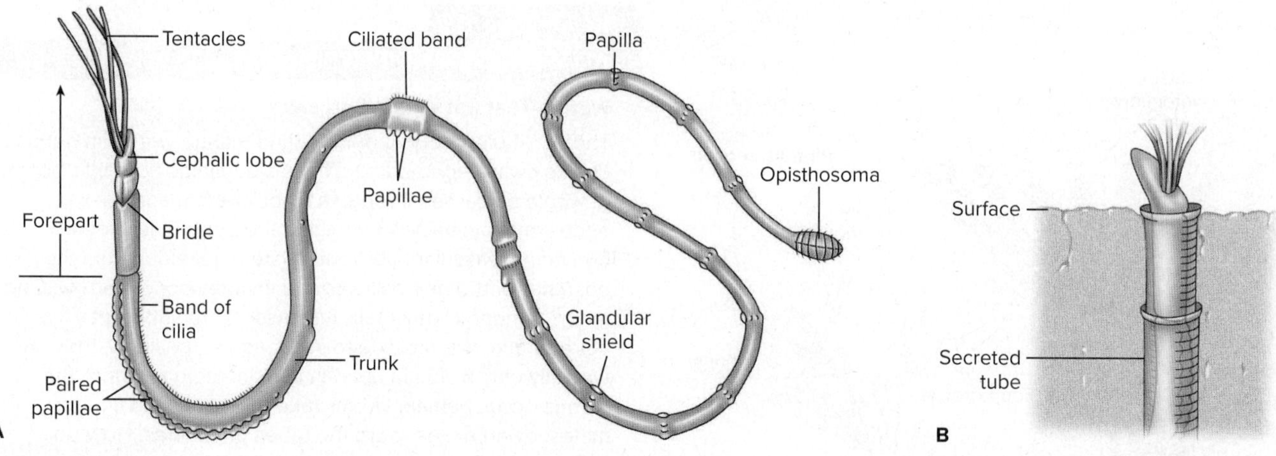

figure 11.13

Diagram of a typical siboglinid. **A,** External features. The body, in life, is much more elongated than shown in this diagram. **B,** Position in its tube.

Siboglinids have a well-developed, closed, blood vascular system. Their photoreceptor cells are very similar to those of clitellates. Sexes are separate.

Key Theme 11.3

ADAPTATION AND PHYSIOLOGY

Housing Bacterial Symbionts

Among the most amazing animals in the deep-water, Pacific rift communities (see Key Theme 2.4) are giant siboglinids, *Riftia pachyptila*. They measure up to 1.5 m in length and 4 cm in diameter. Trophosomes of other siboglinids are confined to the posterior part of the trunk, which is buried in sulfide-rich sediments, but the trophosome of *Riftia* occupies most of its large trunk. It has access to a much larger supply of hydrogen sulfide, enough to nourish its large body, in the effluent of hydrothermal vents.

11.10 Echiuridae

Members of former phylum Echiura (ek-ē-yur´ə) (Gr. *echis*, viper, + *oura*, tail) are now placed within Sedentaria as family Echiuridae. Echiuridae consists of about 140 species of marine worms that burrow into mud or sand or live in empty snail shells, sand dollar tests, or rocky crevices. They live in all oceans, most commonly in littoral zones of warm waters. They vary in length from a few millimeters to 40–50 cm.

Echiurans are cylindrical and somewhat sausage-shaped. Anterior to the mouth is a flattened, extensible proboscis, which cannot be retracted into the trunk. Echiurans are often called "spoon worms" because of the shape of the contracted proboscis in some of them (figure 11.14). The proboscis has a ciliated groove leading to the mouth. There is a complete gut with a posterior anus. While the animal lies buried, its proboscis can extend out over the mud for exploration and deposit feeding (figure 11.15). A different feeding method is used by *Urechis* (Gr. *oura*, tail, + *echis*, viper); it secretes a mucous net in a U-shaped burrow through which it pumps water and strains food particles. *Urechis* is sometimes called the "fat innkeeper" because characteristic species of commensals live with it in its burrow, including species of crab, fish, clam, and scale worm.

Echiurans, with the exception of *Urechis*, have a closed circulatory system with a contractile vessel; most have one to three pairs of nephridia (some have many pairs), and all have a

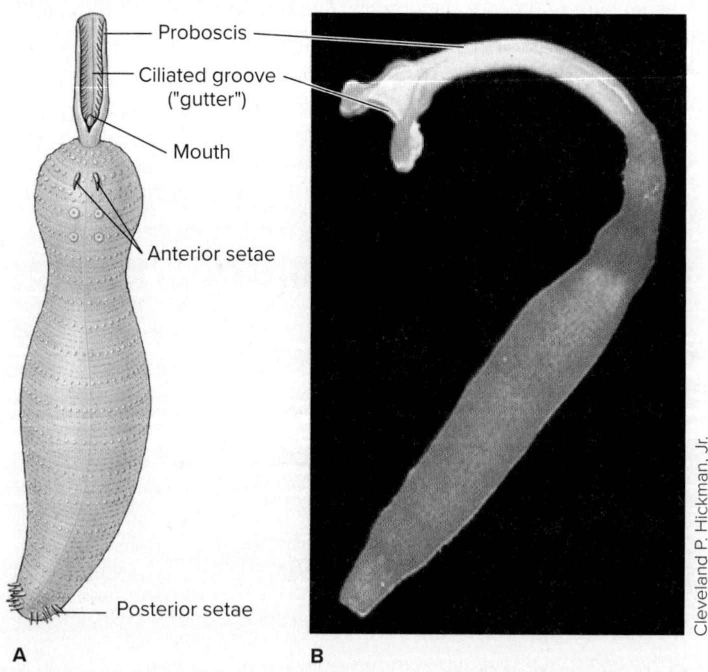

figure 11.14

A, *Echiurus*, an echiuran common on both Atlantic and Pacific coasts of North America. **B,** *Anelassorhynchus*, an echiuran of the tropical Pacific. The shape of their proboscis lends them the common name of "spoon worms."

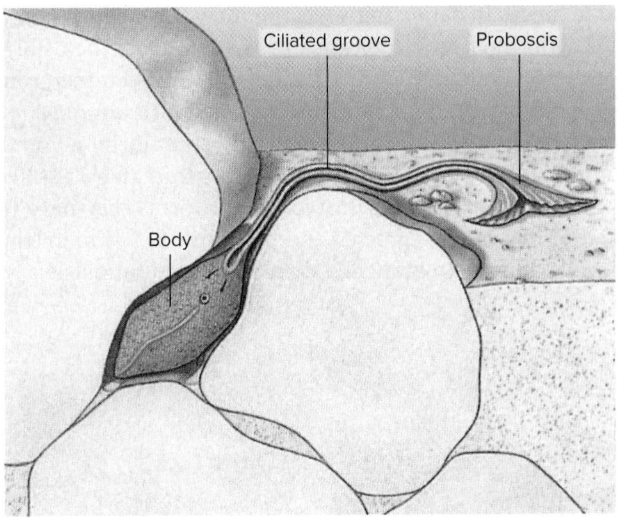

figure 11.15

Bonellia, an echiuran, is a detritus feeder. Lying in its burrow, it explores the surface with its long proboscis, which gathers organic particles and carries them along a ciliated groove to its mouth.

nerve ring and ventral nerve cord. A pair of anal sacs arises from the rectum and opens into the coelom; they are probably respiratory in function and possibly accessory nephridial organs. Early cleavage and trochophore stages are similar to those of other annelids.

Echiurans are not segmented, but they presumably had a segmented ancestor. Many species retain the ancestral paired epidermal setae.

Key Theme 11.4

ADAPTATION AND PHYSIOLOGY

Sex Determination

In some species of echiurans, sexual dimorphism is pronounced, with females much larger than males. *Bonellia* has an extreme sexual dimorphism, and sex is determined in a very interesting way. At first, free-swimming larvae are sexually undifferentiated. Then those that come into contact with the proboscis of a female become tiny males (1–3 mm long) that migrate to the female uterus. About 20 males are usually found in a single female. Larvae that do not contact a female proboscis metamorphose into females. The stimulus for development into males is apparently a hormone produced by the female proboscis.

11.11 Order Clitellata

Clitellata (figure 11.1) contains annelids with an oligochaete body plan, such as earthworms (Family Lumbricidae), as well as leeches (Family Hirudinidae). Group members share a unique reproductive structure called a **clitellum**. The clitellum is a ring of secretory cells in the epidermis that appears on the worm's exterior as a fat band around the body about one-third of the body length from the anterior end. The clitellum is always visible in clitellates outside of Hirudinidae, but it appears only during the reproductive season in leeches. Clitellates lack parapodia, presumably due to an evolutionary loss. Clitellates are all hermaphroditic (monoecious) and exhibit direct development: young develop inside a cocoon secreted by the clitellum, so no trochophore larva is visible. Small worms emerge from cocoons.

Clitellates with an oligochaete body plan do not form a monophyletic group. There are more than 3000 species of clitellates that are not leeches, but we depict only a few branches in figure 11.1. The term oligochaete means "few long hairs," but their setae may be long or short, straight or curved, blunt or needlelike, or arranged singly or in bundles. These worms occur in a great variety of sizes and habitats. Most are terrestrial or freshwater forms, some are parasitic, and a few live in marine or brackish water. Both freshwater and marine clitellates are important food sources for fishes. They include the earthworms and many species that live in freshwater.

Most freshwater forms (figure 11.16) are smaller than earthworms; in general, aquatic forms have more conspicuous, longer setae. They are also more mobile than earthworms and tend to have better-developed sense organs. They are generally benthic forms that creep across a substrate or burrow into soft mud. Their chief foods are algae and detritus, which worms may gather by extending a mucus-coated pharynx. Burrowers swallow mud and digest the organic material. Some, such as *Aeolosoma*, are ciliary feeders that use currents produced by cilia at the anterior end of the body to sweep food particles into their mouth (figure 11.16B). Some aquatic forms have gills, often as long, slender projections from the body surface. Others have ciliated anal gills (figure 11.16D), which they extend

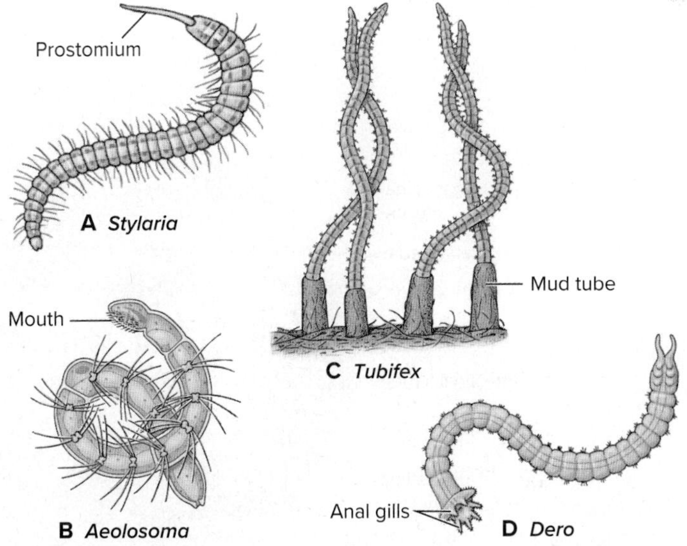

figure 11.16

Some freshwater annelids. **A,** *Stylaria* has its prostomium drawn out into a long snout. **B,** *Aeolosoma* uses cilia around its mouth to sweep in food particles, and it buds off new individuals asexually. **C,** *Tubifex* lives head down in long tubes. **D,** *Dero* has ciliated anal gills.

from their tubes and use to keep water moving. Most forms respire through their skin as do earthworms.

The main features of an clitellate body plan are described with reference to an earthworm. The circulatory system and excretory structures described in earthworms are typical of annelids in general, but the digestive and nervous systems have aspects specific to animals in a terrestrial habitat.

Earthworms ("night crawlers") burrow in moist, rich soil, emerging at night to feed on surface detritus and vegetation and to breed. In damp, rainy weather they stay near the surface, often with their mouth or anus protruding from their burrow. In very dry weather, they may burrow several feet underground, coil in a slime chamber, and become dormant. *Lumbricus terrestris* (L. *lumbricum,* earthworm), the form commonly studied in school laboratories, is approximately 12–30 cm long (figure 11.17). Giant tropical earthworms may have from 150 to 250 or more segments and may grow to 3–4 m in length. They usually live in branched, interconnected tunnels.

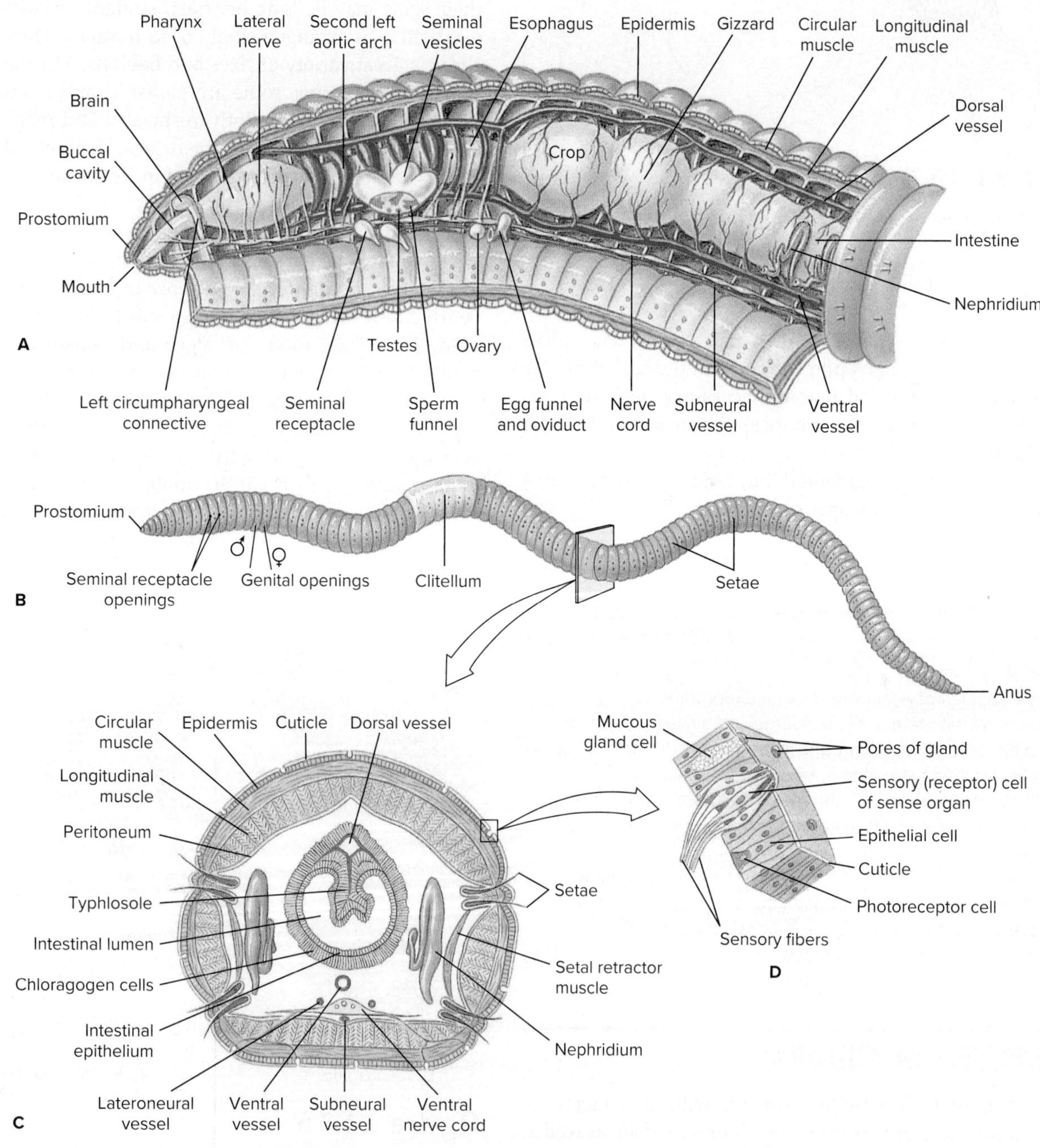

figure 11.17

Earthworm anatomy, as illustrated by *Lumbricus terrestris*. **A,** Internal structure of anterior portion of worm. **B,** External features, lateral view. **C,** Generalized transverse section through region posterior to clitellum. **D,** Portion of epidermis showing sensory, glandular, and epithelial cells.

11.12 Form and Function in Clitellata

In earthworms, a fleshy prostomium overhangs the mouth at the anterior end, with the anus on the terminal end (figure 11.17B). In most earthworms, each segment bears four pairs of chitinous setae (figure 11.17C), although in some each segment may have up to 100 or more setae. Each seta is a bristlelike rod set in a sac within the body wall and moved by tiny muscles (figure 11.18). The setae project through small pores in the cuticle to the outside. During locomotion and burrowing, setae anchor parts of the body to prevent slipping. Earthworms move by peristaltic movement. Contraction of circular muscles in the anterior end lengthens the body, thus pushing the anterior end forward where it is anchored by setae; contractions of longitudinal muscles then shorten the body, pulling the posterior end forward. As these waves of contraction pass along the entire body, it gradually moves forward.

The food of earthworms is mainly decayed organic matter and bits of vegetation drawn in by the muscular **pharynx** (see figure 11.17A). As in other annelids, the digestive tract is unsegmented and extends the length of the worm. The upper digestive tract has distinct regions, including the crop, where food is stored prior to grinding in the gizzard (figure 11.17A). The intestine has a U-shaped fold, the **typhlosole,** that increases the surface area for nutrient absorption. In the typhlosole and around the intestine are **chloragogen cells,** which synthesize glycogen and fat and can break free to distribute these nutrients through the coelom. Chloragogen cells also serve an excretory function.

Annelids have a double transport system—the coelomic fluid and the circulatory system. Both carry food, wastes, and respiratory gases in varying degrees. The blood is in a closed system of blood vessels, including capillary systems in the tissues. There are five main longitudinal trunks, of which the dorsal blood vessel is the main pumping organ (see figure 11.17A and C). Their blood contains colorless ameboid cells and a dissolved respiratory pigment, **hemoglobin.** The blood of other annelids may have respiratory pigments other than hemoglobin.

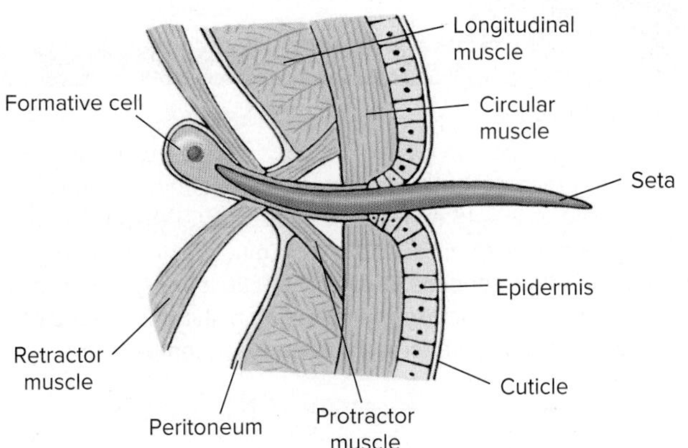

figure 11.18

A seta with its muscle attachments, showing relation to adjacent structures. Setae lost by wear and tear are replaced by new ones, which develop from formative cells.

Key Theme 11.5
ECOLOGY

Earthworms Benefit Soil

Aristotle called earthworms the "intestines of the soil." Some 22 centuries later, Charles Darwin published his observations in his classic, *The Formation of Vegetable Mould Through the Action of Worms.* He showed how worms enrich soil by bringing subsoil to the surface and mixing it with topsoil. An earthworm can ingest its own weight in soil every 24 hours, and Darwin estimated that from 10 to 18 tons of dry earth per acre pass through their intestines annually, thus bringing up potassium and phosphorus from the subsoil and also adding to the soil nitrogenous products from their own metabolism. Earthworms expose the mold to the air and sift it into small particles. They also drag leaves, twigs, and organic substances into their burrows, closer to the roots of plants. Their activities are important in aerating the soil. Darwin's views were at odds with those of his contemporaries, who thought earthworms were harmful to plants. Later research has amply confirmed Darwin's findings, and many countries now practice earthworm management.

The organs of excretion are **nephridia,** a pair of which occurs in each segment except the first three and the last one. Each one occupies parts of two successive segments (figure 11.19). A ciliated funnel, called a nephrostome, lies just anterior to an intersegmental septum and leads by a small ciliated tubule through the septum into the segment behind, where it connects with the main part of the nephridium. This part of the

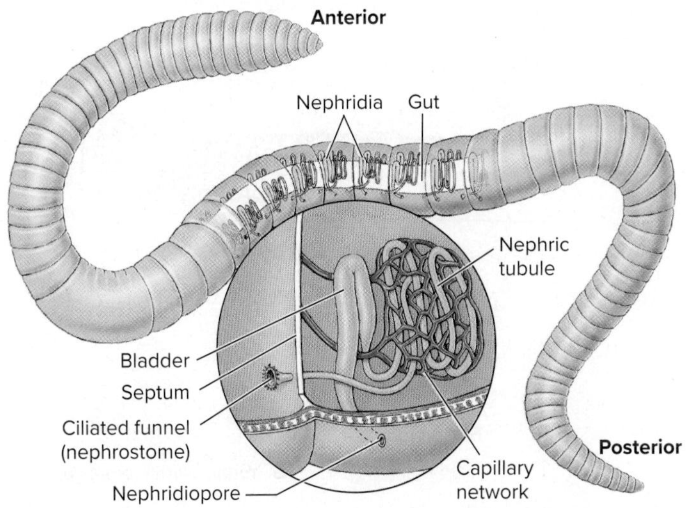

figure 11.19

Nephridium of an earthworm. Wastes are drawn into the ciliated nephrostome in one segment, then passed through loops of the nephridium, and expelled through the nephridiopore of the next segment.

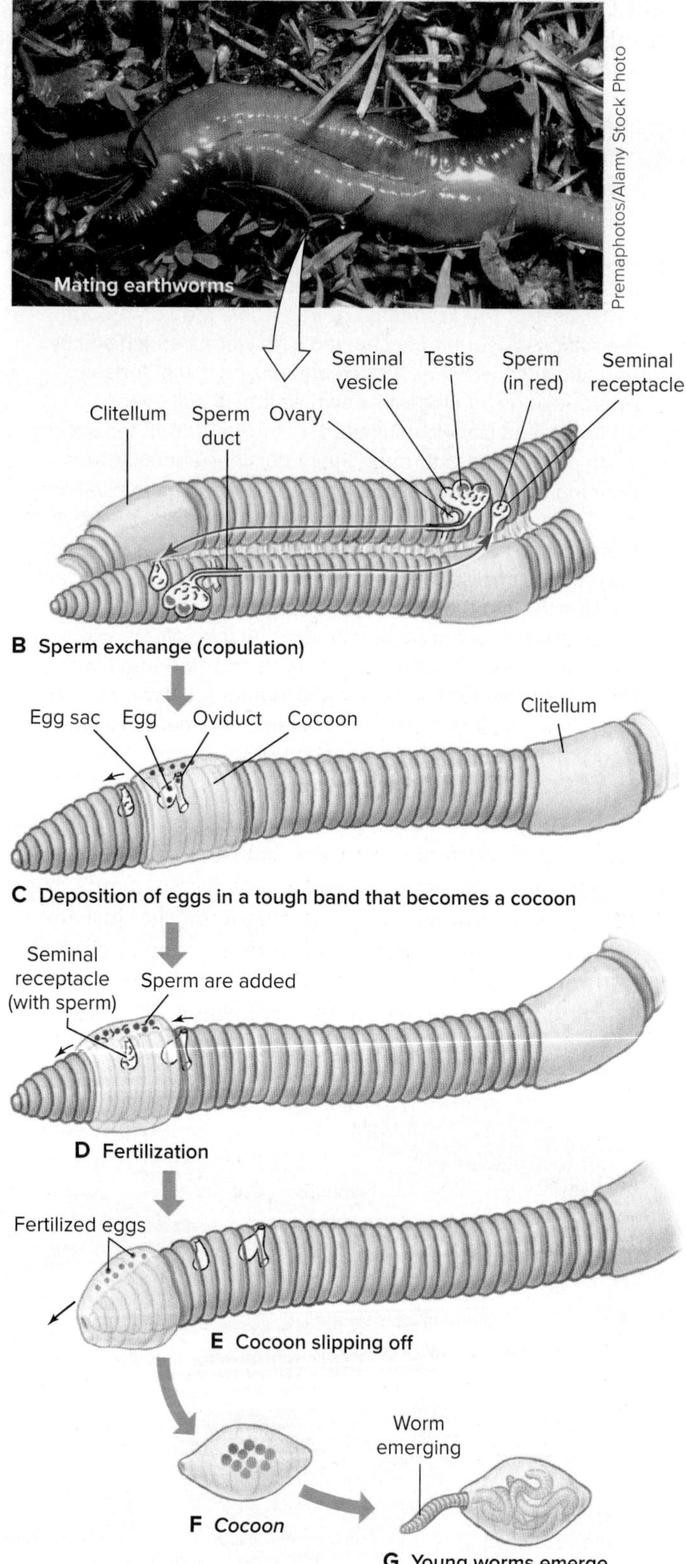

figure 11.20

Earthworm copulation and formation of egg cocoons. **A,** Two earthworms in copulation. Their anterior ends point in opposite directions as their ventral surfaces are held together by mucous bands secreted by the clitella. **B,** Mutual insemination; sperm from genital pore (segment 15) pass along seminal grooves to seminal receptacles (segments 9 and 10) of each mate. **C** and **D,** After the worms separate, the clitellum secretes first a mucous tube and then a tough band that forms a cocoon. The developing cocoon passes forward to receive eggs from oviducts and sperm from seminal receptacles. **E,** As the cocoon slips off over the anterior end, its ends close and seal. **F,** Cocoon is deposited near burrow entrance. **G,** Young worms emerge in 2–3 weeks.

nephridium contains several loops of increasing size, which finally terminate in a bladderlike structure leading to an aperture, or nephridiopore, which opens to the outside near the ventral row of setae. Cilia draw fluid from the coelom into the nephrostome and tubule. In the tubule, water and salts are resorbed, forming a dilute urine that discharges to the outside through the nephridiopore.

The nervous system is typical of all annelids. A pair of cerebral ganglia, which form the animal's brain, lies above the pharynx joined to a ventral nerve cord by a pair of connectives around the pharynx (see figure 11.17A). The ventral nerve cord bears a pair of ganglia in each segment, giving off segmental nerves containing both sensory and motor fibers. For rapid escape movements, most annelids have one or more very large axons commonly called giant axons, or giant fibers, in the ventral nerve cord. Speed of conduction in these giant nerve fibers is much greater than that in small axons.

Earthworms are hermaphroditic and exchange sperm during copulation, which usually occurs at night. When mating, worms extend their anterior ends from their burrows and bring their ventral surfaces together (figure 11.20). They are held together by mucus secreted by each worm's **clitellum** and by special ventral setae, which penetrate each other's bodies in the regions of contact. Sperm are discharged and travel to the seminal receptacles of the other worm in its seminal grooves. After sperm exchange, the worms separate. Each worm secretes around its clitellum first a mucous tube and then a tough, chitinlike band that forms a **cocoon** (figure 11.20). The cocoon slides forward along the body. As it moves, eggs from the oviducts, albumin from skin glands, and sperm from the mate (stored in the seminal receptacles) are poured into it. Fertilization of eggs now occurs within the cocoon. When the cocoon slides off the head end of the worm, its ends close, producing a lemon-shaped body. Embryonic development occurs within the cocoon, and the form that hatches from the egg is a young worm similar to the adult. It does not develop a clitellum until it is sexually mature.

Key Theme 11.6

ADAPTATION AND PHYSIOLOGY

Giant Nerve Fibers

In the dorsal median giant axon of *Lumbricus,* which is 90–160 μm in diameter, speed of conduction has been estimated at 20–45 m/s, several times faster than in ordinary neurons of this species. This is also much faster than in giant axons of other annelids, probably because the giant axons in earthworms are enclosed in myelinated sheaths. Speed of conduction may be altered by changes in temperature.

11.13 Family Hirudinidae

Leeches, numbering over 500 species, occur predominantly in freshwater habitats, but a few are marine, and some have even adapted to terrestrial life in moist, warm areas. Most leeches are between 2 and 6 cm in length, but some are smaller, and some reach 20 cm or more (figure 11.21). They are found in a variety of patterns and colors—black, brown, red, or olive green. They are usually flattened dorsoventrally.

Leeches are highly modified annelids with an oligochaete body plan (see figure 11.1). It is possible to infer some steps in leech evolution from the morphology of members of Acanthobdellida, for example, *Acanthobdella* has five anterior coelomic compartments separated by septa; septa have disappeared in all other leeches.

Members of clade Hirudinea, sometimes called "true leeches," have a fixed number of segments, usually 34, and typically have both an anterior and a posterior sucker (figure 11.22). They have no parapodia and lack septa. Their coelom has become filled with connective tissue and muscle, substantially reducing its effectiveness as a hydrostatic skeleton.

figure 11.21

The world's largest leech, *Haementeria ghilianii,* on the arm of Dr. Roy K. Sawyer, who found it in French Guiana, South America.

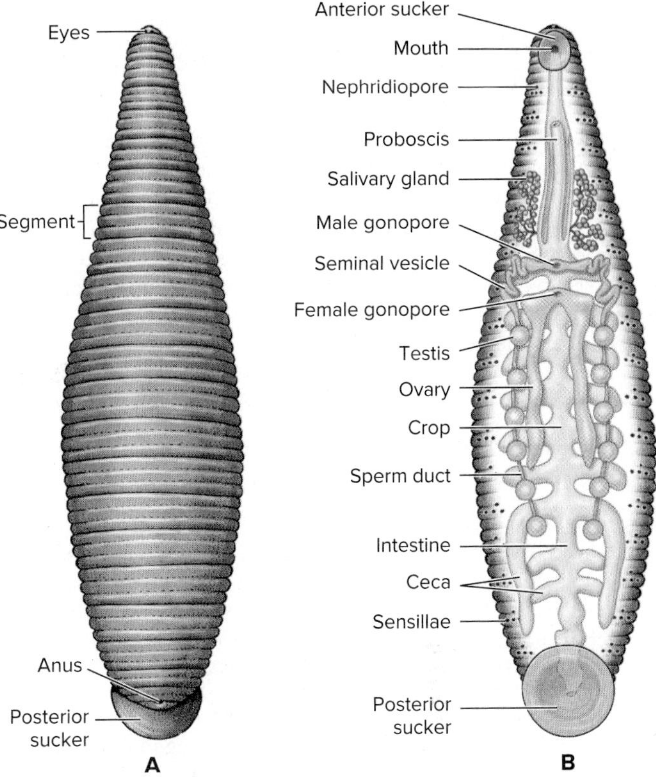

figure 11.22

Structure of a leech, *Placobdella.* **A,** External appearance, dorsal view. **B,** Internal structure, ventral view.

Many leeches live as carnivores on small invertebrates; some are temporary parasites, sucking blood from vertebrates; and some are permanent parasites, never leaving their host. Most leeches have a muscular, protrusible proboscis or a muscular pharynx with three jaws armed with teeth. They feed on body juices of their prey, penetrating its surface with their proboscis or jaws and sucking fluids with their powerful, muscular pharynx. Bloodsucking leeches secrete an anticoagulant in their saliva. Predatory leeches feed frequently, but those that feed on the blood of vertebrates consume large meals (up to several times their body weight) and digest their food slowly. The slow digestion of their meals results from an absence of amylases, lipase, or endopeptidases in their gut secretions. In fact, they appear to depend mostly on bacteria in their gut for digestion of a blood meal.

Key Theme 11.7

ADAPTATION AND PHYSIOLOGY

Attracting Leeches

Leeches are highly sensitive to stimuli associated with the presence of a prey or host. They are attracted by—and will attempt to attach to—an object smeared with appropriate host substances, such as fish scales, oil secretions, or sweat. Those that feed on the blood of mammals are attracted by warmth; terrestrial haemadipsid leeches of the tropics will converge on a person standing in one place.

Taxonomy of Phylum Annelida

Annelids are wormlike forms sharing a segmented ancestor with paired epidermal setae and grooved palps, biramous parapodia. Phylogenetic studies using molecular characters led to a major taxonomic revision. Former classes Polychaeta and Oligochaeta are not monophyletic, but the terms polychaete and oligochaete remain in use as descriptive of particular body plans. Former phyla Pogonophora, Echiura, and Sipuncula were subsumed within Annelida. Most annelids are placed in subphylum Pleistoannelida, but five groups of annelids lie outside this subphylum; Sipuncula and Chaetopteridae are discussed in this chapter, but omitted from the list below.

Subphylum Pleistoannelida Marine, freshwater, and terrestrial annelids, most with segmented bodies.

Class Errantia Freely moving polychaetes. Mostly marine; head distinct and bearing eyes and tentacles; most segments with parapodia (lateral appendages) bearing tufts of many setae; clitellum absent; sexes usually separate; gonads transitory; asexual budding in some; trochophore larva usually present. Examples: *Nereis, Aphrodita, Glycera.*

Class Sedentaria Sedentary annelids including those with polychaete and oligochaete body plans living in tubes and burrows, as well as members of order Clitellata. Examples outside of Clitellata having a polychaete body plan: *Arenicola, Amphitrite,* and Family Siboglinidae (*Osedax, Riftia*); examples having an unsegmented body plan: Family Echiuridae (*Urechis, Bonellia*). We describe only two subgroups, Order Clitellata and family Hirudinidae.

Order Clitellata. Annelids with a clitellum at some phase of the life cycle; conspicuous segmentation; number of segments variable; setae few per segment; no parapodia; head absent; coelom spacious and usually divided by intersegmental septa; hermaphroditic; development direct, no larva; chiefly terrestrial and freshwater. Examples outside of Family Hirudinidae: *Lumbricus, Stylaria, Aeolosoma, Tubifex.*

Family Hirudinidae (hir′u-din′i-dā) (L. *hirudo*, leech, + *-ida*, pl. suffix): leeches. Body with fixed number of segments (normally 34; 15 or 27 in some groups) with many annuli; oral and posterior suckers usually present; clitellum present; no parapodia; setae generally absent; coelom closely packed with connective tissue and muscle; development direct; hermaphroditic; terrestrial, freshwater, and marine. Examples: *Hirudo, Placobdella, Macrobdella.*

Leeches are hermaphroditic but practice cross-fertilization during copulation. A penis or hypodermic impregnation transfers sperm to the mating partner. Leeches have a clitellum, but it is evident only during the breeding season. After copulation, the clitellum secretes a cocoon that receives eggs and sperm. Cocoons are buried underwater in mud, attached to submerged objects, or placed in damp soil (in the case of terrestrial species). Development is similar to that of oligochaetes.

Key Theme 11.8
HUMAN CONNECTIONS

Modern Medicinal Leeches

For centuries, "medicinal leeches" *(Hirudo medicinalis*; figure 11.23) were used for blood letting because of the mistaken idea that body disorders and fevers were caused by an excess of blood. A 10–12 cm long leech can extend to a much greater length when distended with blood, and the amount of blood it can suck is considerable. Leech collecting and leech culture in ponds were practiced in Europe on a commercial scale during the nineteenth century. Wordsworth's poem "The Leech-Gatherer" was based on this use of leeches.

Leeches are once again being used medicinally. When fingers or toes are severed, microsurgeons can often reconnect arteries but not the more delicate veins. Leeches are used to relieve congestion until veins can grow back into a healing digit.

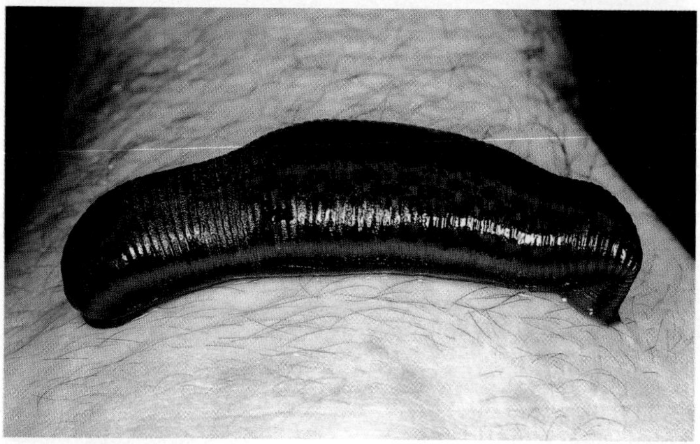

Cleveland P. Hickman, Jr.

figure 11.23

Hirudo medicinalis feeding on blood from a human arm.

11.14 Phylogeny and Adaptive Diversification

Phylogeny

Three worm groups, previously considered separate phyla, are now annelids. Phylogenetic analysis clearly supports inclusion of the pogonophorans, echiurans, and sipunculans within Annelida. All pogonophorans and some echiurans possess the paired epidermal setae characteristic of annelids, but setae are absent in sipunculans. Segments, another annelid character, are visible in parts of the pogonophoran body, but have been completely lost in echiurans and sipunculans. However, there are some serially repeated structures in echiuran larvae, such as nerve cord ganglia and mucus glands, and adults of some species have multiple nephridia.

The annelid fossil record begins in the early Cambrian period with forms that have a polychaete body plan. Parapodia with upper and lower clusters of setae like those found in modern forms (see figure 11.2D) are visible. The fossil record supports an ancestral annelid with a polychaete body plan followed by successive derivation of the oligochaete and clitellate body plans. However, the fossil record of sipunculans starts at about 520 million years ago, so the first loss of segmentation and setae occurred early in annelid evolution.

Clitellates are diagnosed by presence of a clitellum. Unlike other clitellates, leeches display a clitellum only during the reproductive season.

The origins of a metameric (segmented) body are puzzling. True metamerism occurs in annelids, arthropods, and chordates. The placement of the annelids and arthropods in Protostomia and of the chordates in Deuterostomia makes it unlikely that segmentation is homologous among these three taxa because such an ancient origin for segmentation would require that many taxa later lost segmentation. However, one hypothesis suggests that genes within ancient bilaterian genetic regulatory networks involved in determining the anterior-posterior body axis, such as *Notch* and *Wnt*, were secondarily recruited to the development of body segments. In the three segmental lineages, this recruitment occurred independently and somewhat differently, but common genetic elements were used.

Within the protostomes, annelids are placed in clade Lophotrochozoa, whereas arthropods are in clade Ecdysozoa. In both clades, most phyla are not segmented, again making it unlikely that members of these two phyla inherited a segmented body plan from a common ancestor because many evolutionary losses are required.

The selective advantage of a segmented body for annelids appears to lie in the efficiency of burrowing made possible by shape change in individual coelomic compartments of the hydrostatic skeleton. However, this explanation cannot be extended to the arthropods because, as Chapter 13 describes, the rigid exoskeleton of the arthropods prohibits shape change among segments, and the coelom is small in comparison to that of annelids. Clearly, there is much to learn about metamerism.

Adaptive Diversification

Annelids are very diverse in morphology and lifestyle. A basic adaptive feature in the evolution of annelids is their septal arrangement, producing fluid-filled coelomic compartments. Fluid pressure in these compartments is used as a hydrostatic skeleton in precise movements such as burrowing and swimming. Powerful circular and longitudinal muscles can flex, shorten, and lengthen the body. The basic body structure lends itself to great modification. In Errantia, the parapodia have been adapted in many ways and for a variety of functions, chiefly locomotion and respiration.

Feeding adaptations show great variation, from the sucking pharynx of some clitellates and the chitinous jaws of carnivorous polychaetes to the specialized tentacles and radioles of particle feeders. The evolution of a trophosome to house the chemoautotrophic bacteria that provide nutrients to siboglinids is an adaptation to deep-sea life.

In leeches, many adaptations, such as suckers, cutting jaws, pumping pharynx, distensible gut, and the secretion of anticoagulants, are related to their predatory and blood-sucking habits.

■ SUMMARY

Sections	Key Concepts
11.1 Phylum Annelida	• Phylum Annelida is a large, cosmopolitan group whose members inhabit marine, freshwater, and terrestrial habitats. • Annelids feed in diverse ways, including particle feeding, predation, and a reliance on bacterial symbionts. • Annelids are an important part of aquatic and terrestrial food webs. • Burrowing annelids contribute to the health of sediments.
11.2 Taxa within Annelida	• Modern phylogenies are based on morphological and molecular data and often revise our understanding of evolutionary relationships. • The traditional classes Polychaeta and Oligochaeta have been abolished, but the terms remain in use to describe particular body plans. • Several types of worms, previously placed outside Annelida, have been added to the phylum in recent years. These are sipunculans, siboglinids, and echiurans. • Annelida now encompasses Chaetopteridae, Sipuncula, three other groups not discussed here, and the subphylum Pleistoannelida. • Within Pleistoannelida, there are two large clades: Errantia and Sedentaria. • Within Sedentaria, we discuss a large clade, Clitellata, and three other taxa: Siboglinidae, Sabellidae, and Echiuridae. • Within Clitellata are a number of worms with an oligochaete body plan, including earthworms, as well as the leeches (Hirudinidae).
11.3 General Annelid Body Plan	• Typical annelids have a two-part head, but the head can be variously modified for particle collection, predation, and other modes of feeding. • Metamerism, the division of the body into a series of similar segments, each containing a repeated arrangement of many organs and systems, is the most important structural innovation for the phylum. • The coelom is highly developed and works with the septal arrangement of fluid-filled compartments and a well-developed body-wall musculature to make an effective hydrostatic skeleton for burrowing and swimming. • A one-way, regionalized, digestive system is generally present. • The nervous system is well developed.
11.4 Chaetopteridae	• Chaetopterids are tube-dwellers with regionalized bodies. • They collect food on mucous nets.
11.5 Sipuncula	• Sipunculans are not metameric and lack setae. • Commonly called "peanut worms," they burrow in sand or mud. • Sipunculans were placed within Annelida using molecular data.
11.6 Pleistoannelida	• Pleistoannelida is subphylum of Annelida. • It comprises two large clades: Errantia and Sedentaria.
11.7 Class Errantia	• Errantiate worms have a polychaete body plan; most are marine; many are predators. • Parapodia are well-developed, as are the sensory and nervous systems. • Sexual reproduction often occurs via a body form called an epitoke.
11.8 Class Sedentaria	• Sedentary worms have polychaete or oligochaete body plans. • Sedentary worms comprise tube-dwellers, burrowers, and leeches. • Head structures in sedentary annelids are often modified to collect food and function in respiration.
11.9 Siboglinidae (Pogonophorans)	• Siboglinids were once called pogonophorans and vestimentiferans. • These "beardworms" are part of the deep ocean rift-vent community. • They have neither mouths nor guts, but rely on bacterial symbionts for nutrition. • They have highly modified bodies with three distinct regions.
11.10 Echiuridae	• Echiurans, also called spoon worms, are marine animals. • They burrow or live in tubes, shells, or crevices. • They are not segmented, but do possess paired epidermal setae.

Sections	Key Concepts
11.11 Order Clitellata	• Clitellates are marine, freshwater, and terrestrial worms with a diagnostic structure, the clitellum, used in reproduction. • In many the clitellum is always visible, but in leeches it is not apparent outside the reproductive period.
11.12 Form and Function in Clitellata	• Clitellates include earthworms and many freshwater forms; they have a small number of setae per segment and no parapodia. • They have a closed circulatory system, with a dorsal blood vessel serving as the main pumping organ. • Paired nephridia occur in most segments. • Earthworms possess the typical annelid nervous system: dorsal cerebral ganglia connected to a double, ventral nerve cord with segmental ganglia running the length of the worm. • Clitellates are hermaphroditic and practice cross-fertilization. • A clitellum secretes mucus to surround the worms during copulation. It also secretes a cocoon to receive eggs and sperm and provide a moist location for fertilization and embryo development.
11.13 Family Hirudinidae	• Leeches (family Hirudinidae) are mostly freshwater, although a few are marine and a few are terrestrial. • They are primarily fluid-feeders, often consuming blood. Many are predators, some are temporary ectoparasites, and a few are permanent ectoparasites.
11.14 Phylogeny and Adaptive Diversification	• The ancestral annelid had a metameric (segmented) body with paired epidermal setae, but these features are not present in all annelid taxa. • The fossil record includes both typical and non-segmented annelids. • Segmented bodies function well for burrowing animals. • Segmented bodies occur in annelids, arthropods, and chordates, but this body type is thought to have evolved independently in each lineage.

Review Questions

1. What characteristics of phylum Annelida distinguish it from other phyla?
2. How are members of Clitellata distinguished from other annelids?
3. Describe the annelid body plan, including the body wall, segments, coelom and its compartments, and coelomic lining.
4. Explain how the hydrostatic skeleton of annelids helps them burrow. How is burrowing efficiency increased by metamerism?
5. Describe three ways that various errantiate annelids obtain food, and contrast these with food gathering in Sedentaria.
6. Define each of the following: prostomium, peristomium, radioles, parapodium, setae.
7. Explain the functions of each of the following in earthworms: crop, gizzard, typhlosole, chloragogen cells.
8. Which group of annelids has the best-developed sensory structures? How does the extent of sensory perception match lifestyle and habitat?
9. Describe the functions of the clitellum and the cocoon.
10. How are freshwater clitellates generally different from earthworms?
11. Describe how leeches obtain food.
12. Describe the ways in which reproduction and development changed as annelids colonized freshwater and terrestrial habitats. Why might selection have favored certain features such as monoecious reproduction and the use of a cocoon?
13. What is the largest siboglinid known? Where is it found, and how is it nourished?
14. What features of echiurans are shared with other annelids?
15. Where does a sipunculan live, and how does it collect food?
16. Which annelids rely on bacterial symbionts for their nutrition?
17. Compare and contrast the body plans of sedentary and errantiate polychaetes and give a functional explanation for the differences you observe.

For Further Thought Review the extent of segmentation in typical annelids and in siboglinids, echiurans, and sipunculans. Which morphological features must change to produce an unsegmented body cavity? Is there evidence of transitional stages for these features?

12

Smaller Ecdysozoans

LEARNING OBJECTIVES

Readers will be able to:

12.1 Describe the body plan and habits of nematodes with special reference to the life cycles and hosts of parasitic taxa.

12.2 Describe the body plan of a nematomorph and provide one example of its life cycle and host use.

12.3 Identify a loriciferan and describe its habitat and lifestyle.

12.4 Identify a kinorhynch and describe its habitat and lifestyle.

12.5 Identify a priapulid and describe its habitat and lifestyle.

12.6 Identify the taxa within Panarthropoda and describe the structure, function, and formation of the characteristic hemocoel.

12.7 Describe the body plan of an onychophoran with special reference to adaptations for feeding and respiration.

12.8 Identify a tardigrade and describe its habitat and lifestyle, including the use of cryptobiosis; explain how an understanding of cryptobiosis is useful to humans.

12.9 Contrast the evolutionary relationships among ecdysozoans supported by shared developmental, morphological, and molecular characters.

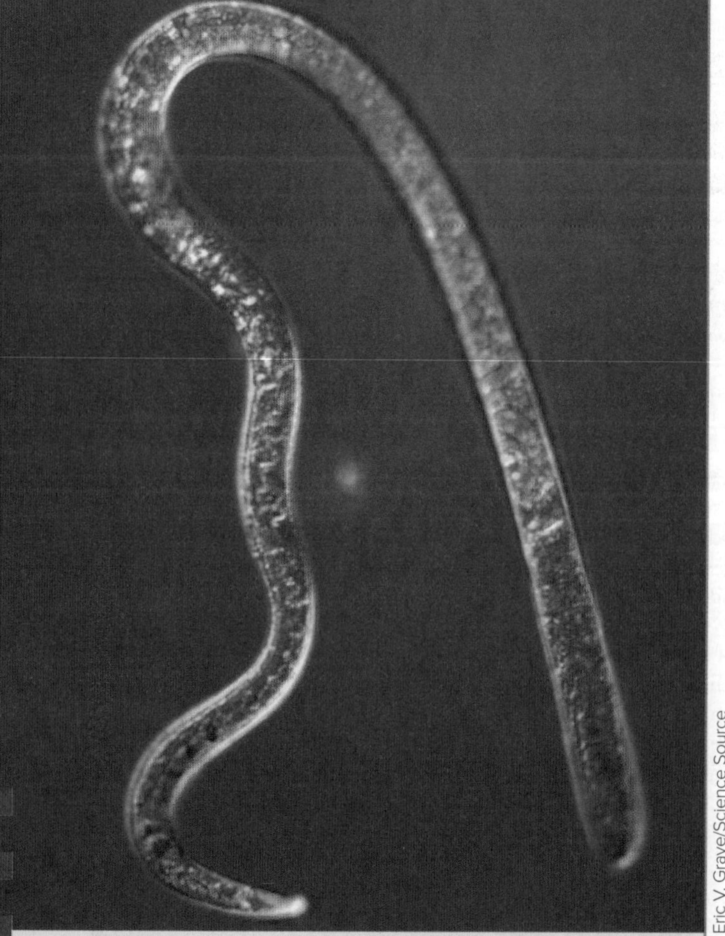

Eric V. Grave/Science Source

Male *Trichinella spiralis*, a nematode.

A World of Nematodes

Without any doubt, nematodes are the most important smaller ecdysozoan animals in terms of both numbers and their impact on humans. Nematodes are abundant over most of the world, and yet most people are only occasionally aware of them as parasites of humans or of pets. We are not aware of the billions of these worms in the soil, in ocean and freshwater habitats, in plants, and in all kinds of other animals. Their dramatic abundance moved N. A. Cobb to write in 1914:

> If all the matter in the universe except the nematodes were swept away, our world would still be dimly recognizable, and if, as disembodied spirits, we could then investigate it, we should find its mountains, hills, vales, rivers, lakes and oceans represented by a thin film of nematodes. The location of towns would be decipherable, since for every massing of human beings there would be a corresponding massing of certain nematodes. Trees would still stand in ghostly rows representing our streets and highways. The location of the various plants and animals would still be decipherable, and, had we sufficient knowledge, in many cases even their species could be determined by an examination of their erstwhile nematode parasites.[1]

[1]N. A. Cobb. 1914. *Yearbook of the United States Department of Agriculture, 1914.*

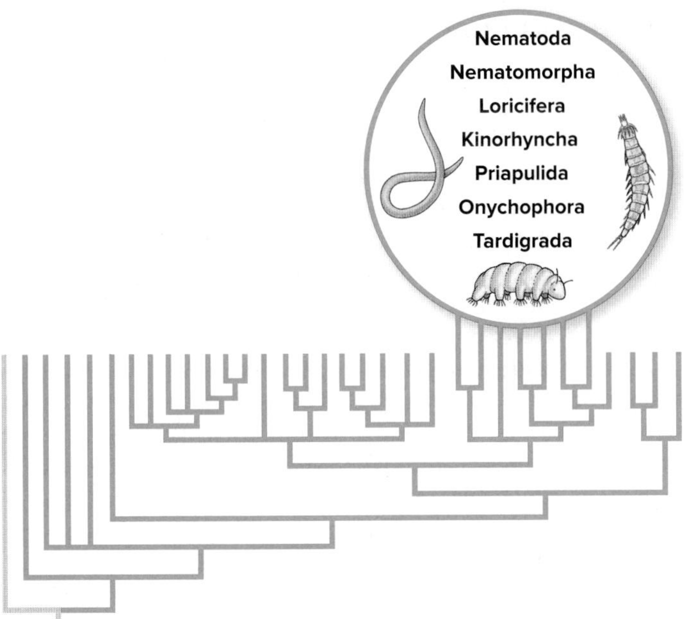

Protostome animals include flatworms, roundworms, molluscs, annelids, and arthropods, among many other taxa (see the Cladogram of All Animal Phyla on the inside front cover). Many protostomes, such as annelids, roundworms, and arthropods, possess a **cuticle,** a nonliving external layer secreted by the epidermis. A firm cuticle surrounding the body wall, like that present in roundworms and arthropods, restricts growth. As the body increases in size, the cuticle is molted and the outer layer shed, a process called **ecdysis.**

Protostome phyla are divided between two large clades: Lophotrochozoa and Ecdysozoa. Ecdysozoa (figure 12.1) comprises those taxa that molt the cuticle as they grow. Where it has been studied, molting is regulated by the hormone **ecdysone;** biologists assume that a homologous set of biochemical steps regulates molting among all ecdysozoans. Ecdysozoan taxa, other than loriciferans, were first united as a clade in phylogenies based on molecular characters.

As with lophotrochozoan phyla, ecdysozoans do not share a common body plan. Members of Nematoda, Nematomorpha,

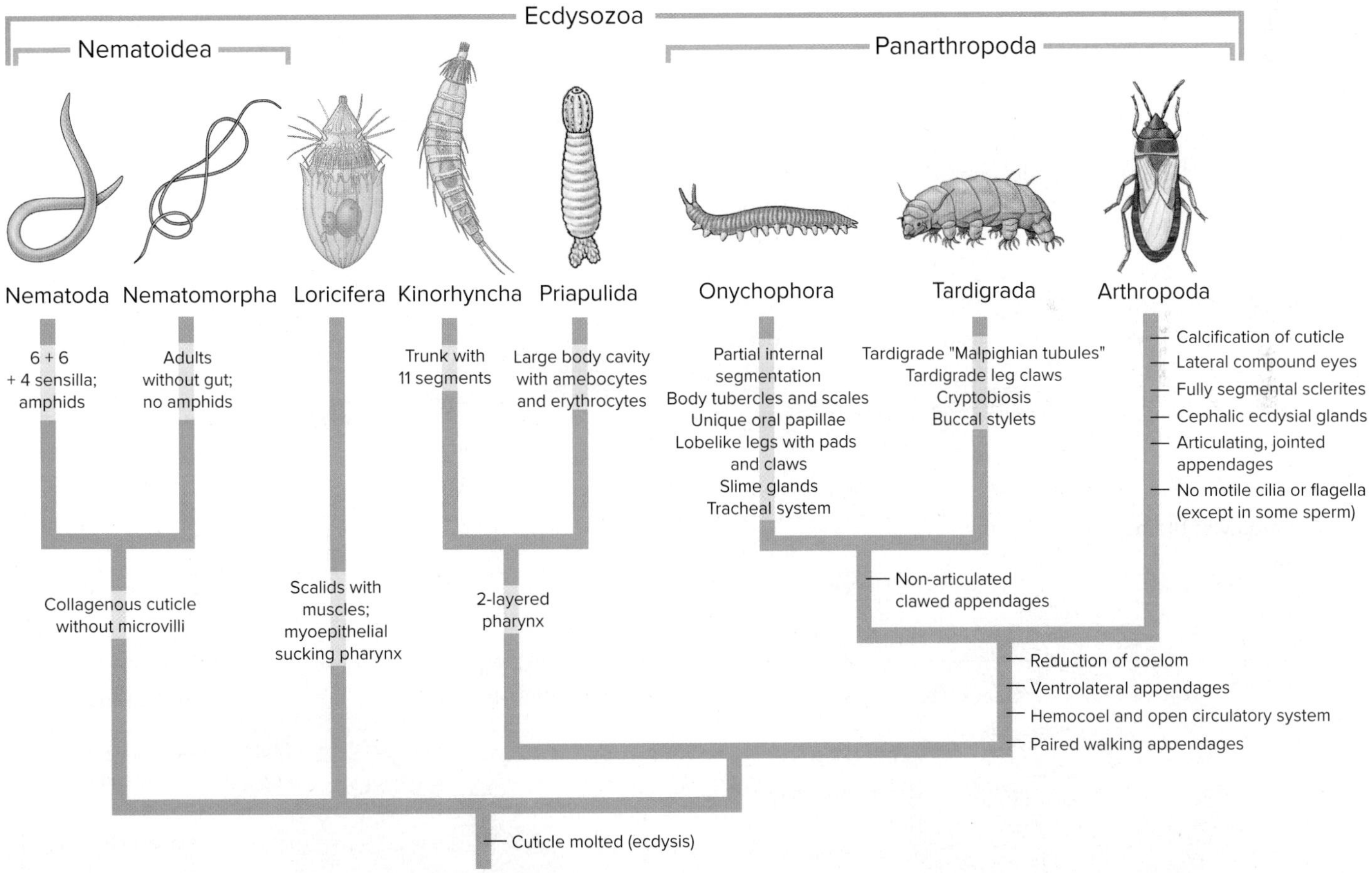

figure 12.1

Cladogram depicting one hypothesis for relationships among ecdysozoan phyla. Characters shown are subsets of those in Nielsen (1995), Neuhaus and Higgins (2002), and Brusca and Brusca (2003); the nematode character "6 + 6 + 4 sensilla" refers to the anterior rings of sensory papillae. Segmentation occurs in Arthropoda and Kinorhyncha and partially in Onychophora. Despite the external appearance of segmentation in Priapulida and Tardigrada, these phyla do not have segmented internal structures. This cladogram depicts three separate origins of segmentation within Ecdysozoa, although this hypothesis is tentative pending further phylogenetic analyses. Segmentation has arisen separately in Lophotrochozoa and Deuterostomia.

and Kinorhyncha have pseudocoelomate bodies. Members of Priapulida have not been carefully studied, but they are assumed to be pseudocoelomate. The pseudocoelom is used as a hydrostatic skeleton in nematodes, kinorhynchs, and perhaps in priapulids. Within Loricifera, species apparently vary in body plan: some are described as pseudocoelomate and others appear acoelomate. Members of clade Panarthropoda have coelomate bodies, but their coeloms are small. Panarthropoda is an enormous group of animals containing three phyla: Onychophora, Tardigrada, and Arthropoda.

Arthropoda, the largest phylum in terms of numbers of described species, forms the subject of Chapter 13. The present chapter describes all other ecdysozoan phyla.

12.1 Phylum Nematoda: Roundworms

Nematodes (Gr., *nematos*, thread) occur in nearly every conceivable kind of ecological niche. Approximately 30,000 species have been named, but most estimates predict well over one million species exist. They live in the sea, in freshwater, and in soil, from polar regions to tropics, and from mountaintops to the depths of seas. Good topsoil may contain billions of nematodes per acre. Nematodes also parasitize virtually every type of animal and many plants. The effects of nematode infestation on crops, domestic animals, and humans make this phylum one of the most important of all parasitic animal groups.

Form and Function

Most nematodes are less than 5 cm in length, and many are microscopic, but some parasitic nematodes are over a meter in length.

Use of the pseudocoelom as a hydrostatic skeleton is a conspicuous feature of nematodes, and we can best understand much of their functional morphology in the context of the high hydrostatic pressure within the pseudocoelom.

The outer body covering is a relatively thick, noncellular cuticle, secreted by the underlying epidermis (**hypodermis**). A cuticle is of great functional importance, serving to contain the high hydrostatic pressure exerted by fluid in the pseudocoelom. Layers of the cuticle are primarily composed of **collagen,** a structural protein present throughout the animals and abundant in vertebrate connective tissue. Crisscrossing fibers form three of the layers, conferring some longitudinal elasticity on the worm but severely limiting its capacity for lateral expansion.

Beneath the hypodermis is a layer of longitudinal muscles. The muscles are arranged in four bands, or quadrants, separated by four epidermal cords that project inward to the pseudocoelom (figure 12.2). There are no circular muscles in the body wall. Another unusual feature of body-wall muscles in nematodes is that the muscles extend processes to synapse with nerve cords, rather than nerves extending an axon to synapse with muscle.

The fluid-filled pseudocoelom, in which internal organs lie, constitutes a hydrostatic skeleton. Many invertebrates use hydrostatic skeletons, which lend support by transmitting the force of

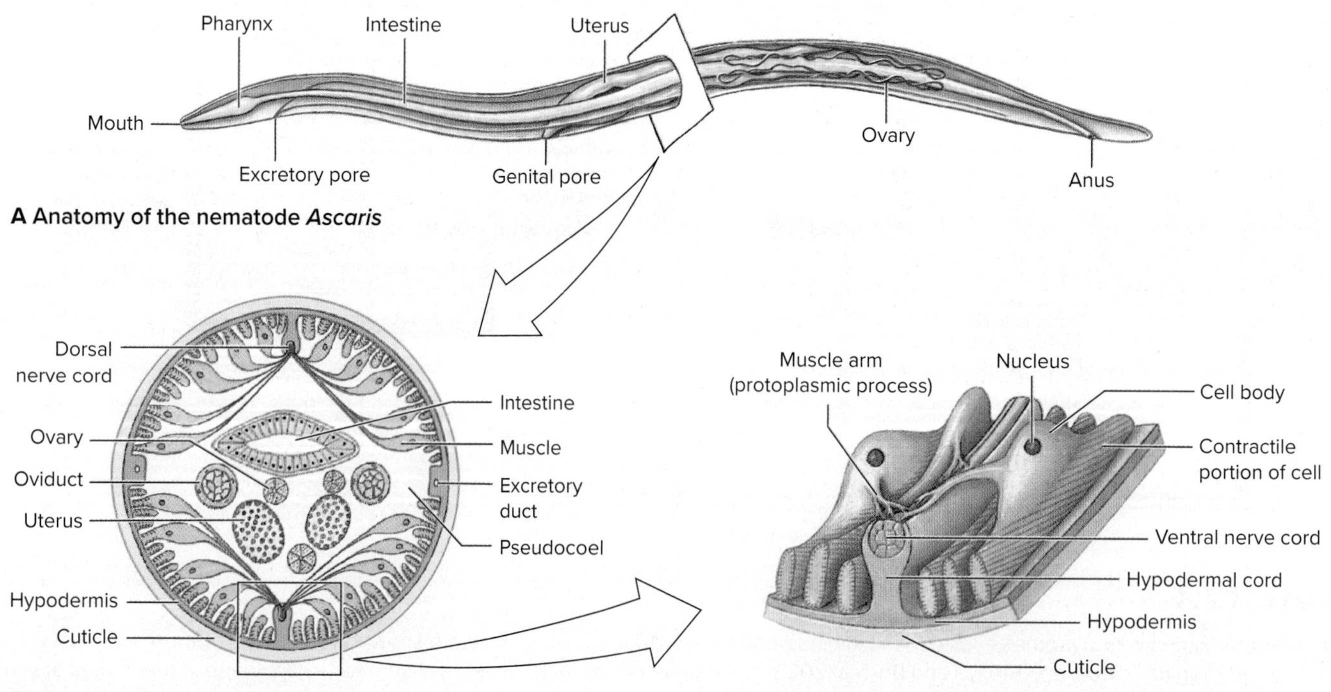

figure 12.2

A, An *Ascaris* female has two ovaries and uteri, which open to the outside by a common genital pore. **B,** Cross section. **C,** Relationship of muscle cells to hypodermis and hypodermal cord.

CHARACTERISTICS of Phylum Nematoda

1. Unique sensory **amphids** or phasmids.
2. Marine, freshwater, and terrestrial.
3. Free-living and parasitic.
4. Body bilaterally symmetrical.
5. Triploblastic body.
6. **Pseudocoelom** functions as a hydrostatic skeleton.
7. Nonliving, flexible cuticle is molted.
8. Digestive system complete; muscular pharynx is triradiate in cross section.
9. Body wall has longitudinal muscles only.
10. Ring of nerve tissue with ganglia around pharynx, dorsal and ventral nerve cords.
11. Sensory system includes papillae and setae for touch; anterior amphids and posterior phasmids are likely chemoreceptors.
12. Parthenogenesis occurs rarely.
13. Sexes separate in most; fertilization is internal, but sperm are ameboid, not flagellated; cleavage pattern unique.
14. Excretory system may include gland cells and canals, opens via excretory pore.
15. No respiratory or circulatory systems.

muscle contraction to the enclosed, noncompressible fluid. Normally, muscles are arranged antagonistically so that, as movement is effected in one direction by contraction of one group of muscles, movement in the opposite direction is effected by contraction of an antagonistic set of muscles. However, nematodes do not have circular body-wall muscles to antagonize the longitudinal muscles; therefore, the cuticle must serve that function. As muscles on one side of the body contract, they compress the cuticle on that side, and the force of the contraction is transmitted (by the fluid in the pseudocoelom) to the other side of the nematode, stretching the cuticle on that side. This compression and stretching of the cuticle serve to antagonize the muscle and provide the forces that return the body to resting position when the muscles relax; this action produces the characteristic thrashing motion when nematodes move through fluids. However, nematode movements are much more directed when they move through sediments or between small objects. An increase in the efficiency of the nematode muscular system can be achieved only by an increase in hydrostatic pressure. Consequently, the hydrostatic pressure in a nematode's pseudocoelom is much higher than is usually found in animals that have both hydrostatic skeletons and antagonistic muscle groups.

The gut tube of a nematode comprises a mouth, a muscular **pharynx,** a long, nonmuscular intestine, a short rectum, and a terminal anus (figure 12.2). The cylindrical pharynx has radial muscles that insert on the cuticular lining of its lumen and on a basement membrane at its periphery. At rest, the lumen is closed. When muscles in the anterior of the pharynx contract, they open its lumen and suck food inside. Relaxation of muscles anterior to the food mass closes the pharyngeal lumen and forces the food posteriorly toward the intestine. The intestine is only one cell layer thick and has no muscles. Food matter moves posteriorly in the intestine by body movements and as additional food is passed into the intestine from the pharynx. Defecation is accomplished by muscles that simply pull the anus open, and pseudocoelomic pressure surrounding the rectum provides expulsive force.

Adults of many parasitic nematodes have an anaerobic energy metabolism; a Krebs cycle and cytochrome system characteristic of aerobic metabolism are absent. They derive energy through glycolysis and some additional electron-transport sequences. Interestingly, some free-living nematodes and free-living stages of parasitic nematodes are obligate aerobes (require oxygen) and have a Krebs cycle and a cytochrome system.

A ring of nerve tissue and ganglia around the pharynx gives rise to small nerves to the anterior end and to two nerve cords, one dorsal and one ventral. Some sensory organs around the lips and around the posterior end are rather elaborate.

Key Theme 12.1

SCIENCE EXPLAINED

C. *Elegans*, a Spectacular Model System

In 1963, Sydney Brenner initiated some extremely fruitful research by studying a free-living nematode, *Caenorhabditis elegans.* This small worm has now become one of the most important experimental models in biology. The origin and lineages of all 959 cells in its body have been traced from zygote to adult, and the complete "wiring diagram" of its nervous system is known—all of the neurons and all of the connections between them. The nematode's genome has been completely mapped, and scientists have sequenced its entire genome of 3 million bases containing 19,820 genes. Many basic discoveries about gene function, such as how genes encode the proteins essential for programmed cell death, have been and will be made using *C. elegans.*

Most nematodes are dioecious. Males are smaller than females, and their posterior end usually bears a pair of copulatory spicules (figure 12.3). Fertilization is internal, and shelled zygotes or embryos are stored in the uterus until deposition. After embryonic development, a juvenile worm hatches from an egg. There are four juvenile stages, each separated by a shedding of the cuticle. Many parasitic nematodes have

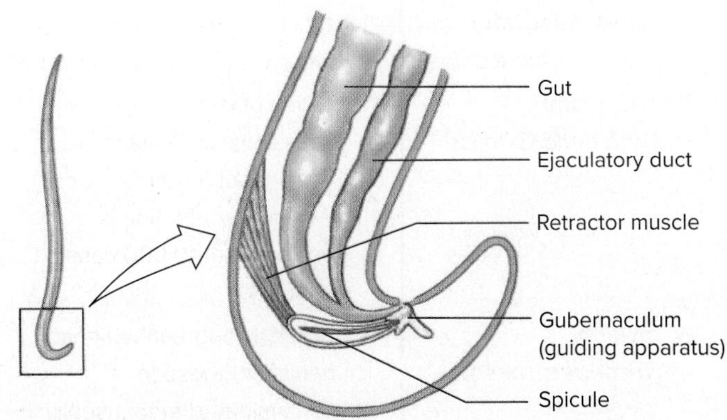

figure 12.3

Posterior end of a male nematode.

free-living juvenile stages, and others require an intermediate host to complete their life cycles.

Key Theme 12.2
ADAPTATION AND PHYSIOLOGY

Mating Under Pressure

The copulatory spicules of male nematodes are not true intromittent organs, since they do not conduct sperm, but are another adaptation to high internal hydrostatic pressure. Spicules must hold the vulva of a female open while ejaculatory muscles in the male reproductive tract overcome the hydrostatic pressure in the female and rapidly inject sperm into her reproductive tract. Nematode spermatozoa are unique among animals in having no flagellum or acrosome. Ameboid sperm crawl using lamellipodia, but the underlying cytoskeleton relies on a major sperm protein, MSP, for extension, rather than actin filaments.

Representative Nematode Parasites

Nearly all vertebrates and many invertebrates are parasitized by nematodes, a number of which are very important pathogens of humans and domestic animals. A few nematodes are common in humans in North America (table 12.1), but they and many others usually abound in tropical countries. We have space to mention only a few, but readers also may recall reading about rat lung worm in Chapter 10 (see Key Theme 10.4).

Table 12.1 Common Parasitic Nematodes of Humans in North America

Common and scientific names	Means of infection; prevalence in humans
Hookworm *(Ancylostoma duodenale* and *Necator americanus)*	Contact with soil infested with juveniles that burrow into skin; occurs in the southern United States, but less prevalent than previously. An estimated 650 million people are infected worldwide.
Pinworm *(Enterobius vermicularis)*	Inhalation of dust containing ova and by contamination with fingers; most common worm parasite in the United States
Intestinal roundworm *(Ascaris lumbricoides)*	Ingestion of ova containing embryos in contaminated food
Trichina worm *(Trichinella spiralis)*	Ingestion of infected muscle; occasional in humans throughout North America. Worldwide infection is estimated at 10,000 cases annually.
Whipworm *(Trichuris trichiura)*	Acquired through unhygienic habits or ingestion of contaminated food; usually common wherever *Ascaris* is found

Ascaris lumbricoides: Large Roundworm of Humans

The large human roundworm, *Ascaris lumbricoides* (Gr. *askaris,* intestinal worm), is one of the most common worm parasites of humans (figure 12.4A). It occurs in warm, humid regions of the earth, and about one billion people worldwide are likely to have this worm. It is now uncommon in the United States. *Ascaris suum,* a parasite of pigs, is morphologically similar and was long considered the same species (figure 12.4B). Adult *Ascaris* worms live in the small intestine of their host. *Ascaris* females may reach 30 cm in length and can produce 200,000 eggs a day. Fertilized eggs leave the host's body with the feces and are amazingly tolerant to dessication or lack of oxygen, but are killed by direct sunlight and high temperatures. In the absence of a sewage system, defecation onto soil allows the life cycle to continue. Given suitable soil conditions, embryos develop into infective juveniles within two weeks. Shelled

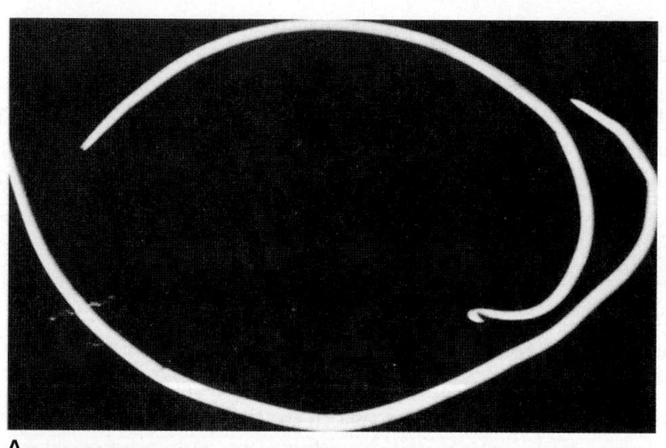

A

Cleveland P. Hickman, Jr.

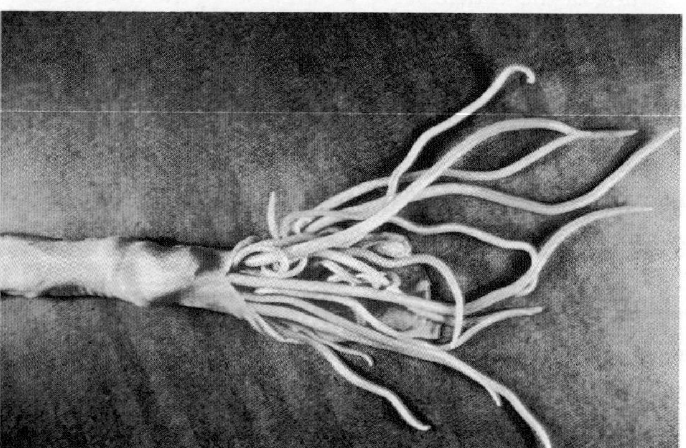

B

Larry S. Roberts

figure 12.4

A, Intestinal roundworm *Ascaris lumbricoides,* male and female. Male *(top)* is smaller and has a characteristic sharp kink in the end of its tail. The females of this large nematode may be over 30 cm long. **B,** Intestine of a pig, nearly completely blocked by *Ascaris suum.* Such heavy infections are also fairly common with *A. lumbricoides* in humans.

juveniles can remain viable for many months or even years in soil, so they may be transmitted to a new host by consumption of contaminated vegetables or ingestion of soil. Tiny juveniles hatch from the shelled eggs once they are swallowed by a host. They burrow through the host's intestinal wall, entering the blood or lymph and then traveling to the heart. From the heart, they migrate to the lungs, moving from the alveoli to the trachea. If abundant, they may cause a serious pneumonia at this stage. The worms, still juveniles, pass from the trachea to the pharynx where they are swallowed, bringing them back into the stomach to complete development. Two months after being ingested as shelled juveniles, the worms feed on intestinal contents and mature. In the intestine, worms cause abdominal symptoms and allergic reactions, and in large numbers, they may cause intestinal blockage. Parasitism by *Ascaris* is rarely fatal unless the intestine is blocked by a heavy infestation. Perforation of the intestine with resultant peritonitis is not uncommon, and wandering worms may occasionally emerge from the anus or throat or may enter the trachea or eustachian tubes and middle ears.

Key Theme 12.3

ADAPTATION AND PHYSIOLOGY

Puppies Parasitized Before Birth

Certain ascarids are common in wild and domestic animals. Species of *Toxocara*, for example, infect dogs and cats. Their life cycle is generally similar to that of *Ascaris*, but juveniles often do not complete their tissue migration in adult dogs, remaining in the host's body in a stage of arrested development. Pregnancy in the female dog, however, stimulates juveniles to wander, and they infect the embryos in the uterus. Puppies are then born with worms. These ascarids also survive in humans but do not complete their development, leading to an occasionally serious condition in children called visceral larva migrans. This condition is a good reason for pet owners to practice hygienic disposal of canine wastes!

Hookworms

Hookworms are so named because their anterior end curves dorsally, suggesting a hook. The most common species is *Necator americanus* (L. *necator*, killer), whose females are up to 11 mm long. Males can reach 9 mm in length. Large plates in the mouth (figure 12.5A & B) cut into the intestinal mucosa of the host, where they suck blood and pump it through their intestine, partially digesting it and absorbing the nutrients. They suck much more blood than they need for food, and heavy infections cause anemia in the host. Hookworm disease in children may retard mental and physical growth and cause a general loss of energy. Effective drug treatments, such as albendazole or mebendazole, are available.

Shelled embryos leave the host's body in feces, and juveniles hatch in the soil, where they live on bacteria. When human skin contacts infested soil, the juveniles burrow through the skin to the blood, reaching the lungs and finally the intestine in a manner similar to that described for *Ascaris*.

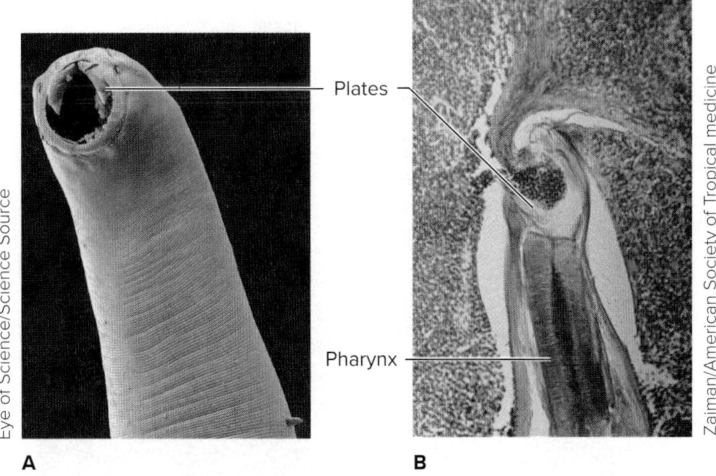

figure 12.5

A, Mouth of hookworm displaying cutting plates. **B,** Section through anterior end of hookworm attached to dog intestine. Note cutting plates pinching off mucosa from which the thick muscular pharynx sucks blood. Esophageal glands secrete anticoagulant to prevent blood from clotting.

Trichina Worm

Trichinella spiralis (Gr. *trichinos*, of hair, + *-ella*, diminutive) is one of the species of tiny nematodes responsible for the potentially lethal disease trichinosis, also called trichinellosis. Adult worms burrow in the mucosa of the host's small intestine, where females produce living young. Juveniles penetrate blood vessels and are carried throughout the body, where they occupy almost any tissue or body space. Eventually, they penetrate skeletal muscle cells, becoming one of the largest known intracellular parasites. Juveniles cause astonishing redirection of gene expression in their host cell, which loses its striations and becomes a **nurse cell** that nourishes the worm (figure 12.6). When meat containing live juveniles is swallowed, the worms are liberated into the host's intestine, where they mature.

Trichinella spp. can infect a wide variety of mammals in addition to humans, including pigs, rats, cats, and dogs. Pigs become infected by eating pork scraps infested with juveniles or by eating infected rats. *Trichinella spiralis* has four sibling species, which differ in geographic distribution, infectivity to different host species, and freezing resistance. Heavy infections may cause death, but lighter infections are much more common. The parasite is now rare in commercially produced pork, but recent cases of infection come from wild boar or from hogs raised in close proximity to wild animals. In northern California, 30 people were infected in 2008 after consuming black bear meat; 12 cases occurred in California in 2017 following consumption of raw pork from privately raised pigs. Effective drug treatments for infected individuals are available.

Pinworms

Pinworms, *Enterobius vermicularis* (Gr. *enteron*, intestine, + *bios*, life), cause relatively little disease, but they are the most

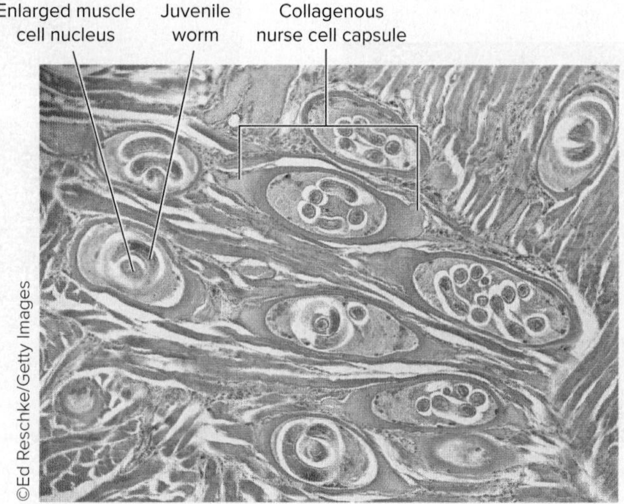

figure 12.6

Section of a human muscle infected with the trichina worm, *Trichinella spiralis.* Juveniles lie within muscle cells that the worms have induced to transform into nurse cells (commonly called cysts). An inflammatory reaction is evident around the nurse cells. Juveniles may live 10–20 years, and nurse cells eventually may calcify.

common worm parasite in the United States, occurring in an estimated 30% of children and 16% of adults. Adult parasites (figure 12.7A) occupy the large intestine and cecum. Females reach about 12 mm in length and migrate to the anal region at night to lay their eggs (figure 12.7B). Scratching the resultant itch effectively contaminates hands and bedclothes. Eggs develop rapidly and become infective within 6 hours at body temperature. After they are swallowed, they hatch in the duodenum, and the worms mature in the large intestine.

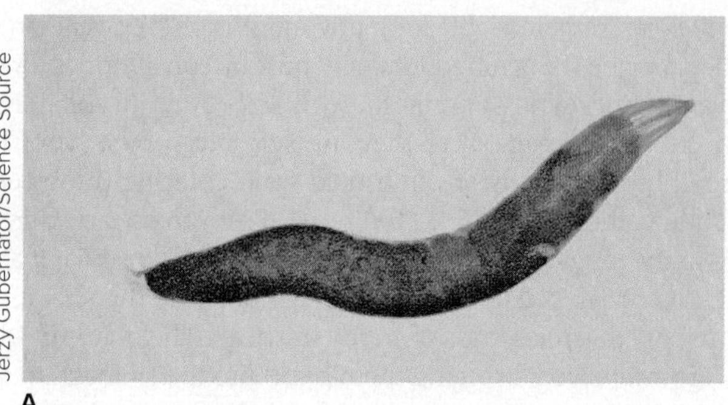

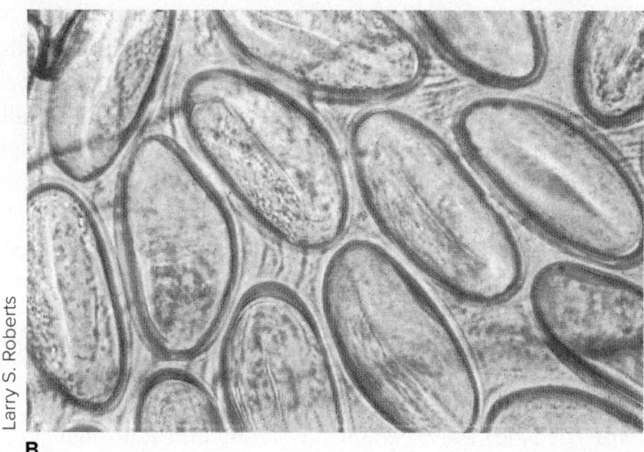

figure 12.7

Pinworms, *Enterobius vermicularis.* **A,** Female worm from human large intestine (slightly flattened in preparation), magnified about 10 times. **B,** Group of shelled juveniles of pinworms, which are usually discharged at night around the anus of the host, who may get fingernails and clothing contaminated by scratching during sleep. This may be the most common and widespread of all human worm parasites.

Key Theme 12.4

HUMAN CONNECTIONS

Pinworm Infection

Most intestinal roundworm infections are usually diagnosed by examining a small bit of feces under a microscope and finding characteristic shelled embryos or juveniles. However, pinworm eggs are not often found in feces because females deposit them on the skin around the anus at night. For pinworms, the "Scotch tape method" is more effective. The sticky side of cellulose tape is applied around the anus to collect the shelled embryos, and then the tape is placed on a glass slide and examined under a microscope. Several drugs are effective against this parasite, but all members of a family should be treated at the same time, since the worm easily spreads throughout a household.

Members of this order of nematodes have **haplodiploidy,** a characteristic shared with a few other animal groups, notably many hymenopteran insects (see section 13.7: Social Behavior). Males are haploid and are produced parthenogenetically; females are diploid and arise from fertilized eggs.

Filarial Worms

At least eight species of filarial nematodes infect humans, and some cause major diseases. Some 120 million people in tropical countries are infected with *Wuchereria bancrofti* (named for Otto Wucherer), *Brugia malayi* (named for S. L. Brug), or *B. timori,* placing these species among the scourges of humanity. The worms infect the lymphatic system, and females may be as long as 100 mm. Disease symptoms include inflammation and obstruction of the lymphatic system. Female worms release tiny living microfilariae into the blood and lymph of the host. These microfilariae are then ingested by mosquitos as the insects feed, and they develop in mosquitos to the infective stage. When the mosquito again feeds on a human, the microfilariae escape from the mosquito and penetrate the wound made by the mosquito bite.

Long and repeated exposure to filarial worms can lead to elephantiasis, a condition marked by excessive growth of connective tissue and enormous swelling of affected parts, such as the scrotum, legs, arms, and more rarely, the vulva and breasts (figure 12.8). Drugs that kill microfilariae in the

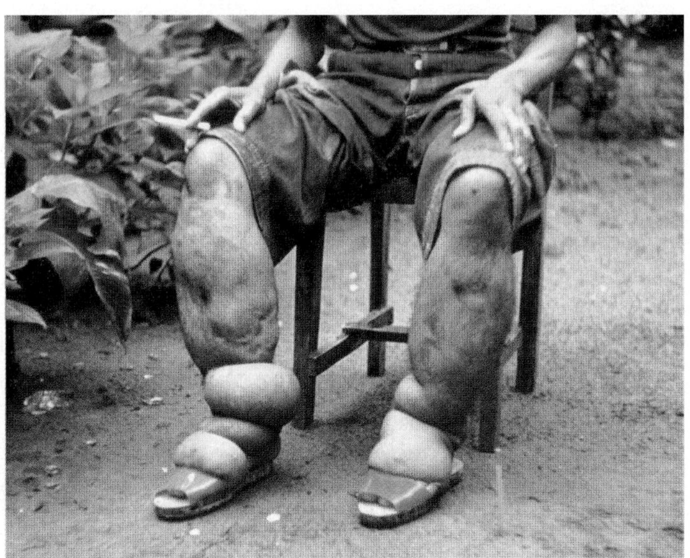

CDC

figure 12.8

Elephantiasis of legs caused by adult filarial worms of *Wuchereria bancrofti*, which live in lymph passages and block the flow of lymph. A mosquito ingests tiny juveniles, called microfilariae, with its blood meal. Microfilariae then develop to the infective stage and can be transmitted by the mosquito to a new host.

Nancy Nehring/Getty Images

figure 12.9

Dirofilaria immitis in a dog's heart. This nematode is a major menace to the health of dogs in North America. Adults live in the heart, and juveniles circulate in the blood, where they are ingested and transmitted by mosquitos.

host are available and effective, but there is limited treatment for the swelling that results from long-term infection. Avoiding mosquito bites is a key element of disease control, along with drug treatments for entire communities. Some drug companies donate drugs as part of a worldwide campaign to eradicate this disease.

Another filarial worm causes river blindness (onchocerciasis) and is carried by blackflies. It infects more than 20 million people in parts of Africa, the Mideast, Central America, and South America (2017 data). Diseases caused by filarial worms fall into the Centers for Disease Control (CDC) category "Neglected Tropical Diseases" (NTDs). Although largely eliminated in the developed world, these diseases caused enormous suffering in the world's poorest countries, where people often could not afford to pay for treatment. However, global attention to these diseases permitted large-scale administration of effective drugs. Four countries have eliminated onchocerciasis and three more no longer require mass drug administration. Eradication of some NTDs is possible; for example, human infection with the nematode causing Guinea-worm disease has decreased from 3.5 million cases worldwide circa 1985 to 54 cases in 2019. This is due to availability of drinking water filtered to remove the tiny arthropod transmitting the disease.

The most common filarial worm in the United States is probably the dog heartworm, *Dirofilaria immitis* (figure 12.9). Carried by mosquitos, it also can infect other canids, cats, ferrets, sea lions, and occasionally humans. Along the Atlantic and Gulf Coast states and northward along the Mississippi River throughout the midwestern states, prevalence in dogs reaches 45%. It occurs in other states at a lower prevalence. This worm causes a very serious disease among dogs, and no responsible owner should fail to provide "heartworm pills" for a dog during mosquito season.

12.2 Phylum Nematomorpha

The popular name for the Nematomorpha (nem′a-to-mor′fa) (Gr. *nema, nematos,* thread, + *morphē,* form) is "horsehair worms," based on an old superstition that the worms arise from horsehairs that happen to fall into water, and indeed the worms do resemble hairs from a horse's tail. They range from 0.5 to 3 mm in diameter and may reach 1 m in length. They were long included in phylum Nematoda because both groups have a cuticle, epidermal cords, longitudinal muscles only, and a similar nervous system pattern. Several studies indicate that nematomorphs are the sister taxon to nematodes; the two groups are united in clade Nematoidea.

About 320 species of horsehair worms have been named. Worldwide in distribution, horsehair worms are free-living as adults and parasitic in arthropods as juveniles (figure 12.10). Early larval forms of some species have a striking resemblance to priapulids. Adults have a vestigial digestive tract and absorb organic molecules through the vestigial gut and body wall, much as do juveniles. They can live almost anywhere in wet or moist surroundings if oxygen is adequate.

Life cycles of nematomorphs are poorly known. In the cosmopolitan genus, *Gordius* (named for an ancient king who tied an intricate knot), juveniles may encyst on vegetation likely to be eaten by a grasshopper or other arthropod. Gordiid larval stages also have hooks or stylets that may be used to bore into

figure 12.10

A horsehair worm, or nematomorph, *Paragordius tricuspidatus*, emerges from the body of a European cricket, *Nemobius sylvestris*. Nematomorphs are very long and very thin. Their pharynx is usually a solid cord of cells and appears nonfunctional. *Paragordius*, whose pharynx opens through to the intestine, is unusual in this respect and also in the possession of a photosensory organ ("eye").

a host, perhaps via the integument or the gut lining. In other cases, the gordiid may infect the host via its drinking water. Larvae encyst in the host; in some cases, it seems that development continues after the first host is eaten by a second host. In the marine nematomorph, *Nectonema* (Gr. *nektos*, swimming, + *nema*, a thread), juveniles occur in hermit crabs and other crabs.

After several months in the hemocoel (see section 12.6) of an arthropod host, juveniles complete a single molt and emerge into water as mature adults. If the host is a terrestrial insect, the parasite somehow stimulates the host to seek water. Worms do not emerge from the host unless water is nearby.

12.3 Phylum Loricifera

Loricifera (L. *lorica*, corselet, + *fero*, to bear) is a recently described phylum of animals (1983). Its members occupy spaces between grains of marine gravel, to which they cling tightly. Although loriciferans were described from specimens collected off the coast of France, they are apparently widely distributed. About 100 extant species are described, but the 2017 discovery of Cambrian loriciferan fossils indicates that this is a very ancient group. Loriciferans have oral styles and scalids similar to those of the kinorhynchs, and the entire forepart of the body can be retracted into the circular lorica (figure 12.11). Loriciferans are dioecious, and their life cycles contain multiple stages. Larvae molt as they grow.

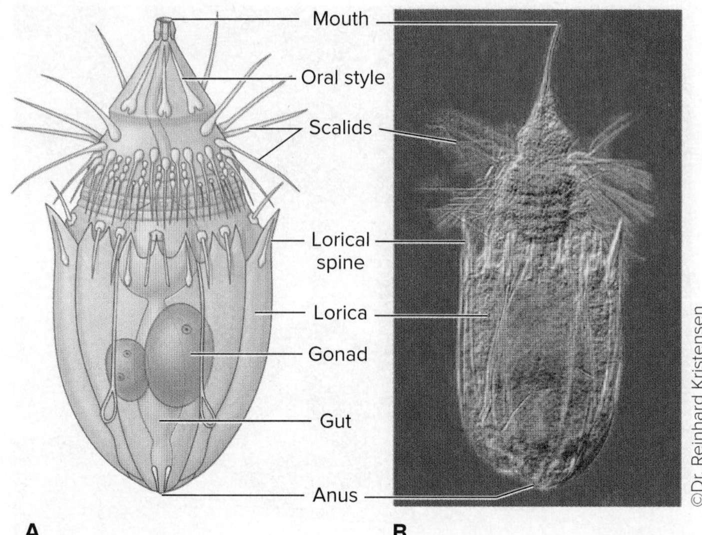

figure 12.11

A, Dorsal view of adult loriciferan, *Nanaloricus mysticus*, showing internal features. **B,** Live animal, 0.3 mm.

12.4 Phylum Kinorhyncha

There are only 179 described species of Kinorhyncha (Gr. *kineo*, to move, + *rhynchos*, beak or snout). These tiny marine worms, usually less than 1 mm long, live in mud or sandy mud. They have no external cilia, but are covered with a cuticle that is

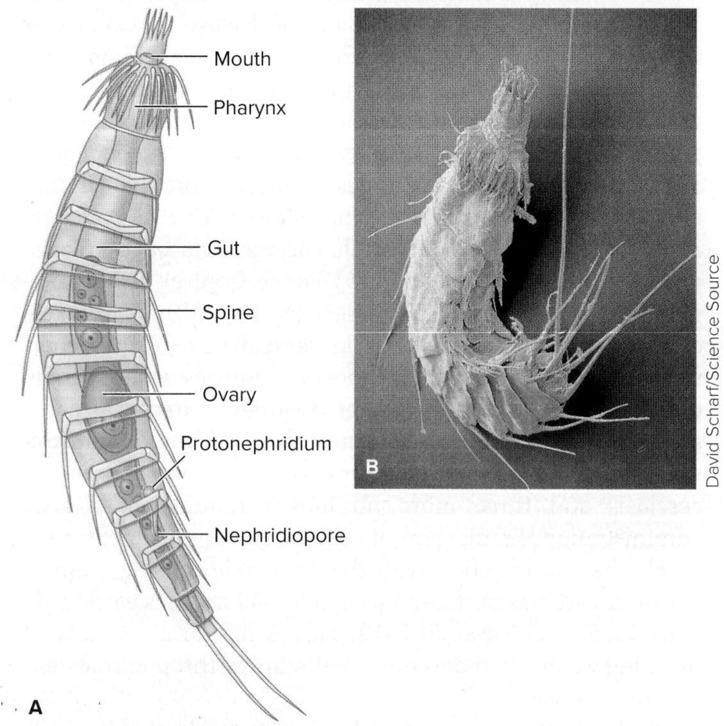

figure 12.12

A, *Echinoderes*, a kinorhynch, is a minute marine worm. Segmentation is superficial. The head with its circle of spines is retractile. **B,** Colored scanning electron micrograph (SEM) of the kinorhynch *Antigomonas* sp.

molted during growth. The body is divided into a head, a neck, and a trunk with 11 external segments (figure 12.12). Kinorhynchs burrow into mud by extending the head, anchoring it by its recurved spines **(scalids),** and drawing their body forward until their head is retracted. They feed on organic sediment in mud, and some feed on diatoms. Kinorhynchs are dioecious.

12.5 Phylum Priapulida

Priapulida (pri´a-pyu´li-da) (Gr. *priapos,* phallus, + *-ida,* pl. suffix) is a small group (only 16 species) of marine worms found chiefly in colder water in both hemispheres. Their cylindrical bodies are rarely more than 12–15 cm long. Most are burrowing predaceous animals that usually orient themselves upright in mud or sand with their mouth at the surface.

Long considered pseudocoelomate, priapulids were mistakenly judged coelomate when nuclei were found in membranes lining the body cavity, the membranes thus supposedly representing a peritoneum. However, electron microscopy showed that the nuclei of their muscle cells were peripheral, and that the muscles secreted an extracellular membrane. The muscle nuclei and extracellular membrane gave the mistaken appearance of an epithelial lining.

The body includes a retractable introvert, a trunk, and usually one or two caudal appendages (figure 12.13). The eversible introvert usually ends with rows of curved spines around the mouth; it is used to sample the surroundings as well as to capture small, soft-bodied prey. Priapulids are not metameric. A chitinous cuticle, molted periodically, covers the body.

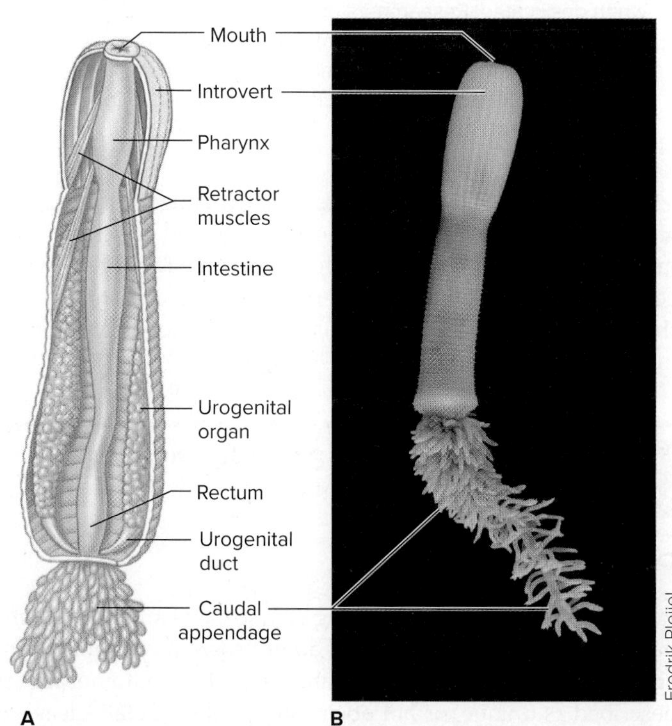

figure 12.13

A, Major internal structures of *Priapulus.* **B,** *Priapulus caudatus.*

Priapulids have no circulatory system, but coelomocytes in their body fluids contain the respiratory pigment hemerythrin. A nerve ring and ventral cord with nerves occurs. There is a protonephridial tubule that serves also as a gonoduct; the anus and urogenital pores open at the end of the trunk.

Sexes are separate, fertilization is external, and embryogenesis is only poorly known.

12.6 Clade Panarthropoda

Panarthropoda contains Arthropoda and two allied phyla, Onychophora and Tardigrada (see figure 12.1). In onychophorans and arthropods, a coelom develops by schizocoely, but coelom formation has been described as enterocoelic in tardigrades. In all three phyla, the main coelomic cavity later fuses with the blastocoel to form a new cavity called a **hemocoel,** or mixocoel. The hemocoel is lined by an extracellular matrix, not the mesodermal peritoneum that originally lined the coelom. Blood from the open circulatory system enters the hemocoel and surrounds the internal organs. There is no separation of extracellular fluid into blood plasma and lymph, so the combined fluid is often called **hemolymph.** A muscular heart is present, but tubular blood vessels occur in only one part of the body; blood enters and leaves the hemocoel through blood vessels. Small coelomic cavities surround a few organs in other parts of the body.

12.7 Phylum Onychophora

Members of phylum Onychophora (on-i-kof´o-ra) (Gr. *onyx,* claw, + *pherō,* to bear) are called "velvet worms" or "walking worms." About 200 species of these caterpillar-like animals, 0.5–15 cm long, occupy rain forests and other tropical and semitropical leafy habitats.

The fossil record of onychophorans describes a 500-million-year history. They were originally marine animals and were probably far more common than they are now. Extant species are all terrestrial and are extremely retiring, being active only at night or when the air is nearly saturated with moisture. Most are predators, feeding on insects, snails, and worms that they entangle in slime extruded from slime glands. Once immobilized by the slime, prey are digested by salivary enzymes and only the liquid portion is consumed.

Onychophorans are covered by a soft cuticle containing chitin and protein. They molt the cuticle in patches. Their head (figure 12.14A) bears a pair of flexible antennae with pigment-cup ocelli at the base. Their wormlike bodies are carried on 14–43 pairs of stumpy, unjointed legs (figure 12.14), each ending with a flexible pad and two claws. Onychophorans are air breathers, using a system of tubes that brings in air from pores scattered over their body. This **tracheal system,** although similar to that of arthropods (see figure 13.45), probably evolved independently. They have segmentally arranged nephridia, a hemocoel, and a muscular body wall.

Nearly all onychophorans are dioecious. In some species, there is a placental attachment between mother and young, and young are born as juveniles (viviparous); others have young

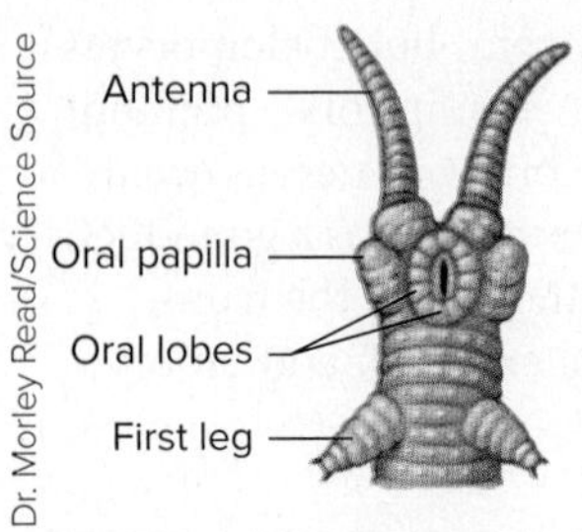

figure 12.14

Peripatus, a caterpillar-like onychophoran, has characteristics in common with both annelids and arthropods. **A,** Ventral view of head showing the region around the mouth. **B,** *Peripatus* in its natural habitat.

that develop in the uterus without attachment (ovoviviparous). Two Australian genera are oviparous and lay shell-covered eggs in moist places.

12.8 Phylum Tardigrada

Tardigrada (tar-di-grā´ da) (L. *tardus,* slow, + *gradus,* step), or "water bears," are minute forms usually less than a millimeter in length. About 1150 species have been described. Some live in freshwater and marine habitats, but most species live on land, occupying a film of water surrounding mosses and lichens.

The tardigrade body bears eight short, unjointed legs, each with claws (figure 12.15). Unable to swim, the animal creeps slowly, clinging to the substrate with its claws. There is a wide range of body shapes. A pair of sharp stylets and a sucking pharynx are adaptations for piercing and sucking plant cells or small prey, such as nematodes and rotifers.

A body covering of nonchitinous cuticle is molted several times during the life cycle. As in arthropods, muscle fibers are attached to the cuticular exoskeleton, and the body cavity is a hemocoel.

The nervous system resembles that of annelids and is surprisingly complex. Some species have a pair of eyespots. Circulatory and respiratory organs are lacking.

One of the most intriguing features of terrestrial tardigrades is their capacity to enter a state of suspended animation, called cryptobiosis, during which metabolism is virtually imperceptible. Under gradual drying conditions, they reduce the water content of their body from 85% to only 3%, movement ceases, and the body becomes barrel shaped. In a cryptobiotic state, tardigrades can withstand harsh environmental conditions—temperature extremes, ionizing radiation, oxygen deficiency, and so on—and many survive for years. Activity resumes when moisture is again available.

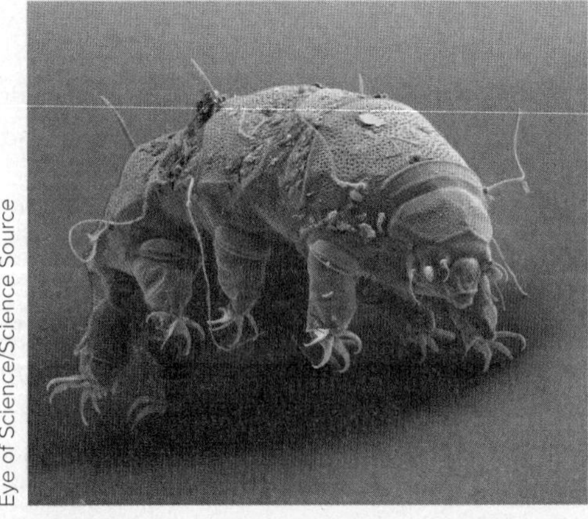

figure 12.15

Color enhanced scanning electron micrograph (SEM) of an active water bear (tardigrade), *Echiniscus granulatus*. This specimen, from Tubingen, Germany, feeds on moss cells. Tardigrades live in aquatic and semi-aquatic habitats, as well as in some hot springs or deep-water zones with high pressures.

Key Theme 12.5
HUMAN CONNECTIONS

Understanding Cryptobiosis

Tardigrades dehydrate almost completely in a process called cryptobiosis. Other animals that dehydrate, such as nematodes and brine shrimp, produce a sugar called trehalose, which preserves their cellular content and permits successful rehydration. Researchers use trehalose to improve the storage periods for blood platelets and food stuffs. Biologists expected tardigrades to use trehalose in cryptobiosis, but we now know that they use a different chemical solution. Tardigrades rely on tardigrade-specific intrinsically disordered proteins. Tardigrade-specific, intrinsically disordered proteins (TDPs) have variable three-dimensional structures and become rigid and glassy when desiccated. Researchers presume that this rigidity prevents vulnerable proteins from unfolding or disintegrating, so they hope to apply TDPs to enhance storage times for drugs and vaccines, and perhaps to reduce the need for refrigerated storage.

Tardigrades may reproduce by parthenogenesis or sexually. Females may deposit their eggs in the old cuticle as they molt or attach them to a substrate. Embryonic formation of the coelom is enterocoelous, a deuterostome characteristic. Nevertheless, molecular and morphological data strongly support membership in Panarthropoda (see figure 12.1).

12.9 Phylogeny and Adaptive Diversification

Phylogeny

Evolutionary relationships among ecdysozoans are not well understood. Members of this clade do not share a common cleavage pattern. Cleavage in nematodes and nematomorphs is described as unique, or not obviously spiral or radial. Cleavage in priapulids is somewhat similar to radial cleavage. Cleavage has not been studied in kinorhynchs, loriciferans, or tardigrades. In onychophoran eggs containing large amounts of yolk, the cytoplasm does not cleave, but nuclei divide in a manner similar to that in insect development (see section 3.3: Syncytial Specification). In onychophoran eggs with little yolk, the cleavage pattern varies, appearing spiral in some taxa and radial in others.

In many phylogenetic analyses, roundworms, phylum Nematoda, are united with horsehair worms, phylum Nematomorpha, in clade Nematoidea (see figure 12.1) where the sister taxa share a collagenous cuticle and unpaired dorsal and ventral nerve cords. Phylum Kinorhyncha is the sister taxon to phylum Priapulida, based on a shared two-layered pharynx. Kinorhynchs have mouthparts (oral styles on a noninversible mouth cone) similar to those of loriciferans, but loriciferans also share some morphological features with larval nematomorphs and with priapulids. Clade Scalidophora, comprising kinorhynchs, priapulids, and loriciferans, emerges in some, but not all, phylogenies based on molecular data.

Clade Panarthropoda unites tardigrades, velvet worms, and arthropods. Velvet worms, phylum Onychophora, are shown as the sister taxon to tardigrades, with this pair as the sister taxon of arthropods, based on mitochondrial genome sequence data; but phylogenies based on shared mouthparts place tardigrades as sister to the arthropods, and this pair as sister to the onychophorans. Tardigrades and arthropods share two morphological features: arthropod-type setae and muscles that insert on the cuticle.

Reconstructing the evolutionary history of life is a fascinating pursuit, but developmental and morphological information is absent for many taxa, especially for those very small animals living in obscure habitats—for example, the spaces between sand. What is known about the morphology of the ancestral ecdysozoan? Shared features of modern taxa predict that the cuticle extended inward at the mouth to line the pharynx. Most ecdysozoans have cuticular teeth in the pharynx and hardened spines or plates around the mouth, but this is not true of modern onychophorans. However, a recent reconstruction of a fossil onychophoran, *Hallucigenia sparsa*, did identify these elements, making it more likely that they were present in the ancestral ecdysozoan.

Adaptive Diversification

Certainly the most impressive adaptive diversification in this group of phyla is by nematodes. They are the most numerous in terms of both individuals and species, and they have adapted to almost every habitat available to animal life. Their basic pseudocoelomate body plan, with the cuticle, hydrostatic skeleton, and longitudinal muscles, has proved generalized and plastic enough to adapt to an enormous variety of physical conditions. Free-living lines gave rise to parasitic forms on at least several occasions, and virtually all potential hosts have been exploited. All types of life cycles occur: from simple and direct to complex, with intermediate hosts; from normal dioecious reproduction to parthenogenesis, hermaphroditism, and alternation of free-living and parasitic generations. A major factor contributing to the evolutionary opportunism of the nematodes is their extraordinary capacity to survive suboptimal conditions—the capacity to arrest development occurs in many free-living and animal parasitic species. Many free-living and plant parasitic species possess the ability to undergo cryptobiosis (survival in harsh conditions by assuming a very low metabolic rate).

SUMMARY

Sections	Key Concepts
12.1 Phylum Nematoda: Roundworms	• Nematodes occur in almost every habitat and exhibit free-living, symbiotic, and parasitic lifestyles. • Only 30,000 species are described currently, but estimates suggest there may be one million species alive today. • Nematodes are roughly cylindrical, tapering at the ends, and covered with a tough, secreted cuticle. Their body-wall muscles are longitudinal only, and to function well in locomotion, such an arrangement must enclose a volume of fluid in the pseudocoel at high hydrostatic pressure. • The internal high hydrostatic pressure has a profound effect on most of their other physiological functions, for example, ingestion of food, egestion of feces, excretion, copulation, and others. • Most nematodes are dioecious, and there are four juvenile stages, each separated by a molt of the cuticle. • Many nematodes are free-living in soil and aquatic habitats. Almost all animals and many plants have nematode parasites; infection may be direct or by animal vectors. • Humans are infected with hookworms, pinworms, intestinal roundworms, trichina worms, filarial worms, lung worms, and Guinea worms, among many others. Modern medicine has treatments for many of these infections.
12.2 Phylum Nematomorpha	• Nematomorphs, called horsehair worms, superficially resemble nematodes. Juvenile stages in the life cycle are parasitic in arthropods, but the aquatic adult stage is free-living. • Nematomorph juveniles undergo a single molt.

(continued)

Sections	Key Concepts
12.3 Phylum Loricifera	• Loricifera is a small phylum of tiny pseudocoelomate animals that live between grains of marine gravel. • They are dioecious with multi-stage life cycles; they molt the cuticle.
12.4 Phylum Kinorhyncha	• Kinorhyncha is a small phylum of tiny, marine pseudocoelomates that live in sand or mud. • They have an externally segmented body; they molt the cuticle.
12.5 Phylum Priapulida	• Priapulids are marine burrowing worms reaching up to 15 cm in length. • Priapulids may be predators or deposit-feeders. • They have a chitinous cuticle that is molted periodically.
12.6 Clade Panarthropoda	• Clade Panarthropoda comprises onychophorans, tardigrades, and arthropods, but sister-taxon relationships are not yet clear. • These animals have an open circulatory system with a hemocoel.
12.7 Phylum Onychophora	• Onychophora, often called velvet worms, are predatory, caterpillar-like animals found in humid, mostly tropical habitats. • They use a series of unjointed, clawed appendages to crawl; nephridia are segmentally arranged. • Onychophorans molt the cuticle in patches.
12.8 Phylum Tardigrada	• Tardigrades are minute animals, mostly terrestrial, living in the water film that surrounds mosses and lichens. • They have eight unjointed legs and a nonchitinous cuticle that is molted several times. • They may undergo cryptobiosis, withstanding extremely adverse conditions for long periods in a dehydrated state. • An understanding of cryptobiosis has been applied to preservation of blood products, foods, and medicines.
12.9 Phylogeny and Adaptive Diversification	• Taxa covered in this chapter possess a range of body plans and developmental patterns, but molecular data group them with phylum Arthropoda as a clade. • Clade Panarthropoda unites tardigrades, velvet worms (onychophorans), and arthropods. • Clade Nematoidea unites nematodes and nematomorphs. • Phyla Kinorhyncha and Priapulida are united by sharing a two-layered pharynx. • The remaining relationships among taxa within Ecdysozoa are not yet clear. • Nematodes are the most abundant and species-rich group of the non-arthropod ecdysozoans. Their highly adaptable body plan permits them to occupy nearly every habitat on earth and to be successful as free-living and parasitic organisms.

Review Questions

1. A skeleton is a supportive structure. Explain how a hydrostatic skeleton supports an animal.
2. What feature of body-wall muscles in nematodes requires a high hydrostatic pressure in the pseudocoelomic fluid for efficient function?
3. Explain how high pseudocoelomic pressure affects feeding and defecation in nematodes. How are ameboid sperm an adaptation to high hydrostatic pressure in the pseudocoelom?
4. Explain the interaction of cuticle, body-wall muscles, and pseudocoelomic fluid in the locomotion of nematodes.
5. Outline the life cycle of each of the following: *Ascaris lumbricoides*, hookworm, *Enterobius vermicularis*, *Trichinella spiralis*, *Wuchereria bancrofti*.
6. Where in a human body are adults of each species in question 5 found?
7. Explain how river blindness or Guinea-worm disease could be eradicated.
8. Where are juveniles and adults of nematomorphs found?
9. What are the approximate lengths of loriciferans, priapulids, and kinorhynchs?
10. Give the main distinguishing characteristics of Onychophora and Tardigrada.
11. What do the members of each of the aforementioned groups eat, and where do they live?
12. What is the survival value of cryptobiosis in tardigrades?
13. How is a hemocoel different from a true coelom?
14. In what sense is a hemocoel part of the circulatory system?
15. Describe the two major protostome clades, and give a defining feature for each.

For Further Thought The comments of N. A. Cobb quoted in the opening essay for this chapter make clear just how successful are nematodes. Do any characteristics of nematodes explain their much greater abundance than the other ecdysozoan phyla covered in this chapter?

13

Arthropods

LEARNING OBJECTIVES

Readers will be able to:

13.1 Describe six aspects of the arthropod body plan associated with ecological success and high diversity; explain the evolutionary history of arthropods according to the mandibulate hypothesis.

13.2 Identify a trilobite and describe its habitat, lifestyle, and time periods of greatest abundance.

13.3 Identify the key features of chelicerates; compare and contrast the three chelicerate classes and distinguish horseshoe crabs, sea spiders, and arachnids. Within class Arachnida, distinguish spiders, scorpions, ticks, and mites; explain which animals are agents of disease.

13.4 Identify the myriapod body plan; compare and contrast the morphology and lifestyles of centipedes and millipedes.

13.5 Describe the crustacean body plan including segments, appendages, and circulatory, respiratory, and excretory systems. Briefly describe how crustaceans feed, reproduce, and molt.

13.6 Identify members of the crustacean subgroups Oligostraca and Vericrustacea, focusing on the most common types such as copepods, barnacles, and decapods.

13.7 Identify the key features of hexapods; briefly describe how insects fly, feed, respire, sense the world, reproduce, and molt. Compare and contrast direct and indirect development and explain how metamorphosis is accomplished.

13.8 Describe the various ways that insects defend themselves and communicate; explain insect social behavior. Discuss the ways in which insects enhance human life, as well as areas in which they cause harm. Describe common methods for controlling insect populations.

13.9 Describe the hypothetical body plan of an ancestral arthropod and propose a mechanism for the evolution of tagmata. Explain why biramous and uniramous appendages may not be useful characters for phylogenetic inference. Discuss the evolutionary relationship between Crustacea and Hexapoda, covering any areas of controversy.

Digital Images Studio/Shutterstock

This brilliantly colored jewel bug (*Chrysocoris stollii*) is a beetle in order Coleoptera.

A Winning Combination

Insect species far outnumber all other species of animals combined. It has been estimated that there are 200 million insects for every single human alive today! Insects have an unmatched ability to adapt to all land environments and virtually all climates, as well as freshwater and shoreline habitats due to their versatile exoskeleton, metamerism, and a host of other features described in this chapter.

Humans have a long history of interaction with insects. Some compete with us for access to important food plants—locust populations cover 16 million square kilometers in 30 African countries at low abundance and cover 29 million square kilometers of land in 60 countries in episodic outbreaks. Other insects feed on crops, spread diseases across forests, or transmit illnesses such as

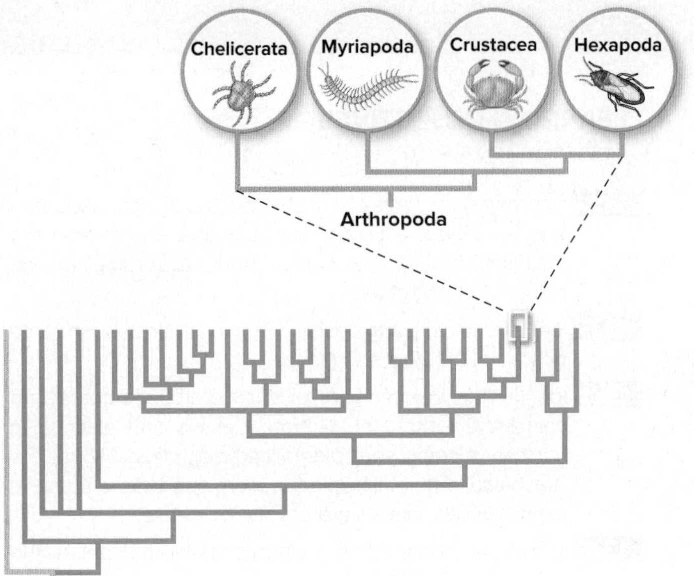

A B

figure 13.1

Fossils of early arthropods. **A,** Dorsal view of a trilobite fossil from Barrandien in Czech republic. **B,** Eurypterid fossil from the Silurian period, *Eurypterus remipes*.

malaria or Lyme disease. A fungus causing Dutch elm disease, mainly transmitted by European bark beetles, has virtually obliterated American elm trees in North America.

Nonetheless, insects are essential for human survival and well-being. We rely on them to pollinate most of our crops, and they play a central role in both terrestrial and aquatic food webs. A 2006 study estimated that pollination, pest control, contributions to animal food webs, and other ecosystem services performed by insects have an annual monetary value of $57 billion for the United States. These benefits are in jeopardy: a study of changes in insect abundance at various sites around the world concluded that up to 40 percent of known insect species were in decline in 2019. While climate change is an important potential driver of this decline, especially in tropical regions, there are other factors that play a role, such as habitat loss, land conversion to intensive agriculture, urban expansion, widespread pesticide use, and the spread of invasive species. The global decline in honeybees was due in part to human distribution of bees infected with *Varroa* mites, which carry the deformed wing virus, DWV (see section 13.8 and Key Theme 13.10). Although many practices that are detrimental to insects are common throughout the world, these practices can be reduced or discontinued through sustained efforts; we can reconsider pesticide use, reduce pollutants, conserve native vegetation, and maintain nonurban landscapes, to name only a few of the actions proposed in many nations in the wake of data on insect decline. Our world would be unrecognizable without insects, so vast improvements are needed in stewardship.

Phylum Arthropoda (ar-thro-pod´a) (Gr. *arthron,* joint, + *podos,* foot) comprises spiders, scorpions, ticks, mites, crustaceans, millipedes, centipedes, insects, and some smaller groups. Its rich fossil record extends back to the mid-Cambrian period and includes many extinct taxa (figure 13.1).

Arthropods are members of clade Panarthropoda (see section 12.6). They are coelomate protostomes with well-developed organ systems. The arthropod cuticle is strengthened with chitin and sclerotized protein, and sometimes with calcium salts, creating an exoskeleton. Arthropods are conspicuously segmented; their ancestral body pattern is a linear series of similar segments, each with a pair of jointed appendages. Often, various segments are combined or fused into functional groups, called **tagmata** (sing., **tagma**). The head and thorax are two such tagmata. The evolutionary process of fusing segments into body regions is called **tagmatization**. Appendages are frequently differentiated and specialized for walking, swimming, flying, or eating. Few arthropods exceed 60 cm in length, and most are far smaller. The largest is a Japanese crab (*Macrocheira kaempferi*) with a span of approximately 3.7 m; the smallest is a parasitic mite less than 0.1 mm long.

13.1 Why Are Arthropods So Diverse and Abundant?

In diversity of ecological distribution, arthropods have no rivals. Arthropods occur in all types of environments, from low ocean depths to very high altitudes and from the tropics far into both the north and south polar regions. Some species are adapted for life on land or in fresh, brackish, and marine waters; others live in or on plants and other animals. Most species use flight to move among their favored habitats.

Although all feeding types—carnivorous, omnivorous, and herbivorous—occur in this vast group, most are herbivorous. Most aquatic arthropods depend on algae for their nourishment, and most land forms live chiefly on plants. Many are parasites.

Particular aspects of arthropod morphology and physiology have helped them achieve their amazing degree of diversity:

1. **Versatile exoskeleton.** The arthropod cuticle is secreted by the underlying epidermis; it must protect the body without restricting movement.

   The cuticle consists of an outer, relatively thin epicuticle and an inner, thicker procuticle (figure 13.2).

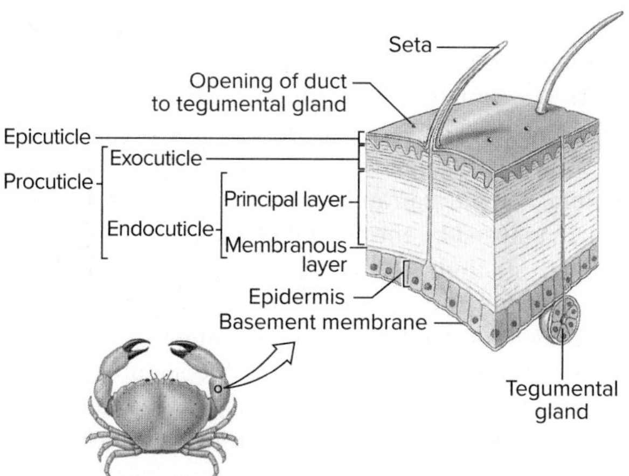

figure 13.2

Structure of crustacean cuticle.

Both the epicuticle and the procuticle are composed of several layers. The epicuticle is made of protein and often lipids. The protein is stabilized and sclerotized, adding further protection. The procuticle is divided into exocuticle, which is secreted before a molt, and endocuticle, which is secreted after molting. Both layers of the procuticle contain **chitin** bound with protein. Chitin is a tough, resistant, nitrogenous polysaccharide that is insoluble in water, alkalies, and weak acids. Thus, the procuticle is not only flexible and lightweight but also affords protection, particularly against dehydration. In most crustaceans, some areas of the procuticle are also impregnated with calcium salts, which reduce flexibility. In the hard shells of lobsters and crabs, for instance, this calcification is extreme.

The cuticle may be soft and permeable or may form a veritable coat of armor. Between body segments and between segments of appendages, it is thin and flexible, permitting free movement of joints. In crustaceans and insects, the cuticle grows inward to create sites for muscle attachment. It may also line the foregut and hindgut, line and support the tubes that bring air into the body (tracheae), and be adapted for a variety of other purposes.

The nonexpansible cuticular exoskeleton imposes important conditions on body size. To increase in size, an arthropod must shed its outer covering at intervals and produce a larger one—a process called **molting.** The process of molting ends in the shedding of skin, called **ecdysis**. Arthropods molt many times before reaching adulthood, and some continue to molt after that. Much of an arthropod's physiology centers on molting, particularly in young animals—preparation, molting itself, and then all of the processes that must be completed in the postmolt period. More details of the molting process are given for crustaceans (see section 13.5: Ecdysis, and figure 13.20) and for insects (see section 13.7: Metamorphosis and Growth).

2. **Segmentation and appendages for more efficient locomotion.** Typically, each arthropod segment has a pair of jointed appendages, but this arrangement is often modified, with both segments and appendages specialized for a range of functions. The jointed appendages are equipped with sensory hairs and are variously modified for sensory functions, food handling, and swift, efficient walking or swimming. Each section of an appendage functions as a hollow tube moved by muscles, which insert on the inside.
3. **Air piped directly to cells.** Most land arthropods have a highly efficient tracheal system of air tubes, which delivers oxygen directly to tissues and cells and makes a high metabolic rate possible. Tracheae open to the outside through openings called spiracles. Tracheae are not present in aquatic arthropods; they breathe mainly by gills.

CHARACTERISTICS of Phylum Arthropoda

1. **Jointed appendages;** ancestrally, one pair to each segment, but number often reduced; appendages often modified for specialized functions.
2. Living in marine, freshwater, and terrestrial habitats; many capable of flight.
3. Free-living and parasitic taxa.
4. Bilateral symmetry; **segmented body** divided into functional groups called **tagmata:** head and trunk; head, thorax, and abdomen; or cephalothorax and abdomen; definite head.
5. Triploblastic body.
6. **Small coelom** in adult; body cavity consists mostly of hemocoel (spaces in the tissues) filled with blood.
7. **Cuticular exoskeleton;** contains protein, lipid, **chitin,** and often calcium carbonate secreted by underlying epidermis and shed (molted) at intervals.
8. **Complete digestive system;** mouthparts modified from ancestral appendages and adapted for different methods of feeding; gut tube shows great specialization by having, in various arthropods, chitinous teeth, compartments, and gastric ossicles.
9. **Complex muscular system** attached to exoskeleton; **striated muscles** for rapid actions, smooth muscles for visceral organs; no cilia.
10. **Nervous system** with dorsal brain connected by a ring around the gullet to a double nerve chain of ventral ganglia; fused ganglia in some species.
11. Well-developed sensory organs; behavioral patterns much more complex than those of most invertebrates, with more **social organization.**
12. Parthenogenesis in some taxa.
13. **Sexes usually separate,** with paired reproductive organs and ducts; usually internal fertilization; oviparous, ovoviviparous, or viviparous; often with **metamorphosis.**
14. Paired excretory glands called **coxal, antennal,** or **maxillary glands** present in some; others have excretory organs called **Malpighian tubules.**
15. Respiration by **body surface, gills, tracheae** (air tubes), or **book lungs.**
16. **Open circulatory system,** with dorsal **contractile heart,** arteries, and hemocoel (see section 12.6).

4. Highly **developed sensory organs.** Sensory organs show great variety, from compound (mosaic) eyes to organs of touch, smell, hearing, balance, and chemical reception. Arthropods are keenly alert to environmental stimuli.
5. **Complex behavior patterns.** Arthropods exceed most other invertebrates in the complexity and organization of their activities. Innate (unlearned) behavior controls much of what they do, but many arthropods also demonstrate learned behaviors.
6. **Trophic breadth through metamorphosis.** Many arthropods pass through metamorphic changes, including a larval form quite different from the adult structure. Larval forms often differ from adults in the habitats they occupy and foods they consume, permitting a single species to exploit diverse resources.

Arthropoda is so astonishingly diverse that zoologists have argued over whether all its members belong within a single phylum. Current evidence supports Arthropoda as a monophyletic group. Formerly, arthropods were divided among four subphyla: the extinct Trilobita and three extant taxa Chelicerata, Crustacea, and Uniramia. The subphylum Uniramia was diagnosed by the shared presence of **uniramous** appendages, with a single branch (figure 13.3, *top*), and included insects and myriapods (centipedes, millipedes, and a few less well-known forms). Uniramia is not supported by molecular data, so we divide uniramians into two subphyla: Myriapoda and Hexapoda. Hexapoda contains class Insecta and one other very small class. Arthropod subphyla are relatively easy to distinguish based on the number of tagmata and the types of appendages present.

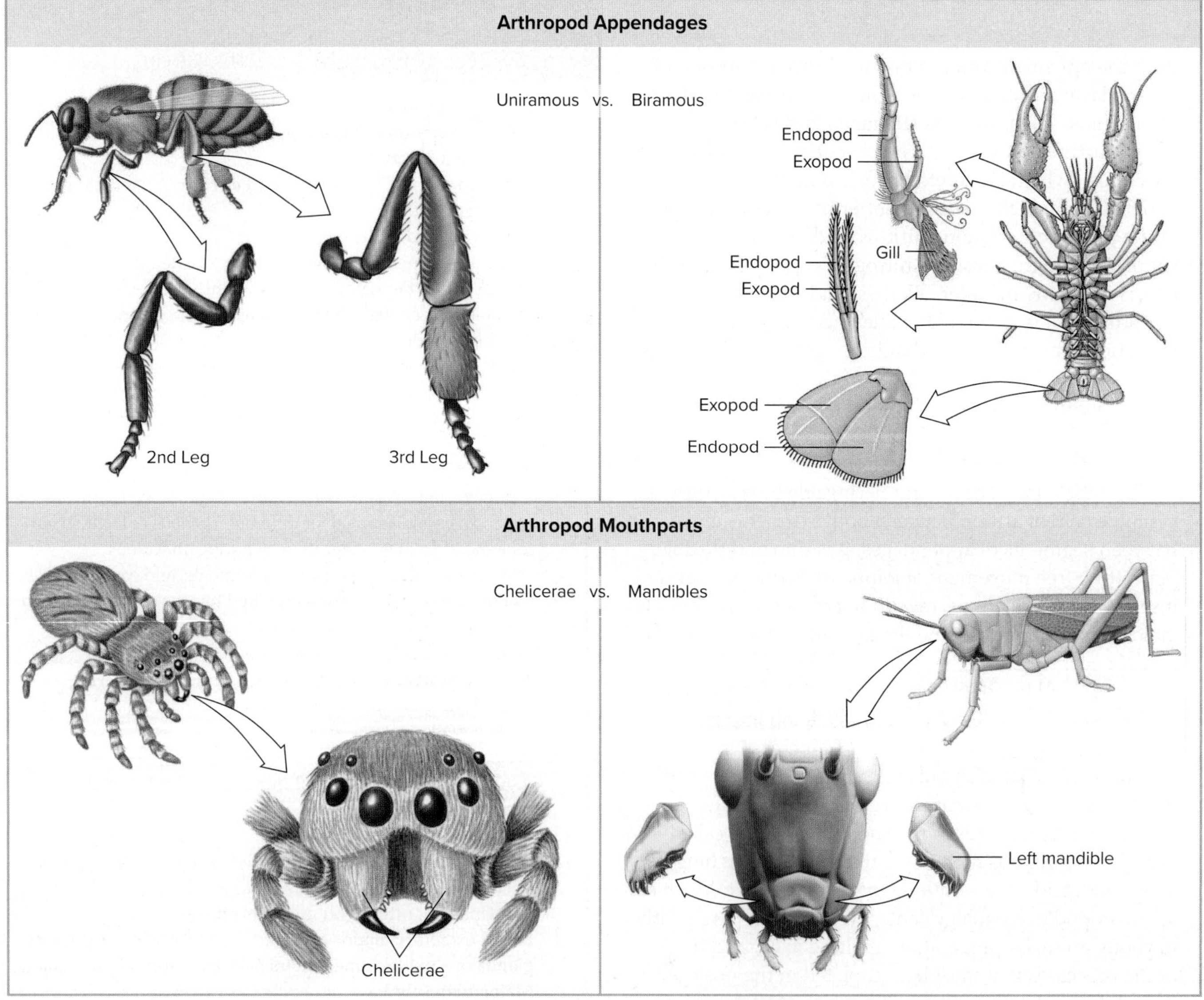

figure 13.3

Two important arthropod characters: Appendages may be uniramous (honeybee leg) or biramous (lobster limb); mouthparts may include chelicerae (spider) or mandibles (grasshopper). Note that the presence or absence of gills is unrelated to appendage form.

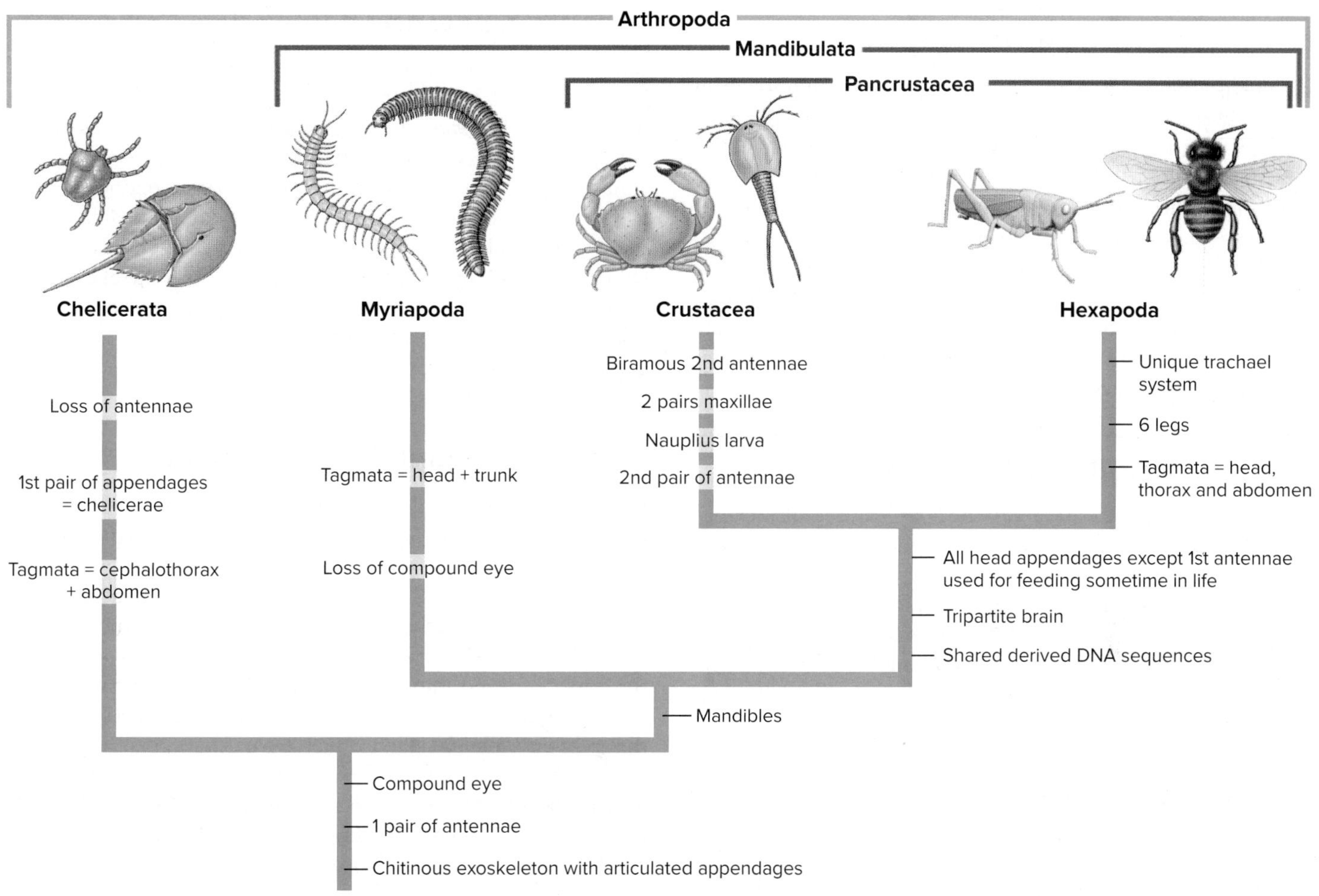

figure 13.4

Cladogram of arthropods showing probable relationships of the four extant subphyla. Only a few synapomorphies are included here. Crustaceans and hexapods are shown as sister taxa, in Pancrustacea. A sister taxon relationship between Pancrustacea and Chelicerata is based on shared possession of mandibles and data from molecular phylogenies.

Relationships among the subphyla are controversial. One hypothesis proposes that all arthropods possessing a particular mouthpart, called a **mandible,** form a single clade, Mandibulata. This clade includes members of Myriapoda, Hexapoda, and Crustacea. Arthropods that do not have mandibles possess **chelicerae** (figure 13.3, *bottom*), as exemplified by spiders. Thus, according to the "mandibulate hypothesis," myriapods, hexapods, and crustaceans are more closely related to each other than any of them are to chelicerates. Critics of the mandibulate hypothesis argue that the mandibles in each group are so different from each other that they could not be homologous. The mandibles of crustaceans are multijointed, with chewing or biting surfaces on the mandible bases (called a gnathobasic mandible), whereas those of myriapods and hexapods have a single joint with the biting surface on the distal edge (called an entire-limb mandible). There are also some differences in the muscles controlling the two types. Proponents of the mandibulate hypothesis respond that the 550-million-year history of the mandibulates makes possible the evolution of diverse mandibles from an ancestral type.

An alternative to the "mandibulate hypothesis" is the "myriochelata hypothesis" under which myriapods and chelicerates are sister taxa. In this latter hypothesis, the clade comprising myriapods and chelicerates together is the sister taxon of a group containing crustaceans and hexapods.

We assume that subphylum Trilobita is the sister taxon to a group containing all other known arthropods, although one phylogeny indicates that trilobites are the sister taxon to the Chelicerata. This phylogeny supports the mandibulate hypothesis[1] as shown in figure 13.4. We depict subphylum Crustacea as the sister taxon of subphylum Hexapoda. Evidence that hexapods and crustaceans form a clade emerged from several phylogenetic studies using molecular characters; these studies prompted a reevaluation of the morphological characters in members of both taxa. We unite subphylum Crustacea with subphylum Hexapoda in clade Pancrustacea.

[1] Legg, D. A., M. D. Sutton, and G. D. Edgecombe. 2013. Arthropod fossil data increase congruence of morphological and molecular phylogenies. *Nature Communications* **4:**2485.

13.2 Subphylum Trilobita

Trilobites (see figure 13.1A) likely evolved well before the Cambrian period. They have been extinct some 250 million years, but were abundant during the Cambrian and Ordovician periods. Their name denotes the trilobed shape of the body, caused by a pair of longitudinal grooves. They were largely bottom-dwellers, probably scavengers. They ranged in size from 2 to 67 cm.

Trilobites have an extensive fossil record due to preservation of their exoskeletons. They can be identified at the species level and, when the time range of a species is known, trilobite species are important indicators for biostratigraphy, the determination of relative age of geological strata.

13.3 Subphylum Chelicerata

Chelicerate arthropods are a very ancient group that includes eurypterids (extinct), horseshoe crabs, spiders, ticks and mites, scorpions, sea spiders, and others. They are characterized by the presence of two tagmata and six pairs of cephalothoracic appendages that include a pair of **chelicerae** (see figure 13.3), a pair of **pedipalps,** and **four pairs of walking legs.** They have no mandibles and no antennae. Most chelicerates liquefy their prey before consumption.

Class Merostomata

Subclass Eurypterida

Eurypterids, or giant water scorpions (see figure 13.1B), lived 200–500 million years ago, and some were perhaps the largest of all arthropods, reaching a length of 3 m. They shared some features with marine horseshoe crabs (figure 13.5) and scorpions, their terrestrial counterparts.

Subclass Xiphosurida: Horseshoe Crabs

Xiphosurids are an ancient marine group that dates from the Cambrian period. Only three genera (five species) survive today. *Limulus* (L. *limus,* sidelong, askew) lives in shallow water along the North American Atlantic coast; it is externally very similar to fossil forms dating from the Triassic period. Horseshoe crabs have an unsegmented, horseshoe-shaped **carapace** (hard dorsal shield) and a broad abdomen, which has a long, spinelike **telson,** or tailpiece (figure 13.5). On some abdominal appendages, **book gills** (flat, leaflike gills) are exposed. Horseshoe crabs swim slowly by means of their abdominal plates and can walk on their walking legs. They feed at night on worms and small molluscs and are harmless to humans.

Class Pycnogonida: Sea Spiders

Pycnogonids, commonly called sea spiders, move on four pairs of long, thin walking legs, drinking juices from hydroids and soft-bodied animals with their large suctorial proboscis (figure 13.6). Their odd appearance is enhanced by a much reduced abdomen attached to an elongated cephalothorax. The small abdomen is completely occupied by the digestive tract, so the gonads extend into the legs. Males often use a pair of legs called **ovigers** to carry the egg masses. They are cryptic animals that rely on sensory setae for tactile information-gathering, although they have four eyes on raised tubercles that together afford a 360-degree view of their surroundings. Most pycnogonids are only a few millimeters long; there are about 1000 species and they are common in all oceans.

Class Arachnida

Arachnids (Gr. *arachnē,* spider) are numerous and diverse, with over 100,000 species described so far. They include spiders,

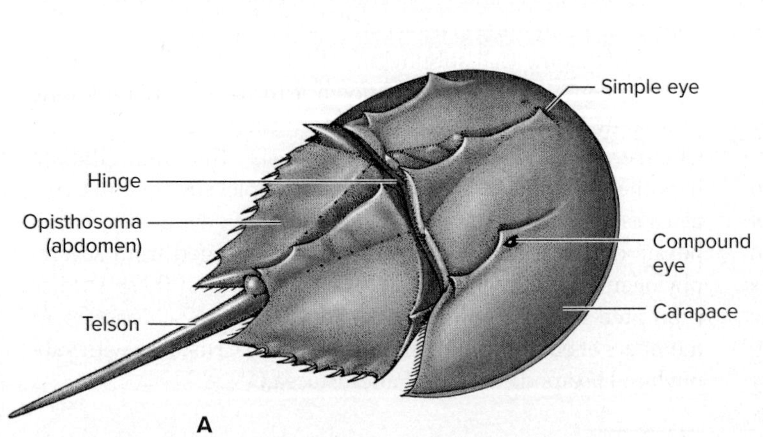

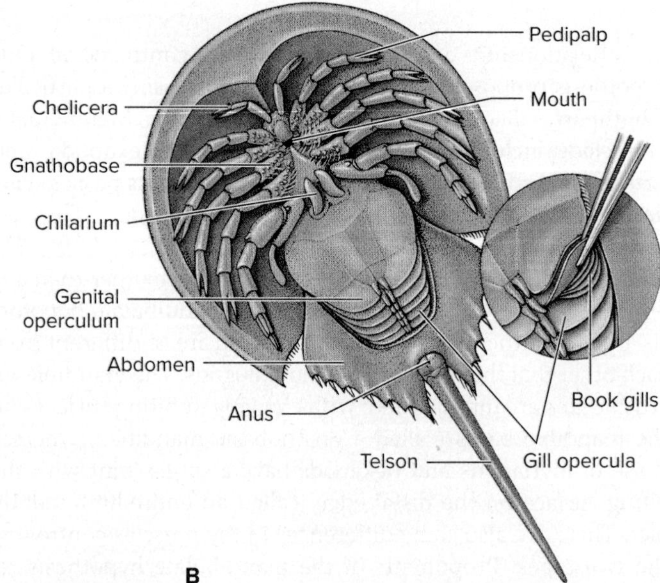

figure 13.5

A, Dorsal view of horseshoe crab *Limulus* (class Merostomata). Individuals grow to 0.5 m in length. **B,** Ventral view of a female.

scorpions, pseudoscorpions, whip scorpions, ticks, mites, harvestmen (daddy longlegs), and others. The arachnid tagmata are a cephalothorax and an abdomen.

Order Araneae: Spiders

Spiders are a large group of 40,000 species, worldwide in distribution. The cephalothorax and abdomen show no external segmentation, and the tagmata are joined by a narrow, waistlike pedicel (figure 13.7).

All spiders are predaceous and feed largely on insects (figure 13.8). Their chelicerae function as fangs and bear ducts from their venom glands, with which they effectively dispatch their prey. Some spiders chase their prey, others ambush them, and many trap them in a net of silk. After a spider seizes its prey with its chelicerae and injects venom, it liquefies the prey's tissues with a digestive fluid and draws the resulting broth into the stomach. Spiders with teeth at the bases of their chelicerae crush or chew prey, and enzymes from their mouth further aid digestion. Many spiders provision their young with previously captured prey.

Spiders breathe by **book lungs** or **tracheae** or both. Book lungs, which are unique to spiders, consist of many parallel air pockets extending into a blood-filled chamber (see figure 13.7C). Air enters the chamber through a slit in the body wall. Tracheae are a system of air tubes that carry air directly to tissues from openings called **spiracles.** The tracheae of spiders are similar to those of insects (see figure 13.45), but much less extensive.

Spiders and insects have a unique excretory system of **Malpighian tubules** (see figure 13.7C) that work in conjunction with specialized rectal glands. Potassium, other solutes, and waste materials are secreted into the tubules, which drain the fluid, or "urine," into the intestine. Rectal glands reabsorb most of the potassium and water, leaving behind wastes such as uric acid. Through this cycling of water and potassium, species living in dry environments conserve body fluids, producing a nearly dry mixture of urine and feces. Many spiders also have coxal glands, which are modified nephridia that open at the coxa, or base, of the first and third walking legs.

Spiders usually have eight simple eyes, each provided with a lens, optic rods, and a retina (see figure 13.7B). Chiefly, the eyes perceive moving objects, but some, such as those of the hunting and jumping spiders, may form images. Because vision is usually poor, a spider's awareness of its environment depends especially on its hairlike sensory setae. Every seta on its surface is useful in communicating some information about the spider's surroundings, air currents, or changing tensions in its web. By sensing vibrations of its web, a spider can judge the size and activity of its entangled prey or receive a message tapped on a silk thread by a prospective mate.

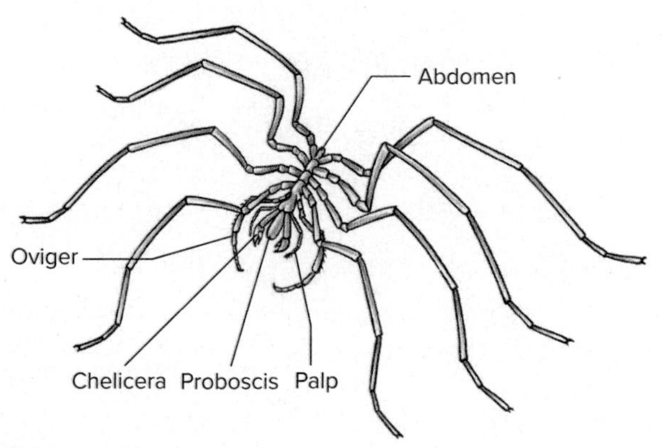

figure 13.6

A pycnogonid, *Nymphon* sp. In this genus, all anterior appendages (chelicerae, palps, and ovigers) are present in both sexes, although ovigers are often not present in females of other genera.

Web-Spinning Habits The ability to spin silk is important in the life of a spider, and in some other arachnids. A spider's silk is a protein fiber of great tensile strength. Two or three pairs of spinnerets containing hundreds of microscopic tubes

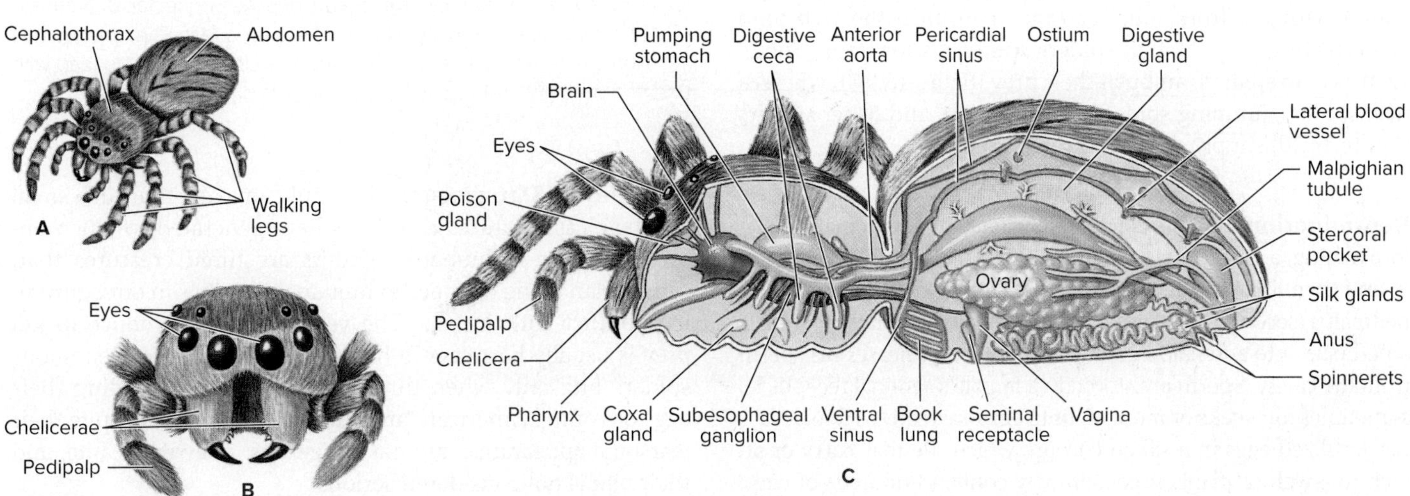

figure 13.7

A, External anatomy of a jumping spider. **B,** Anterior view of head. **C,** Internal anatomy of a female spider.

Michael Lustbader/Science Source

figure 13.8

A grasshopper, snared and helpless in the web of a golden garden spider (*Argiope aurantia*), is wrapped in silk while still alive. If the spider is not hungry, its prize is saved for a later meal.

connect to special abdominal silk glands (see figure 13.7C). A protein secretion emitted as a liquid hardens on contact with air to form a silk thread. Spiders' silk threads are stronger than steel threads of the same diameter and are probably second in tensional strength only to fused quartz fibers. The threads can stretch one-fifth of their length before breaking. Spider silk is not homologous to the commercially valuable silk produced by silkmoth larvae, which is used by the textile industry.

The spider web used for trapping insects is familiar to most people. The webs of some species consist of merely a few strands of silk radiating out from a spider's burrow or place of retreat. Other species spin beautiful, geometric orb webs. Besides making webs, spiders use silk threads to line their nests; form sperm webs or egg sacs; build draglines; make bridge lines, warning threads, molting threads, attachment discs, or nursery webs; or wrap their prey securely (figure 13.8). Net-casting spiders hold small rectangular webs in front of their bodies with the front four legs and then push the web onto unsuspecting prey. Not all spiders spin webs for traps. Some, such as crab spiders, ambush their prey (figure 13.9A), whereas wolf spiders, jumping spiders (figure 13.9B), and fisher spiders (figure 13.10) simply chase and catch their prey.

Reproduction A courtship ritual usually precedes mating. Before mating, a male spins a small web, deposits a drop of sperm on it, and then lifts the package and stores it in special cavities of his pedipalps (second pair of appendages). When he mates, he inserts a pedipalp into a female's genital opening and releases his sperm from the cavity. Sperm are stored in a female's seminal receptacle, sometimes for weeks or months, until eggs are ready. A female lays her fertilized eggs in a silken cocoon, which she may carry or attach to a web or plant. A cocoon may contain hundreds of eggs, which hatch in approximately 2 weeks. The young usually remain in their egg sac for a few weeks and molt once before leaving it. Several molts occur before adulthood.

James Carmichael, Jr./NHPA

A

James Carmichael, Jr./NHPA

B

figure 13.9

A, A red-banded crab spider, *Misumenoides formosipes*, awaits its insect prey. Its coloration matches the petals among which it lies, thus deceiving insects that visit the flowers in search of pollen or nectar. **B,** A female regal jumping spider, *Phidippus regius*, sits on a leaf. This species has excellent vision and stalks an insect until it is close enough to leap with unerring precision, fixing its chelicerae into its prey.

Are Spiders Dangerous? It is truly amazing that such small and helpless creatures as spiders have generated so much unreasoned fear in humans. Spiders are timid creatures that, rather than being enemies to humans, are allies in our continuing conflict with insects. The venom spiders produce to kill prey is usually harmless to humans. Even the most venomous spiders bite only when threatened or when defending their eggs or young. American tarantulas (figure 13.11), despite their fearsome appearance, are not dangerous. They rarely bite, and their bite is not considered serious.

Two spider genera in the United States can inflict severe or even fatal bites: *Latrodectus* (L. *latro,* robber, + *dektes,* biter) and *Loxosceles* (Gr. *loxos,* crooked, + *skelos,* leg). The most important

John Serrao/Science Source

figure 13.10

An Okefenokee fishing spider, *Dolomedes okefenokensis*, has captured a fish. This handsome spider pulls its paralyzed victim from the water, pumps in digestive enzymes, and then sucks out the predigested contents.

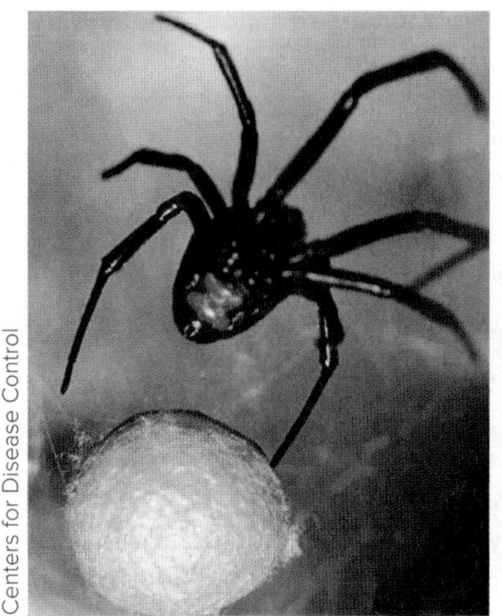

Centers for Disease Control

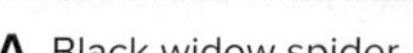

A Black widow spider

Gregory G. Dimijian, M.D./Science Source

B Brown recluse spider

figure 13.12

A, *Latrodectus mactans* next to her egg sac. Note the orange "hourglass" on the ventral side of her abdomen. **B,** *Loxosceles reclusa* is a small venomous spider. Note the small, violin-shaped marking on its cephalothorax. The venom is hemolytic and dangerous.

D. Kucharski K. Kucharska/Shutterstock

figure 13.11

Mexican redrump tarantula *Brachypelma vagans*.

species are *Latrodectus mactans,* or the black widows, and *Loxosceles reclusa,* or the brown recluse. Black widows are moderate to small in size and shiny black, with a bright orange or red "hourglass" on the underside of their abdomen (figure 13.12A). Their venom is neurotoxic, acting on the nervous system. About four or five of each 1000 bites reported are fatal. Black widows occur in the southern and western regions of the United States.

Brown recluse spiders, which are smaller than black widows, are brown, and bear a violin-shaped dorsal stripe on their cephalothorax (figure 13.12B). Their venom is hemolytic rather than neurotoxic, destroying the tissues and skin surrounding a bite. Their bite can be mild to serious and rarely fatal. Brown recluses occur in the midwest and southern regions of the United States.

Dangerous spiders in other parts of the world include funnel-web spiders (*Atrax robustus*) in Australia and certain ctenid spiders in South America, such as *Phoneutria fera.* In contrast to most spiders, these are quite aggressive.

Order Scorpionida: Scorpions

Although scorpions are more common in tropical and subtropical regions, some occur in temperate zones. Scorpions are generally reclusive, hiding in burrows or under objects by day and feeding at night. They feed largely on insects and spiders, which they seize with chelate pedipalps and rip with jawlike chelicerae.

A scorpion's body consists of a rather short cephalothorax, which bears appendages and from one to six pairs of eyes, and a clearly segmented abdomen without appendages. The abdomen is divided into a broader preabdomen and a tail-like postabdomen, which ends in a stinging apparatus used to inject venom. The venom of most species is not harmful to humans, although that of certain species of *Androctonus* in Africa and *Centruroides* in Mexico, Arizona, and New Mexico can be fatal.

Scorpions bear well-developed young, which their mother carries on her back until after their first molt (figure 13.13A).

Order Solifugae: Sun or Camel Spiders

Solpugids, sometimes also called solfugids and by such common names as sun, camel, or wind spiders, are nonvenomous arachnids that shred prey with their large chelicerae (figure 13.13B). They are often less than 1 cm long, but some species approach 15 cm. They are common in tropical and subtropical deserts in America, the Middle East, Africa, and Asia.

Order Opiliones: Harvestmen

Harvestmen, often called "daddy longlegs," are common in the United States and other parts of the world (figure 13.13C). They are easily distinguished from spiders by a broad joining of their abdomen and cephalothorax without the constriction of a pedicel, and by the external segmentation of their abdomen. They have four pairs of long, spindly legs and, without apparent ill effect, can cast off one or more legs if they are grasped by a predator (or human hand). The ends of their chelicerae are pincerlike, and they feed much more as scavengers than do most arachnids.

Order Acari: Ticks and Mites

Acarines differ from all other arachnids in having their cephalothorax and abdomen completely fused, with no external division or segmentation (figure 13.14A). Their mouthparts are on a little anterior projection, or **capitulum.** They occur almost everywhere—in both fresh and salt water, on vegetation, on the ground, and parasitic on vertebrates and invertebrates. Over 50,000 species have been described, many of which are important to humans, but this is probably only a fraction of the species that exist.

Many species of mites are entirely free-living. *Dermatophagoides farinae* (Gr. *dermatos,* skin, + *phago,* to eat, + *eidos,* likeness of form) (figure 13.14B) and related species are denizens of house dust all over the world, sometimes causing allergies and dermatoses. Some mites are marine, but most aquatic species live in fresh water. They have long, hairlike

Karl H. Switak/Science Source
A

Gladden Williams Will/AGE Fotostock
B

Cleveland P. Hickman, Jr.
C

figure 13.13

A, An emperor scorpion (order Scorpionida), *Paninus imperator,* with young, which stay with their mother until their first molt. **B,** A camel spider (order Solifugae). **C,** Two harvestmen (order Opiliones). Harvestmen run rapidly on their stiltlike legs. They are especially noticeable during the harvesting season; hence the common name.

James Gathany;William Nicholson/Center for Disease Control
A

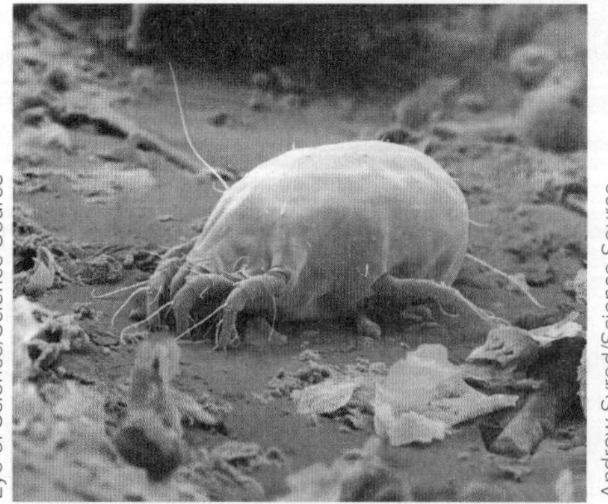
Eye of Science/Science Source
B Color enhanced scanning electron micrograph (SEM) of a house dust mite

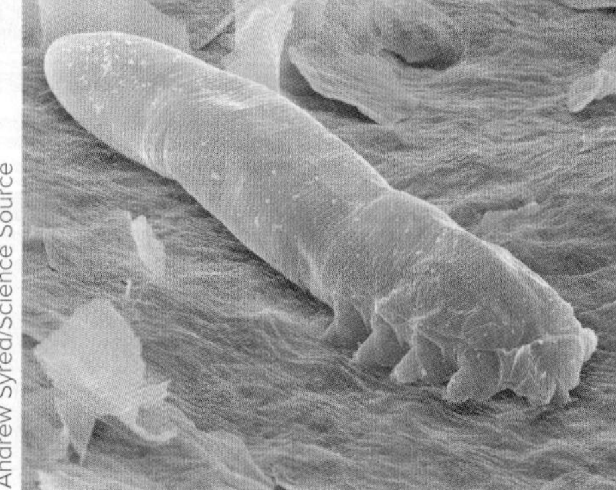
Andrew Syred/Science Source
C Colored scanning electron micrograph (SEM) of a follicle mite, *Demodex follicuorum*

figure 13.14

Ticks and mites are arachnids that often affect humans. **A,** The western blacklegged tick, *Ixodes pacificus* (order Acari), is a vector for the bacterium *Borrelia burgdorferi*, which causes Lyme disease. **B,** Dust mites, *Dermatophagoides pteronyssinus*, often trigger allergic reactions. **C,** The human follicle mite, *Demodex follicuorum*, is tiny (100–400 μm) and lives in follicles, particularly around the nose and eyes. Its prevalence ranges from about 20% in persons 20 years of age or younger to nearly 100% in the aged.

setae on their legs for swimming, and their larvae may be parasitic on aquatic invertebrates. Their sheer abundance gives acarines ecological importance, but many acarines have direct effects on our food supply and health as well. Spider mites (family Tetranychidae) are serious agricultural pests on fruit trees, cotton, clover, and many other plants; they damage leaves by sucking the contents from individual cells. Larvae of genus *Trombicula* are called chiggers or redbugs. They feed on dermal tissues of terrestrial vertebrates, including humans, and cause an irritating dermatitis; some species of chiggers transmit the disease Asiatic scrub typhus. Hair-follicle mites, *Demodex* (figure 13.14C), are apparently nonpathogenic in humans; they infect most of us although we are unaware of them. Other species of *Demodex* and other genera of mites cause mange in domestic animals.

The *Varroa* mite was implicated in the dramatic loss of honeybees that began in the United States in 2006 (see discussion of Colony Collapse Disorder, section 13.8: Insects and Human Welfare). The mite carries "deformed wing virus" or DWV. Global reemergence of this virus is linked to viral evolution, new hosts for the mite, and human movement of infected bees.

Key Theme 13.1

HUMAN CONNECTIONS

Why Chigger Bites Itch

The inflamed welt and intense itching that follow a chigger bite are not caused by a chigger burrowing into the skin, as is popularly believed. Rather, the chigger bites through skin with its chelicerae and injects a salivary secretion containing powerful enzymes that liquefy skin cells. Human skin responds defensively by forming a hardened tube that the larva uses as a sort of drinking straw and through which it gorges itself with host cells and fluid. Scratching usually removes the chigger but leaves the tube, which remains a source of irritation for several days.

Ticks are usually larger than mites. They pierce the skin of vertebrates and suck blood until their bodies become enormously distended; then they drop off and digest their meal. After molting, they are ready for another meal. In addition to disease conditions that they themselves cause, ticks are among the world's premier disease vectors, ranking second only to mosquitoes. They carry a greater variety of infectious agents than any other arthropod; such agents include unicellular eukaryotes, and rickettsial, viral, bacterial, and fungal organisms. Species of *Ixodes* carry the most common arthropod-borne infection in the United States, Lyme disease (see Key Theme 13.2). Species of *Dermacentor* and other ticks transmit Rocky Mountain spotted fever, a misnamed disease since most cases occur in the eastern United States. *Dermacentor* also transmits tularemia and agents of several other diseases. Texas cattle fever, also called red-water fever, is caused by a unicellular eukaryotic parasite transmitted by the cattle tick *Boophilus annulatus.*

Key Theme 13.2

HUMAN CONNECTIONS

Lyme Disease

In the 1970s, people in the town of Lyme, Connecticut, experienced an epidemic of arthritis-like symptoms. Subsequently called Lyme disease, it is caused by a bacterium and carried by ticks of the genus *Ixodes*. Thousands of cases are reported each year in Europe and North America, and other cases have been reported from Japan, Australia, and South Africa. Many people bitten by infected ticks recover spontaneously or do not suffer any ill effects. Others, if not treated at an early stage, develop a chronic, disabling disease. Lyme disease is now the leading arthropod-borne disease in the United States.

13.4 Subphylum Myriapoda

The term myriapod (Gr. *myrias,* a myriad, + *podos,* foot) denotes several classes that share a pattern of two tagmata—head and trunk—with paired appendages on most or all trunk segments. Myriapods include Chilopoda (centipedes), Diplopoda (millipedes), Pauropoda (pauropods), and Symphyla (symphylans). We describe only the chilopods and diplopods.

The head of myriapods has one pair of **antennae**. It also has **mandibles** (see figure 13.3, *bottom*) and two pairs of **maxillae** (one pair of maxillae in millipedes). The legs are all uniramous.

Respiratory exchange occurs through body surface and tracheal systems. In some species, aquatic juvenile stages may have gills.

Class Chilopoda: Centipedes

Centipedes are active predators with a preference for moist places such as under logs or stones, where they feed largely on earthworms and insects. Their bodies are somewhat flattened dorsoventrally, and they may contain from a few to 177 segments (figure 13.15). Each segment, except the one behind the head and the last two, bears one pair of appendages. Those of the first body segment (maxillipeds) are modified to form venom claws, which centipedes use to kill their prey. Most species are harmless to humans.

The centipede head bears a pair of eyes, each consisting of a group of ocelli (simple eyes). Respiration is by tracheal tubes with a pair of spiracles in each segment. Sexes are separate, and all species are oviparous. Young are similar to adults. Familiar genera are the common house centipedes *Scutigera,* with 15 pairs of legs, and *Scolopendra* (figure 13.15), with 21 pairs of legs.

Class Diplopoda: Millipedes

Diplopods, or "double-footed" arthropods, are commonly called millipedes, which means "thousand feet," although they do not literally have a thousand legs (figure 13.16). Their cylindrical bodies contain 25–100 segments. The four thoracic segments bear only one pair of legs each, but the abdominal

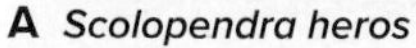

A *Scolopendra heros*

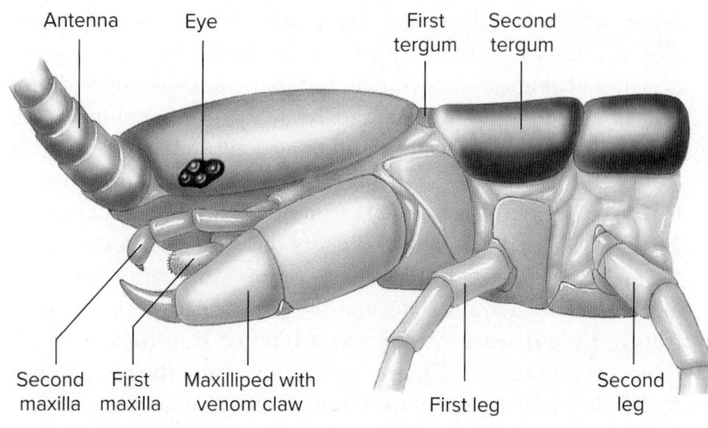

B Lateral view of the head of a centipede

figure 13.15

A, A centipede, *Scolopendra* (class Chilopoda), sits atop a lichen-covered rock. It occurs in the central United States from Texas to Colorado and may be more than 15 cm long. Most segments have one pair of appendages each. The first segment bears a pair of venom claws, which in some species can inflict serious wounds. Centipedes are carnivorous. **B,** The centipede head has eyes, antennae, and two pairs of maxillae.

A Mating millipedes

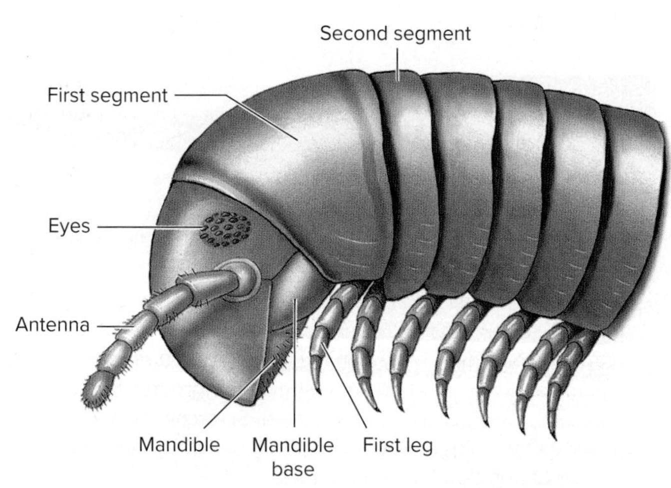

B Head of a millipede

figure 13.16

A, A pair of millipedes from South Africa. **B,** The millipede head has eyes, antennae, and chewing mandibles.

segments each have two pairs, a condition that may have evolved from fusion of segments. Two pairs of spiracles occur on each abdominal segment, each opening into an air chamber that gives rise to tracheal tubes.

Millipedes are less active than centipedes and are generally herbivorous, living on decayed plant and animal matter and sometimes living plants. They prefer dark, moist places under stones and logs. Females lay eggs in a nest and guard them carefully. Larval forms have only one pair of legs per segment. Many millipedes protect themselves from predation by secreting toxic or repellent fluids from special glands along the sides of their body. Some chemical defenses are cyanide-based.

13.5 Subphylum Crustacea

The 67,000 or more species of Crustacea (L. *crusta,* shell) include lobsters, crayfishes, shrimp, crabs, water fleas, copepods, and barnacles. Crustaceans compose the only arthropod subphylum that is primarily aquatic; they are mainly marine, but many freshwater and a few terrestrial species are known. Most are free-living, but some are sessile, commensal, or parasitic. Crustaceans are often very important components of aquatic ecosystems, and several have considerable economic importance.

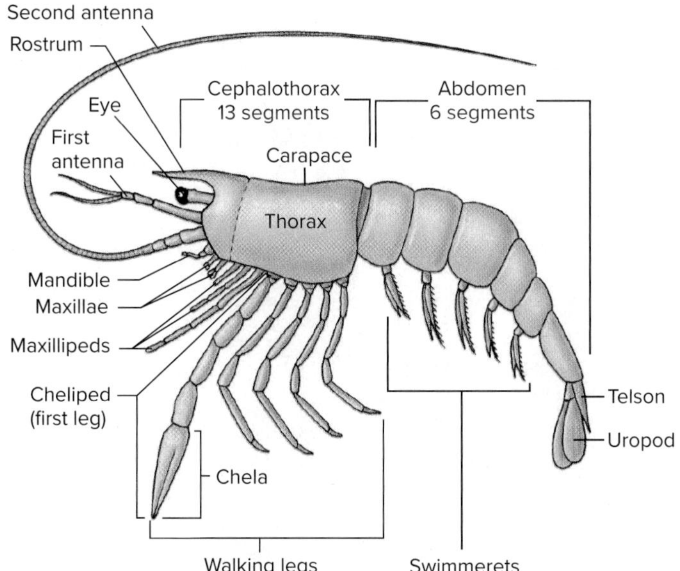

figure 13.17

Archetypical plan of Malacostraca. Note that maxillae and maxillipeds have been separated diagrammatically to illustrate the general plan. Typically, in living animals, only the third maxilliped is visible externally. In order Decapoda, the carapace covers the cephalothorax, as shown here. The head region is beneath the carapace to the left of the dashed line.

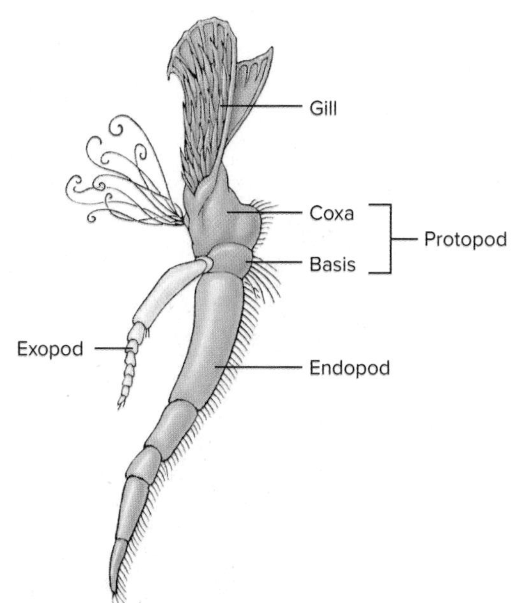

figure 13.18

Parts of a biramous crustacean appendage (third maxilliped of a crayfish). The two branches of the appendage are the exopod and the endopod; both extend from the protopod.

Crustaceans are the only arthropods with two pairs of **antennae** (figure 13.17). In addition to antennae and **mandibles,** they have two pairs of **maxillae** on the head, followed by a pair of appendages on each body segment (although appendages on some segments are absent in some groups). All appendages, except perhaps the first antennae (antennules), were ancestrally **biramous** (see figure 13.3, *top*), and at least some appendages of all present-day adults show that condition. Organs specialized for respiration, if present, are in the form of gills. Malpighian tubules are absent.

Ancestral crustaceans typically had 60 segments or more, but most modern forms have between 16 and 20 segments and are tagmatized. The major tagmata are head, thorax, and abdomen, but these are not homologous throughout the subphylum (or even within some classes) because of variations in exactly which segments are fused to create particular tagmata—for example, many crustaceans have a cephalothorax in place of a head and thorax.

In many crustaceans, the dorsal cuticle of the head extends posteriorly and around the sides of the animal to cover or fuse with some or all of the thoracic and abdominal segments. This covering is called a **carapace.** In some groups, the carapace forms clamshell-like valves that cover most or all of the body. In decapods (including lobsters, shrimp, crabs, and others), the carapace covers the entire cephalothorax but not the abdomen. Crustacea has recently expanded to include the wormlike pentastomids; these parasites previously were considered a phylum of animals, but are now placed in Crustacea, near the fish lice, subclass Branchiura.

Form and Function

Appendages

Some modifications of crustacean appendages are illustrated by those of crayfishes and lobsters (class Malacostraca, order Decapoda, see figures 13.17 and 13.19). Swimmerets, or abdominal appendages, retain the ancestral biramous condition. Such an appendage consists of inner and outer branches, called the **endopod** and **exopod,** which are attached to one or more basal segments collectively called a **protopod** (figure 13.18).

There are many modifications of this plan. In the ancestral character state for crustaceans, all trunk appendages are rather similar in structure and adapted for swimming. The evolutionary trend, shown in crayfishes, has been toward a reduced number of appendages and an increased variety of modifications that fit the appendages for many functions. Some appendages are foliaceous (flat and leaflike), as are maxillae; some are biramous, as are swimmerets, maxillipeds, uropods, and antennae; and others have lost one branch and are secondarily uniramous, as are walking legs.

In crayfishes, the first three pairs of thoracic appendages, called **maxillipeds,** serve along with the two pairs of maxillae as food handlers; the other five pairs of appendages are lengthened and strengthened for walking and defense (figure 13.19). The first pair of walking legs, called **chelipeds,** is enlarged with a strong claw, or **chela,** for defense. Abdominal swimmerets serve not only for locomotion; in males the first pair, called gonopods, are modified for copulation, and in females they all serve as a nursery for attached eggs and young. The last pair of appendages, called **uropods,** are wide and serve as paddles for swift backward movements; with the telson, they form a protective device for eggs or young on the swimmerets.

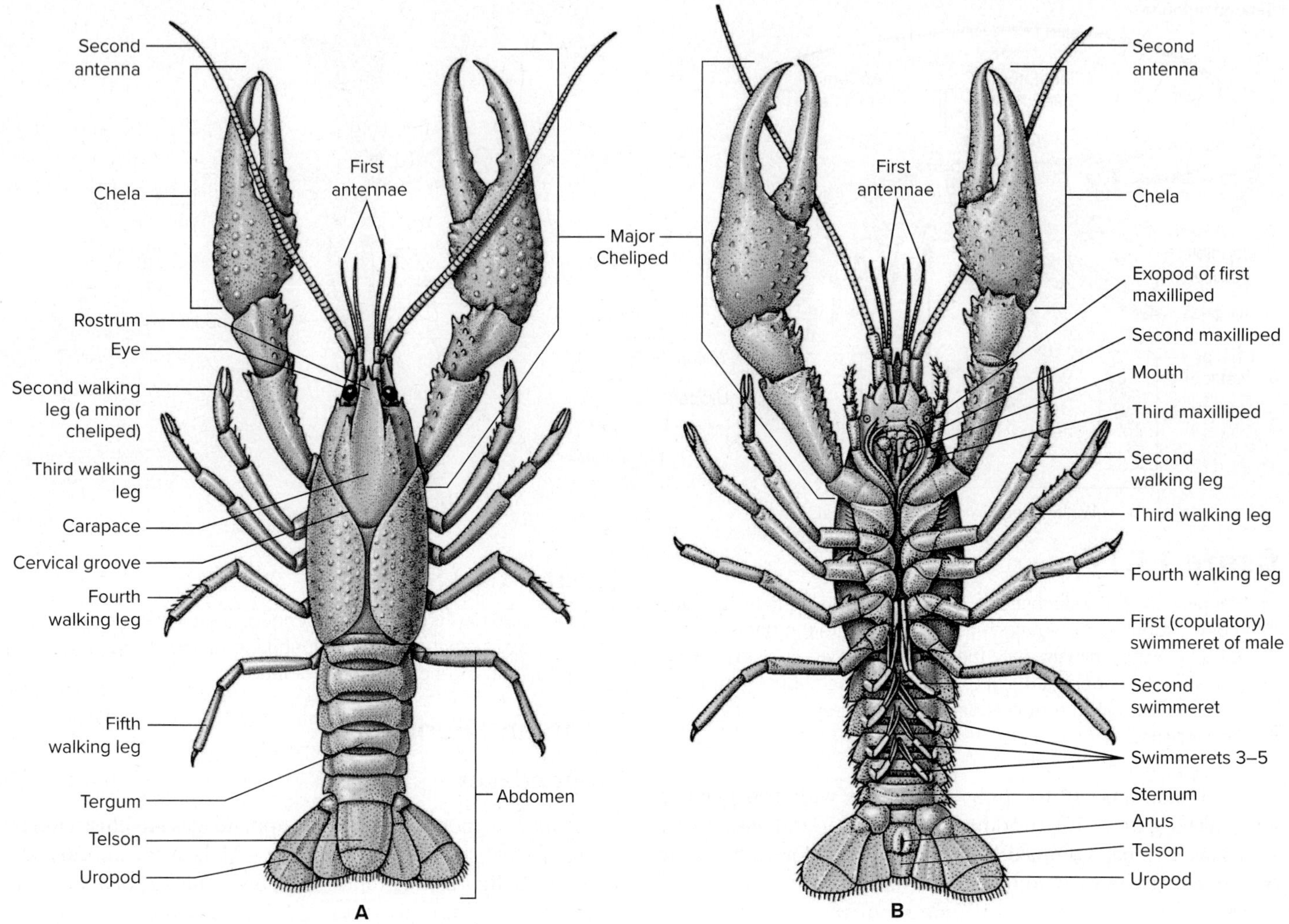

figure 13.19

External structure of crayfishes. **A,** Dorsal view. **B,** Ventral view.

Key Theme 13.3

SCIENCE EXPLAINED

Naming Appendages

Zoologists have not adopted a uniform terminology to describe crustacean appendages. We have adopted one system, but at least two systems are in wide use. For example, alternative terms to those used in this book are exopodite, endopodite, basipodite, and coxopodite (see figure 13.18). The first and second pairs of antennae may be called antennules and antennae (see figure 13.17), and the first and second maxillae are often called maxillules and maxillae. A rose by any other name....

Ecdysis

Ecdysis (Gr. *ekdysis*, to strip off) permits an increase in body size despite a restrictive exoskeleton in crustaceans and other arthropods. The periodic shedding of old cuticle and formation of a larger new one occurs most frequently during larval stages and less often as an animal reaches adulthood. Although actual shedding of the cuticle (ecdysis) is periodic, the molting process and preparations for it, involving storage of reserves and changes in the integument, are a continuous process lasting most of an animal's life.

During each premolt period, the old cuticle becomes thinner as inorganic salts are withdrawn from it and stored in tissues. Other reserves, both organic and inorganic, also accumulate and are stored. The underlying epidermis begins to grow by cell division; it secretes first a new inner layer of epicuticle and then enzymes that digest away the inner layers of old endocuticle (figure 13.20). Gradually, a new cuticle forms inside the degenerating old one. Finally, actual ecdysis occurs as the old cuticle ruptures, usually along the middorsal line, and the animal backs out (figure 13.21). The animal swells with air or water to stretch the new, larger cuticle to its full size. The animal can grow without molting until it fills this larger cuticle. During the postmolt period, the cuticle thickens, its outer layer hardens by tanning, and its inner layer uses salvaged inorganic

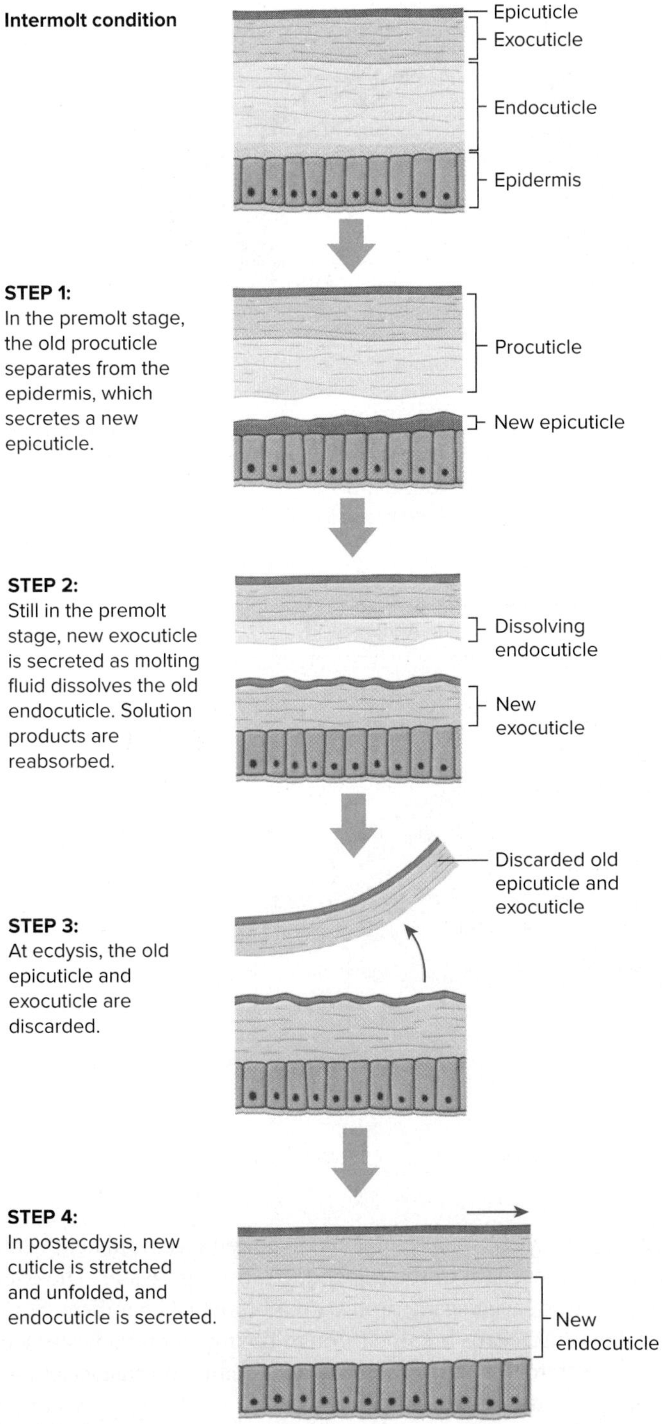

Intermolt condition

Epicuticle
Exocuticle
Endocuticle
Epidermis

STEP 1:
In the premolt stage, the old procuticle separates from the epidermis, which secretes a new epicuticle.

Procuticle
New epicuticle

STEP 2:
Still in the premolt stage, new exocuticle is secreted as molting fluid dissolves the old endocuticle. Solution products are reabsorbed.

Dissolving endocuticle
New exocuticle

STEP 3:
At ecdysis, the old epicuticle and exocuticle are discarded.

Discarded old epicuticle and exocuticle

STEP 4:
In postecdysis, new cuticle is stretched and unfolded, and endocuticle is secreted.

New endocuticle

figure 13.20

Cuticle secretion and reabsorption in ecdysis.

salts and other constituents for strengthening. Usually an animal is very reclusive during its postmolt period when its defenseless condition makes it particularly vulnerable to predation.

That ecdysis is under hormonal control has been demonstrated in both crustaceans and insects, but the process is often initiated by a stimulus perceived by the central nervous system. In decapods, the stimulus decreases production of a molt-inhibiting hormone from neurosecretory cells (see Key Theme 13.4) in the **X-organ** of the eyestalk. The sinus gland, also in the eyestalk, releases the hormone. When the level of molt-inhibiting hormone drops, **Y-organs** near the mandibles produce molting hormone. This hormone initiates processes leading to premolt. Y-organs are homologous to the prothoracic glands of insects, which produce ecdysone.

Key Theme 13.4

ADAPTATION AND PHYSIOLOGY

What Are Neurosecretory Cells?

Neurosecretory cells are modified nerve cells that secrete hormones. They are widespread in invertebrates and also occur in vertebrates. Cells in the vertebrate hypothalamus and in the posterior pituitary are good examples.

Other Endocrine Functions

Pigments in special branched cells **(chromatophores)** in the epidermis produce body color in crustaceans. Chromatophores change color by concentrating pigment granules in the center of each cell, which causes a lightening effect, or by dispersing pigment throughout each cell, which causes darkening. Neurosecretory cells in the eyestalk control pigment behavior. Neurosecretory hormones also control pigment in the eyes for light and dark adaptation, and other neurosecretory hormones control rate and amplitude of heartbeat.

Androgenic glands, which are not neurosecretory, occur in male malacostracans, and their secretion stimulates the expression of male sexual characteristics. If androgenic glands are artificially implanted in a female, her ovaries transform to testes and begin to produce sperm, and her appendages begin to acquire male characteristics at the next molt.

Feeding Habits

Feeding habits and adaptations for feeding vary greatly among crustaceans. Many forms can shift from one type of feeding to another depending on environment and food availability, but fundamentally all use the same set of mouthparts. Maxillipeds hold and manipulate food; mandibles and maxillae shred food and place it in the mouth. In predators, the walking legs, particularly chelipeds, serve in food capture.

Many crustaceans, both large and small, are predatory, and some have interesting adaptations for killing prey. One shrimplike form, *Lygiosquilla,* has on one of its walking legs a specialized digit that can be drawn into a groove and released suddenly to pierce passing prey. Pistol shrimp, *Alpheus,* have one enormously enlarged chela that can be cocked like the hammer of a gun and snapped shut at great speed, forming a cavitation bubble that implodes with a snap sufficient to stun its prey.

The food of crustaceans ranges from plankton, detritus, and bacteria, used by **suspension feeders;** to larvae, worms,

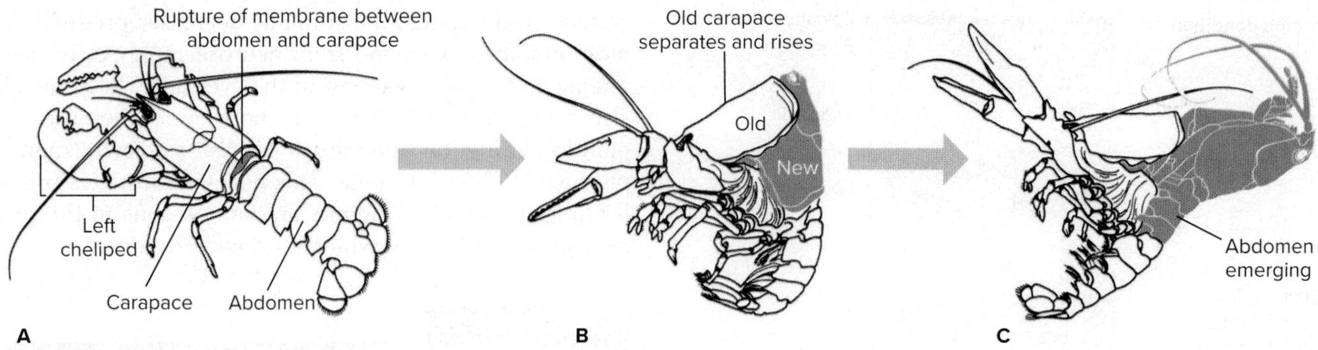

figure 13.21

Molting sequence in a lobster, *Homarus americanus.* **A,** The membrane between the carapace and the abdomen ruptures, and the carapace begins a slow elevation. This step may take up to 2 hours. **B, C,** Head, thorax, and finally the abdomen are withdrawn. This process usually takes no more than 15 minutes. Immediately after ecdysis, chelipeds are desiccated and the body is very soft. The lobster now begins rapid absorption of water so that within 12 hours its body increases about 20% in length and 50% in weight. Water will be replaced by tissue in succeeding weeks.

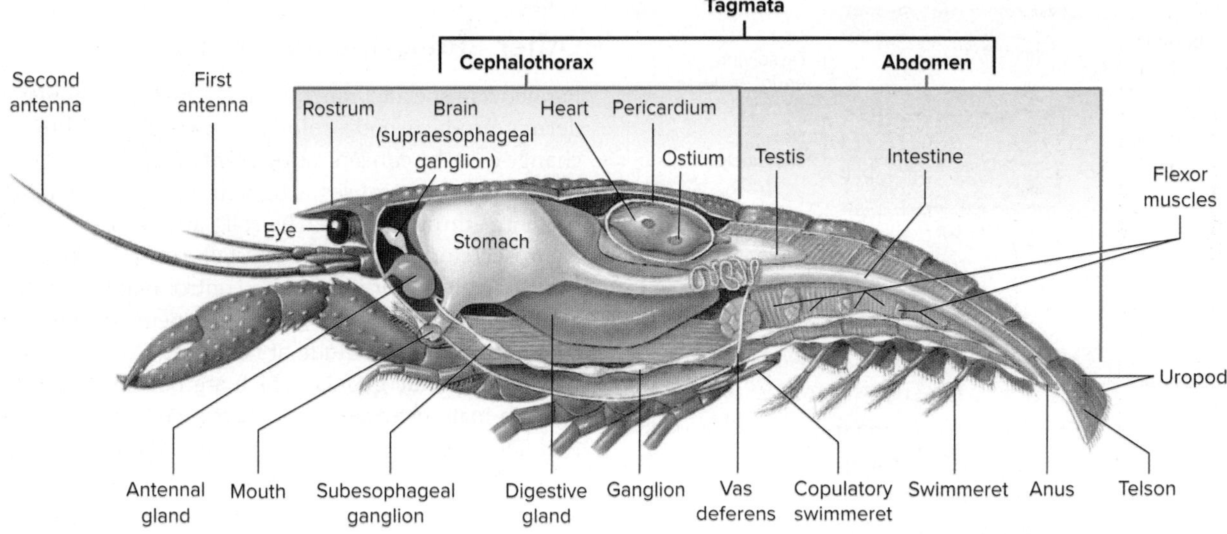

figure 13.22

Internal structure of a male crayfish.

crustaceans, snails, and fishes, used by predators; to dead animal and plant matter, used by scavengers. Suspension feeders, such as fairy shrimps, water fleas, and barnacles, use their legs, which bear a thick fringe of setae, to create water currents that sweep food particles through the setae, where they are captured. Mud shrimps, *Upogebia,* use long setae on their first two pairs of thoracic appendages to strain food material from water circulated through their burrow by movements of their swimmerets.

Crayfishes have a two-part stomach. The first contains a gastric mill in which food, already shredded by the mandibles, is ground further by three calcareous teeth into particles fine enough to pass through a filter of setae in the second part of the stomach; food particles then pass into the intestine for chemical digestion.

Respiration, Excretion, and Circulation

The gills of crustaceans vary in shape. They may be treelike, leaflike, or filamentous, and all are provided with blood vessels or sinuses. They are usually attached to appendages and kept ventilated by movement of the appendages through water. The overlapping carapace usually protects the branchial chambers. Some smaller crustaceans breathe through their general body surface.

Excretory and osmoregulatory organs in crustaceans are paired glands located in the head, with excretory pores opening at the base of either antennae or maxillae; the glands are thus called **antennal glands** or maxillary glands, respectively (figure 13.22). The antennal glands of decapods are also called **green glands** because of their color in life. They resemble the coxal glands of chelicerates. Waste products are mostly ammonia with some urea and uric acid. Some wastes diffuse through the gills as well as through the excretory glands.

Circulation, as in other arthropods, is an open system consisting of a heart, either compact or tubular, and arteries, which transport blood to different areas of the hemocoel. Some smaller crustaceans lack a heart. An open circulatory system depends less on heartbeats for circulation than does a closed system because movement of organs and limbs circulates blood more effectively in open sinuses than in capillaries. Blood may contain as respiratory pigments either hemocyanin or hemoglobin (hemocyanin in decapods), and it has the property of clotting to prevent loss of blood in minor injuries.

Nervous and Sensory Systems

A cerebral ganglion above the esophagus sends nerves to the anterior sense organs and connects to a subesophageal ganglion by a pair of connectives around the esophagus. A double ventral nerve cord has a ganglion in each segment that sends nerves to the viscera, appendages, and muscles (figure 13.22). Giant fiber systems are common among crustaceans.

Sensory organs are well developed. There are two types of eyes—a median (or nauplius) eye and compound eyes. A median eye usually consists of a group of three pigment cups containing retinal cells, and it may or may not have a lens. Median eyes occur in nauplius larvae and in some adult forms, and they may be an adult's only eye, as in copepods.

Most crustaceans have compound eyes similar to insect eyes. In crabs and crayfishes, the eyes are on the ends of movable eyestalks (figure 13.22). Compound eyes are precise instruments, different from vertebrate eyes, and yet especially adept at detecting motion; they can analyze polarized light. The convex corneal surface provides a wide visual field, particularly in stalked eyes where the surface may cover an arc of 200 degrees or more.

Compound eyes are composed of many tapering units called **ommatidia** set close together (figure 13.23). Facets, or corneal surfaces, of the ommatidia give the eye surface the appearance of a fine mosaic. Most crustacean eyes are adapted either to bright or to dim light, depending on their diurnal or nocturnal habits, but some are able, by means of screening pigments, to adapt somewhat to both bright and dim light. The number of ommatidia varies from a dozen or two in some small crustaceans to 15,000 or more in a large lobster. By comparison, some insects have approximately 30,000.

Other sensory organs include statocysts, tactile setae on the cuticle of most of the body, and chemosensitive setae, especially on antennae, antennules, and mouthparts.

Reproduction and Life Cycles

Most crustaceans have separate sexes, and numerous specializations for copulation occur among different groups. Almost all barnacles are monoecious but generally practice cross-fertilization. In some ostracods, males are scarce, and reproduction is usually parthenogenetic. Most crustaceans brood their eggs in some manner—branchiopods and barnacles have special brood chambers, copepods have egg sacs attached to the sides of their abdomen (see figure 13.25C), and malacostracans usually carry eggs and young attached to their appendages.

A crayfish hatchling is a tiny juvenile similar in form to the adult and has a complete set of appendages and segments. However, most crustaceans produce larvae that undergo a series of changes, either gradual or abrupt over a series of molts, to assume adult form (metamorphosis). The ancestral larva of crustaceans is a **nauplius** (pl., **nauplii**) (figure 13.24). It has an

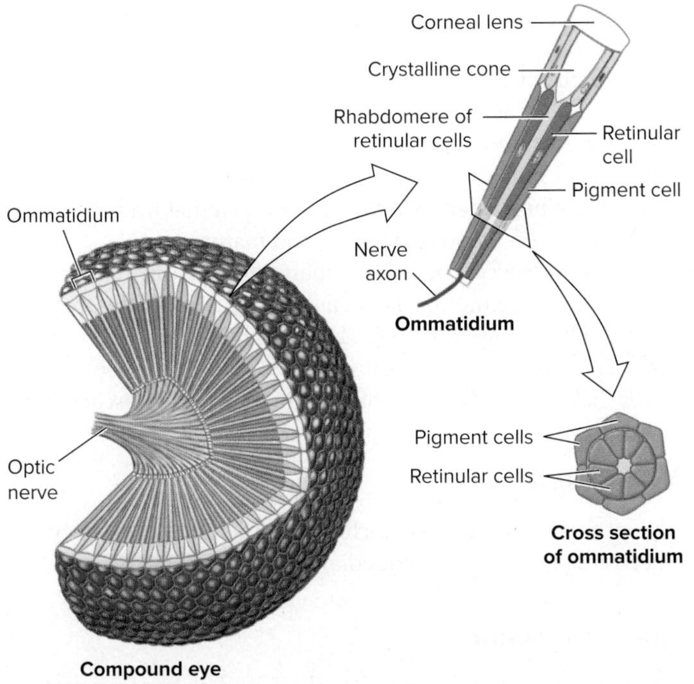

figure 13.23

Compound eye of an insect. *Right*, A single ommatidium is shown enlarged.

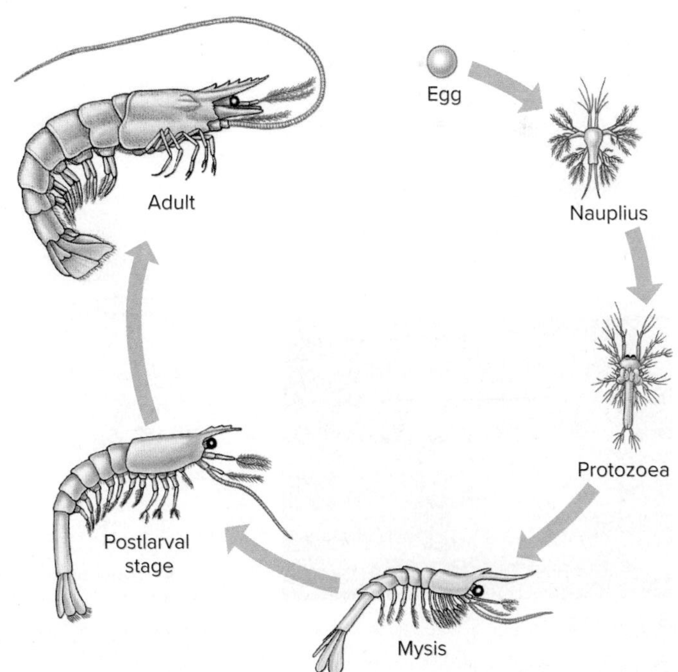

figure 13.24

Life cycle of a Gulf shrimp, *Penaeus*. Penaeids spawn at depths of 40–90 m. Young larval forms are planktonic and move inshore to water of lower salinity to develop as juveniles. Older shrimp return to deeper water offshore.

unsegmented body, a frontal eye, and three pairs of appendages, representing the two pairs of antennae and the mandibles. The developmental stages and postlarvae of different groups of Crustacea are varied and have special names.

13.6 Taxa within Subphylum Crustacea

Classification within Crustacea is in flux because traditional classes and subclasses are not supported by molecular phylogenies. The group Vericrustacea contains most of the well-known crustaceans, but members of Oligostraca will be familiar to some readers. Oligostraca comprises the Ostracoda, Branchiura, and Pentastomida.

Oligostraca

Members of **Ostracoda** (os-tra´kōd-ə) (Gr. *ostrakodes,* having a shell) are enclosed in a bivalved carapace and resemble tiny clams, 0.25–8 mm long (figure 13.25A). Ostracods show considerable fusion of trunk segments, and their thoracic appendages are reduced to two or none. Most ostracods crawl or burrow in marine or freshwater sediments. They scavenge food, feed on detritus, or collect suspended particles from the water.

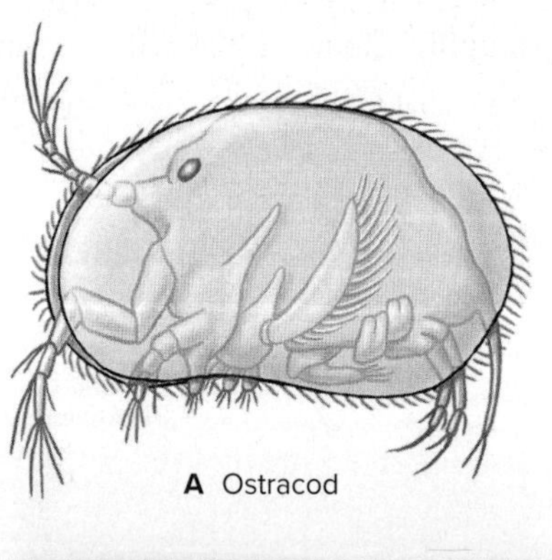
A Ostracod

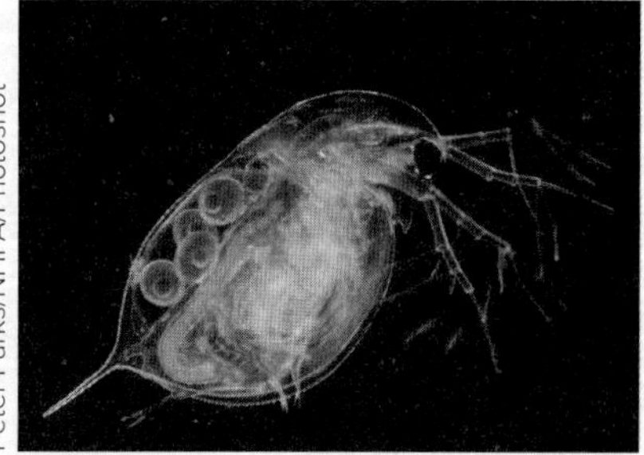
Peter Parks/NHPA/Photoshot

B *Daphnia*

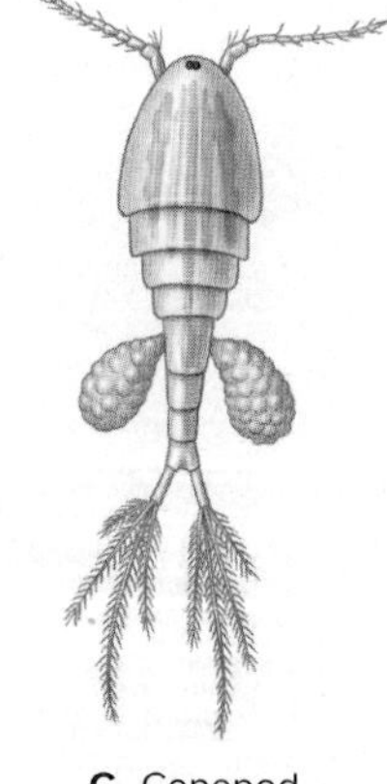
C Copepod

figure 13.25

A, An ostracod (Ostracoda). **B,** A water flea, *Daphnia* (Diplostraca, Cladocera), with eggs in the dorsal brood pouch. These tiny forms occur in great numbers in northern lakes and are an important component of the food chain leading to fishes. **C,** A copepod with attached ovisacs (Copepoda).

Branchiura (bran-kē-u´ra) (Gr. *branchia,* gills, + *ura,* tail) is a small group of primarily fish parasites that, despite its name, has no gills (figure 13.26A). Members of this group are usually between 5 and 10 mm long and parasitize marine or freshwater fishes. They typically have a broad, shieldlike carapace, compound eyes, four biramous thoracic appendages for swimming, and a short, unsegmented abdomen. The second maxillae have become modified as suction cups (figure 13.26B).

Pentastomida (pen-ta-stom´i-da) (Gr. *pente,* five, + *stoma,* mouth), comprising the tongue worms, consists of about 130 species of wormlike parasites of the respiratory system of vertebrates. Adult pentastomids live mostly in the lungs of reptiles, but one species, *Linguatula serrata* (Gr. *lingua,* tongue), lives in the nasopharynx of canines and felines (and occasionally humans). Although more common in tropical areas, they also occur in North America, Europe, and Australia.

Adult pentastomida range from 1 to 13 cm in length. Transverse rings give their bodies a segmented appearance (figure 13.26C and D). Their body is covered with a nonchitinous cuticle that is molted periodically during larval stages. The anterior end may bear five short protuberances (hence the name Pentastomida). Four of these bear chitinous claws, and the fifth bears the mouth (figure 13.26E). The digestive system is simple and adapted for sucking blood from the host. The nervous system, similar to that of other arthropods, has paired ganglia along the ventral nerve cord. The only sense organs appear to be papillae. There are no circulatory, excretory, or respiratory organs.

Sexes are separate, and females are usually larger than males. A female may produce several million eggs, which pass up the trachea of the host, are swallowed, and exit with feces. Larvae hatch as oval, tailed creatures with four stumpy legs. Most pentastomid life cycles require an intermediate vertebrate host such as a fish, a reptile, or, rarely, a mammal, that is eaten by the definitive vertebrate host. After ingestion by an intermediate host, larvae penetrate the intestine, migrate randomly in the body, and finally metamorphose into nymphs. After growth and several molts, a nymph finally becomes encapsulated and dormant. When eaten by a final host, a juvenile finds its way to a lung, feeds on blood and tissue, and matures.

Several species have been found encysted in humans, the most common being *Armillifer armillatus* (L. *armilla,* ring, bracelet, + *fero,* to bear) (figure 13.26C), but usually they cause few symptoms. *Linguatula serrata* is a cause of nasopharyngeal pentastomiasis, or "halzoun," a disease of humans in the Middle East and India.

Vericrustacea

The group Vericrustacea includes Branchiopoda, Copepoda, Thecostraca (formerly Cirripedia), and Malacostraca.

Branchiopoda

Members of Branchiopoda (branke-ä´pōd-ə) have several ancestral characteristics. Branchiopods have small first antennae and second maxillae. Their legs are flattened and leaflike (phyllopodia) and are the chief respiratory organs (hence, the

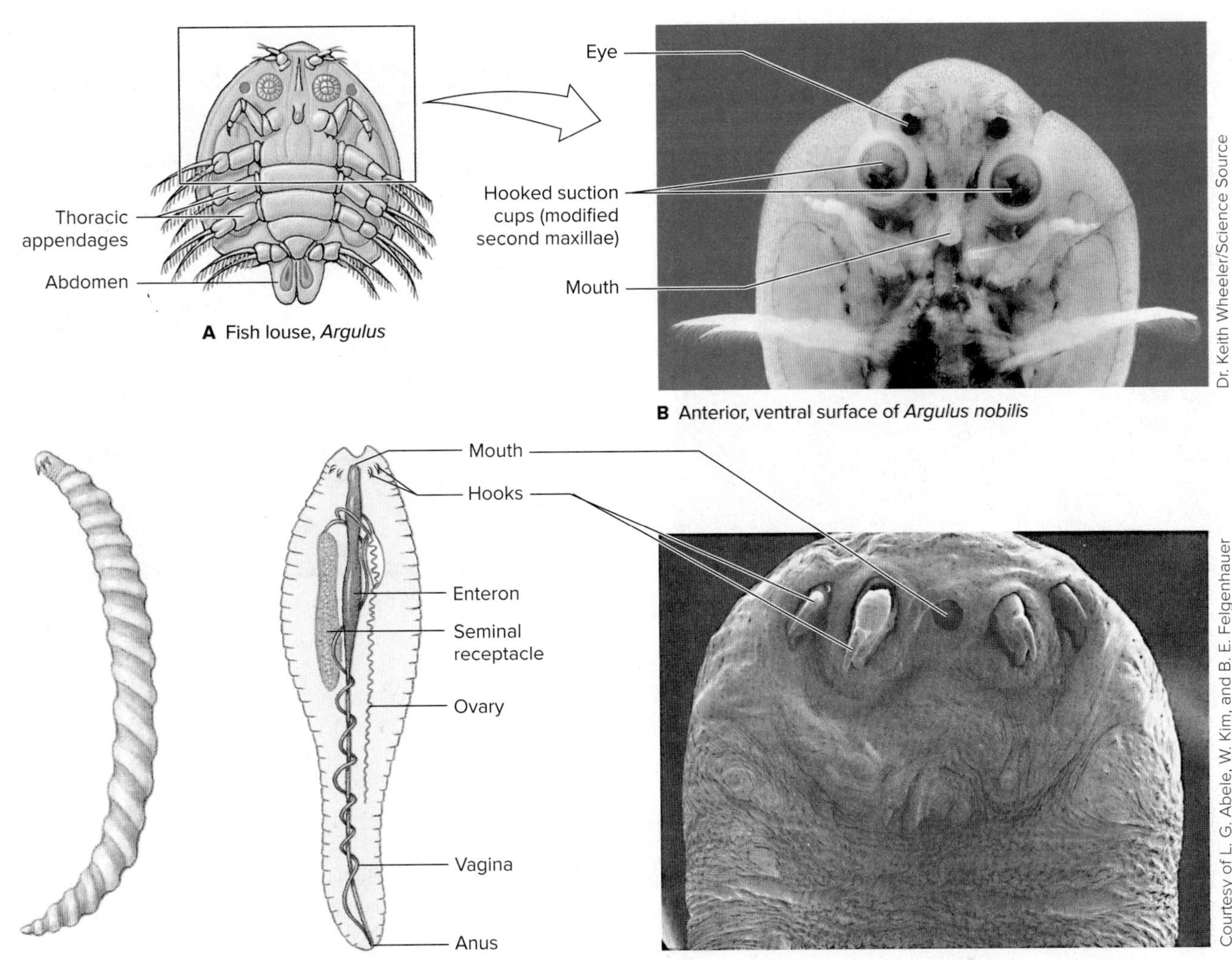

A Fish louse, *Argulus*

B Anterior, ventral surface of *Argulus nobilis*

C *Armillifer*

D *Linguatula*

E Anterior portion of *Porocephalus crotali*

figure 13.26

A, Fish lice, in subclass Branchiura, are closely related to pentastomids (**C** and **D**). Compare a photo of the anterior region of a fish louse (**B**) with an SEM of the anterior region of a pentostomid (**E**) to see morphological similarities.

name branchiopods). Most branchiopods also use their legs in suspension feeding, and groups other than cladocerans use their legs for locomotion as well.

Three orders are recognized: **Anostraca** (fairy shrimp and brine shrimp), which lack a carapace; **Notostraca** (tadpole shrimp such as *Triops*), whose carapace forms a large dorsal shield covering most trunk segments; and **Diplostraca**. Diplostraca contains two groups: Conchostraca (clam shrimp such as *Lynceus*), whose carapace is bivalved and usually encloses the entire body; and Cladocera (water fleas such as *Daphnia,* see figure 13.25B), with a carapace typically covering the entire body but not the head. The most important and diverse taxon is Cladocera, a large component of freshwater zooplankton.

Copepoda

Copepoda (ko-pe´-pōd-ə) (Gr. *kōpē,* oar, + *podos,* foot) is an important group of Crustacea, second only to Malacostraca in number of species. Copepods are small (usually a few millimeters or less in length), rather elongate, tapering toward the posterior end, lacking a carapace, and retaining a simple, median, nauplius eye in adults (see figure 13.25C). They have four pairs of rather flattened, biramous, thoracic swimming appendages, and a fifth, reduced pair. Their abdomen bears no legs. Free-living copepods occur in planktonic and benthic habitats. Planktonic species are essential members of marine and freshwater food webs, often dominating the primary consumer level (herbivore) in aquatic communities. Many symbiotic species are known, and parasitic forms may be so highly modified as adults (and may depart so far from the description just given) that they can hardly be recognized as arthropods.

Thecostraca

Thecostraca, formerly called **Cirripedia** (sir-i-ped´i-a) (L. *cirrus,* curl of hair, + *pedis,* foot), includes typical barnacles, which are

A Acorn barnacles, *Balanus balanoides*

B Gooseneck barnacles, *Lepas sp.*

figure 13.27

A, Sessile barnacles are found on rocks along the Pacific Coast of North America. **B,** Stalked barnacles extend their feeding legs, or cirri. Barnacles attach themselves to a variety of firm substrates, including rocks, pilings, and boat bottoms.

usually enclosed in a shell of calcareous plates, as well as three smaller orders of burrowing or parasitic forms. Barnacles are sessile as adults and may be attached to their substrate directly, as with acorn barnacles (figure 13.27A), or by a stalk, as with gooseneck barnacles (figure 13.27B). Typically, a carapace (mantle) surrounds their body and secretes a shell of calcareous plates. Their head is reduced, they have no abdomen, and their thoracic legs are long, many-jointed cirri with hairlike setae. The cirri are extended through an opening between the calcareous plates to filter from water small particles on which the animal feeds (figure 13.27B).

Key Theme 13.5

HUMAN CONNECTIONS

Barnacles Slow Ships

Barnacles frequently foul ship bottoms by settling and growing there. There may be so many barnacles that the speed of a ship is reduced 30%–40%, requiring expensive drydocking while the barnacles are removed. Vast sums of money have been spent on paints and coatings designed to prevent the settlement and attachment of barnacle larvae. Some paints contain metals such as copper, and other coatings incorporate biocides. International bans on certain coatings are now in place to prevent the spreading of toxic chemicals across ocean ecosystems.

Ships are not the only moving homes for barnacles: They may live atop whales, where they also negatively impact the streamlining of their hosts.

Malacostraca

Malacostraca (mal′a-kos′tra-ka) (Gr. *malakos,* soft, + *ostrakon,* shell) is the largest crustacean subgroup and shows great diversity. We mention only 4 of its 12–13 orders. The trunk of malacostracans usually has eight thoracic and six abdominal segments, each with a pair of appendages. There are many marine and freshwater species.

Isopoda (i-so′ pod-ə) (Gr. *isos,* equal, + *podos,* foot) are commonly dorsoventrally flattened, lack a carapace, and have sessile compound eyes. Their abdominal appendages bear gills. Common land forms are sow bugs or pill bugs (*Porcellio* and *Armadillidium,* figure 13.28A), which live under stones and in other damp places. *Asellus* is common in fresh water, and *Ligia* is abundant on sea beaches and rocky shores. Some isopods are parasites of other crustaceans or of fish (figure 13.29).

Members of **Amphipoda** (am-fi′-pod-ə) (Gr. *amphis,* on both sides, + *podos,* foot) resemble isopods in having sessile compound eyes and no carapace. However, they are usually compressed laterally, and their gills are in the thoracic position, as in other malacostracans. The many marine amphipods (figure 13.30) include beach fleas (*Orchestia*), and there are also numerous freshwater species.

Euphausiacea (yu-faws′i-a′se-a) (Gr. *eu,* well, + *phausi,* shining bright, + *acea,* L. suffix, pertaining to) is a group of only about 90 species that constitute the important oceanic plankton called "krill." Individuals are about 3–6 cm long (figure 13.31) and commonly occur in great oceanic swarms, where they are eaten by baleen whales, penguins, seals, and many fishes.

figure 13.28

A, Four pill bugs, *Armadillidium vulgare* (order Isopoda), common terrestrial forms. **B,** Freshwater sow bug, *Caecidotea* sp., an aquatic isopod.

figure 13.29

An isopod parasite (*Anilocra* sp., order Isopoda, class Malacostraca) on a coney (***Cephalopholis*** *fulva*) inhabiting a Caribbean coral reef.

figure 13.30

Marine amphipods. **A,** Free-swimming amphipod. **B,** Skeleton shrimp, such as *Caprella linearis* from the Russian Arctic, resemble praying mantids. **C,** *Phronima,* a marine pelagic amphipod, normally occupies the tunic of a salp (subphylum Urochordata, see Chapter 15). Swimming by means of its abdominal swimmerets, which protrude from the opening of the barrel-shaped tunic, the amphipod (order Amphipoda, class Malacostraca) maneuvers to catch its prey. In this photo, only the amphipod is visible and the salp tunic is not shown.

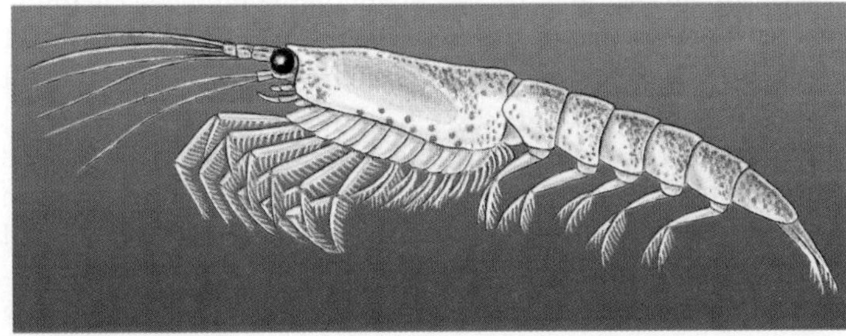

figure 13.31

Meganyctiphanes, order Euphausiacea, "northern krill."

Decapoda (deca′-pod-ə) (Gr. *deka,* ten, + *podos,* foot) have five pairs of walking legs of which the first is often modified to form pincers **(chelae)** (see figures 13.17 and 13.19). These are lobsters, crayfishes (see figure 13.19), shrimps (see figure 13.24), and crabs, the largest of the crustaceans (figure 13.32). True crabs differ from others in having a broader carapace and a much reduced abdomen (figure 13.32A and C). Familiar examples are fiddler crabs, *Uca,* which burrow in sand just below high-tide level (figure 13.32C); decorator crabs, which cover their carapaces with sponges and sea anemones for camouflage; and spider crabs, such as *Libinia.* Hermit crabs

Cleveland P. Hickman, Jr.

A

Dynamic Graphics Group / IT Stock Free / Alamy

B

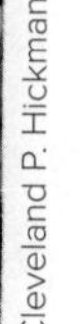

Cleveland P. Hickman, Jr.

C

Larry S. Roberts

D

Larry S. Roberts

E

figure 13.32

Decapod crustaceans. **A,** A bright orange tropical rock crab, *Grapsus grapsus,* is a conspicuous exception to the rule that most crabs exhibit cryptic coloration. **B,** A hermit crab, which has a soft abdominal exoskeleton, occupies and carries a snail shell into which it can withdraw for protection. **C,** A male fiddler crab, *Uca* sp., uses its enlarged cheliped in territorial displays and in threat and combat. **D,** A red night shrimp, *Rhynchocinetes rigens,* prowls caves and overhangs of coral reefs, but only at night. **E,** Spiny lobster *Panulirus argus* (order Decapoda, class Malacostraca).

(figure 13.32B) have become adapted to living in snail shells; their abdomen, which lacks a hard exoskeleton, is protected by the snail shell.

13.7 Subphylum Hexapoda

The subphylum Hexapoda is named for **six legs** present in its members. All legs are **uniramous.** Hexapods have **three tagmata**—head, thorax, and abdomen—with appendages on the head and thorax. Abdominal appendages are greatly reduced in size or absent. The subphylum has two classes: Entognatha, a small group whose members have the bases of mouthparts enclosed within the head capsule, and Insecta, an enormous group whose members have the bases of mouthparts visible outside the head capsule (hence, ectognathous mouthparts).

Winged insects are called pterygotes, and wingless insects are called apterygotes. The wingless order Thysanura forms the living sister taxon to all other insects. Insect wings evolved in a common ancestor of the latter clade (figure 13.33). Thysanurans are called primitively wingless to distinguish them from orders whose members do not have wings now, but whose ancestors were winged.

Class Insecta: Insects

Insects are the most numerous and diverse of all groups of arthropods (figure 13.34). There are more species of insects than species in all the other classes of animals combined. The number of insect species named has been estimated at close to 1 million, with thousands, perhaps millions, of other species yet to be discovered and classified.

There are more than 30 orders of insects. Some of the most common ones, such as Diptera and Hymenoptera, are briefly outlined in Taxonomy of Subphylum Hexapoda, placed near the end of section 13.8.

It is difficult to appreciate the great significance of this extensive group and its role in the biological network of animal life. The study of insects **(entomology)** occupies the time

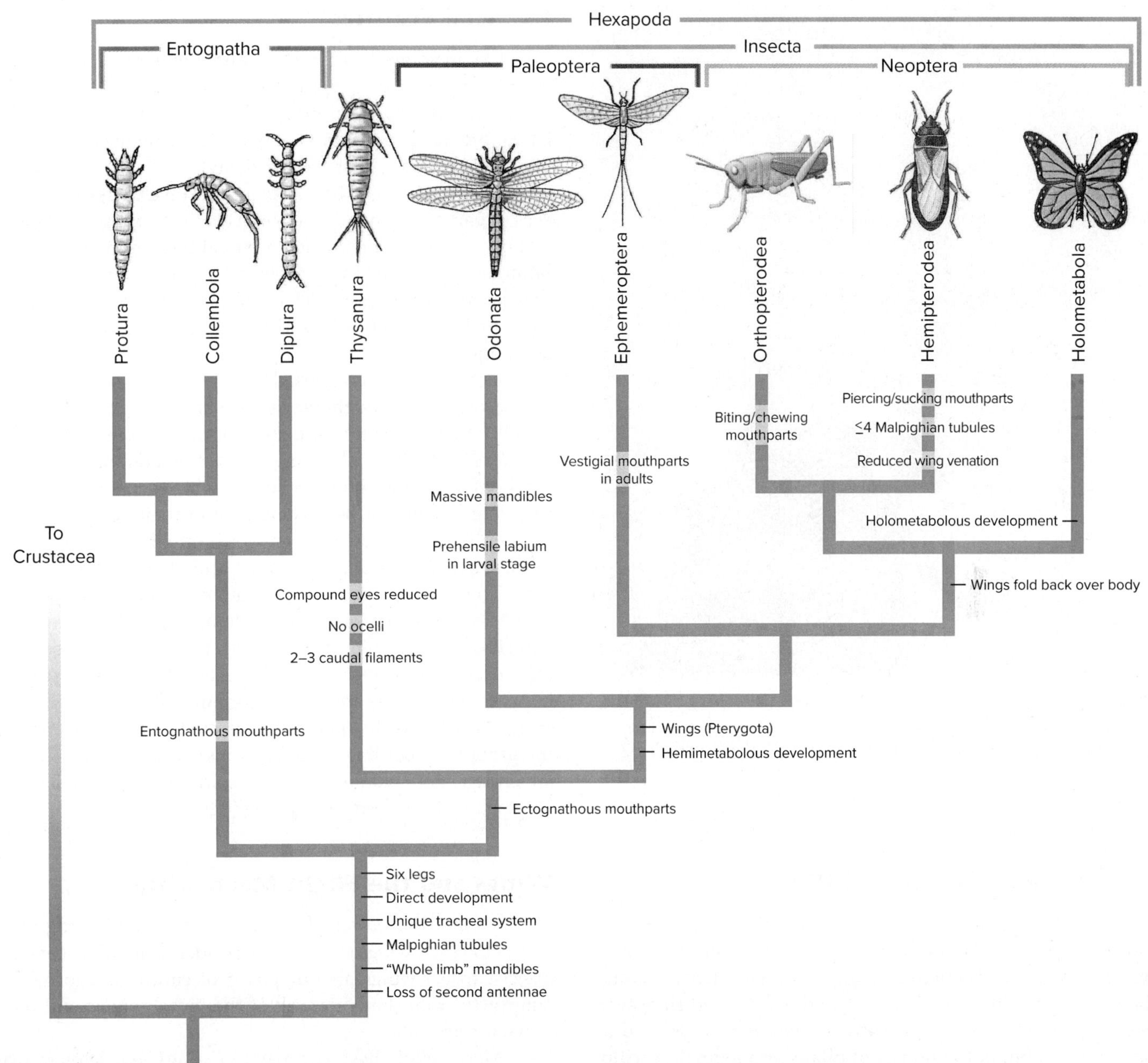

figure 13.33

Cladogram showing hypothetical relationships among hexapods. Many synapomorphies have been omitted. Orders Protura, Collembola, and Diplura are entognathous. These orders, plus Thysanura, originated before the earliest winged ancestors. Orders Odonata and Ephemeroptera form Paleoptera, where wings are outspread. The remaining orders have wings that can fold back over the abdomen (Neoptera). Superorder Orthopterodea includes orders Orthoptera, Blattodea, Phasmatodea, Mantodea, Mantophasmatodea, Isoptera, Plecoptera, Embiidina, and Dermaptera. Hemipterodea includes orders Zoraptera, Psocoptera, Hemiptera, Thysanoptera, and Phthiraptera; and superorder Holometabola encompasses all holometabolous orders. Somewhat different relationships have been proposed from a phyogenomic data set (see Misof et. al., 2014).

and resources of thousands of skilled researchers across the world. The struggle between humans and insect pests seems to be endless, and yet paradoxically, insects are so interwoven into the economy of nature that humans would not thrive without them.

Insects differ from other arthropods in having **three pairs of legs** and usually **two pairs of wings** on the thoracic region of the body (figure 13.35), although some have one pair of wings or none. In size, insects range from less than 1 mm to 20 cm in length, most being less than 2.5 cm long.

figure 13.34

Pie diagram indicating the relative numbers of insect species compared to the rest of the animal kingdom and unicellular eukaryotic groups.

Distribution and Adaptability

Insects occupy practically all habitats that can support life, but relatively few are marine. They are common in brackish water, in salt marshes, and on sandy beaches. They are abundant in fresh water, soils, forests, and plants, and they even occur in deserts and wastelands, on mountaintops, and as parasites in and on the bodies of plants and animals, including other insects.

Their wide distribution is made possible by their powers of flight and their highly adaptable nature. Many insects can easily surmount barriers that are impassable to other animals. Their small size and well-protected eggs allow them to be carried great distances by wind, water, and other animals.

Much of insects' success is due to the adaptive qualities of their cuticular exoskeleton, as is true of other arthropods as well. However, the great exploitation of terrestrial environments by insects has been made possible by an array of adaptations. For example, to minimize evaporative water loss, their epicuticle has both a waxy layer and a varnish layer, and they can close their spiracles. Insects extract maximal fluid from food and fecal material, and many can retain water produced in oxidative metabolism. Many can enter a resting stage (diapause) and lie dormant during inhospitable conditions.

External Features

Insect tagmata are head, thorax, and abdomen. The cuticle of each body segment is typically composed of four plates **(sclerites),** a dorsal notum **(tergum),** a ventral **sternum,** and a pair of lateral **pleura.** The pleura of abdominal segments are membranous rather than sclerotized (hardened).

The head usually bears a pair of relatively large compound eyes, a pair of antennae, and three ocelli. Mouthparts typically include a **labrum,** a pair each of **mandibles** and **maxillae,** a **labium,** and a tonguelike hypopharynx. The type of mouthparts that an insect possesses determines how it feeds. We discuss some of these modifications in the section titled "Internal Form and Function."

The thorax is composed of three segments: prothorax, mesothorax, and metathorax, each bearing a pair of legs (figure 13.35B). In most insects, the mesothorax and metathorax each bear a pair of wings. The wings consist of a double membrane that contains veins of thicker cuticle, which serve to strengthen the wing. Although these veins vary in their patterns among different species, they often are constant within a species and serve as one means of classification and identification.

The legs of insects are often modified for special purposes. Terrestrial forms have walking legs with terminal pads and claws, as in beetles, for example. These pads may be sticky for walking upside down, as in house flies. The hindlegs of grasshoppers and crickets are adapted for jumping (figure 13.35B). In mole crickets, the first pair of legs is modified for burrowing in the ground. Water bugs and many beetles have paddle-shaped appendages for swimming. For grasping prey, the forelegs of a praying mantis are long and strong (figure 13.36).

Wings and the Flight Mechanism

Insects share the power of flight with birds and flying mammals. However, wings evolved independently in birds, bats, and insects. Insect wings are composed of cuticle and formed by outgrowth from the body wall of the mesothoracic and metathoracic segments.

Most insects have two pairs of wings, but Diptera (true flies, figure 13.37) have only one pair, the hindwings being represented by a pair of small **halteres** (sing., **halter;** balancer) that vibrate and are responsible for equilibrium during flight. Males in order Strepsiptera have only a hind pair of wings and an anterior pair of halteres. Males of scale insects also have one pair of wings but no halteres. Some insects are wingless. Ants and termites, for example, have wings only on males and on females during certain periods; workers (females) are always wingless. Lice, bedbugs, and fleas are always wingless.

Wings may be thin and membranous, as in flies and many others; thick and stiff, as in the forewings of beetles; parchment-like, as in the forewings of grasshoppers; covered with fine scales, as in butterflies and moths; or hairlike structures, as in caddisflies.

Wing movements are controlled by a complex of thoracic muscles. Direct flight muscles are attached to a part of the wing itself. Indirect flight muscles are not attached to the wing and

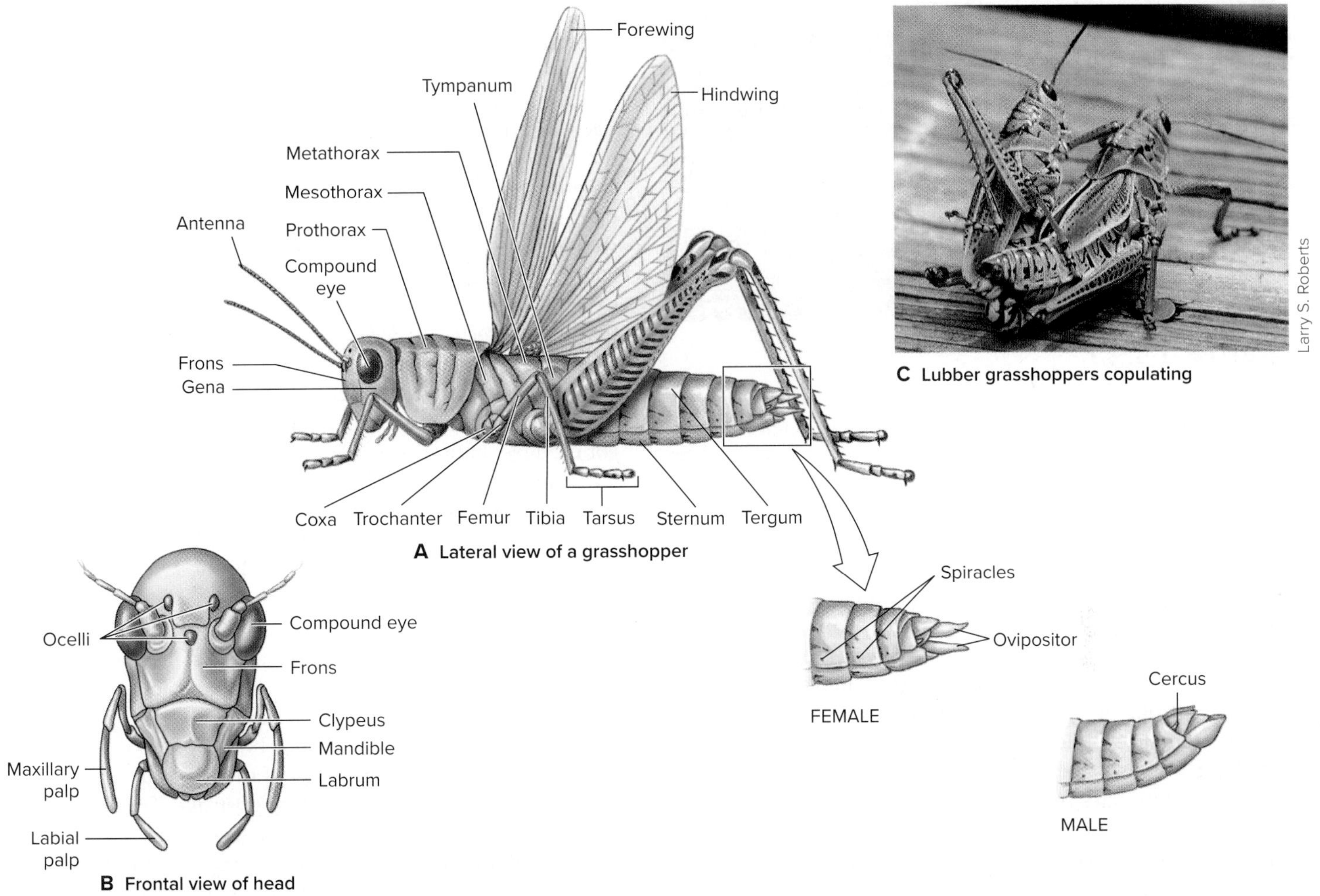

Forewing
Tympanum
Hindwing
Metathorax
Mesothorax
Antenna
Prothorax
Compound eye
Frons
Gena
Coxa
Trochanter
Femur
Tibia
Tarsus
Sternum
Tergum

A **Lateral view of a grasshopper**

Spiracles
Ovipositor
FEMALE
Cercus
MALE

Ocelli
Compound eye
Frons
Clypeus
Mandible
Maxillary palp
Labrum
Labial palp

B **Frontal view of head**

Larry S. Roberts

C **Lubber grasshoppers copulating**

figure 13.35

A, External features of a female grasshopper. The terminal segment of a male with external genitalia is shown in inset. **B,** Grasshoppers have multiple eyes, antennae, chewing mandibles, and palps. **C,** A pair of grasshoppers, *Romalea guttata* (order Orthoptera).

Simon Murrell / Alamy Stock Photo

A

NHPA/A.N.T. Photo Library

B

figure 13.36

A, Praying mantis, *Mantis religiosa* (order Mantodea), feeding on crickets. **B,** South African female praying mantis, *Miomantis caffra,* laying eggs.

McPhoto/SHU/INSADCO Photography/Alamy

figure 13.37

House fly, *Musca domestica* (order Diptera). House flies can become contaminated with over 100 human pathogens which may be transferred to human and animal food by direct contact, regurgitated food, and feces.

cause wing movement by altering the shape of the thorax. The wing is hinged at the thoracic tergum and also slightly laterally on a pleural process, which acts as a fulcrum (figure 13.38). In all insects, the upstroke of a wing is effected by contracting indirect muscles that pull the tergum down toward the sternum. Locusts, dragonflies, and cockroaches (figure 13.38A) accomplish the downstroke by contracting direct muscles attached to the wings lateral to the pleural fulcrum. In flies, bees, and midges, all flight muscles are indirect. The downstroke occurs when sternotergal muscles relax and longitudinal muscles of the thorax arch the tergum (figure 13.38B), pulling the tergal articulations upward relative to the pleura. The downstroke in beetles and grasshoppers involves both direct and indirect muscles.

Contraction of flight muscles is governed by two basic types of neural control: synchronous and asynchronous. Larger insects such as dragonflies and butterflies have synchronous muscles, in which a single volley of nerve impulses stimulates a muscle contraction and thus one wing stroke. Asynchronous muscles occur in other insects. Their mechanism of action is complex and depends on storage of potential energy in resilient parts of the thoracic cuticle. As one set of muscles contracts (moving the wing in one direction), the cuticle changes shape. When these muscles relax, the cuticle rebounds, stretching the antagonistic set of muscles passively. These muscles then actively contract (moving the wing in the other direction). Because the muscle contractions are not phase-related to nervous stimulation, only occasional nerve impulses are necessary to keep the muscles responsive to alternating stretch activation. Thus, extremely rapid wing beats are possible. For example, butterflies (with synchronous muscles) may beat as few as four times per second. Insects with asynchronous muscles, such as flies and bees, may vibrate at 100 beats per second or more. Fruit flies, *Drosophila* (Gr. *drosos,* dew, + *philos,* loving), can fly at 300 beats per second, and midges have been clocked at more than 1000 beats per second!

Obviously, flying entails more than simple flapping of wings; a forward thrust is necessary. As the indirect flight muscles alternate rhythmically to raise and lower the wings, the direct flight muscles alter the angle of the wings so that they act as lifting airfoils during both upstroke and downstroke, twisting the leading edge of the wings downward during downstroke and upward during upstroke. This modulation produces a figure-eight movement (see figure 13.38C) that aids in spilling air from the trailing edges of the wings. The quality of the forward thrust depends, of course, on several

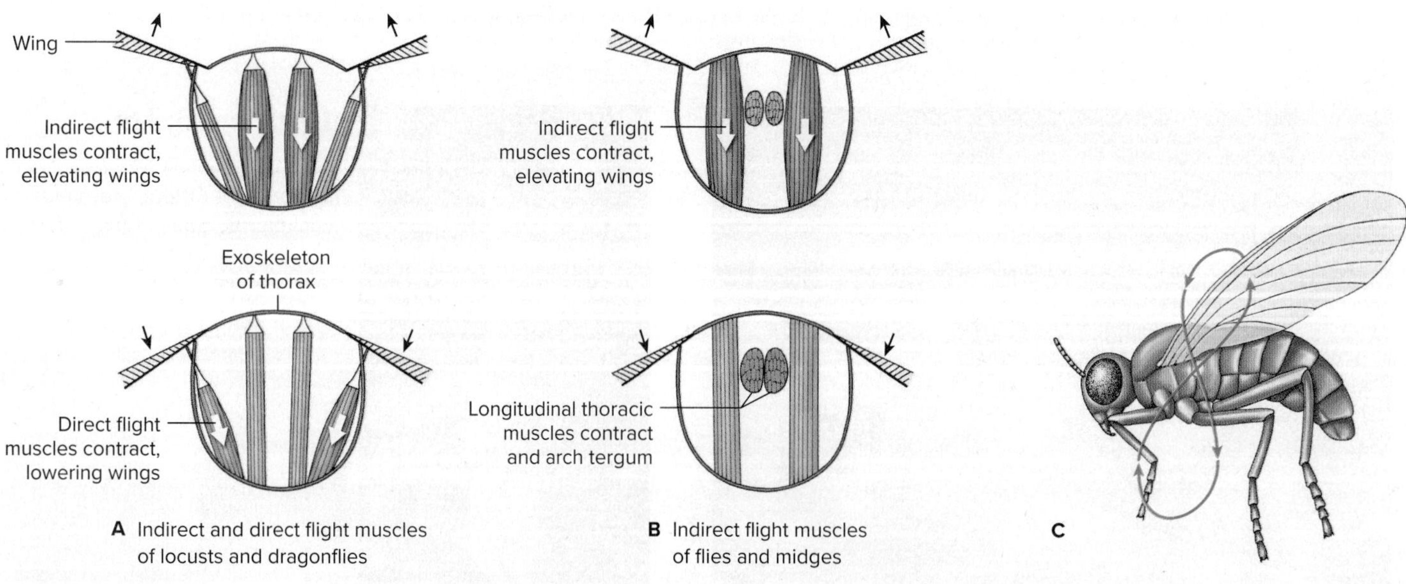

figure 13.38

A, Flight muscles of insects such as locusts, dragon flies, and cockroaches, in which upstroke is by indirect muscles and downstroke is by direct muscles. **B,** In insects such as flies, bees, and midges, both upstroke and downstroke are by indirect muscles. **C,** The figure-eight path followed by the wing of a flying insect during the upstroke and downstroke.

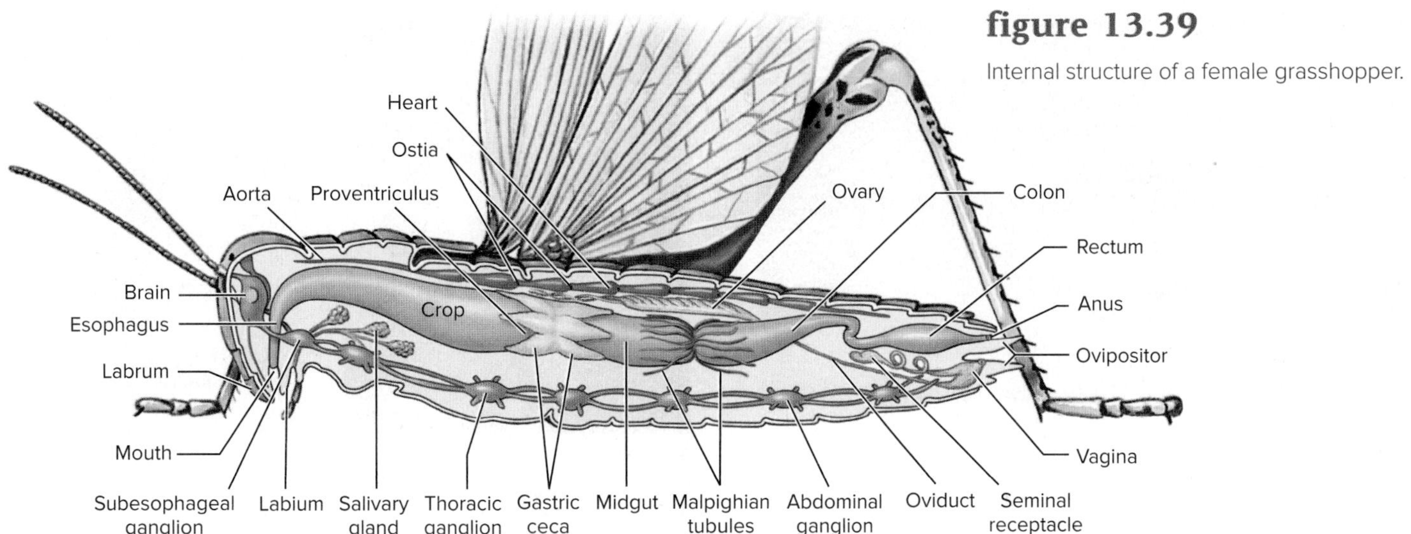

figure 13.39

Internal structure of a female grasshopper.

factors, such as variations in wing venation, how much the wings are tilted, and how they are tapered.

Flight speeds vary. The fastest flyers usually have narrow, fast-moving wings with a pronounced tilt and a strong figure-eight component. Sphinx moths and horse flies are said to achieve approximately 48 km (30 miles) per hour, and dragonflies approximately 40 km (25 miles) per hour. Some insects are capable of long, continuous flights. Migrating monarch butterflies, *Danaus plexippus* (Gr. after Danaus, mythical king of Arabia) (see figure 13.47), travel south for hundreds of miles in the fall, flying at a speed of approximately 10 km (6 miles) per hour.

Internal Form and Function

Nutrition The digestive system (figure 13.39) consists of a foregut (mouth with salivary glands, esophagus, **crop** for storage, and **proventriculus** for grinding), midgut (stomach and gastric ceca), and hindgut (intestine, rectum, and anus). The foregut and hindgut are lined with cuticle, so absorption of food is confined largely to the midgut, although some absorption may occur in all sections. Most insects feed on plant juices and plant tissues, a feeding habit described as **phytophagous.** Some insects feed on specific plants; others, such as grasshoppers, can eat almost any plant. Caterpillars of many moths and butterflies eat the foliage of only certain plants. Certain species of ants and termites cultivate fungus gardens as a source of food.

Many beetles and the larvae of many insects live on dead animals, a feeding habit called **saprophagous.** A number of insects are **predaceous,** catching and eating other insects as well as other types of animals.

Many insects are parasitic as adults or as larvae, and in some cases, both juveniles and adults are parasites. For example, fleas (figure 13.40) live on the blood of mammals as adults, but their larvae are free-living scavengers. Lice (figures 13.41 and 13.42) are parasitic throughout their life cycle. Many parasitic insects are themselves parasitized by other insects, a condition called **hyperparasitism.**

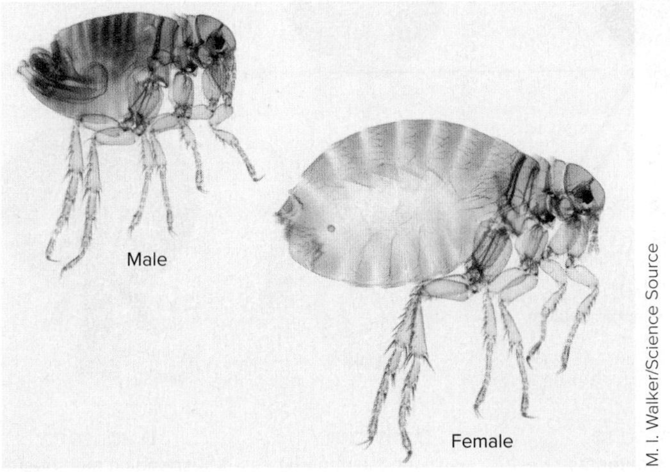

Male

Female

M. I. Walker/Science Source

figure 13.40

Male (upper image) and female (lower image) human fleas, *Pulex irritans* (order Siphonaptera).

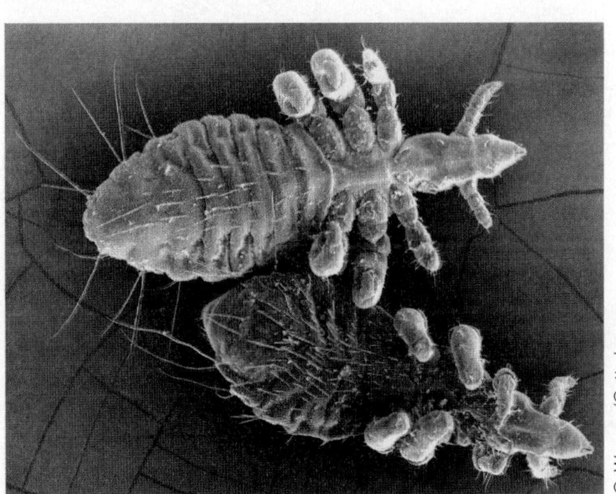

figure 13.41

Linognathus vituli, the long-nosed cattle louse, attaches to hairs on cattle in nature. These animals are photographed in the laboratory.

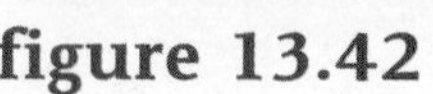

figure 13.42

The head and body louse of humans, *Pediculus humanus* (order Phthiraptera, suborder Anoplura), feeding.

figure 13.43

A, Larval stage of the tomato hornworm, *Manduca sexta* (order Lepidoptera). The more than 100 species of North American sphinx moths are strong fliers and mostly nocturnal feeders. Their larvae, called hornworms because of the large fleshy posterior spine, are often pests of tomatoes, tobacco, and other plants. **B,** Hornworm parasitized by a tiny wasp, *Apanteles,* which laid its eggs inside the caterpillar. The wasp larvae have emerged, and their pupae are on the caterpillar's skin. Young wasps emerge in 5–10 days, and the caterpillar usually dies.

Larvae of many varieties of wasps live inside the bodies of spiders or other insects (figure 13.43), consuming their hosts and eventually killing them. Parasites which ultimately kill their host are called **parasitoids**. Parasitoid insects are enormously important in controlling other insect populations.

The feeding habits of insects are determined to some extent by their mouthparts, which are highly specialized for each type of feeding. Biting and chewing mouthparts, such as those of grasshoppers and many herbivorous insects, are adapted for seizing and crushing food (figure 13.44). The mandibles of chewing insects are strong, toothed plates with edges that can bite or tear while the maxillae hold the food and pass it toward the mouth. Enzymes secreted by the salivary glands add chemical action to the chewing process.

Sucking mouthparts are greatly varied. House flies and fruit flies have no mandibles; their labium is modified into two soft lobes containing many small tubules that absorb liquids with a capillary action, much as do the holes of a commercial sponge (figure 13.44D; see figure 13.37). Horse flies, however, not only soak up surface liquids but also bite into skin with their slender, tapering mandibles and then collect blood. Mosquitoes combine piercing by means of needlelike stylets and sucking through a food channel (figure 13.44B). In honeybees, the labium forms a flexible and contractile "tongue" covered with

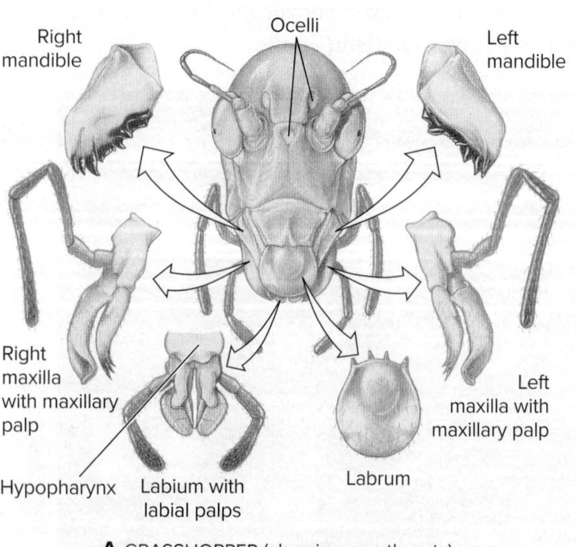

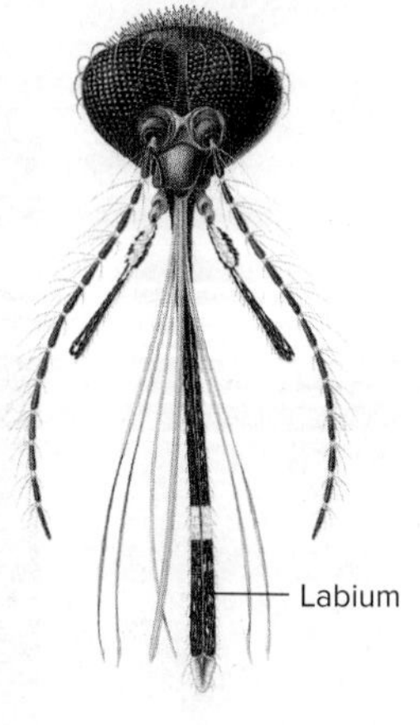

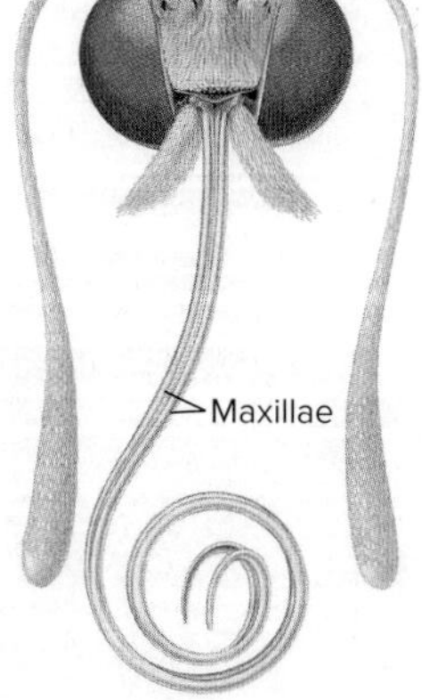

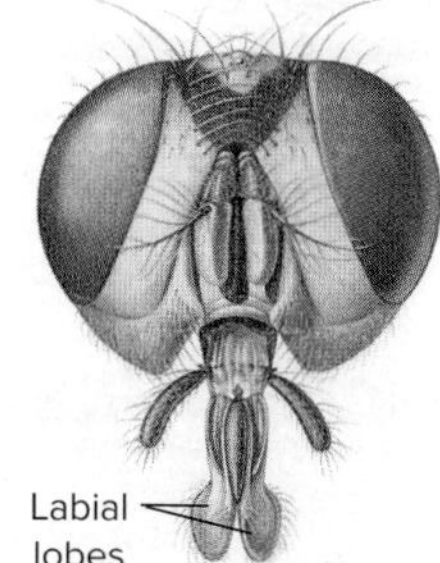

figure 13.44

Four types of insect mouthparts.

many hairs. When a bee plunges its proboscis into nectar, the tip of the tongue bends upward and moves back and forth rapidly. Liquid enters the tube by capillarity and is drawn up the tube continuously by a pumping pharynx. In butterflies and moths, mandibles are usually absent, and maxillae are modified into a long, sucking proboscis (figure 13.44C) for drawing nectar from flowers. At rest, the proboscis is coiled into a flat spiral. In feeding, it extends, and pharyngeal muscles pump fluid.

Circulation A tubular heart in the pericardial cavity (see figure 13.39) moves **hemolymph,** consisting of plasma and amebocytes, forward through the only blood vessel, a dorsal aorta. The heartbeat is a peristaltic wave. Accessory pulsatory organs help move hemolymph into the wings and legs, and flow is also facilitated by various body movements. Hemolymph apparently has little role in oxygen transport; rather, it distributes substances such as molting hormones and nutrients throughout the body. Oxygen transport is done by the **tracheal system,** an unusual system of tubes that pipe air directly to each cell.

Gas Exchange Terrestrial animals obtain oxygen from the air as it dissolves across a wet membrane. Maintaining a wet membrane in a terrestrial environment is difficult, so gas exchange typically occurs in an internal cavity. The twin goals of an efficient respiratory system are to permit rapid oxygen–carbon dioxide exchange and to restrict water loss. In insects, this is the function of the tracheal system, an extensive network of thin-walled tubes that branch into every part of the body (figure 13.45). Tracheal trunks open to the outside by paired spiracles, usually two on the thorax and seven or eight on the abdomen. A spiracle may be a simple hole in the integument, as in primitively wingless insects, but it more often has a valve or other closing mechanism that decreases water loss. Evolution of such a device must have been very important in enabling insects to move into drier habitats.

Tracheae are composed of a single layer of cells and are lined with cuticle that is shed, along with the outer cuticle, during molts. Spiral thickenings of the cuticle, called **taenidia,** support the tracheae and prevent their collapse. Tracheae branch into smaller tubes, ending in very fine, fluid-filled tubules called **tracheoles** (not lined with cuticle), which branch into a fine network over the cells. Scarcely any living cell is located more than a few micrometers away from a tracheole. In fact, the ends of some tracheoles actually indent the membranes of the cells that they supply, so that they terminate close to mitochondria. The tracheal system affords an efficient system of transport without the use of oxygen-carrying pigments in hemolymph.

In some very small insects, gas transport occurs entirely by diffusion along a concentration gradient. As oxygen is used, a partial vacuum develops in the tracheae, drawing air inward through the spiracles. Larger or more active insects employ a ventilation device for moving air into and out of the tubes. Usually muscular movements in the abdomen perform the pumping action that draws air in or expels it.

The tracheal system is primarily adapted for breathing air, but many insects (nymphs, larvae, and adults) live in water. In small, soft-bodied aquatic nymphs, gas exchange may occur by

figure 13.45

A and **B,** Components of the tracheal system. Air sacs and tracheoles not shown. **C,** Tracheal components visible through a transparent cuticle.

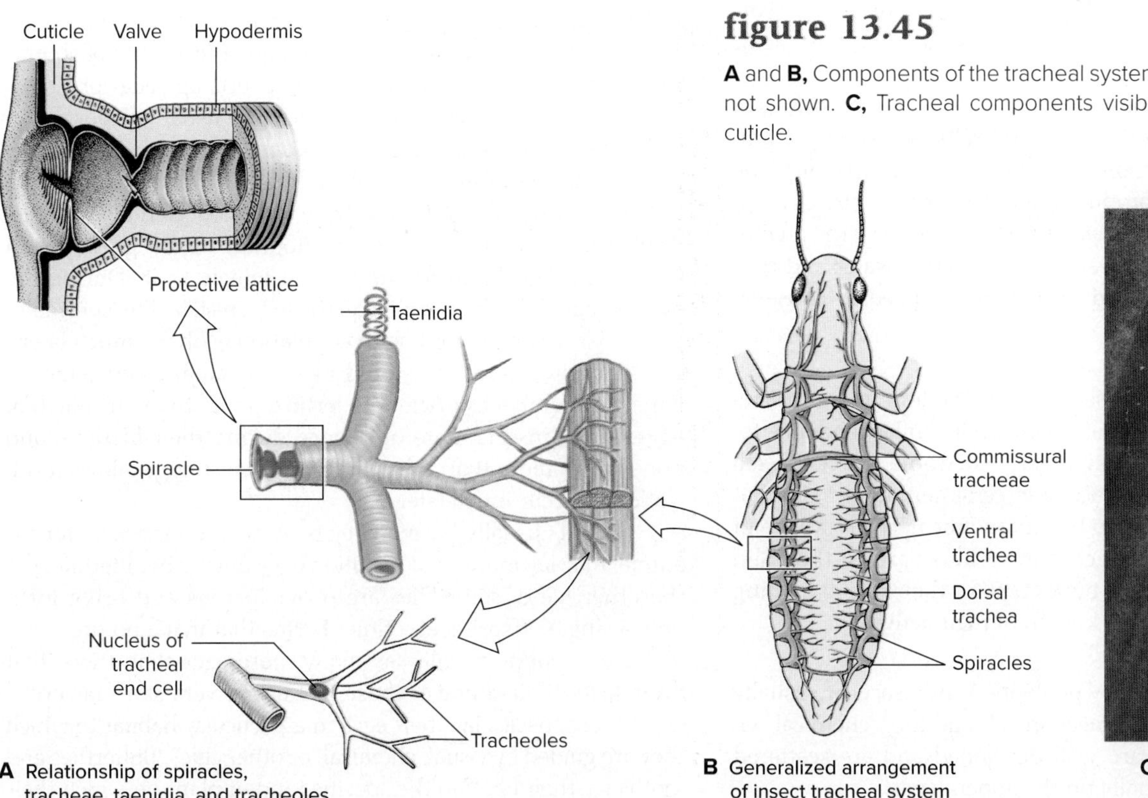

A Relationship of spiracles, tracheae, taenidia, and tracheoles

B Generalized arrangement of insect tracheal system

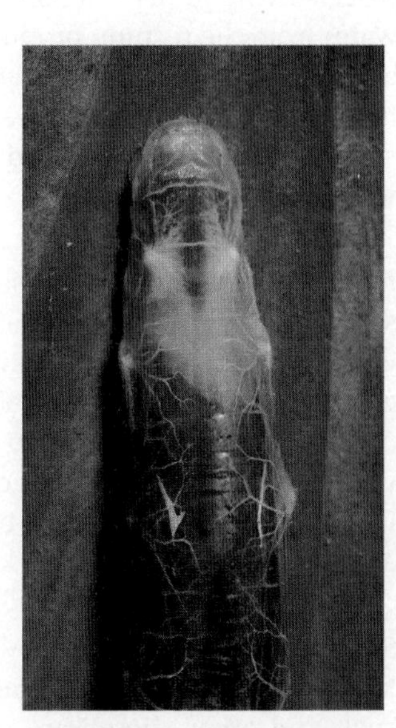

C Spiracles and tracheae of a caterpiller

diffusion through the body wall, usually into and out of a tracheal network just under the integument. The aquatic nymphs of stoneflies and mayflies are equipped with tracheal gills, which are thin extensions of the body wall containing a rich tracheal supply. The gills of dragonfly nymphs are ridges in the rectum (rectal gills) where gas exchange occurs as water enters and leaves.

Key Theme 13.6

ADAPTATION AND PHYSIOLOGY

Breathing Air Under Water

Although diving beetles, *Dytiscus* (Gr. dytikos, able to swim), can fly, they are excellent swimmers and spend most of their life in water. They use an "artificial gill" in the form of a bubble of air held under the first pair of wings. The bubble is kept stable by a layer of hairs on top of the abdomen and is in contact with the spiracles on the abdomen. Oxygen from the bubble diffuses into the tracheae and is replaced by diffusion of oxygen from the surrounding water. However, nitrogen from the bubble diffuses into the water, slowly decreasing the size of the bubble; therefore, diving beetles must surface every few hours to replace the air. Mosquito larvae are not good swimmers but live just below the surface, putting out short breathing tubes like snorkels to the surface for air. Spreading oil on the water, a favorite method of mosquito control, clogs the tracheae with oil and suffocates the larvae. "Rattailed maggots" of syrphid flies have an extensible tail that can stretch as much as 15 cm to the water surface.

Excretion and Water Balance Malpighian tubules (see figure 13.39) occur in most insects. As in spiders (see figure 13.7C), Malpighian tubules are very efficient, both as excretory organs and as a means of conserving body fluids—an important factor in the success of terrestrial animals.

Because water requirements vary among different types of insects, this ability to cycle water and salts is very important. Insects living in dry environments may resorb nearly all of the water from the rectum, producing a nearly dry mixture of urine and feces. Leaf-feeding insects take in and excrete large quantities of fluid. Freshwater larvae need to excrete water and conserve salts. Insects that feed on dry grains need to conserve water and to excrete salt.

Nervous System The insect nervous system in general resembles that of larger crustaceans, with a similar tendency toward fusion of ganglia (see figure 13.39). Some insects have a giant fiber system. There is also a visceral nervous system that corresponds in function with the autonomic nervous system of vertebrates. Neurosecretory cells in various parts of the brain have an endocrine function, but except for their role in molting and metamorphosis, little is known of their activity.

Sense Organs The sensory perceptions of insects are usually keen. Organs receptive to mechanical, auditory, chemical, visual, and other stimuli are well developed and are scattered over the body, but especially on the appendages.

Photoreceptors include both ocelli and compound eyes. Compound eyes are large and constructed of ommatidia, as are those of crustaceans (see figure 13.23). Apparently, the visual acuity of insect eyes is much lower than that of human eyes, but most flying insects rate much higher than humans in flicker-fusion tests. Flickers of light become fused in human eyes at a frequency of 45–55 per second, but in bees and blow flies they do not fuse until 200–300 per second. This would be an advantage in analyzing a fast-changing landscape. Most insects have three ocelli on their head, and they also have dermal light receptors on their body surface.

Insects may detect sounds by means of sensitive, hairlike **sensilla** or by relying on tympanic organs sensitive to sonic or ultrasonic sound. Sensilla are modifications in the cuticular surface for reception of sensory stimuli other than light and are supplied with one or more neurons. Tympanic organs, found in grasshoppers (see figure 13.35B), crickets, cicadas, butterflies, and moths, involve a number of sensory cells extending to a thin tympanic membrane that encloses an air space in which vibrations can be detected.

Chemoreceptive sensilla, which are peglike setae, are especially abundant on the antennae, mouthparts, or legs. Mechanical stimuli, such as contact pressure, vibrations, and tension changes in the cuticle, are detected by sensilla or by sensory cells in the epidermis. Insects also sense temperature, humidity, body position (proprioception), and gravity.

Reproduction Sexes are separate in insects, and fertilization is usually internal. Insects have various means of attracting mates. Female moths emit a chemical (pheromone) that males can detect for a great distance—several miles from females. Fireflies use flashes of light; other insects find each other by means of sounds or color signals and by various kinds of courtship behavior.

At the time of copulation, sperm are usually deposited in the female's vagina (see figure 13.35A). In some orders, sperm are encased in spermatophores that may be transferred at copulation or deposited on the substrate to be collected by a female. A male silverfish deposits a spermatophore on the ground, and then spins signal threads to guide a female to it. During the evolutionary transition from aquatic to terrestrial life, spermatophores were widely used, with copulation evolving much later.

Usually sperm are stored in the seminal receptacle of a female in numbers sufficient to fertilize more than one batch of eggs. Many insects mate only once during their lifetime, and none mates more than a few times. Sperm storage allows fertilization to occur much later.

Insects usually lay many eggs. A queen honeybee, for example, may lay more than 1 million eggs during her lifetime. On the other hand, some flies are ovoviviparous and bring forth only a single offspring at a time. Forms that make no provision for care of young usually lay many more eggs than those that provide for their young or those that have a very short life cycle.

Most species lay their eggs in a particular habitat to which they are guided by visual, chemical, or other clues. Butterflies and moths lay their eggs on the specific kind of plant on which their

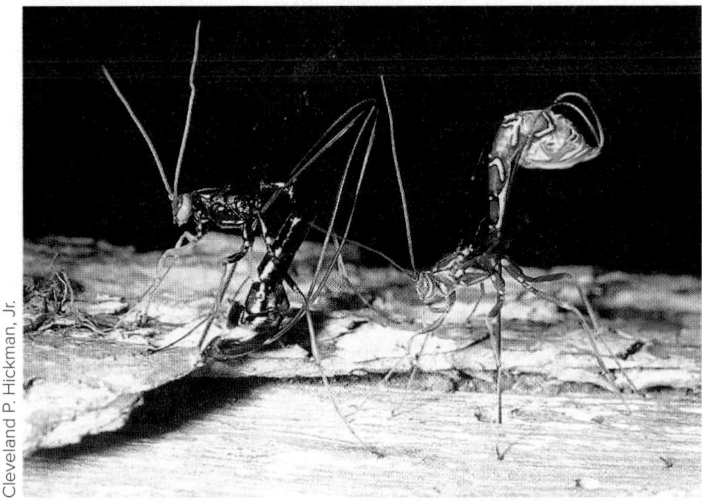

figure 13.46

An ichneumon wasp with the end of the abdomen raised to thrust her long ovipositor into wood. She can bore 13 mm or more into wood to find a tunnel made by the larva of a wood wasp or wood-boring beetle. After she lays her eggs in the larva, the wood-boring beetle larva becomes host for the ichneumon larvae. Other ichneumon species attack spiders, moths, flies, crickets, caterpillars, and other insects.

caterpillars must feed. A tiger moth may look for a pigweed, a sphinx moth for a tomato or tobacco plant, and a monarch butterfly for a milkweed plant. Insects whose immature stages are aquatic lay their eggs in water. A tiny braconid wasp lays her eggs on a caterpillar of the sphinx moth where they will feed and pupate in tiny white cocoons (see figure 13.43B). An ichneumon wasp, with unerring accuracy, seeks out a certain kind of larva in which her young will live as internal parasites. Her long ovipositor may have to penetrate 1–2 cm of wood to deposit her eggs in the larva of a wood wasp or a wood-boring beetle (figure 13.46).

Metamorphosis and Growth

The early development of insects occurs within the eggshell, and hatching young escape in various ways. During postembryonic development, most insects change in form; they undergo **metamorphosis.** A number of molts are necessary during the growth period, and each stage between molts is called an **instar.** Although many animals undergo metamorphosis, insects illustrate it more dramatically than does any other group. The transformation of a caterpillar into a beautiful moth or butterfly is indeed an astonishing morphological change.

Approximately 88% of insects undergo **holometabolous** (complete) **metamorphosis** (Gr. *holo,* complete, + *metabolē,* change), which separates the physiological processes of growth **(larva)** from those of differentiation **(pupa)** and reproduction **(adult).** Each stage functions efficiently without competition with other stages, because the larvae and adults often live in entirely different surroundings and eat different foods. The wormlike larvae, which usually have chewing mouthparts, have various common names, such as caterpillars, maggots,

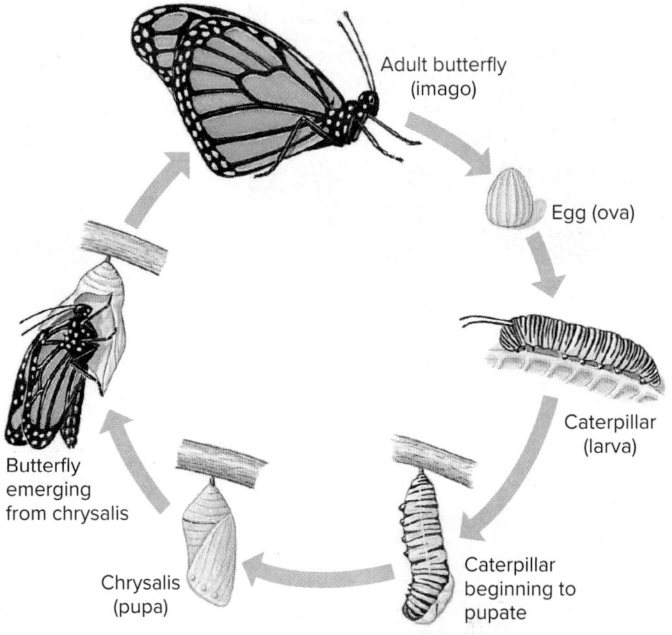

figure 13.47

Holometabolous (complete) metamorphosis in a butterfly, *Danaus plexippus* (order Lepidoptera). Eggs hatch to produce the first of several larval instars. The last larval instar molts to become a pupa. At the pupal molt, an adult emerges.

bagworms, fuzzy worms, and grubs. After a series of instars, a larva forms a case or cocoon around itself and becomes a pupa, or chrysalis, a nonfeeding stage in which many insects pass the winter. When the final molt occurs, a full-grown adult emerges (figure 13.47), pale and with wrinkled wings. In a short time, the wings expand and harden, and the insect flies for the first time. Adults undergo no further molting.

Some insects undergo **hemimetabolous** (gradual, or incomplete) **metamorphosis** (Gr. *hemi,* half, + *metabolē* change) (figure 13.48). These include bugs, scale insects, lice, and grasshoppers, which have terrestrial young, and mayflies,

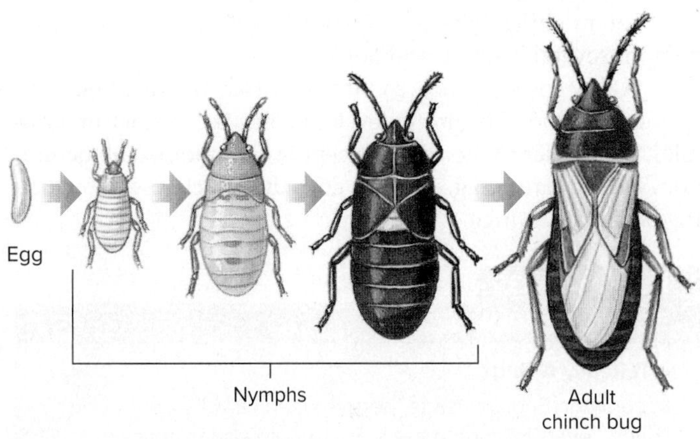

figure 13.48

Life history of a hemimetabolous insect.

Cleveland P. Hickman, Jr.
A

MyLoupe/Getty Images
B

NHPA/Roger Tidman
C

figure 13.49

A, A stonefly, *Perla* sp. (order Plecoptera). **B,** A ten-spot dragonfly, *Libellula pulchella* (order Odonata). **C,** Nymph (juvenile) of an Emperor dragonfly, *Anax imperator*, in its underwater habitat. Both stoneflies and dragonflies have aquatic larvae that undergo gradual metamorphosis.

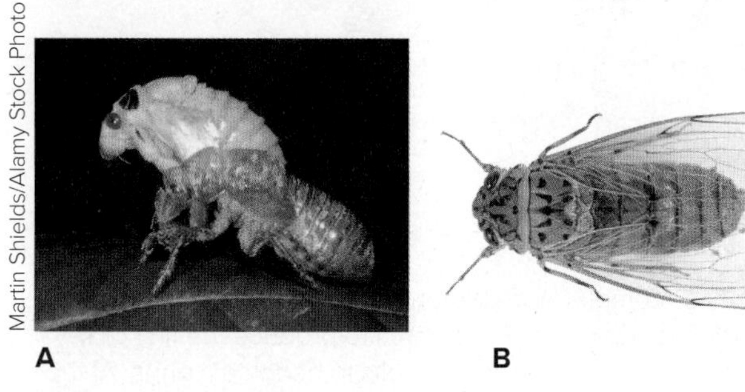
Martin Shields/Alamy Stock Photo
Ingram Publishing/Fotosearch
A B

figure 13.50

A, Ecdysis in a cicada, *Magicicada septendecium* (order Hemiptera, suborder Auchenorrhyncha). The old cuticle splits along a dorsal midline due to increased blood pressure and air forced into the thorax by muscle contraction. The emerging insect is pale, and its new cuticle is soft. The wings expand as hemolymph is pumped into veins, and the insect enlarges by taking in air. **B,** An adult *cicada.*

stoneflies (figure 13.49A), and dragonflies (figure 13.49B), which lay their eggs in water. The young are called **nymphs** or simply juveniles (figure 13.49C), and their wings develop externally as budlike outgrowths in the early instars and increase in size as the animal grows by successive molts and becomes a winged adult (figure 13.50). Aquatic nymphs have tracheal gills or other modifications for aquatic life. The stages are egg, nymph (several instars), and adult.

A few insects, such as silverfish (see figure 13.60) and springtails, undergo direct development. The young, or juveniles, are similar to the adults except in size and sexual maturation. The stages are egg, juvenile, and adult. Such insects include the primitively wingless insects.

Key Theme 13.7
SCIENCE EXPLAINED

Is It Really A Bug?

In common English usage, people often call all insects "bugs," even extending the word to such nonanimals as bacteria, viruses, and glitches in computer programs. Biologically speaking, however, a bug is a member of order Hemiptera, suborder Heteroptera, and nothing else.

Hormones control and regulate metamorphosis in insects. Three major endocrine organs guide development through juvenile instars and the eventual emergence of adults. These organs and the hormones that they produce are the brain (ecdysiotropin), the **ecdysial** or prothoracic glands (ecdysone), and the **corpora allata** (juvenile hormone). Hormonal control of molting and metamorphosis is the same in holometabolous and hemimetabolous insects.

Diapause Many animals can enter a state of dormancy during adverse conditions, and the life cycle of many insects includes a long dormant period during which external climatic conditions are too harsh for normal activity. Most insects enter such a stage facultatively when some environmental factor, such as temperature, becomes unfavorable, and the state continues until conditions again become favorable.

However, some species experience a prolonged arrest of growth that is internally programmed and usually seasonal. This type of dormancy is called **diapause** (di´a-poz) (Gr. *dia,* through, dividing into two parts, + *pausis,* a stopping), and it is an important adaptation to adverse environmental conditions. Diapause is usually triggered by some external signal, such as shortening day length. Diapause always occurs at the end of an active growth stage of the molting cycle so that, when the diapause period is over, the insect is ready for another molt.

13.8 Insect Interactions

Insects' keen sensory perceptions make them extremely responsive to many stimuli. The stimuli may be internal (physiological) or external (environmental), and the responses are governed by both the physiological state of the animal and the pattern of nerve pathways involved. Many responses are simple, such as orientation toward or away from a stimulus as occurs when a moth flies toward light, a cockroach avoids light, or carrion flies follow the odor of dead flesh.

James Hager/robertharding/Getty Images

figure 13.51

Two dung beetles (order Coleoptera) atop a ball of dung, Serengeti National Park, Tanzania, Africa. The beetles chew off a bit of dung, roll it into a ball, and then roll it to where they bury it in soil. They then lay their eggs in the ball, and the larvae feed on the dung.

Behavior and Communication

Much insect behavior, however, is not a simple matter of orientation but involves a complex series of responses. A pair of dung beetles chew off a bit of dung, roll it into a ball, and roll the ball laboriously to where they intend to bury it after laying their eggs in it (figure 13.51). A female cicada slits the bark of a twig and then lays an egg in each of the slits. A female potter wasp (*Eumenes*) scoops up clay into pellets, carries them one by one to her building site, and fashions them into dainty, narrow-necked clay pots, into each of which she lays an egg. Then she hunts and paralyzes a number of caterpillars, pokes them into the opening of a pot, and closes the opening with clay. Each egg, in its own protective pot, hatches to find food awaiting.

Key Theme 13.8

ADAPTATION AND PHYSIOLOGY

Insects Learn!

Some insects can memorize and perform in sequence tasks involving multiple signals in various sensory areas. Worker honeybees have been trained to walk through mazes that involve five turns in sequence, using such clues as the color of a marker, the distance between two spots, or the angle of a turn. The same is true of ants. Workers of one species of *Formica* learned a six-point maze at a rate only two or three times slower than that of laboratory rats. Foraging trips of ants and bees often wind and loop in a circuitous route, but once the forager has found food, the return trip is relatively direct.

Much insect behavior is "innate," meaning that entire sequences of actions apparently have been programmed. However, learning also takes place. A potter wasp, for example, must learn where she has left her pots if she is to return to fill them with caterpillars one at a time. Social insects, which have been studied extensively, are capable of most of the basic forms of learning used by mammals. Bees presented with a test where pulling a string results in a food reward showed a surprising capacity to learn. A few bees were able to innovate and obtain the reward without training, but many bees could be trained to pull the string to obtain food. Other bees who watched the trained bees learned the behavior through observation. Perhaps the most exciting result of these experiments was the cultural transmission of the new skill. Once a trained bee was added to a hive, other bees learned; this learning spread even if the original trained bee was no longer present.

Insects communicate with other members of their species by chemical, visual, auditory, and tactile signals. Chemical signals take the form of **pheromones,** which are substances secreted by one individual that affect the behavior or physiological processes of another individual. Examples of pheromones include sex attractants, releasers of certain behavior patterns, trail markers, alarm signals, and territorial markers. Like hormones, pheromones are effective in minute quantities. Social insects, such as bees, ants, wasps, and termites, can recognize a nestmate—or an alien in the nest—by means of identification pheromones. Pheromones determine caste in termites, and to some extent in ants and bees. In fact, pheromones are probably a primary integrating force in populations of social insects. Many insect pheromones have been extracted and chemically identified.

Sound production and sound reception (phonoproduction and phonoreception) in insects have been studied extensively, and although not all insects have a sense of hearing, those that do use sounds as warning devices, to advertise territorial claims, or for courtship. For example, the sounds of crickets and grasshoppers function in both courtship and aggression. Male crickets scrape the modified edges of their forewings together to produce their characteristic chirping. The long, drawn-out sound of male cicadas, a call to attract females, is produced by vibrating membranes in a pair of organs located on the ventral side of the basal abdominal segment.

Insects practice many forms of tactile communication, such as tapping, stroking, grasping, and antennae touching, which evoke varying responses, including recognition, recruitment, and alarm. Certain kinds of flies, springtails, and beetles manufacture their own visual signals in the form of **bioluminescence.** The best-known luminescent beetles are fireflies, or lightning bugs (which are neither flies nor bugs, but beetles), which use a flash of light to locate a prospective mate. Each species has its own characteristic flashing rhythm produced on the ventral side of the last abdominal segments. Females flash an answer to the species-specific pattern to attract males. This interesting "love call" has been adopted by species of *Photuris* (figure 13.52), which prey on male fireflies of other species they attract.

figure 13.52

A lightning "bug" is actually a beetle in the genus *Photuris*, family Lampyridae. A female emits flashes of light to attract members of the opposite sex for mating. However, females of certain species prey on males of other species. They flash in a pattern characteristic of the other species, but then consume the males arriving to mate.

Social Behavior In terms of being organized into social groups, insects rank very high in the animal kingdom, and cooperation within such complex groups depends heavily on chemical and tactile communication. However, not all social communities are complex. Some community groups are temporary and uncoordinated, such as hibernating associations of carpenter bees or feeding gatherings of aphids (figure 13.53). Some insects are coordinated for only brief periods, such as the tent caterpillars, *Malacosoma,* that join in building a home web and a feeding net. Still, all these are open communities with social behavior.

In true societies of some orders, such as Hymenoptera (honeybees and ants) and Isoptera (termites), a complex social life occurs. Such societies are closed. They involve all stages of the life cycle, communities are usually permanent, all activities are collective, and reciprocal communication occurs. Division of labor is highly efficient. Such a society is essentially a family group in which the mother or perhaps both parents remain with their young, sharing the duties of the group in a cooperative manner. The society usually demonstrates polymorphism, or **caste** differentiation.

Honeybees have one of the most complex social organizations in the insect world. Instead of lasting one season, their organization continues for a more or less indefinite period. As many as 60,000–70,000 honeybees may live in a single hive. Of these, there are three castes: a single, sexually mature female, or **queen**; a few hundred drones, which are sexually mature males; and thousands of workers, which are sexually inactive genetic females (figure 13.54).

Workers tend the young, secrete wax with which they build the six-sided cells of the honeycomb, gather nectar from flowers, manufacture honey, collect pollen, and ventilate and guard the hive. One drone, or sometimes more, fertilizes the queen during the mating flight, at which time enough sperm are stored in her seminal receptacle to last for her lifetime.

figure 13.53

At center, a *Myrmica* ant (order Hymenoptera) tends a group of aphids (order Hemiptera, suborder Sternorrhyncha). Ants in genus *Formica* appear at the bottom of the photo also tending aphids. The aphids feed copiously on plant juices and excrete the excess as a clear liquid rich in carbohydrates ("honey-dew"), which ants cherish as a food.

figure 13.54

Queen bee surrounded by her court. The queen is the only egg layer in the colony. The attendants, attracted by her pheromones, constantly lick her body. As food is transferred from these bees to others, the queen's presence is communicated throughout the colony.

Castes are determined partly by fertilization and partly by what is fed to the larvae. Drones develop parthenogenetically from unfertilized eggs (and consequently are haploid); queens and workers develop from fertilized eggs (and thus are diploid; see haplodiploidy, section 12.1: Pinworms). Female larvae that will become queens are fed royal jelly, a secretion from the salivary glands of nurse workers. Royal jelly differs from the "worker jelly" fed to ordinary larvae, but the components in it that are essential for queen determination have not yet been identified. Honey and pollen are added to the worker diet about the third day of larval life. Pheromones in "queen substance," which is produced by the queen's mandibular glands, prevent female workers from maturing sexually. Workers produce royal jelly only when the level of "queen substance" pheromone in the colony drops. This change occurs when the queen becomes too old, dies, or is removed. Then workers start enlarging a larval cell and feeding a larva royal jelly that produces a new queen.

Honeybees have evolved an efficient system of communication by which, through certain body movements, their scouts inform workers of the location and quantity of food sources.

Termite colonies contain several castes, consisting of fertile individuals, both males and females, and sterile individuals (figure 13.55). Some fertile individuals may have wings and leave the colony, mate, lose their wings, and start a new colony as king and **queen.** Wingless fertile individuals may, under certain conditions, substitute for the king or queen. Sterile members are wingless and become workers and soldiers. Soldiers, which have large heads and mandibles, defend the colony. As in bees and ants, extrinsic factors cause caste differentiation. Reproductive individuals and soldiers secrete inhibiting pheromones that pass throughout the colony to nymphs via a mutual feeding process called **trophallaxis,** so that they become sterile workers. Workers also produce pheromones, and if the level of "worker substance" or "soldier substance" falls, as might happen after an attack by marauding predators, for example, the next generation produces compensating proportions of the appropriate caste.

George D. Lepp/Science Source

A

Kjell Sandved/Science Source

B

figure 13.55

A, Five workers of the western subterranean termite, *Reticulitermes hesperus* (order Isoptera). Workers are wingless immature animals that tend the nest and care for the young. **B,** The termite queen becomes a distended egg-laying machine. A queen (*Microcerotermes* from Darwin, Australia) and several workers are shown.

Ants also have highly organized societies. Superficially, they resemble termites, but they are quite different (belong to a different order) and can be distinguished easily. In contrast to termites, ants are usually dark in color, are hard bodied, and have a constriction posterior to their first abdominal segment.

Key Theme 13.9

SCIENCE EXPLAINED

Describing Insect Societies

Entomologists describe insect societies by borrowing terms commonly used to describe human societies: queen, king, royal, soldier, worker, and caste. Such terms can be misleading by implying a correspondence between human and insect societies that does not exist. For example, in human societies "queen" suggests a position of political power that has no correspondence to the reproductive role of the "queen" of a bee colony. Confusion of these terms led a famous population geneticist, Ronald Fisher, to argue that human societies could achieve greater stability by emulating insect societies, specifically by concentrating reproduction among members of the upper classes. This argument is now considered an embarrassment to his otherwise highly regarded and influential book, *The Genetical Theory of Natural Selection* (1930).

In ant colonies, males die soon after mating, and the queen either starts her own new colony or joins an established colony and does the egg laying. Sterile females are wingless workers and soldiers that do the work of the colony—gather food, care for young, and protect the colony. Many larger colonies may have two or three types of individuals within each caste.

Ants have evolved some striking patterns of "economic" behavior, such as making slaves, farming fungi, herding "ant cows" (aphids or other hemipterans, see figure 13.53), sewing their nests together with silk (figure 13.56), and using tools.

Insects and Human Welfare

Beneficial Insects Although some think of insects primarily as pests, humanity would have great difficulty surviving if all insects were suddenly to disappear. Insects are necessary for cross-fertilization (pollination) of many crops. Bees pollinate over $14 billion worth of food crops per year in the United States alone, and this value does not include pollination of forage crops for livestock or pollination by other insects. In addition, some insects produce useful materials—for example, honey and beeswax from bees, silk from silkworms, and shellac from a wax secreted by lac insects.

Bee populations are in decline in many parts of the world, including the United States. Bee populations have fluctuated in ancient and modern times, but the disappearance of entire colonies that began around 2006 has had a significant

Larry S. Roberts

figure 13.56

A weaver ant nest in Australia.

impact on agriculture. In Colony Collapse Disorder, worker bees disappear and leave the queen stranded with little or no care for her offspring. The workers likely fall victim to the "deformed wing virus" (DWV), which is carried by *Varroa* mites and probably spread by human traffic in bee colonies (see section 13.3: Order Acari and associated references by Villalobos, and by Wilfert et al.).

Key Theme 13.10
EVOLUTION

A Behavioral Defense Against *Varroa* Mites and Colony Collapse Disorder

Four populations of honeybees (in France, Norway, and Sweden) have independently evolved a way to improve survivorship of developing worker bees. The mite parasites infect developing workers in brood chambers, as well as adult bees. In the four populations, workers uncap the brood chambers of developing worker bees and then recap them. Changes in humidity and other factors during the uncapped phase harm the mites, but recapping protects the developing worker bee. This behavior represents a form of "social immunity," a phenomenon occurring in eusocial insects when nestmates cooperate to reduce parasite success. The parallel evolution of this behavior in four distinct populations offers hope for bees worldwide.

Very early in their evolution, insects and flowering plants developed mutual adaptations that have functioned to each other's advantage. Insects exploit flowers for food, and flowers exploit insects for pollination. Each floral petal and sepal arrangement is correlated with the sensory adjustment of certain pollinating insects. Among these mutual adaptations are amazing allurements, traps, specialized structures, and precise timing.

Many predaceous insects, such as tiger beetles, aphid lions, ant lions, praying mantids, and lady beetles, destroy harmful insects (figure 13.57A and B). Some insects control harmful ones by parasitizing them or by laying their eggs where their young, when hatched, may devour the host (figure 13.57C). Dead animals are quickly consumed by maggots hatched from eggs laid on carcasses.

U.S. Department of Agriculture (USDA)

A

Scott Bauer/U.S. Department of Agriculture

B

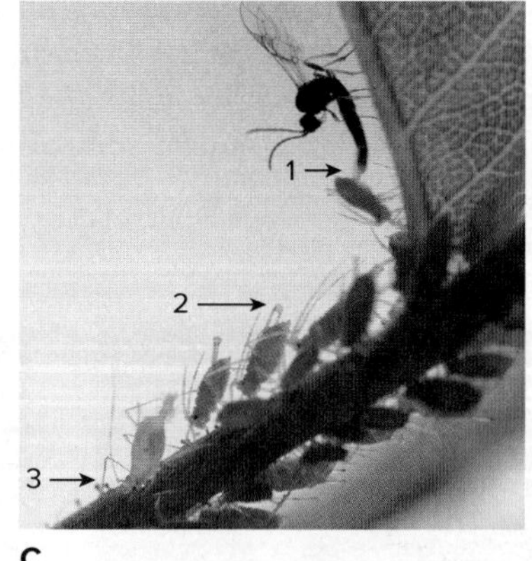

Fran Keller

C

figure 13.57

Some beneficial insects. **A,** A predaceous spined soldier bug (order Hemiptera) feeds on a larva of the Mexican bean beetle. Note the sucking proboscis of the bug. Mexican bean beetles are pests of snap beans and soybeans. **B,** A lady beetle ("ladybug," order Coleoptera). Adults (and larvae of most species) feed voraciously on plant pests such as mites, aphids, scale insects, and thrips. **C,** A parasitic braconid wasp (*Aphidius sp.*, order Hymenoptera) pierces an aphid (arrow 1) to lay her eggs inside. Nearby aphids release alarm pheromones (arrow 2). Aphids, such as *Macrosiphum rosae* shown here, damage plants by feeding on phloem; droplets of phloem from aphid feeding tubes on a rose bush are visible (arrow 3). The green aphid in the lower left is giving birth to live offspring.

A

B

C

figure 13.58

Insect pests. **A,** Japanese beetles, *Popillia japonica* (order Coleoptera), are serious pests of fruit trees and ornamental shrubs. They were introduced into the United States from Japan in 1917. **B,** Citrus mealybugs, *Planococcus citri* (order Hemiptera, suborder Sternorrhyncha). Many mealybugs are pests of commercially valuable plants. **C,** Corn ear worms, *Heliothis zea* (order Lepidoptera). An even more serious pest of corn is the infamous corn borer, an import from Europe in 1908 or 1909.

Insects and their larvae serve as an important source of food for many birds, fish, and other animals.

Harmful Insects Harmful insects include those that eat and destroy plants and fruits; examples are grasshoppers, chinch bugs, corn borers, boll weevils, grain weevils, San Jose scale, and scores of others (figure 13.58). Practically every cultivated crop has several insect pests. Humans expend enormous resources in all agricultural activities, in forestry, and in the food industry to counter insects and the damage they engender. Outbreaks of bark beetles or defoliators, such as spruce budworms and gypsy moths, have generated tremendous economic losses and have had a major impact on the composition of forests in the United States. Gypsy moths, introduced into the United States in 1869 in an ill-advised attempt to breed a better silkworm, have spread throughout the Northeast as far south as Virginia and as far west as Minnesota. In outbreak years, they defoliate oak forests; for example, in 1981, they defoliated 13 million acres in 17 northeastern states.

Lice, blood-sucking flies, warble flies, bot flies, and many others attack humans or domestic animals or both. Malaria (see figure 5.23), carried by the *Anopheles* mosquito (figure 13.59), is still one of the world's major diseases; mosquitoes also transmit yellow fever, lymphatic filariasis, West Nile virus (see Key Theme 13.11) and most recently Zika virus. Fleas carry plague, which at times in history has eradicated significant portions of the human population. House flies are vectors of typhoid, as are lice for typhus fever; tsetse flies carry African sleeping sickness (see section 5.5: Euglenozoa); and certain blood-sucking bugs, *Rhodnius* and related genera, transmit Chagas disease (see sction 5.5: Euglenozoa).

Tremendous destruction of food, clothing, and property is caused by weevils, cockroaches, ants, clothes moths, termites, and carpet beetles. Not the least of the insect pests are bed bugs, *Cimex,* blood-sucking hemipterous insects originally contracted by humans from bats that shared their caves early in human evolution. Bedbug infestation is increasing throughout the developed world, for unknown reasons. Proposed explanations include increased importation by travelers and reduced use of insecticides.

Key Theme 13.11

HUMAN CONNECTIONS

West Nile Virus

West Nile virus, a disease agent spread by mosquitoes, affects mammals and birds throughout the world. First identified in Uganda in 1937, it spread to North America in 1999. Birds serve as a reservoir for the virus: a bird bitten by an infected mosquito plays host to the virus for 1–4 days, during which the virus can be picked up by other mosquitoes and spread to new hosts. Human response to infection varies, with about 80% showing no symptoms, close to 20% exhibiting flu-like symptoms such as fever and achiness, and less than 1% developing potentially fatal encephalitis or other possibly permanent neurological effects. Preventing mosquito bites is the best way to avoid infection, so studies of mosquito behavior are useful. For example, researchers wonder whether mosquitoes that initially feed on infected birds tend to choose another bird for the next bite, or simply feed opportunistically on any available animal. Understanding virus transmission makes possible mathematical models to predict how and when the disease will spread.

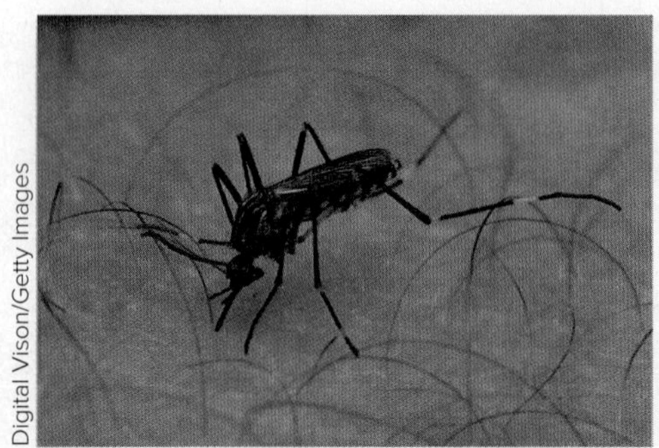

figure 13.59

A mosquito in genus *Aedes* (order Diptera) from South Africa. Members of this genus spread dengue fever and yellow fever, among other illnesses.

Control of Insects We now recognize that insects are an integral part of the ecological communities to which they belong. Without insects, food chains would be disturbed, many of our favorite birds would disappear, and the biological cycles by which dead animal and plant matter disintegrates and returns to enrich the soil would be seriously impeded. The beneficial roles of insects in our environment are often overlooked, and in our zeal to control the pests, we spray the landscape indiscriminately with extremely effective "broad-spectrum" insecticides that eradicate good, as well as harmful, insects. We know that many chemical insecticides persist in the environment and accumulate as residues in the bodies of animals higher in the food chain. Honeybees, on which we rely for pollination, are especially susceptible to insecticides. Furthermore, many insects have developed resistance to commonly used insecticides.

Methods of control other than chemical insecticides have been under intense investigation, experimentation, and development. Economics, concern for the environment, and consumer demand are causing farmers across the United States to use alternative methods to control insect pests, rather than depending strictly on chemicals.

Several types of biological controls have been developed and are under investigation. All of these areas present problems but also show great potential. One method uses bacterial, viral, and fungal pathogens. A bacterium, *Bacillus thuringiensis,* is quite effective in controlling lepidopteran pests (cabbage looper, imported cabbage worm, tomato worm, and gypsy moth). Other strains of *B. thuringiensis* attack insects in other orders, and the species diversity of target insects is being widened by genetic engineering techniques. Genes coding for the toxin produced by *B. thuringiensis* (Bt) have also been introduced into the DNA of the plants themselves, which makes the plants resistant to insect attack, while the Bt is harmless to humans. Many of our grain crops, especially corn, now possess genes that express proteins toxic to specific insect pests, thereby obviating the need for pesticides. However, some insects have now evolved resistance to the Bt toxin. Among the most promising methods for reducing the ability of mosquitoes to spread diseases such as dengue fever involves infecting mosquitoes with a bacterium, *Wolbachia,* that successfully competes with other viruses in the mosquitoes (see www.worldmosquitoprogram.org/en/work/wolbachia-method). Field tests of this method are underway now in several parts of the world.

Taxonomy of Subphylum Hexapoda

Hexapods are divided into orders chiefly by morphology including developmental features. Entomologists are not unanimous on the names of the orders or on the contents of each order. Some tend to combine and others to divide groups. However, the following synopsis of major orders is rather widely accepted.

Class Entognatha

Order Protura (pro-tu´ra) (Gr.*protos,* first, + *oura,* tail). Minute (1–1.5 mm); no eyes or antennae; appendages on abdomen as well as thorax; live in soil and dark, humid places; slight, gradual metamorphosis.

Order Diplura (dip-lu´ra) (Gr. *diploos,* double, + *oura,* tail): **japygids.** Usually less than 10 mm; pale, eyeless; have a pair of long terminal filaments or a pair of caudal forceps; live in damp humus or rotting logs; direct development.

Order Collembola (col-lem´bo-la) (Gr. *kolla,* glue, + *embolon,* peg, wedge): **springtails** and **snowfleas.** Small (5 mm or less); no eyes; respiration by trachea or body surface; use a springing organ folded under the abdomen for leaping; abundant in soil; sometimes swarm on pond surface film or on snowbanks in spring; direct development.

Class Insecta

Order Thysanura (thy-sa-nu´ra) (Gr. *thysanos,* tassel, + *oura,* tail): **silverfish** (figure 13.60) and **bristletails.** Small to medium size; large eyes; long antennae; three long terminal cerci; live under stones and leaves and around human habitations; wingless; direct development.

Order Ephemeroptera (e-fem-er-op´ter-a) (Gr. *ephēmeros,* lasting but a day, + *pteron,* wing): **mayflies** (figure 13.61). Wings membranous; forewings larger than hindwings; adult mouthparts vestigial; nymphs aquatic, with lateral tracheal gills; hemimetabolous development.

Order Odonata (o-do-na´ta) (Gr. *odontos,* tooth, + *ata,* characterized by): **dragonflies** (see figure 13.49B) and **damselflies.** Large; membranous wings are long, narrow, net-veined, and similar in size; long and slender body; aquatic nymphs have aquatic gills and prehensile labium for capturing prey; hemimetabolous development.

Order Orthoptera (or-thop´ter-a) (Gr. *orthos,* straight, + *pteron* wing): **grasshoppers** (see figure 13.35), **locusts, crickets,** and **katydids.** Wings, when present, have forewings thickened and hindwings folded like a fan under forewings; chewing mouthparts.

(continued)

figure 13.60

Silverfish, *Lepisma saccharina* (order Thysanura), is often found in homes.

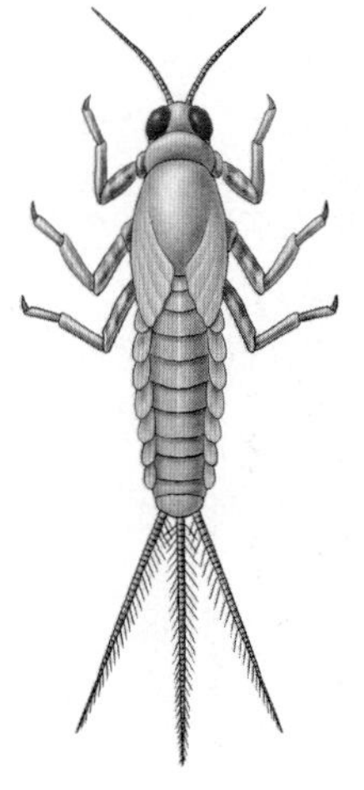

figure 13.61

Mayfly (order Ephemeroptera). **A,** Nymph. **B,** Adult mayfly, *Ephemera danica*.

Order Blattodea (blə-´tō-dēə) (L. *blatta,* cockroach, + Gr. *eidos,* form, + *ea,* characterized by): **cockroaches.** Common insects in tropical arcas; often infest houses in northern areas; oval, flattened bodies may exceed 5 cm in length; tarsi with 5 segments; wings typically present, often reduced.

Order Phasmatodea (faz-mə-´tō-dēə) (Gr. *phasma,* apparition, + *eidos* form, + *ea,* characterized by): **walkingsticks** and **leaf insects.** Bodies elongated and sticklike or flattened and laterally expanded; herbivorous, tropical forms may be very large (up to 30 cm).

Order Mantodea (man-´tō-dēə) (Gr. *mantis,* soothsayer, + *eidos,* form, + *ea,* characterized by): **mantids** (see figure 13.36). Bodies elongated with raptorial front legs; predatory; may reach 10 cm in length.

Order Mantophasmatodea (man-´tō-faz-mə-´tō-dēə) (an amalgamation of the order names for praying mantids [Mantodea] and walkingsticks [Phasmatodea]): **gladiators.** Secondarily wingless; chewing mouthparts; resemble a combination of a praying mantis and a walkingstick; nocturnal predators on insects and spiders; described in 2002; rare, found in Africa; 6–8 species.

Order Dermaptera (der-map´ter-a) (Gr. *derma,* skin, + *pteron,* wing): **earwigs.** Very short forewings; large and membranous hindwings folded under forewings when at rest; chewing mouthparts; forceps-like cerci.

Order Plecoptera (ple-kop´ter-a) (Gr. *plekein,* to twist, + *pteron,* wing): **stoneflies** (see figure 13.49A). Membranous wings; larger and fanlike hindwings; aquatic nymph with tufts of tracheal gills.

Order Isoptera (i-sop´ter-a) (Gr. *isos,* equal, + *pteron,* wing): **termites** (see figure 13.55). Small; membranous, narrow wings similar in size with few veins; wings shed at maturity; erroneously called "white ants"; distinguishable from true ants by broad union of thorax and abdomen; complex social organization; hemimetabolous development.

Order Embiidina (em-bē-ə´ di-nə) (Gr. *embios,* lively, + *eidos,* form, + *ina* resembling): **webspinners.** Small; male wings membranous, narrow, and similar in size; wingless females; chewing mouthparts; colonial; make silk-lined channels in tropical soil.

Order Psocoptera (so-cop´ter-a) (Gr. *psoco,* rub away, + *pteron,* wing) **(Corrodentia): psocids, book lice,** and **bark lice.** Body usually small, may be as large as 10 mm; membranous, narrow wings with few veins, usually held rooflike over abdomen when at rest; some wingless species; found in books, bark, bird nests, on foliage.

Order Zoraptera (zo-rap´ter-a) (Gr. *zōros,* pure, + *apteryos,* wingless): **zorapterans.** As large as 2.5 mm; membranous, narrow wings usually shed at maturity; colonial and termite-like.

Order Phthiraptera (th¯ı-rap´ter-a) (Gr. *phteir,* louse, + *apteros,* wingless): **lice.** Wingless ectoparasites adapted for clinging to warm-blooded hosts. **Sucking lice** (see figure 13.42) in former order Anoplura now constitute suborder Anoplura, mouthparts adapted for piercing and sucking, includes head lice, crab lice, and body lice. **Chewing lice** (see figure 13.41) in former order Mallophaga now divided among three suborders.

Order Thysanoptera (thy-sa-nop´ter-a) (Gr. *thysanos,* tassel, + *pteron,* wing): **thrips.** Length 0.5–5 mm (a few longer); wings, if present, long, very narrow, with few veins, and fringed with long hairs; sucking mouthparts; destructive plant-eaters, but some feed on insects.

Order Hemiptera (he-mip´ter-a) (Gr. *hemi,* half, + *pteron,* wing). Members have unique mouthparts specialized for piercing and sucking. Hemiptera is divided into three suborders: Heteroptera, Auchenorrhyncha, and Sternorrhyncha. Heteroptera contains **true bugs;** size 2–100 mm; wings present or absent; forewings with basal portion thickened and partly sclerotized; apical portion membranous; hindwings membranous; at rest, wings held flat over abdomen; many with odorous scent glands; includes water scorpions, water striders, bedbugs, squash bugs, assassin bugs, chinch bugs, stink bugs, soldier bugs (see figure 13.57A), plant bugs, lace bugs, and many others. Auchenorrhyncha contains **hoppers** (figure 13.62) and **cicadas** (see figure 13.50); four wings typical if wings are present. Sternorrhyncha contains **whiteflies, psyllids, aphids, mealybugs** (see figure 13.58B), and **scale insects;** four wings

(*continued*)

(continued)

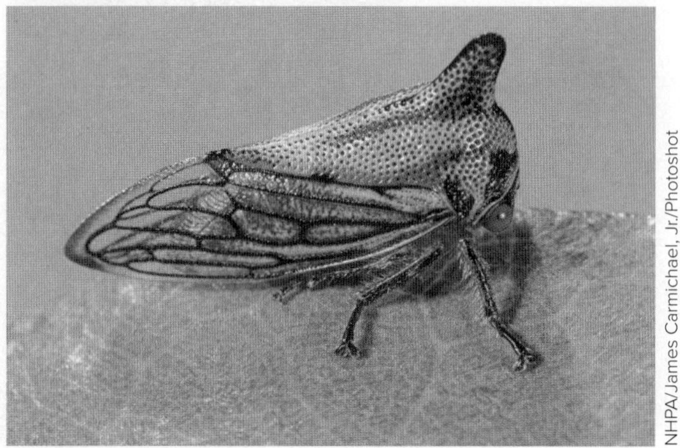

figure 13.62

An oak treehopper, *Platycotis vittata* (order Hemiptera, suborder Auchenorrhyncha).

typical if wings are present; often have complex life histories; many species are plant pests.

Order Neuroptera (neu-rop´ter-a) (Gr. *neuron,* nerve, + *pteron,* wing): **dobsonflies, ant lions** (figure 13.63), and **lacewings.** Medium to large size; similar, membranous wings with many cross veins; chewing mouthparts; dobsonflies have greatly enlarged mandibles in males and aquatic larvae; ant lion larvae (doodlebugs) make craters in sand to trap ants; holometabolous development.

Order Coleoptera (ko-le-op´ter-a) (Gr. *koleos,* sheath, + *pteron,* wing): **beetles** (see figure 13.58A and B), **fireflies** (see figure 13.52), and **weevils.** The largest order of animals; forewings (elytra) thick, hard, opaque; membranous hindwings folded under forewings at rest; mouthparts for biting and chewing; includes ground beetles, carrion beetles, whirligig beetles, darkling beetles, stag beetles, dung beetles (see figure 13.51), diving beetles, boll weevils, fireflies, lady beetles (ladybugs), others; holometabolous development.

Order Strepsiptera (strep-sip´ter-a) (Gr. *strepsis,* a turning, + *pteron,* wing): **stylops** or **twisted wing parasites.** Females wingless, without eyes or antennae; males have vestigial forewings and fan-shaped hindwings; females and larvae parasites of bees, wasps, and other insects.

Order Mecoptera (me-kop´ter-a) (Gr. *mekos* length, + *pteron,* wing): **scorpionflies.** Small to medium size; wings long, slender, with many veins; at rest, wings held rooflike over back; males have scorpion-like clasping organ at end of abdomen; carnivorous; live in most woodlands.

Order Lepidoptera (lep-i-dop´ter-a) (Gr. *lepidos,* scale, + *pteron,* wing): **butterflies** and **moths** (see figure 13.47 and figure 13.58C [larval form]). Membranous wings covered with overlapping scales, wings coupled at base; mouthpart is a sucking tube, coiled when not in use; larvae (caterpillars) have chewing mandibles for eating plants, stubby prolegs on the abdomen, and silk glands for spinning cocoons; antennae knobbed in butterflies and usually plumed in moths; holometabolous development.

figure 13.63

Adult ant lion, *Myrmeleon formicarius* (order Neuroptera).

Order Diptera (dip´ter-a) (Gr. *dis,* two, + *pteron,* wing): **true flies.** Single pair of wings, membranous and narrow; hindwings reduced to inconspicuous balancers (halteres); sucking mouthparts or adapted for sponging, lapping, or piercing; legless larvae called maggots or, when aquatic, wigglers; include crane flies, mosquitoes (see figure 13.59), moth flies, midges, fruit flies, flesh flies, house flies (see figure 13.37), horse flies, bot flies, blow flies, gnats, and many others; holometabolous development.

Order Trichoptera (tri-kop´ter-a) (Gr. *trichos,* hair, + *pteron,* wing): **caddisflies.** Small, soft-bodied; wings well-veined and hairy, folded rooflike over hairy body; chewing mouthparts; aquatic larvae construct cases of leaves, sand, gravel, bits of shell, or plant matter, bound together with secreted silk or cement; some make silk feeding nets attached to rocks in streams; holometabolous development.

Order Siphonaptera (si-fon-ap´ter-a) (Gr. *siphon,* a siphon, + *apteros,* wingless): **fleas** (see figure 13.40). Small; wingless; bodies laterally compressed; legs adapted for leaping; no eyes; ectoparasitic on birds and mammals; larvae legless and scavengers; holometabolous development.

Order Hymenoptera (hi-men-op´ter-a) (Gr. *hymen,* membrane, + *pteron,* wing): **ants, bees** (see figure 13.54), and **wasps** (see figure 13.46). Very small to large; membranous, narrow wings coupled distally; subordinate hindwings; mouthparts for lapping liquids and biting; ovipositor sometimes modified into stinger, piercer, or saw; both social and solitary species; most larvae legless, blind, and maggot-like; holometabolous development.

Introduction of natural predators or parasites of the insect pests has had some success. In the United States, vedalia beetles from Australia help control the cottony-cushion scale on citrus plants, and numerous instances of control by using insect parasites have been recorded. However, the introduction of exotic species to control insect pests may have unexpected negative consequences, and should be done with caution.

Another approach to biological control is to interfere with the reproduction or behavior of insect pests by sterilizing males or by using naturally occurring organic compounds that act as hormones or pheromones. Such research, although very promising, is slow because of our limited understanding of insect behavior and the problems of isolating and identifying complex compounds that an insect produces in such minute amounts. Nevertheless, pheromones may play an important role in biological pest control in the future.

A systems approach called integrated pest management is practiced with many crops. This approach involves integrating all possible, practical techniques to contain pest infestations at a tolerable level—for example, cultural techniques (resistant plant varieties, crop rotation, tillage techniques, timing of sowing, planting, or harvesting, and others), use of biological controls, and sparing use of insecticides.

13.9 Phylogeny and Adaptive Diversification

Phylogeny

Biologists hypothesize that the ancestral arthropod had a segmented body with one pair of appendages per segment. During evolution, adjacent segments fused to make body regions (tagmata). How many segments contributed to a head in each group of arthropods? A partial answer to this question comes from studies of *Hox* genes that are highly correlated with specific anterior–posterior differentiation of body parts in bilateral animals. *Hox* gene studies indicate that the first five segments, at least, fused to form the head tagma in all four extant subphyla. It is surprising to find the same pattern of fusion in chelicerates as in other subphyla because a head is not immediately obvious in a chelicerate. Spider bodies have two tagmata: prosoma, or cephalothorax, and opisthosoma, or abdomen. Is the head part of the prosoma? *Hox* gene comparisons indicate that the entire prosoma corresponds to the head of other arthropods.

Another controversial area of arthropod biology for which genetic studies have proved helpful involves the evolution and antiquity of uniramous and biramous appendages. Hexapods and myriapods have uniramous appendages, but trilobites and some crustaceans have biramous appendages. If the ancestral appendage were biramous, then the switch to uniramous appendages might have occurred in one lineage whose descendants now carry this trait. Such reasoning led biologists to group hexapods with myriapods, but phylogenies using molecular characters repeatedly placed hexapods with crustaceans. Is it likely that the uniramous limb evolved more than once? This question would be more easily answered if the genetic basis of limb structure were understood. It is now known that modulation in expression of the *Distal-less (Dll)* gene determines the location of distal ends of arthropod limbs. In each primordial (embryonic) biramous appendage, the gene product of *Dll* can be observed in two groups of cells, each of which becomes a branch of the limb. In a uniramous limb primordium, there is only one such group of cells, and in primordia of phyllopodous limbs (as in class Branchiopoda), there are as many groups expressing *Dll* as there are limb branches. Gene expression can be modified within a lineage, so the number of limb branches is not likely to be an homologous character.

The best explanation for the similarities between crustaceans and insects, such as the basic structure of their ommatidia,

Taxonomy of Phylum Arthropoda

Subphylum Trilobita (tri´lo-bi´ta) (Gr. *tri-*, three, + *lobos*, lobe): **trilobites.** All extinct forms; Cambrian to Permian periods; body divided by two longitudinal furrows into three lobes; distinct head, thorax, and abdomen; biramous appendages.

Subphylum Chelicerata (ke-lis´e-ra´ta) (Gr. *chēle*, claw, + *keratos*, a horn): **eurypterids, horseshoe crabs, spiders,** and **ticks.** First pair of appendages modified to form chelicerae; pair of pedipalps and four pairs of legs; no antennae, no mandibles; cephalothorax and abdomen often with segments fused.

Class Merostomata (mer´o-sto´ma-ta) (Gr. *meros*, thigh, + *stomatos*, mouth): **aquatic chelicerates.** Cephalothorax and abdomen; compound lateral eyes; appendages with gills; sharp telson; **subclasses Eurypterida** (all extinct) and **Xiphosurida,** the horseshoe crabs.

Class Pycnogonida (pik´no-gon´i-da) (Gr. *pyknos*, compact, + *gonia*, knee, angle): **sea spiders.** Small (3–4 mm), but some reach 500 mm; body chiefly cephalothorax; tiny abdomen; usually four pairs of long walking legs (some with five or six pairs); one pair of subsidiary legs (ovigers) for egg bearing; mouth on long proboscis; four simple eyes; no respiratory or excretory system. Example: *Pycnogonum.*

Class Arachnida (ar-ack´ni-da) (Gr. *arachnē*, spider): **scorpions, spiders, mites, ticks,** and **harvestmen.** Four pairs of legs; segmented or unsegmented abdomen with or without appendages and generally distinct from cephalothorax;

(continued)

(*continued*)

respiration by gills, tracheae, or book lungs; excretion by Malpighian tubules or coxal glands; dorsal bilobed brain connected to ventral ganglionic mass with nerves; simple eyes; sexes separate; chiefly oviparous; no true metamorphosis. Examples: *Argiope, Centruroides.*

Subphylum Crustacea (crus-ta´she-a) (L. *crusta,* shell, + *acea,* group suffix): **crustaceans.** Mostly aquatic, with gills; cephalothorax usually with dorsal carapace; biramous appendages, modified for various functions; head appendages consist of two pairs of antennae, one pair of mandibles, and two pairs of maxillae; sexes usually separate; ancestral development pattern with nauplius stage.

Oligostraca

Ostracoda (ästrə' kōdə) (Gr. *ostrakodes,* having a shell). **Ostracods**. Bivalve carapace entirely encloses body; body unsegmented or indistinctly segmented; no more than two pairs of trunk appendages. Examples: *Cypris, Cypridina, Gigantocypris.*

Branchiura (bran-'kēyurə) (Gr. *branchia,* gills, + *ura,* tail). **Fish lice**. Body oval, head and most of trunk covered by flattened carapace, incompletely fused to first thoracic segment; thorax with four pairs of appendages, biramous; abdomen unsegmented, bilobed; eyes compound; antennae and antennules reduced; maxillules often forming suctoral discs. Examples: *Argulus, Chonopeltis.*

Pentastomida (pen-ta-stōm'i-da) (Gr. *pente,* five, + *stoma,* mouth). **Pentastomids**. Wormlike unsegmented body with five short anterior protuberances, four bear claws and the fifth bears the sucking mouth. Examples: *Armillifer, Linguatula.*

Vericrustacea

Branchiopoda (bran-kēä'pōdə) (Gr. *branchia,* gills, + *pous, podos,* foot). Phyllopodia; carapace present or absent; no maxillipeds; antennules reduced; compound eyes present; no abdominal appendages; maxillae reduced.

Anostraca (ənästrəkə) (Gr. *an-,* prefix meaning without, + *ostrakon,* shell). **Fairy shrimp** and **brine shrimp**. No carapace; no abdominal appendages; uniramous antennae. Examples: *Artemia, Branchinecta*

Notostraca (nōtästrəkə) (Gr. *nōtos,* the back, + *ostrakon,* shell). **Tadpole shrimp**. Carapace forming large dorsal shield; abdominal appendages present, reduced posteriorly; antennae vestigial. Examples: *Triops, Lepidurus.*

Diplostraca (diplōstrəkə) (Gr. *diploos,* double, + *ostrakon,* shell). **Water fleas** (cladocerans) and **clam shrimps** (conchostracans). Carapace folded, usually enclosing trunk but not head (cladocerans) or enclosing entire body (conchostracans); biramous antennae. Examples: *Daphnia, Leptodora, Lynceus.*

Copepoda (kōpe-'pədə) (Gr. *kōpē,* oar, + *pous, podos,* foot). **Copepods**. No carapace; thorax typically of seven segments, of which first and sometimes second fuse with head to form cephalothorax; antennules uniramous; antennae bi- or uniramous; four to five pairs swimming legs; parasitic forms often highly modified. Examples: *Cyclops, Diaptomus, Calanus, Ergasilus, Lernaea, Salmincola, Caligus.*

Thecostraca (Cirripedia) (sirə-'pēdēə) (L. *cirrus,* curl of hair, + *pes, pedis,* foot). **Barnacles**. Sessile or parasitic as adults; head reduced and abdomen rudimentary; paired compound eyes absent; body segmentation indistinct; usually hermaphroditic; in free-living forms carapace becomes mantle, which secretes calcareous plates; antennules become organs of attachment, then disappear. Examples: *Balanus, Policipes, Sacculina.*

Malacostraca (malə-'kä-strəkə) (Gr. *malakos,* soft, + *ostrakon,* shell)**: isopods, amphipods, krill,** and **decapods**. Usually with eight segments in thorax and six plus telson in abdomen; all segments with appendages; antennules often biramous; first one to three thoracic appendages often maxillipeds; carapace covering head and part or all of thorax, sometimes absent; gills usually thoracic epipods. Examples: *Armadillidium, Gammarus, Megacytiphanes, Grapsus, Homarus, Panulirus.*

Subphylum Myriapoda (mir-ē-a´pod-ə) (Gr. *myrias,* a myriad, + *podos,* foot): **myriapods.** All appendages uniramous; head appendages consist of one pair of antennae, one pair of mandibles, and one or two pairs of maxillae.

Class Diplopoda (dip´lō-pod-ə) (Gr. *diploos,* double, + *pədos,* foot): **millipedes.** Subcylindrical body; head with short antennae and simple eyes; body with variable number of segments; short legs, usually two pairs of legs to a segment; separate sexes. Examples: *Julus, Spirobolus.*

Class Chilopoda (ki´-lō-pod-ə) (Gr. *cheilos,* lip, + *podos,* foot): **centipedes.** Dorsoventrally flattened body; variable number of segments, each with one pair of legs; one pair of long antennae; separate sexes. Examples: *Cermatia, Lithobius, Geophilus.*

Class Pauropoda (pāu-rō-pod-ə) (Gr. *pauros,* small, + *podos,* foot): **pauropods.** Minute (1–1.5 mm), cylindrical body consisting of double segments and bearing nine or ten pairs of legs; no eyes. Example: *Pauropus.*

Class Symphyla (sim´fi-la) (Gr. *syn,* together, + *phylon,* tribe): **garden centipedes.** Slender (1–8 mm) with long, threadlike antennae; body consists of 15–23 segments with 10–12 pairs of legs; no eyes. Example: *Scutigerella.*

Subphylum Hexapoda (hek´sa-pod-ə) (Gr. *hex,* six, + *podos,* foot): **hexapods.**Body has distinct head, thorax, and abdomen; pair of antennae; mouthparts modified for different food habits; head composed of six fused segments; thorax has three segments; abdomen has variable number of segments, usually 11 somites; thorax has two pairs of wings (sometimes one pair or none) and three pairs of jointed legs; separate sexes; usually oviparous; gradual or abrupt metamorphosis.

Class Entognatha (en ´tog-na-tha) (Gr. *entos,* within, inside + *gnathos,* jaw): **entognaths.** Base of mouthparts lies within head capsule; mandibles have one articulation. Example: *Entomobrya.*

Class Insecta (in-sek ´ta) (L. *insectus,* cut into): **insects.** Bases of mouthparts exposed and exiting head capsule; mandibles generally have two regions of articulation. Examples: *Drosophila, Bombus, Anopheles* (insect orders listed in Taxonomy of Subphylum Hexapoda, placed near the end of section 13.8.).

is that these two taxa form a clade within Arthropoda. We depict crustaceans and hexapods as sister taxa (see figure 13.4), but some evidence suggests that hexapods arose from *within* the crustacean group, making Crustacea paraphyletic unless it includes hexapods. Insects may be the descendants of freshwater branchiopods. Crustacea now includes members of the former phylum Pentastomida as a subclass. The taxon Pancrustacea includes crustaceans and hexapods.

Phylogenetic placement of subphylum Myriapoda is controversial. Molecular phylogenies supported both the "mandibulata" and "myriochelata" hypotheses. However, a phylogeny incorporating data from fossils and molecular characters in living taxa supports the "mandibulata" hypothesis as shown in figure 13.4.

Evolution within hexapods involved specialization of the first three postcephalic body segments (somites) to become locomotor segments (thorax) and a loss or reduction of appendages on the rest of the body (abdomen). The wingless orders traditionally have been regarded as having the most ancestral characteristics. Three wingless orders (Diplura, Collembola, and Protura) have their mandibles and first maxillae located deeply in pouches in the head, a condition called **entognathy.** All other insects are **ectognathous,** including the wingless order Thysanura. Ectognathous insects do not have their mandibles and maxillae in pouches, and they share other synapomorphies. Entognathous and ectognathous insects form sister groups, and Thysanura diverged from a common ancestor of ectognathous insects before the advent of flight, which unites the remaining ectognathous orders (see figure 13.33).

The earliest phylogenetic split among winged insects separates three taxa that differ in their ability to flex their wings (see figure 13.33). Two of these (Odonata and Ephemeroptera) have outspread wings. The other taxon, which has wings that can fold back over the abdomen, branched into three groups: one with hemimetabolous metamorphosis and chewing mouthparts (Orthopterodea); one group with hemimetabolous metamorphosis and usually sucking mouthparts (Hemipterodea); and a group with holometabolous metamorphosis (Holometabola).

Adaptive Diversification

The adaptive trend in arthropods has been toward tagmatization of the body by differentiation or fusion of segments, giving rise to such tagmata as head and trunk; head, thorax, and abdomen; or cephalothorax (fused head and thorax) and abdomen. A series of similar appendages, one pair on each trunk segment, is the ancestral character state, still retained by some crustaceans and by myriapods. Derived forms include appendages specialized for specific functions and some appendages lost entirely.

Much of the amazing diversity in arthropods seems to have developed because of modification and specialization of their cuticular exoskeleton and their jointed appendages, thus producing a wide variety of locomotor and feeding adaptations. Whether in the area of habitat, feeding adaptations, means of locomotion, reproduction, or general mode of living, the adaptive achievements of arthropods are truly remarkable.

■ SUMMARY

Sections	Key Concepts
13.1 Why are Arthropods So Diverse and Abundant?	• Arthropoda is the largest, most abundant, and diverse phylum of animals. Arthropods occur in virtually all habitats capable of supporting life. • Arthropods are segmented, coelomate, ecdysozoan protostomes with well-developed organ systems. Most show marked tagmatization. • Perhaps more than any other single factor, prevalence of arthropods is explained by adaptations made possible by their cuticular exoskeleton and small size. Other important elements are jointed appendages, tracheal respiration, efficient sensory organs, complex behavior, and metamorphosis. • According to the mandibulate hypothesis, Crustacea and Hexapoda are sister taxa forming the group Pancrustacea. • Pancrustacea is the sister taxon of Myriapoda, together forming the group Mandibulata. • Chelicerata is the sister taxon to Mandibulata. Chelicerate mouthparts differ significantly from those in Mandibulata.
13.2 Subphylum Trilobita	• Trilobites were common animals in the Cambrian and Ordovician periods. They went extinct at the end of the Permian period. • Trilobites were named for three lobes on the body that are separated by longitudinal grooves. • The presence of trilobite species is useful for biostratigraphy.

(*continued*)

Sections	Key Concepts
13.3 Subphylum Chelicerata	• Members of subphylum Chelicerata have no antennae, and their main feeding appendages are chelicerae. • They have a pair of pedipalps (which may be similar to the walking legs) and four pairs of walking legs. • Class Merostomata includes the extinct eurypterids and the ancient, although still extant, horseshoe crabs. • Class Pycnogonida comprises the sea spiders—odd little animals with a large suctorial proboscis and vestigial abdomen. • The great majority of living chelicerates are in class Arachnida. This group comprises spiders (order Araneae), scorpions (order Scorpiones), harvestmen (order Opiliones), ticks and mites (order Acari), and others. • Spiders are predaceous, and their chelicerae are provided with venom glands for paralyzing or killing prey. They breathe by book lungs, tracheae, or both. • Most spiders spin silk, which they use for a variety of purposes, including webs for trapping prey. Spider silk is not homologous to the commercially valuable silk produced by certain moth larvae. • Distinctive characters of scorpions are their large, claw-like pedipalps and their clearly segmented abdomen, which bears a terminal stinging apparatus. • Harvestmen have small, ovoid bodies with very long, slender legs. Their abdomen is segmented and broadly joined to their cephalothorax. • In ticks and mites, the cephalothorax and abdomen are completely fused, and mouthparts are borne on an anterior capitulum. Like spiders, some mites can spin silk. • Ticks are the most numerous arachnids; some are important carriers of disease such as Lyme disease and Rocky Mountain Spotted Fever, and others are serious plant pests.
13.4 Subphylum Myriapoda	• Members of subphylum Myriapoda have a head followed by a series of trunk segments. • The most familiar myriapods are predatory centipedes and herbivorous millipedes.
13.5 Subphylum Crustacea	• Crustacea is a large, primarily aquatic subphylum. • Crustaceans have a pair of mandibles, two pairs of antennae, and two pairs of maxillae. Their tagmata are a head and trunk or a head, thorax, and abdomen. Many have a carapace. Crustacean appendages are ancestrally biramous. • Arthropods must periodically cast off their old cuticle (ecdysis) and grow in size before the newly secreted cuticle hardens. Premolt and postmolt periods are hormonally controlled, as are several other processes, such as change in body color and expression of sexual characteristics. • Feeding habits vary greatly in Crustacea; there are predators, scavengers, suspension feeders, and parasites. Some even obtain nutrition from chemosynthetic bacteria. • Respiration is through the body surface or by gills, and excretory organs take the form of maxillary or antennal glands. • Circulation, as in other arthropods, is through an open system of sinuses (hemocoel), and a dorsal, tubular heart is the chief pumping organ. • Crustaceans have a double ventral nerve cord, statocysts, and tactile sensory setae. Most crustaceans have compound eyes composed of units called ommatidia.
13.6 Taxa within Subphylum Crustacea	• There are two major groups of crustaceans: Oligostraca and Vericrustacea. We discuss only some group members. • Members of Oligostraca discussed are Ostracoda, Branchiura, and Pentastomida. • Branchiura contains fish lice. Closely related to fish lice are tongue worms (former members of phylum Pentastomida); they are parasitic in the lungs and nasal cavities of vertebrates. • Members of Vericrustacea discussed are Branchiopoda, Copepoda, Thecostraca, and Malacostraca. • Branchiopoda is characterized by phyllopodia and contains, among others, order Diplostraca, ecologically important as zooplankton. • Copepods lack a carapace and abdominal appendages. They are abundant and are among the most important of the primary consumers in many freshwater and marine ecosystems. Many are parasitic. • Most members of Thecostraca (barnacles) are sessile as adults, secrete a shell of calcareous plates, and filter-feed by means of their thoracic appendages • Malacostraca is a diverse and familiar group whose important orders are Isopoda, Amphipoda, Euphausiacea, and Decapoda. All have both abdominal and thoracic appendages. • Isopods lack a carapace and are usually dorsoventrally flattened. • Amphipods lack a carapace but are usually laterally flattened. • Euphausiaceans are important oceanic plankton called krill. • Decapods include crabs, shrimps, lobsters, crayfishes, and others; they have five pairs of walking legs (including chelipeds) on their thorax.

Sections	Key Concepts
13.7 Subphylum Hexapoda	• Hexapoda comprises classes Entognatha and Insecta. • In entognathans, the base of the mouthparts is inside the head capsule, whereas in insects the base of the mouthparts extends out of the head capsule. • Insecta is the largest class of the world's largest phylum. • Insects have three tagmata: head, thorax, and abdomen. • Most insects bear two pairs of wings on their thorax, although some have one pair and some are primitively or secondarily wingless. • The evolutionary success of insects in terrestrial habitats is largely explained by features such as wings (present in most), waterproofing of their cuticle and other mechanisms to minimize water loss, and the ability to become dormant during adverse conditions. • Wing movements in some insects are controlled by direct flight muscles, which insert directly on the base of the wings in the thorax, whereas others have indirect flight muscles, which move the wings by changing the shape of the thorax. • Feeding habits vary greatly among insects, and there is an enormous variety of specialization of mouthparts reflecting the particular feeding habits of a given insect. • Insects breathe by means of a tracheal system, which is a system of tubes that brings oxygen directly to each cell. Tracheae open by spiracles on the thorax and abdomen. • There is an open circulatory system. • Excretory organs are Malpighian tubules, located in the hemocoel. • Insects perceive the surrounding world visually, chemically, and via mechano-sensory structures. • Sexes are separate in insects, and fertilization is usually internal. • Almost all insects undergo metamorphosis during development. • In hemimetabolous (incomplete) metamorphosis, larval instars are called immatures or nymphs and resemble small adults in many ways. Sexually mature adults emerge at the last molt. • In holometabolous (complete) metamorphosis, larvae are often caterpillar or maggot-like and do not resemble adults. The last larval molt gives rise to a nonfeeding stage (pupa). A winged adult (sexually mature) emerges at the final pupal molt.
13.8 Insect Interactions	• Insects have excellent sensory mechanisms. They respond to environmental information with complex behaviors such as cutting and rolling dung or building tiny clay pots in which to lay eggs. • Insects are capable of learning; they learn efficient routes to food or other resources. • Insects communicate chemically, through sounds, by producing visual signals, and through touch. • Social behavior can be extremely complex. In eusocial colonies, castes are created through chemical control during larval development. • Insects are important to human welfare, particularly because they pollinate food and forage crop plants. • Insects control populations of other, harmful insects by predation and parasitism. • Insects serve as food for other animals and are key players in many food webs. • Many insects are harmful to human interests because they damage crops, food, forests, clothing, and property. • Many insects are carriers of important diseases affecting humans and domestic animals. • Insects are controlled though chemical pesticides as well as biological control agents such as bacteria, other insects, or use of sterilization to reduce population sizes.
13.9 Phylogeny and Adaptive Diversification	• The ancestral arthropod is hypothesized to have had a segmented body with one pair of appendages per segment. • Tagmatization, the evolutionary fusion of segments to make body regions called tagma, has led to bodies with two or three tagma. • Work with the *Distal-less (Dll)* gene led to the discovery that uniramous and biramous appendages are modulated by expression of a single gene and not likely to be a strong phylogenetic character. • Hexapods and crustaceans are depicted as sister taxa, but some evidence suggests that hexapods arose from within Crustacea. • Winged insects are divided among three groups based on the ability to flex the wings. • Arthropods have adapted to life in almost every part of the world, on land, in freshwater, and in the sea, and feed in a myriad of ways. Their flexible body plan and hardened cuticle make possible this diversity. Adaptive diversity and the numbers of both species and individuals in Insecta are enormous.

Review Questions

1. List some characteristics of arthropods that clearly distinguish them from Annelida.
2. Name the subphyla of arthropods, and give a few examples of each.
3. Much of the success of arthropods has been attributed to their cuticle. Why? Describe some other factors that probably contributed to their success.
4. What is a trilobite?
5. What appendages characterize chelicerates?
6. Briefly describe the appearance of each of the following: eurypterids, horseshoe crabs, pycnogonids.
7. Describe the mechanism of each of the following with respect to spiders: feeding, excretion, sensory reception, webspinning, reproduction.
8. Distinguish each of the following orders from each other: Araneae, Scorpionida, Opiliones, Acari.
9. People fear spiders and scorpions, but ticks and mites are far more important medically and economically. Why? Give examples.
10. The only venomous crustacean known is a small, blind, cave-dwelling animal called a remipedian. Which other arthropods use venoms?
11. What features of pentastomids support their current phylogenetic position as close relatives of fish lice (branchiurans)?
12. Distinguish among Ostracoda, Copepoda, and Thecostraca.
13. Copepods sometimes are called "insects of the sea" because marine planktonic copepods probably are the most abundant animals in the world. What is their ecological importance?
14. Define each of the following: swimmeret, maxilliped, cheliped, nauplius.
15. Describe molting in Crustacea, including the action of hormones.
16. Explain the mechanism of each of the following with respect to crustaceans: feeding, respiration, excretion, circulation, sensory reception, reproduction.
17. Distinguish the following from each other: Diplopoda, Chilopoda, Insecta.
18. Explain why wings powered by indirect flight muscles can beat much more rapidly than those powered by direct flight muscles.
19. What different modes of feeding do insects use, and how are these reflected in their mouthparts?
20. Describe each of the following with respect to insects: respiration, excretion and water balance, sensory reception, reproduction.
21. Explain the difference between holometabolous and hemimetabolous metamorphosis in insects, including the stages of each.
22. Describe insect learning and explain cultural transmission of learning in honeybees.
23. What castes occur in honeybees and in termites, and what is the function of each? What is trophallaxis?
24. Describe several ways in which insects benefit humans and several ways in which they are detrimental.
25. For the past 50 or more years, people have relied on toxic insecticides for control of harmful insects. What problems have arisen as a result? What are the alternatives? Describe integrated pest management and the use of *Wolbachia* to control viruses spread by mosquitoes.
26. We hypothesize that the earliest insects were wingless, making lack of wings the ancestral condition. Does winglessness characterize a hexapod class? In what sense is it useful for classification?
27. What is Colony Collapse Disorder, why is it of concern at a worldwide level, and how might honeybees be protected?

For Further Thought Filter feeding is a very common way to collect food from water, but this method is rarely used on land. Why might web-spinning spiders be considered filter feeders?

14

Echinoderms and Hemichordates

LEARNING OBJECTIVES

Readers will be able to:

14.1 Identify the members of clade Ambulacraria and describe the features that unite its members.

14.2 Describe the functional anatomy of the five echinoderm groups: sea stars, brittle stars, sea urchins, sea cucumbers, and crinoids.

14.3 Describe the symmetry of the earliest echinoderms and explain the lifestyle and associated selection factors favoring radial symmetry in the ancestors of modern forms.

14.4 Describe the functional anatomy of the two hemichordate types: acorn worms and pterobranchs.

Corey Hochachka/Design Pics

Sea stars (*Pisaster ochraceus*) surrounded by green algae at low tide.

A Design to Puzzle the Zoologist

Libbie Hyman, a distinguished American zoologist, once described echinoderms as a "noble group especially designed to puzzle the zoologist." With a combination of characteristics that should delight the most avid reader of science fiction, echinoderms would seem to confirm Lord Byron's observation that

> Tis strange—but true;
> for truth is always strange;
> Stranger than fiction.

Some of the earliest fossil echinoderms were bilaterally symmetrical. This, in conjunction with their bilaterally symmetrical larvae, indicates that the current radial symmetry is derived. Despite the limitations of radial symmetry for free-moving animals, echinoderms are among the most abundant organisms in some marine habitats.

A key feature of echinoderms is the unique water-vascular system that uses hydraulic pressure to operate a multitude of tiny tube feet used in food gathering and in locomotion. Dermal ossicles can fuse together to invest echinoderms in endoskeletal armor or can be reduced to microscopic plates in soft-bodied forms. Many echinoderms have miniature jawlike pincers (pedicellariae) scattered on their body, often stalked and sometimes equipped with venomous glands.

This constellation of characteristics is unique in the animal kingdom. Perhaps strangest of all is that the closest living relatives of chordates, a group that includes humans, are the echinoderms and wormlike hemichordates, and not the more cephalized arthropods.

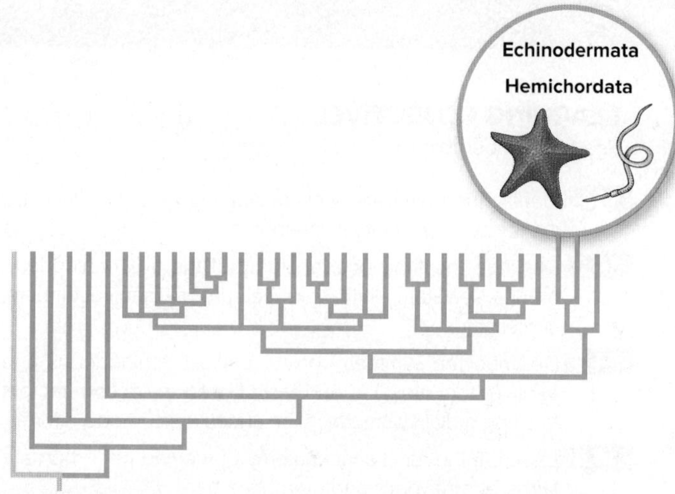

Triploblastic animals are divided into two large clades: Protostomia and Deuterostomia. These clades are diagnosed by a combination of molecular and morphological features, with many morphological differences visible in early development. Classical developmental characters associated with protostomes are spiral mosaic cleavage, formation of the mouth from the embryonic blastopore (protostomy), and formation of a coelom by **schizocoely**, when a coelom is present (see figure 3.13). Classical deuterostome developmental features include radial regulative cleavage, formation of the mouth from a second opening (deuterostomy), and a true coelom, formed by **enterocoely**.

Recent analyses also highlight the importance of the pharyngeal gill-slit complex as a synapomorphy for Deuterostomia. The deuterostome ancestor is now envisioned as a filter-feeder, ingesting small food particles transported via ciliary movement of water into its mouth and out its slits.

Members of some phyla possess all the developmental characters in each suite: marine annelids and molluscs are typical protostomes, and echinoderms are typical deuterostomes. However, readers of the preceding chapters know that some taxa, such as the lophophorates and chaetognaths, have a mix of protostome and deuterostome features. In this chapter, we discuss two deuterostome phyla, Echinodermata and Hemichordata. These two phyla comprise the clade Ambulacraria (figure 14.1).

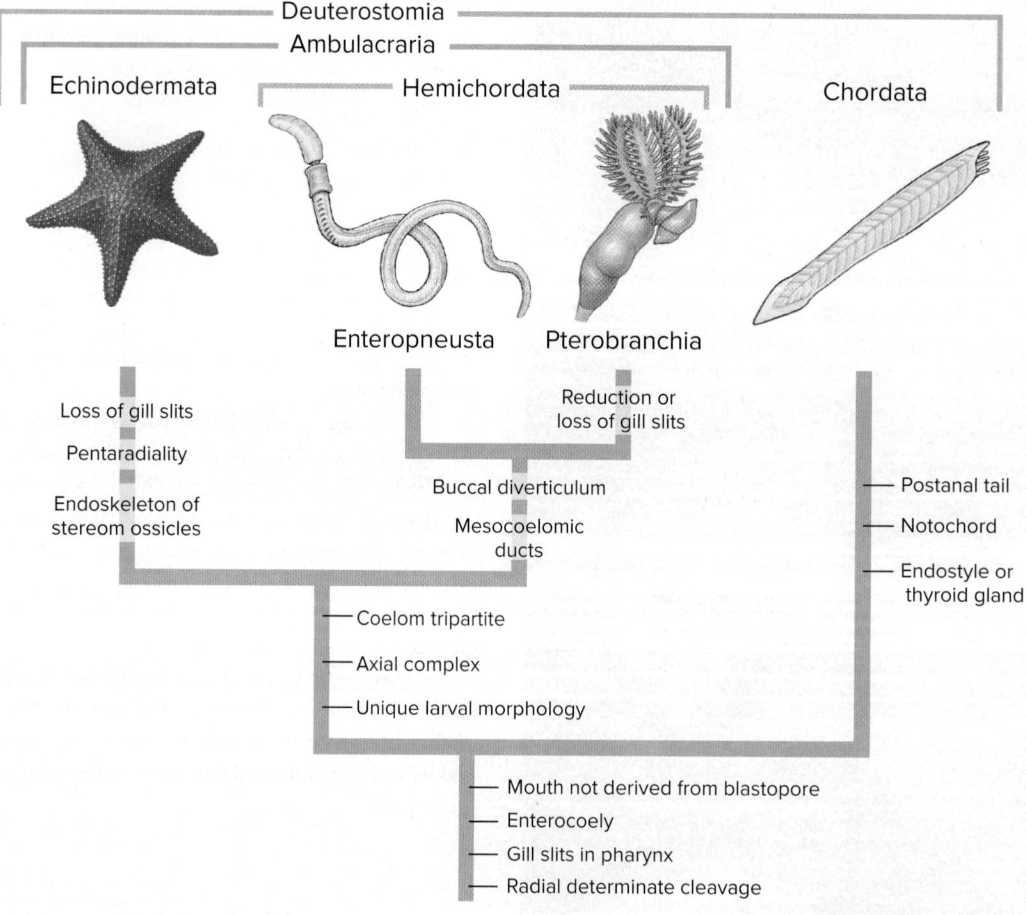

figure 14.1

Cladogram showing hypothetical relationships among deuterostome phyla. Based on this hypothesis, the ancestral deuterostome was a marine form with gill slits. Gill slits apparently were present in stem echinoderms, but lost in modern echinoderms.

14.1 Clade Ambulacraria

Ambulacraria contains two deuterostome phyla: Echinodermata and Hemichordata (see figure 14.1). Echinoderms, including sea stars and sea urchins, are familiar to many people, but hemichordates, including acorn worms and pterobranchs, are much less familiar.

The relationship of hemichordates to other deuterostome phyla has been puzzling. With chordates, they share pharyngeal slits. With echinoderms, they share a similar ciliated larva, a three-part coelom, and a kind of filtration structure called an axial complex. Phylogenetic analyses of 18S rDNA sequences and *Hox* genes support a close relationship between echinoderms and hemichordates. Nearly all zoologists now agree that echinoderms and hemichordates form a clade, Ambulacraria, which is the sister taxon to chordates (see figure 14.1).

The Ambulacraria hypothesis suggests gill slits in the pharynx are an ancestral character for deuterostomes. Gill slits are presumed lost in some pterobranch hemichordates and in all living echinoderm groups, but apparently were present in some of the earliest, extinct echinoderms.

14.2 Phylum Echinodermata

Members of Echinodermata (ē-kī′nō-der′mä-tä) (L. *echinatus*, prickly, + Gr. *derma*, skin) are marine forms and include sea stars (also called starfishes), brittle stars, sea urchins, sea cucumbers, and sea lilies. They have a combination of characteristics found in no other phylum: (1) a calcareous endoskeleton of large plates or small ossicles, (2) a water-vascular system, (3) pedicellariae, (4) dermal branchiae, and (5) basic pentaradial symmetry. Echinoderms have an excellent fossil record, which extends back to at least the early Cambrian period, but reconstruction of the early evolution of echinoderms has been difficult. It seems clear that they descend from bilateral ancestors because their larvae are bilateral but become radially symmetrical later in development. Recent fossils of benthic, bilaterally symmetrical adult echinoderms from the middle Cambrian period further support the idea that pentaradial symmetry is derived. Many early, fossil echinoderms were sessile (attached) forms with radial symmetry (figure 14.2). Sessile forms (all in the Class Crinoidea) still survive, but most living forms are free-moving.

Ecological Relationships

Echinoderms are all marine; they have no ability to osmoregulate and therefore rarely inhabit brackish waters. Virtually all are benthic as adults, inhabiting all oceans of the world and all depths, from intertidal to abyssal regions.

Some sea stars (figure 14.3) are particle feeders, but many are predators, feeding particularly on sedentary or sessile prey. Brittle stars (see figure 14.11A) are the most active echinoderms, moving by their arms. They may be scavengers, browsers, or deposit or filter feeders. Some are commensal in large sponges. Compared to other echinoderms, sea cucumbers (see figure 14.18) are greatly extended in the oral-aboral axis and are oriented with that axis more or less parallel to their substrate and lying on one side. Most are suspension or deposit feeders. "Regular" sea urchins (see figure 14.14), which are radially symmetrical, prefer hard substrates and feed chiefly on algae. "Irregular" urchins (sand dollars and heart urchins) (see figure 14.15), which are secondarily bilateral, are usually found on sand and feed on detritus. Crinoids (see figures 14.21 and 14.22) stretch their arms outward and up like a flower's petals, anchor themselves to the substrate, and feed on plankton and suspended particles.

Class Asteroidea: Sea Stars

Sea stars, often called starfishes, demonstrate basic features of echinoderm structure and function. We cover them first, and then comment on major differences among other echinoderm groups.

Sea stars are familiar along shorelines, where large numbers sometimes aggregate on rocks. There are about 1500 living species. They also live on muddy or sandy substrates on the ocean floor and among coral reefs. They are often brightly colored and range in size from a centimeter to about a meter in greatest diameter.

Form and Function

External Features Reflecting their pentaradial symmetry, sea stars typically have five tapering arms (rays), but there may be more (figure 14.3). These merge gradually with the central disc (figure 14.4). The mouth is on the oral surface; the side opposite the mouth is called the aboral surface.

A band from which **tube feet (podia)** project, called an **ambulacrum** (pl., **ambulacra**, L. *ambulacrum*, an alley), extends from the mouth along the oral sides of each arm to its tip. An ambulacral groove occurs along the middle of each ambulacrum and contains rows of tube feet, which are bordered by movable spines (figure 14.4). Viewed from the oral side, a large radial nerve can be seen in the center of each ambulacral groove (figure 14.5C), between the rows of tube feet. This nerve is very superficially located, covered only by thin epidermis. Under the nerve is an extension of the coelom and the radial canal of the water-vascular system (figure 14.5C). In all other classes of living echinoderms except crinoids, ossicles or other dermal tissue cover these structures; thus, the ambulacral grooves in most asteroids and crinoids are *open*, and those of the other groups are *closed*.

The aboral surface is usually rough and spiny, although spines of many species are flattened, so that the surface appears smooth. Around the bases of spines in many sea stars are groups of minute, pincerlike **pedicellariae** (ped-i-cel-ar′e-ē),

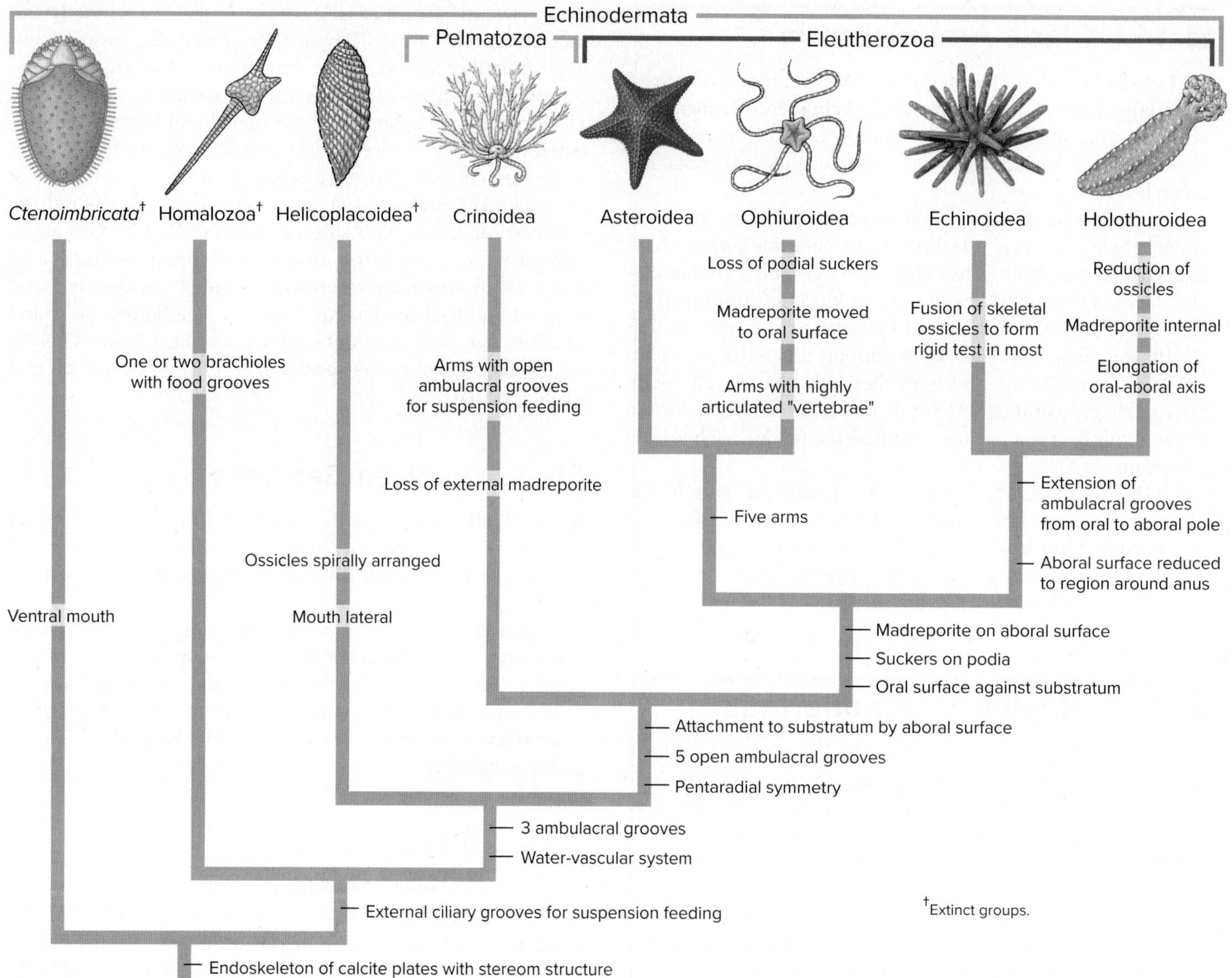

figure 14.2

Cladogram showing hypothetical relationships among echinoderm groups. The extinct Cambrian *Ctenoimbricata* was bilateral, with a stereom endoskeleton of calcite ossicles and a downward-directed mouth, presumably for deposit feeding. The extinct Homalozoa (carpoids) were asymmetrical, mostly sessile forms. The extinct helicoplacoids were radial, with three ambulacral grooves that wound around their bodies in spiral fashion. Pentaradial symmetry was an adaptation to sessile existence and is a synapomorphy for modern echinoderms. The cladogram presented here depicts ophiuroids in a clade with asteroids, with five arms as a synapomorphy. An alternate scenario unites Ophiuroidea, Echinoidea, and Holothuroidea in a clade, with closed ambulacral grooves as a synapomorphy.

bearing tiny jaws manipulated by muscles (figure 14.6). These jaws keep the body surface free of debris, protect the papulae, and sometimes aid in food capture. **Papulae** (pa-pū-lē) (also called **dermal branchiae** or skin gills) are soft, delicate projections of the coelomic cavity, covered only with epidermis and lined internally with peritoneum; they extend outward through spaces between ossicles (see figure 14.5C). Papulae greatly increase the surface area available for respiratory gas exchange. Also on the aboral side are an inconspicuous **anus** and a circular **madreporite** (see figure 14.4A), a calcareous sieve leading to the water-vascular system.

Endoskeleton and Coelom Beneath the epidermis of sea stars is a mesodermal endoskeleton of small calcareous plates, or **ossicles**, bound together with connective tissue. This connective tissue is an unusual form of mutable collagen, called **catch collagen**, that can quickly change from a "liquid" to a "solid" form. This characteristic allows echinoderms to hold various postures without muscular effort. From these ossicles project spines and tubercles that form the spiny surface. Ossicles are penetrated by a meshwork of spaces, usually filled with fibers and dermal cells. This internal meshwork structure is described as **stereom** (see figure 14.19) and is unique to echinoderms.

figure 14.3

Some sea stars (class Asteroidea) from the Pacific. **A,** Pincushion star, *Culcita novaeguineae*, preys on coral polyps and also eats other small organisms and detritus. **B,** Crown-of-thorns star, *Acanthaster planci*, a major coral predator (see Key Theme 14.1), can have 10–20 arms. **C,** Necklace star, *Fromia monilis*, browses encrusting organisms. **D,** Granulated sea star, *Choriaster granulatus*, scavenges dead animals on shallow Pacific reefs.

figure 14.4

External anatomy of a sea star, an asteroid. **A,** Aboral view. **B,** Oral view.

figure 14.5

A, Internal anatomy of a sea star. **B,** Water-vascular system. (Polian vesicles are not present in *Asterias*.) **C,** Cross section of an arm at level of gonads, illustrating open ambulacral groove.

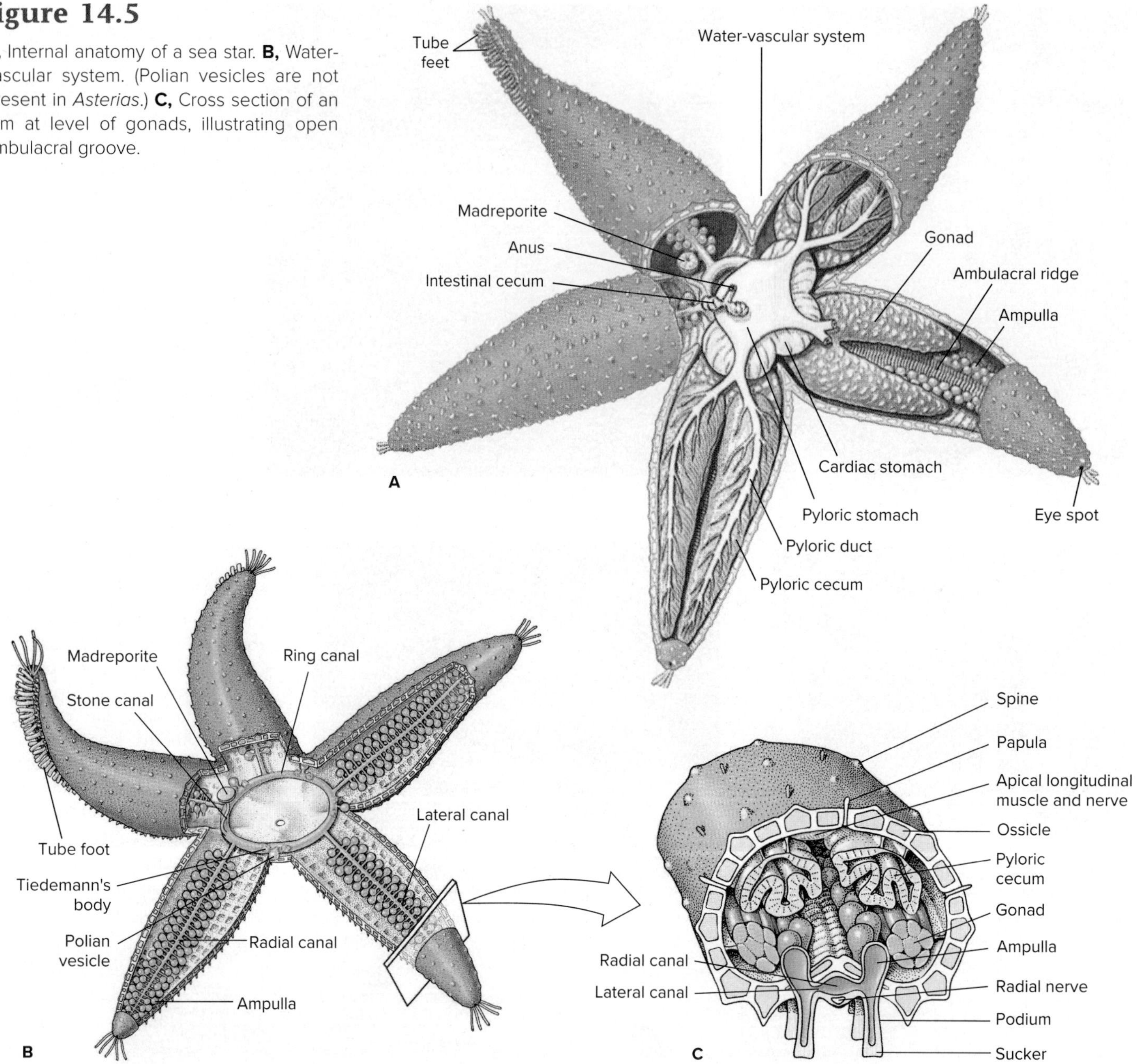

Coelomic compartments of larval echinoderms give rise to several structures in adults, one of which is a spacious body **coelom** filled with fluid. Coelomic fluid circulates around the body cavity and into the papulae, propelled by cilia on the peritoneal lining. Exchange of respiratory gases and excretion of nitrogenous waste, principally ammonia, occur by diffusion through the thin walls of papulae and tube feet.

Water-Vascular System The water-vascular system is another coelomic compartment and is unique to echinoderms. It is a set of canals and specialized tube feet that, together with the dermal ossicles, forms a hydraulic system. In sea stars, the primary functions of the water-vascular system are locomotion and food gathering, in addition to respiration and excretion.

Structurally, the water-vascular system opens to the outside through small pores in the madreporite. The madreporite of asteroids is on the aboral surface (see figure 14.4A), and leads into a **stone canal**, which descends toward a **ring canal** around the mouth (see figure 14.5B). **Radial canals** diverge from the ring canal, one into the ambulacral groove of each ray. **Polian vesicles** also attach to the ring canal of most asteroids (but not *Asterias*) and apparently serve as fluid reservoirs for the water-vascular system.

A series of small **lateral canals**, each with a one-way valve, connects the radial canal to the cylindrical podia (tube feet), along the sides of the ambulacral groove in each ray. Each podium is a hollow, muscular tube, the inner end of which is a muscular sac, or **ampulla**, which lies within the body coelom

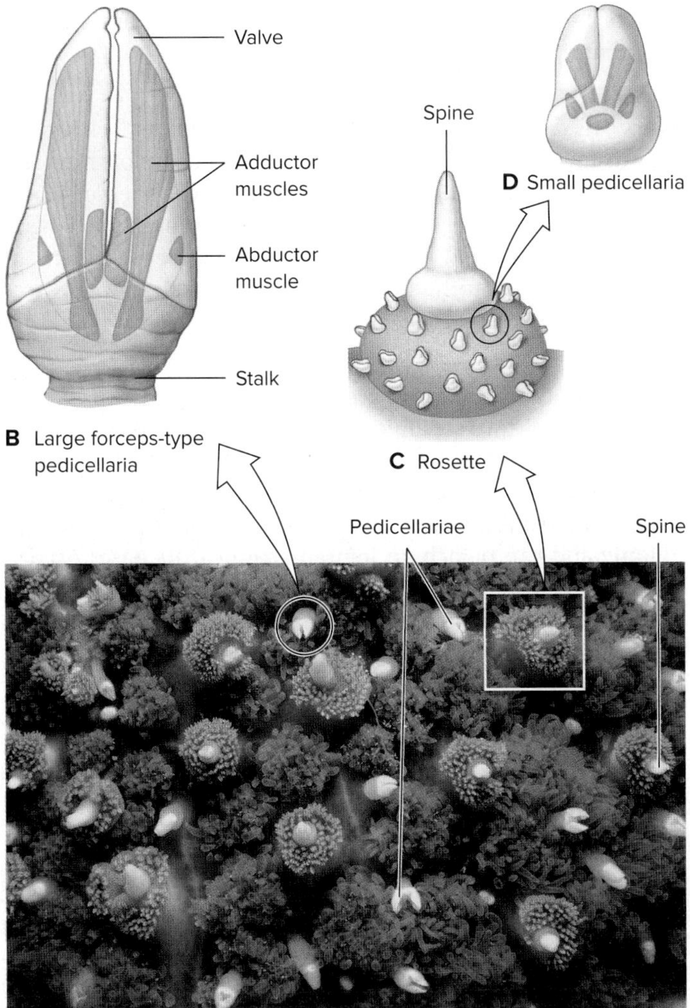

Sunflower Sea Star.

A Close-up view of the aboral surface of the sea star *Pycnopodia helianthoides.*

figure 14.6

Pedicellariae of sea stars. **A,** Close-up view of the aboral surface of the sea star *Pycnopodia helianthoides*. Note the large pedicellariae, as well as grops of small pedicellariae around the spines. Many thin-walled papulae (gray) can be seen. **B,** Forceps-type pedicellaria with associated muscles. **C,** A rosette of small pedicellariae are shown at the base of a spine; note the relative size of the spine. **D,** Structure of a small pedicellaria.

(see figure 14.5), and the outer end of which usually bears a sucker. Podia pass to the outside between ossicles in the ambulacral groove.

The water-vascular system operates hydraulically as an effective locomotor mechanism. Valves in the lateral canals prevent backflow of fluid into the radial canals. Tube feet have in their walls connective tissue that maintains the cylinder at a relatively constant diameter. Contraction of muscles in an ampulla forces fluid into the tube foot, extending it. Conversely, contraction of longitudinal muscles in the tube foot retracts the podium, forcing fluid back into the ampulla. Contraction of muscles in one side of the tube foot bends the organ toward that side. Small muscles at the end of the tube foot can raise the middle of the disclike end, creating a suction-cup effect when the end is applied to a firm substrate. By combining mucous adhesion with suction, a single tube foot can exert a pull equal to 0.25 to 0.3 newtons. Coordinated action of all or many of the tube feet is sufficient to draw a sea star up a vertical surface or over rocks. On a soft surface, such as mud or sand, suckers are ineffective (numerous sand-dwelling species have no suckers), so the tube feet are employed as legs.

Feeding and Digestive System The mouth of a sea star leads into a two-part stomach located in the central disc (see figure 14.5A). The large, lower cardiac stomach can be everted through the mouth during feeding. The smaller upper pyloric stomach connects with pyloric ceca (digestive glands) located in the arms. Digestion is mostly extracellular, occurring in pyloric ceca. A short intestine leads from the stomach to the inconspicuous anus on the aboral side. Some species lack an intestine and anus.

Many sea stars are relatively unselective carnivores, feeding on molluscs, crustaceans, polychaetes, echinoderms, other invertebrates, and rarely small fishes, but some show particular preferences. Some select brittle stars, sea urchins, or sand dollars, swallow their prey whole, and later regurgitate undigestible ossicles and spines. Others attack other sea stars, and if they are small compared to their prey, they may attack and begin eating at the end of one of the prey's arms.

Some asteroids feed heavily on molluscs, and *Asterias* is a significant predator on commercially important mussels and oysters. When feeding on a bivalve, a sea star wraps around its prey, attaching its podia to the valves, and then exerts a steady pull, using its feet in relays. A force of some 12.75 newtons can thus be exerted. In half an hour or so, the adductor muscles of the bivalve fatigue and relax. With a very small gap available, the star inserts its soft, everted stomach into the space between the valves, wraps it around the soft parts of the bivalve, and secretes digestive enzymes to begin digestion. After feeding, the sea star draws its stomach back inside its body by contracting the stomach muscles and relaxing the body-wall muscles.

Some sea stars feed on small particles, either entirely or in addition to carnivorous feeding. Tiny plankton or other organic particles coming in contact with a sea star's surface are carried by epidermal cilia to ambulacral grooves and then to the mouth.

Key Theme 14.1
ECOLOGY

Crown-of-thorns Starfish as Coral Predators

Numbers of crown-of-thorns starfish (see figure 14.3B) appear to have increased in recent decades, damaging large areas of coral reef in the Pacific Ocean. Crown-of-thorns stars (*Acanthaster planci* [Gr. *akantha*, thorn, + *asteros*, star]) feed on coral polyps. When their numbers are low, these sea stars kill only a few corals (usually the faster-growing ones), increasing reef diversity. However, at high densities they kill nearly all corals. Increasing outbreaks of these sea stars appear to be linked to nutrient-rich runoff, which increases the plankton on which the larval sea stars feed.

CHARACTERISTICS of Phylum Echinodermata

1. Unique **water-vascular system** of coelomic origin extends from body surface as a series of tentacle-like projections (**podia** or **tube feet**) protracted by an increase of fluid pressure within them; opening to exterior (**madreporite** or **hydropore**) usually present.
2. Live in marine habitats.
3. Free-living taxa.
4. Body unsegmented (nonmetameric) with **pentaradial symmetry**; body rounded, cylindrical, or star-shaped, with five or more radiating areas, or **ambulacra**, alternating with interambulacral areas; no head.
5. Triploblastic.
6. Coelom enterocoelous and extensive, forming perivisceral cavity and cavity of water-vascular system; coelomic fluids circulated by peritoneal cilia.
7. **Endoskeleton** of **dermal calcareous ossicles** with **spines** or of calcareous **spicules** in dermis; covered by epidermis (ciliated in most); **pedicellariae** (in some).
8. Digestive system usually complete; axial or coiled; anus absent in ophiuroids.
9. Skeletal elements connected by ligaments of mutable collagenous tissue under neural control, ligaments can be "locked" into rigid posture or relaxed to allow free movement; locomotion by **tube feet**, which project from **ambulacra**, by movement of spines, or by movement of arms, which project from the body's central disc.
10. Nervous system with circumoral ring and radial nerves; usually two or three systems of networks located at different levels in the body, varying in degree of development according to group.
11. **No brain**; few specialized sensory organs; sensory system of tactile and chemoreceptors, podia, terminal tentacles, photoreceptors, and statocysts.
12. Autotomy and regeneration of lost parts conspicuous; asexual reproduction by fragmentation in some.
13. Sexes separate (except a few hermaphroditic) with large gonads, single in holothuroids but multiple in most; simple ducts, with no elaborate copulatory apparatus or secondary sexual structures; fertilization usually external; eggs brooded in some; development through **free-swimming, bilateral, larval stages** (some with direct development); metamorphosis to radial adult or subadult form; radial cleavage and regulative development.
14. **Excretory organs absent.**
15. Respiration by **papulae, tube feet, respiratory tree** (holothuroids), and **bursae** (ophiuroids).
16. Blood-vascular system (**hemal system**) much reduced, playing little if any role in circulation, and surrounded by extensions of coelom (**perihemal sinuses**).

Hemal System The so-called hemal system is composed of tissue strands enclosing unlined channels and is itself enclosed in another coelomic compartment, the perihemal channels. The hemal system is not well developed in asteroids, and in all echinoderms its function is unclear. It has little use in circulation of body fluids; some research suggests it plays a role in distributing digested nutrients.

Nervous and Sensory System Echinoderms lack a brain and distinct ganglia. The nervous system consists of three subsystems, each formed by a nerve ring and radial nerves placed at different levels in the disc and arms. An epidermal nerve plexus, or nerve net, connects the systems. Sense organs include an eyespot at the tip of each arm and sensory cells scattered over the epidermis.

Reproductive System, Regeneration, and Autotomy Most sea stars have separate sexes. A pair of gonads lies in each interradial space (see figure 14.5A), and fertilization is external.

Echinoderms can regenerate lost parts. Sea star arms can regenerate readily, even if all are lost. Sea stars also exhibit **autotomy** and can detach an injured arm near its base. An arm may take months to regenerate.

If a removed arm contains a part of the central disc (about one-fifth), the arm can regenerate a complete new sea star! In former times, fishermen dispatched sea stars collected from their oyster beds by chopping them in half with a hatchet—a futile activity. Some sea stars reproduce asexually under normal conditions by cleaving the central disc, each part regenerating the rest of the disc and missing arms (see figure 14.7).

Development Some species produce benthic egg masses in which juveniles develop. Other species brood their eggs, either under the oral side of the animal or in specialized aboral structures, and development is direct. However, most sea stars produce free-swimming larvae.

Early embryogenesis shows a typical ancestral deuterostome pattern. Gastrulation is by invagination, and the anterior end of the archenteron pinches off to become a coelomic cavity,

Larry S. Roberts

figure 14.7

The Pacific sea star, *Echinaster luzonicus*, can reproduce itself by splitting across the disc, and then regenerating missing arms. The one shown here has evidently regenerated six arms from the longer one at top left.

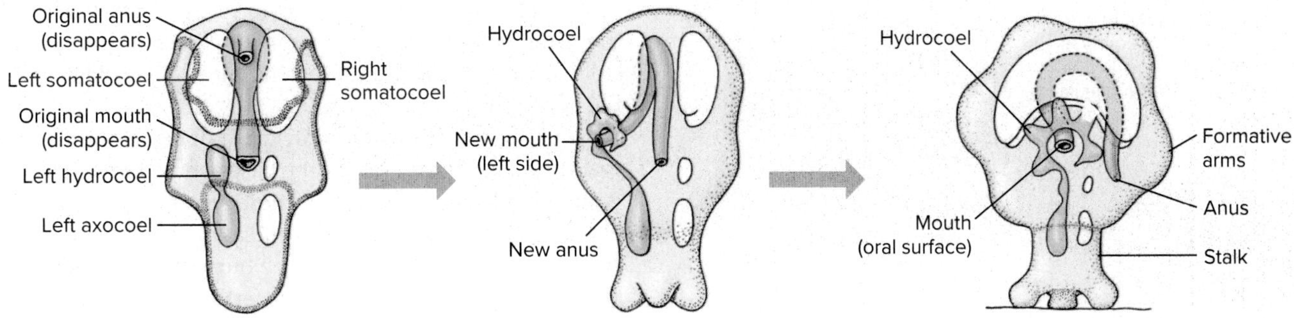

figure 14.8

Asteroid metamorphosis. The left somatocoel becomes the oral coelom, and the right somatocoel becomes the aboral coelom. The left hydrocoel becomes the water-vascular system, and the left axocoel the stone canal and perihemal channels. The right axocoel and hydrocoel disappear.

which expands in a U shape to fill the blastocoel. Each leg of the U, at the posterior end, constricts to become a separate vesicle, and these eventually give rise to the main coelomic compartments of the body (metacoels, called **somatocoels** in echinoderms) (figure 14.8). The anterior portions of the U undergo subdivision to form protocoels and mesocoels (called **axocoels** and **hydrocoels** in echinoderms). The left hydrocoel becomes the water-vascular system, and the left axocoel gives rise to the stone canal and perihemal channels. The right axocoel and hydrocoel disappear. The free-swimming larva, called a **bipinnaria** (figure 14.9A), has cilia arranged in bands. These ciliated tracts become extended into larval arms. The larva grows three adhesive arms and a sucker at its anterior end and is then called a **brachiolaria** (figure 14.9B). It attaches to the substrate with a temporary stalk and undergoes metamorphosis.

Metamorphosis involves dramatic reorganization of a bilateral larva into a radial juvenile. The anteroposterior axis of the larva is lost. *The larval left side becomes the oral surface, and the larval right side becomes the aboral surface* (see figure 14.8). Correspondingly, the larval mouth and anus disappear, and a new mouth and anus form on what were originally the left and right sides, respectively. As internal reorganization proceeds, short, stubby arms and the first podia appear. The animal then detaches from its stalk and begins life as a young sea star.

Sea Daisies Strange little (less than 1 cm diameter) disc-shaped animals, sometimes called sea daisies (figure 14.10), were discovered in water more than 1000 meters deep off New Zealand. The three species in this group are pentaradial but have no arms, and their tube feet are around the periphery of the disc, rather than along the ambulacral areas. Their water-vascular system includes two concentric ring canals; the outer ring may represent radial canals because podia arise from it. A hydropore, homologous to the madreporite, connects the inner ring canal to the aboral surface. Although they were originally described (1986) as a new class of echinoderms (Concentricycloidea), phylogenetic analysis of rDNA places them within Asteroidea.

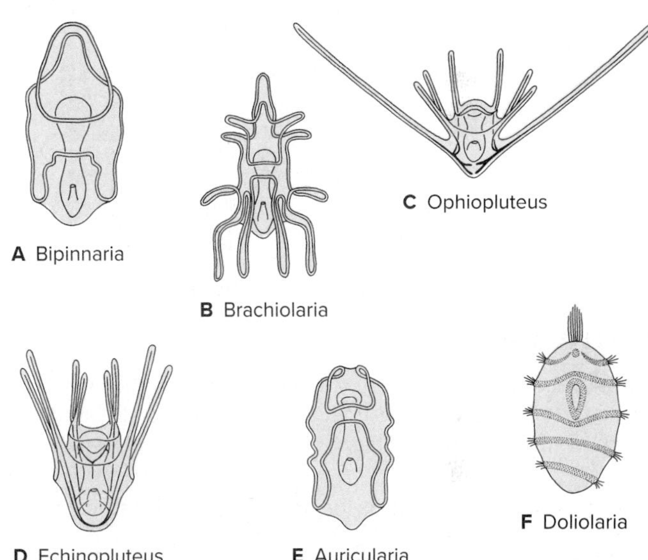

figure 14.9

Larvae of echinoderms. **A,** Bipinnaria of asteroids. **B,** Brachiolaria of asteroids. **C,** Ophiopluteus of ophiuroids. **D,** Echinopluteus of echinoids. **E,** Auricularia of holothuroids. **F,** Doliolaria of crinoids.

Class Ophiuroidea: Brittle Stars and Basket Stars

Brittle stars are the largest group of echinoderms, with over 2000 extant species, and they are probably the most abundant also. They abound in all types of benthic marine habitats, even carpeting the abyssal sea floor in many areas.

Apart from their typical possession of five arms, brittle stars are surprisingly different from asteroids. The arms of brittle stars are slender and sharply set off from the central disc (figure 14.11). They have no pedicellariae or papulae, and their ambulacral grooves are closed and covered with arm ossicles. Their tube feet, which are without suckers, aid in feeding but are of limited use in locomotion. In contrast to asteroids, the

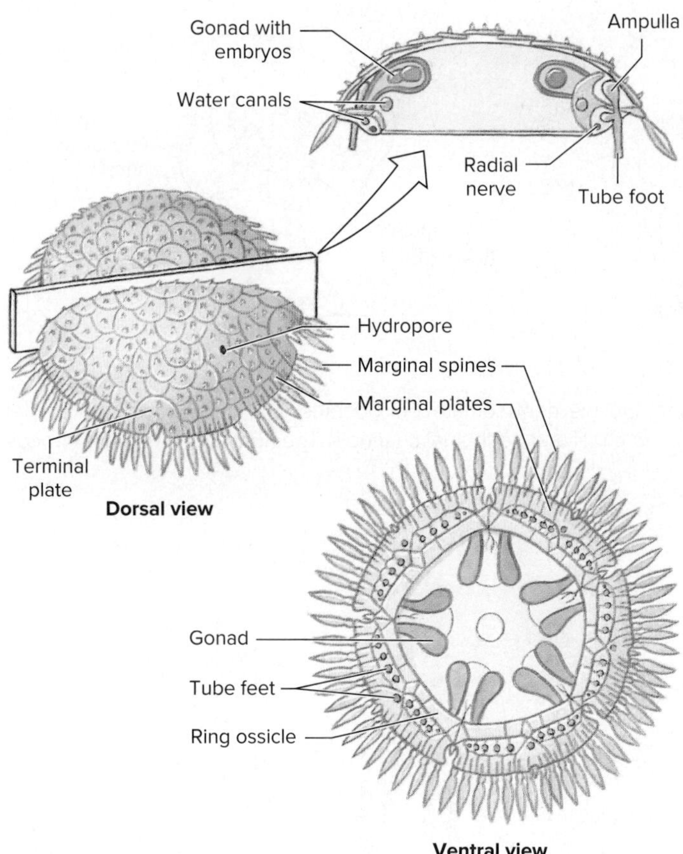

figure 14.10

Sea daisies, *Xyloplax* (class Asteroidea), are peculiar little disc-shaped echinoderms. In contrast to all other echinoderms, their tube feet lie around the margin instead of being distributed along ambulacral areas.

madreporite of ophiuroids is located on the oral surface, on one of the oral-shield ossicles (figure 14.12).

Each jointed arm consists of a column of articulated ossicles connected by muscles and covered by plates. Locomotion is by arm movement.

Five movable plates that serve as jaws surround the mouth (figure 14.12). There is no anus. Their skin is leathery, with dermal plates and spines arranged in characteristic patterns.

Visceral organs are confined to the central disc, because the arms are too slender to contain them (figure 14.13). The stomach is saclike, and there is no intestine. Indigestible material is cast out of the mouth.

Five pairs of invaginations called **bursae** open toward the oral surface by bursal slits at the bases of the arms (figure 14.13). Water circulates in and out of these sacs for exchange of gases. On the coelomic wall of each bursa are small **gonads** that discharge their ripe sex cells into the bursa. Gametes pass through the bursal slits into the water for fertilization. Sexes are usually separate; a few ophiuroids are hermaphroditic. Most species produce a free-swimming larva called an **ophiopluteus**, and its ciliated bands extend onto delicate, beautiful arms (see figure 14.9C). During metamorphosis to the juvenile stage, there is no temporarily attached phase as occurs in asteroids. Water-vascular, nervous, and hemal systems are similar to those of sea stars.

Brittle stars tend to be secretive, living on hard substrates where no light penetrates. They are often negatively phototropic and insinuate themselves into small crevices between rocks, becoming more active at night. They are commonly fully exposed in the permanent darkness of deep seas. Ophiuroids consume a variety of small particles, either deposit or suspension feeding. Podia are important in transferring food to the mouth. Some brittle stars extend their arms into the water and catch suspended particles in mucous strands between arm spines. Basket stars (see figure 14.11B) perch themselves on corals, extending their branched arms to capture small plankton.

Regeneration and autotomy are even more pronounced in brittle stars than in sea stars. Many seem very fragile, releasing an arm or even part of the disc at the slightest provocation. Some can reproduce asexually by cleaving the disc; each new individual then regenerates missing parts.

Class Echinoidea: Sea Urchins, Sand Dollars, and Heart Urchins

There are about 950 living species of echinoids, which have a compact body enclosed in an endoskeletal **test**, or shell. Dermal ossicles, which have become closely fitting plates, form the test.

figure 14.11

A, Brittle star, *Ophiothrix suensonii* (class Ophiuroidea) on a red rope sponge in Belize. Brittle stars do not use their tube feet for locomotion but can move rapidly (for an echinoderm) by means of their arms. **B,** Oral view of a giant basket star, *Astrophyton muricatum*. Basket stars extend their many-branched arms to filter-feed, usually at night.

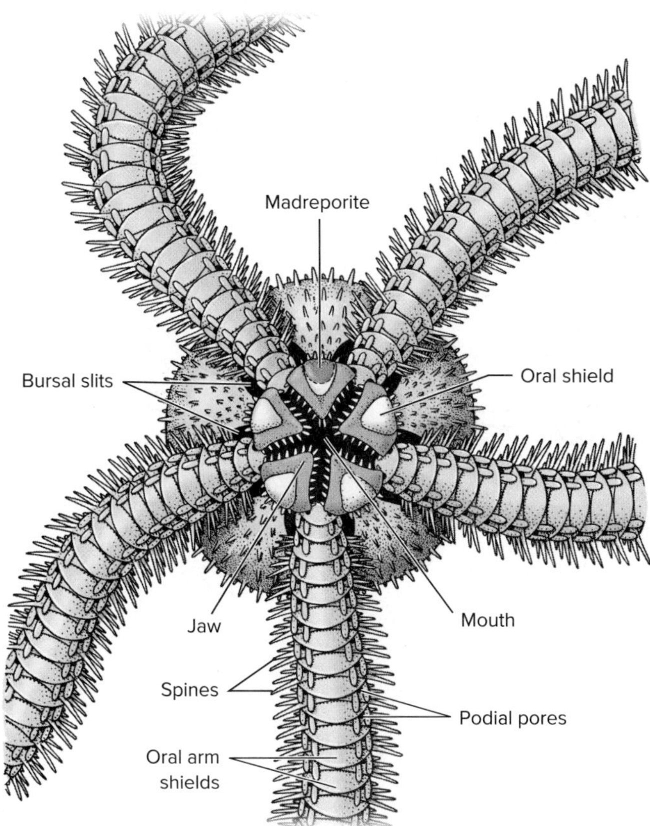

figure 14.12

Oral view of spiny brittle star, *Ophiothrix*.

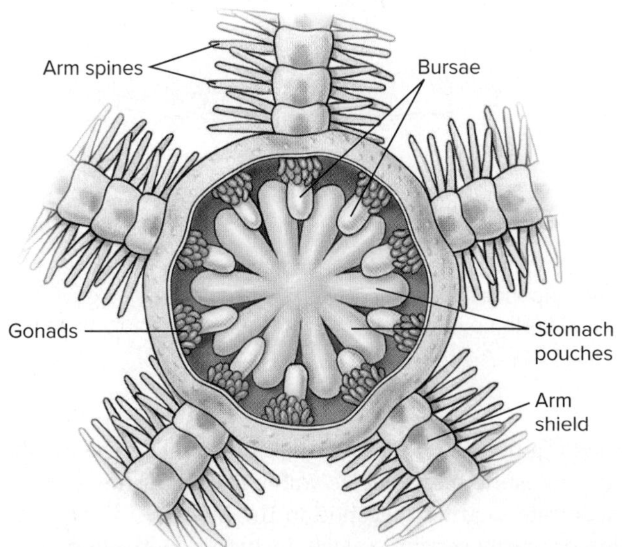

figure 14.13

Ophiuroid with aboral disc wall cut away to show principal internal structures. Bursae are fluid-filled sacs in which water constantly circulates for respiration. They also serve as brood chambers. Only bases of arms are shown.

figure 14.14

Purple sea urchin, *Strongylocentrotus purpuratus* (class Echinoidea), is common along the Pacific Coast of North America where wave action is heavy.

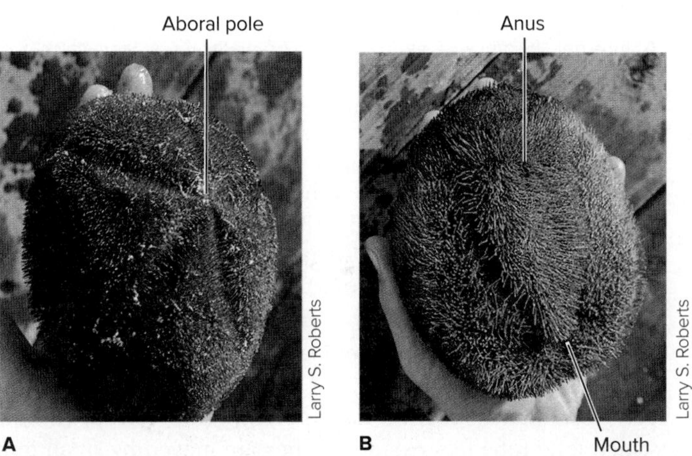

figure 14.15

An irregular echinoid, *Meoma*, one of the largest heart urchins (test up to 18 cm). *Meoma* occurs in the West Indies and from the Gulf of California to the Galápagos Islands. **A,** Aboral view. **B,** Oral view. Note the curved mouth at the anterior end and the anus at the posterior end.

Echinoids lack arms, but their tests bear the typical pentaramous plan of echinoderms in their five ambulacral areas. The most notable modification of the ancestral body plan is that the oral surface, which bears the tube feet and faces the substrate in sea stars, has expanded around to the aboral side, so that ambulacral areas extend upward to an area close to the anus **(periproct).** Most living species of sea urchins are "regular;" they have a hemispherical shape, radial symmetry, and medium to long spines (figure 14.14). Sand dollars and heart urchins (figure 14.15) are "irregular" because members of their orders have become secondarily bilateral; their spines are usually short. Regular urchins move by means of their tube feet, with

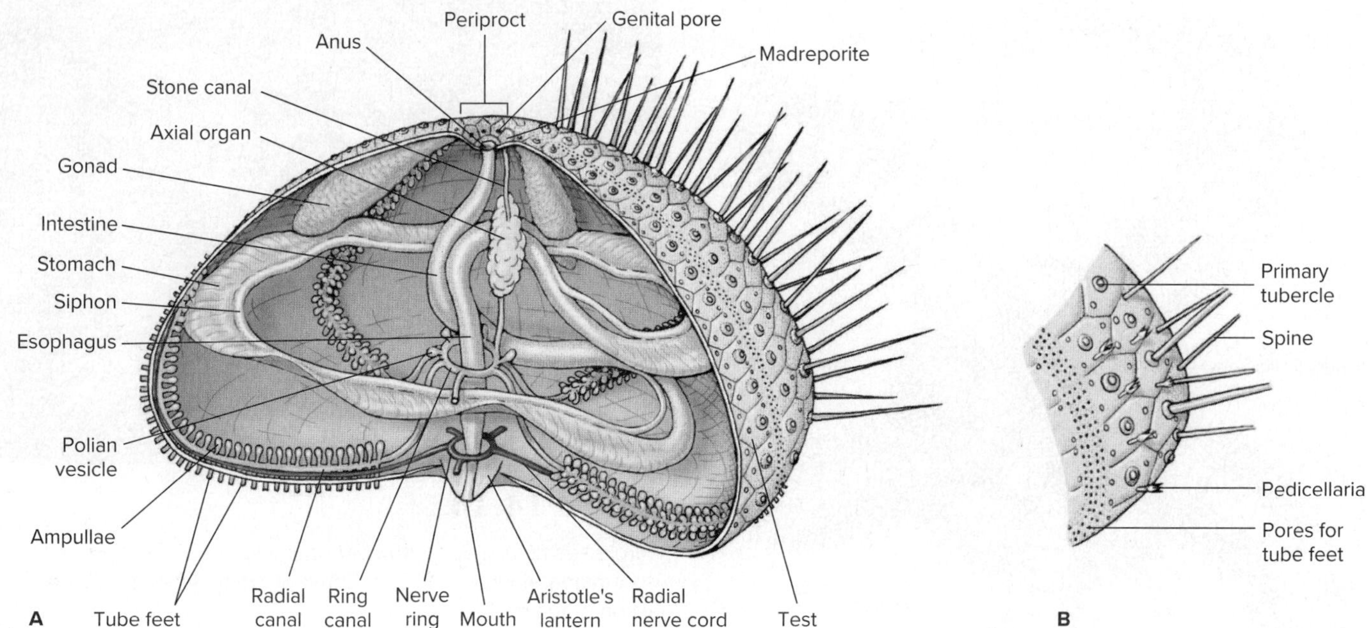

figure 14.16

A, Internal structure of a sea urchin. The water-vascular system is shown in tan. **B,** Detail of a portion of the endoskeleton.

some assistance from their spines, and irregular urchins move chiefly by their spines. Some echinoids are quite colorful.

Echinoids have wide distribution in all seas, from intertidal regions to deep oceans. Regular urchins often prefer rocky or hard substrates, whereas sand dollars and heart urchins burrow into a sandy substrate.

An echinoid test is a compact skeleton of 10 double rows of plates that bear movable, stiff spines (figure 14.16). The five pairs of ambulacral rows have pores (figure 14.16B) through which long tube feet extend. Spines are moved by small muscles around their bases.

There are several kinds of pedicellariae, the most common of which have three jaws and are mounted on long stalks. Pedicellariae of many species bear venom glands, and their toxin paralyzes small prey.

Five converging teeth surround the mouth of regular urchins and sand dollars. In some sea urchins, branched gills (modified podia) encircle the region around the mouth. An anus, genital pores, and a madreporite are located aborally in the periproct region (figure 14.16). The mouth of sand dollars is located at about the center of the oral side, but their anus has shifted to the margin or even to the oral side of the disc, so that an anteroposterior axis and bilateral symmetry can be recognized. Bilateral symmetry is even more accentuated in heart urchins, with their anus near the posterior on their oral side and their mouth moved away from the oral pole toward the anterior (see figure 14.15).

Inside the test (figure 14.16) are a coiled digestive system and a complex chewing apparatus (in regular urchins and in sand dollars), called **Aristotle's lantern**, to which teeth are attached (figure 14.17). A ciliated siphon connects the esophagus to the intestine and enables water to bypass the stomach to concentrate food for digestion in the intestine. Typical sea urchins primarily consume algae, including kelp, supplemented by a wide variety of small invertebrates. Sand dollars collect fine particles on ciliated tracts or with podia. When abundant, sea urchins can eliminate macroalgae from ecosystems, creating sea urchin "barrens."

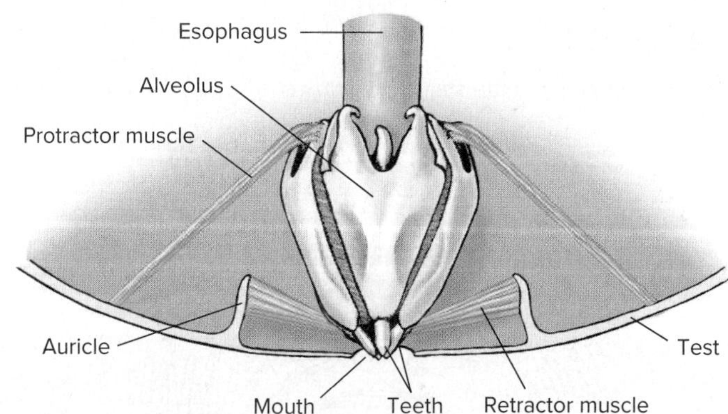

figure 14.17

Aristotle's lantern, a complex mechanism used by sea urchins for masticating their food. Five pairs of retractor muscles draw the lantern and teeth up into the test; five pairs of protractors push the lantern down and expose the teeth.

Hemal and nervous systems are similar to those of asteroids. Ambulacral grooves are closed, and radial canals of the

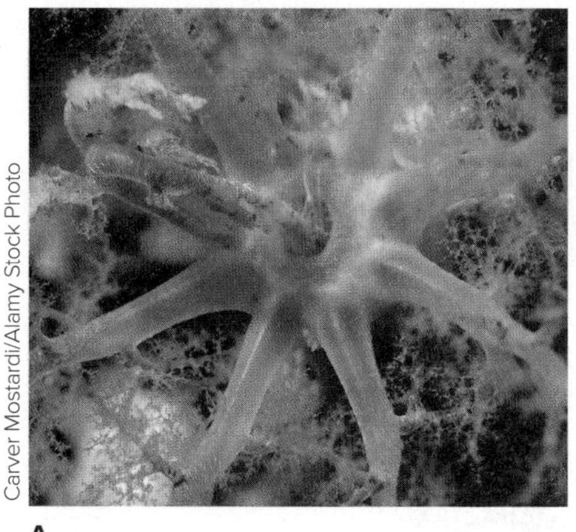

A

B

C

figure 14.18

Sea cucumber (class Holothuroidea). **A,** *Eupentata quinquesemita* extends its tentacles to collect particulate matter in the water, then puts them one by one into its mouth and cleans the food from them. **B,** *Parastichopus californicus* uses moplike tentacles for deposit-feeding on the sea floor. **C,** *Bohadschia argus* expels its Cuvierian tubules, modified parts of its respiratory tree, when it is disturbed. These sticky strands, containing a toxin, discourage potential predators.

water-vascular system run just beneath the test, one in each ambulacral radius (see figure 14.16).

Sexes are separate, and both eggs and sperm are shed into the sea for external fertilization. **Echinopluteus** larvae (see figure 14.9D) may live a planktonic existence for several months and then metamorphose quickly into young urchins. Sea urchins have been used extensively as models in studies of development.

Class Holothuroidea: Sea Cucumbers

In a phylum characterized by odd animals, class Holothuroidea contains members that both structurally and physiologically are among the strangest. There are about 1150 living species of holothuroids. These animals, commonly called sea cucumbers, have a remarkable resemblance to the vegetable after which they are named (figure 14.18). Compared to other echinoderms, sea cucumbers are greatly elongated in their oral-aboral axis, and ossicles are much reduced in most (figure 14.19); consequently, the body wall is usually leathery. Many species crawl along the sea floor, using ventral tube feet and body wall muscles. Others live beneath rocks, and some are burrowers. Despite a prominent anterior end, cephalization is absent.

Because sea cucumbers typically lie on one side, called a sole, their tube feet usually are well developed only in the three strips of ambulacra in contact with the substrate. Tube feet on the dorsal side are reduced and may serve a sensory role (see figure 14.18B). Thus, a secondary bilaterality is present, albeit of different origin from that of irregular urchins. Oral tentacles consist of 10 to 30 retractile, modified tube feet around the mouth. Locomotion in burrowing species, which often lack podia, is primarily accomplished by contraction of well-developed circular and longitudinal muscles in the body wall.

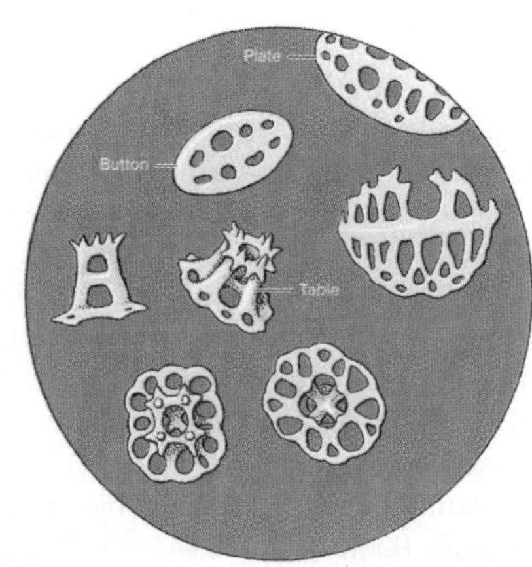

figure 14.19

Ossicles of sea cucumbers are usually microscopic bodies buried in a leathery dermis. They can be extracted from this tissue with commercial bleach and are important taxonomic characteristics. The ossicles shown here, called tables, buttons, and plates, are from the sea cucumber *Holothuria difficilis*. They illustrate the meshwork (stereom) structure observed in ossicles of all echinoderms at some stage in their development. (×250)

The coelomic cavity is spacious and filled with fluid, serving as a hydrostatic skeleton. A **respiratory tree** (figure 14.20), composed of two long, many-branched tubes, empties into the **cloaca**. Water is pumped in and out by the muscular cloaca. The respiratory tree serves for both respiration and excretion and is not present in any other group of living echinoderms. Gas exchange also occurs through the skin and tube feet. The water-vascular system is peculiar in that the madreporite lies free in the coelom.

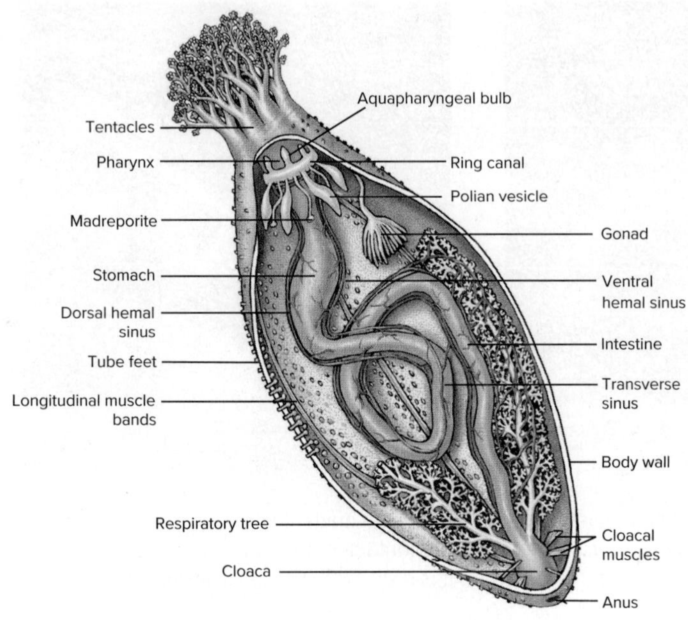

figure 14.20

Internal anatomy of a sea cucumber, *Sclerodactyla*. Hemal system is shown in red.

Sexes are usually separate, but some sea cucumbers are hermaphroditic. Among echinoderms, only sea cucumbers have a single gonad. Fertilization is external, and the free-swimming larva is called an **auricularia** (see figure 14.9E).

Most sedentary species trap suspended food particles in the mucus of their outstretched oral tentacles; others are deposit feeders, collecting fine organic matter in their tentacles. They then stuff the tentacles into their pharynx, one by one, ingesting captured food. As deposit feeders, sea cucumbers are very successful, forming up to 90% of the biomass on the surface of some deep-sea floors.

Some species of sea cucumbers have a peculiar power of what appears to be self-mutilation but in reality is a defense mechanism. When disturbed or subjected to unfavorable conditions, they discharge **Cuvierian tubules**, which are attached to the posterior part of the respiratory tree (see figure 14.18C). These tubules, which sometimes contain toxins, become long and sticky after expulsion and can entangle a predator. Some species also discharge the digestive tract, respiratory tree, or gonads. All of these structures can be regenerated.

Class Crinoidea: Sea Lilies and Feather Stars

Crinoids include about 625 living species. As fossil records reveal, crinoids were once far more numerous than they are now. They differ from other echinoderms by being attached to a substrate during a substantial part of their lives. Sea lilies have a flower-shaped body at the tip of an attached stalk (figure 14.21). Feather stars have long, many-branched arms, and adults are free-moving, although they may remain in the same spot for long periods (figure 14.22). They can move by using the cirri to crawl or by sweeping their long, feather arms to swim. Many crinoids are deep-water forms, but feather stars may inhabit shallow waters, especially in Indo-Pacific and West Indian–Caribbean regions, where the largest numbers of species occur.

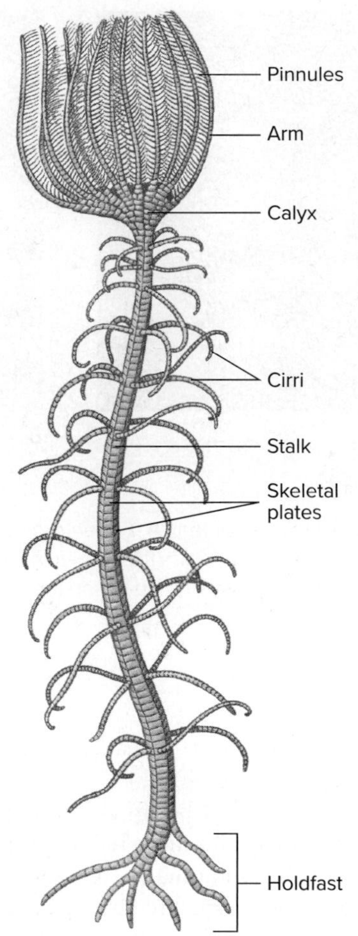

figure 14.21

A sea lily (stalked crinoid). Modern crinoid stalks rarely exceed 60 cm, but fossil forms were as long as 20 m.

The crinoid body is covered with a leathery skin containing calcareous plates. Epidermis is poorly developed. Five flexible arms branch to form many more arms, each with many lateral pinnules arranged like barbs on a feather (see figure 14.21). Sessile forms have a long, jointed **stalk** attached to the aboral side of the body **(calyx)**. This stalk is composed of plates, appears jointed, and may bear jointed, flexible appendages called **cirri**. Cirri at the base of the stalk form a holdfast used to anchor the animal to the substrate. Madreporite, spines, and pedicellariae are absent.

The upper (oral) surface bears a mouth and anus. With the aid of tube feet and mucous strands, crinoids capture small organisms and transfer them to **ambulacral grooves** at the bases of the arms. Their ambulacral grooves, which are open and ciliated, carry food to the mouth. The **water-vascular**

Larry S. Roberts

figure 14.22

Comantheria briareus is a feather star that occupies Pacific coral reefs. It extends its arms into the water to catch food particles both during the day and at night.

system of crinoids follows the basic echinoderm plan. Sense organs are scanty and simple.

Sexes are separate, and gonads are simple. **Doliolaria** larvae (see figure 14.9F) are free-swimming for a time before they attach and metamorphose.

14.3 Phylogeny and Adaptive Diversification of Echinoderms

Phylogeny

Echinoderms left an extensive fossil record and evolved about 26 anatomically distinct body forms in 20 currently recognized classes. Most of these became extinct by the end of the Paleozoic era, and only five classes survive today. Based on their bilateral larvae and newly discovered bilateral fossil species, such as *Ctenoimbricata* (see figure 14.2), it appears that ancestral echinoderms were bilateral deposit feeders with a calcite, stereom endoskeleton (see figure 14.2).

Bilateral symmetry was lost early in echinoderm evolution. This change is seen in the extinct homalozoans, also called carpoids, which were asymmetrical, mostly sessile forms. The first radial echinoderms were the spindle-like, extinct helioplacoids (see figure 14.2). Helioplacoids were triradial, with three, spirally-arranged ambulacra bearing pits for tube feet, and a mouth on the side of their body.

Attachment to a substrate by their aboral surface would have selected for radial symmetry, explaining the origin of subphylum Pelmatozoa, whose living members are crinoids. An ancestor that became free-moving and applied its oral surface to the substrate would have given rise to subphylum Eleutherozoa.

Phylogeny within Eleutherozoa is controversial. Most investigators agree that echinoids and holothuroids form a clade, but opinions diverge on the relationship of ophiuroids and asteroids. Figure 14.2 illustrates the currently favored view that asteroids and ophiuroids form a clade, with independent evolution of closed ambulacral grooves in ophiuroids and the echinoid-holothuroid clade.

Adaptive Diversification

Diversification of echinoderms has been limited by their most important characters: radial symmetry, water-vascular system, and dermal endoskeleton. If their ancestors had a brain and specialized sense organs, these were lost in adoption of radial symmetry. Only now are gene expression studies beginning to help researchers identify the anterior-posterior axis in adult echinoderms. The best evidence currently available suggests that the oral surface is anterior and the aboral surface is posterior. Under this hypothesis, the arms represent lateral growth zones.

Echinoderms have diversified within the benthic habitat. There are large numbers of creeping forms with filter-feeding, deposit-feeding, scavenging, and herbivorous habits, and very rare pelagic forms. In this light, the relative success of asteroids as predators is impressive and probably attributable to the extent to which they have exploited the hydraulic mechanism of their tube feet.

14.4 Phylum Hemichordata

Hemichordata (hem′ē-kor-dä′tə) (Gr. *hemi*, half, + *chorda*, string, cord) are marine animals formerly considered a subphylum of chordates, based on their possession of gill slits and a rudimentary notochord. However, the hemichordate "notochord" is really an evagination of their mouth cavity and not homologous with the chordate notochord. Hemichordates have a three-part coelom typical of deuterostomes.

Hemichordates are wormlike bottom-dwellers, living usually in shallow waters. Some colonial species live in secreted tubes. Most are sedentary or sessile. They are widely distributed, but their secretive habits and fragile bodies make collecting them difficult.

Members of class Enteropneusta (Gr. *enteron*, intestine, + *pneus*, breathing) (acorn worms) range from 20 mm to 2.5 m in length. Members of class Pterobranchia (Gr. *pteron*, wing, + *branchia*, gills) (pterobranchs) are smaller, usually 1 to 7 mm, not including their stalk. About 90 species of enteropneusts and 30 species of pterobranchs are recognized.

Class Enteropneusta

Enteropneusts, or acorn worms (figure 14.23), are sluggish, wormlike animals that live in burrows or under stones, usually in mud or sand flats of intertidal zones.

The mucus-covered body is divided into a tonguelike **proboscis**, a short collar, and a long trunk.

Taxonomy of Phylum Echinodermata

There are about 7000 living and 20,000 extinct or fossil species of Echinodermata. The traditional classification placed all the free-moving forms that were oriented with their oral side down in subphylum Eleutherozoa, containing most living species. The other subphylum, Pelmatozoa, contained mostly forms with their stems and oral side up; most extinct classes and living Crinoidea belong to this group. Although alternative schemes have strong supporters, cladistic analysis provides evidence that the two traditional subphyla are monophyletic. This listing includes only groups with living members.

Subphylum Pelmatozoa (pel-ma'tō-zō-ə) (Gr. *pelmatos*, a stalk, + *zoōn*, animal). Body in form of cup or calyx, borne on aboral stalk during part or all of life; oral surface directed upward; open ambulacral grooves; madreporite absent; both mouth and anus on oral surface; several fossil classes plus living Crinoidea.

Class Crinoidea (cri-noi'dē-ə) (Gr. *krinon*, lily, + *eidos*, form): **sea lilies** and **feather stars**. Five arms branching at base and bearing pinnules; ciliated ambulacral grooves on oral surface with tentacle-like tube feet for food gathering; spines and pedicellariae absent. Examples: *Antedon, Comantheria*.

Subphylum Eleutherozoa (e-lū'ther-ō-zō'ə) (Gr. *eleutheros*, free, not bound, + *zoōn*, animal). Body form star-shaped, globular, discoidal, or cylindrical; oral surface directed toward substrate or oral-aboral axis parallel to substrate.

Class Asteroidea (as'ter-oi'dē-ə) (Gr. *aster*, star, + *eidos*, form): **sea stars (starfish)**. Star-shaped, with arms not sharply demarcated from the central disc; ambulacral grooves open, with tube feet on oral side; tube feet often with suckers; anus and madreporite aboral; pedicellariae present. **Sea daisies**, included in this group, lack arms, have a disc-shaped body, and have a ring of suckerless podia near the body margin. Examples: *Asterias, Pisaster, Xyloplax*.

Class Ophiuroidea (o'fē-ur-oi'dē-ə) (Gr. *ophis*, snake, + *oura*, tail, + *eidos*, form): **brittle stars** and **basket stars**. Star-shaped, with arms sharply demarcated from central disc; ambulacral grooves closed, covered by ossicles; tube feet without suckers and not used for locomotion; pedicellariae absent. Examples: *Ophiothrix, Astrophyton*.

Class Echinoidea (ek'i-noi'dē-ə) (Gr. *echinos*, sea urchin, hedgehog, + *eidos*, form): **sea urchins**, **sea biscuits**, and **sand dollars**. Globular or disc-shaped, with no arms; compact skeleton or test with closely fitting plates; movable spines; ambulacral grooves closed; tube feet often with suckers; pedicellariae present. Examples: *Arbacia, Strongylocentrotus, Meoma*.

Class Holothuroidea (ho'lo-thur-oi'dē-ə) (Gr. *holothourion*, sea cucumber, + *eidos*, form): **sea cucumbers**. Cylindrical, with no arms; spines absent; microscopic ossicles embedded in muscular body wall; anus present; ambulacral grooves closed; tube feet with suckers; oral tentacles (modified tube feet); pedicellariae absent; madreporite internal. Examples: *Sclerodactyla, Parastichopus, Cucumaria*.

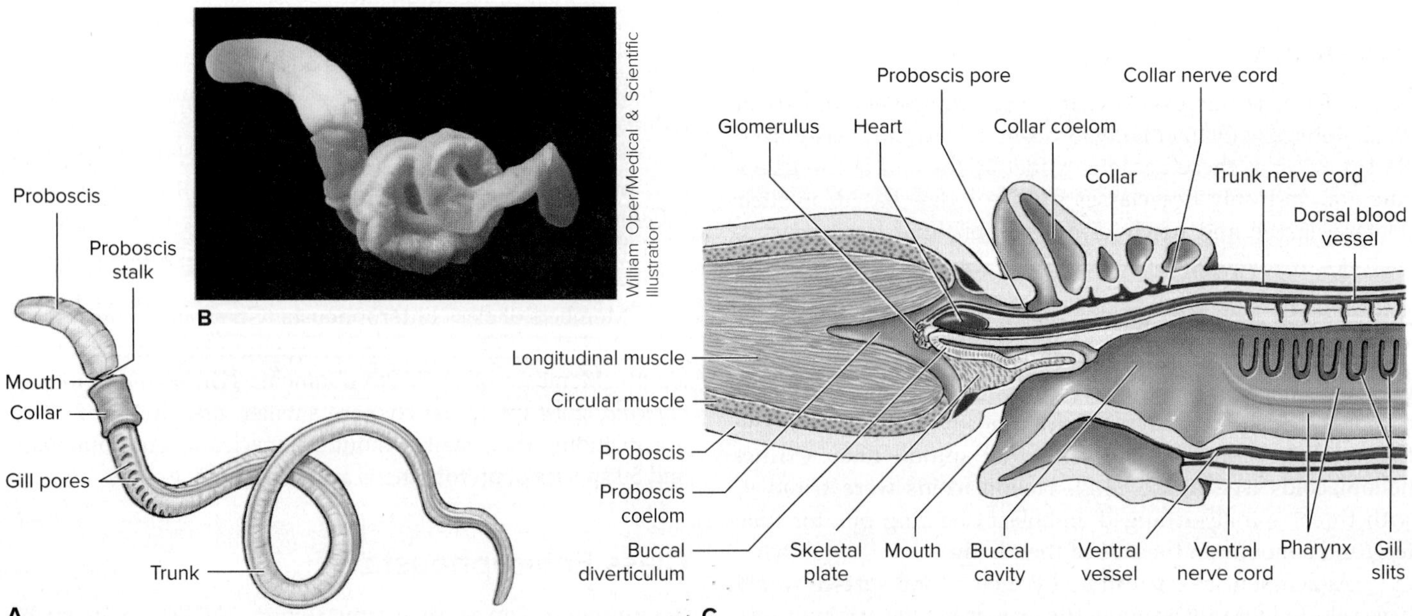

figure 14.23

An acorn worm, *Saccoglossus kovalevskii* (Hemichordata, class Enteropneusta). **A,** External lateral view. **B,** Live animal. **C,** Longitudinal section through anterior end.

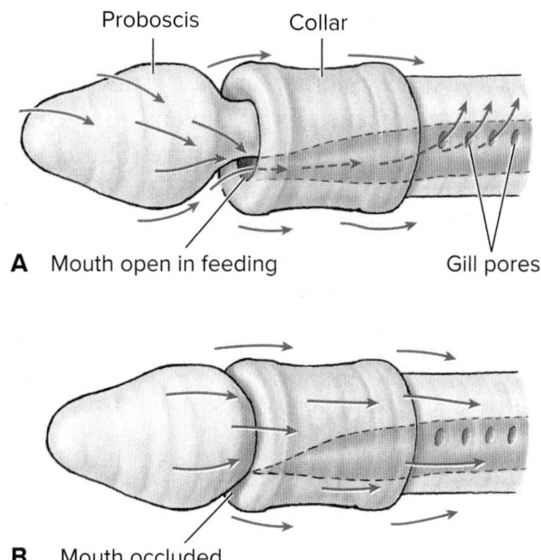

figure 14.24

Food currents of an acorn worm. **A,** Side view with mouth open, showing the direction of currents created by cilia on proboscis and collar. Food particles are directed toward the mouth and digestive tract. Rejected particles move outside of the collar. Water leaves through gill pores. **B,** When the mouth is occluded, all particles are rejected and passed onto the collar. Nonburrowing and some burrowing hemichordates use this feeding method.

Many acorn worms are **deposit feeders**, extracting the organic components of sediments. Others are **suspension feeders**, capturing plankton and detritus. The muscular proboscis is the active part of the animal, collecting food in mucous strands on its surface. Captured particles are transported by ciliary action to the groove at the edge of the collar and then to the mouth (figure 14.24). The food-laden mucus continues to be directed by cilia along the ventral part of the pharynx and esophagus to the intestine. A row of gill pores extends dorsolaterally on each side of the trunk just behind the collar (figure 14.24). These gill pores open from a series of densely ciliated gill chambers that in turn connect with a series of U-shaped gill slits in the sides of the pharynx (see figure 14.23C). Acorn worms have no gills, but some respiratory gas exchange occurs in the vascular branchial epithelium, as well as on the body surface. The gill slits allow removal of excess water from the pharynx, but they are not used as a filter-feeding structure.

In the posterior end of the proboscis is a small coelomic sac (protocoel) into which extends a buccal diverticulum, a slender, blind pouch of the gut that was formerly considered a notochord. Above this buccal diverticulum is the heart, which receives blood carried forward from a dorsal blood vessel. Blood pumped from the heart flows into a network of blood sinuses called a glomerulus, which has an excretory function, and then flows posteriorly through a ventral blood vessel, passing through extensive sinuses in the body wall and gut. Acorn worms have a ventral nerve cord and a larger dorsal nerve cord that is hollow in some species. Sexes are separate,

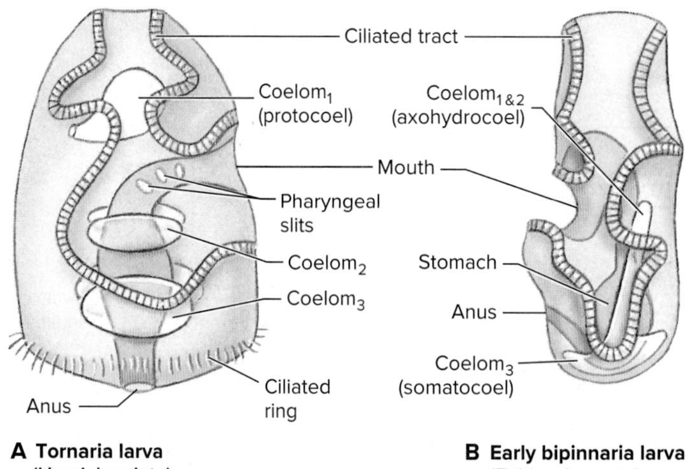

figure 14.25

Comparison of **A,** a hemichordate tornaria with **B,** an echinoderm bipinnaria.

and fertilization is external. In some species, a ciliated tornaria larva develops that closely resembles the echinoderm bipinnaria (figure 14.25).

CHARACTERISTICS of Phylum Hemichordata

1. Body divided into **proboscis**, **collar**, and **trunk; buccal diverticulum** in posterior part of proboscis.
2. Enteropneusta free-moving and of burrowing habits; pterobranchs sessile, mostly colonial, living in secreted tubes.
3. Free-living.
4. Bilaterally symmetrical, soft bodied; wormlike or short and compact with stalk for attachment.
5. Triploblastic.
6. Single coelomic pouch in proboscis, but paired pouches in collar and trunk.
7. Ciliated epidermis.
8. Digestive system complete.
9. Longitudinal and circular muscles in body wall in some.
10. A subepidermal nerve plexus thickened to form dorsal and ventral nerve cords, with a **connective** ring in the collar; some species have hollow **dorsal nerve cord.**
11. Sensory neurons in proboscis likely function in chemoreception.
12. Colonies form by asexual budding in pterobranchs; asexual reproduction by fragmentation in enteropneusts.
13. Sexes separate in Enteropneusta, with gonads projecting into body cavity; tornaria larva in some Enteropneusta.
14. A single **glomerulus** connected to blood vessels may have excretory function and is considered a metanephridium.
15. Respiratory system of **gill slits** (few or none in pterobranchs) connects the pharynx with external environment.
16. Circulatory system is composed of dorsal and ventral vessels and dorsal heart.

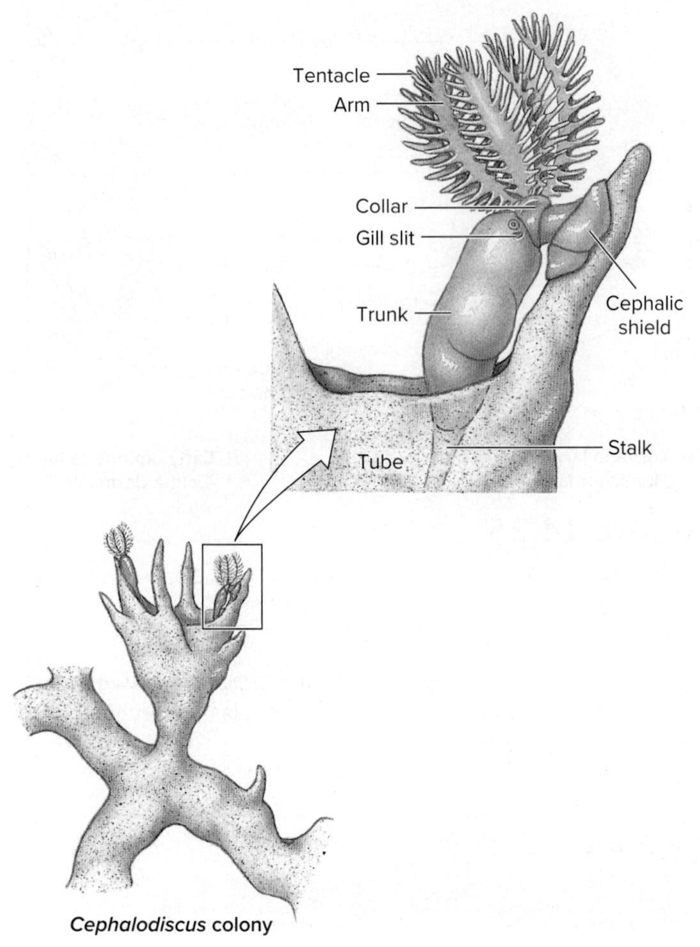

Class Pterobranchia

Pterobranchs are small colonial animals. They bear arms with tentacles containing extensions of the coelomic compartment of the mesosome, similar to a lophophore (figure 14.26). Food captured in mucus on the crown of tentacles is transported with ciliated grooves to the mouth. The digestive tract is U-shaped, with the anus near the mouth. A dorsal, hollow nerve cord is absent, but otherwise their nervous system is similar to that of acorn worms. One genus, *Cephalodiscus*, has a single pair of gill slits (figure 14.26). Some species are dioecious, and others are monoecious; asexual reproduction occurs by budding.

figure 14.26

Cephalodiscus, a pterobranch hemichordate. These tiny (5–7 mm) forms live in tubes in which they can move freely. Ciliated tentacles and arms direct currents of food and water toward the mouth.

■ SUMMARY

Sections	Key Concepts
14.1 Clade Ambulacraria	• Within Deuterostomia, Ambulacraria unites echinoderms and hemichordates on the basis of similar larval forms, a three-part coelom, and a shared axial complex. • Pharyngeal gill slits, present in chordates, some hemichordates, and fossil echinoderms, are now considered a synapomorphy of deuterostomes.
14.2 Phylum Echinodermata	• Echinoderms are an important marine group sharply distinguished from other animal phyla. They have pentaradial symmetry, but were derived from bilateral ancestors. • Members of Echinodermata possess many typical characteristics of Deuterostomia, but the water-vascular system is a synapomorphy for echinoderms. • The echinoderm water-vascular system is an elaborate hydraulic system derived embryonically from one of their coelomic compartments. Along the ambulacral areas, branches of the water-vascular system (tube feet) are important in locomotion, food gathering, respiration, and excretion. • Sea stars (class Asteroidea) usually have five arms, which merge gradually with the central disc. Like other echinoderms, they have no head and few specialized sensory organs. Their mouth is directed toward the substratum. They have stereom dermal ossicles, respiratory papulae, and open ambulacral grooves. Many sea stars have pedicellariae. Sexes are separate, with a bilateral, free-swimming larva that becomes a motile sea star.

Sections	Key Concepts
	• Sea daises (former Class Concentricycloidea) are an enigmatic group now placed with class Asteroidea. They are circular in shape and have marginal tube feet. • Arms of brittle stars (class Ophiuroidea) are slender and sharply set off from the central disc. Ophiuroids have no pedicellariae. They crawl by means of arm movements, and their tube feet, which have no suckers, function in food gathering. • For most sea urchins (class Echinoidea), dermal ossicles fit together closely, forming a hard, spherical test around which lie five ambulacral areas. Ambulacral areas are closed, and there are no arms. Sea urchins move by means of tube feet or by their spines. Some echinoids (sand dollars and heart urchins) have returned to adult bilateral symmetry. • In sea cucumbers (class Holothuroidea), dermal ossicles are very small and loosely connected, so the body wall is soft. Their ambulacral areas are closed. Sea cucumbers are greatly elongated in the oral-aboral axis and lie on their side. Because certain ambulacral areas are against the substrate, sea cucumbers have undergone some return to bilateral symmetry. Tube feet around the mouth are modified into tentacles, with which they feed. Some species can eject Cuvierian tubules (attached to the respiratory tree) or other internal organs as a defense mechanism. • Sea lilies and feather stars (class Crinoidea) are the only group of living echinoderms, other than asteroids, with open ambulacral groups. They are sedentary, muco-ciliary particle feeders with their oral side up.
14.3 Phylogeny and Adaptive Diversification of Echinoderms	• The earliest echinoderms probably were bilaterally symmetrical, as indicated by newly discovered fossil forms. • Fossil evidence suggests they evolved through a sessile stage that became radially symmetrical and then gave rise to free-moving forms.
14.4 Phylum Hemichordata	• Hemichordate bodies have three parts (proboscis, collar, trunk) and are coelomate. • Acorn worms (class Enteropneusta) are suspension-feeding and deposit-feeding forms. Food is captured using mucus and cilia on the proboscis. • Members of class Pterobranchia are colonial tube-dwellers that also use mucus and cilia to feed, but they capture food on tentacles.

Review Questions

1. What constellation of characteristics possessed by echinoderms occurs in no other phylum?
2. How do we know that echinoderms were derived from an ancestor with bilateral symmetry?
3. What evidence suggests that echinoderms and hemichordates form a monophyletic group?
4. What is an ambulacrum, and what is the difference between open and closed ambulacral grooves?
5. Explain the anatomy and function of the water-vascular system of a sea star.
6. Distinguish the following groups of echinoderms from one another: Asteroidea, Ophiuroidea, Echinoidea, Holothuroidea, Crinoidea.
7. Match these groups with *all* correct answers in the lettered choices.

   _____ Crinoidea
   _____ Ophiuroidea
   _____ Holothuroidea
   _____ Asteroidea
   _____ Echinoidea

   a. Closed ambulacral grooves
   b. Oral surface generally upward
   c. With arms
   d. Without arms
   e. Approximately globular or discshaped
   f. Elongated in oral-aboral axis
   g. Have pedicellariae
   h. Madreporite internal
   i. Madreporite on oral plate
8. Define the following: pedicellariae, madreporite, respiratory tree, Aristotle's lantern, papulae, Cuvierian tubules.
9. Give three examples of how echinoderms are important to humans.
10. Distinguish Enteropneusta from Pterobranchia.

For Further Thought Echinoderms are abundant off the rocky west coast of North America, where thick "forests" of kelp grow. How might a change in sea urchin or sea star populations affect these communities?

15

Vertebrate Beginnings: *The Chordates*

LEARNING OBJECTIVES

Readers will be able to:

15.1 Describe the evolutionary origin of chordates.

15.2 List and explain the significance of the five diagnostic characteristics of chordates.

15.3 Describe the functional anatomy of urochordates.

15.4 Describe the functional anatomy of cephalochordates, distinguishing it from that of urochordates.

15.5 Briefly describe characteristics of vertebrates that distinguish them from other chordates. Summarize the evolutionary changes from earliest fossil chordates to ostracoderms and to gnathostomes.

Natural Visions/Alamy Stock Photo

An amphioxus in feeding posture.

It's a Long Way from Amphioxus

Along the coasts of many continents, half buried in sand on the sea floor, lives a small, fishlike, translucent animal quietly filtering organic particles from seawater. Inconspicuous, of no commercial value, and largely unknown, this creature is nonetheless one of the famous animals of classical zoology. It is amphioxus, an animal that wonderfully exhibits the five diagnostic characteristics of the phylum Chordata: (1) a dorsal hollow nerve cord overlying; (2) a supportive notochord; (3) pharyngeal pouches or slits; (4) an endostyle for filter feeding; and (5) a postanal tail for propulsion. Amphioxus is an animal that might have been designed by a zoologist for the classroom. During the nineteenth century, with interest in vertebrate ancestry running high, many zoologists thought amphioxus closely resembled the earliest vertebrates. Its exalted position was later acknowledged by Philip Pope in a poem sung to the tune of "Tipperary." It ends with the refrain:

It's a long way from amphioxus
It's a long way to us,
It's a long way from amphioxus
To the meanest human cuss.
Well, it's good-bye to fins and gill slits
And it's welcome lungs and hair,
It's a long, long way from amphioxus
But we all came from there.

Amphioxus's place in the sun did not endure. Absence of a head, together with several specialized features, suggests to today's zoologists that amphioxus represents an early departure from vertebrate ancestry. We are a very long way indeed from amphioxus. Nevertheless, amphioxus probably resembles the chordate condition immediately preceding the origin of vertebrates more closely than does any other living animal.

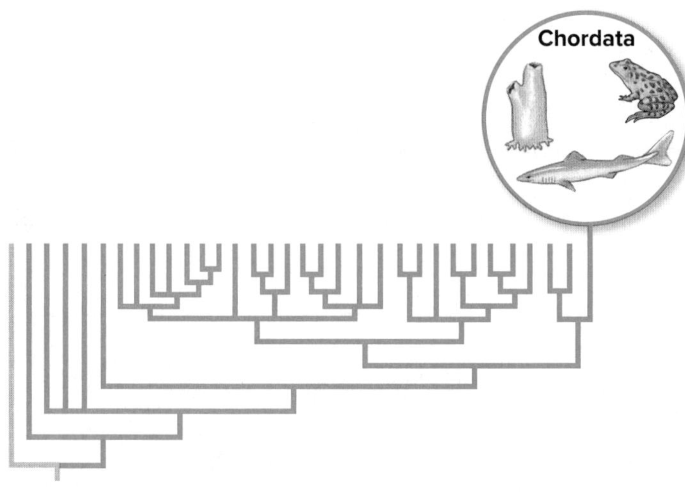

15.1 Chordate Evolution and Classification

The animals most familiar to most people belong to the phylum Chordata. Humans are members and share the characteristic from which the phylum derives its name—the **notochord** (Gr. *nōton*, back, + L. *chorda*, cord) (figure 15.1). All members of the phylum possess this structure, either restricted to early development or present throughout life. The notochord is a rodlike, semirigid body of fluid-filled cells enclosed by a fibrous sheath, which extends, in most cases, the length of the body just ventral to the central nervous system. Thus, the notochord is a hydrostatic organ, similar to the hydrostatic skeletons of nematodes (Chapter 12). Its primary purpose is to stiffen the body, providing skeletal scaffolding for the attachment of swimming muscles.

The structural plan of chordates retains many features that occur in nonchordate invertebrates, such as bilateral symmetry, anteroposterior axis, coelom, tube-within-a-tube arrangement, metamerism, and cephalization.

Chordates are deuterostomes and share with other members of that clade several important characteristics: radial cleavage (see figure 3.10), an anus derived from the first embryonic opening (blastopore) and a mouth derived from an opening of secondary origin, and a coelom primitively formed by fusion of enterocoelous pouches (although in most vertebrates coelom formation is schizocoelous, but independently derived, as an accommodation for their large yolks). These uniquely shared characteristics indicate a natural unity among the deuterostomes.

Traditional and Cladistic Classification of the Chordates

Traditional Linnaean classification of chordates provides a convenient way to indicate the taxa included in each major group. However, in cladistic usage, some of the traditional taxa, such as Agnatha and Reptilia, are no longer recognized or have been redefined. Such taxa do not satisfy the cladistic requirement that only **monophyletic** groups, those groups containing all known descendants of a single common ancestor, are taxonomically valid. For example, the animals traditionally called reptiles (turtles, lizards, snakes, and crocodilians), form a **paraphyletic** group because this group does not contain all descendants of their most recent common ancestor (see figure 4.5). The common ancestor of reptiles as traditionally recognized is also an ancestor of birds (figure 15.2). Reptilia can be considered a clade if birds are included with those animals traditionally called reptiles.

The cladogram of chordates (figure 15.2) shows a nested hierarchy of taxa grouped by their shared derived characters. These characters may be morphological, physiological, embryological, behavioral, chromosomal, or molecular. By contrast, the branches of a phylogenetic tree (figure 15.3) are intended to represent real lineages that occurred in the evolutionary past. Geological information regarding ages of lineages is added to the information from the cladogram to generate a phylogenetic tree for the same taxa.

In our treatment of chordates, we have used a cladistic classification because such usage is necessary to reconstruct the evolution of characters in chordates. At present, because of numerous nested levels of clades, particularly class and subclass for vertebrates, use of Linnaean ranks is incompatible with some taxa. For example, traditional Linnaean classification

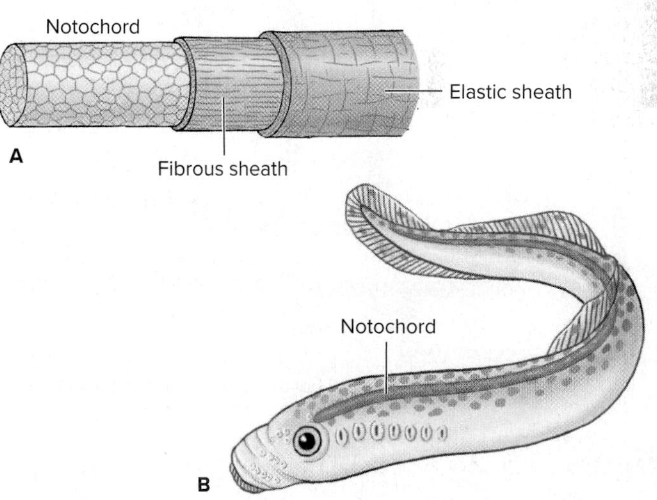

figure 15.1

A, Structure of the notochord and its surrounding sheaths. Cells of the notochord proper are thick walled, pressed together closely, and filled with semifluid. Stiffness is caused mainly by the turgidity of cells with large, fluid-filled vacuoles, enclosed within a connective tissue sheath. This type of endoskeleton is characteristic of all chordates at some stage of life. The notochord provides longitudinal stiffening of the main body axis and a base for trunk muscles. **B,** In hagfishes and lampreys, the notochord persists throughout life, but in other vertebrates it is largely replaced by vertebrae. In mammals, remnants are found in intervertebral discs.

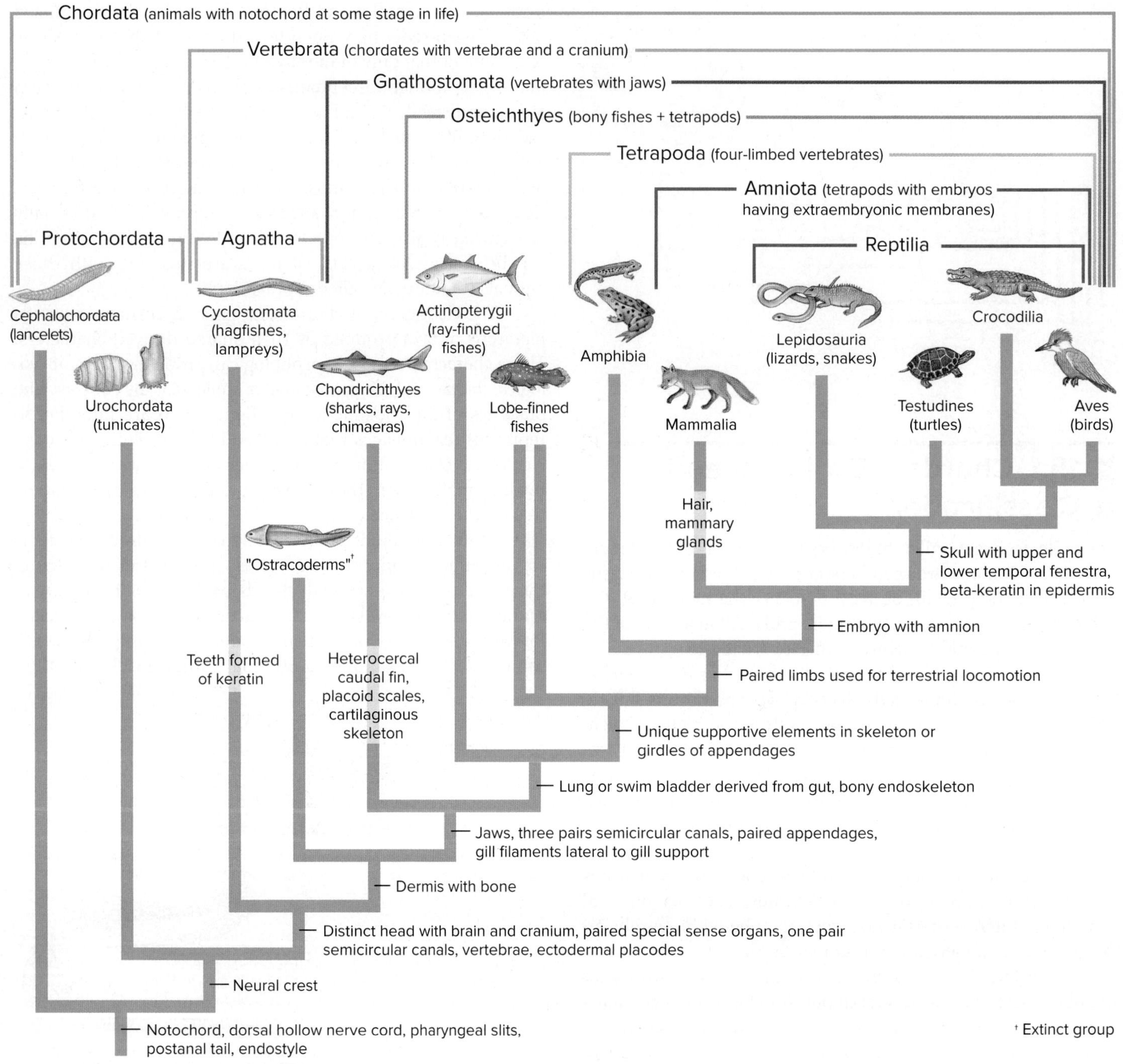

figure 15.2

Cladogram of phylum Chordata, showing probable relationships of the monophyletic groups composing the phylum. Nesting brackets across the top of the cladogram identify monophyletic groupings within the phylum. The lower brackets identify the traditional groupings Protochordata and Agnatha, paraphyletic groups that are not recognized in cladistic treatments, but are shown because of historical use.

ranks Actinoptergyii (ray-finned bony fishes), Reptilia, Mammalia, and Aves all as classes. However, the cladistic taxa Reptilia and Mammalia are actually subgroups of Actinopterygii, and Aves is a subgroup of Reptilia. Continued use of the traditional Linnaean system does not accurately depict the evolutionary relationships of these groups.

Several traditional divisions of phylum Chordata used in Linnaean classifications are shown in table 15.1. A fundamental separation is Protochordata, chordates lacking vertebrae, from Vertebrata. The vertebrates may be variously subdivided into groups based on shared characteristics. Two such subdivisions shown in table 15.1 are (1) Agnatha, vertebrates lacking

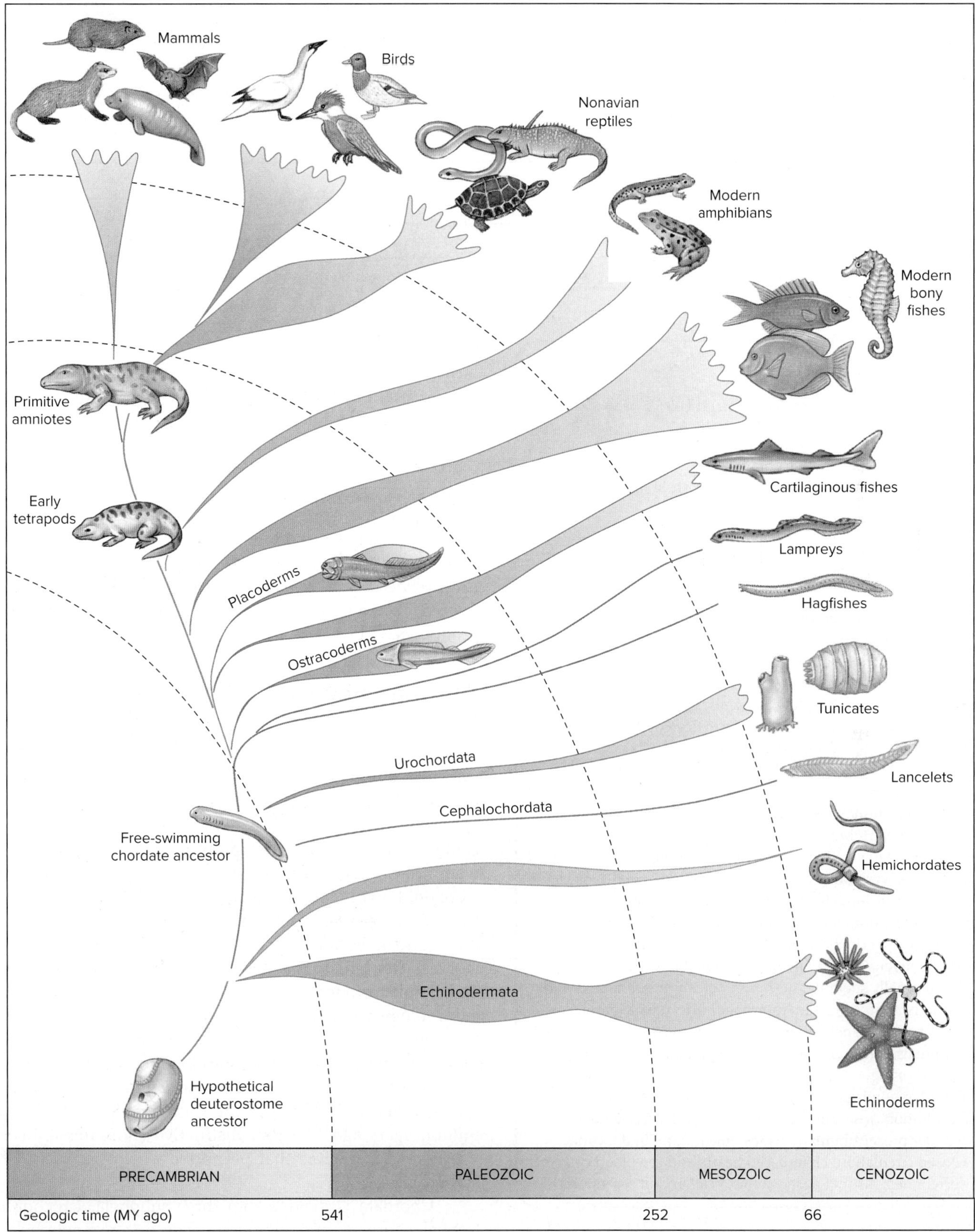

figure 15.3

Phylogenetic tree of the chordates, suggesting their probable origin and relationships. Other schemes have been suggested and are possible. The relative abundance in numbers of species of each group through geological time, as indicated by the fossil record, is indicated here by the bulging and thinning of that group's line of descent.

Table 15.1 Traditional Divisions of the Phylum Chordata

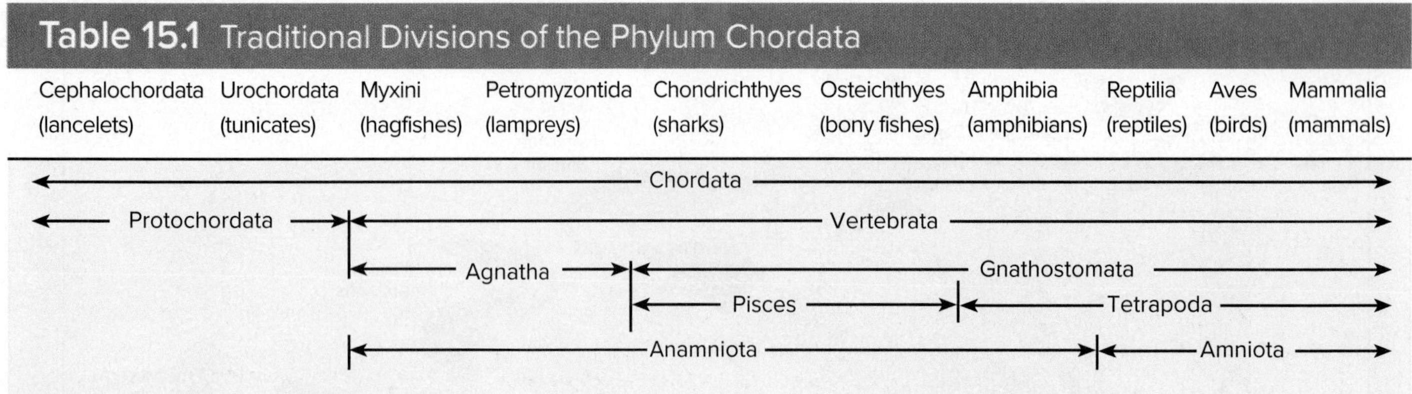

CHARACTERISTICS of Phylum Chordata

1. **Postanal tail; notochord; endostyle or thyroid gland; cartilage and bone** in vertebrates.
2. Living in marine, freshwater, and terrestrial habitats; many capable of flight.
3. Free-living, but a very few fishes are ectoparasitic.
4. Bilateral symmetry; segmented, but segmentation inconspicuous in many.
5. Triploblastic.
6. **Coelom well developed.**
7. Epidermis present in all; dermis in vertebrates; keratinized or bony structures often present in vertebrate integument; glands often diverse and abundant in vertebrates.
8. Digestive system complete; muscular gut in vertebrates; **pharyngeal pouches** present early in development, erupting to outside as gill slits in aquatic forms.
9. Smooth, skeletal, and cardiac muscle tissue present; segmented myomeres in fishes and amphibians.
10. **Nerve cord hollow and dorsal;** distinct, **three-lobed brain** present in vertebrates.
11. Protochordates with simple, unpaired photoreceptors and statocysts; vertebrates with well-developed paired sensory organs for vision, chemoreception, hearing, balance, electroreception, and vibration sensitivity.
12. Asexual reproduction uncommon but does occur by parthenogenesis in a few fishes, amphibians, and lizards.
13. Sexes usually separate; hermaphroditism in sea squirts and some fishes; fertilization internal or external; oviparous or viviparous; distinct larval stage in some; crocodilians, birds, mammals, and some fishes and amphibians with parental care of young.
14. Paired, glomerular kidneys and ducts in vertebrates.
15. Respiration primarily via gills, lungs, and skin; swim bladder present in many fishes, functioning in buoyancy.
16. Closed circulation; **chambered hearts** and red blood cells in vertebrates; distinct aortic arches in all except sea squirts.

jaws (hagfishes, lampreys, ostracoderms, and other jawless fossil fishes), and Gnathostomata, vertebrates having jaws (all other vertebrates) and (2) Amniota, vertebrates whose embryos develop within a fluid-filled sac, the amnion ("reptiles," birds, and mammals), and Anamniota, vertebrates lacking this adaptation (fishes and amphibians). Gnathostomata in turn can be subdivided into Pisces, jawed fishes, bearing fin-like appendages, and Tetrapoda (Gr. *tetras*, four, + *podos*, foot), vertebrates with appendages, if any, in the form of limbs. Note that several of these groupings are paraphyletic (Protochordata, Agnatha, Anamniota, Pisces) and consequently are not accepted in cladistic classifications (although living agnathans form a monophyletic group). Accepted monophyletic taxa are shown at the top of the cladogram in figure 15.2 as a nested hierarchy of increasingly more inclusive groupings.

Ancestry and Evolution

Since the mid-nineteenth century when Darwin's theory of common descent became the focal point for recognizing relationships among groups of living organisms, zoologists have debated the question of chordate origins. Zoologists at first speculated that chordates evolved within the protostome clade (annelids and arthropods) but rejected such ideas when they realized that supposed morphological similarities were not homologous. Early in the twentieth century, further theorizing became rooted in developmental patterns of animals, and it became apparent that chordates are deuterostomes. Deuterostomia, which also includes echinoderms and hemichordates (see figure 14.1), has several important embryological features, as well as shared gene sequences, that clearly separate it from Protostomia and establish its monophyly. Thus, deuterostomes likely arose in ancient Precambrian seas, probably as benthic, bilaterally symmetrical forms. Later, at the base of the Cambrian period some 540 million years ago, the first distinctive chordates appeared, possessing characters that permitted a relatively active lifestyle (figure 15.3).

Chordata is divided into three subphyla, Urochordata, Cephalochordata, and Vertebrata. The first two subphyla are covered in this chapter; their members are informally called **protochordates**. Vertebrata is introduced in this chapter, and its subgroups are covered in detail in Chapters 16–20.

Key Theme 15.1
EVOLUTION

Analogy and Homology

Most early efforts to determine chordate relationships were based on similarities due to analogy rather than homology (see section 4.3: Taxonomic Characters and Reconstruction of Phylogeny). Analogous structures perform similar functions but have different origins (such as the wings of birds and butterflies). Homologous structures, on the other hand, share a common origin but can look quite different (at least superficially) and often perform different functions. For example, all vertebrate forelimbs are homologous because they are derived from a pentadactyl limb of the same ancestor, even though they may be modified as differently as a human's arm and a bird's wing. Homologous structures share a genetic heritage; analogous structures do not. Obviously, only homologous similarities reveal common ancestry.

15.2 Five Chordate Characteristics

Five diagnostic characteristics that set chordates apart from all other phyla are a **notochord,** a **dorsal hollow nerve cord, pharyngeal slits or pouches,** an **endostyle,** and a **postanal tail.** These characteristics are always found at some embryonic stage, although they may change or disappear in later stages of life. All but pharyngeal pouches or slits are unique to chordates; hemichordates also have pharyngeal slits, and these are presumed ancestral to deuterostomes. We include pharyngeal slits as a chordate character because they are particularly distinctive in chordates and an important part of the chordate body plan.

Notochord

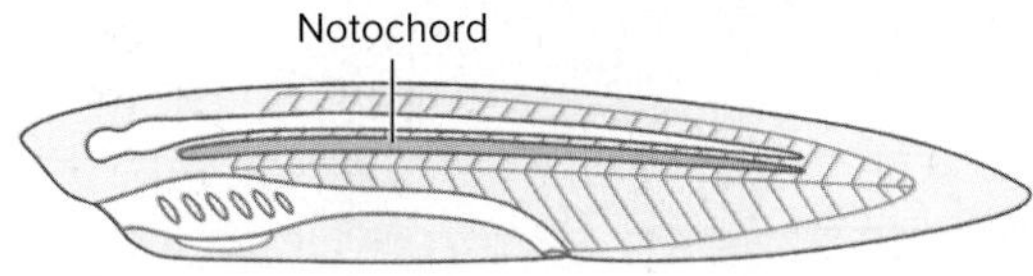

The notochord is a flexible rod extending the length of the body (see figure 15.1); it is the first part of the endoskeleton to appear in the embryo. The notochord is a hydrostatic organ, but unlike nematodes, which contain fluid in a single, large cavity, the fluid in a notochord is contained in cell vacuoles or in compartments between cells, all enclosed in a fibrous sheath. Muscles attach to the notochord, and because it can bend laterally without shortening, it permits undulatory movements of the body. In amphioxus and in jawless vertebrates, the notochord persists throughout life (see figure 15.1), but in nearly all jawed vertebrates it is at least partly replaced by a series of cartilaginous or bony vertebrae.

Dorsal Tubular Nerve Cord

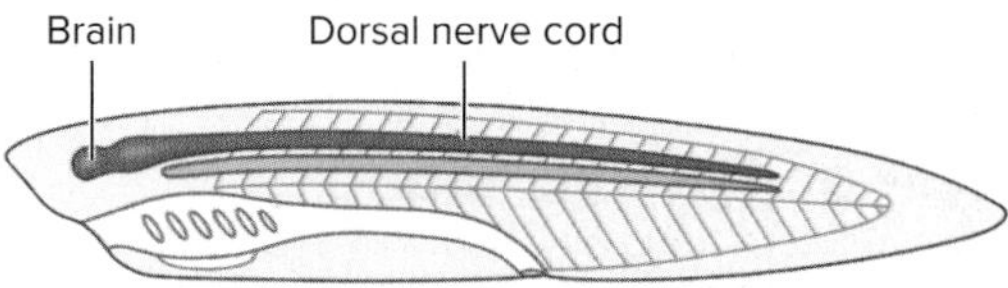

In most invertebrate phyla that have a nerve cord, it is ventral to the digestive tract and is solid, but in chordates the single cord is dorsal to the digestive tract and notochord and is a tube (although the hollow center may be nearly obliterated during growth). The anterior end becomes enlarged to form the brain in vertebrates. The hollow cord is produced in the embryo by infolding of ectodermal cells on the dorsal side of the body above the notochord. Among vertebrates, the nerve cord passes through the protective neural arches of the vertebrae, and the brain is surrounded by a bony or cartilaginous cranium.

Key Theme 15.2
GENETICS AND DEVELOPMENT

Upside-Down Chordates

Chordates and other bilaterians use the same genes, *Bmp* and *Chordin*, to establish their dorsoventral axis. However, chordates have apparently undergone a body inversion during their evolution, resulting in their characteristic dorsal nerve cords.

Pharyngeal Pouches or Slits

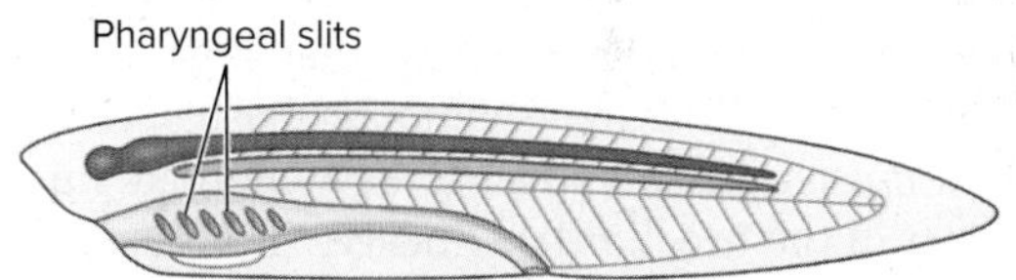

Pharyngeal slits are openings that lead from the pharyngeal cavity to the outside. They are formed by the inpocketing of the outside ectoderm (pharyngeal grooves) and the evagination, or outpocketing, of the endodermal lining of the pharynx (pharyngeal pouches). In aquatic chordates, the two pockets break through the pharyngeal cavity where they meet to form the pharyngeal slits. In amniotes, some pockets do not break through the pharyngeal cavity, so only pouches are formed instead of slits. In tetrapod (four-limbed) vertebrates, pharyngeal pouches give rise to several different structures, including the Eustachian tube, middle ear cavity, tonsils, and parathyroid glands.

The perforated pharynx evolved as a filter-feeding apparatus and is used as such in protochordates. Water with suspended food particles is drawn by ciliary action through the mouth and flows out through pharyngeal slits, where food is trapped in mucus. In vertebrates, ciliary action is replaced by muscular pharyngeal contractions that drive water through the pharynx. The addition of a capillary network and thin, gas-permeable walls in the pharyngeal arches led to the development of internal gills, completing conversion of the pharynx from a filter-feeding apparatus in protochordates to a respiratory organ in aquatic vertebrates.

Endostyle or Thyroid Gland

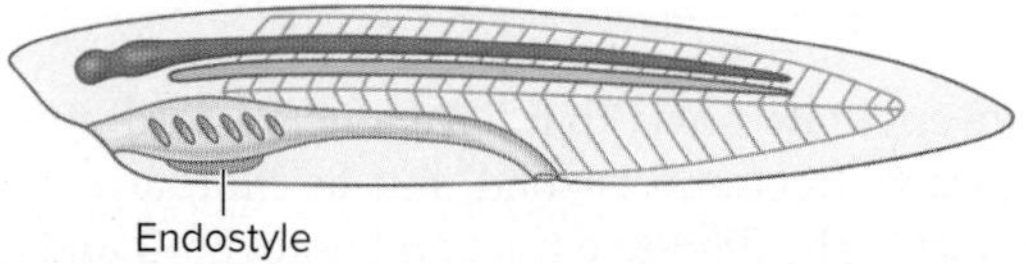

Until recently, the endostyle was not recognized as a chordate character. However, it or its derivative, the thyroid gland, occurs in all chordates, and in no other animals. The endostyle, located in the pharyngeal floor, secretes mucus that traps small food particles brought into the pharyngeal cavity. An endostyle occurs in protochordates and lamprey larvae. Some cells in the endostyle secrete iodinated proteins. These cells are homologous with the iodinated-hormone-secreting thyroid gland of adult lampreys and all other vertebrates. In protochordates and lamprey larvae an endostyle and a perforated pharynx work together to create an efficient filter-feeding apparatus.

Postanal Tail

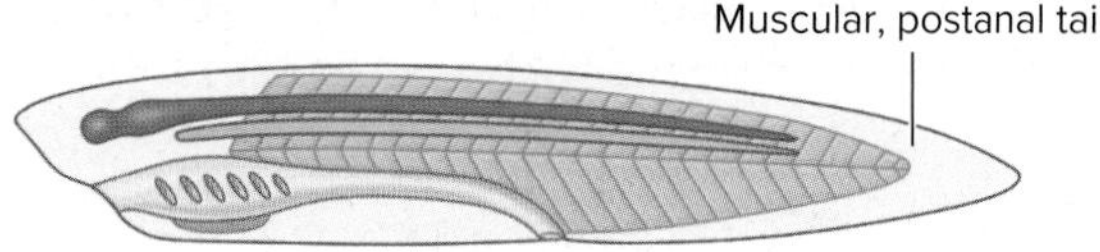

A postanal tail, together with somatic musculature and the stiffening notochord, provides the motility that larval tunicates and amphioxus need for their free-swimming existence. **Myomeres** (segmented muscles) also appear to be a chordate innovation, but were lost in the urochordate lineage. As a structure added to the body behind the anus, the tail clearly has evolved specifically for propulsion in water. Its efficiency is later increased in fishes with the addition of fins. A tail is evident in humans only as a vestige (the coccyx, a series of small vertebrae at the end of the spinal column), but most other mammals have a waggable tail as adults.

15.3 Subphylum Urochordata

The urochordates ("tail-chordates"), more commonly called tunicates, include about 3000 species. They live in all seas, from near shoreline to great depths. Most are sessile as adults, although some are free-living. The name "tunicate" describes the usually tough, nonliving **tunic,** or test, that surrounds the animal and contains cellulose (figure 15.4). As adults, tunicates are highly specialized chordates, for in most species only the larval form, which resembles a microscopic tadpole, bears all the chordate diagnostic characteristics. During metamorphosis, the notochord and the tail disappear, while the dorsal nerve cord is reduced to a single ganglion.

Urochordata is divided into three classes–**Ascidiacea** (Gr. *askiolion,* little bag), **Appendicularia** (L. *larva,* ghost), and **Thaliacea** (Gr. *thalia,* luxuriance). Members of Ascidiacea, commonly called ascidians, or sea squirts, are by far the most common and best known. Ascidians may be solitary, colonial, or compound. All but a few ascidian species are sessile, attached to rocks or other hard substances such as pilings or the bottoms of ships. In some areas, they are among the most abundant of intertidal animals.

Solitary or colonial ascidians are usually spherical or cylindrical forms. Lining the tunic is an inner membrane, the mantle. On the outside are two projections: an **incurrent siphon** and an **excurrent siphon** (figure 15.4). Water enters the incurrent siphon and passes into the **pharynx**. On the midventral side of the pharynx is a groove, the **endostyle,** which is ciliated and secretes a sheet of mucus. As cilia carry the mucous sheet across the inner surface of the pharynx to the dorsal side, it traps small food particles from water passing through slits in the wall of the **pharynx**. The mucus with its entrapped food is formed into a rope and passed posteriorly to the esophagus. The water, now largely cleared of food particles, is driven by cilia into the atrial cavity and finally out the excurrent siphon. Nutrients are absorbed in the midgut, and indigestible wastes are discharged from the anus, located near the excurrent siphon.

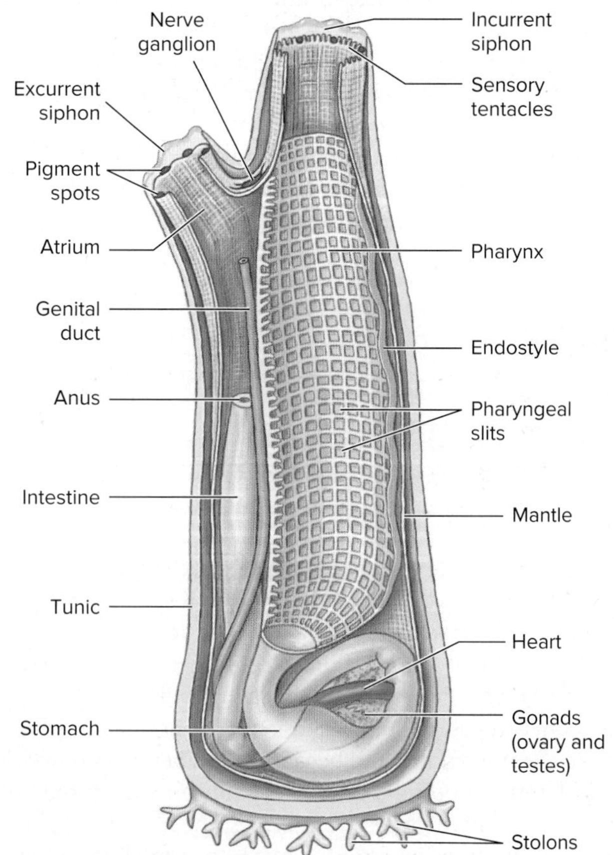

figure 15.4

Structure of a common tunicate, *Ciona.*

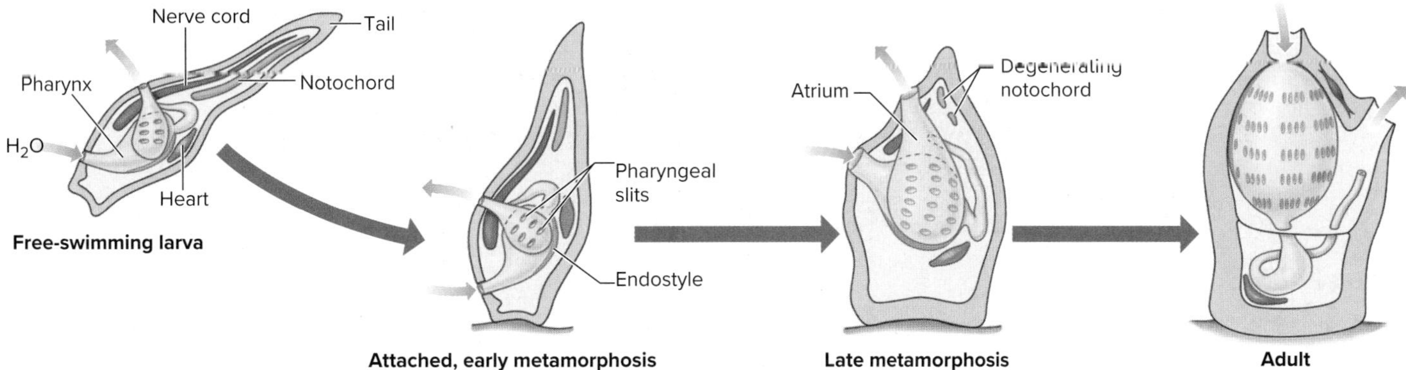

figure 15.5

Metamorphosis of a solitary ascidian from a free-swimming larval stage.

The circulatory system consists of a ventral heart near the stomach and two large vessels, one on either side of the heart. An odd feature, found in no other chordate, is that the heart first drives the blood in one direction for a few beats, then pauses, reverses, and drives the blood in the opposite direction. The nervous system of adults is restricted to a nerve ganglion and a few nerves that lie on the dorsal side of the pharynx. Sea squirts are hermaphroditic. Gametes are carried out of the excurrent siphon into the surrounding water, where fertilization occurs.

Of the five diagnostic characteristics of chordates, adult sea squirts have only two: pharyngeal slits and an endostyle. However, the larval forms reveal the secret of their true relationship. The tiny "tadpole" larva (figure 15.5) is an elongate, transparent form with all five diagnostic chordate characteristics: a notochord, a dorsal hollow nerve cord, a propulsive postanal tail, and a large pharynx with an endostyle and gill slits. The larva does not feed but swims for several hours or days before fastening vertically by adhesive papillae to a solid object. It then metamorphoses to become a sessile adult.

The remaining two classes of Urochordata—Appendicularia and Thaliacea—are mostly small, transparent animals of the open sea (figure 15.6). Appendicularians, also called larvaceans, are small, tadpole-shaped forms resembling the larval stage of ascidians. An appendicularian builds and inhabits a hollow, transparent sphere of mucus. Inside the sphere, feeding filters trap and ingest tiny plankton. Thaliaceans, which include salps (figure 15.6), are barrel- or lemon-shaped forms surrounded by circular muscle bands. Ocean currents transport them as part of the plankton. Many have luminous organs and emit a brilliant light at night.

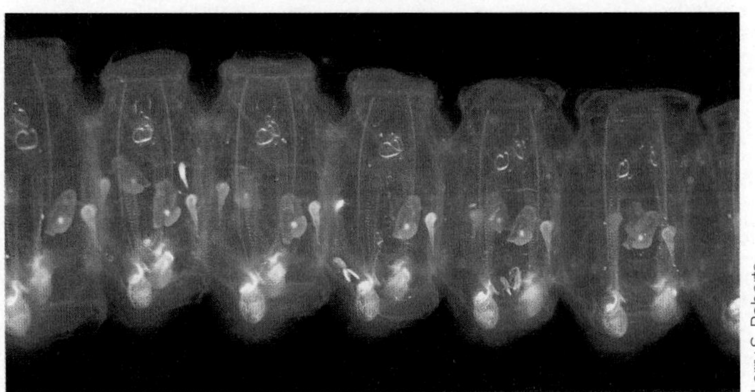
Larry S. Roberts

figure 15.6

Salps. The transparent individuals of this delicate, planktonic species are grouped in a chain. Visible within each individual is an opaque gonad, an opaque gut, and a long, serrated gill bar. Class Thaliacea.

15.4 Subphylum Cephalochordata

Cephalochordates are lancelets: slender, laterally compressed, translucent animals about 3–7 cm in length (figure 15.7) that inhabit sandy sediments of coastal waters around the world. Lancelets originally bore the generic name *Amphioxus* (Gr. *amphi*, both ends, + *oxys*, sharp), but later the genus reverted to the older name *Branchiostoma* (Gr. *branchia*, gills, + *stoma*, mouth). Amphioxus is still used, however, as a common name for all the approximately 32 species in this diminutive subphylum. Five species of amphioxus occur in North American coastal waters. Despite the name Cephalochordata, cephalization is only moderately developed in amphioxus. Their nerve cord is not expanded anteriorly into a distinct brain, and they lack the paired special sense organs that appear in the distinctive head of vertebrates.

An amphioxus spends most of its time partly buried in sediments, with only its anterior end exposed (see photo at the beginning of this chapter). During feeding, water enters the mouth, driven by cilia in the buccal cavity, and then passes through numerous pharyngeal slits, where mucus secreted by the endostyle traps food, which, is then moved by cilia into the gut. Here, the smallest food particles are separated from the mucus and pass into the **hepatic cecum,** where they are phagocytized and digested intracellularly. As in tunicates, filtered water passes first into an **atrium,** and then leaves the body by an **atriopore** (equivalent to the excurrent siphon of tunicates).

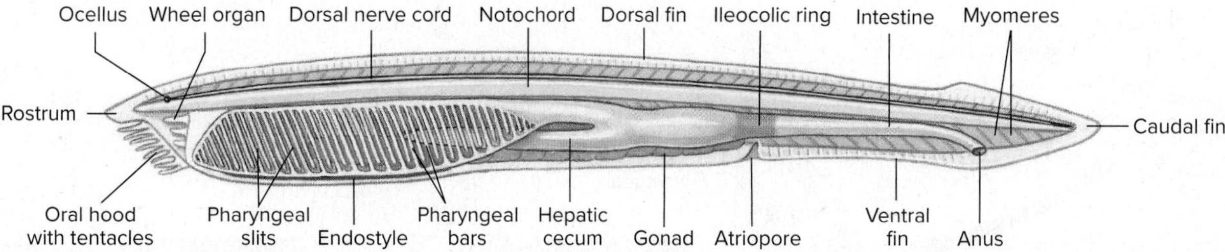

figure 15.7

Amphioxus. This filter-feeding cephalochordate possesses the five diagnostic chordate characteristics (notochord, dorsal hollow nerve cord, pharyngeal slits, endostyle, and postanal tail).

The closed circulatory system is remarkably similar to that of fishes, although there is no heart. Blood is pumped forward in the **ventral aorta** by peristaltic-like contractions of the vessel wall, and then passes upward through branchial arteries (aortic arches) in pharyngeal arches to paired **dorsal aortas**. From there, blood is distributed to the body tissues by capillaries and then collected in veins, which return it to the ventral aorta. Lacking both erythrocytes and hemoglobin, the blood of a lancelet is thought to circulate nutrients but not respiratory gases. There are no gills specialized for respiration in the pharynx; gas exchange occurs over the surface of the body, especially across the lining of the atrium.

The nervous system is centered around a hollow nerve cord lying above the notochord. Sense organs are simple, including an anterior, unpaired **ocellus** that functions as a photoreceptor. There is no distinctive brain.

Sexes are separate. Gametes are released in the atrium, and then pass through the atriopore to the outside, where fertilization occurs. Larvae hatch soon after the eggs are fertilized and gradually assume the shape of adults.

In addition to the five chordate diagnostic characteristics, amphioxus possesses several structural features that resemble the vertebrate plan. Among these are a hepatic cecum, which is similar to the vertebrate liver and pancreas in function, **segmented trunk musculature (myomeres)**, and the basic circulatory pattern of vertebrates.

15.5 Subphylum Vertebrata

The third subphylum of chordates is the large and diverse Vertebrata, the subject of Chapters 16–20. This monophyletic group shares the five diagnostic chordate characteristics with the other two subphyla, but it has novel characters that the others do not share.

In this section we first present characteristics of vertebrates that distinguish them from "protochordates" (amphioxus and tunicates). Second, we discuss the evolutionary origin of vertebrates, including the earliest well-known vertebrates, the ostracoderms. Finally we discuss the evolution of two important characters common to many vertebrates, jaws and paired fins, and two groups of fossil fishes with these adaptations. Fish groups with living members are discussed in Chapter 16.

Adaptations That Have Guided Early Vertebrate Evolution

The earliest vertebrates were substantially larger and considerably more active than the protochordates. Modifications of the skeleton and muscles permitted increased speed and mobility. The higher activity level and size of vertebrates also required structures specialized for locating, capturing, and digesting food as well as adaptations that support a high metabolic rate.

Invertebrates have one cluster of *Hox* genes, while most vertebrates have four clusters (see figure 3.8), suggesting two *Hox* duplication events occurred early during vertebrate evolution. Some researchers have proposed that duplications in homeotic genes, genes that control body patterning, are linked to the increased morphological complexity of vertebrates.

Musculoskeletal Modifications

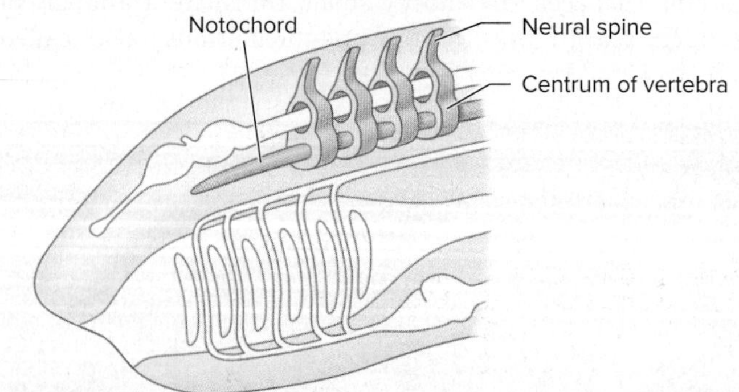

The endoskeleton of vertebrates permits almost unlimited body size, with much greater economy of building materials than does the exoskeleton of arthropods. The most massive animals on earth are vertebrates. In most vertebrates, disclike centra replace the notochord, and dorsal projections called **neural spines** are present on the vertebrae, providing more area for attachment of segmented muscles.

The endoskeleton probably was composed initially of cartilage but later of bone. Cartilage, with its fast growth and flexibility, is ideal for constructing the first skeletal framework of all vertebrate embryos. The endoskeleton of living hagfishes, lampreys, sharks, and their kin, and even that of some "bony" fishes, such as sturgeons, is mostly cartilage. The extracellular matrix of bone tissue is of collagen fibers and the mineral hydroxyapatite, which is mostly formed from calcium and phosphate ions. Bone may have been adaptive in early vertebrates in several ways. Plates of bone in the skin of ostracoderms and other ancient fishes certainly provided protection from predators. Bone is structurally stronger than cartilage, making it ideal for muscle attachment in areas of high mechanical stress. One of the most interesting ideas is that bone's original function was mineral storage and homeostasis. Phosphorus and calcium are used for many physiological processes and are in particularly high demand in organisms with high metabolic rates.

Some vertebrates possess an extensive exoskeleton (one that develops from the skin). Most of the earliest fishes, including the ostracoderms (see figure 15.11) and placoderms (see figure 15.12), were partly covered by a bony dermal armor. This armor is modified as scales in later fishes. Most terrestrial vertebrates are protected with keratinized structures that develop from the epidermis, such as reptilian scales, hair, and feathers.

Physiology Upgrade

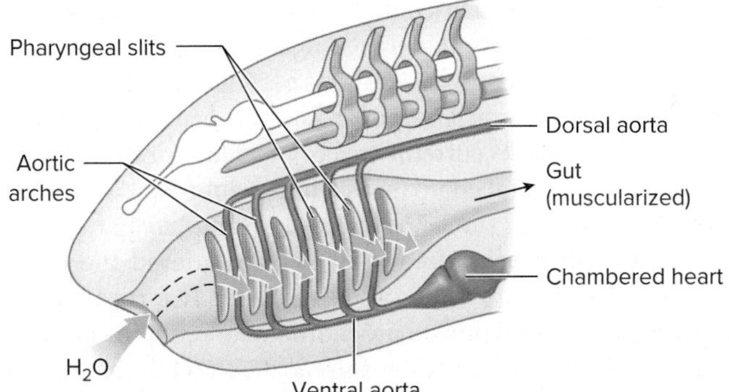

Vertebrate digestive, respiratory, and circulatory systems are modified to meet an increased metabolic demand. The perforated pharynx evolved as a filter-feeding device in early chordates, using cilia and mucus to move water and to trap small, suspended food particles. In vertebrates, the addition of muscles to the pharynx created a powerful pump for moving water. With the origin of highly vascularized gills, the function of the pharynx shifted primarily to gas exchange. Changes in the gut, including a shift from movement of food by ciliary action to muscular action and the addition of accessory digestive glands, the liver and pancreas, managed the increased amount of food ingested. A ventral, chambered heart and erythrocytes with hemoglobin enhanced transportation of respiratory gases, nutrients, and other substances.

New Head, Brain, and Sensory Systems

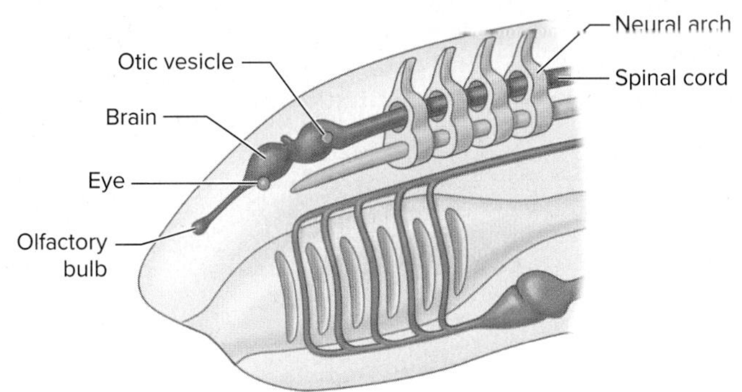

When vertebrate ancestors shifted from sessile filter feeding to active predation, new sensory, motor, and integrative controls became essential for locating and capturing larger prey. The anterior end of the nerve cord became enlarged as a **tripartite brain** (forebrain, midbrain, and hindbrain). Paired special sense organs adapted for distance reception evolved. These included eyes with lenses and inverted retinas; pressure receptors, such as paired inner ears designed for equilibrium and sound reception; chemical receptors, including taste and exquisitely sensitive olfactory organs; lateral-line receptors for detecting water vibrations; and electroreceptors for detecting electrical currents that signal prey.

Neural Crest and Ectodermal Placodes

Development of the vertebrate head and special sense organs resulted from two embryonic tissues nearly unique to vertebrates: **neural crest** and **ectodermal placodes**. The neural crest, a population of ectodermal cells lying along the length of the embryonic neural tube, contributes to the formation of most of the cranium, the pharyngeal skeleton, tooth dentine, some endocrine glands, and other structures. Ectodermal placodes (Gr. *placo*, plate) are platelike thickenings of the epidermis. These give rise to the olfactory epithelium, the lens of the eye, inner-ear epithelium, lateral-line mechanoreceptors, and electroreceptors. Thus, the vertebrate head, with its complex sensory structures located adjacent to the mouth (later equipped with prey-capturing jaws), stemmed from new embryonic tissues.

The Search for the Ancestral Vertebrate Stock

Most of the early Paleozoic vertebrate fossils, the jawless ostracoderms (see figure 15.11), share many novel features of organ system development with living vertebrates. These organ systems must have originated in an early vertebrate or invertebrate chordate lineage. Fossil invertebrate chordates are rare and known primarily from two places—the well-known middle-Cambrian Burgess Shale of Canada (see figure 1.12) and the early-Cambrian fossil beds of Chengjiang and Haikou, China. *Pikaia*, a ribbon-shaped, somewhat fishlike creature about 5 cm

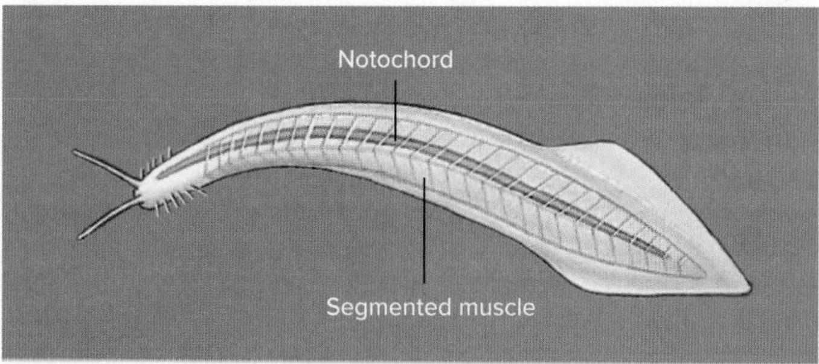

figure 15.8

Pikaia, an early chordate from the Burgess Shale of British Columbia, Canada.

in length, was discovered in the Burgess Shale (figure 15.8). The presence of myomeres and a notochord clearly identifies *Pikaia* as a chordate. The superficial resemblance of *Pikaia* to living amphioxus suggests that it may be an early cephalochordate.

Haikouella lanceolata, a small, fishlike creature discovered in China, possesses several characters that clearly identify it as a chordate, including notochord, pharynx, and dorsal nerve cord. Additional fossil features, interpreted as pharyngeal muscles, paired eyes, and a distinct brain, are characteristic of vertebrates (figure 15.9). However, it is not a vertebrate, because the fossils lack evidence of several diagnostic vertebrate traits, including a cranium and a distinct telencephalon (anterior region of the forebrain). *Haikouella* has been hypothesized to be the sister taxon of vertebrates, although this is not universally accepted. Despite recent fossil discoveries of early chordates, many speculations regarding vertebrate ancestry have focused on the living protochordates, in part because they are better known than are the fossil forms.

Chordate Evolution and the Position of Amphioxus

Some chordates–sea squirts–are sedentary, whereas others–amphioxus and vertebrates–are active and mobile. Which was

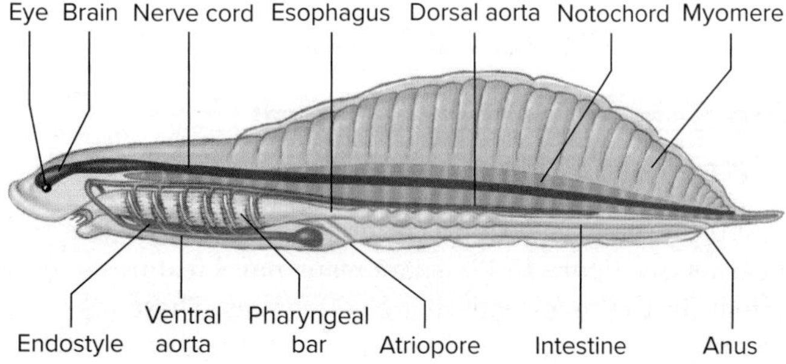

figure 15.9

Haikouella, a chordate with several vertebrate features from the early Cambrian shales of Haikou, China. It has been hypothesized that *Haikouella* is the sister group of vertebrates.

the ancestral form? In 1928 Walter Garstang suggested the ancestral chordate was a sedentary filter feeder, like adult ascidian tunicates. Garstang hypothesized that the vertebrate ancestor lost the ability to metamorphose into a sessile adult, instead developing gonads and reproducing in the larval form. Garstang called this process **paedomorphosis** (Gr. *pais*, child, + *morphe*, form), a term that describes the evolutionary retention of juvenile or larval traits in the adult body. Paedomorphosis is a well-known phenomenon in several different animal groups (paedomorphosis in amphibians is described in section 17.4).

Key Theme 15.3
EVOLUTION

Paedomorphosis

Paedomorphosis, the displacement of ancestral larval or juvenile features into a descendant adult, can be produced by two different evolutionary-developmental processes: neoteny and progenesis. In neoteny, the growth rate of body form is slowed so that the animal does not attain the ancestral adult form when it reaches maturity. Progenesis is precocious maturation of gonads in a larval (or juvenile) body that then stops growing and never attains an adult body form.

Recently collected genetic and developmental evidence allows testing of Garstang's hypothesis. Numerous phylogenetic reconstructions, along with fossil evidence, place cephalochordates as the sister taxon to a clade consisting of urochordates plus vertebrates, suggesting that cephalochordates and vertebrates retain the ancestral chordate condition and that the sessile ascidians represent a derived condition (see figure 15.2). In addition, incipient neural crest tissue recently has been identified in urochordates, supporting a sister-group relationship with vertebrates. Thus, most zoologists currently reject Garstang's hypothesis and envision the ancestral chordate as a free-swimming creature (see figure 15.3), perhaps similar to the modern amphioxus.

Although urochordates are now considered the closest living relatives of vertebrates, their sessile body form is evolved in a lineage ancestral only to urochordates and not in vertebrate ancestry. Most zoologists view amphioxus as retaining much of the prevertebrate and ancestral chordate body structure. Cephalochordates share several characters with vertebrates that are absent from tunicates, including segmented musculature, dorsal and ventral aortas, and branchial or aortic arches. Because of amphioxus' similarity to the presumed archetypal vertebrate form, it remains a popular reference organism in studies of early vertebrate evolution.

The Earliest Vertebrates

The earliest known vertebrate fossils, are small, mostly soft-bodied forms from the early Cambrian period, 530–500 million years ago: *Myllokunmingia* and *Haikouichthys* from the Chengjiang beds in China and *Metaspriggina* from the Burgess Shale. These fishlike fossils show many vertebrate characteristics, including paired eyes and other sense organs, complex swimming muscles, prominent gills, and what have been interpreted as rudimentary vertebrae.

For decades, geologists have used strange, microscopic, toothlike fossils called **conodonts** (Gr. *kōnos*, cone, + *odontos*, tooth) to date Paleozoic marine sediments without having any idea what kind of creature originally possessed these elements. The discovery in the early 1980s of fossils of complete conodont animals changed this situation. With their phosphatized toothlike elements, myomeres, and paired eye and otic capsules, conodonts clearly belong to the vertebrate clade (figure 15.10).

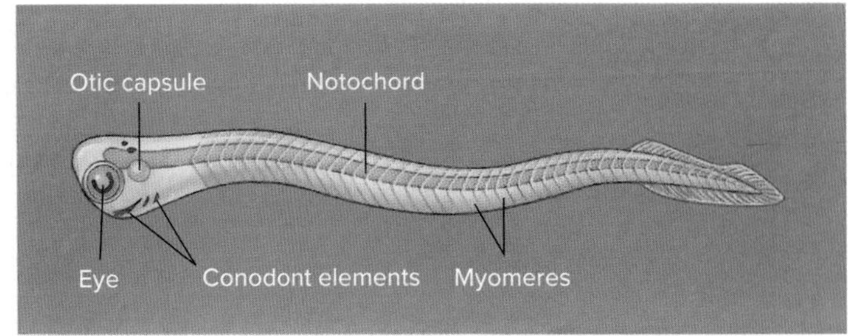

figure 15.10

Restoration of a conodont. Conodonts superficially resembled amphioxus, but they possessed a much greater degree of cephalization (paired eyes and otic [ear] capsules) and bonelike mineralized elements—all indicating that conodonts were probably vertebrates. Conodont elements are thought to be part of a food-handling apparatus.

Much better known are the **ostracoderms** (os-trak'o-derms) (Gr. *ostraco*, shell + *derma*, skin) from the Ordovician to Devonian periods, 440–360 mya. Ostacoderms do not form a monophyletic group, but were similar in morphology. These jawless fishes were armored with bone in their dermis. Most lacked paired fins (pectoral and pelvic), but some had pectoral fins (figure 15.11). Ostracoderms were small animals, usually less than 30 cm in length, but larger than the first chordates, such as *Pikaia* (see figure 15.8), *Haikouella* (see figure 15.9), and *Metaspriggina*. Due to their bony armor, ostracoderms are well-represented in the fossil record.

One of the earliest ostracoderm groups was the **heterostracans** (figure 15.11), peculiar-looking fishes with several large bony plates on the head and a hypocercal caudal fin (with a large lower lobe) and lacking all paired fins. Appearing later, in the Devonian period, were the **osteostracans**. They were covered with heavy plates of bone in their dermis, including a single-piece head shield. Unlike most other ostracoderms, osteostracans had pectoral fins, an innovation that improved swimming efficiency by stabilizing the body during swimming and providing better maneuverability. A third group, the **anaspids**, were more streamlined than other ostracoderms, and had little dermal armor.

Cephalization was prominent in ostracoderms; examination of the internal features of the cranium of one group, the osteostracans, revealed a sophisticated nervous system and sense organs. Ostracoderms had circular or slit-like mouth openings. Although many likely fed by filtering small particles from the sea floor, some were probably predators on soft-bodied invertebrates. Unlike the ciliary, filter-feeding amphioxus and tunicates, ostracoderms had an impressive,

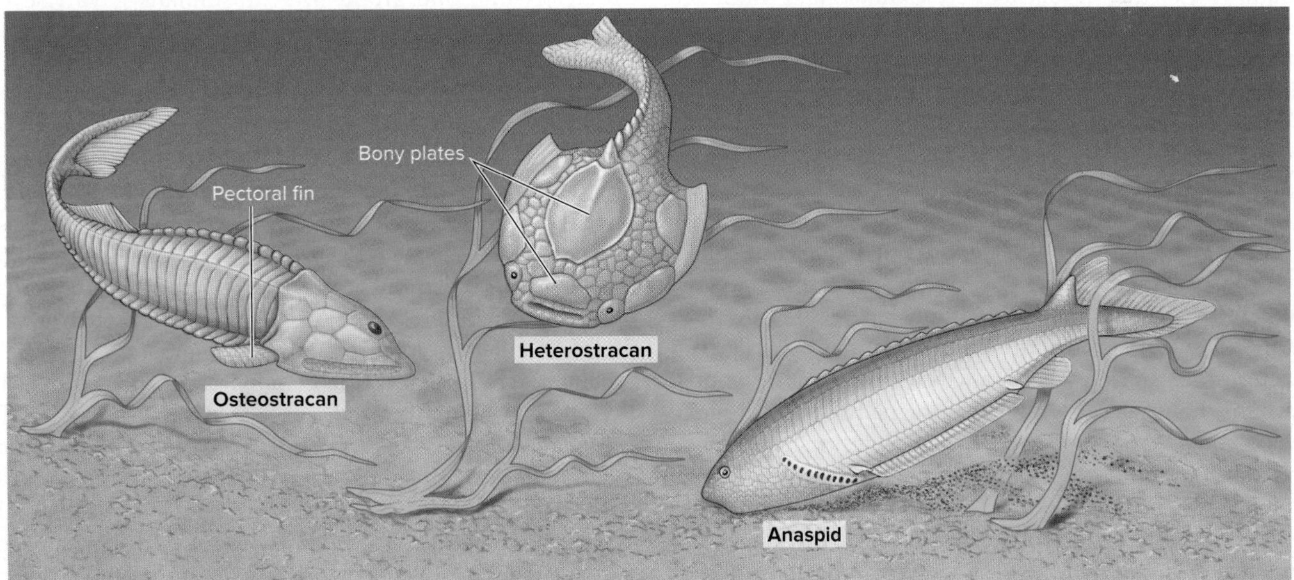

figure 15.11

Three ostracoderms, jawless fishes of Silurian and Devonian times. They are shown as they might have appeared while searching for food on the floor of a Devonian sea. These fishes, 10-15 cm in length, employed a strong pharyngeal pump to circulate water rather than relying on the much more limiting mode of ciliary action used by their ancestors (presumably resembling amphioxus in this way). Shown are *Hemicyclaspis* (osteostracan), *Drepanaspis* (heterostracan), and *Pharyngolepis* (anaspid).

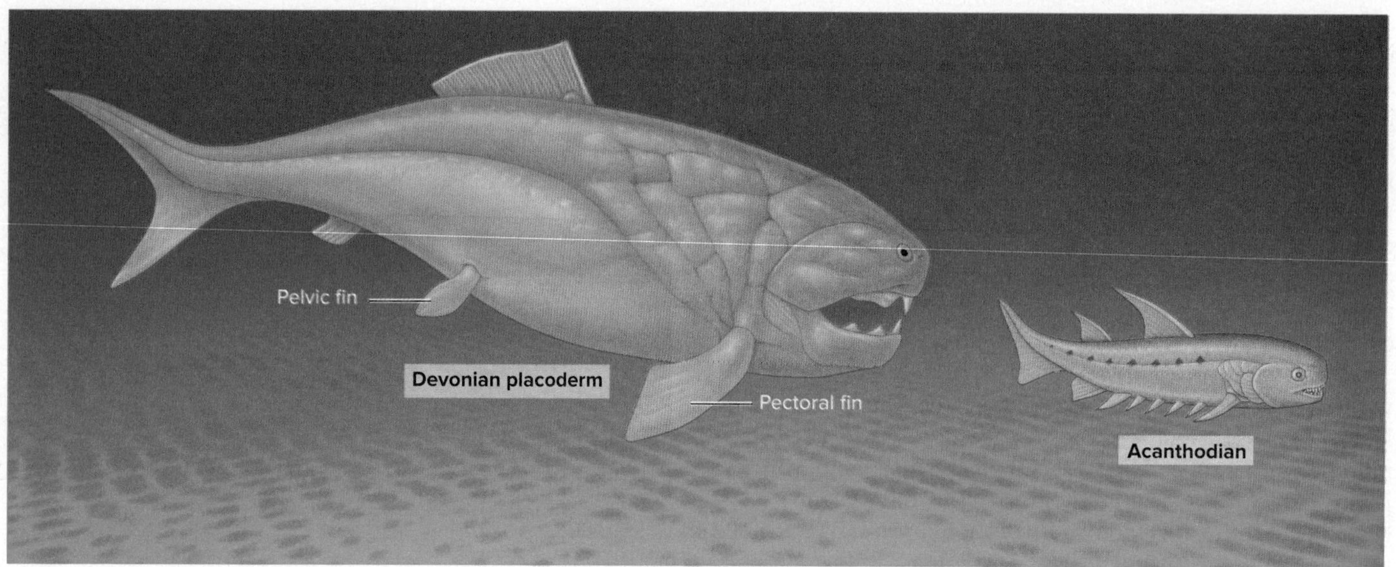

figure 15.12

Early jawed fishes of the Devonian period, 400 million years ago. Placoderms were diverse in morphology, but most had bony plates on the front of the body. The largest placoderms, such as *Dunkleosteus* (shown here), were predators, reaching over 6 m in length. Acanthodians (*Climatius* is shown here) carried less armor than placoderms and had large, anteriorly placed eyes and prominent spines on paired fins. Most were marine but several species lived in fresh water. Both groups had paired pectoral and pelvic fins, usually absent in jawless fishes.

muscle-powered pharyngeal skeleton, which could pump water past the gills and suck-up prey.

For about 90 million years, the only fishes in the seas were jawless fishes. Then, in the Silurian and Devonian periods, ostracoderms were joined by fishes that had an important innovation—jaws. Ostracoderms were common and continued to diversify alongside jawed fishes for another 40 million years before beginning a slow decline. All ostracoderms went extinct by the end of the Devonian period, leaving the world's seas dominated by the jawed fishes.

Early Jawed Vertebrates

All jawed vertebrates, whether extinct or living, are collectively called **gnathostomes** ("jaw mouth") in contrast to jawless vertebrates, the **agnathans** ("without jaw"). Gnathostomes are a monophyletic group; the presence of jaws is a derived character state shared by all jawed fishes and tetrapods. Agnathans, however, are defined principally by the absence of jaws, a character that is not unique to jawless fishes because jaws are lacking in vertebrate ancestors. Thus, "Agnatha" is paraphyletic.

The first jawed vertebrates were the heavily armored **placoderms** (Gr. *plax*, plate, + *derma*, skin) and the lightly armored **acanthodians** (figure 15.12). Both groups first appeared in the fossil record in the Silurian period, about 430 million years ago, and were especially diverse in the Devonian period. Placoderms evolved a great variety of forms, some very large (one was 6 m in length!). They were armored fishes covered with diamond-shaped scales or large plates of bone. **Acanthodians** had large, anteriorly set eyes and fins with large spines, with bone limited to a few places on the skin of the head and base of the fins. Both groups are probably paraphyletic (see figure 16.2). Recent studies suggest one group of placoderms gave rise to all other gnathostomes, while one group of acanthodians gave rise to the sharks and rays. Otherwise, the placoderms and acanthodians became extinct at the end of the Devonian and Permian periods, respectively.

The origin of **jaws** was one of the most important events in vertebrate evolution. The utility of jaws is obvious: they allow predation on large and active forms of food not available to jawless vertebrates. How did jaws evolve? Jawless fishes have paired, serially repeated cartilaginous arches that support the pharynx, and, with attached muscles, pump water across the gills (figure 15.13). Ample evidence suggests that jaws arose from an anterior arch surrounding the mouth, called the mandibular arch. In sharks, the mandibles and gill arches both develop from neural crest tissue, have a similar form, and have similar associated muscles and cranial nerves. How did this mandibular arch change from a function of gill support and ventilation to one of feeding as jaws? Expansion of this arch and evolution of new, associated muscles may have first assisted gill ventilation, perhaps to meet the increasing metabolic demands of early vertebrates or to enhance suction feeding. Once enlarged and equipped with extra muscles, the first pharyngeal arch easily could have been modified to serve as jaws.

An additional feature characteristic of all gnathostomes is the presence of **paired pectoral** and **pelvic appendages** in the form of fins or limbs. These likely originated as stabilizers

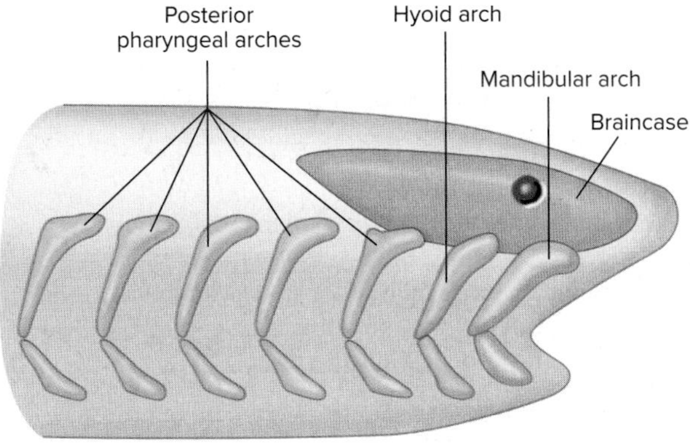

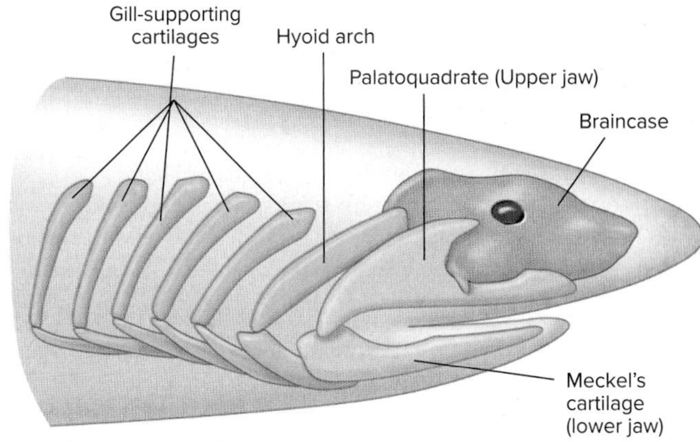

figure 15.13

How vertebrates got their jaws. **A,** Hypothetical, ancestral, jawless vertebrate with serially repeated pharyngeal arches. **B,** Extinct, Carboniferous shark. In gnathostomes, jaws form from the first (mandibular) arch, the hyoid arch forms from the second arch, and gill supports form from the posterior arches.

to check yaw, pitch, and roll generated during active swimming. According to the fin-fold hypothesis, paired fins arose from paired continuous, ventrolateral folds or fin-forming zones. The addition of skeletal supports in the fins enhanced the fins' ability to provide stability during swimming. These paired folds might have been simliar to the metapleural folds of amphioxus or the paired flaps of *Myllokunmingia* and anaspids (figure 15.11). Distinct pectoral and pelvic fins arose from the fin folds, permitting better control of steering during swimming. In one fish lineage, the muscle and skeletal supports in the paired fins became strengthened, adapting them for locomotion on land as limbs.

Living Fishes

Chapter 16 covers the living fishes. Two small groups, hagfishes and lampreys, are the sole surviving agnathans. Arising from the early gnathostomes were two clades that diversified into large groups: Chondrichthyes, the cartilaginous fishes, and Osteichthyes, the bony fishes and tetrapods. Both jaws and paired fins were major innovations in vertebrate evolution, among the most important reasons for the subsequent major radiations of vertebrates that produced the modern fishes and all tetrapods, including you, the reader of this book.

Taxonomy of Living Members of Phylum Chordata

The following classification represents a cladistic approach to describing the relationships within phylum Chordata. Linnaean ranks are omitted for taxa within Vertebrata, because a cladistic classification system for all vertebrates using such ranks has not been established. Nested relationships are shown by indenting.

Subphylum Urochordata (yur'ō-kor-dä'tə) (Gr. *oura*, tail, + L. *chorda*, cord, + *ata*, characterized by) **(Tunicata): tunicates**. Notochord and nerve cord in free-swimming larva only; ascidian adults sessile, encased in tunic. 3000 species.

Subphylum Cephalochordata (sef'ə-lō-kor-dä'tə) (Gr. *kephale-*, head, + L. *chorda*, cord): **lancelets (amphioxus)**. Notochord and nerve cord found along entire length of body and persist throughout life; fishlike in form. 32 species.

Subphylum Vertebrata (ver'tē-brä'tə) (L. *vertebratus*, backboned): **vertebrates**. Bony or cartilaginous cranium surrounding tripartite brain; well-developed head with paired sense organs, usually with vertebrae; heart with multiple chambers; muscularized digestive tract; paired kidneys.

Cyclostomata (sīi'klō-stō-mä'tä) (Gr. *cyclos, circle, stoma*, mouth): **hagfishes, lampreys**. Without true jaws or paired appendages; teeth of keratin.

Myxini (mik-sē'nē) (Gr. *myxa*, slime): **hagfishes**. Four pairs of tentacles around mouth; buccal funnel absent; 1–16 pairs of gill openings; slime glands present; vertebrae vestigial. About 78 species.

Petromyzontida (pet'trō-mī-zon'ti'də) (Gr. *petros*, stone, + *myzon*, sucking): **lampreys**. Buccal funnel with keratinized teeth, nasal sac not connected to pharynx; vertebrae present only as neural arches. 41 species.

(*continued*)

(*continued*)

Gnathostomata (na'thō-stō'mă-tə) (Gr. *gnathos*, jaw, + *stoma*, mouth): **jawed fishes, tetrapods**. With jaws and (usually) paired appendages.

Chondrichthyes (kon-drik'thē-ēz) (Gr. *chondros*, cartilage, + *ichthys*, a fish): **sharks, skates, rays, chimaeras**. Cartilaginous skeleton, intestine with spiral valve; claspers present in males; no swim bladder. About 1200 species.

Actinopterygii (ak'ti-nop'te-rij'ē-ī) (Gr. *aktis*, ray, + *pteryx*, fin, wing): **ray-finned fishes.** Bony skeleton; single gill opening covered by operculum; paired fins supported primarily by dermal rays; appendage musculature within body; swim bladder, if present, mainly for buoyancy atrium and ventricle not divided. About 29,650 species.

Sarcopterygii (sär-cop-te-rij'ē-ī) (Gr. *sarkos*, flesh, + *pteryx*, fin, wing): **lobe-finned fishes, tetrapods.** Bony skeleton; paired appendages with sturdy internal bones and musculature within appendage; atrium and ventricle at least partly divided.

Dipnoi (dip'nōi) (Gr. *di*, double + *pnoi*, breathing): **lungfishes.** Freshwater habitat; platelike, crushing teeth; caudal fin joined with anal and dorsal fins; swim bladder modified as one or two lungs. 6 species.

Actinistia (ak'tin-is-tē-ə) (Gr. *actino*, ray + *istia*, a superlative ending): **coelacanths.** Marine habitat; three-lobed caudal fin; swim bladder filled with fat. 2 species.

Amphibia (am-fib'ē-ə) (Gr. *amphi*, both or double, + *bios*, life): **amphibians.** Ectothermic tetrapods; respiration by lungs, gills, or skin; development through larval stage; skin moist, containing mucous glands and lacking scales. About 7900 species.

Reptilia (rep-til'ē-ə) (L. *repere*, to creep): **birds, turtles, lizards, snakes, crocodilians.** Tetrapods with an amniotic membrane in embryos; lungs ventilated by aspiration; skin dry. Birds comprise about 10,500 species of endothermic reptiles with front limbs modified as wings; body covered with feathers. The remainder of reptiles (turtles, lizards, snakes, tuataras, and crocodilians) are about 9500 species of ectothermic tetrapods covered by epidermal scales.

Mammalia (ma-mā'lē-ə) (L. *mamma*, breast): **mammals.** Endothermic tetrapods with mammary glands; amniotic membrane in embryos; body more or less covered with hair; brain large, with neocortex; three middle ear bones. About 5400 species.

SUMMARY

Sections	Key Concepts
15.1 Chordate Evolution and Classification	• Chordates are named for the notochord, a fluid-filled skeletal rod that provides an anchor for swimming muscles. • Chordates are deuterostomes, sharing several developmental traits with echinoderms and hemichordates, including radial cleavage and the anus formed from the blastopore. • Chordata is composed of three subphyla: Urochordata, Cephalochordata, and Vertebrata. • The chordate ancestor was likely a free-swimming, filter-feeding creature.
15.2 Five Chordate Characteristics	• Chordates have five diagnostic characters that distinguish them from all other animal phyla: (1) notochord; (2) dorsal, hollow nerve cord; (3) pharyngeal pouches or slits; (4) muscular, postanal tail; (5) endostyle. These characters are unique to chordates except for pharyngeal slits, which are found in simple form in a few other deuterostomes.
15.3 Subphylum Urochordata	• Members of subphylum Urochordata are commonly called tunicates, sea squirts, and salps. • Most are sedentary as adults, with a swimming larval stage. The swimming larval stage has all five diagnostic chordate characters, but adults possess only pharyngeal slits and an endostyle of the five. • They are covered with a tough tunic and filter feed by circulating water via siphons and trapping food in their netlike pharynx.
15.4 Subphylum Cephalochordata	• Members of Cephalochordata, commonly called amphioxus, are small, fishlike creatures that partly bury in the substrate and filter feed using their pharynx. • Amphioxus, with their fins and segmented muscles (myomeres), are stronger swimmers than urochordates. • Unlike urochordates, cephalochordates retain all five diagnostic chordate characters throughout life.

Sections	Key Concepts
15.5 Subphylum Vertebrata	• Vertebrates are characterized by vertebrae (reduced in some), a well-developed head, large size, and a high degree of mobility. • Vertebrates have an endoskeleton of bone and cartilage, providing a sturdy framework for attachment of muscles and a store for calcium and phosphorus. • High metabolic rates are supported with a number of structures, including a muscular pharynx, equipped with high-surface-area gills. Oxygen picked up at the gills is circulated in the blood with a chambered heart, another distinctive vertebrate trait. • The complex sensory systems and distinct brain provide vertebrates with superb information about their environment. • The earliest chordates, *Pikaia* and *Haikouella,* were soft-bodied, fish-like forms, but lacked a cranium and vertebrae. • Bone is abundant in skin of the first well-known vertebrates, the ostracoderms, a group of extinct jawless fishes. • Jaws and paired, pectoral and pelvic fins both first appeared in two extinct fish groups, placoderms and acanthodians. The evolution of jaws and paired fins were important evolutionary events allowing greater food handling and precise swimming movements, respectively.

Review Questions

1. What characteristics are shared by the deuterostome phyla that indicate a monophyletic group of interrelated animals?
2. Explain how use of a cladistic classification for vertebrates changes the major important regroupings of traditional vertebrate taxa (refer to figure 15.2). Why are Agnatha and Reptilia, as traditionally recognized, inconsistent with cladistic principles?
3. Name five diagnostic characteristics shared by all chordates, and explain the function of each.
4. In debating the question of chordate origins, zoologists eventually agreed that chordates must have evolved within the deuterostome assemblage rather than from a protostome group as earlier argued. What embryological evidence supports this view?
5. Offer a description of an adult tunicate that would identify it as a chordate, yet distinguish it from any other chordate group.
6. Both sea squirts (urochordates) and lancelets (cephalochordates) are filter-feeding organisms. Describe the filter-feeding apparatus of a sea squirt, and explain in what ways its mode of feeding is similar to, and different from, that of amphioxus.
7. Explain why it is necessary to know the life history of a tunicate in order to understand why tunicates are chordates.
8. List three groups of adaptations that guided vertebrate evolution, and explain how each has contributed to the success of vertebrates.
9. In 1928, Walter Garstang hypothesized that tunicates resemble the ancestral stock of the vertebrates. Explain this hypothesis and evaluate its validity based on recent phylogenetic and developmental data.
10. What are the phylogenetic placements of *Haikouella* and *Pikaia*, and what evidence supports their respective placements?
11. Distinguish between ostracoderms and placoderms. What important evolutionary adaptations first appeared in each group? What are conodonts?
12. What is the currently favored explanation for the evolution of vertebrate jaws?

For Further Thought Urochordates, instead of amphioxus, are now considered to be the sister group of vertebrates. Why is amphioxus, instead of urochordates, still considered a better model for the vertebrate ancestral body plan?

16

Fishes

LEARNING OBJECTIVES

Readers will be able to:

16.1 Identify the five major groups of living fishes and diagram their phylogenetic relationship.

16.2 Compare and contrast the anatomy and ecology of hagfishes and lampreys.

16.3 Describe the functional anatomy of sharks, rays, and chimaeras, emphasizing the sensory and reproductive systems.

16.4 Compare the two major clades containing bony fishes.

16.5 Describe, using appropriate terminology, how fishes (1) swim, (2) maintain buoyancy, (3) osmoregulate, (4) obtain oxygen, and (5) reproduce. Explain why fishes migrate, comparing anadromy and catadromy.

Larry S. Roberts

Grey snappers, *Lutjanus griseus*, in the Florida Keys.

What Is a Fish?

In common (and especially older) usage, the term "fish" denotes a mixed assortment of water-dwelling animals. We speak of jellyfish, cuttlefish, starfish, crayfish, and shellfish, knowing full well that when we use the word "fish" in such combinations, we are not referring to a true fish. In earlier times, even biologists did not make such a distinction. Sixteenth century natural historians classified seals, whales, amphibians, crocodiles, and even hippopotamuses, as well as a host of aquatic invertebrates, as fishes. Later biologists were more discriminating, eliminating first the invertebrates and then the amphibians, reptiles, and mammals from the narrowing concept of a fish. Today we recognize a fish as an aquatic vertebrate with gills; appendages, if present, in the form of fins; and usually a skin with scales of dermal origin.

Even this modern concept of the term "fish" is used for convenience, not as a taxonomic unit. Fishes do not form a monophyletic group, because the ancestor of land vertebrates (tetrapods) is found within one group of fishes (the sarcopterygians). Thus, fishes can be defined as all vertebrates that are not tetrapods. The world's fishes have diversified to produce over 31,000 living species—about as many as all other species of vertebrates combined—with adaptations for almost every conceivable aquatic environment. No other animal group matches their domination of the world's seas, lakes, and streams.

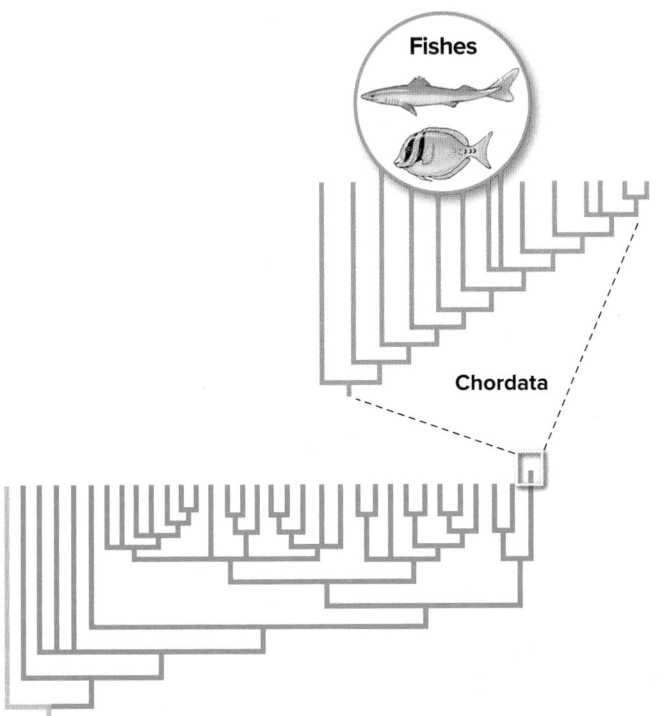

The body of a fish is streamlined for movement through water. Suspended in a medium that is 800 times more dense than air, a trout or pike can remain motionless, varying its neutral buoyancy by adding or removing air from its swim bladder. It may dart forward or at angles, using its fins as brakes and tilting rudders. With excellent organs for salt and water exchange, fishes can finely tune their body fluid composition in their freshwater or seawater environment. Their gills are the most effective respiratory devices in the animal kingdom for extracting oxygen from a medium that contains less than 1/20 as much oxygen as air. Fishes have keen olfactory and visual senses and a unique lateral-line system, which has exquisite sensitivity to water currents and vibrations. Thus, in mastering the physical problems of their element, early fishes evolved a basic body plan and a set of physiological strategies that both shaped and constrained the evolution of their descendants.

Key Theme 16.1

SCIENCE EXPLAINED

Fish or Fishes?

The use of *fishes* as the plural form of *fish* may sound odd to most people accustomed to using *fish* in both the singular and the plural. *Fish* refers to one or more individuals of the same species; *fishes* refer to more than one species.

16.1 Ancestry and Relationships of Major Groups of Fishes

Fishes are a vast array of distantly related, gill-breathing, aquatic vertebrates with fins, comprising about half of the approximately 64,000 vertebrate species. The five major groups of living fishes are (1) hagfishes, (2) lampreys, (3) cartilaginous fishes, (4) ray-finned fishes, and (5) lobe-finned fishes.

The first fishes appeared in the early Paleozoic era, about 550 million years ago (hypotheses concerning vertebrate origins are discussed in Chapter 15). The earliest fishes were a paraphyletic assemblage of jawless fishes, traditionally called **agnathans,** which included ostracoderms (see figure 15.11).

The agnathans include the living **hagfishes** and **lampreys,** fishes adapted as scavengers or parasites. Hagfishes and lampreys form the clade Cyclostomata. Hagfishes and lampreys superficially look much alike, but they diverged from each other at least 450 million years ago and are placed in different clades, Myxini and Petromyzontida, respectively.

All remaining living fishes have paired appendages and jaws and are included, along with tetrapods (land vertebrates), in the monophyletic group **Gnathostomata** (the evolution of jaws and paired fins is discussed in section 15.5: Early Jawed Vertebrates). Gnathostomes appear in the fossil record in the late Silurian period with fully formed jaws, and no intermediate forms between agnathans and gnathostomes are known. By the Devonian period, the "age of fishes," several distinct groups of jawed fishes were well represented. One group, the **placoderms** (see figure 15.12), is a paraphyletic assemblage of armored fishes that went extinct by the end of the Devonian period, but includes a lineage ancestral to all remaining gnathostomes. **Cartilaginous fishes**, forming clade Chondrichthyes (sharks, rays, and chimaeras), lost the heavy dermal armor of the early jawed fishes and adopted cartilage for the endoskeleton. Most are active predators with sharklike or raylike body forms that changed only slightly over the ages.

The other two groups of gnathostomes, **acanthodians** (see figure 15.12) and **Osteichthyes** (bony fishes and tetrapods), were abundant and diverse in the Devonian period. There are two distinct clades of Osteichthyes. Actinopterygii consists of the **ray-finned fishes**, which include nearly all of our familiar bony fishes. The clade Sarcopterygii comprises the **lobe-finned fishes** and the tetrapods. The lobe-finned fishes are represented today by lungfishes and coelacanths—meager remnants of diverse lineages that flourished in the Devonian period (figure 16.1). A taxonomy of major fish taxa appears at the end of this chapter.

16.2 Living Jawless Fishes: Cyclostomata

Living jawless fishes include about 119 species divided between two clades: Myxini (hagfishes) and Petromyzontida (lampreys). Members of each group lack jaws, internal ossification, scales, and paired fins, and both groups share porelike gill openings and an eel-like body form.

Based on their morphological similarity, early zoologists united these two groups under the name **Cyclostomata.** This grouping fell out of favor by the 1990s, when cladistic analysis of morphological characters placed lampreys with gnathostomes, making Cyclostomata paraphyletic. Numerous

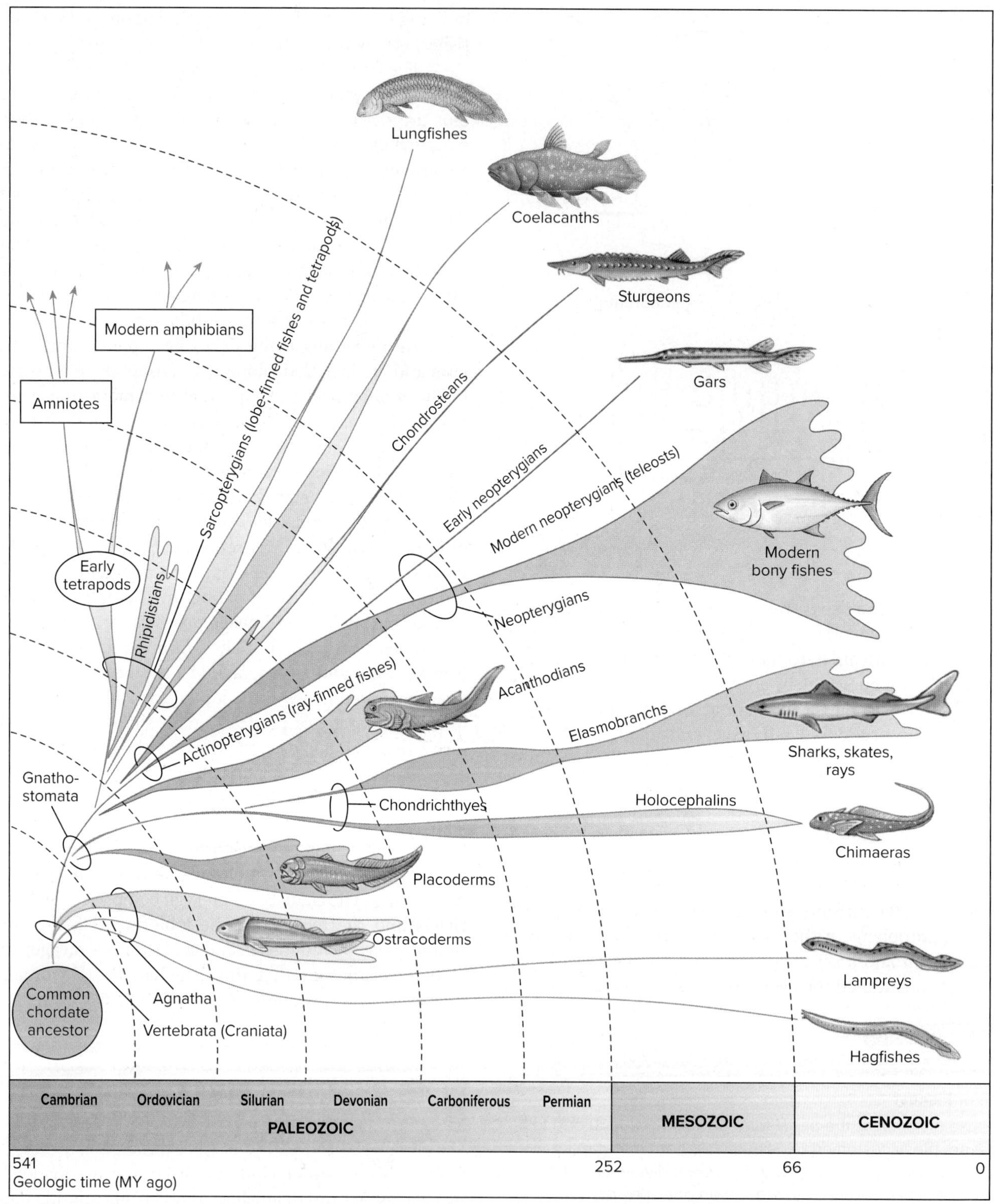

figure 16.1

Graphic representation of the family tree of fishes, showing the evolution of major groups through geological time. Numerous lineages of extinct fishes are not shown. Widened areas in the lines of descent indicate periods of adaptive diversification and the relative number of species in each group. The lobe-finned fishes (sarcopterygians), for example, flourished in the Devonian period, but declined and are today represented by only six species (lungfishes and coelacanths). Sharks and rays diversified during the Carboniferous period, declined in the Permian period, and then diversified again in the Mesozoic era. Johnny-come-latelies in fish evolution are the spectacularly diverse modern fishes, or teleosts, which include most living fishes.

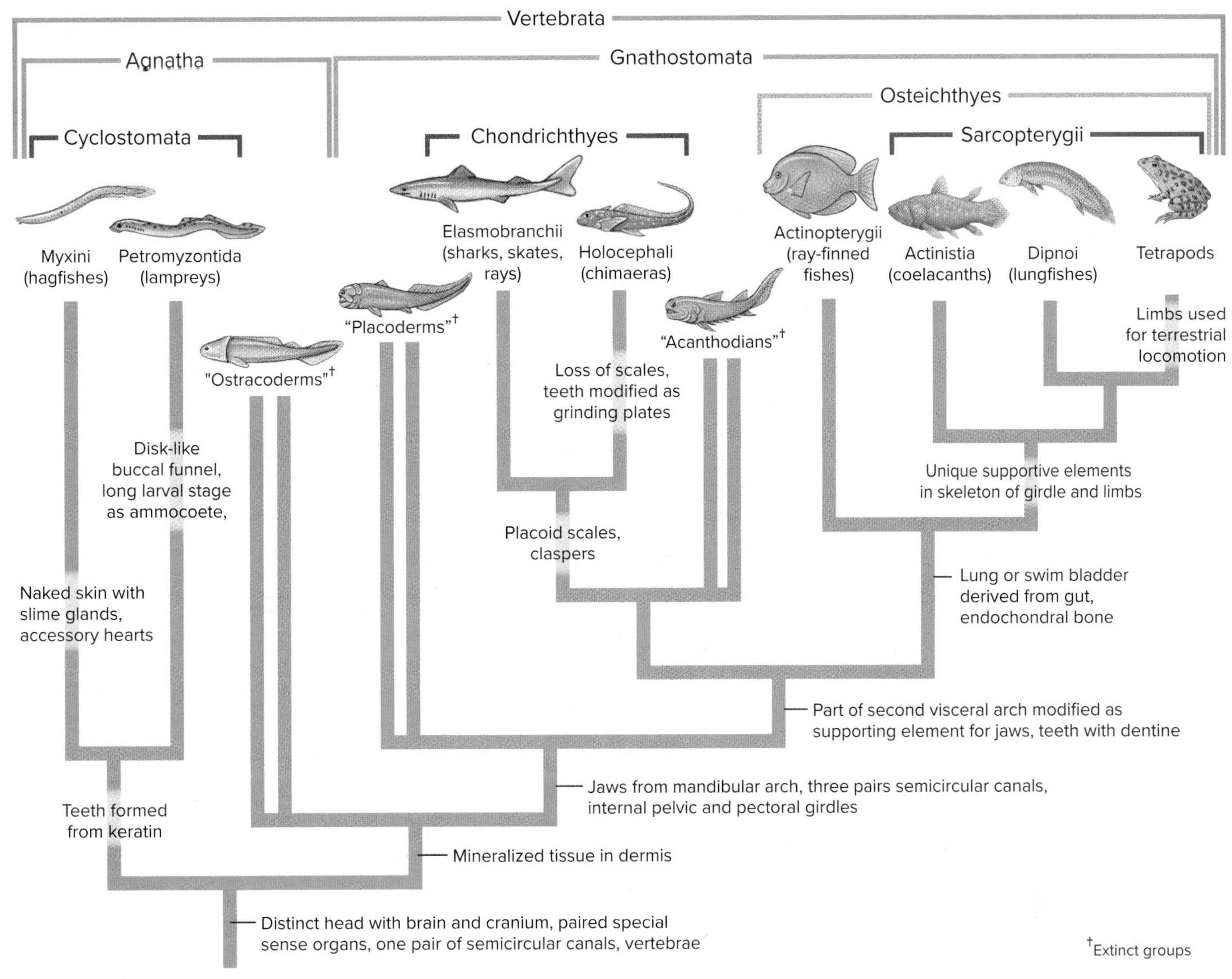

figure 16.2

Cladogram of the fishes, showing the probable relationships of major fish taxa. Agnatha is paraphyletic with respect to gnathostomes. The placement of acanthodians is unclear; we show them as paraphyletic with respect to Chondrichthyes.

characters, including a cerebellum, eye lens, and apparently, vertebrae, are shared in lampreys and gnathostomes, but absent in hagfishes. However, subsequent phylogenetic analyses of molecular data consistently support the Cyclostomata grouping of hagfishes and lampreys, suggesting that the apparently "simple" morphology of hagfishes represents considerable degeneracy (loss of characters) rather than an ancestral vertebrate condition. In 2006, Kinya Ota and Shigeru Kuratani developed a method to rear embryos of hagfishes, and numerous papers on hagfish development and evolution soon followed. These studies also support the Cyclostomata hypothesis and are providing a wealth of new information about early vertebrate evolution. We now know that vertebrae, formerly considered absent in hagfishes, appear as rudimentary structures in some hagfish embryos, suggesting vertebrae were present in the common ancestor of all living vertebrates! We treat Cyclostomata as a monophyletic group of vertebrates, containing lampreys and hagfishes as sole living members (figure 16.2).

Hagfishes: Myxini

Hagfishes are an entirely marine group that feeds on annelids, molluscs, crustaceans, and dead or dying animals. There are about 78 species of hagfishes, of which the best known in North America are the Atlantic hagfish, *Myxine glutinosa* (Gr. *myxa,* slime) (figure 16.3), and the Pacific hagfish, *Eptatretus stoutii* (N.L., *ept,* Gr. *hepta,* seven, + *tretos,* perforated). Although almost completely blind, hagfishes are quickly attracted to food, especially dead or dying fishes, by their keenly developed

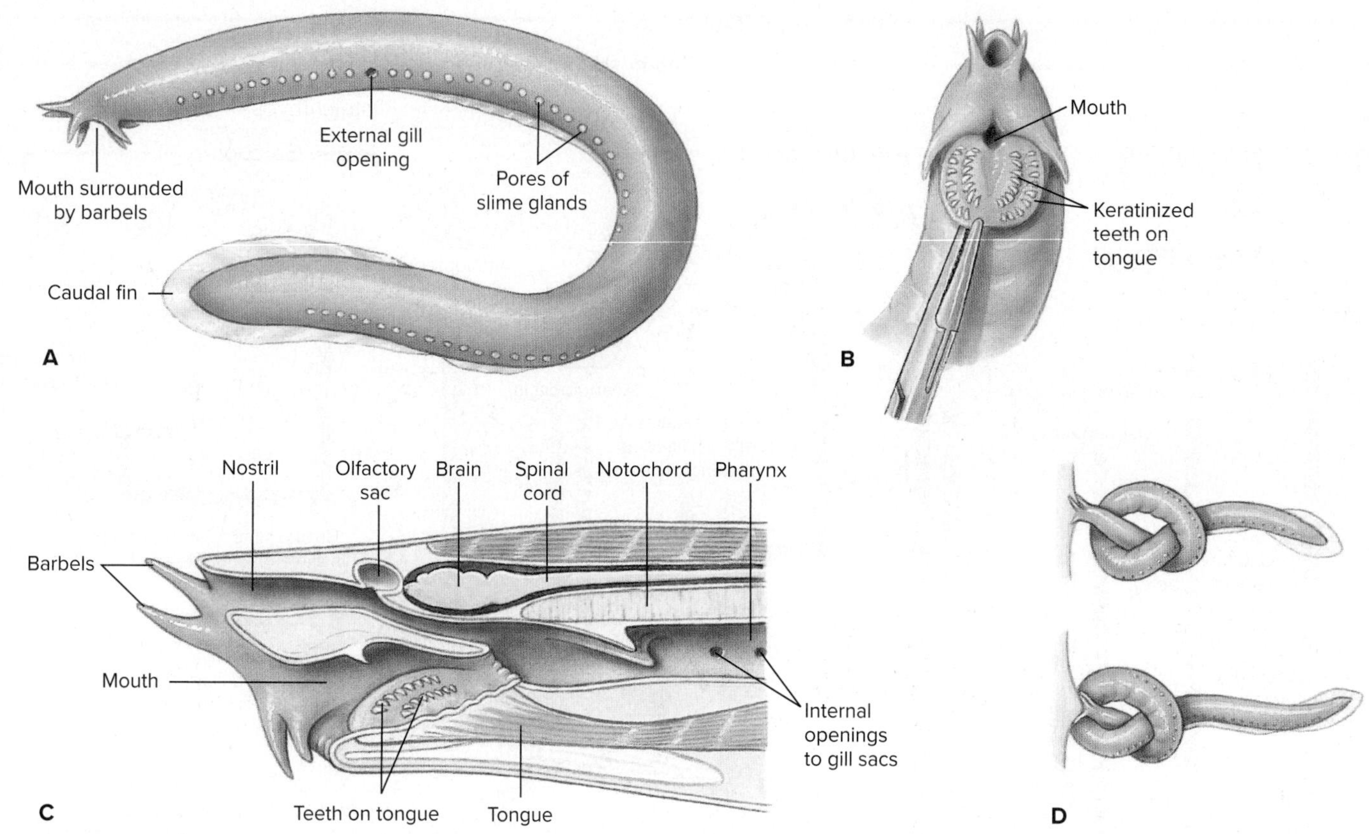

figure 16.3

Atlantic hagfish, *Myxine glutinosa* (Myxini). **A,** External anatomy. **B,** Ventral view of head, showing keratinized teeth used to grasp food during feeding. **C,** Sagittal section of head region (note retracted position of rasping tongue and internal openings into a row of gill sacs). **D,** Hagfish knotting, showing how it obtains leverage to remove flesh from prey.

senses of smell and touch. A whale that dies and falls to the bottom of the ocean attracts thousands of hagfishes, who feed on the carcass for several years. Using two toothed, keratinized plates on its tongue that fold together in a pincerlike action, the hagfish rasps bits of flesh from its prey. For extra leverage, the hagfish often ties a knot in its tail and then passes the knot forward along its body until it is pressed securely against the side of its prey (figure 16.3D).

Unlike the body fluids of other vertebrates, those of hagfishes are in osmotic equilibrium with seawater, as are those of most marine invertebrates. Hagfishes have several other anatomical and physiological peculiarities, including a low-pressure circulatory system served by three accessory hearts in addition to the main heart positioned behind the gills. Hagfishes also are renowned for their ability to generate enormous quantities of slime.

Much of the reproductive biology of hagfishes remains a mystery. Reproduction rarely has been documented; in these cases, females produced small numbers of large, yolky eggs, which take at least 5 months to hatch. There is no larval stage. Almost nothing is known of spawning times, places, and behaviors or age at maturity.

Key Theme 16.2

HUMAN CONNECTIONS

Hagfishes, from Trash to Treasure

While the strange features of hagfishes fascinate many people, hagfishes have not endeared themselves to commercial fishermen. In earlier days, when commercial fishing was done mainly by gill nets and set lines, hagfishes often bit into the bodies of captured fishes and devoured the contents, leaving behind a useless sack of skin and bones. But as large and efficient trawls came into use, hagfishes ceased to be important pests. The commercial fishing industry "turned the tables" and began targeting hagfishes as a source of leather for golf bags and boots. Fishing pressure has been so intense that some species have greatly declined.

Lampreys: Petromyzontida

Of the 41 described species of lampreys distributed around the world, one of the best known is the destructive, 1-m long sea lamprey, *Petromyzon marinus,* which occurs on both sides of the

CHARACTERISTICS of Fish Groups

Cyclostomata (Myxini and Petromyzontida)	Chondrichthyes	Actinopterygii	Sarcopterygii (fish members only)
1. Body slender, eel-like; **no paired appendages**	Body fusiform or dorsoventrally compressed; caudal fin **heterocercal** (sharks and rays) or diphycercal (chimaeras) (see figure 16.13); **paired pectoral** and **pelvic fins** supported by cartilaginous rays	Caudal fin heterocercal (ancestral condition) or **homocercal**; paired pectoral and pelvic fins usually present, supported by **bony rays**; muscles controlling fin movements within trunk	Caudal fin heterocercal (fossil forms) or **diphycercal**; paired pectoral and pelvic fins usually present, supported by **stout bones and bony rays**; muscles controlling fin movements within fin
2 **Skin naked** (no scales)	Skin with **placoid scales** (see figure 16.12) of dermal origin or **naked**	Skin with **ganoid** (ancestral condition), **cycloid, or ctenoid scales** of dermal origin or naked	Skin with elasmoid scales (in living species) with dense bone and some dentine
3. **Fibrous and cartilaginous skeleton**; notochord persistent; **vertebrate reduced or absent**	**Skeleton cartilaginous**; notochord persistent but reduced; **vertebrae** distinct	**Skeleton of bone**; notochord usually absent; **vertebrae** distinct	**Skeleton of bone**; notochord absent or nearly so; **vertebrae** distinct
4. **Jaws absent**; mouth with keratinized plates (hagfishes) or teeth (lampreys); no distinctive stomach	**Jaws present** with **polyphyodont teeth**; stomach large (absent in chimaeras); intestine with **spiral valve** (see figure 16.6); liver often large and oil filled	**Jaws present**, usually with **enameloid, polyphyodont teeth**; spiral valve present (ancestral condition) or absent	**Jaws present**; teeth as enamel-covered crushing plates in lungfishes; intestine with spiral valve
5. Brain small, but distinct; 10 pairs of cranial nerves	Brain well developed; 10 pairs of cranial nerves	Brain well developed, but relatively small; 10 pairs of cranial nerves	Brain well developed, but relatively small; 10 pairs of cranial nerves
6. Eyes poorly developed (hagfishes) or moderately developed (lampreys); **one pair** (hagfishes) or **two pairs** (lampreys) of **semicircular canals**	Senses of smell, vibration reception (lateral line), vision and **electroreception** well developed; **three pairs of semicircular canals**	Senses of vision, hearing, smell, and vibration reception usually well developed; **three pairs of semicircular canals**	Senses of vision, hearing, and smell usually well developed, but highly variable; **three pairs of semicircular canals**
7. Sexes separate; external fertilization	Sexes separate; **internal fertilization with claspers**	Sexes usually separate; some hermaphroditic; some reproduce asexually by parthenogenesis; fertilization usually external, but internal in some	Sexes separate; many hermaphroditic; fertilization external (lungfishes) or internal (coelacanths)
8. Large yolky eggs and no larval stage in hagfishes; small eggs and long larval stage (**ammocoete**) in lampreys	Oviparous or viviparous; embryos of viviparous species nourished by **placenta, yolk sac (ovoviviparity), or cannibalism**; no larval stage	Oviparous or viviparous; embryos of viviparous species nourished by placenta or yolk sac (ovoviviparity); larval stage often greatly different from adult	Oviparous (lungfishes) or ovoviviparous (coelacanths)
9. Excretory system of **pronephric and mesonephric** (hagfishes) or **opisthonephric** (lampreys) kidneys; kidneys drain via archinephric duct to cloaca; **ammonia** main nitrogenous waste	Excretory system of **opisthonephric kidneys**, which drain via **archinephric duct** to cloaca**; high concentration of urea and trimethylamine oxide in blood**; **rectal gland** present	Excretory system of **opisthonephric kidneys**, which drain via archinephric duct to cloaca; **ammonia** usually main nitrogenous waste	Excretory system of **opisthonephric kidneys**, which drain via archinephric duct to cloaca; **ammonia** and **urea** usually main nitrogenous wastes
10. Hagfishes with 5-16 pairs of gills; lampreys with 7 pairs of gills	**Five to seven pairs of gills** leading to gill slits in rays and sharks or covered by operculum in chimaeras; **no swim bladder or lung**	**Gills covered by bony operculum**; **swim bladder present**, usually functioning for buoyancy, sometimes used for respiration	**Gills covered by bony operculum**; **swim bladder present**, used primarily in respiration (fat filled in coelacanths)
11. Heart with a sinus venosus, atrium, and ventricle; **single circulation**; accessory hearts in hagfishes; nucleated red blood cells	Heart with a sinus venosus, atrium, ventricle, and conus arteriosus; **single circulation**; nucleated red blood cells	Heart with a sinus venosus, atrium, and ventricle; **single circulation**; nucleated red blood cells	Heart with a sinus venosus, two atria, and partly divided ventricle; **pulmonary and systemic circuits incompletely separated**; nucleated red blood cells

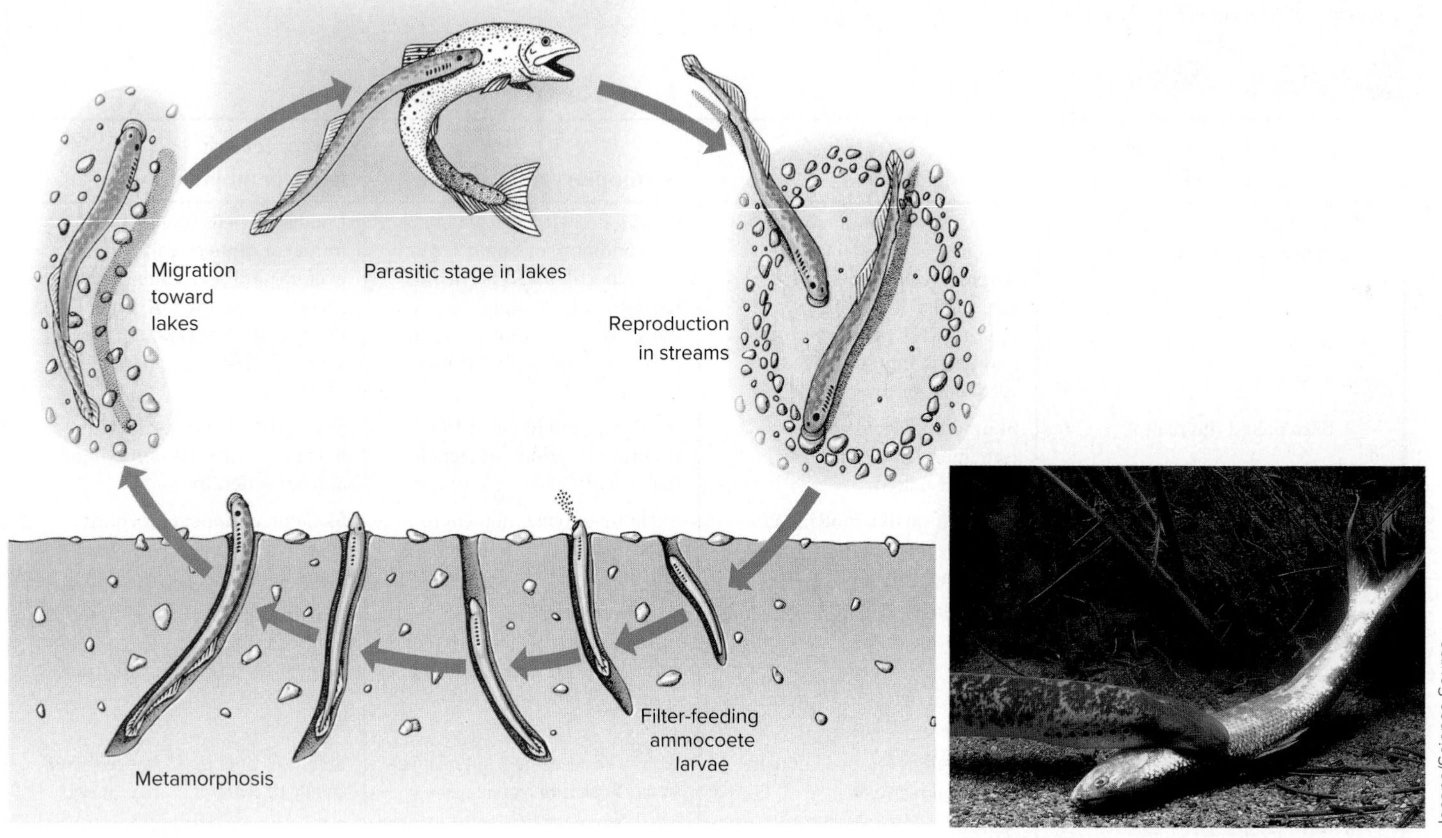

A Life cycle of sea lamprey

B Sea lamprey, *Petromyzon marinus*, feeding on fish

figure 16.4

Sea lamprey, *Petromyzon marinus*. **A,** Life cycle of the "landlocked" form; **B,** feeding on body fluids of a dying fish.

Atlantic Ocean (figure 16.4). The name *Petromyzon* (Gr. *petros,* stone, + *myzon,* sucking) refers to the lamprey's habit of grasping a stone with its mouth to hold its position in a current. There are 24 species of lampreys in North America, of which about half are parasitic; the rest, often called "brook lampreys," are species that never feed after metamorphosis and die soon after spawning.

All lampreys ascend freshwater streams to spawn in the spring in shallow, gravel-filled areas. Males begin building nests and are joined later by females. Using their oral discs to lift stones and pebbles and vigorous body vibrations to sweep away light debris, they form an oval depression. As a female sheds eggs into the nest, a male fertilizes them. The sticky eggs adhere to pebbles in the nest and become lightly covered with sand. Adults die soon after spawning.

Eggs hatch in about 2 weeks, releasing small larvae called **ammocoetes** (figure 16.4). When the larvae reach about 1 cm long, they leave the nest and then burrow into sand, where they filter-feed on small organisms and fine organic matter. The larvae live from 3 to 7 years and then rapidly metamorphose into adults.

Parasitic lampreys either migrate to the sea, if marine, or remain in fresh water, where they attach themselves by their suckerlike mouth to fish and use their sharp, keratinized teeth to rasp through flesh and consume body fluids or flesh. To promote the flow of blood, a lamprey injects an anticoagulant into the wound. When gorged, the lamprey releases its hold but leaves the fish with a large wound that can be fatal. Parasitic freshwater adults live 1–2 years before spawning and then die; marine forms live 2–3 years.

Nonparasitic lampreys do not feed after emerging as adults, and their digestive tract degenerates. Within a few months, they spawn and die.

Invasion of the Great Lakes by the sea lamprey (see figure 16.4) devastated fisheries. No lampreys were present in the Great Lakes upstream of Niagara Falls until the Welland Ship Canal was deepened in the 1910s, allowing lampreys to bypass the Falls. Moving first through Lake Erie and then to Lakes Huron, Michigan, and Superior, sea lampreys, combined with the effects of overfishing, caused the total collapse of a multimillion-dollar lake trout fishery in the early 1950s. Other, less valuable fish species were attacked and destroyed in turn. After reaching peak abundance in 1951 in Lakes Huron and Michigan and in 1961 in Lake Superior, sea lampreys began to decline, due to depletion of their food supply and to expensive control measures (mainly chemical larvicides placed in selected spawning streams). Lake trout, aided by a restocking program, are now recovering, but wounding rates are still high in some lakes.

16.3 Cartilaginous Fishes: Chondrichthyes

There are about 1200 living species in the clade Chondrichthyes, an ancient group that appeared in the Devonian period. Although they are a much smaller and less diverse assemblage than bony fishes, their impressive combination of well-developed sense organs, powerful jaws and swimming musculature, and predaceous habits ensures them a secure and lasting place in the aquatic community. One of their distinctive features is a cartilaginous skeleton. Their skeletons are extensively calcified, but bone is entirely absent, although phosphatized minerals, characteristic of bone, are retained in scales, teeth, and spines. Chondrichthyes is subdivided into two clades: Elasmobranchii, the sharks, skates, and rays; and Holocephali, the chimaeras.

Sharks, Skates, and Rays: Elasmobranchii

Sharks, which include about 45% of the approximately 1150 species in Elasmobranchii, are typically predaceous fishes with five to seven pairs of gill slits and gills and (usually) a spiracle behind each eye. Larger sharks, such as the massive (but harmless) plankton-feeding whale shark, can reach 15 m in length, the largest of all fishes. Spiny dogfish sharks, widely used in zoological laboratories, rarely exceed 1 m (figures 16.5 and 16.6).

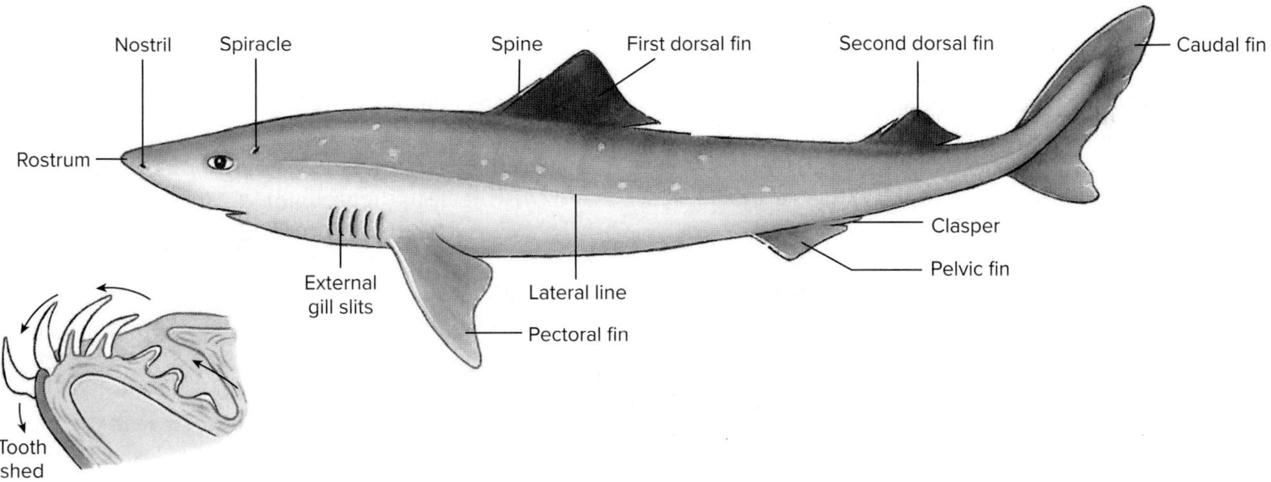

figure 16.5

Male spiny dogfish shark, *Squalus acanthias*. *Inset:* Section of lower jaw shows new teeth developing inside the jaw. These move forward to replace lost teeth. The rate of replacement varies among species.

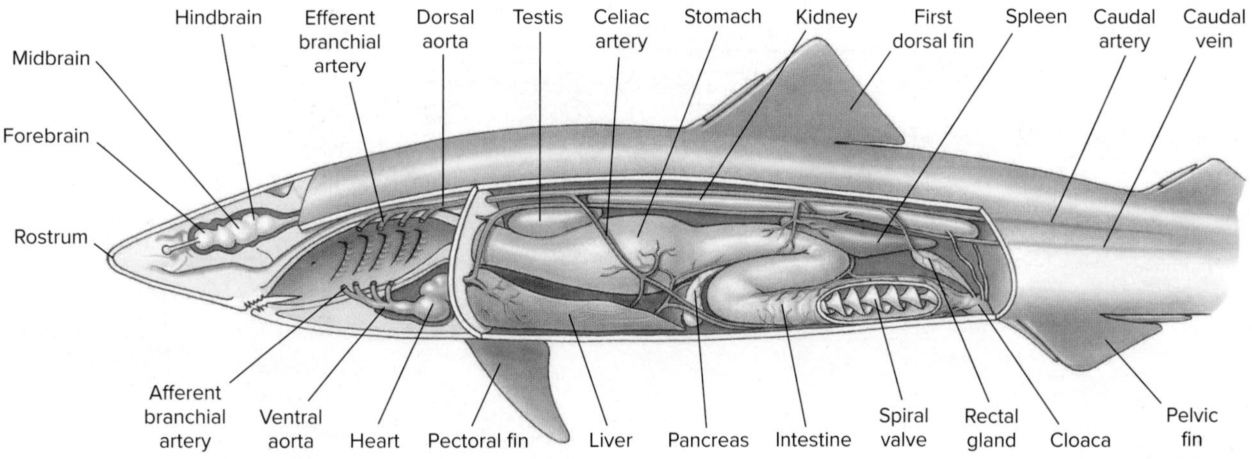

figure 16.6

Internal anatomy of a male spiny dogfish shark, *Squalus acanthias*.

Although to many people sharks have a sinister appearance and a fearsome reputation, they are at the same time among the most gracefully streamlined of all fishes (figure 16.5). Sharks are heavier than water and will sink if not swimming forward. The asymmetrical **heterocercal tail,** in which the vertebral column turns upward and extends into the dorsal lobe of the tail (see figure 16.13), provides thrust and some lift as it sweeps back and forth in the water.

Key Theme 16.3

HUMAN CONNECTIONS

Shark Finning

The worldwide shark fishery is experiencing unprecedented pressure, driven, in part, by the high demand of shark fins for shark-fin soup, an Asian delicacy. Rapid declines in shark populations led to the outlawing of "finning" in a few countries, including the United States and Canada. Following public campaigns from environmental groups, consumption of shark fin soup in China and Hong Kong substantially declined. However, worldwide commercial harvest of sharks remains high, and about one-fourth of all shark species are jeopardized, according the IUCN. Contributing to the threatened collapse of shark fisheries worldwide is the low fecundity of sharks and the many years most sharks require to reach sexual maturity; some species take as long as 35 years.

Sharks are well equipped for their predatory life. Their tough, leathery skin is covered with numerous dermal **placoid scales** (see figure 16.12) that are modified anteriorly to form replaceable rows of teeth in both jaws (figure 16.7). Placoid scales reduce the turbulence of water flowing over the body surface during swimming. Sharks use their keen sense of smell to guide them to food. They can also locate prey from long distances by sensing low-frequency vibrations with mechanoreceptors in the **lateral-line system.** This system is composed of special receptor organs (**neuromasts**) in interconnected tubes and pores extending along the sides of the body and over the head (figure 16.8). At close range, a shark switches to vision as its primary method of tracking prey. During the final stage of the attack, sharks are guided to their prey by the bioelectric fields that surround all animals. Electroreceptors, the **ampullae of Lorenzini,** are located primarily on the shark's head (see figure 16.7).

All chondrichthyans have internal fertilization; sperm is introduced into the female reproductive tract by a modified portion of the male's pelvic fin called a **clasper** (see figure 16.5). Some sharks and all skates (figure 16.9) are **oviparous,** laying eggs soon after fertilization. Embryos are nourished from the

Wolfgang Pölzer/Alamy Stock Photo

figure 16.7

Head of a sand tiger shark, *Carcharias taurus*. Note the series of successional teeth. Also visible below the eye are ampullae of Lorenzini, visible as tiny dark pits

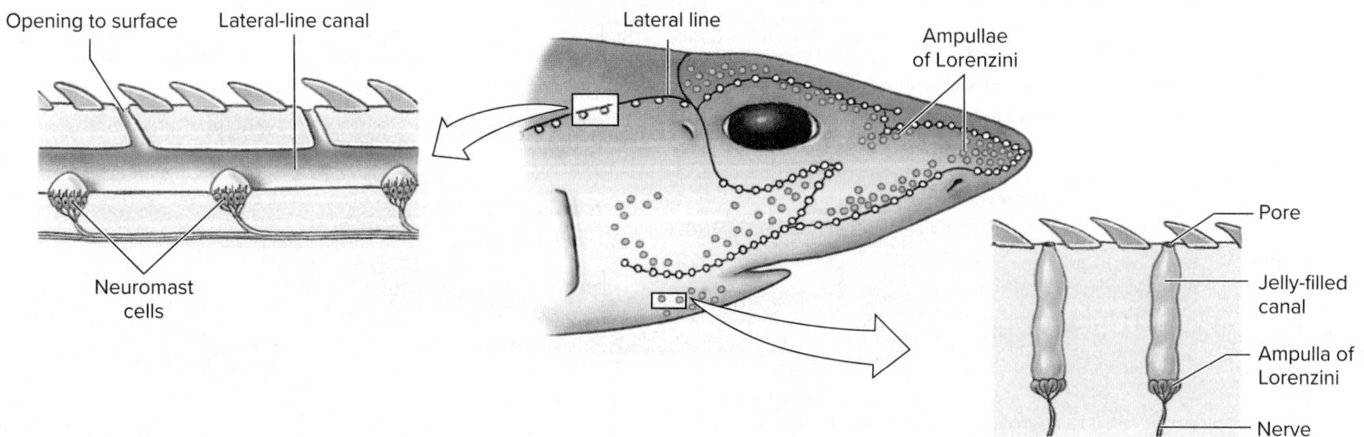

figure 16.8

Sensory canals and receptors in a shark. Ampullae of Lorenzini respond to weak electrical fields. Lateral-line sensors, called neuromast cells, are sensitive to disturbances in the water, enabling a shark to detect nearby objects by reflected waves in the water.

A

B

Jeff Rotman/Alamy Stock Photo

Comstock Images/PictureQuest

figure 16.9

Skates and rays are specialized for life on the sea floor. Both **A,** the clearnose skate, *Raja eglanteria*, and **B,** the bluespotted ribbontail ray, *Taeniura lymma*, are flattened dorsoventrally and move by undulations of winglike pectoral fins.

egg yolk for a long period—up to 2 years in one species— before hatching. Rays and many sharks are **viviparous,** retaining developing young in the uterus, and giving birth to fully formed young. One kind of viviparity is **ovoviviparity,** also called yolk-sac viviparity or lecithotrophic viviparity, in which the young are nourished only by the contents of their yolk sacs; another is **placental viviparity,** in which the young are nourished via a placenta. The embryos of a few species, including sand tiger sharks (see figure 16.7), consume siblings or eggs while in the uterus. Regardless of the initial amount of maternal support, all parental care ends once eggs are laid or young are born.

More than half of all elasmobranchs are rays, a group that includes skates, stingrays, electric rays, and manta rays. Rays are distinguished by their dorsoventrally flattened bodies and enlarged pectoral fins, which they move in a wavelike fashion to propel themselves (figure 16.9). Gill openings are on the underside of the head, and the large **spiracles** are on top. Respiratory water enters through these spiracles to prevent clogging of the gills, because the mouth is often buried in sand. Teeth are adapted for crushing prey: molluscs, crustaceans, and an occasional small fish.

Stingrays have a slender, whiplike tail armed with one or more saw-toothed spines that can inflict painful wounds to a would-be predator. Electric rays have on both sides of their head powerful electrical organs, which produce sufficient voltage to stun prey.

Chimaeras: Holocephali

The approximately 48 species of chimaeras (kī-mer′uz; L. monster) (figure 16.10), are remnants of a group that diverged from the earliest shark lineage, which originated at least 380 million years ago. Fossil chimaeras first appeared in the Mississippian period, reached their zenith in the Cretaceous and early Paleogene periods (120 million to 50 million years ago), and then declined. Anatomically, they have several features linking them to elasmobranchs, but they possess a suite of unique characters, too. Instead of distinct teeth, their jaws bear large, flat plates and they have a single pair of gill openings. Their food is a mixed diet of molluscs, echinoderms, crustaceans, and fishes. They are not commercial species and are seldom caught. Their strange appearance is responsible for their common names of chimaera and ratfish, but in life they are beautifully colored with a pearly iridescence.

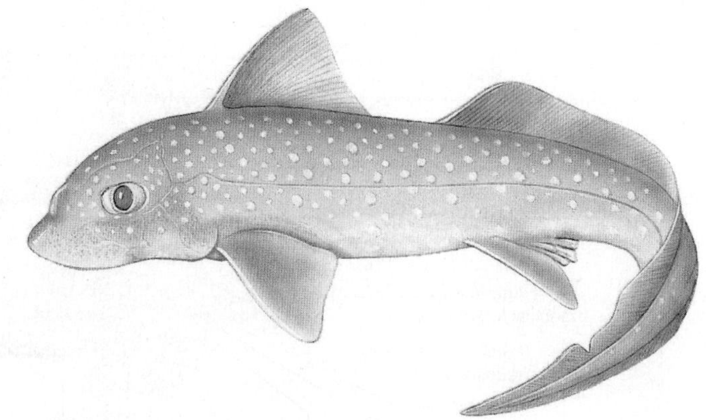

figure 16.10

A spotted ratfish, *Hydrolagus colliei*, of the North American west coast. This species is one of the most handsome chimaeras, which tend to be bizarre in appearance.

16.4 Bony Fishes and Tetrapods: Osteichthyes

Origin, Evolution, and Diversity

In the Silurian period, a lineage of fishes with bony endoskeletons gave rise to a clade (**Osteichthyes**) of vertebrates that contains 96% of living fishes and all tetrapods. Fishes of this clade are called "bony fishes." Bony fishes and tetrapods are united by **endochondral bone** (bone that replaces cartilage developmentally), lungs or a swim bladder derived from the gut, and several cranial and dental characters. The traditional use of "Osteichthyes" included only fishes, making it not monophyletic. Here we treat it as a clade by also including tetrapods (see figure 16.2).

By the middle-Devonian period, bony fishes already had diversified extensively into two major clades, with adaptations that fitted them for every aquatic habitat except the most inhospitable. One of these clades, the ray-finned fishes (Actinopterygii), includes teleosts (figure 16.11), the most species-rich clade of living vertebrates. A second clade (Sarcopterygii) is represented today by eight fishlike species, the lungfishes and coelacanths (see figures 16.19 and 16.20), and the land vertebrates (tetrapods).

Several key adaptations contributed to the diversification of bony fishes. They have an **operculum** over the gill composed of bony plates and attached to a series of muscles, which pump water across the gills (see figure 16.24). A gas-filled pouch branching from the esophagus provides an additional means of gas exchange in oxygen-poor waters and an efficient means of achieving neutral buoyancy. In fishes that use these pouches primarily for gas exchange, the pouches are called lungs, while in fishes that use these pouches primarily for buoyancy, these pouches are called **swim bladders** (figure 16.11). Progressive specialization of jaw musculature and skeletal elements involved in suction feeding is another key feature of bony fish evolution.

Ray-Finned Fishes: Actinopterygii

Ray-finned fishes are an enormous assemblage containing our familiar bony fishes—over 31,000 species. The ancestral forms that swam in Devonian waters were small and heavily armored, with thick **ganoid scales** (figure 16.12) and heterocercal caudal fins (figure 16.13). In contrast to reptiles, which have keratinized, epidermal scales (see figure 18.4), fishes have bony, dermal scales (figure 16.14).

From those earliest ray-finned fishes arose several clades. Bichirs, in the clade Cladistia, have lungs, heavy ganoid scales, and other characteristics similar to those of their bony-fish ancestors (figure 16.15A). The 13 species of bichirs live in freshwaters of Africa. A second clade is the chondrosteans (Gr. *chondros,* cartilage, + *osteon,* bone), represented by 27 species of freshwater and anadromous sturgeons and paddlefishes (figure 16.15B and C).

The third major group to emerge from the early ray-finned stock were **neopterygians** (Gr. *neos,* new, + *pteryx,* fin). Neopterygians appeared in the late Permian period and diversified extensively during the Mesozoic era, when one lineage gave rise to modern bony fishes, the teleosts. The living non-teleost neopterygians are the bowfin *Amia* (Gr., tunalike fish), which inhabits shallow, weedy waters of the Great Lakes and the Mississippi River basin, and the gars *Lepisosteus* (Gr. *lepidos,* scale, + *osteon,* bone) and *Atractosteus* (Gr. *atraktos,* spindle, + *osteon,* bone), both native to eastern and southern North America

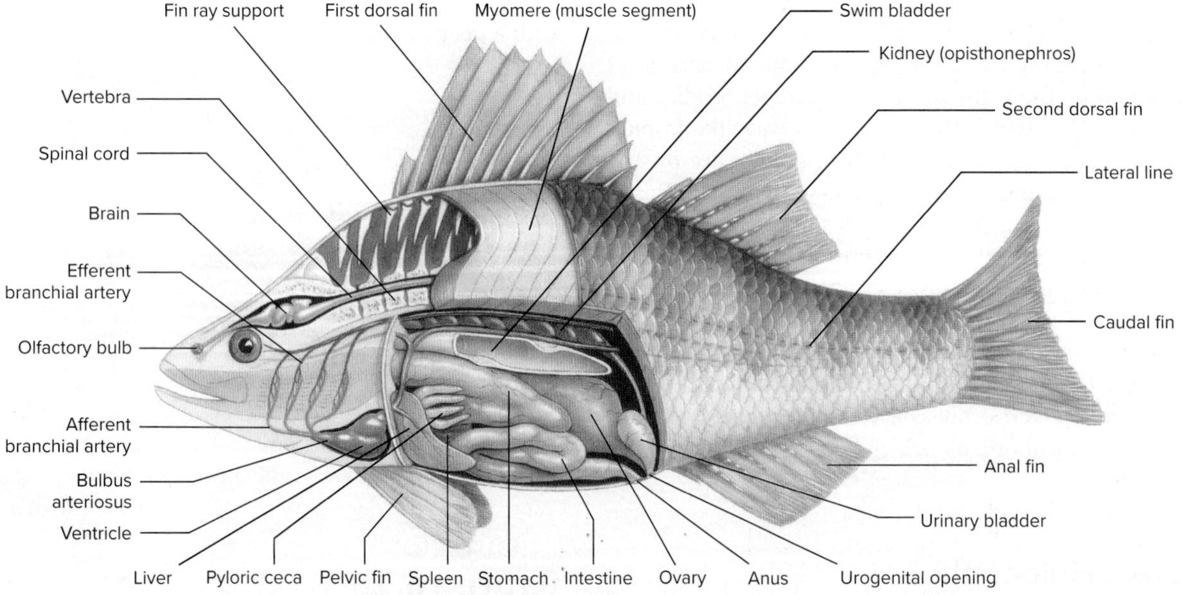

figure 16.11

Anatomy of a yellow perch, *Perca flavescens*, a freshwater teleost fish.

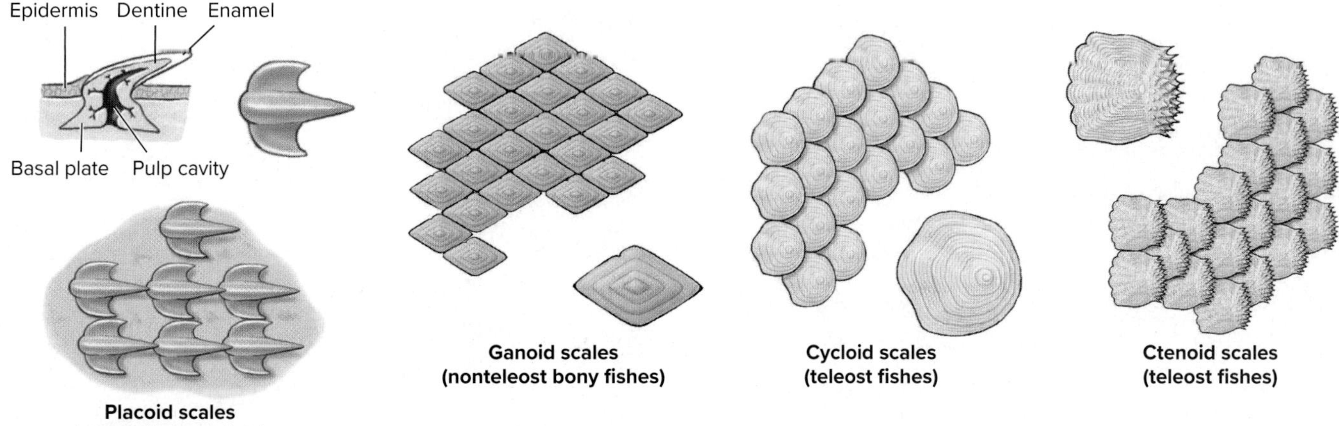

figure 16.12

Types of fish scales, shown with anterior to the left. Placoid scales are small, conical, toothlike structures characteristic of Chondrichthyes. Diamond-shaped ganoid scales, present in early bony fishes and living gars, are composed of layers of silvery enamel (ganoin) on the upper surface and bone on the lower. Teleosts have either cycloid or ctenoid scales that contain bone, but are thin and flexible and arranged in overlapping rows.

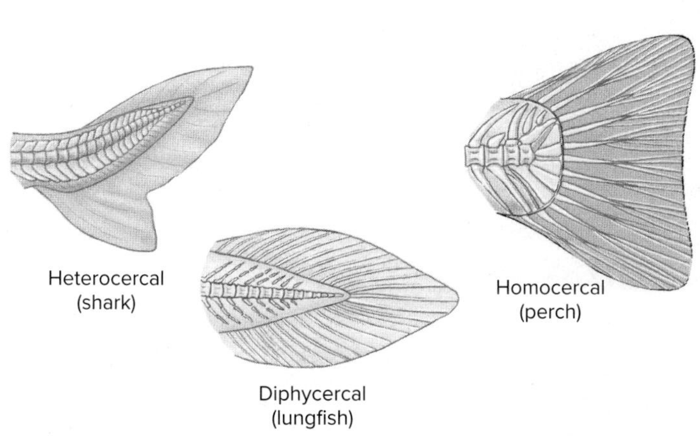

figure 16.13

Types of caudal fins among fishes.

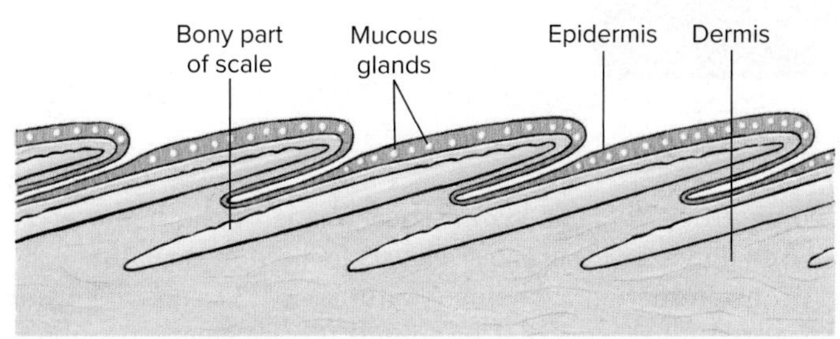

figure 16.14

Section through the skin of a bony fish, showing the overlapping scales (yellow). The scales lie in the dermis and are covered by epidermis.

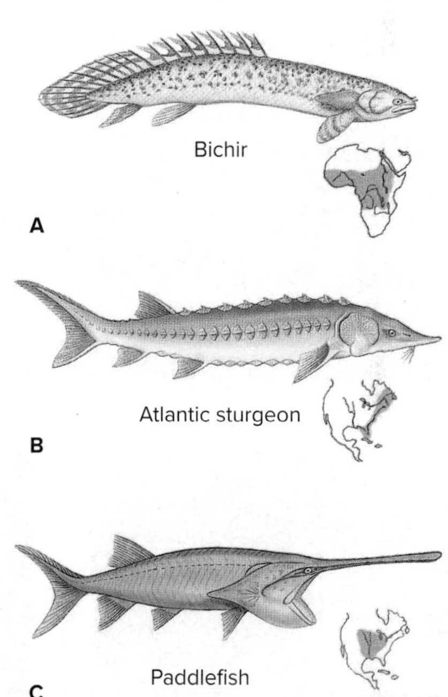

figure 16.15

Nonteleost ray-finned fishes of class Actinopterygii. **A,** Bichir, *Polypterus bichir*, of equatorial West Africa is a nocturnal predator. **B,** Atlantic sturgeon, *Acipenser oxyrhynchus* (now uncommon), inhabits Atlantic coastal rivers. **C,** Paddlefish, *Polyodon spathula*, of the Mississippi River basin reaches 2 m and 80 kg.

(figure 16.16). Gars are large, ambush predators with elongate bodies and jaws filled with needlelike teeth.

Teleosts (Gr. *teleos,* complete, + *osteon,* bone), the modern bony fishes, form the largest clade of neopterygians. Teleost diversity is astounding, with almost 30,000 described species representing about 96% of all living fishes and almost half of all vertebrates (figure 16.17). They display an amazing array of

figure 16.16

Nonteleost neopterygian fishes. **A,** Bowfin, *Amia calva*. **B,** Longnose gar, *Lepisosteus osseus*. Both species frequent slow-moving streams and swamps of eastern North America, where they can suspend motionless in the water, ready to snatch passing fishes.

figure 16.17

Diversity among teleosts. **A,** This striped marlin, *Kajikia audax*, chasing sardines near the Baja California peninsula, is one of the fastest teleosts. **B,** Mudskippers, *Periopthalmus*, make extensive excursions on land to graze on algae and capture insects; they build nests in which the young hatch and are guarded by the mother. **C,** Protective coloration of the lionfish, *Pterois volitans*, advises caution; the dorsal spines are venomous. **D,** The sucking disc on the sharksucker, *Echeneis naucrates*, is a modification of the dorsal fin.

morphological form and size, occupy nearly every aquatic habitat on Earth, and may even make excursions onto land, as do mudskippers (figure 16.17B).

Several morphological trends in the teleost lineage allowed it to diversify into a truly incredible variety of habitats and forms. The heavy dermal armor of the earliest ray-finned fishes was replaced by light, thin, flexible **cycloid** and **ctenoid scales** (see figure 16.12). Increased mobility and speed permitted by loss of heavy armor improved predator avoidance and feeding efficiency. The symmetrical shape of the **homocercal** tail (see figure 16.13) of most teleosts focused musculature contractions on the tail, allowing greater speed. Various fins were elaborated into forms that permitted precise steering, camouflage, protection, attachment, or social communication (figure 16.17C and D). Teleost lineages demonstrate an increasingly fine control of gas resorption and secretion in the swim bladder, improving control of buoyancy. Nearly all teleosts capture food by **suction feeding,** where rapid expansion of the orobranchial cavity creates low pressure in the cavity, drawing in water and prey. Changes in jaw suspension allowed teleosts to use this method more effectively than did their ancestors. With so many innovations, teleosts have become the most diverse of fishes.

Lobe-Finned Fishes: Sarcopterygii

The ancestor of tetrapods is found within a group of extinct sarcopterygian fishes called **rhipidistians,** which included several lineages that flourished in shallow waters in the late Paleozoic era. The evolution of tetrapods from rhipidistians is discussed in Chapter 17.

All early sarcopterygians had lungs as well as gills, and a tail of the **heterocercal** type. However, during the Paleozoic era, the orientation of the vertebral column changed so that the tail became symmetrical and **diphycercal** (see figure 16.13). These fishes had powerful jaws; heavy, enameled scales; and strong, fleshy, paired lobed fins that they may have used to maneuver through shallow, debris-filled aquatic habitats. The sarcopterygian clade today includes only eight fish species: six species of lungfishes and two species of coelacanths (figures 16.18 and 16.19).

Of the three surviving genera of **lungfishes,** most similar to early forms is *Neoceratodus* (Gr. *neos,* new, + *keratos,* horn, + *odes,* form), the Australian lungfish, which may attain a length of 1.5 m. This lungfish, unlike its living relatives, normally relies on gill respiration and cannot survive long out of water. The South American lungfish, *Lepidosiren* (L. *lepido, scaly,* + *siren,* mythical mermaid), and the African lungfishes, *Protopterus* (Gr. *protos,* first, + *pteron,* wing) (figure 16.18), can live out of water for long periods of time. They breathe air with lungs, which are highly-folded, vascularized swim bladders. *Protopterus* lives in African streams and ponds that are baked hard by the hot tropical sun during the dry season. The fish burrows at the approach of the dry season and secretes a copious slime that mixes with mud to form a hard cocoon in which it remains dormant until rain returns.

Coelacanths also arose in the Devonian period, diversified somewhat, and reached their peak of diversity in the Mesozoic era. At the end of that era, they nearly disappeared but left one remarkable surviving genus, *Latimeria* (figure 16.19). Because the last coelacanths were believed to have become extinct 70 million years ago, the scientific world was astonished when a coelacanth was found on a trawl off the coast of South Africa in 1938. An intensive search to

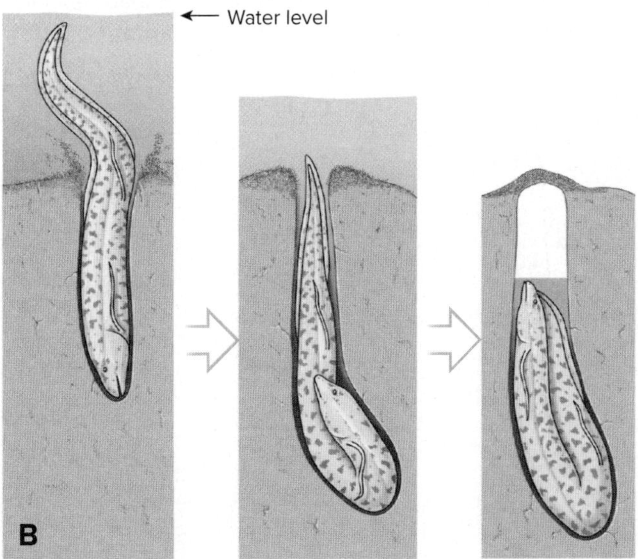

figure 16.18

A, An African lungfish, *Protopterus annectens*, of Sarcopterygii. **B,** An African lungfish burrows into the sediments of a pool as the water dries. There it secretes a mucous cocoon and enters a state of dormancy (aestivation) for several months to 4 years, when rains return water to its pool.

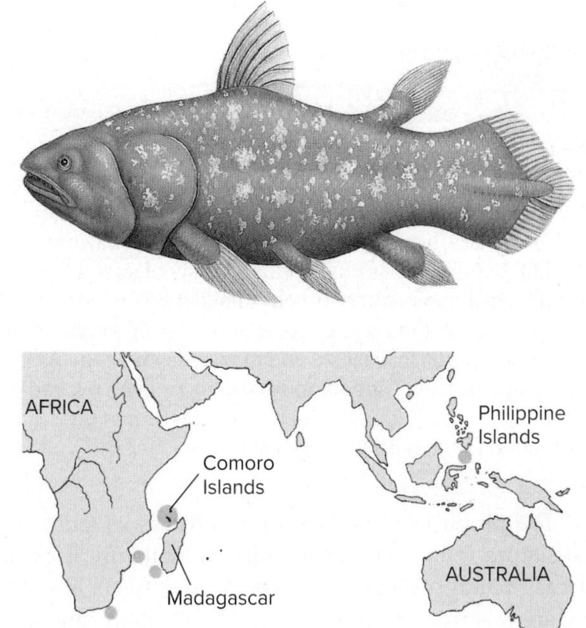

figure 16.19

The coelacanth genus *Latimeria* is a surviving marine relict of a group of lobe-finned fishes that flourished some 350 million years ago.

locate more specimens was successful off the coast of the Comoro Islands. There, fishermen occasionally catch them at great depths with hand lines, providing specimens for research. This was assumed to be the only population of *Latimeria* until 1998, when the scientific world was again surprised by the capture of a new species of coelacanth in Indonesia, 10,000 km from the Comoros!

The tail of coelacanths is diphycercal (see figure 16.13) but possesses a small lobe between the upper and lower caudal lobes, producing a three-pronged structure (figure 16.19). Their swim bladders are fat-filled and not used for respiration.

Coelacanths are a deep metallic blue with irregular white or brassy flecks, providing camouflage against the dark lava-cave reefs that they inhabit. Young are born fully formed after hatching internally from eggs 9 cm in diameter–the largest among bony fishes.

16.5 Structural and Functional Adaptations of Fishes

Locomotion in Water

To the human eye, some fishes appear capable of swimming at extremely high speeds, but our judgment is unconsciously tempered by our own experience that water is a highly resistant medium through which to move. Most fishes, such as a trout or a minnow, can swim maximally about 10 body lengths per second, obviously an impressive performance by human standards. Yet, when these speeds are translated into kilometers per hour, we realize that a 30 cm (1 foot) trout can swim only about 10.4 km (6.5 miles) per hour. As a general rule, the larger the fish, the faster it can swim.

Key Theme 16.4

ADAPTATION AND PHYSIOLOGY

Fast Fishes

Measuring fish cruising speeds is best done in a "fish wheel," a large, ring-shaped channel filled with water that is turned at a speed equal and opposite to that of the fish. Much more difficult to measure are the sudden bursts of speed that most fish can make to capture prey or to avoid being captured. A hooked bluefin tuna was once "clocked" at 66 km per hour (41 mph); swordfish, sailfish, and marlin may be capable of incredible bursts of speed approaching, or even exceeding, 110 km per hour (68 mph). Such high speeds can be sustained for no more than 1–5 seconds.

The propulsive mechanism of a fish is its trunk and tail musculature. The axial, locomotory musculature is composed of zigzag bands, called **myomeres.** Muscle fibers in each myomere are relatively short and connect the tough connective tissue partitions that separate each myomere from the next. On the surface, myomeres take the shape of a W lying on its side (figure 16.20), but internally the bands are complexly folded and nested so that the pull of each myomere extends over several vertebrae. This arrangement produces more power and finer control of movement, since many myomeres are involved in bending a given segment of the body.

Understanding how fishes swim can be approached by studying the motion of a very flexible fish such as an eel (figure 16.21, *left*). The movement is serpentine, not unlike that of a snake, with waves of contraction moving backward along the body by alternate contraction of the myomeres on either side. The anterior end of the body bends less than the posterior end, so that each undulation increases in amplitude as it travels along the body. While undulations move backward, bending of the body pushes laterally against the water, producing a **reactive force** that is directed forward, but at an angle. It can be analyzed as having two components: **thrust,** which is used to overcome drag and propels the fish forward, and **lateral force,** which tends to make the fish's head "yaw," or deviate from the course in the same direction as the tail. This side-to-side head movement is very obvious in a swimming

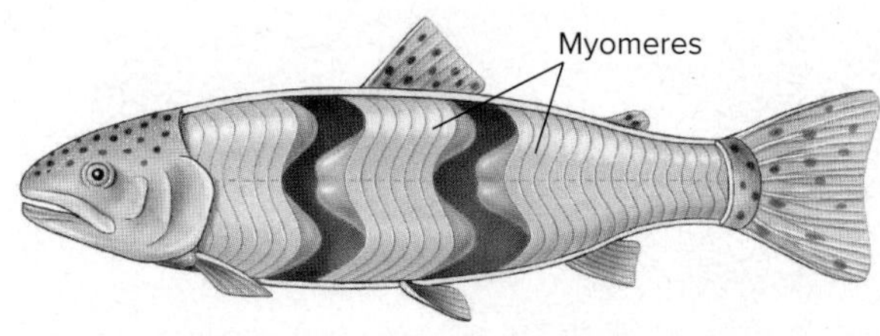

figure 16.20

Trunk musculature of a teleost fish, partly dissected to show the internal arrangement of the muscle bands (myomeres). The myomeres are folded into a complex, nested grouping, an arrangement that favors stronger and more controlled swimming.

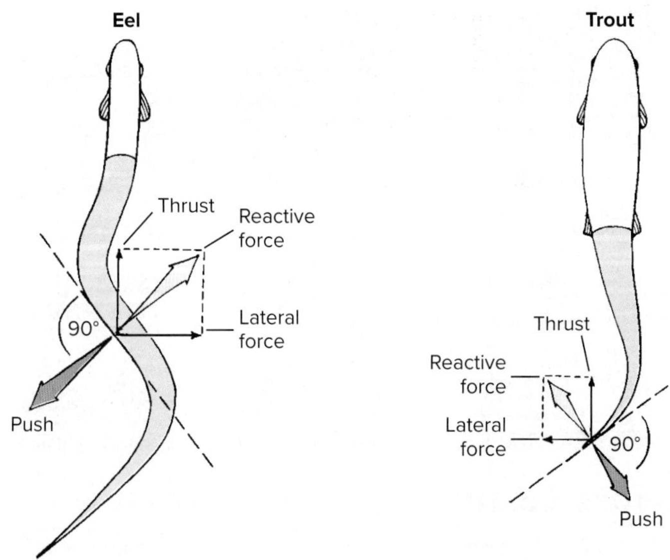

figure 16.21

Movements of swimming fishes, showing the forces developed by an eel-shaped fish and a spindle-shaped fish.

eel, but many fishes have a large, rigid head with enough surface resistance to minimize yaw.

The movement of an eel is reasonably efficient at low speed, but its body shape generates too much frictional drag for rapid swimming. Fishes that swim rapidly, such as trout, are less flexible and limit body undulations mostly to the caudal region (figure 16.21, *right*). Muscle force generated in the muscle mass is transferred through tendons to the caudal peduncle and caudal fin, where thrust is generated. This form of swimming reaches its highest development in tunas, whose bodies barely flex. Virtually all thrust is derived from powerful beats of the caudal fin (figure 16.22). Many fast oceanic fishes, such as marlin (see figure 16.17A), swordfish, amberjacks, and mackerel, have swept-back caudal fins shaped much like a sickle. Such fins are the aquatic counterpart of the high-aspect wings of the swiftest birds (see figure 19.16). Swimming is the most economical form of vertebrate locomotion, largely because aquatic animals are almost perfectly supported by their medium and expend little energy to overcome the force of gravity.

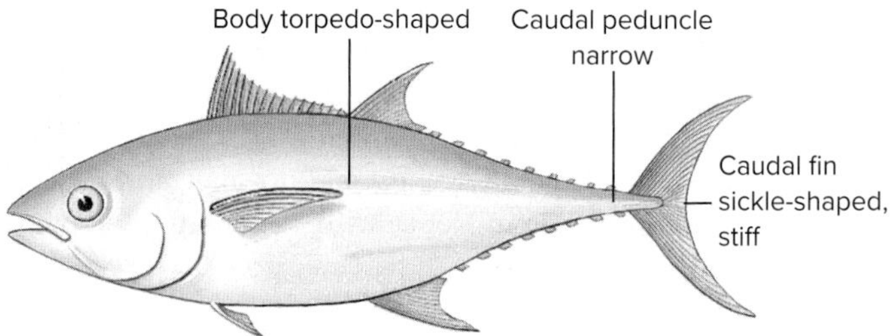

figure 16.22

Bluefin tuna, *Thunnus thynnus*, showing adaptations for fast swimming. Powerful trunk muscles pull on the slender caudal peduncle. Because the body does not bend, all of the thrust comes from beats of the stiff, sickle-shaped caudal fin.

Key Theme 16.5
ADAPTATION AND PHYSIOLOGY

Endothermic Fishes

The body temperature of most fishes is the same as their environment, because any heat generated internally is lost quickly into the surrounding water. However, some fishes, such as tunas and mako sharks, maintain a high temperature in their swimming muscles and viscera—as much as 10°C warmer than the surrounding water. Marlins and other billfishes elevate the temperature of their brain and retina. Research by F. G. Carey and others explains how these fishes accomplish this kind of thermoregulation, called **regional endothermy.** Heat is generated as a by-product of various activities, including digestion and swimming, or for billfishes, by a specialized heat-generating organ beneath the brain. This heat is conserved with a **rete mirabile,** a parallel bundle of blood vessels arranged to provide a countercurrent flow of blood. High temperatures apparently promote powerful swimming and enhance digestive and nervous system activity. Fishes with regional endothermy are the fastest in the world.

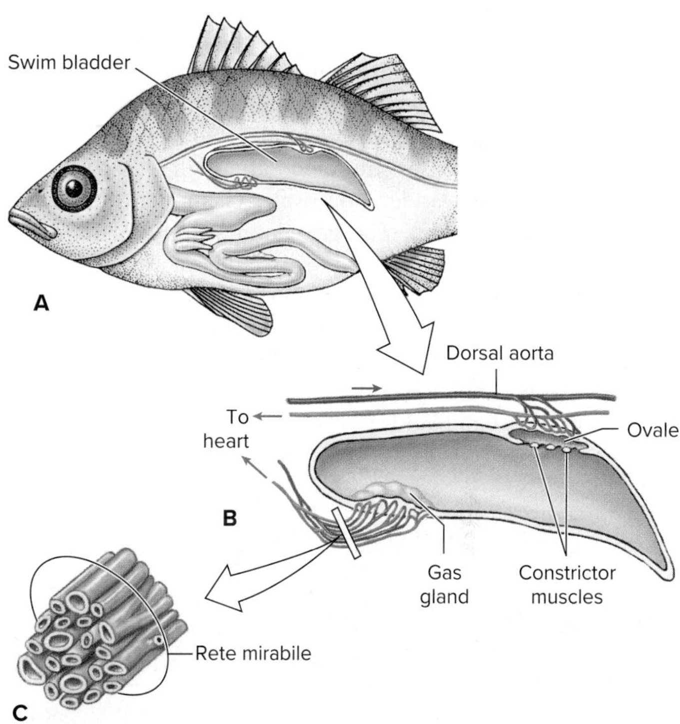

figure 16.23

A, Swim bladder of a teleost fish. The swim bladder lies in the coelom just beneath the vertebral column. **B,** Gas (primarily oxygen) diffuses into the swim bladder at the gas gland. Gas from the blood is moved into the gas gland by the rete mirabile, a complex array of tightly packed capillaries that act as a countercurrent multiplier to increase oxygen concentration. The arrangement of venous and arterial capillaries in the rete is shown in **C,** To release gas during ascent, a muscular valve opens, allowing gas to enter the ovale, and then to diffuse into the blood.

Neutral Buoyancy and the Swim Bladder

All fishes are slightly heavier than water because their skeletons and other tissues contain heavy elements that are present only in trace amounts in natural waters. To keep from sinking, sharks, which lack a swim bladder, must always keep moving forward in the water. The asymmetrical (heterocercal) tail of a shark provides lift as it sweeps through the water, and the broad head and flat pectoral fins (see figure 16.5) act as angled planes to provide additional lift. Sharks are also aided in buoyancy by having very large livers containing a lipid called **squalene** with a density of only 0.86 grams per milliliter. The liver thus acts like a large sack of buoyant oil that helps compensate for the shark's heavy body.

By far the most efficient device for adjusting buoyancy is a gas-filled space. The **swim bladder** serves this purpose in bony fishes (figure 16.23). It arose from the paired lungs of early Devonian fishes. Swim bladders are present in most pelagic bony fishes but are absent from tunas, most abyssal fishes, and most bottom-dwellers, such as flounders and sculpins. Although its name suggests otherwise, a swim bladder is not used to swim.

By adjusting the volume of gas (primarily oxygen) in its swim bladder, a fish can achieve neutral buoyancy and remain suspended indefinitely at any depth with no muscular effort. If a fish swims to a greater depth, the greater pressure exerted by the surrounding water compresses the gas in the swim bladder, so that the fish becomes less buoyant and begins to sink. Gas must be added to the swim bladder to establish a new equilibrium buoyancy. When a fish swims upward, gas in the bladder expands because of the reduced surrounding water pressure, making the fish more buoyant. Unless gas is removed, the fish continues to ascend with increasing speed as the swim bladder continues to expand.

Gas is removed from the swim bladder in two ways. Some fishes (trout, for example) have a pneumatic duct that connects the swim bladder to the esophagus, through which they can expel air. Most teleosts have lost the pneumatic duct and exchange air in the swim bladder with the blood. Gas exits the swim bladder and diffuses into blood at a highly vascularized region of the swim bladder, the **ovale**. Gas enters the swim bladder at the **gas gland.** The gas gland secretes lactic acid, causing oxygen to be released from hemoglobin, into the plasma of the blood. A remarkable network of parallel blood capillaries, the **rete mirabile,** supplies the gas gland and acts as a countercurrent multiplier (figure 16.23). The rete allows oxygen to reach high concentrations in the gas gland and to diffuse into the swim bladder.

The amazing effectiveness of this device is exemplified by a fish living at a depth of 2400 m (8000 feet). To keep the bladder inflated at that depth, the gas inside (mostly oxygen) must have a pressure exceeding 240 atmospheres, which is much greater than the pressure in a fully charged steel gas cylinder. Yet the oxygen pressure in the fish's blood cannot exceed 0.2 atmosphere, in equilibrium with the oxygen pressure in the atmosphere at the sea surface.

Respiration

Fish gills are composed of thin filaments, each covered with a thin epidermal membrane that is folded repeatedly into platelike **lamellae** (figure 16.24). These lamellae are richly supplied with blood vessels. The gills are located inside a pharyngeal cavity and are covered with a moveable bony flap, the **operculum.** This arrangement protects the delicate gill filaments, streamlines the body, and makes possible a pumping system for moving water through the mouth, across the gills, and out the gill slits. Instead of opercular flaps as in bony fishes, elasmobranchs have a series of **gill slits** (see figure 16.5) out of which the water flows. In both elasmobranchs and bony fishes, the branchial mechanism is arranged to pump water continuously and smoothly over the gills, although to an observer it appears that fish breathing is pulsatile.

The direction of water flow is opposite that of blood flow (countercurrent flow), the best arrangement for extracting the greatest possible amount of oxygen from water. Some bony fishes can remove as much as 85% of the dissolved oxygen from

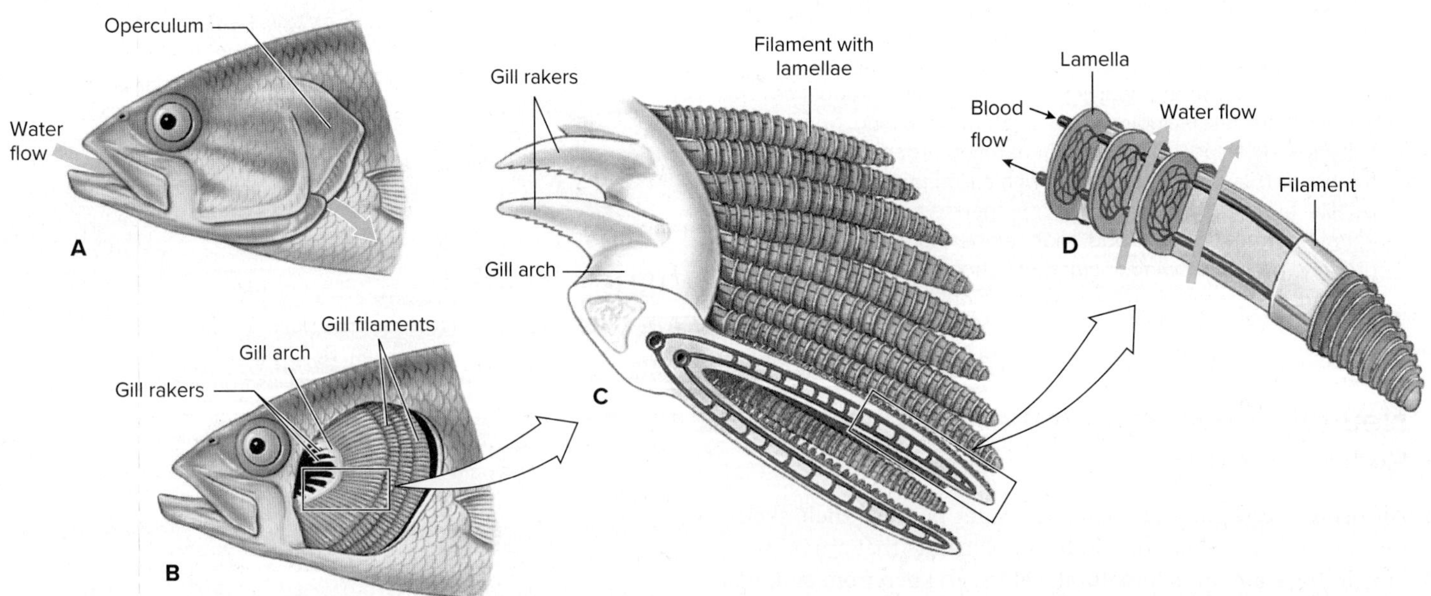

figure 16.24

Gills of a fish. Muscles attached to the operculum, **A,** pump water over gills and out the gill slit. The bony, protective flap covering the gills (operculum) has been removed, **B,** to reveal the branchial chamber containing the gills. There are four gill arches on each side, each bearing numerous filaments. **C,** A portion of a gill arch shows gill rakers that project forward to strain out food and debris, and gill filaments that project to the rear. **D,** A single gill filament is dissected to show the blood capillaries within the platelike lamellae. Direction of water flow (*blue arrows*) is opposite the direction of blood flow.

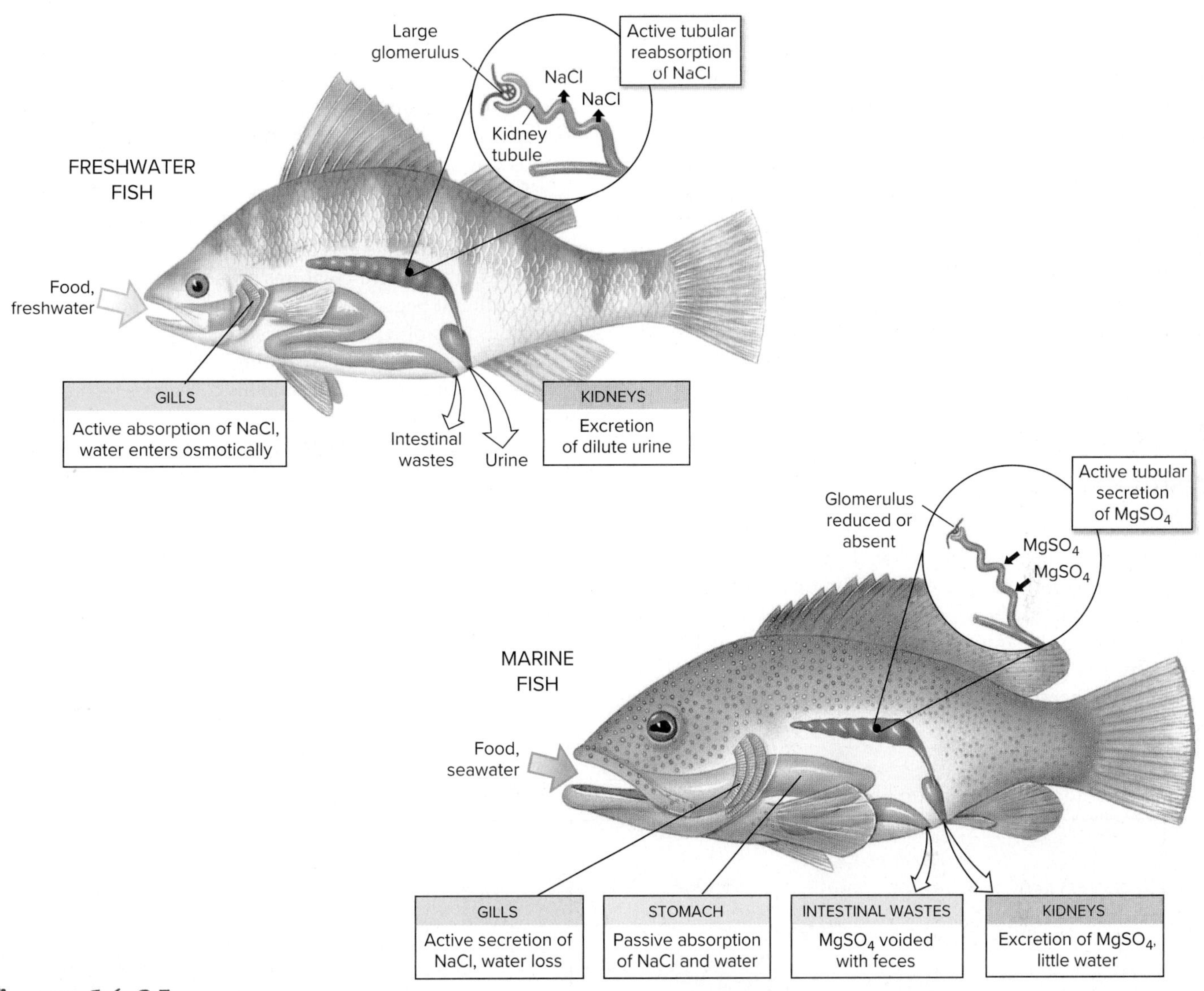

figure 16.25

Osmotic regulation in freshwater and marine bony fishes. *Top*: A freshwater fish maintains osmotic and ionic balance in its dilute environment by actively absorbing sodium chloride (NaCl) across the gills (some salt is gained with food). To flush out excess water that constantly enters the body, the glomerular kidneys produce a dilute urine by reabsorbing sodium chloride. *Bottom*: A marine fish must drink seawater to replace water lost osmotically to its salty environment. Sodium chloride and water are absorbed from the stomach. Excess sodium chloride is actively transported outward by the gills. Divalent sea salts, mostly magnesium sulfate ($MgSO_4$), are eliminated with feces and secreted by the kidneys.

water passing over their gills. Very active fishes, such as herring and mackerel, can obtain sufficient water for their high oxygen demands only by swimming forward continuously to force water into their open mouth and across their gills. This process is called ram ventilation.

Osmotic Regulation

Fresh water is an extremely dilute medium with a salt concentration (0.001–0.005 gram moles per liter [M]) much below that of the blood of freshwater fishes (0.2–0.3 M). Water therefore tends to enter their bodies osmotically, and salt is lost by diffusion outward. Although the scaled and mucus-covered body surface is almost totally impermeable to water, water gain and salt loss do occur across thin membranes of the gills. Freshwater fishes are **hyperosmotic regulators** with several defenses against these problems (figure 16.25, *top*). First, excess water is pumped out by the kidneys, which are capable of forming very dilute urine. Second, special **salt-absorbing cells** located in the gill epithelium actively move salt ions, principally sodium and chloride, from water to the blood. This absorption, together with salt present in the fish's food, replaces diffusive salt loss. These mechanisms are so efficient that a freshwater fish devotes only a small part of its total energy expenditure to maintaining osmotic balance.

Marine fishes are **hypoosmotic regulators** that encounter a completely different problem. Having a much lower blood salt concentration (0.3–0.4 M) than the seawater around them (about 1 M), they tend to lose water and to gain salt. A marine teleost quite literally risks drying out, much like a desert mammal deprived of water. To compensate for water loss, a marine teleost drinks seawater (figure 16.25, *bottom*). Excess salt accompanying the seawater is eliminated in multiple ways. Major sea salt ions (sodium, chloride, and potassium) are carried by the blood to the gills, where they are expelled by special salt-secretory cells. The remaining sea salt ions, mostly magnesium, sulfate, and calcium, are voided with feces or excreted by the kidneys.

Migration

Freshwater Eels

For centuries, naturalists had been puzzled by the life history of freshwater eels, *Anguilla* (an-gwil'ə) (L. eel), a common and commercially important species inhabiting coastal streams of the North Atlantic. Eels are **catadromous** (Gr. *kata,* down, + *dromos,* running), meaning they spend most of their lives in fresh water but migrate to the sea to spawn. Each fall, people saw large numbers of eels swimming down rivers toward the sea, but no adults ever returned. Each spring, countless numbers of young eels, called "elvers" (figure 16.26), each about the size of a wooden matchstick, appeared in coastal rivers and began swimming upstream. Beyond the assumption that eels must spawn somewhere at sea, the location of their breeding grounds was unknown.

The first clue was provided by two Italian scientists, Grassi and Calandruccio, who in 1896 discovered that elvers were advanced juvenile eels and that true larval eels were tiny, leaf-shaped, transparent creatures called **leptocephali.** In 1905, Johann Schmidt began a systematic study of eel biology, examining thousands of leptocephali caught in plankton nets from many areas of the Atlantic. By noting where larvae in different stages were captured, Schmidt and his colleagues reconstructed the spawning migration.

When adult eels leave the coastal streams of Europe and North America, they swim to the Sargasso Sea, a vast area of warm oceanic water southeast of Bermuda. Here, at depths of 300 m or more, the eels spawn and die. Minute larvae then begin an incredible journey back to the streams of Europe and North America (figure 16.26). Because the Sargasso Sea is much closer to the American coastline, American eel larvae make their journey in only about 1 year, compared to 3 years for European eel larvae. Males typically remain in the brackish water of coastal rivers, while females migrate as far as

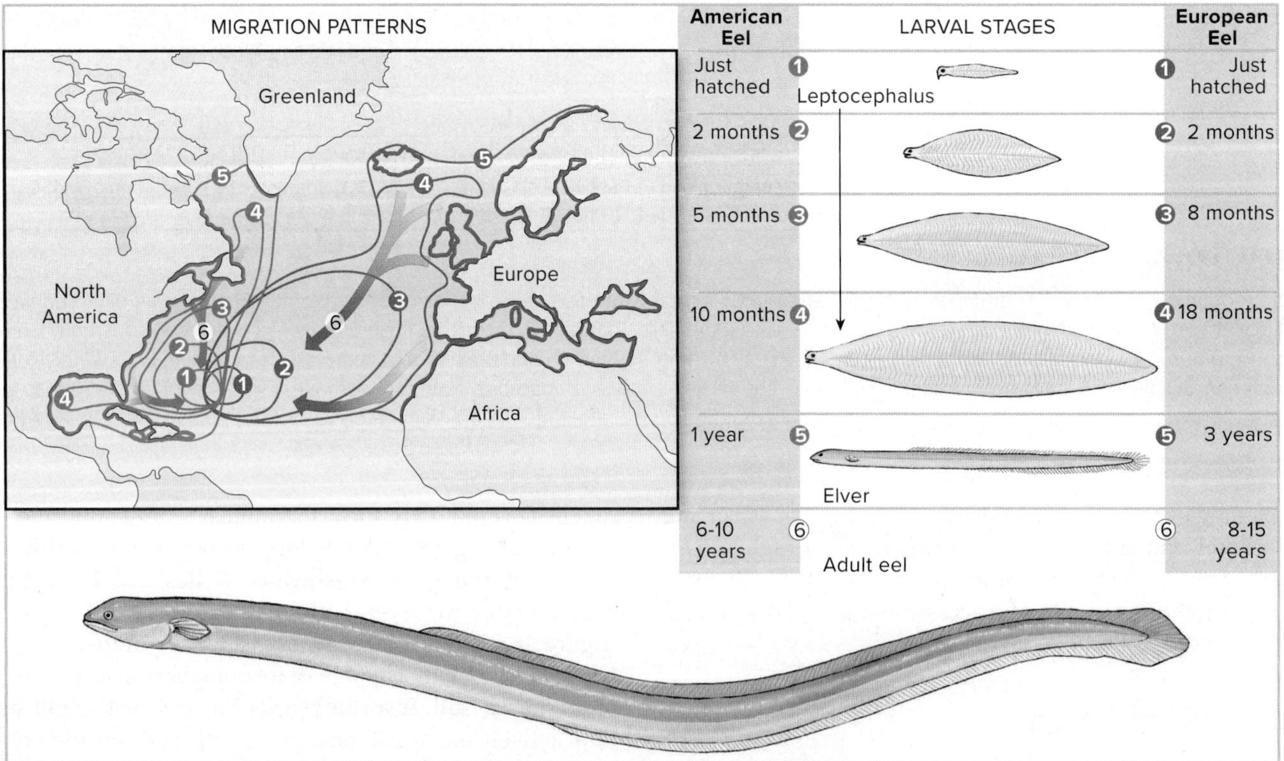

figure 16.26

Life histories of American eels, *Anguilla rostrata*, and European eels, *Anguilla anguilla*. Migration patterns of American eels are shown in blue, while those of European eels are shown in red. Circled numbers refer to stages of development. American eels complete their larval metamorphosis and sea journey in 1 year, while European eels take 3 years to complete their much longer journey.

several hundred kilometers upstream. After 6–15 years of growth, females, now 1 m long, return to the ocean to join the smaller males in the journey back to the spawning grounds in the Sargasso Sea.

Homing Salmon

The life history of salmon is as remarkable as that of freshwater eels and certainly has received far more popular attention. Salmon are **anadromous** (Gr. *anadromous,* running upward), spending their adult lives at sea but returning to freshwater streams to spawn. Atlantic salmon (*Salmo salar*) can make repeated upstream spawning runs. The seven Pacific salmon (*Oncorhynchus*) species (sockeye, coho, pink, Chinook, chum, steelhead, and masu) each make a single spawning run (figures 16.27 and 16.28), after which they die.

The virtually infallible homing instinct of the Pacific species is legendary. After migrating downstream as a smolt (a juvenile stage; figure 16.28), a sockeye salmon travels many hundreds of kilometers over the Pacific for nearly 4 years, grows to 2–5 kg in weight, and then returns almost unerringly to spawn in the headwaters of its parent stream. Some straying does occur and is an important means of increasing gene flow and populating new streams.

Experiments show that homing salmon are guided upstream by the characteristic odor of their parent stream. When salmon finally reach the spawning beds of their parents (where they themselves were hatched), they spawn and die. The following spring, newly hatched fry transform into smolts before and during the downstream migration. At this time they are imprinted with the distinctive odor of the stream, which is apparently a mosaic of compounds released by the characteristic vegetation and soil in the watershed of the natal stream. They also seem to imprint on odors of other streams that they pass while migrating downriver and use these odors in reverse sequence as a map during the upriver migration as returning adults.

Design Pics Inc/Alamy Stock Photo

figure 16.27

Sockeye salmon, *Oncorhynchus nerka*, spawning in Power Creek, Alaska. Females build nests called rudds, using their caudal fins to excavate shallow depressions in gravel riffles. After a female deposits her eggs, a male, distinguished by hooked jaws and large teeth, fertilizes the eggs. The female covers the eggs with gravel, and the young begin their development.

Key Theme 16.6

HUMAN CONNECTIONS

Salmons and Dams

Salmon runs along the Pacific coast of North America have been devastated by a lethal combination of increased siltation from logging, pollution, and, especially, the presence of more than 50 hydroelectric dams, which obstruct upstream migration of adult salmon and kill downstream migrants as they pass through the dams' power-generating turbines. The result is that the annual run of wild salmon is today less than 10% of the 10–16 million fish that ascended the rivers 150 years ago. Some hope comes from dam removals, which are occurring at an increasing rate in the United States. The "new" streams formed are quickly recovering. Only days after removal of a 62-m dam on the Elwha River in Washington in 2012, Chinook salmon were detected moving upstream of the old dam.

Reproduction

In a group as diverse as fishes, it is no surprise to find extraordinary variations on the basic theme of sexual reproduction. Most fishes favor a simple theme: they are **dioecious,** with external fertilization and external development of their eggs and embryos (oviparity). However, as tropical fish enthusiasts are well aware, the ever-popular ovoviviparous guppies and mollies of home aquaria develop in the ovarian cavity of the mother and then are born (figure 16.29). As described earlier in this chapter (see section 16.3: Sharks, Skates, and Rays: Elasmobranchii), some viviparous sharks develop a kind of placental attachment through which the young are nourished during gestation.

Oviparity is the most common mode of reproduction in fishes. Many marine fishes are extraordinarily prolific egg producers. Males and females come together in great schools and, without elaborate courtship behavior, release vast numbers of gametes into the water to drift with currents. A large female cod may release 4–6 million eggs at a single spawning. Less than one in a million will survive the numerous perils of the ocean to reach reproductive maturity.

Unlike the minute, semibuoyant, transparent eggs of pelagic marine teleosts, the eggs of many near-shore and bottom-dwelling (benthic) species are larger, yolky, nonbuoyant, and adhesive. Some bury their eggs, some attach them to vegetation, some deposit them in nests, and others even incubate them in their mouths (figure 16.30). Many benthic spawners guard their eggs. Intruders expecting an easy meal of eggs may be met with a vivid and often belligerent display by the guard, which is almost always male.

Freshwater fishes usually produce nonbuoyant eggs. Some, such as yellow perch, provide no parental care and

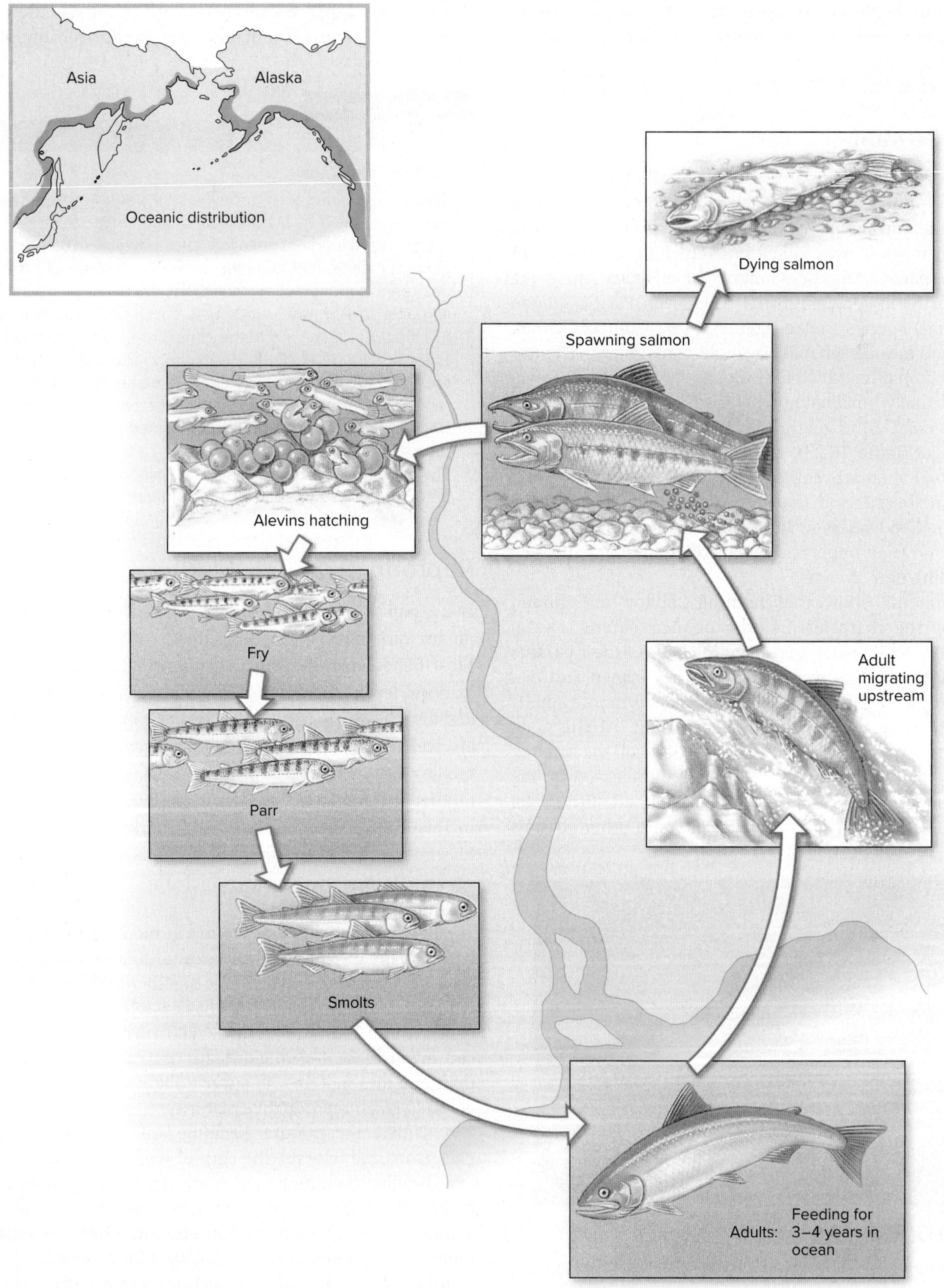

figure 16.28

Life history of Pacific salmon, *Oncorhynchus*.

figure 16.29

Eastern mosquitofish, *Gambusia holbrooki*, giving birth. This ovoviviparous species has been introduced worldwide for mosquito control.

figure 16.30

Male banded jawfish, *Opistognathus macrognathus*, orally brooding its eggs. The male retrieves the female's eggs and incubates them until they hatch. During brief periods when the jawfish is feeding, the eggs are left in the burrow.

simply scatter their myriads of eggs among weeds or along the sediment. Freshwater fishes that do provide egg care, such as catfishes, produce fewer, larger eggs that have a better chance for survival.

Elaborate preliminaries to mating are the rule for freshwater fishes. A female Pacific salmon, for example, performs a ritualized mating "dance" with her breeding partner after arriving at the spawning bed in a fast-flowing, gravel-bottomed stream (see figure 16.27). She then turns on her side and scoops out a depression with her tail. As the eggs are laid by the female, they are fertilized by the male. After the female covers the eggs with gravel, the exhausted fish dies.

Soon after an egg of an oviparous species is fertilized, it absorbs water, and the outer layer hardens. Cleavage follows, and a blastoderm forms astride a relatively large yolk mass. Many fishes hatch as larvae, carrying a semitransparent sac of yolk, which provides their food supply until the mouth and digestive tract have developed and the larvae can feed on their own. After a period of growth, a larva undergoes a metamorphosis, especially dramatic in many marine species, including eels (see figure 16.26). Body shape is refashioned, fin and color patterns change, and the larva becomes a juvenile bearing the unmistakable definitive body form of its species.

Taxonomy of Living Fishes

The following taxonomy primarily follows that of Nelson (2016). The probable relationships of these traditional groupings, together with the major extinct groups of fishes, are shown in the cladogram in figure 16.2. Because of the difficulty of determining relationships among the numerous living and fossil species, we can appreciate why fish classification has undergone, and will continue to undergo, continuous revision. As in the previous chapter, we do not show Linnaean ranks below subphylum. Nested hierarchies of monophyletic groups are shown by indenting.

Subphylum Vertebrata

Cyclostomata (sī'klō-stō-mä'tä) (Gr. *cyclos,* circle, + *stoma,* mouth). Without true jaws or paired appendages; teeth of keratin; vertebrae reduced or absent in adults.

Myxini (mik-sin'ē) (Gr. *myxa,* slime): **hagfishes.** Mouth with four pairs of tentacles; buccal funnel absent; 1–16 pairs of external gill openings; vertebrae rudimentary, only in embryos; slime glands present. Examples: *Myxine, Epaptretus;* about 78 species, marine.

Petromyzontida (pet'trō-mī-zon'ti-də) (Gr. *petros,* stone, + *myzon,* sucking): **lampreys.** Mouth surrounded by keratinized teeth but no barbels; buccal funnel present; seven pairs of external gill openings; vertebrae present only as neural arches. Examples: *Petromyzon, Ichthyomyzon, Lampetra;* 41 species, freshwater and anadromous.

Gnathostomata (na'thō-stō'mä-tə) (Gr. *gnathos,* jaw, + *stoma,* mouth). Jaws present; paired appendages present (secondarily lost in a few forms); three pairs of semicircular canals; notochord partly or completely replaced by vertebral centra.

Chondrichthyes (kon-drik'thē-ēz') (Gr. *chondros,* cartilage, + *ichthys,* fish): **cartilaginous fishes.** Cartilaginous skeleton; teeth not fused to jaws and usually replaced; no swim bladder; intestine with spiral valve; claspers present in males.

Elasmobranchii (ē-laz'mō-brānk'ē-ī) (Gr. *elasmos,* plated, + *branchia,* gills): **sharks, skates,** and **rays.** Placoid scales or derivatives (scutes and spines) usually present; five to seven gill arches and gill slits in separate

(continued)

(*continued*)

clefts along pharynx; upper jaw not fused to cranium. Examples: *Squalus, Raja, Sphyrna;* about 1150 species, mostly marine.

Holocephali (hō-lō-sef′ə-lī) (Gr. *holos,* entire, + *kephalē,* head): **chimaeras** and **ratfishes.** Scales absent; four gill slits covered by operculum; jaws with tooth plates; accessory clasping organ (tentaculum) in males; upper jaw fused to cranium. Examples: *Chimaera, Hydrolagus;* 48 species, marine.

Actinopterygii (ak′ti-nop′te-rij′ē-ī) (Gr. *aktis,* ray, + *pteryx,* fin, wing): **ray-finned fishes.** Skeleton ossified; gills covered by bony operculum; paired fins supported primarily by dermal rays; fin musculature within body; swim bladder mainly for buoyancy, if present; atrium and ventricle not divided; teeth with enameloid covering.

Cladistia (clə-dis′tē-ə) (Gr. *cladi,* branch): **bichirs.** Rhombic ganoid scales; lungs; spiracle present; dorsal fin consisting of 5–18 finlets. Examples: *Polypterus;* about 16 species, freshwater.

Chondrostei (kon-dros′tē-ī) (Gr. *chondros,* cartilage, + *osteon,* bone): **paddlefishes, sturgeons.** Skeleton primarily cartilage; caudal fin heterocercal; large scutes or tiny ganoid scales present. Examples: *Polyodon, Acipenser;* 27 species, freshwater and anadromous.

Neopterygii (nē′op-te-rij′ē-ī) (Gr. *neo,* new, + *pteryx,* fin, wing): **gars, bowfin,** and **teleosts.** Skeleton primarily bone; caudal fin usually homocercal; scales cycloid, ctenoid, absent, or rarely, ganoid. Examples: *Amia, Lepisosteus, Anguilla, Oncorhynchus, Perca;* about 29,600 species, nearly all aquatic habitats.

Sarcopterygii (sär-cop-te-rij′-ē-ī) (Gr. *sarkos,* flesh, + *pteryx,* fin, wing): **lobe-finned fishes, tetrapods.** Fishlike members with skeleton ossified; gills covered by bony operculum; paired fins with sturdy internal skeleton and musculature within appendage; diphycercal tail; usually with lungs; atrium and ventricle at least partly divided; teeth with enamel covering. Fish examples: *Latimeria* (coelacanths); *Neoceratodus, Lepidosiren, Protopterus* (lungfishes); 8 fish species, marine and freshwater.

SUMMARY

Sections	Key Concepts
16.1 Ancestry and Relationships of Major Groups of Fishes	• Fishes are a non-monophyletic group of gill-breathing aquatic vertebrates with fins for appendages. • There are five major, mutually exclusive, clades containing living fishes: Myxini (hagfishes), Petromyzontida (lampreys), Chondrichthyes (cartilaginous fishes), Actinopterygii (ray-finned fishes), and Sarcopterygii (lobe-finned fishes).
16.2 Living Jawless Fishes: Cyclostomata	• The hagfishes and lampreys are jawless, eel-like fishes without paired fins that form the clade Cyclostomata. • Hagfishes are marine fishes, producing slime as a defense. They feed on invertebrates and carcasses using keratinized teeth in their mouths. • Lampreys are anadromous or freshwater. They have a filter-feeding larval stage called ammocoetes. The adults either do not feed or feed on other fishes with their disc-like mouths.
16.3 Cartilaginous Fishes: Chondrichthyes	• Chondrichthyes include sharks, rays, and chimaeras. • They are characterized by strong jaws, paired fins, a cartilaginous skeleton, and well-developed sensory organs, including a lateral line to detect water vibrations and electroreceptors to detect the electrical field of their prey. • All are predators and have teeth with hard enamel. • Reproduction in sharks and rays is diverse, although all use internal fertilization. Some are oviparous (egg-laying) and some are viviparous, with development occurring in the mother's uterus. The nutritional needs of viviparous embryos can be solely supplied by their yolk sac (ovoviviparity) or by a placental connection with the mother.
16.4 Bony Fishes and Tetrapods: Osteichthyes	• The fishlike members of Osteichthyes have bony endoskeletons and are divided into Sarcopterygii and Actinopterygii. • Sarcopterygii includes lobe-finned fishes (lungfishes and coelacanths) and tetrapods (terrestrial vertebrates). • Actinopterygii is composed of the ray-finned fishes, most of which are the teleosts, a large group characterized by adaptations that enhance precise movements and suction feeding. • The earliest bony fishes had heavy, often ganoid scales, whereas later fishes (teleosts) have lighter, cycloid or ctenoid scales.

Sections	Key Concepts
16.5 Structural and Functional Adaptations of Fishes	• Most fishes swim by undulatory contractions of the body muscles, generating thrust against the surrounding water. Eel-like fishes oscillate the whole body, but faster fishes, such as trout and tuna, limit the motion to the caudal (tail) region of the body. • Bony fishes achieve neutral buoyancy in water with a gas-filled swim bladder. Gas (oxygen) is added to the swim bladder from the bloodstream at the gas gland, and gas is removed from the swim bladder into the bloodstream at the ovale. Sharks lack swim bladders, but achieve near-neutral buoyancy with lightweight oils in their body. • Gills of bony fishes, having efficient countercurrent flow between water and blood, facilitate high rates of oxygen exchange. • In bony fishes, osmotic regulation primarily occurs in the gills and kidneys. Marine fishes, which are hypoosmotic relative to seawater, drink water and pump out ions at the gills and kidneys. Freshwater fishes, which are hyperosmotic relative to surrounding water, pump in ions at the gills and excrete dilute urine. • Oviparity without parental care is the most common reproductive mode, although viviparity is common in some groups, and some fishes guard their eggs or young. • Catadromous fishes (such as eels) migrate from freshwater to spawn in saltwater, whereas anadromous fishes (such as salmon) migrate from saltwater or lakes to spawn in freshwater streams.

Review Questions

1. Provide a brief description of fishes, citing characteristics that distinguish them from all other animals.
2. What evidence suggests that hagfishes and lampreys form a monophyletic group?
3. Describe feeding behavior in hagfishes and lampreys. How do they differ?
4. Describe the life cycle of sea lampreys, *Petromyzon marinus,* and the history of their invasion of the Great Lakes.
5. In what ways are sharks well equipped for a predatory life habit?
6. What function does the lateral-line system serve? Where are receptors located?
7. Explain how bony fishes differ from sharks and rays in the following systems or features: skeleton, scales, buoyancy, respiration, and reproduction.
8. Match the ray-finned fishes in the right column with the group to which each belongs in the left column:

___ Chondrosteans	a. Perch
___ Nonteleost neopterygians	b. Sturgeon
	c. Gar
___ Teleosts	d. Salmon
	e. Paddlefish
	f. Bowfin

9. Make a cladogram that includes the following groups of fishes: chimaeras, hagfishes, lampreys, salmon, sharks, sturgeons. Add the following synapomorphies to the diagram: claspers, endochondral bone, fins with stout bones, jaws, keratinized teeth, and vertebrae.
10. List four characteristics of teleosts that contributed to their incredible evolutionary diversity.
11. What morphological characteristics distinguish lobe-finned fishes?
12. Explain how lungfishes are adapted to survive out of water.
13. Describe discoveries of living coelacanths. What is the evolutionary significance of the group to which they belong?
14. Compare the swimming movements of eels with those of trout, and explain why the latter are more efficient for rapid locomotion.
15. Sharks and bony fishes approach or achieve neutral buoyancy in different ways. Describe the methods of each group. Why must a teleost adjust the gas volume in its swim bladder when it swims upward or downward? How is gas volume adjusted?
16. What is meant by "countercurrent flow" as it applies to fish gills?
17. Compare the osmotic problem and the mechanism of osmotic regulation in freshwater and marine bony fishes.
18. Describe the life cycle of freshwater eels.
19. How do adult Pacific salmon find their way back to their parent stream to spawn?
20. What mode of reproduction in fishes is described by each of the following terms: oviparity, ovoviviparity, placental viviparity?
21. Reproduction in marine pelagic fishes and in benthic freshwater fishes is distinctively different. How and why do they differ?

For Further Thought Regarding osmoregulation, what physiological and behavioral changes would occur as a fish migrates from a freshwater stream to the ocean?

17

The Early Tetrapods and Modern Amphibians

LEARNING OBJECTIVES

Readers will be able to:

17.1 Describe the best hypotheses for explaining from key fossils how the ancestral tetrapod body plan first evolved in the Paleozoic Era.

17.2 Describe the life-history characteristics considered ancestral for living amphibians.

17.3 Explain the major characteristics of caecilians, amphibian order Gymnophiona.

17.4 Explain the major characteristics of salamanders, amphibian order Urodela, including life cycles, respiration, and paedomorphosis.

17.5 Explain the major characteristics of frogs, amphibian order Anura, including habitat, distribution, and life history.

Jason Edwards/National Geographic/Getty Images

A frog tadpole undergoing metamorphosis.

From Water to Land in Ontogeny and Phylogeny

The chorus of frogs beside a pond in spring announces the start of a new life cycle. Mating frogs produce masses of eggs, which soon hatch into limbless, gill-breathing tadpole larvae that feed and grow. Then a remarkable transformation occurs. Hindlimbs appear and gradually lengthen. The tail shortens and eventually disappears. Larval teeth and gills are lost. Eyelids develop. The forelimbs emerge. In a matter of weeks, the aquatic tadpole has completed its metamorphosis to an adult, land-dwelling frog.

By contrast, the evolutionary transition from water to land occurred not in weeks but over millions of years. A lengthy series of alterations cumulatively fitted vertebrate bodies for life on land. The origin of land vertebrates is no less remarkable, and it is unlikely to happen again because well-established competitors would exclude a transitional form.

Amphibians include the only living vertebrates that have a transition from water to land in both their ontogeny (organismal development) and their phylogeny (evolutionary history). After some 350 million years of evolution, amphibians remain quasiterrestrial, hovering between aquatic and land environments. Even the amphibians best adapted for a terrestrial existence cannot stray far from moist conditions. Many, however, have evolved ways to keep their eggs out of open water where the larvae would encounter enemies.

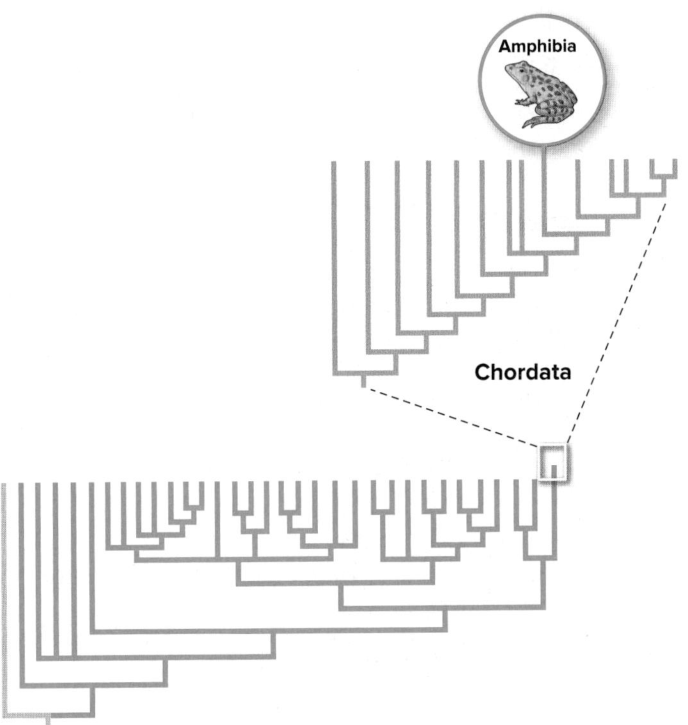

Life on land is a major theme of the remaining vertebrate groups, which form a clade called the superclass Tetrapoda (from the Greek term for "four feet"). Amphibians and amniotes (including nonavian reptiles, birds, and mammals) are the two major extant branches of **tetrapod** phylogeny, which originates in the Devonian period. Amphibians are ectothermic, relying on environmental context rather than metabolic energy to regulate body temperature. Many amphibians depend on freshwater streams or pools for their reproduction. Many tetrapod lineages have lost one or both of the two pairs of limbs that give the group its name; amphibian examples include sirens and caecilians (section 17.3). In this chapter, we review the origins of terrestrial vertebrates and discuss the amphibian branch in detail. We discuss the major amniote groups in Chapters 18–20.

Movement from water to land is perhaps the most dramatic event in animal evolution, because land is physically hazardous. Life originated in water. Animal bodies are mostly water, and all cellular activities occur in water. Nevertheless, organisms invaded land, carrying their watery composition with them. Vascular plants, pulmonate snails, and tracheate arthropods made this transition much earlier than did vertebrates, and thus generated a source of food for terrestrial vertebrates. Although the invasion of land required modification of almost every organ system, aquatic and terrestrial vertebrates retain many structural and functional similarities. Today, the transition between aquatic and terrestrial vertebrates occurs in many living amphibians that make this transition during their life histories.

When moving from water to land, animals encounter important physical differences in environment: (1) oxygen content, (2) fluid density, (3) temperature regulation, and (4) habitat diversity. Oxygen is at least 20 times more abundant in air, and it diffuses much more rapidly through air than through water. It is thus readily available to animals that possess lungs and/or a skin surface suitable for respiratory gas exchange. Compared to water, air has 1000 times less buoyant density and 50 times less viscosity. It therefore provides relatively little support against gravity, requiring terrestrial animals to develop strong limbs and to remodel their skeletons to gain adequate structural support. Air fluctuates in temperature more readily than does water, and terrestrial environments therefore experience harsh and unpredictable cycles of freezing, thawing, drying, and flooding. Terrestrial animals require behavioral and physiological strategies to protect themselves from thermal extremes.

17.1 Devonian Origin of Tetrapods

By the Devonian period, beginning about 416 million years ago, bony fishes had diversified to include many freshwater forms. An important combination of characteristics that evolved originally in aquatic habitats provided access to terrestrial habitats. An air-filled cavity, which could be used as a swim bladder, and paired internal nares (nostrils; evolutionary origin shown on figure 17.1), serving chemoreception, connected to the pharynx. On land, this combination of structures would draw oxygen-rich air through the nares and into the air-filled cavity, whose moist surface would exchange oxygen and carbon dioxide with body fluids. The bony elements of paired fins, modified for support and movement on underwater surfaces (evolutionary origin shown in figure 17.2), gained sufficient strength to support and to move the body on land.

The internal nares, air-filled cavity, and paired limbs of an aquatic tetrapod ancestor therefore made possible evolution of terrestrial breathing and support. The air-filled cavity illustrates the evolutionary principle of exaptation (see section 1.4: Natural Selection), in which a structure that has evolved by natural selection in an initial utility or role is later recruited or "coopted" in a new role. Note that the air-filled cavities are homologous, with *lung* indicating that the structure's primary role is air breathing, as it is in terrestrial forms, and *swim bladder* denoting that the structure serves primarily in providing buoyancy during swimming in aquatic animals. Zoologists continue to debate the controversial question of whether the lung or swim bladder was the original role of the air-filled cavity.

Freshwater habitats are inherently unstable, being prone to evaporation or depletion of the dissolved oxygen needed to support vertebrate life. It is therefore not surprising that multiple fish groups, given a combination of structures that could be coopted for terrestrial breathing and locomotion, evolved some degree of terrestriality. Some fishes supplement the low oxygen in stagnant waters by taking air breaths or holding internal air bubbles, and terrestriality elaborates this tendency. Mudskippers and lungfish are two familiar examples of evolution of terrestriality by fishes; however, only one such transition, occurring in the early Devonian period, provided the ancestral lineage of all tetrapod vertebrates. This lineage ultimately evolved the

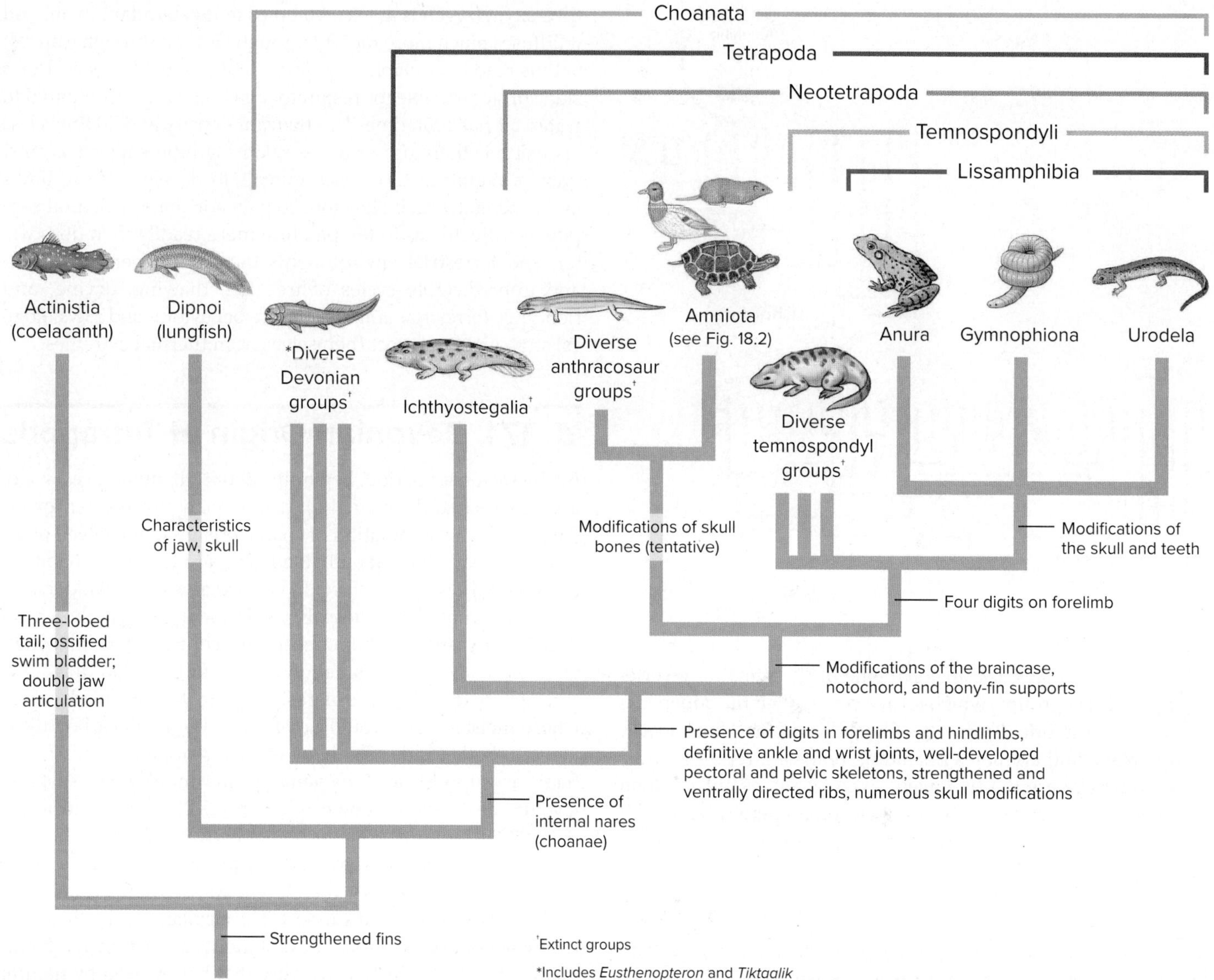

figure 17.1

Tentative cladogram of the Tetrapoda with emphasis on descent of the amphibians. Especially controversial are the relationships of major tetrapod groups (Amniota, Temnospondyli, and diverse early tetrapod groups) and outgroups (Actinistia, Dipneusti, extinct Devonian groups). All aspects of this cladogram are controversial, however, including relationships of the Lissamphibia. Extinct Devonian groups include the fossils *Eusthenopteron*, *Tiktaalik*, and ichthyostegalians (including *Acanthostega and Ichthyostega*).

characteristic tetrapod adaptations to air breathing, including increased vascularization of the air-filled cavity with a rich capillary network to form an efficient lung, and a **double circulation** directing deoxygenated blood into the lungs for oxygenation, and oxygenated blood from the lungs to other body tissues.

Tetrapods evolved limbs in an ancestral aquatic habitat during the Devonian period prior to their evolutionary movement onto land. Although fish fins at first appear very different from the jointed limbs of tetrapods, an examination of the bony elements of the paired fins of the lobe-finned fishes reveals their homology with amphibian limbs. In *Eusthenopteron,* a Devonian lobe-fin living approximately 385 million years ago, we recognize an upper arm bone (humerus) and two forearm bones (radius and ulna) as well as other elements that we homologize with the wrist bones of tetrapods (figure 17.2). *Eusthenopteron* could push itself through the bottom mud of pools with its fins, but it could not walk upright because backward and forward movement of the fins was limited to about 20–25 degrees. The fossil genus *Tiktaalik,* which lived approximately 375 million years ago, is morphologically intermediate between lobe-finned fishes and tetrapods. *Tiktaalik* probably inhabited shallow, oxygen-depleted streams or swamps, using its limbs to support the body while placing its snout above water to breathe air. This form also might have traversed land.

Acanthostega, an early tetrapod that lived approximately 365 million years ago, had well-formed tetrapod limbs with clearly formed digits on both fore- and hindlimbs, but it was clearly an aquatically adapted form whose limbs were too weakly

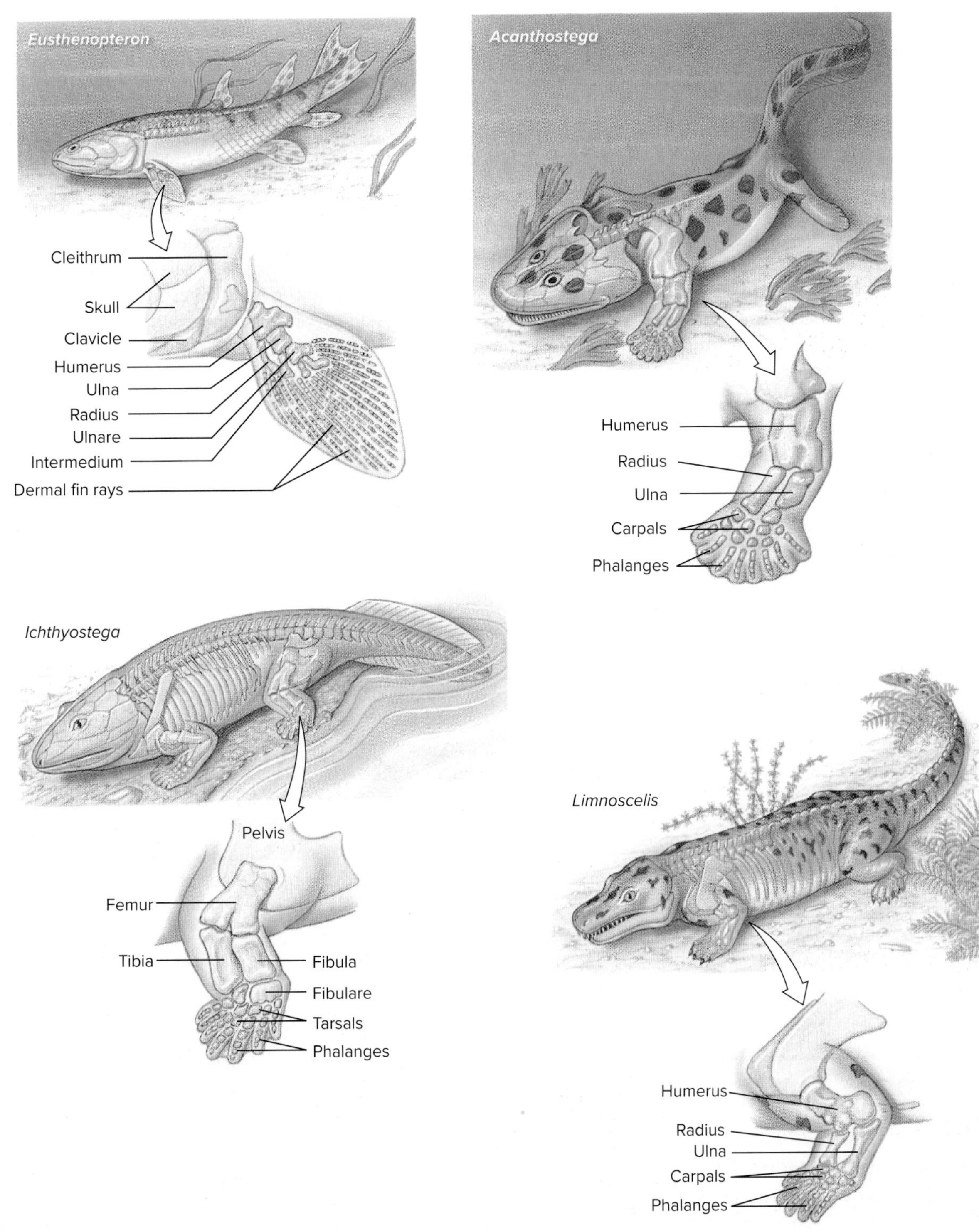

figure 17.2

Evolution of tetrapod limbs. The limbs of tetrapods evolved from the fins of Paleozoic fishes. *Eusthenopteron,* a late Devonian lobe-finned fish, had paired muscular fins supported by bony elements that foreshadowed the bones of tetrapod limbs. The anterior fin contained an upper arm bone (humerus), two forearm bones (radius and ulna), and smaller elements homologous to the wrist bones of tetrapods. As typical of fishes, the pectoral girdle, consisting of the cleithrum, clavicle, and other bones, was firmly attached to the skull. In *Acanthostega,* one of the earliest known Devonian tetrapods (appearing about 365 million years ago), dermal fin rays of the anterior appendage were replaced by eight fully evolved fingers. *Acanthostega* was probably exclusively aquatic because its limbs were too weak for travel on land. *Ichthyostega,* a contemporary of *Acanthostega,* had fully formed tetrapod limbs and must have been able to walk on land. The hindlimb bore seven toes (the number of forelimb digits is unknown). *Limnoscelis,* an anthracosaur of the Carboniferous period (about 300 million years ago), had five digits on both fore- and hindlimbs, the basic pentadactyl model that became the tetrapod standard.

constructed for walking on land. *Ichthyostega,* (Gr. *ichthys,* fish, + *stegē,* roof or covering, in reference to the roof of the skull, which was shaped like that of a fish) another fossil genus approximately 365 million years old, had a fully developed shoulder girdle, bulky limb bones, and well-developed muscles. *Acanthostega* and *Ichthyostega* reveal that the earliest tetrapods had more than five digits per limb and that the five-digit pattern more characteristic of living forms was stabilized later in tetrapod evolution.

Ichthyostega possessed several adaptations, in addition to jointed limbs, that equipped it for life on land (figure 17.2): a stronger backbone and associated muscles that support the body in air, new muscles that elevate the head, strengthened shoulder and hip girdles, a protective rib cage, a modified ear structure that detects airborne sounds, a foreshortened skull, and a lengthened snout. *Ichthyostega* nonetheless resembled aquatic forms in retaining a tail complete with fin rays and in having opercular (gill-covering) bones.

Key Theme 17.1

SCIENCE EXPLAINED

Ichthyostega

Bones of *Ichthyostega,* the most thoroughly studied of all early tetrapods, were first discovered on an east Greenland mountainside in 1897 by Swedish scientists looking for three explorers lost two years earlier during an ill-fated attempt to reach the North Pole by hot-air balloon. Later expeditions by Gunnar Säve-Söderberg uncovered skulls of *Ichthyostega,* but before he could examine the skulls, Säve-Söderberg died at age 38. After Swedish paleontologists returned to the Greenland site, where they found the remainder of *Ichthyostega*'s skeleton, Erik Jarvik, one of Säve-Söderberg's assistants, devoted his life's work to producing the detailed description of *Ichthyostega* available today.

Evolutionary relationships of early tetrapod groups are still controversial. We present a tentative cladogram (see figure 17.1), which almost certainly will be revised as new data are reported.

17.2 Temnospondyls and Modern Amphibians

Many extinct lineages plus the **Lissamphibia,** which contains the modern amphibians, are called **temnospondyls** (see figures 17.1 and 17.3). Temnospondyls first appeared early in the Carboniferous period and diversified to form more than 25 extinct taxonomic families. Among temnospondyls, only the living amphibians survived the end-Cretaceous mass extinction. Hypotheses that living amphibians arose within temnospondyls and that all temnospondyls are closer to amphibians than to amniotes are not universally accepted. Some paleontologists consider extinct temnospondyls to form a clade outside living tetrapods. Nonetheless, close resemblance of tooth structure of some temnospondyls to those of frogs and salamanders supports origin of modern amphibians within temnospondyls. This is the arrangement shown in figure 17.1, with temnospondyl lineages all placed closer to amphibians than to amniotes. This group generally has only four digits on the forelimb rather than the five digits typical of most tetrapods.

Lissamphibians likely arose during the Carboniferous period and later diversified, probably by the late Triassic period (early Mesozoic era), to produce ancestors of the three major groups of amphibians alive today: **frogs** (Anura or Salientia), **salamanders** (Caudata or Urodela), and **caecilians** (Apoda or Gymnophiona). The earliest fossils attributable to modern amphibians appear in the middle Permian period.

Two additional generally recognized but nonetheless controversial groupings of Carboniferous and Permian tetrapods, **lepospondyls** and **anthracosaurs,** are judged from skull structure to be closer to amniotes than to temnospondyls (figure 17.1). We cover the amniote branch of tetrapod phylogeny in Chapters 18–20.

The three living amphibian orders comprise more than 7900 species. Most share general adaptations to life on land, including skeletal strengthening.

In the ancestral amphibian life history, eggs are aquatic, and they hatch to produce an aquatic larval form that breathes using gills. A metamorphosis follows in which gills are lost. Metamorphosed amphibians use cutaneous respiration on land, exchanging gases with the surrounding air across their moist skin. Presence of lungs is the ancestral condition in amphibians; lungs occur throughout larval life and at metamorphosis become active for breathing air. Absence of lungs is known only for single species in caecilians and in frogs, although many species of salamanders are lungless. Gas exchange across the skin is the primary means of respiration in the many terrestrial salamanders that lack lungs. Many amphibians retain this general pattern, with some important exceptions. Some salamanders do not undergo a complete metamorphosis and retain a permanently aquatic, larval morphology throughout life. Some caecilians, some frogs, and other salamanders live entirely on land and lack the aquatic larval phase completely. Both alternatives are evolutionarily derived conditions. Some frogs, salamanders, and caecilians that undergo a complete metamorphic life cycle nonetheless remain in water as adults rather than moving onto land during their metamorphosis.

Even the most terrestrial amphibians remain dependent on very moist environments. Their skin is thin, and it requires moisture for protection against desiccation in air. Amphibians also require moderately cool environments. As **ectotherms,** their body temperature matches environmental temperatures, greatly restricting where they can live. Cool and wet environments are especially important for reproduction. Eggs must be shed directly into water or onto moist terrestrial surfaces to avoid desiccation.

17.3 Caecilians: Order Gymnophiona (Apoda)

The order Gymnophiona (jim'no-fī'o-na) (Gr. *gymnos,* naked, + *ophineos,* of a snake) contains over 200 species of elongate, limbless, burrowing creatures commonly called **caecilians**

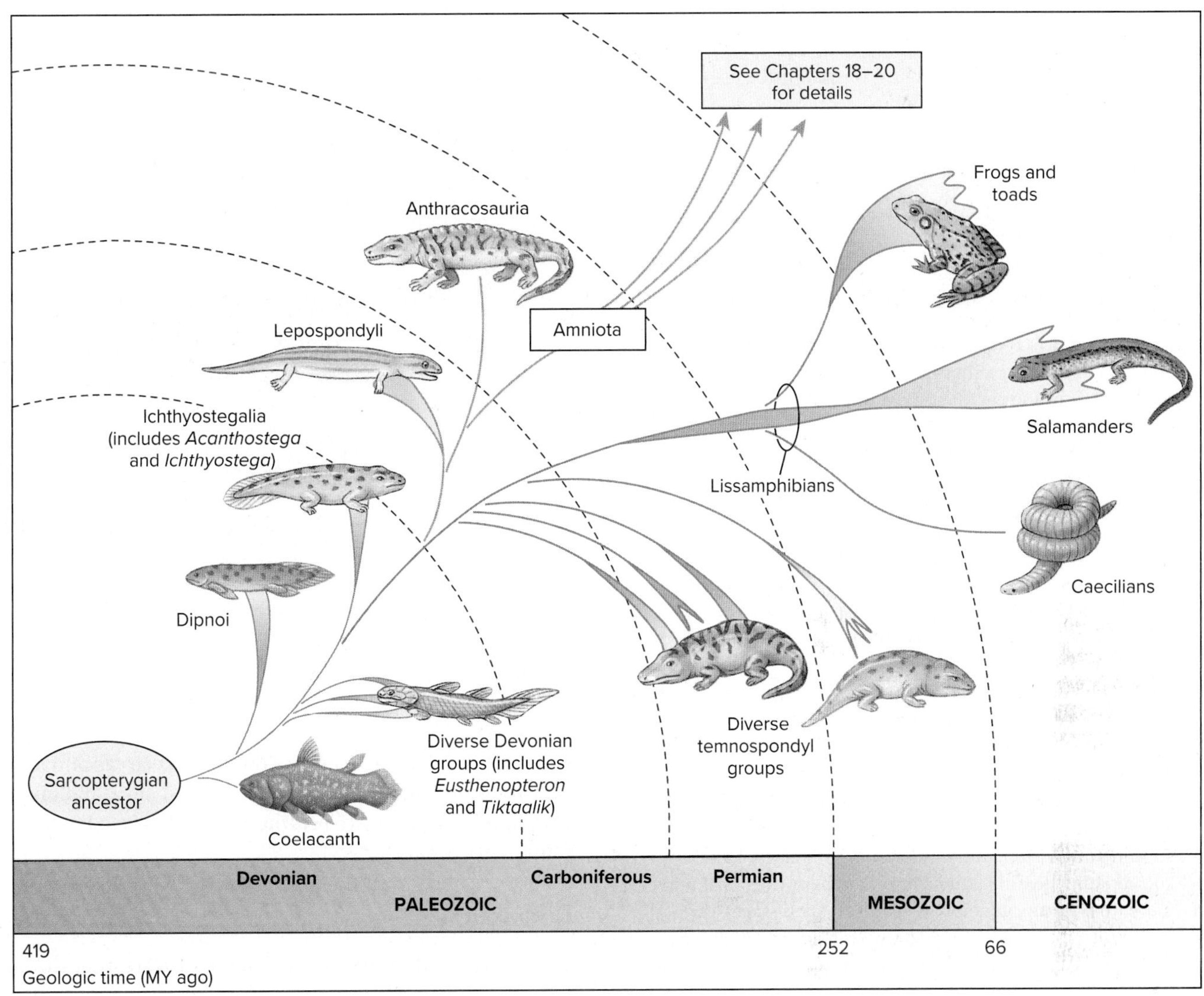

figure 17.3

Early tetrapod evolution and the descent of amphibians. The tetrapods share most recent common ancestry with the extinct Devonian groups shown; of living groups, tetrapods are most closely related to lungfishes. Amphibians share most recent common ancestry with the diverse temnospondyls of the Carboniferous and Permian periods of the Paleozoic era, and the Triassic period of the Mesozoic era.

(figure 17.4). They occur in tropical forests of South America (their principal home), Africa, India, and Southeast Asia, and many species are aquatic. They possess a long, slender body, many vertebrae, long ribs, no limbs, and a terminal anus; some have small dermal scales in the skin. Eyes are small, and most species are totally blind as adults. Their food is mostly earthworms and small insects, which they find underground. Caecilians represent one of several cases in tetrapods of a complete evolutionary loss of limbs.

Fertilization of eggs occurs inside the female's body, and the male has a protrusible copulatory organ. Caecilians usually deposit eggs in moist ground near water. In some species, the female parent carefully guards developing eggs in folds of her body. Some species have aquatic larvae with a tail fin, an open gill slit, and external gills; larval development in other species occurs within the egg. Viviparity is common in some caecilians, with the embryos obtaining nourishment by eating the wall of the oviduct.

17.4 Salamanders: Order Urodela (Caudata)

Order Urodela (Gr. *oura,* tail, + *delos,* evident) comprises tailed amphibians, over 700 species of salamanders. Salamanders occur in almost all northern temperate regions of the world, and they are abundant and diverse in North America. Salamanders also inhabit tropical areas of Central America and northern South America. Salamanders are typically small; most of the common North American salamanders are less than 15 cm long. Some aquatic forms are considerably longer, and Japanese giant salamanders sometimes exceed 1.5 m in length.

Most salamanders have limbs set at right angles to the trunk, with forelimbs and hindlimbs of approximately equal size. In some aquatic and burrowing forms, limbs are rudimentary or absent.

CHARACTERISTICS of Modern Amphibians

1. **Limbs usually four (quadrupedal)** in two pairs with associated shoulder/hip girdle, although some salamanders have forelimbs only and caecilians have no limbs; no true nails; feet often webbed; **forelimb usually with four digits** and hindlimb with five. **Skeleton mostly bony** with varying numbers of vertebrae; salamanders usually have a distinct head, neck, trunk and tail; adult frogs have a fused head and trunk and usually no tail; caecilians have an elongated trunk not strongly demarcated from the head and a terminal anus; **ectothermic.**
2. **Skin smooth, moist and glandular;** integument modified for **cutaneous respiration;** pigment cells (chromatophores) common and varied; **granular glands** secrete defensive compounds.
3. Skull relatively light, less ossified, flattened in profile, and with fewer bones than other vertebrates.
4. Mouth usually large with small teeth in upper or both jaws and on the vomer/palate.
5. **Tripartite brain** includes forebrain (telencephalon) coordinating olfaction, midbrain (mesencephalon) coordinating vision, and hindbrain (rhombencephalon) coordinating hearing and balance; ten pairs of cranial nerves.
6. Ear with **tympanic membrane** (eardrum) and **stapes** (columella) for transmitting vibrations to inner ear; for vision in air, cornea rather than lens is principal refractive surface for bending light; **eyelids** and **lachrymal glands** protect and wash eyes; paired **internal nostrils** open into a nasal cavity lined with **olfactory epithelium** at anterior part of mouth cavity.
7. Separate sexes; fertilization mostly external in frogs and toads but internal via a spermatophore in most salamanders and caecilians; predominantly oviparous, but some ovoviviparous or viviparous.
8. **Eggs moderately yolky (mesolecithal) with jellylike membrane coverings;** aquatic larva often present with metamorphosis to a more terrestrial adult form.
9. Excretory system of paired mesonephric or opisthonephric kidneys; urea main nitrogenous waste.
10. Respiration by skin and in some forms by gills and/or lungs; presence of gills and lungs varies among species and by developmental state of some species; forms with aquatic larvae lose gills at metamorphosis in frogs; many salamanders retain gills and an aquatic habitat throughout life; paired nostrils enable breathing in lung-breathing forms; vocal cords present between lungs and vocal sacs especially in frogs.
11. **Heart with a sinus venosus,** two atria, one ventricle, a conus arteriosus; **double circulation** through the heart in which pulmonary arteries and veins supply lungs (when present) and return oxygenated blood to heart; skin abundantly supplied with blood vessels.

A

B

Mark Smith/Science Source

figure 17.4

A, Female caecilian coiled around eggs in a burrow. **B,** Pink-head caecilian (*Herpele multiplicata*), native to western Africa.

Salamanders are carnivorous both as larvae and as adults, preying on worms, small arthropods, and small molluscs. Like all amphibians, they are **ectothermic** and have a low metabolic rate.

Life Cycles

Some salamanders are either aquatic or terrestrial throughout their life cycle, but the ancestral life cycle is metamorphic, having aquatic larvae and terrestrial adults that occupy moist places under stones and rotten logs. Eggs of most salamanders are fertilized internally; a female recovers in her vent (common opening for anus and reproductive system) a packet of sperm (**spermatophore**) deposited by a male on a leaf or stick (figure 17.5). Aquatic species lay eggs in clusters or stringy masses in water. Their eggs hatch to produce an aquatic larva having external gills and a finlike tail. Completely terrestrial species deposit eggs in small, grapelike clusters under logs or in excavations in soft moist earth, and many species guard their eggs (figure 17.6). Terrestrial species undergo **direct development.** They bypass the larval stage and hatch as miniature versions of their parents. A particularly complex life cycle occurs in some American newts, whose aquatic larvae metamorphose to form terrestrial juveniles that later metamorphose again to produce secondarily aquatic, breeding adults (figure 17.7).

Respiration

At various stages of their life history, salamanders may have external gills, lungs, both, or neither of these structures. They also share the general amphibian condition of having extensive

figure 17.5

Courtship and sperm transfer in the pygmy salamander, *Desmognathus wrighti.* After judging the female's receptivity by the presence of her chin on his tail base, the male deposits a spermatophore on the ground, and then moves forward a few paces. **A,** The white mass of the sperm atop a gelatinous base is visible at the level of the female's forelimb. The male moves ahead, the female following until the spermatophore is at the level of her vent. **B,** The female has recovered the sperm mass in her vent, while the male arches his tail, tilting the female upward and presumably facilitating recovery of the sperm mass. The female later uses sperm stored in her body to fertilize eggs internally before laying them.

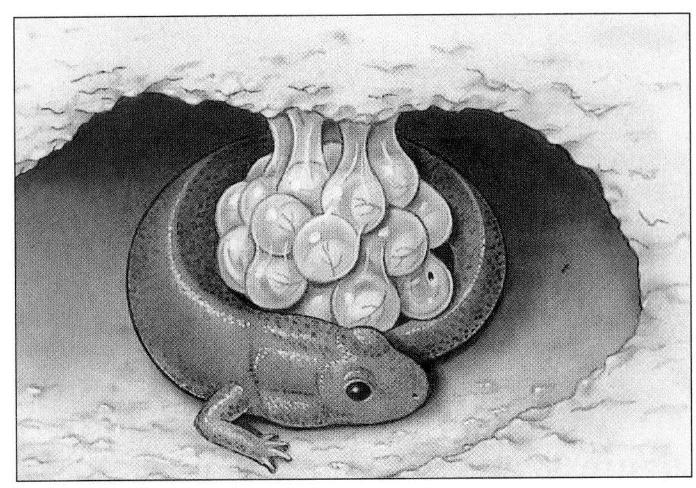

figure 17.6

Female dusky salamander (*Desmognathus* sp.) attending eggs. Many salamanders care for their eggs, which includes rotating the eggs and protecting them from fungal infections and predation by various arthropods and other salamanders.

vascular nets in their skin for respiratory exchange of oxygen and carbon dioxide with the external environment (called "cutaneous respiration"). Salamanders that have an aquatic larval stage hatch with gills, but lose them later if a metamorphosis occurs. Several diverse lineages of salamanders have evolved permanently aquatic forms that fail to complete metamorphosis and retain their gills and finlike tail throughout life (figure 17.8). Lungs, the most widespread respiratory organ of terrestrial vertebrates, are present from birth in the salamanders that have them, and they become the primary means of respiration following metamorphosis. Amphiumas, while having a completely aquatic life history, nonetheless lose their gills before adulthood and then breathe primarily by lungs, raising their nostrils above the water surface to get air.

In contrast to amphiumas, all species of the large family Plethodontidae (see figures 17.5, 17.6, and 17.9) are lungless,

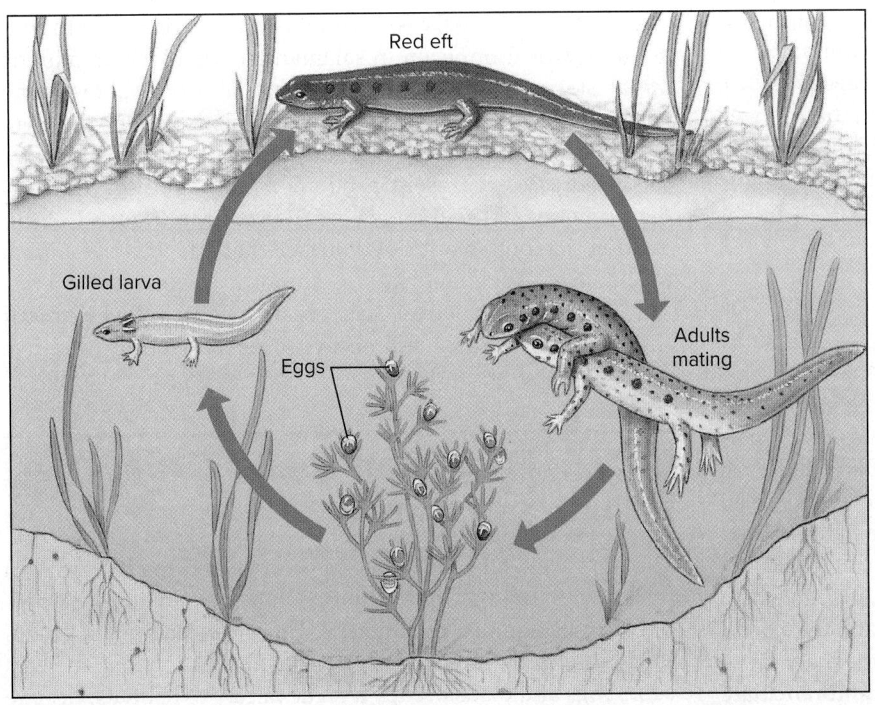

figure 17.7

Life history of the red-spotted newt, *Notophthalmus viridescens,* of the family Salamandridae. In many habitats, the aquatic larva metamorphoses into a brightly colored "red eft" stage, which remains on land from one to three years before transforming into a secondarily aquatic adult.

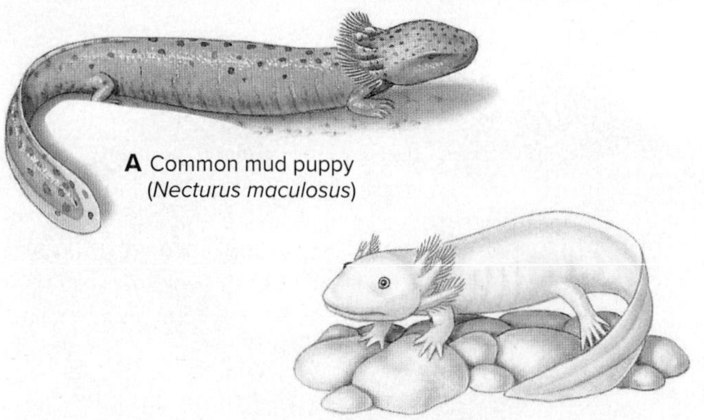

figure 17.8

Permanently-gilled, aquatic salamanders. **A,** The mud puppy (*Necturus* sp.) and **B,** the axolotl (*Ambystoma mexicanum*). Some other species of *Ambystoma* are facultatively metamorphic; they may remain permanently gilled or, should their pond habitat evaporate, metamorphose to a terrestrial form that loses its gills and breathes by lungs. The axolotl shown is an albino form commonly used in laboratory experiments but uncommon in natural populations. Evolutionary change from a metamorphosing ancestor to a nonmetamorphosing, permanently gilled descendant illustrates paedomorphosis (juvenile ancestral characters retained by adult descendants).

Cleveland P. Hickman, Jr.

figure 17.9

Longtail salamander, *Eurycea longicauda*, a common plethodontid salamander.

and many of these species are entirely terrestrial. Efficiency of cutaneous respiration is increased by penetration of a capillary network into the epidermis or by thinning of the epidermis over superficial dermal capillaries. Cutaneous respiration is supplemented by air pumped into the mouth, where respiratory gases are exchanged across the vascularized membranes of the buccal (mouth) cavity in a process called buccopharyngeal breathing. Lungless plethodontids probably originated in streams, where lungs would have been disadvantageous by providing excess buoyancy, and where water is so cool and well oxygenated that cutaneous respiration suffices. It is odd that the most completely terrestrial group of salamanders is one that lacks lungs.

Paedomorphosis

Whereas most salamanders complete their development by metamorphosis to the adult body form, some species reach sexual maturity while retaining their gills, aquatic lifestyle, and other larval characteristics. This condition illustrates **paedomorphosis** (Gr. "child form"), defined as evolution of an adult form that resembles an ancestral juvenile. Some characteristics of an ancestral adult morphology are consequently eliminated. Examples of such nonmetamorphic, permanently gilled species are mud puppies of the genus *Necturus* (see figure 17.8A), which inhabit the muddy substrates of ponds and lakes, and the **axolotl** of Mexico (see figure 17.8B). These species never metamorphose under natural conditions, and are called **perennibranchiate** ("permanently gilled"). Experimental treatment with a thyroid hormone can induce metamorphosis in axolotls but not mud puppies.

Other species of salamanders that reach sexual maturity with larval morphology metamorphose to terrestrial forms under certain environmental conditions. We find good examples in *Ambystoma tigrinum* and related species from North America. They typically inhabit small ponds that can disappear through evaporation in dry weather. When ponds evaporate, the aquatic form metamorphoses to a terrestrial form, losing its gills and developing lungs. It then can traverse land to find new sources of water in which to live and to reproduce.

Paedomorphosis is an important means of evolutionary diversification, even in salamanders that lack an aquatic larval stage. For example, paedomorphosis enables some species of the tropical plethodontid genus *Bolitoglossa* to climb forest vegetation. The highly webbed feet of *Bolitoglossa rufescens* represent paedomorphic evolution in which development of the digits is greatly reduced (figure 17.10B), making the foot an adhesive surface that attaches the animal to banana trees and other smooth leaves and stems. The ancestral structure has greater digital development (figure 17.10A and C) and occurs in related species that live mainly on the ground.

17.5 Frogs and Toads: Order Anura (Salientia)

The approximately 7000 species of frogs and toads that form the order Anura (Gr. *an*, without, + *oura*, tail) are for most people the most familiar amphibians. The Anura are an old group, known from the early Triassic period, almost 250 million years ago. Frogs and toads occupy a great variety of habitats, despite

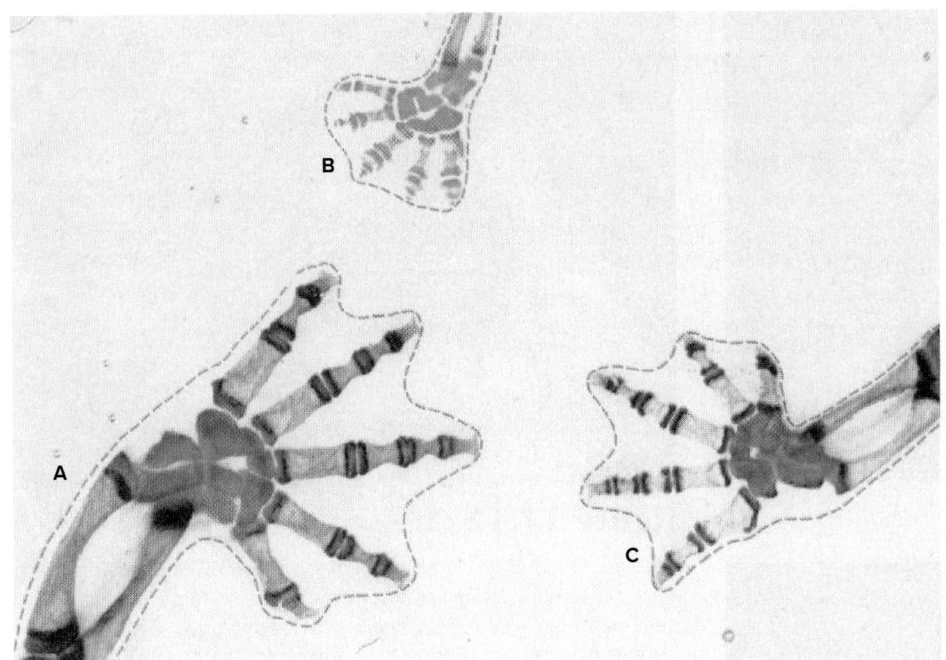

figure 17.10

Foot structure of representatives of three different species of the tropical plethodontid salamander genus *Bolitoglossa.* These specimens have been treated chemically to clear the skin and muscles and to stain the bone red/pink and the cartilage blue. The species having the most fully ossified and distinct digits **(A, C)** live primarily on the forest floor. The species having the padlike foot caused by restricted digital growth **(B)** climbs smooth leaves and stems using the foot surface to produce suction or adhesion for attachment. The padlike foot evolved by paedomorphosis; it was derived evolutionarily by truncating development of the body, which prevents full digital development.

their aquatic mode of reproduction and water-permeable skin, which prevent them from wandering too far from sources of water, and their ectothermy, which bars them from polar and subarctic habitats. The name of the order, Anura, denotes an obvious group characteristic, the absence of tails in adults (although all pass through a tailed stage during embryonic development and often as tadpole larvae). Frogs and toads are specialized for jumping by simultaneous extensor thrusts of the hindlimbs, as suggested by the alternative order name, Salientia, which means leaping.

We see in the appearance and life habit of their larvae further distinctions between Anura and Urodela. Eggs of most frogs hatch into a tadpole having a long, finned tail, both internal and external gills, no legs, specialized mouthparts for herbivorous feeding (some tadpoles and all salamander larvae are carnivorous), and a highly specialized internal anatomy. They bear little resemblance to adult frogs. Metamorphosis of a frog tadpole to an adult frog is thus a striking transformation. The permanently gilled larval condition never occurs in frogs and toads as it does in many salamanders.

Taxonomists recognize 56 families of frogs and toads. The best-known frog families in North America are Ranidae, which contains most of our familiar frogs (figure 17.11A), and Hylidae, the tree frogs (figure 17.11B). True toads, belonging to family Bufonidae, have short legs, stout bodies, and thick skins usually with prominent warts (figure 17.12). However, the term "toad" sometimes denotes thick-skinned, terrestrial members of several other families.

The largest anuran is the West African *Conraua goliath,* which is more than 30 cm long in head and body length (figure 17.13). This giant eats animals as big as rats and ducks. The smallest frogs recorded are *Eleutherodactylus iberia* and *Psyllophryne didactyla,* measuring less than 1 cm in length; they are also the smallest known tetrapods. These tiny frogs, which can be more than covered by a dime, live respectively in Cuba and in the Brazilian rain forest. The largest American frog is the bullfrog, *Lithobates catesbeianus* (see figure 17.11A), which reaches a head and body length of 20 cm.

Habitats and Distribution

Probably the most familiar frogs are species of the ranid genera *Lithobates, Pelophylax,* and *Rana* (Gr. frog), which collectively occupy most temperate and tropical regions of the world. They usually live near water, although some, such as the wood frog, *L. sylvaticus,* spend most of their time on damp forest floors. The larger bullfrogs, *L. catesbeianus,* and the green frogs, *L. clamitans,* are nearly always in or near permanent water or swamps. Leopard frogs, *L. pipiens* and related species, occur in nearly every state and Canadian province and are the most widespread of all North American frogs. The northern leopard frog, *L. pipiens,* is the species most commonly used in biology laboratories and for classical electrophysiological research.

Most larger frogs are solitary except during breeding season. During breeding periods, most of them, especially males, are very noisy. Each male usually occupies a particular perch near water, where he may remain for hours or even days, trying to attract a female to that spot. At times, frogs are mainly silent, and their presence is not detected until they are disturbed. When they enter the water, they dart swiftly to the bottom of the pool, where they kick the substrate to conceal themselves in a cloud of muddy water. In swimming, they hold the forelimbs near the body and kick backward with their webbed hindlimbs, which propel them forward. When they surface to breathe, only the head and foreparts are exposed, and they conceal themselves behind any available vegetation.

T. Kitchin & V. Hurst/NHPA/Photoshot

A

Cleveland P. Hickman, Jr.

figure 17.12

American toad, *Bufo americanus* of the family Bufonidae. This principally nocturnal yet familiar amphibian feeds on large numbers of insect pests as well as on snails and earthworms. The warty skin contains numerous glands that produce a surprisingly poisonous milky fluid, giving the toad excellent protection from a variety of potential predators.

Cleveland P. Hickman, Jr.

B

figure 17.11

Two common North American frogs. **A,** Bullfrog, *Lithobates catesbeianus,* largest American frog and mainstay of the frog-leg epicurean market (family Ranidae). **B,** Green tree frog, *Hyla cinerea,* a common inhabitant of swamps of the southeastern United States (family Hylidae). Note the adhesive pads on the feet.

NHPA/Daniel Heuclin

figure 17.13

Conraua goliath (family Petropedetidae) of West Africa, the world's largest frog. This specimen weighed 3.3 kg (approximately 7-1/2 pounds).

During winter months, most frogs hibernate in the oxygen-rich water of pools and streams. Their life processes reach a very low ebb during hibernation, sustained by diffusion of oxygen across the skin and oxidation of glycogen and fat stored in their bodies during the spring and summer. More terrestrial frogs, such as tree frogs, hibernate in humus of the forest floor. They tolerate low temperatures, and many actually survive prolonged freezing of all extracellular fluid, representing 35% of their body water. Such frost-tolerant frogs

Key Theme 17.2

ECOLOGY

Amphibian Declines

Amphibian populations have declined in various parts of the world, although many species thrive. No single explanation fits all declines, although loss of habitat and epidemic infections by chytrid fungi predominate. In some populations, changes are simply random fluctuations caused by episodic droughts. Frog and toad eggs exposed on the surfaces of ponds are especially sensitive to damaging ultraviolet radiation. Climatic changes that reduce water depth at oviposition sites increase the ultraviolet exposure of embryos and make them more susceptible to fungal infection. Declines in population sometimes are accompanied by an increased incidence of malformed individuals, such as frogs with extra limbs. Malformed limbs are often associated with infection by trematodes (see section 8.3: Class Trematoda).

Declines in some amphibian populations are caused by other amphibians. For example, an exotic frog introduced into southern California thrives in its new American home. African clawed frogs, *Xenopus laevis* (figure 17.14), are voracious, aggressive, primarily aquatic frogs that rapidly displace native frogs and fishes from waterways. This species was introduced into North America in the 1940s when it was used extensively in human pregnancy tests. A sample of a woman's urine was injected into the back of a female *Xenopus* frog; hormones in the urine of a pregnant woman would cause the frog to lay eggs 12-24 hours following the injection. The injection did not damage the frog, which could be reused for the same procedure. Some hospitals dumped surplus frogs into nearby streams, where these prolific breeders flourished and are now considered pests. Similar results occurred when giant toads, *Rhinella marina* (up to 23 cm in length), were introduced to Queensland, Australia, to control agricultural pests. They rapidly spread, producing numerous ecological changes, including displacement of native anurans.

Cleveland P. Hickman, Jr.

figure 17.14

African clawed frog, *Xenopus laevis*. The claws, an unusual feature in frogs, are on the hind feet. This frog has been introduced into California, where it is considered a serious pest.

prepare for winter by accumulating glucose and glycerol in their body fluids, which protects tissues from ice-crystal formation.

Adult frogs have numerous enemies, including snakes, aquatic birds, turtles, raccoons, and humans; fish prey upon tadpoles, only a few of which survive to maturity. Some adult frogs defend themselves by feigning death. Most anurans can inflate their lungs to make themselves difficult for a predator to swallow. When disturbed along the margin of a pond or brook, a frog often remains quite still. When it senses danger, it jumps, not always into the water where enemies may lurk, but into grassy cover on the bank. When held, a frog might cease its struggles for an instant to put its captor off guard and then suddenly leap, voiding its urine. A frog's best protection is its ability to leap and, in some species, to use poison glands. Many frogs and toads in the tropics and subtropics are aggressive, jumping and biting at predators. The brightly colored poison-dart frogs of the family Dendrobatidae use potent toxins for defense.

Life History

Because frogs and toads are ectothermic, they breed, feed, and grow only during warm seasons, often in a predictable annual cycle. With warming spring temperatures and rainfall, males call vociferously to attract females. After a brief courtship, mated pairs enter the water where the male clasps the female in a process called **amplexus** (figure 17.15). As the

Cleveland P. Hickman, Jr.

figure 17.15

A male green frog, *Hyla cinerea,* clasps a larger female during the breeding season in a South Carolina swamp. Clasping (amplexus) lasts until the female deposits her eggs, which are fertilized externally (outside the body). Like most tree frogs, these can change color rapidly; the male here, normally green, has darkened during amplexus.

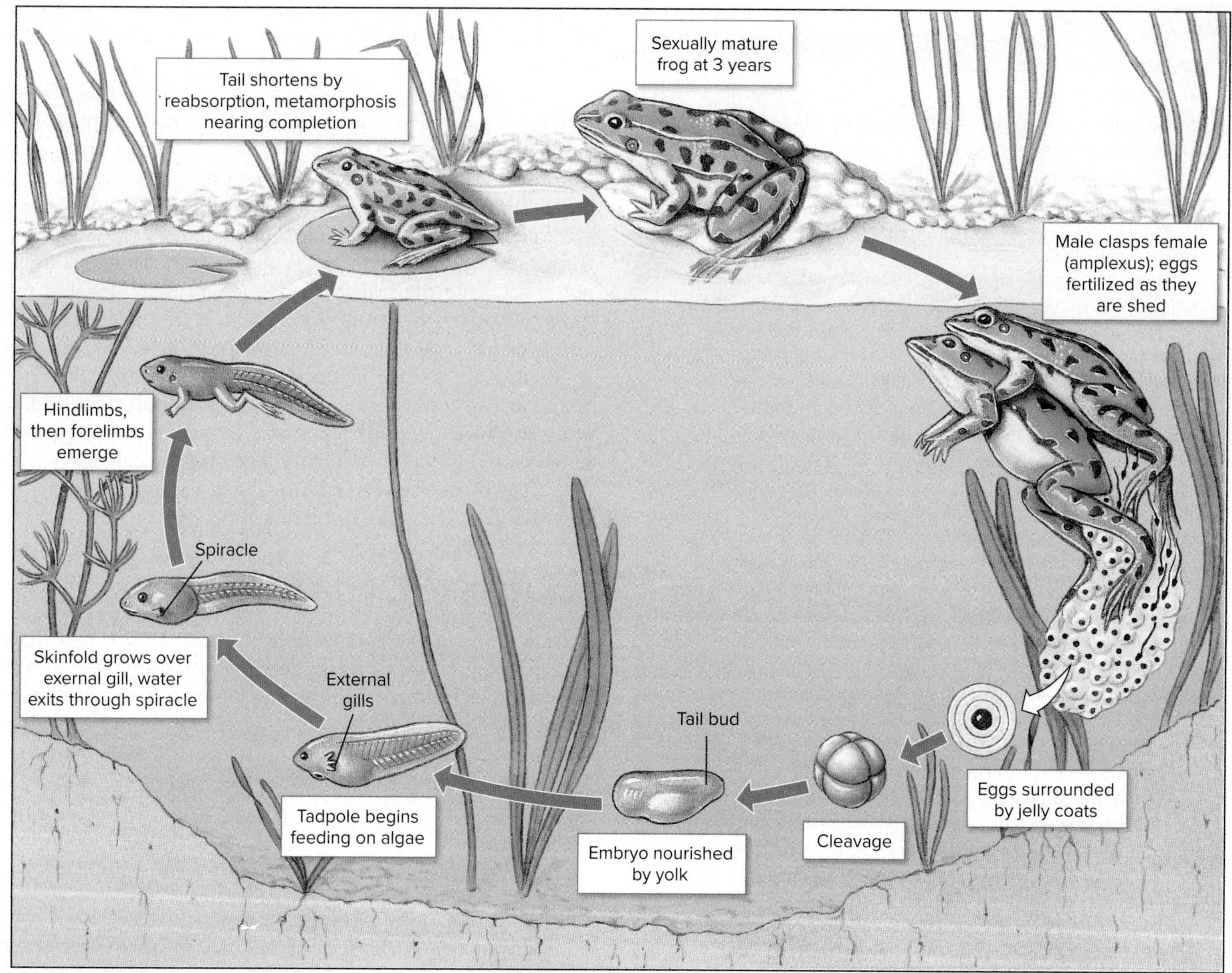

figure 17.16

Life cycle of a leopard frog, *Lithobates pipiens*. Cleavage and embryonic stages occur within the jelly-coated eggs within the egg masses as shown at the right side of the figure. The cleavage and embryonic stages are shown free and disproportionately large to reveal structural detail.

female lays eggs in a gelatinous mass, the male discharges seminal fluid containing sperm over the eggs to fertilize them. After fertilization, the jelly layers absorb water and swell (figure 17.16). Eggs are laid in large masses, often anchored to vegetation, and then abandoned by the parents. The eggs begin development immediately. Within a few days, the embryos have developed into tiny tadpoles visible through the translucent jelly layers surrounding them (figure 17.16). The

Taxonomy of Class Amphibia

Order Gymnophiona (jim′no-fī′ō-na) (Gr. *gymnos,* naked, + *ophioneos,* of a snake) **(Apoda): caecilians.** Body elongate; limbs and limb girdle absent; mesodermal scales present in skin of some; tail short or absent; 95–285 vertebrae; pantropical, 10 families, 32 genera, approximately 210 species.

Order Urodela (ūr′uh-dēl′ə) (Gr. *oura,* tail, + *delos,* evident) **(Caudata): salamanders.** Body with head, trunk, and tail; no scales; usually two pairs of equal limbs; 10–60 vertebrae; predominantly Holarctic; 9 living families, 68 genera, approximately 725 species.

Order Anura (uh-nūr′ə) (Gr. *an,* without, + *oura,* tail) **(Salientia): frogs** and **toads.** Head and trunk fused; no tail; no scales; two pairs of limbs; large mouth; lungs; 6–10 vertebrae including urostyle (coccyx); cosmopolitan, predominantly tropical; 56 families; 451 genera; approximately 7000 species.

embryonic tadpoles grow and develop for several weeks before hatching from their eggs to become free-living. If the eggs were laid in a temporary pond or puddle, tadpoles race against time to complete development before the habitat dries.

At hatching, a tadpole has a distinct head and body with a compressed tail. The mouth is on the ventral side of the head and has keratinized jaws for scraping vegetation from objects for food. Behind the mouth is a ventral adhesive disc for clinging to objects. In front of the mouth are two deep pits, which later become nostrils. Swellings occur on each side of the head, and these later become external gills. There are three pairs of external gills, which later transform into internal gills that become covered with a flap of skin (the operculum) on each side. On the right side, the operculum completely fuses with the body wall, but on the left side a small opening, the **spiracle** (L. *spiraculum,* airhole), permits water to exit after entering the mouth and passing the internal gills. The hindlimbs appear first, whereas the forelimbs are hidden for a time by folds of the operculum. The tail is absorbed, the intestine becomes much shorter, the mouth undergoes a transformation into the adult condition, lungs develop, and the gills are absorbed. Leopard frogs usually complete metamorphosis within three months; a bullfrog takes much longer to complete this process.

The life cycle just described is typical of most temperate-zone anurans but is only one of many alternative patterns in tropical anurans. Some remarkable reproductive strategies of tropical anurans are illustrated in figure 17.17. Some species lay their eggs in foam masses that float on the surface of the water; some deposit their eggs on leaves overhanging ponds and streams into which the emerging tadpoles drop; some lay eggs in damp burrows; and others place their eggs in water-trapped tree cavities or in water-filled chambers of some bromeliads (epiphytic plants in the tropical forest canopy).

While most frogs abandon their eggs, some tend their eggs. Marsupial frogs carry their developing eggs in a pouch on the back (figure 17.17A). In Surinam frogs (figure 17.17B), the male and female do backward somersaults during mating, and eggs and sperm slide into the space between the mating pair. The male presses the fertilized eggs into the female's back, which develops a spongy incubating layer that eventually sloughs once young are hatched. Hatchling poison-dart frogs squirm onto the parent's back to be carried for varying lengths of time (figure 17.16C). Tadpoles of Darwin's frog develop into froglets in the protection of their father's vocal pouch

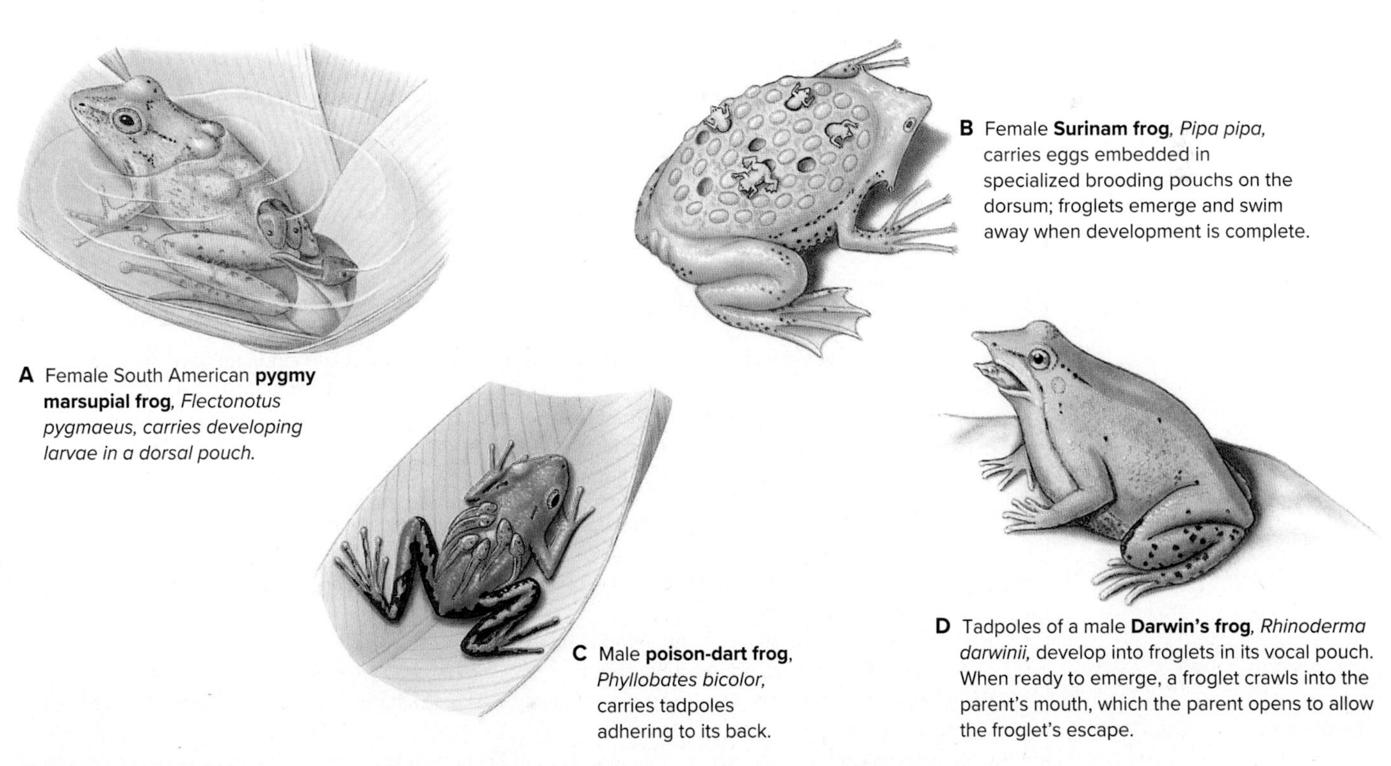

figure 17.17

Unusual reproductive strategies of anurans. **A,** A female South American pygmy marsupial frog, *Flectonotus pygmaeus*, carries developing larvae in a dorsal pouch. **B,** A female Surinam frog carries eggs embedded in specialized brooding pouches on the dorsum; froglets emerge and swim away when development is complete. **C,** A male poison-dart frog, *Phyllobates bicolor,* carries tadpoles adhering to its back. **D,** Tadpoles of Darwin's frog, *Rhinoderma darwinii,* develop into froglets in their male parent's vocal pouch. When the froglets are ready to emerge, their father opens his mouth to let them escape.

(figure 17.17D), whereas Australian gastric-brooding frogs develop in their mother's stomach.

Although most frogs develop through a larval stage (the tadpole), many tropical frogs have evolved direct development. In direct development, no free-living tadpole stage occurs, and the froglet that emerges is a miniature replica of the adult. In the species-rich tropical genus *Eleutherodactylus,* mating occurs on land, and eggs hatch directly into froglets; the aquatic larval stage is eliminated, freeing these frogs from an obligatory association with pools or streams. A Puerto Rican species, *E. jasperi,* has evolved internal fertilization and live birth of offspring.

SUMMARY

Sections	Key Concepts
17.1 Devonian Origin of Tetrapods	• Internal nares, an air-filled cavity, and paired limbs of an aquatic tetrapod ancestor made possible the evolution of terrestrial breathing and support. • Air-filled cavities called "lungs" and "swim bladders" are homologous, with "lungs" denoting the structure's role for air breathing in lungfishes and tetrapods. • *Eusthenopteron,* a Devonian lobe-finned fish that lived approximately 385 million years ago, had an upper arm bone (humerus) and two forearm bones (radius and ulna) as well as other elements that we homologize with wrist bones of tetrapods. • The fossil genus *Tiktaalik,* which lived approximately 375 million years ago, is morphologically intermediate between lobe-finned fishes and tetrapods. • *Acanthostega,* one of the earliest-known Devonian tetrapods, had well-formed tetrapod limbs with clearly formed digits on both forelimbs and hindlimbs.
17.2 Temnospondyls and Modern Amphibians	• Temnospondyls first appeared early in the Carboniferous period and diversified to produce more than 25 extinct taxonomic families, including ancestors of modern amphibians. This group typically has only four digits on the forelimb rather than the five digits typical of most tetrapods. Modern amphibians are the only temnospondyl group to have survived the end-Cretaceous mass extinction. • Modern amphibians, also called Lissamphibia, probably originated in the late Carboniferous period and diversified, probably by the early Triassic period, to produce ancestors of the three major groups of living amphibians: frogs, salamanders, and caecilians. • In the ancestral amphibian life cycle, aquatic eggs hatch to produce an aquatic larval form that respires using gills. At metamorphosis, gills are lost, and metamorphosed amphibians use cutaneous respiration on land; many have lungs, which exist throughout larval life and are activated for breathing air at metamorphosis. • Some salamanders lack a complete metamorphosis and retain a permanently aquatic, larval morphology throughout life. • Some caecilians, some frogs, and some other salamanders live entirely on land and have no aquatic larval phase. • Even the most terrestrial amphibians remain dependent on very moist environments because their skin is thin and requires moisture for protection against desiccation. • To avoid desiccation, eggs must be shed directly into water or onto moist terrestrial surfaces.
17.3 Caecilians: Order Gymnophiona (Apoda)	• Order Gymnophiona contains approximately 210 species of elongate, limbless, burrowing creatures called caecilians. • Caecilians possess a long, slender body, small dermal scales in the skin of some, many vertebrae, long ribs, no limbs, and a terminal anus. • Eyes are small, and most species are blind as adults. • Special sensory tentacles occur on the snout. • Fertilization of eggs occurs inside the female's body, and males have a protrusible copulatory organ for delivering sperm to the eggs. • Caecilians usually deposit their eggs in moist ground near water. • Some species have aquatic larvae; larval development in other species occurs within the egg. • Viviparity also occurs in some caecilians, with embryos obtaining nourishment by eating the wall of the oviduct.

Sections	Key Concepts
17.4 Salamanders: Order Urodela (Caudata)	• Order Urodela comprises tailed amphibians, approximately 725 species of salamanders. • Some salamanders are aquatic or terrestrial throughout their entire life, but the ancestral condition is metamorphic, having aquatic larvae and terrestrial adults that occupy moist places under stones and rotten logs. • Eggs of most salamanders are fertilized internally; a female acquires and stores in her cloaca a packet of sperm (spermatophore) deposited by a male on a leaf or stick. • Salamanders share the general amphibian condition of having in their skin extensive vascular nets for respiratory exchange of oxygen and carbon dioxide. • At various stages of their life history, salamanders also may have external gills, lungs, both, or neither gills nor lungs. • Cutaneous respiration is supplemented by pumping air through the mouth, where respiratory gases are exchanged across the vascularized membranes of the buccal (mouth) cavity (buccopharyngeal breathing). • Some plethodontids are completely terrestrial and they are the only vertebrates to have neither lungs nor gills at any stage of their life history. • A phylogenetic trend in salamander evolution is for descendants to retain into adulthood features that occurred only in pre-adult stages of their ancestors. Some characteristics of ancestral adult morphology are eliminated. This condition is called paedomorphosis. • The most dramatic paedomorphosis occurs in perennibranchiate species, which become sexually mature while retaining their gills, aquatic life habit, and other larval characteristics. Mud puppies of genus *Necturus*, which inhabit submerged substrates in ponds and lakes, are an extreme example.
17.5 Frogs and Toads: Order Anura (Salientia)	• Approximately 7000 species of frogs and toads compose order Anura, typically lacking tails in adults, although all have a tailed stage during embryonic or larval development. • Adult anurans are specialized for jumping by simultaneous extensor thrusts of the hindlimbs. • In spring, males call vociferously to attract females. • When their eggs are mature, females enter water and males clasp them in a process called amplexus, fertilizing eggs leaving the female's body. • Fully terrestrial frogs mate on land, and eggs hatch directly into froglets; the aquatic larval stage is eliminated, freeing these frogs from an obligatory association with pools or streams. • Frogs are carnivorous as adults, but tadpoles are usually herbivorous, feeding on pond algae and other vegetable matter.

Review Questions

1. How did the characteristic differences between aquatic and terrestrial environments influence the early evolution of tetrapods?
2. Describe the different modes of respiration used by amphibians. What paradox do the amphiumas and terrestrial plethodontids present regarding the association of lungs with life on land?
3. Evolution of the tetrapod limb was one of the most important advances in vertebrate history. Describe the inferred sequence of its evolution.
4. Compare the general life-history patterns of salamanders with those of frogs. Which group shows a greater variety of evolutionary changes from the ancestral biphasic amphibian life cycle?
5. Give the literal meaning of the name Gymnophiona. What animals are in this amphibian order? Describe their appearance and habitats.
6. What are the literal meanings of the order names Urodela and Anura? What major features distinguish members of these two orders from each other?
7. Describe the breeding behavior of a typical woodland salamander.
8. How is paedomorphosis important to evolutionary diversification of salamanders?
9. Briefly describe the reproductive behavior of frogs. In what important ways do frogs and salamanders differ in their reproduction?
10. Amphibians are tied to water for reproduction. How have some frogs evolved to provide a moist environment for larval development using the parent's body?

For Further Thought On the nineteenth-century notion of a "scale of nature," living amphibians were considered remnants of archaic terrestrial vertebrates largely superseded by "higher" forms including birds and mammals. Nonetheless, amphibian species are often more abundant and of longer evolutionary duration than avian or mammalian species. In what ways are amphibians unusually well suited for evolutionary persistence?

18

Amniote Origins and Nonavian Reptiles

LEARNING OBJECTIVES

Readers will be able to:

18.1 Describe the early evolution and major adaptations of amniotes, and explain changes in the use of the term "reptiles."

18.2 Describe the morphology and behavior of turtles, lizards, snakes, tuataras, crocodilians, and dinosaurs.

Zigmund Leszczynski/age fotostock

Hatching Madagascar day gecko, *Phelsuma madagascariensis.*

Enclosing an Aquatic Habitat

Amphibians, with their well-developed limbs, redesigned sensory and respiratory systems, and modifications of the postcranial skeleton for supporting the body in air, have made a notable conquest of land. However, their shell-less eggs, thin, moist skin, and gilled larvae keep their development hazardously tied to freshwater or moist terrestrial habitats. An ancestor of a clade containing turtles, lizards, snakes, tuataras, crocodilians, birds, and mammals evolved an egg better adapted for dry terrestrial conditions. This shelled egg enclosed the formerly aquatic early stages of development. In fact, the "pond-dwelling" stages were not eliminated but enclosed within a series of extraembryonic membranes that provided complete support for embryonic development. One membrane, the amnion, encloses a fluid-filled cavity, an aquatic environment for the developing embryo. Another membranous sac, the allantois, serves both as a respiratory surface and as a chamber for storing nitrogenous wastes. Enclosing these membranes is a third membrane, the chorion, through which oxygen and carbon dioxide freely pass. Finally, surrounding and protecting everything is a porous shell.

With the last ties to aquatic reproduction severed, conquest of land by vertebrates was ensured. Paleozoic tetrapods that developed this reproductive pattern and their descendants are called Amniota, named after the innermost of the three extraembryonic membranes, the amnion. Before the end of the Paleozoic era, amniotes diversified into multiple lineages that gave rise to turtles, snakes, lizards, tuataras, crocodilians, birds, and mammals.

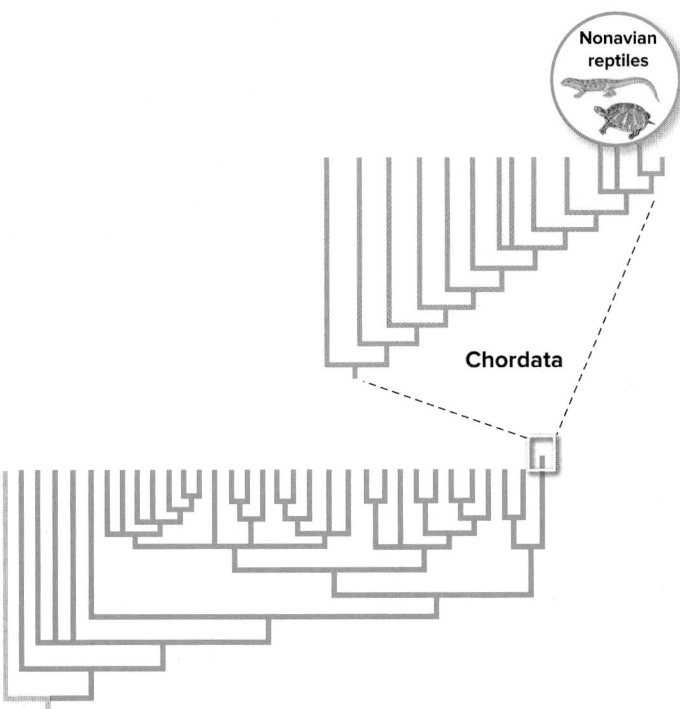

The clade Reptilia (rep-til´ē-ə) (L. *repto,* to creep) includes birds (Aves), turtles, lizards, snakes, tuataras, and crocodilians, in addition to numerous extinct groups. Living nonavian reptiles include nearly 9500 species (approximately 320 species in the United States and Canada) occupying a great variety of aquatic and terrestrial habitats, in many of which they are diverse and abundant. Nevertheless, reptiles are often remembered for what they once were, especially the awe-inspiring dinosaurs, reptiles of giant stature that diversified during the Mesozoic era. The "age of reptiles" in the Mesozoic era lasted for 165 million years, finally ending at the close of the Cretaceous period, when many reptilian lineages became extinct.

Reptiles (including birds) are members of the clade Amniota, as are mammals. In section 18.1, we discuss the origin of amniotes, their diversification into various groups, their adaptations for life on dry land, and the taxonomy of Reptilia. Section 18.2 covers the four living groups of nonavian reptiles. Birds and mammals are discussed in Chapters 19 and 20, respectively.

18.1 Origin and Early Evolution of Amniotes

As mentioned in the prologue to this chapter, amniotes are a monophyletic group that appeared and diversified in the late Paleozoic era. Most zoologists agree that amniotes are most closely related to anthracosaurs, a group of **anamniotes** (vertebrates lacking an amnion) of the early Carboniferous period. Early diversification of amniotes produced three patterns of holes (fenestrae) in the temporal region of the skull. **Anapsid** (Gr. *an,* without, + *apsis,* arch) skulls have no openings in the temporal area of the skull behind the **orbit** (opening in the skull for the eye); thus, the temporal region of the skull is completely roofed by dermal bones (see figure 18.2). This skull morphology occurs in the earliest amniotes, and in one living group, the turtles, although the anapsid condition in turtles likely evolved secondarily, from ancestors having temporal fenestrae. Two other amniote clades, Diapsida and Synapsida, represent separate evolutionary derivations from the ancestral anapsid condition.

A **diapsid** (Gr. *di,* double, + *apsis,* arch) skull has two temporal openings: one pair located low over the cheeks, and a second pair positioned above the lower pair, in the roof of the skull, and separated from the first by a bony arch (see figure 18.2). Diapsid skulls characterize birds and all amniotes traditionally considered "reptiles," except turtles (figure 18.1). In many living diapsids (lizards, snakes, and birds), one or both of the bony arches and openings have been lost, perhaps associated with skull kinesis (see figure 18.9). The earliest diapsids gave rise to five morphologically distinct clades. The **lepidosaurs** include lizards, snakes, and tuataras. The **archosaurs** include dinosaurs, pterosaurs, crocodilians, and birds. A third, smaller clade, the **sauropterygians,** includes several extinct aquatic groups, the most familiar of which are the large, long-necked plesiosaurs (figure 18.1). **Ichthyosaurs,** represented by extinct, aquatic, dolphinlike forms (figure 18.1), form a fourth clade of diapsids. Placement of the last clade, the **turtles,** with diapsids has been controversial, in part because of the "anapsid" form of their skull. Turtle skulls lack temporal fenestrae and were hypothesized to be the only living descendants of parareptiles, an early anapsid group. However, other morphological and genetic evidence published over the past 20 years places turtles within the diapsid clade, suggesting the two pairs of temporal fenestrae characteristic of diapsids were lost early in turtle evolution.

The third skull fenestration condition is **synapsid** (Gr. *syn,* together, + *apsis,* arch), characterized by a single pair of temporal openings located low on the cheeks and bordered by a bony arch (figure 18.2). The synapsid condition occurs in a clade that includes mammals and their extinct relatives, the therapsids and pelycosaurs (see figure 18.1).

What was the functional significance of the temporal openings in early amniotes? In living forms, these openings are occupied by large muscles that elevate (adduct) the lower jaw. Expansion of jaw-elevating musculature might reflect a shift from suction feeding in aquatic vertebrates (see section 16.4 Ray-Finned Fishes: Actinopterygii) to a terrestrial feeding method that required larger muscles to produce more static pressure, for such activities as nipping plant material with the anterior teeth or grinding food with the posterior teeth.

Adaptations of Amniotes

Derived characters of amniotes include an amniotic egg, rib ventilation, and desiccation-resistant skin, in addition to skeletal characteristics of the head, shoulder, and ankle. These and

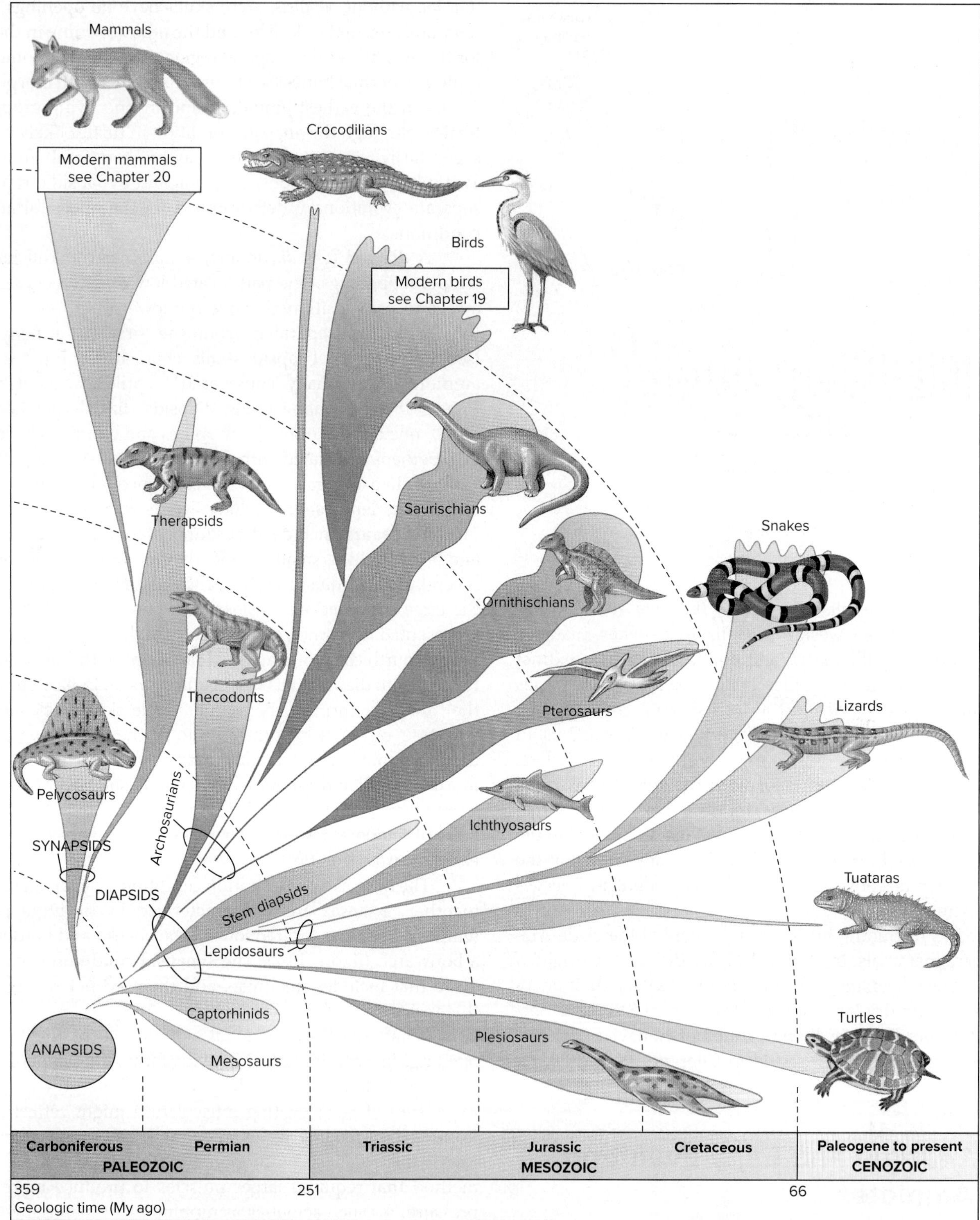

figure 18.1

Evolution of amniotes. The earliest amniotes evolved an amniotic egg, which allowed amniotes to exploit drier habitats than their ancestors. The living amniotes, which include turtles, lizards, snakes, tuataras, crocodilians, birds, and mammals, evolved from a lineage of small, lizardlike forms that retained the anapsid skull pattern of early, anamniote tetrapods. One lineage that descended from the early amniotes had a synapsid skull pattern and gave rise to mammals. Birds, squamates, and crocodilians have a diapsid skull pattern. Turtles have an anapsid skull, although turtles probably evolved from a diapsid ancestor.

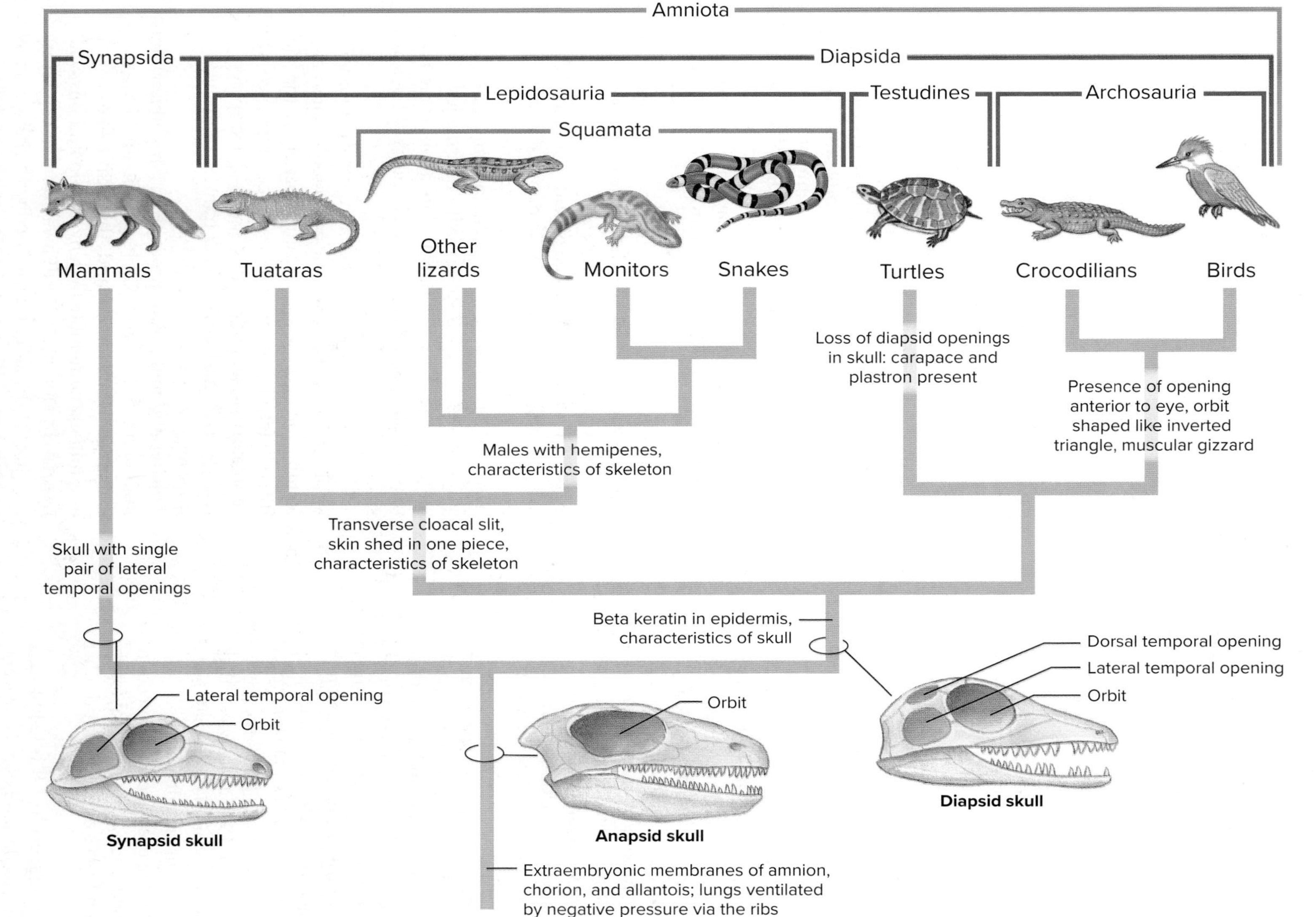

figure 18.2

Cladogram of living Amniota, showing monophyletic groups. The skulls represent the ancestral anapsid, synapsid, and diapsid conditions. Skulls of modern diapsids and synapsids are often highly modified by loss or fusion of skull bones that obscures the ancestral condition. Representative skulls for anapsids are *Nyctiphruretus* of the upper Permian period; for diapsids, *Youngina* of the upper Permian period; for synapsids, *Aerosaurus*, a pelycosaur of the lower Permian period. The relationship of turtles to other reptiles is controversial, although most recent genetic evidence indicates a sister-taxon relationship with archosaurs, as shown here. Reptilia, as currently recognized, is equivalent to Diapsida, as shown in this cladogram.

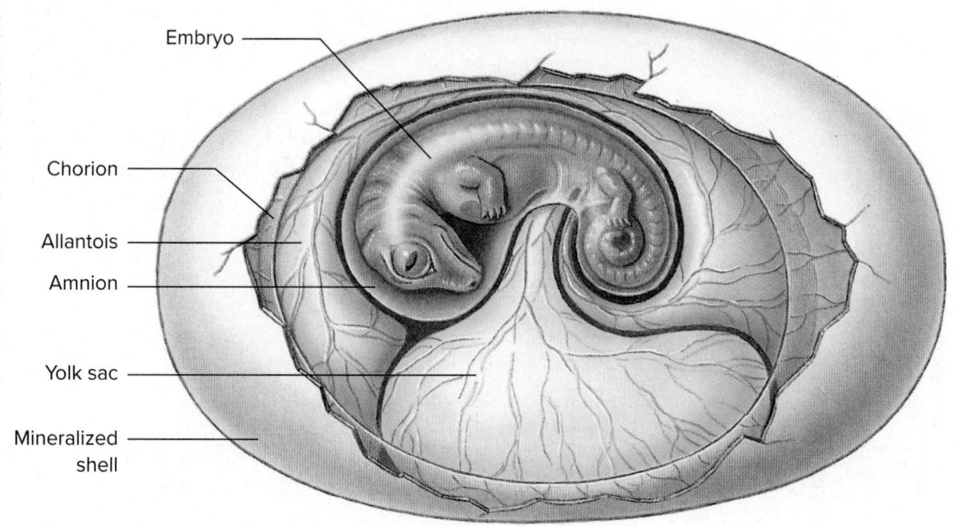

figure 18.3

Amniotic egg. The embryo develops within the amnion and is cushioned by amniotic fluid. Food is provided by yolk from the yolk sac, and metabolic wastes are deposited within the allantois. As development proceeds, the allantois fuses with the chorion, a membrane lying against the inner surface of the shell; both membranes are supplied with blood vessels that assist in the exchange of oxygen and carbon dioxide across the porous shell. Because this kind of egg is an enclosed, self-contained system, it is often called a "cleidoic" egg (Gr. *kleidoun*, to lock in).

other morphological and physiological characters allowed amniotes to be more energetic and to be more successful at exploiting dry, terrestrial habitats than anamniotes (especially modern amphibians).

1. **Amniotic egg.** All amniotes have eggs with four extraembryonic membranes, the **amnion, allantois, chorion**, and **yolk sac** (figure 18.3). The amnion encloses the embryo in fluid, cushioning the embryo and providing an aqueous medium for growth. Metabolic wastes are stored in a sac formed by the allantois. The chorion surrounds the entire contents of the egg, and like the allantois, is highly vascularized. Thus, the chorion and allantois form an efficient respiratory organ for removing carbon dioxide and acquiring oxygen. Most amniotic eggs are surrounded by a mineralized but often flexible shell, although many lizards and snakes and most mammals lack shelled eggs. The shell forms an important mechanical support, and especially for birds, a semipermeable barrier, which allows passage of gases but limits water loss. Like eggs of anamniotes, amniotic eggs have a yolk sac for nutrient storage, although the yolk sac tends to be larger in amniotes. In marsupial and placental mammals, the yolk sac does not store yolk, but it may form a temporary or persistent yolk-sac placenta to transfer nutrients, gases, and wastes between the embryo and mother. For many amniote species, egg or embryonic development occurs in a female's reproductive tract, providing even greater protection from predators and dehydration, and potential for the mother to manage the embryo's nutritional and other physiological needs.

   How did the amniotic egg evolve? It is tempting to think of the amniotic egg as *the* land egg. However, many amphibians lay eggs on land and many amniotic eggs, such as those of turtles, must be buried in wet soil or deposited in areas of high humidity. Still, amniotic eggs can be laid in places too dry for any amphibian; clearly, the evolution of amniotic eggs was a major factor in the success of tetrapods on land. Perhaps a more important advantage of the amniotic egg was that it permitted development of a larger, faster-growing embryo. Anamniote eggs are supported mainly by a thick jellylike layer. This jelly layer is inadequate to support large eggs and limits movement of oxygen into the eggs. One hypothesis suggests that a first step in the evolution of the amniotic egg was replacement of the jelly layer with a shell, which provided better support and movement of oxygen. Furthermore, calcium deposited in the shell can be dissolved and absorbed by the growing embryo, to provide one of the raw materials needed for skeleton construction. This hypothesis is supported by physiological studies, which show that embryos of species with the smallest amniotic eggs have a metabolic rate about three times that of embryos of anamniotes with eggs of the same size.

   All amniotes lack gilled larvae and have internal fertilization. Shelled amniotic eggs require internal fertilization because sperm cannot penetrate the shell. Internal fertilization in amniotes is accomplished with a copulatory organ (hemepenis or penis), except for tuataras and most birds, which transfer sperm by pressing their cloacas together.
2. **Thicker and more waterproof skin.** Amphibians must maintain a thin, moist skin to permit effective gas exchange. However, on land, this skin makes amphibians vulnerable to dehydration and physical trauma. In amniotes, a shift away from a respiratory function of the skin is associated with changes in skin morphology. Although the skin varies widely in structure among living amniotes and anamniote tetrapods, amniote skin tends to be much thicker, more

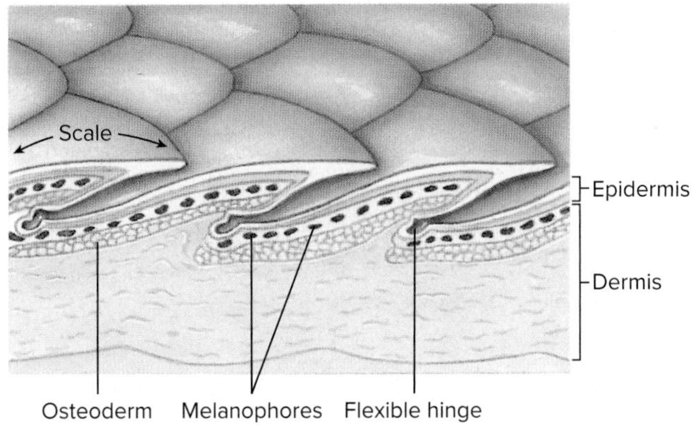

figure 18.4

Section of the skin of a reptile showing overlapping, keratinized scales in the epidermis and bony osteoderms in the dermis.

keratinized, and less permeable to water. A wide variety of structures composed of **keratin,** such as scales (figure 18.4), hair, feathers, and claws, project from amniote skin. Keratin gives the skin protection from physical trauma, and hydrophobic lipids in the skin limit water loss through the skin. Unique to clade Reptilia (birds and nonavian reptiles), the epidermis contains a hard form of keratin called **beta keratin.** The characteristic scales of nonavian reptiles, formed mostly of beta keratin, provide protection against wear in terrestrial environments. The epidermal scales of reptiles are not homologous to the scales of fishes, which are mostly bony, dermal structures (see figure 16.14).

In crocodilians, scales remain throughout life, growing gradually to replace wear. In lizards and snakes, new keratinized epidermis grows beneath the old, which is then shed at intervals. Turtles add new layers of keratin under the old layers of the platelike scutes, which are modified scales. Crocodiles and many lizards (skinks, for example) possess bony plates called **osteoderms** (figure 18.4) located in the dermis, beneath the keratinized scales. The dermis has **chromatophores**, color-bearing cells that give many lizards and snakes their colorful hues. This layer, unfortunately for their bearers, is converted into alligator and snakeskin leather, esteemed for expensive handbags and shoes.

Keratin and lipids limit the skin's ability to exchange respiratory gases—so, unlike most amphibians, few amniotes use their skin as a primary respiratory organ. Amniote gas exchange occurs primarily in the lungs.

3. **Rib ventilation of the lungs.** Compared to amphibians, amniote lungs are larger (figure 18.5), have much more surface area, and are ventilated by a different mechanism. These changes reflect both the increased metabolic demands of amniotes and the reduced ability of amniote skin to serve as a gas-exchange organ.

   Amphibians, like air-breathing fishes, fill their lungs by *pushing* air from the oral and pharyngeal cavities into the lungs (buccal pumping). In contrast, amniotes *draw* air into their lungs (**aspiration**) by expanding the thoracic cavity using costal (rib) muscles or pulling the liver (with other muscles) posterior. Although lungs are the primary gas exchange organ for nearly all amniotes, other areas of the body can be used. Many aquatic turtles supplement pulmonary respiration with respiration at their pharynx or cloaca, and most gas exchange in sea snakes occurs across their skin.

4. **Stronger jaws.** Most fish jaws are adapted for suction and quick closure, but once prey is seized, little static force can be applied. Suction feeding is not possible with terrestrial vertebrates, and the skeleton and muscles of the jaws of early tetrapods became adapted to seize prey. In amniotes, expansion of the jaw musculature, often into temporal openings (see figure 18.2) or notches, provided better mechanical advantage. In contrast to fishes, the tongue of tetrapods is muscular and mobile, functioning to move food in the mouth for mastication and swallowing.

5. **High-pressure cardiovascular systems.** All amniotes have functionally separate pulmonary and systemic circulation: the right side of the heart receives deoxygenated blood and the left side of the heart receives oxygenated blood. In mammals, birds, and crocodilians, complete separation of these circuits in the heart is accomplished with left and right ventricles. Other reptiles have a single ventricle incompletely partitioned into multiple chambers (figure 18.6). Even in these species, flow patterns within the heart limit mixing of oxygen-rich blood and oxygen-poor blood; all amniotes therefore have functionally separate circulations. This separation of circuits provides higher systemic blood pressure in amniotes; fishes and amphibians typically have systemic pressures of 15–40 mm Hg, compared with about 80 mm Hg in monitors (a group of large, active lizards, see figure 18.14). Higher pressure is adaptive for active, terrestrial organisms because of their higher metabolic needs and because the heart must pump blood against gravity.

   Is the incomplete separation of hearts of many nonavian reptiles merely a transitional stage on the evolutionary path to an "advanced" heart of mammals and birds? No, incomplete separation is adaptive for these vertebrates, as it allows blood to bypass the lungs when pulmonary respiration is not occurring (for example, diving or estivation). Thus, during lung-bypass, energy is not expended pumping blood through the capillary beds of the lungs.

6. **Water-conserving nitrogen excretion.** Most amphibians excrete their metabolic waste primarily as ammonia or urea. Ammonia is toxic at relatively low concentrations

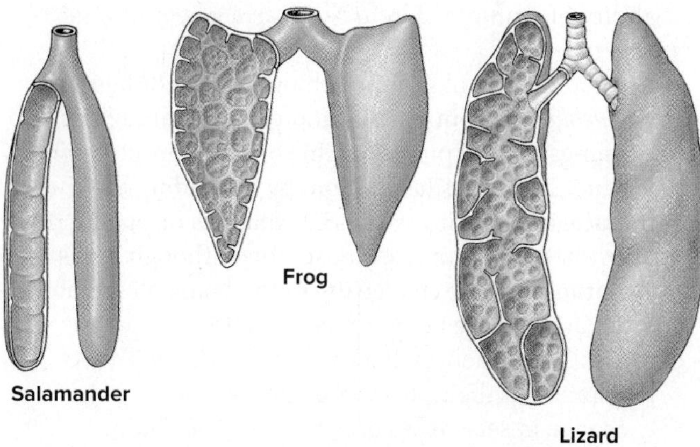

figure 18.5

Amphibians (salamander and frog) have lungs with simple inner linings, while reptiles (lizard) have lungs with inner foldings, creating a more efficient gas-exchange surface between air spaces and blood.

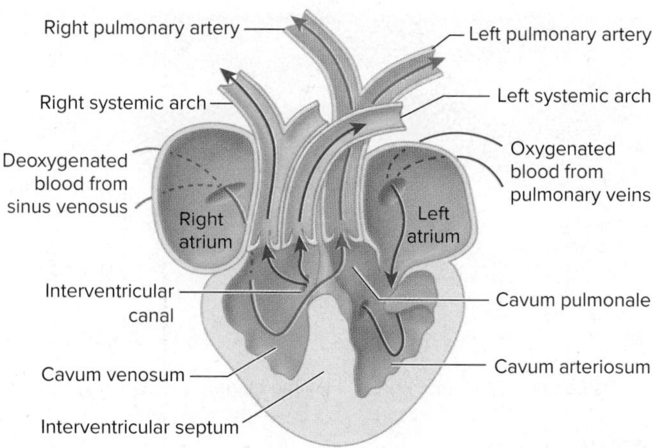

figure 18.6

Route of blood flow through a turtle or squamate heart. Atria are completely separated and the ventricle is incompletely partitioned into three cavities: cavum venosum, cavum pulmonale, and cavum arteriosum. When lung respiration is occurring, little mixing of pulmonary and systemic blood occurs. When the lungs are not in use, the pulmonary circuit can be bypassed by sending blood from the cavum venosum directly to the systemic arches.

and must be removed in a dilute solution. Because of the water required for excretion of ammonia, it is not adaptive for vertebrates occupying dry, terrestrial habitats. Mammals excrete their nitrogenous waste as urea, which is concentrated in their kidneys, reducing water loss through excretion. Birds and nonavian reptiles excrete their nitrogenous wastes as concentrated uric acid. Birds and nonavian reptiles have limited abilities to concentrate their urine in the kidneys, so their urinary bladder receives dilute urine. Water (and many salts) are resorbed in the bladder, and "urine" is voided as a semisolid mass of uric acid.

7. **Expanded brain and sensory organs.** The brain has a relatively large cerebrum and cerebellum in all amniotes, but especially so in birds and mammals. Enlargement of the cerebrum is correlated with integration of sensory information and control of muscles during locomotion. Nonavian reptiles and birds have particularly good vision, processed in the optic lobe (figure 18.5), and many species display brilliant coloration. Olfaction is highly developed in mammals, snakes, and lizards, many of which supplement detection of odors with vomeronasal organs, specialized olfactory chambers in the roof of the mouth (see section 18.2 Snakes: Serpentes).

Changes in Traditional Classification of Reptiles

With increasing use of cladistic methodology in zoology and its insistence on hierarchical arrangement of monophyletic groups (see section 4.4: Phylogenetic Systematics/Cladistics), important changes have been made in the classification of amniotes. As traditionally defined, "reptiles" are snakes, lizards, tuataras, crocodilians, and turtles, as well as several extinct groups, such as dinosaurs, plesiosaurs, pterosaurs, and many early amniotes,

CHARACTERISTICS of Nonavian Reptiles

1. Two paired limbs, usually with five toes each; limbs vestigial or absent in many; **ectothermic.**
2. Body covering of **keratinized epidermal scales** and sometimes bony dermal plates; integument with few glands.
3. Skull with **one occipital condyle** (bony bump that connects to the first vertebra); lower jaws of several bones; distinct atlas and axis; usually two sacral vertebrae.
4. Teeth **polyphyodont** (replaced many times) or absent (turtles); when present, teeth usually **homodont** (all similar in function and shape), with a single point; **gizzard** in crocodilians.
5. Brain moderately well developed with expanded cerebrum; 12 pairs of cranial nerves.
6. Eyes with color vision in some; snakes and some lizards with highly developed chemoreception using olfactory epithelia and **Jacobson's organ;** some snakes with **heat-sensitive pit organs;** middle ear with a single bone.
7. Usually separate sexes, but some lizards reproduce asexually by parthenogenesis; internal fertilization; copulatory organ a **penis, hemipenes,** or (rarely) absent; sex determined by chromosomes or by environment.
8. Fetal membranes of **amnion, chorion, and allantois;** oviparous or viviparous; eggs with leathery or calcareous shells; embryos of viviparous species nourished by **placenta** or **yolk sac** (ovoviviparity); parental care absent, except in crocodilians; no larval stage.
9. Excretory system of **metanephric kidneys** and ureters that open into a cloaca; **uric acid** usually main nitrogenous waste.
10. Lungs filled by **aspiration** (negative ventilation); **no gills;** some supplement gas exchange with cloaca, skin, or pharynx.
11. Heart with a sinus venosus, two atria, and ventricles incompletely divided into three cavities; crocodilian heart with sinus venosus, two atria, and two ventricles; **pulmonary and systemic circuits incompletely separated;** nucleated red blood cells.

The Mesozoic World of Dinosaurs

In 1842, when the English anatomist Richard Owen coined the term *dinosaur* ("fearfully great lizard") to describe fossil Mesozoic reptiles of gigantic size, only three poorly known dinosaur genera were distinguished. New and marvelous fossil discoveries quickly followed, and by 1887 zoologists were able to distinguish two groups of dinosaurs based on differences in the structure of the pelvic girdles. The Saurischia ("lizard-hipped") had a simple, three-pronged pelvis with hip bones arranged much as they are in other nonavian reptiles. The Ornithischia ("bird-hipped") had a somewhat more complex pelvis. The ilium and ischium were arranged similarly in ornithischians and saurischians, but the ornithischian pubis was a narrow, rod-shaped bone with anteriorly and posteriorly directed processes lying alongside the ischium. Oddly, while the ornithischian pelvis, as the name suggests, was similar to that of birds, birds are of the saurischian clade. New evidence suggests that Saurischia is not monophyletic, although this finding is not universally accepted.

Dinosaurs and their living descendants, the birds, are archosaurs ("ruling lizards"), a group that includes crocodilians and pterosaurs (see figure 18.1). As traditionally recognized, dinosaurs are a paraphyletic group because they do not include birds.

From among the various archosaurian diversifications of the Triassic period there emerged a lineage with limbs drawn under the body, which provided an upright posture. This lineage gave rise to the earliest dinosaurs of the late Triassic period. *Herrerasaurus,* a bipedal dinosaur from Argentina, has one of the most distinctive characteristics of dinosaurs: walking upright on pillarlike legs, rather than on legs splayed outward like those of amphibians and nonavian reptiles. This arrangement allowed the legs to support the great weight of the body while providing an efficient and rapid stride.

Two groups of saurischian dinosaurs are recognized based on differences in feeding habits and locomotion: the carnivorous and bipedal theropods, and the herbivorous and quadrupedal sauropods. *Coelophysis* was an early theropod with a body form typical of all theropods: powerful hindlimbs with three-toed feet; a long, heavy, counterbalancing tail; slender, grasping forelimbs; a flexible neck; and a large head with jaws armed with daggerlike teeth. Large predators such as *Allosaurus,* common during the Jurassic period, were replaced by even more massively built carnivores of the Cretaceous period, such as *Tyrannosaurus,* which reached a length of 12.2 m (40 ft), stood nearly 6 m high, and weighed more than 7200 kg (8 tons). Not all predatory saurischians were massive; several were swift and nimble, such as *Velociraptor* ("speedy predator") of the late Cretaceous period.

Herbivorous saurischians, the quadrupedal sauropods, appeared in the late Triassic period. Although early sauropods were small- and medium-sized dinosaurs, those of the Jurassic and Cretaceous periods attained gigantic proportions and were the largest terrestrial vertebrates ever to have lived. *Brachiosaurus* reached 25 m (82 ft) in length and may have weighed more than 30,000 kg (33 tons). Even larger sauropods have been discovered; *Argentinosaurus* was 40 m (132 ft) long and weighed at least 80,000 kg. With long necks and long front legs, the sauropods were the first vertebrates adapted to feed on trees.

The second group of dinosaurs, the Ornithischia, were all herbivorous. Although more varied in appearance than saurischians, the ornithischians are united by several derived skeletal features that indicate common ancestry. The huge back-plated *Stegosaurus* of the Jurassic period is a well-known example of armored ornithischians, which formed two of the five major groups of ornithischians. Even more shielded with bony plates than stegosaurs were the heavily built ankylosaurs, "armored tanks" of the dinosaur world. As the Jurassic period gave way to the Cretaceous period, several groups of unarmored ornithischians appeared, although many bore impressive horns. The steady increase in ornithiscian diversity in the Cretaceous period paralleled a concurrent gradual decline in giant sauropods, which had flourished in the Jurassic period. *Triceratops* is representative of horned dinosaurs that were common in the late Cretaceous period. Even more prominent in the late Cretaceous period were the hadrosaurs, such as *Parasaurolophus,* which probably lived in large herds. Many hadrosaurs had skulls elaborated with crests that probably functioned as vocal resonators to produce species-specific calls. The fifth group, the bipedal pachycephalosaurs of the late Cretaceous period, had thick skulls possibly used in fighting.

Dinosaurs likely provided considerably more complex parental care than did most other nonavian reptilians. Because both crocodilians (see figure 18.27) and birds, members of the clade Archosauria, share complex parental care, it is likely that dinosaurs exhibited similar behavior. Fossil nests of dinosaurs have been discovered for several groups. In one case, a fossil adult of the small theropod *Oviraptor* was found with a nest of eggs. Originally it was hypothesized that the adult was a predator on the eggs (*Oviraptor* means "egg seizer"), but later, an embryo in a similar egg was found and identified as *Oviraptor,* indicating that the adult was probably with its own eggs! Young of *Maiasaura* (a hadrosaur) found in a nest were too large to be hatchlings and had considerable wear on their teeth, suggesting that the young had remained in the nest and possibly been fed by adults during part of their early life.

Sixty-six million years ago, the last Mesozoic dinosaurs became extinct, leaving birds and crocodilians as the only surviving lineages of archosaurs. The demise of dinosaurs coincided with a large asteroid impact on the Yucatán peninsula that would have produced worldwide environmental upheaval. Although the impact event is generally accepted as the primary cause of the extinctions, other events, including a massive volcanic eruption in the Deccan Plateau of India and lowering of sea level, have been suggested to have contributed to the demise of dinosaurs and other animals. We continue to be fascinated by the awe-inspiring, often staggeringly large creatures that dominated the Mesozoic era for 165 million years—an incomprehensibly long period of time. Today, inspired by clues from fossils and footprints from a lost world, scientists continue to piece together the puzzle of how the various dinosaur groups arose, behaved, and diversified.

(*continued*)

SAURISCHIANS
ORNITHISCHIANS
66
MY ago
CRETACEOUS
145
MY ago
JURASSIC
201
MY ago
TRIASSIC
252
MY ago
Titanosaurus
12 m (40 ft)
Velociraptor
1.8 m (6 ft)
Parasaurolophus
10 m (33 ft)
Triceratops
9 m (30 ft)
Brachiosaurus
25 m (82 ft)
Allosaurus
11 m (35 ft)
Stegosaurus
9 m (30 ft)
Ilium
Ischium
Pubis
Coelophysis
3 m (10 ft)
Ilium
Pubis
Ischium
Herrerasaurus 4 m (13 ft)
One of the oldest known
dinosaurs, currently thought
to be a basal saurischian.

including "mammal-like reptiles," (extinct amniotes with a synapsid skull that are closer to mammals than to other amniotes [see figure 20.2]). Thus, "reptiles" as traditionally defined form a **paraphyletic** group because they do not include all descendants (birds and mammals) of their most recent ancestor. Reptilia is made monophyletic by removal of synapsids and inclusion of birds (see figure 18.2).

Crocodilians and birds are sister groups; they share a unique common ancestor. In other words, crocodilians and birds belong to a monophyletic group apart from other reptiles and, according to the rules of cladistics, should be assigned to a clade that separates them from the remaining reptiles. This clade is in fact recognized; it is Archosauria (see figures 18.1 and 18.2), a grouping that also includes the extinct dinosaurs and pterosaurs. Archosaurs, along with lepidosaurs (tuataras, lizards, and snakes) and turtles, form a monophyletic group called Reptilia. Here we use Reptilia in a cladistic fashion to include living amniote groups traditionally termed "reptiles" together with birds and all extinct groups more closely related to these than to mammals. The term "nonavian reptiles" is used informally to denote a paraphyletic group that includes the living turtles, lizards, snakes, tuataras, and crocodilians, and a number of extinct groups, including plesiosaurs, ichthyosaurs, pterosaurs, and dinosaurs. The four clades (often given the rank of "order" in Linnaean classifications) of living nonavian reptiles are the subject of the remainder of the chapter: (1) Testudines, turtles; (2) Squamata, lizards and snakes; (3) Sphenodonta, tuataras; (4) Crocodilia, crocodilians.

18.2 Characteristics and Natural History of Reptilian Groups

Turtles: Testudines

Turtles appear in the fossil record in the Triassic period some 240 million years ago. The earliest turtles had teeth and a reduced shell, but otherwise were similar to those of today. Lacking teeth, a modern turtle's jaws have tough keratinized plates for gripping food (figure 18.7). Turtles are enclosed in shells consisting of a dorsal **carapace** (Fr., from Sp. *carapacho*, covering) and a ventral **plastron** (Fr., breastplate). The shell is bone, overlain by enlarged scales of keratin. Like a medieval coat of armor, the shell offers protection for the head and appendages, which most turtles can retract into it (figure 18.8). Because its ribs are fused to the shell, a turtle cannot expand its chest to breathe. Instead, turtles employ certain abdominal and pectoral muscles as a "diaphragm." Air is drawn in by contracting limb flank muscles to make the body cavity larger. Exhalation is also active, accomplished by drawing the limbs back into the shell, thus compressing the viscera and forcing air out of the lungs.

Key Theme 18.1
SCIENCE EXPLAINED

Turtles, Terrapins, or Tortoises?

The terms "turtle," "tortoise," and "terrapin" are applied variously to different members of the turtle order. In North American usage, they are all correctly called turtles. The term "tortoise" is frequently applied to land turtles, especially large forms. British usage of the terms is different: "tortoise" is the inclusive term, whereas "turtle" is applied only to the aquatic members.

Cleveland P. Hickman, Jr.

figure 18.7

Common snapping turtle, *Chelydra serpentina*, showing the absence of teeth. Instead, the jaw edges are covered with a keratinized plate.

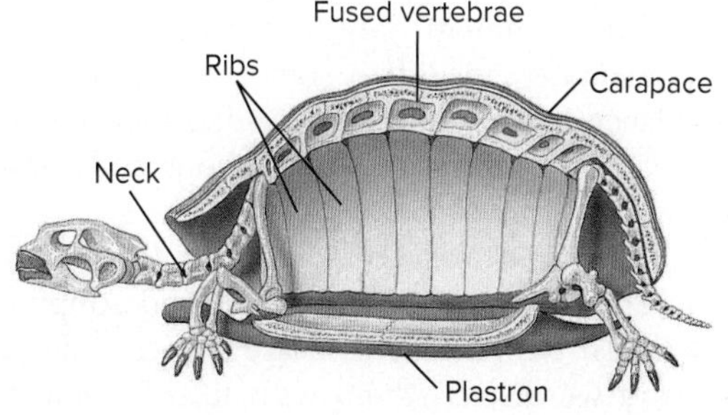

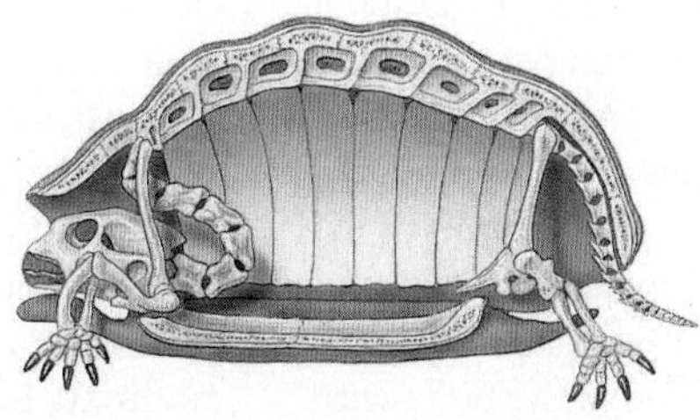

figure 18.8

Skeleton and shell of a turtle, showing fusion of vertebrae and ribs with the carapace. The long, flexible neck allows many turtles to withdraw their heads inside the shells for protection.

figure 18.9

Green sea turtle, *Chelonia mydas*. Young green sea turtles are omnivorous, but adults are strictly herbivorous, subsisting on marine grasses and algae. Although widely distributed in warm oceans, they are uncommon and considered critically endangered. Like other sea turtles, they return to land only to deposit their eggs.

figure 18.10

Mating Galápagos tortoises, *Geochelone elaphantopus*. The male has a concave plastron that fits over the highly convex carapace of the female, helping to provide stability during mating. Males utter a roaring sound during mating, the only time they are known to emit vocalizations.

Sound perception is poor in turtles, and most turtles are mute. Compensating for poor hearing are good senses of smell and color vision. Turtles are oviparous, and fertilization is internal. All turtles, even marine forms, bury their shelled, amniotic eggs in the ground. An interesting feature of turtle reproduction is that in some turtle families, as in all crocodilians and some lizards, nest temperature determines the sex of the hatchlings. In turtles, low temperatures during incubation produce males, and high temperatures produce females.

Marine turtles, buoyed by their aquatic environment, can reach great size. Leatherbacks are the largest, attaining a length of 2 m and a weight of 900 kg. Green sea turtles, so named because of their greenish body fat (colored from their herbivorous diet), may exceed 300 kg, although most are smaller (figure 18.9). Some land tortoises weigh several hundred kilograms, such as the giant tortoises of the Galápagos Islands (figure 18.10) that so intrigued Darwin during his visit there in 1835. Most tortoises move rather slowly; one hour of determined trudging carries a large Galápagos tortoise approximately 300 m. Low metabolism may explain the longevity of turtles; some live for more than 150 years.

Lizards and Snakes: Squamata

Squamates are highly diverse, comprising 95% of known species of living nonavian reptiles. Lizards appeared in the fossil record in the Jurassic period, but they did not diversify until the Cretaceous period of the Mesozoic era when the dinosaurs were at the climax of their diversity. Snakes appeared during the late Jurassic period, probably evolving from a group of lizards whose descendants include the Gila monster, iguanas, and monitor lizards (see figure 18.2). Two specializations in particular characterize snakes: (1) extreme elongation of the body with accompanying displacement and rearrangement of internal organs and (2) skull specializations for swallowing large prey.

Skulls of squamates are modified from the ancestral diapsid condition by loss of bone ventral and posterior to the lower temporal opening. This modification allowed evolution in most lizards and snakes of a **kinetic skull** having movable joints (figure 18.11). These joints, located at the posterior of the quadrate and pterygoid bones, palate, and roof of the skull, allow the snout to be tilted. Specialized mobility of the skull enables squamates to seize and to manipulate their prey; it also increases the effective closing force of the jaw musculature. Such exceptional skull mobility is considered a major factor in the diversification of lizards and snakes.

Squamates copulate with a paired structure called a **hemipenis** (pl. hemipenes) which is held inside the cloaca when not in use. Most squamates are **oviparous** species that lay their shelled, ellipitical-shaped eggs beneath logs or rocks, or in holes in the ground. The remainder are **viviparous**, giving birth to fully-developed young. Many of these, including geckos, monitors, and nearly all pit vipers, are **ovoviviparous**, in which the embryos receive nourishment only from their yolk sacs. Other viviparous squamates, including some skinks, nourish their young via a placenta. Most species abandon their young or eggs immediately, but a few, including pythons and some skinks, protect their eggs from predators.

Traditionally, Squamata was divided into three subgroups: Lacertilia (lizards), Serpentes (snakes), and Amphisbaenia (amphisbaenians). Amphisbaenians are now considered modified lizards, and we discuss them with lizards. Snakes

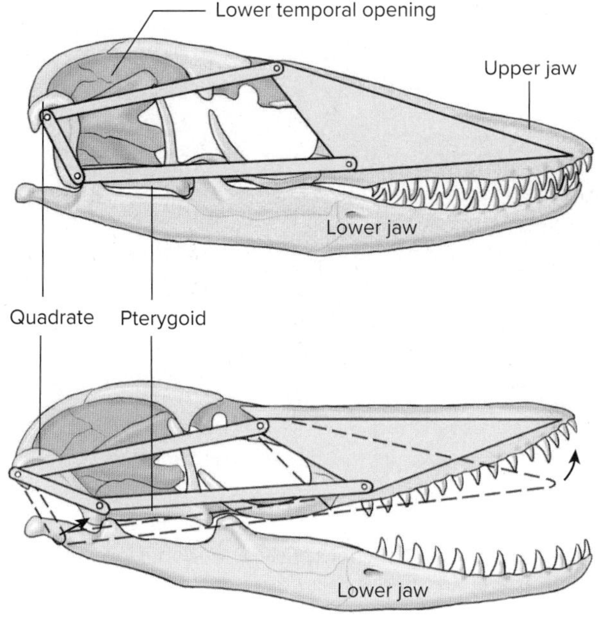

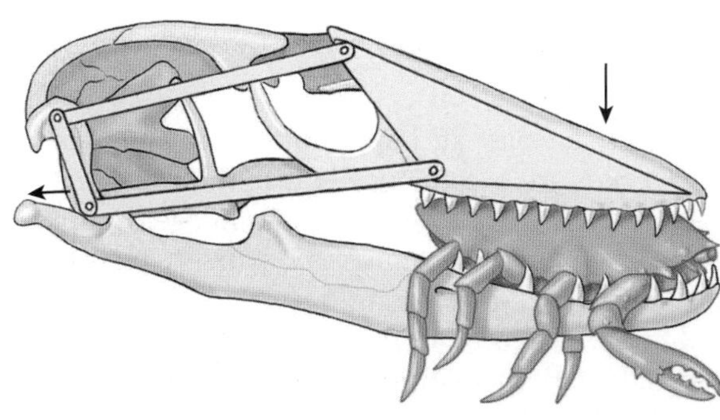

figure 18.11

Kinetic diapsid skull of a monitor lizard, *Varanus*, showing joints that allow the snout and upper jaw to move on the rest of the skull. The quadrate bone can move at its dorsal end and ventrally at both the lower jaw and the pterygoid bone. The front part of the skull is also movable, allowing the snout to be raised or depressed to ensure parallel bite force between the jaws. Note that the lower temporal opening is very large, with no lower border. The upper temporal opening lies dorsal and medial to the lower temporal opening and is not visible in this drawing.

John Mitchell/Science Source

figure 18.12

The tokay, *Gekko gecko*, of Southeast Asia has a true voice and is named after its strident, repeated *to-kay, to-kay* call.

Cleveland P. Hickman, Jr.

figure 18.13

A large male marine iguana, *Amblyrhynchus cristatus*, of the Galápagos Islands, feeding underwater on algae. This is the only marine lizard in the world. It has special salt-removing glands in the eye orbits and long claws that enable it to cling to the bottom while feeding on small red and green algae, its principal diet. It may dive to depths exceeding 10 m (33 feet) and remain submerged for more than 30 minutes.

form a monophyletic group that evolved within a lizard subgroup, thus making "Lacertilia" paraphyletic. Here we use "lizard" and "Lacertilia" to describe squamates that are not snakes.

Lizards: "Lacertilia"

Lizards are an extremely diverse group that includes terrestrial, burrowing, aquatic, arboreal, and aerial members. Among the more familiar groups in this varied suborder are **geckos** (figure 18.12), small, agile, mostly nocturnal forms with adhesive toe pads that enable them to walk upside down and on vertical surfaces; **iguanas,** often brightly colored New World lizards with ornamental crests, frills, and throat fans, including the remarkable marine iguana of the Galápagos Islands (figure 18.13); **skinks,** which have elongate bodies, an armor of tight-fitting osteoderms, and reduced limbs in many species; **monitors,** large, active predators that include the largest lizard, the Komodo dragon, *Varanus komodoensis* (figure 18.14); and **chameleons,** a group of arboreal lizards, mostly of Africa and Madagascar. Chameleons are entertaining creatures that catch insects with a sticky-tipped tongue that can be flicked accurately

Susan K./Flickr Open/Getty Images

figure 18.14

The Komodo dragon, *Varanus komodoensis*, is the largest lizard, reaching up to 3 m in length. It feeds on pigs, deer, and carrion, and like other monitors, is mildly venomous.

CathyKeifer/iStock/Getty Images

figure 18.15

A chameleon snares a cricket. After cautiously edging close to its target, the chameleon suddenly lunges forward, anchoring its tail and feet to the branch. A split second later, it launches its sticky-tipped, foot-long tongue to trap the prey. The eyes of this panther chameleon, *Fucifer pardalis*, are swiveled forward to provide binocular vision and excellent depth perception.

James Gerholdt/Getty Images

figure 18.16

A European glass lizard or sheltopusik, *Pseudopus apodus*. This legless lizard is stiff and has an extremely long, fragile tail that readily fractures when the animal is struck or seized. Most specimens, such as this one, have only a partly regenerated tip to replace a much longer tail previously lost. Glass lizards can be readily distinguished from snakes by the deep, flexible groove running along each side of the body. They mainly feed on arthropods and other invertebrates.

and rapidly to a distance greater than the length of their body (figure 18.15). Most lizards have four limbs and relatively short bodies, but in many, the limbs are reduced, and a few, such as glass lizards (figure 18.16), are completely limbless.

Most lizards have movable eyelids, whereas a snake's eyes are permanently covered with a transparent cap. Lizards have keen vision for daylight (retinas rich in both cones and rods), although one group, the nocturnal geckos, has retinas composed entirely of rods. Most lizards have an external ear opening that snakes lack. However, as with other nonavian reptiles, hearing does not play an important role in the lives of most lizards. Geckos are exceptions because the males are strongly vocal (to announce territory and to discourage the approach of other males), and they must, of course, hear their own vocalizations.

Many lizards occupy the world's hot and arid regions, aided by adaptations that make desert life possible. Lipids in their thick skin minimize water loss. Little water is lost in their urine because they primarily excrete uric acid, as do other groups that are successful in arid habitats (birds, insects, and pulmonate snails). Some, such as Gila monsters of the southwestern United States deserts, store fat in their tails, and then use the fat during droughts to produce energy and metabolic water (figure 18.17).

Lizards, like nearly all nonavian reptiles, are **ectothermic,** adjusting their body temperature by moving among different microclimates. Because cold climates provide few opportunities for ectotherms to raise their body temperature, relatively few nonavian reptile species live in these habitats. Ectotherms use considerably less energy than do endotherms; therefore, nonavian reptiles are successful in ecosystems with low productivity and warm climates, such as tropical deserts, dry, open forests, and grasslands. Thus, ectothermy is not an "inferior" characteristic of nonavian reptiles, but rather a successful strategy for coping with specific environmental challenges.

The amphisbaenians, or "worm lizards," are lizards highly specialized for a **fossorial** (burrowing) life. Amphisbaenia means "double walk," in reference to their peculiar ability to move backward nearly as effectively as forward. Amphisbaenians have elongate, cylindrical bodies of nearly uniform diameter, and most lack any trace of external limbs (figure 18.18).

figure 18.17

A Gila monster, *Heloderma suspectum*, of southwestern United States desert regions. It and its relative, the Mexican beaded lizard, are the only highly venomous lizards. These brightly colored lizards feed principally on birds' eggs, nesting birds, mammals, and insects. Unlike venomous snakes, the Gila monster secretes venom from glands in its lower jaw. The chewing bite is painful to humans but seldom fatal.

figure 18.18

An amphisbaenian, or "worm lizard." Amphisbaenians burrow, using their solid skull as a digging tool. The species pictured, the Iberian worm lizard, *Blanus cinereus*, occurs in Portugal and Spain.

Their eyes are usually hidden underneath the skin, and they have no external ear openings. Their skull is solidly built and either conical or spade-shaped, which aids in tunneling through the soil. The skin is formed into numerous, independently moving rings that can grip the soil, creating a movement not unlike that of earthworms. Amphisbaenians have an extensive distribution in South America and tropical Africa; one species occurs in Florida.

figure 18.19

The great mobility of a snake's jaws and cranial bones is evident in this snake, *Dasypeltis inornata*, swallowing an egg.

Snakes: Serpentes

Snakes are a monophyletic group of squamates. They are limbless and usually lack both pectoral and pelvic girdles (the latter persists as a vestige in pythons, boas, and a few other snakes). The numerous vertebrae of snakes, shorter and wider than those of most tetrapods, permit quick lateral undulations through grass and over rough terrain. Ribs increase rigidity of the vertebral column, providing more resistance to lateral stresses.

The skull of snakes is more kinetic than that of lizards, enabling snakes to swallow much larger prey. The two halves of the lower jaw of snakes are joined only by muscles and skin, permitting them to spread widely apart (figure 18.19). Snakes differ from lizards in having no movable eyelids and no external ear openings. Most snakes have relatively poor vision, but arboreal snakes possess excellent binocular vision, which is useful for tracking prey through branches where scent trails are difficult to follow (figure 18.20). Snakes' internal ears are mainly sensitive to sounds in a limited range of low frequency (100–700 Hz). Snakes are also quite sensitive to vibrations conducted through the ground.

Nevertheless, most snakes employ chemical senses rather than vision or vibration to hunt their prey. In addition to the usual olfactory areas in the nose, which are not well developed, snakes have a pair of pitlike **Jacobson's organs** (vomeronasal organs) in the roof of the mouth. These organs are lined with a richly innervated olfactory epithelium. The **forked tongue,** flicked through the air, collects scent molecules (figure 18.21). The tongue is then drawn past Jacobson's organs and information is transmitted to the brain, where scents are identified.

Boids (pythons and boas) and pit vipers have special heat-sensitive **pit organs** on their heads, located between their nostrils and eyes (figure 18.22). They are exceedingly sensitive to

figure 18.20

A parrot snake, *Leptophis ahaetulla*. The slender body of this Central American tree snake is an adaptation for traveling among branches.

figure 18.21

A copperhead, *Agkistrodon contortrix*, flicks its tongue to smell its surroundings. Scent particles trapped on the tongue's surface are transferred to Jacobson's organs, olfactory organs in the roof of the mouth. Note the heat-sensitive pit organ between the nostril and eye.

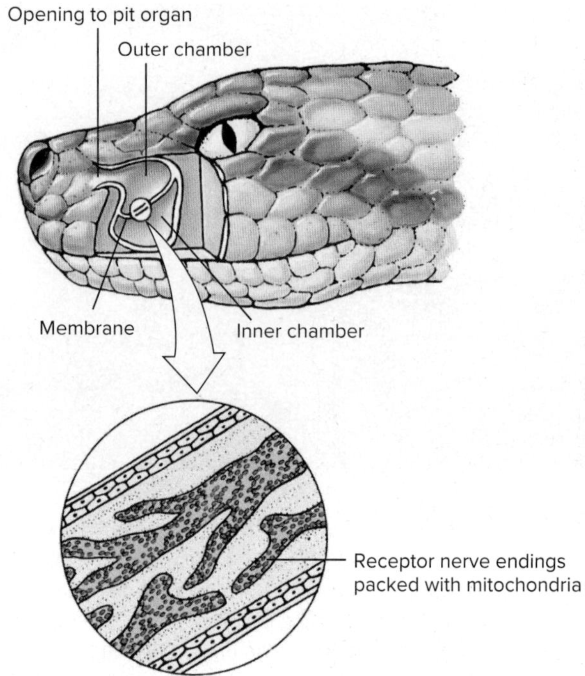

figure 18.22

Pit organ of a rattlesnake, a pit viper. Cutaway shows the location of a deep membrane that divides the pit into inner and outer chambers. Heat-sensitive nerve endings are concentrated in the membrane.

figure 18.23

Nonvenomous African house snake, *Boaedon fuluginosus*, constricting a mouse before swallowing it.

radiant energy (long-wave infrared) and can distinguish temperature differences smaller than 0.003°C from a radiating surface. The pits are used to track warm prey and to aim strikes, which are as effective in total darkness as in daylight.

Most snakes capture their prey by grabbing it with their mouth and swallowing it while it is still alive. Swallowing a struggling, kicking, biting animal is dangerous, so most snakes that swallow prey alive specialize on small prey, such as worms, insects, fish, frogs, and, less frequently, small mammals. Many of these snakes, which may be quite fast, locate prey by actively foraging. Snakes that first kill their prey by constriction (figure 18.23) usually specialize on larger, often mammalian, prey. The largest constrictors can kill and swallow prey as large as deer, leopards, and crocodilians. However, because muscle rearrangements that permit constricting also reduce speed of travel, most constrictors ambush their prey.

Other snakes kill their prey by injecting it with venom. Less than 20% of all snakes are venomous, although venomous species outnumber nonvenomous species by 4 to 1 in Australia. Venomous snakes are usually divided into five families, based in part on type of fangs.

Vipers (family Viperidae) have large, movable, tubular fangs at the front of the mouth (figure 18.24). The fangs lie in a membranous sheath when the mouth is closed. When a viper strikes, a special muscle and bone lever system erects the fangs as the mouth opens (figure 18.24). Fangs are driven into prey by the thrust, and venom is injected into the wound through a canal in the fangs. A viper immediately releases its prey after the bite and follows it until it is paralyzed or dead. Vipers of the Old World, including puff adders, saw-scaled vipers, and Russell's vipers, lack facial heat-sensing pits. Snakes of subfamily Crotalinae within family Viperidae are called **pit vipers** because they possess pit organs on their heads (see figure 18.22). All best-known North American venomous snakes are pit vipers, including rattlesnakes, cottonmouths, and copperheads (see figure 18.21). Each year in the United States approximately 3500 bites from pit vipers are reported, causing only about 5 deaths.

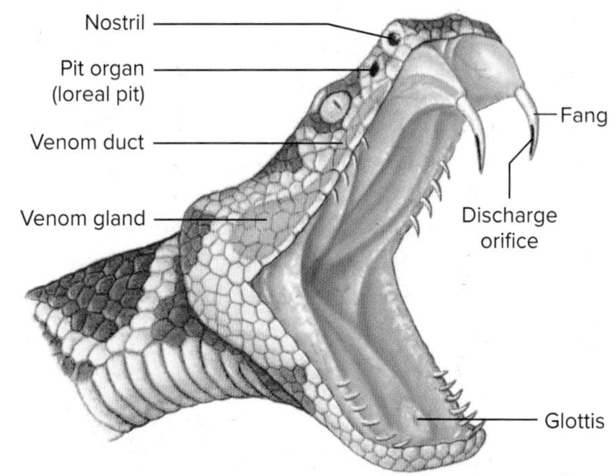

figure 18.24

Head of a rattlesnake showing the venom apparatus. The venom gland, a modified salivary gland, is connected by a duct to the hollow fang.

Key Theme 18.2

HUMAN CONNECTIONS

Pythons on the Loose

The Burmese python, *Python bivittatus*, is rare and declining in southeast Asia, where it is native, but has become an invasive species in south Florida. This large (up to 5 m) constrictor was imported into the United States by the tens of thousands, where they were sold mostly as pets. Accidental escapes from breeding facilities and pet owners intentionally releasing their difficult-to-care-for (and dangerous) pets produced a reproducing population in Everglades National Park by about 2006. A 2011 study documented declines of 88–100% in the native midsize mammals, such as raccoons, rabbits, and opossums, that these snakes eat. The python's range expansion to the Florida Keys jeopardizes a number of endangered native species, such as the Key Largo wood rat. Through intensive efforts, thousands of Burmese pythons have been removed from south Florida, but the population continues to grow.

Mark Carwardine/Getty Images

figure 18.25

Monocled cobra, *Naja kaouthia*. Cobras erect the front part of the body when startled and as a threat display. Although a cobra's strike range is limited, all cobras are dangerous because of the extreme toxicity of their venom.

A second family of venomous snakes (Elapidae) has short, permanently erect fangs in the front of the mouth and includes cobras (figure 18.25), mambas, coral snakes, and kraits. The highly venomous sea snakes are usually placed in a third family (Hydrophiidae). Mole vipers and stiletto snakes (Atractaspididae) are small, fossorial, venomous species that vary in fang type. The large family Colubridae, which contains most familiar nonvenomous snakes, also includes numerous mildly venomous species and a few, such as the boomslang and savanna twigsnake of Africa, that are very dangerous and have caused human fatalities. Venomous colubrids have fixed fangs in the rear of the mouth.

Even the saliva of harmless snakes possesses limited toxic qualities, which is likely the ancestral state from which highly toxic venom evolved. Although sea snakes and the Australian inland taipan have perhaps the most toxic of snake venoms, several larger snakes are more dangerous, because they inject more venom. Black mambas are large and fast, and inject a large amount of fast-acting toxin during a bite. This is one of the most feared snakes in Africa, as untreated bites are almost always fatal, sometimes in as little as 1 hour. It is estimated worldwide that at least 20,000 (possibly up to 90,000) people die from snakebites each year. Most deaths occur in India, Pakistan, Bangladesh, and nearby countries where poorly shod people frequently encounter venomous snakes and do not get immediate medical attention following a snake bite. The snakes primarily responsible for deaths in these areas are Russell's viper, saw-scaled vipers, and several species of cobras.

Tuataras: Sphenodonta

Sphenodonta is represented by a single living species of the genus *Sphenodon* (Gr. *sphenos,* wedge, + *odontos,* tooth) of New Zealand (figure 18.26). Tuataras are lizardlike forms, measuring up to 80 cm long, that live in burrows they often share with sea birds called petrels. They are sole survivors of the sphenodontid lineage that diversified modestly during the early Mesozoic but declined toward the end of that era. Tuataras were once widespread throughout the two main islands of New Zealand but are now restricted to small islets of Cook Strait and off the northeast coast of North Island. Loss of tuatara populations on the main islands of New Zealand was caused by humans intentionally or accidentally introducing nonnative animals, including cats, dogs, rats, and goats, which preyed upon tuataras and their eggs or destroyed their habitat. Tuataras are particularly vulnerable because they have slow growth and low reproductive rates.

Tuataras have one of the slowest reproduction rates of reptiles: they take 10–20 years to become sexually mature and typically produce eggs, which take seven months to hatch, only once every four years. They are slow-growing animals with long lives; one captive male, named "Henry," is thought to be over 100 years old, and successfully reproduced in 2009.

Tuataras have captured the interest of zoologists because of numerous features that are almost identical to those of early Mesozoic diapsids from 200 million years ago. These features include a diapsid skull with two pairs of temporal openings, and a prominent median parietal "third eye." Parietal eyes are not used for vision, but instead detect changes in light intensity, important in regulating daily and seasonal rhythms of behavior.

Asa Thoresen/Science Source

figure 18.26

Tuatara, *Sphenodon punctatus*. This "living fossil" has on top of its head a well-developed parietal eye with retina, lens, and nervous connections to the brain. Although covered with scales, this third eye is sensitive to light. Tuataras are mostly restricted to certain islands off the coast of New Zealand.

Alligators, Caimans, Crocodiles, and Gharials: Crocodilia

Modern crocodilians and birds are the only survivors of the archosaurian lineage that gave rise to the great Mesozoic diversification of dinosaurs and their kin. Crocodilians differ little in anatomy from crocodilians of the early Mesozoic era. Having remained mostly unchanged for nearly 200 million years, crocodilians face an uncertain future in a world dominated by humans.

All crocodilians have an elongate, robust, well-reinforced skull and massive jaw musculature arranged to provide a wide gape and rapid, powerful closure. Teeth are set in sockets, a type of dentition that was typical of Mesozoic archosaurs as well as the earliest birds. Another adaptation, found in no other vertebrate except mammals, is a complete secondary palate. This feature allows a crocodilian to breathe when its mouth is filled with water or food (or both).

Estuarine crocodiles (*Crocodylus porosus*) of southern Asia and Nile crocodiles (*C. niloticus*) of Africa (figure 18.27A) grow to great size (adults weighing 1000 kg have been reported) and are swift and aggressive. Crocodiles are known to attack animals as large as cattle, deer, and people. Alligators (figure 18.27B) are less aggressive than these crocodiles and far less dangerous to people. In the United States, *Alligator mississipiensis* is the only species of alligator; *Crocodylus acutus,* restricted to extreme southern Florida, is the only species of crocodile.

Crocodilians are oviparous. The female usually lays 20–90 eggs in a mass of dead vegetation or buries them in sand, and then guards them. Unlike other nonavian reptiles, crocodilians provide extensive parental care. The mother hears vocalizations from hatching young and responds by opening the nest to allow the hatchlings to escape. Young are guarded by their mother for

A

B

figure 18.27

Crocodilians. **A,** Nile crocodile, *Crocodylus niloticus*, basking. The lower jaw tooth fits *outside* the slender upper jaw; alligators lack this feature. **B,** American alligator, *Alligator mississipiensis*, an increasingly noticeable resident of rivers, bayous, and swamps of the southeastern United States.

two years or more after hatching. Although the young are capable of catching their own food immediately after hatching, they also feed on small pieces that fall from food the mother is eating. As with many turtles and some lizards, the incubation temperature of the eggs determines the sex ratio of the offspring. However, unlike turtles, low nest temperatures produce only females, whereas high nest temperatures produce only males.

Taxonomy of Living Nonavian Reptiles

The following cladistic taxonomy lists the living members of the clade Reptilia, except for Aves (birds), a clade within Archosauria that is the subject of Chapter 19. Relationships of turtles to other diapsids are controversial.

Diapsida (dī-a p′si-də) (Gr. *di,* double, + *apsis,* arch): **diapsids.** Amniotes having a skull with two pairs of temporal openings in ancestral state.

Testudines (tes-tū′di-nēz) (L. *testudo,* tortoise) **(Chelonia): turtles.** Body in a bony case of dorsal carapace and ventral plastron; jaws with keratinized plates instead of teeth; vertebrae and ribs fused to overlying carapace; temporal openings lost; about 325 species.

Lepidosauria (le-pi-dō-sor′ē-ə) (Gr. *lepidos,* scale, + *sauros,* lizard). Characterized by sprawling posture; no bipedal specializations; diapsid skull often modified by loss of one or both temporal arches; transverse cloacal slit; skin shed in one piece.

Squamata (skwä-mā′tə) (L. *squamatus,* scaly, + *ata,* characterized by): **snakes and lizards.** Skin of keratinized epidermal scales or plates, which are shed; quadrate movable; skull kinetic (except in amphisbaenians); vertebrae usually concave in front; paired copulatory organs; about 5810 lizard species and about 3370 snake species.

Sphenodonta (sfē′nō-don′tə) (Gr. *sphen,* wedge, + *odontos,* tooth): **tuataras.** Skull retains both pairs of temporal openings; vertebrae biconcave; quadrate immovable; parietal eye prominent; one species.

Archosauria (ärk′ō-sor′ē-ə) (Gr. *archon,* ruling, + *sauros,* lizard). Orbit shaped like an upside-down triangle; anteorbital fenestra (opening in the skull anterior to the orbit) and gizzard present; ventricle fully divided; parental care of young; many forms bipedal.

Crocodilia (krok′ə-dil′ē-ə) (L. *crocodilus,* crocodile): **alligators, caimans, crocodiles, and gharials.** Skull elongate and massive; nares terminal; secondary palate present; vertebrae usually concave in front; forelimbs usually with five digits; hindlimbs with four digits; quadrate immovable; 25 species.

SUMMARY

Sections	Key Concepts
18.1 Origin and Early Evolution of Amniotes	• Amniotes diverged from a group of early tetrapods during the late Paleozoic era, about 300 million years ago, and diversified to occupy a range of aquatic and terrestrial habitats. • Before the end of the Paleozoic era, amniotes diversified to form three groups distinguished by skull structure: anapsids, which lack temporal fenestrae; synapsids, which have one pair of temporal fenestrae; and diapsids, which have two pairs of temporal fenestrae. • Mammals evolved from early synapsids. Early diapsids gave rise to all living nonavian reptiles (lizards, snakes, tuataras, crocodilians, and tentatively, turtles) and to birds. No living amniotes retain the Paleozoic anapsid condition. • One clade of diapsids, the archosaurs, underwent a great worldwide diversification during the Mesozoic era into large and morphologically diverse forms, including the ichthyosaurs, plesiosaurs, pterosaurs, and dinosaurs. Although most of these lineages are extinct, descendants of some dinosaurs survived the great extinction at the end of the Cretaceous period, and underwent their own diversification as birds. • The two major groups of dinosaurs, Saurischia and Ornithischia, differ in their hip structure. At least some dinosaurs cared for their young, perhaps in a manner similar to that of crocodilians and birds. Dinosaurs appeared in the early Mesozoic era and were the dominant large vertebrates until all became extinct at the end of the Mesozoic era. • "Reptilia," as traditionally conceived, is paraphyletic because it includes extinct synapsid groups more closely related to mammals and excludes birds. Removal of extinct synapsid amniotes and inclusion of birds in Reptilia makes it a monophyletic group. Reptilia includes four clades with living members: Testudines (turtles), Squamata (lizards and snakes), Sphenodonta (tuataras), and Archosauria (crocodiles and birds). • The success of amniotes as terrestrial vertebrates can be attributed to several adaptations, including the amniotic egg. The amniotic egg, with its shell and four extraembryonic membranes—the amnion, allantois, chorion, and yolk sac—permits rapid development of embryos in terrestrial environments. Additional amniote adaptations that support occupation of dry environments and a relatively active lifestyle include a thick, water-resistant skin, excretion of urea or uric acid, high-surface-area lungs ventilated by trunk muscles, expanded jaw musculature, and an efficient cardiovascular system.
18.2 Characteristics and Natural History of Reptilian Groups	• Turtles (Testudines), with their distinctive shells, have changed little in anatomy since the Triassic period. Turtles are a small group of long-lived terrestrial, semiaquatic, and aquatic species. They lack teeth, and instead bear keratinized plates on their jaws. All are oviparous, and all, including marine forms, bury their eggs on land. • Lizards and snakes (Squamata) represent 95% of all living nonavian reptiles. Both groups are diverse and successful, particularly in hot climates. Most lizards and snakes are oviparous, although viviparity is not uncommon, especially in cooler climates. Lizards and snakes are ectothermic, regulating their temperature by moving among different microenvironments. Lizards are distinguished from snakes by having united lower-jaw halves, moveable eyelids, and external ear openings. Amphisbaenians are a small group of tropical lizards specialized for burrowing. They have ringed, usually limbless bodies and a solid skull. • Snakes, which evolved from one group of lizards, are characterized by elongate, limbless bodies and a highly kinetic skull with several moveable joints that allow them to swallow prey several times larger than the snake's diameter. Most snakes rely on chemical senses, including Jacobson's organ, to hunt prey, rather than on visual and auditory senses. Two groups of snakes (pit vipers and boids) have unique infrared-sensing organs for tracking prey. Some snakes swallow their prey alive; others kill their prey by constriction or with venom. Different groups of venomous snakes are distinguished by the anatomy of their fangs. • The tuatara of New Zealand (Sphenodonta) is the sole survivor of a group that otherwise disappeared 100 million years ago. It bears several features, including two pairs of temporal openings and a prominent parietal eye, that are almost identical to those of early Mesozoic diapsids. These rare reptiles are particularly vulnerable because they have slow growth and low reproductive rates. • Crocodiles, alligators, caimans, and gharials (Crocodilia) are members of the clade Archosauria, which also includes the extinct dinosaurs and the living birds. Crocodilians have several adaptations for a carnivorous, semiaquatic life, including a massive skull with powerful jaws, and a secondary palate. They have the most complex parental care of any living nonavian reptile.

Review Questions

1. What are the four membranes associated with amniotic eggs? What is the function of each of these membranes?
2. How do the skin and respiratory systems of amniotes differ from those of their early tetrapod ancestors?
3. Amniotes are divided into three groups based on their skull morphologies. What are these three groups, and how do the skulls differ? Which living amniotes, if any, originated from each of these three groups?
4. Why are "reptiles," as traditionally defined, a paraphyletic group? How has cladistic taxonomy revised Reptilia to make it monophyletic?
5. Describe ways in which amniotes are more functionally or structurally suited for terrestriality than were the earliest tetrapods.
6. Describe the principal structural features of turtles that distinguish them from other nonavian reptiles.
7. Distinguish ornithiscian and saurischian dinosaurs, based on their hip anatomy. Was a dinosaur's style of parental care more like a lizard's or a crocodilian's?
8. How might nest temperature affect egg development in turtles? In crocodilians?
9. What is meant by a "kinetic" skull, and what benefit does it confer? How are snakes able to eat large prey?
10. Most lizards and snakes are oviparous, but some are ovoviviparous or have placental viviparity. Compare these methods of reproduction in squamates.
11. In what ways are the special senses of snakes similar to those of lizards, and in what ways have they evolved for specialized feeding strategies?
12. What are amphisbaenians? What morphological adaptations do they have for burrowing?
13. What is the function of Jacobson's organ in snakes?
14. What is the function of the "pit" in pit vipers?
15. What are the differences in structure or location of fangs in a rattlesnake, a cobra, and an African boomslang?
16. Why are tuataras (*Sphenodon*) of special interest to biologists? Why are they rare?
17. From which diapsid subgroup have crocodilians descended? What other major fossil and living vertebrate groups belong to this same subgroup? In what structural and behavioral ways do crocodilians differ from other living nonavian reptiles?
18. How do crocodilians breathe when their mouths are full of food?

For Further Thought How could changes in the environment affect populations of species with temperature-dependent sex determination?

19

Birds

LEARNING OBJECTIVES

Readers will be able to:

19.1 Describe the early evolution of birds.

19.2 Identify how the feathers and organ systems of birds are adapted for flight.

19.3 Explain how birds generate lift and thrust to fly.

19.4 Explain how birds navigate in migration, and how migration can be adaptive.

19.5 Describe different mating systems of birds.

19.6 Identify the major human threats to bird populations.

Steve Whiston/Moment/Getty Images

Snow geese, *Chen caerulescens*, during migration.

Long Trip to a Summer Home

Almost every human has dreamed of rising free from the earth, in bird-like flight. Dreams of flying are not necessary for birds, for they are masters of flight. Some birds use this power to make long seasonal migrations. Moving between southern wintering regions and northern summer breeding regions with long days and an abundance of insects provides parents with ample food to rear their young. Predators of birds are not so abundant in the far North, and a brief once-a-year appearance of vulnerable young birds does not encourage buildup of predator populations. Migration also vastly increases the amount of space available for breeding and reduces aggressive territorial behavior. Finally, migration favors homeostasis—the balancing of internal physiological processes—by allowing birds to avoid climatic extremes.

The migratory pageant inspires wonder, and the physiological mechanisms of migration are equally challenging to researchers. What determines the timing of migration, and what induces each bird to store sufficient fuel for the journey? How did the sometimes difficult migratory routes originate, and what cues do birds use to navigate? How does instinct drive the migratory waves in spring and fall, carrying most birds successfully to their northern nests, while countless others fail and die, winnowed by the ever-challenging task?

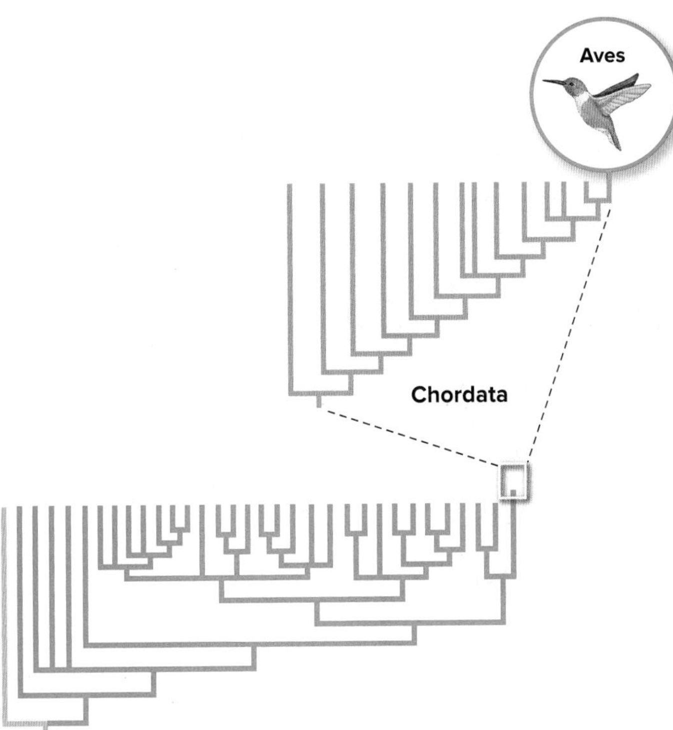

Of the vertebrates, birds (Aves, ā´vēz; L. pl. of *avis*, bird) are the most noticeable, the most melodious, and, many think, the most beautiful. With more than 10,500 species distributed nearly worldwide, birds outnumber any other vertebrate group except fishes. Birds inhabit forests and deserts, mountains and prairies, and all oceans. Four species are known to have visited the North Pole, and one, a skua, was seen at the South Pole. Some birds live in total darkness in caves, finding their way by echolocation, and others dive to depths greater than 45 m to prey on aquatic life.

The single unique feature that distinguishes birds from other living animals is their feathers. If an animal has feathers, it is a bird; if it lacks feathers, it is not a bird. However, we note that feathers were not so diagnostic in the past; some theropod dinosaurs had feathers.

There is great uniformity of structure among birds. Despite approximately 150 million years of evolution, during which they proliferated and adapted to specialized ways of life, we have no difficulty recognizing a living bird. In addition to feathers, all birds have forelimbs modified into wings (although not always used for flight); all have hindlimbs adapted for walking, swimming, or perching; all have keratinized beaks; and all lay eggs. The reason for this great structural and functional uniformity is that birds evolved into flying machines, which forces them to retain these diagnostic characteristics.

A bird's entire anatomy is designed around flight. An airborne life for a large vertebrate is a highly demanding evolutionary challenge. A bird must, of course, have wings for lift and propulsion. Bones must be light and yet serve as a rigid airframe. The respiratory system must be highly efficient to meet the intense metabolic demands of flight. A bird must have a rapid and efficient digestive system to process an energy-rich diet, and it must have a high-pressure circulatory system. Above all, birds must have a finely tuned nervous system and acute senses, especially superb vision, to handle the complex demands of headfirst, high-velocity flight.

19.1 Origin and Relationships

Approximately 147 million years ago, a flying animal died and settled to the bottom of a shallow marine lagoon in what is now Bavaria, Germany. It was rapidly covered with fine silt and eventually fossilized. There it remained until discovered in 1861 by a workman splitting slate in a limestone quarry. The fossil was approximately the size of a crow, with a skull not unlike that of modern birds except that the beaklike jaws bore small, bony teeth set in sockets like those of dinosaurs (figure 19.1). The skeleton was decidedly reptilian, with a long bony tail, clawed fingers, and abdominal ribs. It might have been classified as a theropod dinosaur except that it carried an unmistakable imprint of feathers, those marvels of biological engineering that only birds possess. Named *Archaeopteryx lithographica* (är-kē-op´ter-iks lith-ō-graf´i-kə; Gr., "ancient wing inscribed in stone"), the fossil was an especially fortunate discovery because it demonstrated beyond reasonable doubt the phylogenetic relatedness of birds and theropod dinosaurs.

Zoologists had long recognized the similarity of birds to other reptiles. The distinguished English zoologist Thomas Henry Huxley (see Key Theme 1.3) was so impressed with these affinities that he called birds "glorified reptiles" and classified them with a group of dinosaurs called theropods (see "The Mesozoic World of Dinosaurs" in Chapter 18). Theropod dinosaurs share many derived characters with birds, the most obvious of which is an elongate, mobile, S-shaped neck (figures 19.2 and 19.3).

In particular, a group of feathered theropods, called dromaeosaurs, shares many features, such as a furcula (fused clavicles), with birds. A phylogenetic approach to classification groups birds with several bipedal dinosaur lineages in the clade Theropoda (figure 19.3). In this view, dinosaurs are not extinct—they are with us today as birds!

The early evolution of birds from their nonavian, dinosaur ancestors illustrates **mosaic evolution.** The numerous characters of modern birds (Aves) did not appear all at once, but rather they appeared one at a time, so that transitional forms bear a mixture of ancestral and derived traits, as does *Archaeopteryx* (see figure 19.5). Birds (Avialae), as used here, are theropods with feathered wings capable of powered flight, and include *Archaeopteryx* and several other extinct Mesozoic forms, along with modern birds (Aves).

Feathers preceded both birds and flight. Much of the evidence documenting the early evolution of birds comes from numerous, recently unearthed fossils from late Jurassic and early Cretaceous deposits in Liaoning Province, China. These spectacular fossils include some with filaments, such as *Sinosauropteryx,* and some with feathers, such as *Caudipteryx* (figure 19.3). The filaments were hollow structures, similar to

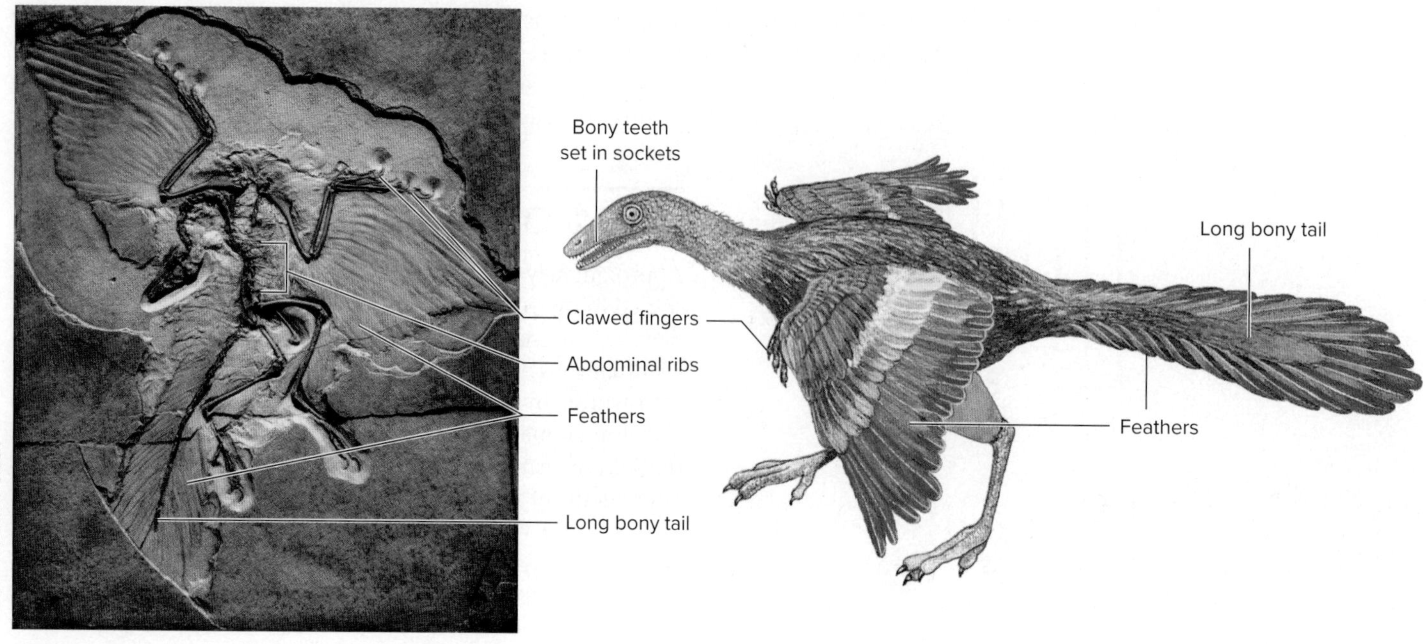

figure 19.1

Archaeopteryx, a 147-million-year-old relative of modern birds. **A,** Cast of the second and most nearly perfect fossil of *Archaeopteryx*, which was discovered in a Bavarian stone quarry. Twelve specimens of *Archaeopteryx* have been discovered, the most recent one in 2011. **B,** Reconstruction of *Archaeopteryx*.

an early developmental stage of modern feathers (figure 19.4). *Caudipteryx* could not fly, however, because it had short forelimbs and symmetrical, vaned feathers (the flight feathers of modern flying birds are asymmetrical). Clearly, these early filaments and feathers served a different purpose, perhaps providing thermoregulation or **crypsis,** or use in courtship displays. *Archaeopteryx* and later birds bear asymmetrical feathers and elongated forelimbs, **exaptations** for powered flight.

In the Cretaceous period, birds diversified extensively, with various forms feeding on fishes, invertebrates, leaves, or seeds. These Cretaceous birds, with their short tail, keeled sternum, and anteriorly-placed body mass, would have looked much more bird-like than their ancestors. However, unlike modern birds (Aves), most of the Cretaceous birds had teeth. Aves is the clade containing the last common ancestor of all living birds and its descendants. Diversification of modern birds occurred rapidly in the late Cretaceous and Paleogene periods, producing the incredible diversity of familiar living birds (see figure 19.2).

Living birds (Neonithes) are divided into two groups: (1) **Paleognathae** (Gr. *palaios,* ancient, + *gnathos,* jaw), large, flightless, ostrichlike birds and kiwis, often called **ratite** birds, that have a flat sternum with poorly developed pectoral muscles and (2) **Neognathae** (Gr. *neos,* new, + *gnathos,* jaw), all other birds, nearly all of which are flying birds that have a keeled sternum to which powerful flight muscles attach. There are several flightless neognathus birds, some of which lack a keeled sternum. Flightlessness has appeared independently among many groups of birds; the fossil record reveals flightless wrens, pigeons, parrots, cranes, ducks, auks, and even a

CHARACTERISTICS of Aves

1. **Neck elongate and S-shaped;** forelimbs modified as **wings; endothermic.**
2. Epidermal covering of **feathers** and **leg scales;** thin epidermis and dermis; no sweat glands; oil gland at base of tail.
3. Skull with **one occipital condyle;** many **bones with air cavities;** ribs with strengthening, uncinate processes; tail short, caudal vertebrate reduced to a **pygostyle;** bones of pelvic girdle fused to form a **synsacrum; sternum usually large and keeled.**
4. **No teeth;** each jaw covered with a keratinized sheath, forming a **beak; gizzard** present.
5. Brain well developed, with **large optic lobes and cerebellum;** 12 pairs of cranial nerves.
6. Eyes large, with **pecten** (see figure 19.12); middle ear with a single bone.
7. Separate sexes; internal fertilization; copulatory organ (penis) only in paleognathids, ducks, geese, and a few others; **females with functional left ovary and oviduct only;** sex determined by chromosomes (female is heterogametic).
8. Fetal membranes of **amnion, chorion, and allantois; oviparous; eggs with much yolk and hard, calcareous shells;** extensive parental care of young.
9. Excretory system of **metanephric kidneys** and ureters that open into cloaca; uric acid main nitrogenous waste.
10. Lungs of **parabronchi** with **continuous air flow; syrinx** (voice box) present; **air sacs** among visceral organs and skeleton.
11. Heart with two atria and two ventricles; **separate pulmonary and systemic circuits; persistent right aortic arch;** nucleated red blood cells.

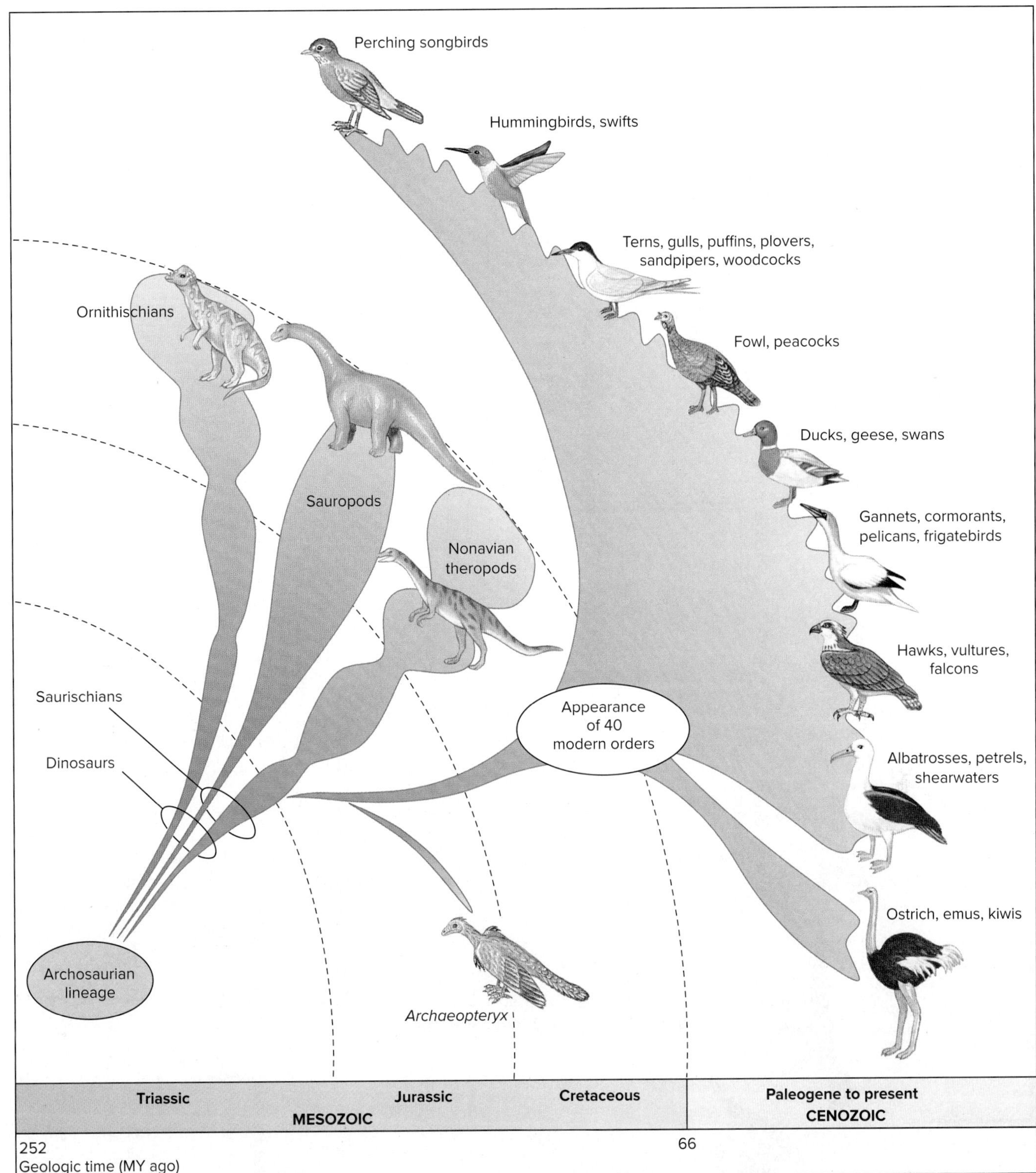

figure 19.2

Evolution of modern birds. Of 40 living bird orders, 9 of the more important are shown. The earliest known bird, *Archaeopteryx,* lived in the upper Jurassic period about 147 million years ago. *Archaeopteryx* uniquely shares many specialized aspects of its skeleton with the smaller theropod dinosaurs and is considered to have evolved within the theropod clade. Evolution of modern bird orders occurred rapidly during the Cretaceous and early Paleogene periods.

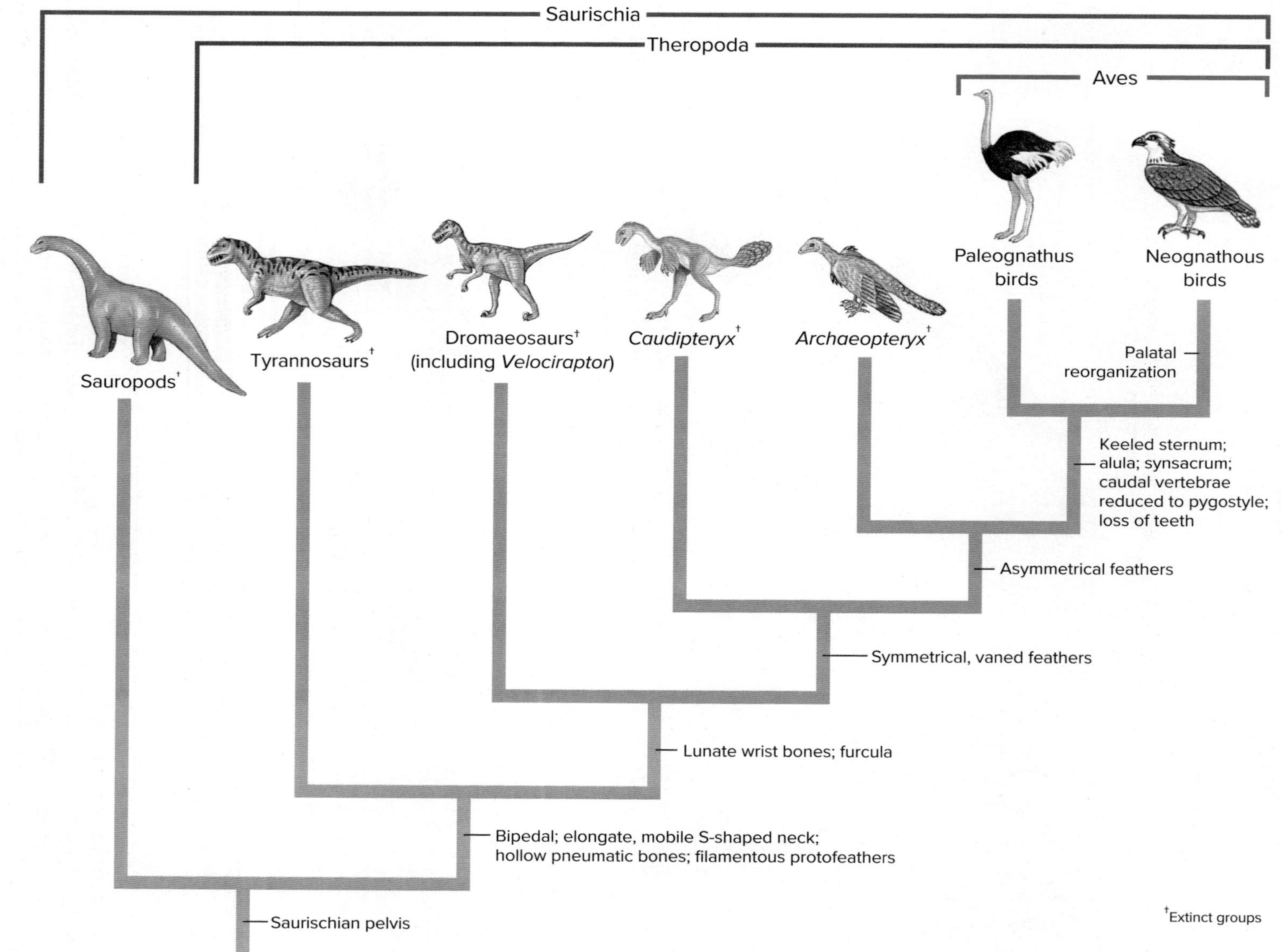

figure 19.3

Cladogram of Saurischia, illustrating the relationship of several taxa to modern birds. Shown are a few of the shared derived characters, mostly related to flight. The ornithischians are the sister group to the saurischians, and all are members of the clade Archosauria (see figures 18.1 and 18.2).

flightless owl. Penguins are flightless although they use their wings to "fly" through water (see figure 4.6). Flightlessness has almost always evolved on islands lacking terrestrial predators. Flightless birds living on continents today are the large paleognathids (ostrich, rhea, cassowary, emu), which can run fast enough to escape predators. Ostriches can run 70 km (43 miles) per hour, faster than any other bipedal animal. The evolution of ratites is discussed in figure 1.19.

Key Theme 19.1

EVOLUTION

Giant Extinct Birds of Islands

The bodies of flightless birds are dramatically redesigned because of the lack of restrictions for flight. The keel of the sternum is lost, and heavy flight muscles (as much as 17% of the body weight of flying birds) disappear. Because body weight is no longer a restriction, flightless birds tend to become large. The largest living birds, ostriches, emus, and cassowaries, are flightless; they can weigh over 50 kg and reach 2 m in height. Several extinct flightless birds were enormous: the giant moas of New Zealand weighed more than 225 kg (500 pounds), and the elephant birds of Madagascar, the largest birds that ever lived, probably weighed nearly 450 kg (about 1000 pounds) and stood nearly 3 m tall.

19.2 Structural and Functional Adaptations for Flight

Just as an airplane must be designed and built according to rigid aerodynamic specifications to fly, so too must birds meet stringent structural requirements. Flight by humans became possible when we developed an internal combustion engine and learned how to reduce the weight-to-power ratio to a critical point. Birds accomplished flight millions of years ago. Unlike airplanes, birds also must feed themselves, convert food into metabolic energy, escape predators, repair their own injuries, maintain a constant body temperature, and reproduce.

Feathers

Feathers are very lightweight and yet possess remarkable toughness and tensile strength. Most bird feathers are **contour feathers,** vaned feathers that cover and streamline a bird's body. A contour feather consists of a hollow **quill** emerging from a skin follicle, and a **shaft,** which is a continuation of the quill and bears numerous **barbs** (figure 19.4). Barbs are arranged in closely parallel fashion and spread diagonally outward from both sides of the central shaft to form a flat, expansive, webbed surface, the **vane.** There may be several hundred barbs in a vane.

Through a microscope, each barb appears to be a miniature replica of the feather, with numerous parallel filaments called **barbules** set in each side of the barb and spreading laterally from it. There may be 600 barbules on each side of a barb, adding up to more than 1 million barbules for the feather. Barbules of one barb overlap the barbules of a neighboring barb in a herringbone pattern and are held together with great tenacity by tiny hooks. Should two adjoining barbs become separated—and considerable force is needed to pull the vane apart—they are instantly zipped together again by drawing the feather through the fingertips. Birds do this preening with their beaks.

Different types of bird feathers serve different functions. **Contour feathers** (figure 19.4E) give the bird its outward form and are the type we have already described. Contour feathers that extend beyond the body and are used in flight are called **flight feathers. Filoplume feathers** (figure 19.4G) are hairlike, degenerate feathers; each is a weak shaft with a tuft of short barbs at the tip. They are the "hairs" of a plucked fowl. They have no known function. **Down feathers** (figure 19.4H) are soft tufts without a prominent rachis, hidden beneath contour feathers. They are soft because their barbules lack hooks. Down feathers are especially abundant on the breast and abdomen of water birds and on young quail and grouse and function principally to conserve heat.

Like a reptile's scale, to which it is homologous, a feather develops from an epidermal elevation overlying a nourishing dermal core. However, rather than flattening like a scale, a feather bud rolls into a cylinder and sinks into the follicle from which it is growing. During growth, pigments are added to epidermal cells. As the feather enlarges and nears the end of its growth, the soft shaft and barbs are transformed into hard structures by deposition of **keratin.** The protective sheath splits apart, allowing the end of a feather to protrude and barbs to unfold.

When fully grown, a feather, like mammalian hair, is a dead structure. Shedding, or molting, of feathers is a highly orderly process; feathers are discarded gradually, which avoids appearance of bare spots. Flight and tail feathers are lost in exact pairs, one from each side, maintaining balance. Replacements emerge before the next pair is lost, and most birds can continue to fly unimpaired during the molting period; however, many water birds (ducks, geese, loons, and others) lose all their primary feathers at once and are grounded during the molt. Many prepare for molting by moving to isolated bodies of water where they can find food and more easily escape enemies. Nearly all birds molt at least once a year, usually in late summer after nesting season.

Skeleton

A major structural requirement for flight is a light, yet sturdy skeleton (figure 19.5A). Compared with the earliest known bird, *Archaeopteryx* (figure 19.5B), bones of modern birds are phenomenally light, delicate, and laced with air cavities. Such pneumatized bones (figure 19.6) are nevertheless strong. The skeleton of a frigate bird with a 2.1 m (7-foot) wingspan weighs only 114 g (4 ounces), less than the weight of all its feathers.

As archosaurs, birds are diapsids (see figure 18.2). However, skulls of modern birds are so specialized that it is difficult

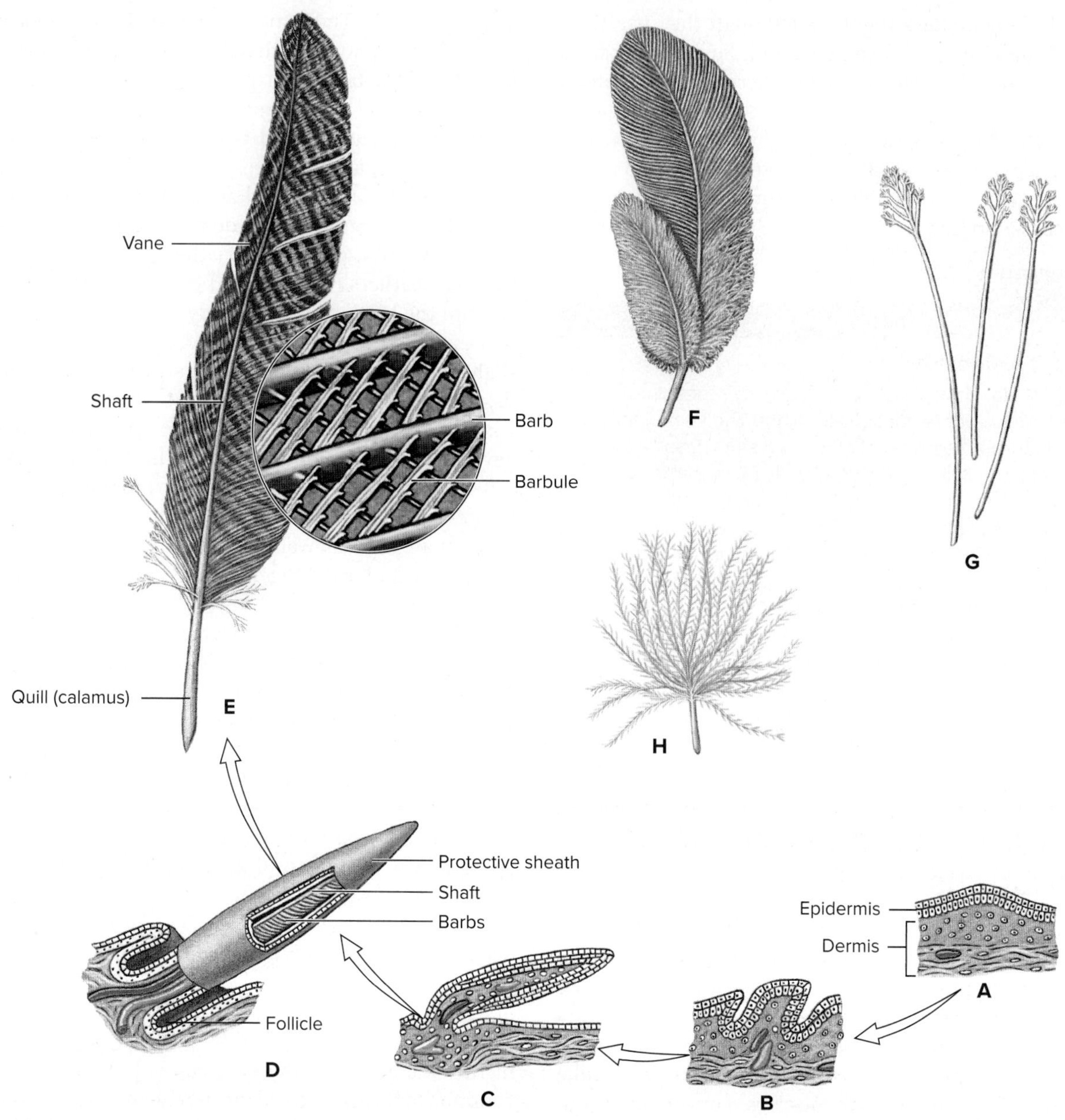

figure 19.4

Types of bird feathers and their development. **A** to **E,** Successive stages in development of a vaned, or contour, feather. Growth occurs within a protective sheath, **D,** that splits open when growth is complete, allowing the mature feather to spread flat. **F** to **H,** Other feather types, including a pheasant feather with aftershaft, **F;** filoplumes, **G;** and down feathers, **H.**

to see any trace of the original diapsid condition. A bird's skull is built lightly and mostly fused into one piece. A pigeon's skull weighs only 0.21% of its body weight; by comparison, a rat's skull weighs 1.25% of its body weight. The braincase and orbits are large in bird skulls to accommodate a bulging brain and large eyes needed for coordination of flight muscles and superior vision.

In *Archaeopteryx,* both jaws contained teeth set in sockets, an archosaurian characteristic. Modern birds are completely toothless, having instead a keratinous beak molded around the bony jaws. The mandible is a complex of several bones hinged to provide a double-jointed action, which permits the mouth to gape widely. Most birds have kinetic skulls (see figure 18.11 for a kinetic lizard skull) with a flexible attachment between upper jaw and skull.

The most distinctive feature of the vertebral column is its rigidity. Most vertebrae, except the **cervicals** (neck vertebrae), are fused together, and along with the pelvic girdle form a stiff but light framework to support the legs and to provide rigidity for flight. To assist in this rigidity, ribs are braced against each other with uncinate processes (see figure 19.5A). Except in flightless birds, the sternum bears a large, thin keel that provides an attachment for powerful flight muscles. Fused clavicles form an elastic **furcula** (the "wishbone" of a turkey or chicken)

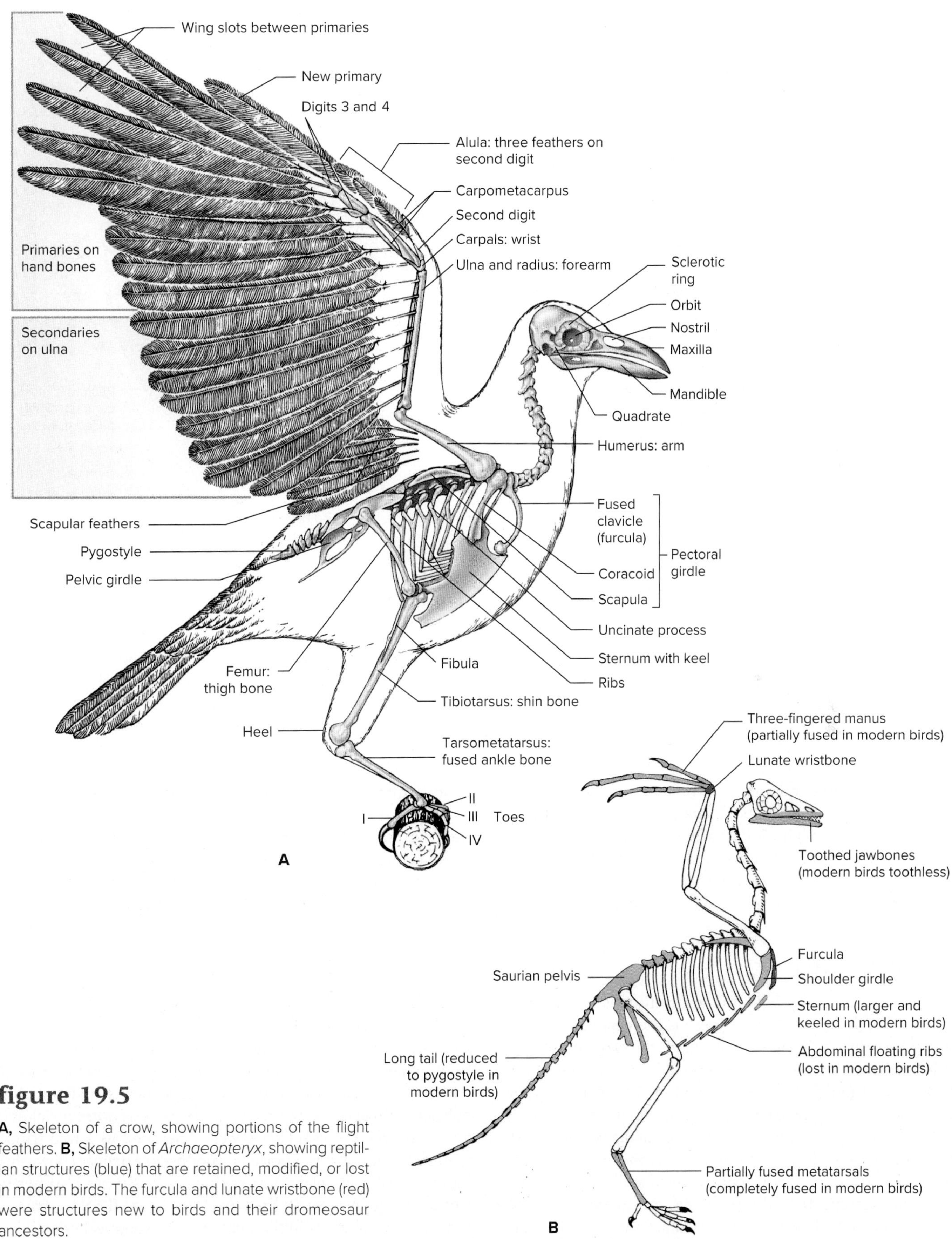

figure 19.5

A, Skeleton of a crow, showing portions of the flight feathers. **B,** Skeleton of *Archaeopteryx*, showing reptilian structures (blue) that are retained, modified, or lost in modern birds. The furcula and lunate wristbone (red) were structures new to birds and their dromeosaur ancestors.

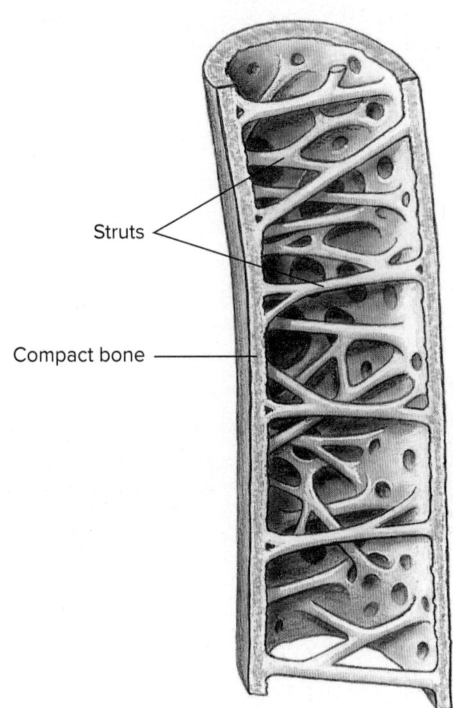

figure 19.6

Hollow wing bone of a songbird, showing stiffening struts and air spaces that replace bone marrow. Such pneumatized bones are remarkably light and strong.

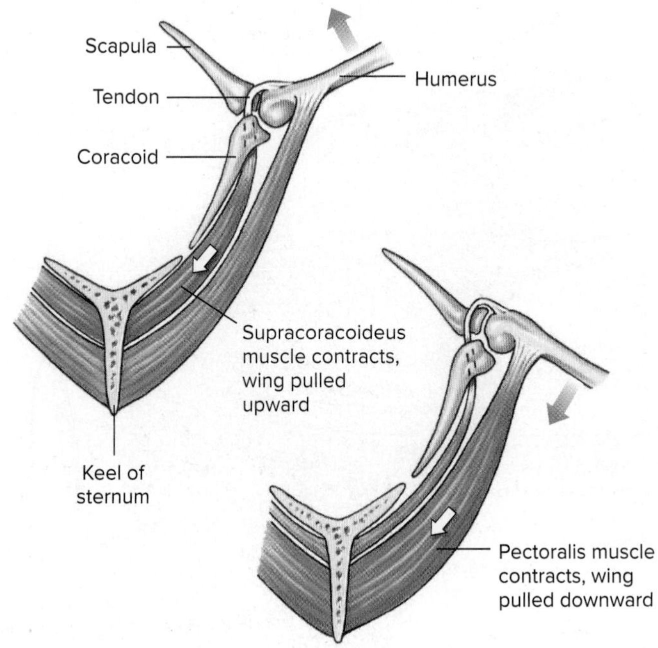

figure 19.7

Flight muscles of a bird are arranged to keep the center of gravity low in the body. Both major flight muscles are anchored to the sternum keel. Contraction of the pectoralis muscle pulls the wing downward. Then, as the pectoralis relaxes, the supracoracoideus muscle contracts and, acting as a pulley system, pulls the wing upward.

that apparently stores energy as it flexes during wing beats. The asymmetrical flight feathers and distinct furcula of *Archaeopteryx,* along with the anatomy of its brain and inner ear, suggest it had some flight abilities. However, it would have been a weak flier, because its small sternum offered relatively little area for flight muscle attachment (see figure 19.5B).

Bones of the forelimbs are highly modified for flight. They are reduced in number, and several are fused together. Despite these alterations for flight, a bird wing is clearly a rearrangement of the basic vertebrate tetrapod limb from which it arose (see figure 17.1), and all the elements—arm, forearm, wrist, and fingers—are represented in modified form (see figure 19.5A).

Muscular System

Locomotor muscles of wings are relatively massive to meet demands of flight. The largest of these is the **pectoralis,** which depresses the wings in flight. Its antagonist is the **supracoracoideus** muscle, which raises the wing (figure 19.7). Surprisingly, perhaps, this latter muscle is not located on the backbone (anyone who has been served the back of a chicken knows that it offers little meat) but is positioned under the pectoralis on the breast. It is attached by a tendon to the upper side of the humerus of the wing so that it pulls from below by an ingenious "rope-and-pulley" arrangement. Both the pectoralis and the supracoracoideus are anchored to the keel of the sternum. Positioning the main muscle mass low in the body improves aerodynamic stability.

From the main leg muscle mass in the thigh, thin but strong tendons extend downward through sleevelike sheaths to the toes. Consequently, the feet are nearly devoid of muscles, explaining the thin, delicate appearance of a bird's leg. This arrangement places the main muscle mass near a bird's center of gravity and at the same time allows great agility to the slender, lightweight feet. Because the feet are composed mostly of bone, tendon, and tough, scaly skin, they are highly resistant to damage from freezing. When a bird perches on a branch, an ingenious toe-locking mechanism (figure 19.8) is activated, which prevents the bird from falling off its perch when asleep. The same mechanism causes the talons of a hawk or owl to sink deeply into its prey as the legs bend under the impact of the strike.

Digestive System

Beaks of birds are strongly adapted to specialized food habits—from generalized types, such as the strong, pointed beaks of crows and ravens, to highly specialized beaks in flamingos, pelicans, and avocets (figure 19.9). The beak of a woodpecker is a straight, hard, chisel-like device. Anchored to a tree trunk, with its tail serving as a brace, a woodpecker delivers powerful, rapid blows to excavate nest cavities or to expose burrows of wood-boring insects. It then uses its long, flexible, barbed tongue to seek insects in their galleries. A woodpecker's skull is especially thick to absorb shock.

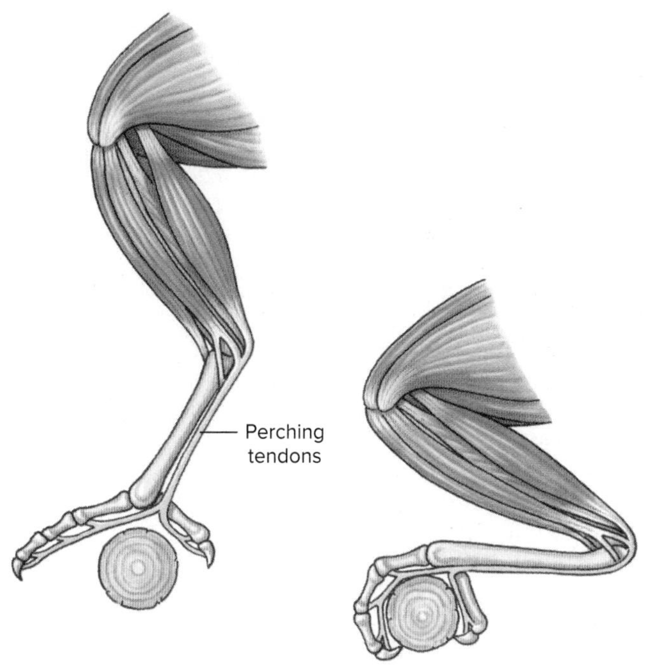

figure 19.8

Perching mechanism of a bird. When a bird settles on a branch, tendons automatically tighten, closing the toes around the perch.

Birds process an energy-rich diet rapidly and thoroughly with efficient digestive equipment. A shrike can digest a mouse in 3 hours, and berries pass completely through the digestive tract of a thrush in just 30 minutes. Although many animal foods find their way into diets of birds, insects constitute by far the largest component. Because birds lack teeth, foods that require grinding are reduced in the gizzard. Many birds have an enlargement **(crop)** at the lower end of the esophagus, which serves as a storage chamber. In pigeons, doves, and some parrots, the crop not only stores food but also produces a lipid- and protein-rich "milk," composed of sloughed epithelial cells of the crop lining. For a few days after hatching, the young are fed regurgitated crop milk by both parents.

The stomach proper consists of a **proventriculus,** which secretes gastric juice, and a muscular **gizzard,** a region specialized for grinding food. To assist in grinding food, birds swallow gritty objects or pebbles, which lodge in the gizzard. A turkey's gizzard is especially powerful and can break up acorns, hickory nuts, and walnuts. Certain birds of prey, such as owls, form pellets of indigestible materials, mainly bones and fur, in the proventriculus and eject them through the mouth. At the junction of the small intestine with the colon are paired **ceca;** these are well developed in herbivorous birds in which they serve as fermentation chambers. The terminal part of the digestive system is the **cloaca,** which also receives reproductive ducts and ureters.

Circulatory System

The general plan of circulation in birds is not greatly different from that of mammals, although it evolved independently. Their four-chambered heart is large with strong ventricular walls. Birds share with mammals a complete separation of

figure 19.9

Some beaks of birds showing a variety of adaptations.

respiratory and systemic circulations. Their heartbeat is extremely fast, and as in mammals an inverse relationship occurs between heart rate and body weight. For example, a turkey has a heart rate at rest of about 93 beats per minute, a chicken has a rate of 250 beats per minute, and a black-capped chickadee has a rate of 500 beats per minute when asleep, which may increase to a phenomenal 1000 beats per minute during exercise. Blood pressure in birds is roughly equivalent to that in mammals of similar size. Birds' blood contains nucleated, biconvex erythrocytes, different from the enucleate, biconcave erythroctes of mammals.

Respiratory System

The respiratory system of birds differs radically from that of non-avian reptiles and mammals and is marvelously adapted for meeting the high metabolic demands of flight. In birds, the finest branches of the bronchi, rather than ending in saclike alveoli as in mammals, are tubelike **parabronchi** through which air flows continuously. The parabronchi form the lungs of birds. Also unique is the extensive system of nine interconnecting air sacs that are located in pairs in the thorax and abdomen and even extend by tiny tubes into the centers of long bones (figure 19.10A). Air sacs connect to the lungs in such a way that most of the inspired air bypasses the lungs and flows directly into the posterior air sacs, which serve as reservoirs for fresh air. On expiration, this oxygenated air is passed through the lungs. A second respiratory cycle draws air from the lungs to the anterior air sacs and then to the outside. Thus, it takes two respiratory cycles for a single breath of air to pass through the respiratory system (figure 19.10B). The advantage of such a system is that an almost continuous stream of oxygenated air is passed through a system of richly vascularized parabronchi. In contrast, the lungs of other tetrapods operate as dead-end sacs; thus fresh air mixes with old air that remains in the lungs between breaths.

Key Theme 19.2

ADAPTATION AND PHYSIOLOGY

High-Altitude Flying

The remarkable efficiency of a bird's respiratory system is emphasized by bar-headed geese, *Anser indicus*, that routinely migrate over the Himalayan mountains at altitudes up to 7300 m (24,000 ft), under conditions that are hypoxic to humans. The geese reach these altitudes in less than a day, without the acclimatization that is absolutely essential for humans even to approach the upper reaches of the Himalayas.

Excretory System

Urine is formed in the relatively large, paired kidneys. Urine passes by way of tubes called ureters to the **cloaca.** There is no urinary bladder.

Like other reptiles, birds excrete their nitrogenous wastes as uric acid. In shelled eggs, all excretory products must remain within the eggshell with the growing embryo. Uric acid crystallizes from solution and can be stored harmlessly in the allantois (see figure 18.3). Because of uric acid's low solubility, a bird can excrete 1 g of uric acid in only 1.5–3 ml of water, whereas a mammal can require 60 ml of water to excrete 1 g of urea. Concentration of uric acid occurs almost entirely in the cloaca, where it is combined with fecal matter, and the water is reabsorbed. Bird kidneys are much less efficient than those of mammals in removing salts, especially sodium, potassium, and chloride. The kidneys of most birds concentrate salts only

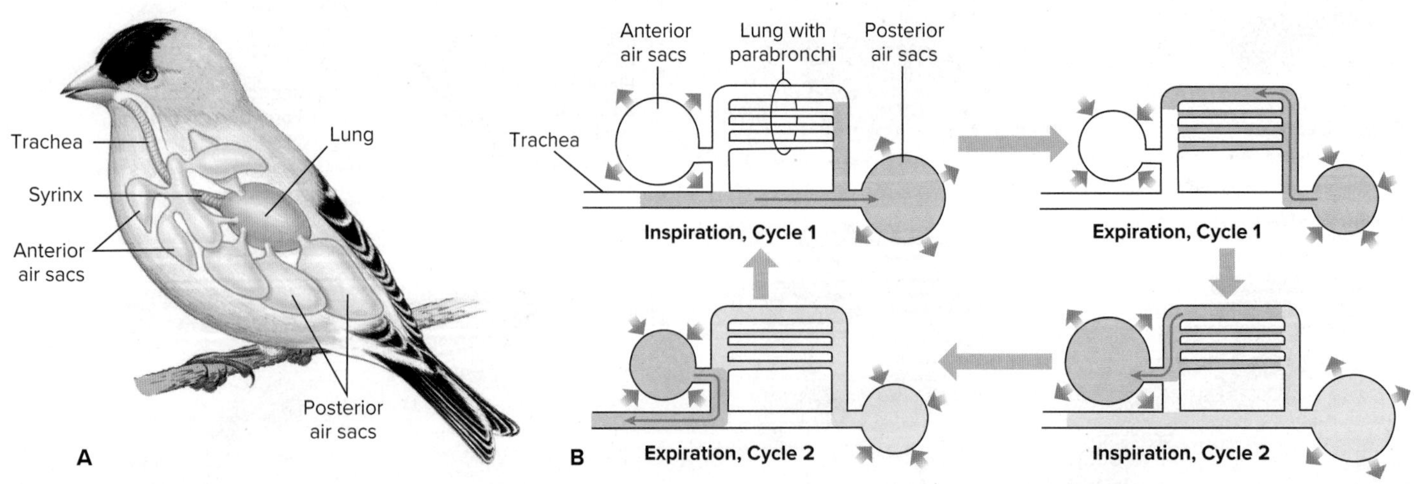

figure 19.10

Respiratory system of a bird. **A,** Lungs and air sacs. One side of the bilateral air sac system is shown. **B,** Movement of a single volume of air through a bird's respiratory system. Two full respiratory cycles are required to move air through the system. The first breath of air is shown in dark blue, while the second breath of air is shown in light blue. Note that air actually moves continually through the parabronchi, but we depict individual phases to make air flow easier to follow.

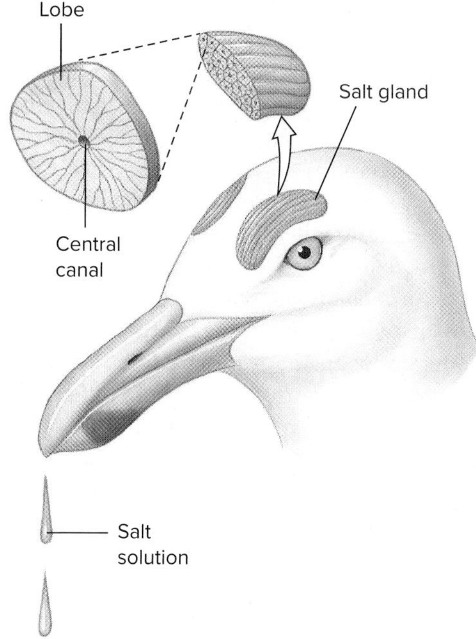

figure 19.11

Salt glands of a marine bird (gull). One salt gland is located above each eye. Each gland consists of several lobes arranged in parallel. One lobe is shown in cross section, much enlarged. Salt is secreted into many radially arranged tubules, and then flows into a central canal that leads into the nose.

slightly greater than that of blood, but most mammals concentrate salts 4–8 times that of blood, and some desert rodents concentrate salts to nearly 25 times that of blood.

To compensate for the weak solute-concentrating ability of the kidneys, some birds, especially marine birds, use additional methods to excrete salts gained from the food they eat and the seawater they drink. **Salt glands,** one located above each eye (figure 19.11), excrete a highly concentrated solution of sodium chloride, up to twice the concentration of seawater. The salt solution runs out the internal or external nostrils, giving gulls, petrels, and other sea birds a perpetually runny nose.

Nervous and Sensory System

The design of a bird's nervous and sensory system reflects the complex problems of flight and a highly visible existence, in which the bird must gather food, mate, defend territory, incubate and rear young, and correctly distinguish friend from foe. Its brain has well-developed **cerebral hemispheres, cerebellum,** and **optic lobes.** The **cerebral cortex**—chief coordinating center of a mammalian brain—is thin, unfissured, and poorly developed in birds. The core of the cerebrum, the **dorsal ventricular ridge,** has enlarged into the principal integrative center, controlling such activities as eating, singing, flying, and complex reproductive behavior. Relatively intelligent birds, such as crows and parrots, have larger cerebral hemispheres than do less intelligent birds, such as chickens and pigeons. The cerebellum, much larger in birds than in other reptiles, coordinates muscle position, equilibrium, and visual information used in movement and balance. The optic lobes, laterally bulging structures of the midbrain, organize visual information.

The senses of smell and taste are poor in some birds, but they are relatively well developed in many other birds, such as carnivorous birds, flightless birds, oceanic birds, and waterfowl. Birds have good hearing and superb vision, the keenest in the animal kingdom. The organ of hearing, the **cochlea,** is much shorter than the coiled mammalian cochlea, and yet birds can hear roughly the same range of sound frequencies as humans. Actually, a bird's ear far surpasses that of humans in its capacity to distinguish differences in intensities and to respond to rapid fluctuations in pitch.

A bird's eye resembles that of other vertebrates in gross structure but is relatively larger, less spherical, and almost immobile; instead of turning their eyes, birds turn their heads with their long flexible necks to scan the visual field. The light-sensitive **retina** (figure 19.12) is generously equipped with rods (for dim light vision) and cones (for good acuity and color vision). Cones predominate in diurnal (day-active) birds, and rods are more numerous in nocturnal birds. A distinctive feature of a bird's eye is the **pecten,** a highly vascularized organ attached to the retina and jutting into the vitreous humor (figure 19.12). The pecten is thought to provide nutrients and oxygen to the eye.

The **fovea,** a region of keenest vision on the retina, is in a deep pit (in birds of prey and some others), which requires the bird to focus exactly on the subject. Many birds, moreover, have two sensitive spots (foveae) on the retina (figure 19.12): a central one for sharp monocular views and a posterior one for binocular vision. The visual acuity of a hawk is about eight times that of humans (enabling a hawk to see clearly a crouching rabbit 2 km away), and an owl's ability to see in dim light is more than 10 times that of a human.

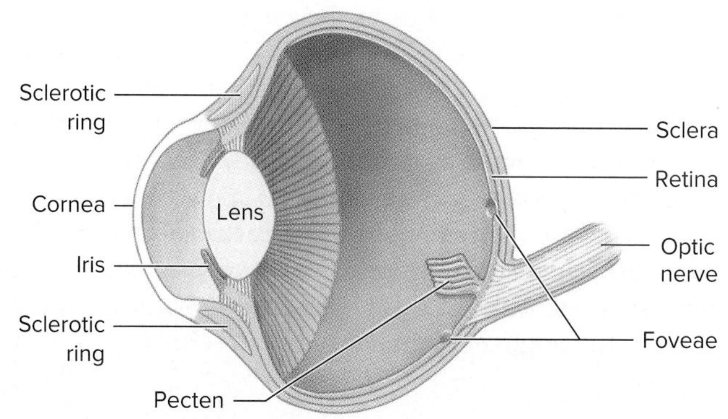

figure 19.12

A hawk's eye has all the structural components of a mammalian eye, plus a pleated structure, the pecten, thought to provide nourishment to the retina. The extraordinarily keen vision of hawks is attributed to the extreme density of cone cells in the foveae: 1.5 million per fovea compared to 0.2 million for humans.

Key Theme 19.3
ADAPTATION AND PHYSIOLOGY

Ultraviolet Vision

Many birds can see at ultraviolet wavelengths, enabling them to view environmental features inaccessible to humans but accessible to insects (such as flowers with ultraviolet-reflecting "nectar guides" that attract pollinating insects). Several species of ducks, hummingbirds, kingfishers, and passerines (songbirds) can see at near ultraviolet (UV) wavelengths down to 370 nm (human eyes filter out ultraviolet light below 400 nm). For what purpose do birds use their UV sensitivity? Some, such as hummingbirds, may be attracted to nectar-guiding flowers, but for others, the benefit derived from UV sensitivity is unknown.

19.3 Flight

Flight allows birds to capture airborne prey, to escape predators, and to migrate between areas offering seasonally available resources or favorable climates. How did flight evolve?

Two competing hypotheses of the origin of bird flight have been offered: (1) birds began to fly either by climbing to a high place and gliding down or (2) by flapping their wings to launch into the air from the ground. The first hypothesis, termed arboreal, or "trees down," has been long favored. Proponents of this view envision an arboreal ancestor of *Archaeopteryx* gliding from tree to tree, or perhaps "pouncing" on prey below using wings to control its attack. Modifications permitting lift and powered flight would be highly advantageous for this kind of life. Indeed, there are many arboreal squirrels and lizards that use gliding to move among trees. A weakness of this hypothesis is that few feathered dromaeosaurs were arboreal.

Proponents of the cursorial, or "ground-up," hypothesis suggest that the feathered wings of bipedal, ground-dwelling ancestors of the first flying birds may have been used as snares to capture insects or to refine aerodynamic control during leaps to capture flying insects. Thus, as the wings became larger, they would have been capable of powered flight. However, a ground launch requires working against gravity rather than enlisting its help! No living gliders launch from the ground. Bats cannot launch upward from the ground; instead, they must take to flight from an elevated surface. A slightly more convincing scenario is suggested by studies of chukar partridge chicks, which use wingbeats to assist running up steep inclines. Although the evidence is weighted toward the arboreal hypothesis, the debate about the origin of flight has not been settled.

Aerodynamics of Flight

To become airbone, a bird must generate lift forces greater than its own mass and to move forward it must generate thrust to move against resistive forces of drag. It uses its wings to do both. A wing is streamlined in cross section, with a slightly concave lower surface **(cambered)** with small, tight-fitting feathers where the leading edge meets the air. Air slips smoothly over the wing, creating lift with minimal drag. In general, the outer part of the wing, the modified hand bones with the attached **primary feathers** (see figure 19.5A), provides the thrust needed to move the bird forward against the resistive forces of friction. The inner, highly cambered, part of the wing, with the **secondary feathers** and associated forearm, has less vertical motion than the outer wing, and mainly acts as an airfoil, generating lift.

Wings create lift by deflecting air downward and creating higher air pressure below the wing than above the wing. A flying bird holds its wings at an angle, so that the leading edge of the wing is higher than the trailing edge (figure 19.13A). Air moving past the wing, from front to back, deflects downward. In accordance with Newton's Third Law (for every action there is an equal but opposite reaction), the wing (and the bird) deflects upward. You can demonstrate this by moving your hand, held at an angle, through a pool of water. Lift also is created by the wing's camber or angle, which creates a low-pressure area on the top of the wing, where air moves the fastest (Bernoulli's principle). Thrust is generated when a bird's outer wing moves down and slightly forward, rotating so that the leading edge is below the trailing edge. As the wing moves downward, the primaries bite through the air like a propeller, displacing the air backward and propelling the bird forward. The wing folds slightly during the upstroke, and it returns to its original position with a minimum of drag (figure 19.14). Birds that hover have an especially powerful upstroke and orient their wing motions in a way that produces lift, but not thrust (figure 19.15).

The morphology of a bird's wings is related to the bird's typical flight speed. At high flight speeds sufficient lift is generated so that the wing is held at a low angle of attack (see figure 19.13A), creating less drag. As speed decreases, lift can be increased by increasing the angle of attack, but lift-destroying turbulence can appear (see figure 19.13B), causing stalling. Turbulence can be prevented by **wing slots,** which direct a layer of rapidly moving air across the upper wing surface (see figure 19.13C). Wing slots are formed by an **alula,** a group of small feathers on the thumb (see figure 19.5) or gaps between the primary feathers. Eddies of air that form at the tips of wings, called **wing-tip vortexes,** are particularly problematic at high speeds because they create drag (see figure 19.13D). This drag is reduced in long, slender wings with pointed tips **(high-aspect ratio wings)**.

Basic Forms of Bird Wings

Bird wings vary in size and form because successful exploitation of different habitats has imposed special aerodynamic requirements. Four types of bird wings are recognized.

Elliptical Wings

Birds such as warblers, sparrows, doves, woodpeckers, and magpies, which must maneuver in forested or brushy habitats,

Air moves faster here, and exerts less pressure on wing

α

Angle of attack

Air flowing past the wing is deflected downward, displacing the wing upward

A Air flow around wing

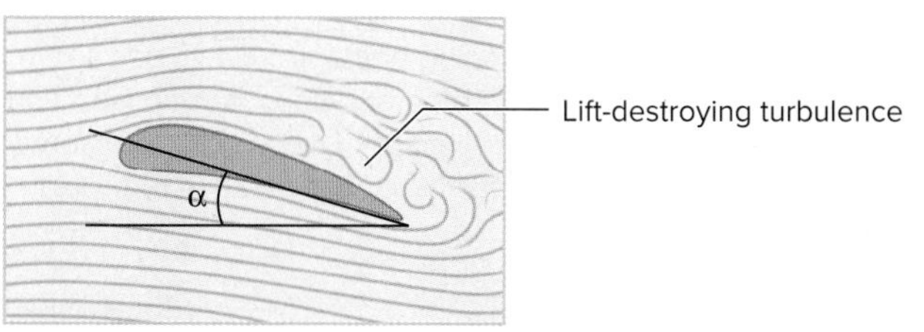

B Stalling at low speed

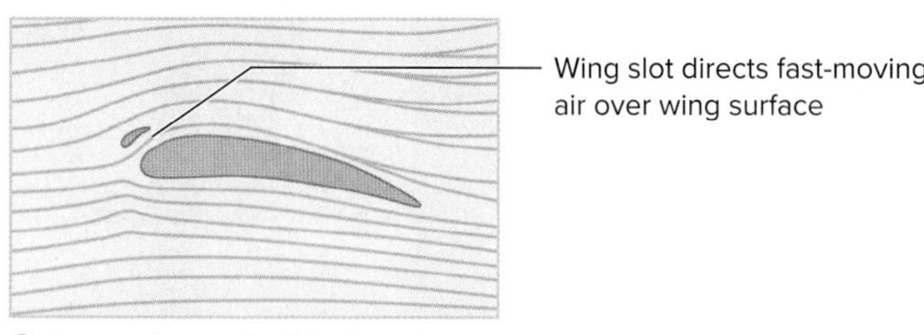

C Preventing stall with wing slots

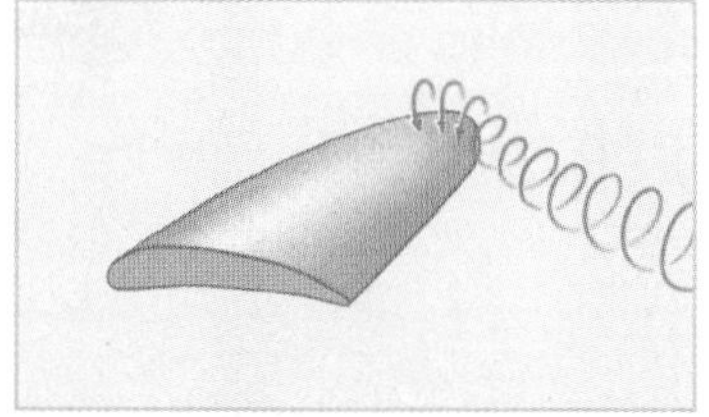

D Formation of wing tip vortex

figure 19.13

A, Air patterns formed by the airfoil, or wing, moving from right to left. **B,** At low speed, the angle of attack (α) must increase to maintain lift, but this increases the threat of stalling. **C,** Low-speed stalling can be prevented with wing slots. **D,** Wing tip vortex, a turbulence that tends to develop at high speeds, reduces flight efficiency. The effect is reduced in slender wings that sweep back and taper to a tip.

have elliptical wings (figure 19.16A). This type has a **low-aspect ratio** (ratio of length to average width). Wings of the highly maneuverable British Spitfire fighter plane of World War II fame conformed closely to the outline of a sparrow's wing. Elliptical wings have both an alula and gaps between the primary feathers; this arrangement helps to prevent stalling during sharp turns, low-speed flight, and frequent landing and takeoff. Each separated primary feather behaves as a narrow wing with a high angle of attack, providing high lift at low speed. The high maneuverability of elliptical wings is exemplified by the tiny chickadee, which can change course within 0.03 of a second.

High-Speed Wings

Birds that feed during flight, such as swallows, falcons, and swifts, or those that make long migrations, such as plovers, sandpipers, terns, and gulls (figure 19.16B), have wings that sweep back and taper to a slender tip. They are rather flat in section, have a **high-aspect ratio,** and lack wing-tip slotting. Sweepback and wide separation of wing tips reduce drag from wing-tip vortices. This type of wing is aerodynamically efficient for high-speed flight but cannot easily keep a bird airborne at low speeds, except for hummingbirds, which rapidly move their wings in a specialized way to hover (see figure 19.15). The fastest birds, such as sandpipers, clocked at 175 km (109 miles) per hour, belong to this group.

Active Soaring Wings

Oceanic soaring birds, including albatrosses, shearwaters, and gannets (figure 19.16C), also have high-aspect ratio wings, shaped like those of sailplanes. Such long, narrow wings lack slots and are adapted for **active (dynamic) soaring.** Active soaring can be done only over seas with strong, reliable winds, and it exploits different wind speeds near the ocean surface (slow) and well above the surface (fast). Passive soaring, by contrast, occurs when a bird uses rising columns of warm air for lift, as described in the next section. A bird that uses dynamic soaring begins a downwind glide from an elevated position, gaining speed as it descends. Near the surface of the ocean, it turns into

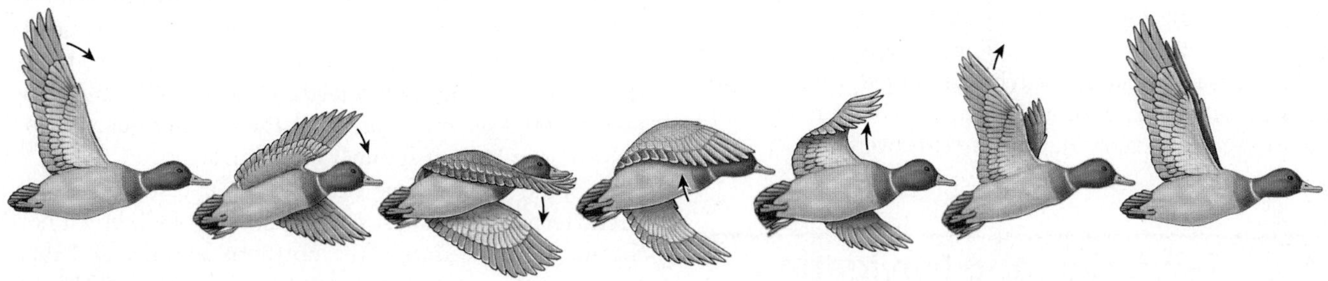

figure 19.14

In the normal flapping flight of strong fliers such as ducks, the wings sweep downward and forward fully extended. Thrust is provided by primary feathers at the wing tips. To begin the upbeat, the wing is bent, bringing it upward and backward. The wing then extends, ready for the next downbeat.

figure 19.15

The secret of a hummingbird's ability to change direction instantly, or to hang motionless in the air while sipping nectar from a flower, lies in its wing structure. The wing is nearly rigid, but hinged at the shoulder by a swivel joint and powered by a supracoracoideus muscle (see figure 19.7) that is unusually large for the bird's size. When hovering, the wing moves in a sculling motion. The leading edge of the wing moves forward on the forward stroke, and then swivels nearly 180 degrees at the shoulder to move backward on the backstroke. The effect is to provide lift without propulsion on **both** forward and backward strokes.

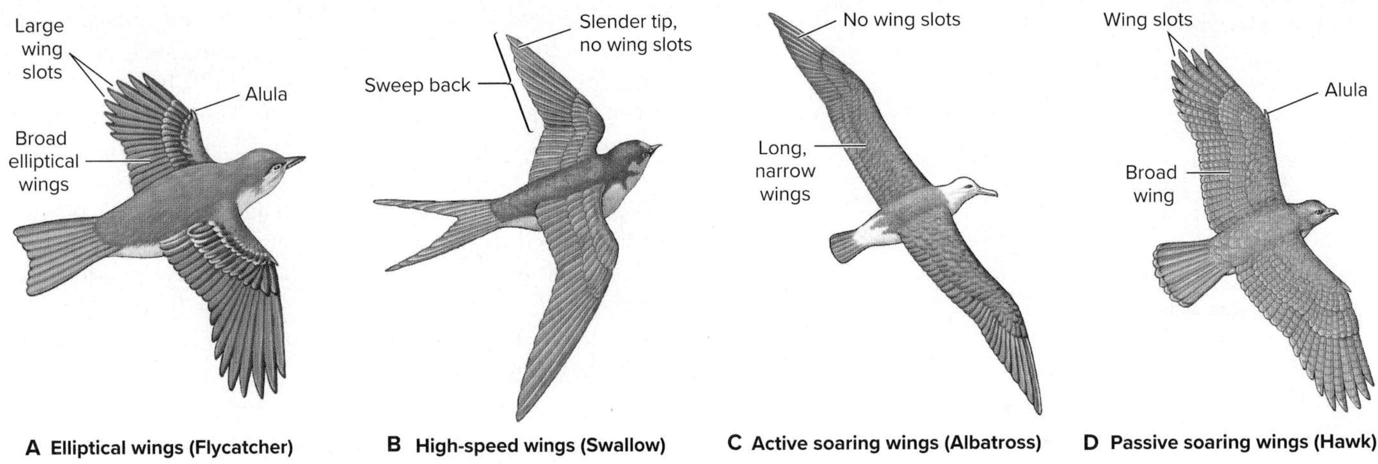

figure 19.16

Four basic forms of bird wings.

the wind and rises into stronger winds. Although its velocity relative to the ocean slows, the strong winds over its wings provide the lift to keep it aloft.

Passive Soaring Wings

Hawks, vultures, eagles, owls, and ospreys (figure 19.16D)—predators that carry heavy loads—have large wings with high camber, promoting high lift. Their wings have alulas and primary-feather slots, preventing stalling at low speeds. Many of these birds are land soarers, with broad, slotted wings that provide the sensitive response and maneuverability required for static soaring in the capricious air currents over land.

19.4 Migration and Navigation

We described advantages of migration in the prologue to this chapter. Not all birds migrate, of course, but most North American and European species do, and the biannual journeys of some are truly extraordinary undertakings.

Migration Routes

Most migratory birds have well-established routes trending north and south. Because most migratory birds breed in the Northern Hemisphere, where most of the earth's landmass is concentrated, most birds migrate south for the northern winter and north to nest in the northern summer (figure 19.17). Some complete their migratory routes in a very short time. Bar-tailed godwits, *Limosa lapponica,* fly non-stop, 11,000 km from Alaska to New Zealand, relying on large stores of body fat to fuel their nine-day journey. Others, however, make a leisurely trip, often stopping along the way to feed. Some warblers are known to take 50–60 days to migrate from their winter quarters in Central America to their summer breeding areas in Canada.

Some species have extremely long-distance migrations. Arctic terns, the greatest globe spanners of all, breed north of the Arctic Circle during the northern summer and then migrate to Antarctic regions for the northern winter. This species is also known to take a circuitous route in migrations from North America, flying to the coastlines of Europe and Africa and then to winter quarters, a trip that may exceed 18,000 km (11,200 miles).

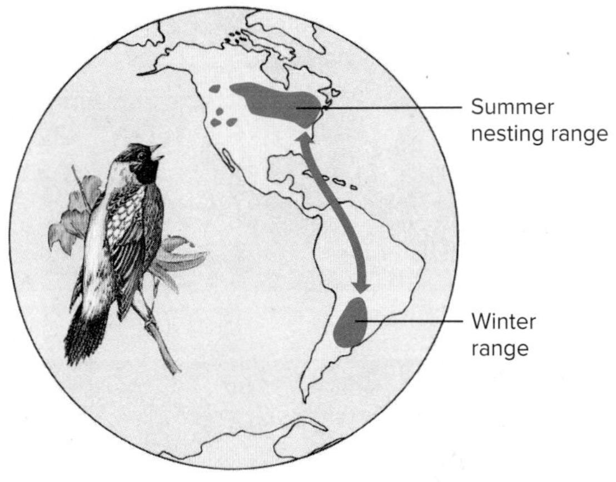

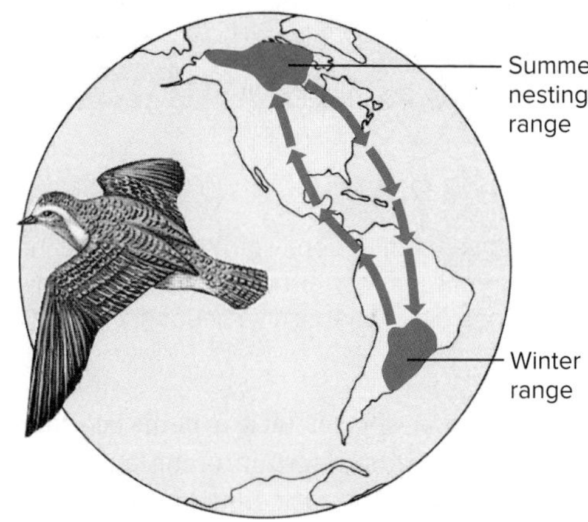

figure 19.17

Migrations of bobolinks, *Dolichonyx oryzivorus,* and American golden plovers, *Pluvialis dominica.* Bobolinks commute 22,500 km (14,000 miles) each year between their nesting sites in North America and their range in Argentina, where they spend the northern winters, a phenomenal feat for such a small bird. Although the breeding range has extended to colonies in western areas, these birds take no short-cuts but adhere to the ancestral seaboard route. American golden plovers fly a loop migration, striking out across the Atlantic in their southward autumnal migration but returning in the spring by way of Central America and the Mississippi Valley because ecological conditions are more favorable at that time.

Many small songbirds, such as warblers, vireos, thrushes, flycatchers, and sparrows, as well as shorebirds, such as sandpipers and plovers, also make great migratory treks (figure 19.17). Migratory birds that nest in Europe or Central Asia spend the northern winter in Africa.

Stimulus for Migration

People have known for centuries that the onset of reproductive cycles in birds is closely related to season. It has been demonstrated that lengthening days of late winter and early spring stimulate development of gonads and accumulation of fat—both important internal changes that predispose birds to migrate northward. Increasing day length stimulates the anterior lobe of the pituitary into activity. Release of pituitary gonadotropic hormone in turn sets in motion a complex series of physiological and behavioral changes, stimulating gonadal growth, fat deposition, migration, courtship and mating behavior, and care of the young.

Direction Finding in Migration

Numerous experiments suggest most birds navigate chiefly by sight. Birds recognize topographical landmarks and follow familiar migratory routes—a behavior assisted by flock migration, during which the navigational resources and experience of older birds can be pooled. In addition to visual navigation, birds use a variety of orientation cues. Birds have a highly accurate sense of time. Numerous studies support a hypothesis that birds can navigate by the earth's magnetic field. In the early 1970s, W. T. Keeton showed that flight bearings of homing pigeons were significantly disturbed by magnets attached to the birds' heads, or by minor fluctuations in the geomagnetic field. Deposits of a magnetic substance called magnetite (Fe_3O_4) have been discovered in the beaks of pigeons. Recent experiments showed that a pigeon could discriminate between the presence and absence of a magnetic anomaly, but not when its upper beak was anesthetized, nor when its trigeminal nerve, which innervates the upper beak, was severed.

Experiments by German ornithologists G. Kramer and E. Sauer and American ornithologist S. Emlen demonstrated convincingly that birds can navigate by celestial cues: the sun by day and the stars by night. Using special circular cages, Kramer concluded that birds maintain compass direction by referring to the sun, regardless of the time of day (figure 19.18). This process is called **sun-azimuth orientation** (Arabic *azimuth,* compass bearing of the sun). Sauer's and Emlen's ingenious planetarium experiments, in which birds that were exposed to a night sky that could be experimentally rotated, also strongly suggest that some birds, probably many, are able to navigate by the North Star axis around which the constellations appear to rotate.

19.5 Social Behavior and Reproduction

The adage says "birds of a feather flock together," and many birds are indeed highly social creatures. Especially during the breeding season, sea birds gather, often in enormous colonies, to nest and to rear young. Land birds, with some conspicuous exceptions (such as starlings and rooks), tend to be less gregarious than sea birds during breeding and to seek isolation for rearing their brood. Nonetheless, species that covet separation from their kind during breeding may aggregate for migration or feeding. Togetherness offers advantages: mutual protection from enemies, greater ease in finding mates, less opportunity

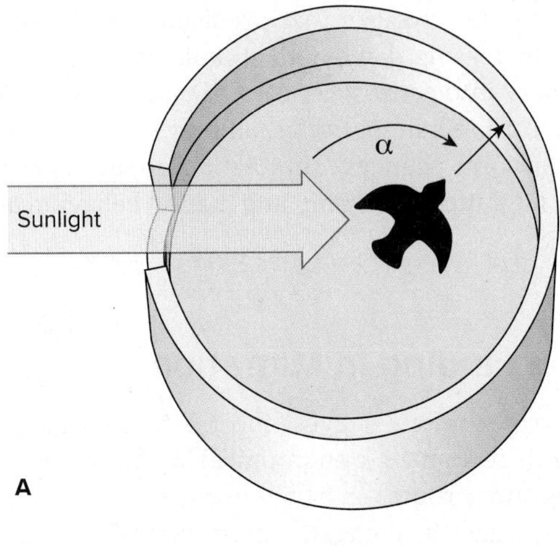

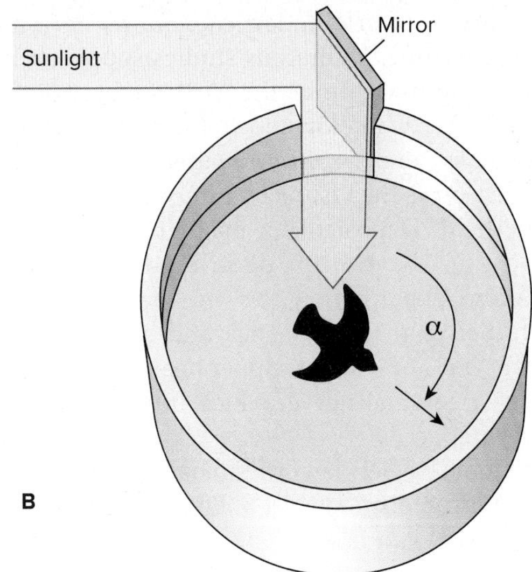

figure 19.18

Gustav Kramer's experiments with sun-compass navigation in starlings. **A,** In a windowed, circular cage, the bird fluttered to align itself in the direction it would normally follow if it were free. **B,** When the true angle of the sun is deflected with a mirror, the bird maintains the same relative position to the sun. This shows that these birds use the sun as a compass. The bird orients correctly throughout the day, changing its orientation to the sun as the sun moves across the sky.

imageBROKER/Alamy Stock Photo

Dorit Bar-Zakay/Flickr Open/Getty Images

figure 19.19

Cooperative feeding behavior by white pelicans, *Pelecanus onocrotalus.* **A,** Pelicans form a horseshoe to drive fish together. **B,** Then they plunge simultaneously to scoop fish in their huge beaks. Order Pelecaniformes.

for individual straying during migration, and mass huddling for protection against low night temperatures during migration. Certain species, such as pelicans (figure 19.19), use cooperative behavior to feed. At no time are the highly organized social interactions of birds more evident than during the breeding season, as they establish territorial claims, select mates, build nests, incubate and hatch their eggs, and rear their young.

Reproductive System

During most of the year, testes of male birds are tiny, bean-shaped bodies, but during the breeding season they enlarge greatly, to as much as 300 times their nonbreeding size. Because males of most species lack a penis, copulation is a matter of bringing cloacal surfaces into contact, usually while the male stands on the back of the female (figure 19.20).

In females of most birds, only the left ovary and oviduct develop (figure 19.21); those on the right dwindle to vestigial

Cleveland P. Hickman, Jr.

figure 19.20

Copulation in waved albatrosses, *Diomeda irrorata.* In most bird species, males lack a penis. A male passes sperm by standing on the back of a female and pressing his cloaca against that of the female. Order Procellariiformes.

structures (loss of one ovary might be an adaptation for reducing weight). Eggs discharged from the ovary enter the oviduct, where fertilization occurs. Several hours later, while eggs are passing down the oviduct, **albumin,** or egg white, from special glands is added to them; farther down the oviduct, shell membrane, shell, and shell pigments are secreted around the egg. Sperm remain alive in the female oviduct for many days after a single mating.

Key Theme 19.4

HUMAN CONNECTIONS

DDT and Thinning of Egg Shells

Upper-level carnivores, species at the top of the food chain, are vulnerable to decline through biomagnification of toxins. After World War II, DDT was widely used to control insect populations, especially mosquitoes, which transmit malaria, in the United States. Brown pelicans, hawks, eagles, ospreys, and other raptorial birds consumed prey containing DDT, which then became concentrated in the birds' bodies. A primary effect on these birds was thinning of eggshells, possibly because DDT (or its metabolite, DDE) interferes with movement of calcium from blood into the shell glands. The fragile eggs often broke before hatching and these birds were in great decline by the middle of the 20th century. Public outcry led to banning of DDT use in 1972, in part championed by Rachel Carson's *Silent Spring,* which warned of the dangers of DDT. Subsequently, most raptor populations rebounded, including bald eagles, leading to their removal from the U.S. Fish and Wildlife Service's list of threatened and endangered wildlife. However, DDT is persistent in the environment and continues to affect some birds. In 2013, shell thinning was still a problem in certain California condor populations, which feed on sea lions that forage in the highly contaminated Palo Verdes Shelf.

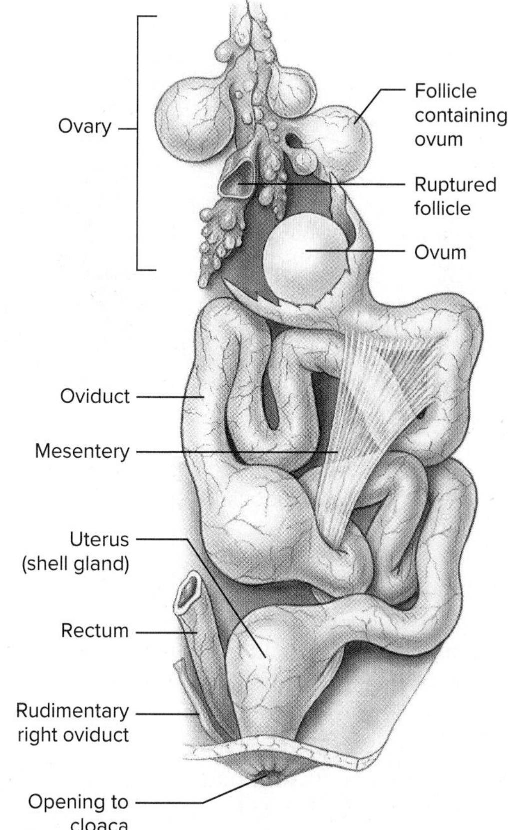

figure 19.21

Reproductive system of a female bird. For most birds, only the left ovary and reproductive tract are functional. The corresponding structures on the right dwindle to vestiges.

Mating Systems

Two types of mating systems are **monogamy,** in which an individual has only one mate, and **polygamy,** in which an individual has more than one mate during a breeding period. Monogamy is rare in most animal groups, but it is common in birds; more than 90% are monogamous. A few bird species, such as swans and geese, choose partners for life and often remain together throughout the year. Seasonal monogamy is more common because most migrant birds pair only during the breeding season, living independent lives the rest of the year and perhaps choosing a different mate the next breeding season.

One reason monogamy is much more common among birds than among mammals is perhaps because male and female birds are equally adept at most aspects of parental care. Because female mammals gestate the young and feed them by lactation, they provide important kinds of parental care that males cannot. Female and male birds can alternate care of the nest and young, which permits one parent to be at the nest at all times. For some species, a female remains on the nest for months at a time and is fed by the male. This constant attendance to the nest may be particularly important in species that would experience high loss of eggs or young to predators or rival birds if a nest were left unguarded. For many bird species, the high demands on a male to care for the young or his mate preclude the establishment of nests with additional females.

Most birds exhibit a form of monogamy called **social monogamy,** in which male-female pair-bonds are formed and both sexes cooperate in rearing of the young. However, either member of a pair may mate with an individual that is not its partner. Recent DNA analyses have shown that most songbird species frequently are "unfaithful," engaging in extra-pair copulations. As a result, nests of many of these monogamous species contain a sizeable portion (30% or more) of young with fathers other than the attendant male. Why do individuals engage in extra-pair copulations? One possible benefit of extra-pair copulations is increased genetic diversity of the young. Second, extra-pair copulations permit a male to increase the number of sired young, increasing his reproductive success. Third, by mating with an individual of better genetic quality than the present mate, fitness of offspring could be increased. This can be particularly beneficial for females, who cannot easily increase the number of their offspring because the large size of a bird's egg

Jeannie Stafford/USFS

figure 19.22

Dominant male sage grouse, *Centrocercus urophasianus*, by a hen that has been attracted by his "booming" display. Order Galliformes.

limits the number of eggs (and offspring) she can produce. Thus, extra-pair copulations allow males to increase the number of their offspring and could allow females to improve the genetic quality of their offspring. Genetic diversity of offspring is increased by extra-pair copulation for both males and females.

The most common form of polygamy in birds is **polygyny** ("many females"), in which a male has more than one female mate. In many species of grouse, males gather in a collective display ground, or **lek,** which is divided into individual territories, each vigorously defended by a displaying male (figure 19.22). There is nothing of value in a lek to a female; instead, females judge males based on their physical appearance and athleticism of their displays, presumably an indicator of their genetic quality. Usually a dominant male and several subordinate males occur in a lek. Competition among males for females is intense, but females appear to choose the dominant male for mating because, presumably, social rank correlates with genetic quality.

Polyandry ("many males"), in which a female mates with several males and the male incubates the eggs, is relatively rare in birds. It is practiced by several shorebirds, including spotted sandpipers. Female spotted sandpipers defend territories and mate with multiple males. Males incubate eggs within the female's territory and provide most parental care. This unusual reproductive strategy may be in response to high predation on spotted sandpiper nests.

Nesting and Care of Young

A few birds simply lay eggs on bare ground or rocks, but most build some form of nest in which to rear their young. Some birds build elaborate nests, such as the pendant nests constructed by orioles, the delicate lichen-covered mud nests of hummingbirds (figure 19.23) and flycatchers, the chimney-shaped mud nests of cliff swallows, the floating nests of red-necked grebes, and the huge sand and vegetation mounds of Australian brush turkeys. Nest parasites, such as brown-headed cowbirds and European cuckoos, build no nests at all but simply lay their eggs in the nests of birds smaller than themselves. When their eggs hatch, the foster parents care for the cowbird young, which outcompete the host's own hatchlings.

Richard R. Hansen/Science Source

figure 19.23

Anna's hummingbird, *Calypte anna*, feeding its young in its nest of plant fibers, bound together with spiderwebs and camouflaged with lichens. A female builds the nest, incubates two pea-sized eggs, and rears the young with no assistance from a male. Order Apodiformes.

Newly hatched birds are of two types: **altricial** and **precocial** (figure 19.24). Altricial young, which are naked and unable to walk or see at birth, remain in the nest for a week or more. Parents of altricial young must supply food to their young almost constantly, for young birds can eat more than their weight each day (see figure 19.23). Precocial young, including quail, fowl, ducks, and most water birds, are covered with down when hatched and can run or swim as soon as their plumage is dry. However, most precocial young, even those able to leave the nest soon after hatching, are still fed or protected from predators by their parents for some time. Some young are not easily categorized as altricial or precocial because they are intermediate in development at hatching. For example, gulls and terns are born covered with down and with their eyes open, but are unable to leave the nest for some time.

Although it may seem that precocial chicks have all the advantages, with their greater ability to find food and escape predation, altricial chicks have an advantage of their own. Altricial chicks grow faster, perhaps due to the higher growth potential of immature tissue.

19.6 Humans and Bird Populations

Activities of people may cause spectacular changes in bird distribution. Both common starlings (figure 19.25) and house sparrows have been accidentally or deliberately introduced into

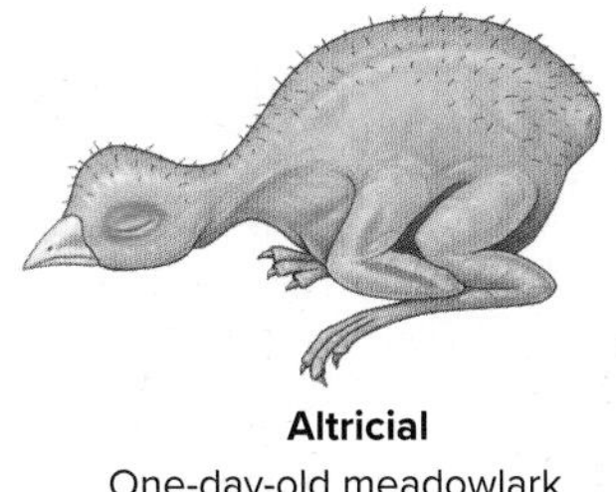

Altricial
One-day-old meadowlark

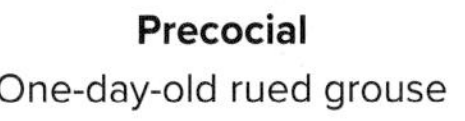

Precocial
One-day-old rued grouse

figure 19.24

Comparison of altricial and precocial young. Altricial American robin and meadowlark hatchlings are nearly naked, blind, and helpless. Precocial ruffed grouse and black-necked stilt hatchlings are covered with down, alert, and strong-legged. Black-necked stilts are able to run and swim within two hours after hatching.

numerous countries, to become the two most abundant bird species on earth, with the exception of domestic fowl.

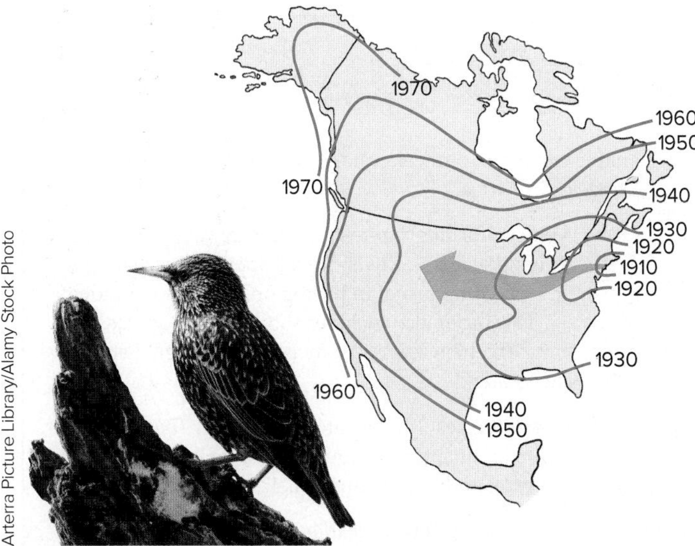

figure 19.25

Colonization of North America by common starlings, *Sturnus vulgaris*, after the introduction of 60 birds into Central Park in New York City in 1890. There are now an estimated 150 million starlings in the United States alone, testimony to the great reproductive potential of birds. Starlings are omnivorous, eating mostly insects in spring and summer and shifting to wild fruits in the fall. Order Passeriformes.

Humans also are responsible for the extinction of many bird species. More than 140 species of birds have, since 1681, followed the last dodo to extinction. Many were victims of changes in their habitat or competition with introduced species. Overhunting contributed to the extinction of some species, among them passenger pigeons, which 150 years ago darkened skies over North America in numbers estimated in the billions (figure 19.26).

Today, game birds are a well-managed renewable resource in the United States and Canada, and while hunters kill millions of game birds each year, none of the bird species legally hunted are endangered. Hunting interests, by acquiring large areas of wetlands for migratory bird refuges and sanctuaries, have contributed to the recovery of both game and nongame birds.

The 2018 State of the World's Birds report shows that 40% of the world's birds are in decline. The most important reasons for decline are habitat-associated, in the form of intensification and expansion of agriculture and deforestation. Agricultural lands now occupy 38% of the Earth's land surface; expansion is particularly rapid in the tropics. Intensification of agriculture, enabled by use of pesticides and fertilizers, has deprived ground-nesting birds of fields that formerly were left fallow.

The rapid loss of tropical forests, approximately 158,000 km^2 per year, an area almost the size of Washington state, is depriving hundreds of species of neotropical migrant birds of their wintering homes. Fragmentation of forests exposes nests of forest-dwelling birds to nest predators such as raccoons and opossums as well as nest parasites such as brown-headed cowbirds. Additional threats include invasive alien species,

figure 19.26

Sport-shooting of passenger pigeons in Louisiana during the nineteenth century, before establishment of state and federal hunting regulations. In addition to clearing of the hardwood forests that served as nesting habitats, relentless sport and market hunting eventually dropped the population to levels too low to sustain colonial breeding. The last passenger pigeon died in captivity in 1914.

hunting (especially in Europe, Africa, and Asia), and climate change.

Some of our favorite pets contribute to the decline in birds. A 2013 study estimates that each year 2.4 billion birds are killed by cats in the United States alone. Feral cats are thought to be responsible for about 70% of the deaths.

Some birds, such as robins, house sparrows, and starlings, can accommodate various changes, and may even thrive on them, but for most birds the changes are adverse. Unless we take leadership in managing our natural resources wisely, we soon could be facing the "silent spring" that Rachel Carson envisioned in 1962.

Key Theme 19.5

HUMAN CONNECTIONS

Lead Poisoning of Birds

Lead poisoning of waterfowl is a side effect of hunting and fishing. Before federal regulations in 1991 requiring the use of nonlead shot for all inland and coastal waterfowl hunting, shotguns scattered more than 3000 tons of lead each year in the United States alone. When waterfowl eat the pellets (which they mistake for seeds or grist), the pellets are ground and eroded in their gizzards, facilitating absorption of lead into their blood. Lead poisoning paralyzes or weakens birds, leading to death by starvation. The primary threat to inland populations of endangered California condors is consuming lead shot when feeding off a carcass. Copper shot is required in areas with condors, but illegal use of lead shot continues. Although poisoning of birds from lead shot has declined, lead sinkers used in fishing still poison large amounts of waterfowl. Recently, several states have banned lead sinkers, requiring anglers to use nontoxic alternatives.

Taxonomy of Living Members of Aves

The clade Aves contains about 10,500 species distributed among 40 orders of living birds. Understanding the relationships of living birds, and consequently placing them in a classification, has been difficult because of the apparent rapid diversification of birds in the Cretaceous and Paleogene periods. Relationships among bird groups are still being evaluated, leading to refinements in classification schemes. We present a classification of 22 selected bird orders, which follows the 2015 International Ornithologists Union World Bird List.

Aves (L. *avis*, bird)

Paleognathae (Gr. *palaios*, ancient, + *gnathos*, jaw). Modern birds with ancestral archosaurian palate. Ratites, which include ostriches, rheas, cassowaries, emus, and kiwis (with unkeeled sternum), and tinamous (with keeled sternum).

Order Struthioniformes (strū´thē-on-i-for´mēz) (L. *struthio*, ostrich, + *forma*, form): **ostriches.** The ostrich, *Struthio camelus* (figure 19.27), of Africa, is the largest living bird, which reaches up to 2.4 m tall and 145 kg. The feet have only two toes of unequal size covered with pads, which enable the birds to travel rapidly over sandy ground. 2 species in Africa.

Order Apterygiformes (ap-te-rij´i-for´mēz) (Gr. *a*, without + *pteryg*, wing + form): **kiwis.** Kiwis, about the size of a domestic fowl, are unusual in having only the mere vestige of a wing. This order also includes the extinct, flightless moas, some of which reached 2 m at the shoulder. 5 species, all in New Zealand.

Order Tinamiformes (tin-am´i-for´mēz) (N.L. *Tinamus*, type genus, + form): **tinamous.** Ground-dwelling, grouselike birds of Central and South America. 47 species.

Neognathae (Gr. *neos*, new, + *gnathos*, jaw). Modern birds with flexible palate.

Order Anseriformes (an´ser-i-for´mēz) (L. *anser*, goose, + form): **swans, geese, ducks.** Members of this order have broad beaks with filtering ridges at their margins, a foot web restricted to the front toes, and a long sternum with a low keel. 177 species, worldwide.

figure 19.27

Ostrich, *Struthio camelus*, of Africa, the largest of all living birds. Order Struthioniformes.

figure 19.28

Greater flamingos, *Phoenicopterus ruber*, on an alkaline lake in East Africa. Order Phoenicopteriformes.

Order Galliformes (gal′li-for′mēz) (L. *gallus*, cock, + form): **quail, grouse, pheasants, ptarmigans, turkeys, domestic fowl.** Chickenlike ground-nesting herbivores with strong beaks and heavy feet. The bobwhite quail, *Colinus virginianus*, occurs across the eastern half of the United States. The ruffed grouse, *Bonasa umbellus*, occupies the same region, but in woods instead of the open pastures and grain fields the bobwhite frequents. 298 species, worldwide.

Order Sphenisciformes (sfē-nis′i-for′mēz) (Gr. *Sphēniskos*, dim. of *sphen*, wedge, from the shortness of the wings, + form): **penguins.** Web-footed marine swimmers of southern seas from Antarctica north to the Galápagos Islands. Although penguins have a keeled sternum, they use their wings as paddles for swimming rather than for flight. 18 species.

Order Phoenicopteriformes (fē′ni-cop-ter′i-for′mēz) (Gr *phoenico*, reddish-purple, + *pter*, wing, + form): **flamingos** (figure 19.28). Large, colorful, wading birds that use lamellae in their beaks to strain zooplankton from the water. 6 species.

Order Procellariiformes (prō-sel-lar′ē-i-for′mēz) (L. *procella*, tempest, + form): **albatrosses, petrels, fulmars, shearwaters.** All are marine birds with hooked beaks and tubular nostrils. In wingspan (more than 3.6 m in some), albatrosses are the largest flying birds. 143 species, worldwide.

Order Pelecaniformes (pel-e-can′i-for′mēz) (Gr. *pelekan*, pelican, + form): **pelicans, ibises,** and **herons.** Mostly colonial fish-eaters inhabiting coasts, lakes, wetlands, and streams. 118 species, worldwide distribution, especially in the tropics.

Order Suliformes (sū′li-for′mēz) (Ice. *sul*, gannet + form): **frigatebirds, gannets, boobies, cormorants.** Medium to large diving seabirds that feed mostly on fishes. 60 species, worldwide.

Order Accipitriformes (as-sip′i-tri-for′mez) (L. *accipiter*, hawk + form): **vultures, condors, eagles, buzzards, hawks.** Most are diurnal birds of prey with keen vision and sharp, curved talons. 265 species, worldwide.

Order Falconiformes (fal′ko-ni-for′mēz) (L. *falco*, falcon, + form): **falcons.** Very fast birds of prey that primarily eat other birds. The peregrine falcon, *Falco peregrinus*, dives at speeds up to 320 km/h (200 mph). 67 species, worldwide.

Order Charadriiformes (kä-ra-drē′i-for′mēz) (Gr. *charadri*, curlew (a sandpiper), + form): **gulls** (figure 19.29), **plovers, sandpipers, terns, woodcocks, snipes, avocets, phalaropes, skuas, skimmers, auks, puffins.** All are shorebirds. They are strong fliers and are usually colonial. 384 species, worldwide.

Order Columbiformes (kō-lum′bē-i-for′mēz) (L. *columba*, dove, + form): **pigeons, doves.** All have short necks, short legs, and a short, slender beak. The flightless dodo, *Raphus cucullatus*, of the Mauritius Islands became extinct in 1681. 334 species, worldwide.

(*continued*)

(continued)

Cleveland P. Hickman, Jr.

figure 19.29

Laughing gulls, *Larus atricilla*, in flight. Order Charadriiformes.

Order Psittaciformes (sit´ta-sē´-for´mēz) (L. *psittacus*, parrot, + form): **parrots, parakeets.** Birds with hinged and movable upper beak and a fleshy tongue. 395 species, pantropical.

Order Cuculiformes (ku-ku´li-for´mēz) (L. *cuculus*, cuckoo, + form): **cuckoos, roadrunners.** European cuckoos, *Cuculus canorus*, lay their eggs in nests of smaller birds, which rear the young cuckoos. American cuckoos usually rear their own young. 149 species, worldwide.

Order Strigiformes (strij´i-for´mēz) (L. *strix*, screech owl, + form): **owls.** Nocturnal predators with large eyes, powerful beaks and feet, and silent flight. 240 species, worldwide.

Order Caprimulgiformes (kap-ri-mul´ji-for´mēz) (L. *caprimulgus*, goatsucker, + form): **nightjars, nighthawks.** Night and twilight feeders with small, weak legs and wide mouths fringed with bristles. Whippoorwills, *Antrostomus vociferus*, are common in the woods of the eastern states, and common nighthawks, *Chordeiles minor*, are often seen and heard during evening flights. 122 species, worldwide.

Order Apodiformes (ə-pōd´i-for´mēz) (Gr. *apous*, footless, + form): **swifts, hummingbirds.** Small birds with short legs and rapid wingbeat. The familiar chimney swift, *Chaetura pelagia*, fastens its nest in chimneys by means of saliva. Most species of hummingbirds occur in the tropics, but there are 24 species in the United States, of which only one, the ruby-throated hummingbird, commonly occurs in the eastern part of the country. 469 species, worldwide.

Order Coraciiformes (kä-rā´sē´i´for´mēz) (N.L. *coracii* from Gr. *korakias*, a kind of raven, + form): **kingfishers, bee-eaters.** Birds that have strong, prominent beaks and nest in cavities. In the eastern half of the United States, belted kingfishers, *Megaceryle alcyon*, are common along waterways. 156 species, worldwide.

Order Piciformes (pis´i-for´mēz) (L. *picus*, woodpecker, + form): **woodpeckers, toucans, puffbirds, honeyguides.** Birds with highly specialized beaks and two toes extending forward and two backward. All nest in cavities. The largest species in North America is the pileated woodpecker, usually found in mature forests. 443 species, worldwide.

Order Passeriformes (pas-ser´i-for´mēz) (L. *passer*, sparrow, + form): **perching songbirds.** This is the largest order of birds, containing 140 families and 60% of all birds. Most have a highly developed syrinx (voice box). Their feet are adapted for perching on thin stems and twigs. The young are altricial. To this order belong many birds with beautiful songs, such as thrushes, warblers, finches (figure 19.30), mockingbirds, meadowlarks, sparrows, vireos, chickadees, and hosts of others. Others of this order, such as swallows, magpies, starlings, crows, ravens, jays, nuthatches, and creepers, have no songs worthy of the name. 6390 species, worldwide.

Cleveland P. Hickman, Jr.

figure 19.30

Ground finch, *Geospiza fuliginosa*, one of the famous Darwin's finches of the Galápagos Islands. Order Passeriformes.

SUMMARY

Sections	Key Concepts
19.1 Origin and Relationships	• The more than 10,500 species of living birds are feathered, egg-laying, endothermic vertebrates with forelimbs modified as wings. • *Archaeopteryx*, a well-known fossil species from the Jurassic period of the Mesozoic era, has the ancestral diapsid traits of a long tail and teeth, but has well-developed wings and asymmetrical (flight) feathers, which identify it as a bird (Avialae). Its morphology clearly places birds within a group of bipedal, otherwise extinct dinosaurs called theropods. • Feathers or filaments, present in many theropods, preceded flight. Their original function in bird ancestors probably was for thermoregulation or as displays. • Diversification of modern birds (Aves) into the 40 orders living today occurred primarily in the Cretaceous and Paleogene periods. • Flightlessness in birds is not uncommon and evolved independently in several bird orders, usually on islands where terrestrial predators are absent. All flightless birds are derived from flying ancestors.
19.2 Structural and Functional Adaptations for Flight	• Adaptations of birds for flight are of two basic kinds: those reducing body weight and those promoting more power for flight. • Feathers, made of keratin, combine lightness with strength, water repellency, and high insulative value. • Body weight is further reduced by elimination of some bones, fusion of others (also providing structural stability in flight), and hollow, air-filled spaces in many bones. The light, keratinized beak, which replaces the heavy jaws and teeth of nonavian reptiles, serves as a bird's hand and mouth and is variously adapted for different feeding habits. • Adaptations that provide power for flight include a high metabolic rate and body temperature coupled with an energy-rich diet, a highly efficient respiratory system consisting of air sacs arranged to provide a constant, one-way flow of air through the lungs, powerful flight muscles arranged to place muscle mass near the bird's center of gravity, and an efficient, high-pressure circulation. • Birds have keen eyesight (due to their large eyes and nutrient-supplying pecten), good hearing, and a brain adapted for fine coordination of flight muscles. The kidneys produce uric acid as the principal nitrogenous waste.
19.3 Flight	• Birds generate lift during flight when air passes angled or curved (cambered) wings. The downward-deflected air pushes the wing (and bird) upward. • Birds fly by applying the same aerodynamic principles as an airplane and using similar equipment: wings for lift and propulsion, a tail for steering and landing control, and wing slots for control at the slow speed of take-offs and landings. • Bird wing anatomy reflects the flight style of birds. In general, slower-flying, highly maneuverable birds have short, rounded wings with high camber and prominent slots. The fastest birds have long, pointed wings with flat profiles and no slots.
19.4 Migration and Navigation	• Bird migration refers to regular movements between summer nesting places and wintering regions. Because land mass is concentrated in the northern hemisphere, most migratory birds move north in spring and south in fall. • Birds use many cues to navigate during migration, including using geographic landmarks, the position of the sun or stars, and the Earth's magnetic field.
19.5 Social Behavior and Reproduction	• The highly developed social behavior of birds is manifested in vivid courtship displays, mate selection, territorial behavior, and incubation of eggs and care of young. • Most birds are monogamous, in which a male and female form a pair-bond during the reproductive season. Many of these are only socially monogamous—pairing is maintained, but mating is not restricted to their partner. • Some birds practice polygyny, where one male mates with multiple females; or polyandry, where one female mates with multiple males. • Most birds build a nest to incubate their eggs. For some species, only the mother cares for the eggs and young, but in many species, both parents participate in care. • Young hatch at various levels of development: altricial young are naked and helpless, while precocial young are feathered and able to walk and feed themselves.
19.6 Humans and Bird Populations	• Some introduced bird species, including the common starling and house sparrow, have greatly increased in population size, becoming the most abundant wild birds in North America. • Many birds have experienced great declines in populations, or even become extinct. Primary causes of bird declines include loss of habitat as nesting and overwintering areas are converted to agricultural lands or urbanized areas, competition or predation from invasive species, predation from feral cats, and overhunting.

Review Questions

1. Explain the significance of the discovery of *Archaeopteryx*. Why did this fossil demonstrate beyond reasonable doubt that birds are grouped phylogenetically with dinosaurs?
2. The special adaptations of birds contribute to two essentials for flight: more power and less weight. Explain how each of the following contributes to one or both of these two essentials: feathers, skeleton, muscle distribution, digestive system, circulatory system, respiratory system, excretory system, reproductive system.
3. How do marine birds rid themselves of excess salt?
4. In what ways are a bird's ears, eyes, and brain specialized for the demands of flight?
5. Explain how a bird's wing produces lift. What design features help prevent stalling at low flight speeds? What features help to reduce drag?
6. Describe four basic forms of bird wings. How does wing shape correlate with flight speed and maneuverability?
7. What advantages does seasonal migration provide for birds?
8. Describe different navigational resources birds may use in long-distance migration.
9. What are some advantages of social aggregation among birds?
10. More than 90% of all bird species are monogamous. Explain why monogamy is much more common among birds than among mammals.
11. Briefly describe examples of polygyny and polyandry among birds.
12. Why might a "monogamous" bird seek extra-pair copulations?
13. Define altricial and precocial as they relate to birds.
14. Offer some examples of how human activities have affected bird populations.

For Further Thought Reproductive behavior and strategies are better known in birds than in any other vertebrate group. Why?

20

Mammals

LEARNING OBJECTIVES

Readers will be able to:

20.1 Describe the evolution of mammals from their amniote ancestors.

20.2 Describe (1) the integument, including glands, of mammals, including adaptations for endothermy; (2) mammalian adaptations for an insectivorous, carnivorous, herbivorous, or omnivorous diet; (3) the reproductive biology of monotremes, marsupials, and placentals; and (4) how mammals deal with the challenges of winter by migration or hibernation.

20.3 Describe the evolution of humans.

Mint Images/Art Wolfe/Getty Images RF

Juvenile grizzly bear, *Ursus arctos horribilis*.

The Tell-Tale Hair

Hair evolved in a common ancestor of all mammals and has been retained to varying degrees in all species descended from that ancestor. Hair is therefore diagnostic for mammals; except in some pathological conditions, all mammals have hair at some point in their lives, and hair occurs in no other living organisms. Even those living mammals apparently without hair, such as whales, usually have a few hairs on their bodies. Mammalian hair has undergone numerous adaptive modifications for diverse uses. Mammals use hair for concealment, behavioral signaling, waterproofing, and buoyancy; their hair may serve as sensitive vibrissae on their snouts or as prickly quills. Perhaps the most important use of their hair is thermal insulation, helping to maintain a high, constant body temperature in all climates, and thus support a high level of activity.

Mammals are among the most active animals, exhibiting speed and endurance in aquatic, aerial, and terrestrial habitats. They maintain this activity in nearly all environmental conditions, including cool nights, baking deserts, frigid polar seas, and icy winters. Although hair is perhaps the most obvious feature of mammals, many other adaptations support their diversification. These adaptations include mammary glands for nourishing newborns, a large brain with a unique covering of the cerebrum (the neocortex), a diaphragm for efficient ventilation of lungs, and a highly developed sense of smell. Most mammals have an intrauterine, vascular placenta for feeding the embryo, specialized teeth and jaw musculature for processing food, and an upright gait for rapid and efficient locomotion.

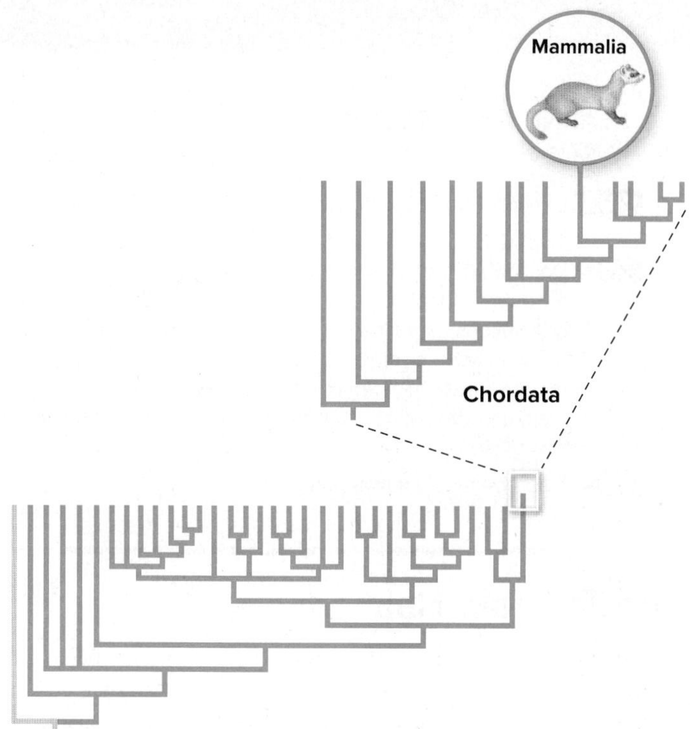

Mammals occupy almost every environment on earth that supports life. Although not a large group (about 5700 species, compared with more than 10,000 species of birds, 32,000 species of fishes, and 1,100,000 species of insects), clade Mammalia (ma-ma´lē-ə; L. *mamma,* breast) is among the most biologically differentiated groups in the animal kingdom. Mammals are exceedingly diverse in size, shape, form, and function. They range in size from Kitti's hognosed bat, weighing only 2 g, to blue whales, exceeding 170 metric tons.

More than any other animals, mammals are targets of human activity. We have domesticated numerous mammals for food and clothing, as beasts of burden, and as pets. We use millions of mammals each year in biomedical research. We have introduced alien mammals into new habitats and extinguished mammalian populations from their native habitats. In 2019, 692 species of mammals were listed as "critically endangered" or "endangered" by the International Union for Conservation of Nature and Natural Resources (IUCN); especially jeopardized are rodents, bats, cetaceans, cats, and primates. Because our welfare continues to be closely tied to that of mammals, we must strive to preserve their natural populations and environments.

20.1 Origin and Evolution of Mammals

The evolutionary descent of mammals from their earliest amniote ancestors is perhaps the most fully documented transition in vertebrate history. From the fossil record, we can trace the evolution over 150 million years of endothermic, furry mammals from their small, ectothermic, hairless ancestors. Mammals and their closest extinct relatives have a pair of openings in the temporal region of the skull associated with attachment of mandibular muscles. This condition identifies them as **synapsids,** one of three major groups of amniotes that diversified during the late Paleozoic era (see figure 18.2).

The earliest synapsids included diverse herbivorous and carnivorous forms collectively called **pelycosaurs** (figures 20.1 and 20.2). Pelycosaurs share a general outward resemblance to lizards, but this resemblance is misleading. Pelycosaurs are not closely related to lizards, which are diapsids (see figure 18.2), nor are they a monophyletic group. From one group of early carnivorous synapsids arose the **therapsids** (figure 20.2), the only synapsid group to survive beyond the Paleozoic era. Therapsids evolved an efficient erect gait with upright limbs positioned beneath the body rather than sprawled to the side, as in lizards and early pelycosaurs. Elevation of the body reduced stabilty, and the muscular coordination center of the brain, the cerebellum, assumed an expanded role. Changes in the morphology of the skull and jaw-closing muscles increased feeding efficiency. Therapsids diversified into numerous herbivorous and carnivorous forms, but most early forms disappeared during a great extinction at the end of the Permian period. Previously, pelycosaurs and early therapsids were called "mammal-like reptiles," but use of this term is inappropriate, because they are not part of the clade Reptilia (see section 18.1: Changes in Traditional Classification of Reptiles).

One therapsid group to survive into the Mesozoic era was the **cynodonts** (figures 20.1 and 20.2). Cynodonts evolved several features that supported a high metabolic rate: powerful and specialized jaw musculature, permitting a stronger bite; **heterodont** teeth, permitting better food processing and use of more diverse foods; **turbinate bones** in the nasal cavity, aiding retention of body heat (figure 20.3); and a **secondary palate** (figure 20.3), enabling breathing while holding prey in its mouth or chewing food. The secondary palate would be important to subsequent mammalian evolution by permitting the young to breathe while suckling. Loss of lumbar ribs in cynodonts is correlated with the evolution of a **diaphragm** and also may have provided greater dorsoventral flexibility to the spinal column.

The earliest mammals of the late Triassic period were small, mouse- or shrew-sized animals with enlarged crania, redesigned jaws, and a new type of dentition, called **diphyodont,** in which teeth are replaced only once (deciduous and permanent teeth). This condition contrasts with the ancestral amniote pattern of continual tooth replacement throughout life (**polyphyodont**). Two bones, the articular and the quadrate, which previously served as the jaw joint, were reduced in size and relocated to the middle ear, becoming the malleus and incus, respectively. A new jaw joint formed between the dentary and squamosal (temporal) bones; this joint is the defining character for fossil mammals.

The earliest mammals were almost certainly endothermic, although their body temperatures would have been lower than those of modern placental (Eutherian) mammals. Hair was essential for insulation, and the presence of hair implies that

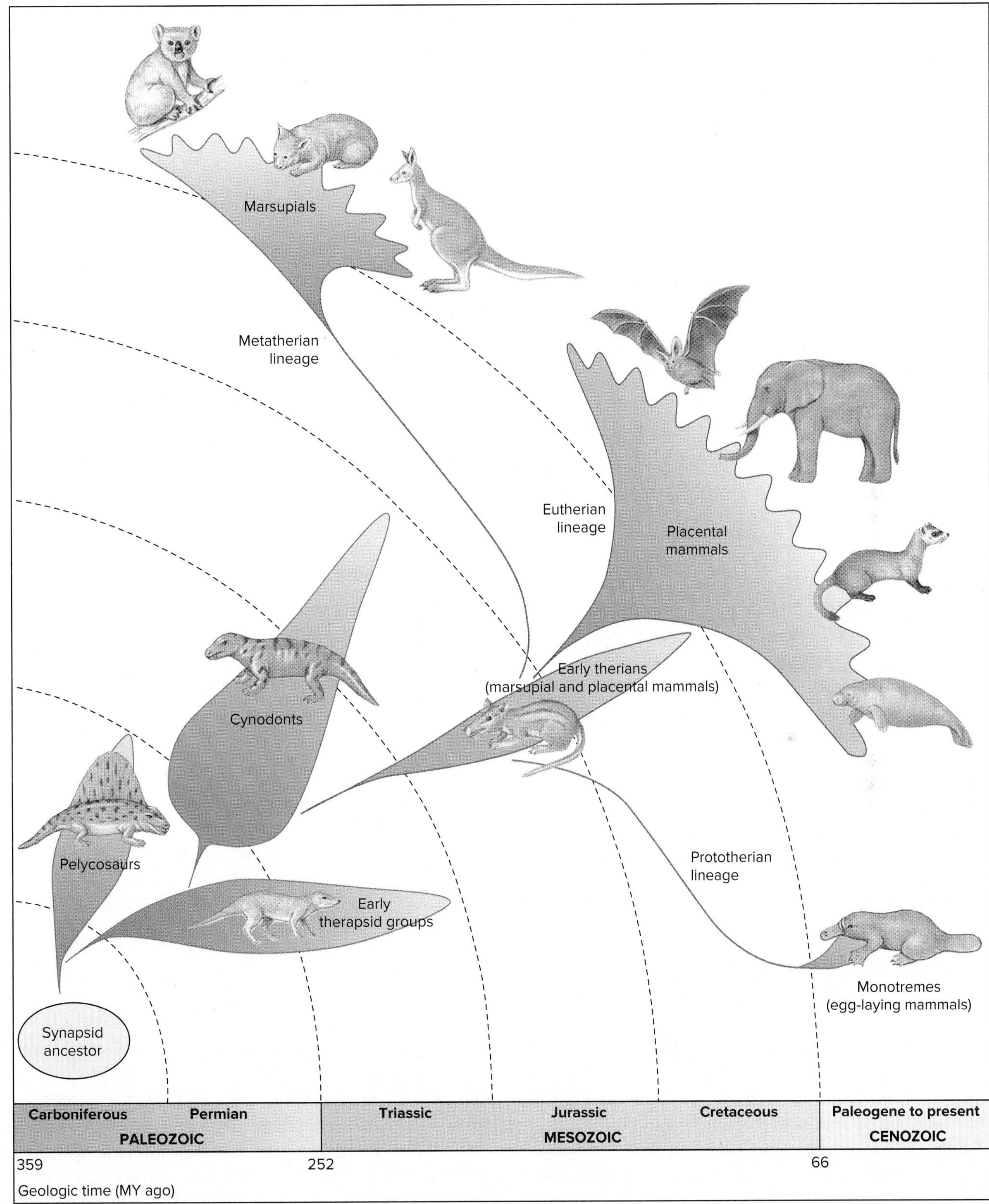

figure 20.1

Evolution of major groups of synapsids. The synapsids, characterized by lateral temporal openings in the skull, began with pelycosaurs, early amniotes of the Permian period. Pelycosaurs diversified extensively and then were replaced by the therapsids. One lineage of therapsids, the cynodonts, gave rise in the Triassic period to early mammals. Fossil evidence indicates that all three groups of living mammals—monotremes, marsupials, and placentals—are derived from the same cynodont lineage. The great diversification of modern placental orders occurred during the Paleogene period.

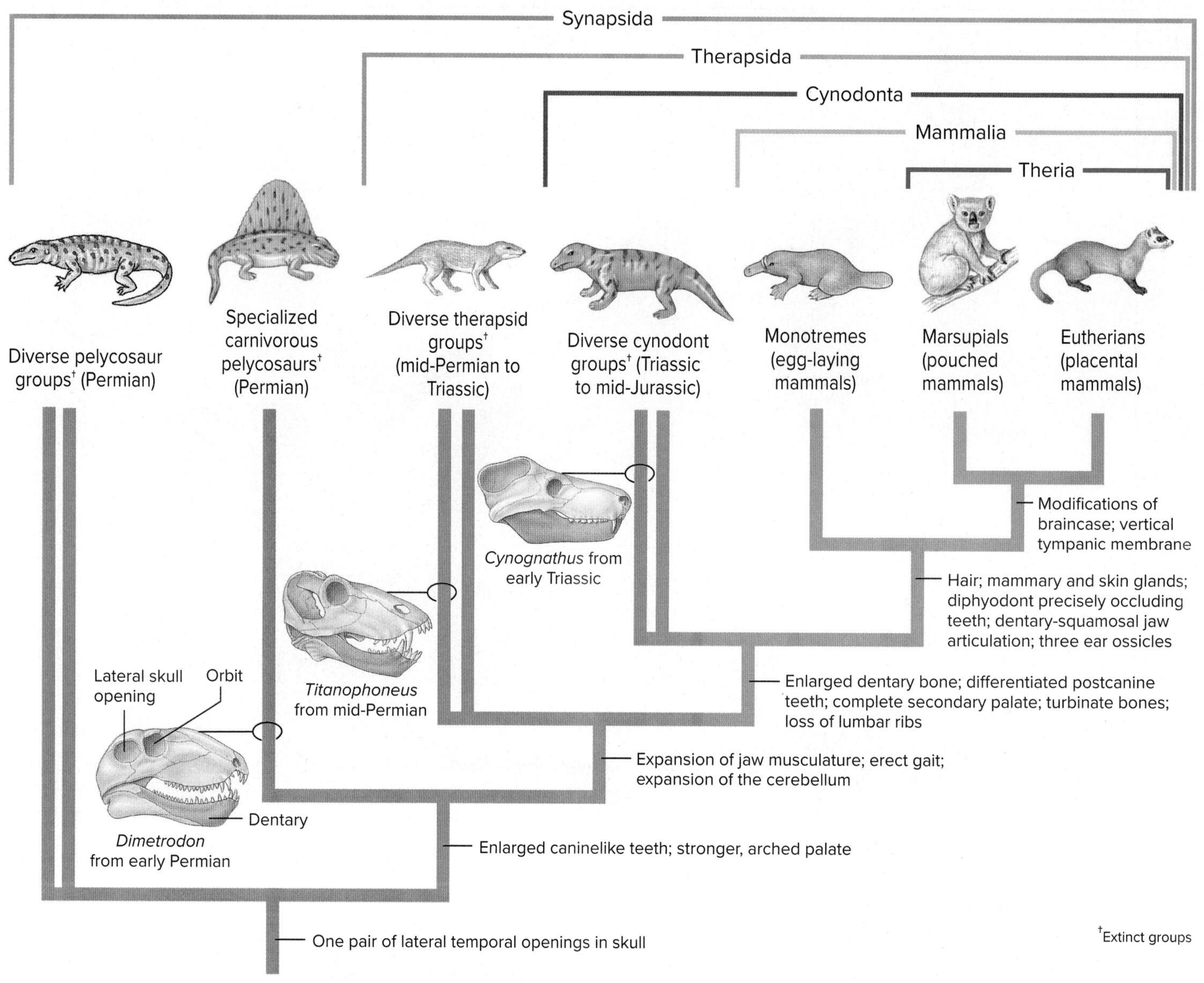

figure 20.2

Abbreviated cladogram of synapsids, emphasizing origins of important characteristics of mammals. The skulls show progressive increase in size of the dentary bone relative to other bones in the lower jaw and increasing heterodonty.

sebaceous and sweat glands must have evolved at this time to condition hair and to facilitate thermoregulation. The fossil record is silent on the appearance of mammary glands, but they must have evolved before the end of the Triassic period.

Oddly, early mammals of the Late Triassic period, having developed nearly all novel attributes of modern mammals, had to wait for another 150 million years before they could achieve their great diversity. While dinosaurs became diverse and abundant, all nonmammalian synapsid groups became extinct. Mammals survived mostly as small, shrew-sized, probably nocturnal creatures. Then, in the Jurassic and Cretaceous periods, mammals rapidly diversified into numerous forms, including catlike carnivores, otterlike aquatic species, raccoon-sized omnivores, and gliders. Mammalian diversification in the Cretaceous period paralleled insect and flowering plant diversification; mammals' diverse and precisely occluding dentition likely allowed them to exploit these new and changing food resources. The success of early mammal diversification likely was strongly enhanced by extensive maternal care, including feeding of the young by **lactation.** In the subsequent Cenozoic era, following the dinosaurs' demise, mammals rapidly diversified into the forms familiar to us today.

There are 29 orders of living mammals: one order of **monotremes** (Prototheria), seven orders of **marsupials** (Metatheria), and 21 orders of **placentals** (Eutheria). A taxonomy of selected orders is at the end of this chapter.

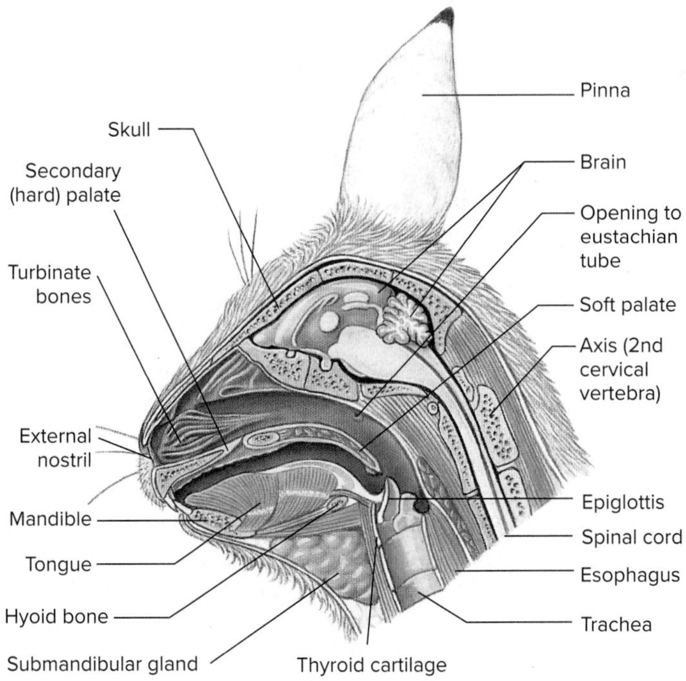

figure 20.3

Sagittal section of the head of a rabbit. The secondary palate, which consists of bony (hard) and nonbony (soft) regions, separates the routes of air (dorsal) and food (ventral).

20.2 Structural and Functional Adaptations of Mammals

Integument and Its Derivatives

Mammalian skin and its associated structures distinguish mammals as a group. In general, skin is thicker in mammals than in other vertebrates, although as in all vertebrates it is composed of **epidermis** and **dermis** (figure 20.4). The epidermis is thinner where it is well protected by hair, but in places that are subject to much contact and use, such as the palms or soles, its outer layers become thickened with **keratin,** a fibrous protein that also constitutes nails, claws, hooves, and hair.

Hair

Hair is especially characteristic of mammals, although humans are not very hairy creatures, and hair in whales is reduced to a few sensory bristles on the snout. A hair grows from a hair follicle that, although epidermal in origin, is sunk into the dermis of the skin (figure 20.4). A hair grows continuously by rapid proliferation of cells in the follicle. As a hair shaft is pushed upward, new cells are carried away from their source of nourishment and die, filled with keratin.

Mammals characteristically have two kinds of hair forming their **pelage** (fur coat): (1) dense and soft **underhair** for insulation, and (2) coarse and longer **guard hair** for protection against wear and to provide coloration. Underhair traps a layer of insulating air. In aquatic mammals, such as fur seals, otters, and beavers, underhair is so dense that it is almost impossible to wet. In water, guard hairs become wet and adhere to each other, forming a protective blanket over the underhair (figure 20.5).

When a hair reaches a certain length, it stops growing. Normally it remains in its follicle until a new growth starts, whereupon it falls out. Most mammals experience periodic molts of the entire coat. In humans, hair is shed and replaced throughout life (although replacement is not assured, as evidenced by balding males).

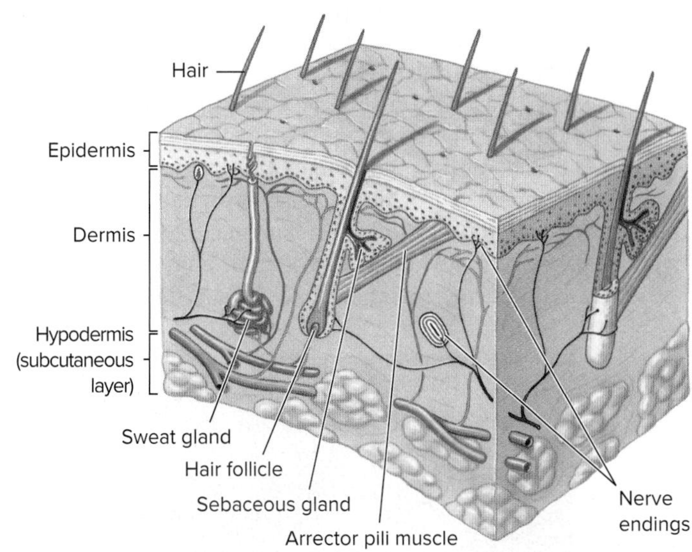

figure 20.4

Structure of human skin (epidermis and dermis) and hypodermis, showing hair and glands.

figure 20.5

American beaver, *Castor canadensis,* feeding on tree bark. This second-largest rodent (the South American capybara is larger) has a heavy waterproof pelage consisting of long, tough guard hairs overlying the thick, silky underhair so valued in the fur trade. Order Rodentia, family Castoridae.

Key Theme 20.1

ADAPTATION AND PHYSIOLOGY

Hairs are Not All the Same

A hair is more than a strand of keratin. It consists of three layers: the medulla or pith in the center of the hair, the cortex with pigment granules next to the medulla, and the outer cuticle composed of overlapping scales. Hair of different mammals shows a considerable range of structure. It may be deficient in cortex, like the brittle hair of deer, or it may be deficient in medulla, like the hollow, air-filled hairs of wolverines. Hairs of rabbits and some others are scaled to interlock when pressed together. Curly hair, such as that of sheep, grows from curved follicles.

Some mammals, such as foxes and seals, shed their coat every summer. Most mammals have two annual molts, one in spring and one in fall. Summer coats are always much thinner than winter coats and in some mammals may be a different color. Several northern mustelid carnivores, such as weasels, have white winter coats and brown summer coats. It was once hypothesized that the white pelage of arctic animals conserved body heat by reducing radiation loss; in fact, dark and white pelages radiate heat equally well. The white winter pelage of arctic animals is simply camouflage in a land of snow. The snowshoe hare of North America has three annual molts: the white winter coat is replaced by a brownish-gray summer coat, and this is replaced in autumn by a grayer coat, which is soon shed to reveal the white winter coat beneath (figure 20.6).

Most mammals have somber colors that disguise their presence. Often a species is marked with "salt-and-pepper" coloration or a disruptive pattern that helps make it inconspicuous in its natural surroundings. Examples are the spots of leopards and fawns and the stripes of tigers. Skunks advertise their presence with conspicuous warning coloration.

The hair of mammals has become modified to serve many purposes. Bristles of hogs, spines of porcupines and their kin, and vibrissae on the snouts of most mammals are examples. **Vibrissae,** commonly called "whiskers," are really sensory hairs that provide a tactile sense to many mammals. The slightest movement of a vibrissa generates impulses in sensory nerve endings that travel to special sensory areas in the brain. Vibrissae are especially long in nocturnal and burrowing animals.

Porcupines, hedgehogs, echidnas, and a few other mammals have developed an effective and dangerous spiny armor. When cornered, the common North American porcupine turns its back toward its attacker and lashes out with its barbed tail. The lightly attached quills break off at their bases when they enter the skin and, aided by backward-pointing hooks on the tips, work deeply into tissues. Dogs are frequent victims (figure 20.7), but fishers, wolverines, and bobcats can flip the porcupine onto its back to expose vulnerable underparts.

Alan and Sandy Carey/Getty Images

A

Howard Sandler/Shutterstock

B

figure 20.6

Snowshoe hare, *Lepus americanus,* in **A,** brown summer coat, and **B,** white winter coat. In winter, extra hair growth on the hind feet broadens the animal's support in snow. Snowshoe hares are common residents of the taiga and are an important food for lynxes, foxes, and other carnivores. Order Lagomorpha, family Leporidae.

Horns and Antlers

Several kinds of horns or hornlike structures occur in mammals. **True horns** of the family Bovidae (antelopes, sheep, and cattle) are hollow sheaths of keratinized epidermis that embrace a core of bone arising from the skull. True horns are not shed, usually are not branched (although they may be greatly curved), grow continuously, and occur in both sexes.

Kevin Mingora/AP Photo

figure 20.7

Dogs are frequent victims of a porcupine's impressive quills. Unless removed (usually by a veterinarian), quills continue to work their way deeper into the flesh, causing great distress and sometimes death.

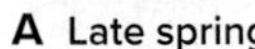

figure 20.8

Annual growth of male deer antlers. **A,** Antlers begin to grow in late spring, stimulated by pituitary gonadotropins. **B,** Bone grows very rapidly until halted by a rapid rise in testosterone production by the testes. **C,** The skin (velvet) dies and rubs off. **D,** Testosterone levels peak during the fall breeding season. The antlers are shed in January as testosterone levels subside.

Antlers of the deer family Cervidae are branched and composed of solid bone when mature. During their annual spring growth, antlers develop beneath a covering of highly vascular soft skin called **velvet** (figure 20.8). When growth of antlers is complete just before the breeding season, blood vessels constrict, and a male removes the velvet by rubbing its antlers against trees. Antlers, which usually occur only in males, are shed after the breeding season. New buds appear a few months later to begin the next set of antlers. For several years, each new pair of antlers is larger and more elaborate than the previous set. Annual growth of antlers places a strain on mineral metabolism because during the growing season an older moose or elk must accumulate about 5.5 kg (12 lb) of calcium salts from its vegetable diet.

Horns of the pronghorn, *Antilocapra americana,* are similar to the true horns of bovids, except that the keratinized portion is forked and shed annually. Giraffe horns are similar to antlers but retain their integumentary covering and are not shed. Rhinoceros horn consists of hairlike, keratinized filaments that arise from dermal papillae cemented together but not attached to the skull.

Key Theme 20.2

HUMAN CONNECTIONS

Rhino Horn Trade

Trade in rhino products, especially the horn, has pushed Asian and African rhinos to the brink of extinction. Rhino horn was used in traditional Chinese medicine to treat fever, gout, and other ailments. Such supposed medicinal values are totally without pharmacological basis. Until recently the principal use of rhino horns was to fashion handles for ceremonial daggers called jambiyas in the Middle East. Between 1970 and 1997, horns from 22,350 rhinos were imported into northern Yemen alone. Thanks to education efforts, use of rhino horn in Yemen has nearly ceased; the destination of most illegal rhino horn is China and Vietnam. An international ban prohibiting the trade of rhino horn has reduced, but not eliminated, the now illegal trade, and populations continue to suffer.

Glands

Of all vertebrates, mammals have the greatest variety of integumentary glands. Most fall into one of four classes: sweat, scent, sebaceous, or mammary. All are derivatives of epidermis.

CHARACTERISTICS of Mammalia

1. Fleshy external ears **(pinna); endothermic.**
2. Body mostly covered with **hair,** but reduced coverage in some; **sweat, scent, and sebaceous glands** present; skin underlain by a thick layer of fat.
3. Skull with **two occipital condyles;** lower jaw a **single enlarged bone (dentary);** jaw joint between squamosal and dentary bones; **seven cervical vertebrae** (except in manatees and some sloths); **pelvic bones fused.**
4. Mouth with **diphyodont teeth;** teeth **heterodont** in most.
5. Brain highly developed, especially **cerebral cortex;** 12 pairs of cranial nerves.
6. Olfactory sense highly developed; middle ear with **three bones** (malleus, incus, and stapes).
7. Separate sexes; internal fertilization; copulatory organ a **penis;** testes usually in a **scrotum;** sex determined by chromosomes (male is heterogametic).
8. Embryonic membranes of **amnion, chorion, and allantois;** most **viviparous** with embryos developing in a **uterus** with **placental attachment,** except monotremes, which are **oviparous;** young nourished by **milk from mammary glands.**
9. Excretory system of **metanephric kidneys** and ureters that usually open into a bladder; urea main nitrogenous waste.
10. Lungs with high surface area from **alveoli** and ventilated by **aspiration;** larynx present; **secondary palate** separates air and food passages (see figure 20.3); muscular **diaphragm** ventilates lungs; convoluted **turbinate bones** in the nasal cavity warm and moisten inspired air.
11. Heart with two atria and two ventricles; **separate pulmonary and systemic circuits; persistent left aortic arch;** nonnucleated, biconcave red blood cells.

Sweat glands are tubular, highly coiled glands that occur over much of the body surface in most mammals (see figure 20.4). They are absent in other vertebrates. There are two kinds of sweat glands: eccrine and apocrine. **Eccrine glands** secrete a watery fluid that, if evaporated on the skin's surface, draws heat away from the skin and cools it. Eccrine glands occur in hairless regions, especially foot pads, in most mammals, although in horses and most primates they are scattered over the body. **Apocrine glands** are larger than eccrine glands and have longer and more convoluted ducts. They always open into a hair follicle or where a hair once was. Apocrine glands develop near sexual puberty and are restricted (in humans) to the axillae (armpits), genital regions, breasts, and external auditory canals. In contrast to the watery secretions of eccrine glands, apocrine secretions are milky fluids, whitish or yellow in color, that dry on the skin to form a film. Apocrine glands are not involved in heat regulation. Their activity is correlated with the reproductive cycle.

Scent glands occur in nearly all mammals. Their location and functions vary greatly. They are used for communication with members of the same species, for marking territorial boundaries, for warning, or for defense. Scent-producing glands are located in orbital, metatarsal, and interdigital regions; at the base of the tail; and in the penis. The most odoriferous of all glands are those of skunks, which open by ducts into the anus; their secretions can be discharged forcefully for 2–3 m. During mating season, many mammals produce strong scents for attracting the opposite sex. Human scents mainly arise from apocrine glands in the axillary region. However, we tend to dislike our own scent, a concern that has stimulated a lucrative deodorant industry to produce an endless array of soaps and odor-masking concoctions.

Sebaceous glands are usually associated with hair follicles (see figure 20.4), although some are free and open directly onto the surface. Gland cells are secreted in their entirety and are continuosly renewed by cell division. These cells become distended with a fatty accumulation, then die, and are expelled as a greasy mixture called **sebum** into the hair follicle. Called a "polite fat" because it does not turn rancid, sebum keeps skin and hair pliable and glossy. Most mammals have sebaceous glands over their entire body; in humans, they are most numerous in the scalp and on the face.

Mammary glands, for which mammals are named, occur on all female mammals and in a rudimentary form on all male mammals. They develop from downward-growing epidermal cells that form a branching system of ducts, surrounded by milk-producing cells. Mammary glands grow rapidly during the late stage of pregnancy, in response to hormones (estrogen and progesterone). The number of mammary glands varies from 2 in primates to about 20 in some marsupials. In most mammals, milk is secreted from mammary glands via nipples, but monotremes lack nipples and simply secrete milk onto the fur of the mother's belly where it is lapped by the young. Release of milk to a suckling infant is called **lactation** (figure 20.9).

Milk contains a mixture of carbohydrates, fats, and proteins. Cow's milk is watery, with only about 3.5% fat and 3.5% protein. In contrast, seal and whale milk has as much as 60% fat and 11% protein. This rich milk allows juveniles to grow rapidly and develop thick adipose layers, adaptive in the cold seas they inhabit.

figure 20.9

A female Galápagos sea lion, *Zalophis wollebaeki*, produces a single pup, which feeds solely on its mother's fat- and protein-rich milk for 5–6 months. Thereafter, its diet is supplemented by fish, but it continues to consume milk until 11 months of age, when it is weaned. Order Carnivora, family Otariidae.

Feeding young by lactation is one of the most important adaptations of mammals. With lactation, young do not depend on their own ability to forage. Instead, nutrients ingested by the mother can be stored, as fat, and used to make milk. Lactation protects against uncertain and variable food supplies, and its (usually) constant supply ensures rapid growth.

Olfaction and the Cerebral Cortex

Mammals have an array of highly developed special senses, including vision, hearing, smell, and taste, all of which combine to provide a detailed picture of the world around them. The sense of smell (olfaction) is especially well developed in and of great social importance to most mammals. This link to olfaction is associated with the abundance and number of kinds of mammalian skin glands: sweat, sebaceous, and scent (see section 20.2: Glands), which communicate considerable information. Olfaction is amazingly acute in mammals. Even humans, not celebrated for their sense of smell, can discriminate about 10,000 different odors.

Olfactory receptors are distributed on the surfaces of the nasal cavity, as olfactory endings of the olfactory bulb of the brain (figure 20.10). Some mammals supplement their olfactory sense with vomeronasal organs, also located in the nasal cavity. Olfactory information is processed in an anterior region of the **cerebral cortex.** The cerebral cortex, also called the neocortex in mammals, is the outermost layer of gray matter in the cerebrum. Unlike other vertebrates, the cortex of mammals usually

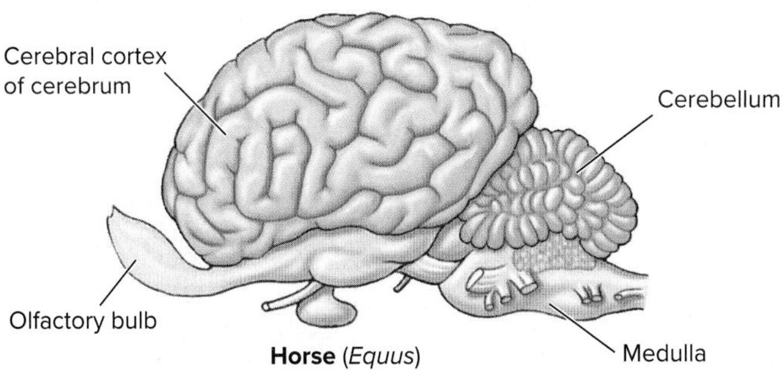

figure 20.10

Brain of a horse (anterior to left). Olfactory information detected by nerve endings of the olfactory bulb is relayed to the cerebral cortex for processing. The cerebral cortex is the outer layer of the cerebrum, highly folded into ridges and grooves. It receives nearly all sensory information, sends out directions for muscle contraction, and performs higher functions, such as memory, learning, and communication.

is highly folded. The cortex receives nearly all sensory information, controls voluntary muscle movements, and performs integrative functions such as memory, judgment, and reasoning. The cerebrum is much larger in mammals than in other vertebrates; this expansion in size is linked, in part, to the complex social systems of mammals.

Key Theme 20.3
ADAPTATIONS AND PHYSIOLOGY

The Sweet Smell of Research Success

Much of our understanding of the olfactory system is due to the research of Linda Buck and Richard Axel, who discovered a large gene family of about 1000 olfactory genes, comprising about 3% of the human genome. Each of these genes is associated with one of the 1000 different receptors on the olfactory epithelium, with each receptor responding to a select few odorants, sending its information to the olfactory bulb. This research earned Buck and Axel the 2004 Nobel Prize in Physiology or Medicine.

Food and Feeding

Mammals exploit an enormous variety of food sources; some mammals require highly specialized diets, whereas others are opportunistic feeders that thrive on diversified diets. A mammal's adaptations for finding, capturing, chewing, swallowing, and digesting food are inextricably linked with its diet.

Teeth, perhaps more than any other single physical characteristic, reveal the life habit of a mammal (figure 20.11). All mammals have teeth (with few exceptions), and their modifications are correlated with what a mammal eats.

As mammals evolved during the Mesozoic era, major changes occurred in teeth and jaws. Unlike the uniform **homodont** dentition of the first synapsids, mammalian teeth became differentiated to perform specialized functions such as cutting, seizing, gnawing, grinding, and crushing. Teeth differentiated in this manner are called **heterodont.** Mammalian teeth are differentiated into four types: **incisors (I),** with simple crowns and sharp edges, used mainly for snipping; **canines (C),** with long conical crowns, specialized for piercing; and **premolars (PM)** and **molars (M),** with compressed crowns and one or more cusps, suited for shearing, crushing, or grinding. The ancestral tooth formula, which expresses the number of each tooth type in one-half of the upper and lower jaw, was I 3/3, C 1/1, PM 4/4, M 3/3 = 44. Shrews, some omnivores, and some carnivores come closest to this ancestral pattern (figure 20.11).

Unlike most other vertebrates, mammals do not continuously replace their teeth throughout their lives. Most mammals grow just two sets of teeth: a temporary set, called **deciduous teeth,** or milk teeth, is replaced by a permanent set when the skull has grown large enough to accommodate a full set. Only incisors, canines, and premolars are deciduous; molars are never replaced and the single permanent set must last a lifetime.

Feeding Specializations

The feeding, or trophic, apparatus of a mammal—teeth and jaws, tongue, and digestive tract—is adapted to the animal's particular feeding habits. Mammals are customarily divided among four basic trophic categories—insectivores, carnivores, omnivores, and herbivores—but many other feeding specializations have evolved in mammals, and feeding habits of some mammals defy exact classification. The principal feeding specializations of mammals are shown in figure 20.11.

Insectivorous mammals, such as shrews, moles, anteaters, and most bats, feed on a variety of insects and other small invertebrates. Most insectivorous mammals have teeth with pointed cusps, permitting them to puncture the exoskeleton or skin of their prey. Some large insectivores, such as anteaters and pangolins, completely lack teeth (figure 20.11).

Herbivorous mammals, which feed on grasses and other vegetation, are of two main groups: (1) **browsers** and **grazers,** including ungulates (hooved mammals, including horses, deer, antelope, cattle, sheep, and goats); and (2) gnawers, including many rodents, rabbits, and hares. In herbivores, canines are absent or reduced in size, whereas molars and premolars, which are adapted for grinding, are broad, ridged, and usually high crowned. Rodents have chisel-sharp incisors that grow throughout life and must be worn away to keep pace with their continual growth (figure 20.11).

Herbivores have a number of adaptations for processing their fibrous diet of plant food. **Cellulose,** the structural carbohydrate of plants, is composed of long chains of glucose molecules and therefore is a potentially nutritious food resource. However, no vertebrates synthesize cellulose-splitting enzymes **(cellulases).** Instead, herbivorous vertebrates harbor anaerobic bacteria and ciliates (see section 5.5: Ciliophora) that produce cellulase in fermentation chambers in their gut. Here, **fermentation** refers to a process in which ATP is generated, without oxygen, by enzymatic breakdown

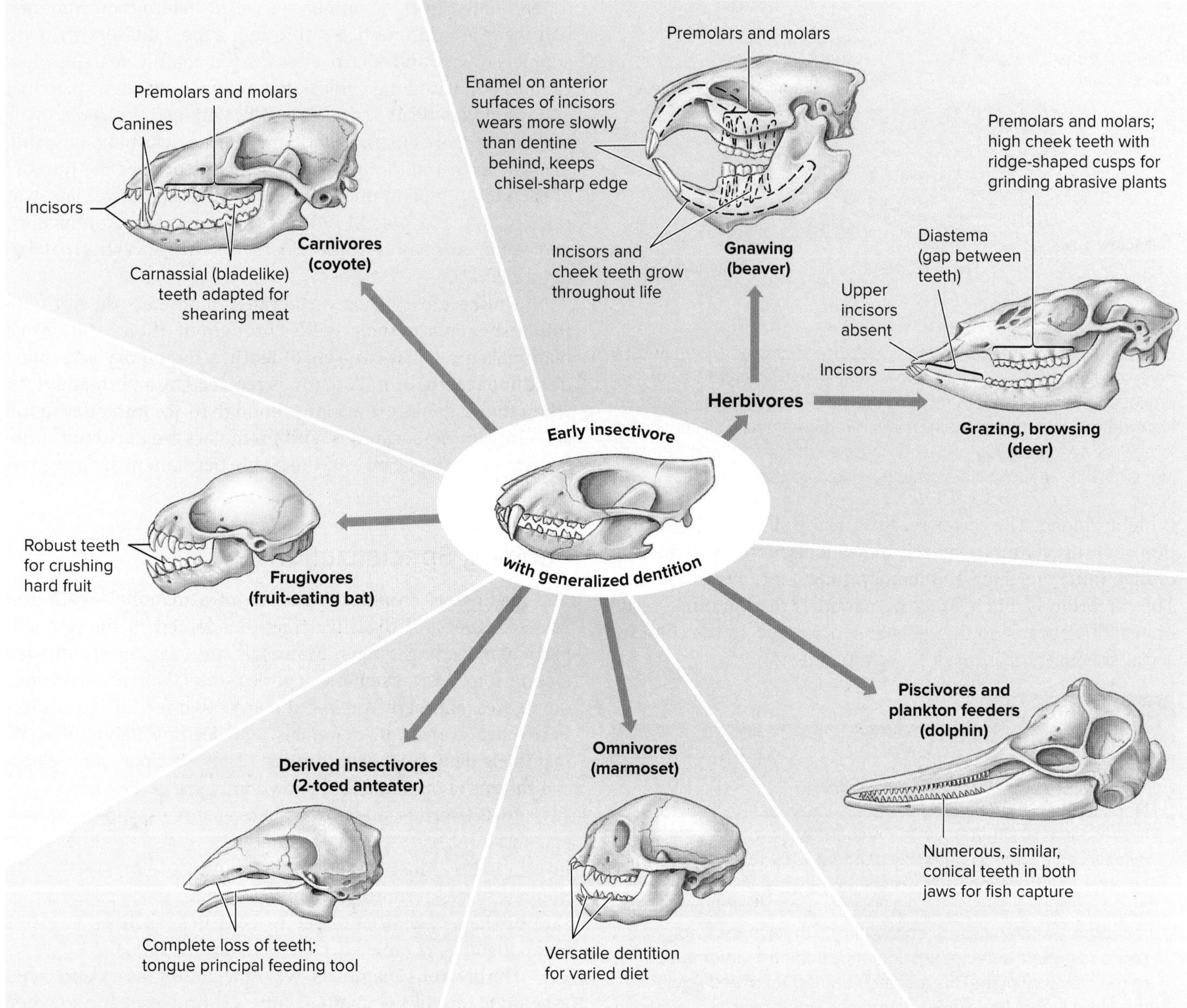

figure 20.11

Feeding specializations of major trophic groups of placental (eutherian) mammals. Early placentals were insectivores; all other types are descended from them.

of cellulose. Simple carbohydrates, proteins, and lipids produced by the microorganisms can be absorbed by the host, and the host can digest the microorganisms as well.

Fermentation in some herbivores, such as rabbits, horses, elephants, some primates, and many rodents, occurs primarily in the colon and in a spacious side pocket, or diverticulum, called a **cecum** (figure 20.12). Although some absorption occurs in the colon and cecum, most fermentation occurs posterior to the small intestine, where nutrients are absorbed, so many nutrients are lost in feces. Rabbits and many rodents eat their fecal pellets **(coprophagy),** giving food a second pass through the gut to extract additional nutrients.

Ruminants (cattle, bison, buffalo, goats, antelopes, sheep, deer, giraffes, and okapis) have a huge four-chambered stomach (figure 20.12). As a ruminant feeds, food passes down the esophagus to the **rumen,** where it is digested by microorganisms and then formed into small balls of cud. The ruminant returns cud to its mouth, where it is deliberately chewed at length to crush the fiber. Swallowed again, the food returns to the rumen, where the cellulolytic microorganisms continue fermentation. The pulp passes to the **reticulum,** and then to the **omasum,** where water-soluble food and microbial products are absorbed. The remainder proceeds to the **abomasum** ("true" acid stomach) and small

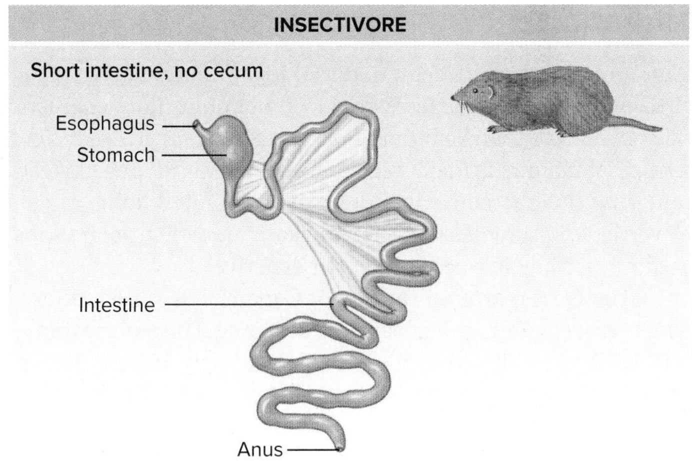

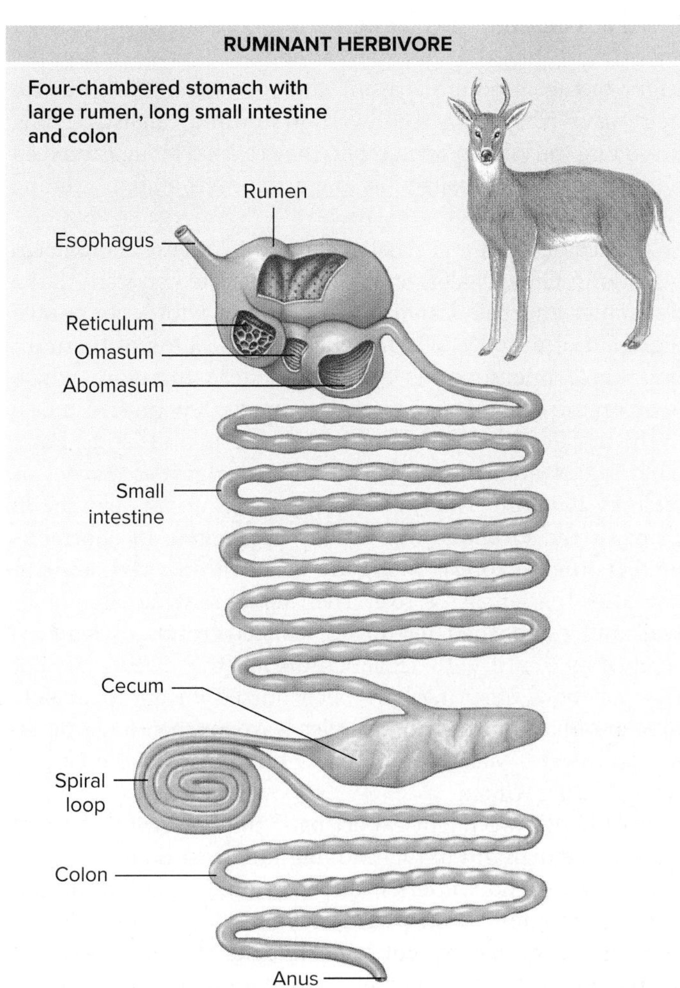

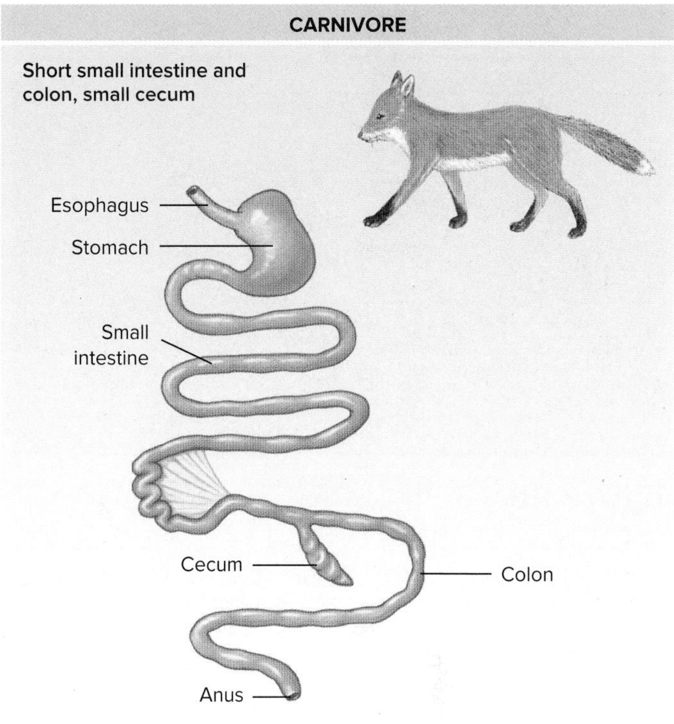

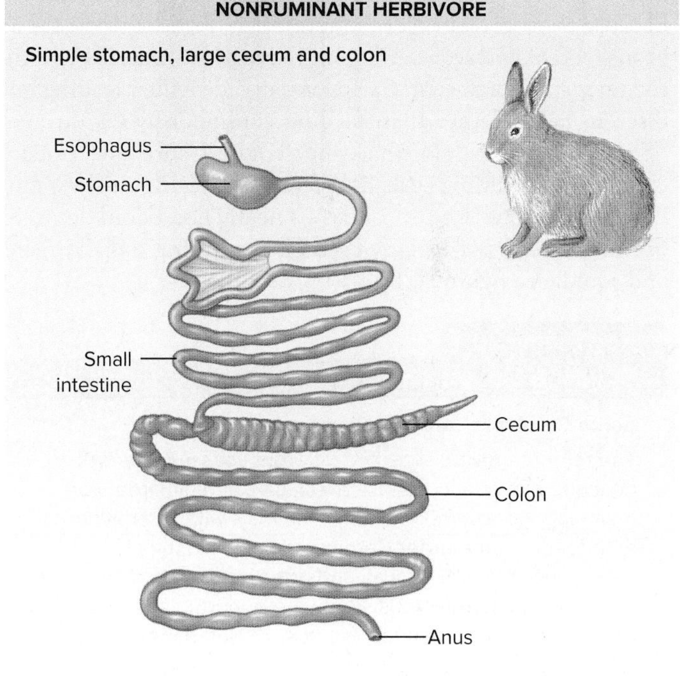

figure 20.12

Digestive systems of mammals, showing different morphology with different diets.

intestine, where proteolytic enzymes are secreted and normal digestion occurs.

Herbivores generally have large, long digestive tracts and must eat a considerable amount of plant food to survive. An African elephant weighing 6 tons must consume 135–150 kg (300–400 pounds) of rough fodder each day to obtain sufficient nourishment.

Carnivorous mammals, including foxes, dogs, weasels, wolverines, fishers, and cats, feed mainly on other vertebrates. Carnivores are well equipped with long canine teeth and

Miguel Sanz/Flickr Open/Getty Images

figure 20.13

Lions, *Panthera leo*, eating a wildebeest. Lacking stamina for a long chase, lions stalk prey and then charge suddenly, surprising their prey. After gorging themselves, lions sleep and rest for periods as long as one week before eating again. Order Carnivora, family Felidae.

powerful clawed limbs for killing their prey (figure 20.13). Their premolars and molars often are bladelike, and used in a scissors-like way to cut muscle and tendon (see figure 20.11). Because their protein diet is more easily digested than the fibrous food of herbivores, their digestive tract is shorter, and the cecum is small or absent. Carnivores organize their feeding into discrete meals rather than feeding continuously (as do most herbivores) and therefore have much more leisure time. Aquatic carnivores, including seals and cetaceans, feed on fishes, molluscs, and crustaceans. Most have a nearly homodont dentition of conical teeth, particularly useful for capturing slippery fishes and squid (see figure 20.11).

Key Theme 20.4

SCIENCE EXPLAINED

Some Carnivores are Not Carnivores

The term "carnivore" has two different uses in mammals: to describe diet and to denote specific taxonomic orders of mammals. For example, not all mammals with a carnivorous diet belong to the order Carnivora (many marsupials and cetaceans are carnivorous), and not all members of the order Carnivora are carnivorous. Many are opportunistic feeders, and some, such as pandas, are strict vegetarians.

Omnivorous mammals, which use both plants and animals for food, include pigs, raccoons, many rodents, bears, and most primates, including humans. Omnivores have a versatile dentition, with broad, rounded molars for crushing food. Their cecum is usually poorly developed.

Most mammals are notably opportunistic in their diets; many are difficult to categorize in a feeding guild. For example, foxes, often regarded as carnivores, usually feed on rodents, rabbits, and birds, but consume fruit, grains, nuts, and insects when their normal food is scarce.

Migration

Migration is a more difficult undertaking for mammals than for birds or fishes, because terrestrial locomotion is more energetically expensive than swimming or flying. Not surprisingly, few terrestrial mammals make regular seasonal migrations, instead centering their activities in a defined and limited home range. Nevertheless, some striking terrestrial mammalian migrations occur, especially in northern North America.

The barren-ground caribou of Canada and Alaska undertakes direct and purposeful mass migrations spanning 160–1100 km (100–700 miles) twice annually (figure 20.14). From winter ranges in boreal forests (taiga), they migrate rapidly in late winter and spring to calving ranges on the barren grounds (tundra). Calves are born in mid-June. As summer progresses, caribou are increasingly harassed by warble and nostril flies that bore into their flesh, by mosquitos that drink their blood (estimated at a liter per caribou each week during the height of the mosquito season), and by wolves that prey on their calves. They move southward in July and August, feeding little along the way. In September, they reach the taiga and feed there almost continuously on low ground vegetation. Mating (rut) occurs in October.

Oceanic seals and whales make the longest mammalian migrations. Gray whales, for example, migrate between Alaska in summer and Baja California, Mexico, in winter, an annual migration of over 18,000 km (11,250 miles). One of the most remarkable migrations is that of northern fur seals, which breed on the Pribilof Islands approximately 300 km (185 miles) off the coast of Alaska and north of the Aleutian Islands. From wintering grounds off southern California, females journey as much as 2800 km (1740 miles) across open ocean, arriving in spring at the Pribilofs, where they congregate in enormous numbers (figure 20.15). Young are born within a few hours or days after the females arrive. Then males, having already arrived and established territories, collect groups of females, which they guard with vigilance during the mating period. After newborn offspring have been nursed for approximately three months, females and juveniles leave for their long migration southward. Males do not follow but remain in the Gulf of Alaska during winter.

Although we might expect bats, the only winged mammals, to use their gift of flight to migrate, most do not migrate and instead spend winters in hibernation. Those that do migrate, like the little brown bat (see figure 20.17), travel only a few hundred kilometers, but some, like the Mexican free-tailed bat, travel over 1500 km on their journey from the United States to Central America.

Flight and Echolocation

Many mammals scamper through trees with amazing agility; some can glide from tree to tree, and one group, bats, has full flight. Gliding and flying evolved independently in several groups of mammals, including marsupials, rodents, flying lemurs, and bats. Flying squirrels (figure 20.16) actually glide

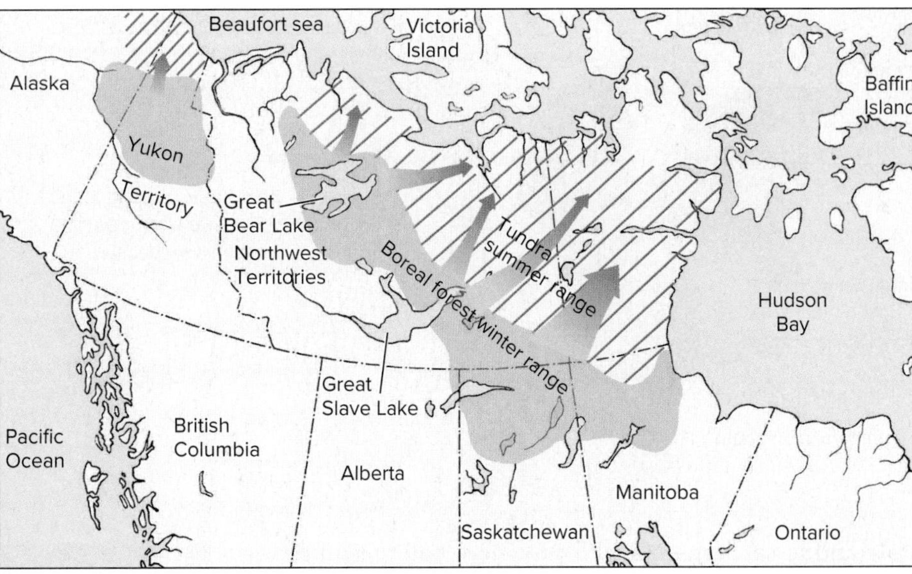

figure 20.14

Barren-ground caribou, *Rangifer tarandus,* of Canada and Alaska. **A,** Adult male caribou in autumn pelage and antlers in velvet. **B,** Summer and winter ranges of some major caribou herds in Canada and Alaska. (Other herds not shown occur on Baffin Island and in western and central Alaska.) The principal spring migration routes are indicated by arrows; routes vary considerably from year to year. The same species is known as reindeer in Europe. Order Artiodactyla, family Cervidae.

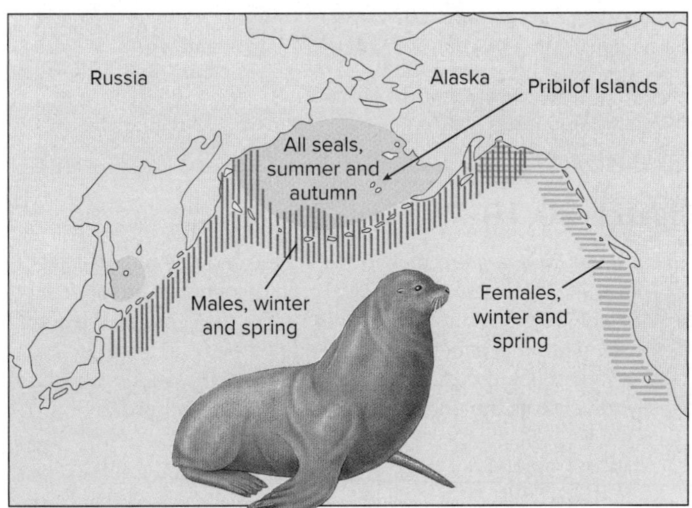

figure 20.15

Annual migrations of northern fur seals, *Callorhinus ursinus,* showing the separate wintering grounds of males and females. Both males and females of the large Pribilof population migrate in early summer to the Pribilof Islands, where females give birth and then mate with males. Order Carnivora, family Otariidae.

figure 20.16

Southern flying squirrel, *Glaucomys volans,* gliding in for a landing. The area of the undersurface is nearly tripled in area when the gliding skin is spread. Glides of 40–50 m are possible. Good maneuverability during flight is achieved by adjusting the position of the gliding skin and flattened tail. Flying squirrels are nocturnal and have superb night vision. Order Rodentia, family Sciuridae.

rather than fly, using the gliding skin (patagium) that extends from the sides of their body. Even bats must initiate their flight by gliding from a vertical surface or ledge; most bats cannot fly upward from the ground (vampire bats are an exception).

Bats are mostly nocturnal or crepuscular (active at twilight), and thus hold a niche unoccupied by most birds. Their achievement is attributed to two features: flight and capacity to navigate by echolocation. Together, these adaptations enable bats to avoid obstacles in absolute darkness, to

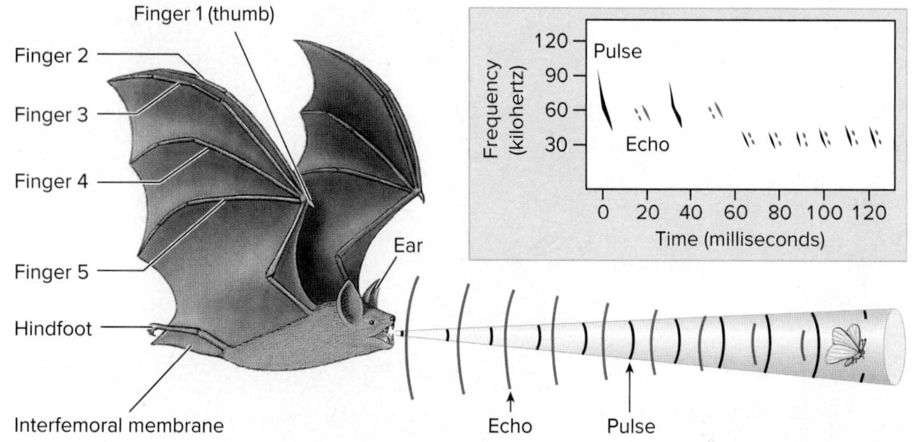

figure 20.17

Echolocation of an insect by a little brown bat, *Myotis lucifugus*. Frequency-modulated pulses are directed in a narrow beam from the bat's mouth. As the bat nears its prey, it emits shorter, lower signals at a faster rate. Order Chiroptera, family Vespertilionidae.

locate and to catch insects with precision, and to find their way deep into caves (a habitat largely unexploited by other mammals and birds) where they sleep during daytime hours.

When in flight, a bat emits short pulses 5–10 msec in duration in a narrow directed beam from its mouth or nose (figure 20.17). Each pulse is frequency modulated; it is highest at the beginning, up to 100,000 Hz (hertz, cycles per second), and sweeps down to perhaps 30,000 Hz at the end. Sounds at this frequency are ultrasonic to human ears, which have an upper limit of about 20,000 Hz. When bats search for prey, they produce about 10 pulses per second. If prey is detected, the rate increases rapidly up to 200 pulses per second in the final phase of approach and capture. Pulses are spaced so that the echo of each is received before the next pulse is emitted, an adaptation that prevents jamming. Since transmission-to-reception time decreases as a bat approaches an object, the bat can increase pulse frequency to obtain more information about an object. Pulse length is shortened as the bat nears an object. Some nocturnal moths can detect a bat's ultrasonic pulses and avoid predation.

External ears of bats are large, like hearing trumpets, and shaped variously in different species. A bat builds a mental image of its surroundings from echo scanning that approaches the visual resolution of a diurnal animal's eyes.

Some bats, including the approximately 185 species of Old World fruit bats, lack echolocation abilities. Even so, most are primarily nocturnal, using large eyes and olfaction to find their meals of fruits, flowers, and nectar.

figure 20.18

African lions, *Panthera leo*, mating. During the short period a female is receptive, she may mate repeatedly. Three or four cubs are born after gestation of 100 days. Once the mother introduces cubs into the pride, they are treated with affection by both adult males and females. Once mature, male cubs are driven from the pride, but maturing females usually remain with the pride. Order Carnivora, family Felidae.

Reproduction

Most mammals have definite mating seasons, usually in winter or spring and timed so that birth and the rearing of young occur at the most favorable time of the year. Many male mammals can copulate at any time, but females are receptive only during a relatively brief period of the **estrous cycle** called **estrus** or heat (figure 20.18).

There are three different patterns of reproduction in mammals, represented by monotremes, marsupials, and placental mammals. **Monotremes** are egg-laying (oviparous) mammals. The duck-billed platypus, has one breeding season each year. Embryos develop for 10–12 days in the uterus, where they are nourished by yolk supplies deposited prior to ovulation and by secretions from the mother. A thin, leathery shell is secreted around the embryos before the eggs are laid. The platypus lays its eggs in a burrow, where they hatch in a relatively underdeveloped state after about 12 days. After hatching, the young feed on milk produced by the mother's mammary glands. Because monotremes have no nipples, the young lap milk secreted onto the belly fur of the mother.

Marsupials are pouched, viviparous mammals that exhibit a second pattern of reproduction. Although only eutherians are called "placental mammals," marsupials do have a transient type of placenta called a **choriovitteline** (yolk sac)

placenta. Embryos (blastocysts) of a marsupial are first encapsulated by shell membranes and float freely for several days in uterine fluid. After "hatching" from the shell membranes, embryos of most marsupials do not implant, or "take root," in the uterus as they would in eutherians, but erode shallow depressions in the uterine wall in which they lie and absorb nutrient secretions from the mucosa through the vascularized yolk sac. **Gestation** (the intrauterine period of development) is brief in marsupials, and therefore all marsupials give birth to tiny young that are effectively still embryos, both anatomically and physiologically (figure 20.19). However, early birth is followed by a prolonged interval of lactation and parental care (figure 20.20).

The third pattern of reproduction is that of viviparous **placental mammals,** eutherians. In placentals, the reproductive investment is in prolonged gestation—unlike marsupials, in which the reproductive investment is in prolonged lactation (figure 20.20). Like marsupials, most placental embryos initially receive nutrients through a choriovitteline placenta. The choriovitteline placenta may persist, as in mice, but usually is transitory and quickly replaced by a **chorioallantoic placenta,** formed from the chorionic and allantoic membranes (see figure 18.3). In eutherians, both types of placentae are extensively vascularized. The numerous capillaries of the placenta and the mother's uterus are close, but do not connect, and allow nutrients, respiratory gases, wastes, and other molecules to diffuse between the maternal and embryonic circulation. Length of gestation is longer in placentals than in marsupials, and generally increases with body size. For example, mice have a gestation period of 21 days; rabbits and hares, 30–36 days; cats and dogs, 60 days; cattle, 280 days; and elephants, 22 months (the longest). Important exceptions include baleen whales, the largest mammals, whose gestation period is only 12 months, and small bats no larger than mice, whose gestation period extends 4–5 months. The condition of the young at birth also varies. Most antelopes bear **precocial** young that are well-furred, eyes open, and able to run. Newborn mice, however, are **altricial:** blind, naked, and helpless.

figure 20.19

Virginia opossums, *Didelphis virginiana*, fastened to nipples in their mother's pouch. When born after a gestation period of only 12 days, they are the size of honey bees. They remain attached to the nipples for 50–60 days. Order Didelphimorphia, family Didelphidae.

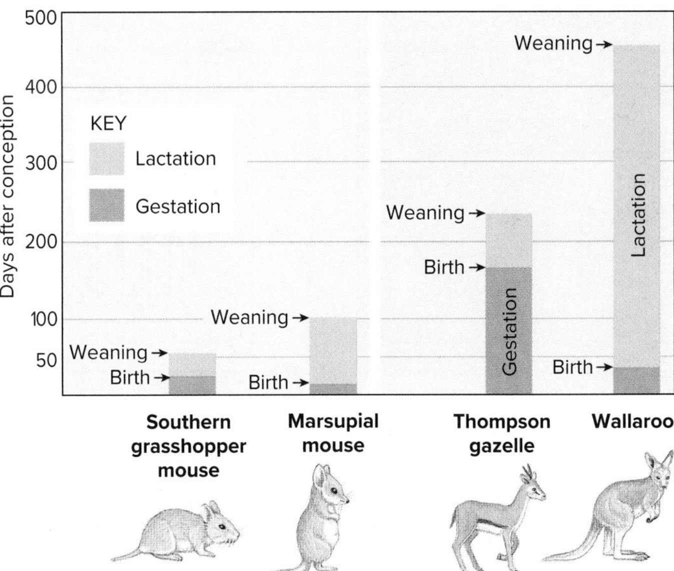

figure 20.20

Comparison of gestation and lactation periods between matched pairs of ecologically similar species of marsupial and placental mammals. The graph shows that marsupials have shorter intervals of gestation and longer intervals of lactation than do similar species of placentals.

Key Theme 20.5

GENETICS AND DEVELOPMENT

Delayed Implantation

Delayed implantation lengthens the gestation period of many mammals. The blastocyst remains dormant while its implantation in the uterine wall is postponed for periods of a few weeks to several months. For many mammals (for example, bears, seals, weasels, some bats, and a few deer), delayed implantation extends gestation so that the young are born at a time of year best for their survival.

Torpor

Endotherms, animals that regulate their body temperature using a metabolic process, have a number of adaptations, such as hair and subcutaneous fat (see figures 20.4 and 20.5) that reduce their energetic costs in cold environments. Still, it is energetically expensive to maintain a high body temperature, especially for small mammals. Some mammals migrate to warmer climates (see figures 20.14 and 20.15), but most cold-inhabiting mammals do not. It is not surprising, then, that many small mammals abandon homeostasis, allowing their body temperature to drop below normal levels.

NPS / Jacob W. Frank

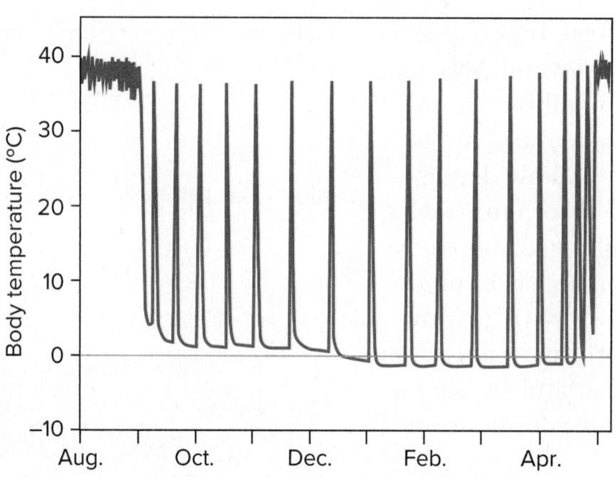

figure 20.21

Hibernation in the Arctic ground squirrel (*Urocitellus parryii*) in Alaska. During the winter, body temperature decreases to about 0° C, although there are frequent, brief episodes of arousal. The function of these episodes is unclear. Order Rodentia, family Sciuridae.

Torpor refers to a state of reduced activity and decreased metabolic rate. Many small mammals (body sizes less than 8 kg) survive the cold northern winter, with its scarcity of food, by entering into a deep state of torpor called **hibernation**. True hibernators include ground squirrels, chipmunks, marmots, and some bats, among others. During hibernation, body temperature cools to near-ambient temperature and metabolic rate decreases to a fraction of normal (figure 20.21). In ground squirrels, for example, the respiratory rate decreases from a normal rate of 100 breaths per minute to 1–3 breaths per minute, and the heart rate decreases from 250 to 5 beats per minute. Feeding does not occur; instead, hibernators use considerable fat reserves accumulated prior to hibernation.

Large mammals, such as bears and badgers, can enter a state of torpor in the winter characterized by a modest drop in body temperature (3–7° C drop in black bears). This sometimes is called "shallow torpor" to distinguish it from the deep torpor of rodents, but many researchers now use hibernation for all forms of winter torpor, emphasizing the common decrease in metabolic rate. Some very small mammals, especially bats, allow their body temperature to fall on a daily basis during their period of sleep or inactivity. This **daily torpor** achieves energy-saving benefits similar to those of other forms of torpor.

20.3 Human Evolution

Darwin devoted an entire book, *The Descent of Man and Selection in Relation to Sex* (1871), largely to human evolution. The idea that humans share common descent with apes and other animals was repugnant to the Victorian world, which responded with predictable outrage (see figure 1.18). When Darwin's views were first debated, few human fossils had been unearthed, but the current accumulation of fossil evidence has strongly vindicated Darwin's hypothesis that humans descend from other primates.

Primate Diversity

The earliest primate was probably a small, nocturnal animal similar in appearance to a tree shrew. This ancestral primate lineage split into two lineages, one of which gave rise to lemurs (figure 20.22) and lorises, and the other to tarsiers, monkeys (figure 20.23), and apes (figure 20.24). Most primates are arboreal (tree-dwelling), which is the probable ancestral lifestyle for the group. All have long arms with a large range of motion, essential for active animals moving through the trees. Grasping hands and feet, in contrast to the clawed feet of squirrels and other rodents, enable primates to grip limbs, to hang from branches, to seize and to manipulate food, and most significantly, to use tools. Primates have highly developed special senses, especially acute, binocular vision, and proper coordination of limb and finger muscles to assist their active arboreal life. Of course, sense organs are no better than the brain at processing sensory information. A large **cerebral cortex** supports timing, judgment of distance, and perception of the environment.

Three clades of monkeys and apes are recognized: (1) New World monkeys of the Americas, including howler monkeys, spider monkeys (see figure 20.23A), and tamarins;

Daniel Hernanz Ramos/Moment/Getty Images

figure 20.22

Ring-tailed lemurs, *Lemur catta*, like all living lemurs, only occur on Madagascar. Order Primates, family Lemuridae.

A

B

figure 20.23

Monkeys. **A,** Geoffroy's spider monkey, *Ateles geoffroyi*, order Primates, family Atelidae, is a New World monkey. **B,** Olive baboons, *Papio homadryas*, order Primates, family Cercopithecidae, are Old World monkeys.

figure 20.24

Western lowland gorillas, *Gorilla gorilla gorilla*, order Primates, family Hominidae, are apes.

(2) Old World monkeys of Africa and Asia, including baboons (see figure 20.23B), mandrills, and colobus monkeys; and (3) apes (figure 20.24). Old World monkeys differ from New World monkeys in lacking a grasping tail, while having close-set nostrils, better opposable and grasping thumbs, and only two premolars in each jaw half.

Apes, which differ from Old World monkeys in having a larger cerebrum and loss of a tail, include gibbons, orangutans, gorillas, chimpanzees, bonobos, and humans. Except for gibbons, all apes are in the family Hominidae and are here called **hominids.** Chimpanzees and bonobos, *Pan,* form the living sister group to humans (see figure 4.7). All fossil hominid species that are phylogenetically closer to living humans than to chimpanzees and bonobos are called here **humans** or **hominins.** The earliest known fossils of apes are from 23-million-year-old rocks in eastern Africa.

The First Humans and Bipedalism

Trends in the evolution of skeletal differences between humans and other hominids often are associated with changes in diet and posture. Human jaws are less robust and bear smaller canines than those of other hominids, reflecting a more omnivorous diet. Position of the **foramen magnum** (a hole in the skull through which the spinal cord passes) shifted to directly underneath the braincase in humans, indicative of bipedalism and upright posture. Bipedalism provided a better view of the landscape and freed the hands for using tools, defense, carrying young, and gathering food.

Genetic evidence suggests that humans diverged from chimpanzees about 10–7 million years ago. Fossil evidence of humans from this period is sparse and controversial. In 2001 the desert sands of Chad yielded a remarkably complete skull of a hominid, *Sahelanthropus tchadensis,* dated at 7–6 million years ago (figure 20.25). Although its brain was no larger than that of a chimpanzee (320–380 cm^3), its relatively small canine teeth and ventral position of the foramen magnum suggest that the skull could be human.

The earliest well-known human is *Ardipithecus ramidus* from Ethiopia, dated at about 4.4 million years ago (figure 20.25). Until recently this species was known only from teeth, but in 2009 numerous additional fossils were described, including a 45% complete skeleton, named "Ardi." *Ardipithecus* was about 120 cm tall and bipedal, although it retained many ancestral

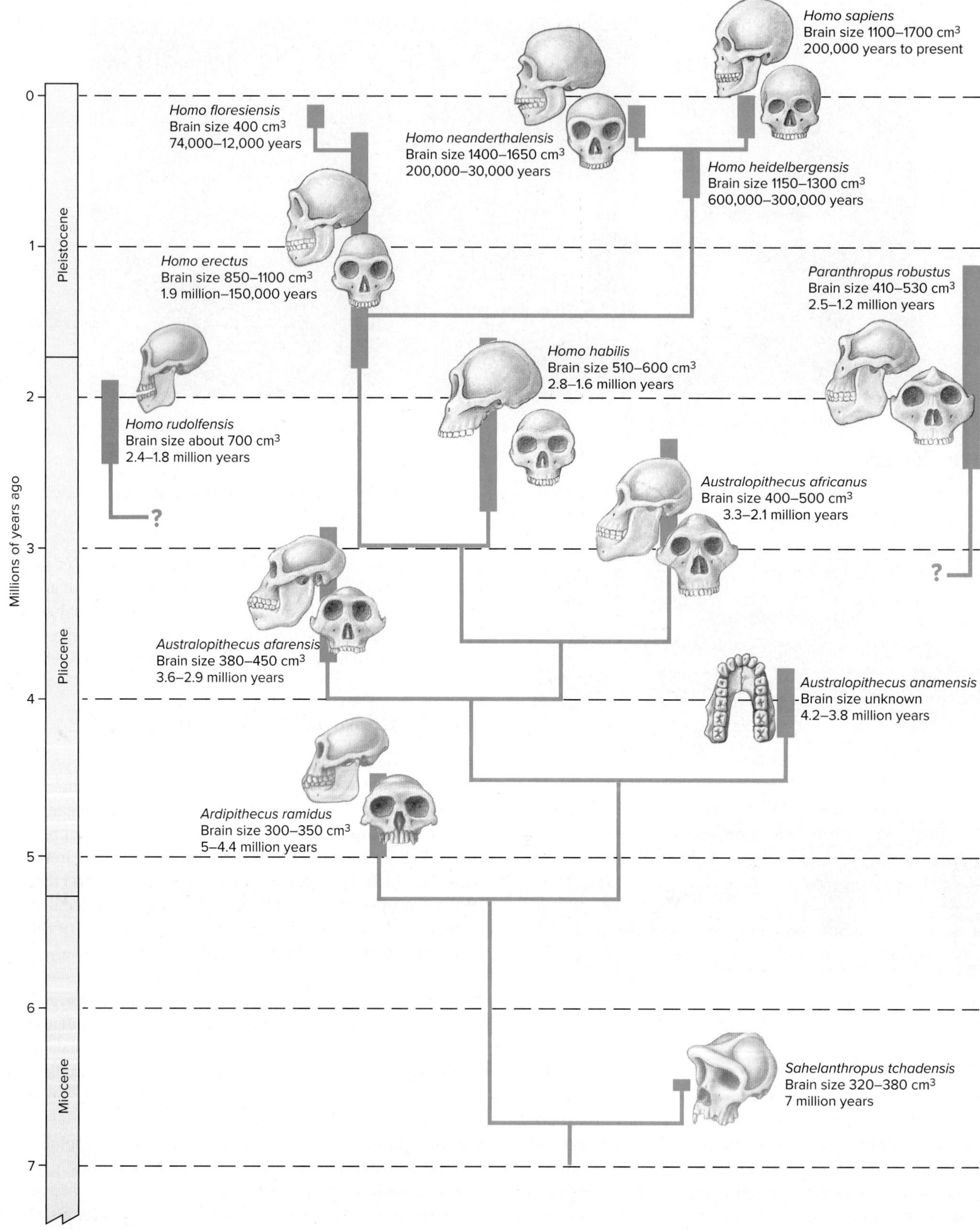

figure 20.25

Evolution of hominins. Shown is one possible phylogenetic reconstruction. Many relationships are controversial; for example, some divide *H. erectus* and *H. heidelbergensis* into several species, while others consider each of these to be single species lineages. Note that in this reconstruction *Australopithecus* and *Homo erectus* are paraphyletic. Thick vertical lines denote known durations of fossil forms in the stratigraphic record.

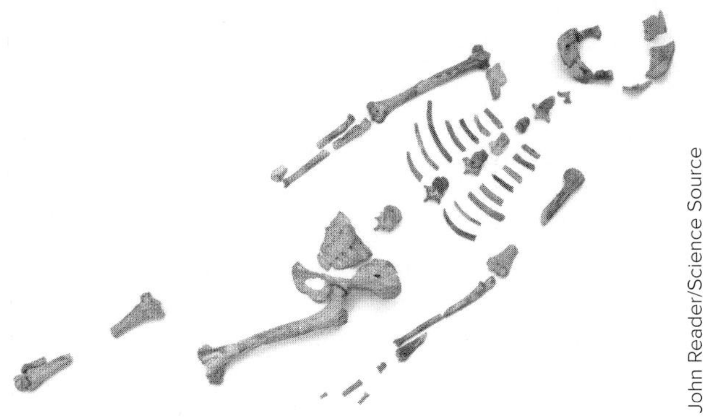

figure 20.26

Lucy (*Australopithecus afarensis*), one of the most nearly complete skeletons of an early human ever found. Lucy is dated at 3.2 million years old. Numerous additional fossils of this species have been found, including complete skulls in 1992 and 2000.

adaptations for arboreal life, including long arms, fingers, and toes. Both *Sahelanthropus* and *Ardipithecus* were probably woodland dwellers, based on associated fossil vertebrates and invertebrates from the same deposits. Earlier hypotheses suggested that bipedalism arose as an adaptation to drying African environments, as woodlands were replaced with open savannas. However, because formation of the savannas of Africa did not occur until about 3 million years ago, it is now clear that bipedalism first arose in woodland-dwelling hominins.

Another celebrated early human fossil is a 40% complete skeleton of a female *Australopithecus afarensis* (figure 20.26). Unearthed in 1974 and named "Lucy" by its discoverer, Donald Johanson, *A. afarensis* was a short, bipedal human with a brain size slightly larger than that of a chimpanzee (figure 20.27). This species was 1–1.5 m tall, and its teeth suggest it was primarily a fruit-eater, but also ate meat. The numerous fossils of *A. afarensis* date from 3.7 to 3 million years ago.

An explosion of australopithecine fossil finds during the last five decades documents at least eight species. Most of the early forms are considered gracile australopithecines because of their comparatively light build, especially in skull and teeth (although all were more robust than modern humans). The oldest of these, *Australopithecus anamensis,* lived 4.2–3.9 million years ago, while the youngest, *A. sediba,* lived about 2 million years ago.

Coexisting with the earliest species of *Homo* and the latest species of gracile *Australopithecus* were at least three robust australopithecines, including *Paranthropus robustus* (see figure 20.25). The "robust" australopithecines, which lived from 2.5 to 1.2 million years ago, were characterized by prominent skull crests, heavy jaws, and large back molars. They had a broad diet, consuming coarse seeds, roots, and nuts in addition to the fruit, soft plants, and invertebrates typical of other early hominins. They form an extinct branch in hominid evolution and are not part of our own ancestry.

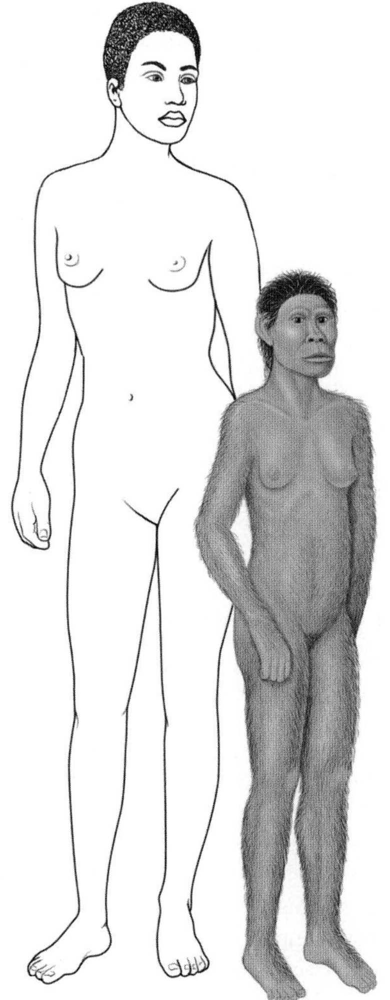

figure 20.27

A reconstruction of the appearance of Lucy (*right*) compared with a modern human (*left*).

Early *Homo:* Tool Making and Migration Out of Africa

The earliest *Homo* fossils are dated about 2.8 million years ago. Because the human fossil record during this time is sparse, the evolution of *Homo* is unclear, although most researchers consider *Homo habilis* ("handy man") to be representative of the earliest *Homo*. This species was similar in form to *Australopithecus,* about 111–135 cm tall (3.7–4.5 feet), with long arms and short legs. However, it had a larger brain (500–600 cm^3) than did *Australopithecus,* and may have used it in a novel cultural feature—the making of stone tools. This capacity to make stone tools may be the defining feature of genus *Homo*. Chimpanzees and bonobos do not make stone tools, and they cannot be taught to make simple stone tools, despite efforts by patient researchers. *Homo habilis* shared the early Pleistocene drying African landscape with many other hominids, including species of *Paranthropus, Australopithecus,* and other species of *Homo*. Although the expansion of savannas is no longer linked to the origin of bipedalism, perhaps it contributed to the spread of savanna-dwelling *Homo*.

About 1.9 million years ago, *Homo erectus* appeared, a large hominid standing 150–190 cm (5–6 feet) tall, with a low but distinct forehead and strong brow ridges. This species is known from Africa and Eurasia; those from Africa sometimes are considered to be a different species, *H. ergaster*. The cranial capacity of *H. erectus* ranged from 850 cm^3 (early forms) to 1100 cm^3 (late forms), between that of *H. habilis* and modern humans (see figure 20.25). The technology of *H. erectus* is characterized by more advanced tools and the control and use of fire, as indicated by charcoal deposits. It spread to southern Europe and to Asia as far east as China and Java, where it survived until about 150,000 years ago. One of the most unusual hominid finds was announced in 2004: *Homo floresiensis*, a species only 1 m tall, from the island of Flores, Indonesia. This species diverged from *H. erectus* and became extinct only about 13,000 years ago.

Modern Humans

Modern humans diverged from African *H. erectus* about 800,000 years ago. These humans, now assigned to *H. heidelbergensis*, spread throughout Africa, Europe, and the Middle East. About 200,000 years ago, when long glacial conditions predominated, *Homo heidelbergensis* was replaced by two hominins, *H. neanderthalensis* in Europe and *H. sapiens* in Africa. **Neanderthals** (*H. neanderthalensis*) occupied most of Europe and the Middle East. They had a brain size slightly larger than that of modern humans and made more sophisticated stone tools than did *H. erectus*. Their robust, heavily muscled bodies allowed them to survive the cold climates of the Ice Age and to hunt large Pleistocene mammals.

Homo sapiens, as presently defined, arose in Africa about 200,000 years ago. About 39,000 years ago Neanderthals disappeared, about 3000 years after *H. sapiens* spread into Europe and Asia. The first modern humans were tall people with a culture different from that of Neanderthals. Implement crafting developed rapidly, and human culture became enriched with aesthetics, artistry, and sophisticated language.

What happened to the Neanderthals? Direct evidence is lacking, but most anthropologists venture that *H. sapiens*, with its technological and cultural advantages, outcompeted other humans. Recent evidence suggests some interbreeding with *H. sapiens* about 50,000 years ago (and possibly later, in Asia, with a group of early humans known as Denisovans). This apparently occurred after *H. sapiens* migrated from Africa, because Neanderthal genes are present in living Asians and Europeans, but absent in Africans. Because of this interbreeding, some researchers consider *H. erectus* and its descendants a single species lineage.

In closing our discussion of human evolution, it is important to note that human evolution was characterized by the same patterns and processes evident in the evolution of other animal groups, including speciation and extinction. At present there is only one human species alive, an unusual situation when one considers that three to five species of humans have been present during much of the past 4 million years (see figure 20.25).

We are unique, with a nongenetic cultural evolution that provides constant feedback between past and future experience. Our symbolic languages, capacity for conceptual thought, knowledge of our history, and ability to manipulate our environment emerge from this nongenetic cultural endowment. Finally, we owe much of our cultural and intellectual achievements to our arboreal ancestry, which gave us binocular vision, superb visuotactile discrimination, and manipulative use of our hands. If horses (with one toe instead of five fingers) had human mental capacity, could they have accomplished what humans have?

Taxonomy of Living Mammalian Orders

This taxonomy follows that of Wilson and Reeder (2005). Of the 29 recognized mammalian orders, three smaller marsupial orders and ten smaller placental orders are omitted.

Mammalia

Prototheria (prō′tō-thir′ē-ə) (Gr. *prōtos*, first, + *thēr*, wild animal). Monotreme mammals.

Order Monotremata (mon′ō-trē-mä′tä) (Gr. *monos*, single, + *trēma*, hole): **egg-laying (oviparous) mammals: duck-billed platypus, echidnas.** Five species in this order from Australia, Tasmania, and New Guinea. Echidnas, *Tachyglossus* and *Zaglossus*, have long, narrow snouts adapted for feeding on ants, their chief food.

Theria (thir′ē-a) (Gr. *thēr*, wild animal).

Metatheria (met′ə-thir′ē-ə) (Gr. *meta*, after, + *thēr*, wild animal). Marsupial mammals.

Order Didelphimorphia (di′del-fī-mor′fē-ə) (Gr. *di*, two, + *delphi*, uterus, + *morph*, form): **American opossums.** These mammals, like other marsupials, are characterized by an abdominal pouch, or marsupium, in which they rear their young. Most species occur in Central and South America, but one species, the Virginia opossum, is widespread in North America; 87 species.

Order Dasyuormorphia (das-ē-yur′ō-mor′fē-ə) (Gr. *dasy,* hairy, + *uro,* tail, + *morph,* form): **Australian carnivorous mammals.** In addition to a number of larger carnivores, this order includes a number of marsupial "mice," all of which are carnivorous. Confined to Australia, Tasmania, and New Guinea; 71 species.

Order Peramelemorphia (per′ə-mel-e-mor′fē-ə) (Gr. *per,* pouch, + *mel,* badger, + *morph,* form): **bandicoots.** Like placentals, members of this group have a chorioallantoic placenta and a relatively high rate of reproduction. Confined to Australia, Tasmania, and New Guinea; 22 species.

Order Diprotodontia (di′prō-tō-don′-tē-ə) (Gr. *di,* two, + *pro,* front, + *odont,* tooth): **koala, wombats, possums, wallabies, kangaroos.** Diverse marsupial group containing some of the largest and most familiar marsupials. Present in Australia, Tasmania, New Guinea, and many islands of the East Indies; 143 species.

Eutheria (yū-ther′ē-ə) (Gr. *eu,* true, + *thēr,* wild animal). Placental mammals.

Order Proboscidea (prō′bo-sid′ē-ə) (Gr. *proboskis,* elephant's trunk, from *pro,* before, + *boskein,* to feed): **elephants.** Largest of living land animals, with two upper incisors elongated as tusks and well-developed molar teeth. Three extant species: Indian elephant, with relatively small ears, and two species of African elephants, with large ears.

Order Cingulata (sin′gyū-lä-tə) (L. *cingul,* belt): **armadillos** (figure 20.28). Insectivorous mammals with small, peglike teeth and beltlike bands of armor. Inhabit South and Central America; the nine-banded armadillo is expanding its range northward in the United States; 21 species.

Order Primates (prī-mā′tēz or prī′māts) (L. *prima,* first): **prosimians, monkeys, apes.** First among animals in brain development with especially large cerebral cortex. Most species are arboreal, with large eyes, binocular vision, grasping hands, and five digits (usually with flat nails) on both forelimbs and hindlimbs; two suborders; 376 species.

Suborder Strepsirrhini (strep′sər-rī-nē) (Gr. *strepsō,* to turn, twist, + *rhinos,* nose): **lemurs, aye-ayes, lorises, pottos, bush babies.** Seven families of arboreal primates concentrated on Madagascar, but with species in Africa, Southeast Asia, and Malay peninsula. All have a wet, naked region surrounding comma-shaped nostrils, long nonprehensile tail, and second toe provided with a claw; 88 species.

Suborder Haplorhini (hap′lō-rī-nē) (Gr. *haploos,* single, simple, + *rhinos,* nose): **tarsiers, marmosets, New and Old World monkeys, apes.** Six families, all except tarsiers, are in the clade Anthropoidea. Haplorhine primates have dry, hairy noses, ringed nostrils, and differences in skull morphology that distinguish them from strepsirrhine primates; 288 species.

Joel Zatz/Alamy Stock Photo

figure 20.28

Nine-banded armadillo, *Dasypus novemcintus.* During the day this nocturnal species occupies long tunnels, which it digs using powerful, clawed forelimbs. Order Cingulata, family Dasypodidae.

Cleveland P. Hickman, Jr.

figure 20.29

A collared pika, *Ochotona collaris,* atop a rockslide in Alaska. This little rat-sized mammal does not hibernate but prepares for winter by storing dried plants beneath boulders. Pikas face an uncertain future because their alpine tundras are shrinking due to warming climates. Order Lagomorpha, family Ochotonidae.

Order Lagomorpha (lag′ō-mor′fə) (Gr. *lagos,* hare, + *morphē,* form): **rabbits, hares, pikas** (figure 20.29). Dentition resembling that of rodents but with four upper incisors rather than two; 92 species.

Order Rodentia (rō-den′tē-ə) (L. *rodere,* to gnaw): **gnawing mammals: squirrels, mice, rats, marmots.** Most numerous of all mammals both in numbers and species. Characterized

(continued)

(*continued*)

figure 20.30

A northern shorttail shrew, *Blarina brevicauda*. This tiny but fierce mammal, with a prodigious appetite for insects, mice, snails, and worms, spends most of its time underground and is seldom seen. Shrews resemble the insectivorous ancestors of placental mammals. Order Soricomorpha, family Soricidae.

figure 20.31

Wolverine, *Gulo gulo*, in Yellowstone National Park. Wolverines, the largest members of the weasel family, are residents in boreal forests and alpine and tundra habitats in North America, Europe, and Asia. Order Carnivora, family Mustelidae.

by two pairs of chisel-like incisors that grow continually and are adapted for gnawing; 2277 species.

Order Soricomorpha (sor'i-cō-mor'fə) (L. *soric* shrew, + *morph*, form): **shrews** (figure 20.30), **moles.** Small, sharp-snouted animals that feed principally on invertebrates; 428 species.

Order Chiroptera (kī-rop'ter-ə) (Gr. *cheir,* hand, + *pteron,* wing): **bats.** Flying mammals with forelimbs modified into wings. Most are nocturnal insect-eaters and navigate by echolocation, but fruit bats are herbivorous and all but one species lack echolocation abilities; 1116 species.

Order Carnivora (car-niv'or-ə) (L. *caro,* flesh, + *vorare,* to devour): **flesh-eating mammals: dogs, wolves, cats, bears, weasels** (figure 20.31)**, seals, sea lions, walruses.** All except the giant panda have predatory habits, and their teeth are especially adapted for grabbing and killing prey and consuming flesh; worldwide but Australia has only seals and the dingo (*Canis*); 286 species.

Order Perissodactyla (pe-ris'sō-dak'til-ə) (Gr. *perissos,* odd, + *dactylos,* toe): **odd-toed hoofed mammals: horses, asses, zebras, tapirs, rhinoceroses.** Mammals with an odd number (one or three) of toes, each with a keratinized hoof

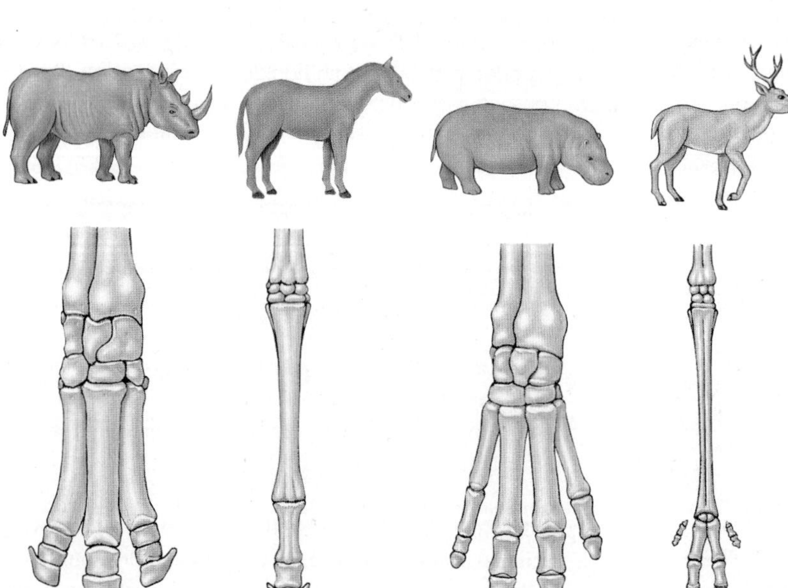

figure 20.32

Odd-toed and even-toed ungulates. Rhinoceroses and horses (order Perissodactyla) are odd-toed; hippopotamuses and deer (order Artiodactyla) are even-toed. Lighter, faster mammals run on only one or two toes.

(figure 20.32). All are herbivorous. Both Perissodactyla and Artiodactyla, often called **ungulates,** or hoofed mammals, with teeth adapted for grinding plants; 17 species.

Order Artiodactyla (är′tē-ō-dak′til-ə) (Gr. *artios,* even, + *daktylos,* toe): **even-toed hoofed mammals: swine, camels, deer, hippopotamuses, antelopes, cattle, sheep, goats.** Each toe sheathed in a keratinized hoof. Most species have two toes, although hippopotamuses and some others have four (figure 20.32). Many, such as cattle, deer, and antelopes, have horns or antlers. Most are ruminants; 240 species.

Order Cetacea (se-tā′shē-ə) (L. *cetus,* whale): **whales** (figure 20.33), **dolphins, porpoises.** Anterior limbs of cetaceans are modified into broad flippers; posterior limbs are absent. Nostrils are represented by a single or double blowhole on top of the head. Hairs are limited to a few on the muzzle, skin glands are absent except for mammary glands and those of the eye, and there is no external ear; 84 species.

figure 20.33

Humpback whale, *Megaptera novaeangliae*, breaching. Among the most acrobatic of whales, humpbacks appear to breach to stun fish schools or to communicate information to other pod members. Order Cetacea, family Balaenopteridae.

SUMMARY

Sections	Key Concepts
20.1 Origin and Evolution of Mammals	• Mammals are endothermic vertebrates whose bodies have hair and who nurse their young with milk. • The approximately 5700 species of mammals descend from the synapsid lineage of amniotes that arose during the Carboniferous period of the Paleozoic era. Their evolution is traced from pelycosaurs of the Permian period to therapsids of the late Permian and Triassic periods. One group of therapsids, the cynodonts, gave rise during the Triassic period to mammals. • Mammalian diversification began during the Cretaceous period: by the Eocene epoch of the Cenozoic era, more modern orders had appeared. • Mammal evolution is characterized by changes that resulted in heterodont teeth, limbs oriented underneath the body, a secondary palate, three middle ear bones, and a jaw joint formed by the dentary and temporal bones. • Living mammals are in three clades: Monotremes (egg-laying mammals), Metatheria (marsupials), and Eutheria (placentals).
20.2 Structural and Functional Adaptations of Mammals	• Mammals are named for the glandular milk-secreting organs of females (rudimentary in males), a unique adaptation that, combined with prolonged parental care, buffers infants from demands of foraging for themselves and eases the transition to adulthood. • Hair, the integumentary outgrowth that covers most mammals, provides mechanical protection, thermal insulation, protective coloration, and waterproofing. • Many hoofed herbivorous mammals have horns (primarily composed of keratin) or antlers (composed of bone). • Mammal skin is rich in glands: sweat glands that function in evaporative cooling, scent glands used in social interactions, and sebaceous glands that secrete conditioning oil. • The olfactory sense of mammals is highly developed, functioning in social interactions and finding food. Mammals have large brains compared to other vertebrates and have a distinctive outer layer of the cerebrum, the cerebral cortex. • Placental mammals have deciduous teeth that are replaced by permanent teeth (diphyodont condition). Mammals have a diverse (heterodont) dentition, including four kinds of teeth–incisors, canines, premolars, and molars. These are highly modified in different mammals for specialized feeding tasks, or they may be absent.

(continued)

Sections	Key Concepts
	• Mammals' food habits strongly influence their body form and physiology. Insectivorous mammals have pointed teeth for piercing and crushing insects and other small invertebrates. Herbivorous mammals have specialized molars and premolars for grinding cellulose and silica-rich plants; they also have specialized regions of the gut for harboring microorganisms that digest cellulose. Carnivorous mammals have specialized jaw muscles and teeth for killing and removing flesh from their prey, which are mainly herbivorous mammals. Omnivorous mammals feed on both plant and animal foods and have a variety of tooth types. • Some marine, territorial, and aerial mammals migrate: migrations of some, such as whales, fur seals, and caribou, are extensive. Migrations usually are made toward climatic conditions favorable for finding food, mating, or rearing young. • The mammals with true flight, bats, are mainly nocturnal and thus avoid direct competition with birds. Most employ ultrasonic echolocation to navigate and to feed in darkness. • Monotremes, of the Australian region, are the only mammals that retain the ancestral trait of egg-laying. After hatching, the young are nourished with their mother's milk. All other mammals (therians) are viviparous. • Embryos of marsupials have brief gestation periods, are born underdeveloped, and complete their early growth in the mother's pouch, nourished by milk. • The remaining mammals are eutherians ("placentals"), mammals that develop a sophisticated placental attachment between mother and embryo through which an embryo is nourished for a prolonged period. • Many northern mammals survive the cold winters by hibernating, in which body temperatures and metabolic rates are greatly reduced.
20.3 Human Evolution	• Humans are primates, a mammalian group that descended from a shrew-like ancestor. The common ancestor of all modern primates was arboreal and had grasping fingers and forward-facing eyes capable of binocular vision. • Primates diversified to form two groups: (1) lemurs and lorises; and (2) tarsiers, monkeys, and apes (including humans). Chimpanzees and bonobos together form the sister group of humans. • The earliest humans appeared in Africa 7 to 10 million years ago and gave rise to several genera of early humans, including *Ardipithecus, Australopithecus,* and *Paranthropus,* which persisted for about 3 million years. These early humans were shorter and smaller-brained than modern humans, but were bipedal. • *Homo habilis,* the first maker of stone tools, appeared about 2.8 million years ago, and coexisted with other early humans. • *Homo erectus* appeared about 1.9 million years ago and spread throughout Africa, Europe, and Asia. It was eventually replaced by Neanderthals, *Homo neanderthalensis,* and modern humans, *Homo sapiens.*

Review Questions

1. Describe the evolution of mammals, tracing their ancestry from early synapsids to true mammals. How would you distinguish pelycosaurs, early therapsids, cynodonts, and mammals?
2. Describe structural and functional adaptations in early amniotes that foreshadowed the mammalian body plan. Which mammalian attributes were especially important to successful diversification of mammals?
3. Hair is hypothesized to have evolved in therapsids as an adaptation for insulation, but modern mammals have adapted hair for several other purposes. Describe these.
4. What is distinctive about each of the following: horns of bovids, antlers of deer, and horns of rhinos? Describe the growth cycle of antlers.
5. Describe the location and principal function(s) of each of the following skin glands: sweat glands, scent glands, sebaceous glands, and mammary glands.
6. Define the terms "diphyodont" and "heterodont," and explain how both terms apply to mammalian dentition.
7. Based on the function of the cerebral cortex, hypothesize why the cerebral cortex is fairly smooth in shrews, but highly folded in primates.
8. Compare the food habits of insectivores, herbivores, carnivores, and omnivores. List common names of some mammals belonging to each group.

9. Most herbivorous mammals depend on cellulose as their main energy source, and yet no mammal synthesizes cellulose-splitting enzymes. How are the digestive tracts of mammals specialized for symbiotic digestion of cellulose?
10. How does fermentation differ between horses and cattle?
11. Describe the annual migrations of barren-ground caribou and fur seals.
12. Explain what is distinctive about the life habit and mode of navigation in bats.
13. Describe and distinguish patterns of reproduction in monotremes, marsupials, and placental mammals. What aspects of mammalian reproduction occur in *all* mammals but in no other vertebrates?
14. Describe the changes that occur in a ground squirrel during hibernation.
15. What anatomical characteristics distinguish primates from other mammals?
16. What role do the fossils named "Ardi" and "Lucy" play in reconstruction of human evolutionary history?
17. In what ways do the genera *Australopithecus* and *Homo* differ?
18. When did the different species of *Homo* appear, and how did their cultures differ?

For Further Thought Many zoologists consider mammalian teeth, rather than endothermy, hair, or lactation, the characteristic most responsible for the success of mammals. Explain why this could be true.

General References

These references pertain to groups covered in more than one chapter. They include a number of very valuable field manuals that aid in identification, as well as general texts.

Brusca, R. C., W. Moore, and S. M. Shuster. 2016. Invertebrates, ed. 3. Sunderland, Massachusetts, Sinauer Associates, Inc. *Invertebrate text organized around the bauplan ("body plan") concept—structural range, architectural limits, and functional aspects of a design—for each phylum. Includes cladistic analysis of phylogeny for most groups.*

Cracraft, J., and M. J. Donoghue (eds.). 2004. Assembling the tree of life. New York, Oxford University Press. *This volume summarizes much of the current work on the Tree of Life project, an attempt to uncover the evolutionary relationships for all living taxa. The early chapters of the book discuss the relevance of the Tree of Life project for society. The overview and introductory sections are very readable, and these are paired with results of recent phylogenetic analyses for all domains of life.*

Foelix, R. F. 2011. ed 3. Biology of spiders. New York, Oxford University Press. *Attractive comprehensive book with extensive references; of interest to amateurs and professionals.*

Gosner, K. L. 1999. A field guide to the Atlantic seashore: invertebrates and seaweeds of the Atlantic coast from the Bay of Fundy to Cape Hatteras. The Peterson Field Guide Series. Boston, Houghton Mifflin Harcourt. *A helpful aid for students of invertebrates that are found along the northeastern coast of the United States.*

Halstead, B. W. 1992. Dangerous aquatic animals of the world: a color atlas: with prevention, first aid, and emergency treatment procedures. Princeton, New Jersey, Darwin Press. *Excellent coverage of threats to humans by dangerous aquatic animals and how to avoid fateful encounters.*

Humann, P., N. DeLoach, and L. Wilk. 2002. Reef creature identification. Florida, Caribbean, Bahamas. Third Edition. Jacksonville, Florida, New World Publications. *Excellent field guide to aid identification of Atlantic reef invertebrates, except corals.*

Hyman, L. H. 1940–1967. The invertebrates, 6 vols. New York, McGraw-Hill Book Company. *Informative discussions on the phylogenies of most invertebrates are treated in this outstanding series of monographs. Volume I contains a discussion of the colonial hypothesis of the origin of metazoa, and volume 2 contains a discussion of the origin of bilateral animals, body cavities, and metamerism.*

Kaplan, E. H. 1999. A field guide to southeastern and Caribbean seashores: Cape Hatteras to the Gulf Coast, Florida, and the Caribbean. Second Edition A. Peterson Field Guide Series. Boston, Houghton Mifflin Harcourt. *More than just a field guide, this comprehensive book is filled with information on the biology of seashore animals: complements Gosner's field guide, which covers animals north of Cape Hatteras.*

Kardong, K. V. 2018. Vertebrates: comparative anatomy, function, evolution, ed. 8. New York, McGraw-Hill. *Evolution of vertebrate structure.*

Kozloff, E. N. 1996. Marine invertebrates of the Pacific Northwest. Seattle, University of Washington Press. *Contains keys for many marine groups.*

Lane, R. P., and H. W. Crosskey. 1993. Medically important insects and arachnids. London, Chapman and Hall. *The most up-to-date medical entomology text available.*

Mebs, D. 2002. Venomous and poisonous animals. Stuttgart, Germany, Medpharm Scientific Publishers. *Fascinating coverage of animals dangerous to humans. Highlights include color photos of every species and medical treatment descriptions.*

Morris, R. H., D. P. Abbott, and E. C. Haderlie. 1980. Intertidal invertebrates of California. Stanford, Stanford University Press. *An essential reference on the most important invertebrates of the intertidal zone in California. Contains 900 color photographs.*

Nielsen, C. 2012. Animal evolution: interrelationships of the living phyla. Third Edition. New York, Oxford University Press. *Cladistic analysis of molecular genetic data and morphology is used to develop sister-group relationships of the living animals. An advanced but essential reference.*

Palumbi, S. R., and A. R. Palumbi. 2014. The extreme life of the sea. Princeton, New Jersey, Princeton University Press. *An entertaining account of the extraordinary animals and environments contained within the world's oceans.*

Pechenik, J. A. 2014. Biology of the invertebrates, ed. 7. Dubuque, Iowa, McGraw-Hill Companies, Inc. *Up-to-date text, including recognition of superphyla Ecdysozoa and Lophotrochozoa.*

Pough, F. H., and C. M. Janis. 2018. Vertebrate life, ed. 10. Upper Saddle River, New Jersey, Prentice Hall. *Vertebrate morphology, physiology, ecology, and behavior cast in a cladistic framework.*

Ricketts, E. F., J. Calvin, and J. W. Hedgpeth (revised by D. W. Phillips). 1985. Between Pacific tides, ed. 5. Stanford, Stanford University Press. *A revision of a classic work in marine biology. It stresses the habits and habitats of the Pacific coast invertebrates, and the illustrations are revealing. It includes an excellent, annotated systematic index and bibliography.*

Roberts, L. S., J. Janovy, Jr., and S. Nadler. 2012. Gerald D. Schmidt and Larry S. Roberts' Foundations of parasitology, ed. 9. Dubuque, Iowa, McGraw-Hill Publishers. *Highly readable and up-to-date account of parasitic protozoa, worms, and arthropods.*

Rothschild, S. B. 2004. Beachcomber's guide to Gulf Coast marine life. Third Edition. Houston, Texas, Gulf Publishing Company. *Coverage arranged by habitats. No keys, but common forms that occur near shore can be identified.*

Ruppert, E. E., R. S. Fox, and R. D. Barnes. 2004. Invertebrate zoology, ed. 7. Florence, Brooks/Cole Publishing. *Authoritative, detailed coverage of invertebrate phyla.*

Smith, D. G. 2001. Pennak's freshwater invertebrates of the United States, ed. 4. New York, John Wiley & Sons, Inc. *Contains keys for identification of freshwater invertebrates with brief accounts of each group. Indispensable for freshwater biologists.*

Smith, D. L., and K. B. Johnson. 1996. A guide to marine coastal plankton and marine invertebrate larvae, ed. 2. Dubuque, Iowa, Kendall/Hunt Publishing Company. *Valuable manual for identification of marine plankton, which is usually not covered in most field guides.*

Tudge, C. 2002. The variety of life; a survey and celebration of all the creatures that have ever lived. New York, Oxford University Press. *This well-written book might be called a celebration of systematics–the discipline, as the author says, that introduces us to the creatures themselves. A fine introduction that explains cladistics is followed by chapters that describe . . . almost everything!*

Valentine, J. V. 2004. On the origin of phyla. Chicago, University of Chicago Press. *A thorough coverage of paleontological, morphological, and molecular evidence on relationships among the animal phyla and the evolutionary origins of animal body plans.*

Willmer, P., G. Stone, and I. A. Johnston. 2004. Environmental physiology of animals, ed. 2. Malden, Massachusetts, Wiley-Blackwell Publishers. *A comparative treatment of the physiological ecology of animals.*

Glossary

This glossary lists definitions, pronunciations, and derivations of the most important recurrent technical terms, units, and names (excluding taxa) used in this text.

A

abiotic (ā′bī-ät′ik) (Gr. *a,* without, + *biōtos,* life, livable) Characterized by the absence of life.

abomasum (ab′ō-mā′səm) (L. *ab,* from, + *omasum,* paunch) Fourth and last chamber of the stomach of ruminant mammals.

aboral (ab-o′rəl) (L. *ab,* from, + *os,* mouth) Region of an animal opposite the mouth.

acanthodians (a′kan-thō′dē-əns) (Gr. *akantha,* prickly, thorny) Group of jawed fishes, characterized by large spines in their fins, from Lower Silurian to Lower Permian.

acclimatization (ə-klī′mə-tə-zā-shən) (L. *ad,* to, + Gr. *klima,* climate) Gradual physiological adjustment of an organism in response to relatively long-lasting environmental changes.

acoelomate (ā-sēl′ə-māt′) (Gr. *a,* not, + *koilōma,* cavity) Without a coelom, as in flatworms and proboscis worms.

acontium (ə-kän′chē-əm), pl. **acontia** (Gr. *akontion,* dart) Threadlike structure bearing nematocysts located on mesentery of sea anemone.

actin filament (ak′-tin fi′-lə-ment) A component of the eukaryotic cytoskeleton generally functioning in cell movement, also called a microfilament.

adaptation (L. *adaptatus,* fitted) Anatomical structure, physiological process, or behavioral trait that evolved by natural selection and improves an organism's ability to survive and leave descendants.

adaptive radiation Evolutionary diversification that produces numerous ecologically disparate lineages from a single ancestral one, especially when this diversification occurs within a short interval of geological time.

adaptive zone A characteristic reaction and mutual relationship between environment and organism ("way of life") demonstrated by a group of evolutionarily related organisms.

adductor (ə-duk′tər) (L. *ad,* to, + *ducere,* to lead) A muscle that draws a part toward a median axis, or a muscle that draws the two valves of a mollusc shell together.

adipose (ad′ə-pōs) (L. *adeps,* fat) Fatty tissue; fatty.

aerobic (ā-rō′bik) (Gr. *aēr,* air, + *bios,* life) Oxygen-dependent form of respiration.

afferent (af′ə-rənt) (L. *ad,* to, + *ferre,* to bear) Adjective meaning leading or bearing toward some organ—for example, nerves conducting impulses toward the brain or blood vessels carrying blood toward an organ; opposed to efferent.

age structure An accounting of the ages of individuals in a population at a particular time and place.

agnathan (ag-nā′-thən) (Gr. *a,* without, + *gnathos,* jaw) A jawless fish of the paraphyletic superclass Agnatha of the phylum Chordata.

alate (ā′lāt) (L. *alatus,* wing) Winged.

albumin (al-bū′mən) (L. *albuman,* white of egg) Any of a large class of simple proteins that are important constituents of vertebrate blood plasma and tissue fluids and also present in milk, whites of eggs, and other animal substances.

allantois (ə-lan′tois) (Gr. *allas,* sausage, + *eidos,* form) One of the extraembryonic membranes of the amniotes that functions in respiration and excretion in birds and nonavian reptiles and plays an important role in the development of the placenta in most mammals.

allele (ə-lēl′) (Gr. *allēlōn,* of one another) Alternative forms of genes coding for the same trait, and situated at the same locus in homologous chromosomes.

allelic frequency An estimation of the proportion of gametes produced in a population (gene pool) that contains a particular allelic form of a particular gene.

allopatric (Gr. *allos,* other, + *patra,* native land) In separate and mutually exclusive geographical regions.

allopatric speciation The hypothesis that new species are formed by dividing an ancestral species into geographically isolated subpopulations that evolve **reproductive barriers** between them through independent evolutionary divergence from their common ancestor.

altricial (al-tri′shəl) (L. *altrices,* nourishers) Referring to young animals (especially birds) having the young hatched in an immature, dependent condition.

alula (al′yə-lə) (L. dim of *ala,* wing) The first digit or thumb of a bird's wing, much reduced in size.

alveoli (al-vē′ə-lī) Pockets or spaces bounded by membrane or epithelium.

alveolin (al-vē′ə-lin) A member of a family of proteins, structurally similar to intermediate filaments, that connect to the cytoskeleton in Alveolates; proposed as an alveolate synapomorphy.

ambulacra (am′byə-lak′rə) (L. *ambulare,* to walk) In echinoderms, radiating grooves where podia of the water-vascular system characteristically project outside the organism.

ameboid movement (ə-mē′boid) (Gr. *amoibē,* change, + *oid,* like) Cellular locomotion by protrusion of cytoplasm to form pseudopodia.

amensalism (ā-men′səl-iz′əm) An asymmetric competitive interaction between two species in an ecological community in which only one of the species is affected.

amictic (ā-mik′tic) (Gr. *a,* without, + *miktos,* mixed or blended) Pertaining to female rotifers, which produce only diploid eggs that cannot be fertilized, or to the eggs produced by such females; contrasts with **mictic.**

ammocoetes (am-ə-sēt′ēz) (Gr. *ammos,* sand, + *koitē,* bed) The filter-feeding larval stage of lampreys.

amnion (am′nē-än) (Gr. *amnion,* membrane around the fetus) The innermost of the extraembryonic membranes forming a fluid-filled sac around the embryo in amniotes.

amniote (am′nē-ōt) Having an amnion; as a noun, an animal that develops an amnion in embryonic life; refers collectively to nonavian reptiles, birds, and mammals. Adj., **amniotic.**

amphid (am′fəd) (Gr. *amphidea,* anything that is bound around) One of a pair of anterior sense organs in certain nematodes.

amplexus (am-plek′səs) (L., embrace) The embrace of frogs or toads in which males fertililze eggs as they leave the female's body.

ampulla (am-pūl′ə) (L., flask) Membranous vesicle; dilation at one end of each semicircular canal containing sensory epithelium; muscular vesicle above the tube foot in water-vascular system of echinoderms.

ampullae of Lorenzini (*Lorenzini,* 17th century physician and ichthyologist) Tiny, gel-filled canals that function as electroreceptors on the skin surface of members of Chondrichthyes.

anadromous (an-ad′rə-məs) (Gr. *anadromos,* running upward) Refers to fishes that migrate up streams from the sea to spawn.

anaerobic (an′ə-rō′bik) (Gr. *an,* not, + *aēr,* air, + *bios,* life) Not dependent on oxygen for respiration.

analogy (L. *analogous,* ratio) Similarity of function but not of origin.

anamniote (an′am-nē-ōt) A vertebrate that lacks an amniotic membrane around the embryo. Includes fishes and amphibians.

anapsids (ə-nap′sәds) (Gr. *an,* without, + *apsis,* arch) Extinct amniotes in which the skull lacks temporal openings.

ancestral character state The condition of a taxonomic character inferred to have been present in the most recent common ancestor of a taxonomic group being studied cladistically.

ancoracyst (an-kor′-a-sist) (Gr. ancora, anchor + kystis, bladder) A newly discovered type of extrusome that may function to immobilize prey, discovered in the unicellular eukaryote *Ancoracysta twista.*

androgenic gland (an′drō-jen′ək) (Gr. *anēr,* male, + *gennaein,* to produce) Gland in Crustacea that causes development of male characteristics.

annulus (an′yəl-əs) (L., ring) Any ringlike structure, such as superficial rings on leeches.

antenna (L., sail yard) A sensory appendage on the head of arthropods, or the second pair of two such pairs of structures in crustaceans.

antennal gland Excretory gland of Crustacea located in the antennal metamere.

anterior (L. comparative of *ante,* before) The head of an organism or (as the adjective) toward that end.

anthropoid (an′thrə-poyd) (Gr. *anthrōpos,* man, + *eidos,* form) Resembling humans, especially the great apes.

aperture (ap′ər-chər) (L. *apertura* from *aperire,* to uncover) An opening; the opening into the first whorl of a gastropod shell.

apical (ā′pə-kl) (L. *apex,* tip) Pertaining to the tip or apex.

apical complex A certain combination of organelles found in the protozoan phylum Apicomplexa.

apicoplast (ap′ə-kə-plast) A non-photosynthetic chloroplast derivative in apicomplexan unicellular eukaryotes such as the malarial parasite *Plasmodium.*

apocrine (ap′ə-krən) (Gr. *apo,* away, + *krinein,* to separate) Applies to a type of mammalian sweat gland that produces a viscous secretion by breaking off a part of the cytoplasm of secreting cells.

apodeme (ap′-ə-dēm) A protrusion on the inside of the cuticle of certain arthropods (crustaceans and insects) to which the muscles attach.

apopyle (ap′ə-pīl) (Gr. *apo,* away from, + *pylē,* gate) In sponges, opening of the radial canal into the spongocoel.

aposematic (ap-ə-si-mat′ik) A conspicuous condition that serves as a warning; for example, the bright colors of a Monarch butterfly's wings warn potential predators that the butterfly is distasteful.

arboreal (är-bōr′ē-əl) (L. *arbor,* tree) Living in trees.

archaeocyte (ar′kē-ə-sīt) (Gr. *archaios,* beginning, + *kytos,* hollow vessel) Ameboid cells of varied function in sponges.

archenteron (ärk-en′tə-rän) (Gr. *archē,* beginning, + *eneron,* gut) The main cavity of an embryo in the gastrula stage; it is lined with endoderm and represents the future digestive cavity.

archosaur (är′kə-sor) (Gr. *archōn,* ruling, + *sauros,* lizard) A clade of diapsid vertebrates that includes the living crocodiles and birds and the extinct pterosaurs and dinosaurs.

Aristotle's lantern Masticating apparatus of some sea urchins.

artiodactyl (är′tē-ō-dak′təl) (Gr. *artios,* even, + *daktylos,* toe) One of an order of mammals with two or four digits on each foot.

asconoid (Gr. *askos,* bladder) Simplest form of sponges, with canals leading directly from the outside to the interior.

asexual Without distinct sexual organs; not involving formation of gametes.

aspiration (as′pir-ā-shun) Drawing a fluid into a cavity by suction, as when air is drawn into lungs of amniotes.

asymmetric competition See **amensalism.**

atoke (ā′tōk) (Gr. *a,* without, + *tokos,* offspring) Anterior, nonreproductive part of a marine polychaete, as distinct from the posterior, reproductive part (epitoke) during the breeding season.

atrium (ā′trē-əm) (L. *atrium,* vestibule) One of the chambers of the heart. Also, the tympanic cavity of the ear. Also, the large cavity containing the pharynx in tunicates and cephalochordates.

auricle (aw′ri-kl) (L. *auricula,* dim. of *auris,* ear) One of the less muscular chambers of the heart; atrium; the external ear, or pinna; any earlike lobe or process.

auricularia (ə-rik′u-lar′ē-ə) (L. *auricula,* a small ear) Type of larva found in Holothuroidea.

autogamy (aw-täg′ə-mē) (Gr. *autos,* self, + *gamos,* marriage) Condition in which the gametic nuclei produced by meiosis fuse within the same organism that produced them to restore the diploid number.

autotomy (aw-tät′ə-mē) (Gr. *autos,* self, + *tomos,* a cutting) Detachment of a part of the body by the organism itself.

autotroph (aw′tō-trōf) (Gr. *autos,* self, + *trophos,* feeder) Organism that makes its organic nutrients from inorganic raw materials.

autotrophic nutrition (Gr. *autos,* self, + *trophia,* denoting nutrition) Nutrition characterized by the ability to use simple inorganic substances to synthesize more complex organic compounds, as in green plants and some bacteria.

average effect A quantitative-genetic parameter estimating the incremental contribution of each copy of a particular allele to the mean value of a particular organismal phenotype (such as height or weight) in the population being studied. Average effect is calculated from measurements of the population frequencies of all genotypes containing the allele, and the mean deviations of each genotypic class from the mean value of the phenotype in the population as a whole.

axial (L. *axis,* axle) Relating to the axis, or stem; on or along the axis.

axocoel (aks′ə-sēl) (Gr. *axon,* axle, + *koilos,* hollow) Anterior coelomic compartment in echinoderms; corresponds to protocoel.

axolotl (ak′sə-lot′l) (Nahuatl *atl,* water, + *xolotl,* doll, servant, spirit) Salamanders of the species *Ambystoma mexicanum,* which do not metamorphose and which retain aquatic larval characteristics throughout adulthood.

axon (ak′sän) (Gr. *axōn*) Elongate extension of a neuron that conducts impulses away from the cell body and toward the synaptic terminals.

axoneme (aks′ə-nēm) (L. *axis,* axle, + Gr. *nēma,* thread) The microtubules in a cilium or flagellum, usually arranged as a circlet of nine pairs enclosing one central pair; also, the microtubules of an axopodium.

axopodium (ak′sə-pō′dē-əm) (Gr. *axon,* an axis, + *podion,* small foot) Long, slender, more or less permanent pseudopodium in certain amebas. (Also called an **axopod.**)

axostyle (aks′ō-stīl) Tubelike organelle in some flagellate protozoa, extending from the area of the kinetosomes to the posterior end, where it often protrudes.

B

basal body Also called kinetosome or blepharoplast, a cylinder of nine triplets of microtubules found at the base of a flagellum or cilium; same structure as a centriole.

basal disc Aboral attachment site on a cnidarian polyp.

basis, basipodite (bā′səs, bā-si′pō-dīt) (Gr. *basis*, base, + *pous, podos*, foot) The distal or second joint of the protopod of a crustacean appendage.

benthos (ben′thōs) (Gr., depth of the sea) Organisms that live on the floor of a sea or lake. Adj., **benthic.** Also, the submerged substrate itself.

beta keratin A hardened form of keratin that makes up feathers and scales of birds and other reptiles.

Bilateria (bī′lə-tir′-ē-ə) (L. *bi-*, two, + *latus*, side) Bilaterally symmetrical animals.

binary fission Mode of asexual reproduction in which the animal splits into two approximately equal offspring.

binomial nomenclature The Linnean system of naming species in which the first word is the name of the genus (first letter capitalized) and the second word is the specific epithet (uncapitalized), usually an adjective modifying the name of the genus. Both of these words are italicized.

biogenetic law A statement postulating a characteristic relationship between **ontogeny** and **phylogeny.** Examples include Haeckel's law of **recapitulation** and Von Baer's law that general characteristics (those shared by many species) appear earlier in ontogeny than more restricted ones; neither of these statements is universally true.

biogeochemical cycle A description of the flow of elementary matter, such as carbon or phosphorus, through the component parts of an ecosystem and its **abiotic** environment, including the amount of an element present at the various stages of a **food web.**

biological species concept A reproductive community of populations (reproductively isolated from others) that occupies a specific niche in nature.

bioluminescence Method of light production by living organisms; usually, certain proteins (luciferins), in the presence of oxygen and an enzyme (luciferase), are converted to oxyluciferins with liberation of light.

biomass (Gr. *bios*, life, + *maza*, lump or mass) The weight of total living organisms or of a species population per unit of area.

biome (bī′ōm) (Gr. *bios*, life, + *ōma*, abstract group suffix) Complex of plant and animal communities characterized by climatic and soil conditions; the largest ecological unit.

biosphere (Gr. *bios*, life, + *sphaira*, globe) That part of earth containing living organisms.

biotic (bī-ät′ik) (Gr. *biōtos*, life, livable) Of or relating to life.

bipinnaria (L. *bi*, double, + *pinna*, wing, + *aria*, like or connected with) Free-swimming, ciliated, bilateral larva of the asteroid echinoderms; develops into the brachiolaria larva.

biradial symmetry Type of radial symmetry in which only two planes passing through the oral-aboral axis yield mirror images because some structure is paired.

biramous (bī-rā m′əs) (L. *bi*, double, + *ramus*, a branch) Adjective describing appendages with two distinct branches; contrasts with uniramous, unbranched.

blastocoel (blas′tə-sēl) (Gr. *blastos*, germ, + *koilos*, hollow) Cavity of the blastula.

blastomere (Gr. *blastos*, germ, + *meros*, part) An early cleavage cell.

blastopore (Gr. *blastos*, germ, + *poros*, passage, pore) External opening of the archenteron in the gastrula.

blastula (Gr. *blastos*, germ, + *ula*, dim) Early embryological state of many animals; consists of a hollow mass of cells.

book gill Respiratory structure of aquatic chelicerates (Arthropoda) in which many thin, blood-filled gills are layered like the pages of a book. Gas exchange occurs as seawater passes between each pair of gills.

book lung Respiratory structure of terrestrial chelicerates (Arthropoda) in which many thin-walled air pockets extend into a blood-filled chamber in the abdomen.

BP Before the present.

brachial (brāk′ē-əl) (L. *brachium*, forearm) Referring to the arm.

branchial (brank′ē-əl) (Gr. *branchia*, gills) Referring to gills.

brown bodies Remnants of the lophophore and digestive tract of a degenerating adult ectoproct left behind in the chamber as a new lophophore and digestive tract are formed.

buccal (buk′əl) (L. *bucca*, cheek) Referring to the mouth cavity.

buccal pumping A method of ventilation in which air is pushed from the mouth into the lungs, as occurs in amphibians and air-breathing fishes.

budding Reproduction in which the offspring arises as a small outgrowth from the parent. Failure of the offspring to separate from the parent leads to colony formation.

bursa, pl. **bursae** (M.L. *bursa*, pouch, purse made of skin) A saclike cavity. In ophiuroid echinoderms, pouches opening at bases of arms and functioning in respiration and reproduction (genitorespiratory bursae).

C

caecilian (si-sil′-yən) Denotes any member of the amphibian order Gymnophiona (also called Apoda).

calyx (kā′-liks) (L., bud cup of a flower) Any of various cup-shaped zoological structures.

capitulum (ka-pi′chə-ləm) (L., small head) Term applied to small, headlike structures of various organisms, including projection from body of ticks and mites carrying mouthparts.

carapace (kar′ə-pās) (F. from Sp. *carapacho*, shell) Shieldlike plate covering the cephalothorax of certain crustaceans; dorsal part of the shell of a turtle.

carinate (kar′ə-nāt) (L. *carina*, keel) Having a keel; in particular, the flying birds with a keeled sternum for the insertion of flight muscles; contrasts with **ratite.**

carnivore (kar′nə-vōr′) (L. *carnivorus*, flesh-eating) A mammal of the order Carnivora. Also, any organism that eats animals. Adj., **carnivorous.**

carrying capacity The maximum number of individuals of a particular species that can persist under specified environmental conditions.

cartilage (L. *cartilago;* akin to L. *cratis*, wickerwork) Translucent elastic tissue that forms most of the skeleton of embryos, very young vertebrates, and adult cartilaginous fishes, such as sharks and rays; in adult amniotes, much of it is converted into bone.

caste (kast) (L. *castus*, pure, separated) One of the polymorphic forms within an insect society; each caste has its specific duties, such as queen, worker, soldier, and others.

catadromous (kə-tad′rə-məs) (Gr. *kata*, down, + *dromos*, a running) Refers to fishes that migrate from fresh water to the ocean to spawn.

catastrophic species selection Differential survival among species during a time of mass extinction based on character variation that permits some species but not others to withstand severe environmental disturbances, such as those caused by an asteroid impact.

caudal (käd′l) (L. *cauda*, tail) Constituting, belonging to, or relating to a tail.

cecum, pl. ceca (sē′kəm) (L. *caecus*, blind) A blind pouch at the beginning of the large intestine; any similar pouch.

cellulase (sel′ū-lās) (L. *cella*, small room) Enzyme that cleaves cellulose; synthesized only by bacteria and some protists.

cellulose (sel′ū-lōs) (L. *cella*, small room) Chief polysaccharide constituent of the cell wall of green plants and some fungi; an insoluble carbohydrate $(C_6H_{10}O_5)_n$ that is converted to glucose by hydrolysis.

centriole (sen′trē-ōl) (Gr. *kentron*, center of a circle, + L. *ola*, small) Minute cytoplasmic organelle usually found in the centrosome and considered the active division center of the animal cell; organizes spindle fibers during mitosis and meiosis. Same structure as basal body or kinetosome.

cephalization (sef′ə-li-zā′shən) (Gr. *kephalē*, head) The evolutionary process by which specialization, particularly of the sensory organs and appendages, became localized in the head end of animals.

cephalothorax (sef′ə-lä-thō′raks) (Gr. *kephalē*, head, + thorax) A body division found in many Arachnida and higher Crustacea in which the head is fused with some or all of the thoracic segments.

cerata (sə-ra′tə) (Gr. *keras,* a horn, bow) Dorsal processes on some nudibranchs for gas exchange.

cercaria (ser-kar′ē-ə) (Gr. *kerkos,* tail, + L. *aria,* like or connected with) Tadpolelike juveniles of trematodes (flukes).

cerebellum (ser-ə-bel′əm) (L. *cereb,* brain) A region of the hindbrain that processes spatial orientation information and coordinates muscle movement.

cerebral cortex (L. *cerebro,* brain) The outer (superficial) layer of the cerebrum, the largest part of the forebrain. In mammals, it is uniquely arranged in six layers and is also known as the neocortex.

cervical (sər′və-kəl) (L. *cervix,* neck) Relating to a neck; usually refers to vertebrae in the neck region, or to the neck (cervix) of a mammalian uterus.

character (kar′ik-tər) A component of phenotype (including specific molecular, morphological, behavioral, or other features) used by systematists to diagnose species or higher taxa, or to evaluate phylogenetic relationships among different species or higher taxa, or relationships among populations within a species.

chela (kēl′ə) (Gr. *chēlē,* claw) Pincerlike claw.

chelicera (kə-lis′ə-rə), pl. **chelicerae** (Gr. *chēlē,* claw, + *keras,* horn) One of a pair of the most anterior head appendages on members of subphylum Chelicerata.

chelipeds (kēl′ə-peds) (Gr. *chēlē, claw,* + L. *pes,* foot) Pincerlike first pair of legs in most decapod crustaceans; specialized for seizing and crushing.

chemoautotroph (kē-mō-aw′tō-trōf) (Gr. *chemeia,* transmutation, + *autos,* self, + *trophos,* feeder) An organism utilizing inorganic compounds as a source of energy.

chitin (kī′tən) (Fr. *chitine,* from Gr. *chitōn,* tunic) A hard substance that forms part of the cuticle of arthropods and occurs sparingly in certain other invertebrates; a nitrogenous polysaccharide insoluble in water, alcohol, dilute acids, and the digestive juices of most animals.

chloragogen cells (klōr′ə-gog-ən) (Gr. *chlōros,* light green, + *agōgos,* a leading, a guide) Modified peritoneal cells, greenish or brownish, clustered around the digestive tract of certain annelids; apparently they aid in elimination of nitrogenous wastes and in food transport.

chlorophyll (klō′rə-fil) (Gr. *chlōros,* light green, + *phyllōn,* leaf) Green photosynthetic pigment found in cyanobacteria, plants, and some flagellates.

chloroplast (Gr. *chlōros,* light green, + *plastos,* molded) An organelle in plant cells that contains chlorophyll and is the physical location of photosynthesis; occurs also in photosynthetic unicellular eukaryotes and *Volvox.*

choanoblast (kō-an′ə-blast) (Gr. *choanē,* funnel, + *blastos,* germ) One of several cellular elements within the syncytial tissue of a hexactinellid sponge.

choanocyte (kō-an′ə-sīt) (Gr. *choanē,* funnel, + *kytos,* hollow vessel) One of the flagellate collar cells that line cavities and canals of sponges.

choanoflagellate (kō-an′ə-fla-jel′āt) Any member of a protozoan clade having a single flagellum surrounded by a column of microvilli; some form colonies, and all are included within the larger clade of opisthokonts.

chorioallantoic placenta (kō′rē-ō-al′an-tō′-ic) (Gr. *chorion,* skin, + *allas,* sausage) Type of placenta that occurs in placental mammals and some marsupials; a structure modified from the embryonic chorionic and allantoic membranes through which materials are exchanged between the embryo and the mother.

chorion (kō′rē-on) (Gr. *chorion,* skin) The outer layer of the double membrane that surrounds the embryo of reptiles, birds, and mammals; in mammals, it contributes to the placenta.

choriovitelline placenta (kor′ē-ō-vi′təl-ən) (Gr. *chorion,* skin, + *vittel,* yolk of an egg) An often transitory placenta that forms during the early developmental stages of marsupials and placental mammals. Also called the "yolk sac placenta," it is formed from the yolk sac and chorionic membrane of the embryo.

chromatophore (krō-mat′ə-fōr) (Gr. *chrōma,* color, + *herein,* to bear) Pigment cell, usually in the dermis, in which usually the pigment can be dispersed or concentrated.

chromosomal theory of inheritance The well-established principle, initially proposed by Sutton and Boveri in 1903–1904, that nuclear chromosomes are the physical bearers of genetic material in eukaryotic organisms. It is the foundation for modern evolutionary genetics.

chrysalis (kris′ə-lis) (L., from Gr. *chrysos,* gold) The pupal stage of a butterfly.

cilium (sil′ē-əm), pl. **cilia** (L., eyelid) A hairlike, vibratile organellar process found on many animal cells. Cilia may be used for moving particles along the cell surface or, in ciliated unicellular forms, for locomotion. Some cilia are sensory, as are those in the inner ear and lateral-line organs.

cirrus (sir′əs) (L., curl) Hairlike tuft on an insect appendage; locomotor organelle of fused cilia; male copulatory organ of some invertebrates. Cirri can refer to tufts of fused cilia in members of the Ciliophora.

clade (klād) (Gr. *klados,* branch) A taxon or other group consisting of a particular ancestral lineage and all of its descendants, forming a distinct branch on a cladogram or phylogenetic tree.

cladistics (klad-is′təks) (Gr. *klados,* branch, sprout) A system of arranging taxa by analysis of primitive and derived characteristics so that the arrangement reflects phylogenetic relationships.

cladogram (klād′ə-gram) (Gr. *klados,* branch, + *gramma,* letter) A branching diagram showing the pattern of sharing of evolutionarily derived characters among species or higher taxa.

clasper Digitiform projection on the medial side of the pelvic fins of male chondrichthians and some placoderms; used as an intromittent organ to transfer sperm to the female reproductive tract.

cleavage An orderly sequence of cell divisions where the zygote divides into two cells, the two cells divide to make four cells, the four make eight cells, and so on until there are hundreds of cells in an embryo.

clitellum (kli-tel′əm) (L. *clitellae,* pack-saddle) Thickened, saddlelike portion of certain midbody segments of many oligochaetes and leeches.

cloaca (klō-ā′kə) (L., sewer) Posterior chamber of the digestive tract in many vertebrates, receiving feces and urogenital products. In certain invertebrates, a terminal portion of the digestive tract that serves also as a respiratory, excretory, or reproductive duct.

cloning (klō′ning) Production of genetically identical organisms by asexual reproduction.

cnida (nī′də) (Gr. *knidē,* nettle) Stinging or adhesive organelles formed within cnidocytes in phylum Cnidaria; nematocysts are a common type.

cnidoblast (nī′də′-blast) (Gr. *knidē,* nettle, + Gr. *blastos,* germ) A cnidocyte is called a cnidoblast during the time when a cnida is forming within it.

cnidocil (nī′dō-sil) (Gr. *knidē,* nettle,+ L. *cilium,* hair) Triggerlike spine on a nematocyst.

cnidocyte (nī′dō-sīt) (Gr. *knidē,* nettle, + *kytos,* hollow vessel) Modified interstitial cell that holds the cnida.

coccidian (kok-sid′ē-ən) (Gr. *kokkis,* kernel, grain) Intracellular protozoan parasite belonging to a class within phylum Apicomplexa; the organism causing malaria is an example.

cochlea (kōk′lē-ə) (L., snail, from Gr. *kochlos,* a shellfish) A tubular cavity of the inner ear containing the essential organs of hearing; occurs in crocodiles, birds, and mammals; spirally coiled in mammals.

cocoon (kə-kün′) (Fr. *cocon,* shell) Protective covering of a resting or developmental stage, sometimes including its contents; for example, the cocoon of a moth or the protective covering for the developing embryos in some annelids.

coelenteron (sē-len′tər-on) (Gr. *koilos,* hollow, + *enteron,* intestine) Internal cavity of a cnidarian; gastrovascular cavity; archenteron.

coelom (sē′lōm) (Gr. *koilōma,* cavity) The body cavity in triploblastic animals, lined with mesodermal peritoneum.

coelomate (sē′lō m-āt) Animals that possess a **coelom;** also called eucoelomate.
coelomoduct (sē-lō′mə-dukt) (Gr. *koilos,* hollow, + *ductus,* a leading) Duct that carries gametes or excretory products (or both) from the coelom to the exterior.
cohesion species concept The most inclusive population of individuals having the potential for phenotypic cohesion through intrinsic cohesion mechanisms; a refinement of the **evolutionary species concept** emphasizing population-genetic processes.
cohort (kō′hort) All organisms of a population born within a specified time interval.
collagen (käl′ə-jən) (Gr. *kolla,* glue, + *genos,* descent) Tough, fibrous protein occurring in vertebrates as the chief constituent of collagenous connective tissue; also occurs in invertebrates—for example, the cuticle of nematodes.
collar bodies Extensions of choanoblasts bearing flagellated collars in hexactinellid sponges.
collar cells Cells having a single flagellum surrounded by a ring of microvilli. Sponge choanocytes are collar cells, as are choanoflagellates, but collar cells also occur outside these taxa.
collencyte (käl′ən-sīt) (Gr. *kolla,* glue, + *kytos,* hollow vessel) Type of cell in sponges that secretes fibrillar collagen.
colloblast (käl′ə-blast) (Gr. *kolla,* glue, + *blastos,* germ) A glue-secreting cell on the tentacles of ctenophores.
comb plate One of the plates of fused cilia arranged in rows for ctenophore locomotion.
commensalism (kə-men′səl-iz′əm) (L. *com,* together with, + *mensa,* table) A relationship in which one individual lives close to or on another and benefits, and the host is unaffected; often symbiotic.
common descent Darwin's theory that all forms of life are derived from a shared ancestral population through a branching of evolutionary lineages.
community (L. *communitas,* community, fellowship) See **ecological community** and **reproductive community.**
comparative biochemistry Studies of the structures of biological macromolecules, especially proteins and nucleic acids, and their variation within and among species to reveal homologies of macromolecular structure.
comparative cytology or **karyology** Studies of the structures of chromosomes within and among species to reveal homologies of chromosomal structure.
comparative method Use of patterns of similarity and dissimilarity among species or populations to test hypotheses of character **homology** and to infer phylogenetic relationships; use of phylogeny to examine evolutionary processes and history.
comparative morphology Studies of organismal form and its variation within and among species to reveal homologies of organismal characters.
competition Some degree of overlap in ecological niches of two populations in the same community, such that both depend on the same food source, shelter, or other resources, and negatively affect each other's survival.
competitive exclusion An ecological principle stating that two species whose niches are very similar cannot coexist indefinitely in the same community; one species is driven to extinction by competition between them.
conditional specification During embryonic cleavage, diffusion of molecules from neighboring cells provides positional information to specify cell fate.
conjugation (kon′jū-gā′shən) (L. *conjugare,* to yoke together) Temporary union of two ciliate protozoa while they are exchanging chromatin material and undergoing nuclear phenomena leading to binary fission. Also, formation of cytoplasmic bridges between bacteria for transfer of plasmids.
conodont (kōn′ə-dänt) (Gr. *kōnos,* cone, + *odontos,* tooth) Toothlike element from a Paleozoic animal now considered an early marine chordate.
conspecific (L. *com,* together, + species) Of the same species.
consumer (kən-sū′-mer) Organism whose energy and matter are acquired by eating other organisms, which may be **primary producers, herbivores,** or **carnivores.**
contractile vacuole Clear, fluid-filled cell vacuole in protozoa and a few animals; collects water and releases it to the outside in a cyclical manner, for osmoregulation and some excretion.
control That part of a scientific experiment to which the experimental variable is not applied, but similar to the experimental group in all other respects.
convexity The property of a taxonomic group that a path can be drawn between any two members on a cladogram or phylogenetic tree without leaving the group. **Monophyletic** and **paraphyletic** groups are convex, whereas polyphyletic groups are not.
coprophagy (ko-prō′fā′-jē) (Gr. *kopros,* dung, + *phagein,* to eat) Feeding on dung or excrement as a normal behavior among animals; reingestion of feces.
copulation (Fr., from L. *copulare,* to couple) Sexual union to facilitate fertilization of eggs by sperm.
coral bleaching Corals become white and brittle after expelling their mutualist zooxanthellae in response to increased ocean temperatures.
corneum (kor′nē-əm) (L. *corneus,* horny) Outermost epithelial layer of dead, keratinized cells. Also called stratum corneum.
corona (kə-rō′nə) (L., crown) Head or upper portion of a structure; ciliated disc on anterior end of rotifers.
corpora allata (kor′pə-rə əl-la′tə) (L. *corpus,* body, + *allatum,* aided) Endocrine glands in insects that produce juvenile hormone.
cortex (kor′teks) (L., bark) The outer layer of a structure.
cosmopolitan Denotes a species that has a very large geographic distribution, such as the human species. Contrasts with endemic.
coxa, coxopodite (kox′ə, kox′-ə pō′-dīt) (L. *coxa,* hip, + Gr. *pous, podos,* foot) The proximal joint of an insect or arachnid leg; in crustaceans, the proximal joint of the protopod.
crepuscular (krə-pus′kyū-lar) (L. *crepusculum,* twilight) Active at dawn and dusk, at twilight levels.
Cretaceous extinction A **mass extinction** that occurred 65 million years ago in which 76% of existing species, including all dinosaurs, became extinct, marking the end of the Mesozoic era.
crop Region of the esophagus specialized for storing food.
crypsis (krip′sis) (Gr. *crypt,* hidden) The ability to avoid being detected, by using camouflage, burrowing, or mimicry. Adj. cryptic.
crystalline style Rod containing digestive enzymes present in the stomach of a bivalve.
ctene (tēn) (Gr. *kteis, ktenos,* comb) Fused cilia that form a comb plate in members of phylum Ctenophora.
ctenidia (te-ni′dē-ə) (Gr. *kteis,* comb) Comblike structures, especially gills of molluscs; also applied to comb plates of Ctenophora.
ctenoid scales (tēn′oid) (Gr. *kteis, ktenos,* comb) Thin, overlapping dermal scales of teleost fishes; exposed posterior margins have fine, toothlike spines.
cuticle (kū′ti-kəl) (L. *cutis,* skin) Protective, noncellular, organic layer secreted by the external epithelium (hypodermis) of many invertebrates. In vertebrates, the term refers to the epidermis or outer skin.
Cuvierian tubules (*Cuvier,* 19th-century French comparative vertebrate anatomist) Sticky, often toxic elongate internal organs of holothurians expelled to entangle potential predators; these can be regenerated.
cyanobacteria (sī-an′ō-bak-ter′ē-ə) (Gr. *kyanos,* a dark-blue substance, + *bakterion,* dim. of *baktron,* a staff) Photosynthetic prokaryotes, also called blue-green algae, cyanophytes.
cycloid scales (sī′klȯid) (Gr. *kyklos,* circle) Thin, overlapping dermal scales of teleost fishes; posterior margins are smooth.
cynodonts (sin′ə-dänts) (Gr. *kynodōn,* canine tooth) Group of mammal-like carnivorous synapsids of the Upper Permian and Triassic periods.
cyst (sist) (Gr. *kystis,* a bladder, pouch) Resistant, quiescent stage of an organism, usually with a secreted wall.
cysticercus (sis′tə-ser′kəs) (Gr. *kystis,* bladder, + *kerkos,* tail) Type of juvenile tapeworm having an invaginated and introverted scolex contained in a fluid-filled bladder.
cystid (sis′tid) (Gr. *kystis,* bladder) In an ectoproct, the dead secreted outer parts plus the adherent underlying living layers.

cytopharynx (Gr. *kytos,* hollow vessel, + *pharynx,* throat) Short, tubular gullet in ciliate unicellular eukaryotes.

cytoplasm (sī'tə-plasm) (Gr. *kytos,* hollow vessel, + *plasma,* mold) The living matter of the cell, excluding the nucleus.

cytoplasmic specification During embryonic cleavage, molecules within the cytoplasm of each cell specify cell fate, same as autonomous specification.

cytoproct (sī'tə-prokt) (Gr. *kytos,* hollow vessel, + *prōktos,* anus) Site on a protozoan where indigestible matter is expelled.

cytopyge (sī'tə-pīj) (Gr. *kytos,* hollow vessel, + *pyge,* rump or buttocks) In some protozoa, localized site for expulsion of wastes.

cytostome (sī'tə-stōm) (Gr. *kytos,* hollow vessel, + *stoma,* mouth) The cell mouth in many protozoa.

D

dactylozooid (dak-til'ə-zō-id) (Gr. *dakos,* bite, sting, + *tylos,* knob, + *zōon,* animal) Polyp of a colonial hydroid specialized for defense or killing food.

Darwinism Theory of evolution emphasizing common descent of all living organisms, gradual change, multiplication of species, and natural selection.

data, sing. **datum** (Gr. *dateomai,* to divide, cut in pieces) The results of a scientific experiment, or descriptive observations, upon which a conclusion is based.

deciduous (də-sij'ü-wəs) (L. *deciere,* to fall off) Shed at the end of a growing period.

decomposer (dē'-kəm-pō'zər) **Consumer** that breaks organic matter into soluble components available to plants at the base of the food web; most are bacteria or fungi.

deduction (L. *deductus,* led apart, split, separated) Reasoning from the general to the particular, from given premises to their necessary conclusion.

definitive host The host in which sexual reproduction of a symbiont occurs; if no sexual reproduction, then the host in which the symbiont becomes mature and reproduces; contrasts with **intermediate host**.

deme (dēm) (Gr., populace) A local population of closely related animals.

demographic transition A change in the characteristic age structure and rate of growth of a population; specifically, a change in human populations coincident with industrialization from having a higher proportion of young individuals and a net increase in population size to a more even distribution of old and young individuals and a more stable population size.

demography (də-mäg'rə-fē) (Gr. *demos,* people, + *graphy*) The properties of the rate of growth and the age structure of populations.

density-dependent Biotic environmental factors, such as predators and parasites, whose effects on a population vary according to the number of organisms in the population.

density-independent Abiotic environmental factors, such as fires, floods, and temperature changes, whose effects on a population are unaffected by the number of organisms in the population.

deposit feeders Aquatic organisms that consume detritus and small organisms in submerged sediments.

derived character state Condition of a taxonomic character inferred by cladistic analysis to have arisen within a taxon being examined cladistically rather than having been inherited from the most recent common ancestor of all members of the taxon.

dermal (Gr. *derma,* skin) Pertaining to the skin; cutaneous.

dermal branchia See **papula.**

dermal ostia (Gr. *derma,* skin, + L. *ostium,* door) Incurrent pores in a sponge.

dermis The inner, sensitive mesodermal layer of skin; corium.

determinate cleavage Type of cleavage, usually spiral, in which the fate of the blastomeres is determined very early in development; see **mosaic development.**

detorsion (L. *de,* down, from, + *torquere,* to twist) Developmental process in post-veliger stage of certain gastropod molluscs whereby orientation of internal organs resulting from torsion is altered to resemble a state prior to torsion; anus and mantle cavity are posterior after detorsion.

detritus (də-trī'tus) (L., that which is rubbed or worn away) Any fine particulate debris of organic or inorganic origin.

Deuterostomia (dū'tə-rō-stō'mē-ə) (Gr. *deuteros,* second, secondary, + *stoma,* mouth) A group of higher phyla in which cleavage is indeterminate and ancestrally radial. The endomesoderm is enterocoelous, and the mouth is derived away from the blastopore. Includes Echinodermata, Chordata, and Hemichordata; contrasts with **Protostomia.**

diapause (dī'ə-pawz) (Gr. *diapausis,* pause) Period of arrested development in the life cycle of insects and certain other animals in which physiological activity is very low and the animal is highly resistant to unfavorable external conditions.

diaphragm (dī'ə-fram) (Gr. *dia,* separate, + *phragm,* partition) A sheetlike muscle that separates the thoracic and abdominal cavities of mammals. Contraction of this muscle draws air into the lungs.

diapsids (dī-ap'sədz) (Gr. *di,* two, + *apsis,* arch) Amniotes in which the skull bears two pairs of temporal openings; includes nonavian reptiles and birds. The temporal openings are evolutionarily lost in turtles.

dictyosome (dik'tē-ə-sōm) (Gr. *diction,* to throw, + *sōma,* body) A part of the secretory system of endoplasmic reticulum in unicellular eukaryotes; also called Golgi bodies.

diffusion (L. *diffusion,* dispersion) Movement of particles or molecules from an area of high concentration of the particles or molecules to an area of lower concentration.

digitigrade (dij'ə-tə-grād) (L. *digitus,* finger, toe, + *gradus,* step, degree) Walking on the digits, with the heel of the foot raised; contrasts with **plantigrade** in describing mammalian locomotion.

dimorphism (dī-mor'fizm) (Gr. *di,* two, + *morphē,* form) Existence within a species of two distinct forms according to color, sex, size, organ structure, or behavior. Occurrence of two kinds of zooids in a colonial organism.

dioecious (dı-esh'əs) (Gr. *di,* two, + *oikos,* house) Having male and female organs in separate individuals.

diphycercal (dif'i-sər'kəl) (Gr. *diphyēs,* twofold, + *kerkos,* tail) A tail that tapers to a point, as in lungfishes; vertebral column extends to tip without upturning.

diphyodont (dī-fi'ə-dänt) (Gr. *diphyēs,* twofold, + *odous,* tooth) Having successive sets of deciduous and permanent teeth.

diploblastic (di'plə-blas'tək) (Gr. *diploos,* double, + *blastos,* bud) Organism with two germ layers, endoderm and ectoderm.

diploid (dip'lȯid) (Gr. *diploos,* double, + *eidos,* form) Having the somatic (double, or 2n) number of chromosomes or twice the number characteristic of a gamete of a given species.

direct development A postnatal ontogeny primarily featuring growth in size rather than a major change in body shape or organs present; a life history that lacks **metamorphosis.**

directional selection **Natural selection** that favors one extreme value of a continuously varying trait and disfavors other values.

disruptive selection **Natural selection** that simultaneously favors two different extreme values of a continuously varying trait but disfavors intermediate values.

distal (dis'təl) Farther from the center of the body than a reference point.

DNA barcoding Technique for identifying organisms to species using sequence information of a standard gene present in all animals. The mitochondrial gene encoding cytochrome *c* oxidase I (*COI*) is often used.

domain (dō-mān') An informal taxonomic rank above the Linnean kingdom; Archaea, Bacteria, and Eucarya are ranked as domains.

dorsal (dor'səl) (L. *dorsum,* back) Toward the back, or upper surface, of an animal.

double circulation A blood-transport system having a distinct pulmonary circuit of blood vessels separate from the circuit of blood vessels serving the remainder of the body.

dual-gland adhesive organ Organs in the epidermis of most turbellarians, with three cell types: viscid and releasing gland cells and anchor cells.

E

eccrine (ek′rən) (Gr. *ek,* out of, + *krinein,* to separate) Applies to a type of mammalian sweat gland that produces a watery secretion.

ecdysiotropin (ek-dī′-sē-ə-trō′pən) (Gr. *ekdysis,* to strip off, escape + *tropos,* a turn, change) Hormone secreted in brain of insects that stimulates prothoracic gland to secrete molting hormone. Prothoracicotropic hormone; brain hormone.

ecdysis (ek-dī′-sis) (Gr. *ekdysis,* to strip off, escape) Shedding of outer cuticular layer; molting, as in insects or crustaceans.

ecdysone (ek-dī′sōn) (Gr. *ekdysis,* to strip off) Molting hormone of arthropods; stimulates growth and ecdysis; produced by prothoracic glands in insects and Y-organs in crustaceans.

ecdysozoan protostome (ek-dī′-sō-zō′ən prō′tə-stōm) (Gr. *ekdysis,* to strip off, escape, + *zōon,* animal. Gr. *protos,* first, + *stoma,* mouth) Any member of a clade within Protostomia whose members shed the cuticle as they grow; includes arthropods, nematodes, and several smaller phyla.

ECM (extracellular matrix) A layer composed of proteoglycans, linking proteins, and fibrous proteins, such as collagen; the layer is secreted by cells and has structural and functional roles (e.g., cell orientation, adhesion, communication, filtering of materials); the matrix is sometimes called a basal lamina or a basal membrane.

ecological character displacement Differences in morphology or behavior within a species caused by competition with another species; characteristics typical of one species differ according to whether the other species is present in or absent from a local community.

ecological community An assemblage of species that are associated in a common area and interact with each other in a self-regulating relationship.

ecological pyramid A quantitative measurement of a **food web** in terms of amount of **biomass,** numbers of **organisms,** or energy at each of the different **trophic** levels present **(producers, herbivores,** first-level **carnivores,** higher-level carnivores).

ecology (Gr. *oikos,* house, + *logos,* discourse) Part of biology that concerns the relationship between organisms and their environment.

ecosystem (ek′ō-sis-təm) (eco[logy] from Gr. *oikos,* house, + system) An ecological unit comprising both the biotic communities and the nonliving (abiotic) environment, which interact to produce a stable system.

ectoderm (ek′tō-derm) (Gr. *ektos,* outside, + *derma,* skin) Outer layer of cells of an early embryo (gastrula stage); one of the germ layers, also sometimes used to include tissues derived from ectoderm.

ectognathous (ek′tō-nā′thəs) (Gr. *ektos,* outside, without, + *gnathos,* jaw) Derived character shared by most insects in which mandibles and maxillae are not in pouches.

ectolecithal (ek′tō-les′ə-thəl) (Gr. *ektos,* ouside, + *ekithos,* yolk) Yolk for nutrition of the embryo contributed by cells that are separate from the egg cell and are combined with the zygote by envelopment within the eggshell.

ectomesoderm In a developing animal, a middle layer of cells derived from ectoderm, in contrast to the more typical derivation from endoderm.

ectoparasite (ek′tō-par′ə-sīt) **Parasite** that resides on the outside surface of its host organism; contrasts with **endoparasite.**

ectoplasm (ek′tō-pla-zm) (Gr. *ektos,* outside, + *plasma,* form) The cortex of a cell or that part of cytoplasm just under the cell surface; contrasts with **endoplasm.**

ectothermic (ek′tō-therm′ic) (Gr. *ektos,* outside, + *thermē,* heat) Having a body temperature derived from heat acquired from the environment; contrasts with **endothermic.**

efferent (ef′ə-rənt) (L. *ex,* out, + *ferre,* to bear) Leading or conveying away from some organ—for example, nerve impulses conducted away from the brain, or blood conveyed away from an organ; contrasts with **afferent.**

egestion (ē-jes′chən) (L. *egestus,* to discharge) Act of dispelling indigestible or waste matter from the body by any normal route.

elephantiasis (el-ə-fən-tī′ə-səs) Disfiguring condition caused by chronic infection with filarial worms *Wuchereria bancrofti* and *Brugia malayi.*

Eltonian pyramid An **ecological pyramid** showing numbers of organisms at each of the **trophic** levels.

embryogenesis (em′brē-ō-jen′ə-səs) (Gr. *embryon,* embryo, + *genesis,* origin) The origin and development of the embryo; embryogeny.

encystment Process of **cyst** formation.

endemic (en-dem′ik) (Gr. *en,* in, + *demos,* populace) Geographically restricted to a certain region or country; native to a restricted area; not introduced.

endoderm (en′də-dərm) (Gr. *endon,* within, + *derma,* skin) Innermost germ layer of an embryo, forming the primitive gut; also may refer to tissues derived from endoderm.

endognathous (en′dō-nā-thəs) (Gr. *endon,* within, + *gnathous,* jaw) Ancestral character in insects (orders Diplura, Collembola, and Protura) in which the mandibles and maxillae are located in pouches.

endolecithal (en′dō-les′ə-thəl) (Gr. *endon,* within, + *lekithos,* yolk) Yolk for nutrition of the embryo incorporated into the egg cell itself.

endoparasite (en′dō-par′ə-sīt) Parasite that resides inside the body of its host organism; contrasts with **ectoparasite.**

endoplasm (en′dō-pla-zm) (Gr. *endon,* within, + *plasma,* mold or form) The portion of cytoplasm that immediately surrounds the nucleus.

endopod, endopodite (en′dō-päd, en-dop′ə-dīt) (Gr. *endon,* within, + *pous, podos,* foot) Medial branch of a biramous crustacean appendage.

endoskeleton (Gr. *endon,* within, + *skeletos,* hard) Skeleton or supporting framework within the living tissues of an organism; contrasts with **exoskeleton.**

endostyle (en′dō-stīl) (Gr. *endon,* within, + *stylos,* a pillar) Mucus-secreting, ciliated groove(s) in the floor of the pharynx of tunicates, cephalochordates, and larval lampreys; used for accumulating and moving food particles to the stomach.

endosymbiosis A symbiosis in which the symbiont lives inside its host; origin of eukaryotes, in which one prokaryote (symbiont) came to live inside another prokaryote (host), and symbionts eventually became organelles, such as mitochondria, of the host.

endothermic (en′dō-therm′ik) (Gr. *endon,* within, + *thermē,* heat) Having a body temperature determined by heat derived from an animal's own oxidative metabolism; contrasts with **ectothermic.**

energy budget Economic analysis of the energy used by an organism, partitioned into **gross productivity, net productivity,** and **respiration.**

enterocoel (en′tər-ō-sēl′) (Gr. *enteron,* gut, + *koilos,* hollow) Type of coelom formed by the outpouching of a mesodermal sac from the endoderm of the primitive gut.

enteron (en′tə-rän) (Gr., intestine) The digestive cavity.

entognathy (en′tə-nā-thē) Bases of mouthparts enclosed within the head capsule in insect orders Collembola, Diplura, and Protura.

entomology (en′tə-mol′ə-jē) (Gr. *entoma,* an insect, + *logos,* discourse) Study of insects.

ephyra (ef′ə-rə) (Gr. *Ephyra,* Greek city) Juvenile medusa budded from a strobilating polyp in class Scyphozoa, phylum Cnidaria.

epidermis (ep′ə-dər′məs) (Gr. *epi,* on, upon, + *derma,* skin) The outer, nonvascular layer of skin of ectodermal origin; in invertebrates, a single layer of ectodermal epithelium.

epipod, epipodite (ep'ə-päd, e-pi-pō'-dīt) (Gr. *epi,* on, upon, + *pous, podos,* foot) Lateral process on the protopod of a crustacean appendage, often modified as a gill.

epistome (ep'i-stōm) (Gr. *epi,* on, upon, + *stoma,* mouth) Flap over the mouth in some lophophorates bearing the protocoel.

epithelium (ep'i-thē'lē-um) (Gr. *epi,* on, upon, + *thēlē,* nipple) Cellular tissue covering a free surface or lining a tube or cavity.

epitoke (ep'i-tōk) (Gr. *epitokos,* fruitful) Posterior part of a marine polychaete when swollen with developing gonads during the breeding season; contrasts with **atoke.**

erythrocyte (ə-rith'rō-sīt) (Gr. *erythros,* red, + *kytos,* hollow vessel) Red blood cell; has hemoglobin to carry oxygen from lungs or gills to tissues. During formation in mammals, erythrocytes lose their nuclei, whereas those of other vertebrates retain the nuclei.

estivation (es'tə-vā'shen) (L. *aestivates,* to spend the summer) A state of dormancy during the summer when temperatures are high, food is scarce and/or dehydration threatens. Metabolism and breathing rate decrease.

estrous cycle Periodic episodes of estrus, or "heat," when females of most mammalian species become sexually receptive.

estrus (es'trəs) (L. *oestrus,* gadfly, frenzy) The period of heat, or rut, especially of a female mammal during ovulation. Associated with maximum sexual receptivity.

eukaryotic, eucaryotic (ū'ka-rē-ot'ik) (Gr. *eu,* good, true, + *karyon,* nut,kernel) Organisms whose cells characteristically contain a membrane-bound nucleus or nuclei; contrasts with **prokaryotic.**

eumetazoan (ü'-met-ə-zō'ən) (Gr. *eu,* good, true, + *meta,* after, + *zōon,* animal) Any multicellular animal with distinct germ layers that form true tissues; animals beyond the cellular grade of organization.

euryhaline (ū-rə-hā'līn) (Gr. *eurys,* broad, + *hals,* salt) Able to tolerate wide ranges of saltwater concentrations.

euryphagous (yə-rif'ə-gəs) (Gr. *eurys,* broad, + *phagein,* to eat) Eating a large variety of foods.

eurytopic (ū-rē-tōp'ik) (Gr. *eurys,* broad, + *topos,* place) Describes an organism with a wide environmental range.

eutely (ū'tē-lē) (Gr. *euteia,* thrift) Condition of a body composed of a constant number of cells or nuclei in all adult members of a species, as in rotifers, acanthocephalans, and nematodes.

evagination (ē-vaj-ə-nā'shən) (L. *e,* out, + *vagina,* sheath) An outpocketing from a hollow structure.

evolution (L. *evolvere,* to unfold) All changes in the characteristics and diversity of life on earth throughout its history.

evolutionary sciences Empirical investigation of ultimate causes in biology using the comparative method.

evolutionary species concept A single lineage of ancestral-descendant populations that maintains its identity from other such lineages and has its own evolutionary tendencies and historical fate; differs from the **biological species concept** by explicitly including a time dimension and including asexual lineages.

evolutionary taxonomy System of classification, formalized by George Gaylord Simpson, that groups species into Linnean higher taxa representing a hierarchy of distinct adaptive zones; such taxa may be **monophyletic** or **paraphyletic** but not **polyphyletic.**

exaptation (ek-sap'-tā'shən) Evolutionary cooption of an organismal or molecular character for a biological role unrelated to the character's evolutionary origin. Bird feathers are considered an exaptation for flight because they originated prior to avian flight but provided utility for flight following its origin; contrasts with **adaptation;** bird feathers are considered an adaptation for the biological role of thermoregulation.

exopod, exopodite (ex'ə-päd, ex-äp'ō-dīt) (Gr. *exō,* outside, + *pous, podos,* foot) Lateral branch of a biramous crustacean appendage.

exoskeleton (ek'sō-skel'ə-tən) (Gr. *exō,* outside, + *skeletos,* hard) Supporting structure secreted by ectoderm or epidermis; external, not enveloped by living tissue, as opposed to **endoskeleton**; in vertebrates, a supporting structure formed within the integument.

experiment (L. *experiri,* to try) A trial made to support or to disprove a hypothesis.

experimental method General procedure for testing hypotheses by predicting how a biological system will respond to a disturbance, making the disturbance under controlled conditions, and then comparing the observed results with the predicted ones.

experimental sciences Empirical investigation of proximate causes in biology using the **experimental method.**

extrinsic factor An environmental variable that influences the biological properties of a population, such as observed number of individuals or rate of growth.

extrusome (eks'trə-sōm) (L. *extrusus,* driven out, + *soma,* body) Any membrane-bound organelle used to extrude something from a cell.

eyelid (ī'-lid) A thin surface of skin and muscle that can be closed to protect an eye from light, abrasion, and/or desiccation. Appears in many but not all terrestrial vertebrates.

F

fermentation (L. *fermentum,* ferment) Enzymatic breakdown of an organic molecule in the absence of oxygen, releasing energy in the process. In some mammalian digestive tracts, symbiotic microbes metabolize cellulose via the process of fermentation.

filopodium (fi'lo-pō'dē-əm) (L. *filum,* thread, + Gr. *pous, podos,* foot) Type of pseudopodium that is very slender and may branch but does not rejoin to form a mesh.

filter feeding Any feeding process by which particulate food is filtered from water in which it is suspended.

fission (L. *fissio,* a splitting) Asexual reproduction by division of the body into two or more parts.

fitness Degree of adjustment and suitability for a particular environment. Genetic fitness is the relative contribution of a genotype to the next generation; organisms with high genetic fitness are those favored by natural selection.

flagellum (flə-jel'əm), pl. **flagella** (L., a whip) Whiplike organelle of locomotion.

flame cell Specialized hollow excretory or osmoregulatory structure composed of one or several small cells containing a tuft of flagella (the "flame") and situated at the end of a minute tubule; connected tubules ultimately open to the outside. See **protonephridium.**

fluke (O.E. *flōc,* flatfish) Member of platyhelminth classes Trematoda or Monogenera. Also, certain of the flatfishes (order Pleuronectiformes).

food chain Movement of energy from plant compounds to organisms that eat plants, then to other organisms that eat the plant feeders, and possibly further through a linear series of organisms that feed and are then eaten by others. Food chains connect and branch to form **food webs.**

food vacuole A digestive organelle in the cell.

food web An analysis relating species in an ecological community according to how they acquire nutrition, such as by fixing atmospheric carbon **(producers),** consuming producers **(herbivores),** consuming herbivores (first-level **carnivores**), or consuming carnivores (higher-level carnivores).

fossil (fos'əl) Any remains or impression of an organism from a past geological age that has been preserved by natural processes, usually by mineralization in the earth's crust.

fossorial (fä-sōr'ē-əl) (L. *fossor,* digger) Characterized by digging or burrowing.

fouling Contamination of feeding or respiratory areas of an organism by excrement, sediment, or other matter. Also, accumulation of sessile marine organisms on the hull of a boat or ship so as to impede its movement through water.

fovea (fō′vē-ə) (L., a small pit) Small pit or depression, especially the fovea centralis, a small, rodless pit in the retina of some vertebrates; a point of acute vision.

frog (fräg′) Denotes any member of the amphibian order Anura (also called Salientia).

frontal plane Plane parallel to the main axis of the body and at right angles to the sagittal plane.

fundamental niche A variety of roles potentially performed by an organism or population in an ecological community; limits on such roles are set by the intrinsic biological attributes of an organism or population. See also **niche** and **realized niche.**

furcula (fur′cū-la) (L. *furc*, fork) Fused clavicles of birds and some dinosaurs.

funnel The tube from which a jet of water exits the mantle cavity of a cephalopod mollusc.

G

gamete (ga′mēt, gə-mēt′) (Gr. *gamos,* marriage) A mature haploid sex cell; usually male and female gametes can be distinguished. An egg or a sperm.

gametic meiosis Meiosis that occurs during formation of the gametes, as in humans and other metazoa.

gametocyte (ga-mēt′ə-sīt) (Gr. *gametēs,* spouse, + *kytos,* hollow vessel) The mother cell of a gamete; an immature gamete.

ganglion (gang′lē-ən), pl. **ganglia** (Gr., little tumor) Aggregation of nerve tissue containing nerve cells.

ganoid scales (ga′noyd) (Gr. *ganos,* brightness) Thick, bony, rhombic scales of some bony fishes; not overlapping.

gastrocoel (gas′trō-sēl) (Gr.*gastēr,* stomach, + *koilos,* hollow) Embryonic cavity forming in gastrulation that becomes the adult gut; also called an archenteron.

gastrodermis (gas′trō-dər′mis) (Gr. *gastēr,* stomach, + *derma,* skin) Lining of the digestive cavity of cnidarians.

gastrovascular cavity (Gr. *gastēr,* stomach, + L. *vasculum,* small vessel) Body cavity in certain lower invertebrates that functions in both digestion and circulation and has a single opening serving as both mouth and anus.

gastrozooid (gas′trō-zō-id) (Gr. *gastēr,* stomach, + *zoon,* animal) The feeding polyp of a hydroid; a hydranth.

gastrula (gas′trə-lə) (Gr. *gastēr,* stomach, + L. *ula,* dim.) Embryonic stage, usually cap- or sac-shaped, with walls of two layers of cells surrounding a cavity (archenteron) with one opening (blastopore).

gemmule (je′mūl) (L. *gemma,* bud, + *ula,* dim.) Asexual, cystlike reproductive unit in freshwater sponges; formed in summer or autumn and capable of overwintering.

gene (Gr. *genos,* descent) The part of a chromosome that is the hereditary determiner and is transmitted from one generation to another. Specifically, a gene is a nucleic acid sequence (usually DNA) that encodes a functional polypeptide or RNA sequence.

general lineage concept Claim by Kevin de Queiroz that the primary definition of the species category is that it is a segment of a population **lineage**, and that all contrasting species concepts (**biological species concept, cohesion species concept, evolutionary species concept, phylogenetic species concept**) differ from each other only by secondary or incidental properties necessary to identify species taxa.

gene pool Collection of all the alleles of all the genes in a population.

genetic drift Change in allelic frequencies by chance processes in the evolution of animals. In small populations, one allele may drift to fixation, becoming the only representative of that gene locus.

genotype (jēn′ə-tīp) (Gr. *genos,* offspring, + *typos,* form) The genetic constitution, expressed and latent, of an organism; the total set of genes present in the cells of an organism; contrasts with **phenotype.**

genus (jē′nus), pl. **genera** (L., race) Group of related species with the taxonomic rank between family and species.

germinative zone The site immediately following the scolex on the body of a mature tapeworm where new proglottids are produced.

germ layer In an animal embryo, one of three basic layers (ectoderm, endoderm, mesoderm) from which the various organs and tissues arise in the multicellular animal.

germovitellarium (jer′mə-vit-əl-ar′ē-əm) (L. *germen,* a bud, offshoot, + *vitellus,* yolk) Closely associated ovary (germarium) and yolk-producing structure (vitellarium) in rotifers.

germ plasm The germ cells of an organism, as distinct from the somatoplasm; the hereditary material (genes) of the germ cells.

gestation (je-stā′shən) (L. *gestare,* to bear) Period during which offspring are carried in the uterus.

gizzard (giz′erd) A modified stomach with thick muscular walls that grinds food, often with the aid of stones.

glideosome (glī′-dē-ō-sōm) A unique structure in apicomplexan parasites that binds the membrane of the parasite cell to that of the host cell to permit parasite entry without cell rupture.

glochidium (glō-kid′ē-əm) (Gr. *glochis,* point, + *idion,* dimin. suffix) Bivalved larval stage of freshwater mussels.

glycogen (glī′kə-jən) (Gr. *glykys,* sweet, + *genes,* produced) A polysaccharide constituting the principal form in which carbohydrate is stored in animals; animal starch.

gnathobase (nath′ə-bās′) (Gr. *gnathos,* jaw, base) Median basic process on certain appendages in some arthropods, usually used for biting or crushing food.

gnathostomes (nath′ə-stōmz) (Gr. *gnathos,* jaw, + *stoma,* mouth) Vertebrates with jaws.

gonad (gō′nad) (N.L. *gonas,* a primary sex organ) Organ that produces gametes (ovary in the female and testis in the male).

gonangium (gō-nan′jē-əm) (N.L. *gonas,* primary sex organ, + *angeion,* dimin. of vessel) Reproductive zooid of hydroid colony (Cnidaria).

gonoduct (gän′ə-dukt) (Gr. *gonos,* seed, progeny, + duct) Duct leading from a gonad to the exterior.

gonophore (gon′ə-for) (Gr. *gonos,* seed, progeny, + *phoros,* bearer) Sexual reproductive structure developing from reduced medusae in some hydrozoans; it may be retained on the colony or released.

gonopore (gän′ə-pōr) (Gr. *gonos,* seed, progeny, + *poros,* an opening) Genital pore in many invertebrates.

grade (L. *gradus,* step) Level of organismal complexity or adaptive zone characteristic of a taxonomic group.

gradualism (graj′ə-wal-iz′əm) A component of Darwin's evolutionary theory postulating that evolution occurs by the temporal accumulation of small, incremental changes by populations, usually across very long periods of geological time; it opposes claims that evolution can occur by mutations having large and discontinuous phenotypic effects.

granular glands Integumentary structures of modern amphibians associated with secretion of defensive compounds.

green gland Excretory gland of certain Crustacea; the antennal gland.

gregarine (gre-ga-rin′) (L. *gregarious,* belonging to a herd or flock) Protozoan parasites belonging to class Gregarinea within phylum Apicomplexa; these organisms infect the guts or body cavities of invertebrates.

gross productivity Measurement of the total energy assimilated by an organism.

ground substance The matrix in which connective tissue fibers are embedded.

growth rate The proportion by which a population changes in numbers of individuals at a given time by reproduction and possibly by immigration.

guild (gild) (M.E. *gilde,* payment, tribute) Species of a local community that partition resources through character displacement to avoid niche overlap and competition, such as Galápagos finch communities whose component species differ in beak size for specializing on different-sized seeds.

gynecophoric canal (gī′nə-kə-fōr′ik) (Gr. *gynē,* woman, + *pherein,* to carry) Groove in male schistosomes (certain trematodes) that carries the female.

H

habitat (L. *habitare,* to dwell) The place where an organism normally lives or where individuals of a population live.

habitat fragmentation Emergence of geographic barriers that disrupt a formerly continuous geographic distribution of populations of a species. Evolutionary rates of species formation and species extinction are increased by this occurrence.

halter (hal′tər), pl. **halteres** (hal-ti′rēz) (Gr., leap) In Diptera, small, club-shaped structure on each side of the metathorax representing the hind wings; thought to be sense organ for balancing; also called a balancer.

haplodiploidy (Gr. *haploos,* single, + *diploos,* double, + *eidos,* form) Reproduction in which haploid males are produced parthenogenetically and diploid females develop from fertilized eggs.

haploid (Gr. *haploos,* single) The reduced, or n, number of chromosomes, typical of gametes, as opposed to the diploid, or 2n, number found in somatic cells. In certain groups, some mature organisms have a haploid number of chromosomes.

Hardy-Weinberg equilibrium Mathematical demonstration that the Mendelian hereditary process does not change the populational frequencies of alleles or genotypes across generations, and that change in allelic or genotypic frequencies requires factors such as natural selection, genetic drift in finite populations, recurring mutation, migration of individuals among populations, and nonrandom mating.

hemal system (hē′məl) (Gr. *haima,* blood) System of small vessels in echinoderms; function unknown.

hemimetabolous (he′mē-mə-ta′bə-ləs) (Gr. *hēmi,* half, + *metabolē,* change) Refers to gradual metamorphosis during development of insects without a pupal stage; contrasts with **holometabolous**.

hemipenis (he′mē-pē-nis) pl. **hemipenes** (Gr. *hemi,* half + L. *penis*) One of a pair of copulatory organs of male squamates (snakes and lizards), held inside the cloaca when not in use.

hemocoel (hē′mə-sēl) (Gr. *haima,* blood, + *koilos,* hollow) Main body cavity of arthropods; may be subdivided into sinuses, through which blood flows.

hemoglobin (Gr. *haima,* blood, + L. *globulus,* globule) Iron-containing respiratory pigment occurring in vertebrate red blood cells and in blood plasma of many invertebrates; a compound of an iron porphyrin heme and globin proteins.

hemolymph (hē′mə-limf) (Gr. *haima,* blood, + L. *lympha,* water) Fluid composed of blood plasma and lymph residing in the coelom or hemocoel of invertebrates having open circulatory systems.

herbivore ([h]erb′ə-vōr′) (L. *herba,* green crop, + *vorare,* to devour) Any organism subsisting on plants. Adj., **herbivorous.** Feeding on plants is called **herbivory.**

hermaphrodite (hər-maf′rə-dīt) (Gr. *hermaphroditos,* containing both sexes; from Greek mythology, Hermaphroditos, son of Hermes and Aphrodite) An organism with both male and female functional reproductive organs. **Hermaphroditism** often denotes an aberration in species that normally have separate male and female individuals; **monoecism** implies that this is the normal condition for the species.

heterocercal (het′ər-o-sər′kəl) (Gr. *heteros,* different, + *kerkos,* tail) In some fishes, a tail with the upper lobe larger than the lower, and the end of the vertebral column somewhat upturned in the upper lobe, as in sharks; contrasts with **homocercal**.

heterochrony (het′ə-rō-krōn-ē) (Gr. *heteros,* different, + *chronos,* time) Evolutionary change in the relative time of appearance or rate of development of characteristics from ancestor to descendant.

heterodont (het′ə-rō-dänt) (Gr. *heteros,* different, + *odous,* tooth) Having teeth differentiated into incisors, canines, and molars for different purposes in mammals; contrasts with **homodont**.

heterolobosea (het′ə-rō-lo-bō′sē-ə) (Gr. *heteros,* other, different, + *lobos,* lobe) Protozoan clade in which most members can assume both ameboid and flagellate forms.

heterostracans (het′ə-rō-strā′-kanz) (Gr. *beteros,* different, + *ostrakon,* shell) Group of extinct fishes with dermal armor and no jaws or paired fins; known from the Ordovician to Devonian periods.

heterotroph (het′ə-rō-trōf) (Gr. *heteros,* different, + *trophos,* feeder) Organism that obtains both organic and inorganic raw materials from the environment in order to live; includes most animals and microorganisms and those plants that do not have photosynthesis.

heterozygous Describes an organism in which homologous chromosomes contain different allelic forms (often dominant and recessive) of a gene; derived from a zygote formed by union of gametes of dissimilar allelic constitution.

hexamerous (hek-sam′ər-əs) (Gr. *hex,* six, + *meros,* part) Six parts; specifically, symmetry based on six or multiples thereof.

hibernation (L. *hibernus,* wintry) Condition, especially of mammals, of passing the winter in a torpid state in which the body temperature drops nearly to freezing and the metabolism drops close to zero.

hierarchical system Scheme that arranges organisms into a series of taxa of increasing inclusiveness, as illustrated by Linnean classification.

histology (hi-stäl′-ə-jē) (Gr. *histos,* web, tissue, + *logos,* discourse) Study of the microscopic anatomy of tissues.

holometabolous (hō′lō-mə-ta′bə-ləs) (Gr. *holo,* complete, + *metabolē,* change) Complete metamorphosis during insect development from larval to pupal to adult stages; contrasts with **hemimetabolous**.

holophytic nutrition (hō-lō-fit′ik) (Gr. *holo,* whole, + *phyt,* plant) Occurs in green plants and certain protozoa and involves synthesis of carbohydrates from carbon dioxide and water in the presence of light, chlorophyll, and certain enzymes.

holozoic nutrition (hō-lō-zō′ik) (Gr. *holo,* whole, + *zoikos,* of animals) Type of nutrition involving ingestion of liquid or solid organic food particles.

homeobox (hō′mē-ō-box) (Gr. *homolos,* like, resembling, + L. *buxus,* boxtree [used in the sense of enclosed, contained]) A highly conserved 180-base-pair sequence found in homeotic genes, regulatory sequences of protein-coding genes that regulate development.

homeothermic (hō′mē-ō-thər′mik) (Gr. *homeo,* alike, + *thermē,* heat), or **homeothermy** Having a nearly uniform body temperature, regulated independent of the environmental temperature; "warm-blooded," **endothermic**.

homeotic genes (hō-mē-ät′ik) (Gr. *homolos,* like, resembling) Genes, identified through mutations, that give developmental identity to specific body segments.

home range The area over which an animal ranges in its activities. Unlike territories, home ranges are not defended.

hominid (häm′ə-nid) (L. *homo, hominis,* man) A member of the family Hominidae, which includes chimpanzees, gorillas, humans, orangutans, and extinct forms descended from their most recent common ancestor.

hominoid (häm′ə-noyd) Relating to the Hominoidea, a superfamily of primates to which the great apes and humans belong.

homocercal (hō′mə-ser′kəl) (Gr. *homos,* same, common, + *kerkos,* tail) A tail with the upper and lower lobes symmetrical and the vertebral column ending near the middle of the base, as in most teleost fishes; contrasts with **heterocercal**.

homodont (hō′mō-dänt) (Gr. *homos,* same, + *odous,* tooth) Having all teeth similar in form; contrasts in mammals with **heterodont**.

homology (hō-mäl′ə-jē) (Gr. *homologos,* agreeing) Similarity of parts or organs of different organisms caused by evolutionary derivation from a corresponding part or organ in a common ancestor, and usually having a similar embryonic origin. May also refer to a matching pair of chromosomes. Serial homology is the correspondence in the same individual of repeated structures having the same origin and development, such as the appendages of arthropods. Adj., **homologous.**

homoplasy (hō′mō-plā′sē) Phenotypic similarity among characteristics of different species or populations (including molecular, morphological, behavioral, or other features) that does not accurately represent patterns of common evolutionary descent (+ nonhomologous similarity); it is produced by evolutionary parallelism, convergence, and/or reversal, and is revealed by incongruence among different characters on a cladogram or phylogenetic tree.

hyaline (hī′ə-lən) (Gr. *hyalos,* glass) Adj., glassy, translucent. Noun, a clear, glassy, structureless material occurring in cartilage, vitreous bodies, mucin, and glycogen, for example.

hydatid cyst (hī-da′təd) (Gr. *hydatis,* watery vesicle) Type of cyst formed by juveniles of certain tapeworms (*Echinococcus*) in their vertebrate hosts.

hydranth (hī′dranth) (Gr. *hydōr,* water, + *anthos,* flower) Nutritive zooid of hydroid colony.

hydrocoel (hī′drə-sēl) (Gr. *hydōr,* water, + *koilos,* hollow) Second or middle coelomic compartment in echinoderms; left hydrocoel gives rise to water-vascular system.

hydrocoral (Gr. *hydōr,* water, + *korallion,* coral) Certain members of the cnidarian class Hydrozoa that secrete calcium carbonate, resembling true corals.

hydrogenosomes (hī-drə-jen′ə-sōmz) Small organelles in certain anaerobic protozoa that produce molecular hydrogen as an end product of energy metabolism.

hydroid The polyp form of a cnidarian as distinguished from the medusa form. Formerly, hydrozoans of the discontinued, paraphyletic order Hydroida.

hydrostatic skeleton A mass of fluid or plastic parenchyma enclosed within a muscular wall to provide the support necessary for antagonistic muscle action; for example, parenchyma in acoelomates and perivisceral fluids in pseudocoelomates serve as hydrostatic skeletons.

hydrothermal vent A submarine hot spring; seawater seeping through the sea floor is heated by magma and expelled back into the sea through a hydrothermal vent.

hyperosmotic (hī′pər-oz-mo′-tik) (Gr. *hyper,* over, + *ōsmos,* impulse) Describes a solution that contains a greater concentration of dissolved particles than another solution to which it is compared; gains water through a selectively permeable membrane from a solution containing fewer particles; contrasts with **hypoosmotic.**

hyperparasitism (hī′pər-par′ə-si-ti′zm) A parasite itself parasitized by another parasite.

hypodermis (hī′pə-dər′mis) (Gr. *hypo,* under, + L. *dermis,* skin) The cellular layer lying beneath and secreting the cuticle of annelids, arthropods, and certain other invertebrates.

hypoosmotic (hī′pō-oz-mo′-tik) (Gr. *hypo,* under, + *ōsmos,* impulse) Describes a solution that contains a lesser concentration of dissolved particles than another solution to which it is compared; loses water through a selectively permeable membrane from a solution containing more particles; contrasts with **hyperosmotic.**

hypostome (hī′pə-stōm) (Gr. *hypo,* under, + *stoma,* mouth) Name applied to a structure in various invertebrates (such as mites and ticks), located at the posterior or ventral area of the mouth; elevation supporting the mouth of a hydrozoan.

hypothesis (hī′po-thə′-sis) (Gr. *hypothesis,* foundation, supposition) A statement or proposition that can be tested by observation or experiment.

hypothetico-deductive method The central procedure of scientific inquiry in which a postulate is advanced to explain a natural phenomenon and then subjected to observational or experimental testing that potentially could reject the postulate.

I

ichthyosaur (ik′thē-ō-sor) (Gr. *ichthyo,* fish, + *saur,* lizard) Aquatic, Mesozoic reptiles characterized by a porpoiselike body, but having a vertical tail and large eyes.

immediate cause See **proximate cause.**

inbreeding The tendency among members of a population to mate preferentially with close relatives.

indigenous (in-dij′ə-nəs) (L. *indigna,* native) Pertains to organisms that are native to a particular region; not introduced.

induction (L. *inducere, inductum,* to lead) Reasoning from the particular to the general; deriving a general statement (hypothesis) based on individual observations. In embryology, the alteration of cell fates caused by interaction with neighboring cells.

infraciliature (in′frə-sil′ē-ə-chər) (L. *infra,* below, + *cilia,* eyelashes) The organelles just below the cilia in ciliate protozoa.

inheritance of acquired characteristics The discredited Lamarckian notion that organisms, by striving to meet the demands of their environments, obtain new adaptations and pass them by heredity to their offspring.

instar (inz′tär) (L., form) Stage in the life of an insect or other arthropod between molts.

integument (in-teg′ū-mənt) (L. *integumentum,* covering) An external covering or enveloping layer.

intermediary meiosis Meiosis that occurs neither during gamete formation nor immediately after zygote formation, resulting in both haploid and diploid generations, such as in foraminiferan protozoa.

intermediate filament A component of the cytoskeleton intermediate in size between actin filaments (microfilaments) and microtubules.

intermediate host A host in which some development of a symbiont occurs, but in which maturation and sexual reproduction do not occur; contrasts with **definitive host**.

internal nostrils Palatal structures connecting the nasal cavity and throat in lungfishes and tetrapod vertebrates; used for olfaction and/or breathing when the mouth is closed.

interstitial (in′tər-sti′shəl) (L. *inter,* among, + *sistere,* to stand) Situated in the interstices or spaces between structures such as cells, organs, or grains of sand.

intracellular (in-trə-sel′yə-lər) (L. *intra,* inside, + *cellula,* chamber) Occurring within a body cell or within body cells.

intrinsic growth rate Exponential growth rate of a population; the difference between the density-independent components of the birth and death rates of a natural population with stable age distribution.

intrinsic rate of increase See **intrinsic growth rate.**

introvert (L. *intro,* inward, + *vertere,* to turn) The anterior narrow portion that can be withdrawn (introverted) into the trunk of a sipunculid worm.

iteroparity (i′tər-o-pā′ri-tē′) A life history in which individual organisms of a population normally reproduce more than one time before dying; contrasts with **semelparity.**

J

Jacobson's organ (*Jacobson,* 19th-century Danish surgeon and anatomist) Chemosensory organ in the roof of the mouth of many terrestrial vertebrates; the tongue transfers pheromones and food-derived chemicals to this organ, also called the vomeronasal organ.

juvenile hormone Hormone produced by the corpora allata of insects; among its effects are maintenance of larval or nymphal characteristics during development.

K

karyology (ker′-ē-ol′ə-jē) See **comparative cytology**

keratin (ker′ə-tən) (Gr. *kera,* horn, + *in,* suffix of proteins) A scleroprotein found in epidermal tissues and modified into hard structures such as horns, hair, and nails.

keystone species Species (typically a predator) whose removal leads to reduced species diversity within the community.

kinetoplast (kī-nēt'ə-plast) (Gr. *kinētos,* moving, + *plastos,* molded, formed) Cellular organelle that functions in association with a kinetosome at the base of a flagellum; presumed to be derived from a mitochondrion.

kinetosome (kin-et'ə-sōm) (Gr. *kinētos,* moving, + *sōma,* body) The self-duplicating granule at the base of a flagellum or cilium; similar to centriole; also called a basal body or blepharoplast.

L

labium (lā'bē-əm) (L., a lip) The lower lip of an insect formed by fusion of the second pair of maxillae. Also part of human female external genitalia.

labrum (lā'brəm) (L., a lip) The upper lip of insects and crustaceans situated above or in front of the mandibles; also denotes the outer lip of a gastropod shell.

labyrinthodont (lab'ə-rin'thə-dänt) (Gr. *labyrinthos,* labyrinth, + *odous, odontos,* tooth) Group of Paleozoic amphibians containing the temnospondyls and the anthracosaurs.

lachrymal glands (lak'rə-məl) (L. *lacrimia,* tear) Structures in terrestrial vertebrates that secrete tears to lubricate the eyes.

lactation (lak-tā'shən) (L. *lacteus,* milk) The production of milk from mammary glands.

lacunar system Netlike set of circulatory canals filled with fluid in an acanthocephalan.

Lamarckism Hypothesis, as expounded by Jean-Baptiste de Lamarck, of evolution by acquisition during an organism's lifetime of characteristics that are transmitted to offspring.

lamella (lə-mel'ə) (L. dim. of *lamina,* plate) One of the two plates forming a gill in a bivalve mollusc. One of the thin layers of bone laid concentrically around an osteon (Haversian canal). Any thin, platelike structure.

larva (lar'və), pl. **larvae** (L., a ghost) An immature stage that is quite different from the adult.

lateral (L. *latus,* the side, flank) Of or pertaining to the side of an animal; a *bilateral* animal has two sides.

lateral-line system Sensory organ that detects water vibrations; consists of neuromast organs in canals and grooves on the head and sides of the body of fishes and some amphibians.

lek (lek) (Sw., play, game) Area where animals assemble for communal courtship display and mating.

lemniscus (lem-nis'kəs) (L., ribbon) One of a pair of internal projections of the epidermis from the neck region of Acanthocephala, functioning in fluid control in the protrusion and invagination of the proboscis.

lepidosaurs (lep'ə-dō-sors) (L. *lepidos,* scale, + *sauros,* lizard) A group of diapsid reptiles that appeared in the Permian period and that includes the modern snakes, lizards, and tuataras.

leptocephalus (lep'tə-sef'ə-ləs), pl. **leptocephali** (Gr. *leptos,* thin, + *kephalē,* head) Transparent, ribbonlike migratory larva of the European or American eel.

limiting resource A particular source of nutrition, energy, or living space whose scarcity is causally associated with a population having fewer individuals than otherwise expected in a particular environment.

lineage (lin'əj) An unbranched succession of ancestral-descendant populations in evolutionary time. Lineages related to others through branching of ancestral lineages form a phylogenetic tree. In molecular evolution, a succession of ancestral-descendant DNA molecules traced through an organismal genealogy or phylogeny.

lobopodium (lō'bə-pō'dē-əm) (Gr. *lobos,* lobe, + *pous, podos,* foot) Blunt, lobelike pseudopodium.

lobosea (lə-bō'sē-ə) (Gr. *lobos,* lobe) A protozoan clade comprising amebas with lobopodia.

lophophore (lōf'ə-fōr) (Gr. *lophos,* crest, + *phoros,* bearing) Tentacle-bearing ridge or arm within which is an extension of the coelomic cavity in lophophorate animals (ectoprocts, brachiopods, and phoronids).

lophotrochozoan protostome (lō'fō-trō'kō-zō'ən) (Gr. *lophos,* crest, + *trochos,* wheel, + *zōon,* animal) Any member of a clade within Protostomia whose members generally possess either a trochophore larva or a lophophore; examples are annelids, molluscs, and ectoprocts.

lorica (lor'ə-kə) (L. *lorica,* corselet) A secreted, protective covering, as in phylum Loricifera.

lymph (limf) (L. *lympha,* water) The interstitial (intercellular) fluid in the body; also, the fluid in the lymphatic space.

M

macroevolution (L. *makros,* long, large, + *evolvere,* to unfold) Evolutionary change on a grand scale, encompassing the origin of novel designs, evolutionary trends, adaptive radiation, and mass extinction.

macrogamete (mak'rə-gam'ēt) (Gr. *makros,* long, large, + *gamos,* marriage) The larger of the two gamete types in a heterogametic species, considered the female gamete.

macronucleus (ma'krō-nū'klē-əs) (Gr. *makros,* long, large, + *nucleus,* kernel) The larger of the two kinds of nuclei in ciliate protozoa; controls all cell function except reproduction.

madreporite (ma'drə-pōr'īt) (Fr. *madrépore,* reef-building coral, + *ite,* suffix for some body parts) Sievelike structure that is the intake of the water-vascular system of echinoderms.

malaria (mə-lar'-ē-ə) (It. *malaria,* bad air) Disease marked by periodic chills, fever, anemia, and other symptoms; caused by *Plasmodium* spp.

Malpighian tubules (mal-pig'ē-ən) (Marcello Malpighi, Italian anatomist, 1628–94) Blind tubules opening into the hindgut of nearly all insects and some myriapods and arachnids and functioning primarily as excretory organs.

mandible (L. *mandibula,* jaw) One of the lower jaw bones in vertebrates; one of the head appendages in arthropods.

mantle Soft extension of the body wall in certain invertebrates—for example, brachiopods and molluscs—which usually secretes a shell; thin body wall of tunicates.

manubrium (mə-nü'brē-əm) (L., handle) The portion projecting from the oral side of a jelly medusa, bearing the mouth; oral cone; presternum or anterior part of sternum; handlelike part of malleus of ear.

marsupial (mär-sü'pē-əl) (Gr. *marsypion,* little pouch) One of the pouched mammals of the subclass Metatheria.

mass extinction Relatively short interval of geological time in which a large portion (75–95%) of existing species or higher taxa are eliminated nearly simultaneously.

mastax (mas'təx) (Gr., jaws) Pharyngeal mill of rotifers.

matrix (mā'triks) (L. *mater,* mother) The extracellular substance of a tissue, or that part of a tissue into which an organ or process is set.

maxilla (mak-sil'ə) (L. dim. of *mala,* jaw) One of the upper jawbones in vertebrates; one of the head appendages in arthropods.

maxilliped (mak-sil'ə-ped) (L. *maxilla,* jaw, + *pes,* foot) One of the pairs of head appendages located just posterior to the maxilla in crustaceans; a thoracic appendage that has become incorporated into the feeding mouthparts.

medial (mē'dē-əl) Situated, or occurring, in the middle.

medulla (mə-dül'ə) (L., marrow) The inner portion of an organ in contrast to the cortex, or outer portion. Also, hindbrain.

medusa (mə-dü'-sə) (Gr. mythology, female monster with snake-entwined hair) A sea jelly, or the free-swimming stage that reproduces sexually in the life cycle of cnidarians.

megafauna (me′-gə -faw′-nə) Large (>44 kg) terrestrial animals, especially in reference to the large fauna of the Pleistocene epoch.

Mehlis glands (me′ləs) Glands of uncertain function surrounding the junction of yolk duct, oviduct, and uterus in trematodes and cestodes.

meiosis (mī-ō′səs) (Gr. from *meioun*, to make small) The nuclear changes by means of which the chromosomes are reduced from the diploid to the haploid number; in animals, usually occurs in the last two divisions in the formation of the mature egg or sperm.

melanin (mel′ə-nin) (Gr. *melas*, black) Black or dark-brown pigment found in plant or animal structures.

membranelle Tiny, membrane-like structure; may be formed by fused cilia.

merozoite (me′rə-zō′īt) (Gr. *meros*, part, + *zōon*, animal) A very small trophozoite at the stage just after cytokinesis has been completed in multiple fission of a protozoan.

mesenchyme (me′zn-kīm) (Gr. *mesos*, middle, + *enchyma*, infusion) Embryonic connective tissue; irregular or amebocytic cells often embedded in gelatinous matrix.

mesocoel (mēz′ō-sēl) (Gr. *mesos*, middle, + *koilos*, hollow) Middle body coelomic compartment in some deuterostomes, anterior in lophophorates; corresponds to hydrocoel in echinoderms.

mesoderm (me′zə-dərm) (Gr. *mesos*, middle, + *derma*, skin) The third germ layer, formed in the gastrula between the ectoderm and endoderm; gives rise to connective tissues, muscle, urogenital and vascular systems, and the peritoneum.

mesoglea (mez′ō-glē′ə) (Gr. *mesos*, middle, + *glia*, glue) The layer of jellylike or cement material between the epidermis and gastrodermis in cnidarians and ctenophores.

mesohyl (me′zō-hil) (Gr. *mesos*, middle, + *hyle*, a wood) Gelatinous matrix surrounding sponge cells; also called mesoglea, mesenchyme.

mesonephros (me-zō-nef′rōs) (Gr. *mesos*, middle, + *nephros*, kidney) The middle of three pairs of embryonic renal organs in vertebrates. Functional kidney of embryonic amniotes; its collecting duct is a Wolffian duct. Adj., **mesonephric.**

mesosome (mez′ə-sōm) (Gr. *mesos*, middle, + *sōma*, body) The portion of the body in lophophorates and some deuterostomes that contains the mesocoel.

metacercaria (me′tə-sər-ka′rē-ə) (Gr. *meta*, after, + *kerkos*, tail, + L. *aria*, connected with) Fluke juvenile (cercaria) that has lost its tail and become encysted.

metacoel (met′ə-sēl) (Gr. *meta*, after, + *koilos*, hollow) Posterior coelomic compartment in some deuterostomes and lophophorates; corresponds to somatocoel in echinoderms.

metamere (met′ə-mēr) (Gr. *meta*, after, + *meros*, part) A repeated body unit along the longitudinal axis of an animal; a somite or segment.

metamerism (mə-ta′-mə-ri′zəm) (Gr. *meta*, between, after, + *meros*, part) Being composed of serially repeated parts (metameres); serial segmentation.

metamorphosis (Gr. *meta*, after, + *morphē*, form, + *osis*, state of) Sharp change in form during postembryonic development— for example, tadpole to frog, or larval insect to adult.

metanephridium (me′tə-nə-fri′di-əm) (Gr. *meta*, after, + *nephros*, kidney) Type of tubular nephridium with the inner open end draining the coelom and the outer open end discharging to the exterior.

metanephros (me′tə-ne′frōs) (Gr. *meta*, between, among, after, + *nephros*, kidney) Vertebrate kidney formed from the most posterior of three embryonic regions capable of forming renal organs; functional kidney of adult amniotes; drained by ureter. Adj., **metanephric.**

metapleural folds (me-tə′-plə-rəl fōldz) Paired, elongate ventrolateral fin-like flaps on amphioxus (Subclass Cephalochordata).

metapopulation A large population that comprises numerous semi-autonomous subpopulations, termed demes, with some limited movement of individuals among demes. Demes of a metapopulation are often geographically distinct. Patterns of movement of individuals among versus within demes constitute **metapopulation dynamics**.

metasome (met′ə-sōm) (Gr. *meta*, after, behind, + *sōma*, body) The portion of the body in lophophorates and some deuterostomes that contains the metacoel.

metazoan (met-ə-zō′ ə) (Gr. *meta*, after, + *zōon*, animal) Animal; contrasts with **protozoan**.

microevolution (mī′krō-ev-ə-lü-shən) (L. *mikros*, small, + *evolvere*, to unfold) A change in the gene pool of a population across generations.

microfilariae (mīk′rə-fil-ar′ē-ē) (Gr. *mikros*, small, + L. *filum*, a thread) Born alive, partially developed juvenile filarial worms (phylum Nematoda).

microgamete (mīk′rə-ga′-mēt) (Gr. *mikros*, small, + *gamos*, marriage) The smaller of the two gamete types in a heterogametic species, considered the male gamete.

microhabitat selection When species having similar ecological niches occur in the same community, they partition shared resources by specializing on different aspects of the shared resource, such as insectivorous birds favoring different characteristic foraging sites.

micron (μ) (mī′-krän) (Gr. neuter of *mikros*, small) One-thousandth of a millimeter; about 1/25,000 of an inch. Now largely replaced by micrometer (μ).

microneme (mī′krə-nēm) (Gr. *mikros*, small, + *nēma*, thread) One type of structure forming the apical complex in phylum Apicomplexa; slender and elongate, leading to the anterior; thought to function in host-cell penetration.

micronucleus The smaller of two nuclei in ciliated protozoa; controls the reproductive functions of these organisms.

microsporidian (mī′krō-spo-rid′ē-ən) (Gr. *micros*, small, + *spora*, seed, + *idion*, dim. suffix) Any member of a unicellular eukaryote clade comprising intracellular parasites with a distinctive spore morphology.

microthrix See **microvillus.**

microtubule (Gr. *mikros*, small, + L. *tubule*, pipe) Long, tubular cytoskeletal element with an outside diameter of 20 to 27 nm. Microtubules influence cell shape and play important roles during cell division.

microvillus (Gr. *mikros*, small, + L. *villus*, shaggy hair) Narrow, cylindrical cytoplasmic projection from epithelial cells; microvilli form the brush border of several types of epithelial cells. Also, microvilli with unusual structure cover the surface of cestode tegument (also called **microthrix** [pl. **microtriches**]).

mictic (mik′tik) (Gr. *miktos*, mixed or blended) Pertaining to haploid eggs of rotifers or the females that lay such eggs.

mimic (mim′ik) (Gr. *mimicus*, imitator) Species whose morphological or behavioral characteristics copy those of another species because those characteristics deter shared predators.

mimicry Evolution by natural selection of similar forms by different species, such as the sharing of warning signals that discourage common predators. In Batesian mimicry, a species tasteful to a predator deceptively copies warning signals of a distasteful species. In Müllerian mimicry, two or more distasteful species evolve common warning signals to avoid a common predator.

miracidium (mīr′ə-sid′ē-əm) (Gr. *meirakidion*, youthful person) Minute, ciliated larval stage in the life of flukes.

mitochondrion (mīt′ō-kän′drē-ən) (Gr. *mitos*, a thread, + *chondrion*, dim. of *chondros*, corn, grain) Cellular organelle in which aerobic metabolism occurs.

mitosis (mī-tō′səs) (Gr. *mitos*, thread, + *osis*, state of) Nuclear division in which chromosomal material is equally divided both qualitatively and quantitatively between the two resulting nuclei; ordinary cell division.

mitosome A highly modified mitochondrion that does not produce energy; most lack genetic material.

mixotrophy (mik′sō-trō′fē) (Gr. *mikso*, mixed + *trophia*, denoting nutrition) A nutritional strategy where autotrophy and heterotrophy are used by one organism to acquire organic carbon and/or other elements, such as N, P, or Fe.

model (mod′l) (Fr. *modèle*, pattern) A species whose morphological or behavioral characteristics are copied by another species because those characteristics deter shared predators.

modular (moj′ə-lər) Describes the structure of a colony of genetically identical organisms that are physically associated and produced asexually by cloning.

molting Shedding of the outer cuticular layer; see **ecdysis.**

monoecious (mə-nē′shəs) (Gr.*monos,* single, + *oikos,* house) Having both male and female gonads in the same organism, usually denoting that this is the typical condition of a species; hermaphroditic.

monogamy (mə-näg′ə-mē) (Gr. *monos,* single, + *gamos,* marriage) The condition of having a single mate at any one time. Adj., **monogamous.**

monophyly (män′ō-fī′lē) (Gr. *monos,* single, + *phyle,* tribe) Condition occurring when a taxon or other group of organisms contains the most recent common ancestor of the group and all of its descendants. Adj., **monophyletic.**

monotreme (mä′nō-trēm) (Gr. *monos,* single, + *trēma,* hole) Egg-laying mammal of the order Monotremata.

morphogenesis (mor′fə-je′nə-səs) (Gr. *morphē,* form, + *genesis,* origin) Development of the architectural features of organisms; formation and differentiation of tissues and organs.

morphology (Gr. *morphē,* form, + *logos,* discourse) The science of structure. Includes cytology, the study of cell structure; histology, the study of tissue structure; and anatomy, the study of gross structure.

mosaic development Embryonic development characterized by independent differentiation of each part of the embryo; cytoplasmic specification.

mosaic evolution The concept that evolutionary change takes place in some characters first, then others later, rather than all characters changing simultaneously. For example, the transitional fossil *Archaeopteryx* has wings and feathers nearly identical to modern birds, but retains teeth and a long tail basically unchanged from its ancestors.

MROs Mitochondria-related organelles of three general types, including hydrogenosomes and mitosomes.

MTOCs Microtubule-organizing centers of several different types where free tubulin molecules come together to form strands.

mucus (mū′kəs) (L. *mucus,* nasal mucus) Viscid, slippery secretion rich in mucins produced by secretory cells such as those in mucous membranes. Adj., **mucous.**

multiple fission Mode of asexual reproduction in some protistans in which the nuclei divide more than once before cytokinesis occurs.

multiplication of species The Darwinian theory that the evolutionary process generates new species through a branching of evolutionary lineages derived from an ancestral species.

mutation (mū-tā′shən) (L. *mutare,* to change) Stable and abrupt change of a gene; the heritable modification of a character.

mutualism (mü′chə-wə-li′zəm) (L. *mutuus,* lent, borrowed, reciprocal) Type of interaction in which two different species derive benefit from their association, sometimes necessary for both species to survive; often symbiotic.

myocyte (mī′ə-sīt) (Gr. *mys,* muscle, + *kytos,* hollow vessel) Contractile cell (pinacocyte) in sponges.

myofibril (Gr. *mys,* muscle, + L. dim. of *fibra,* fiber) Contractile filament within muscle or muscle fiber.

myomere (mī′ə-mēr) (Gr. *mys,* muscle, + *meros,* part) A muscle segment of successive segmental trunk musculature.

myotome (mī′ə-tōm) (Gr. *mys,* muscle, + *tomos,* cutting) A voluntary muscle segment in cephalochordates and vertebrates; that part of a somite destined to form muscles; the muscle group innervated by a single spinal nerve.

N

nacre (nā′kər) (F., mother-of-pearl) Innermost, lustrous layer of a mollusc shell, secreted by mantle epithelium. Adj., **nacreous.**

nares (na′rēz), sing. **naris** (L., nostrils) Openings into the nasal cavity, both internally and externally, in the head of a vertebrate.

natural selection The interactions between organismal character variation and the environment that cause differences in rates of survival and reproduction among varying organisms in a population; leads to evolutionary change if variation is heritable.

nauplius (naw′plē-əs) (L., a kind of shellfish) Free-swimming microscopic larval stage of certain crustaceans, with three pairs of appendages (antennules, antennae, and mandibles) and a median eye. Characteristic of ostracods, copepods, barnacles, and some others.

nekton (nek′tən) (Gr. neuter of *nēktos,* swimming) Term for actively swimming organisms, essentially independent of wave and current action; contrasts with **plankton.**

nematocyst (ne-mat′ə-sist′) (Gr. *nēma,* thread, + *kystis,* bladder) Stinging organelle of cnidarians.

neo-Darwinism (nē′ō′där′wə-niz′əm) A modified version of Darwin's evolutionary theory that eliminates elements of the Lamarckian inheritance of acquired characteristics and pangenesis that were present in Darwin's formulation; this theory originated with August Weismann in the late nineteenth century and, after incorporating Mendelian genetic principles, has become the currently favored version of Darwinian evolutionary theory.

neopterygian (nē-äp′tə-rij′ē-ən) (Gr. *neos,* new, + *pteryx,* fin) Any of a large group of bony fishes that includes most modern species.

nephridium (nə-frid′ē-əm) (Gr. *nephridios,* of the kidney) One of the segmentally arranged, paired excretory tubules of many invertebrates, notably the annelids. In a broad sense, any tubule specialized for excretion and/or osmoregulation; with an external opening and with or without an internal opening.

nephron (ne′frän) (Gr. *nephros,* kidney) Functional unit of a vertebrate kidney comprising Bowman's capsule, an enclosed glomerulus, and the attached uriniferous tubule.

nephrostome (nef′rə-stōm) (Gr. *nephros,* kidney, + *stoma,* mouth) Ciliated, funnel-shaped opening of a nephridium.

nested hierarchy Ordering of species into a series of increasingly more inclusive clades according to the taxonomic distribution of synapomorphies.

net productivity The energy stored by an organism, equal to the energy assimilated **(gross productivity)** minus the energy used for metabolic maintenance **(respiration).**

neural crest Populations of ectodermally derived embryonic cells that differentiate into many skeletal, neural, and sensory structures; unique to vertebrates.

neural spine A dorsal projection of a vertebra that functions as a muscle attachment site.

neuroglia (nur′ə-glē′ə) (Gr. *neuron,* nerve, + *glia,* glue) Tissue supporting and filling the spaces between the nerve cells of the central nervous system.

neuromast (Gr. *neuron,* sinew, nerve, + *mastos,* knoll) Cluster of sense cells on or near the surface of a fish or amphibian that is sensitive to water movements.

neuron (Gr., nerve) A nerve cell.

neuropodium (nur′ə-pō′dē-əm) (Gr. *neuron,* nerve, + *pous, podos,* foot) Lobe of parapodium nearer the ventral side in polychaete annelids.

neurosecretory cell (nur′ə-sə-krē′to-rē) Any cell (neuron) of the nervous system that produces a hormone.

niche (nich′) The role of an organism, population, or species in an ecological community comprising its usage of resources, its unique way of life, and its relationship to other biotic and abiotic factors.

niche overlap A comparison of two species that quantifies the proportion of each species' resources also utilized by the other species.

node (nōd′) Point in a cladogram or phylogenetic tree where two or more branches join to form a clade. In a phylogenetic tree, a node represents the most recent common ancestor of all lineages descended from it. Sometimes, taxa at the tip branches are called terminal nodes, and the nodes as defined here are called internal nodes.

notochord (nō'tə-kord') (Gr. *nōtos,* back, + *chorda,* cord) Elongated cartilaginous cellular cord, enclosed in a sheath, that forms the primitive axial skeleton of chordate embryos, adult cephalochordates, and jawless vertebrates.

notopodium (nō'tə-pō'dē-əm) (Gr. *nōtos,* back, + *pous, podos,* foot) Lobe of parapodium nearer the dorsal side in polychaete annelids.

nucleolus (nü-klē'ə-ləs) (dim. of L. *nucleus,* kernel) Deeply staining body within the nucleus of a cell and containing ribosomal RNA; nucleoli are specialized portions of certain chromosomes that carry multiple copies of the genes encoding ribosomal RNA and where ribosomal RNA is actively synthesized.

nucleoplasm (nü'klē-ə-pla'zəm) (L. *nucleus,* kernel, + Gr. *plasma,* mold) Protoplasm of nucleus, as distinguished from cytoplasm.

nucleus (nü'klē-əs) (L. *nucleus,* a little nut, the kernel) The organelle in eukaryotes that contains the chromatin and is bounded by a double membrane (nuclear envelope).

nurse cells Single cells or layers of cells surrounding or adjacent to other cells or structures for which the nurse cells provide nutrients or other molecules (for example, for insect oocytes or *Trichinella* spp. juveniles).

nymph (L. *nympha,* nymph, bride) Immature stage (following hatching) of a hemimetabolous insect that lacks a pupal stage.

O

ocellus (ō-sel'əs) (L. dim. of *oculus,* eye) A simple eye or eyespot in many types of invertebrates.

octomerous (ok-tom'ər-əs) (Gr. *oct,* eight, + *meros,* part) Eight parts; specifically, symmetry based on eight.

odontophore (ō-don'tə-for') (Gr. *odous,* tooth, + *pherein,* to carry) Tooth-bearing organ in molluscs, including the radula, radular sac, muscles, and cartilages.

olfactory epithelium A specialized chemosensory surface tissue inside the nasal cavities of aquatic and terrestrial vertebrates.

omasum (ō-ma'səm) (L., paunch) The third compartment of the stomach of a ruminant mammal.

ommatidium (ä'mə-tid'ē-əm) (Gr. *omma,* eye, + *idium,* small) One of the optical units of the compound eye of arthropods.

omnivore (äm'nə-vōr) (L. *omnis,* all, + *vorare,* to devour) Animal whose diet includes a variety of animal and plant material.

oncosphere (än'kō-sfiər) (Gr. *onkinos,* a hook, + *sphaira,* ball) Rounded larva common to all cestodes; bears hooks.

ontogeny (on-to'jə-nē) (Gr. *ontos,* being, + *geneia,* act of being born, from *genēs,* born) The course of development of an individual from egg to senescence.

oocyst (ō'ə-sist) (Gr. *ōion,* egg, + *kystis,* bladder) Cyst formed around zygote of malarial parasite and related organisms.

oocyte (ō'ə-sīt) (Gr. *ōion,* egg, + *kytos,* hollow) Stage in formation of ovum, just preceding first meiotic division (primary oocyte) or just following first meiotic division (secondary oocyte).

ookinete (ō-ə-kī'nēt) (Gr. *ōion,* egg, + *kinein,* to move) Motile zygote of malarial parasites.

operculum (ō-per'kū-ləm) (L., cover) The gill cover in bony fishes; keratinized plate in some snails.

opisthaptor (ō'pəs-thap'tər) (Gr. *opisthen,* behind, + *haptein,* to fasten) Posterior attachment organ of a monogenetic trematode.

opisthokont (ō-pis'thō-kont) (G. *opisthen,* behind, + *kontos,* a pole) Any member of the eukaryotic clade comprising fungi, microsporidians, choanoflagellates, and animals; if present, flagellated cells possess a single posterior flagellum.

opisthonephros (ō-pis'-thō-nef'-rō s) (Gr. *opisth,* back + *nephros,* kidney) A kidney that develops from the middle and posterior portions of the nephrogenic region of vertebrates and is drained by the Wolffian duct or accessory ducts. Functional kidney of most adult anamniotes (fishes and amphibians). Adj., **opisthonephric.**

oral disc The end of a cnidarian polyp bearing the mouth.

oral lobe Flaplike extension of the mouth of a scyphozoan medusa that aids in feeding.

organelle (Gr. *organon,* tool, organ, + L. *ella,* dimin. suffix) Specialized part of a cell; a subcellular structure that performs functions analogous to organs of multicellular animals.

organism (or'-gə-niz'-əm) A biological individual composed of one or more cells, tissues, and/or organs whose parts are interdependent in producing a collective physiological system. Organisms of the same species may form **populations.**

osculum (os'kū-ləm) (L. *osculum,* a little mouth) Excurrent opening in a sponge.

osmoregulation Maintenance of proper internal salt and water concentrations in a cell or in the body of a living organism; active regulation of internal osmotic pressure.

osmosis (oz-mō'sis) (Gr. *ōsmos,* act of pushing, impulse) A flow of solvent (usually water) through a semipermeable membrane.

osmotroph (oz'mə-trōf) (Gr. *ōsmos,* a thrusting, impulse, + *trophē,* to eat) Heterotrophic organism that absorbs dissolved nutrients.

osphradium (os-frā'dē-əm) (Gr. *osphradion,* strong smell) Sense organ in aquatic snails and bivalves that tests incoming water.

ossicles (L. *ossiculum,* small bone) Small separate pieces of echinoderm endoskeleton. Also, tiny bones of the middle ear of vertebrates.

osteoderm (os'tē-ə-dərm') (Gr. *osteon,* bone, + derma, skin) A bony, dermal plate located under and supporting an epidermal scale.

osteostracans (os-tē-os'trə-kəns) (Gr. *osteon,* bone, + *ostrakon,* shell) Group of jawless, extinct fishes with dermal armor and pectoral fins from the Silurian and Devonian periods.

ostium (L., door) Opening.

ostracoderm (os-trak'ō-derm) (Gr. *ostrakon,* shell, + *derma,* skin) A paraphyletic group of extinct, jawless fishes with dermal armor known from the late Cambrian to Devonian periods.

otolith (ōt'ə-lith') (Gr. *ous, otos,* ear, + *lithos,* stone) Calcareous concretions in the membranous labyrinth of the inner ear of lower vertebrates or in the auditory organ of certain invertebrates.

outgroup In phylogenetic systematic studies, a species or group of species closely related to but not included within a taxon whose phylogeny is being studied, and used to polarize variation of characters and to root the phylogenetic tree.

outgroup comparison Method for determining the polarity of a character in cladistic analysis of a taxonomic group. Character states found within the group being studied are judged ancestral if they occur also in related taxa outside the study group (+ outgroups); character states that occur only within the taxon being studied but not in outgroups are judged to have been derived evolutionarily within the group being studied.

oviger (ō'vi-jər) (L. *ovum,* egg, + *gerere,* to bear) Leg that carries eggs in pycnogonids.

oviparity (ō'və-pa'rət-ē) (L. *ovum,* egg, + *parere,* to bring forth) Reproduction in which eggs are released by the female; development of offspring occurs outside the maternal body. Adj., **oviparous** (ō-vip'ə-rəs).

ovipositor (ō've-poz'ət-ər) (L. *ovum,* egg, + *positor,* builder, placer, + *or,* suffix denoting agent or doer) In many female insects, a structure at the posterior end of the abdomen for laying eggs.

ovoviviparity (ō'vo-vī-və-par'ə-tē) (L. *ovum,* egg, + *vivere,* to live, + *parere,* to bring forth) A reproductive mode in which young develop (often in an egg) in the uterus, but rely only on their yolk sacs for nutrition, with no placental connection. Also called lecithotropic viviparity. Adj., **ovoviviparous** (ō-vo-vī-vip'ə-rəs).

ovum (L. *ovum,* egg) Mature female germ cell (egg).

P

paedomorphosis (pē-dō-mor'fə-səs) (Gr. *pais,* child, + *morphē,* form) Displacement of ancestral juvenile features to later stages of the ontogeny of descendants.

pangenesis (pan-jen′ə-sis) (Gr. *pan,* all, + *genesis,* descent) Darwin's discredited hypothesis that hereditary factors are produced by individual body parts according to the organism's use of those parts, and collected in the germ cells for transmission to offspring.

papilla (pə-pil′ə), pl. **papillae** (L., nipple) Small, nipplelike projection. A vascular process that nourishes the root of a hair, feather, or developing tooth.

papula (pa′pü-lə), pl. **papulae** (L., pimple) Respiratory processes on skin of sea stars; also, pustules on skin. Also called **dermal branchia.**

parabasal apparatus Cellular organelles similar to Golgi bodies, presumed to function as part of the secretory system in endoplasmic reticulum.

parabronchi (par-ə-bron′kī) (Gr. *para,* beside, + *bronchos,* windpipe) Fine air-conduction pathways of a bird lung.

paraphyly (per′ə-fī′lē) (Gr. *para,* before, + *phyle,* tribe) The condition that a taxon or other group of organisms contains the most recent common ancestor of all members of the group but excludes some descendants of that ancestor; shows **convexity.** Adj., **paraphyletic.**

parapodium (pe′rə-pō′dē-əm) (Gr. *para,* beside, + *pous, podos,* foot) One of the paired lateral processes on each side of most segments in polychaete annelids; variously modified for locomotion, respiration, or feeding.

parasite (pər′ə-sīt) Organism that lives physically on or in, and at the expense of, another organism.

parasitism (per′ə-sit′iz-əm) (Gr. *parasitos,*from *para,* beside, + *sitos,* food) The condition of an organism living in or on another organism (host) at whose expense the parasite is maintained; destructive symbiosis.

parasitoid (per′ə-si′-tȯid) Organism that is a typical parasite early in its development but that finally kills the host during or at the completion of development; refers to many insect parasites of other insects.

parenchyma (pə-ren′kə-mə) (Gr., anything poured in beside) In simpler animals, a spongy mass of vacuolated mesenchyme cells filling spaces between viscera, muscles, or epithelia; in some, the cells are cell bodies of muscle cells. Also, the specialized tissue of an organ as distinguished from the supporting connective tissue.

parenchymula (pə′rən-kīm′yə-lə) (Gr. *para,* beside, + *enchyma,* infusion) Flagellated, solid-bodied larva of some sponges.

parietal (pä-rī′-ə-təl) (L. *paries,* wall) Something next to, or forming part of,a wall of a structure.

parsimony (pär′sə-mō′nē′) (L. *parsus,* to spare) A general methodological principle that the simplest hypothesis capable of explaining observations is the best working hypothesis and should be tested first before investigating more complex hypotheses. In phylogenetic systematics, this principle involves using the phylogenetic tree that requires the smallest amount of evolutionary change as the best working hypothesis of phylogenetic relationships.

parthenogenesis (pär′thə-nō-gen′ə-sis) (Gr. *parthenos,* virgin, + L. from Gr. *genesis,* origin) Unisexual reproduction involving the production of young by females not fertilized by males; common in rotifers, cladocerans, aphids, bees, ants, and wasps. A parthenogenetic egg may be diploid or haploid.

pecten (pek′-tən) (L., comb) Any of several types of comblike structures on various organisms; for example, a pigmented, vascular, and comblike process that projects into the vitreous humor from the retina at a point of entrance of the optic nerve in the eyes of all birds and many reptiles.

pectoral (pek′tə-rəl) (L. *pectoralis,* from *pectus,* the breast) Of or pertaining to the breast or chest, to the pectoral girdle, or to a pair of keratinized shields of the plastron of certain turtles.

pedalium (pə-dal′ē-əm) (Gr. *pedalion,* a prop, rudder) The flattened, bladelike base of a tentacle or group of tentacles in the cnidarian class Cubozoa.

pedal laceration Asexual reproduction in sea anemones; a form of fission.

pedicel (ped′ə-sel) (L. *pediculus,* little foot) A small or short stalk or stem. In insects, the second segment of an antenna or the waist of an ant.

pedicellaria (ped′ə-sə-ler′ē-ə) (L. *pediculus,* little foot, + *aria,* like or connected with) One of many minute, pincerlike organs on the surface of certain echinoderms.

pedipalps (ped′ə-palps′) (L. *pes, pedis,* foot, + *palpus,* stroking, caress) Second pair of appendages of arachnids.

peduncle (pē-dun′kəl) (L. *pedunculus,* dim. of *pes,* foot) A stalk. Also, a band of white matter joining different parts of the brain.

pelage (pel′ij) (Fr., fur) Hairy covering of mammals.

pelagic (pə-laj′ik) (Gr. *pelagos,* the open sea) Occupying or moving through water rather than the underlying substrate; contrasts with **benthic.**

pellicle (pel′ə-kəl) (L. *pellicula,* dim. of *pelis,* skin) Thin, translucent, secreted envelope covering many protozoa.

pelycosaur (pel′ə-kō-sor) (Gr. *pelyx,* basin, + *sauros,* lizard) Any of a group of Permian synapsids characterized by homodont dentition and sprawling limbs.

pen A flattened, flexible internal support in a squid; a remnant of the ancestral shell.

pentadactyl (pen-tə-dak′təl) (Gr. *pente,* five, + *daktylos,* finger) Having five digits, or five fingerlike parts, to the hand or foot.

perennibranchiate (pə-rən′ə-brank′ē-ət) (L. *perennis,* throughout the year, + Gr. *branchia,* gills) Having permanent gills, relating especially to certain paedomorphic salamanders.

periostracum (pe-rē-äs′trə-kəm) (Gr. *peri,* around, + *ostrakon,* shell) Outer keratinized layer of a mollusc shell.

peripheral (pə-ri′fər-əl) (Gr. *peripherein,* to move around) Structure or location distant from center, near outer boundaries.

periproct (per′ə-präkt) (Gr. *peri,* around, + *prōktos,* anus) Region of aboral plates around the anus of echinoids.

perisarc (per′ə-särk) (Gr. *peri,* around, + *sarx,* flesh) Sheath covering the stalk and branches of a hydroid.

perissodactyl (pə-ris′ə-dak′təl) (Gr. *perissos,* odd, + *daktylos,* finger, toe) Pertaining to an order of ungulate mammals with an odd number of digits.

peristomium (per′ə-stō′mē-əm) (Gr. *peri,* around, + *stoma,* mouth) One of two parts forming the annelid head; it bears the mouth.

peritoneum (per′ə-tə-nē′əm) (Gr. *peritonaios,* stretched around) The membrane that lines the coelom and covers the coelomic viscera.

Permian extinction A **mass extinction** that occurred 245 million years ago in which 96% of existing species became extinct, marking the end of the Paleozoic era.

perpetual change The most basic theory of evolution, stating that the living world is neither constant nor cycling, but is always undergoing irreversible modification through time.

phagocyte (fag′ə-sīt) (Gr. *phagein,* to eat, + *kytos,* hollow vessel) Any cell that engulfs and devours microorganisms or other particles.

phagocytosis (fag′ə-sī-tō′səs) (Gr. *phagein,* to eat, + *kytos,* hollow vessel) The engulfment of a particle by a phagocyte or a protozoan.

phagosome (fa′gə-sōm) (Gr. *phagein,* to eat, + *sōma,* body) Membrane-bound vessel in cytoplasm containing food material engulfed by phagocytosis.

phagotroph (fag′ə-trōf) (Gr. *phagein,* to eat, + *trophē,* food) Heterotrophic organism that ingests solid particles for food.

pharyngeal pouches Paired evaginations, or pockets, that develop on the inside of the pharynx in chordates. In fishes and protochordates, they open to the outside of the body as pharyngeal slits.

pharynx (far′inks), pl. **pharynges** (Gr. *pharynx,* gullet) The part of the digestive tract between the mouth cavity and the esophagus that, in chordates, is common to both the digestive and the respiratory tracts.

phenetic taxonomy (fə-ne′tik) (Gr. *phaneros,* visible, evident) Uses overall similarity to classify organisms into taxa; contrasts with classification based explicitly on reconstruction of phylogeny.

phenotype (fē′nə-tīp) (Gr. *phainein,* to show) The visible or expressed characteristics of an organism; influenced by the genotype in interaction with environmental conditions.

pheromone (fer′ə-mōn) (Gr. *pherein,* to carry, + *hormōn,* exciting, stirring up) Chemical substance released by one organism that influences the behavior or physiological processes of another organism.

photoautotroph (fō-tō-aw′tō-trōf) (Gr. *phōtos,* light, + *autos,* self, + *trophos,* feeder) Organism requiring light as a source of energy for making organic nutrients from inorganic raw materials.

photosynthesis (fō-tō-sin′thə-sis) (Gr. *phōs,* light, + *synthesis,* action or putting together) The synthesis of carbohydrates from carbon dioxide and water in chlorophyll-containing cells exposed to light.

phototaxis (fō-tō-tak′sis) (Gr. *phōtos,* light, + *taxis,* arranging, order) A taxis in which light is the orienting stimulus. Involuntary tendency for an organism to turn toward (positive) or away from (negative) light.

phototrophs (fō-tō-trōfs) (Gr. *phōtos,* light, + *trophē,* nourishment) Organisms capable of using CO_2 in the presence of light as a source of metabolic energy.

phyletic gradualism A model of evolution in which morphological evolutionary change is continuous and incremental and occurs mainly within unbranched species or lineages over long periods of geological time; contrasts with **punctuated equilibrium.**

phylogenetic species concept An irreducible (basal) cluster of organisms, diagnosably distinct from other such clusters, and within which there is a parental pattern of ancestry and descent.

phylogenetic systematics See **cladistics.**

phylogenetic tree A diagram whose branches represent evolutionary lineages; depicts the common descent of species or higher taxa.

phylogeny (fī′läj′ə-nē) (Gr. *phylon,* tribe, race, + *geneia,* origin) The origin and diversification of any taxon, or the evolutionary history of its origin and diversification, usually presented as a dendrogram.

phylum (fī′ləm), pl. **phyla** (N.L. from Gr. *phylon,* race, tribe) A chief taxonomic category, between kingdom and class, into which are grouped organisms of common descent that share a fundamental pattern of organization.

physiology (L. *physiologia,* natural science) Branch of biology covering the organic processes and phenomena of an organism or any of its parts or a particular body process.

phytophagous (fi-täf′ə-gəs) (Gr. *phyton,* plant, + *phagein,* to eat) Feeding on plants.

pinacocyte (pin′a-kō-sīt′) (Gr. *pinax,* tablet, + *kytos,* hollow vessel) Flattened cells comprising dermal epithelium in sponges.

pinna (pin′ə) (L., feather, sharp point) The external ear. Also, a feather, wing, fin, or similar part.

pinocytosis (pin′o-sī-tō′sis, pīn′o-sī-tō′sis) (Gr. *pinein,* to drink, + *kytos,* hollow vessel,+ *osis,* condition) Acquisition of fluid by a cell in which the plasma membrane invaginates and pinches off to form small vesicles.

placenta (plə-sen′tə) (L., flat cake) The vascular structure, both embryonic and maternal, through which a mammalian embryo and fetus are nourished while in the uterus; also used to denote analogous but nonhomologous structures in some elasmobranchs, bony fishes, and squamates.

placoderms (plak′ə-dərmz) (Gr. *plax,* plate, + *derma,* skin) A group of heavily armored jawed fishes of the Lower Devonian to Lower Carboniferous periods.

placoid scale (pla′kȯid) (Gr. *plax, plakos,* tablet, plate) Type of scale found in cartilaginous fishes, composed of a basal plate of dentine embedded in the skin and a backward-pointing spine tipped with enamel.

plankton (plank′tən) (Gr. neuter of *planktos,* wandering) The passively floating animal and plant life of a body of water; contrasts with **nekton.**

plantigrade (plan′tə-grād′) (L. *planta,* sole, + *gradus,* step, degree) Pertaining to mammals that walk on the whole surface of the foot (for example, humans and bears); contrasts with **digitigrade.**

planula (plan′yə-lə) (N.L. dim. from L. *planus,* flat) Free-swimming, ciliated larval type of cnidarians; usually flattened and ovoid, with an outer layer of ectodermal cells and an inner mass of endodermal cells.

plasma membrane (plaz′mə) (Gr. *plasma,* a form, mold) Living, external, limiting, protoplasmic structure that regulates the exchange of nutrients across the cell surface.

plasmid (plaz′məd) (Gr. *plasma,* a form, mold) Small circle of DNA that a bacterium may carry in addition to its genomic DNA.

plastron (plas′trən) (Fr. *plastron,* breast plate) Ventral body shield of turtles; structure in corresponding position in certain arthropods; thin film of gas retained by epicuticle hairs of aquatic insects.

pleura (plü′rə) (Gr., side, rib) The membrane that lines each half of the thorax and covers the lungs.

pneumostome (nü′-mo-stōm) An external opening to the internal modified mantle cavity that functions as a lung in terrestrial gastropods.

podium (pō′dē-əm) (Gr. *pous, podos,* foot) A footlike structure–for example, the tube foot of echinoderms.

poikilothermic (pȯi-ki′lə-thər′mik) (Gr. *poikilos,* variable, + thermal) Pertaining to animals whose body temperature is variable and fluctuates with that of the environment; cold-blooded; contrasts with **homeothermic.**

polarity (Gr. *polos,* axis) In systematics, the ordering of alternative states of a taxonomic character from ancestral to successively derived conditions in an evolutionary transformation series. In developmental biology, the tendency for the axis of an ovum to orient corresponding to the axis of the mother. Also, condition of having opposite poles; differential distribution of gradation along an axis.

Polian vesicles (pō′lē-ən) (From G. S. Poli, 1746–1825, Italian naturalist) Vesicles opening into a ring canal in most asteroids and holothuroids.

polyandry (pol′ē-an′drē) (Gr. *polys,* many, + *anēr,* man) Having more than one male mate at one time.

polygamy (pə-lig′ə-mē) (Gr. *polys,* many, + *gamos,* marriage) Having more than one mate at one time.

polygyny (pə-lij′ə-nē) (Gr. *polys,* many, + *gynē,* woman) Having more than one female mate at one time.

polymorphism (pä′lē-mor′fi-zəm) (Gr. *polys,* many, + *morphē,* form) The presence in a species of more than one structural type of individual; genetic variation in a population.

polyp (päl′əp) (Fr. *polype,* octopus, from L. *polypus,* many-footed) The sessile stage in the life cycle of cnidarians.

polyphyletic (pä′lē-fī-let′-ik) (Gr. *polys,* many, + *phylon,* tribe) Derived from more than one ancestral source; contrasts with monophyletic and paraphyletic.

polyphyly (pä′lē-fī′lē) (Gr. *polys,* full,+ *phylon,* tribe) The condition that a taxon or other group of organisms does not contain the most recent common ancestor of all members of the group, implying that it has multiple evolutionary origins; such groups are not valid as formal taxa and are recognized as such only through error.

polyphyodont (pä-lē-fī′ə-dänt) (Gr. *polyphyes,* manifold, + *odous,* tooth) Having several sets of teeth in succession.

polypide (pä′lē-pīd) (L. *polypus,* polyp) An individual or zooid in a colony, specifically in ectroprocts, that has a lophophore, digestive tract, muscles, and nerve centers.

polytypic (pol-ē-tip′-ik) Denotes a species that contains two or more taxonomically designated subspecies; taxonomic recognition of subspecies is controversial and rejected by many taxonomists.

population (L. *populus,* people) A group of organisms of the same species inhabiting a specific geographical locality.

porocyte (pō′rə-sīt) (Gr. *porus,* passage, pore, + *kytos,* hollow vessel) Type of cell in asconoid sponges through which water enters the spongocoel.

portal system (L. *porta,* gate) System of large veins beginning and ending with a bed of capillaries; for example, the hepatic portal and renal portal systems in vertebrates.

positive assortative mating Tendency of an individual to mate preferentially with others whose phenotypes are similar to its own.

posterior (L., latter) Situated at or toward the rear of the body; in bilateral forms, the end of the main body axis opposite the head.

preadaptation The possession of a trait that coincidentally predisposes an organism for survival in an environment different from those encountered in its evolutionary history.

precocial (prē-kō´shəl) (L. *praecoquere,* to ripen beforehand) Referring (especially) to birds whose young are covered with down and are able to walk when newly hatched.

predaceous, predacious (prē-dā´shəs) (L. *praedator,* a plunderer; *praeda,* prey) Living by killing and consuming other animals; predatory.

predation (prə-dā´shən) Interaction between species in an ecological community in which members of one species (prey) serve as food for another species (**predator**).

predator (pred´ə-tər) (L. *praedator,* a plunderer; *praeda,* prey) Organism that preys on other organisms for its food.

prehensile (prē-hen´səl) (L. *prehendere,* to seize) Adapted for grasping.

primary producer Species whose members begin **productivity** by acquiring energy and matter from **abiotic** sources, as for example when plants synthesize sugars from water and carbon dioxide using solar energy (see **photosynthesis**).

primate (prī-māt) (L. *primus,* first) Any mammal of the order Primates, which includes the tarsiers, lemurs, marmosets, monkeys, apes, and humans.

primitive (L. *primus,* first) Primordial; ancient; little evolved; characteristics closely approximating those possessed by early ancestral types.

proboscis (prō-bäs´əs) (Gr. *pro,* before, + *boskein,* feed) A snout or trunk. Also, a tubular sucking or feeding organ with the mouth at the end, as in planarians, leeches, and insects. Also, the sensory and defensive organ at the anterior end of certain invertebrates.

producers (L. *producere,* to bring forth) Organisms, such as plants, able to produce their own food from inorganic substances.

production In ecology, the energy accumulated by an organism that becomes incorporated into new biomass.

productivity (prō´duk-tiv´-ət-ē) Property of a biological system measured by the amount of energy and/or materials that it incorporates.

proglottid (prō-glät´əd) (Gr. *proglōttis,* tongue tip, from *pro,* before, + *glōtta,* tongue, + *id,* suffix) Portion of a tapeworm containing a set of reproductive organs; usually corresponds to a segment.

prokaryotic, procaryotic (pro-kar´ē-ät´ik) (Gr. *pro,* before, + *karyon,* kernel, nut) Not having a membrane-bound nucleus or nuclei. Prokaryotic cells characterize Bacteria and Archaea.

pronephros (prō-nef´rōs) (Gr. *pro,* before, + *nephros,* kidney) Most anterior of three pairs of embryonic renal organs of vertebrates; functional only in adult hagfishes and larval fishes and amphibians; vestigial in mammalian embryos. Adj., **pronephric.**

prosimian (prō-sim´ē-ən) (Gr. *pro,* before, + L. *simia,* ape) Any member of a group of arboreal primates including lemurs, tarsiers, and lorises but excluding monkeys, apes, and humans.

prosopyle (prōs´-ə-pīl) (Gr. *prosō,* forward, + *pylē,* gate) Connections between the incurrent and radial canals in some sponges.

prostomium (prō-stō´mē-əm) (Gr. *pro,* before, + *stoma,* mouth) In most annelids and some molluscs, that part of the head located in front of the mouth.

protein (prō´tēn, prō´tē-ən) (Gr. *protein,* from *proteios,* primary) A macromolecule consisting of carbon, hydrogen, oxygen, and nitrogen and usually containing sulfur; composed of chains of amino acids joined by peptide bonds; present in all cells.

protein polymorphism Occurrences of allelic variants in the amino acid sequences of proteins within a population or species. Separation of allelic variants by protein electrophoresis provided an early method for quantifying the amount of genetic variation present in natural populations.

prothoracic glands Glands in the prothorax of insects that secrete the hormone ecdysone.

prothoracicotropic hormone See **ecdysiotropin.**

protist (prō´-tist) (Gr. *prōtos,* first) A member of the discontinued, paraphyletic kingdom Protista, generally considered to include the protozoa and eukaryotic algae.

protocoel (prō-tō-sēl) (Gr. *prōtos,* first, + *koilos,* hollow) The anterior coelomic compartment in some deuterostomes; corresponds to the axocoel in echinoderms.

protocooperation Mutually beneficial interaction between organisms in which the interaction is not physiologically necessary to the survival of either.

protochordate (prō´tō-kōr´dāt) (Gr. *protos,* first, + *chorda,* cord) A member of the phylum Chordata that is not a vertebrate. Living protochordates are in the subphyla Cephalochordata and Urochordata.

protonephridium (prō-tō-nə-frid´ē-əm) (Gr. *prōtos,* first, + *nephros,* kidney) Primitive osmoregulatory or excretory organ consisting of a tubule terminating internally with a flame bulb or solenocyte; the unit of a flame bulb system.

proton pump Active transport of hydrogen ions (protons) across an inner mitochondrial membrane during cellular respiration.

protopod, protopodite (prō´-tō-päd, prō´-tō-pō-dīt) (Gr. *prōtos,* first, + *pous, podos,* foot) Basal portion of crustacean appendage, containing coxa and basis.

protostome (prō´tō-stōm) (Gr. *protos,* first, + *stoma,* mouth) A member of the group Protostomia. Protostome taxa have recently been divided into ecdysozoan protostomes and lophotrochozoan protostomes.

Protostomia (prō-tō-stō´mē-ə) (Gr. *prōtos,* first, + *stoma,* mouth) A clade in which cleavage is determinate, the mouth is derived at or near the blastopore, and a coelom (when present) is formed by proliferation of mesodermal bands that later split (schizocoely). Phyla within the clade are divided between two subgroups: Lophotrochozoa and Ecdysozoa. Lophotrochozoans, exemplified by annelids and molluscs, share spiral cleavage and derive the mesoderm from a particular blastomere (called 4d). Ecdysozoans (arthropods and related taxa) have a unique cleavage pattern and do not form mesoderm from the 4d cell. Contrasts with **Deuterostomia.**

protozoan A common name often used to denote unicellular eukaryotes; contrasts with "metazoan" (+ animal), and also excludes fungi and plants.

proventriculus (pro´ven-trik´ū-ləs) (L. *pro,* before, + *ventriculum,* ventricle) In birds, the glandular stomach between the crop and the gizzard. In insects, a muscular dilation of the foregut armed internally with chitinous teeth.

proximal (L. *proximus,* nearest) Situated toward or near the point of attachment; opposite of distal, distant.

proximate cause (L. *proximus,* nearest, + *causa*) The factors that underlie the functioning of a biological system at a particular place and time, including those responsible for metabolic, physiological, and behavioral functions at the molecular, cellular, organismal, and population levels. Immediate cause.

pseudocoelom (sü´də-sē-lōm) (Gr. *pseudēs,* false, + *koilos,* hollow) Body cavity not lined with peritoneum and not a part of the blood or digestive systems; embryonically derived from the blastocoel.

pseudopodium (sü´də-pō´dē-əm) (Gr. *pseudēs,* false, + *podion,* small foot, + *eidos,* form) A temporary cytoplasmic protrusion extended from a protozoan or ameboid cell and serving for locomotion or for engulfing food.

punctuated equilibrium Model of evolution in which morphological evolutionary change is discontinuous, being associated primarily with discrete, geologically instantaneous events of speciation that lead to phylogenetic branching; morphological evolutionary stasis characterizes species between episodes of speciation; contrasts with **phyletic gradualism.**

pupa (pū′pə) (L., girl, doll, puppet) Inactive quiescent state of the holometabolous insects. It follows the larval stages and precedes the adult stage.

pygidium (pə-jid′ē-əm) (Gr. *pyge,* rump, buttocks, + *idion,* dim. ending) Posterior region of a segmented animal bearing the anus.

pygostyle (pī′gō stīl) (Gr. *pygo,* rump + *styl,* piller) A bone at the end of a bird's vertebral column, formed from fused caudal vertebrae.

Q

queen In entomology, a reproductive female in a colony of social insects such as bees, ants, and termites; distinguished from workers, nonreproductive females, and soldiers.

R

radial canals Canals along the ambulacra radiating from the ring canal of echinoderms. Also, choanocyte-lined canals in syconoid sponges.

radial cleavage Embryonic development in which early cleavage planes are symmetrical to the polar axis, each blastomere of one tier lying directly above the corresponding blastomere of the next layer; see **indeterminate cleavage**.

radial symmetry A morphological condition in which the parts of an animal are arranged concentrically around an oral-aboral axis, and more than one imaginary plane through this axis yields halves that are mirror images of each other.

radiole (rā′dē-ōl) (L. *radiolus,* dim. of *radius,* ray, spoke of a wheel) Featherlike structure extending from head of some sedentary polychaetes; used in feeding on suspended particles.

radula (ra′jə-lə) (L., scraper) Rasping tongue found in most molluscs.

ratite (ra′tīt) (L. *ratis,* raft) Referring to birds having an unkeeled sternum; contrasts with **carinate.**

realized niche The role actually performed by an organism or population in its ecological community at a particular time and place as constrained by both its intrinsic biological attributes and its particular environmental conditions. See also **niche** and **fundamental niche.**

recapitulation Summarizing or repeating; hypothesis that an individual repeats its phylogenetic history in its development.

redia (rē′dē-ə), pl. **rediae** (rē′dē-ē) (from Francesco Redi, 1626–97, Italian biologist) Larval stage in the life cycle of flukes; it is produced by a sporocyst larva, and in turn gives rise to many cercariae.

regulative development Progressive determination and restriction of initially totipotent embryonic material.

reproductive barriers (L. *re,* + *producere,* to lead forward; M.F. *barriere,* bar) The factors that prevent one sexually propagating population from interbreeding and exchanging genes with another population.

reproductive community A general criterion for the species category shared to some degree by all formal species concepts is that species constitute a reproductively bounded population or lineage of populations that does not freely merge with others in nature.

resource (rē′so(ə)rs) Available source of nutrition, energy, or space in which to live.

resource partitioning To coexist in the same habitat, two or more species specialize on different portions of a shared resource; for example, different warbler species forage in different characteristic parts of a conifer tree when they are part of the same ecological community.

respiration (L. *respiratio,* breathing) Gas interchange between an organism and its surrounding medium. In the cell, the release of energy by the oxidation of food molecules.

rete mirabile (rē′tē mə-rab′ə-lē) (L., wonderful net) Network of small blood vessels so arranged that the incoming blood runs countercurrent to the outgoing blood and thus makes possible efficient exchange between the two bloodstreams. Such a mechanism serves to maintain the high concentration of gases in the fish swim bladder.

reticulopodia (rə-tik′ū-lə-pō′dē-ə) (L. *reticulum,* dim. of *rete,* net, + *podos, pous,* foot) Pseudopodia that branch and rejoin extensively.

reticulum (rə-tik′yə-ləm) (L. *rete,* dim. *reticulum,* a net) Second stomach of ruminants; a netlike structure.

retina (ret′ən-ə) (L. *rete,* dim of reticulum, a net) The posterior sensory cell layer of the eye. Sensory cells in this layer (rods and/or cones) receive light signals and transmit them to the brain, where images are formed.

retortamonad (rə-tort′ə-mō′nad) (L. *retro,* bend backward, + *monas,* single) Any member of a protozoan clade composed of certain heterotrophic flagellates.

rhabdite (rab′dīt) (Gr. *rhabdos,* rod) Rodlike structures in the cells of the epidermis or underlying parenchyma in certain turbellarians. They are discharged in mucous secretions.

rheoreceptor (rē′ō-rē-cep′tor) (Gr. *rheos,* a flowing, + L. *receptus,* accept) Sensory organ of aquatic animals that responds to water current.

rhinophore (rī′nə-fōr) (Gr. *rhis,* nose, + *pherein,* to carry) Chemoreceptive tentacles in some molluscs (opisthobranch gastropods).

rhipidistian (rip-ə-dis′tē-ən) (Gr. *rhipis,* fan, + *histion,* sail, web) Member of a group of Paleozoic lobe-finned fishes.

rhizopodia (rī′zə-pō′dē-ə) (Gr. *rhiza,* root, + *podos,* foot) Branched filamentous pseudopodia made by some amebas.

rhopalium (rō-pā′lē-əm) (N.L. from Gr. *rhopalon,* a club) One of the marginal, club-shaped sense organs of certain sea jellies; tentaculocyst.

rhoptries (rō′p′trēz) (Gr. *thopalon,* club, + *tryō,* to rub, wear out) Club-shaped bodies in Apicomplexa forming one of the structures of the apical complex; open at anterior and apparently functioning in penetration of the host cell.

rhynchocoel (rink′ō-sēl) (Gr. *rhynchos,* snout, + *koilos,* hollow) In nemertines, the dorsal tubular cavity that contains the inverted proboscis. It has no opening to the outside.

rostrum (ros′trəm) (L., ship's beak) A snoutlike projection on the head.

rumen (rü′mən) (L., cud) The large first compartment of the stomach of ruminant mammals. Fermentation of cellulose by microorganisms occurs in it.

ruminant (rüm′ə-nənt) (L. *ruminare,* to chew the cud) Cud-chewing artiodactyl mammals with a complex, four-chambered stomach.

S

sagittal plane (saj′ə-təl) (L. *sagitta,* arrow) Pertaining to the median anteroposterior plane that divides a bilaterally symmetrical organism into right and left halves.

salamander (sal′-ə-man′-der) Denotes any member of the amphibian order Urodela (also called Caudata, sometimes with a different content of fossil forms).

saprophagous (sə-prof′ə-gəs) (Gr. *sapros,* rotten, + *phagos,* from *phagein,* to eat) Feeding on decaying matter; saprobic; saprozoic.

saprozoic nutrition (sap′rə-zō′ik) (Gr. *sapros,* rotten, + *zōon,* animal) Animal nutrition by absorption of dissolved salts and simple organic nutrients from surrounding medium; also refers to feeding on decaying matter.

sauropterygians (so-rop′tə-rij′ē-əns) (Gr. *sauros,* lizard, + *pteryginos,* winged) Mesozoic marine reptiles often with paddle-like limbs and elongate necks; includes plesiosaurs, pliosaurs, and placodonts.

scalids (skā-lədz) (Gr. *skalis,* hoe, mattock) Recurved spines on the head of kinorhynchs.

schistosomiasis (shis′tō-sō-mī′-ə-sis) (Gr. *schistos,* divided, + *soma,* body, + *iasis,* a diseased condition) Infection with blood flukes of the genus *Schistosoma.*

schizocoel (skiz′ə-sēl) (Gr. *schizo,* from *schizein,* to split, + *koilos,* hollow) A coelom formed by the splitting of embryonic mesoderm. Noun, **schizocoelomate,** an animal with a schizocoel, such as an arthropod or mollusc. Adj., **schizocoelous.**

schizocoelous mesoderm formation (skiz′ō-sē-ləs) Embryonic formation of the mesoderm as cords of cells between ectoderm and endoderm; splitting of these cords produces the coelomic space.

schizogony (skə-zo′gə-nē) (Gr. *schizein,* to split, + *gonos,* seed) Multiple asexual fission.

sclerite (skle′rīt) (Gr. *sklēros,* hard) A hard, chitinous or calcareous plate or spicule; one of the plates forming the exoskeleton of arthropods, especially insects.

scleroblast (skler′ə-blast) (Gr. *sklēros,* hard, + *blastos,* germ) Amebocyte specialized to secrete a spicule; occurs in sponges.

sclerocyte (skler′ə-sīt) (Gr. *sklēros,* hard, + *kytos,* hollow vessel) Amebocyte in sponges that secretes spicules.

sclerotin (skler′-ə-tən) (Gr. *sklērotēs,* hardness) Insoluble, tanned protein permeating the cuticle of arthropods.

sclerotization (skle′rə-tə-zā′shən) Hardening of the cuticle of arthropods by the formation of stabilizing cross linkages between peptide chains of adjacent protein molecules.

scolex (skō′leks) (Gr. *skōlex,* worm, grub) The holdfast, or so-called head, of a tapeworm; bears suckers and, in some, hooks; posterior to it, new proglottids are differentiated.

scyphistoma (sī-fis′tə-mə) (Gr. *skyphos,* cup, + *stoma,* mouth) The polyp form of a scyphozoan.

sebaceous (sə-bāsh′əs) (L. *sebaceus,* made of tallow) Type of mammalian epidermal gland that produces a fatty substance.

sebum (sē′bəm) (L., grease, tallow) Oily secretion of the sebaceous glands of the skin.

secondary palate A plate of bone and other connective tissue separating the oral cavity and nasal cavity, present in mammals and crocodilians.

sedentary (sed′ən-ter-ē) Stationary, sitting, inactive; staying in one place.

segmentation Division of the body into discrete segments or metameres; also called **metamerism.**

semelparity (se′məl-pā′ri-tē′) A life history in which individual organisms of a population normally reproduce only one time before dying, although numerous offspring may be produced at the time of reproduction; contrasts with **iteroparity.**

sensillum, pl. **sensilla** (sen-si′ləm) (L., *sensus,* sense) A small sense organ, especially in arthropods.

septum, pl. **septa** (L., fence) A wall between two cavities.

serial homology See **homology.**

serosa (sə-rō′sə) (N.L. from L. *serum,* serum) The outer embryonic membrane of birds and reptiles; chorion. Also, the peritoneal lining of the body cavity.

serous (sir′əs) (L. *serum,* serum) Watery, resembling serum; applied to glands, tissue, cells, or fluid.

serum (sir′əm) (L., whey, serum) The liquid that separates from the blood after coagulation; blood plasma from which fibrinogen has been removed. Also, the clear portion of a biological fluid separated from its particulate elements.

sessile (ses′əl) (L. *sessilis,* low, dwarf) Attached at the base; fixed to one spot.

seta (sēt′ə), pl. **setae** (sē′tē) (L. *bristle*) Needlelike chitinous structure of the integument of annelids, arthropods, and others.

sex ratio The proportion of males versus females in a population at a particular time and place.

sexual selection Charles Darwin's theory that a struggle for mates exists and that characteristics favorable for mating may prevail through reproductive success even if they are not advantageous in the struggle for survival.

siliceous (si-li′shəs) (L. *silex,* flint) Containing silica, glassy.

simian (sim′ē-ən) (L. *simia,* ape) Pertaining to monkeys or apes.

sink deme A subpopulation (deme) whose members are drawn disproportionately from other subpopulations of the same species (see **metapopulation dynamics**); for example, a deme occupying an environmentally unstable area whose members are periodically destroyed by climatic changes and then replenished by colonists from other demes when favorable conditions are restored.

sinus (sī′nəs) (L., curve) Cavity or space in tissues or in bone.

siphon (sī′-fən) Tube for directing water flow.

siphonoglyph (sī′fan′ə-glif′) (Gr. *siphōn,* reed, tube, siphon, + *glyphē,* carving) Ciliated furrow in the gullet of sea anemones.

siphuncle (sī′fun-kəl) (L. *siphunculus,* small tube) Cord of tissue running through the shell of a nautiloid, connecting all chambers with the animal's body.

sister taxon (="sister group") The relationship between a pair of species or higher taxa that are each other's closest phylogenetic relatives.

social monogamy A living relationship between an adult male and female. Although both sexes typically cooperate in rearing young, sexual activity and reproduction may involve other individuals.

solenia (sə-len′ē-ə) (Gr. *solen,* pipe) Channels through the material connecting the polyps in an octocorallian colony (phylum Cnidaria).

soma (sō′mə) (Gr., body) The whole of an organism except the germ cells (germ plasm).

somatic (sō-mat′ik) (Gr. *sōma,* body) Refers to the body—for example, somatic cells in contrast to germ cells.

somatocoel (sə-mat′ə-sēl) (Gr. *sōma,* the body, + *koilos,* hollow) Posterior coelomic compartment of echinoderms; left somatocoel gives rise to oral coelom, and right somatocoel becomes aboral coelom.

somite (sō′mīt) (Gr. *sōma,* body) One of the blocklike masses of mesoderm arranged segmentally (metamerically) in a longitudinal series beside the digestive tube of the embryo; metamere.

sorting Differential survival and reproduction among varying individuals; often confused with natural selection, which is one possible cause of sorting.

source deme A stable subpopulation (deme) that serves differentially as a source of colonists for establishing, joining, or replacing other such subpopulations of the same species (see **metapopulation dynamics**); for example, a deme inhabiting an environmentally stable area whose members routinely establish transitory populations in environmentally unstable nearby areas.

speciation (spē′sē-ā′shən) (L. *species,* kind) The evolutionary process or event by which new species arise.

species (spē′shez, spē′sēz) sing. and pl. (L., particular kind) A group of interbreeding individuals of common ancestry that are reproductively isolated from all other such groups; a taxonomic unit ranking below a genus and designated by a binomial name consisting of its genus and the species name.

species diversity The number of different species that coexist at a given time and place to form an ecological community.

species epithet The second, uncapitalized word in the binomial name of a species. It is usually an adjective modifying the first word, the genus into which the species is placed.

species richness The number of different **species** that coexist at a given time and place to form an **ecological community.**

species selection Differential rates of speciation and/or extinction among varying evolutionary lineages caused by interactions among species-level characteristics and the environment.

spermatheca (spər′mə-thē′kə) (Gr. *sperma,* seed, + *thēkē,* case) A sac in the female reproductive organs for the reception and storage of sperm.

spermatophore (spər-ma′-tə-fōr′) (Gr. *sperma, spermatos,* seed, + *pherein,* to bear) Capsule or packet enclosing sperm; produced by males of several invertebrate groups and a few vertebrates.

spicule (spi′kyul) (L. dim. of *spica,* point) One of the minute calcareous or siliceous skeletal bodies found in sponges, radiolarians, soft corals, and sea cucumbers.

spiracle (spi′rə-kəl) (L. *spiraculum,* from *spirare,* to breathe) External opening of a trachea in arthropods. One of a pair of openings on the head of elasmobranchs for passage of water. Exhalant aperture of tadpole gill chamber.

spiral cleavage Type of early embryonic cleavage in which cleavage planes are diagonal to the polar axis, and unequal cells are produced by the alternate clockwise and counterclockwise cleavage around the axis of polarity; see **determinate cleavage.**

spongin (spun′jin) (L. *spongia,* sponge) Fibrous, collagenous material forming the skeletal network of some sponges.

spongocoel (spun′jō-sēl) (Gr. *spongos,* sponge, + *koilos,* hollow) Central cavity in sponges.

spongocyte (spun′jō-sīt) (Gr. *spongos,* sponge, + *kytos,* hollow vessel) A cell in sponges that secretes spongin.

sporocyst (spō′rə-sist) (Gr. *sporos,* seed, + *kystis,* pouch) Larval stage in the life cycle of flukes; it originates from a miracidium.

sporogony (spor-og′ə-nē) (Gr. *sporos,* seed, + *gonos,* birth) Multiple fission to produce sporozoites after zygote formation.

sporozoite (spo′rə-zō′īt) (Gr. *sporos,* seed, + *zōon,* animal, + *ite,* suffix for body part) Stage in the life history of many sporozoan protozoa; released from oocysts.

stabilizing selection Natural selection that favors average values of a continuously varying trait and disfavors extreme values.

stapes (stā′pēz) (L., stirrup) Stirrup-shaped, innermost bone of the middle ear of mammals; the only ear bone of other tetrapods; homologous to the hyomandibula of fishes.

statoblast (sta′-tō-blast) (Gr. *statos,* standing, fixed, + *blastos,* germ) Asexually produced overwintering stage of many freshwater ectoprocts.

statocyst (Gr. *statos,* standing, + *kystis,* bladder) Sense organs of equilibrium; a fluid-filled cellular cyst containing one or more granules (statoliths) used to sense direction of gravity.

stenohaline (sten-ə-ha′līn, -lən) (Gr. *stenos,* narrow, + *hals,* salt) Pertaining to aquatic organisms that have restricted tolerance to changes in environmental saltwater concentration.

stenophagous (stə-näf′ə-gəs) (Gr. *stenos,* narrow, + *phagein,* to eat) Eating few kinds of foods.

stereom (ster′ē-ōm) (Gr. *stereos,* solid, hard, firm) Meshwork structure of endoskeletal ossicles of echinoderms.

sternum (ster′nəm) (L., breastbone) Ventral plate of an arthropod body segment; breastbone of vertebrates.

stigma (Gr. *stigma,* mark, tattoo mark) Eyespot in certain protozoa. Spiracle of certain terrestrial arthropods.

stolon (stō′lən) (L. *stolō, stolonis,* a shoot, or sucker of a plant) Rootlike extension of the body wall giving rise to buds that may develop into new zooids, thus forming a compound animal in which the zooids remain united by the stolon. Occurs in some colonial anthozoans, hydrozoans, ectoprocts, and ascidians.

stoma (stō′mə) (Gr., mouth) A mouthlike opening.

strobila (strō′bə-lə) (Gr. *strobilē,* lint plug like a pine cone [*strobilos*]) Scyphozoan jelly polyp with a stack of ephyrae atop it. Also, the chain of proglottids of a tapeworm.

strobilation (strō′bi-lā′shən) (Gr. *strobilos,* a pine cone) Repeated linear budding of individuals, as in scyphozoan ephyrae (phylum Cnidaria), or sets of reproductive organs in tapeworms (phylum Platyhelminthes).

stroma (strō′mə) (Gr. *stroma,* bedding) Supporting connective tissue framework of an animal organ; filmy framework of red blood corpuscles and certain cells.

subnivean (səb-ni′vē-ən) (L. *sub,* under, below, + *nivis,* snow) Applied to environments beneath snow, which acts as an insulator against colder atmospheric temperature.

survivorship (sər-vī′vər-ship) The proportion of individuals of a cohort or population that persist from one point in their life history, such as birth, to another one, such as reproductive maturity or a specified age.

suspension feeders Aquatic organisms that collect suspended food particles from the surrounding water; particles may be filtered or taken by other methods.

swim bladder Gas-filled sac of many bony fishes used in buoyancy and, in some cases, respiratory gas exchange.

sycon (sī′kon) (Gr. *sykon,* fig) Type of canal system in certain sponges. Sometimes called syconoid.

symbiogenesis (sim′-bī-ō-je′nəs-əs) An evolutionary process in which two organisms form an association leading to mutual dependency; one organism eventually becomes an organelle within the other, as in the origin of mitochondria and chloroplasts.

symbiosis (sim-bī-ōs′əs, sim′bē-ōs′əs) (Gr. *syn,* with, + *bios,* life) The living together of two different species in an intimate relationship. The symbiont always benefits; the host may benefit (mutualism), be unaffected (commensalism), or be harmed (parasitism).

synapomorphy (sin-ap′ə-mōr′fē) (Gr. *syn,* together with, + *apo,* of, + *morphē,* form) Shared, evolutionarily derived character states that are used to recover patterns of common descent among two or more species.

synapsids (si-nap′sədz) (Gr. *synapsis,* contact, union) An amniote lineage comprising the mammals and their closest fossil relatives, having a skull with a single pair of temporal openings.

syncytial specification During embryonic cleavage, molecules that specify cell fate diffuse within the cytoplasm of a single multinucleate cell.

syncytium (sin-sish′ē-əm) (Gr. *syn,* with, + *kytos,* hollow vessel) A multinucleated cell. Adj., **syncytial**.

syngamy (sin′gə-mē) (Gr. *syn,* with, + *gamos,* marriage) Fertilization of one gamete with another individual gamete to form a zygote; found in most animals that have sexual reproduction.

synsacrum (sin-sa′krəm) (Gr. *syn,* together + sacrum) A bone of birds, consisting of fused lumbar, sacral, and caudal vertebrae.

syrinx (sir′inks) (Gr., shepherd's pipe) The vocal organ of birds located at the base of the trachea.

systematics (sis-tə-mat′iks) Science of taxonomy and reconstruction of phylogeny.

systematization (sis-tə-mə-tī-zā′shən) Construction of a hierarchical taxonomy that conveys the structure of common evolutionary descent among species. Each recognized taxon comprises an ancestral population lineage and all of its descendants.

T

tactile (tak′til) (L. *tactilis,* able to be touched, from *tangere,* to touch) Pertaining to touch.

taenidia (tə′nid′-ē-ə) (Gr. *tainia,* ribbon) Spiral thickenings of the cuticle that support tracheae (phylum Arthropoda).

tagma, pl. **tagmata** (Gr. *tagma,* arrangement, order, row) Compound body section of an arthropod resulting from embryonic fusion of two or more segments; for example, head, thorax, abdomen.

tagmatization, tagmosis Organization of the arthropod body into tagmata.

taxon (tak′son), pl. **taxa** (Gr. *taxis,* order, arrangement) Any taxonomic group or entity.

taxonomy (tak-sän′ə-mē) (Gr. *taxis,* order, arrangement, + *nomas,* law) Study of the principles of systematic ordering and naming of organisms.

tegument (teg′ū-ment) (L. *tegumentum,* from tegere, to cover) An integument; specifically, the external covering in cestodes and trematodes.

teleology (tel′ē-äl′ə-jē) (Gr. *telos,* end, + L. *logia,* study of, from Gr. *logos,* word) The philosophical view that natural events are goal-directed and preordained, as opposed to the scientific view of mechanical determinism.

teleost (tē′lē-ost) A clade of advanced ray-finned fishes characterized by a homocercal caudal fin.

telson (tel′sən) (Gr., extremity) Posterior projection of the last body segment in many crustaceans.

temnospondyls (tem-nō-spän′dəls) (Gr. *temnō,* to cut, + *spondylos,* vertebra) A large group of early tetrapods that lived from the Carboniferous to the Triassic.

tergum (ter′gəm) (L., back) Dorsal part of an arthropod body segment.

territory (L. *territorium,* from *terra,* earth) Restricted area preempted by an animal or pair of animals, usually for breeding purposes, and guarded from other individuals of the same species.

test (L. *testa,* shell) A shell or hardened outer covering.

tetrapods (te′trə-pods) (Gr. *tetras,* four, + *pous, podos,* foot) Four-limbed vertebrates; the group includes amphibians, reptiles, birds, and mammals, including taxa that have lost limbs, such as caecilians and snakes.

theory A scientific hypothesis or set of related hypotheses that offers very powerful explanations for a wide variety of related phenomena and serves to organize the scientific investigation of those phenomena.

therapsid (thə-rap′sid) (Gr. *theraps,* an attendant) Extinct Mesozoic synapsid amniotes from which mammals evolved.

thoracic (thō-ra′sək) (L. *thōrax,* chest) Pertaining to the thorax or chest.

Tiedemann's bodies (tēd′ə-mənz) (from F. Tiedemann, German anatomist) Four or five pairs of pouchlike bodies attached to the ring canal of sea stars, apparently functioning in production of coelomocytes.

tissue (ti′shü) (M.E. *tissu,* tissue) An aggregation of cells and cell products organized to perform a common function.

tornaria (tor-na′re-ə) (Gr. *tornos,* compass, circle, wheel) A hemichordate (acorn worm) larva that closely resembles bipinnaria larvae of sea stars.

torpor (tor′pur) (L. *torpere,* numbed) A state of reduced physiological activity, typically characterized by reduced metabolic rate and temperature. Seasonal torpor is called **hibernation** in the winter and aestivation in the summer. Torpor occurring in bouts of less than 24 hours is referred to as daily torpor.

torsion (L. *torquere,* to twist) A twisting phenomenon in gastropod development that alters the position of the visceral and pallial organs by 180 degrees.

toxicyst (tox′i-sist) (Gr. *toxikon,* poison, + *kystis,* bladder) Structures of predatory ciliate protozoa that, on stimulation, expel a poison to subdue prey.

trabecular reticulum (trə-bek′yə-lər rə-tik′yə-ləm) An extensive syncytial tissue in hexactinellid sponges, bearing choanoblasts and collar bodies and forming flagellated chambers.

trachea (trā′kē-ə) (M.L., windpipe) The windpipe. Also, any of the air tubes of insects.

tracheal system (trā k′ē-əl) (M.L. *trachia,* windpipe) A network of thin-walled tubes that branch throughout the entire body of terrestrial insects; used for respiration.

tracheole (trā k′ē-ōl) (L. *trachia,* windpipe) Fine branches of the tracheal system, filled with fluid, but not shed at ecdysis.

transcription factor A protein that controls the rate at which messenger RNA is copied from a DNA template. Transcription factors typically bind to a specific DNA sequence and to one another to control tissue-specific patterns of gene expression during animal development.

transverse plane (L. *transversus,* across) A plane or section that lies or passes across a body or structure dividing it into cephalic and caudal pieces.

trend A directional change in the characteristic features or patterns of diversity in a group of organisms when viewed over long periods of evolutionary time in the fossil record.

trichinosis (trik-ən-o′səs) Disease caused by infection with the nematode *Trichinella* spp.

trichocyst (trik′ə-sist) (Gr. *thrix,* hair, + *kystis,* bladder) Saclike protrusible organelle in the ectoplasm of ciliates, which discharges as a threadlike weapon of defense.

triploblastic (trip′lō-blas′tik) (Gr. *triploos,* triple, + *blastos,* germ) Pertaining to animals in which the embryo has three primary germ layers—ectoderm, mesoderm, and endoderm.

trochophore (trō k′ə-fōr) (Gr. *trochos,* wheel, + *pherein,* to bear) A free-swimming ciliated marine larva characteristic of most molluscs and certain ectoprocts, brachiopods, and marine worms; an ovoid or pyriform body with a preoral circlet of cilia and sometimes a secondary circlet behind the mouth.

trophallaxis (tröf′ə-lak′səs) (Gr. *trophē,* food, + *allaxis,* barter, exchange) Exchange of food between young and adults, especially certain social insects.

trophic (trō′fək) (Gr. *trophē,* food) Pertaining to nutrition.

trophic level Position of a species in a **food web,** such as producer, herbivore, first-level carnivore, or higher-level carnivore.

trophosome (trō f′ə-sō m) (Gr. *trophē,* food, + *soma,* body) Organ in pogonophorans bearing mutualistic bacteria; derived from midgut.

trophozoite (trō f′ə-zō′ī t) (Gr. *trophē,* food, + *zōon,* animal) Adult stage in the life cycle of a protozoan in which it is actively absorbing nourishment.

tube feet (podia) Numerous small, muscular, fluid-filled tubes projecting from an echinoderm; part of water-vascular system; used in locomotion, clinging, food handling, and respiration.

tubulin (tü′bū-lən) (L. *tubulus,* small tube, + *in,* belonging to) Globular protein forming the hollow cylinder of microtubules.

tunic (L. *tunica,* tunic, coat) In tunicates, a cuticular, cellulose-containing covering of the body secreted by the underlying body wall.

turbinates (tər′bin-ā ts) (L. *turbin,* whirling) Highly convoluted bones covered in mucous membrane in the nasal cavity of endotherms that serve to reduce heat and water lost during respiration.

tympanic membrance (tim-pan′ik) (Gr. *tympanon,* drum) The surface that separates the outer and middle ear; also called the eardrum.

typhlosole (tif′lə-sō l′) (Gr. *typhlos,* blind, + *sōlēn,* channel, pipe) A longitudinal fold projecting into the intestine in certain invertebrates such as the earthworm.

typological species concept The discredited, pre-Darwinian notion that species are classes defined by the presence of fixed, unchanging characters (+ "essence") shared by all members.

U

ultimate cause (L. *ultimatus,* last, + *causa*) The evolutionary factors responsible for the origin, state of being, or role of a biological system.

umbo (um′bō), pl. **umbones** (um-bō′nēz) (L., boss of a shield) One of the prominences on either side of the hinge region in a bivalve mollusc shell. Also, the "beak" of a brachiopod shell.

undulating membrane Membranous structure on a protozoan associated with a flagellum; on other protozoa, may be formed from fused cilia.

ungulate (un′gū-lət) (L. *ungula,* hoof) Hooved. Noun, any hooved mammal.

uniformitarianism (ū′nə-for′mə-ter′ē-ə-niz′əm) Methodological assumptions that the laws of chemistry and physics have remained constant throughout the history of the earth, and that past geological events occurred by processes that can be observed today.

uniramous (ü′nə-rā m′əs) (L. *unus,* one, + *ramus,* a branch) Adjective describing unbranched appendages (phylum Arthropoda).

unitary (ü′nə-ter′ē) Describes the structure of a population in which reproduction is strictly sexual and each organism is genetically distinct from others.

uropod (ū′rə-pod) (Gr. *oura,* tail, + *pous, podos,* foot) Posteriormost appendage of many crustaceans.

V

vacuole (vak′yə-wō l) (L. *vacuus,* empty, + Fr. *ole;* dimin. suffix) A membrane-bound, fluid-filled space in a cell.

valve (L. *valva,* leaf of a double door) One of the two shells of a typical bivalve mollusc or brachiopod.

variation (L. *varius,* various) Differences among individuals of a group or species that cannot be ascribed to age, sex, or position in the life cycle.
velarium (və-la′rē-əm) (L. *velum,* veil, covering) Shelflike extension of the subumbrellar edge in cubozoans (phylum Cnidaria).
veliger (vēl′ə-jər) (L. *velum,* veil, covering) Larval form of certain molluscs; develops from the trochophore and has the beginning of a foot, mantle, and shell.
velum (vē′ləm) (L., veil, covering) A membrane on the subumbrellar surface of jellies of class Hydrozoa. Also, a ciliated swimming organ of the veliger larva.
ventral (ven′trəl) (L. *venter,* belly) Situated on the lower or abdominal surface.
vestige (ves′tij) (L. *vestigium,* footprint) A rudimentary organ that may have been well developed in some ancestor or in the embryo.
vibrissa (vī-bris′ə), pl. **vibrissae** (L., nostril-hair) Stiff hairs that grow from the nostrils or other parts of the face of many mammals and that serve as tactile organs; "whiskers."
villus (vil′əs), pl. **villi** (L., tuft of hair) Small fingerlike, vascular process on the wall of the small intestine. Also, one of the branching, vascular processes on the embryonic portion of the placenta.
viscera (vis′ər-ə) (L. pl. of *viscus,* internal organ) Internal organs in the body cavity.
visceral (vis′ər-əl) Pertaining to viscera.
viviparity (vī′və-par′ə-tē) (L. *vivus,* alive, + *parere,* to bring forth) Reproduction in which eggs develop within the female body, which supplies nutritional aid; occurs in therian mammals, many non-avian reptiles, and some fishes; offspring are born as juveniles. Adj., **viviparous** (vī-vip′ə-rəs).

W

water-vascular system System of fluid-filled closed tubes and ducts peculiar to echinoderms; used to move tentacles and tube feet that serve variously for clinging, food handling, locomotion, and respiration.
weir (wēr) (Old English *wer,* a fence placed in a stream to catch fish) Interlocking extensions of a flame cell and a collecting tubule cell in some protonephridia.

X

X-organ Neurosecretory organ in eyestalk of crustaceans that secretes molt-inhibiting hormone.

Y

Y-organ Gland in the antennal or maxillary segment of some crustaceans that secretes molting hormone.
yolk sac (yōlk sak) A membranous sac on the outside of many embryos or larvae of animals, filled with nutrients that support the animal's growth.

Z

zoecium (zō-ē′shē-əm) (Gr. *zōon,* animal, + *oikos,* house) Cuticular sheath or shell of Ectoprocta; also spelled **zooecium.**
zooid (zō-oid) (Gr. *zōon,* animal) An individual member of a colony of animals, such as colonial cnidarians and ectoprocts.
zooxanthella (zō′ə-zan-thəl′ə) (Gr. *zōon,* animal, + *xanthos,* yellow) A golden-brown dinoflagellate living in the tissues of many types of marine invertebrates; green algae in these tissues are called zoochlorellae.
zygote (Gr. *zygōtos,* yoked) A fertilized egg.
zygotic meiosis Meiosis that occurs within the first few divisions after zygote formation; thus, all stages in the life cycle other than the zygote are haploid.

Index

Note: Page numbers followed by *b*, *f*, and *t* indicate boxes, figures, and tables, respectively.

A

B

C

D

E

F

I

J

K

L

M

N

Q

R

S

U

V

W

X